I0485806

The Language of The Mathematical Proof

Logical Reasoning & Mathematics

Edited by Paul F. Kisak

Contents

Chapter 1

The Mathematical Proof

In mathematics, a **proof** is a deductive argument for a mathematical statement. In the argument, other previously established statements, such as theorems, can be used. In principle, a proof can be traced back to self-evident or assumed statements, known as axioms.[2][3][4] Proofs are examples of deductive reasoning and are distinguished from inductive or empirical arguments; a proof must demonstrate that a statement is always true (occasionally by listing *all* possible cases and showing that it holds in each), rather than enumerate many confirmatory cases. An unproven statement that is believed true is known as a conjecture.

Proofs employ logic but usually include some amount of natural language which usually admits some ambiguity. In fact, the vast majority of proofs in written mathematics can be considered as applications of rigorous informal logic. Purely formal proofs, written in symbolic language instead of natural language, are considered in proof theory. The distinction between formal and informal proofs has led to much examination of current and historical mathematical practice, quasi-empiricism in mathematics, and so-called folk mathematics (in both senses of that term). The philosophy of mathematics is concerned with the role of language and logic in proofs, and mathematics as a language.

1.1 History and etymology

The word "proof" comes from the Latin *probare* meaning "to test". Related modern words are the English "probe", "probation", and "probability", the Spanish *probar* (to smell or taste, or (lesser use) touch or test),[5] Italian *provare* (to try), and the German *probieren* (to try). The early use of "probity" was in the presentation of legal evidence. A person of authority, such as a nobleman, was said to have probity, whereby the evidence was by his relative authority, which outweighed empirical testimony.[6]

Plausibility arguments using heuristic devices such as pictures and analogies preceded strict mathematical proof.[7] It is likely that the idea of demonstrating a conclusion first arose in connection with geometry, which originally meant the same as "land measurement".[8] The development of mathematical proof is primarily the product of ancient Greek mathematics, and one of its greatest achievements. Thales (624–546 BCE) proved some theorems in geometry. Eudoxus (408–355 BCE) and Theaetetus (417–369 BCE) formulated theorems but did not prove them. Aristotle (384–322 BCE) said definitions should describe the concept being defined in terms of other concepts already known. Mathematical proofs were revolutionized by Euclid (300 BCE), who introduced the axiomatic method still in use today, starting with undefined terms and axioms (propositions regarding the undefined terms assumed to be selfevidently true from the Greek "axios" meaning "something worthy"), and used these to prove theorems using deductive logic. His book, the *Elements*, was read by anyone who was considered educated in the West until the middle of the 20th century.[9] In addition to the familiar theorems of geometry, such as the Pythagorean theorem, the *Elements* includes a proof that the square root of two is irrational and that there are infinitely many prime numbers.

Further advances took place in medieval Islamic mathematics. While earlier Greek proofs were largely geometric demonstrations, the development of arithmetic and algebra by Islamic mathematicians allowed more general proofs that no longer depended on geometry. In the 10th century CE, the Iraqi mathematician Al-Hashimi provided general proofs for numbers (rather than geometric demonstrations) as he considered multiplication, division, etc. for "lines." He used this method to provide a proof of the existence of irrational numbers.[10] An inductive proof for arithmetic sequences was introduced in the *Al-Fakhri* (1000) by Al-Karaji, who used it to prove the binomial theorem and properties of Pascal's triangle. Alhazen also developed the method of proof by contradiction, as the first attempt at proving the Euclidean parallel postulate.[11]

Modern proof theory treats proofs as inductively defined data structures. There is no longer an assumption that axioms are "true" in any sense; this allows for parallel mathematical theories built on alternate sets of axioms (see Axiomatic set theory and Non-Euclidean geometry for examples).

1.2 Nature and purpose

As practised, a proof is expressed in natural language and is a rigorous argument intended to convince the audience of the truth of a statement. The standard of rigor is not absolute and has varied throughout history. A proof can be presented differently depending on the intended audience. In order to gain acceptance, a proof has to meet communal statements of rigor; an argument considered vague or incomplete may be rejected.

The concept of a proof is formalized in the field of mathematical logic.[12] A formal proof is written in a formal language instead of a natural language. A formal proof is defined as sequence of formulas in a formal language, in which each formula is a logical consequence of preceding formulas. Having a definition of formal proof makes the

concept of proof amenable to study. Indeed, the field of proof theory studies formal proofs and their properties, for example, the property that a statement has a formal proof. An application of proof theory is to show that certain undecidable statements are not provable.
The definition of a formal proof is intended to capture the concept of proofs as written in the practice of mathematics. The soundness of this definition amounts to the belief that a published proof can, in principle, be converted into a formal proof. However, outside the field of automated proof assistants, this is rarely done in practice. A classic question in philosophy asks whether mathematical proofs are analytic or synthetic. Kant, who introduced the analyticsynthetic distinction, believed mathematical proofs are synthetic.
Proofs may be viewed as aesthetic objects, admired for their mathematical beauty. The mathematician Paul Erdős was known for describing proofs he found particularly elegant as coming from "The Book", a hypothetical tome containing the most beautiful method(s) of proving each theorem. The book *Proofs from THE BOOK*, published in

2003, is devoted to presenting 32 proofs its editors find particularly pleasing.

1.3 Methods

1.3.1 Direct proof

Main article: Direct proof
In direct proof, the conclusion is established by logically combining the axioms, definitions, and earlier theorems.[13]
For example, direct proof can be used to establish that the sum of two even integers is always even:
Consider two even integers x and y. Since they are even, they can be written as $x = 2a$ and $y = 2b$, respectively, for integers a and b. Then the sum $x + y = 2a + 2b = 2(a+b)$. Therefore $x+y$ has 2 as a factor and, by definition, is even. Hence the sum of any two even integers is even.
This proof uses the definition of even integers, the integer properties of closure under addition and multiplication, and distributivity.

1.3.2 Proof by mathematical induction

Main article: Mathematical induction
Mathematical induction is not a form of inductive reasoning. In proof by mathematical induction, a single "base case" is proved, and an "induction rule" is proved, which establishes that a certain case implies the next case. Applying the induction rule repeatedly, starting from the independently proved base case, proves many, often infinitely many, other cases.[14] Since the base case is true, the infinity of other cases must also be true, even if all of them cannot be proved directly because of their infinite number. A subset of induction is infinite descent. Infinite descent can be used to prove the irrationality of the square root of two.
A common application of proof by mathematical induction is to prove that a property known to hold for one number holds for all natural numbers:[15] Let **N** = {1,2,3,4,...} be the set of natural numbers, and $P(n)$ be a mathematical statement involving the natural number n belonging to **N** such that
_ **(i)** $P(1)$ is true, i.e., $P(n)$ is true for $n = 1$.
_ **(ii)** $P(n+1)$ is true whenever $P(n)$ is true, i.e., $P(n)$ is true implies that $P(n+1)$ is true.
_ **Then $P(n)$ is true for all natural numbers n.**
For example, we can prove by induction that all integers of the form $2n + 1$ are odd:
(i) For $n = 1$, $2n + 1 = 2(1) + 1 = 3$, and 3 is odd. Thus $P(1)$ is true.
(ii) For $2n + 1$ for some n, $2(n+1) + 1 = (2n+1) + 2$. If $2n + 1$ is odd, then $(2n+1) + 2$ must also be odd, because adding 2 to an odd number results in an odd number. So $P(n+1)$ is true if $P(n)$ is true.
Thus $2n + 1$ is odd, for all natural numbers n.
It is common for the phrase "proof by induction" to be used for a "proof by mathematical induction".[16]

1.3.3 Proof by contraposition

Main article: Contraposition
Proof by contraposition infers the conclusion "if p then q" from the premise "if *not q* then *not p*". The statement "if *not q* then *not p*" is called the contrapositive of the statement "if p then q". For example, contraposition can be used to establish that, given an integer x, if x^2 is even, then x is even:

Suppose x is not even. Then x is odd. The product of two odd numbers is odd, hence $x^2 = x{\cdot}x$ is odd.
Thus x^2 is not even.

1.3.4 Proof by contradiction

Main article: Proof by contradiction

In proof by contradiction (also known as *reductio ad absurdum*, Latin for "by reduction to the absurd"), it is shown that if some statement were true, a logical contradiction occurs, hence the statement must be false. A famous example of proof by contradiction shows that

*p*2 is an irrational number:

Suppose that

*p*2 were a rational number, so by definition

*p*2 = *ab* where *a* and *b* are non-zero integers

with no common factor. Thus, *bp*2 = *a* . Squaring both sides yields 2*b*2 = *a*2. Since 2 divides the left hand side, 2 must also divide the right hand side (as they are equal and both integers). So *a*2 is even, which implies that *a* must also be even. So we can write *a* = 2*c*, where *c* is also an integer. Substitution into the original equation yields 2*b*2 = (2*c*)2 = 4*c*2. Dividing both sides by 2 yields *b*2 = 2*c*2. But then, by the same argument as before, 2 divides *b*2, so *b* must be even. However, if *a* and *b* are both even, they share a factor, namely 2. This contradicts our assumption, so we are forced to conclude that

*p*2 is an irrational number.

1.3.5 Proof by construction

Main article: Proof by construction

Proof by construction, or proof by example, is the construction of a concrete example with a property to show that something having that property exists. Joseph Liouville, for instance, proved the existence of transcendental numbers by constructing an explicit example. It can also be used to construct a counterexample to disprove a proposition that all elements have a certain property.

1.3.6 Proof by exhaustion

Main article: Proof by exhaustion

In proof by exhaustion, the conclusion is established by dividing it into a finite number of cases and proving each one separately. The number of cases sometimes can become very large. For example, the first proof of the four color theorem was a proof by exhaustion with 1,936 cases. This proof was controversial because the majority of the cases were checked by a computer program, not by hand. The shortest known proof of the four color theorem as of 2011 still has over 600 cases.

1.3.7 Probabilistic proof

Main article: Probabilistic method

A probabilistic proof is one in which an example is shown to exist, with certainty, by using methods of probability theory. Probabilistic proof, like proof by construction, is one of many ways to show existence theorems.

This is not to be confused with an argument that a theorem is 'probably' true, a 'plausibility argument'. The work on the Collatz conjecture shows how far plausibility is from genuine proof.[17]

1.3.8 Combinatorial proof

Main article: Combinatorial proof

1.4. UNDECIDABLE STATEMENTS 5

A combinatorial proof establishes the equivalence of different expressions by showing that they count the same object in different ways. Often a bijection between two sets is used to show that the expressions for their two sizes are equal. Alternatively, a double counting argument provides two different expressions for the size of a single set, again showing that the two expressions are equal.

1.3.9 Nonconstructive proof

Main article: Nonconstructive proof

A nonconstructive proof establishes that a mathematical object with a certain property exists without explaining how such an object can be found. Often, this takes the form of a proof by contradiction in which the nonexistence of the object is proven to be impossible. In contrast, a constructive proof establishes that a particular object exists by providing a method of finding it. A famous example of a nonconstructive proof shows that there exist two irrational numbers *a* and *b* such that *ab* is a rational number:

Either

*p*2*p*2 is a rational number and we are done (take *a* = *b* =*p*2),

or *p*2*p*2 is irrational so we can write *a* =*p*2*p*2 and *b* =*p*2 .

This then gives (*p*2*p*2)*p*2=*p*22= 2 , which is thus a rational of the form *ab:*

1.3.10 Statistical proofs in pure mathematics

Main article: Statistical proof
The expression “statistical proof” may be used technically or colloquially in areas of pure mathematics, such as involving cryptography, chaotic series, and probabilistic or analytic number theory.[18][19][20] It is less commonly used to refer to a mathematical proof in the branch of mathematics known as mathematical statistics. See also "Statistical proof using data" section below.

1.3.11 Computer-assisted proofs

Main article: Computer-assisted proof
Until the twentieth century it was assumed that any proof could, in principle, be checked by a competent mathematician to confirm its validity.[7] However, computers are now used both to prove theorems and to carry out calculations that are too long for any human or team of humans to check; the first proof of the four color theorem is an example of a computer-assisted proof. Some mathematicians are concerned that the possibility of an error in a computer program or a run-time error in its calculations calls the validity of such computer-assisted proofs into question. In practice, the chances of an error invalidating a computer-assisted proof can be reduced by incorporating redundancy and self-checks into calculations, and by developing multiple independent approaches and programs. Errors can never be completely ruled out in case of verification of a proof by humans either, especially if the proof contains natural language and requires deep mathematical insight.

1.4 Undecidable statements

A statement that is neither provable nor disprovable from a set of axioms is called undecidable (from those axioms). One example is the parallel postulate, which is neither provable nor refutable from the remaining axioms of Euclidean geometry.
Mathematicians have shown there are many statements that are neither provable nor disprovable in Zermelo-Fraenkel set theory with the axiom of choice (ZFC), the standard system of set theory in mathematics (assuming that ZFC is consistent); see list of statements undecidable in ZFC.

Gödel’s (first) incompleteness theorem shows that many axiom systems of mathematical interest will have undecidable statements.

1.5 Heuristic mathematics and experimental mathematics

Main article: Experimental mathematics
While early mathematicians such as Eudoxus of Cnidus did not use proofs, from Euclid to the foundational mathematics developments of the late 19th and 20th centuries, proofs were an essential part of mathematics.[21] With the increase in computing power in the 1960s, significant work began to be done investigating mathematical objects outside of the proof-theorem framework,[22] in experimental mathematics. Early pioneers of these methods intended the work ultimately to be embedded in a classical proof-theorem framework, e.g. the early development of fractal geometry,[23] which was ultimately so embedded.

1.6 Related concepts

1.6.1 Visual proof

Although not a formal proof, a visual demonstration of a mathematical theorem is sometimes called a "proof without words". The left-hand picture below is an example of a historic visual proof of the Pythagorean theorem in the case of the (3,4,5) triangle.
_ Visual proof for the (3, 4, 5) triangle as in the Chou Pei Suan Ching 500–200 BC.
_ Animated visual proof for the Pythagorean theorem by rearrangement.
_ A second animated proof of the Pythagorean theorem.
Some illusory visual proofs, such as the missing square puzzle, can be constructed in a way which appear to prove a supposed mathematical fact but only do so under the presence of tiny errors (for example, supposedly straight lines which actually bend slightly) which are unnoticeable until the entire picture is closely examined, with lengths and angles precisely measured or calculated.

1.6.2 Elementary proof

Main article: Elementary proof
An elementary proof is a proof which only uses basic techniques. More specifically, the term is used in number theory

to refer to proofs that make no use of complex analysis. For some time it was thought that certain theorems, like the prime number theorem, could only be proved using "higher" mathematics. However, over time, many of these results have been reproved using only elementary techniques.

1.6.3 Two-column proof

A particular way of organising a proof using two parallel columns is often used in elementary geometry classes in the United States.[24] The proof is written as a series of lines in two columns. In each line, the left-hand column contains a proposition, while the right-hand column contains a brief explanation of how the corresponding proposition in the left-hand column is either an axiom, a hypothesis, or can be logically derived from previous propositions. The left-hand column is typically headed "Statements" and the right-hand column is typically headed "Reasons".[25]

1.6. RELATED CONCEPTS 7

A two-column proof published in 1913

1.6.4 Colloquial use of "mathematical proof"

The expression "mathematical proof" is used by lay people to refer to using mathematical methods or arguing with mathematical objects, such as numbers, to demonstrate something about everyday life, or when data used in an argument is numerical. It is sometimes also used to mean a "statistical proof" (below), especially when used to argue from data.

1.6.5 Statistical proof using data

Main article: Statistical proof

"Statistical proof" from data refers to the application of statistics, data analysis, or Bayesian analysis to infer propositions regarding the probability of data. While *using* mathematical proof to establish theorems in statistics, it is usually not a mathematical proof in that the *assumptions* from which probability statements are derived require empirical evidence from outside mathematics to verify. In physics, in addition to statistical methods, "statistical proof" can refer to the specialized *mathematical methods of physics* applied to analyze data in a particle physics experiment or observational study in cosmology. "Statistical proof" may also refer to raw data or a convincing diagram involving data, such as scatter plots, when the data or diagram is adequately convincing without further analysis.

8 *CHAPTER 1. MATHEMATICAL PROOF*

1.6.6 Inductive logic proofs and Bayesian analysis

Main articles: Inductive logic and Bayesian analysis

Proofs using inductive logic, while considered mathematical in nature, seek to establish propositions with a degree of certainty, which acts in a similar manner to probability, and may be less than one certainty. Bayesian analysis establishes assertions as to the degree of a person's subjective belief. Inductive logic should not be confused with mathematical induction.

1.6.7 Proofs as mental objects

Main articles: Psychologism and Language of thought

Psychologism views mathematical proofs as psychological or mental objects. Mathematician philosophers, such as Leibniz, Frege, and Carnap have attempted to develop a semantics for what they considered to be the language of thought, whereby standards of mathematical proof might be applied to empirical science.

1.6.8 Influence of mathematical proof methods outside mathematics

Philosopher-mathematicians such as Spinoza have attempted to formulate philosophical arguments in an axiomatic manner, whereby mathematical proof standards could be applied to argumentation in general philosophy. Other mathematician-philosophers have tried to use standards of mathematical proof and reason, without empiricism, to arrive at statements outside of mathematics, but having the certainty of propositions deduced in a mathematical proof, such as Descarte's *cogito* argument.

1.7 Ending a proof

Main article: Q.E.D.

Sometimes, the abbreviation *"Q.E.D."* is written to indicate the end of a proof. This abbreviation stands for *"Quod Erat Demonstrandum"*, which is Latin for *"that which was to be demonstrated"*. A more common alternative is to use a square or a rectangle, such as □ or ■, known as a "tombstone" or "halmos" after its eponym Paul Halmos. Often, "which was to be shown" is verbally stated when writing "QED", "□", or "■" in an oral presentation on a board.

1.8 See also

_ Automated theorem proving
_ Invalid proof
_ List of incomplete proofs
_ List of long proofs
_ List of mathematical proofs
_ Nonconstructive proof
_ Proof by intimidation
_ Termination analysis
_ *What the Tortoise Said to Achilles*
1.9. REFERENCES 9

1.9 References

[1] Bill Casselman. "One of the Oldest Extant Diagrams from Euclid". University of British Columbia. Retrieved 2008-09-26.
[2] Clapham, C. and Nicholson, JN. *The Concise Oxford Dictionary of Mathematics, Fourth edition.* "A statement whose truth is either to be taken as self-evident or to be assumed. Certain areas of mathematics involve choosing a set of axioms and discovering what results can be derived from them, providing proofs for the theorems that are obtained."
[3] Cupillari, Antonella. *The Nuts and Bolts of Proofs*. Academic Press, 2001. Page 3.
[4] Gossett, Eric. *Discrete Mathematics with Proof.* John Wiley and Sons, 2009. Definition 3.1 page 86. ISBN 0-470-45793-7
[5] New Shorter Oxford English Dictionary, 1993, OUP, Oxford.
[6] The Emergence of Probability, Ian Hacking
[7] The History and Concept of Mathematical Proof, Steven G. Krantz. 1. February 5, 2007
[8] Kneale, p. 2
[9] Howard Eves, *An Introduction to the History of Mathematics*, Saunders, 1990, ISBN 0-03-029558-0 p. 141: "No work, except The Bible, has been more widely used...."
[10] Matvievskaya, Galina (1987), *The Theory of Quadratic Irrationals in Medieval Oriental Mathematics*, *Annals of the New York Academy of Sciences* **500**: 253–277 [260], doi:10.1111/j.1749-6632.1987.tb37206.x
[11] Eder, Michelle (2000), *Views of Euclid's Parallel Postulate in Ancient Greece and in Medieval Islam*, Rutgers University, retrieved 2008-01-23
[12] Buss, 1997, p. 3
[13] Cupillari, page 20.
[14] Cupillari, page 46.
[15] Examples of simple proofs by mathematical induction for all natural numbers
[16] Proof by induction, University of Warwick Glossary of Mathematical Terminology
[17] While most mathematicians do not think that probabilistic evidence ever counts as a genuine mathematical proof, a few mathematicians and philosophers have argued that at least some types of probabilistic evidence (such as Rabin's probabilistic algorithm for testing primality) are as good as genuine mathematical proofs. See, for example, Davis, Philip J. (1972), "Fidelity in Mathematical Discourse: Is One and One Really Two?" *American Mathematical Monthly* 79:252-63. Fallis, Don (1997), "The Epistemic Status of Probabilistic Proof." *Journal of Philosophy* 94:165-86.
[18] "in number theory and commutative algebra... in particular the statistical proof of the lemma."
[19] "Whether constant π (i.e., pi) is normal is a confusing problem without any strict theoretical demonstration except for some **statistical** proof"" (Derogatory use.)
[20] "these observations suggest a statistical proof of Goldbach's conjecture with very quickly vanishing probability of failure for large E"
[21] "*What to do with the pictures? Two thoughts surfaced: the first was that they were unpublishable in the standard way, there were no theorems only very suggestive pictures. They furnished convincing evidence for many conjectures and lures to further exploration, but theorems were coins of the realm ant the conventions of that day dictated that journals only published theorems*", David Mumford, Caroline Series and David Wright, Indra's Pearls, 2002
[22] "*Mandelbrot, working at the IBM Research Laboratory, did some computer simulations for these sets on the reasonable assumption that, if you wanted to prove something, it might be helpful to know the answer ahead of time.*"A Note on the History of Fractals,
[23] "*... brought home again to Benoit [Mandelbrot] that there was a 'mathematics of the eye', that visualization of a problem was as valid a method as any for finding a solution. Amazingly, he found himself alone with this conjecture. The teaching of mathematics in France was dominated by a handful of dogmatic mathematicians hiding behind the pseudonym 'Bourbaki'...*
", Introducing Fractal Geometry, Nigel Lesmoir-Gordon
[24] Patricio G. Herbst, Establishing a Custom of Proving in American School Geometry: Evolution of the Two-Column Proof in the Early Twentieth Century, Educational Studies in Mathematics, Vol. 49, No. 3 (2002), pp. 283-312,
[25] Introduction to the Two-Column Proof, Carol Fisher
10 *CHAPTER 1. MATHEMATICAL PROOF*

1.10 Sources

_ Pólya, G. (1954), *Mathematics and Plausible Reasoning*, Princeton University Press.

_ Fallis, Don (2002), *What Do Mathematicians Want? Probabilistic Proofs and the Epistemic Goals of Mathematicians, Logique et Analyse* **45**: 373–388.
_ Franklin, J.; Daoud, A. (2011), *Proof in Mathematics: An Introduction*, Kew Books, ISBN 0-646-54509-4.
_ Solow, D. (2004), *How to Read and Do Proofs: An Introduction to Mathematical Thought Processes*, Wiley, ISBN 0-471-68058-3.
_ Velleman, D. (2006), *How to Prove It: A Structured Approach*, Cambridge University Press, ISBN 0-521-67599-5.

1.11 External links

_ Hazewinkel, Michiel, ed. (2001), "Proof theory", *Encyclopedia of Mathematics*, Springer, ISBN 978-1-55608-010-4
_ What are mathematical proofs and why they are important?
_ 2πix.com: Logic Part of a series of articles covering mathematics and logic.
_ How To Write Proofs by Larry W. Cusick
_ How to Write a Proof by Leslie Lamport, and the motivation of proposing such a hierarchical proof style.
_ Proofs in Mathematics: Simple, Charming and Fallacious
_ *The Seventeen Provers of the World*, ed. by Freek Wiedijk, foreword by Dana S. Scott, Lecture Notes in Computer Science 3600, Springer, 2006, ISBN 3-540-30704-4. Contains formalized versions of the proof that
p
2 is irrational in several automated proof systems.
_ What is Proof? Thoughts on proofs and proving.
_ ProofWiki.org A wiki compendium of mathematical proofs.
_ planetmath.org A wiki style encyclopedia of proofs
_ A lesson about proofs, in a course from Wikiversity
_ The role and function of proof by Michael de Villiers
_ Developing understanding of different roles of proof by Michael de Villiers
_ "Book of Proof" by Richard Hammack 2009 Part of the Open Textbook Initiative. Provides an introduction to mathematical proofs.
_ *On proof and progress in mathematics*. Thurston, William P. 1994.

Chapter 2

Proposition

This article is about the term in logic and philosophy. For other uses, see Proposition (disambiguation).
Not to be confused with preposition.
The term ***proposition*** has a broad use in contemporary philosophy. It is used to refer to some or all of the following: the primary bearers of truth-value, the objects of belief and other "propositional attitudes" (i.e., what is believed, doubted, etc.), the referents of that-clauses and the meanings of declarative sentences. Propositions are the sharable objects of attitudes and the primary bearers of truth and falsity. This stipulation rules out certain candidates for propositions, including thought- and utterance-tokens which are not sharable, and concrete events or facts, which cannot be false.[1]

2.1 Historical usage

2.1.1 By Aristotle

Aristotelian logic identifies a proposition as a sentence which affirms or denies a predicate of a subject. An Aristotelian proposition may take the form "All men are mortal" or "Socrates is a man." In the first example the subject is "All men" and the predicate "are mortal." In the second example the subject is "Socrates" and the predicate is "is a man."

2.1.2 By the logical positivists

Often propositions are related to closed sentences to distinguish them from what is expressed by an open sentence. In this sense, propositions are "statements" that are truth-bearers. This conception of a proposition was supported by the philosophical school of logical positivism.
Some philosophers argue that some (or all) kinds of speech or actions besides the declarative ones also have propositional content. For example, yes–no questions present propositions, being inquiries into the truth value of them. On the other hand, some signs can be declarative assertions of propositions without forming a sentence nor even being linguistic, e.g. traffic signs convey definite meaning which is either true or false.
Propositions are also spoken of as the content of beliefs and similar intentional attitudes such as desires, preferences, and hopes. For example, "I desire *that I have a new car*," or "I wonder *whether it will snow*" (or, whether it is the case that "it will snow"). Desire, belief, and so on, are thus called propositional attitudes when they take this sort of content.

2.1.3 By Russell

Bertrand Russell held that propositions were structured entities with objects and properties as constituents. Wittgenstein held that a proposition is the set of possible worlds/states of affairs in which it is true. One important difference between these views is that on the Russellian account, two propositions that are true in all the same states of affairs
11
12 *CHAPTER 2. PROPOSITION*
can still be differentiated. For instance, the proposition that two plus two equals four is distinct on a Russellian account from three plus three equals six. If propositions are sets of possible worlds, however, then all mathematical truths (and all other necessary truths) are the same set (the set of all possible worlds).

2.2 Relation to the mind

In relation to the mind, propositions are discussed primarily as they fit into propositional attitudes. Propositional attitudes are simply attitudes characteristic of folk psychology (belief, desire, etc.) that one can take toward a proposition (e.g. 'it is raining,' 'snow is white,' etc.). In English, propositions usually follow folk psychological attitudes by a "that clause" (e.g. "Jane believes *that* it is raining"). In philosophy of mind and psychology, mental states are often taken to primarily consist in propositional attitudes. The propositions are usually said to be the "mental content" of the attitude. For example, if Jane has a mental state of believing that it is raining, her mental content is the proposition 'it is raining.' Furthermore, since such mental states are *about* something (namely propositions), they are said to be intentional mental states. Philosophical debates surrounding propositions as they relate to propositional attitudes have also recently centered on whether they are internal or external to the agent or whether they are mind-dependent or mind-independent entities (see the entry on internalism and externalism in philosophy of mind).

2.3 Treatment in logic

As noted above, in Aristotelian logic a proposition is a particular kind of sentence, one which affirms or denies a predicate of a subject. Aristotelian propositions take forms like "All men are mortal" and "Socrates is a man." Propositions show up in formal logic as objects of a formal language. A formal language begins with different types of symbols. These types can include variables, operators, function symbols, predicate (or relation) symbols, quantifiers, and propositional constants. (Grouping symbols are often added for convenience in using the language but do not play a logical role.) Symbols are concatenated together according to recursive rules in order to construct strings to which truth-values will be assigned. The rules specify how the operators, function and predicate symbols, and quantifiers are to be concatenated with other strings. A proposition is then a string with a specific form. The form that a proposition takes depends on the type of logic.

The type of logic called propositional, sentential, or statement logic includes only operators and propositional constants as symbols in its language. The propositions in this language are propositional constants, which are considered atomic propositions, and composite propositions, which are composed by recursively applying operators to propositions. *Application* here is simply a short way of saying that the corresponding concatenation rule has been applied.

The types of logics called predicate, quantificational, or *n*-order logic include variables, operators, predicate and function symbols, and quantifiers as symbols in their languages. The propositions in these logics are more complex. First, terms must be defined. A term is (i) a variable or (ii) a function symbol applied to the number of terms required by the function symbol's arity. For example, if + is a binary function symbol and x, y, and z are variables, then $x+(y+z)$ is a term, which might be written with the symbols in various orders. A proposition is (i) a predicate symbol applied to the number of terms required by its arity, (ii) an operator applied to the number of propositions required by its arity, or (iii) a quantifier applied to a proposition. For example, if = is a binary predicate symbol and $\forall$ is a quantifier, then $\forall x,y,z\ [(x = y) \rightarrow (x+z = y+z)]$ is a proposition. This more complex structure of propositions allows these logics to make finer distinctions between inferences, i.e., to have greater expressive power.

In this context, propositions are also called sentences, statements, statement forms, formulas, and well-formed formulas, though these terms are usually not synonymous within a single text. This definition treats propositions as syntactic objects, as opposed to semantic or mental objects. That is, propositions in this sense are meaningless, formal, abstract objects. They are assigned meaning and truth-values by mappings called interpretations and valuations, respectively.

2.4 Objections to propositions

Attempts to provide a workable definition of proposition include
Two meaningful declarative sentences express the same proposition if and only if they mean the same thing.
2.5. SEE ALSO 13
thus defining *proposition* in terms of synonymity. For example, "Snow is white" (in English) and "Schnee ist weiß" (in German) are different sentences, but they say the same thing, so they express the same proposition.
Two meaningful declarative sentence-tokens express the same proposition if and only if they mean the same thing.
Unfortunately, the above definition has the result that two sentences/sentence-tokens which have the same meaning and thus express the same proposition, could have different truth-values, e.g. "I am Spartacus" said by Spartacus and said by John Smith; and e.g. "It is Wednesday" said on a Wednesday and on a Thursday.
A number of philosophers and linguists claim that all definitions of a proposition are too vague to be useful. For them, it is just a misleading concept that should be removed from philosophy and semantics. W.V. Quine maintained that the indeterminacy of translation prevented any meaningful discussion of propositions, and that they should be discarded in favor of sentences.[2] Strawson advocated the use of the term "statement".

2.5 See also

_ Main contention
_ Proposition usage in modern mathematical papers.

2.6 References

[1] "Propositions (Stanford Encyclopedia of Philosophy)". Plato.stanford.edu. Retrieved 2014-06-23.
[2] Quine W.V. *Philosophy of Logic*, Prentice-Hall NJ USA: 1970, pp 1-14

2.7 External links

_ Stanford Encyclopedia of Philosophy articles on:
_ Propositions, by Matthew McGrath
_ Singular Propositions, by Greg Fitch
_ Structured Propositions, by Jeffrey C. King

Chapter 3

Argument-deduction-proof distinctions

Argument-deduction-proof distinctions originated with logic itself.[1] Naturally, the terminology evolved.

3.1 Argument

An **argument**, more fully a premise-conclusion argument, is a two-part system composed of premises and conclusion. An argument is *valid* if and only if its conclusion is a consequence of its premises. Every premise set has infinitely many consequences each giving rise to a valid argument. Some consequences are obviously so but most are not: most are hidden consequences. Most valid arguments are not yet known to be valid. To determine validity in non-obvious cases deductive reasoning is required. There is no deductive reasoning in an argument *per se*; such must come from the outside.
Every argument's premises are conclusions of other arguments. Every argument's conclusion is a premise of other arguments. The word *constituent* may be used for either a premise or conclusion.In the context of this article and in most classical contexts, all candidates for consideration as argument constituents fall under the category of truthbearer: propositions, statements, sentences, judgments, etc.

3.2 Deduction

A **deduction** is a three-part system composed of premises, a conclusion, and chain of intermediates — steps of reasoning showing that its conclusion is a consequence of its premises. The reasoning in a deduction is by definition cogent. Such reasoning itself, or the chain of intermediates representing it, has also been called an argument, more fully a deductive argument. In many cases, an argument can be known to be valid by means of a deduction of its conclusion from its premises but non-deductive methods such as Venn diagrams and other graphic procedures have been proposed.

3.3 Proof

A **proof** is a deduction whose premises are known truths. A proof of the Pythagorean theorem is a deduction that might use several premises — axioms, postulates, and definitions — and contain dozens of intermediate steps. As Alfred Tarski famously emphasized in accord with Aristotle, truths can be known by proof but proofs presuppose truths not known by proof.

3.4 Comparison

Premise-conclusion arguments do not require or produce either knowledge of validity or knowledge of truth. Premise sets may be chosen arbitrarily and conclusions may be chosen arbitrarily. Deductions require knowing how to reason but they do not require knowledge of truth of their premises. Deductions produce knowledge of the validity of

arguments but ordinarily they do not produce knowledge of the truth of their conclusions. Proofs require knowledge of the truth of their premises, they require knowledge of deductive reasoning, and they produce knowledge of their conclusions.

3.5 Context

Modern logicians disagree concerning the nature of argument constituents.Quine devotes the first chapter of *Philosophy of Logic* to this issue.[2] Historians have not even been able to agree on what Aristotle took as constituents.[3]
Argument-deduction-proof distinctions are inseparable from what have been called the *consequence-deducibility* distinction and the *truth-and-consequence conception of proof*.[1] Variations among argument-deduction-proof distinctions are not all terminological.

Logician Alonzo Church[4] never used the word *argument* in the above sense and had no synonym. Moreover, Church never explained that deduction is the process of producing knowledge of consequence and it never used the common noun *deduction* for an application of the deduction process. His primary focus in discussing proof was "conviction" produced by generation of chains of logical truths—not the much more widely applicable and more familiar general process of demonstration as found in pre-Aristotelian geometry and discussed by Aristotle.[1] He did discuss deductions in the above sense but not by that name: he called them awkwardly "proofs from premises" — an expression he coined for the purpose.

The absence of argument-deduction-proof distinctions is entirely consonant with Church's avowed Platonistic logicism. Following Dummett's insightful remarks[5] about Frege, which — *mutatis mutandis* — apply even more to Church, it might be possible to explain the today-surprising absence.

3.6 References

[1] Corcoran, John (2009). "Aristotle's Demonstrative Logic". *History and Philosophy of Logic* **30**: 1–20. doi:10.1080/01445340802228362.
[2] WILLARD QUINE, *Philosophy of logic*, Harvard, 1970/1986.
[3] JOHN CORCORAN, Aristotle's syllogistic premises. Bulletin of Symbolic Logic. 18 (2012) 300–1.
[4] Church, Alonzo (1956). *Introduction to Mathematical Logic*. Princeton University Press.|isbn=9780691029061
[5] Dummett, Michael (1973). *Frege: Philosophy of Language*. Harvard University Press. pp. 432ff.

Chapter 4

Theorem

For the Italian film, see Teorema (film).

In mathematics, a **theorem** is a statement that has been proven on the basis of previously established statements, such as other theorems—and generally accepted statements, such as axioms. The proof of a mathematical theorem is a logical argument for the theorem statement given in accord with the rules of a deductive system. The proof of a theorem is often interpreted as justification of the truth of the theorem statement. In light of the requirement that theorems be proved, the concept of a theorem is fundamentally *deductive*, in contrast to the notion of a scientific theory, which is *empirical*.[2]

Many mathematical theorems are conditional statements. In this case, the proof deduces the conclusion from the hypotheses. In light of the interpretation of proof as justification of truth, the conclusion is often viewed as a necessary consequence of the hypotheses, namely, that the conclusion is true in case the hypotheses are true, without any further assumptions. However, the conditional could be interpreted differently in certain deductive systems, depending on

the meanings assigned to the derivation rules and the conditional symbol.
Although they can be written in a completely symbolic form, for example, within the propositional calculus, theorems are often expressed in a natural language such as English. The same is true of proofs, which are often expressed as logically organized and clearly worded informal arguments, intended to convince readers of the truth of the statement of the theorem beyond any doubt, and from which a formal symbolic proof can in principle be constructed. Such arguments are typically easier to check than purely symbolic ones—indeed, many mathematicians would express a preference for a proof that not only demonstrates the validity of a theorem, but also explains in some way *why* it is obviously true. In some cases, a picture alone may be sufficient to prove a theorem. Because theorems lie at the core of mathematics, they are also central to its aesthetics. Theorems are often described as being “trivial”, or “difficult”, or “deep”, or even “beautiful”. These subjective judgments vary not only from person to person, but also with time: for example, as a proof is simplified or better understood, a theorem that was once difficult may become trivial. On the other hand, a deep theorem may be simply stated, but its proof may involve surprising and subtle connections between disparate areas of mathematics. Fermat’s Last Theorem is a particularly well-known example of such a theorem.

4.1 *Informal account of theorems*

Logically, many theorems are of the form of an indicative conditional: *if A, then B*. Such a theorem does not assert *B*, only that *B* is a necessary consequence of *A*. In this case *A* is called the **hypothesis** of the theorem (note that “hypothesis” here is something very different from a conjecture) and *B* the **conclusion** (formally, *A* and *B* are termed the *antecedent* and *consequent*). The theorem “If *n* is an even natural number then *n*/2 is a natural number” is a typical example in which the hypothesis is "*n* is an even natural number” and the conclusion is "*n*/2 is also a natural number”.
To be proven, a theorem must be expressible as a precise, formal statement. Nevertheless, theorems are usually expressed in natural language rather than in a completely symbolic form, with the intention that the reader can produce a formal statement from the informal one.
It is common in mathematics to choose a number of hypotheses within a given language and declare that the theory consists of all statements provable from these hypotheses. These hypothesis form the foundational basis of the theory and are called axioms or postulates. The field of mathematics known as proof theory studies formal languages, axioms
16
4.2. PROVABILITY AND THEOREMHOOD 17
and the structure of proofs.
Some theorems are “trivial”, in the sense that they follow from definitions, axioms, and other theorems in obvious ways and do not contain any surprising insights. Some, on the other hand, may be called “deep”, because their proofs may be long and difficult, involve areas of mathematics superficially distinct from the statement of the theorem itself, or show surprising connections between disparate areas of mathematics.[3] A theorem might be simple to state and yet be deep. An excellent example is Fermat’s Last Theorem, and there are many other examples of simple yet deep theorems in number theory and combinatorics, among other areas.
Other theorems have a known proof that cannot easily be written down. The most prominent examples are the four color theorem and the Kepler conjecture. Both of these theorems are only known to be true by reducing them to a computational search that is then verified by a computer program. Initially, many mathematicians did not accept this form of proof, but it has become more widely accepted. The mathematician Doron Zeilberger has even gone so far as to claim that these are possibly the only nontrivial results that mathematicians have ever proved.[4] Many mathematical theorems can be reduced to more straightforward computation, including polynomial identities, trigonometric identities and hypergeometric identities.[5]

4.2 Provability and theoremhood

To establish a mathematical statement as a theorem, a proof is required, that is, a line of reasoning from axioms in the system (and other, already established theorems) to the given statement must be demonstrated. However, the proof is usually considered as separate from the theorem statement. Although more than one proof may be known for a single theorem, only one proof is required to establish the status of a statement as a theorem. The Pythagorean theorem and the law of quadratic reciprocity are contenders for the title of theorem with the greatest number of distinct proofs.

4.3 Relation with scientific theories

Theorems in mathematics and theories in science are fundamentally different in their epistemology. A scientific theory cannot be proven; its key attribute is that it is falsifiable, that is, it makes predictions about the natural world that are testable by experiments. Any disagreement between prediction and experiment demonstrates the incorrectness of the scientific theory, or at least limits its accuracy or domain of validity. Mathematical theorems, on the other hand, are purely abstract formal statements: the proof of a theorem cannot involve experiments or other empirical evidence in the same way such evidence is used to support scientific theories.

Nonetheless, there is some degree of empiricism and data collection involved in the discovery of mathematical theorems. By establishing a pattern, sometimes with the use of a powerful computer, mathematicians may have an idea of what to prove, and in some cases even a plan for how to set about doing the proof. For example, the Collatz conjecture has been verified for start values up to about 2.88×10^{18}. The Riemann hypothesis has been verified for the first 10 trillion zeroes of the zeta function. Neither of these statements is considered proven.

Such evidence does not constitute proof. For example, the Mertens conjecture is a statement about natural numbers that is now known to be false, but no explicit counterexample (i.e., a natural number n for which the Mertens function $M(n)$ equals or exceeds the square root of n) is known: all numbers less than 10^{14} have the Mertens property, and the smallest number that does not have this property is only known to be less than the exponential of 1.59×10^{40}, which is approximately 10 to the power 4.3×10^{39}. Since the number of particles in the universe is generally considered less than 10 to the power 100 (a googol), there is no hope to find an explicit counterexample by exhaustive search.

Note that the word "theory" also exists in mathematics, to denote a body of mathematical axioms, definitions and theorems, as in, for example, group theory. There are also "theorems" in science, particularly physics, and in engineering, but they often have statements and proofs in which physical assumptions and intuition play an important role; the physical axioms on which such "theorems" are based are themselves falsifiable.

4.4 Terminology

A number of different terms for mathematical statements exist, these terms indicate the role statements play in a particular subject. The distinction between different terms is sometimes rather arbitrary and the usage of some terms has evolved over time.

- An **axiom** or **postulate** is a statement that is accepted without proof and regarded as fundamental to a subject. Historically these have been regarded as "self-evident", but more recently they are considered assumptions that characterize the subject of study. In classical geometry, axioms are general statements while postulates are statements about geometrical objects.[6] A definition is also accepted without proof since it simply gives the meaning of a word or phrase in terms of known concepts.
- A **proposition** is a generic term for a theorem of no particular importance. This term sometimes connotes a statement with a simple proof, while the term **theorem** is usually reserved for the most important results or those with long or difficult proofs. In classical geometry, a proposition may be a construction that satisfies given requirements; for example, Proposition 1 in Book I of Euclid's elements is the construction of an equilateral triangle.[7]
- A **lemma** is a "helping theorem", a proposition with little applicability except that it forms part of the proof of a larger theorem. In some cases, as the relative importance of different theorems becomes more clear, what was once considered a lemma is now considered a theorem, though the word "lemma" remains in the name. Examples include Gauss's lemma, Zorn's lemma, and the Fundamental lemma.
- A **corollary** is a proposition that follows with little or no proof from one other theorem or definition.[8]
- A **converse** of a theorem is a statement formed by interchanging what is given in a theorem and what is to be proved. For example, the isosceles triangle theorem states that if two sides of a triangle are equal then two angles are equal. In the converse, the given (that two sides are equal) and what is to be proved (that two angles are equal) are swapped, so the converse is the statement that if two angles of a triangle are equal then two sides are equal. In this example, the converse can be proven as another theorem, but this is often not the case. For example, the converse to the theorem that two right angles are equal angles is the statement that two equal angles must be right angles, and this is clearly not always the case.[9]

There are other terms, less commonly used, that are conventionally attached to proven statements, so that certain theorems are referred to by historical or customary names. For examples:

- **Identity**, used for theorems that state an equality between two mathematical expressions. Examples include Euler's formula and Vandermonde's identity.
- **Rule**, used for certain theorems such as Bayes' rule and Cramer's rule, that establish useful formulas.
- **Law**. Examples include the law of large numbers, the law of cosines, and Kolmogorov's zero-one law.[10]
- **Principle**. Examples include Harnack's principle, the least upper bound principle, and the pigeonhole principle.

A few well-known theorems have even more idiosyncratic names. The **division algorithm** (see Euclidean division) is a theorem expressing the outcome of division in the natural numbers and more general rings. The **Bézout's identity** is a theorem asserting that the greatest common divisor of two numbers may be written as a linear combination of these numbers. The **Banach–Tarski paradox** is a theorem in measure theory that is paradoxical in the sense that it contradicts common intuitions about volume in three-dimensional space.

An unproven statement that is believed true is called a **conjecture** (or sometimes a **hypothesis**, but with a different meaning from the one discussed above). To be considered a conjecture, a statement must usually be proposed publicly, at which point the name of the proponent may be attached to the conjecture, as with Goldbach's conjecture. Other famous conjectures include the Collatz conjecture and the Riemann hypothesis. On the other hand, Fermat's last theorem has always been known by that name, even before it was proven; it was never known as "Fermat's conjecture".

4.5 Layout

A theorem and its proof are typically laid out as follows:
Theorem (name of person who proved it and year of discovery, proof or publication).
Statement of theorem (sometimes called the proposition*).*
Proof.
Description of proof.
End mark.
4.6. LORE 19
The end of the proof may be signalled by the letters Q.E.D. (*quod erat demonstrandum*) or by one of the tombstone marks "□" or "■" meaning "End of Proof", introduced by Paul Halmos following their usage in magazine articles.
The exact style depends on the author or publication. Many publications provide instructions or macros for typesetting in the house style.
It is common for a theorem to be preceded by definitions describing the exact meaning of the terms used in the theorem. It is also common for a theorem to be preceded by a number of propositions or lemmas which are then used in the proof. However, lemmas are sometimes embedded in the proof of a theorem, either with nested proofs, or with their proofs presented after the proof of the theorem.
Corollaries to a theorem are either presented between the theorem and the proof, or directly after the proof. Sometimes, corollaries have proofs of their own that explain why they follow from the theorem.

4.6 Lore

It has been estimated that over a quarter of a million theorems are proved every year.[11]
The well-known aphorism, "A mathematician is a device for turning coffee into theorems", is probably due to Alfréd Rényi, although it is often attributed to Rényi's colleague Paul Erdős (and Rényi may have been thinking of Erdős), who was famous for the many theorems he produced, the number of his collaborations, and his coffee drinking.[12]
The classification of finite simple groups is regarded by some to be the longest proof of a theorem. It comprises tens of thousands of pages in 500 journal articles by some 100 authors. These papers are together believed to give a complete proof, and several ongoing projects hope to shorten and simplify this proof.[13] Another theorem of this type is the Four color theorem whose computer generated proof is too long for a human to read. It is certainly the longest known proof of a theorem whose statement can be easily understood by a layman.

4.7 Theorems in logic

Logic, especially in the field of proof theory, considers theorems as statements (called **formulas** or **well formed formulas**) of a formal language. The statements of the language are strings of symbols and may be broadly divided into nonsense and well-formed formulas. A set of **deduction rules**, also called **transformation rules** or rules of inference, must be provided. These deduction rules tell exactly when a formula can be derived from a set of premises.
The set of well-formed formulas may be broadly divided into theorems and non-theorems. However, according to Hofstadter, a formal system often simply defines all its well-formed formula as theorems.[14]
Different sets of derivation rules give rise to different interpretations of what it means for an expression to be a theorem. Some derivation rules and formal languages are intended to capture mathematical reasoning; the most common examples use first-order logic. Other deductive systems describe term rewriting, such as the reduction rules for λ calculus.
The definition of theorems as elements of a formal language allows for results in proof theory that study the structure of formal proofs and the structure of provable formulas. The most famous result is Gödel's incompleteness theorem; by representing theorems about basic number theory as expressions in a formal language, and then representing this language within number theory itself, Gödel constructed examples of statements that are neither provable nor disprovable from axiomatizations of number theory.
A theorem may be expressed in a formal language (or "formalized"). A formal theorem is the purely formal analogue of a theorem. In general, a formal theorem is a type of well-formed formula that satisfies certain logical and syntactic conditions. The notation S is often used to indicate that S is a theorem.
Formal theorems consist of formulas of a formal language and the transformation rules of a formal system. Specifically, a formal theorem is always the last formula of a derivation in some formal system each formula of which is a logical consequence of the formulas that came before it in the derivation. The initially accepted formulas in the derivation are called its **axioms**, and are the basis on which the theorem is derived. A set of theorems is called a **theory**.
What makes formal theorems useful and of interest is that they can be interpreted as true propositions and their derivations may be interpreted as a proof of the truth of the resulting expression. A set of formal theorems may be
20 *CHAPTER 4. THEOREM*
referred to as a **formal theory**. A theorem whose interpretation is a true statement about a formal system is called a

metatheorem.

4.7.1 Syntax and semantics

Main articles: Syntax (logic) and Formal semantics (logic)
The concept of a formal theorem is fundamentally syntactic, in contrast to the notion of a *true proposition,* which introduces semantics. Different deductive systems can yield other interpretations, depending on the presumptions of the derivation rules (i.e. belief, justification or other modalities). The soundness of a formal system depends on whether or not all of its theorems are also validities. A validity is a formula that is true under any possible interpretation, e.g. in classical propositional logic validities are tautologies. A formal system is considered semantically complete when all of its tautologies are also theorems.

4.7.2 Derivation of a theorem

Main article: Formal proof
The notion of a theorem is very closely connected to its formal proof (also called a "derivation"). To illustrate how derivations are done, we will work in a very simplified formal system. Let us call ours *FS* Its alphabet consists only of two symbols { **A**, **B** } and its formation rule for formulas is:
FS
The single axiom of *FS* is:
ABBA.
The only rule of inference (transformation rule) for *FS* is:
Any occurrence of "**A**" in a theorem may be replaced by an occurrence of the string "**AB**" and the result is a theorem.
Theorems in *FS* are defined as those formulae that have a derivation ending with that formula. For example
1. **ABBA** (Given as axiom)
2. **ABBBA** (by applying the transformation rule)
3. **ABBBAB** (by applying the transformation rule)
is a derivation. Therefore "**ABBBAB**" is a theorem of *FS :* The notion of truth (or falsity) cannot be applied to the formula "**ABBBAB**" until an interpretation is given to its symbols. Thus in this example, the formula does not yet represent a proposition, but is merely an empty abstraction.
Two metatheorems of *FS* are:
Every theorem begins with "**A**".
Every theorem has exactly two "**A**"s.

4.7.3 Interpretation of a formal theorem

Main article: Interpretation (logic)
4.8. SEE ALSO 21

4.7.4 Theorems and theories

Main articles: Theory and Theory (mathematical logic)

4.8 See also

_ Inference
_ List of theorems
_ Toy theorem
_ Metamath – a language for developing strictly formalized mathematical definitions and proofs accompanied by a proof checker for this language and a growing database of thousands of proved theorems.

4.9 Notes

[1] For full text of 2nd edition of 1940, see Elisha Scott Loomis. "The Pythagorean proposition: its demonstrations analyzed and classified, and bibliography of sources for data of the four kinds of proofs". *Education Resources Information Center*. Institute of Education Sciences (IES) of the U.S. Department of Education. Retrieved 2010-09-26. Originally published in 1940 and reprinted in 1968 by National Council of Teachers of Mathematics.
[2] However, both theorems and theories are investigations. See Heath 1897 Introduction, The terminology of Archimedes, p. clxxxii:"theorem (θεὼρνμα) from θεωρεῖν to investigate"
[3] Weisstein, Eric W., "Deep Theorem", *MathWorld.*
[4] Doron Zeilberger. "Opinion 51".
[5] Petkovsek et al. 1996.
[6] Wentworth, G.; Smith, D.E. (1913). "Art. 46, 47". *Plane Geometry*. Ginn & Co.
[7] Wentworth & Smith Art. 50

[8] Wentworth & Smith Art. 51
[9] Follows Wentworth & Smith Art. 79
[10] The word *law* can also refer to an axiom, a rule of inference, or, in probability theory, a probability distribution.
[11] Hoffman 1998, p. 204.
[12] Hoffman 1998, p. 7.
[13] An enormous theorem: the classification of finite simple groups, Richard Elwes, Plus Magazine, Issue 41 December 2006.
[14] Hofstadter 1980

4.10 References

_ Heath, Sir Thomas Little (1897). *The works of Archimedes*. Dover. Retrieved 2009-11-15.
_ Hoffman, P. (1998). *The Man Who Loved Only Numbers: The Story of Paul Erdős and the Search for Mathematical Truth*. Hyperion, New York. ISBN 1-85702-829-5.
_ Hofstadter, Douglas (1979). *Gödel, Escher, Bach: An Eternal Golden Braid*. Basic Books.
_ Hunter, Geofrfrey (1996) [1973]. *Metalogic: An Introduction to the Metatheory of Standard First Order Logic*. University of California Press. ISBN 0-520-02356-0.
_ Mates, Benson (1972). *Elementary Logic*. Oxford University Press. ISBN 0-19-501491-X.
_ Petkovsek, Marko; Wilf, Herbert; Zeilberger, Doron (1996). *A = B*. A.K. Peters, Wellesley, Massachusetts. ISBN 1-56881-063-6.

4.11 External links

_ Weisstein, Eric W., "Theorem", *MathWorld*.
_ Theorem of the Day

A planar map with five colors such that no two regions with the same color meet. It can actually be colored in this way with only four colors. The four color theorem states that such colorings are possible for any planar map, but every known proof involves a computational search that is too long to check by hand.

The Collatz conjecture: one way to illustrate its complexity is to extend the iteration from the natural numbers to the complex numbers. The result is a fractal, which (in accordance with universality) resembles the Mandelbrot set.

Symbols and strings of symbols Well-formed formulas Theorems

This diagram shows the syntactic entities that can be constructed from formal languages. The symbols and strings of symbols may be broadly divided into nonsense and well-formed formulas. A formal language can be thought of as identical to the set of its well-formed formulas. The set of well-formed formulas may be broadly divided into theorems and non-theorems.

Chapter 5

Axiom

This article is about logical propositions. For other uses, see Axiom (disambiguation).
"Postulation" redirects here. For the term in algebraic geometry, see Postulation (algebraic geometry).
An **axiom** or **postulate** is a premise or starting point of reasoning. As classically conceived, an axiom is a premise so evident as to be accepted as true without controversy.[1] The word comes from the Greek *axíōma* (ἀξίωμα) 'that which is thought worthy or fit' or 'that which commends itself as evident.'[2][3] As used in modern logic, an axiom is simply a premise or starting point for reasoning.[4] Axioms define and delimit the realm of analysis; the relative truth of an axiom is taken for granted within the particular domain of analysis, and serves as a starting point for deducing and inferring other relative truths. No explicit view regarding the absolute truth of axioms is ever taken in the context of modern mathematics, as such a thing is considered to be an irrelevant and impossible contradiction in terms.
In mathematics, the term *axiom* is used in two related but distinguishable senses: "logical axioms" and "non-logical axioms". Logical axioms are usually statements that are taken to be true within the system of logic they define (e.g., (A and B) implies A), while non-logical axioms (e.g., $a + b = b + a$) are actually defining properties for the domain of a specific mathematical theory (such as arithmetic). When used in the latter sense, "axiom," "postulate", and "assumption" may be used interchangeably. In general, a non-logical axiom is not a self-evident truth, but rather a formal logical expression used in deduction to build a mathematical theory. As modern mathematics admits multiple, equally "true" systems of logic, precisely the same thing must be said for logical axioms - they both define and are

specific to the particular system of logic that is being invoked. To axiomatize a system of knowledge is to show that its claims can be derived from a small, well-understood set of sentences (the axioms). There are typically multiple ways to axiomatize a given mathematical domain.
In both senses, an axiom is any mathematical statement that serves as a starting point from which other statements are logically derived. Within the system they define, axioms (unless redundant) cannot be derived by principles of deduction, nor are they demonstrable by mathematical proofs, simply because they are starting points; there is nothing else from which they logically follow otherwise they would be classified as theorems. However, an axiom in one system may be a theorem in another, and vice versa.

5.1 Etymology

The word "axiom" comes from the Greek word ἀξίωμα (*axioma*), a verbal noun from the verb ἀξιόειν (*axioein*), meaning "to deem worthy", but also "to require", which in turn comes from ἄξιος (*axios*), meaning "being in balance", and hence "having (the same) value (as)", "worthy", "proper". Among the ancient Greek philosophers an axiom was a claim which could be seen to be true without any need for proof.
The root meaning of the word 'postulate' is to 'demand'; for instance, Euclid demands of us that we agree that some things can be done, e.g. any two points can be joined by a straight line, etc.[5]
Ancient geometers maintained some distinction between axioms and postulates. While commenting Euclid's books Proclus remarks that "Geminus held that this [4th] Postulate should not be classed as a postulate but as an axiom, since it does not, like the first three Postulates, assert the possibility of some construction but expresses an essential property".[6] Boethius translated 'postulate' as *petitio* and called the axioms *notiones communes* but in later manuscripts this usage was not always strictly kept.
27
28 *CHAPTER 5. AXIOM*

5.2 Historical development

5.2.1 Early Greeks

The logico-deductive method whereby conclusions (new knowledge) follow from premises (old knowledge) through the application of sound arguments (syllogisms, rules of inference), was developed by the ancient Greeks, and has become the core principle of modern mathematics. Tautologies excluded, nothing can be deduced if nothing is assumed. Axioms and postulates are the basic assumptions underlying a given body of deductive knowledge. They are accepted without demonstration. All other assertions (theorems, if we are talking about mathematics) must be proven with the aid of these basic assumptions. However, the interpretation of mathematical knowledge has changed from ancient times to the modern, and consequently the terms *axiom* and *postulate* hold a slightly different meaning for the present day mathematician, than they did for Aristotle and Euclid.
The ancient Greeks considered geometry as just one of several sciences, and held the theorems of geometry on par with scientific facts. As such, they developed and used the logico-deductive method as a means of avoiding error, and for structuring and communicating knowledge. Aristotle's posterior analytics is a definitive exposition of the classical view.
An "axiom", in classical terminology, referred to a self-evident assumption common to many branches of science. A good example would be the assertion that
When an equal amount is taken from equals, an equal amount results.
At the foundation of the various sciences lay certain additional hypotheses which were accepted without proof. Such a hypothesis was termed a *postulate.* While the axioms were common to many sciences, the postulates of each particular science were different. Their validity had to be established by means of real-world experience. Indeed, Aristotle warns that the content of a science cannot be successfully communicated, if the learner is in doubt about the truth of the postulates.[7]
The classical approach is well-illustrated by Euclid's Elements, where a list of postulates is given (common-sensical geometric facts drawn from our experience), followed by a list of "common notions" (very basic, self-evident assertions).
Postulates
1. It is possible to draw a straight line from any point to any other point.
2. It is possible to extend a line segment continuously in both directions.
3. It is possible to describe a circle with any center and any radius.
4. It is true that all right angles are equal to one another.
5. ("Parallel postulate") It is true that, if a straight line falling on two straight lines make the interior angles on the same side less than two right angles, the two straight lines, if produced indefinitely, intersect on that side on which are the angles less than the two right angles.

Common notions

1. Things which are equal to the same thing are also equal to one another.
2. If equals are added to equals, the wholes are equal.
3. If equals are subtracted from equals, the remainders are equal.
4. Things which coincide with one another are equal to one another.
5. The whole is greater than the part.

5.2.2 Modern development

A lesson learned by mathematics in the last 150 years is that it is useful to strip the meaning away from the mathematical assertions (axioms, postulates, propositions, theorems) and definitions. One must concede the need for primitive

notions, or undefined terms or concepts, in any study. Such abstraction or formalization makes mathematical knowledge more general, capable of multiple different meanings, and therefore useful in multiple contexts. Alessandro Padoa, Mario Pieri, and Giuseppe Peano were pioneers in this movement.

Structuralist mathematics goes further, and develops theories and axioms (e.g. field theory, group theory, topology, vector spaces) without *any* particular application in mind. The distinction between an "axiom" and a "postulate" disappears. The postulates of Euclid are profitably motivated by saying that they lead to a great wealth of geometric facts. The truth of these complicated facts rests on the acceptance of the basic hypotheses. However, by throwing out Euclid's fifth postulate we get theories that have meaning in wider contexts, hyperbolic geometry for example. We must simply be prepared to use labels like "line" and "parallel" with greater flexibility. The development of hyperbolic geometry taught mathematicians that postulates should be regarded as purely formal statements, and not as facts based on experience.

When mathematicians employ the field axioms, the intentions are even more abstract. The propositions of field theory do not concern any one particular application; the mathematician now works in complete abstraction. There are many examples of fields; field theory gives correct knowledge about them all.

It is not correct to say that the axioms of field theory are "propositions that are regarded as true without proof." Rather, the field axioms are a set of constraints. If any given system of addition and multiplication satisfies these constraints, then one is in a position to instantly know a great deal of extra information about this system.

Modern mathematics formalizes its foundations to such an extent that mathematical theories can be regarded as mathematical objects, and mathematics itself can be regarded as a branch of logic. Frege, Russell, Poincaré, Hilbert, and Gödel are some of the key figures in this development.

In the modern understanding, a set of axioms is any collection of formally stated assertions from which other formally stated assertions follow by the application of certain well-defined rules. In this view, logic becomes just another formal system. A set of axioms should be consistent; it should be impossible to derive a contradiction from the axiom. A set of axioms should also be non-redundant; an assertion that can be deduced from other axioms need not be regarded as an axiom.

It was the early hope of modern logicians that various branches of mathematics, perhaps all of mathematics, could be derived from a consistent collection of basic axioms. An early success of the formalist program was Hilbert's formalization of Euclidean geometry, and the related demonstration of the consistency of those axioms.

In a wider context, there was an attempt to base all of mathematics on Cantor's set theory. Here the emergence of Russell's paradox, and similar antinomies of naïve set theory raised the possibility that any such system could turn out to be inconsistent.

The formalist project suffered a decisive setback, when in 1931 Gödel showed that it is possible, for any sufficiently large set of axioms (Peano's axioms, for example) to construct a statement whose truth is independent of that set of axioms. As a corollary, Gödel proved that the consistency of a theory like Peano arithmetic is an unprovable assertion within the scope of that theory.

It is reasonable to believe in the consistency of Peano arithmetic because it is satisfied by the system of natural numbers, an infinite but intuitively accessible formal system. However, at present, there is no known way of demonstrating the consistency of the modern Zermelo–Fraenkel axioms for set theory. Furthermore, using techniques of forcing (Cohen) one can show that the continuum hypothesis (Cantor) is independent of the Zermelo–Fraenkel axioms. Thus, even this very general set of axioms cannot be regarded as the definitive foundation for mathematics.

5.2.3 Other sciences

Axioms play a key role not only in mathematics, but also in other sciences, notably in theoretical physics. In particular, the monumental work of Isaac Newton is essentially based on Euclid's axioms, augmented by a postulate on the nonrelation of spacetime and the physics taking place in it at any moment.

In 1905, Newton's axioms were replaced by those of Albert Einstein's special relativity, and later on by those of general relativity.

Another paper of Albert Einstein and coworkers (see EPR paradox), almost immediately contradicted by Niels Bohr, concerned the interpretation of quantum mechanics. This was in 1935. According to Bohr, this new theory should be

probabilistic, whereas according to Einstein it should be deterministic. Notably, the underlying quantum mechanical theory, i.e. the set of "theorems" derived by it, seemed to be identical. Einstein even assumed that it would be sufficient to add to quantum mechanics "hidden variables" to enforce determinism. However, thirty years later, in
30 *CHAPTER 5. AXIOM*
1964, John Bell found a theorem, involving complicated optical correlations (see Bell inequalities), which yielded measurably different results using Einstein's axioms compared to using Bohr's axioms. And it took roughly another twenty years until an experiment of Alain Aspect got results in favour of Bohr's axioms, not Einstein's. (Bohr's axioms are simply: The theory should be probabilistic in the sense of the Copenhagen interpretation.)
As a consequence, it is not necessary to explicitly cite Einstein's axioms, the more so since they concern subtle points on the "reality" and "locality" of experiments.
Regardless, the role of axioms in mathematics and in the above-mentioned sciences is different. In mathematics one neither "proves" nor "disproves" an axiom for a set of theorems; the point is simply that in the conceptual realm identified by the axioms, the theorems logically follow. In contrast, in physics a comparison with experiments always makes sense, since a falsified physical theory needs modification.

5.3 Mathematical logic

In the field of mathematical logic, a clear distinction is made between two notions of axioms: *logical* and *non-logical* (somewhat similar to the ancient distinction between "axioms" and "postulates" respectively).

5.3.1 Logical axioms

These are certain formulas in a formal language that are universally valid, that is, formulas that are satisfied by every assignment of values. Usually one takes as logical axioms *at least* some minimal set of tautologies that is sufficient for proving all tautologies in the language; in the case of predicate logic more logical axioms than that are required, in order to prove logical truths that are not tautologies in the strict sense.
Examples
Propositional logic In propositional logic it is common to take as logical axioms all formulae of the following forms, where ϕ , _ , and can be any formulae of the language and where the included primitive connectives are only " *:* " for negation of the immediately following proposition and " *!* " for implication from antecedent to consequent propositions:
1. ϕ *!* (*!* ϕ)
2. (ϕ *!* (*!* _)) *!* ((ϕ *!*) *!* (ϕ *!* _))
3. (*:*ϕ *!* *:*) *!* (*!* ϕ)*:*
Each of these patterns is an *axiom schema*, a rule for generating an infinite number of axioms. For example, if *A* , *B* , and *C* are propositional variables, then *A ! (B ! A)* and *(A ! :B) ! (C ! (A ! :B))* are both instances of axiom schema 1, and hence are axioms. It can be shown that with only these three axiom schemata and *modus ponens*, one can prove all tautologies of the propositional calculus. It can also be shown that no pair of these schemata is sufficient for proving all tautologies with *modus ponens*.
Other axiom schemas involving the same or different sets of primitive connectives can be alternatively constructed.[8]
These axiom schemata are also used in the predicate calculus, but additional logical axioms are needed to include a quantifier in the calculus.[9]
First-order logic Axiom of Equality. Let L be a first-order language. For each variable *x* , the formula
$x = x$
is universally valid.
This means that, for any variable symbol *x ;* the formula $x = x$ can be regarded as an axiom. Also, in this example, for this not to fall into vagueness and a never-ending series of "primitive notions", either a precise notion of what we mean by $x = x$ (or, for that matter, "to be equal") has to be well established first, or a purely formal and syntactical

5.3. MATHEMATICAL LOGIC 31

usage of the symbol = has to be enforced, only regarding it as a string and only a string of symbols, and mathematical logic does indeed do that.
Another, more interesting example axiom scheme, is that which provides us with what is known as **Universal Instantiation**:
Axiom scheme for Universal Instantiation. Given a formula ϕ in a first-order language L , a variable *x* and a term *t* that is substitutable for *x* in ϕ , the formula
8x ϕ *!* ϕ*x*
t
is universally valid.
Where the symbol ϕ*x*
t stands for the formula ϕ with the term *t* substituted for *x* . (See Substitution of variables.) In informal terms, this example allows us to state that, if we know that a certain property *P* holds for every *x* and that *t* stands for a particular object in our structure, then we should be able to claim *P*(*t*) . Again, *we are claiming that*

the formula 8xϕ ! ϕx
t is valid, that is, we must be able to give a "proof" of this fact, or more properly speaking, a *metaproof*. Actually, these examples are *metatheorems* of our theory of mathematical logic since we are dealing with the very concept of *proof* itself. Aside from this, we can also have **Existential Generalization**:
Axiom scheme for Existential Generalization. Given a formula ϕ in a first-order language L , a variable x and a term t that is substitutable for x in ϕ , the formula
ϕx
t
! 9x ϕ
is universally valid.

5.3.2 Non-logical axioms

Non-logical axioms are formulas that play the role of theory-specific assumptions. Reasoning about two different structures, for example the natural numbers and the integers, may involve the same logical axioms; the non-logical axioms aim to capture what is special about a particular structure (or set of structures, such as groups). Thus nonlogical axioms, unlike logical axioms, are not *tautologies*. Another name for a non-logical axiom is *postulate*.[10]
Almost every modern mathematical theory starts from a given set of non-logical axioms, and it was thought that in principle every theory could be axiomatized in this way and formalized down to the bare language of logical formulas.
Non-logical axioms are often simply referred to as *axioms* in mathematical discourse. This does not mean that it is claimed that they are true in some absolute sense. For example, in some groups, the group operation is commutative, and this can be asserted with the introduction of an additional axiom, but without this axiom we can do quite well developing (the more general) group theory, and we can even take its negation as an axiom for the study of noncommutative groups.
Thus, an *axiom* is an elementary basis for a formal logic system that together with the rules of inference define a **deductive system**.

Examples
This section gives examples of mathematical theories that are developed entirely from a set of non-logical axioms (axioms, henceforth). A rigorous treatment of any of these topics begins with a specification of these axioms.
Basic theories, such as arithmetic, real analysis and complex analysis are often introduced non-axiomatically, but implicitly or explicitly there is generally an assumption that the axioms being used are the axioms of Zermelo–Fraenkel set theory with choice, abbreviated ZFC, or some very similar system of axiomatic set theory like Von Neumann–Bernays–Gödel set theory, a conservative extension of ZFC. Sometimes slightly stronger theories such as Morse-Kelley set theory or set theory with a strongly inaccessible cardinal allowing the use of a Grothendieck universe are used, but in fact most mathematicians can actually prove all they need in systems weaker than ZFC, such as second-order arithmetic.
The study of topology in mathematics extends all over through point set topology, algebraic topology, differential topology, and all the related paraphernalia, such as homology theory, homotopy theory. The development of *abstract algebra* brought with itself group theory, rings, fields, and Galois theory.
32 *CHAPTER 5. AXIOM*
This list could be expanded to include most fields of mathematics, including measure theory, ergodic theory, probability, representation theory, and differential geometry.
Combinatorics is an example of a field of mathematics which does not, in general, follow the axiomatic method.

Arithmetic The Peano axioms are the most widely used *axiomatization* of first-order arithmetic. They are a set of axioms strong enough to prove many important facts about number theory and they allowed Gödel to establish his famous second incompleteness theorem.[11]
We have a language L*NT* = *f*0; *Sg* where 0 is a constant symbol and *S* is a unary function and the following axioms:
1. *8x::(Sx* = 0)
2. *8x:8y:(Sx* = *Sy ! x* = *y*)
3. ((*ϕ*(0) ^ *8x:* (*ϕ*(*x*) *! ϕ*(*Sx*))) *! 8x:ϕ*(*x*) for any L*NT* formula *ϕ* with one free variable.
The standard structure is N = *h*N*;* 0*; Si* where N is the set of natural numbers, *S* is the successor function and 0 is naturally interpreted as the number 0.

Euclidean geometry Probably the oldest, and most famous, list of axioms are the 4 + 1 Euclid's postulates of plane geometry. The axioms are referred to as "4 + 1" because for nearly two millennia the fifth (parallel) postulate ("through a point outside a line there is exactly one parallel") was suspected of being derivable from the first four. Ultimately, the fifth postulate was found to be independent of the first four. Indeed, one can assume that exactly one parallel through a point outside a line exists, or that infinitely many exist. This choice gives us two alternative forms of geometry in which the interior angles of a triangle add up to exactly 180 degrees or less, respectively, and are known as Euclidean and hyperbolic geometries. If one also removes the second postulate ("a line can be extended indefinitely") then elliptic geometry arises, where there is no parallel through a point outside a line, and in which the interior angles of a triangle add up to more than 180 degrees.

Real analysis The object of study is the real numbers. The real numbers are uniquely picked out (up to isomorphism) by the properties of a *Dedekind complete ordered field*, meaning that any nonempty set of real numbers with an upper bound has a least upper bound. However, expressing these properties as axioms requires use of second-order logic. The Löwenheim-Skolem theorems tell us that if we restrict ourselves to first-order logic, any axiom system for the reals admits other models, including both models that are smaller than the reals and models that are larger. Some of the latter are studied in non-standard analysis.

5.3.3 Role in mathematical logic

Deductive systems and completeness

A **deductive system** consists of a set _ of logical axioms, a set _ of non-logical axioms, and a set *f*(□; ϕ)*g* of *rules of inference*. A desirable property of a deductive system is that it be **complete**. A system is said to be complete if, for all formulas ϕ ,

if _ *j*= ϕ then _ ⊢ ϕ

that is, for any statement that is a *logical consequence* of _ there actually exists a *deduction* of the statement from _ . This is sometimes expressed as "everything that is true is provable", but it must be understood that "true" here means "made true by the set of axioms", and not, for example, "true in the intended interpretation". Gödel's completeness theorem establishes the completeness of a certain commonly used type of deductive system.

Note that "completeness" has a different meaning here than it does in the context of Gödel's first incompleteness theorem, which states that no *recursive*, *consistent* set of non-logical axioms_ of the Theory of Arithmetic is *complete*, in the sense that there will always exist an arithmetic statement ϕ such that neither ϕ nor :ϕ can be proved from the given set of axioms.

There is thus, on the one hand, the notion of *completeness of a deductive system* and on the other hand that of *completeness of a set of non-logical axioms*. The completeness theorem and the incompleteness theorem, despite their names, do not contradict one another.

5.4. SEE ALSO 33

5.3.4 Further discussion

Early mathematicians regarded axiomatic geometry as a model of physical space, and obviously there could only be one such model. The idea that alternative mathematical systems might exist was very troubling to mathematicians of the 19th century and the developers of systems such as Boolean algebra made elaborate efforts to derive them from traditional arithmetic. Galois showed just before his untimely death that these efforts were largely wasted. Ultimately, the abstract parallels between algebraic systems were seen to be more important than the details and modern algebra was born. In the modern view axioms may be any set of formulas, as long as they are not known to be inconsistent.

5.4 See also

_ Axiomatic system
_ Dogma
_ List of axioms
_ Model theory
_ Regulæ Juris
_ Theorem

5.5 References

[1] "A proposition that commends itself to general acceptance; a well-established or universally conceded principle; a maxim, rule, law" axiom, n., definition 1a. *Oxford English Dictionary* Online, accessed 2012-04-28. Cf. Aristotle, *Posterior Analytics* I.2.72a18-b4.

[2] Cf. axiom, n., etymology. *Oxford English Dictionary*, accessed 2012-04-28.

[3] Oxford American College Dictionary: "n. a statement or proposition that is regarded as being established, accepted, or self-evidently true. ORIGIN: late 15th cent.: ultimately from Greek axiōma 'what is thought fitting,' from axios 'worthy.' http://www.highbeam.com/doc/1O997-axiom.html (subscription required)

[4] "A proposition (whether true or false)" axiom, n., definition 2. *Oxford English Dictionary* Online, accessed 2012-04-28.

[5] Wolff, P. Breakthroughs in Mathematics, 1963, New York: New American Library, pp 47–8

[6] Heath, T. 1956. The Thirteen Books of Euclid's Elements. New York: Dover. *p200*

[7] Aristotle, Metaphysics Bk IV, Chapter 3, 1005b "Physics also is a kind of Wisdom, but it is not the first kind. – And the attempts of some of those who discuss the terms on which truth should be accepted, are due to want of training in logic; for they should know these things already when they come to a special study, and not be inquiring into them while they are listening to lectures on it." W.D. Ross translation, in The Basic Works of Aristotle, ed. Richard McKeon, (Random House, New York, 1941)|date=June 2011
[8] Mendelson, "6. Other Axiomatizations" of Ch. 1
[9] Mendelson, "3. First-Order Theories" of Ch. 2
[10] Mendelson, "3. First-Order Theories: Proper Axioms" of Ch. 2
[11] Mendelson, "5. The Fixed Point Theorem. Gödel's Incompleteness Theorem" of Ch. 2

5.6 Further reading

_ Mendelson, Elliot (1987). *Introduction to mathematical logic.* Belmont, California: Wadsworth & Brooks. ISBN 0-534-06624-0

5.7 External links

_ Axiom at PhilPapers
_ Axiom at PlanetMath.org.
_ *Metamath* axioms page

Chapter 6

Deductive reasoning

Deductive reasoning, also **deductive logic** or **logical deduction** or, informally, **"top-down" logic**,[1] is the process of reasoning from one or more statements (premises) to reach a logically certain conclusion.[2]
Deductive reasoning links premises with conclusions. If all premises are true, the terms are clear, and the rules of deductive logic are followed, then the conclusion reached is necessarily true.
Deductive reasoning (top-down logic) contrasts with inductive reasoning (bottom-up logic) in the following way: In deductive reasoning, a conclusion is reached reductively by applying general rules that hold over the entirety of a closed domain of discourse, narrowing the range under consideration until only the conclusion is left. In inductive reasoning, the conclusion is reached by generalizing or extrapolating from initial information. As a result, induction can be used even in an open domain, one where there is epistemic uncertainty. Note, however, that the inductive reasoning mentioned here is not the same as induction used in mathematical proofs – mathematical induction is actually a form of deductive reasoning.

6.1 Simple example

An example of a deductive argument:
1. All men are mortal.
2. Socrates is a man.
3. Therefore, Socrates is mortal.
The first premise states that all objects classified as "men" have the attribute "mortal". The second premise states that "Socrates" is classified as a "man" – a member of the set "men". The conclusion then states that "Socrates" must be "mortal" because he inherits this attribute from his classification as a "man".

6.2 Law of detachment

Main article: Modus ponens
The law of detachment (also known as **affirming the antecedent** and **Modus ponens**) is the first form of deductive reasoning. A single conditional statement is made, and a hypothesis (P) is stated. The conclusion (Q) is then deduced from the statement and the hypothesis. The most basic form is listed below:
1. $P \rightarrow Q$ (conditional statement)
2. P (hypothesis stated)
3. Q (conclusion deduced)

In deductive reasoning, we can conclude Q from P by using the law of detachment.[3] However, if the conclusion (Q) is given instead of the hypothesis (P) then there is no definitive conclusion.
The following is an example of an argument using the law of detachment in the form of an if-then statement:
1. If an angle satisfies $90° < A < 180°$, then A is an obtuse angle.

2. A = 120°.
3. A is an obtuse angle.
Since the measurement of angle A is greater than 90° and less than 180°, we can deduce that A is an obtuse angle.

6.3 Law of syllogism

The law of syllogism takes two conditional statements and forms a conclusion by combining the hypothesis of one statement with the conclusion of another. Here is the general form:
1. $P \rightarrow Q$
2. $Q \rightarrow R$
3. Therefore, $P \rightarrow R$.
The following is an example:
1. If Larry is sick, then he will be absent.
2. If Larry is absent, then he will miss his classwork.
3. Therefore, if Larry is sick, then he will miss his classwork.
We deduced the final statement by combining the hypothesis of the first statement with the conclusion of the second statement. We also allow that this could be a false statement. This is an example of the Transitive Property in mathematics. The Transitive Property is sometimes phrased in this form:
1. A = B.
2. B = C.
3. Therefore A = C.

6.4 Law of contrapositive

Main article: Modus tollens
The law of contrapositive states that, in a conditional, if the conclusion is false, then the hypothesis must be false also. The general form is the following:
1. $P \rightarrow Q$.
2. $\sim Q$.
3. Therefore we can conclude $\sim P$.
The following are examples:
1. If it is raining, then there are clouds in the sky.
2. There are no clouds in the sky.
3. Thus, it is not raining.
6.5. VALIDITY AND SOUNDNESS 37

6.5 Validity and soundness

Deductive arguments are evaluated in terms of their *validity* and *soundness*.
An argument is valid if it is impossible for its premises to be true while its conclusion is false. In other words, the conclusion must be true if the premises are true. An argument can be valid even though the premises are false.
An argument is sound if it is valid and the premises are true.
It is possible to have a deductive argument that is logically valid but is not sound. Fallacious arguments often take that form.
The following is an example of an argument that is valid, but not sound:
1. Everyone who eats carrots is a quarterback.
2. John eats carrots.
3. Therefore, John is a quarterback.
The example's first premise is false – there are people who eat carrots and are not quarterbacks – but the conclusion must be true, so long as the premises are true (i.e. it is impossible for the premises to be true and the conclusion false). Therefore the argument is valid, but not sound. Generalizations are often used to make invalid arguments, such as "everyone who eats carrots is a quarterback." Not everyone who eats carrots is a quarterback, thus proving the flaw of such arguments.
In this example, the first statement uses categorical reasoning, saying that all carrot-eaters are definitely quarterbacks. This theory of deductive reasoning – also known as term logic – was developed by Aristotle, but was superseded by propositional (sentential) logic and predicate logic.
Deductive reasoning can be contrasted with inductive reasoning, in regards to validity and soundness. In cases of inductive reasoning, even though the premises are true and the argument is "valid", it is possible for the conclusion to be false (determined to be false with a counterexample or other means).

6.6 Education

Deductive reasoning is generally thought of as a skill that develops without any formal teaching or training. As a result of this belief, deductive reasoning skills are not taught in secondary schools, where students are expected to use reasoning more often and at a higher level.[4] It is in high school, for example, that students have an abrupt introduction to mathematical proofs – which rely heavily on deductive reasoning.[4]

6.7 See also

_ Argument (logic)
_ Logic
_ Mathematical logic
_ Abductive reasoning
_ Analogical reasoning
_ Correspondence theory of truth
_ Defeasible reasoning
_ Decision making
_ Decision theory
_ Fallacy

_ Fault Tree Analysis
_ Geometry
_ Hypothetico-deductive method
_ Inquiry
_ Mathematical induction
_ Inductive reasoning
_ Inference
_ Logical consequence
_ Natural deduction
_ Propositional calculus
_ Retroductive reasoning
_ Scientific method
_ Theory of justification
_ Soundness
_ Syllogism

6.8 References

[1] Deduction & Induction, Research Methods Knowledge Base
[2] Sternberg, R. J. (2009). *Cognitive Psychology*. Belmont, CA: Wadsworth. p. 578. ISBN 978-0-495-50629-4.
[3] Guide to Logic
[4] Stylianides, G. J.; Stylianides (2008). "A. J.". *Mathematical Thinking and Learning* **10** (2): 103–133. doi:10.1080/10986060701854425.

6.9 Further reading

_ Vincent F. Hendricks, *Thought 2 Talk: A Crash Course in Reflection and Expression*, New York: Automatic Press / VIP, 2005, ISBN 87-991013-7-8
_ Philip Johnson-Laird, Ruth M. J. Byrne, *Deduction*, Psychology Press 1991, ISBN 978-0-86377-149-1jiii
_ Zarefsky, David, *Argumentation: The Study of Effective Reasoning Parts I and II*, The Teaching Company 2002
_ Bullemore, Thomas, * The Pragmatic Problem of Induction.

6.10 External links

_ Deductive reasoning at PhilPapers
_ Deductive reasoning at the Indiana Philosophy Ontology Project
_ Deductive reasoning entry in the *Internet Encyclopedia of Philosophy*

Chapter 7

Inductive reasoning

"Inductive inference" redirects here. For the technique in mathematical proof, see Mathematical induction. For the theory introduced by Ray Solomonoff, see Solomonoff's theory of inductive inference.

Inductive reasoning (as opposed to *deductive* reasoning) is reasoning in which the premises seek to supply strong evidence for (not absolute proof of) the truth of the conclusion. While the conclusion of a deductive argument is supposed to be certain, the truth of the conclusion of an inductive argument is supposed to be *probable*, based upon the evidence given.[1]
The philosophical definition of inductive reasoning is more nuanced than simple progression from particular/individual instances to broader generalizations. Rather, the premises of an inductive logical argument indicate some degree of support (inductive probability) for the conclusion but do not entail it; that is, they suggest truth but do not ensure it. In this manner, there is the possibility of moving from general statements to individual instances (for example, statistical syllogisms, discussed below).
Many dictionaries define inductive reasoning as reasoning that derives general principles from specific observations, though some sources disagree with this usage.[2]

7.1 Description

Inductive reasoning is inherently uncertain. It only deals in degrees to which, given the premises, the conclusion is *credible* according to some theory of evidence. Examples include a many-valued logic, Dempster–Shafer theory, or probability theory with rules for inference such as Bayes' rule. Unlike deductive reasoning, it does not rely on universals holding over a closed domain of discourse to draw conclusions, so it can be applicable even in cases of epistemic uncertainty (technical issues with this may arise however; for example, the second axiom of probability is a closed-world assumption).[3]
An example of an inductive argument:
100% of biological life forms that we know of depend on liquid water to exist.
Therefore, if we discover a new biological life form it will probably depend on liquid water to exist.
This argument could have been made every time a new biological life form was found, and would have been correct every time; however, it is still possible that in the future a biological life form not requiring water could be discovered.
As a result, the argument may be stated less formally as:
All biological life forms that we know of depend on liquid water to exist.
All biological life probably depends on liquid water to exist.
39
40 *CHAPTER 7. INDUCTIVE REASONING*

7.2 Inductive vs. deductive reasoning

Unlike deductive arguments, inductive reasoning allows for the possibility that the conclusion is false, even if all of the premises are true.[4] Instead of being valid or invalid, inductive arguments are either *strong* or *weak*, which describes how *probable* it is that the conclusion is true.[5]
A classical example of an **incorrect** inductive argument was presented by John Vickers:
All of the swans we have seen are white.
Therefore, all swans are white.
Note that this definition of *inductive* reasoning excludes mathematical induction, which is a form of *deductive* reasoning.

7.3 Criticism

Main article: Problem of induction
Inductive reasoning has been criticized by thinkers as diverse as Sextus Empiricus[6] and Karl Popper.[7]
The classic philosophical treatment of the problem of induction was given by the Scottish philosopher David Hume. Although the use of inductive reasoning demonstrates considerable success, its application has been questionable. Recognizing this, Hume highlighted the fact that our mind draws uncertain conclusions from relatively limited experiences. In deduction, the truth value of the conclusion is based on the truth of the premise. In induction, however, the dependence on the premise is always uncertain. As an example, let's assume "all ravens are black." The fact that there are numerous black ravens supports the assumption. However, the assumption becomes inconsistent with the fact that there are white ravens. Therefore, the general rule of "all ravens are black" is inconsistent with the existence of the white raven. Hume further argued that it is impossible to justify inductive reasoning: specifically, that it cannot be justified deductively, so our only option is to justify it inductively. Since this is circular he concluded that our use of induction is unjustifiable with the help of "Hume's Fork".[8]
However, Hume then stated that even if induction were proved unreliable, we would still have to rely on it. So instead of a position of severe skepticism, Hume advocated a practical skepticism based on common sense, where the inevitability of induction is accepted.[9]

7.3.1 Biases

Inductive reasoning is also known as hypothesis construction because any conclusions made are based on current knowledge and predictions. As with deductive arguments, biases can distort the proper application of inductive argument, thereby preventing the reasoner from forming the most logical conclusion based on the clues. Examples of these biases include the availability heuristic, confirmation bias, and the predictable-world bias.

The availability heuristic causes the reasoner to depend primarily upon information that is readily available to him/her. People have a tendency to rely on information that is easily accessible in the world around them. For example, in surveys, when people are asked to estimate the percentage of people who died from various causes, most respondents would choose the causes that have been most prevalent in the media such as terrorism, and murders, and airplane accidents rather than causes such as disease and traffic accidents, which have been technically "less accessible" to the individual since they are not emphasized as heavily in the world around him/her.

The confirmation bias is based on the natural tendency to confirm rather than to deny a current hypothesis. Research has demonstrated that people are inclined to seek solutions to problems that are more consistent with known hypotheses rather than attempt to refute those hypotheses. Often, in experiments, subjects will ask questions that seek answers that fit established hypotheses, thus confirming these hypotheses. For example, if it is hypothesized that Sally is a sociable individual, subjects will naturally seek to confirm the premise by asking questions that would produce answers confirming that Sally is in fact a sociable individual.

The predictable-world bias revolves around the inclination to perceive order where it has not been proved to exist, either at all or at a particular level of abstraction. Gambling, for example, is one of the most popular examples of

predictable-world bias. Gamblers often begin to think that they see simple and obvious patterns in the outcomes and, therefore, believe that they are able to predict outcomes based upon what they have witnessed. In reality, however, the outcomes of these games are difficult to predict and highly complex in nature. However, in general, people tend to seek some type of simplistic order to explain or justify their beliefs and experiences, and it is often difficult for them to realise that their perceptions of order may be entirely different from the truth.[10]

7.4 Types

7.4.1 Generalization

A generalization (more accurately, an *inductive generalization*) proceeds from a premise about a sample to a conclusion about the population.

The proportion Q of the sample has attribute A.

Therefore:

The proportion Q of the population has attribute A.

Example

There are 20 balls—either black or white—in an urn. To estimate their respective numbers, you draw a sample of four balls and find that three are black and one is white. A good inductive generalization would be that there are 15 black, and five white, balls in the urn.

How much the premises support the conclusion depends upon (a) the number in the sample group, (b) the number in the population, and (c) the degree to which the sample represents the population (which may be achieved by taking a random sample). The hasty generalization and the biased sample are generalization fallacies.

7.4.2 Statistical syllogism

Main article: Statistical syllogism

A statistical syllogism proceeds from a generalization to a conclusion about an individual.

A proportion Q of population P has attribute A.

An individual X is a member of P.

Therefore:

There is a probability which corresponds to Q that X has A.

The proportion in the first premise would be something like "3/5ths of", "all", "few", etc. Two dicto simpliciter fallacies can occur in statistical syllogisms: "accident" and "converse accident".

7.4.3 Simple induction

Simple induction proceeds from a premise about a sample group to a conclusion about another individual.

Proportion Q of the known instances of population P has attribute A.

Individual I is another member of P.

Therefore:

There is a probability corresponding to Q that I has A.

This is a combination of a generalization and a statistical syllogism, where the conclusion of the generalization is also the first premise of the statistical syllogism.

Argument from analogy
Main article: Argument from analogy
The process of analogical inference involves noting the shared properties of two or more things, and from this basis inferring that they also share some further property:[11]
P and Q are similar in respect to properties a, b, and c.
Object P has been observed to have further property x.
Therefore, Q probably has property x also.
Analogical reasoning is very frequent in common sense, science, philosophy and the humanities, but sometimes it is accepted only as an auxiliary method. A refined approach is case-based reasoning.[12]

7.4.4 Causal inference

A causal inference draws a conclusion about a causal connection based on the conditions of the occurrence of an effect. Premises about the correlation of two things can indicate a causal relationship between them, but additional factors must be confirmed to establish the exact form of the causal relationship.

7.4.5 Prediction

A prediction draws a conclusion about a future individual from a past sample.
Proportion Q of observed members of group G have had attribute A.
Therefore:
There is a probability corresponding to Q that other members of group G will have attribute A when next observed.

7.5 Bayesian inference

As a logic of induction rather than a theory of belief, Bayesian inference does not determine which beliefs are *a priori* rational, but rather determines how we should rationally change the beliefs we have when presented with evidence. We begin by committing to a prior probability for a hypothesis based on logic or previous experience, and when faced with evidence, we adjust the strength of our belief in that hypothesis in a precise manner using Bayesian logic.

7.6 Inductive inference

Around 1960, Ray Solomonoff founded the theory of universal inductive inference, the theory of prediction based on observations; for example, predicting the next symbol based upon a given series of symbols. This is a formal inductive framework that combines algorithmic information theory with the Bayesian framework. Universal inductive inference is based on solid philosophical foundations[13] and can be considered as a mathematically formalized Occam's razor. Fundamental ingredients of the theory are the concepts of algorithmic probability and Kolmogorov complexity.

7.7 See also

_ Abductive reasoning
_ Algorithmic information theory
_ Algorithmic probability

_ Analogy
_ Bayesian probability
_ Counterinduction
_ Deductive reasoning
_ Explanation
_ Failure mode and effects analysis
_ Falsifiability
_ Grammar induction
_ Inductive inference
_ Inductive logic programming
_ Inductive probability
_ Inductive programming
_ Inductive reasoning aptitude
_ Inquiry
_ Kolmogorov complexity
_ Lateral thinking
_ Laurence Jonathan Cohen
_ Logic
_ Logical positivism

_ Machine learning
_ Mathematical induction
_ Mill's Methods
_ Minimum description length
_ Minimum message length
_ Open world assumption
_ Raven paradox
_ Recursive Bayesian estimation
_ Retroduction
_ Solomonoff's theory of inductive inference
_ Statistical inference
_ Stephen Toulmin
_ Universal artificial intelligence

7.8 References

[1] Copi, I. M., Cohen, C., & Flage, D. E. (2007). Essentials of logic (2nd ed.). Upper Saddle River, NJ: Pearson Education, Inc.
[2] *Deductive and Inductive Arguments*, *Internet Encyclopedia of Philosophy*, "Some dictionaries define "deduction" as reasoning from the general to specific and "induction" as reasoning from the specific to the general. While this usage is still sometimes found even in philosophical and mathematical contexts, for the most part, it is outdated."
[3] Bart Kosko, *Fuzziness vs. Probability*, International Journal of General Systems, vol. 17, no. 1, pp. 211-240, 1990.
[4] John Vickers. The Problem of Induction. The Stanford Encyclopedia of Philosophy.
[5] Herms, D. "Logical Basis of Hypothesis Testing in Scientific Research" (pdf).
[6] Sextus Empiricus, *Outlines Of Pyrrhonism*. Trans. R.G. Bury, Harvard University Press, Cambridge, Massachusetts, 1933, p. 283.
[7] Karl R. Popper, David W. Miller. "A proof of the impossibility of inductive probability." *Nature* 302 (1983), 687–688.
[8] Vickers, John. "The Problem of Induction" (Section 2). *Stanford Encyclopedia of Philosophy*. 21 June 2010
[9] Vickers, John. "The Problem of Induction" (Section 2.1). *Stanford Encyclopedia of Philosophy*. 21 June 2010.
[10] Gray, Peter. Psychology. New York: Worth, 2011. Print.
[11] Baronett, Stan (2008). *Logic*. Upper Saddle River, NJ: Pearson Prentice Hall. pp. 321–325.
[12] For more information on inferences by analogy, see Juthe, 2005.
[13] Samuel Rathmanner and Marcus Hutter. A philosophical treatise of universal induction. Entropy, 13(6):1076–1136, 2011

7.9 Further reading

_ Cushan, Anna-Marie (1983/2014). *Investigation into Facts and Values: Groundwork for a theory of moral conflict resolution*. [Thesis, Melbourne University], Ondwelle Publications (online): Melbourne.
_ Herms, D. "Logical Basis of Hypothesis Testing in Scientific Research" (PDF).
_ Kemerling, G. (27 October 2001). "Causal Reasoning".
_ Holland, J. H.; Holyoak, K. J.; Nisbett, R. E.; Thagard, P. R. (1989). *Induction: Processes of Inference, Learning, and Discovery*. Cambridge, MA, USA: MIT Press. ISBN 0-262-58096-9.
_ Holyoak, K.; Morrison, R. (2005). *The Cambridge Handbook of Thinking and Reasoning*. New York: Cambridge University Press. ISBN 978-0-521-82417-0.

7.10 External links

_ Confirmation and Induction entry in the *Internet Encyclopedia of Philosophy*
_ Inductive Logic entry in the *Stanford Encyclopedia of Philosophy*
_ Inductive reasoning at PhilPapers
_ Inductive reasoning at the Indiana Philosophy Ontology Project
_ *Four Varieties of Inductive Argument* from the Department of Philosophy, University of North Carolina at Greensboro.
_ *Properties of Inductive Reasoning* PDF (166 KiB), a psychological review by Evan Heit of the University of California, Merced.
_ *The Mind, Limber* An article which employs the film The Big Lebowski to explain the value of inductive reasoning.
_ The Pragmatic Problem of Induction, by Thomas Bullemore

Chapter 8

Empirical evidence

"Empirical" redirects here. For the jazz ensemble, see Empirical (jazz band).
"A posteriori" redirects here. For other uses, see A posteriori (disambiguation).

Empirical evidence (also **empirical data**, **sense experience**, **empirical knowledge**, or the **a posteriori**) is a source of knowledge acquired by means of observation or experimentation.[1] The term comes from the Greek word for experience, Εμπειρία (empeiría).

Empirical evidence is information that justifies a belief in the truth or falsity of a claim. In the empiricist view, one can claim to have knowledge only when one has a true belief based on empirical evidence. This stands in contrast to the rationalist view under which reason or reflection alone is considered to be evidence for the truth or falsity of some propositions.[2] The senses are the primary source of empirical evidence. Although other sources of evidence, such as memory, and the testimony of others ultimately trace back to some sensory experience, they are considered to be secondary, or indirect.[2]

In another sense, empirical evidence may be synonymous with the outcome of an experiment. In this sense, an empirical result is a unified confirmation. In this context, the term *semi-empirical* is used for qualifying theoretical methods which use in part basic axioms or postulated scientific laws and experimental results. Such methods are opposed to theoretical *ab initio* methods which are purely deductive and based on first principles.

In science, empirical evidence is required for a hypothesis to gain acceptance in the scientific community. Normally, this validation is achieved by the scientific method of hypothesis commitment, experimental design, peer review, adversarial review, reproduction of results, conference presentation and journal publication. This requires rigorous communication of hypothesis (usually expressed in mathematics), experimental constraints and controls (expressed necessarily in terms of standard experimental apparatus), and a common understanding of measurement.

Statements and arguments depending on empirical evidence are often referred to as **a posteriori** ("from the later") as distinguished from *a priori* ("from the earlier"). (See A priori and a posteriori). *A priori* knowledge or justification is independent of experience (for example "All bachelors are unmarried"); whereas *a posteriori* knowledge or justification is dependent on experience or empirical evidence (for example "Some bachelors are very happy"). The notion of the distinction between *a priori* and *a posteriori* as tantamount to the distinction between empirical and non-empirical knowledge comes from Kant's Critique of Pure Reason.[3]

The standard positivist view of empirically acquired information has been that observation, experience, and experiment serve as neutral arbiters between competing theories. However, since the 1960s, a persistent critique most associated with Thomas Kuhn,[4] has argued that these methods are influenced by prior beliefs and experiences. Consequently it cannot be expected that two scientists when observing, experiencing, or experimenting on the same event will make the same theory-neutral observations. The role of observation as a theory-neutral arbiter may not be possible. Theory-dependence of observation means that, even if there were agreed methods of inference and interpretation, scientists may still disagree on the nature of empirical data.[5]

8.1 See also

_ Anecdotal evidence
45
46 *CHAPTER 8. EMPIRICAL EVIDENCE*
_ Empirical distribution function
_ Empirical formula
_ Empirical measure
_ Empirical research (more on the scientific usage)
_ Phenomenology (science)
_ Scientific evidence
_ Scientific method
_ Theory

8.2 Footnotes

[1] Pickett 2006, p. 585
[2] Feldman 2001, p. 293
[3] Craig 2005, p. 1
[4] Kuhn 1970
[5] Bird 2013

8.3 References

_ Bird, Alexander (2013). "4.2 Perception, Observational Incommensurability, and World-Change". In Zalta, Edward N. "Thomas Kuhn". *Stanford Encyclopedia of Philosophy*. Retrieved 25 January 2012.

_ Craig, Edward (2005). "a posteriori". *The Shorter Routledge Encyclopedia of Philosophy*. Routledge. ISBN 9780415324953.
_ Feldman, Richard (2001) [1999]. "Evidence". In Audi, Robert. *The Cambridge Dictionary of Philosophy* (2nd ed.). Cambridge, UK: Cambridge University Press. pp. 293–294. ISBN 978-0521637220.
_ Kuhn, Thomas S. (1970) [1962]. *The Structure of Scientific Revolutions* (2nd ed.). Chicago: University of Chicago Press. ISBN 978-0226458045.
_ Pickett, Joseph P., ed. (2011). "Empirical". *The American Heritage Dictionary of the English Language* (5th ed.). Houghton Mifflin. ISBN 978-0-547-04101-8.

8.4 External links

_ The dictionary definition of empirical evidence at Wiktionary
_ A Priori and A Posteriori entry in the *Internet Encyclopedia of Philosophy*

Chapter 9

Logic

This article is about reasoning and its study. For other uses, see Logic (disambiguation).
Logic (from the Ancient Greek: λογική, *logike*)[1] is the use and study of valid reasoning.[2][3] The study of logic features most prominently in the subjects of philosophy, mathematics, and computer science.
Logic was studied in several ancient civilizations, including India,[4] China,[5] Persia and Greece. In the West, logic was established as a formal discipline by Aristotle, who gave it a fundamental place in philosophy. The study of logic was part of the classical trivium, which also included grammar and rhetoric. Logic was further extended by Al-Farabi who categorized it into two separate groups (idea and proof). Later, Avicenna revived the study of logic and developed relationship between temporalis and the implication. In the East, logic was developed by Buddhists and Jains.
Logic is often divided into three parts: inductive reasoning, abductive reasoning, and deductive reasoning.

9.1 The study of logic

The concept of logical form is central to logic, it being held that the validity of an argument is determined by its logical form, not by its content. Traditional Aristotelian syllogistic logic and modern symbolic logic are examples of formal logics.
_ **Informal logic** is the study of natural language arguments. The study of fallacies is an especially important branch of informal logic. The dialogues of Plato[6] are good examples of informal logic.
_ **Formal logic** is the study of inference with purely formal content. An inference possesses a *purely formal content* if it can be expressed as a particular application of a wholly abstract rule, that is, a rule that is not about any particular thing or property. The works of Aristotle contain the earliest known formal study of logic. Modern formal logic follows and expands on Aristotle.[7] In many definitions of logic, logical inference and inference with purely formal content are the same. This does not render the notion of informal logic vacuous, because no formal logic captures all of the nuances of natural language.
_ **Symbolic logic** is the study of symbolic abstractions that capture the formal features of logical inference.[8][9] Symbolic logic is often divided into two branches: propositional logic and predicate logic.
_ **Mathematical logic** is an extension of symbolic logic into other areas, in particular to the study of model theory, proof theory, set theory, and recursion theory.

9.1.1 Logical form

Main article: Logical form
Logic is generally considered **formal** when it analyzes and represents the *form* of any valid argument type. The form of an argument is displayed by representing its sentences in the formal grammar and symbolism of a logical language
47
48 *CHAPTER 9. LOGIC*
to make its content usable in formal inference. If one considers the notion of form too philosophically loaded, one could say that formalizing simply means translating English sentences into the language of logic.
This is called showing the *logical form* of the argument. It is necessary because indicative sentences of ordinary language show a considerable variety of form and complexity that makes their use in inference impractical. It requires, first, ignoring those grammatical features irrelevant to logic (such as gender and declension, if the argument is in Latin), replacing conjunctions irrelevant to logic (such as "but") with logical conjunctions like "and" and replacing ambiguous, or alternative logical expressions ("any", "every", etc.) with expressions of a standard type (such as "all", or the universal quantifier ∀).

Second, certain parts of the sentence must be replaced with schematic letters. Thus, for example, the expression "all As are Bs" shows the logical form common to the sentences "all men are mortals", "all cats are carnivores", "all Greeks are philosophers", and so on.

That the concept of form is fundamental to logic was already recognized in ancient times. Aristotle uses variable letters to represent valid inferences in *Prior Analytics*, leading Jan Łukasiewicz to say that the introduction of variables was "one of Aristotle's greatest inventions".[10] According to the followers of Aristotle (such as Ammonius), only the logical principles stated in schematic terms belong to logic, not those given in concrete terms. The concrete terms "man", "mortal", etc., are analogous to the substitution values of the schematic placeholders *A*, *B*, *C*, which were called the "matter" (Greek *hyle*) of the inference.

The fundamental difference between modern formal logic and traditional, or Aristotelian logic, lies in their differing analysis of the logical form of the sentences they treat.

_ In the traditional view, the form of the sentence consists of (1) a subject (e.g., "man") plus a sign of quantity ("all" or "some" or "no"); (2) the copula, which is of the form "is" or "is not"; (3) a predicate (e.g., "mortal"). Thus: all men are mortal. The logical constants such as "all", "no" and so on, plus sentential connectives such as "and" and "or" were called "syncategorematic" terms (from the Greek *kategorei* – to predicate, and *syn* – together with). This is a fixed scheme, where each judgment has an identified quantity and copula, determining the logical form of the sentence.

_ According to the modern view, the fundamental form of a simple sentence is given by a recursive schema, involving logical connectives, such as a quantifier with its bound variable, which are joined by juxtaposition to other sentences, which in turn may have logical structure.

_ The modern view is more complex, since a single judgement of Aristotle's system involves two or more logical connectives. For example, the sentence "All men are mortal" involves, in term logic, two non-logical terms "is a man" (here *M*) and "is mortal" (here *D*): the sentence is given by the judgement A(M,D). In predicate logic, the sentence involves the same two non-logical concepts, here analyzed as $m(x)$ and $d(x)$, and the sentence is given by *8x:(m(x) ! d(x))* , involving the logical connectives for universal quantification and implication.

_ But equally, the modern view is more powerful. Medieval logicians recognized the problem of multiple generality, where Aristotelian logic is unable to satisfactorily render such sentences as "Some guys have all the luck", because both quantities "all" and "some" may be relevant in an inference, but the fixed scheme that Aristotle used allows only one to govern the inference. Just as linguists recognize recursive structure in natural languages, it appears that logic needs recursive structure.

9.1.2 Deductive and inductive reasoning, and abductive inference

Deductive reasoning concerns what follows necessarily from given premises (if *a*, then *b*). However, inductive reasoning—the process of deriving a reliable generalization from observations—has sometimes been included in the study of logic. Similarly, it is important to distinguish deductive validity and inductive validity (called "cogency"). An inference is deductively valid if and only if there is no possible situation in which all the premises are true but the conclusion false. An inductive argument can be neither valid nor invalid; its premises give only some degree of probability, but not certainty, to its conclusion.

The notion of deductive validity can be rigorously stated for systems of formal logic in terms of the well-understood notions of semantics. Inductive validity on the other hand requires us to define a reliable generalization of some set of observations. The task of providing this definition may be approached in various ways, some less formal than others; some of these definitions may use mathematical models of probability. For the most part this discussion of logic deals only with deductive logic.

Abduction[11] is a form of logical inference that goes from observation to a hypothesis that accounts for the reliable data (observation) and seeks to explain relevant evidence. The American philosopher Charles Sanders Peirce (1839–1914) first introduced the term as "guessing".[12] Peirce said that to *abduce* a hypothetical explanation *a* from an observed surprising circumstance *b* is to surmise that *a* may be true because then *b* would be a matter of course.[13] Thus, to abduce *a* from *b* involves determining that *a* is sufficient (or nearly sufficient), but not necessary, for *b* .

9.1.3 Consistency, validity, soundness, and completeness

Among the important properties that logical systems can have:

_ **Consistency**, which means that no theorem of the system contradicts another.[14]

_ **Validity**, which means that the system's rules of proof never allow a false inference from true premises. A logical system has the property of soundness when the logical system has the property of validity and uses only premises that prove true (or, in the case of axioms, are true by definition).[14]

_ **Completeness**, of a logical system, which means that if a formula is true, it can be proven (if it is true, it is a theorem of the system).

_ **Soundness**, the term soundness has multiple separate meanings, which creates a bit of confusion throughout the literature. Most commonly, soundness refers to logical systems, which means that if some formula can be

proven in a system, then it is true in the relevant model/structure (if A is a theorem, it is true). This is the converse of completeness. A distinct, peripheral use of soundness refers to arguments, which means that the premises of a valid argument are true in the actual world.
Some logical systems do not have all four properties. As an example, Kurt Gödel's incompleteness theorems show that sufficiently complex formal systems of arithmetic cannot be consistent and complete;[9] however, first-order predicate logics not extended by specific axioms to be arithmetic formal systems with equality can be complete and consistent.[15]

9.1.4 Rival conceptions of logic

Main article: Definitions of logic
Logic arose (see below) from a concern with correctness of argumentation. Modern logicians usually wish to ensure that logic studies just those arguments that arise from appropriately general forms of inference. For example, Thomas Hofweber writes in the Stanford Encyclopedia of Philosophy that logic "does not, however, cover good reasoning as a whole. That is the job of the theory of rationality. Rather it deals with inferences whose validity can be traced back to the formal features of the representations that are involved in that inference, be they linguistic, mental, or other representations".[16]
By contrast, Immanuel Kant argued that logic should be conceived as the science of judgement, an idea taken up in Gottlob Frege's logical and philosophical work. But Frege's work is ambiguous in the sense that it is both concerned with the "laws of thought" as well as with the "laws of truth", i.e. it both treats logic in the context of a theory of the mind, and treats logic as the study of abstract formal structures.

9.2 History

Main article: History of logic
In Europe, logic was first developed by Aristotle.[17] Aristotelian logic became widely accepted in science and mathematics and remained in wide use in the West until the early 19th century.[18] Aristotle's system of logic was responsible for the introduction of hypothetical syllogism,[19] temporal modal logic,[20][21] and inductive logic,[22] as well as influential terms such as terms, predicables, syllogisms and propositions. In Europe during the later medieval period, major efforts were made to show that Aristotle's ideas were compatible with Christian faith. During the High Middle Ages, logic became a main focus of philosophers, who would engage in critical logical analyses of philosophical arguments, often using variations of the methodology of scholasticism. In 1323, William of Ockham's influential
50 *CHAPTER 9. LOGIC*
Aristotle, 384–322 BCE.
Summa Logicae was released. By the 18th century, the structured approach to arguments had degenerated and fallen out of favour, as depicted in Holberg's satirical play *Erasmus Montanus*.
The Chinese logical philosopher Gongsun Long (c. 325–250 BCE) proposed the paradox "One and one cannot become two, since neither becomes two."[23] In China, the tradition of scholarly investigation into logic, however, was repressed by the Qin dynasty following the legalist philosophy of Han Feizi.
In India, innovations in the scholastic school, called Nyaya, continued from ancient times into the early 18th century
9.3. TYPES OF LOGIC 51
with the Navya-Nyaya school. By the 16th century, it developed theories resembling modern logic, such as Gottlob Frege's "distinction between sense and reference of proper names" and his "definition of number", as well as the theory of "restrictive conditions for universals" anticipating some of the developments in modern set theory.[24] Since 1824, Indian logic attracted the attention of many Western scholars, and has had an influence on important 19thcentury logicians such as Charles Babbage, Augustus De Morgan, and George Boole.[25] In the 20th century, Western philosophers like Stanislaw Schayer and Klaus Glashoff have explored Indian logic more extensively.
The syllogistic logic developed by Aristotle predominated in the West until the mid-19th century, when interest in the foundations of mathematics stimulated the development of symbolic logic (now called mathematical logic). In 1854, George Boole published *An Investigation of the Laws of Thought on Which are Founded the Mathematical Theories of Logic and Probabilities*, introducing symbolic logic and the principles of what is now known as Boolean logic. In 1879, Gottlob Frege published *Begriffsschrift*, which inaugurated modern logic with the invention of quantifier notation. From 1910 to 1913, Alfred North Whitehead and Bertrand Russell published *Principia Mathematica*[8] on the foundations of mathematics, attempting to derive mathematical truths from axioms and inference rules in symbolic logic. In 1931, Gödel raised serious problems with the foundationalist program and logic ceased to focus on such issues.
The development of logic since Frege, Russell, and Wittgenstein had a profound influence on the practice of philosophy and the perceived nature of philosophical problems (see Analytic philosophy), and Philosophy of mathematics.
Logic, especially sentential logic, is implemented in computer logic circuits and is fundamental to computer science. Logic is commonly taught by university philosophy departments, often as a compulsory discipline.

9.3 Types of logic

9.3.1 Syllogistic logic

Main article: Aristotelian logic

The *Organon* was Aristotle's body of work on logic, with the *Prior Analytics* constituting the first explicit work in formal logic, introducing the syllogistic.[26] The parts of syllogistic logic, also known by the name term logic, are the analysis of the judgements into propositions consisting of two terms that are related by one of a fixed number of relations, and the expression of inferences by means of syllogisms that consist of two propositions sharing a common term as premise, and a conclusion that is a proposition involving the two unrelated terms from the premises.

Aristotle's work was regarded in classical times and from medieval times in Europe and the Middle East as the very picture of a fully worked out system. However, it was not alone: the Stoics proposed a system of propositional logic that was studied by medieval logicians. Also, the problem of multiple generality was recognized in medieval times. Nonetheless, problems with syllogistic logic were not seen as being in need of revolutionary solutions.

Today, some academics claim that Aristotle's system is generally seen as having little more than historical value (though there is some current interest in extending term logics), regarded as made obsolete by the advent of propositional logic and the predicate calculus. Others use Aristotle in argumentation theory to help develop and critically question argumentation schemes that are used in artificial intelligence and legal arguments.

9.3.2 Propositional logic (sentential logic)

Main article: Propositional calculus

A propositional calculus or logic (also a sentential calculus) is a formal system in which formulae representing propositions can be formed by combining atomic propositions using logical connectives, and in which a system of formal proof rules establishes certain formulae as "theorems".

9.3.3 Predicate logic

Main article: Predicate logic

Predicate logic is the generic term for symbolic formal systems such as first-order logic, second-order logic, manysorted logic, and infinitary logic.

Predicate logic provides an account of quantifiers general enough to express a wide set of arguments occurring in natural language. Aristotelian syllogistic logic specifies a small number of forms that the relevant part of the involved judgements may take. Predicate logic allows sentences to be analysed into subject and argument in several additional ways—allowing predicate logic to solve the problem of multiple generality that had perplexed medieval logicians.

The development of predicate logic is usually attributed to Gottlob Frege, who is also credited as one of the founders of analytical philosophy, but the formulation of predicate logic most often used today is the first-order logic presented in Principles of Mathematical Logic by David Hilbert and Wilhelm Ackermann in 1928. The analytical generality of predicate logic allowed the formalization of mathematics, drove the investigation of set theory, and allowed the development of Alfred Tarski's approach to model theory. It provides the foundation of modern mathematical logic.

Frege's original system of predicate logic was second-order, rather than first-order. Second-order logic is most prominently defended (against the criticism of Willard Van Orman Quine and others) by George Boolos and Stewart Shapiro.

9.3.4 Modal logic

Main article: Modal logic

In languages, modality deals with the phenomenon that sub-parts of a sentence may have their semantics modified by special verbs or modal particles. For example, "*We go to the games*" can be modified to give "*We should go to the games*", and "*We can go to the games*" and perhaps "*We will go to the games*". More abstractly, we might say that modality affects the circumstances in which we take an assertion to be satisfied.

Aristotle's logic is in large parts concerned with the theory of non-modalized logic. Although, there are passages in his work, such as the famous sea-battle argument in *De Interpretatione* § 9, that are now seen as anticipations of modal logic and its connection with potentiality and time, the earliest formal system of modal logic was developed by Avicenna, whom ultimately developed a theory of "temporally modalized" syllogistic.[27]

While the study of necessity and possibility remained important to philosophers, little logical innovation happened until the landmark investigations of Clarence Irving Lewis in 1918, who formulated a family of rival axiomatizations of the alethic modalities. His work unleashed a torrent of new work on the topic, expanding the kinds of modality treated to include deontic logic and epistemic logic. The seminal work of Arthur Prior applied the same formal language to treat temporal logic and paved the way for the marriage of the two subjects. Saul Kripke discovered (contemporaneously with rivals) his theory of frame semantics, which revolutionized the formal technology available to modal logicians and gave a new graph-theoretic way of looking at modality that has driven many applications in computational linguistics and computer science, such as dynamic logic.

9.3.5 Informal reasoning

Main article: Informal logic

The motivation for the study of logic in ancient times was clear: it is so that one may learn to distinguish good from bad arguments, and so become more effective in argument and oratory, and perhaps also to become a better person. Half of the works of Aristotle's Organon treat inference as it occurs in an informal setting, side by side with the development of the syllogistic, and in the Aristotelian school, these informal works on logic were seen as complementary to Aristotle's treatment of rhetoric.

This ancient motivation is still alive, although it no longer takes centre stage in the picture of logic; typically dialectical logic forms the heart of a course in critical thinking, a compulsory course at many universities.

Argumentation theory is the study and research of informal logic, fallacies, and critical questions as they relate to every day and practical situations. Specific types of dialogue can be analyzed and questioned to reveal premises, conclusions, and fallacies. Argumentation theory is now applied in artificial intelligence and law.

9.3.6 Mathematical logic

Main article: Mathematical logic

Mathematical logic really refers to two distinct areas of research: the first is the application of the techniques of formal logic to mathematics and mathematical reasoning, and the second, in the other direction, the application of mathematical techniques to the representation and analysis of formal logic.[28]

The earliest use of mathematics and geometry in relation to logic and philosophy goes back to the ancient Greeks such as Euclid, Plato, and Aristotle.[29] Many other ancient and medieval philosophers applied mathematical ideas and methods to their philosophical claims.[30]

One of the boldest attempts to apply logic to mathematics was undoubtedly the logicism pioneered by philosopherlogicians such as Gottlob Frege and Bertrand Russell: the idea was that mathematical theories were logical tautologies, and the programme was to show this by means to a reduction of mathematics to logic.[8] The various attempts to carry this out met with a series of failures, from the crippling of Frege's project in his *Grundgesetze* by Russell's paradox, to the defeat of Hilbert's program by Gödel's incompleteness theorems.

Both the statement of Hilbert's program and its refutation by Gödel depended upon their work establishing the second area of mathematical logic, the application of mathematics to logic in the form of proof theory.[31] Despite the negative nature of the incompleteness theorems, Gödel's completeness theorem, a result in model theory and another application of mathematics to logic, can be understood as showing how close logicism came to being true: every rigorously defined mathematical theory can be exactly captured by a first-order logical theory; Frege's proof calculus is enough to *describe* the whole of mathematics, though not *equivalent* to it. Thus we see how complementary the two areas of mathematical logic have been.

If proof theory and model theory have been the foundation of mathematical logic, they have been but two of the four pillars of the subject. Set theory originated in the study of the infinite by Georg Cantor, and it has been the source of many of the most challenging and important issues in mathematical logic, from Cantor's theorem, through the status of the Axiom of Choice and the question of the independence of the continuum hypothesis, to the modern debate on large cardinal axioms.

Recursion theory captures the idea of computation in logical and arithmetic terms; its most classical achievements are the undecidability of the Entscheidungsproblem by Alan Turing, and his presentation of the Church–Turing thesis.[32] Today recursion theory is mostly concerned with the more refined problem of complexity classes—when is a problem efficiently solvable?—and the classification of degrees of unsolvability.[33]

9.3.7 Philosophical logic

Main article: Philosophical logic

Philosophical logic deals with formal descriptions of ordinary, non-specialist ("natural") language. Most philosophers assume that the bulk of everyday reasoning can be captured in logic if a method or methods to translate ordinary language into that logic can be found. Philosophical logic is essentially a continuation of the traditional discipline called "logic" before the invention of mathematical logic. Philosophical logic has a much greater concern with the connection between natural language and logic. As a result, philosophical logicians have contributed a great deal to the development of non-standard logics (e.g. free logics, tense logics) as well as various extensions of classical logic (e.g. modal logics) and non-standard semantics for such logics (e.g. Kripke's supervaluationism in the semantics of logic).

Logic and the philosophy of language are closely related. Philosophy of language has to do with the study of how our language engages and interacts with our thinking. Logic has an immediate impact on other areas of study. Studying logic and the relationship between logic and ordinary speech can help a person better structure his own arguments and critique the arguments of others. Many popular arguments are filled with errors because so many people are untrained in logic and unaware of how to formulate an argument correctly.

9.3.8 Computational logic

Main article: Logic in computer science

Logic cut to the heart of computer science as it emerged as a discipline: Alan Turing's work on the *Entscheidungsproblem* followed from Kurt Gödel's work on the incompleteness theorems. The notion of the general purpose computer that came from this work was of fundamental importance to the designers of the computer machinery in the 1940s.

In the 1950s and 1960s, researchers predicted that when human knowledge could be expressed using logic with mathematical notation, it would be possible to create a machine that reasons, or artificial intelligence. This was more difficult than expected because of the complexity of human reasoning. In logic programming, a program consists of a set of axioms and rules. Logic programming systems such as Prolog compute the consequences of the axioms and rules in order to answer a query.

Today, logic is extensively applied in the fields of Artificial Intelligence, and Computer Science, and these fields provide a rich source of problems in formal and informal logic. Argumentation theory is one good example of how logic is being applied to artificial intelligence. The ACM Computing Classification System in particular regards:

_ Section F.3 on Logics and meanings of programs and F.4 on Mathematical logic and formal languages as part of the theory of computer science: this work covers formal semantics of programming languages, as well as work of formal methods such as Hoare logic;

_ Boolean logic as fundamental to computer hardware: particularly, the system's section B.2 on Arithmetic and logic structures, relating to operatives AND, NOT, and OR;

_ Many fundamental logical formalisms are essential to section I.2 on artificial intelligence, for example modal logic and default logic in Knowledge representation formalisms and methods, Horn clauses in logic programming, and description logic.

Furthermore, computers can be used as tools for logicians. For example, in symbolic logic and mathematical logic, proofs by humans can be computer-assisted. Using automated theorem proving the machines can find and check proofs, as well as work with proofs too lengthy to write out by hand.

9.3.9 Bivalence and the law of the excluded middle

Main article: Principle of bivalence

The logics discussed above are all "bivalent" or “two-valued"; that is, they are most naturally understood as dividing propositions into true and false propositions. Non-classical logics are those systems that reject bivalence.

Hegel developed his own dialectic logic that extended Kant's transcendental logic but also brought it back to ground by assuring us that “neither in heaven nor in earth, neither in the world of mind nor of nature, is there anywhere such an abstract 'either–or' as the understanding maintains. Whatever exists is concrete, with difference and opposition in itself”.[34]

In 1910, Nicolai A. Vasiliev extended the law of excluded middle and the law of contradiction and proposed the law of excluded fourth and logic tolerant to contradiction.[35] In the early 20th century Jan Łukasiewicz investigated the extension of the traditional true/false values to include a third value, “possible”, so inventing ternary logic, the first multi-valued logic.

Logics such as fuzzy logic have since been devised with an infinite number of “degrees of truth”, represented by a real number between 0 and 1.[36]

Intuitionistic logic was proposed by L.E.J. Brouwer as the correct logic for reasoning about mathematics, based upon his rejection of the law of the excluded middle as part of his intuitionism. Brouwer rejected formalization in mathematics, but his student Arend Heyting studied intuitionistic logic formally, as did Gerhard Gentzen. Intuitionistic logic is of great interest to computer scientists, as it is a constructive logic and can be applied for extracting verified programs from proofs.

Modal logic is not truth conditional, and so it has often been proposed as a non-classical logic. However, modal logic is normally formalized with the principle of the excluded middle, and its relational semantics is bivalent, so this inclusion is disputable.

9.3.10 “Is logic empirical?"

Main article: Is logic empirical?

What is the epistemological status of the laws of logic? What sort of argument is appropriate for criticizing purported principles of logic? In an influential paper entitled “Is logic empirical?"[37] Hilary Putnam, building on a suggestion of W. V. Quine, argued that in general the facts of propositional logic have a similar epistemological status as facts about the physical universe, for example as the laws of mechanics or of general relativity, and in particular that what physicists have learned about quantum mechanics provides a compelling case for abandoning certain familiar principles of classical logic: if we want to be realists about the physical phenomena described by quantum theory, then we should abandon the principle of distributivity, substituting for classical logic the quantum logic proposed by Garrett Birkhoff and John von Neumann.[38]

Another paper of the same name by Sir Michael Dummett argues that Putnam's desire for realism mandates the law of distributivity.[39] Distributivity of logic is essential for the realist's understanding of how propositions are true of the world in just the same way as he has argued the principle of bivalence is. In this way, the question, "Is logic empirical?" can be seen to lead naturally into the fundamental controversy in metaphysics on realism versus anti-realism.

9.3.11 Implication: strict or material?

Main article: Paradox of entailment

It is obvious that the notion of implication formalized in classical logic does not comfortably translate into natural language by means of "if ... then ...", due to a number of problems called the paradoxes of material implication.

The first class of paradoxes involves counterfactuals, such as *If the moon is made of green cheese, then 2+2=5*, which are puzzling because natural language does not support the principle of explosion. Eliminating this class of paradoxes was the reason for C. I. Lewis's formulation of strict implication, which eventually led to more radically revisionist logics such as relevance logic.

The second class of paradoxes involves redundant premises, falsely suggesting that we know the succedent because of the antecedent: thus "if that man gets elected, granny will die" is materially true since granny is mortal, regardless of the man's election prospects. Such sentences violate the Gricean maxim of relevance, and can be modelled by logics that reject the principle of monotonicity of entailment, such as relevance logic.

9.3.12 Tolerating the impossible

Main article: Paraconsistent logic

Hegel was deeply critical of any simplified notion of the Law of Non-Contradiction. It was based on Leibniz's idea that this law of logic also requires a sufficient ground to specify from what point of view (or time) one says that something cannot contradict itself. A building, for example, both moves and does not move; the ground for the first is our solar system and for the second the earth. In Hegelian dialectic, the law of non-contradiction, of identity, itself relies upon difference and so is not independently assertable.

Closely related to questions arising from the paradoxes of implication comes the suggestion that logic ought to tolerate inconsistency. Relevance logic and paraconsistent logic are the most important approaches here, though the concerns are different: a key consequence of classical logic and some of its rivals, such as intuitionistic logic, is that they respect the principle of explosion, which means that the logic collapses if it is capable of deriving a contradiction. Graham Priest, the main proponent of dialetheism, has argued for paraconsistency on the grounds that there are in fact, true contradictions.[40]

9.3.13 Rejection of logical truth

The philosophical vein of various kinds of skepticism contains many kinds of doubt and rejection of the various bases on which logic rests, such as the idea of logical form, correct inference, or meaning, typically leading to the conclusion

that there are no logical truths. Observe that this is opposite to the usual views in philosophical skepticism, where logic directs skeptical enquiry to doubt received wisdoms, as in the work of Sextus Empiricus.

Friedrich Nietzsche provides a strong example of the rejection of the usual basis of logic: his radical rejection of idealization led him to reject truth as a "... mobile army of metaphors, metonyms, and anthropomorphisms—in short ... metaphors which are worn out and without sensuous power; coins which have lost their pictures and now matter only as metal, no longer as coins."[41] His rejection of truth did not lead him to reject the idea of either inference or logic completely, but rather suggested that "logic [came] into existence in man's head [out] of illogic, whose realm originally must have been immense. Innumerable beings who made inferences in a way different from ours perished".[42] Thus there is the idea that logical inference has a use as a tool for human survival, but that its existence does not support the existence of truth, nor does it have a reality beyond the instrumental: "Logic, too, also rests on assumptions that do not correspond to anything in the real world".[43]

This position held by Nietzsche however, has come under extreme scrutiny for several reasons. He fails to demonstrate the validity of his claims and merely asserts them rhetorically. Although, since he is criticising the established criteria of validity, this does not undermine his position for one could argue that the demonstration of validity provided in the name of logic was just as rhetorically based. Some philosophers, such as Jürgen Habermas, claim his position is self-refuting—and accuse Nietzsche of not even having a coherent perspective, let alone a theory of knowledge.[44] Again, it is unclear if this is a decisive critique for the criteria of coherency and consistent theory are exactly what is under question. Georg Lukács, in his book *The Destruction of Reason*, asserts that, "Were we to study Nietzsche's statements in this area from a logico-philosophical angle, we would be confronted by a dizzy chaos of the most lurid assertions, arbitrary and violently incompatible."[45] Still, in this respect his "theory" would be a much better depicition of a confused and chaotic reality than any consistent and compatible theory. Bertrand Russell described Nietzsche's irrational claims with "He is fond of expressing himself paradoxically and with a view to shocking conventional readers" in his book *A History of Western Philosophy*.[46]

9.4 See also

_ Digital electronics (also known as *digital logic* or logic gates)
_ Fallacies
_ List of logic journals
_ Logic puzzle
_ Logic symbols
_ Mathematics
_ List of mathematics articles
_ Outline of mathematics
_ Metalogic
_ Outline of logic
_ Philosophy
_ List of philosophy topics
_ Outline of philosophy
_ Reason
_ *Straight and Crooked Thinking* (book)
_ Table of logic symbols
_ Truth
_ Vector logic

9.5 Notes and references

[1] "possessed of reason, intellectual, dialectical, argumentative", also related to λόγος (*logos*), "word, thought, idea, argument, account, reason, or principle" (Liddell & Scott 1999; Online Etymology Dictionary 2001).
[2] Richard Henry Popkin; Avrum Stroll (1 July 1993). *Philosophy Made Simple*. Random House Digital, Inc. p. 238. ISBN 978-0-385-42533-9. Retrieved 5 March 2012.
[3] Jacquette, D. (2002). *A Companion to Philosophical Logic*. Wiley Online Library. p. 2.
[4] For example, Nyaya (syllogistic recursion) dates back 1900 years.
[5] Mohists and the school of Names date back at 2200 years.
[6] Plato (1976). Buchanan, Scott, ed. *The Portable Plato*. Penguin. ISBN 0-14-015040-4.
[7] Aristotle (2001). "Posterior Analytics". In Mckeon, Richard. *The Basic Works*. Modern Library. ISBN 0-375-75799-6.
[8] Whitehead, Alfred North; Russell, Bertrand (1967). *Principia Mathematica to *56*. Cambridge University Press. ISBN 0-521-62606-4.
[9] For a more modern treatment, see Hamilton, A. G. (1980). *Logic for Mathematicians*. Cambridge University Press. ISBN 0-521-29291-3.
[10] Łukasiewicz, Jan (1957). *Aristotle's syllogistic from the standpoint of modern formal logic* (2nd ed.). Oxford University Press. p. 7. ISBN 978-0-19-824144-7.
[11] _ Magnani, L. "Abduction, Reason, and Science: Processes of Discovery and Explanation". *Kluwer Academic Plenum Publishers, New York, 2001*. xvii þ 205 pages. Hard cover, ISBN 0-306-46514-0.
_ R. Josephson, J. & G. Josephson, S. "Abductive Inference: Computation, Philosophy, Technology" *Cambridge University Press, New York & Cambridge (U.K.)*. viii þ 306 pages. Hard cover (1994), ISBN 0-521-43461-0, Paperback (1996), ISBN 0-521-57545-1.
_ Bunt, H. & Black, W. "Abduction, Belief and Context in Dialogue: Studies in Computational Pragmatics" *(Natural Language Processing, 1.) John Benjamins, Amsterdam & Philadelphia, 2000*. vi þ 471 pages. Hard cover, ISBN 90-272-4983-0 (Europe),
1-58619-794-2 (U.S.)
[12] Peirce, C. S.
_ "On the Logic of drawing History from Ancient Documents especially from Testimonies" (1901), *Collected Papers* v. 7, paragraph 219.
_ "PAP" ["Prolegomena to an Apology for Pragmatism"], MS 293 c. 1906, *New Elements of Mathematics* v. 4, pp. 319-320.
_ A Letter to F. A. Woods (1913), *Collected Papers* v. 8, paragraphs 385-388.
(See under "Abduction" and "Retroduction" at *Commens Dictionary of Peirce's Terms*.)
[13] Peirce, C. S. (1903), Harvard lectures on pragmatism, *Collected Papers* v. 5, paragraphs 188–189.
[14] Bergmann, Merrie; Moor, James; Nelson, Jack (2009). *The Logic Book* (Fifth ed.). New York, NY: McGraw-Hill. ISBN 978-0-07-353563-0.
[15] Mendelson, Elliott (1964). "Quantification Theory: Completeness Theorems". *Introduction to Mathematical Logic*. Van Nostrand. ISBN 0-412-80830-7.
[16] Hofweber, T. (2004). "Logic and Ontology". In Zalta, Edward N. *Stanford Encyclopedia of Philosophy*.
[17] E.g., Kline (1972, p.53) wrote "A major achievement of Aristotle was the founding of the science of logic".

[18] "Aristotle", MTU Department of Chemistry.
[19] Jonathan Lear (1986). "*Aristotle and Logical Theory*". Cambridge University Press. p.34. ISBN 0-521-31178-0
[20] Simo Knuuttila (1981). "*Reforging the great chain of being: studies of the history of modal theories*". Springer Science & Business. p.71. ISBN 90-277-1125-9
[21] Michael Fisher, Dov M. Gabbay, Lluís Vila (2005). "*Handbook of temporal reasoning in artificial intelligence*". Elsevier. p.119. ISBN 0-444-51493-7

[22] Harold Joseph Berman (1983). "*Law and revolution: the formation of the Western legal tradition*". Harvard University Press. p.133. ISBN 0-674-51776-8
[23] The four Catuşkoţi logical divisions are formally very close to the four opposed propositions of the Greek tetralemma, which in turn are analogous to the four truth values of modern relevance logic Cf. Belnap (1977); Jayatilleke, K. N., (1967, The logic of four alternatives, in *Philosophy East and West*, University of Hawaii Press).
[24] Kisor Kumar Chakrabarti (June 1976). "Some Comparisons Between Frege's Logic and Navya-Nyaya Logic". *Philosophy and Phenomenological Research* (International Phenomenological Society) **36** (4): 554–563. doi:10.2307/2106873. JSTOR 2106873. "This paper consists of three parts. The first part deals with Frege's distinction between sense and reference of proper names and a similar distinction in Navya-Nyaya logic. In the second part we have compared Frege's definition of number to the Navya-Nyaya definition of number. In the third part we have shown how the study of the so-called 'restrictive conditions for universals' in Navya-Nyaya logic anticipated some of the developments of modern set theory."
[25] Jonardon Ganeri (2001). *Indian logic: a reader*. Routledge. pp. vii, 5, 7. ISBN 0-7007-1306-9.
[26] "Aristotle". *Encyclopædia Britannica.*
[27] "History of logic: Arabic logic". Encyclopædia Britannica.
[28] Stolyar, Abram A. (1983). *Introduction to Elementary Mathematical Logic*. Dover Publications. p. 3. ISBN 0-486-64561-4.
[29] Barnes, Jonathan (1995). *The Cambridge Companion to Aristotle*. Cambridge University Press. p. 27. ISBN 0-521-42294-9.
[30] Aristotle (1989). *Prior Analytics*. Hackett Publishing Co. p. 115. ISBN 978-0-87220-064-7.
[31] Mendelson, Elliott (1964). "Formal Number Theory: Gödel's Incompleteness Theorem". *Introduction to Mathematical Logic*. Monterey, Calif.: Wadsworth & Brooks/Cole Advanced Books & Software. OCLC 13580200.
[32] Brookshear, J. Glenn (1989). "Computability: Foundations of Recursive Function Theory". *Theory of computation: formal languages, automata, and complexity*. Redwood City, Calif.: Benjamin/Cummings Pub. Co. ISBN 0-8053-0143-7.
[33] Brookshear, J. Glenn (1989). "Complexity". *Theory of computation: formal languages, automata, and complexity*. Redwood City, Calif.: Benjamin/Cummings Pub. Co. ISBN 0-8053-0143-7.
[34] Hegel, G. W. F (1971) [1817]. *Philosophy of Mind*. Encyclopedia of the Philosophical Sciences. trans. William Wallace. Oxford: Clarendon Press. p. 174. ISBN 0-19-875014-5.
[35] Joseph E. Brenner (3 August 2008). *Logic in Reality*. Springer. pp. 28–30. ISBN 978-1-4020-8374-7. Retrieved 9 April 2012.
[36] Hájek, Petr (2006). "Fuzzy Logic". In Zalta, Edward N.. *Stanford Encyclopedia of Philosophy.*
[37] Putnam, H. (1969). "Is Logic Empirical?". *Boston Studies in the Philosophy of Science* **5**.
[38] Birkhoff, G.; von Neumann, J. (1936). "The Logic of Quantum Mechanics". *Annals of Mathematics* (Annals of Mathematics) **37** (4): 823–843. doi:10.2307/1968621. JSTOR 1968621.
[39] Dummett, M. (1978). "Is Logic Empirical?". *Truth and Other Enigmas*. ISBN 0-674-91076-1.
[40] Priest, Graham (2008). "Dialetheism". In Zalta, Edward N.. *Stanford Encyclopedia of Philosophy.*
[41] Nietzsche, 1873, On Truth and Lies in a Nonmoral Sense.
[42] Nietzsche, 1882, *The Gay Science*.
[43] Nietzsche, 1878, *Human, All Too Human*
[44] Babette Babich, Habermas, Nietzsche, and Critical Theory
[45] Georg Lukács. "The Destruction of Reason by Georg Lukács 1952". Marxists.org. Retrieved 2013-06-16.
[46] Russell, Bertrand (1945), *A History of Western Philosophy And Its Connection with Political and Social Circumstances from the Earliest Times to the Present Day*, Simon and Schuster, p. 762

9.6 Bibliography

_ Nuel Belnap, (1977). "A useful four-valued logic". In Dunn & Eppstein, *Modern uses of multiple-valued logic*. Reidel: Boston.
_ Józef Maria Bocheński (1959). *A précis of mathematical logic*. Translated from the French and German editions by Otto Bird. D. Reidel, Dordrecht, South Holland.
_ Józef Maria Bocheński, (1970). *A history of formal logic*. 2nd Edition. Translated and edited from the German edition by Ivo Thomas. Chelsea Publishing, New York.
_ Brookshear, J. Glenn (1989). *Theory of computation: formal languages, automata, and complexity*. Redwood City, Calif.: Benjamin/Cummings Pub. Co. ISBN 0-8053-0143-7.
_ Cohen, R.S, and Wartofsky, M.W. (1974). *Logical and Epistemological Studies in Contemporary Physics*. Boston Studies in the Philosophy of Science. D. Reidel Publishing Company: Dordrecht, Netherlands. ISBN 90-277-0377-9.
_ Finkelstein, D. (1969). "Matter, Space, and Logic". in R.S. Cohen and M.W. Wartofsky (eds. 1974).

_ Gabbay, D.M., and Guenthner, F. (eds., 2001–2005). *Handbook of Philosophical Logic*. 13 vols., 2nd edition. Kluwer Publishers: Dordrecht.
_ Hilbert, D., and Ackermann, W, (1928). *Grundzüge der theoretischen Logik* (*Principles of Mathematical Logic*). Springer-Verlag. OCLC 2085765
_ Susan Haack, (1996). *Deviant Logic, Fuzzy Logic: Beyond the Formalism*, University of Chicago Press.
_ Hodges, W., (2001). *Logic. An introduction to Elementary Logic*, Penguin Books.
_ Hofweber, T., (2004), Logic and Ontology. *Stanford Encyclopedia of Philosophy*. Edward N. Zalta (ed.).
_ Hughes, R.I.G., (1993, ed.). *A Philosophical Companion to First-Order Logic*. Hackett Publishing.
_ Kline, Morris (1972). *Mathematical Thought From Ancient to Modern Times*. Oxford University Press. ISBN 0-19-506135-7.
_ Kneale, William, and Kneale, Martha, (1962). *The Development of Logic*. Oxford University Press, London, UK.
_ Liddell, Henry George; Scott, Robert. "Logikos". *A Greek-English Lexicon*. Perseus Project. Retrieved 8 May 2009.
_ Mendelson, Elliott, (1964). *Introduction to Mathematical Logic*. Wadsworth & Brooks/Cole Advanced Books & Software: Monterey, Calif. OCLC 13580200
_ Harper, Robert (2001). "Logic". *Online Etymology Dictionary*. Retrieved 8 May 2009.
_ Smith, B., (1989). "Logic and the Sachverhalt". *The Monist* 72(1):52–69.
_ Whitehead, Alfred North and Bertrand Russell, (1910). *Principia Mathematica*. Cambridge University Press: Cambridge, England. OCLC 1041146

9.7 External links

_ Logic at PhilPapers
_ Logic at the Indiana Philosophy Ontology Project
_ Logic entry in the *Internet Encyclopedia of Philosophy*
_ Hazewinkel, Michiel, ed. (2001), "Logical calculus", *Encyclopedia of Mathematics*, Springer, ISBN 978-1-55608-010-4

_ An Outline for Verbal Logic
_ Introductions and tutorials
_ An Introduction to Philosophical Logic, by Paul Newall, aimed at beginners.
_ forall x: an introduction to formal logic, by P.D. Magnus, covers sentential and quantified logic.
_ Logic Self-Taught: A Workbook (originally prepared for on-line logic instruction).
_ Nicholas Rescher. (1964). *Introduction to Logic*, St. Martin's Press.
_ Essays
_ "Symbolic Logic" and "The Game of Logic", Lewis Carroll, 1896.
_ Math & Logic: The history of formal mathematical, logical, linguistic and methodological ideas. In *The Dictionary of the History of Ideas.*
_ Online Tools
_ Interactive Syllogistic Machine A web based syllogistic machine for exploring fallacies, figures, terms, and modes of syllogisms.
_ Reference material
_ Translation Tips, by Peter Suber, for translating from English into logical notation.
_ Ontology and History of Logic. An Introduction with an annotated bibliography.
_ Reading lists
_ The London Philosophy Study Guide offers many suggestions on what to read, depending on the student's familiarity with the subject:
_ Logic & Metaphysics
_ Set Theory and Further Logic
_ Mathematical Logic

Chapter 10

Natural language

This article is about natural language in neuropsychology and linguistics. For natural language in computer systems, see Natural language processing.

In neuropsychology, linguistics and the philosophy of language, a **natural language** or **ordinary language** is any language which arises, unpremeditated, in the brains of human beings. Typically, therefore, these are the languages human beings use to communicate with each other, whether by speech, signing, touch or writing. They are distinguished from constructed and formal languages such as those used to program computers or to study logic.[1]

10.1 Defining natural language

Though the exact definition varies between scholars, natural language can broadly be defined in contrast to artificial or constructed languages (such as computer programming languages and international auxiliary languages) and to other communication systems in nature (such as bees' waggle dance).[2] Definitions of "natural language" also usually state or imply that a "natural" language is one that any cognitively normal human infant is able to learn and whose development has been through use rather than by prescription. An unstandardized language such as African American Vernacular English, for example, is a natural language, whereas a standardized language such as Standard American English is, in part, prescribed.[3]

10.2 Native language learning

Main article: Language acquisition

The learning of one's own native language, typically that of one's parents, normally occurs spontaneously in early human childhood and is biologically, socially and ecologically driven. A crucial role of this process is the ability of humans from an early age to engage in speech repetition and so quickly acquire a spoken vocabulary from the pronunciation of words spoken around them. This together with other aspects of speech involves the neural activity of parts of the human brain such as the Wernicke's and Broca's areas.[4]

There are approximately 7,000 current human languages, and many, if not most seem to share certain properties, leading to the hypothesis of Universal Grammar, as argued by the generative grammar studies of Noam Chomsky and his followers. Recently, it has been demonstrated that a dedicated network in the human brain (crucially involving Broca's area, a portion of the left inferior frontal gyrus), is selectively activated by complex verbal structures (but not simple ones) of those languages that meet the Universal Grammar requirements.[5][6]

While it is clear that there are innate mechanisms that enable the learning of language and define the range of languages that can be learned, it is not clear that these mechanisms in anyway resemble a human language or universal grammar. The study of language acquisition is the domain of psycholinguistics and Chomsky always declined to engage in questions of how his putative language organ, the Language Acquisition Device or Universal Grammar, might have evolved.[7] During a period (the 1970s and 80s) when nativist Transformational Generative Grammar was becoming dominant in Linguistics, and called "Standard Theory", linguists who questioned these tenets were disenfranchised

61

62 *CHAPTER 10. NATURAL LANGUAGE*

and Cognitive Linguistics and Computational Psycholinguistics were born and the more general term Emergentism developed for the anti-nativist view that language is emergent from more fundamental cognitive processes that are not specifically linguistic in nature.

10.3 Origins of natural language

Main article: Origin of language

There is disagreement among anthropologists on when language was first used by humans. Estimates range from about two million (2,000,000) years ago, during the time of *Homo habilis*, to as recently as forty thousand (40,000) years ago, during the time of Cro-Magnon man. However recent evidence suggests modern human language was invented or evolved in Africa prior to the dispersal of humans from Africa around 50,000 years ago. Since all people including the most isolated indigenous groups such as the Andamanese or the Tasmanian aboriginals possess language, then it was presumedly present in the ancestral populations in Africa before the human population split into various groups to inhabit the rest of the world.[8][9]

10.4 Controlled languages

Main article: Controlled natural language

Controlled natural languages are subsets of natural languages whose grammars and dictionaries have been restricted in order to reduce or eliminate both ambiguity and complexity (for instance, by cutting down on rarely used superlative or adverbial forms or irregular verbs). The purpose behind the development and implementation of a controlled natural language typically is to aid non-native speakers of a natural language in understanding it, or to ease computer processing of a natural language. An example of a widely used controlled natural language is Simplified English, which was originally developed for aerospace industry maintenance manuals.

10.5 Constructed languages and international auxiliary languages

Main article: Constructed language

Main article: International auxiliary language
Constructed international auxiliary languages such as Esperanto and Interlingua (even those that have native speakers) are not generally considered natural languages.[10] The problem is that other languages have been used to communicate and evolve in a natural way, while Esperanto was selectively designed by L.L. Zamenhof from natural languages, not grown from the natural fluctuations in vocabulary and syntax. Some natural languages have become naturally "standardized" by children's natural tendency to correct for illogical grammar structures in their parents' language, which can be seen in the development of pidgin languages into creole languages (as explained by Steven Pinker in The Language Instinct), but this is not the case in many languages, including constructed languages such as Esperanto, where strict rules are in place as an attempt to consciously remove such irregularities. The possible exception to this are true native speakers of such languages.[11] More substantive basis for this designation is that the vocabulary, grammar, and orthography of Interlingua are natural; they have been standardized and presented by a linguistic research body, but they predated it and are not themselves considered a product of human invention.[12] Most experts, however, consider Interlingua to be naturalistic rather than natural.[10] Latino Sine Flexione, a second naturalistic auxiliary language, is also naturalistic in content but is no longer widely spoken.[13]

10.6 Modalities

Natural language manifests itself in modalities other than speech. See also nonverbal communication.

10.6.1 Sign languages

A sign language is a language which conveys meaning through visual rather than acoustic patterns—simultaneously combining hand shapes, orientation and movement of the hands, arms or body, and facial expressions to express a speaker's thoughts. Sign languages are natural languages which have developed in Deaf communities, which can include interpreters and friends and families of deaf people as well as people who are deaf or hard of hearing themselves. In contrast, a manually coded language (or signed oral language) is a constructed sign system combining elements of a sign language and an oral language. For example, Signed Exact English (SEE) did not develop naturally in any population, but was "created by a committee of individuals".[14]

10.6.2 Written languages

Main article: Written language
In a sense, written language should be distinguished from natural language. Until recently in the developed world, it was common for many people to be fluent in spoken and yet remain illiterate; this is still the case in poor countries today. Furthermore, natural language acquisition during childhood is largely spontaneous, while literacy must usually be intentionally acquired.[15]

10.7 See also

10.8 Notes

[1] Lyons, John (1991). *Natural Language and Universal Grammar*. New York: Cambridge University Press. pp. 68–70. ISBN 978-0521246965.
[2] Fernandes, Keith (2008). *On the Significance of Speech: How Infants Discover Symbols and Structure*. Ann Arbor, MI: ProQuest. pp. 7–8. ISBN 978-1243524065.
[3] Language: Journal of the Linguistic Society of America
[4] Kendra A. Palmer (2009). "Understanding Human Language: An In-Depth Exploration of the Human Facility for Language". StudentPulse.com. Retrieved 22 August 2012.
[5] A. Moro, M. Tettamanti, D. Perani, C. Donati, S. F. Cappa, F. Fazio "Syntax and the brain: disentangling grammar by selective anomalies", NeuroImage, 13, January 2001, Academic Press, Chicago, pp. 110-118
[6] Musso, M., Moro, A. , Glauche. V., Rijntjes, M., Reichenbach, J., Büchel, C., Weiller, C. "Broca's area and the language instinct," Nature neuroscience, 2003, vol. 6, pp. 774-781.
[7] Piattelli-Palmarini M., Language and Learning: The Debate between Jean Piaget & Noam Chomsky, Routledge and Kegan Paul, 1980.
[8] Early Voices: The Leap to Language nytimes article by Nicholas Wade
[9] http://www.benjamins.com/cgi-bin/t_bookview.cgi?bookid=CELCR%205
[10] Gopsill, F. P., "A historical overview of international languages". In *International languages: A matter for Interlingua*. Sheffield, England: British Interlingua Society, 1990.
[11] Proponents contend that there are 200-2000 native speakers of Esperanto.
[12] Gode, Alexander, *Interlingua-English: A dictionary of the international language*. New York: Storm Publishers, 1951. (Original edition)
[13] Gopsill, F. P., "Naturalistic international languages". In *International languages: A matter for Interlingua*. Sheffield, England: British Interlingua Society, 1990.
[14] Emmorey, Karen. *Language, cognition, and the brain: insights from sign language research* (2001), p. 11.

[15] Pinker, Steven. 1994. The Language Instinct

10.9 References

_ ter Meulen, Alice, 2001, "Logic and Natural Language," in Goble, Lou, ed., *The Blackwell Guide to Philosophical Logic*. Blackwell.

Chapter 11

Informal logic

Informal logic, intuitively, refers to the principles of logic and logical thought outside of a formal setting. However, perhaps because of the "informal" in the title, the precise definition of "informal logic" is a matter of some dispute.[1] Ralph H. Johnson and J. Anthony Blair define informal logic as "a branch of logic whose task is to develop non-formal standards, criteria, procedures for the analysis, interpretation, evaluation, criticism and construction of argumentation."[2] This definition reflects what had been implicit in their practice and what others[3][4][5] were doing in their informal logic texts.

Informal logic is associated with (informal) fallacies, critical thinking, the Thinking Skills Movement[6] and the interdisciplinary inquiry known as argumentation theory. Frans H. van Eemeren writes that the label "informal logic" covers a "collection of normative approaches to the study of reasoning in ordinary language that remain closer to the practice of argumentation than formal logic."[7]

11.1 History

Informal logic as a distinguished enterprise under this name emerged roughly in the late 1970s as a sub-field of philosophy. The naming of the field was preceded by the appearance of a number of textbooks that rejected the symbolic approach to logic on pedagogical grounds as inappropriate and unhelpful for introductory textbooks on logic for a general audience, for example Howard Kahane's *Logic and Contemporary Rhetoric*, subtitled "The Use of Reason in Everyday Life", first published in 1971. Kahane's textbook was described on the notice of his death in the *Proceedings And Addresses of the American Philosophical Association* (2002) as "a text in informal logic, [that] was intended to enable students to cope with the misleading rhetoric one frequently finds in the media and in political discourse. It was organized around a discussion of fallacies, and was meant to be a practical instrument for dealing with the problems of everyday life. [It has] ... gone through many editions; [it is] ... still in print; and the thousands upon thousands of students who have taken courses in which his text [was] ... used can thank Howard for contributing to their ability to dissect arguments and avoid the deceptions of deceitful rhetoric. He tried to put into practice the ideal of discourse that aims at truth rather than merely at persuasion. (Hausman et al. 2002)"[8][9] Other textbooks from the era taking this approach were Michael Scriven's *Reasoning* (Edgepress, 1976) and *Logical Self-Defense* by Ralph Johnson and J. Anthony Blair, first published in 1977.[8] Earlier precursors in this tradition can be considered Monroe Beardsley's *Practical Logic* (1950) and Stephen Toulmin's *The Uses of Argument* (1958).[10]

The field perhaps became recognized under its current name with the *First International Symposium on Informal Logic* held in 1978. Although initially motivated by a new pedagogical approach to undergraduate logic textbooks, the scope of the field was basically defined by a list of 13 problems and issues which Blair and Johnson included as an appendix to their keynote address at this symposium:[8][11]

_ the theory of logical criticism
_ the theory of argument
_ the theory of fallacy
_ the fallacy approach vs. the critical thinking approach

_ the viability of the inductive/deductive dichotomy
_ the ethics of argumentation and logical criticism
_ the problem of assumptions and missing premises
_ the problem of context
_ methods of extracting arguments from context
_ methods of displaying arguments
_ the problem of pedagogy
_ the nature, division and scope of informal logic
_ the relationship of informal logic to other inquiries

David Hitchcock argues that the naming of the field was unfortunate, and that *philosophy of argument* would have been more appropriate. He argues that more undergraduate students in North America study informal logic than any other branch of philosophy, but that as of 2003 informal logic (or philosophy of argument) was not recognized as separate sub-field by the World Congress of Philosophy.[8] Frans H. van Eemeren wrote that "informal logic" is mainly an approach to argumentation advanced by a group of US and Canadian philosophers and largely based on the previous works of Stephen Toulmin and to a lesser extent those of Chaïm Perelman.[7]
Alongside the symposia, since 1983 the journal *Informal Logic* has been the publication of record of the field, with Blair and Johnson as initial editors, with the editorial board now including two other colleagues from the University of Windsor—Christopher Tindale and Hans V. Hansen.[12] Other journals that regularly publish articles on informal logic include *Argumentation* (founded in 1986), *Philosophy and Rhetoric*, *Argumentation and Advocacy* (the journal of the American Forensic Association), and *Inquiry: Critical Thinking Across the Disciplines* (founded in 1988).[13]

11.2 Proposed definitions

Johnson and Blair (2000) proposed the following definition: "Informal logic designates that branch of logic whose task is to develop non-formal$_2$ standards, criteria, procedures for the analysis, interpretation, evaluation, critique and construction of argumentation in everyday discourse." Their meaning of non-formal$_2$ is taken from Barth and Krabbe (1982), which is explained below.
To understand the definition above, one must understand "informal" which takes its meaning in contrast to its counterpart "formal." (This point was not made for a very long time, hence the nature of informal logic remained opaque, even to those involved in it, for a period of time.) Here it is helpful to have recourse[14] to Barth and Krabbe (1982:14f) where they distinguish three senses of the term "form." By "form$_1$," Barth and Krabbe mean the sense of the term which derives from the Platonic idea of form—the ultimate metaphysical unit. Barth and Krabbe claim that most traditional logic is formal in this sense. That is, syllogistic logic is a logic of terms where the terms could naturally be understood as place-holders for Platonic (or Aristotelian) forms. In this first sense of "form," almost all logic is informal (not-formal). Understanding informal logic this way would be much too broad to be useful.
By "form$_2$," Barth and Krabbe mean the form of sentences and statements as these are understood in modern systems of logic. Here validity is the focus: if the premises are true, the conclusion must then also be true. Now validity has to do with the logical form of the statement that makes up the argument. In this sense of "formal," most modern and contemporary logic is "formal." That is, such logics canonize the notion of logical form, and the notion of validity plays the central normative role. In this second sense of form, informal logic is not-formal, because it abandons the notion of logical form as the key to understanding the structure of arguments, and likewise retires validity as normative for the purposes of the evaluation of argument. It seems to many that validity is too stringent a requirement, that there are good arguments in which the conclusion is supported by the premises even though it does not follow necessarily from them (as validity requires). An argument in which the conclusion is thought to be "beyond reasonable doubt, given the premises" is sufficient in law to cause a person to be sentenced to death, even though it does not meet the standard of logical validity. This type of argument, based on accumulation of evidence rather than pure deduction, is called a conductive argument.
By "form$_3$," Barth and Krabbe mean to refer to "procedures which are somehow regulated or regimented, which take place according to some set of rules." Barth and Krabbe say that "we do not defend formality$_3$ of all kinds and under

all circumstances." Rather "we defend the thesis that verbal dialectics must have a certain form (i.e., must proceed according to certain rules) in order that one can speak of the discussion as being won or lost" (19). In this third sense of "form", informal logic can be formal, for there is nothing in the informal logic enterprise that stands opposed to the idea that argumentative discourse should be subject to norms, i.e., subject to rules, criteria, standards or procedures. Informal logic does present standards for the evaluation of argument, procedures for detecting missing premises etc.
Johnson and Blair (2000) noticed a limitation of their own definition, particularly with respect to "everyday discourse", which could indicate that it does not seek to understand specialized, domain-specific arguments made in natural languages. Consequently, they have argued that the crucial divide is between arguments made in formal languages and those made in natural languages.
Fisher and Scriven (1997) proposed a more encompassing definition, seeing informal logic as "the discipline which studies the practice of critical thinking and provides its intellectual spine". By "critical thinking" they understand "skilled and active interpretation and evaluation of observations and communications, information and argumentation."[15]

11.3 Criticisms

Some hold the view that informal logic is not a branch or subdiscipline of logic, or even the view that there cannot be such a thing as informal logic.[16][17][18] Massey criticizes informal logic on the grounds that it has no theory underpinning it. Informal logic, he says, requires detailed classification schemes to organize it, which in other disciplines is provided by the underlying theory. He maintains that there is no method of establishing the invalidity of an argument aside from the formal method, and that the study of fallacies may be of more interest to other disciplines, like psychology, than to philosophy and logic.[16]

11.4 Relation to critical thinking

See also: Critical thinking

Since the 1980s, informal logic has been partnered and even equated,[19] in the minds of many, with critical thinking. The precise definition of "critical thinking" is a subject of much dispute.[20] Critical thinking, as defined by Johnson, is the evaluation of an intellectual product (an argument, an explanation, a theory) in terms of its strengths and weaknesses.[20] While critical thinking will include evaluation of arguments and hence require skills of argumentation including informal logic, critical thinking requires additional abilities not supplied by informal logic, such as the ability to obtain and assess information and to clarify meaning. Also, many believe that critical thinking requires certain dispositions.[21] Understood in this way, "critical thinking" is a broad term for the attitudes and skills that are involved in analyzing and evaluating arguments. The critical thinking movement promotes critical thinking as an educational ideal. The movement emerged with great force in the '80s in North America as part of an ongoing critique of education as regards the thinking skills not being taught.

11.5 Relation to argumentation theory

See also: Argumentation theory

The social, communicative practice of argumentation can and should be distinguished from implication (or entailment)—a relationship between propositions; and from inference—a mental activity typically thought of as the drawing of a conclusion from premises. Informal logic may thus be said to be a logic of argumentation, as distinguished from implication and inference.[22]

Argumentation theory (or the theory of argumentation) has come to be the term that designates the theoretical study of argumentation. This study is interdisciplinary in the sense that no one discipline will be able to provide a complete account. A full appreciation of argumentation requires insights from logic (both formal and informal), rhetoric, communication theory, linguistics, psychology, and, increasingly, computer science. Since the 1970s, there has been significant agreement that there are three basic approaches to argumentation theory: the logical, the rhetorical and the dialectical. According to Wenzel,[23] the logical approach deals with the product, the dialectical with the process,

and the rhetorical with the procedure. Thus, informal logic is one contributor to this inquiry, being most especially concerned with the norms of argument.

11.6 See also

- Argument map
- Informal fallacy
- Inference objection
- Lemma
- Philosophy of language
- Semantics

11.7 Footnotes

[1] See Johnson 1999 for a survey of definitions.

[2] Johnson, Ralph H., and Blair, J. Anthony (1987), "The Current State of Informal Logic", *Informal Logic*, 9(2–3), 147–151. Johnson & Blair added "... in everyday discourse" but in (2000), modified their definition, and broadened the focus now to include the sorts of argument that occurs not just in everyday discourse but also disciplined inquiry—what Weinstein (1990) calls "stylized discourse."

[3] Scriven, 1976

[4] Munson, 1976

[5] Fogelin, 1978

[6] Resnick, 1989

[7] Frans H. van Eemeren (2009). "The Study of Argumentation". In Andrea A. Lunsford, Kirt H. Wilson, Rosa A. Eberly. *The SAGE handbook of rhetorical studies*. SAGE. p. 117. ISBN 978-1-4129-0950-1.

[8] David Hitchcock, Informal logic 25 years later in *Informal Logic at 25: Proceedings of the Windsor Conference (OSSA 2003)*

[9] JSTOR 3218569

[10] Fisher (2004) p. vii

[11] J. Anthony Blair and Ralph H. Johnson (eds.), Informal Logic: The First International Symposium, 3-28. Pt. Reyes, CA: Edgepress

[12] http://ojs.uwindsor.ca/ojs/leddy/index.php/informal_logic/about/editorialTeam

[13] Johnson and Blair (2000), p. 100

[14] As Johnson (1999) does.

[15] Johnson and Blair (2000), p. 95

[16] Massey, 1981

[17] Woods, 1980
[18] Woods, 2000
[19] Johnson (2000) takes the conflation to be part of the Network Problem and holds that settling the issue will require a theory of reasoning.
[20] Johnson, 1992
[21] Ennis, 1987
[22] Johnson, 1999
[23] Wenzel (1990)

11.8 References

_ Barth, E. M., & Krabbe, E. C. W. (Eds.). (1982). From axiom to dialogue: A philosophical study of logics and argumentation. Berlin: Walter De Gruyter.
_ Blair, J. A & Johnson, R.H. (1980). The recent development of informal logic. In J. Anthony Blair and Ralph H. Johnson (Eds.). Informal logic: The first international symposium, (pp. 3–28). Inverness, CA: Edgepress.
_ Ennis, R.H. (1987). A taxonomy of critical thinking dispositions and abilities. In J.B. Baron and R.J. Sternberg (Eds.), Teaching critical thinking skills: Theory and practice, (pp. 9–26). New York: Freeman.
_ Eemeren, F. H. van, & Grootendorst, R. (1992). Argumentation, communication and fallacies. Hillsdale, NJ: Lawrence Erlbaum Associates.
_ Fisher, A. and Scriven, M. (1997). Critical thinking: It's definition and assessment. Point Reyes, CA: Edgepress
_ Fisher, Alec (2004). *The logic of real arguments* (2nd ed.). Cambridge University Press. ISBN 978-0-521-65481-4.
_ Govier, T. (1987). Problems in argument analysis and evaluation. Dordrecht: Foris.
_ Govier, T. (1999). The Philosophy of Argument. Newport News, VA: Vale Press.
_ Groarke, L. (2006). Informal Logic. Stanford Encyclopedia of Philosophy, from http://plato.stanford.edu/entries/logic-informal/
_ Hitchcock, David (2007). "Informal logic and the concept of argument". In Jacquette, Dale. *Philosophy of logic*. Elsevier. ISBN 978-0-444-51541-4. preprint
_ Johnson, R. H. (1992). The problem of defining critical thinking. In S. P. Norris (Ed.), The generalizability of critical thinking (pp. 38–53). New York: Teachers College Press. (Reprinted in Johnson (1996).)
_ Johnson, R. H. (1996). The rise of informal logic. Newport News, VA: Vale Press
_ Johnson, R. H. (1999). The relation between formal and informal logic. *Argumentation*, 13(3) 265-74.
_ Johnson, R. H. (2000). Manifest rationality: A pragmatic theory of argument. Mahwah, NJ: Lawrence Erlbaum Associates.
_ Johnson, R. H. & Blair, J. A. (1987). The current state of informal logic. *Informal Logic* 9, 147-51.
_ Johnson, R. H. & Blair, J. A. (1996). Informal logic and critical thinking. In F. van Eemeren, R. Grootendorst, & F. Snoeck Henkemans (Eds.), Fundamentals of argumentation theory (pp. 383–86). Mahwah, NJ: Lawrence Erlbaum Associates
_ Johnson, R. H. & Blair, J. A. (2002). Informal logic and the reconfiguration of logic. In D. Gabbay, R. H. Johnson, H.-J. Ohlbach and J. Woods (Eds.). Handbook of the logic of argument and inference: The turn towards the practical (pp. 339–396). Elsivier: North Holland.
_ MacFarlane, J. (2005). Logical Constants. Stanford Encyclopedia of Philosophy.
_ Massey, G. (1981). The fallacy behind fallacies. *Midwest Studies of Philosophy*, 6, 489-500.
_ Munson, R. (1976). The way of words: an informal logic. Boston: Houghton Mifflin.
_ Resnick, L. (1987). Education and learning to think. Washington, DC: National Academy Press..
_ Walton, D. N. (1990). What is reasoning? What is an argument? *The Journal of Philosophy*, 87, 399-419.
_ Weinstein, M. (1990) Towards a research agenda for informal logic and critical thinking. *Informal Logic*, 12, 121-143.
_ Wenzel, J. 1990 Three perspectives on argumentation. In R Trapp and J Scheutz, (Eds.), Perspectives on argumentation: Essays in honour of Wayne Brockreide, 9-26 Waveland Press: Prospect Heights, IL
_ Woods, J. (1980). What is informal logic? In J.A. Blair & R. H. Johnson (Eds.), Informal Logic: The First International Symposium (pp. 57–68). Point Reyes, CA: Edgepress.

11.8.1 Special journal issue

The open access issue 20(2) of *Informal Logic* from year 2000 groups a number of papers addressing foundational issues, based on the Panel on Informal Logic that was held at the 1998 World Congress of Philosophy, including:
_ Hitchcock, D. (2000) The significance of informal logic for philosophy. *Informal Logic* 20(2), 129-138.
_ Johnson, R. H. & Blair, J. A. (2000). Informal logic: An overview. *Informal Logic* 20(2): 93-99.
_ Woods, J. (2000). How Philosophical is Informal Logic? *Informal Logic* 20(2): 139-167. 2000

11.8.2 Textbooks

_ Kahane, H. (1971). Logic and contemporary rhetoric:The use of reasoning in everyday life. Belmont: Wadsworth. Still in print as Nancy Cavender; Howard Kahane (2009). *Logic and Contemporary Rhetoric: The Use of Reason in Everyday Life* (11th ed.). Cengage Learning. ISBN 978-0-495-80411-6.
_ Scriven, M. (1976). Reasoning. New York. McGraw Hill.
_ Johnson, R. H. & Blair, J. A. (1977). Logical self-defense. Toronto: McGraw-Hill Ryerson. US Edition. (2006). New York: Idebate Press.
_ Fogelin, R.J. (1978). Understanding arguments: An introduction to informal logic. New York: Harcourt, Brace, Jovanovich. Still in print as Sinnott-Armstrong, Walter; Fogelin, Robert (2010), *Understanding Arguments: An Introduction to Informal Logic* (8th ed.), Belmont, California: Wadsworth Cengage Learning, ISBN 978-0-495-60395-5
_ Stephen N. Thomas (1997). *Practical reasoning in natural language* (4th ed.). Prentice Hall. ISBN 978-0-13-678269-8.
_ Irving M. Copi; Keith Burgess-Jackson (1996). *Informal logic* (3rd ed.). Prentice Hall. ISBN 978-0-13-229048-7.
_ Woods, John, Andrew Irvine and Douglas Walton, 2004. Argument: Critical Thinking, Logic and the Fallacies. Toronto: Prentice Hall
_ Groarke, Leo and Christopher Tindale, 2004. Good Reasoning Matters! (3rd edition). Toronto: Oxford University Press
_ Douglas N. Walton (2008). *Informal logic: a pragmatic approach* (2nd ed.). Cambridge University Press. ISBN 978-0-521-71380-1.
_ Trudy Govier (2009). *A Practical Study of Argument* (7th ed.). Cengage Learning. ISBN 978-0-495-60340-5.

11.9 External links

_ Informal Logic entry by Leo Groarke in the *Stanford Encyclopedia of Philosophy*

Chapter 12

Formal proof

A **formal proof** or **derivation** is a finite sequence of sentences (called well-formed formulas in the case of a formal language) each of which is an axiom or follows from the preceding sentences in the sequence by a rule of inference. The last sentence in the sequence is a theorem of a formal system. The notion of theorem is not in general effective, therefore there may be no method by which we can always find a proof of a given sentence or determine that none exists. The concept of natural deduction is a generalization of the concept of proof.[1]
The theorem is a syntactic consequence of all the well-formed formulas preceding it in the proof. For a well-formed formula to qualify as part of a proof, it must be the result of applying a rule of the deductive apparatus of some formal system to the previous well-formed formulae in the proof sequence.
Formal proofs often are constructed with the help of computers in interactive theorem proving. Significantly, these proofs can be checked automatically, also by computer. Checking formal proofs is usually simple, while the problem of *finding* proofs (automated theorem proving) is usually computationally intractable and/or only semi-decidable, depending upon the formal system in use.

12.1 Background

12.1.1 Formal language

Main article: Formal language
A *formal language* is a set of finite sequences of symbols. Such a language can be defined without reference to any meanings of any of its expressions; it can exist before any interpretation is assigned to it – that is, before it has any meaning. Formal proofs are expressed in some formal language.

12.1.2 Formal grammar

Main articles: Formal grammar and Formation rule
A *formal grammar* (also called *formation rules*) is a precise description of the well-formed formulas of a formal language. It is synonymous with the set of strings over the alphabet of the formal language which constitute well

formed formulas. However, it does not describe their semantics (i.e. what they mean).

12.1.3 Formal systems

Main article: Formal system

A *formal system* (also called a *logical calculus*, or a *logical system*) consists of a formal language together with a deductive apparatus (also called a *deductive system*). The deductive apparatus may consist of a set of transformation

71

72 *CHAPTER 12. FORMAL PROOF*

rules (also called *inference rules*) or a set of axioms, or have both. A formal system is used to derive one expression from one or more other expressions.

12.1.4 Interpretations

Main articles: Formal semantics (logic) and Interpretation (logic)

An *interpretation* of a formal system is the assignment of meanings to the symbols, and truth-values to the sentences of a formal system. The study of interpretations is called formal semantics. *Giving an interpretation* is synonymous with *constructing a model.*

12.2 See also

_ Proof (truth)
_ Mathematical proof
_ Proof theory
_ Axiomatic system

12.3 References

[1] The Cambridge Dictionary of Philosophy, *deduction*

12.4 External links

_ "A Special Issue on Formal Proof". *Notices of the American Mathematical Society*. December 2008.
_ 2πix.com: Logic Part of a series of articles covering mathematics and logic.

Chapter 13
Proof theory

Proof theory is a branch of mathematical logic that represents proofs as formal mathematical objects, facilitating their analysis by mathematical techniques. Proofs are typically presented as inductively-defined data structures such as plain lists, boxed lists, or trees, which are constructed according to the axioms and rules of inference of the logical system. As such, proof theory is syntactic in nature, in contrast to model theory, which is semantic in nature. Together with model theory, axiomatic set theory, and recursion theory, proof theory is one of the so-called *four pillars* of the foundations of mathematics.[1]

Proof theory is important in philosophical logic, where the primary interest is in the idea of a proof-theoretic semantics, an idea which depends upon technical ideas in structural proof theory to be feasible.

13.1 History

Although the formalisation of logic was much advanced by the work of such figures as Gottlob Frege, Giuseppe Peano, Bertrand Russell, and Richard Dedekind, the story of modern proof theory is often seen as being established by David Hilbert, who initiated what is called Hilbert's program in the foundations of mathematics. Kurt Gödel's seminal work on proof theory first advanced, then refuted this program: his completeness theorem initially seemed to bode well for Hilbert's aim of reducing all mathematics to a finitist formal system; then his incompleteness theorems showed that this is unattainable. All of this work was carried out with the proof calculi called the Hilbert systems.

In parallel, the foundations of structural proof theory were being founded. Jan Łukasiewicz suggested in 1926 that one could improve on Hilbert systems as a basis for the axiomatic presentation of logic if one allowed the drawing of conclusions from assumptions in the inference rules of the logic. In response to this Stanisław Jaśkowski (1929) and Gerhard Gentzen (1934) independently provided such systems, called calculi of natural deduction, with Gentzen's approach introducing the idea of symmetry between the grounds for asserting propositions, expressed in introduction rules, and the consequences of accepting propositions in the elimination rules, an idea that has proved very important in proof theory.[2] Gentzen (1934) further introduced the idea of the sequent calculus, a calculus advanced in a similar spirit that better expressed the duality of the logical connectives,[3] and went on to make fundamental advances in the formalisation of intuitionistic logic, and provide the first combinatorial proof of the consistency of Peano arithmetic. Together, the presentation of natural deduction and the sequent calculus introduced the fundamental idea of analytic proof to proof theory.

13.2 Formal and informal proof

Main article: Formal proof

The *informal* proofs of everyday mathematical practice are unlike the *formal* proofs of proof theory. They are rather like high-level sketches that would allow an expert to reconstruct a formal proof at least in principle, given enough time and patience. For most mathematicians, writing a fully formal proof is too pedantic and long-winded to be in common use.

73

74 *CHAPTER 13. PROOF THEORY*

Formal proofs are constructed with the help of computers in interactive theorem proving. Significantly, these proofs can be checked automatically, also by computer. (Checking formal proofs is usually simple, whereas *finding* proofs (automated theorem proving) is generally hard.) An informal proof in the mathematics literature, by contrast, requires weeks of peer review to be checked, and may still contain errors.

13.3 Kinds of proof calculi

The three most well-known styles of proof calculi are:

_ The Hilbert calculi

_ The natural deduction calculi

_ The sequent calculi

Each of these can give a complete and axiomatic formalization of propositional or predicate logic of either the classical or intuitionistic flavour, almost any modal logic, and many substructural logics, such as relevance logic or linear logic. Indeed it is unusual to find a logic that resists being represented in one of these calculi.

13.4 Consistency proofs

Main article: Consistency proof

As previously mentioned, the spur for the mathematical investigation of proofs in formal theories was Hilbert's program. The central idea of this program was that if we could give finitary proofs of consistency for all the sophisticated formal theories needed by mathematicians, then we could ground these theories by means of a metamathematical argument, which shows that all of their purely universal assertions (more technically their provable _0
1 sentences)
are finitarily true; once so grounded we do not care about the non-finitary meaning of their existential theorems, regarding these as pseudo-meaningful stipulations of the existence of ideal entities.

The failure of the program was induced by Kurt Gödel's incompleteness theorems, which showed that any ω-consistent theory that is sufficiently strong to express certain simple arithmetic truths, cannot prove its own consistency, which on Gödel's formulation is a _0
1 sentence.

Much investigation has been carried out on this topic since, which has in particular led to:

_ Refinement of Gödel's result, particularly J. Barkley Rosser's refinement, weakening the above requirement of ω-consistency to simple consistency;

_ Axiomatisation of the core of Gödel's result in terms of a modal language, provability logic;

_ Transfinite iteration of theories, due to Alan Turing and Solomon Feferman;

_ The recent discovery of self-verifying theories, systems strong enough to talk about themselves, but too weak to carry out the diagonal argument that is the key to Gödel's unprovability argument.

13.5 Structural proof theory

Main article: Structural proof theory

Structural proof theory is the subdiscipline of proof theory that studies proof calculi that support a notion of analytic proof. The notion of analytic proof was introduced by Gentzen for the sequent calculus; there the analytic proofs are those that are cut-free. His natural deduction calculus also supports a notion of analytic proof, as shown by Dag Prawitz. The definition is slightly more complex: we say the analytic proofs are the normal forms, which are related to the notion of normal form in term rewriting. More exotic proof calculi such as Jean-Yves Girard's proof nets also support a notion of analytic proof.

13.6. PROOF-THEORETIC SEMANTICS 75

Structural proof theory is connected to type theory by means of the Curry-Howard correspondence, which observes a structural analogy between the process of normalisation in the natural deduction calculus and beta reduction in the typed lambda calculus. This provides the foundation for the intuitionistic type theory developed by Per Martin-Löf, and is often extended to a three way correspondence, the third leg of which are the cartesian closed categories.

13.6 Proof-theoretic semantics

Main articles: proof-theoretic semantics and logical harmony
In linguistics, type-logical grammar, categorial grammar and Montague grammar apply formalisms based on structural proof theory to give a formal natural language semantics.

13.7 Tableau systems

Main article: Method of analytic tableaux
Analytic tableaux apply the central idea of analytic proof from structural proof theory to provide decision procedures and semi-decision procedures for a wide range of logics.

13.8 Ordinal analysis

Main article: Ordinal analysis
Ordinal analysis is a powerful technique for providing combinatorial consistency proofs for theories formalising arithmetic and analysis.

13.9 Logics from proof analysis

Main article: Substructural logic
Several important logics have come from insights into logical structure arising in structural proof theory.

13.10 See also

_ Intermediate logic
_ Model theory
_ Proof (truth)
_ Proof techniques

13.11 Notes

[1] E.g., Wang (1981), pp. 3–4, and Barwise (1978).
[2] Prawitz (2006, p. 98).
[3] Girard, Lafont, and Taylor (1988).

13.12 References

_ J. Avigad, E.H. Reck (2001). "Clarifying the nature of the infinite": the development of metamathematics and proof theory. Carnegie-Mellon Technical Report CMU-PHIL-120.
_ J. Barwise (ed., 1978). Handbook of Mathematical Logic. North-Holland.
_ A. S. Troelstra, H. Schwichtenberg (1996). *Basic Proof Theory*. In series *Cambridge Tracts in Theoretical Computer Science*, Cambridge University Press, ISBN 0-521-77911-1.
_ G. Gentzen (1935/1969). Investigations into logical deduction. In M. E. Szabo, editor, *Collected Papers of Gerhard Gentzen*. North-Holland. Translated by Szabo from "Untersuchungen über das logische Schliessen", Mathematisches Zeitschrift 39: 176-210, 405-431.
_ Hazewinkel, Michiel, ed. (2001), "Proof theory", *Encyclopedia of Mathematics*, Springer, ISBN 978-1-55608-010-4
_ Luis Moreno & Bharath Sriraman (2005).*Structural Stability and Dynamic Geometry: Some Ideas on Situated Proof. International Reviews on Mathematical Education. Vol. 37, no.3, pp. 130–139*
_ Prawitz, Dag (2006) [1965]. *Natural deduction: A proof-theoretical study*. Mineola, New York: Dover Publications. ISBN 978-0-486-44655-4.
_ J. von Plato (2008). The Development of Proof Theory. Stanford Encyclopedia of Philosophy.
_ Wang, Hao (1981). *Popular Lectures on Mathematical Logic*. Van Nostrand Reinhold Company. ISBN 0-442-23109-1.

Chapter 14

Mathematical practice

Mathematical practice is used to distinguish the working practices of professional mathematicians (e.g. selecting theorems to prove, using informal notations to persuade themselves and others that various steps in the final proof can be formalised, and seeking peer review and publication) from the end result of proven and published theorems.

14.1 Quasi-empiricism

This distinction is considered especially important by adherents of quasi-empiricism in mathematics, which denies the possibility of foundations of mathematics and attempts to refocus attention on the ways in which mathematicians arrive at mathematical statements.

14.2 Folk mathematics

If modern mathematical practices are what distinguish modern professional mathematicians from older ideas of folk mathematics. Although such "folk" practices may well include useful formulae or algorithms, they are generally without the accompanying proof discipline.

14.3 Historical tradition

The evolution of mathematical practice was slow, and some contributors to modern mathematics did not follow even the practice of their time, e.g. Pierre de Fermat who was infamous for withholding his proofs, but nonetheless had a vast reputation for correct assertions of results. Likewise there is contrast between the practices of Pythagoras and Euclid. While Euclid was the originator of what we now understand as the published geometric proof, Pythagoras created a closed community and suppressed results; he is even said to have drowned a student in a barrel for revealing the existence of irrational numbers. Modern mathematicians admire Euclid's practices, and usually frown on those of both Fermat and Pythagoras. Nonetheless, all three are considered important contributors to mathematics, despite the variance in method.

One motivation to study mathematical practice is that, despite much work in the 20th century, some still feel that the foundations of mathematics remain unclear and ambiguous. One proposed remedy is to shift focus to some degree onto 'what is meant by a proof', and other such questions of method.

If mathematics has been informally used throughout history, in numerous cultures and continents, then it could be argued that "mathematical practice" is the practice, or use, of mathematics in everyday life. One definition of mathematical practice, as described above, is the "working practices of professional mathematicians." However, another definition, more in keeping with the predominant usage of mathematics, is that mathematical practice is the everyday practice, or use, of math. Whether one is estimating the total cost of their groceries, calculating miles per gallon, or figuring out how many minutes on the treadmill that chocolate éclair will require, math as used by most people relies less on proof than on practicality (i. e., does it answer the question?)

14.4 Teaching practice

Mathematical teaching usually requires the use of several important teaching pedagogies or components. Most GCSE, A-Level and undergraduate mathematics require the following components:

1. Textbooks or lecture notes which display the mathematical material to be covered/taught within the context of the teaching of mathematics. This requires that the mathematical content being taught at the (say) undergraduate level is of a well documented and widely accepted nature that has been unanimously verified as being correct and meaningful within a mathematical context.
2. Workbooks. Usually, in order to ensure that students have an opportunity to learn and test the material that they have learnt, workbooks or question papers enable mathematical understanding to be tested. It is not unknown for exam papers to draw upon questions from such test papers, or to require prerequisite knowledge of such test papers for mathematical progression.
3. Exam papers and standardised (and preferably apolitical) testing methods. Often, within countries such as the US, the UK (and, in all likelihood, China) there are standardised qualifications, examinations and workbooks that form the concrete teaching materials needed for secondary-school and pre-university courses (for example, within the UK, all students are required to sit or take Scottish Highers/Advanced Highers, A-levels or their equivalent in order to ensure that a certain minimal level of mathematical competence in a wide variety of topics has been obtained). Note, however, that at the undergraduate, post-graduate and doctoral levels within these countries, there need not be any standardised process via which mathematicians of differing ability levels can be tested or examined. Other common test formats within the UK and beyond include the BMO (which is a multiple-choice test competition paper used in order to determine the best candidates that are to represent countries within the International Mathematical Olympiad).

14.5 Assessment practice

Assessment practice overlaps with teaching practice in a sense (it is difficult to teach individuals to a certain level of mathematical competence without first having fore-knowledge of their current mathematical abilities).
These test practices sometimes require written exams to be sat (exams in which answers are in actuality written on exam scripts). However, given the usually lofty moral standards by which mathematical assessment has been tauted to have been conducted according to (together with the ease of statistical data interpretation that such test formats are associated with), multiple choice questions are often seen as useful in determining or verifying a given level of mathematical capability.

14.6 Other aspects of mathematical practice

14.7 See also

_ Informal mathematics
_ Foundations of mathematics
_ Philosophy of mathematics

Chapter 15

Quasi-empiricism in mathematics

Quasi-empiricism in mathematics is the attempt in the philosophy of mathematics to direct philosophers' attention to mathematical practice, in particular, relations with physics, social sciences, and computational mathematics, rather than solely to issues in the foundations of mathematics. Of concern to this discussion are several topics: the relationship of empiricism (See Maddy) with mathematics, issues related to realism, the importance of culture, necessity of application, etc.

15.1 Primary arguments

A primary argument with respect to Quasi-empiricism is that whilst mathematics and physics are more frequently being considered as closely linked fields of study, this may reflect human cognitive bias. It is claimed that, despite rigorous application of appropriate empirical methods or mathematical practice in either field, this would nonetheless be insufficient to disprove alternate approaches.
Eugene Wigner (1960)[1] noted that this culture need not be restricted to mathematics, physics, or even humans. He stated further that "The miracle of the appropriateness of the language of mathematics for the formulation of the laws of physics is a wonderful gift which we neither understand nor deserve. We should be grateful for it and hope that it will remain valid in future research and that it will extend, for better or for worse, to our pleasure, even though perhaps also to our bafflement, to wide branches of learning." Wigner used several examples to demonstrate why 'bafflement' is an appropriate description, such as showing how mathematics adds to situational knowledge in ways that are either not possible otherwise or are so outside normal thought to be of little notice. The predictive ability, in the sense of describing potential phenomena prior to observation of such, which can be supported by a mathematical system would be another example.
Following up on Wigner, Richard Hamming (1980) [2] wrote about applications of mathematics as a central theme to this topic and suggested that successful use can trump, sometimes, proof, in the following sense: where a theorem has evident veracity through applicability, later evidence that shows the theorem's proof to be problematic would result more in trying to firm up the theorem rather than in trying to redo the applications or to deny results obtained to date. Hamming had four explanations for the 'effectiveness' that we see with mathematics and definitely saw this topic as worthy of discussion and study.
1) "We see what we look for." Why 'quasi' is apropos in reference to this discussion. 2) "We select the kind of mathematics to use." Our use and modification of mathematics is essentially situational and goal driven. 3) "Science in fact answers comparatively few problems." What still needs to be looked at is a larger set. 4) "The evolution of man provided the model." There may be limits attributable to the human element.
Hilary Putnam (1975) [3] stated that mathematics had accepted informal proofs and proof by authority, and had made and corrected errors all through its history. Also, he stated that Euclid's system of proving geometry theorems was unique to the classical Greeks and did not evolve similarly in other mathematical cultures in China, India, and Arabia.

This and other evidence led many mathematicians to reject the label of Platonists, along with Plato's ontology— which, along with the methods and epistemology of Aristotle, had served as a foundation ontology for the Western

world since its beginnings. A truly international culture of mathematics would, Putnam and others (1983) [4] argued, necessarily be at least 'quasi'-empirical (embracing 'the scientific method' for consensus if not experiment).

Imre Lakatos (1976 - posthumous), [5] who did his original work on this topic for his dissertation (1961, Cambridge), argued for 'Research Programs' as a means to support a basis for mathematics and considered thought experiments as appropriate to mathematical discovery. Lakatos may have been the first to use 'quasi-empiricism' in the context of this subject.

15.1.1 Operational aspects

Recent work that pertains to this topic are several. Chaitin's and Stephen Wolfram's work, though their positions may be considered controversial, apply. Chaitin (1997/2003) [6] suggests an underlying randomness to mathematics and Wolfram (A New Kind of Science, 2002) [7] argues that undecidability may have practical relevance, that is, be more than an abstraction.

Another relevant addition would be the discussions concerning Interactive computation, especially those related to the meaning and use of Turing's model (Church-Turing, TM, etc.).

These works are heavily computational and raise another set of issues. To quote Chaitin (1997/2003): "Now everything has gone topsy-turvy. It's gone topsy-turvy, not because of any philosophical argument, not because of Gödel's results or Turing's results or my own incompleteness results. It's gone topsy-turvy for a very simple reason — the computer!".[6]

The collection of "Undecidables" in Wolfram (A New Kind of Science, 2002) [7] is another example.

Wegner's recent paper [8] suggests that *interactive computation* can help mathematics form a more appropriate framework (empirical) than can be founded with rationalism alone. Related to this argument is that the function (even recursively related ad infinitum) is too simple of a construct to handle the reality of entities that resolve (via computation or some type of analog) n-dimensional (general sense of the word) systems.

15.2 See also

_ Gregory Chaitin
_ Richard Hamming
_ Imre Lakatos
_ Penelope Maddy
_ Charles Sanders Peirce
_ Karl Popper
_ Hilary Putnam
_ Thomas Tymoczko
_ Eugene Wigner
_ Stephen Wolfram
_ Beyond the traditional schools
_ Entscheidungsproblem
_ Foundations of mathematics
_ Interactive computation
_ Philosophy of mathematics
_ Unreasonable Ineffectiveness of Mathematics

15.3 References

[1] Eugene Wigner, 1960, "The Unreasonable Effectiveness of Mathematics in the Natural Sciences," *Communications on Pure and Applied Mathematics 13*:
[2] R. W. Hamming, 1980, The Unreasonable Effectiveness of Mathematics, *The American Mathematical Monthly* Volume 87 Number 2 February 1980
[3] Putnam, Hilary, 1975, *Mind, Language, and Reality. Philosophical Papers, Volume 2*. Cambridge University Press, Cambridge, UK. ISBN 88-459-0257-9
[4] _ Benacerraf, Paul, and Putnam, Hilary (eds), 1983, *Philosophy of Mathematics, Selected Readings*, 1st edition, Prentice–Hall, Englewood Cliffs, NJ, 1964. 2nd edition, Cambridge University Press, Cambridge, UK, 1983
[5] Lakatos, Imre 1976, *Proofs and Refutations*. Cambridge: Cambridge University Press. ISBN 0-521-29038-4
[6] Chaitin, Gregory J., 1997/2003, "*Limits of Mathematics*", Springer-Verlag, New York, NY. ISBN 1-85233-668-4
[7] Wolfram, Stephen, 2002, *A New Kind of Science* (Undecidables), Wolfram Media, Chicago, IL. ISBN 1-57955-008-8
[8] Peter Wegner, Dina Goldin, 2006, *Principles of Problem Solving*. **Communications of the ACM** 49 (2006), pp.27-29

Chapter 16

Mathematical folklore

See also folk theorem for other uses of this expression.
As the term is understood by mathematicians, **folk mathematics** or **mathematical folklore** means theorems, definitions, proofs, or mathematical facts or techniques that are found by investigation and may circulate among mathematicians by word-of-mouth but have not appeared in print, either in books or in scholarly journals. Knowledge of folklore is the coin of the realm of academic mathematics, showing relative insight of investigators.
Quite important at times for researchers are **folk theorems**, which are results known, at least to experts in a field, and considered to have established status, but not published in complete form. Sometimes these are only alluded to in the public literature. An example is a book of exercises, described on the back cover:
This book contains almost 350 exercises in the basics of ring theory. The problems form the 'folklore' of ring theory, and the solutions are given in as much detail as possible.[1]
Another distinct category is **wellknowable** mathematics, a term introduced by John Conway. This consists of matters that are known and factual, but not in active circulation in relation with current research. Both of these concepts are attempts to describe the actual context in which research work is done.
Some people, principally non-mathematicians, use the term *folk mathematics* to refer to the informal mathematics studied in many ethno-cultural studies of mathematics.

16.1 Stories, sayings and jokes

See also: Mathematical joke
Mathematical folklore may also refer to unusual (and possibly apochryphal) stories or jokes involving mathematicians or mathematics that are told verbally in mathematics departments. Compilations include tales collected in G. H. Hardy's *A Mathematician's Apology* and (Krantz 2002); examples include:
_ Galileo dropping weights from the Leaning Tower of Pisa.
_ An apple falling on Isaac Newton's head to inspire his theory of gravitation.
_ The drinking, duel and early death of Galois.
_ Richard Feynman cracking safes in the Manhattan Project.
_ Alfréd Rényi's definition of a mathematician - "*a mathematician is a device for turning coffee into theorems*".
_ The "turtles all the way down" story told by Stephen Hawking.
_ Fermat's lost simple proof.
_ The unwieldy proof and associated controversies of the Four Color Theorem.
82
16.2. SEE ALSO 83

16.2 See also

16.3 Notes

[1] Grigore Calugareau & Peter Hamburg (1998) *Exercises in Basic Ring Theory*, Kluwer,[ISBN 0792349180]

16.4 References

_ Krantz, Steven G. (2002), *Mathematical Apocrypha: Stories & Anecdotes of Mathematicians & the Mathematical*
_ David Harel, "On Folk Theorems", *Communications of the ACM* **23**:7:379-389 (July 1980)

Chapter 17

Philosophy of mathematics

The **philosophy of mathematics** is the branch of philosophy that studies the philosophical assumptions, foundations, and implications of mathematics. The aim of the philosophy of mathematics is to provide an account of the nature and methodology of mathematics and to understand the place of mathematics in people's lives. The logical and structural

nature of mathematics itself makes this study both broad and unique among its philosophical counterparts.
The terms *philosophy of mathematics* and *mathematical philosophy* are frequently used as synonyms.[1] The latter, however, may be used to refer to several other areas of study. One refers to a project of formalizing a philosophical subject matter, say, aesthetics, ethics, logic, metaphysics, or theology, in a purportedly more exact and rigorous form, as for example the labors of scholastic theologians, or the systematic aims of Leibniz and Spinoza. Another refers to the working philosophy of an individual practitioner or a like-minded community of practicing mathematicians. Additionally, some understand the term "mathematical philosophy" to be an allusion to the approach to the foundations of mathematics taken by Bertrand Russell in his books *The Principles of Mathematics* and *Introduction to Mathematical Philosophy*.

17.1 Recurrent themes

Recurrent themes include:

- What is the role of Mankind in developing mathematics?
- What are the sources of mathematical subject matter?
- What is the ontological status of mathematical entities?
- What does it mean to refer to a mathematical object?
- What is the character of a mathematical proposition?
- What is the relation between logic and mathematics?
- What is the role of hermeneutics in mathematics?
- What kinds of inquiry play a role in mathematics?
- What are the objectives of mathematical inquiry?
- What gives mathematics its hold on experience?
- What are the human traits behind mathematics?
- What is mathematical beauty?
- What is the source and nature of mathematical truth?
- What is the relationship between the abstract world of mathematics and the material universe?

17.2 History

The origin of mathematics is subject to argument. Whether the birth of mathematics was a random happening or induced by necessity duly contingent of other subjects, say for example physics, is still a matter of prolific debates. Many thinkers have contributed their ideas concerning the nature of mathematics. Today, some philosophers of mathematics aim to give accounts of this form of inquiry and its products as they stand, while others emphasize a role for themselves that goes beyond simple interpretation to critical analysis. There are traditions of mathematical philosophy in both Western philosophy and Eastern philosophy. Western philosophies of mathematics go as far back as Plato, who studied the ontological status of mathematical objects, and Aristotle, who studied logic and issues related to infinity (actual versus potential).
Greek philosophy on mathematics was strongly influenced by their study of geometry. For example, at one time, the Greeks held the opinion that 1 (one) was not a number, but rather a unit of arbitrary length. A number was defined as a multitude. Therefore 3, for example, represented a certain multitude of units, and was thus not "truly" a number. At another point, a similar argument was made that 2 was not a number but a fundamental notion of a pair. These views come from the heavily geometric straight-edge-and-compass viewpoint of the Greeks: just as lines drawn in a geometric problem are measured in proportion to the first arbitrarily drawn line, so too are the numbers on a number line measured in proportion to the arbitrary first "number" or "one".
These earlier Greek ideas of numbers were later upended by the discovery of the irrationality of the square root of two. Hippasus, a disciple of Pythagoras, showed that the diagonal of a unit square was incommensurable with its (unit-length) edge: in other words he proved there was no existing (rational) number that accurately depicts the proportion of the diagonal of the unit square to its edge. This caused a significant re-evaluation of Greek philosophy of mathematics. According to legend, fellow Pythagoreans were so traumatized by this discovery that they murdered Hippasus to stop him from spreading his heretical idea. Simon Stevin was one of the first in Europe to challenge Greek ideas in the 16th century. Beginning with Leibniz, the focus shifted strongly to the relationship between mathematics and logic. This perspective dominated the philosophy of mathematics through the time of Frege and of Russell, but was brought into question by developments in the late 19th and early 20th centuries.

17.2.1 20th century

A perennial issue in the philosophy of mathematics concerns the relationship between logic and mathematics at their joint foundations. While 20th century philosophers continued to ask the questions mentioned at the outset of this article, the philosophy of mathematics in the 20th century was characterized by a predominant interest in formal logic, set theory, and foundational issues.

It is a profound puzzle that on the one hand mathematical truths seem to have a compelling inevitability, but on the other hand the source of their "truthfulness" remains elusive. Investigations into this issue are known as the foundations of mathematics program.

At the start of the 20th century, philosophers of mathematics were already beginning to divide into various schools of thought about all these questions, broadly distinguished by their pictures of mathematical epistemology and ontology. Three schools, formalism, intuitionism, and logicism, emerged at this time, partly in response to the increasingly widespread worry that mathematics as it stood, and analysis in particular, did not live up to the standards of certainty and rigor that had been taken for granted. Each school addressed the issues that came to the fore at that time, either attempting to resolve them or claiming that mathematics is not entitled to its status as our most trusted knowledge.

Surprising and counter-intuitive developments in formal logic and set theory early in the 20th century led to new questions concerning what was traditionally called the *foundations of mathematics*. As the century unfolded, the initial focus of concern expanded to an open exploration of the fundamental axioms of mathematics, the axiomatic approach having been taken for granted since the time of Euclid around 300 BCE as the natural basis for mathematics. Notions of axiom, proposition and proof, as well as the notion of a proposition being true of a mathematical object (see Assignment (mathematical logic)), were formalized, allowing them to be treated mathematically. The Zermelo–Fraenkel axioms for set theory were formulated which provided a conceptual framework in which much mathematical discourse would be interpreted. In mathematics, as in physics, new and unexpected ideas had arisen and significant changes were coming. With Gödel numbering, propositions could be interpreted as referring to themselves or other propositions, enabling inquiry into the consistency of mathematical theories. This reflective critique in which the theory under review "becomes itself the object of a mathematical study" led Hilbert to call such study *metamathematics* or *proof theory*.[2]

At the middle of the century, a new mathematical theory was created by Samuel Eilenberg and Saunders Mac Lane, known as category theory, and it became a new contender for the natural language of mathematical thinking.[3] As the 20th century progressed, however, philosophical opinions diverged as to just how well-founded were the questions about foundations that were raised at the century's beginning. Hilary Putnam summed up one common view of the situation in the last third of the century by saying:

When philosophy discovers something wrong with science, sometimes science has to be changed—Russell's paradox comes to mind, as does Berkeley's attack on the actual infinitesimal—but more often it is philosophy that has to be changed. I do not think that the difficulties that philosophy finds with classical mathematics today are genuine difficulties; and I think that the philosophical interpretations of mathematics that we are being offered on every hand are wrong, and that "philosophical interpretation" is just what mathematics doesn't need.[4]:169–170

Philosophy of mathematics today proceeds along several different lines of inquiry, by philosophers of mathematics, logicians, and mathematicians, and there are many schools of thought on the subject. The schools are addressed separately in the next section, and their assumptions explained.

17.3 Major themes

17.3.1 Mathematical realism

Mathematical realism, like realism in general, holds that mathematical entities exist independently of the human mind. Thus humans do not invent mathematics, but rather discover it, and any other intelligent beings in the universe would presumably do the same. In this point of view, there is really one sort of mathematics that can be discovered; triangles, for example, are real entities, not the creations of the human mind.

Many working mathematicians have been mathematical realists; they see themselves as discoverers of naturally occurring objects. Examples include Paul Erdős and Kurt Gödel. Gödel believed in an objective mathematical reality that could be perceived in a manner analogous to sense perception. Certain principles (e.g., for any two objects, there is a collection of objects consisting of precisely those two objects) could be directly seen to be true, but the continuum hypothesis conjecture might prove undecidable just on the basis of such principles. Gödel suggested that quasi-empirical methodology could be used to provide sufficient evidence to be able to reasonably assume such a conjecture.

Within realism, there are distinctions depending on what sort of existence one takes mathematical entities to have, and how we know about them. Major forms of mathematical realism include Platonism and empiricism.

17.3.2 Mathematical anti-realism

Mathematical anti-realism generally holds that mathematical statements have truth-values, but that they do not do so by corresponding to a special realm of immaterial or non-empirical entities. Major forms of mathematical anti-realism include Formalism and Fictionalism.

17.4 Contemporary schools of thought

17.4.1 Platonism

Mathematical Platonism is the form of realism that suggests that mathematical entities are abstract, have no spatiotemporal or causal properties, and are eternal and unchanging. This is often claimed to be the view most people have of numbers. The term *Platonism* is used because such a view is seen to parallel Plato's Theory of Forms and a "World of Ideas" (Greek: *eidos* (εἶδος)) described in Plato's Allegory of the cave: the everyday world can only imperfectly approximate an unchanging, ultimate reality. Both *Plato's cave* and *Platonism* have meaningful, not just superficial connections, because Plato's ideas were preceded and probably influenced by the hugely popular *Pythagoreans* of ancient Greece, who believed that the world was, quite literally, generated by numbers.

The major problem of mathematical platonism is this: precisely where and how do the mathematical entities exist, and how do we know about them? Is there a world, completely separate from our physical one, that is occupied by the mathematical entities? How can we gain access to this separate world and discover truths about the entities? One answer might be the Ultimate Ensemble, which is a theory that postulates all structures that exist mathematically also exist physically in their own universe.

Plato spoke of mathematics by:

How do you mean?

I mean, as I was saying, that arithmetic has a very great and elevating effect, compelling the soul to reason about abstract number, and rebelling against the introduction of visible or tangible objects into the argument. You know how steadily the masters of the art repel and ridicule any one who attempts to divide absolute unity when he is calculating, and if you divide, they multiply, taking care that one shall continue one and not become lost in fractions.

That is very true.

Now, suppose a person were to say to them: O my friends, what are these wonderful numbers about which you are reasoning, in which, as you say, there is a unity such as you demand, and each unit is equal, invariable, indivisible, --what would they answer?

—Plato, Chapter 7. "The Republic" (Jowell translation).

In context, chapter 8, of H.D.P. Lee's translation, reports the education of a philosopher contains five mathematical disciplines:

1. mathematics;
2. arithmetic, written in unit fraction "parts" using theoretical unities and abstract numbers;
3. plane geometry and solid geometry also considered the line to be segmented into rational and irrational unit "parts";
4. astronomy
5. harmonics

Translators of the works of Plato rebelled against practical versions of his culture's practical mathematics. However, Plato himself and Greeks had copied 1,500 older Egyptian fraction abstract unities, one being a hekat unity scaled to (64/64) in the Akhmim Wooden Tablet, thereby not getting lost in fractions.

Gödel's Platonism postulates a special kind of mathematical intuition that lets us perceive mathematical objects directly. (This view bears resemblances to many things Husserl said about mathematics, and supports Kant's idea that mathematics is synthetic *a priori*.) Davis and Hersh have suggested in their book *The Mathematical Experience* that most mathematicians act as though they are Platonists, even though, if pressed to defend the position carefully, they may retreat to formalism (see below).

Some mathematicians hold opinions that amount to more nuanced versions of Platonism.

Full-blooded Platonism is a modern variation of Platonism, which is in reaction to the fact that different sets of mathematical entities can be proven to exist depending on the axioms and inference rules employed (for instance, the law of the excluded middle, and the axiom of choice). It holds that all mathematical entities exist, however they may be provable, even if they cannot all be derived from a single consistent set of axioms.

17.4.2 Empiricism

Empiricism is a form of realism that denies that mathematics can be known *a priori* at all. It says that we discover mathematical facts by empirical research, just like facts in any of the other sciences. It is not one of the classical three positions advocated in the early 20th century, but primarily arose in the middle of the century. However, an important early proponent of a view like this was John Stuart Mill. Mill's view was widely criticized, because, according to critics, it makes statements like "2 + 2 = 4" come out as uncertain, contingent truths, which we can only learn by observing instances of two pairs coming together and forming a quartet.

Contemporary mathematical empiricism, formulated by Quine and Putnam, is primarily supported by the indispensability

argument: mathematics is indispensable to all empirical sciences, and if we want to believe in the reality of the phenomena described by the sciences, we ought also believe in the reality of those entities required for this description.
That is, since physics needs to talk about electrons to say why light bulbs behave as they do, then electrons must exist. Since physics needs to talk about numbers in offering any of its explanations, then numbers must exist. In keeping with Quine and Putnam's overall philosophies, this is a naturalistic argument. It argues for the existence of mathematical entities as the best explanation for experience, thus stripping mathematics of being distinct from the other sciences.
Putnam strongly rejected the term "Platonist" as implying an over-specific ontology that was not necessary to mathematical practice in any real sense. He advocated a form of "pure realism" that rejected mystical notions of truth and accepted much quasi-empiricism in mathematics. Putnam was involved in coining the term "pure realism" (see below).
The most important criticism of empirical views of mathematics is approximately the same as that raised against Mill. If mathematics is just as empirical as the other sciences, then this suggests that its results are just as fallible as theirs, and just as contingent. In Mill's case the empirical justification comes directly, while in Quine's case it comes indirectly, through the coherence of our scientific theory as a whole, i.e. consilience after E.O. Wilson. Quine suggests that mathematics seems completely certain because the role it plays in our web of belief is incredibly central, and that it would be extremely difficult for us to revise it, though not impossible.
For a philosophy of mathematics that attempts to overcome some of the shortcomings of Quine and Gödel's approaches by taking aspects of each see Penelope Maddy's *Realism in Mathematics*. Another example of a realist theory is the embodied mind theory (below). For a modern revision of mathematical empiricism see New Empiricism (below).
For experimental evidence suggesting that human infants can do elementary arithmetic, see Brian Butterworth.

17.4.3 Mathematical monism

Max Tegmark's mathematical universe hypothesis goes further than full-blooded Platonism in asserting that not only do all mathematical objects exist, but nothing else does. Tegmark's sole postulate is: *All structures that exist mathematically also exist physically*. That is, in the sense that "in those [worlds] complex enough to contain self-aware substructures [they] will subjectively perceive themselves as existing in a physically 'real' world".[5][6]

17.4.4 Logicism

Logicism is the thesis that mathematics is reducible to logic, and hence nothing but a part of logic.[7]:41 Logicists hold that mathematics can be known *a priori*, but suggest that our knowledge of mathematics is just part of our knowledge of logic in general, and is thus analytic, not requiring any special faculty of mathematical intuition. In this view, logic is the proper foundation of mathematics, and all mathematical statements are necessary logical truths.
Rudolf Carnap (1931) presents the logicist thesis in two parts:[7]
1. The *concepts* of mathematics can be derived from logical concepts through explicit definitions.
2. The *theorems* of mathematics can be derived from logical axioms through purely logical deduction.
Gottlob Frege was the founder of logicism. In his seminal *Die Grundgesetze der Arithmetik* (*Basic Laws of Arithmetic*) he built up arithmetic from a system of logic with a general principle of comprehension, which he called "Basic Law V" (for concepts *F* and *G*, the extension of *F* equals the extension of *G* if and only if for all objects *a*, *Fa* if and only if *Ga*), a principle that he took to be acceptable as part of logic.
Frege's construction was flawed. Russell discovered that Basic Law V is inconsistent (this is Russell's paradox). Frege abandoned his logicist program soon after this, but it was continued by Russell and Whitehead. They attributed the paradox to "vicious circularity" and built up what they called ramified type theory to deal with it. In this system, they were eventually able to build up much of modern mathematics but in an altered, and excessively complex form (for example, there were different natural numbers in each type, and there were infinitely many types). They also had to make several compromises in order to develop so much of mathematics, such as an "axiom of reducibility". Even Russell said that this axiom did not really belong to logic.

Modern logicists (like Bob Hale, Crispin Wright, and perhaps others) have returned to a program closer to Frege's. They have abandoned Basic Law V in favor of abstraction principles such as Hume's principle (the number of objects falling under the concept *F* equals the number of objects falling under the concept *G* if and only if the extension of *F* and the extension of *G* can be put into one-to-one correspondence). Frege required Basic Law V to be able to give an explicit definition of the numbers, but all the properties of numbers can be derived from Hume's principle. This would not have been enough for Frege because (to paraphrase him) it does not exclude the possibility that the number 3 is in fact Julius Caesar. In addition, many of the weakened principles that they have had to adopt to replace Basic Law V no longer seem so obviously analytic, and thus purely logical.

17.4.5 Formalism

Main article: Formalism (mathematics)
Formalism holds that mathematical statements may be thought of as statements about the consequences of certain

string manipulation rules. For example, in the "game" of Euclidean geometry (which is seen as consisting of some strings called "axioms", and some "rules of inference" to generate new strings from given ones), one can prove that the Pythagorean theorem holds (that is, you can generate the string corresponding to the Pythagorean theorem). According to formalism, mathematical truths are not about numbers and sets and triangles and the like—in fact, they aren't "about" anything at all.

Another version of formalism is often known as deductivism. In deductivism, the Pythagorean theorem is not an absolute truth, but a relative one: *if* you assign meaning to the strings in such a way that the rules of the game become true (i.e., true statements are assigned to the axioms and the rules of inference are truth-preserving), *then* you have to accept the theorem, or, rather, the interpretation you have given it must be a true statement. The same is held to be true for all other mathematical statements. Thus, formalism need not mean that mathematics is nothing more than a meaningless symbolic game. It is usually hoped that there exists some interpretation in which the rules of the game hold. (Compare this position to structuralism.) But it does allow the working mathematician to continue in his or her work and leave such problems to the philosopher or scientist. Many formalists would say that in practice, the axiom systems to be studied will be suggested by the demands of science or other areas of mathematics.

A major early proponent of formalism was David Hilbert, whose program was intended to be a complete and consistent axiomatization of all of mathematics. Hilbert aimed to show the consistency of mathematical systems from the assumption that the "finitary arithmetic" (a subsystem of the usual arithmetic of the positive integers, chosen to be philosophically uncontroversial) was consistent. Hilbert's goals of creating a system of mathematics that is both complete and consistent were dealt a fatal blow by the second of Gödel's incompleteness theorems, which states that sufficiently expressive consistent axiom systems can never prove their own consistency. Since any such axiom system would contain the finitary arithmetic as a subsystem, Gödel's theorem implied that it would be impossible to prove the system's consistency relative to that (since it would then prove its own consistency, which Gödel had shown was impossible). Thus, in order to show that any axiomatic system of mathematics is in fact consistent, one needs to first assume the consistency of a system of mathematics that is in a sense stronger than the system to be proven consistent.

Hilbert was initially a deductivist, but, as may be clear from above, he considered certain metamathematical methods to yield intrinsically meaningful results and was a realist with respect to the finitary arithmetic. Later, he held the opinion that there was no other meaningful mathematics whatsoever, regardless of interpretation.

Other formalists, such as Rudolf Carnap, Alfred Tarski, and Haskell Curry, considered mathematics to be the investigation of formal axiom systems. Mathematical logicians study formal systems but are just as often realists as they are formalists.

Formalists are relatively tolerant and inviting to new approaches to logic, non-standard number systems, new set theories etc. The more games we study, the better. However, in all three of these examples, motivation is drawn from existing mathematical or philosophical concerns. The "games" are usually not arbitrary.

The main critique of formalism is that the actual mathematical ideas that occupy mathematicians are far removed from the string manipulation games mentioned above. Formalism is thus silent on the question of which axiom systems ought to be studied, as none is more meaningful than another from a formalistic point of view.

Recently, some formalist mathematicians have proposed that all of our *formal* mathematical knowledge should be systematically encoded in computer-readable formats, so as to facilitate automated proof checking of mathematical proofs and the use of interactive theorem proving in the development of mathematical theories and computer software. Because of their close connection with computer science, this idea is also advocated by mathematical intuitionists and

David Hilbert
constructivists in the "computability" tradition (see below). See QED project for a general overview.

17.4.6 Conventionalism

The French mathematician Henri Poincaré was among the first to articulate a conventionalist view. Poincaré's use of non-Euclidean geometries in his work on differential equations convinced him that Euclidean geometry should not

be regarded as *a priori* truth. He held that axioms in geometry should be chosen for the results they produce, not for their apparent coherence with human intuitions about the physical world.

17.4.7 Psychologism

Psychologism in the philosophy of mathematics is the position that mathematical concepts and/or truths are grounded in, derived from or explained by psychological facts (or laws).

John Stuart Mill seems to have been an advocate of a type of logical psychologism, as were many 19th-century German logicians such as Sigwart and Erdmann as well as a number of psychologists, past and present: for example, Gustave Le Bon. Psychologism was famously criticized by Frege in his *The Foundations of Arithmetic*, and many of his works and essays, including his review of Husserl's *Philosophy of Arithmetic*. Edmund Husserl, in the first volume of his *Logical Investigations*, called "The Prolegomena of Pure Logic", criticized psychologism thoroughly and sought to distance himself from it. The "Prolegomena" is considered a more concise, fair, and thorough refutation of psychologism than the criticisms made by Frege, and also it is considered today by many as being a memorable refutation for its decisive blow to psychologism. Psychologism was also criticized by Charles Sanders Peirce and Maurice Merleau-Ponty.

17.4.8 Intuitionism

Main article: Mathematical intuitionism

In mathematics, intuitionism is a program of methodological reform whose motto is that “there are no non-experienced mathematical truths” (L.E.J. Brouwer). From this springboard, intuitionists seek to reconstruct what they consider to be the corrigible portion of mathematics in accordance with Kantian concepts of being, becoming, intuition, and knowledge. Brouwer, the founder of the movement, held that mathematical objects arise from the *a priori* forms of the volitions that inform the perception of empirical objects.[8]

A major force behind intuitionism was L.E.J. Brouwer, who rejected the usefulness of formalized logic of any sort for mathematics. His student Arend Heyting postulated an intuitionistic logic, different from the classical Aristotelian logic; this logic does not contain the law of the excluded middle and therefore frowns upon proofs by contradiction. The axiom of choice is also rejected in most intuitionistic set theories, though in some versions it is accepted. Important work was later done by Errett Bishop, who managed to prove versions of the most important theorems in real analysis within this framework.

In intuitionism, the term “explicit construction” is not cleanly defined, and that has led to criticisms. Attempts have been made to use the concepts of Turing machine or computable function to fill this gap, leading to the claim that only questions regarding the behavior of finite algorithms are meaningful and should be investigated in mathematics. This has led to the study of the computable numbers, first introduced by Alan Turing. Not surprisingly, then, this approach to mathematics is sometimes associated with theoretical computer science.

Constructivism

Main article: Mathematical constructivism

Like intuitionism, constructivism involves the regulative principle that only mathematical entities which can be explicitly constructed in a certain sense should be admitted to mathematical discourse. In this view, mathematics is an exercise of the human intuition, not a game played with meaningless symbols. Instead, it is about entities that we can create directly through mental activity. In addition, some adherents of these schools reject non-constructive proofs, such as a proof by contradiction.

Finitism

Finitism is an extreme form of constructivism, according to which a mathematical object does not exist unless it can be constructed from natural numbers in a finite number of steps. In her book *Philosophy of Set Theory*, Mary

Tiles characterized those who allow countably infinite objects as classical finitists, and those who deny even countably infinite objects as strict finitists.

The most famous proponent of finitism was Leopold Kronecker,[9] who said:

God created the natural numbers, all else is the work of man.

Ultrafinitism is an even more extreme version of finitism, which rejects not only infinities but finite quantities that cannot feasibly be constructed with available resources.

17.4.9 Structuralism

Main article: Mathematical structuralism

Structuralism is a position holding that mathematical theories describe structures, and that mathematical objects are exhaustively defined by their *places* in such structures, consequently having no intrinsic properties. For instance, it would maintain that all that needs to be known about the number 1 is that it is the first whole number after 0. Likewise all the other whole numbers are defined by their places in a structure, the number line. Other examples of mathematical objects might include lines and planes in geometry, or elements and operations in abstract algebra.

Structuralism is an epistemologically realistic view in that it holds that mathematical statements have an objective truth value. However, its central claim only relates to what *kind* of entity a mathematical object is, not to what kind of *existence* mathematical objects or structures have (not, in other words, to their ontology). The kind of existence mathematical objects have would clearly be dependent on that of the structures in which they are embedded; different sub-varieties of structuralism make different ontological claims in this regard.[10]

The *Ante Rem*, or fully realist, variation of structuralism has a similar ontology to Platonism in that structures are held to have a real but abstract and immaterial existence. As such, it faces the usual problems of explaining the interaction between such abstract structures and flesh-and-blood mathematicians.

In Re, or moderately realistic, structuralism is the equivalent of Aristotelian realism. Structures are held to exist inasmuch as some concrete system exemplifies them. This incurs the usual issues that some perfectly legitimate structures might accidentally happen not to exist, and that a finite physical world might not be “big” enough to accommodate some otherwise legitimate structures.

The *Post Res* or eliminative variant of structuralism is anti-realist about structures in a way that parallels nominalism.

According to this view mathematical *systems* exist, and have structural features in common. If something is true of a structure, it will be true of all systems exemplifying the structure. However, it is merely convenient to talk of structures being "held in common" between systems: they in fact have no independent existence.

17.4.10 Embodied mind theories

Embodied mind theories hold that mathematical thought is a natural outgrowth of the human cognitive apparatus which finds itself in our physical universe. For example, the abstract concept of number springs from the experience of counting discrete objects. It is held that mathematics is not universal and does not exist in any real sense, other than in human brains. Humans construct, but do not discover, mathematics.

With this view, the physical universe can thus be seen as the ultimate foundation of mathematics: it guided the evolution of the brain and later determined which questions this brain would find worthy of investigation. However, the human mind has no special claim on reality or approaches to it built out of math. If such constructs as Euler's identity are true then they are true as a map of the human mind and cognition.

Embodied mind theorists thus explain the effectiveness of mathematics—mathematics was constructed by the brain in order to be effective in this universe.

The most accessible, famous, and infamous treatment of this perspective is *Where Mathematics Comes From*, by George Lakoff and Rafael E. Núñez. In addition, mathematician Keith Devlin has investigated similar concepts with his book *The Math Instinct*, as has neuroscientist Stanislas Dehaene with his book *The Number Sense*. For more on the philosophical ideas that inspired this perspective, see cognitive science of mathematics.

New empiricism

A more recent empiricism returns to the principle of the English empiricists of the 18th and 19th centuries, in particular John Stuart Mill, who asserted that all knowledge comes to us from observation through the senses. This applies not only to matters of fact, but also to "relations of ideas", as Hume called them: the structures of logic which interpret, organize and abstract observations.

To this principle it adds a materialist connection: all the processes of logic which interpret, organize and abstract observations, are physical phenomena which take place in real time and physical space: namely, in the brains of human beings. Abstract objects, such as mathematical objects, are ideas, which in turn exist as electrical and chemical states of the billions of neurons in the human brain.

This second concept is reminiscent of the social constructivist approach, which holds that mathematics is produced by humans rather than being "discovered" from abstract, *a priori* truths. However, it differs sharply from the constructivist implication that humans arbitrarily construct mathematical principles that have no inherent truth but which instead are created on a conveniency basis. On the contrary, new empiricism shows how mathematics, although constructed by humans, follows rules and principles that will be agreed on by all who participate in the process, with the result that everyone practicing mathematics comes up with the same answer—except in those areas where there is philosophical disagreement on the meaning of fundamental concepts. This is because the new empiricism perceives this agreement as being a physical phenomenon, one which is observed by other humans in the same way that other physical phenomena, like the motions of inanimate bodies, or the chemical interaction of various elements, are observed.

Combining the materialist principle with Millisian epistemology evades the principal difficulty with classical empiricism—that all knowledge comes from the senses. That difficulty lies in the observation that mathematical truths based on logical deduction appear to be more certainly true than knowledge of the physical world itself. (The physical world in this case is taken to mean the portion of it lying outside the human brain.)

Kant argued that the structures of logic which organize, interpret and abstract observations were built into the human mind and were true and valid *a priori*. Mill, on the contrary, said that we believe them to be true because we have enough individual instances of their truth to generalize: in his words, "From instances we have observed, we feel warranted in concluding that what we found true in those instances holds in all similar ones, past, present and future, however numerous they may be".[11] Although the psychological or epistemological specifics given by Mill through which we build our logical apparatus may not be completely warranted, his explanation still nonetheless manages to demonstrate that there is no way around Kant's *a priori* logic. To recant Mill's original idea in an empiricist twist: "*Indeed, the very principles of logical deduction are true because we observe that using them leads to true conclusions*", which is itself an *a priori* presupposition.

If all this is true, then where do the world senses come in? The early empiricists all stumbled over this point. Hume asserted that all knowledge comes from the senses, and then gave away the ballgame by excepting abstract propositions, which he called "relations of ideas". These, he said, were absolutely true (although the mathematicians who thought them up, being human, might get them wrong). Mill, on the other hand, tried to deny that abstract ideas exist outside the physical world: all numbers, he said, "must be numbers of something: there are no such things as numbers in the abstract". When we count to eight or add five and three we are really counting spoons or bumblebees. "All things possess quantity", he said, so that propositions concerning numbers are propositions concerning "all things whatever". But then in almost a contradiction of himself he went on to acknowledge that numerical and algebraic expressions are not necessarily attached to real world objects: they "do not excite in our minds ideas of any things in particular". Mill's low reputation as a philosopher of logic, and the low estate of empiricism in the century and a half following him, derives from this failed attempt to link abstract thoughts to the physical world, when it may be more plausibly arguable that abstraction consists precisely of separating the thought from its physical foundations.

The conundrum created by our certainty that abstract deductive propositions, if valid (i.e. if we can "prove" them), are true, exclusive of observation and testing in the physical world, gives rise to a further reflection ... What if thoughts themselves, and the minds that create them, are physical objects, existing only in the physical world?

This would reconcile the contradiction between our belief in the certainty of abstract deductions and the empiricist principle that knowledge comes from observation of individual instances. We know that Euler's equation is true because every time a human mind derives the equation, it gets the same result, unless it has made a mistake, which can be acknowledged and corrected. We observe this phenomenon, and we extrapolate to the general proposition that it is always true.

This applies not only to physical principles, like the law of gravity, but to abstract phenomena that we observe only in human brains: in ours and in those of others.

Aristotelian realism

Main article: Aristotle's theory of universals

Similar to empiricism in emphasizing the relation of mathematics to the real world, Aristotelian realism holds that mathematics studies properties such as symmetry, continuity and order that can be literally realized in the physical world (or in any other world there might be). It contrasts with Platonism in holding that the objects of mathematics, such as numbers, do not exist in an "abstract" world but can be physically realized. For example, the number 4 is realized in the relation between a heap of parrots and the universal "being a parrot" that divides the heap into so many parrots.[12] Aristotelian realism is defended by James Franklin and the Sydney School in the philosophy of mathematics and is close to the view of Penelope Maddy that when an egg carton is opened, a set of three eggs is perceived (that is, a mathematical entity realized in the physical world).[13] A problem for Aristotelian realism is what account to give of higher infinities, which may not be realizable in the physical world.

17.4.11 Fictionalism

Fictionalism in mathematics was brought to fame in 1980 when Hartry Field published *Science Without Numbers*, which rejected and in fact reversed Quine's indispensability argument. Where Quine suggested that mathematics was indispensable for our best scientific theories, and therefore should be accepted as a body of truths talking about independently existing entities, Field suggested that mathematics was dispensable, and therefore should be considered as a body of falsehoods not talking about anything real. He did this by giving a complete axiomatization of Newtonian mechanics that didn't reference numbers or functions at all. He started with the "betweenness" of Hilbert's axioms to characterize space without coordinatizing it, and then added extra relations between points to do the work formerly done by vector fields. Hilbert's geometry is mathematical, because it talks about abstract points, but in Field's theory, these points are the concrete points of physical space, so no special mathematical objects at all are needed.

Having shown how to do science without using numbers, Field proceeded to rehabilitate mathematics as a kind of useful fiction. He showed that mathematical physics is a conservative extension of his non-mathematical physics (that is, every physical fact provable in mathematical physics is already provable from Field's system), so that mathematics is a reliable process whose physical applications are all true, even though its own statements are false. Thus, when doing mathematics, we can see ourselves as telling a sort of story, talking as if numbers existed. For Field, a statement like "2 + 2 = 4" is just as fictitious as "Sherlock Holmes lived at 221B Baker Street"—but both are true according to the relevant fictions.

By this account, there are no metaphysical or epistemological problems special to mathematics. The only worries left are the general worries about non-mathematical physics, and about fiction in general. Field's approach has been very influential, but is widely rejected. This is in part because of the requirement of strong fragments of second-order logic to carry out his reduction, and because the statement of conservativity seems to require quantification over abstract models or deductions.

17.4.12 Social constructivism or social realism

Social constructivism or *social realism* theories see mathematics primarily as a social construct, as a product of culture, subject to correction and change. Like the other sciences, mathematics is viewed as an empirical endeavor whose results are constantly evaluated and may be discarded. However, while on an empiricist view the evaluation is some sort of comparison with "reality", social constructivists emphasize that the direction of mathematical research is dictated by the fashions of the social group performing it or by the needs of the society financing it. However, although such external forces may change the direction of some mathematical research, there are strong internal constraints—the mathematical traditions, methods, problems, meanings and values into which mathematicians are enculturated—that work to conserve the historically defined discipline.

This runs counter to the traditional beliefs of working mathematicians, that mathematics is somehow pure or objective.

But social constructivists argue that mathematics is in fact grounded by much uncertainty: as mathematical practice evolves, the status of previous mathematics is cast into doubt, and is corrected to the degree it is required or desired by the current mathematical community. This can be seen in the development of analysis from reexamination of the calculus of Leibniz and Newton. They argue further that finished mathematics is often accorded too much status, and folk mathematics not enough, due to an overemphasis on axiomatic proof and peer review as practices. However, this might be seen as merely saying that rigorously proven results are overemphasized, and then "look how chaotic

and uncertain the rest of it all is!"

The social nature of mathematics is highlighted in its subcultures. Major discoveries can be made in one branch of mathematics and be relevant to another, yet the relationship goes undiscovered for lack of social contact between mathematicians. Social constructivists argue each speciality forms its own epistemic community and often has great difficulty communicating, or motivating the investigation of unifying conjectures that might relate different areas of mathematics. Social constructivists see the process of "doing mathematics" as actually creating the meaning, while social realists see a deficiency either of human capacity to abstractify, or of human's cognitive bias, or of mathematicians' collective intelligence as preventing the comprehension of a real universe of mathematical objects. Social constructivists sometimes reject the search for foundations of mathematics as bound to fail, as pointless or even meaningless. Some social scientists also argue that mathematics is not real or objective at all, but is affected by racism and ethnocentrism. Some of these ideas are close to postmodernism.

Contributions to this school have been made by Imre Lakatos and Thomas Tymoczko, although it is not clear that either would endorse the title. More recently Paul Ernest has explicitly formulated a social constructivist philosophy of mathematics.[14] Some consider the work of Paul Erdős as a whole to have advanced this view (although he personally rejected it) because of his uniquely broad collaborations, which prompted others to see and study "mathematics as a social activity", e.g., via the Erdős number. Reuben Hersh has also promoted the social view of mathematics, calling it a "humanistic" approach,[15] similar to but not quite the same as that associated with Alvin White;[16] one of Hersh's co-authors, Philip J. Davis, has expressed sympathy for the social view as well.

A criticism of this approach is that it is trivial, based on the trivial observation that mathematics is a human activity. To observe that rigorous proof comes only after unrigorous conjecture, experimentation and speculation is true, but it is trivial and no-one would deny this. So it's a bit of a stretch to characterize a philosophy of mathematics in this way, on something trivially true. The calculus of Leibniz and Newton was reexamined by mathematicians such as Weierstrass in order to rigorously prove the theorems thereof. There is nothing special or interesting about this, as it fits in with the more general trend of unrigorous ideas which are later made rigorous. There needs to be a clear distinction between the objects of study of mathematics and the study of the objects of study of mathematics. The former doesn't seem to change a great deal; the latter is forever in flux. The latter is what the social theory is about, and the former is what Platonism *et al.* are about.

However, this criticism is rejected by supporters of the social constructivist perspective because it misses the point that the very objects of mathematics are social constructs. These objects, it asserts, are primarily semiotic objects existing in the sphere of human culture, sustained by social practices (after Wittgenstein) that utilize physically embodied signs and give rise to intrapersonal (mental) constructs. Social constructivists view the reification of the sphere of human culture into a Platonic realm, or some other heaven-like domain of existence beyond the physical world, a long-standing category error.

17.4.13 Beyond the traditional schools

Rather than focus on narrow debates about the true nature of mathematical truth, or even on practices unique to mathematicians such as the proof, a growing movement from the 1960s to the 1990s began to question the idea of seeking foundations or finding any one right answer to why mathematics works. The starting point for this was Eugene Wigner's famous 1960 paper *The Unreasonable Effectiveness of Mathematics in the Natural Sciences*, in which he argued that the happy coincidence of mathematics and physics being so well matched seemed to be unreasonable and hard to explain.

The embodied-mind or cognitive school and the social school were responses to this challenge, but the debates raised were difficult to confine to those.

Quasi-empiricism

One parallel concern that does not actually challenge the schools directly but instead questions their focus is the notion of quasi-empiricism in mathematics. This grew from the increasingly popular assertion in the late 20th century that no one foundation of mathematics could be ever proven to exist. It is also sometimes called "postmodernism in mathematics" although that term is considered overloaded by some and insulting by others. Quasi-empiricism argues that in doing their research, mathematicians test hypotheses as well as prove theorems. A mathematical argument can transmit falsity from the conclusion to the premises just as well as it can transmit truth from the premises to the conclusion. Quasi-empiricism was developed by Imre Lakatos, inspired by the philosophy of science of Karl Popper.

Lakatos' philosophy of mathematics is sometimes regarded as a kind of social constructivism, but this was not his

intention.
Such methods have always been part of folk mathematics by which great feats of calculation and measurement are sometimes achieved. Indeed, such methods may be the only notion of proof a culture has.
Hilary Putnam has argued that any theory of mathematical realism would include quasi-empirical methods. He proposed that an alien species doing mathematics might well rely on quasi-empirical methods primarily, being willing often to forgo rigorous and axiomatic proofs, and still be doing mathematics—at perhaps a somewhat greater risk of failure of their calculations. He gave a detailed argument for this in *New Directions*.[17]

Popper's "two senses" theory

Realist and constructivist theories are normally taken to be contraries. However, Karl Popper[18] argued that a number statement such as "2 apples + 2 apples = 4 apples" can be taken in two senses. In one sense it is irrefutable and logically true. In the second sense it is factually true and falsifiable. Another way of putting this is to say that a single number statement can express two propositions: one of which can be explained on constructivist lines; the other on realist lines.[19]

Language

Main article: Philosophy of language
Innovations in the philosophy of language during the 20th century renewed interest in whether mathematics is, as is often said, the *language* of science. Although some mathematicians and philosophers would accept the statement "mathematics is a language", linguists believe that the implications of such a statement must be considered. For example, the tools of linguistics are not generally applied to the symbol systems of mathematics, that is, mathematics is studied in a markedly different way than other languages. If mathematics is a language, it is a different type of language than natural languages. Indeed, because of the need for clarity and specificity, the language of mathematics is far more constrained than natural languages studied by linguists. However, the methods developed by Frege and Tarski for the study of mathematical language have been extended greatly by Tarski's student Richard Montague and other linguists working in formal semantics to show that the distinction between mathematical language and natural language may not be as great as it seems.

17.5 Arguments

17.5.1 Indispensability argument for realism

This argument, associated with Willard Quine and Hilary Putnam, is considered by Stephen Yablo to be one of the most challenging arguments in favor of the acceptance of the existence of abstract mathematical entities, such as numbers and sets.[20] The form of the argument is as follows.
1. One must have ontological commitments to *all* entities that are indispensable to the best scientific theories, and to those entities *only* (commonly referred to as "all and only").
2. Mathematical entities are indispensable to the best scientific theories. Therefore,
3. One must have ontological commitments to mathematical entities.[21]
The justification for the first premise is the most controversial. Both Putnam and Quine invoke naturalism to justify the exclusion of all non-scientific entities, and hence to defend the "only" part of "all and only". The assertion that "all" entities postulated in scientific theories, including numbers, should be accepted as real is justified by confirmation holism. Since theories are not confirmed in a piecemeal fashion, but as a whole, there is no justification for excluding any of the entities referred to in well-confirmed theories. This puts the nominalist who wishes to exclude the existence of sets and non-Euclidean geometry, but to include the existence of quarks and other undetectable entities of physics, for example, in a difficult position.[21]

17.5.2 Epistemic argument against realism

The anti-realist "epistemic argument" against Platonism has been made by Paul Benacerraf and Hartry Field. Platonism posits that mathematical objects are *abstract* entities. By general agreement, abstract entities cannot interact causally with concrete, physical entities. ("the truth-values of our mathematical assertions depend on facts involving Platonic entities that reside in a realm outside of space-time"[22]) Whilst our knowledge of concrete, physical objects is based on our ability to perceive them, and therefore to causally interact with them, there is no parallel account of how mathematicians come to have knowledge of abstract objects.[23][24][25] ("An account of mathematical truth ... must be consistent with the possibility of mathematical knowledge."[26]) Another way of making the point is that if the Platonic world were to disappear, it would make no difference to the ability of mathematicians to generate proofs, etc., which is already fully accountable in terms of physical processes in their brains.
Field developed his views into fictionalism. Benacerraf also developed the philosophy of mathematical structuralism, according to which there are no mathematical objects. Nonetheless, some versions of structuralism are compatible with some versions of realism.
The argument hinges on the idea that a satisfactory naturalistic account of thought processes in terms of brain processes

can be given for mathematical reasoning along with everything else. One line of defense is to maintain that this is false, so that mathematical reasoning uses some special intuition that involves contact with the Platonic realm. A modern form of this argument is given by Sir Roger Penrose.[27]

Another line of defense is to maintain that abstract objects are relevant to mathematical reasoning in a way that is non-causal, and not analogous to perception. This argument is developed by Jerrold Katz in his book *Realistic Rationalism*.

A more radical defense is denial of physical reality, i.e. the mathematical universe hypothesis. In that case, a mathematician's knowledge of mathematics is one mathematical object making contact with another.

17.6 Aesthetics

Many practicing mathematicians have been drawn to their subject because of a sense of beauty they perceive in it. One sometimes hears the sentiment that mathematicians would like to leave philosophy to the philosophers and get back to mathematics—where, presumably, the beauty lies.

In his work on the divine proportion, H.E. Huntley relates the feeling of reading and understanding someone else's proof of a theorem of mathematics to that of a viewer of a masterpiece of art—the reader of a proof has a similar sense of exhilaration at understanding as the original author of the proof, much as, he argues, the viewer of a masterpiece has a sense of exhilaration similar to the original painter or sculptor. Indeed, one can study mathematical and scientific writings as literature.

Philip J. Davis and Reuben Hersh have commented that the sense of mathematical beauty is universal amongst practicing mathematicians. By way of example, they provide two proofs of the irrationality of the $\sqrt{2}$. The first is the traditional proof by contradiction, ascribed to Euclid; the second is a more direct proof involving the fundamental theorem of arithmetic that, they argue, gets to the heart of the issue. Davis and Hersh argue that mathematicians find the second proof more aesthetically appealing because it gets closer to the nature of the problem.

Paul Erdős was well known for his notion of a hypothetical "Book" containing the most elegant or beautiful mathematical proofs. There is not universal agreement that a result has one "most elegant" proof; Gregory Chaitin has argued against this idea.

Philosophers have sometimes criticized mathematicians' sense of beauty or elegance as being, at best, vaguely stated. By the same token, however, philosophers of mathematics have sought to characterize what makes one proof more desirable than another when both are logically sound.

Another aspect of aesthetics concerning mathematics is mathematicians' views towards the possible uses of mathematics for purposes deemed unethical or inappropriate. The best-known exposition of this view occurs in G.H. Hardy's book *A Mathematician's Apology*, in which Hardy argues that pure mathematics is superior in beauty to applied mathematics precisely because it cannot be used for war and similar ends. Some later mathematicians have characterized Hardy's views as mildly dated, with the applicability of number theory to modern-day cryptography.

17.7 See also

17.7.1 Related works

17.7.2 Historical topics

_ History and philosophy of science
_ History of mathematics
_ History of philosophy

17.8 Notes

[1] Maziars, Edward A. (1969). "Problems in the Philosophy of Mathematics (Book Review)". *Philosophy of Science* **36** (3): 325.. For example, when Edward Maziars proposes in a 1969 book review *"to distinguish philosophical mathematics (which is primarily a specialised task for a mathematician) from mathematical philosophy (which ordinarily may be the philosopher's metier)"*, he uses the term *mathematical philosophy* as being synonymous with *philosophy of mathematics*.

[2] Kleene, Stephen (1971). *Introduction to Metamathematics*. Amsterdam, Netherlands: North-Holland Publishing Company. p. 5.

[3] Mac Lane, Saunders (1998), *Categories for the Working Mathematician*, 2nd edition, Springer-Verlag, New York, NY.

[4] _ Putnam, Hilary (1967), "Mathematics Without Foundations", *Journal of Philosophy* 64/1, 5-22. Reprinted, pp. 168–184 in W.D. Hart (ed., 1996).

[5] Tegmark, Max (February 2008). "The Mathematical Universe". *Foundations of Physics* **38** (2): 101–150. arXiv:0704.0646. Bibcode:2008FoPh...38..101T. doi:10.1007/s10701-007-9186-9.

[6] Tegmark (1998), p. 1.

[7] Carnap, Rudolf (1931), "Die logizistische Grundlegung der Mathematik", *Erkenntnis* 2, 91-121. Republished, "The Logicist Foundations of Mathematics", E. Putnam and G.J. Massey (trans.), in Benacerraf and Putnam (1964). Reprinted, pp.

41–52 in Benacerraf and Putnam (1983).
[8] Audi, Robert (1999), *The Cambridge Dictionary of Philosophy*, Cambridge University Press, Cambridge, UK, 1995. 2nd edition. Page 542.
[9] From an 1886 lecture at the 'Berliner Naturforscher-Versammlung', according to H. M. Weber's memorial article, as quoted and translated in Gonzalez Cabillon, Julio (2000-02-03). "FOM: What were Kronecker's f.o.m.?". Retrieved 2008-07-19. Gonzalez gives as the sources for the memorial article, the following: 'Weber, H: "Leopold Kronecker", _Jahresberichte der Deutschen Mathematiker Vereinigung_, vol ii (1893) pp 5-31. Cf page 19. See also _Mathematische Annalen_ vol xliii (1893) pp 1-25'.
[10] Brown, James (2008). *Philosophy of Mathematics*. New York: Routledge. ISBN 978-0-415-96047-2.
[11] A System of Logic Ratiocinative and Inductive, The Collected Works of John Stuart Mill published by the University of Toronto Press in 1973. Book II, Chapter vi, Section 2 (Toronto edition 1975, Vol.7, p. 254)
[12] Franklin, James (2014), "An Aristotelian Realist Philosophy of Mathematics", Palgrave Macmillan, Basingstoke; Franklin, James (2011), "Aristotelianism in the philosophy of mathematics," *Studia Neoaristotelica* 8, 3-15.
[13] Maddy, Penelope (1990), *Realism in Mathematics*, Oxford University Press, Oxford, UK.
[14] Ernest, Paul. "Is Mathematics Discovered or Invented?". University of Exeter. Retrieved 2008-12-26.
[15] Hersh, Reuben (February 10, 1997). *What Kind of a Thing is a Number?*. Interview with John Brockman. Edge Foundation. Retrieved 2008-12-26.
[16] "Humanism and Mathematics Education". *Math Forum*. Humanistic Mathematics Network Journal. Retrieved 2008-12-26.
[17] Tymoczko, Thomas (1998), *New Directions in the Philosophy of Mathematics*. ISBN 978-0691034980.
[18] Popper, Karl Raimund (1946) Aristotelian Society Supplementary Volume XX.

[19] Gregory, Frank Hutson (1996) Arithmetic and Reality: A Development of Popper's Ideas. City University of Hong Kong. Republished in Philosophy of Mathematics Education Journal No. 26 (December 2011)
[20] Yablo, S. (November 8, 1998). "A Paradox of Existence".
[21] Putnam, H. *Mathematics, Matter and Method. Philosophical Papers, vol. 1*. Cambridge: Cambridge University Press, 1975. 2nd. ed., 1985.
[22] Field, Hartry, 1989, Realism, Mathematics, and Modality, Oxford: Blackwell, p. 68
[23] "Since abstract objects are outside the nexus of causes and effects, and thus perceptually inaccessible, they cannot be known through their effects on us" Katz, J. *Realistic Rationalism*, p15
[24] ,Philosophy Now: *Mathematical_Knowledge_A_Dilemma Mathematical Knowledge: A dilemma*
[25] Standard Encyclopaedia of Philosophy
[26] Benacceraf, 1973, p409
[27] Review of The Emperor's New Mind

17.9 Further reading

_ Aristotle, "Prior Analytics", Hugh Tredennick (trans.), pp. 181–531 in *Aristotle, Volume 1*, Loeb Classical Library, William Heinemann, London, UK, 1938.
_ Benacerraf, Paul, and Putnam, Hilary (eds., 1983), *Philosophy of Mathematics, Selected Readings*, 1st edition, Prentice-Hall, Englewood Cliffs, NJ, 1964. 2nd edition, Cambridge University Press, Cambridge, UK, 1983.
_ Berkeley, George (1734), *The Analyst; or, a Discourse Addressed to an Infidel Mathematician. Wherein It is examined whether the Object, Principles, and Inferences of the modern Analysis are more distinctly conceived, or more evidently deduced, than Religious Mysteries and Points of Faith*, London & Dublin. Online text, David R. Wilkins (ed.), Eprint.
_ Bourbaki, N. (1994), *Elements of the History of Mathematics*, John Meldrum (trans.), Springer-Verlag, Berlin, Germany.
_ Chandrasekhar, Subrahmanyan (1987), *Truth and Beauty. Aesthetics and Motivations in Science*, University of Chicago Press, Chicago, IL.
_ Colyvan, Mark (2004), "Indispensability Arguments in the Philosophy of Mathematics", *Stanford Encyclopedia of Philosophy*, Edward N. Zalta (ed.), Eprint.
_ Davis, Philip J. and Hersh, Reuben (1981), *The Mathematical Experience*, Mariner Books, New York, NY.
_ Devlin, Keith (2005), *The Math Instinct: Why You're a Mathematical Genius (Along with Lobsters, Birds, Cats, and Dogs)*, Thunder's Mouth Press, New York, NY.
_ Dummett, Michael (1991 a), *Frege, Philosophy of Mathematics*, Harvard University Press, Cambridge, MA.
_ Dummett, Michael (1991 b), *Frege and Other Philosophers*, Oxford University Press, Oxford, UK.
_ Dummett, Michael (1993), *Origins of Analytical Philosophy*, Harvard University Press, Cambridge, MA.
_ Ernest, Paul (1998), *Social Constructivism as a Philosophy of Mathematics*, State University of New York Press, Albany, NY.
_ George, Alexandre (ed., 1994), *Mathematics and Mind*, Oxford University Press, Oxford, UK.
_ Hadamard, Jacques (1949), *The Psychology of Invention in the Mathematical Field*, 1st edition, Princeton University Press, Princeton, NJ. 2nd edition, 1949. Reprinted, Dover Publications, New York, NY, 1954.
_ Hardy, G.H. (1940), *A Mathematician's Apology*, 1st published, 1940. Reprinted, C.P. Snow (foreword), 1967. Reprinted, Cambridge University Press, Cambridge, UK, 1992.
_ Hart, W.D. (ed., 1996), *The Philosophy of Mathematics*, Oxford University Press, Oxford, UK.

_ Hendricks, Vincent F. and Hannes Leitgeb (eds.). *Philosophy of Mathematics: 5 Questions*, New York: Automatic Press / VIP, 2006.
_ Huntley, H.E. (1970), *The Divine Proportion: A Study in Mathematical Beauty*, Dover Publications, New York, NY.
_ Irvine, A., ed (2009), *The Philosophy of Mathematics*, in *Handbook of the Philosophy of Science* series, North-Holland Elsevier, Amsterdam.
_ Klein, Jacob (1968), *Greek Mathematical Thought and the Origin of Algebra*, Eva Brann (trans.), MIT Press, Cambridge, MA, 1968. Reprinted, Dover Publications, Mineola, NY, 1992.
_ Kline, Morris (1959), *Mathematics and the Physical World*, Thomas Y. Crowell Company, New York, NY, 1959. Reprinted, Dover Publications, Mineola, NY, 1981.
_ Kline, Morris (1972), *Mathematical Thought from Ancient to Modern Times*, Oxford University Press, New York, NY.
_ König, Julius (Gyula) (1905), "Über die Grundlagen der Mengenlehre und das Kontinuumproblem", *Mathematische Annalen* 61, 156-160. Reprinted, "On the Foundations of Set Theory and the Continuum Problem", Stefan Bauer-Mengelberg (trans.), pp. 145–149 in Jean van Heijenoort (ed., 1967).
_ Körner, Stephan, *The Philosophy of Mathematics, An Introduction.* Harper Books, 1960.
_ Lakoff, George, and Núñez, Rafael E. (2000), *Where Mathematics Comes From: How the Embodied Mind Brings Mathematics into Being*, Basic Books, New York, NY.
_ Lakatos, Imre 1976 *Proofs and Refutations:The Logic of Mathematical Discovery* (Eds) J. Worrall & E. Zahar Cambridge University Press
_ Lakatos, Imre 1978 *Mathematics, Science and Epistemology: Philosophical Papers* Volume 2 (Eds) J.Worrall & G.Currie Cambridge University Press
_ Lakatos, Imre 1968 *Problems in the Philosophy of Mathematics* North Holland
_ Leibniz, G.W., *Logical Papers* (1666–1690), G.H.R. Parkinson (ed., trans.), Oxford University Press, London, UK, 1966.
_ Maddy, Penelope (1997), *Naturalism in Mathematics*, Oxford University Press, Oxford, UK.
_ Maziarz, Edward A., and Greenwood, Thomas (1995), *Greek Mathematical Philosophy*, Barnes and Noble Books.
_ Mount, Matthew, *Classical Greek Mathematical Philosophy*, .
_ Parsons, Charles (2014). *Philosophy of Mathematics in the Twentieth Century: Selected Essays.* Cambridge, MA: Harvard University Press. ISBN 978-0-674-72806-6.
_ Peirce, Benjamin (1870), "Linear Associative Algebra", § 1. See *American Journal of Mathematics* 4 (1881).
_ Peirce, C.S., *Collected Papers of Charles Sanders Peirce*, vols. 1-6, Charles Hartshorne and Paul Weiss (eds.), vols. 7-8, Arthur W. Burks (ed.), Harvard University Press, Cambridge, MA, 1931 – 1935, 1958. Cited as CP (volume).(paragraph).
_ Peirce, C.S., various pieces on mathematics and logic, many readable online through links at the Charles Sanders Peirce bibliography, especially under Books authored or edited by Peirce, published in his lifetime and the two sections following it.
_ Plato, "The Republic, Volume 1", Paul Shorey (trans.), pp. 1–535 in *Plato, Volume 5*, Loeb Classical Library, William Heinemann, London, UK, 1930.
_ Plato, "The Republic, Volume 2", Paul Shorey (trans.), pp. 1–521 in *Plato, Volume 6*, Loeb Classical Library, William Heinemann, London, UK, 1935.
_ Resnik, Michael D. *Frege and the Philosophy of Mathematics*, Cornell University, 1980.

_ Resnik, Michael (1997), *Mathematics as a Science of Patterns*, Clarendon Press, Oxford, UK, ISBN 978-0-19-825014-2
_ Robinson, Gilbert de B. (1959), *The Foundations of Geometry*, University of Toronto Press, Toronto, Canada, 1940, 1946, 1952, 4th edition 1959.
_ Raymond, Eric S. (1993), "The Utility of Mathematics", Eprint.
_ Smullyan, Raymond M. (1993), *Recursion Theory for Metamathematics*, Oxford University Press, Oxford, UK.
_ Russell, Bertrand (1919), *Introduction to Mathematical Philosophy*, George Allen and Unwin, London, UK. Reprinted, John G. Slater (intro.), Routledge, London, UK, 1993.
_ Shapiro, Stewart (2000), *Thinking About Mathematics: The Philosophy of Mathematics*, Oxford University Press, Oxford, UK
_ Strohmeier, John, and Westbrook, Peter (1999), *Divine Harmony, The Life and Teachings of Pythagoras*, Berkeley Hills Books, Berkeley, CA.
_ Styazhkin, N.I. (1969), *History of Mathematical Logic from Leibniz to Peano*, MIT Press, Cambridge, MA.
_ Tait, William W. (1986), "Truth and Proof: The Platonism of Mathematics", *Synthese* 69 (1986), 341-370. Reprinted, pp. 142–167 in W.D. Hart (ed., 1996).

_ Tarski, A. (1983), *Logic, Semantics, Metamathematics: Papers from 1923 to 1938*, J.H. Woodger (trans.), Oxford University Press, Oxford, UK, 1956. 2nd edition, John Corcoran (ed.), Hackett Publishing, Indianapolis, IN, 1983.
_ Ulam, S.M. (1990), *Analogies Between Analogies: The Mathematical Reports of S.M. Ulam and His Los Alamos Collaborators*, A.R. Bednarek and Françoise Ulam (eds.), University of California Press, Berkeley, CA.
_ van Heijenoort, Jean (ed. 1967), *From Frege To Gödel: A Source Book in Mathematical Logic, 1879-1931*, Harvard University Press, Cambridge, MA.
_ Wigner, Eugene (1960), "The Unreasonable Effectiveness of Mathematics in the Natural Sciences", *Communications on Pure and Applied Mathematics* **13**(1): 1-14. Eprint
_ Wilder, Raymond L. *Mathematics as a Cultural System*, Pergamon, 1980.

17.10 External links

_ Philosophy of mathematics at PhilPapers
_ Philosophy of mathematics at the Indiana Philosophy Ontology Project
_ Philosophy of Mathematics entry by Leon Horsten in the *Stanford Encyclopedia of Philosophy*
_ Philosophy of mathematics entry in the *Internet Encyclopedia of Philosophy*
_ The London Philosophy Study Guide offers many suggestions on what to read, depending on the student's familiarity with the subject:
_ Philosophy of Mathematics
_ Mathematical Logic
_ Set Theory & Further Logic
_ R.B. Jones' philosophy of mathematics page
_ Philosophy of mathematics at DMOZ
_ The Philosophy of Real Mathematics Blog
_ Kaina Stoicheia by C.S. Peirce.

17.10.1 Journals

_ Philosophia Mathematica journal
_ The Philosophy of Mathematics Education Journal homepage

Chapter 18

Language of mathematics

The **language of mathematics** is the system used by mathematicians to communicate mathematical ideas among themselves. This language consists of a substrate of some natural language (for example English) using technical terms and grammatical conventions that are peculiar to mathematical discourse (see Mathematical jargon), supplemented by a highly specialized symbolic notation for mathematical formulas.
Like natural languages in general, discourse using the language of mathematics can employ a scala of registers. Research articles in academic journals use a more formal tone than oral exchanges over a scribbled-upon napkin in the university cafeteria.

18.1 What is a language?

Here are some definitions of language:
_ *a systematic means of communicating by the use of sounds or conventional symbols* WordNet
_ *a system of words used in a particular discipline* WordNet
_ *the code we all use to express ourselves and communicate to others* Speech & Language Therapy Glossary of Terms
_ *a set (finite or infinite) of sentences, each finite in length and constructed out of a finite set of elements* Noam Chomsky.
These definitions describe language in terms of the following components:
_ A vocabulary of symbols or words
_ A grammar consisting of rules of how these symbols may be used
_ A community of people who use and understand these symbols
_ A range of meanings that can be communicated with these symbols
Each of these components is also found in the language of mathematics.

18.2 The vocabulary of mathematics

Mathematical notation has assimilated symbols from many different alphabets and typefaces. It also includes symbols that are specific to mathematics, such as

8 9 ∇^ 1:

103

Mathematical notation is central to the power of modern mathematics. Though the algebra of Al-Khwārizmī did not use such symbols, it solved equations using many more rules than are used today with symbolic notation, and had great difficulty working with multiple variables (which using symbolic notation can simply be called *x; y; z* , etc.). Sometimes formulas cannot be understood without a written or spoken explanation, but often they are sufficient by themselves, and sometimes they are difficult to read aloud or information is lost in the translation to words, as when several parenthetical factors are involved or when a complex structure like a matrix is manipulated.

Like any other profession, mathematics also has its own brand of technical terminology. In some cases, a word in general usage has a different and specific meaning within mathematics—examples are group, ring, field, category, term, and factor. For more examples, see Category:Mathematical terminology.

In other cases, specialist terms have been created which do not exist outside of mathematics—examples are tensor, fractal, functor. Mathematical statements have their own moderately complex taxonomy, being divided into axioms, conjectures, theorems, lemmas and corollaries. And there are stock phrases in mathematics, used with specific meanings, such as "*if and only if*", "*necessary and sufficient*" and "*without loss of generality*". Such phrases are known as mathematical jargon.

The vocabulary of mathematics also has visual elements. Diagrams are used informally on blackboards, as well as more formally in published work. When used appropriately, diagrams display schematic information more easily. Diagrams also help visually and aid intuitive calculations. Sometimes, as in a visual proof, a diagram even serves as complete justification for a proposition. A system of diagram conventions may evolve into a mathematical notation – for example, the Penrose graphical notation for tensor products.

18.3 The grammar of mathematics

The grammar used for mathematical discourse is essentially the grammar of the natural language used as substrate, but with several mathematics-specific peculiarities.

Most notably, the mathematical notation used for formulas has its own grammar, not dependent on a specific natural language, but shared internationally by mathematicians regardless of their mother tongues. This includes the conventions that the formulas are written predominantly left to right, also when the writing system of the substrate language is right-to-left, and that the Latin alphabet is commonly used for simple variables and parameters. A formula such as

$\sin x + a \cos 2x$ _ 0

is understood by Chinese and Israeli mathematicians alike.

Such mathematical formulas can be a part of speech in a natural-language phrase, or even assume the role of a fullfledged sentence. For example, the formula above, an equation, can be considered a sentence or sentential phrase in which the greater than or equal to symbol has the role of a verb. In careful speech, this can be made clear by pronouncing "≥" as "is greater than or equal to", but in an informal context mathematicians may shorten this to "greater or equal" and yet handle this grammatically like a verb. A good example is the book title *Why does E = mc2?*;[1] here, the equals sign has the role of an infinitive.

Mathematical formulas can be *vocalized* (spoken aloud). The vocalization system for formulas has to be learned, and is dependent on the underlying natural language. For example, when using English, the expression "$f(x)$" is conventionally pronounced "eff of eks", where the insertion of the preposition "of" is not suggested by the notation per se. The expression " $\frac{dy}{dx}$ ", on the other hand, is vocalized like "dee-why-dee-eks", with complete omission of the fraction bar, in other contexts often pronounced "over". The book title *Why does E = mc2?* is said aloud as *Why does ee equal em see-squared?*.

Characteristic for mathematical discourse – both formal and informal – is the use of the inclusive first person plural "we" to mean: "the audience (or reader) together with the speaker (or author)".

18.4 The language community of mathematics

Mathematics is used by mathematicians, who form a global community composed of speakers of many languages. It is also used by students of mathematics. As mathematics is a part of primary education in almost all countries, almost

all educated people have some exposure to pure mathematics. There are very few cultural dependencies or barriers in modern mathematics. There are international mathematics competitions, such as the International Mathematical Olympiad, and international co-operation between professional mathematicians is commonplace.

18.5 The meanings of mathematics

Mathematics is used to communicate information about a wide range of different subjects. Here are three broad categories:

_ **Mathematics describes the real world**: many areas of mathematics originated with attempts to describe and solve real world phenomena - from measuring farms (geometry) to falling apples (calculus) to gambling (probability). Mathematics is widely used in modern physics and engineering, and has been hugely successful in helping us to understand more about the universe around us from its largest scales (physical cosmology) to its smallest (quantum mechanics). Indeed, the very success of mathematics in this respect has been a source of puzzlement for some philosophers (see The Unreasonable Effectiveness of Mathematics in the Natural Sciences by Eugene Wigner).

_ **Mathematics describes abstract structures**: on the other hand, there are areas of pure mathematics which deal with abstract structures, which have no known physical counterparts at all. However, it is difficult to give any categorical examples here, as even the most abstract structures can be co-opted as models in some branch of physics (see Calabi-Yau spaces and string theory).

_ **Mathematics describes mathematics**: mathematics can be used reflexively to describe itself—this is an area of mathematics called metamathematics.

Mathematics can communicate a range of meanings that is as wide as (although different from) that of a natural language. As English mathematician R.L.E. Schwarzenberger says:

My own attitude, which I share with many of my colleagues, is simply that mathematics is a language. Like English, or Latin, or Chinese, there are certain concepts for which mathematics is particularly well suited: it would be as foolish to attempt to write a love poem in the language of mathematics as to prove the Fundamental Theorem of Algebra using the English language.

18.6 Alternative views

Some definitions of language, such as early versions of Charles Hockett's "design features" definition, emphasize the spoken nature of language. Mathematics would not qualify as a language under these definitions, as it is primarily a written form of communication (to see why, try reading Maxwell's equations out loud). However, these definitions would also disqualify sign languages, which are now recognized as languages in their own right, independent of spoken language.

Other linguists believe no valid comparison can be made between mathematics and language, because they are simply too different:

Mathematics would appear to be both more and less than a language for while being limited in its linguistic capabilities it also seems to involve a form of thinking that has something in common with art and music.

- Ford & Peat (1988)

18.7 See also

_ Formulario mathematico

_ Linguistics

_ Philosophy of language

18.8 References

[1] Brian Cox; Jeff Forshaw (2010). *Why does E = mc2? (and why should we care?)*. Da Capo Press. ISBN 978-0-306-81876-9.

_ Knight, Isabel F. (1968). *The Geometric Spirit: The Abbe de Condillac and the French Enlightenment*. New Haven: Yale University Press.

_ R. L. E. Schwarzenberger (2000), *The Language of Geometry*, published in *A Mathematical Spectrum Miscellany*, Applied Probability Trust.

_ Alan Ford & F. David Peat (1988), *The Role of Language in Science*, Foundations of Physics Vol 18.

_ Kay O'Halloran, *Mathematical Discourse: Language, Symbolism and Visual Images*, Continuum, 2004. ISBN 0826468578

18.9 External links

_ What is Language

_ *Mathematics and the Language of Nature* - essay by F. David Peat.

_ Mathematical Words: Origins and Sources (John Aldrich, University of Southampton)

_ Communicating in the Language of Mathematics by Dr. David Moursund

Chapter 19

Arithmetic

Arithmetic or **arithmetics** (from the Greek word ἀριθμός, *arithmos* "number") is the oldest[1] and most elementary branch of mathematics. It consists in the study of numbers, especially the properties of the traditional operations between them — addition, subtraction, multiplication and division. Arithmetic is an elementary part of number theory, and number theory is considered to be one of the top-level divisions of modern mathematics, along with algebra, geometry, and analysis. The terms *arithmetic* and *higher arithmetic* were used until the beginning of the 20th century as synonyms for *number theory* and are, sometimes, still used to refer to a wider part of number theory.[2]

19.1 History

The prehistory of arithmetic is limited to a small number of artifacts which may indicate the conception of addition and subtraction, the best-known being the Ishango bone from central Africa, dating from somewhere between 20,000 and 18,000 BC, although its interpretation is disputed.[3]

The earliest written records indicate the Egyptians and Babylonians used all the elementary arithmetic operations as early as 2000 BC. These artifacts do not always reveal the specific process used for solving problems, but the characteristics of the particular numeral system strongly influence the complexity of the methods. The hieroglyphic system for Egyptian numerals, like the later Roman numerals, descended from tally marks used for counting. In both cases, this origin resulted in values that used a decimal base but did not include positional notation. Complex calculations with Roman numerals required the assistance of a counting board or the Roman abacus to obtain the results.

Early number systems that included positional notation were not decimal, including the sexagesimal (base 60) system for Babylonian numerals and the vigesimal (base 20) system that defined Maya numerals. Because of this place-value concept, the ability to reuse the same digits for different values contributed to simpler and more efficient methods of calculation.

The continuous historical development of modern arithmetic starts with the Hellenistic civilization of ancient Greece, although it originated much later than the Babylonian and Egyptian examples. Prior to the works of Euclid around 300 BC, Greek studies in mathematics overlapped with philosophical and mystical beliefs. For example, Nicomachus summarized the viewpoint of the earlier Pythagorean approach to numbers, and their relationships to each other, in his *Introduction to Arithmetic.*

Greek numerals were used by Archimedes, Diophantus and others in a positional notation not very different from ours. Because the ancient Greeks lacked a symbol for zero (until the Hellenistic period), they used three separate sets of symbols. One set for the unit's place, one for the ten's place, and one for the hundred's. Then for the thousand's place they would reuse the symbols for the unit's place, and so on. Their addition algorithm was identical to ours, and their multiplication algorithm was only very slightly different. Their long division algorithm was the same, and the square root algorithm that was once taught in school was known to Archimedes, who may have invented it. He preferred it to Hero's method of successive approximation because, once computed, a digit doesn't change, and the square roots of perfect squares, such as 7485696, terminate immediately as 2736. For numbers with a fractional part, such as 546.934, they used negative powers of 60 instead of negative powers of 10 for the fractional part 0.934.[4]

The ancient Chinese used a similar positional notation. Because they also lacked a symbol for zero, they had one set of symbols for the unit's place, and a second set for the ten's place. For the hundred's place they then reused the

107

108 *CHAPTER 19. ARITHMETIC*

Arithmetic tables for children, Lausanne, 1835

19.1. HISTORY 109

symbols for the unit's place, and so on. Their symbols were based on the ancient counting rods. It is a complicated question to determine exactly when the Chinese started calculating with positional representation, but it was definitely before 400 BC.[5] The Bishop of Syria, Severus Sebokht (650 AD), "Indians possess a method of calculation that no word can praise enough. Their rational system of mathematics, or of their method of calculation. I mean the system using nine symbols."[6]

Leonardo of Pisa (Fibonacci) in 1200 AD wrote in *Liber Abaci* "The method of the Indians (Modus Indoram) surpasses any known method to compute. Its a marvelous method. They do their computations using nine figures and symbol zero".[7]

The gradual development of Hindu–Arabic numerals independently devised the place-value concept and positional notation, which combined the simpler methods for computations with a decimal base and the use of a digit representing 0. This allowed the system to consistently represent both large and small integers. This approach eventually replaced all other systems. In the early 6th century AD, the Indian mathematician Aryabhata incorporated an existing version of this system in his work, and experimented with different notations. In the 7th century, Brahmagupta established the use of 0 as a separate number and determined the results for multiplication, division, addition and subtraction of zero and all other numbers, except for the result of division by 0. His contemporary, the Syriac bishop Severus Sebokht described the excellence of this system as "... valuable methods of calculation which surpass description". The Arabs also learned this new method and called it *hesab*.

Leibniz's Stepped Reckoner was the first calculator that could perform all four arithmetic operations.

Although the Codex Vigilanus described an early form of Arabic numerals (omitting 0) by 976 AD, Fibonacci was primarily responsible for spreading their use throughout Europe after the publication of his book *Liber Abaci* in 1202. He considered the significance of this "new" representation of numbers, which he styled the "Method of the Indians" (Latin *Modus Indorum*), so fundamental that all related mathematical foundations, including the results of Pythagoras and the algorism describing the methods for performing actual calculations, were "almost a mistake" in comparison.
In the Middle Ages, arithmetic was one of the seven liberal arts taught in universities.
The flourishing of algebra in the medieval Islamic world and in Renaissance Europe was an outgrowth of the enormous simplification of computation through decimal notation.
Various types of tools exist to assist in numeric calculations. Examples include slide rules (for multiplication, division, and trigonometry) and nomographs in addition to the electrical calculator.

19.2 Arithmetic operations

See also: Operation (mathematics)
The basic arithmetic operations are addition, subtraction, multiplication and division, although this subject also includes more advanced operations, such as manipulations of percentages, square roots, exponentiation, and logarithmic functions. Arithmetic is performed according to an order of operations. Any set of objects upon which all four arithmetic operations (except division by 0) can be performed, and where these four operations obey the usual laws, is called a field.[8]

19.2.1 Addition (+)

Main article: Addition
Addition is the basic operation of arithmetic. In its simplest form, addition combines two numbers, the *addends* or *terms*, into a single number, the *sum* of the numbers (Such as 2 + 2 = 4 or 3 + 5 = 8).
Adding more than two numbers can be viewed as repeated addition; this procedure is known as summation and includes ways to add infinitely many numbers in an infinite series; repeated addition of the number 1 is the most basic form of counting.
Addition is commutative and associative so the order the terms are added in does not matter. The identity element of addition (the additive identity) is 0, that is, adding 0 to any number yields that same number. Also, the inverse element of addition (the additive inverse) is the opposite of any number, that is, adding the opposite of any number to the number itself yields the additive identity, 0. For example, the opposite of 7 is −7, so 7 + (−7) = 0.
Addition can be given geometrically as in the following example:
If we have two sticks of lengths *2* and *5*, then if we place the sticks one after the other, the length of the stick thus formed is 2 + 5 = 7.

19.2.2 Subtraction (−)

Main article: Subtraction
See also: Method of complements
Subtraction is the inverse of addition. Subtraction finds the *difference* between two numbers, the *minuend* minus the *subtrahend*. If the minuend is larger than the subtrahend, the difference is positive; if the minuend is smaller than the subtrahend, the difference is negative; if they are equal, the difference is 0.
Subtraction is neither commutative nor associative. For that reason, it is often helpful to look at subtraction as addition of the minuend and the opposite of the subtrahend, that is $a - b = a + (-b)$. When written as a sum, all the properties of addition hold.
There are several methods for calculating results, some of which are particularly advantageous to machine calculation. For example, digital computers employ the method of two's complement. Of great importance is the counting up method by which change is made. Suppose an amount P is given to pay the required amount Q, with P greater than Q. Rather than performing the subtraction $P - Q$ and counting out that amount in change, money is counted out starting at Q and continuing until reaching P. Although the amount counted out must equal the result of the subtraction $P - Q$, the subtraction was never really done and the value of $P - Q$ might still be unknown to the change-maker.

19.2.3 Multiplication (× or · or *)

Main article: Multiplication

Multiplication is the second basic operation of arithmetic. Multiplication also combines two numbers into a single number, the *product*. The two original numbers are called the *multiplier* and the *multiplicand*, sometimes both simply

called *factors*.

Multiplication may be viewed as a scaling operation. If the numbers are imagined as lying in a line, multiplication by a number, say x, greater than 1 is the same as stretching everything away from 0 uniformly, in such a way that the number 1 itself is stretched to where x was. Similarly, multiplying by a number less than 1 can be imagined as squeezing towards 0. (Again, in such a way that 1 goes to the multiplicand.)

Multiplication is commutative and associative; further it is distributive over addition and subtraction. The multiplicative identity is 1, that is, multiplying any number by 1 yields that same number. Also, the multiplicative inverse is the reciprocal of any number (except 0; 0 is the only number without a multiplicative inverse), that is, multiplying the reciprocal of any number by the number itself yields the multiplicative identity.

The product of a and b is written as $a \times b$ or $a \cdot b$. When a or b are expressions not written simply with digits, it is also written by simple juxtaposition: ab. In computer programming languages and software packages in which one can only use characters normally found on a keyboard, it is often written with an asterisk: $a * b$.

19.2.4 Division (÷ or /)

Main article: Division (mathematics)

Division is essentially the inverse of multiplication. Division finds the *quotient* of two numbers, the *dividend* divided by the *divisor*. Any dividend divided by 0 is undefined. For distinct positive numbers, if the dividend is larger than the divisor, the quotient is greater than 1, otherwise it is less than 1 (a similar rule applies for negative numbers). The quotient multiplied by the divisor always yields the dividend.

Division is neither commutative nor associative. As it is helpful to look at subtraction as addition, it is helpful to look at division as multiplication of the dividend times the reciprocal of the divisor, that is $a \div b = a \times 1/b$. When written as a product, it obeys all the properties of multiplication.

19.3 Decimal arithmetic

Decimal representation refers exclusively, in common use, to the written numeral system employing arabic numerals as the digits for a radix 10 ("decimal") positional notation; however, any numeral system based on powers of 10, e.g., Greek, Cyrillic, Roman, or Chinese numerals may conceptually be described as "decimal notation" or "decimal representation".

Modern methods for four fundamental operations (addition, subtraction, multiplication and division) were first devised by Brahmagupta of India. This was known during medieval Europe as "Modus Indoram" or Method of the Indians. Positional notation (also known as "place-value notation") refers to the representation or encoding of numbers using the same symbol for the different orders of magnitude (e.g., the "ones place", "tens place", "hundreds place") and, with a radix point, using those same symbols to represent fractions (e.g., the "tenths place", "hundredths place"). For example, 507.36 denotes 5 hundreds (10^2), plus 0 tens (10^1), plus 7 units (10^0), plus 3 tenths (10^{-1}) plus 6 hundredths (10^{-2}).

The concept of 0 as a number comparable to the other basic digits is essential to this notation, as is the concept of 0's use as a placeholder, and as is the definition of multiplication and addition with 0. The use of 0 as a placeholder and, therefore, the use of a positional notation is first attested to in the Jain text from India entitled the *Lokavibhâga*, dated 458 AD and it was only in the early 13th century that these concepts, transmitted via the scholarship of the Arabic world, were introduced into Europe by Fibonacci[9] using the Hindu–Arabic numeral system.

Algorism comprises all of the rules for performing arithmetic computations using this type of written numeral. For example, addition produces the sum of two arbitrary numbers. The result is calculated by the repeated addition of single digits from each number that occupies the same position, proceeding from right to left. An addition table with ten rows and ten columns displays all possible values for each sum. If an individual sum exceeds the value 9, the result is represented with two digits. The rightmost digit is the value for the current position, and the result for the subsequent addition of the digits to the left increases by the value of the second (leftmost) digit, which is always one. This adjustment is termed a *carry* of the value 1.

The process for multiplying two arbitrary numbers is similar to the process for addition. A multiplication table with ten rows and ten columns lists the results for each pair of digits. If an individual product of a pair of digits exceeds

9, the *carry* adjustment increases the result of any subsequent multiplication from digits to the left by a value equal to the second (leftmost) digit, which is any value from 1 to 8 ($9 \times 9 = 81$). Additional steps define the final result. Similar techniques exist for subtraction and division.

The creation of a correct process for multiplication relies on the relationship between values of adjacent digits. The value for any single digit in a numeral depends on its position. Also, each position to the left represents a value ten times larger than the position to the right. In mathematical terms, the exponent for the radix (base) of 10 increases by 1 (to the left) or decreases by 1 (to the right). Therefore, the value for any arbitrary digit is multiplied by a value of the form 10^n with integer *n*. The list of values corresponding to all possible positions for a single digit is written as $\{..., 10^2, 10, 1, 10^{-1}, 10^{-2}, ...\}$.

Repeated multiplication of any value in this list by 10 produces another value in the list. In mathematical terminology, this characteristic is defined as closure, and the previous list is described as **closed under multiplication**. It is the basis for correctly finding the results of multiplication using the previous technique. This outcome is one example of the uses of number theory.

19.4 Compound unit arithmetic

Compound[10] unit arithmetic is the application of arithmetic operations to mixed radix quantities such as feet and inches, gallons and pints, pounds shillings and pence, and so on. Prior to the use of decimal-based systems of money and units of measure, the use of compound unit arithmetic formed a significant part of commerce and industry.

19.4.1 Basic arithmetic operations

The techniques used for compound unit arithmetic were developed over many centuries and are well-documented in many textbooks in many different languages.[11][12][13][14] In addition to the basic arithmetic functions encountered in decimal arithmetic, compound unit arithmetic employs three more functions:

- **Reduction** where a compound quantity is reduced to a single quantity, for example conversion of a distance expressed in yards, feet and inches to one expressed in inches.[15]
- **Expansion**, the inverse function to reduction, is the conversion of a quantity that is expressed as a single unit of measure to a compound unit, such as expanding 24 oz to 1 lb, 8 oz.
- **Normalization** is the conversion of a set of compound units to a standard form – for example rewriting "1 ft 13 in" as "2 ft 1 in".

Knowledge of the relationship between the various units of measure, their multiples and their submultiples forms an essential part of compound unit arithmetic.

19.4.2 Principles of compound unit arithmetic

There are two basic approaches to compound unit arithmetic:

- **Reduction–expansion method** where all the compound unit variables are reduced to single unit variables, the calculation performed and the result expanded back to compound units. This approach is suited for automated calculations. A typical example is the handling of time by Microsoft Excel where all time intervals are processed internally as days and decimal fractions of a day.
- **On-going normalization method** in which each unit is treated separately and the problem is continuously normalized as the solution develops. This approach, which is widely described in classical texts, is best suited for manual calculations. An example of the on-going normalization method as applied to addition is shown below.

A scale calibrated in imperial units with an associated cost display.

19.4.3 Operations in practice

During the 19th and 20th centuries various aids were developed to aid the manipulation of compound units, particularly in commercial applications. The most common aids were mechanical tills which were adapted in countries such as the United Kingdom to accommodate pounds, shillings, pennies and farthings and "Ready Reckoners" – books aimed at traders that catalogued the results of various routine calculations such as the percentages or multiples of various sums of money. One typical booklet[16] that ran to 150 pages tabulated multiples "from one to ten thousand at the various prices from one farthing to one pound".

The cumbersome nature of compound unit arithmetic has been recognized for many years – in 1586, the Flemish mathematician Simon Stevin published a small pamphlet called *De Thiende* ("the tenth")[17] in which he declared that the universal introduction of decimal coinage, measures, and weights to be merely a question of time while in the modern era, many conversion programs, such as that embedded in the calculator supplied as a standard part of the Microsoft Windows 7 operating system display compound units in a reduced decimal format rather than using an expanded format (i.e. "2.5 ft" is displayed rather than "2 ft 6 in").

19.5 Number theory

Main article: Number theory

Until the 19th century, *number theory* was a synonym of "arithmetic". The addressed problems were directly related to the basic operations and concerned primality, divisibility, and the solution of equations in integers, such as Fermat's last theorem. It appeared that most of these problems, although very elementary to state, are very difficult and may not be solved without very deep mathematics involving concepts and methods from many other branches of mathematics. This led to new branches of number theory such as analytic number theory, algebraic number theory, Diophantine geometry and arithmetic algebraic geometry. Wiles' proof of Fermat's Last Theorem is a typical example of the necessity of sophistical methods, which go far beyond the classical methods of arithmetic, for solving problems that can be stated in elementary arithmetic.

19.6 Arithmetic in education

Primary education in mathematics often places a strong focus on algorithms for the arithmetic of natural numbers, integers, fractions, and decimals (using the decimal place-value system). This study is sometimes known as algorism. The difficulty and unmotivated appearance of these algorithms has long led educators to question this curriculum, advocating the early teaching of more central and intuitive mathematical ideas. One notable movement in this direction was the New Math of the 1960s and 1970s, which attempted to teach arithmetic in the spirit of axiomatic development from set theory, an echo of the prevailing trend in higher mathematics.[18]
Also, arithmetic was used by Islamic Scholars in order to teach application of the rulings related to Zakat and Irth. This was done in a book entitled *The Best of Arithmetic* by Abd-al-Fattah-al-Dumyati.[19]
The book begins with the foundations of mathematics and proceeds to its application in the later chapters.

19.7 See also

_ Lists of mathematics topics
_ Mathematics
_ Outline of arithmetic

19.7.1 Related topics

_ Addition of natural numbers

_ Additive inverse
_ Arithmetic coding
_ Arithmetic mean
_ Arithmetic progression
_ Arithmetic properties
_ Associativity
_ Commutativity
_ Distributivity
_ Elementary arithmetic
_ Finite field arithmetic
_ Integer
_ List of important publications in mathematics
_ Mental calculation
_ Number line

19.8 Notes

[1] "Mathematics". Science Clarified. Retrieved 23 October 2012.
[2] Davenport, Harold, *The Higher Arithmetic: An Introduction to the Theory of Numbers* (7th ed.), Cambridge University Press, Cambridge, UK, 1999, ISBN 0-521-63446-6
[3] Rudman, Peter Strom (2007). *How Mathematics Happened: The First 50,000 Years*. Prometheus Books. p. 64. ISBN 978-1-59102-477-4.
[4] *The Works of Archimedes*, Chapter IV, *Arithmetic in Archimedes*, edited by T.L. Heath, Dover Publications Inc, New York, 2002.
[5] Joseph Needham, *Science and Civilization in China*, Vol. 3, page 9, Cambridge University Press, 1959.
[6] Reference: Revue de l'Orient Chretien by François Nau pp.327-338. (1929)
[7] Reference: Sigler, L., "Fibonacci's Liber Abaci", Springer, 2003.
[8] Tapson, Frank (1996). *The Oxford Mathematics Study Dictionary*. Oxford University Press. ISBN 0 19 914551 2.
[9] Leonardo Pisano - page 3: "Contributions to number theory". *Encyclopædia Britannica* Online, 2006. Retrieved 18 September 2006.
[10] Walkingame, Francis (1860). "The Tutor's Companion; or, Complete Practical Arithmetic". Webb, Millington & Co. pp. 24–39.
[11] Palaiseau, JFG (October 1816). *Métrologie universelle, ancienne et moderne: ou rapport des poids et mesures des empires, royaumes, duchés et principautés des quatre parties du monde* [*Universal, ancient and modern metrology: or report of*

weights and measurements of empires, kingdoms, duchies and principalities of all parts of the world] (in French). Bordeaux. Retrieved October 30, 2011.
[12] Jacob de Gelder (1824). *Allereerste Gronden der Cijferkunst* [*Introduction to Numeracy*] (in Dutch). 's Gravenhage and Amsterdam: de Gebroeders van Cleef. pp. 163–176. Retrieved March 2, 2011.
[13] Malaisé, Ferdinand (1842). *Theoretisch-Praktischer Unterricht im Rechnen für die niederen Classen der Regimentsschulen der Königl. Bayer. Infantrie und Cavalerie* [*Theoretical and practical instruction in arithmetic for the lower classes of the Royal Bavarian Infantry and Cavalry School*] (in German). Munich. Retrieved 20 March 2012.
[14] *Encyclopædia Britannica*, Vol I, Edinburgh, 1772, Arithmetick

[15] Walkingame, Francis (1860). "The Tutor's Companion; or, Complete Practical Arithmetic". Webb, Millington & Co. pp. 43–50.
[16] Thomson, J (1824). *The Ready Reckoner in miniature containing accurate table from one to the thousand at the various prices from one farthing to one pound.*. Montreal. Retrieved 25 March 2012.
[17] O'Connor, John J.; Robertson, Edmund F. (January 2004), "Arithmetic", *MacTutor History of Mathematics archive*, University of St Andrews.
[18] Mathematically Correct: Glossary of Terms
[19] Abd-al-Fattah Bin Abd-al-Rahman al-Banna al-Dumyati (1887). "The Best of Arithmetic". *World Digital Library* (in Arabic). Retrieved 30 June 2013.

19.9 References

_ Cunnington, Susan, *The Story of Arithmetic: A Short History of Its Origin and Development*, Swan Sonnenschein, London, 1904
_ Dickson, Leonard Eugene, *History of the Theory of Numbers* (3 volumes), reprints: Carnegie Institute of Washington, Washington, 1932; Chelsea, New York, 1952, 1966
_ Euler, Leonhard, *Elements of Algebra*, Tarquin Press, 2007
_ Fine, Henry Burchard (1858–1928), *The Number System of Algebra Treated Theoretically and Historically*, Leach, Shewell & Sanborn, Boston, 1891
_ Karpinski, Louis Charles (1878–1956), *The History of Arithmetic*, Rand McNally, Chicago, 1925; reprint: Russell & Russell, New York, 1965
_ Ore, Øystein, *Number Theory and Its History*, McGraw–Hill, New York, 1948
_ Weil, André, *Number Theory: An Approach through History*, Birkhauser, Boston, 1984; reviewed: Mathematical Reviews 85c:01004

19.10 External links

_ MathWorld article about arithmetic
_ The New Student's Reference Work/Arithmetic (historical)
_ The Great Calculation According to the Indians, of Maximus Planudes – an early Western work on arithmetic at Convergence
_ P. H. Vander Weyde (1879). "Arithmetic". *The American Cyclopædia.*

Chapter 20

Algebra

"Algebraist" redirects here. For the novel by Iain M. Banks, see The Algebraist.
For beginner's introduction to algebra, see Wikibooks: Algebra.
Algebra (from Arabic *al-jebr* meaning "reunion of broken parts"[1]) is one of the broad parts of mathematics,
The quadratic formula expresses the solution of the degree two equation $ax^2+bx+c=0$ *in terms of its coefficients* $a; b; c$ *, where* a *is not equal to* 0 .
together with number theory, geometry and analysis. In its most general form algebra is the study of symbols and the rules for manipulating symbols[2] and is a unifying thread of all of mathematics.[3] As such, it includes everything from elementary equation solving to the study of abstractions such as groups, rings, and fields. The more basic parts of algebra are called elementary algebra, the more abstract parts are called abstract algebra or modern algebra. Elementary algebra is essential for any study of mathematics, science, or engineering, as well as such applications as medicine and economics. Abstract algebra is a major area in advanced mathematics, studied primarily by professional mathematicians. Much early work in algebra, as the Arabic origin of its name suggests, was done in the Near East, by such mathematicians as Omar Khayyam (1048-1131).[4][5]
Elementary algebra differs from arithmetic in the use of abstractions, such as using letters to stand for numbers that are either unknown or allowed to take on many values.[6] For example, in $x + 2 = 5$ the letter x is unknown, but the

law of inverses can be used to discover its value: $x = 3$. In $E = mc^2$, the letters E and m are variables, and the letter c is a constant. Algebra gives methods for solving equations and expressing formulas that are much easier (for those who know how to use them) than the older method of writing everything out in words.

The word *algebra* is also used in certain specialized ways. A special kind of mathematical object in abstract algebra is called an "algebra", and the word is used, for example, in the phrases linear algebra and algebraic topology (see below).

A mathematician who does research in algebra is called an **algebraist**.

20.1 How to distinguish between different meanings of "algebra"

For historical reasons, the word "algebra" has several related meanings in mathematics, as a single word or with qualifiers. Such a situation, where a single word has many meanings in the same area of mathematics, may be confusing. However the distinction is easier if one recalls that the name of a scientific area is usually singular and without an article and the name of a specific structure requires an article or the plural. Thus we have:

_ As a single word without article, "algebra" names a broad part of mathematics (see below).

_ As a single word with article or in plural, "algebra" denotes a specific mathematical structure. See algebra (ring theory) and algebra over a field. More generally, in universal algebra, it can refer to any structure.

_ With a qualifier, there is the same distinction:

_ Without article, it means a part of algebra, such as linear algebra, elementary algebra (the symbolmanipulation rules taught in elementary courses of mathematics as part of primary and secondary education), or abstract algebra (the study of the algebraic structures for themselves).

_ With an article, it means an instance of some abstract structure, like a Lie algebra or an associative algebra.

_ Frequently both meanings exist for the same qualifier, as in the sentence: *Commutative algebra is the study of commutative rings, which are commutative algebras over the integers.*

20.2 Algebra as a branch of mathematics

Algebra began with computations similar to those of arithmetic, with letters standing for numbers.[6] This allowed proofs of properties that are true no matter which numbers are involved. For example, in the quadratic equation

$$ax^2 + bx + c = 0;$$

$a; b; c$ can be any numbers whatsoever (except that a cannot be 0), and the quadratic formula can be used to quickly and easily find the value of the unknown quantity x .

As it developed, algebra was extended to other non-numerical objects, such as vectors, matrices, and polynomials. Then the structural properties of these non-numerical objects were abstracted to define algebraic structures such as groups, rings, and fields.

Before the 16th century, mathematics was divided into only two subfields, arithmetic and geometry. Even though some methods, which had been developed much earlier, may be considered nowadays as algebra, the emergence of algebra and, soon thereafter, of infinitesimal calculus as subfields of mathematics only dates from 16th or 17th century. From the second half of 19th century on, many new fields of mathematics appeared, most of which made use of both arithmetic and geometry, and almost all of which used algebra.

Today, algebra has grown until it includes many branches of mathematics, as can be seen in the Mathematics Subject Classification[7] where none of the first level areas (two digit entries) is called *algebra.* Today algebra includes section 08-General algebraic systems, 12-Field theory and polynomials, 13-Commutative algebra, 15-Linear and multilinear algebra; matrix theory, 16-Associative rings and algebras, 17-Nonassociative rings and algebras, 18-Category theory; homological algebra, 19-K-theory and 20-Group theory. Algebra is also used extensively in 11-Number theory and 14-Algebraic geometry.

20.3 Etymology

The word *algebra* comes from the Arabic language (ال جبر *al-jabr* "restoration") from the title of the book *Ilm al-jabr wa'l-muḳābala* by al-Khwarizmi. The word entered the English language during Late Middle English from either Spanish, Italian, or Medieval Latin. Algebra originally referred to a surgical procedure, and still is used in that sense in Spanish, while the mathematical meaning was a later development.[8]

20.4 History

Main articles: History of algebra and Timeline of algebra

The start of algebra as an area of mathematics may be dated to the end of 16th century, with François Viète's work.

Until the 19th century, algebra consisted essentially of the theory of equations. In the following, “Prehistory of algebra” is about the results of the theory of equations that precede the emergence of algebra as an area of mathematics.

20.4.1 Early history of algebra

The roots of algebra can be traced to the ancient Babylonians,[9] who developed an advanced arithmetical system with which they were able to do calculations in an algorithmic fashion. The Babylonians developed formulas to calculate solutions for problems typically solved today by using linear equations, quadratic equations, and indeterminate linear equations. By contrast, most Egyptians of this era, as well as Greek and Chinese mathematics in the 1st millennium BC, usually solved such equations by geometric methods, such as those described in the *Rhind Mathematical Papyrus*, Euclid’s *Elements*, and *The Nine Chapters on the Mathematical Art*. The geometric work of the Greeks, typified in the *Elements*, provided the framework for generalizing formulae beyond the solution of particular problems into more general systems of stating and solving equations, although this would not be realized until mathematics developed in medieval Islam.[10]

By the time of Plato, Greek mathematics had undergone a drastic change. The Greeks created a geometric algebra where terms were represented by sides of geometric objects, usually lines, that had letters associated with them.[6]

Diophantus (3rd century AD) was an Alexandrian Greek mathematician and the author of a series of books called *Arithmetica*. These texts deal with solving algebraic equations,[11] and have led, in number theory to the modern notion of Diophantine equation.

Earlier traditions discussed above had a direct influence on Muhammad ibn Mūsā al-Khwārizmī (c. 780–850). He later wrote *The Compendious Book on Calculation by Completion and Balancing*, which established algebra as a mathematical discipline that is independent of geometry and arithmetic.[12]

The Hellenistic mathematicians Hero of Alexandria and Diophantus[13] as well as Indian mathematicians such as Brahmagupta continued the traditions of Egypt and Babylon, though Diophantus’ *Arithmetica* and Brahmagupta’s *Brahmasphutasiddhanta* are on a higher level.[14] For example, the first complete arithmetic solution (including zero and negative solutions) to quadratic equations was described by Brahmagupta in his book *Brahmasphutasiddhanta.* Later, Arabic and Muslim mathematicians developed algebraic methods to a much higher degree of sophistication. Although Diophantus and the Babylonians used mostly special *ad hoc* methods to solve equations, Al-Khwarizmi contribution was fundamental. He solved linear and quadratic equations without algebraic symbolism, negative numbers or zero, thus he has to distinguish several types of equations.[15]

In the context where algebra is identified with the theory of equations, the Greek mathematician Diophantus has traditionally been known as the “father of algebra” but in more recent times there is much debate over whether al-Khwarizmi, who founded the discipline of *al-jabr*, deserves that title instead.[16] Those who support Diophantus point to the fact that the algebra found in *Al-Jabr* is slightly more elementary than the algebra found in *Arithmetica* and that *Arithmetica* is syncopated while *Al-Jabr* is fully rhetorical.[17] Those who support Al-Khwarizmi point to the fact that he introduced the methods of "reduction" and “balancing” (the transposition of subtracted terms to the other side of an equation, that is, the cancellation of like terms on opposite sides of the equation) which the term *al-jabr* originally referred to,[18] and that he gave an exhaustive explanation of solving quadratic equations,[19] supported by geometric proofs, while treating algebra as an independent discipline in its own right.[20] His algebra was also no longer concerned “with a series of problems to be resolved, but an exposition which starts with primitive terms in which the combinations must give all possible prototypes for equations, which henceforward explicitly constitute the true object of study”. He also studied an equation for its own sake and “in a generic manner, insofar as it does not simply emerge in the course of solving a problem, but is specifically called on to define an infinite class of problems”.[21]

The Persian mathematician Omar Khayyam is credited with identifying the foundations of algebraic geometry and found the general geometric solution of the cubic equation. Another Persian mathematician, Sharaf al-Dīn al-Tūsī, found algebraic and numerical solutions to various cases of cubic equations.[22] He also developed the concept of a function.[23] The Indian mathematicians Mahavira and Bhaskara II, the Persian mathematician Al-Karaji,[24] and the Chinese mathematician Zhu Shijie, solved various cases of cubic, quartic, quintic and higher-order polynomial equations using numerical methods. In the 13th century, the solution of a cubic equation by Fibonacci is representative of the beginning of a revival in European algebra. As the Islamic world was declining, the European world was

ascending. And it is here that algebra was further developed.

20.4.2 History of algebra

François Viète's work on new algebra at the close of the 16th century was an important step towards modern algebra. In 1637, René Descartes published *La Géométrie*, inventing analytic geometry and introducing modern algebraic notation. Another key event in the further development of algebra was the general algebraic solution of the cubic and quartic equations, developed in the mid-16th century. The idea of a determinant was developed by Japanese mathematician Kowa Seki in the 17th century, followed independently by Gottfried Leibniz ten years later, for the purpose of solving systems of simultaneous linear equations using matrices. Gabriel Cramer also did some work on matrices and determinants in the 18th century. Permutations were studied by Joseph-Louis Lagrange in his 1770

paper *Réflexions sur la résolution algébrique des équations* devoted to solutions of algebraic equations, in which he introduced Lagrange resolvents. Paolo Ruffini was the first person to develop the theory of permutation groups, and like his predecessors, also in the context of solving algebraic equations.
Abstract algebra was developed in the 19th century, deriving from the interest in solving equations, initially focusing on what is now called Galois theory, and on constructibility issues.[25] George Peacock was the founder of axiomatic thinking in arithmetic and algebra. Augustus De Morgan discovered relation algebra in his *Syllabus of a Proposed System of Logic*. Josiah Willard Gibbs developed an algebra of vectors in three-dimensional space, and Arthur Cayley developed an algebra of matrices (this is a noncommutative algebra).[26]

20.5 Areas of mathematics with the word algebra in their name

Some areas of mathematics that fall under the classification abstract algebra have the word algebra in their name; linear algebra is one example. Others do not: group theory, ring theory, and field theory are examples. In this section, we list some areas of mathematics with the word "algebra" in the name.
_ Elementary algebra, the part of algebra that is usually taught in elementary courses of mathematics.
_ Abstract algebra, in which algebraic structures such as groups, rings and fields are axiomatically defined and investigated.
_ Linear algebra, in which the specific properties of linear equations, vector spaces and matrices are studied.
_ Commutative algebra, the study of commutative rings.
_ Computer algebra, the implementation of algebraic methods as algorithms and computer programs.
_ Homological algebra, the study of algebraic structures that are fundamental to study topological spaces.
_ Universal algebra, in which properties common to all algebraic structures are studied.
_ Algebraic number theory, in which the properties of numbers are studied from an algebraic point of view.
_ Algebraic geometry, a branch of geometry, in its primitive form specifying curves and surfaces as solutions of polynomial equations.
_ Algebraic combinatorics, in which algebraic methods are used to study combinatorial questions.
Many mathematical structures are called **algebras**:
_ Algebra over a field or more generally algebra over a ring.
Many classes of algebras over a field or over a ring have a specific name:
_ Associative algebra
_ Non-associative algebra
_ Lie algebra
_ Hopf algebra

_ C*-algebra
_ Symmetric algebra
_ Exterior algebra
_ Tensor algebra
_ In measure theory,
_ Sigma-algebra
_ Algebra over a set
_ In category theory
_ F-algebra and F-coalgebra
_ T-algebra
_ In logic,
_ Relational algebra: a set of finitary relations that is closed under certain operators.
_ Boolean algebra, a structure abstracting the computation with the truth values *false* and *true*. See also Boolean algebra (structure).
_ Heyting algebra

20.6 Elementary algebra

Main article: Elementary algebra
Elementary algebra is the most basic form of algebra. It is taught to students who are presumed to have no knowledge of mathematics beyond the basic principles of arithmetic. In arithmetic, only numbers and their arithmetical operations (such as +, −, ×, ÷) occur. In algebra, numbers are often represented by symbols called variables (such as a, n, x, y or z). This is useful because:
_ It allows the general formulation of arithmetical laws (such as $a + b = b + a$ for all a and b), and thus is the first step to a systematic exploration of the properties of the real number system.
_ It allows the reference to "unknown" numbers, the formulation of equations and the study of how to solve these. (For instance, "Find a number x such that $3x + 1 = 10$" or going a bit further "Find a number x such that $ax + b = c$". This step leads to the conclusion that it is not the nature of the specific numbers that allows us to solve

it, but that of the operations involved.)
It allows the formulation of functional relationships. (For instance, "If you sell x tickets, then your profit will be $3x - 10$ dollars, or $f(x) = 3x - 10$, where f is the function, and x is the number to which the function is applied".)

20.6.1 Polynomials

Main article: Polynomial
A **polynomial** is an expression that is the sum of a finite number of non-zero terms, each term consisting of the product of a constant and a finite number of variables raised to whole number powers. For example, $x^2 + 2x - 3$ is a polynomial in the single variable x. A **polynomial expression** is an expression that may be rewritten as a polynomial, by using commutativity, associativity and distributivity of addition and multiplication. For example, $(x - 1)(x + 3)$ is a polynomial expression, that, properly speaking, is not a polynomial. A **polynomial function** is a function that is defined by a polynomial, or, equivalently, by a polynomial expression. The two preceding examples define the same polynomial function.
Two important and related problems in algebra are the factorization of polynomials, that is, expressing a given polynomial as a product of other polynomials that can not be factored any further, and the computation of polynomial greatest common divisors. The example polynomial above can be factored as $(x - 1)(x + 3)$. A related class of problems is finding algebraic expressions for the roots of a polynomial in a single variable.

20.6.2 Teaching algebra

See also: Mathematics education
It has been suggested that elementary algebra should be taught as young as eleven years old,[27] though in recent years it is more common for public lessons to begin at the eighth grade level (≈ 13 y.o. ±) in the United States.[28]
Since 1997, Virginia Tech and some other universities have begun using a personalized model of teaching algebra that combines instant feedback from specialized computer software with one-on-one and small group tutoring, which has reduced costs and increased student achievement.[29]

20.7 Abstract algebra

Main articles: Abstract algebra and Algebraic structure
Abstract algebra extends the familiar concepts found in elementary algebra and arithmetic of numbers to more general concepts. Here are listed fundamental concepts in abstract algebra.
Sets: Rather than just considering the different types of numbers, abstract algebra deals with the more general concept of *sets*: a collection of all objects (called elements) selected by property specific for the set. All collections of the familiar types of numbers are sets. Other examples of sets include the set of all two-by-two matrices, the set of all second-degree polynomials ($ax^2 + bx + c$), the set of all two dimensional vectors in the plane, and the various finite groups such as the cyclic groups, which are the groups of integers modulo n. Set theory is a branch of logic and not technically a branch of algebra.
Binary operations: The notion of addition (+) is abstracted to give a *binary operation*, $*$ say. The notion of binary operation is meaningless without the set on which the operation is defined. For two elements a and b in a set S, $a * b$ is another element in the set; this condition is called closure. Addition (+), subtraction (-), multiplication (×), and division (÷) can be binary operations when defined on different sets, as are addition and multiplication of matrices, vectors, and polynomials.
Identity elements: The numbers zero and one are abstracted to give the notion of an *identity element* for an operation. Zero is the identity element for addition and one is the identity element for multiplication. For a general binary operator $*$ the identity element e must satisfy $a * e = a$ and $e * a = a$. This holds for addition as $a + 0 = a$ and $0 + a = a$ and multiplication $a \times 1 = a$ and $1 \times a = a$. Not all sets and operator combinations have an identity element; for example, the set of positive natural numbers (1, 2, 3, ...) has no identity element for addition.
Inverse elements: The negative numbers give rise to the concept of *inverse elements*. For addition, the inverse of a is written $-a$, and for multiplication the inverse is written a^{-1}. A general two-sided inverse element a^{-1} satisfies the property that $a * a^{-1} = 1$ and $a^{-1} * a = 1$.
Associativity: Addition of integers has a property called associativity. That is, the grouping of the numbers to be added does not affect the sum. For example: $(2 + 3) + 4 = 2 + (3 + 4)$. In general, this becomes $(a * b) * c = a * (b * c)$. This property is shared by most binary operations, but not subtraction or division or octonion multiplication.
Commutativity: Addition and multiplication of real numbers are both commutative. That is, the order of the numbers does not affect the result. For example: $2 + 3 = 3 + 2$. In general, this becomes $a * b = b * a$. This property does not hold for all binary operations. For example, matrix multiplication and quaternion multiplication are both non-commutative.

20.7.1 Groups

Main article: Group (mathematics)
See also: Group theory and Examples of groups
Combining the above concepts gives one of the most important structures in mathematics: a **group**. A group is a combination of a set S and a single binary operation $*$, defined in any way you choose, but with the following properties:

_ An identity element e exists, such that for every member a of S, $e * a$ and $a * e$ are both identical to a.
_ Every element has an inverse: for every member a of S, there exists a member a^{-1} such that $a * a^{-1}$ and $a^{-1} * a$ are both identical to the identity element.
_ The operation is associative: if a, b and c are members of S, then $(a * b) * c$ is identical to $a * (b * c)$.

If a group is also commutative—that is, for any two members a and b of S, $a * b$ is identical to $b * a$—then the group is said to be abelian.
For example, the set of integers under the operation of addition is a group. In this group, the identity element is 0 and the inverse of any element a is its negation, $-a$. The associativity requirement is met, because for any integers a, b and c, $(a + b) + c = a + (b + c)$
The nonzero rational numbers form a group under multiplication. Here, the identity element is 1, since $1 \times a = a \times 1 = a$ for any rational number a. The inverse of a is $1/a$, since $a \times 1/a = 1$.
The integers under the multiplication operation, however, do not form a group. This is because, in general, the multiplicative inverse of an integer is not an integer. For example, 4 is an integer, but its multiplicative inverse is ¼, which is not an integer.
The theory of groups is studied in group theory. A major result in this theory is the classification of finite simple groups, mostly published between about 1955 and 1983, which separates the finite simple groups into roughly 30 basic types.
Semigroups, quasigroups, and monoids are structures similar to groups, but more general. They comprise a set and a closed binary operation, but do not necessarily satisfy the other conditions. A semigroup has an *associative* binary operation, but might not have an identity element. A monoid is a semigroup which does have an identity but might not have an inverse for every element. A quasigroup satisfies a requirement that any element can be turned into any other by either a unique left-multiplication or right-multiplication; however the binary operation might not be associative.
All groups are monoids, and all monoids are semigroups.

20.7.2 Rings and fields

Main articles: Ring (mathematics) and Field (mathematics)
See also: Ring theory, Glossary of ring theory, Field theory (mathematics) and Glossary of field theory
Groups just have one binary operation. To fully explain the behaviour of the different types of numbers, structures with two operators need to be studied. The most important of these are rings, and fields.
A **ring** has two binary operations (+) and (×), with × distributive over +. Under the first operator (+) it forms an *abelian group*. Under the second operator (×) it is associative, but it does not need to have identity, or inverse, so division is not required. The additive (+) identity element is written as 0 and the additive inverse of a is written as $-a$.
Distributivity generalises the *distributive law* for numbers. For the integers $(a + b) \times c = a \times c + b \times c$ and $c \times (a + b) = c \times a + c \times b$, and × is said to be *distributive* over +.
The integers are an example of a ring. The integers have additional properties which make it an **integral domain**.
A **field** is a *ring* with the additional property that all the elements excluding 0 form an *abelian group* under ×. The multiplicative (×) identity is written as 1 and the multiplicative inverse of a is written as a^{-1}.
The rational numbers, the real numbers and the complex numbers are all examples of fields.

20.8 See also

_ Outline of algebra
_ Outline of linear algebra
_ Algebra tile

20.9 Notes

[1] "algebra". *Online Etymology Dictionary.*
[2] I. N. Herstein, *Topics in Algebra*, "An algebraic system can be described as a set of objects together with some operations for combining them." p. 1, Ginn and Company, 1964
[3] I. N. Herstein, *Topics in Algebra*, "...it also serves as the unifying thread which interlaces almost all of mathematics." p. 1, Ginn and Company, 1964

[4] https://en.wikipedia.org/wiki/Omar_Khayyám
[5] "Omar Khayyam". *Encyclopedia Britannica*. Retrieved 5 October 2014.
[6] (Boyer 1991, "Europe in the Middle Ages" p. 258) "In the arithmetical theorems in Euclid's *Elements* VII-IX, numbers had been represented by line segments to which letters had been attached, and the geometric proofs in al-Khwarizmi's *Algebra* made use of lettered diagrams; but all coefficients in the equations used in the *Algebra* are specific numbers, whether represented by numerals or written out in words. The idea of generality is implied in al-Khwarizmi's exposition, but he had no scheme for expressing algebraically the general propositions that are so readily available in geometry."
[7] "2010 Mathematics Subject Classification". Retrieved 5 October 2014.
[8] "algebra". *Oxford English Dictionary*. Oxford University Press.
[9] Struik, Dirk J. (1987). *A Concise History of Mathematics*. New York: Dover Publications. ISBN 0-486-60255-9.
[10] Boyer 1991
[11] Cajori, Florian (2010). *A History of Elementary Mathematics – With Hints on Methods of Teaching*. p. 34. ISBN 1-4460-2221-8.
[12] Roshdi Rashed (November 2009). "Al Khwarizmi: The Beginnings of Algebra". Saqi Books. ISBN 0-86356-430-5
[13] "Diophantus, Father of Algebra". Retrieved 5 October 2014.
[14] "History of Algebra". Retrieved 5 October 2014.
[15] Josef W. Meri (2004). *Medieval Islamic Civilization*. Psychology Press. p. 31. ISBN 978-0-415-96690-0. Retrieved 25 November 2012.
[16] Boyer, Carl B. (1991). *A History of Mathematics* (Second ed.). Wiley. pp. 178, 181. ISBN 0-471-54397-7.
[17] Boyer, Carl B. (1991). *A History of Mathematics* (Second ed.). Wiley. p. 228. ISBN 0-471-54397-7.
[18] (Boyer 1991, "The Arabic Hegemony" p. 229) "It is not certain just what the terms *al-jabr* and *muqabalah* mean, but the usual interpretation is similar to that implied in the translation above. The word *al-jabr* presumably meant something like "restoration" or "completion" and seems to refer to the transposition of subtracted terms to the other side of an equation; the word *muqabalah* is said to refer to "reduction" or "balancing" – that is, the cancellation of like terms on opposite sides of the equation."
[19] (Boyer 1991, "The Arabic Hegemony" p. 230) "The six cases of equations given above exhaust all possibilities for linear and quadratic equations having positive root. So systematic and exhaustive was al-Khwarizmi's exposition that his readers must have had little difficulty in mastering the solutions."
[20] Gandz and Saloman (1936), *The sources of al-Khwarizmi's algebra*, Osiris i, p. 263–277: "In a sense, Khwarizmi is more entitled to be called "the father of algebra" than Diophantus because Khwarizmi is the first to teach algebra in an elementary form and for its own sake, Diophantus is primarily concerned with the theory of numbers".
[21] Rashed, R.; Armstrong, Angela (1994). *The Development of Arabic Mathematics*. Springer. pp. 11–2. ISBN 0-7923-2565-6. OCLC 29181926
[22] O'Connor, John J.; Robertson, Edmund F., "Sharaf al-Din al-Muzaffar al-Tusi", *MacTutor History of Mathematics archive*, University of St Andrews.
[23] Victor J. Katz, Bill Barton; Barton, Bill (October 2007). "Stages in the History of Algebra with Implications for Teaching". *Educational Studies in Mathematics* (Springer Netherlands) **66** (2): 185–201 [192]. doi:10.1007/s10649-006-9023-7
[24] (Boyer 1991, "The Arabic Hegemony" p. 239) "Abu'l Wefa was a capable algebraist as well as a trigonometer. ... His successor al-Karkhi evidently used this translation to become an Arabic disciple of Diophantus – but without Diophantine analysis! ... In particular, to al-Karkhi is attributed the first numerical solution of equations of the form $ax_{2n} + bx_n = c$ (only equations with positive roots were considered),"
20.10. REFERENCES 125
[25] "The Origins of Abstract Algebra". University of Hawaii Mathematics Department.
[26] "The Collected Mathematical Papers".Cambridge University Press.
[27] "Hull's Algebra" (pdf). *New York Times*. July 16, 1904. Retrieved September 21, 2012.
[28] Quaid, Libby (September 22, 2008). "Kids misplaced in algebra" (Report). *Associated Press*. Retrieved September 23, 2012.
[29] Hamilton, Reeve (7 September 2012). "THE TEXAS TRIBUNE; U.T.-Arlington Adopts New Way to Tackle Algebra". *The New York Times*. Retrieved 10 September 2012.

20.10 References

_ Boyer, Carl B. (1991), *A History of Mathematics* (Second Edition ed.), John Wiley & Sons, Inc., ISBN 0-471-54397-7
_ Donald R. Hill, *Islamic Science and Engineering* (Edinburgh University Press, 1994).
_ Ziauddin Sardar, Jerry Ravetz, and Borin Van Loon, *Introducing Mathematics* (Totem Books, 1999).
_ George Gheverghese Joseph, *The Crest of the Peacock: Non-European Roots of Mathematics* (Penguin Books, 2000).
_ John J O'Connor and Edmund F Robertson, *History Topics: Algebra Index*. In *MacTutor History of Mathematics archive* (University of St Andrews, 2005).
_ I.N. Herstein: *Topics in Algebra*. ISBN 0-471-02371-X
_ R.B.J.T. Allenby: *Rings, Fields and Groups*. ISBN 0-340-54440-6
_ L. Euler: *Elements of Algebra*, ISBN 978-1-899618-73-6
_ Asimov, Isaac (1961). *Realm of Algebra*. Houghton Mifflin.

20.11 External links

_ Khan Academy: Conceptual videos and worked examples
_ Khan Academy: Origins of Algebra, free online micro lectures
_ Algebrarules.com: An open source resource for learning the fundamentals of Algebra
_ 4000 Years of Algebra, lecture by Robin Wilson, at Gresham College, October 17, 2007 (available for MP3 and MP4 download, as well as a text file).
_ Algebra entry by Vaughan Pratt in the *Stanford Encyclopedia of Philosophy*

A page from Al-Khwārizmī's al-Kitāb al-muḫtaṣar fī ḥisāb al-ğabr wa-l-muqābala

Italian mathematician Girolamo Cardano published the solutions to the cubic and quartic equations in his 1545 book Ars magna.

Algebraic expression notation:
1 – power (exponent)
2 – coefficient
3 – term
4 – operator
5 – constant term
x y c – *variables/constants*

The graph of a polynomial function of degree 3.

Chapter 21

Irrational number

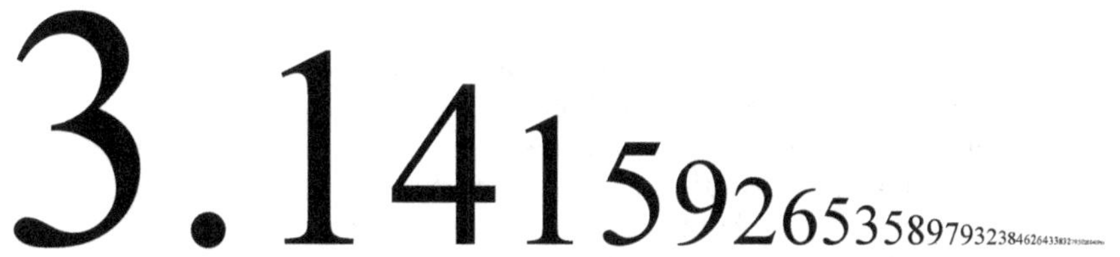

The famous mathematical constant pi (π) is among the best known irrational numbers and is much represented in popular culture

In mathematics, an **irrational number** is any real number that cannot be expressed as a ratio of integers. Informally, this means that an irrational number cannot be represented as a simple fraction. Irrational numbers are those real numbers that cannot be represented as terminating or repeating decimals. As a consequence of Cantor's proof that the real numbers are uncountable (and the rationals countable) it follows that almost all real numbers are irrational.[1]

When the ratio of lengths of two line segments is irrational, the line segments are also described as being *incommensurable*, meaning they share no measure in common.

Perhaps the best-known irrational numbers are: the ratio of a circle's circumference to its diameter π, Euler's number e, the golden ratio φ, and the square root of two √2.[2][3][4]

21.1 History

It has been suggested that the concept of irrationality was implicitly accepted by Indian mathematicians since the 7th century BC, when Manava (c. 750 – 690 BC) believed that the square roots of numbers such as 2 and 61 could not be exactly determined.[5] However, historian Carl Benjamin Boyer writes that "such claims are not well substantiated and unlikely to be true".[6]

21.1.1 Ancient Greece

The first proof of the existence of irrational numbers is usually attributed to a Pythagorean (possibly Hippasus of Metapontum),[7] who probably discovered them while identifying sides of the pentagram.[8] The then-current Pythagorean method would have claimed that there must be some sufficiently small, indivisible unit that could fit evenly into one of these lengths as well as the other. However, Hippasus, in the 5th century BC, was able to deduce that there was in fact no common unit of measure, and that the assertion of such an existence was in fact a contradiction.

He did this by demonstrating that if the hypotenuse of an isosceles right triangle was indeed commensurable with a leg, then that unit of measure must be both odd and even, which is impossible. His reasoning is as follows:
_ Start with an isosceles right triangle with side lengths of integers *a*, *b*, and *c*. The ratio of the hypotenuse to a leg is represented by *c*:*b*.

The number p2 is irrational.
_ Assume *a*, *b*, and *c* are in the smallest possible terms (*i.e.* they have no common factors).
_ By the Pythagorean theorem: $c^2 = a^2+b^2 = b^2+b^2 = 2b^2$. (Since the triangle is isosceles, $a = b$).
_ Since $c^2 = 2b^2$, c^2 is divisible by 2, and therefore even.
_ Since c^2 is even, *c* must be even.
_ Since *c* and *b* have no common factors, and *c* is even, *b* must be odd (if *b* were even, *b* and *c* would have a common factor of 2).
_ Since *c* is even, dividing *c* by 2 yields an integer. Let *y* be this integer ($c = 2y$).
_ Squaring both sides of $c = 2y$ yields $c^2 = (2y)^2$, or $c^2 = 4y^2$.
_ Substituting $4y^2$ for c^2 in the first equation ($c^2 = 2b^2$) gives us $4y^2= 2b^2$.
_ Dividing by 2 yields $2y^2 = b^2$.
_ Since *y* is an integer, and $2y^2 = b^2$, b^2 is divisible by 2, and therefore even.
_ Since b^2 is even, *b* must be even.
_ However, we have already asserted that *b* must be odd, and *b* cannot be both odd and even. This contradiction proves that *c* and *b* cannot both be integers, and thus the existence of a number that cannot be expressed as a ratio of two integers.[9]

Greek mathematicians termed this ratio of incommensurable magnitudes *alogos*, or inexpressible. Hippasus, however, was not lauded for his efforts: according to one legend, he made his discovery while out at sea, and was subsequently thrown overboard by his fellow Pythagoreans "…for having produced an element in the universe which denied the…doctrine that all phenomena in the universe can be reduced to whole numbers and their ratios."[10] Another legend states that Hippasus was merely exiled for this revelation. Whatever the consequence to Hippasus himself, his discovery posed a very serious problem to Pythagorean mathematics, since it shattered the assumption that number and geometry were inseparable–a foundation of their theory.
The discovery of incommensurable ratios was indicative of another problem facing the Greeks: the relation of the discrete to the continuous. Brought into light by Zeno of Elea, who questioned the conception that quantities are discrete and composed of a finite number of units of a given size. Past Greek conceptions dictated that they necessarily must be, for "whole numbers represent discrete objects, and a commensurable ratio represents a relation between two collections of discrete objects."[11] However Zeno found that in fact "[quantities] in general are not discrete collections of units; this is why ratios of incommensurable [quantities] appear….[Q]uantities are, in other words, continuous."[11] What this means is that, contrary to the popular conception of the time, there cannot be an indivisible, smallest unit of measure for any quantity. That in fact, these divisions of quantity must necessarily be infinite. For example, consider a line segment: this segment can be split in half, that half split in half, the half of the half in half, and so on. This process can continue infinitely, for there is always another half to be split. The more times the segment is halved, the closer the unit of measure comes to zero, but it never reaches exactly zero. This is just what Zeno sought to prove. He sought to prove this by formulating four paradoxes, which demonstrated the contradictions inherent in the mathematical thought of the time. While Zeno's paradoxes accurately demonstrated the deficiencies of current mathematical conceptions, they were not regarded as proof of the alternative. In the minds of the Greeks, disproving the validity of one view did not necessarily prove the validity of another, and therefore further investigation had to occur.
The next step was taken by Eudoxus of Cnidus, who formalized a new theory of proportion that took into account commensurable as well as incommensurable quantities. Central to his idea was the distinction between magnitude and number. A magnitude "...was not a number but stood for entities such as line segments, angles, areas, volumes, and time which could vary, as we would say, continuously. Magnitudes were opposed to numbers, which jumped from one value to another, as from 4 to 5."[12] Numbers are composed of some smallest, indivisible unit, whereas magnitudes are infinitely reducible. Because no quantitative values were assigned to magnitudes, Eudoxus was then able to account for both commensurable and incommensurable ratios by defining a ratio in terms of its magnitude, and proportion as an equality between two ratios. By taking quantitative values (numbers) out of the equation, he avoided the trap of having to express an irrational number as a number. "Eudoxus' theory enabled the Greek mathematicians to make tremendous progress in geometry by supplying the necessary logical foundation for incommensurable ratios."[13] Book 10 is dedicated to classification of irrational magnitudes.
As a result of the distinction between number and magnitude, geometry became the only method that could take into account incommensurable ratios. Because previous numerical foundations were still incompatible with the concept of incommensurability, Greek focus shifted away from those numerical conceptions such as algebra and focused almost

exclusively on geometry. In fact, in many cases algebraic conceptions were reformulated into geometrical terms. This may account for why we still conceive of x^2 or x^3 as x squared and x cubed instead of x second power and x third power. Also crucial to Zeno's work with incommensurable magnitudes was the fundamental focus on deductive reasoning that resulted from the foundational shattering of earlier Greek mathematics. The realization that some basic conception within the existing theory was at odds with reality necessitated a complete and thorough investigation of the axioms and assumptions that comprised that theory. Out of this necessity Eudoxus developed his method of exhaustion, a kind of reductio ad absurdum that "...established the deductive organization on the basis of explicit axioms..." as well as "...reinforced the earlier decision to rely on deductive reasoning for proof."[14] This method of exhaustion is the first step in the creation of calculus.

Theodorus of Cyrene proved the irrationality of the surds of whole numbers up to 17, but stopped there probably because the algebra he used couldn't be applied to the square root of 17.[15] It wasn't until Eudoxus developed a theory of proportion that took into account irrational as well as rational ratios that a strong mathematical foundation of irrational numbers was created.[16]

21.1.2 India

Geometrical and mathematical problems involving irrational numbers such as square roots were addressed very early during the Vedic period in India and there are references to such calculations in the *Samhitas*, *Brahmanas* and more notably in the *Sulbha sutras* (800 BC or earlier). (See Bag, Indian Journal of History of Science, 25(1-4), 1990).

It is suggested that Aryabhata (5th century AD) in calculating a value of pi to 5 significant figures, he used the word āsanna (approaching), to mean that not only is this an approximation but that the value is incommensurable (or irrational).

Later, in their treatises, Indian mathematicians wrote on the arithmetic of surds including addition, subtraction, multiplication, rationalization, as well as separation and extraction of square roots. (See Datta, Singh, Indian Journal of History of Science, 28(3), 1993).

Mathematicians like Brahmagupta (in 628 AD) and Bhaskara I (in 629 AD) made contributions in this area as did other mathematicians who followed. In the 12th century Bhaskara II evaluated some of these formulas and critiqued them, identifying their limitations.

During the 14th to 16th centuries, Madhava of Sangamagrama and the Kerala school of astronomy and mathematics discovered the infinite series for several irrational numbers such as π and certain irrational values of trigonometric functions. Jyesthadeva provided proofs for these infinite series in the *Yuktibhāṣā*.[17]

21.1.3 Middle Ages

In the Middle ages, the development of algebra by Muslim mathematicians allowed irrational numbers to be treated as *algebraic objects*.[18] Middle Eastern mathematicians also merged the concepts of "number" and "magnitude" into a more general idea of real numbers, criticized Euclid's idea of ratios, developed the theory of composite ratios, and extended the concept of number to ratios of continuous magnitude.[19] In his commentary on Book 10 of the *Elements*, the Persian mathematician Al-Mahani (d. 874/884) examined and classified quadratic irrationals and cubic irrationals. He provided definitions for rational and irrational magnitudes, which he treated as irrational numbers. He dealt with them freely but explains them in geometric terms as follows:[20]

"It will be a rational (magnitude) when we, for instance, say 10, 12, 3%, 6%, etc., because its value is pronounced and expressed quantitatively. What is not rational is irrational and it is impossible to pronounce and represent its value quantitatively. For example: the roots of numbers such as 10, 15, 20 which are not squares, the sides of numbers which are not cubes *etc.*"

In contrast to Euclid's concept of magnitudes as lines, Al-Mahani considered integers and fractions as rational magnitudes, and square roots and cube roots as irrational magnitudes. He also introduced an arithmetical approach to the concept of irrationality, as he attributes the following to irrational magnitudes:[20]

"their sums or differences, or results of their addition to a rational magnitude, or results of subtracting a magnitude of this kind from an irrational one, or of a rational magnitude from it."

The Egyptian mathematician Abū Kāmil Shujā ibn Aslam (c. 850 – 930) was the first to accept irrational numbers as solutions to quadratic equations or as coefficients in an equation, often in the form of square roots, cube roots and fourth roots.[21] In the 10th century, the Iraqi mathematician Al-Hashimi provided general proofs (rather than geometric demonstrations) for irrational numbers, as he considered multiplication, division, and other arithmetical functions.[22] Iranian mathematician, Abū Ja'far al-Khāzin (900–971) provides a definition of rational and irrational magnitudes, stating that if a definite quantity is:[23]

"contained in a certain given magnitude once or many times, then this (given) magnitude corresponds to a rational number. . . . Each time when this (latter) magnitude comprises a half, or a third, or a quarter of the given magnitude (of the unit), or, compared with (the unit), comprises three, five, or three fifths, it is a rational magnitude. And, in general, each magnitude that corresponds to this magnitude (*i.e.* to the unit), as one number to another, is rational. If, however, a magnitude cannot be represented as a multiple, a part (l/*n*), or parts (*m*/*n*) of a given magnitude, it is irrational, *i.e.* it cannot be expressed other than by means of roots."

Many of these concepts were eventually accepted by European mathematicians sometime after the Latin translations of the 12th century. Al-Hassār, a Moroccan mathematician from Fez specializing in Islamic inheritance jurisprudence during the 12th century, first mentions the use of a fractional bar, where numerators and denominators are separated by a horizontal bar. In his discussion he writes, "..., for example, if you are told to write three-fifths and a third of a fifth, write thus, 3 1
5 3 ." [24] This same fractional notation appears soon after in the work of Leonardo Fibonacci in the 13th century.[25]

134 *CHAPTER 21. IRRATIONAL NUMBER*

21.1.4 Modern period

The 17th century saw imaginary numbers become a powerful tool in the hands of Abraham de Moivre, and especially of Leonhard Euler. The completion of the theory of complex numbers in the 19th century entailed the differentiation of irrationals into algebraic and transcendental numbers, the proof of the existence of transcendental numbers, and the resurgence of the scientific study of the theory of irrationals, largely ignored since Euclid. The year 1872 saw the publication of the theories of Karl Weierstrass (by his pupil Ernst Kossak), Eduard Heine (*Crelle's Journal*, 74), Georg Cantor (Annalen, 5), and Richard Dedekind. Méray had taken in 1869 the same point of departure as Heine, but the theory is generally referred to the year 1872. Weierstrass's method has been completely set forth by Salvatore Pincherle in 1880,[26] and Dedekind's has received additional prominence through the author's later work (1888) and the endorsement by Paul Tannery (1894). Weierstrass, Cantor, and Heine base their theories on infinite series, while Dedekind founds his on the idea of a cut (Schnitt) in the system of real numbers, separating all rational numbers into two groups having certain characteristic properties. The subject has received later contributions at the hands of Weierstrass, Leopold Kronecker (Crelle, 101), and Charles Méray.

Continued fractions, closely related to irrational numbers (and due to Cataldi, 1613), received attention at the hands of Euler, and at the opening of the 19th century were brought into prominence through the writings of Joseph Louis Lagrange. Dirichlet also added to the general theory, as have numerous contributors to the applications of the subject. Johann Heinrich Lambert proved (1761) that π cannot be rational, and that e^n is irrational if n is rational (unless $n = 0$).[27] While Lambert's proof is often called incomplete, modern assessments support it as satisfactory, and in fact for its time it is unusually rigorous. Adrien-Marie Legendre (1794), after introducing the Bessel–Clifford function, provided a proof to show that π^2 is irrational, whence it follows immediately that π is irrational also. The existence of transcendental numbers was first established by Liouville (1844, 1851). Later, Georg Cantor (1873) proved their existence by a different method, that showed that every interval in the reals contains transcendental numbers. Charles Hermite (1873) first proved e transcendental, and Ferdinand von Lindemann (1882), starting from Hermite's conclusions, showed the same for π. Lindemann's proof was much simplified by Weierstrass (1885), still further by David Hilbert (1893), and was finally made elementary by Adolf Hurwitz and Paul Gordan.

21.2 Example proofs

21.2.1 Square roots

The square root of 2 was the first number proved irrational, and that article contains a number of proofs. The golden ratio is another famous quadratic irrational and there is a simple proof of its irrationality in its article. The square roots of all natural numbers which are not perfect squares are irrational and a proof may be found in quadratic irrationals.

21.2.2 General roots

The proof above for the square root of two can be generalized using the fundamental theorem of arithmetic. This asserts that every integer has a unique factorization into primes. Using it we can show that if a rational number is not an integer then no integral power of it can be an integer, as in lowest terms there must be a prime in the denominator that does not divide into the numerator whatever power each is raised to. Therefore if an integer is not an exact kth power of another integer then its kth root is irrational.

21.2.3 Logarithms

Perhaps the numbers most easy to prove irrational are certain logarithms. Here is a proof by contradiction (reductio ad absurdum) that $\log_2 3$ is irrational. Notice that $\log_2 3 \approx 1.58 > 0$.

Assume $\log_2 3$ is rational. For some positive integers m and n, we have

$\log_2 3 = mn$:

It follows that

21.3. TRANSCENDENTAL AND ALGEBRAIC IRRATIONALS 135

$2^{m/n} = 3$
$(2^{m/n})^n = 3^n$
$2^m = 3^n.$

However, the number 2 raised to any positive integer power must be even (because it is divisible by 2) and the number 3 raised to any positive integer power must be odd (since none of its prime factors will be 2). Clearly, an integer cannot be both odd and even at the same time: we have a contradiction. The only assumption we made was that $\log_2 3$ is rational (and so expressible as a quotient of integers m/n with $n \neq 0$). The contradiction means that this assumption must be false, i.e. $\log_2 3$ is irrational, and can never be expressed as a quotient of integers m/n with $n \neq 0$.
Cases such as $\log_{10} 2$ can be treated similarly.

21.3 Transcendental and algebraic irrationals

Almost all irrational numbers are transcendental and all real transcendental numbers are irrational (there are also complex transcendental numbers): the article on transcendental numbers lists several examples. e^r and π^r are irrational if $r \neq 0$ is rational; e^π is irrational.
Another way to construct irrational numbers is as irrational algebraic numbers, i.e. as zeros of polynomials with integer coefficients: start with a polynomial equation
$p(x) = a_n x^n + a_{n-1} x^{n-1} + \cdots + a_1 x + a_0 = 0$
where the coefficients a_i are integers. Suppose you know that there exists some real number x with $p(x) = 0$ (for instance if n is odd and a_n is non-zero, then because of the intermediate value theorem). The only possible rational roots of this polynomial equation are of the form r/s where r is a divisor of a_0 and s is a divisor of a_n; there are only finitely so many such candidates you can check by hand. If neither of them is a root of p, then x must be irrational.
For example, this technique can be used to show that $x = (2^{1/2} + 1)^{1/3}$ is irrational: we have $(x^3 - 1)^2 = 2$ and hence $x^6 - 2x^3 - 1 = 0$, and this latter polynomial does not have any rational roots (the only candidates to check are ± 1).
Because the algebraic numbers form a field, many irrational numbers can be constructed by combining transcendental and algebraic numbers. For example $3\pi + 2$, $\pi + \sqrt{2}$ and $e\sqrt{3}$ are irrational (and even transcendental).

21.4 Decimal expansions

The decimal expansion of an irrational number never repeats or terminates, unlike a rational number. Similarly for binary, octal or hexadecimal expansions, and in general for expansions in every positional notation with natural bases.
To show this, suppose we divide integers n by m (where m is nonzero). When long division is applied to the division of n by m, only m remainders are possible. If 0 appears as a remainder, the decimal expansion terminates. If 0 never occurs, then the algorithm can run at most $m - 1$ steps without using any remainder more than once. After that, a remainder must recur, and then the decimal expansion repeats.
Conversely, suppose we are faced with a repeating decimal, we can prove that it is a fraction of two integers. For example, consider:
$A = 0.7\,162\,162\,162\ldots.$
Here the repitend is 162 and the length of the repitend is 3. First, we multiply by an appropriate power of 10 to move the decimal point to the right so that it is just in front of a repitend. In this example we would multiply by 10 to obtain:
$10A = 7.162\,162\,162\ldots.$

Now we multiply this equation by 10^r where r is the length of the repitend. This has the effect of moving the decimal point to be in front of the "next" repitend. In our example, multiply by 10^3:
$10{,}000A = 7\,162.162\,162\ldots.$
The result of the two multiplications gives two different expressions with exactly the same "decimal portion", that is, the tail end of 10,000A matches the tail end of 10A exactly. Here, both 10,000A and 10A have .162162162 ... at the end.
Therefore, when we subtract the 10A equation from the 10,000A equation, the tail end of 10A cancels out the tail end of 10,000A leaving us with:
$9990A = 7155.$
Then $A = 7155/9990 = 135 \cdot 53/135 \cdot 74 = 53/74.$
(135 is the greatest common divisor of 7155 and 9990). 53/74 is a quotient of integers and therefore a rational number.

21.5 Irrational powers

Dov Jarden gave a simple non-constructive proof that there exist two irrational numbers a and b, such that a^b is rational.[28]
Indeed, if $\sqrt{2}^{\sqrt{2}}$ is rational, then take $a = b = \sqrt{2}$. Otherwise, take a to be the irrational number $\sqrt{2}^{\sqrt{2}}$ and $b = \sqrt{2}$. Then $a^b = (\sqrt{2}^{\sqrt{2}})^{\sqrt{2}} = \sqrt{2}^{\sqrt{2}\cdot\sqrt{2}} = \sqrt{2}^2 = 2$, which is rational.

Although the above argument does not decide between the two cases, the Gelfond–Schneider theorem shows that $\sqrt{2}^{\sqrt{2}}$ is transcendental, hence irrational. This theorem states that if *a* and *b* are both algebraic numbers, and *a* is not equal to 0 or 1, and *b* is not a rational number, then any value of a^b is a transcendental number (there can be more than one value if complex number exponentiation is used).
An example that provides a simple constructive proof is[29]
$(p2)^{\log_{p2} 3} = 3$:
The base of the left side is irrational and the right side is rational, so one must prove that the exponent on the left side, $\log_p$
$_2$ 3 , is irrational. This is so because, by the formula relating logarithms with different bases,
$\log_{p2} 3 = \log_2 3$
$\log_2 p2 = \log_2 3/1/2$
$= 2 \log_2 3$
which we can assume, for the sake of establishing a contradiction, equals a ratio *m/n* of positive integers. Then $\log_2 3 = m/2n$ hence $2^{\log_2 3} = 2^{m/2n}$ hence $3 = 2^{m/2n}$ hence $3^{2n} = 2^m$, which is a contradictory pair of prime factorizations and hence violates the fundamental theorem of arithmetic (unique prime factorization).
A stronger result is the following:[30] Every rational number in the interval $((1/e)^{1/e}; 1)$ can be written either as a^a for some irrational number *a* or as n^n for some natural number *n*. Similarly,[30] every positive rational number can be written either as a^{a^a} for some irrational number *a* or as n^{n^n} for some natural number *n*.

21.6 Open questions

It is not known whether $\pi + e$ or $\pi - e$ is irrational or not. In fact, there is no pair of non-zero integers *m* and *n* for which it is known whether $m\pi + ne$ is irrational or not. Moreover, it is not known whether the set $\{\pi, e\}$ is algebraically independent over **Q**.
It is not known whether πe, π/e, 2^e, e^e, e^{e^e}, π^e, $\pi^{\sqrt{2}}$, ln π, Catalan's constant, or the Euler–Mascheroni gamma constant γ are irrational.[31][32][33] It is not known if n^π or n^e is rational for any positive integer *n*.

21.7 Set of all irrationals

Since the reals form an uncountable set, of which the rationals are a countable subset, the complementary set of irrationals is uncountable.
Under the usual (Euclidean) distance function $d(x, y) = |x - y|$, the real numbers are a metric space and hence also a topological space. Restricting the Euclidean distance function gives the irrationals the structure of a metric space. Since the subspace of irrationals is not closed, the induced metric is not complete. However, being a G-delta set—i.e., a countable intersection of open subsets—in a complete metric space, the space of irrationals is completely metrizable: that is, there is a metric on the irrationals inducing the same topology as the restriction of the Euclidean metric, but with respect to which the irrationals are complete. One can see this without knowing the aforementioned fact about G-delta sets: the continued fraction expansion of an irrational number defines a homeomorphism from the space of irrationals to the space of all sequences of positive integers, which is easily seen to be completely metrizable. Furthermore, the set of all irrationals is a disconnected metrizable space. In fact, the irrationals have a basis of clopen sets so the space is zero-dimensional.

21.8 See also

_ Computable number
_ Dedekind cut
_ Diophantine approximation
_ Golden Ratio
_ *n*th root
_ Proof that e is irrational
_ Proof that π is irrational
_ Square root of 2
_ Square root of 3
_ Square root of 5
_ Transcendental number
_ Trigonometric number

21.9 References

[1] Cantor, Georg (1955) [1915]. Philip Jourdain, ed. *Contributions to the Founding of the Theory of Transfinite Numbers.* New York: Dover. ISBN 978-0-486-60045-1.
[2] The 15 Most Famous Transcendental Numbers. by Clifford A. Pickover. URL retrieved 24 October 2007.

[3] http://www.mathsisfun.com/irrational-numbers.html; URL retrieved 24 October 2007.

[4] Weisstein, Eric W., "Irrational Number", *MathWorld*. URL retrieved 26 October 2007.
[5] T. K. Puttaswamy, "The Accomplishments of Ancient Indian Mathematicians", pp. 411–2, in Selin, Helaine; D'Ambrosio, Ubiratan, eds. (2000). *Mathematics Across Cultures: The History of Non-western Mathematics*. Springer. ISBN 1-4020-0260-2..
[6] Boyer (1991). "China and India". *A History of Mathematics* (2nd ed.). p. 208. "It has been claimed also that the first recognition of incommensurables appears in India during the *Sulbasutra* period, but such claims are not well substantiated. The case for early Hindu awareness of incommensurable magnitudes is rendered most unlikely by the lack of evidence that Indian mathematicians of that period had come to grips with fundamental concepts."
[7] Kurt Von Fritz (1945). "The Discovery of Incommensurability by Hippasus of Metapontum". *The Annals of Mathematics*.
[8] James R. Choike (1980). "The Pentagram and the Discovery of an Irrational Number". *The Two-Year College Mathematics Journal*..
[9] Kline, M. (1990). *Mathematical Thought from Ancient to Modern Times*, Vol. 1. New York: Oxford University Press. (Original work published 1972). p.33.
[10] Kline 1990, p. 32.
[11] Kline 1990, p.34.
[12] Kline 1990, p.48.
[13] Kline 1990, p.49.
[14] Kline 1990, p.50.
[15] Robert L. McCabe (1976). "Theodorus' Irrationality Proofs". *Mathematics Magazine*..
[16] Charles H. Edwards (1982). *The historical development of the calculus*. Springer.
[17] Katz, V. J. (1995), "Ideas of Calculus in Islam and India", *Mathematics Magazine* (Mathematical Association of America) **68** (3): 163–74.
[18] O'Connor, John J.; Robertson, Edmund F., "Arabic mathematics: forgotten brilliance?", *MacTutor History of Mathematics archive*, University of St Andrews..
[19] Matvievskaya, Galina (1987). "The Theory of Quadratic Irrationals in Medieval Oriental Mathematics". *Annals of the New York Academy of Sciences* **500**: 253–277 [254]. doi:10.1111/j.1749-6632.1987.tb37206.x..
[20] Matvievskaya, Galina (1987). "The Theory of Quadratic Irrationals in Medieval Oriental Mathematics". *Annals of the New York Academy of Sciences* **500**: 253–277 [259]. doi:10.1111/j.1749-6632.1987.tb37206.x.
[21] Jacques Sesiano, "Islamic mathematics", p. 148, in Selin, Helaine; D'Ambrosio, Ubiratan (2000). *Mathematics Across Cultures: The History of Non-western Mathematics*. Springer. ISBN 1-4020-0260-2..
[22] Matvievskaya, Galina (1987). "The Theory of Quadratic Irrationals in Medieval Oriental Mathematics". *Annals of the New York Academy of Sciences* **500**: 253–277 [260]. doi:10.1111/j.1749-6632.1987.tb37206.x..
[23] Matvievskaya, Galina (1987). "The Theory of Quadratic Irrationals in Medieval Oriental Mathematics". *Annals of the New York Academy of Sciences* **500**: 253–277 [261]. doi:10.1111/j.1749-6632.1987.tb37206.x..
[24] Cajori, Florian (1928), *A History of Mathematical Notations (Vol.1)*, La Salle, Illinois: The Open Court Publishing Company pg. 269.
[25] (Cajori 1928, pg.89)
[26] Salvatore Pincherle (1880). "Saggio di una introduzione alla teorica delle funzioni analitiche secondo i principi del prof. Weierstrass". *Giornale di Matematiche*.
[27] J. H. Lambert (1761). "Mémoire sur quelques propriétés remarquables des quantités transcendentes circulaires et logarithmiques". *Histoire de l'Académie Royale des Sciences et des Belles-Lettres der Berlin*: 265–276.
[28] George, Alexander; Velleman, Daniel J. (2002). *Philosophies of mathematics*. Blackwell. pp. 3–4. ISBN 0-631-19544-0.
[29] Lord, Nick, "Maths bite: irrational powers of irrational numbers can be rational", *Mathematical Gazette* 92, November 2008, p. 534.
[30] Marshall, Ash J., and Tan, Yiren, "A rational number of the form a^a with a irrational", *Mathematical Gazette* 96, March 2012, pp. 106-109.

[31] Weisstein, Eric W., "Pi", *MathWorld*.
[32] Weisstein, Eric W., "Irrational Number", *MathWorld*.
[33] Some unsolved problems in number theory

21.10 Further reading

_ Adrien-Marie Legendre, *Éléments de Géometrie*, Note IV, (1802), Paris
_ Rolf Wallisser, "On Lambert's proof of the irrationality of π", in *Algebraic Number Theory and Diophantine Analysis*, Franz Halter-Koch and Robert F. Tichy, (2000), Walter de Gruyer

21.11 External links

_ Zeno's Paradoxes and Incommensurability (n.d.). Retrieved April 1, 2008
_ Weisstein, Eric W., "Irrational Number", *MathWorld*.
_ Square root of 2 is irrational

Chapter 22

Mathematical induction

Mathematical induction can be informally illustrated by reference to the sequential effect of falling dominoes.

Mathematical induction is a method of mathematical proof typically used to establish a given statement for all natural numbers. It is a form of direct proof, and it is done in two steps. The first step, known as the **base case**, is to prove the given statement for the first natural number. The second step, known as the **inductive step**, is to prove that the given statement for any one natural number implies the given statement for the next natural number. From these two steps, mathematical induction is the rule from which we infer that the given statement is established for all natural numbers.

The method can be extended to prove statements about more general well-founded structures, such as trees; this generalization, known as structural induction, is used in mathematical logic and computer science. Mathematical induction in this extended sense is closely related to recursion. Mathematical induction, in some form, is the foundation of all correctness proofs for computer programs.[1]

Although its name may suggest otherwise, mathematical induction should not be misconstrued as a form of inductive

140

22.1. HISTORY 141

reasoning (also see Problem of induction). Mathematical induction is an inference rule used in proofs. In mathematics, proofs including those using mathematical induction are examples of deductive reasoning and inductive reasoning is excluded from proofs.[2]

22.1 History

In 370 BC, Plato's Parmenides may have contained an early example of an implicit inductive proof.[3] The earliest implicit traces of mathematical induction can be found in Euclid's[4] proof that the number of primes is infinite and in Bhaskara's "cyclic method".[5] An opposite iterated technique, counting *down* rather than up, is found in the Sorites paradox, where one argued that if 1,000,000 grains of sand formed a heap, and removing one grain from a heap left it a heap, then a single grain of sand (or even no grains) forms a heap.

An implicit proof by mathematical induction for arithmetic sequences was introduced in the *al-Fakhri* written by al-Karaji around 1000 AD, who used it to prove the binomial theorem and properties of Pascal's triangle.

None of these ancient mathematicians, however, explicitly stated the inductive hypothesis. Another similar case (contrary to what Vacca has written, as Freudenthal carefully showed) was that of Francesco Maurolico in his *Arithmeticorum libri duo* (1575), who used the technique to prove that the sum of the first *n* odd integers is n^2. The first explicit formulation of the principle of induction was given by Pascal in his *Traité du triangle arithmétique* (1665).

Another Frenchman, Fermat, made ample use of a related principle, indirect proof by infinite descent. The inductive hypothesis was also employed by the Swiss Jakob Bernoulli, and from then on it became more or less well known.

The modern rigorous and systematic treatment of the principle came only in the 19th century, with George Boole,[6] Augustus de Morgan, Charles Sanders Peirce,[7] Giuseppe Peano, and Richard Dedekind.[5]

22.2 Description

The simplest and most common form of mathematical induction infers that a statement involving a natural number n holds for all values of n. The proof consists of two steps:

1. The **basis** (**base case**): prove that the statement holds for the first natural number n. Usually, $n = 0$ or $n = 1$, rarely, $n = -1$ (although not a natural number, the extension of the natural numbers to -1 is still a well-ordered set).
2. The **inductive step**: prove that, if the statement holds for some natural number n, then the statement holds for $n + 1$.

The hypothesis in the inductive step that the statement holds for some n is called the **induction hypothesis** (or **inductive hypothesis**). To perform the inductive step, one assumes the induction hypothesis and then uses this assumption to prove the statement for $n + 1$.

Whether $n = 0$ or $n = 1$ depends on the definition of the natural numbers. If 0 is considered a natural number, as is common in the fields of combinatorics and mathematical logic, the base case is given by $n = 0$. If, on the other hand, 1 is taken as the first natural number, then the base case is given by $n = 1$.

22.3 Example

Mathematical induction can be used to prove that the following statement, which we will call $P(n)$, holds for all natural numbers n.

0 + 1 + 2 + _ _ _ + n =
$n(n + 1)2$:
$P(n)$ gives a formula for the sum of the natural numbers less than or equal to number n. The proof that $P(n)$ is true for each natural number n proceeds as follows.

Basis: Show that the statement holds for $n = 0$.
$P(0)$ amounts to the statement:
0 =0x(0 + 1)/2:
In the left-hand side of the equation, the only term is 0, and so the left-hand side is simply equal to 0.
In the right-hand side of the equation, $0\cdot(0 + 1)/2 = 0$.
The two sides are equal, so the statement is true for $n = 0$. Thus it has been shown that $P(0)$ holds.
Inductive step: Show that *if* $P(k)$ holds, then also $P(k + 1)$ holds. This can be done as follows.
Assume $P(k)$ holds (for some unspecified value of k). It must then be shown that $P(k + 1)$ holds, that is:
(0 + 1 + 2 + _ _ _ + k) + (k + 1) =
$(k + 1)((k + 1) + 1)$
2
:
Using the induction hypothesis that $P(k)$ holds, the left-hand side can be rewritten to:
$k(k + 1)$
2
+ (k + 1) :
Algebraically:
$k(k + 1)$
2
+ (k + 1) =
$k(k + 1) + 2(k + 1)$
2
=
$(k + 1)(k + 2)$
2
=
$(k + 1)((k + 1) + 1)$
2
thereby showing that indeed $P(k + 1)$ holds.
Since both the basis and the inductive step have been performed, by mathematical induction, the statement $P(n)$ holds for all natural n. Q.E.D.

22.4 Axiom of induction

Mathematical induction as an inference rule can be formalized as a second-order axiom. The *axiom of induction* is, in logical symbols,
8P: [[*P*(0) ^ *8*(*k 2* N)*:* [*P*(*k*) *) P*(*k* + 1)]] *) 8*(*n 2* N)*: P*(*n*)]
where P is any predicate and k and n are both natural numbers.
In words, the basis $P(0)$ and the inductive step (namely, that the inductive hypothesis $P(k)$ implies $P(k + 1)$) together imply that $P(n)$ for any natural number n. The axiom of induction asserts that the validity of inferring that $P(n)$ holds for any natural number n from the basis and the inductive step.
Note that the first quantifier in the axiom ranges over *predicates* rather than over individual numbers. This is a secondorder quantifier, which means that this axiom is stated in second-order logic. Axiomatizing arithmetic induction in first-order logic requires an axiom schema containing a separate axiom for each possible predicate. The article Peano axioms contains further discussion of this issue.

22.5 Heuristic justification

As an inference rule, mathematical induction can be justified as follows. Having proven the base case and the inductive step, then any value can be obtained by performing the inductive step repeatedly. It may be helpful to think of the

domino effect. Consider a half line of dominoes each standing on end, and extending infinitely to the right. Suppose that:
1. The first domino falls right.
2. If a (fixed but arbitrary) domino falls right, then its next neighbor also falls right.
With these assumptions one can conclude (using mathematical induction) that all of the dominoes will fall right.

Mathematical induction, as formalized in the second-order axiom above, works because k is used to represent an *arbitrary* natural number. Then, using the inductive hypothesis, i.e. that $P(k)$ is true, show $P(k+1)$ is also true. This allows us to "carry" the fact that $P(0)$ is true to the fact that $P(1)$ is also true, and carry $P(1)$ to $P(2)$, etc., thus proving $P(n)$ holds for every natural number n.

22.6 Variants

In practice, proofs by induction are often structured differently, depending on the exact nature of the property to be proved.

22.6.1 Starting at some other number

If we want to prove a statement not for all natural numbers but only for all numbers greater than or equal to a certain number b then:

1. Showing that the statement holds when $n = b$.
2. Showing that if the statement holds for $n = m \geq b$ then the same statement also holds for $n = m + 1$.

This can be used, for example, to show that $m \geq 3n$ for $n \geq 3$. A more substantial example is a proof that

$$3n < n! <$$

$$2n \text{ for } n \quad 6:$$

In this way we can prove that $P(n)$ holds for all $n \geq 1$, or even $n \geq -5$. This form of mathematical induction is actually a special case of the previous form because if the statement that we intend to prove is $P(n)$ then proving it with these two rules is equivalent with proving $P(n + b)$ for all natural numbers n with the first two steps.

22.6.2 Building on $n = 2$

In mathematics, many standard functions, including operations such as "+" and relations such as "=", are binary, meaning that they take two arguments. Often these functions possess properties that implicitly extend them to more than two arguments. For example, once addition $a + b$ is defined and is known to satisfy the associativity property ($a + b) + c = a + (b + c)$, then the ternary addition $a + b + c$ makes sense, either as $(a + b) + c$ or as $a + (b + c)$. Similarly, many axioms and theorems in mathematics are stated only for the binary versions of mathematical operations and relations, and implicitly extend to higher-arity versions.

Suppose that we wish to prove a statement about an n-ary operation implicitly defined from a binary operation, using mathematical induction on n. Then it should come as no surprise that the $n = 2$ case carries special weight. Here are some examples.

Example: product rule for the derivative

In this example, the binary operation in question is multiplication (of functions). The usual product rule for the derivative taught in calculus states:

144 *CHAPTER 22. MATHEMATICAL INDUCTION*

(fg)

$= f$

$g + g$

$f:$

or in logarithmic derivative form

(fg)

$/(fg) = f$

$/f + g$

$/g:$

This can be generalized to a product of n functions. One has

$(f_1 f_2 f_3 _ _ _ f_n)$

$= (f$

$_1 f_2 f_3 _ _ _ f_n) + (f_1 f$

$_2 f_3 _ _ _ f_n) + (f_1 f_2 f$

$_3$

$_ _ _ f_n) + _ _ _ + (f_1 f_2 _ _ _ f_{n\square 1} f$

′

n):

or in logarithmic derivative form

$(f_1f_2f_3 \cdots f_n)$

′

$/(f_1f_2f_3 \cdots f_n)$

$= (f$

′

$_1/f_1) + (f$

′

$_2/f_2) + (f$

′

$_3/f_3) + \cdots + (f$

′

$_n/f_n)$:

In each of the n terms of the usual form, just one of the factors is a derivative; the others are not.

When this general fact is proved by mathematical induction, the $n = 0$ case is trivial, $(1)' = 0$(since the empty product is 1, and the empty sum is 0). The $n = 1$ case is also trivial, f'

$_1 = f'$

$_1$: And for each $n \geq 3$, the case is easy to prove from the preceding $n - 1$ case. The real difficulty lies in the $n = 2$ case, which is why that is the one stated in the standard product rule.

An alternative way to look at this is to generalize $f(xy) = f(x) + f(y)$; $f(1) = 0$ (a monoid homomorphism) to

$f($

Π

$x_i) =$

Σ

$f(x_i)$.

Example: Cohen's proof that there is no "horse of a different color"

Main article: All horses are the same color

In this example, the binary relation in question is an equivalence relation applied to horses, such that two horses are equivalent if they are the same color. The argument is essentially identical to the one above, but the crucial $n = 1$ case fails, causing the entire argument to be invalid.

Joel E. Cohen proposed the following argument, which purports to prove by mathematical induction that all horses are of the same color:[8]

_ Basis: If there is only *one* horse, there is only one color.

_ Induction step: Assume as induction hypothesis that within any set of n horses, there is only one color. Now look at any set of $n + 1$ horses. Number them: 1, 2, 3, ..., n, $n + 1$. Consider the sets $\{1, 2, 3, ..., n\}$ and $\{2, 3, 4, ..., n + 1\}$. Each is a set of only n horses, therefore within each there is only one color. But the two sets overlap, so there must be only one color among all $n + 1$ horses.

The basis case $n = 1$ is trivial (as any horse is the same color as itself), and the inductive step is correct in all cases $n > 1$. However, the logic of the inductive step is incorrect for $n = 1$, because the statement that "the two sets overlap" is false (there are only $n + 1 = 2$ horses prior to either removal, and after removal the sets of one horse each do not overlap). Indeed, going from the $n = 1$ case to the $n = 2$ case is clearly the crux of the matter; if one could prove the $n = 2$ case directly without having to infer it from the $n = 1$ case, then all higher cases would follow from the inductive hypothesis.

22.6.3 Induction on more than one counter

It is sometimes desirable to prove a statement involving two natural numbers, n and m, by iterating the induction process. That is, one performs a basis step and an inductive step for n, and in each of those performs a basis step and an inductive step for m. See, for example, the proof of commutativity accompanying *addition of natural numbers*. More complicated arguments involving three or more counters are also possible.

22.6.4 Infinite descent

Main article: Infinite descent

The method of infinite descent was one of Pierre de Fermat's favorites. This method of proof can assume several slightly different forms. For example, it might begin by showing that if a statement is true for a natural number n it must also be true for some smaller natural number m ($m < n$). Using mathematical induction (implicitly) with the inductive hypothesis being that the statement is false for all natural numbers less than or equal to m, we can conclude that the statement cannot be true for any natural number n.

Although this particular form of infinite-descent proof is clearly a mathematical induction, whether one holds all proofs “by infinite descent” to be mathematical inductions depends on how one defines the term “proof by infinite descent.” One might, for example, use the term to apply to proofs in which the well-ordering of the natural numbers is assumed, but not the principle of induction. Such, for example, is the usual proof that 2 has no rational square root (see Infinite descent).

22.6.5 Prefix induction

The most common form of induction requires proving that
(∀k) (P(k) → P(k+1))
or equivalently
(∀k) (P(k-1) → P(k))
whereupon the induction principle “automates” n applications of this inference in getting from P(0) to P(n). This could be called “predecessor induction” because each step proves something about a number from something about that number’s predecessor.
A variant of interest in computational complexity is “prefix induction”, in which one needs to prove
(∀k) (P(k) → P(2k) ∧ P(2k+1))
or equivalently
(∀k) (P(floor(k⁄2)) → P(k))
The induction principle then “automates” log(n) applications of this inference in getting from P(0) to P(n). (It’s called “prefix induction” because each step proves something about a number from something about the “prefix” of that number formed by truncating the low bit of its binary representation.)
If traditional predecessor induction is interpreted computationally as an n-step loop, prefix induction corresponds to a log(n)-step loop, and thus proofs using prefix induction are “more feasibly constructive” than proofs using predecessor induction.
Predecessor induction can trivially simulate prefix induction on the same statement. Prefix induction can simulate predecessor induction, but only at the cost of making the statement more syntactically complex (adding a bounded universal quantifier), so the interesting results relating prefix induction to polynomial-time computation depend on excluding unbounded quantifiers entirely, and limiting the alternation of bounded universal and existential quantifiers allowed in the statement. See [9]
One could take it a step farther to “prefix of prefix induction": one must prove
(∀k) (P(floor(√k)) → P(k))
whereupon the induction principle “automates” log(log(n)) applications of this inference in getting from P(0) to P(n). This form of induction has been used, analogously, to study log-time parallel computation.
146 *CHAPTER 22. MATHEMATICAL INDUCTION*

22.6.6 Complete induction

Another variant, called **complete induction** (or **strong induction** or **course of values induction**), says that in the second step we may assume not only that the statement holds for $n = m$ but also that it is true for **all** n less than or equal to m.
Complete induction is most useful when several instances of the inductive hypothesis are required for each inductive step. For example, complete induction can be used to show that
$F_n =$
φn □ n
φ □
where F_n is the nth Fibonacci number, φ = (1 + √5)/2 (the golden ratio) and ψ = (1 − √5)/2 are the roots of the polynomial $x^2 - x - 1$. By using the fact that $F_{n+2} = F_{n+1} + F_n$ for each $n \in \mathbf{N}$, the identity above can be verified by direct calculation for F_{n+2} if we assume that it already holds for both F_{n+1} and F_n. To complete the proof, the identity must be verified in the two base cases $n = 0$ and $n = 1$.
Another proof by complete induction uses the hypothesis that the statement holds for *all* smaller n more thoroughly. Consider the statement that “every natural number greater than 1 is a product of (one or more) prime numbers", and assume that for a given $m > 1$ it holds for all smaller $n > 1$. If m is prime then it is certainly a product of primes, and if not, then by definition it is a product: $m = n_1 n_2$, where neither of the factors is equal to 1; hence neither is equal to m, and so both are smaller than m. The induction hypothesis now applies to n_1 and n_2, so each one is a product of primes. Then m is a product of products of primes; i.e. a product of primes.
This generalization, complete induction, is equivalent to the ordinary mathematical induction described above. Suppose P(n) is the statement that we intend to prove by complete induction. Let Q(n) mean P(m) holds for all m such that $0 \le m \le n$. Then Q(n) is true for all n if and only if P(n) is true for all n, and a proof of P(n) by complete induction is just the same thing as a proof of Q(n) by (ordinary) induction.
Transfinite induction
Main article: Transfinite induction
The last two steps can be reformulated as one step:

1. Showing that if the statement holds for all $n < m$ then the same statement also holds for $n = m$.
This form of mathematical induction is not only valid for statements about natural numbers, but for statements about elements of any well-founded set, that is, a set with an irreflexive relation < that contains no infinite descending chains. This form of induction, when applied to ordinals (which form a well-ordered and hence well-founded class), is called *transfinite induction*. It is an important proof technique in set theory, topology and other fields.
Proofs by transfinite induction typically distinguish three cases:
1. when m is a minimal element, i.e. there is no element smaller than m
2. when m has a direct predecessor, i.e. the set of elements which are smaller than m has a largest element
3. when m has no direct predecessor, i.e. m is a so-called limit-ordinal
Strictly speaking, it is not necessary in transfinite induction to prove the basis, because it is a vacuous special case of the proposition that if P is true of all $n < m$, then P is true of m. It is vacuously true precisely because there are no values of $n < m$ that could serve as counterexamples.

22.7 Equivalence with the well-ordering principle

The principle of mathematical induction is usually stated as an axiom of the natural numbers; see Peano axioms. However, it can be proved from the well-ordering principle. Indeed, suppose the following:

_ The set of natural numbers is well-ordered.
_ Every natural number is either zero, or n+1 for some natural number n.
_ For any natural number n, n+1 is greater than n.
To derive simple induction from these axioms, we must show that if P(n) is some proposition predicated of n, and if:
_ P(0) holds and
_ whenever P(k) is true then P(k+1) is also true
then P(n) holds for all n.
Proof. Let S be the set of all natural numbers for which P(n) is false. Let us see what happens if we assert that S is nonempty. Well-ordering tells us that S has a least element, say t. Moreover, since P(0) is true, t is not 0. Since every natural number is either zero or some n+1, there is some natural number n such that n+1=t. Now n is less than t, and t is the least element of S. It follows that n is not in S, and so P(n) is true. This means that P(n+1) is true, and so P(t) is true. This is a contradiction, since t was in S. Therefore, S is empty.
It can also be proved that induction, given the other axioms, implies the well-ordering principle.

22.8 See also

_ Combinatorial proof
_ Recursion
_ Recursion (computer science)
_ Structural induction

22.9 Notes

[1] Anderson, Robert B. (1979). *Proving Programs Correct*. New York: John Wiley & Sons. p. 1. ISBN 0471033952.
[2] Suber, Peter. "Mathematical Induction". Earlham College. Retrieved 26 March 2011.
[3] Mathematical Induction: The Basis Step of Verification and Validation in a Modeling and Simulation Course
[4] Proof due to Euclid http://primes.utm.edu/notes/proofs/infinite/euclids.html http://www.mathsisgoodforyou.com/conjecturestheorems/euclidsprimes.htm http://www.hermetic.ch/pns/proof.htm
[5] Cajori (1918), p. 197
"The process of reasoning called "Mathematical Induction" has had several independent origins. It has been traced back to the Swiss Jakob (James) Bernoulli, the Frenchman B. Pascal and P. Fermat, and the Italian F. Maurolycus. [...] By reading a little between the lines one can find traces of mathematical induction still earlier, in the writings of the Hindus and the Greeks, as, for instance, in the "cyclic method" of Bhaskara, and in Euclid's proof that the number of primes is infinite."
[6] "It is sometimes required to prove a theorem which shall be true whenever a certain quantity n which it involves shall be an integer or whole number and the method of proof is usually of the following kind. *1st.* The theorem is proved to be true when n = 1. *2ndly*. It is proved that if the theorem is true when n is a given whole number, it will be true if n is the next greater integer. Hence the theorem is true universally. . .. This species of argument may be termed a continued *sorites*" (Boole circa 1849 *Elementary Treatise on Logic not mathematical* pages 40–41 reprinted in Grattan-Guinness, Ivor and Bornet, Gérard (1997), *George Boole: Selected Manuscripts on Logic and its Philosophy*, Birkhäuser Verlag, Berlin, ISBN 3-7643-5456-9)
[7] _ Peirce, C. S. (1881). "On the Logic of Number". *American Journal of Mathematics* **4** (1–4). pp. 85–95. doi:10.2307/2369151. JSTOR 2369151. MR 1507856. Reprinted (CP 3.252-88), (W 4:299-309).

_ Paul Shields. (1997), "Peirce's Axiomatization of Arithmetic", in Houser *et al.*, eds., *Studies in the Logic of Charles S. Peirce*.

[8] Cohen, Joel E. (1961), *On the nature of mathematical proof, Opus*. Reprinted in *A Random Walk in Science* (R. L. Weber, ed.), Crane, Russak & Co., 1973.
[9] Buss, Samuel (1986). *Bounded Arithmetic*. Naples: Bibliopolis.

22.10 References

Introduction
_ Franklin, J.; A. Daoud (2011). *Proof in Mathematics: An Introduction*. Sydney: Kew Books. ISBN 0-646-54509-4. (Ch. 8.)
_ Hazewinkel, Michiel, ed. (2001), "Mathematical induction", *Encyclopedia of Mathematics*, Springer, ISBN 978-1-55608-010-4
_ Knuth, Donald E. (1997). *The Art of Computer Programming, Volume 1: Fundamental Algorithms* (3rd ed.). Addison-Wesley. ISBN 0-201-89683-4. (Section 1.2.1: Mathematical Induction, pp. 11–21.)
_ Kolmogorov, Andrey N.; Sergei V. Fomin (1975). *Introductory Real Analysis*. Silverman, R. A. (trans., ed.). New York: Dover. ISBN 0-486-61226-0. (Section 3.8: Transfinite induction, pp. 28–29.)
History
_ Acerbi, F. (2000). "Plato: *Parmenides* 149a7-c3. A Proof by Complete Induction?". *Archive for History of Exact Sciences* **55**: 57–76. doi:10.1007/s004070000020.
_ Bussey, W. H. (1917). "The Origin of Mathematical Induction". *The American Mathematical Monthly* **24** (5): 199–207. doi:10.2307/2974308. JSTOR 2974308.
_ Cajori, Florian (1918). "Origin of the Name "Mathematical Induction"". *The American Mathematical Monthly* **25** (5): 197–201. doi:10.2307/2972638. JSTOR 2972638.
_ "Could the Greeks Have Used Mathematical Induction? Did They Use It?". *Physis* **XXXI**: 253–265. 1994. |first1= missing |last1= in Authors list (help)
_ Freudenthal, Hans (1953). "Zur Geschichte der vollständigen Induction". *Archives Internationales d'Histiore des Sciences* **6**: 17–37.
_ Katz, Victor J. (1998). *History of Mathematics: An Introduction*. Addison-Wesley. ISBN 0-321-01618-1.
_ Peirce, C. S. (1881). "On the Logic of Number". *American Journal of Mathematics* **4** (1–4). pp. 85–95. doi:10.2307/2369151. JSTOR 2369151. MR 1507856. Reprinted (CP 3.252-88), (W 4:299-309).
_ Rabinovitch, Nachum L. (1970). "Rabbi Levi Ben Gershon and the origins of mathematical induction". *Archive for History of Exact Sciences* **6** (3): 237–248. doi:10.1007/BF00327237.
_ Rashed, Roshdi (1972). "L'induction mathématique: al-Karajī, as-Samaw'al". *Archive for History of Exact Sciences* (in French) **9** (1): 1–21. doi:10.1007/BF00348537.
_ Shields, Paul (1997). "Peirce's Axiomatization of Arithmetic". In Houser et al. *Studies in the Logic of Charles S. Peirce.*
_ Ungure, S. (1991). "Greek Mathematics and Mathematical Induction". *Physis*. XXVIII: 273–289.
_ Ungure, S. (1994). "Fowling after Induction". *Physis* **XXXI**: 267–272.
_ Vacca, G. (1909). "Maurolycus, the First Discoverer of the Principle of Mathematical Induction". *Bulletin of the American Mathematical Society* **16** (2): 70–73. doi:10.1090/S0002-9904-1909-01860-9.
_ Yadegari, Mohammad (1978). "The Use of Mathematical Induction by Abū Kāmil Shujā' Ibn Aslam (850-930)". *Isis* **69** (2): 259–262. doi:10.1086/352009. JSTOR 230435.

Chapter 23

Arithmetic progression

In mathematics, an **arithmetic progression** (AP) or **arithmetic sequence** is a sequence of numbers such that the difference between the consecutive terms is constant. For instance, the sequence 5, 7, 9, 11, 13, 15 … is an arithmetic progression with common difference of 2.
If the initial term of an arithmetic progression is a_1 and the common difference of successive members is d, then the nth term of the sequence (a_n) is given by:
$a_n = a_1 + (n - 1)d;$
and in general
$a_n = a_m + (n - m)d:$
A finite portion of an arithmetic progression is called a **finite arithmetic progression** and sometimes just called an arithmetic progression. The sum of a finite arithmetic progression is called an **arithmetic series**.
The behavior of the arithmetic progression depends on the common difference d. If the common difference is:
_ Positive, the members (terms) will grow towards positive infinity.

_ Negative, the members (terms) will grow towards negative infinity.

23.1 Sum

This section is about Finite arithmetic series. For Infinite arithmetic series, see Infinite arithmetic series.
Computation of the sum 2 + 5 + 8 + 11 + 14. When the sequence is reversed and added to itself term by term, the resulting sequence has a single repeated value in it, equal to the sum of the first and last numbers (2 + 14 = 16). Thus 16 × 5 = 80 is twice the sum.
The sum of the members of a finite arithmetic progression is called an **arithmetic series**. For example, consider the sum:

2 + 5 + 8 + 11 + 14

This sum can be found quickly by taking the number n of terms being added (here 5), multiplying by the sum of the first and last number in the progression (here 2 + 14 = 16), and dividing by 2:

$$\frac{n(a_1 + a_n)}{2}$$

149

150 *CHAPTER 23. ARITHMETIC PROGRESSION*

In the case above, this gives the equation:

$$2 + 5 + 8 + 11 + 14 = \frac{5(2 + 14)}{2} = \frac{5 \cdot 16}{2} = 40.$$

This formula works for any real numbers a_1 and a_n. For example:

$$\left(-\frac{3}{2}\right) + \left(-\frac{1}{2}\right) + \frac{1}{2} = \frac{3\left(-\frac{3}{2} + \frac{1}{2}\right)}{2} = -\frac{3}{2}.$$

23.1.1 Derivation

To derive the above formula, begin by expressing the arithmetic series in two different ways:

$$S_n = a_1 + (a_1 + d) + (a_1 + 2d) + \cdots + (a_1 + (n - 2)d) + (a_1 + (n - 1)d)$$

$$S_n = (a_n - (n - 1)d) + (a_n - (n - 2)d) + \cdots + (a_n - 2d) + (a_n - d) + a_n.$$

Adding both sides of the two equations, all terms involving d cancel:

$$2S_n = n(a_1 + a_n).$$

Dividing both sides by 2 produces a common form of the equation:

$$S_n = \frac{n}{2}(a_1 + a_n).$$

An alternate form results from re-inserting the substitution: $a_n = a_1 + (n$ □ $1)d$:
$S_n =$
n
2
$[2a_1 + (n$ □ $1)d]$:
Furthermore the mean value of the series can be calculated via: S_n/n :
$n =$
$a_1 + a_n$
2
:
In 499 AD Aryabhata, a prominent mathematician-astronomer from the classical age of Indian mathematics and Indian astronomy, gave this method in the *Aryabhatiya* (section 2.18).

23.2 Product

The product of the members of a finite arithmetic progression with an initial element a_1, common differences d, and n elements in total is determined in a closed expression
$a_1 a_2$ _ _ _ $a_n = d$
a_1
d
$d($
a_1
d
$+ 1)d($
a_1
d
$+ 2)$ _ _ _ $d($
a_1
d
$+ n$ □ $1) = d^n$
$(a_1$
d
$)^n$
$= d^n$ □ $(a_1/d + n)$
□ (a_1/d)
;
where x^n denotes the rising factorial and □ denotes the Gamma function. (Note however that the formula is not valid when a_1/d is a negative integer or zero.)
This is a generalization from the fact that the product of the progression 1 _ 2 _ _ _ _ _ n is given by the factorial $n!$ and that the product

An n member arithmetical progression.
m _ $(m + 1)$ _ $(m + 2)$ _ _ _ _ _ $(n$ □ $2)$ _ $(n$ □ $1)$ _ n
for positive integers m and n is given by
$n!$
$(m$ □ $1)!$
:
Taking the example from above, the product of the terms of the arithmetic progression given by $a_n = 3 + (n$□$1)(5)$ up to the 50th term is
$P_{50} = 5^{50}$ _ □ $(3/5 + 50)$
□ $(3/5)$
_ $3{:}78438$ _ 10^{98}:

23.3 Standard deviation

The standard deviation of any arithmetic progression can be calculated via:
_ $= jdj$
√
$(n$ □ $1)(n + 1)$
12

where n is the number of terms in the progression, and d is the common difference between terms

23.4 See also

_ Arithmetico-geometric sequence
_ Generalized arithmetic progression - is a set of integers constructed as an arithmetic progression is, but allowing several possible differences.
_ Harmonic progression
_ Heronian triangles with sides in arithmetic progression
_ Problems involving arithmetic progressions
_ Utonality

23.5 References

_ Sigler, Laurence E. (trans.) (2002). *Fibonacci's Liber Abaci.* Springer-Verlag. pp. 259–260. ISBN 0-387-95419-8.

23.6 External links

_ Hazewinkel, Michiel, ed. (2001), "Arithmetic series", *Encyclopedia of Mathematics*, Springer, ISBN 978-1-55608-010-4
_ Weisstein, Eric W., "Arithmetic progression", *MathWorld.*
_ Weisstein, Eric W., "Arithmetic series", *MathWorld.*

Chapter 24

Binomial theorem

The binomial coefficients appear as the entries of Pascal's triangle where each entry is the sum of the two above it.

In elementary algebra, the **binomial theorem** describes the algebraic expansion of powers of a binomial, hence it is referred to as **binomial expansion**. According to the theorem, it is possible to expand the power $(x+y)^n$ into a sum involving terms of the form ax^by^c, where the exponents b and c are nonnegative integers with $b+c=n$, and the coefficient a of each term is a specific positive integer depending on n and b. When an exponent is zero, the corresponding power is usually omitted from the term. For example,

$(x+y)^4 = x^4 + 4x^3y + 6x^2y^2 + 4xy^3 + y^4.$

The coefficient a in the term of ax^by^c is known as the binomial coefficient $\binom{n}{b}$ or $\binom{n}{c}$ (the two have the same value). These coefficients for varying n and b can be arranged to form Pascal's triangle. These numbers also arise in combinatorics, where $\binom{n}{b}$ gives the number of different combinations of b elements that can be chosen from an n-element set.

153

154 *CHAPTER 24. BINOMIAL THEOREM*

24.1 History

This formula and the triangular arrangement of the binomial coefficients are often attributed to Blaise Pascal, who described them in the 17th century, but they were known to many mathematicians who preceded him. For instance the 4th century B.C. Greek mathematician Euclid mentioned the special case of the binomial theorem for exponent 2[1][2] as did the 3rd century B.C. Indian mathematician Pingala to higher orders. A more general binomial theorem and the so-called "Pascal's triangle" were known in the 10th century A.D. to Indian mathematician Halayudha.

Arabian mathematician Al-Karaji,[3] in the 11th century was aware of a more general binomial theorem, along with Persian poet and mathematician Omar Khayyam,[4] and in the 13th century to Chinese mathematician Yang Hui, who all derived similar results.[5] Al-Karaji also provided a mathematical proof of both the binomial theorem and Pascal's triangle, using a primitive form of mathematical induction.[3] Sir Isaac Newton is generally credited with the generalised binomial theorem, valid for any rational exponent.[6]

24.2 Statement of the theorem

According to the theorem, it is possible to expand any power of $x + y$ into a sum of the form

$$(x+y)^n = \binom{n}{0}x^n y^0 + \binom{n}{1}x^{n-1}y^1 + \binom{n}{2}x^{n-2}y^2 + \cdots + \binom{n}{n-1}x^1 y^{n-1} + \binom{n}{n}x^0 y^n,$$

where each $\binom{n}{k}$ is a specific positive integer known as a binomial coefficient. This formula is also referred to as the **binomial formula** or the **binomial identity**. Using summation notation, it can be written as

$$(x+y)^n = \sum_{k=0}^{n} \binom{n}{k} x^{n-k} y^k = \sum_{k=0}^{n} \binom{n}{k} x^k y^{n-k}.$$

The final expression follows from the previous one by the symmetry of x and y in the first expression, and by comparison it follows that the sequence of binomial coefficients in the formula is symmetrical.

A simple variant of the binomial formula is obtained by substituting 1 for y, so that it involves only a single variable. In this form, the formula reads

$(1 + x)^n =$

(

$$\binom{n}{0}x^0 + \binom{n}{1}x^1 + \binom{n}{2}x^2 + \cdots + \binom{n}{n-1}x^{n-1} + \binom{n}{n}x^n;$$

or equivalently

$$(1 + x)^n = \sum_{k=0}^{n} \binom{n}{k} x^k.$$

24.3 Examples

The most basic example of the binomial theorem is the formula for the square of $x + y$:

$$(x + y)^2 = x^2 + 2xy + y^2.$$

The binomial coefficients 1, 2, 1 appearing in this expansion correspond to the third row of Pascal's triangle. The coefficients of higher powers of $x + y$ correspond to later rows of the triangle:

Pascal's triangle

$$(x + y)^3 = x^3 + 3x^2y + 3xy^2 + y^3;$$
$$(x + y)^4 = x^4 + 4x^3y + 6x^2y^2 + 4xy^3 + y^4;$$
$$(x + y)^5 = x^5 + 5x^4y + 10x^3y^2 + 10x^2y^3 + 5xy^4 + y^5;$$
$$(x + y)^6 = x^6 + 6x^5y + 15x^4y^2 + 20x^3y^3 + 15x^2y^4 + 6xy^5 + y^6;$$
$$(x + y)^7 = x^7 + 7x^6y + 21x^5y^2 + 35x^4y^3 + 35x^3y^4 + 21x^2y^5 + 7xy^6 + y^7.$$

Notice that

1. the powers of x go down until it reaches 0 ($x^0 = 1$), starting value is n (the n in $(x + y)^n$.)
2. the powers of y go up from 0 ($y^0 = 1$) until it reaches n (also the n in $(x + y)^n$.)
3. the nth row of the Pascal's Triangle will be the coefficients of the expanded binomial. (Note that the top is row 0.)
4. for each line, the number of products (i.e. the sum of the coefficients) is equal to 2^n .
5. for each line, the number of product groups is equal to $n + 1$.

The binomial theorem can be applied to the powers of any binomial. For example,

$$(x + 2)^3 = x^3 + 3x^2(2) + 3x(2)^2 + 2^3$$
$$= x^3 + 6x^2 + 12x + 8.$$

For a binomial involving subtraction, the theorem can be applied as long as the opposite of the second term is used. This has the effect of changing the sign of every other term in the expansion:

$$(x - y)^3 = x^3 - 3x^2y + 3xy^2 - y^3.$$

Another useful example is that of the expansion of the following square roots:

p

$1 + x = 1 + 1$

$2x \square 1$

$8x2 + 1$

$16x3 \square 5$

$128x4 + 7$

$256x5 \square$ _ _ _

p 1

$1 + x$

$= 1 \square 1$

$2x + 3$

$8x2 \square 5$

$16x3 + 35$

$128x4 \square 63$

$256x5 +$ _ _ _

Sometimes it may be useful to expand negative exponents when $jxj < 1$:

$(1 + x)$

$\square 1 =$

1

$1 + x$

$= 1 \square x + x2 \square x3 + x4 \square x5 +$ _ _ _

24.3.1 Geometric explanation

For positive values of a and b, the binomial theorem with $n = 2$ is the geometrically evident fact that a square of side $a + b$ can be cut into a square of side a, a square of side b, and two rectangles with sides a and b. With $n = 3$, the theorem states that a cube of side $a + b$ can be cut into a cube of side a, a cube of side b, three $a \times a \times b$ rectangular boxes, and three $a \times b \times b$ rectangular boxes.

In calculus, this picture also gives a geometric proof of the derivative $(xn)' = nxn\square 1$: [7] if one sets $a = x$ and $b = \Delta x$; interpreting b as an infinitesimal change in a, then this picture shows the infinitesimal change in the volume of an n-dimensional hypercube, $(x + \Delta x)n$; where the coefficient of the linear term (in Δx) is $nxn\square 1$; the area of the n faces, each of dimension $(n \square 1)$:

$(x + \Delta x)n = xn + nxn\square 1\Delta x +$

$(n$

2

$)$

$xn\square 2(\Delta x)2 +$ _ _ _ :

Substituting this into the definition of the derivative via a difference quotient and taking limits means that the higher order terms – $(\Delta x)2$ and higher – become negligible, and yields the formula $(xn)' = nxn\square 1$; interpreted as "the infinitesimal change in volume of an n-cube as side length varies is the area of n of its $(n \square 1)$ -dimensional faces".

If one integrates this picture, which corresponds to applying the fundamental theorem of calculus, one obtains Cavalieri's quadrature formula, the integral

\int

$xn\square 1 \, dx = 1$

nxn – see proof of Cavalieri's quadrature formula for

details.[7]

24.4 The binomial coefficients

Main article: Binomial coefficient

(The coefficients that appear in the binomial expansion are called **binomial coefficients**. These are usually written n

k

$)$

, and pronounced "n choose k".

24.4.1 Formulae

The coefficient of $xn\text{-}kyk$ is given by the formula

$($

n

k

$$) = \frac{n!}{k!\,(n - k)!}$$

which is defined in terms of the factorial function $n!$. Equivalently, this formula can be written

$$\binom{n}{k} = \frac{n(n-1)\cdots(n-k+1)}{k(k-1)\cdots 1} = \prod_{\ell=1}^{k} \frac{n-\ell+1}{\ell} = \prod_{\ell=0}^{k-1} \frac{n-\ell}{k-\ell}$$

with k factors in both the numerator and denominator of the fraction. Note that, although this formula involves a fraction, the binomial coefficient $\binom{n}{k}$ is actually an integer.

24.4.2 Combinatorial interpretation

The binomial coefficient $\binom{n}{k}$ can be interpreted as the number of ways to choose k elements from an n-element set.

This is related to binomials for the following reason: if we write $(x + y)^n$ as a product

$$(x + y)(x + y)(x + y)\cdots(x + y);$$

then, according to the distributive law, there will be one term in the expansion for each choice of either x or y from each of the binomials of the product. For example, there will only be one term x^n, corresponding to choosing x from each binomial. However, there will be several terms of the form $x^{n-2}y^2$, one for each way of choosing exactly two binomials to contribute a y. Therefore, after combining like terms, the coefficient of $x^{n-2}y^2$ will be equal to the number of ways to choose exactly 2 elements from an n-element set.

24.5 Proofs

24.5.1 Combinatorial proof

Example

The coefficient of xy^2 in

$$(x + y)^3 = (x + y)(x + y)(x + y)$$
$$= xxx + xxy + xyx + xyy + yxx + yxy + yyx + yyy$$
$$= x^3 + 3x^2y + 3xy^2 + y^3.$$

equals $\binom{3}{2} = 3$ because there are three x,y strings of length 3 with exactly two y's, namely,

$$xyy;\ yxy;\ yyx;$$

corresponding to the three 2-element subsets of { 1, 2, 3 }, namely,

*f*2; 3*g*; *f*1; 3*g*; *f*1; 2*g*;

where each subset specifies the positions of the y in a corresponding string.

General case

Expanding $(x + y)_n$ yields the sum of the 2_n products of the form $e_1 e_2 \ldots e_n$ where each e_i is x or y. Rearranging factors shows that each product equals $x_{n-k} y_k$ for some k between 0 and n. For a given k, the following are proved equal in succession:

_ the number of copies of $x_{n-k} y_k$ in the expansion
_ the number of n-character x,y strings having y in exactly k positions
_ the number of k-element subsets of $\{ 1, 2, \ldots, n\}$

_

$\binom{n}{k}$

(this is either by definition, or by a short combinatorial argument if one is defining

$\binom{n}{k}$

as $n!$

$k! (n\square k)!$).

This proves the binomial theorem.

24.5.2 Inductive proof

Induction yields another proof of the binomial theorem (1). When $n = 0$, both sides equal 1, since $x_0 = 1$ and

$\binom{0}{0}$

$= 1$

. Now suppose that (1) holds for a given n; we will prove it for $n + 1$. For j, $k \geq 0$, let $[f(x, y)]_{jk}$ denote the coefficient of $x_j y_k$ in the polynomial $f(x, y)$. By the inductive hypothesis, $(x + y)_n$ is a polynomial in x and y such that $[(x + y)_n]_{j,k}$ is

$\binom{n}{k}$

if $j + k = n$, and 0 otherwise. The identity

$(x + y)_{n+1} = x(x + y)_n + y(x + y)_n;$

shows that $(x + y)_{n+1}$ also is a polynomial in x and y, and

$[(x + y)_{n+1}]_{j;k} = [(x + y)_n]_{j\square 1;k} + [(x + y)_n]_{j;k\square 1}.$

If $j + k = n + 1$, then $(j - 1) + k = n$ and $j + (k - 1) = n$, so the right hand side is

$\binom{n}{k}$

$+$

$\binom{n}{k \square 1}$

$=$

$\binom{n + 1}{k}$

$;$

by Pascal's identity. On the other hand, if $j + k \neq n + 1$, then $(j - 1) + k \neq n$ and $j + (k - 1) \neq n$, so we get $0 + 0 = 0$. Thus

$(x + y)_{n+1} =$

$\sum_{k=0}^{n+1}$

$\binom{n+1}{k}$

$x_{n+1\square k} y_k;$

which is the inductive hypothesis with $n + 1$ substituted for n and so completes the inductive step.

24.6 Generalisations

24.6.1 Newton's generalised binomial theorem

Main article: Binomial series

Around 1665, Isaac Newton generalised the formula to allow real exponents other than nonnegative integers, and in fact it can be generalised further, to complex exponents. In this generalisation, the finite sum is replaced by an infinite series. In order to do this one needs to give meaning to binomial coefficients with an arbitrary upper index, which cannot be done using the above formula with factorials; however factoring out $(n - k)!$ from numerator and denominator in that formula, and replacing n by r which now stands for an arbitrary number, one can define

$$\binom{r}{k} = \frac{r(r-1)\cdots(r-k+1)}{k!} = \frac{(r)_k}{k!};$$

where $(\cdot)_k$ is the Pochhammer symbol here standing for a falling factorial. Then, if x and y are real numbers with $|x| > |y|$,[8] and r is any complex number, one has

$$(x+y)^r = \sum_{k=0}^{\infty} \binom{r}{k} x^{r-k} y^k \quad (2)$$

$$= x^r + rx^{r-1}y + \frac{r(r-1)}{2!}x^{r-2}y^2 + \frac{r(r-1)(r-2)}{3!}x^{r-3}y^3 + \cdots.$$

When r is a nonnegative integer, the binomial coefficients for $k > r$ are zero, so (2) specializes to (1), and there are at most $r + 1$ nonzero terms. For other values of r, the series (2) has infinitely many nonzero terms, at least if x and y are nonzero.

This is important when one is working with infinite series and would like to represent them in terms of generalised hypergeometric functions.

Taking $r = -s$ leads to a useful formula:

$$\frac{1}{(1-x)^s} = \sum_{k=0}^{\infty} \binom{s+k-1}{k} x^k \equiv \sum_{k=0}^{\infty} \binom{s+k-1}{s-1}$$

$) x^k.$

Further specializing to $s = 1$ yields the geometric series formula.

Generalisations

Formula (2) can be generalised to the case where x and y are complex numbers. For this version, one should assume $|x| > |y|$[8] and define the powers of $x + y$ and x using a holomorphic branch of log defined on an open disk of radius $|x|$ centered at x.

Formula (2) is valid also for elements x and y of a Banach algebra as long as $xy = yx$, x is invertible, and $\|y/x\| < 1$.

24.6.2 The multinomial theorem

Main article: Multinomial theorem

The binomial theorem can be generalised to include powers of sums with more than two terms. The general version is

$$(x_1 + x_2 + \cdots + x_m)^n = \sum_{k_1+k_2+\cdots+k_m=n} \binom{n}{k_1, k_2, \ldots, k_m} x_1^{k_1} x_2^{k_2} \cdots x_m^{k_m}.$$

where the summation is taken over all sequences of nonnegative integer indices k_1 through k_m such that the sum of all k_i is n. (For each term in the expansion, the exponents must add up to n). The coefficients $\binom{n}{k_1,\cdots,k_m}$ are known as multinomial coefficients, and can be computed by the formula

$$\binom{n}{k_1, k_2, \ldots, k_m} = \frac{n!}{k_1!\, k_2! \cdots k_m!}.$$

Combinatorially, the multinomial coefficient $\binom{n}{k_1,\cdots,k_m}$ counts the number of different ways to partition an n-element set into disjoint subsets of sizes k_1, ..., k_m.

24.6.3 The multi-binomial theorem

It is often useful when working in more dimensions, to deal with products of binomial expressions. By the binomial theorem this is equal to

$$(x_1 + y_1)^{n_1} \cdots (x_d + y_d)^{n_d} = \sum_{k_1=0}^{n_1} \cdots \sum_{k_d=0}^{n_d} \binom{n_1}{k_1} x_1^{k_1}$$

1 $y_{n_1 k_1}$
1 : : :
(
n_d
k_d
)
x_{k_d}
$d\,y_{n_d \square k_d}$
d :

This may be written more concisely, by multi-index notation, as

$(x + y)_ =$
Σ

$\overline{}$
(

$-$

$\overline{}$
)
$x\,_y\,_\square_:$

24.7 Applications

24.7.1 Multiple-angle identities

For the complex numbers the binomial theorem can be combined with De Moivre's formula to yield multiple-angle formulas for the sine and cosine. According to De Moivre's formula,

$\cos(nx) + i\sin(nx) = (\cos x + i\sin x)_n$:

Using the binomial theorem, the expression on the right can be expanded, and then the real and imaginary parts can be taken to yield formulas for $\cos(nx)$ and $\sin(nx)$. For example, since

$(\cos x + i\sin x)_2 = \cos_2 x + 2i\cos x\sin x \square \sin_2 x;$

De Moivre's formula tells us that

$\cos(2x) = \cos_2 x \square \sin_2 x$ and $\sin(2x) = 2\cos x\sin x;$

which are the usual double-angle identities. Similarly, since

$(\cos x + i\sin x)_3 = \cos_3 x + 3i\cos_2 x\sin x \square 3\cos x\sin_2 x \square i\sin_3 x;$

De Moivre's formula yields

$\cos(3x) = \cos_3 x \square 3\cos x\sin_2 x$ and $\sin(3x) = 3\cos_2 x\sin x \square \sin_3 x:$

In general,

$\cos(nx) =$
Σ
$k\,even$
$(\square 1)_{k/2}$
(
n
k
)
$\cos_{n \square k} x \sin_k x$
and
$\sin(nx) =$
Σ
$k\,odd$
$(\square 1)_{(k \square 1)/2}$
(
n
k
)
$\cos_{n \square k} x \sin_k x:$

24.7.2 Series for e

The number e is often defined by the formula

$e = \lim_{n!1}$
(

$\left(1+\frac{1}{n}\right)^n$:

Applying the binomial theorem to this expression yields the usual infinite series for e. In particular:

$$\left(1+\frac{1}{n}\right)^n = 1 + \binom{n}{1}\frac{1}{n} + \binom{n}{2}\frac{1}{n^2} + \binom{n}{3}\frac{1}{n^3} + \cdots + \binom{n}{n}\frac{1}{n^n} :$$

The kth term of this sum is

$$\binom{n}{k}\frac{1}{n^k} = \frac{1}{k!}\cdot\frac{n(n-1)(n-2)\cdots(n-k+1)}{n^k}$$

As $n \to \infty$, the rational expression on the right approaches one, and therefore

$$\lim_{n\to\infty}\binom{n}{k}\frac{1}{n^k} = \frac{1}{k!} :$$

This indicates that e can be written as a series:

$$e = \frac{1}{0!} + \frac{1}{1!} + \frac{1}{2!} + \frac{1}{3!} + \cdots$$

Indeed, since each term of the binomial expansion is an increasing function of n, it follows from the monotone convergence theorem for series that the sum of this infinite series is equal to e.

24.7.3 Derivative of the power function

In finding the derivative of the power function, $f(x) = x^n$, by using the definition of derivative, the expansion of $(x + h)^n$ is employed.

24.7.4 Nth derivative of a product

To indicate the formula for the derivative of order n of the product of two functions, the formula of the binomial theorem is used symbolically.[9]

24.8 The binomial theorem in abstract algebra

Formula (1) is valid more generally for any elements x and y of a semiring satisfying $xy = yx$. The theorem is true even more generally: alternativity suffices in place of associativity.

The binomial theorem can be stated by saying that the polynomial sequence $\{ 1, x, x^2, x^3, \ldots \}$ is of binomial type.

24.9 In popular culture

- The binomial theorem is mentioned in the Major-General's Song in the comic opera The Pirates of Penzance.
- Professor Moriarty is described by Sherlock Holmes as having written a treatise on the binomial theorem.

24.10 See also

- Binomial approximation
- Binomial distribution
- Binomial inverse theorem
- Binomial probability
- Binomial series
- Combination

- Multinomial theorem
- Negative binomial distribution
- Pascal's triangle
- Stirling's approximation

24.11 Notes

[1] Binomial Theorem

[2] The Story of the Binomial Theorem by J. L. Coolidge, *The American Mathematical Monthly* **56**:3 (1949), pp. 147–157

[3] O'Connor, John J.; Robertson, Edmund F., "Abu Bekr ibn Muhammad ibn al-Husayn Al-Karaji", *MacTutor History of Mathematics archive*, University of St Andrews.

[4] Sandler, Stanley (2011). *An Introduction to Applied Statistical Thermodynamics*. Hoboken NJ: John Wiley & Sons, Inc. ISBN 978-0-470-91347-5.

[5] Landau, James A (1999-05-08). "Historia Matematica Mailing List Archive: Re: [HM] Pascal's Triangle" (mailing list email). *Archives of Historia Matematica*. Retrieved 2007-04-13.

[6] Bourbaki: *History of mathematics*

[7] (Barth 2004)

[8] This is to guarantee convergence. Depending on r, the series may also converge sometimes when $|x| = |y|$.

[9] Seely, Robert T.: *Calculus of One and Several Variables*

24.12 References

_ Bag, Amulya Kumar (1966). "Binomial theorem in ancient India". *Indian J. History Sci* **1** (1): 68–74.
_ Barth, Nils R. (November 2004). "Computing Cavalieri's Quadrature Formula by a Symmetry of the *n*-Cube". *The American Mathematical Monthly* (Mathematical Association of America) **111** (9): 811–813. doi:10.2307/4145193. ISSN 0002-9890. JSTOR 4145193, author's copy, further remarks and resources
_ Graham, Ronald; Knuth, Donald; Patashnik, Oren (1994). "(5) Binomial Coefficients". *Concrete Mathematics* (2nd ed.). Addison Wesley. pp. 153–256. ISBN 0-201-55802-5. OCLC 17649857.

24.13 External links

_ Solomentsev, E.D. (2001), "Newton binomial", in Hazewinkel, Michiel, *Encyclopedia of Mathematics*, Springer, ISBN 978-1-55608-010-4
_ Binomial Theorem by Stephen Wolfram, and "Binomial Theorem (Step-by-Step)" by Bruce Colletti and Jeff Bryant, Wolfram Demonstrations Project, 2007.

Chapter 25

Proof by contradiction

See also: Table of logic symbols
In logic, **proof by contradiction** is a form of proof, and more specifically a form of indirect proof, that establishes the truth or validity of a proposition by showing that the proposition's being false would imply a contradiction. Proof by contradiction is also known as **indirect proof**, **apagogical argument**, **proof by assuming the opposite**, and ***reductio ad impossibilem***. It is a particular kind of the more general form of argument known as *reductio ad absurdum*.
G. H. Hardy described proof by contradiction as "one of a mathematician's finest weapons", saying "It is a far finer gambit than any chess gambit: a chess player may offer the sacrifice of a pawn or even a piece, but a mathematician offers the game."[1]

25.1 Examples

25.1.1 Irrationality of the square root of 2

A classic proof by contradiction from mathematics is the proof that the square root of 2 is irrational.[2] If it were rational, it could be expressed as a fraction a/b in lowest terms, where a and b are integers, at least one of which is odd. But if $a/b = \sqrt{2}$, then $a^2 = 2b^2$. Therefore a^2 must be even. Because the square of an odd number is odd, that in turn implies that a is even. This means that b must be odd because a/b is in lowest terms.
On the other hand, if a is even, then a^2 is a multiple of 4. If a^2 is a multiple of 4 and $a^2 = 2b^2$, then $2b^2$ is a multiple of 4, and therefore b^2 is even, and so is b.
So b is odd and even, a contradiction. Therefore the initial assumption—that $\sqrt{2}$ can be expressed as a fraction—must be false.

25.1.2 The length of the hypotenuse

The method of proof by contradiction has also been used to show that for any non-degenerate right triangle, the length of the hypotenuse is less than the sum of the lengths of the two remaining sides.[3] The proof relies on the Pythagorean theorem. Letting c be the length of the hypotenuse and a and b the lengths of the legs, the claim is that $a + b > c$.
The claim is negated to assume that $a + b \leq c$. Squaring both sides results in $(a + b)^2 \leq c^2$ or, equivalently, $a^2 + 2ab + b^2 \leq c^2$. A triangle is non-degenerate if each edge has positive length, so it may be assumed that a and b are greater than 0. Therefore, $a^2 + b^2 < a^2 + 2ab + b^2 \leq c^2$. The transitive relation may be reduced to $a^2 + b^2 < c^2$. It is known from the Pythagorean theorem that $a^2 + b^2 = c^2$. This results in a contradiction since strict inequality and equality are mutually exclusive. The latter was a result of the Pythagorean theorem and the former the assumption that $a + b \leq c$. The contradiction means that it is impossible for both to be true and it is known that the Pythagorean theorem holds. It follows that the assumption that $a + b \leq c$ must be false and hence $a + b > c$, proving the claim.

25.1.3 No least positive rational number

Consider the proposition, *P*: "there is no smallest rational number greater than 0". In a proof by contradiction, we start by assuming the opposite, ¬*P*: that there *is* a smallest rational number, say, *r*.
Now *r*/2 is a rational number greater than 0 and smaller than *r*. (In the above symbolic argument, "*r*/2 is the smallest rational number" would be *Q* and "*r* (which is different from *r*/2) is the smallest rational number" would be ¬*Q*.)
But that contradicts our initial assumption, ¬*P*, that *r* was the *smallest* rational number. So we can conclude that the original proposition, *P*, must be true — "there is no smallest rational number greater than 0".

25.1.4 Other

For other examples, see proof that the square root of 2 is not rational (where indirect proofs different from the above one can be found) and Cantor's diagonal argument.

25.2 In mathematical logic

In mathematical logic, the proof by contradiction is represented as:
If
S [fPg ⊢ F
then
S ⊢ *:P:*
or
If
S [f:Pg ⊢ F
then
S ⊢ *P:*
In the above, *P* is the proposition we wish to prove, and *S* is a set of statements, which are the premises—these could be, for example, the axioms of the theory we are working in, or earlier theorems we can build upon. We consider *P*, or the negation of *P*, in addition to *S*; if this leads to a logical contradiction *F*, then we can conclude that the statements in *S* lead to the negation of *P*, or *P* itself, respectively.
Note that the set-theoretic union, in some contexts closely related to logical disjunction (or), is used here for sets of statements in such a way that it is more related to logical conjunction (and).
A particular kind of indirect proof assumes that some object doesn't exist, and then proves that this would lead to a contradiction; thus, such an object must exist. Although it is quite freely used in mathematical proofs, not every school of mathematical thought accepts this kind of argument as universally valid. See further Nonconstructive proof.

25.3 Notation

Proofs by contradiction sometimes end with the word "Contradiction!". Isaac Barrow and Baermann used the notation Q.E.A., for "*quod est absurdum*" ("which is absurd"), along the lines of Q.E.D., but this notation is rarely used today.[4]
A graphical symbol sometimes used for contradictions is a downwards zigzag arrow "lightning" symbol (U+21AF: ↯), for example in Davey and Priestley.[5] Others sometimes used include a pair of opposing arrows (as*!* or*)(*
), struck-out arrows (↮), a stylized form of hash (such as U+2A33: □), or the "reference mark" (U+203B: ※).[6][7]
The "up tack" symbol (U+22A5: ⊥) used by philosophers and logicians (see contradiction) also appears, but is often avoided due to its usage for orthogonality.

25.4 See also

_ Proof by contrapositive

25.5 References

[1] G. H. Hardy, *A Mathematician's Apology; Cambridge University Press, 1992. ISBN 9780521427067.* p. 94.
[2] Alfield, Peter (16 August 1996). "Why is the square root of 2 irrational?". *Understanding Mathematics, a study guide.* Department of Mathematics, University of Utah. Retrieved 6 February 2013.
[3] Stone, Peter. "Logic, Sets, and Functions: Honors". *Course materials.* pp 14–23: Department of Computer Sciences, The University of Texas at Austin. Retrieved 6 February 2013.
[4] Hartshorne on QED and related
[5] B. Davey and H.A. Priestley, Introduction to lattices and order, Cambridge University Press, 2002.
[6] The Comprehensive LaTeX Symbol List, pg. 20. http://www.ctan.org/tex-archive/info/symbols/comprehensive/symbols-a4.pdf
[7] Gary Hardegree, *Introduction to Modal Logic*, Chapter 2, pg. II–2. http://people.umass.edu/gmhwww/511/pdf/c02.pdf

25.6 Further reading

_ Franklin, James (2011). *Proof in Mathematics: An Introduction.* chapter 6: Kew. ISBN 978-0-646-54509-7.

25.7 External links

_ Proof by Contradiction from Larry W. Cusick's How To Write Proofs

Chapter 26

Euclidean geometry

"Plane geometry" redirects here. For other uses, see Plane geometry (disambiguation).

Euclidean geometry is a mathematical system attributed to the Alexandrian Greek mathematician Euclid, which he described in his textbook on geometry: the *Elements*. Euclid's method consists in assuming a small set of intuitively appealing axioms, and deducing many other propositions (theorems) from these. Although many of Euclid's results had been stated by earlier mathematicians,[1] Euclid was the first to show how these propositions could fit into a comprehensive deductive and logical system.[2] The *Elements* begins with plane geometry, still taught in secondary school as the first axiomatic system and the first examples of formal proof. It goes on to the solid geometry of three dimensions. Much of the *Elements* states results of what are now called algebra and number theory, explained in geometrical language.[3]

For more than two thousand years, the adjective "Euclidean" was unnecessary because no other sort of geometry had been conceived. Euclid's axioms seemed so intuitively obvious (with the possible exception of the parallel postulate) that any theorem proved from them was deemed true in an absolute, often metaphysical, sense. Today, however, many other self-consistent non-Euclidean geometries are known, the first ones having been discovered in the early 19th century. An implication of Albert Einstein's theory of general relativity is that physical space itself is not Euclidean, and Euclidean space is a good approximation for it only where the gravitational field is weak.[4]

Euclidean geometry is an example of synthetic geometry, in that it proceeds logically from axioms to propositions without the use of coordinates. This is in contrast to analytic geometry, which uses coordinates.

26.1 *The Elements*

Main article: Euclid's Elements

The *Elements* are mainly a systematization of earlier knowledge of geometry. Its superiority over earlier treatments was rapidly recognized, with the result that there was little interest in preserving the earlier ones, and they are now nearly all lost.

There are 13 total books in the *Elements*:

Books I–IV and VI discuss plane geometry. Many results about plane figures are proved, e.g., *If a triangle has two equal angles, then the sides subtended by the angles are equal.* The Pythagorean theorem is proved.[5]

Books V and VII–X deal with number theory, with numbers treated geometrically via their representation as line segments with various lengths. Notions such as prime numbers and rational and irrational numbers are introduced. The infinitude of prime numbers is proved.

Books XI–XIII concern solid geometry. A typical result is the 1:3 ratio between the volume of a cone and a cylinder with the same height and base.

Detail from Raphael's The School of Athens *featuring a Greek mathematician – perhaps representing Euclid or Archimedes – using a compass to draw a geometric construction.*

26.1.1 Axioms

Euclidean geometry is an axiomatic system, in which all theorems ("true statements") are derived from a small number of axioms.[6] Near the beginning of the first book of the *Elements*, Euclid gives five postulates (axioms) for plane geometry, stated in terms of constructions (as translated by Thomas Heath):[7]

"Let the following be postulated":

1. "To draw a straight line from any point to any point."
2. "To produce [extend] a finite straight line continuously in a straight line."
3. "To describe a circle with any centre and distance [radius]."

α

β

The parallel postulate: If two lines intersect a third in such a way that the sum of the inner angles on one side is less than two right angles, then the two lines inevitably must intersect each other on that side if extended far enough.

4. "That all right angles are equal to one another."
5. *The parallel postulate*: "That, if a straight line falling on two straight lines make the interior angles on the same side less than two right angles, the two straight lines, if produced indefinitely, meet on that side on which are the angles less than the two right angles."

Although Euclid's statement of the postulates only explicitly asserts the existence of the constructions, they are also taken to be unique.

The *Elements* also include the following five "common notions":

1. Things that are equal to the same thing are also equal to one another (Transitive property of equality).
2. If equals are added to equals, then the wholes are equal (Addition property of equality).
3. If equals are subtracted from equals, then the remainders are equal (Subtraction property of equality).
4. Things that coincide with one another are equal to one another (Reflexive Property).
5. The whole is greater than the part.

26.1.2 Parallel postulate

Main article: Parallel postulate

To the ancients, the parallel postulate seemed less obvious than the others. They were concerned with creating a system which was absolutely rigorous and to them it seemed as if the parallel line postulate should have been able to be proven rather than simply accepted as a fact. It is now known that such a proof is impossible.[8] Euclid himself

seems to have considered it as being qualitatively different from the others, as evidenced by the organization of the *Elements*: the first 28 propositions he presents are those that can be proved without it.

Many alternative axioms can be formulated that have the same logical consequences as the parallel postulate. For example Playfair's axiom states:

In a plane, through a point not on a given straight line, at most one line can be drawn that never meets the given line.

26.2 Methods of proof

Euclidean Geometry is *constructive*. Postulates 1, 2, 3, and 5 assert the existence and uniqueness of certain geometric figures, and these assertions are of a constructive nature: that is, we are not only told that certain things exist, but are also given methods for creating them with no more than a compass and an unmarked straightedge.[9] In this sense, Euclidean geometry is more concrete than many modern axiomatic systems such as set theory, which often assert the existence of objects without saying how to construct them, or even assert the existence of objects that cannot be constructed within the theory.[10] Strictly speaking, the lines on paper are *models* of the objects defined within the formal system, rather than instances of those objects. For example a Euclidean straight line has no width, but any real drawn line will. Though nearly all modern mathematicians consider nonconstructive methods just as sound as constructive ones, Euclid's constructive proofs often supplanted fallacious nonconstructive ones—e.g., some of the Pythagoreans' proofs that involved irrational numbers, which usually required a statement such as "Find the greatest common measure of ..."[11]

Euclid often used proof by contradiction. Euclidean geometry also allows the method of superposition, in which a figure is transferred to another point in space. For example, proposition I.4, side-angle-side congruence of triangles, is proved by moving one of the two triangles so that one of its sides coincides with the other triangle's equal side, and then proving that the other sides coincide as well. Some modern treatments add a sixth postulate, the rigidity of the triangle, which can be used as an alternative to superposition.[12]

26.3 System of measurement and arithmetic

Euclidean geometry has two fundamental types of measurements: angle and distance. The angle scale is absolute, and Euclid uses the right angle as his basic unit, so that, e.g., a 45-degree angle would be referred to as half of a right angle. The distance scale is relative; one arbitrarily picks a line segment with a certain nonzero length as the unit, and other distances are expressed in relation to it. Addition of distances is represented by a construction in which one line segment is copied onto the end of another line segment to extend its length, and similarly for subtraction.

Measurements of area and volume are derived from distances. For example, a rectangle with a width of 3 and a length of 4 has an area that represents the product, 12. Because this geometrical interpretation of multiplication was limited to three dimensions, there was no direct way of interpreting the product of four or more numbers, and Euclid

avoided such products, although they are implied, e.g., in the proof of book IX, proposition 20.
Euclid refers to a pair of lines, or a pair of planar or solid figures, as "equal" (ἴσος) if their lengths, areas, or volumes are equal, and similarly for angles. The stronger term "congruent" refers to the idea that an entire figure is the same size and shape as another figure. Alternatively, two figures are congruent if one can be moved on top of the other so that it matches up with it exactly. (Flipping it over is allowed.) Thus, for example, a 2x6 rectangle and a 3x4 rectangle are equal but not congruent, and the letter R is congruent to its mirror image. Figures that would be congruent except for their differing sizes are referred to as similar. Corresponding angles in a pair of similar shapes are congruent and corresponding sides are in proportion to each other.

26.4 Notation and terminology

26.4.1 Naming of points and figures

Points are customarily named using capital letters of the alphabet. Other figures, such as lines, triangles, or circles, are named by listing a sufficient number of points to pick them out unambiguously from the relevant figure, e.g., triangle

A proof from Euclid's elements that, given a line segment, an equilateral triangle exists that includes the segment as one of its sides. The proof is by construction: an equilateral triangle ABΓ is made by drawing circles Δ and Ε centered on the points A and B, and taking one intersection of the circles as the third vertex of the triangle.
ABC would typically be a triangle with vertices at points A, B, and C.

26.4.2 Complementary and supplementary angles

Angles whose sum is a right angle are called complementary. Complementary angles are formed when a ray shares the same vertex and is pointed in a direction that is in between the two original rays that form the right angle. The

An example of congruence. The two figures on the left are congruent, while the third is similar to them. The last figure is neither. Note that congruences alter some properties, such as location and orientation, but leave others unchanged, like distance and angles. The latter sort of properties are called invariants and studying them is the essence of geometry.
number of rays in between the two original rays is infinite.
Angles whose sum is a straight angle are supplementary. Supplementary angles are formed when a ray shares the same vertex and is pointed in a direction that is in between the two original rays that form the straight angle (180 degree angle). The number of rays in between the two original rays is infinite.

26.4.3 Modern versions of Euclid's notation

In modern terminology, angles would normally be measured in degrees or radians.
Modern school textbooks often define separate figures called lines (infinite), rays (semi-infinite), and line segments (of finite length). Euclid, rather than discussing a ray as an object that extends to infinity in one direction, would normally use locutions such as "if the line is extended to a sufficient length," although he occasionally referred to "infinite lines." A "line" in Euclid could be either straight or curved, and he used the more specific term "straight line" when necessary.

26.5 Some important or well known results

_ The **Pons Asinorum** or **Bridge of Asses theorem** states that in an isosceles triangle, $\alpha = \beta$ and $\gamma = \delta$.
_ The **Triangle Angle Sum theorem** states that the sum of the three angles of any triangle, in this case angles α, β, and γ, will always equal 180 degrees.
_ The **Pythagorean theorem** states that the sum of the areas of the two squares on the legs (a and b) of a right triangle equals the area of the square on the hypotenuse (c).
_ **Thales' theorem** states that if AC is a diameter, then the angle at B is a right angle.

26.5.1 Pons Asinorum

The Bridge of Asses (*Pons Asinorum*) states that *in isosceles triangles the angles at the base equal one another, and, if the equal straight lines are produced further, then the angles under the base equal one another.*[13] Its name may be attributed to its frequent role as the first real test in the *Elements* of the intelligence of the reader and as a bridge to the harder propositions that followed. It might also be so named because of the geometrical figure's resemblance to a steep bridge that only a sure-footed donkey could cross.[14]

26.5.2 Congruence of triangles

Triangles are congruent if they have all three sides equal (SSS), two sides and the angle between them equal (SAS), or two angles and a side equal (ASA) Only the first of these three is actually proven, the other two are postulates,

A

A

S

S

A

S

S

A

S

A

A

S

Congruence of triangles is determined by specifying two sides and the angle between them (SAS), two angles and the side between them (ASA) or two angles and a corresponding adjacent side (AAS). Specifying two sides and an adjacent angle (SSA), however, can yield two distinct possible triangles unless the angle specified is a right angle.

or axioms (i.e., they are assumed to be true) (Book I, propositions 4, 8, and 26). (Triangles with three equal angles (AAA) are similar, but not necessarily congruent. Also, triangles with two equal sides and an adjacent angle are not necessarily equal or congruent.)

26.5.3 Triangle Angle Sum

The sum of the angles of a triangle is equal to a straight angle (180 degrees).[15] This causes an equilateral triangle to have 3 interior angles of 60 degrees. Also, it causes every triangle to have at least 2 acute angles and up to 1 obtuse or right angle.

26.5.4 Pythagorean theorem

The celebrated Pythagorean theorem (book I, proposition 47) states that in any right triangle, the area of the square whose side is the hypotenuse (the side opposite the right angle) is equal to the sum of the areas of the squares whose sides are the two legs (the two sides that meet at a right angle).

26.5.5 Thales' theorem

Thales' theorem, named after Thales of Miletus states that if A, B, and C are points on a circle where the line AC is a diameter of the circle, then the angle ABC is a right angle. Cantor supposed that Thales proved his theorem by means of Euclid Book I, Prop. 32 after the manner of Euclid Book III, Prop. 31.[16] Tradition has it that Thales sacrificed an ox to celebrate this theorem.[17]

26.5.6 Scaling of area and volume

In modern terminology, the area of a plane figure is proportional to the square of any of its linear dimensions, A / L_2

, and the volume of a solid to the cube, V / L^3. Euclid proved these results in various special cases such as the area of a circle[18] and the volume of a parallelepipedal solid.[19] Euclid determined some, but not all, of the relevant constants of proportionality. E.g., it was his successor Archimedes who proved that a sphere has 2/3 the volume of the circumscribing cylinder.[20]

26.6 Applications

Because of Euclidean geometry's fundamental status in mathematics, it would be impossible to give more than a representative sampling of applications here.

_ A surveyor uses a level

_ Sphere packing applies to a stack of oranges.

_ A parabolic mirror brings parallel rays of light to a focus.

As suggested by the etymology of the word, one of the earliest reasons for interest in geometry was surveying,[21] and certain practical results from Euclidean geometry, such as the right-angle property of the 3-4-5 triangle, were used long before they were proved formally.[22] The fundamental types of measurements in Euclidean geometry are distances and angles, and both of these quantities can be measured directly by a surveyor. Historically, distances were often measured by chains such as Gunter's chain, and angles using graduated circles and, later, the theodolite.

An application of Euclidean solid geometry is the determination of packing arrangements, such as the problem of finding the most efficient packing of spheres in n dimensions. This problem has applications in error detection and correction.

Geometric optics uses Euclidean geometry to analyze the focusing of light by lenses and mirrors.

_ Geometry is used in art and architecture.

_ The water tower consists of a cone, a cylinder, and a hemisphere. Its volume can be calculated using solid geometry.

_ Geometry can be used to design origami.

Geometry is used extensively in architecture.

Geometry can be used to design origami. Some classical construction problems of geometry are impossible using compass and straightedge, but can be solved using origami.[23]

26.7 As a description of the structure of space

Euclid believed that his axioms were self-evident statements about physical reality. Euclid's proofs depend upon assumptions perhaps not obvious in Euclid's fundamental axioms,[24] in particular that certain movements of figures do not change their geometrical properties such as the lengths of sides and interior angles, the so-called *Euclidean motions*, which include translations and rotations of figures.[25] Taken as a physical description of space, postulate 2 (extending a line) asserts that space does not have holes or boundaries (in other words, space is homogeneous and unbounded); postulate 4 (equality of right angles) says that space is isotropic and figures may be moved to any location while maintaining congruence; and postulate 5 (the parallel postulate) that space is flat (has no intrinsic curvature).[26]

As discussed in more detail below, Einstein's theory of relativity significantly modifies this view.

The ambiguous character of the axioms as originally formulated by Euclid makes it possible for different commentators to disagree about some of their other implications for the structure of space, such as whether or not it is infinite[27] (see below) and what its topology is. Modern, more rigorous reformulations of the system[28] typically aim for a cleaner separation of these issues. Interpreting Euclid's axioms in the spirit of this more modern approach, axioms 1-4 are consistent with either infinite or finite space (as in elliptic geometry), and all five axioms are consistent with a variety of topologies (e.g., a plane, a cylinder, or a torus for two-dimensional Euclidean geometry).

26.8 Later work

26.8.1 Archimedes and Apollonius

Archimedes (ca. 287 BCE – ca. 212 BCE), a colorful figure about whom many historical anecdotes are recorded, is remembered along with Euclid as one of the greatest of ancient mathematicians. Although the foundations of his work were put in place by Euclid, his work, unlike Euclid's, is believed to have been entirely original.[29] He proved equations for the volumes and areas of various figures in two and three dimensions, and enunciated the Archimedean property of finite numbers.

Apollonius of Perga (ca. 262 BCE–ca. 190 BCE) is mainly known for his investigation of conic sections.

26.8.2 17th century: Descartes

René Descartes (1596–1650) developed analytic geometry, an alternative method for formalizing geometry which focused on turning geometry into algebra.[30] In this approach, a point is represented by its Cartesian (x, y) coordinates, a line is represented by its equation, and so on. In Euclid's original approach, the Pythagorean theorem follows from

Euclid's axioms. In the Cartesian approach, the axioms are the axioms of algebra, and the equation expressing the Pythagorean theorem is then a definition of one of the terms in Euclid's axioms, which are now considered theorems. The equation

$$|PQ| = \sqrt{(p - r)^2 + (q - s)^2}$$

defining the distance between two points $P = (p, q)$ and $Q = (r, s)$ is then known as the *Euclidean metric*, and other metrics define non-Euclidean geometries.

In terms of analytic geometry, the restriction of classical geometry to compass and straightedge constructions means a restriction to first- and second-order equations, e.g., $y = 2x + 1$ (a line), or $x^2 + y^2 = 7$ (a circle).

Also in the 17th century, Girard Desargues, motivated by the theory of perspective, introduced the concept of idealized points, lines, and planes at infinity. The result can be considered as a type of generalized geometry, projective geometry, but it can also be used to produce proofs in ordinary Euclidean geometry in which the number of special cases is reduced.[31]

A sphere has 2/3 the volume and surface area of its circumscribing cylinder. A sphere and cylinder were placed on the tomb of Archimedes at his request.

26.8.3 18th century

Geometers of the 18th century struggled to define the boundaries of the Euclidean system. Many tried in vain to prove the fifth postulate from the first four. By 1763 at least 28 different proofs had been published, but all were found incorrect.[32]

Leading up to this period, geometers also tried to determine what constructions could be accomplished in Euclidean geometry. For example, the problem of trisecting an angle with a compass and straightedge is one that naturally occurs within the theory, since the axioms refer to constructive operations that can be carried out with those tools. However, centuries of efforts failed to find a solution to this problem, until Pierre Wantzel published a proof in 1837 that such a construction was impossible. Other constructions that were proved impossible include doubling the cube and squaring the circle. In the case of doubling the cube, the impossibility of the construction originates from the fact that the compass and straightedge method involve first- and second-order equations, while doubling a cube requires the solution of a third-order equation.

Euler discussed a generalization of Euclidean geometry called affine geometry, which retains the fifth postulate un

René Descartes. Portrait after Frans Hals, 1648.

modified while weakening postulates three and four in a way that eliminates the notions of angle (whence right triangles become meaningless) and of equality of length of line segments in general (whence circles become meaningless) while retaining the notions of parallelism as an equivalence relation between lines, and equality of length of parallel line segments (so line segments continue to have a midpoint).

26.8.4 19th century and non-Euclidean geometry

In the early 19th century, Carnot and Möbius systematically developed the use of signed angles and line segments as a way of simplifying and unifying results.[33]

$\sqrt{\pi}$

r=1

Squaring the circle: the areas of this square and this circle are equal. In 1882, it was proven that this figure cannot be constructed in a finite number of steps with an idealized compass and straightedge.

The century's most significant development in geometry occurred when, around 1830, János Bolyai and Nikolai Ivanovich Lobachevsky separately published work on non-Euclidean geometry, in which the parallel postulate is not valid.[34] Since non-Euclidean geometry is provably relatively consistent with Euclidean geometry, the parallel postulate cannot be proved from the other postulates.

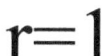

In the 19th century, it was also realized that Euclid's ten axioms and common notions do not suffice to prove all of theorems stated in the *Elements*. For example, Euclid assumed implicitly that any line contains at least two points, but this assumption cannot be proved from the other axioms, and therefore must be an axiom itself. The very first geometric proof in the *Elements,* shown in the figure above, is that any line segment is part of a triangle; Euclid constructs this in the usual way, by drawing circles around both endpoints and taking their intersection as the third vertex. His axioms, however, do not guarantee that the circles actually intersect, because they do not assert the geometrical property of continuity, which in Cartesian terms is equivalent to the completeness property of the real numbers. Starting with Moritz Pasch in 1882, many improved axiomatic systems for geometry have been proposed, the best known being those of Hilbert,[35] George Birkhoff,[36] and Tarski.[37]

26.8.5 20th century and general relativity

Einstein's theory of general relativity shows that the true geometry of spacetime is not Euclidean geometry.[38] For

A disproof of Euclidean geometry as a description of physical space. In a 1919 test of the general theory of relativity, stars (marked with short horizontal lines) were photographed during a solar eclipse. The rays of starlight were bent by the Sun's gravity on their way to the earth. This is interpreted as evidence in favor of Einstein's prediction that gravity would cause deviations from Euclidean geometry.

example, if a triangle is constructed out of three rays of light, then in general the interior angles do not add up to 180 degrees due to gravity. A relatively weak gravitational field, such as the Earth's or the sun's, is represented by a metric that is approximately, but not exactly, Euclidean. Until the 20th century, there was no technology capable of detecting the deviations from Euclidean geometry, but Einstein predicted that such deviations would exist. They were later verified by observations such as the slight bending of starlight by the Sun during a solar eclipse in 1919, and such considerations are now an integral part of the software that runs the GPS system.[39] It is possible to object to this

interpretation of general relativity on the grounds that light rays might be improper physical models of Euclid's lines, or that relativity could be rephrased so as to avoid the geometrical interpretations. However, one of the consequences of Einstein's theory is that there is no possible physical test that can distinguish between a beam of light as a model of a geometrical line and any other physical model. Thus, the only logical possibilities are to accept non-Euclidean geometry as physically real, or to reject the entire notion of physical tests of the axioms of geometry, which can then be imagined as a formal system without any intrinsic real-world meaning.

26.9 Treatment of infinity

26.9.1 Infinite objects

Euclid sometimes distinguished explicitly between "finite lines" (e.g., Postulate 2) and "infinite lines" (book I, proposition 12). However, he typically did not make such distinctions unless they were necessary. The postulates do not explicitly refer to infinite lines, although for example some commentators interpret postulate 3, existence of a circle with any radius, as implying that space is infinite.[27]

The notion of infinitesimal quantities had previously been discussed extensively by the Eleatic School, but nobody had been able to put them on a firm logical basis, with paradoxes such as Zeno's paradox occurring that had not been resolved to universal satisfaction. Euclid used the method of exhaustion rather than infinitesimals.[40]

Later ancient commentators such as Proclus (410–485 CE) treated many questions about infinity as issues demanding proof and, e.g., Proclus claimed to prove the infinite divisibility of a line, based on a proof by contradiction in which he considered the cases of even and odd numbers of points constituting it.[41]

At the turn of the 20th century, Otto Stolz, Paul du Bois-Reymond, Giuseppe Veronese, and others produced controversial work on non-Archimedean models of Euclidean geometry, in which the distance between two points may be infinite or infinitesimal, in the Newton–Leibniz sense.[42] Fifty years later, Abraham Robinson provided a rigorous logical foundation for Veronese's work.[43]

26.9.2 Infinite processes

One reason that the ancients treated the parallel postulate as less certain than the others is that verifying it physically would require us to inspect two lines to check that they never intersected, even at some very distant point, and this inspection could potentially take an infinite amount of time.[44]

The modern formulation of proof by induction was not developed until the 17th century, but some later commentators consider it implicit in some of Euclid's proofs, e.g., the proof of the infinitude of primes.[45]

Supposed paradoxes involving infinite series, such as Zeno's paradox, predated Euclid. Euclid avoided such discussions, giving, for example, the expression for the partial sums of the geometric series in IX.35 without commenting on the possibility of letting the number of terms become infinite.

26.10 Logical basis

See also: Hilbert's axioms, Axiomatic system and Real closed field

26.10.1 Classical logic

Euclid frequently used the method of proof by contradiction, and therefore the traditional presentation of Euclidean geometry assumes classical logic, in which every proposition is either true or false, i.e., for any proposition P, the proposition “P or not P” is automatically true.

26.10.2 Modern standards of rigor

Placing Euclidean geometry on a solid axiomatic basis was a preoccupation of mathematicians for centuries.[46] The role of primitive notions, or undefined concepts, was clearly put forward by Alessandro Padoa of the Peano delegation at the 1900 Paris conference:[46][47]

...when we begin to formulate the theory, we can imagine that the undefined symbols are *completely devoid of meaning* and that the unproved propositions are simply *conditions* imposed upon the undefined symbols.

Then, the *system of ideas* that we have initially chosen is simply *one interpretation* of the undefined symbols; but..this interpretation can be ignored by the reader, who is free to replace it in his mind by *another interpretation*.. that satisfies the conditions...

Logical questions thus become completely independent of *empirical* or *psychological* questions...

The system of undefined symbols can then be regarded as the *abstraction* obtained from the *specialized theories* that result when...the system of undefined symbols is successively replaced by each of the interpretations...

—Padoa, *Essai d'une théorie algébrique des nombre entiers, avec une Introduction logique à une théorie déductive qulelconque*

That is, mathematics is context-independent knowledge within a hierarchical framework. As said by Bertrand Russell:[48]

If our hypothesis is about *anything*, and not about some one or more particular things, then our deductions constitute mathematics. Thus, mathematics may be defined as the subject in which we never know what we are talking about, nor whether what we are saying is true.

—Bertrand Russell, *Mathematics and the metaphysicians*

Such foundational approaches range between foundationalism and formalism.

26.10.3 Axiomatic formulations

Geometry is the science of correct reasoning on incorrect figures.

—George Polyá, *How to Solve It*, p. 208

_ Euclid’s axioms: In his dissertation to Trinity College, Cambridge, Bertrand Russell summarized the changing role of Euclid’s geometry in the minds of philosophers up to that time.[49] It was a conflict between certain knowledge, independent of experiment, and empiricism, requiring experimental input. This issue became clear as it was discovered that the parallel postulate was not necessarily valid and its applicability was an empirical matter, deciding whether the applicable geometry was Euclidean or non-Euclidean.

_ Hilbert’s axioms: Hilbert’s axioms had the goal of identifying a *simple* and *complete* set of *independent* axioms from which the most important geometric theorems could be deduced. The outstanding objectives were to make Euclidean geometry rigorous (avoiding hidden assumptions) and to make clear the ramifications of the parallel postulate.

_ Birkhoff’s axioms: Birkhoff proposed four postulates for Euclidean geometry that can be confirmed experimentally with scale and protractor. This system relies heavily on the properties of the real numbers.[50][51][52]

The notions of *angle* and *distance* become primitive concepts.[53]

_ Tarski’s axioms: Alfred Tarski (1902–1983) and his students defined *elementary* Euclidean geometry as the geometry that can be expressed in first-order logic and does not depend on set theory for its logical basis,[54] in contrast to Hilbert’s axioms, which involve point sets.[55] Tarski proved that his axiomatic formulation of elementary Euclidean geometry is consistent and complete in a certain sense: there is an algorithm that, for every proposition, can be shown either true or false.[37] (This doesn't violate Gödel’s theorem, because Euclidean geometry cannot describe a sufficient amount of arithmetic for the theorem to apply.[56]) This is equivalent to the decidability of real closed fields, of which elementary Euclidean geometry is a model.

26.10.4 Constructive approaches and pedagogy

The process of abstract axiomatization as exemplified by Hilbert’s axioms reduces geometry to theorem proving or predicate logic. In contrast, the Greeks used construction postulates, and emphasized problem solving.[57] For the Greeks, constructions are more primitive than existence propositions, and can be used to prove existence propositions, but not *vice versa*. To describe problem solving adequately requires a richer system of logical concepts.[57] The contrast in approach may be summarized:[58]

_ Axiomatic proof: Proofs are deductive derivations of propositions from primitive premises that are ‘true’ in some sense. The aim is to justify the proposition.

_ Analytic proof: Proofs are non-deductive derivations of hypotheses from problems. The aim is to find hypotheses

capable of giving a solution to the problem. One can argue that Euclid's axioms were arrived upon in this manner. In particular, it is thought that Euclid felt the parallel postulate was forced upon him, as indicated by his reluctance to make use of it,[59] and his arrival upon it by the method of contradiction.[60]
Andrei Nicholaevich Kolmogorov proposed a problem solving basis for geometry.[61][62] This work was a precursor of a modern formulation in terms of constructive type theory.[63] This development has implications for pedagogy as well.[64]

If proof simply follows conviction of truth rather than contributing to its construction and is only experienced as a demonstration of something already known to be true, it is likely to remain meaningless and purposeless in the eyes of students.
—Celia Hoyles, *The curricular shaping of students' approach to proof*

26.11 See also

- Analytic geometry
- Birkhoff's axioms
- Cartesian coordinate system
- Hilbert's axioms
- Incidence geometry
- List of interactive geometry software
- Metric space
- Non-Euclidean geometry
- Ordered geometry
- Parallel postulate
- Type theory

26.11.1 Classical theorems

- Angle bisector theorem
- Butterfly theorem
- Ceva's theorem
- Heron's formula
- Menelaus' theorem
- Nine-point circle
- Pythagorean theorem

26.12 Notes

[1] Eves, vol. 1., p. 19
[2] Eves (1963), vol. 1, p. 10
[3] Eves, p. 19
[4] Misner, Thorne, and Wheeler (1973), p. 47
[5] Euclid, book I, proposition 47
[6] The assumptions of Euclid are discussed from a modern perspective in Harold E. Wolfe (2007). *Introduction to Non-Euclidean Geometry*. Mill Press. p. 9. ISBN 1-4067-1852-1.
[7] tr. Heath, pp. 195–202.
[8] Florence P. Lewis (Jan 1920), *History of the Parallel Postulate, The American Mathematical Monthly* (The American Mathematical Monthly, Vol. 27, No. 1) **27** (1): 16–23, doi:10.2307/2973238, JSTOR 2973238.
[9] Ball, p. 56
[10] Within Euclid's assumptions, it is quite easy to give a formula for area of triangles and squares. However, in a more general context like set theory, it is not as easy to prove that the area of a square is the sum of areas of its pieces, for example. See Lebesgue measure and Banach–Tarski paradox.
[11] Daniel Shanks (2002). *Solved and Unsolved Problems in Number Theory*. American Mathematical Society.
[12] Coxeter, p. 5
[13] Euclid, book I, proposition 5, tr. Heath, p. 251
[14] Ignoring the alleged difficulty of Book I, Proposition 5, Sir Thomas L. Heath mentions another interpretation. This rests on the resemblance of the figure's lower straight lines to a steeply inclined bridge that could be crossed by an ass but not by a horse: "But there is another view (as I have learnt lately) which is more complimentary to the ass. It is that, the figure of the proposition being like that of a trestle bridge, with a ramp at each end which is more practicable the flatter the figure is drawn, the bridge is such that, while a horse could not surmount the ramp, an ass could; in other words, the term is meant to refer to the sure-footedness of the ass rather than to any want of intelligence on his part." (in "Excursis II," volume 1 of Heath's translation of *The Thirteen Books of the Elements*.)
[15] Euclid, book I, proposition 32
[16] Heath, p. 135, Extract of page 135
[17] Heath, p. 318

[18] Euclid, book XII, proposition 2
[19] Euclid, book XI, proposition 33
[20] Ball, p. 66
[21] Ball, p. 5
[22] Eves, vol. 1, p. 5; Mlodinow, p. 7
[23] Tom Hull. "Origami and Geometric Constructions".
[24] Richard J. Trudeau (2008). "Euclid's axioms". *The Non-Euclidean Revolution*. Birkhäuser. pp. 39 *'ff.* ISBN 0-8176-4782-1.
[25] See, for example: Luciano da Fontoura Costa, Roberto Marcondes Cesar (2001). *Shape analysis and classification: theory and practice*. CRC Press. p. 314. ISBN 0-8493-3493-4. and Helmut Pottmann, Johannes Wallner (2010). *Computational Line Geometry*. Springer. p. 60. ISBN 3-642-04017-9. The *group of motions* underlie the metric notions of geometry. See Felix Klein (2004). *Elementary Mathematics from an Advanced Standpoint: Geometry* (Reprint of 1939 Macmillan Company ed.). Courier Dover. p. 167. ISBN 0-486-43481-8.
[26] Roger Penrose (2007). *The Road to Reality: A Complete Guide to the Laws of the Universe*. Vintage Books. p. 29. ISBN 0-679-77631-1.
[27] Heath, p. 200

[28] e.g., Tarski (1951)
[29] Eves, p. 27
[30] Ball, pp. 268ff
[31] Eves (1963)
[32] Hofstadter 1979, p. 91.
[33] Eves (1963), p. 64
[34] Ball, p. 485
[35] _ Howard Eves, 1997 (1958). *Foundations and Fundamental Concepts of Mathematics*. Dover.
[36] Birkhoff, G. D., 1932, "A Set of Postulates for Plane Geometry (Based on Scale and Protractors)," Annals of Mathematics 33.
[37] Tarski (1951)
[38] Misner, Thorne, and Wheeler (1973), p. 191
[39] Rizos, Chris. University of New South Wales. GPS Satellite Signals. 1999.
[40] Ball, p. 31
[41] Heath, p. 268
[42] Giuseppe Veronese, On Non-Archimedean Geometry, 1908. English translation in Real Numbers, Generalizations of the Reals, and Theories of Continua, ed. Philip Ehrlich, Kluwer, 1994.
[43] Robinson, Abraham (1966). Non-standard analysis.
[44] For the assertion that this was the historical reason for the ancients considering the parallel postulate less obvious than the others, see Nagel and Newman 1958, p. 9.
[45] Cajori (1918), p. 197
[46] A detailed discussion can be found in James T. Smith (2000). "Chapter 2: Foundations". *Methods of geometry*. Wiley. pp. 19 *ff.* ISBN 0-471-25183-6.
[47] Société française de philosophie (1900). *Revue de métaphysique et de morale, Volume 8*. Hachette. p. 592.
[48] Bertrand Russell (2000). "Mathematics and the metaphysicians". In James Roy Newman. *The world of mathematics* **3** (Reprint of Simon and Schuster 1956 ed.). Courier Dover Publications. p. 1577. ISBN 0-486-41151-6.
[49] Bertrand Russell (1897). "Introduction". *An essay on the foundations of geometry*. Cambridge University Press.
[50] George David Birkhoff, Ralph Beatley (1999). "Chapter 2: The five fundamental principles". *Basic Geometry* (3rd ed.). AMS Bookstore. pp. 38 *ff.* ISBN 0-8218-2101-6.
[51] James T. Smith. "Chapter 3: Elementary Euclidean Geometry". *Cited work*. pp. 84 *ff.*
[52] Edwin E. Moise (1990). *Elementary geometry from an advanced standpoint* (3rd ed.). Addison–Wesley. ISBN 0-201-50867-2.
[53] John R. Silvester (2001). "§1.4 Hilbert and Birkhoff". *Geometry: ancient and modern*. Oxford University Press. ISBN 0-19-850825-5.
[54] Alfred Tarski (2007). "What is elementary geometry". In Leon Henkin, Patrick Suppes & Alfred Tarski. *Studies in Logic and the Foundations of Mathematics – The Axiomatic Method with Special Reference to Geometry and Physics* (Proceedings of International Symposium at Berkeley 1957–8; Reprint ed.). Brouwer Press. p. 16. ISBN 1-4067-5355-6. "We regard as elementary that part of Euclidean geometry which can be formulated and established without the help of any set-theoretical devices"
[55] Keith Simmons (2009). "Tarski's logic". In Dov M. Gabbay, John Woods. *Logic from Russell to Church*. Elsevier. p. 574. ISBN 0-444-51620-4.
[56] Franzén, Torkel (2005). Gödel's Theorem: An Incomplete Guide to its Use and Abuse. AK Peters. ISBN 1-56881-238-8. Pp. 25–26.

[57] Petri Mäenpää (1999). "From backward reduction to configurational analysis". In Michael Otte, Marco Panza. *Analysis and synthesis in mathematics: history and philosophy*. Springer. p. 210. ISBN 0-7923-4570-3.
[58] Carlo Cellucci (2008). "Why proof? What is proof?". In Rossella Lupacchini, Giovanna Corsi. *Deduction, Computation, Experiment: Exploring the Effectiveness of Proof.* Springer. p. 1. ISBN 88-470-0783-6.
[59] Eric W. Weisstein (2003). "Euclid's postulates". *CRC concise encyclopedia of mathematics* (2nd ed.). CRC Press. p. 942. ISBN 1-58488-347-2.

[60] Deborah J. Bennett (2004). *Logic made easy: how to know when language deceives you.* W. W. Norton & Company. p. 34. ISBN 0-393-05748-8.
[61] AN Kolmogorov, AF Semenovich, RS Cherkasov (1982). *Geometry: A textbook for grades 6–8 of secondary school* [Geometriya. Uchebnoe posobie dlya 6–8 klassov srednie shkoly] (3rd ed.). Moscow: "Prosveshchenie" Publishers. pp. 372–376. A description of the approach, which was based upon geometric transformations, can be found in *Teaching geometry in the USSR* Chernysheva, Firsov, and Teljakovskii
[62] Viktor Vasil'evich Prasolov, Vladimir Mikhaĭlovich Tikhomirov (2001). *Geometry.* AMS Bookstore. p. 198. ISBN 0-8218-2038-9.
[63] Petri Mäenpää (1998). "Analytic program derivation in type theory". In Giovanni Sambin, Jan M. Smith. *Twenty-five years of constructive type theory: proceedings of a congress held in Venice, October 1995.* Oxford University Press. p. 113. ISBN 0-19-850127-7.
[64] Celia Hoyles (Feb 1997). "The curricular shaping of students' approach to proof". *For the Learning of Mathematics* (FLM Publishing Association) **17** (1): 7–16. JSTOR 40248217.

26.13 References

_ Ball, W.W. Rouse (1960). *A Short Account of the History of Mathematics* (4th ed. [Reprint. Original publication: London: Macmillan & Co., 1908] ed.). New York: Dover Publications. pp. 50–62. ISBN 0-486-20630-0.
_ Coxeter, H.S.M. (1961). *Introduction to Geometry.* New York: Wiley.
_ Eves, Howard (1963). *A Survey of Geometry.* Allyn and Bacon.
_ Heath, Thomas L. (1956). *The Thirteen Books of Euclid's Elements* (2nd ed. [Facsimile. Original publication: Cambridge University Press, 1925] ed.). New York: Dover Publications.
(3 vols.): ISBN 0-486-60088-2 (vol. 1), ISBN 0-486-60089-0 (vol. 2), ISBN 0-486-60090-4 (vol. 3).
Heath's authoritative translation of Euclid's Elements plus his extensive historical research and detailed commentary throughout the text.
_ Misner, Thorne, and Wheeler (1973). *Gravitation.* W.H. Freeman.
_ Mlodinow (2001). *Euclid's Window.* The Free Press.
_ Nagel, E. and Newman, J.R. (1958). *Gödel's Proof.* New York University Press.
_ Alfred Tarski (1951) *A Decision Method for Elementary Algebra and Geometry.* Univ. of California Press.

26.14 External links

_ Hazewinkel, Michiel, ed. (2001), "Euclidean geometry", *Encyclopedia of Mathematics*, Springer, ISBN 978-1-55608-010-4
_ Hazewinkel, Michiel, ed. (2001), "Plane trigonometry", *Encyclopedia of Mathematics*, Springer, ISBN 978-1-55608-010-4
_ Kiran Kedlaya, *Geometry Unbound* (a treatment using analytic geometry; PDF format, GFDL licensed)

Chapter 27

Parallel postulate

α

β

If the sum of the interior angles α and β is less than 180°, the two straight lines, produced indefinitely, meet on that side.

In geometry, the **parallel postulate**, also called **Euclid's fifth postulate** because it is the fifth postulate in Euclid's *Elements*, is a distinctive axiom in Euclidean geometry. It states that, in two-dimensional geometry:

If a line segment intersects two straight lines forming two interior angles on the same side that sum to less than two right angles, then the two lines, if extended indefinitely, meet on that side on which the angles sum to less than two right angles.

Euclidean geometry is the study of geometry that satisfies all of Euclid's axioms, *including* the parallel postulate. A geometry where the parallel postulate does not hold is known as a non-Euclidean geometry. Geometry that is independent of Euclid's fifth postulate (i.e., only assumes the modern equivalent of the first four postulates) is known as absolute geometry (or, in other places known as neutral geometry).

186

27.1. EQUIVALENT PROPERTIES 187

27.1 Equivalent properties

Probably the best known equivalent of Euclid's parallel postulate is Playfair's axiom, named after the Scottish mathematician John Playfair, which states:

In a plane, given a line and a point not on it, at most one line parallel to the given line can be drawn through the point.[1]

This axiom is not logically equivalent to the Euclidean parallel postulate since there are geometries in which one is true and the other is not. However, in the presence of the remaining axioms which give Euclidean geometry, each of these can be used to prove the other, so they are equivalent in the context of absolute geometry.[2]

Many other statements equivalent to the parallel postulate have been suggested, some of them appearing at first to be unrelated to parallelism, and some seeming so self-evident that they were unconsciously assumed by people who claimed to have proven the parallel postulate from Euclid's other postulates. This is a summary

1. There is at most one line that can be drawn parallel to another given one through an external point. (Playfair's axiom)
2. The sum of the angles in every triangle is 180° (triangle postulate).
3. There exists a triangle whose angles add up to 180°.
4. The sum of the angles is the same for every triangle.
5. There exists a pair of similar, but not congruent, triangles.
6. Every triangle can be circumscribed.
7. If three angles of a quadrilateral are right angles, then the fourth angle is also a right angle.
8. There exists a quadrilateral in which all angles are right angles.
9. There exists a pair of straight lines that are at constant distance from each other.
10. Two lines that are parallel to the same line are also parallel to each other.
11. In a right-angled triangle, the square of the hypotenuse equals the sum of the squares of the other two sides (Pythagoras' Theorem).[3][4]
12. There is no upper limit to the area of a triangle. (Wallis axiom)[5]
13. The summit angles of the Saccheri quadrilateral are 90°.
14. If a line intersects one of two parallel lines, both of which are coplanar with the original line, then it also intersects the other. (Proclus' axiom)[6]

However, the alternatives which employ the word "parallel" cease appearing so simple when one is obliged to explain which of the four common definitions of "parallel" is meant – constant separation, never meeting, same angles where crossed by *some* third line, or same angles where crossed by *any* third line – since the equivalence of these four is itself one of the unconsciously obvious assumptions equivalent to Euclid's fifth postulate. In the list above, it is always taken to refer to non-intersecting lines. For example, if the word "parallel" in Playfair's axiom is taken to mean 'constant separation' or 'same angles where crossed by any third line', then it is no longer equivalent to Euclid's fifth postulate, and is provable from the first four (the axiom says 'There is at most one line...', which is consistent with there being no such lines). However, if the definition is taken so that parallel lines are lines that do not intersect, or that have some line intersecting them in the same angles, Playfair's axiom is contextually equivalent to Euclid's fifth postulate and is thus logically independent of the first four postulates. Note that the latter two definitions are not equivalent, because in hyperbolic geometry the second definition holds only for ultraparallel lines.

27.2 History

For two thousand years, many attempts were made to prove the parallel postulate using Euclid's first four postulates. The main reason that such a proof was so highly sought after was that, unlike the first four postulates, the parallel postulate isn't self-evident. If the order the postulates were listed in the Elements is significant, it indicates that Euclid included this postulate only when he realised he could not prove it or proceed without it.[7] Many attempts were made to prove the fifth postulate from the other four, many of them being accepted as proofs for long periods of time until the mistake was found. Invariably the mistake was assuming some 'obvious' property which turned out to be equivalent to the fifth postulate (Playfair's axiom). Although known from the time of Proclus, this became known as Playfair's Axiom after John Playfair wrote a famous commentary on Euclid in 1795 in which he proposed replacing Euclid's fifth postulate by his own axiom.

Proclus (410-485) wrote a commentary on *The Elements* where he comments on attempted proofs to deduce the fifth postulate from the other four, in particular he notes that Ptolemy had produced a false 'proof'. Proclus then goes on to give a false proof of his own. However he did give a postulate which is equivalent to the fifth postulate.

Ibn al-Haytham (Alhazen) (965-1039), an Arab mathematician, made an attempt at proving the parallel postulate using a proof by contradiction,[8] in the course of which he introduced the concept of motion and transformation into geometry.[9] He formulated the Lambert quadrilateral, which Boris Abramovich Rozenfeld names the "Ibn al-Haytham–Lambert quadrilateral",[10] and his attempted proof contains elements similar to those found in Lambert quadrilaterals and Playfair's axiom.[11]

Omar Khayyám (1050–1123), a Persian, attempted to prove the fifth postulate from another explicitly given postulate (based on the fourth of the five *principles due to the Philosopher* (Aristotle), namely, "Two convergent straight lines intersect and it is impossible for two convergent straight lines to diverge in the direction in which they converge."[12] He derived some of the earlier results belonging to elliptical geometry and hyperbolic geometry, though his postulate excluded the latter possibility.[13] The Saccheri quadrilateral was also first considered by Omar Khayyám in the late 11th century in Book I of *Explanations of the Difficulties in the Postulates of Euclid*.[10] Unlike many commentators on Euclid before and after him (including Giovanni Girolamo Saccheri), Khayyám was not trying to prove the parallel postulate as such but to derive it from his equivalent postulate. He recognized that three possibilities arose from omitting Euclid's fifth postulate; if two perpendiculars to one line cross another line, judicious choice of the last can make the internal angles where it meets the two perpendiculars equal (it is then parallel to the first line). If those equal internal angles are right angles, we get Euclid's fifth postulate, otherwise, they must be either acute or obtuse. He showed that the acute and obtuse cases led to contradictions using his postulate, but his postulate is now known to be equivalent to the fifth postulate.

Nasir al-Din al-Tusi (1201–1274), in his *Al-risala al-shafiya'an al-shakk fi'l-khutut al-mutawaziya* (*Discussion Which Removes Doubt about Parallel Lines*) (1250), wrote detailed critiques of the parallel postulate and on Khayyám's attempted proof a century earlier. Nasir al-Din attempted to derive a proof by contradiction of the parallel postulate.[14]

He also considered the cases of what are now known as elliptical and hyperbolic geometry, though he ruled out both of them.[13]

Euclidean, elliptical and hyperbolic geometry. The Parallel Postulate is satisfied only for models of Euclidean geometry.

Nasir al-Din's son, Sadr al-Din (sometimes known as "Pseudo-Tusi"), wrote a book on the subject in 1298, based on his father's later thoughts, which presented one of the earliest arguments for a non-Euclidean hypothesis equivalent to

the parallel postulate. "He essentially revised both the Euclidean system of axioms and postulates and the proofs of many propositions from the *Elements*."[14][15] His work was published in Rome in 1594 and was studied by European geometers. This work marked the starting point for Saccheri's work on the subject[14] which opened with a criticism of Sadr al-Din's work and the work of Wallis.[16]

Giordano Vitale (1633-1711), in his book *Euclide restituo* (1680, 1686), used the Khayyam-Saccheri quadrilateral to prove that if three points are equidistant on the base AB and the summit CD, then AB and CD are everywhere equidistant. Girolamo Saccheri (1667-1733) pursued the same line of reasoning more thoroughly, correctly obtaining absurdity from the obtuse case (proceeding, like Euclid, from the implicit assumption that lines can be extended indefinitely and have infinite length), but failing to refute the acute case (although he managed to wrongly persuade himself that he had).

In 1766 Johann Lambert wrote, but did not publish, *Theorie der Parallellinien* in which he attempted, as Saccheri did, to prove the fifth postulate. He worked with a figure that today we call a *Lambert quadrilateral*, a quadrilateral with three right angles (can be considered half of a Saccheri quadrilateral). He quickly eliminated the possibility that the fourth angle is obtuse, as had Saccheri and Khayyám, and then proceeded to prove many theorems under the assumption of an acute angle. Unlike Saccheri, he never felt that he had reached a contradiction with this assumption. He had proved the non-Euclidean result that the sum of the angles in a triangle increases as the area of the triangle decreases, and this led him to speculate on the possibility of a model of the acute case on a sphere of imaginary radius. He did not carry this idea any further.[17]

Where Khayyám and Saccheri had attempted to prove Euclid's fifth by disproving the only possible alternatives, the nineteenth century finally saw mathematicians exploring those alternatives and discovering the logically consistent geometries which result. In 1829, Nikolai Ivanovich Lobachevsky published an account of acute geometry in an obscure Russian journal (later re-published in 1840 in German). In 1831, János Bolyai included, in a book by his father, an appendix describing acute geometry, which, doubtlessly, he had developed independently of Lobachevsky. Carl Friedrich Gauss had also studied the problem, but he did not publish any of his results. Upon hearing of Bolyai's results in a letter from Bolyai's father, Farkas Bolyai, Gauss stated:

"If I commenced by saying that I am unable to praise this work, you would certainly be surprised for a moment. But I cannot say otherwise. To praise it would be to praise myself. Indeed the whole contents of the work, the path taken by your son, the results to which he is led, coincide almost entirely with my meditations, which have occupied my mind partly for the last thirty or thirty-five years."[18]

The resulting geometries were later developed by Lobachevsky, Riemann and Poincaré into hyperbolic geometry (the acute case) and elliptic geometry (the obtuse case). The independence of the parallel postulate from Euclid's other axioms was finally demonstrated by Eugenio Beltrami in 1868.

27.3 Converse of Euclid's parallel postulate

Euclid did not postulate the converse of his fifth postulate, which is one way to distinguish Euclidean geometry from elliptic geometry. The Elements contains the proof of an equivalent statement (Book I, Proposition 27): *If a straight line falling on two straight lines make the alternate angles equal to one another, the straight lines will be parallel to one another.* As De Morgan[19] pointed out, this is logically equivalent to (Book I, Proposition 16). These results do not depend upon the fifth postulate, but they do require the second postulate[20] which is violated in elliptic geometry.

27.4 Criticism

Attempts to logically prove the parallel postulate, rather than the eighth axiom,[21] were criticized by Arthur Schopenhauer. However, the argument used by Schopenhauer was that the postulate is evident by perception, not that it was not a logical consequence of the other axioms.

27.5 See also

_ For more information, see the history of non-Euclidean geometry.

If:

Then:

a + b = 180°

line 1 and line 2 are parallel

a b

line 1 line 2

If the sum of the two interior angles equals 180°, the lines are parallel and will never intersect.

27.6 Notes

[1] Euclid's Parallel Postulate and Playfair's Axiom
[2] Henderson & Taimiņa 2005, pg. 139
[3] Eric W. Weisstein (2003), *CRC concise encyclopedia of mathematics* (2nd ed.), p. 2147, ISBN 1-58488-347-2, "The parallel postulate is equivalent to the *Equidistance postulate*, *Playfair axiom*, *Proclus axiom*, the *Triangle postulate* and the *Pythagorean theorem*."
[4] Alexander R. Pruss (2006), *The principle of sufficient reason: a reassessment*, Cambridge University Press, p. 11, ISBN 0-521-85959-X, "We could include...the parallel postulate and derive the Pythagorean theorem. Or we could instead make the Pythagorean theorem among the other axioms and derive the parallel postulate."
[5] Bogomolny, Alexander. "Euclid's Fifth Postulate". *Cut The Knot*. Retrieved 30 September 2011.
[6] Weisstein, Eric W. "Proclus' Axiom – MathWorld". Retrieved 2009-09-05.
[7] Florence P. Lewis (Jan 1920), *History of the Parallel Postulate*, *The American Mathematical Monthly* (The American Mathematical Monthly, Vol. 27, No. 1) **27** (1): 16–23, doi:10.2307/2973238, JSTOR 2973238.
[8] Katz 1998, pg. 269
[9] Katz 1998, p. 269:
In effect, this method characterized parallel lines as lines always equidistant from one another and also introduced the concept of motion into geometry.
[10] Rozenfeld 1988, p. 65
[11] Smith 1992
[12] Boris A Rosenfeld and Adolf P Youschkevitch (1996), *Geometry*, p.467 in Roshdi Rashed, Régis Morelon (1996), *Encyclopedia of the history of Arabic science*, Routledge, ISBN 0-415-12411-5.
[13] Boris A. Rosenfeld and Adolf P. Youschkevitch (1996), "Geometry", in Roshdi Rashed, ed., *Encyclopedia of the History of Arabic Science*, Vol. 2, p. 447-494 [469], Routledge, London and New York:

"Khayyam's postulate had excluded the case of the hyperbolic geometry whereas al-Tusi's postulate ruled out both the hyperbolic and elliptic geometries."
[14] Katz 1998, pg.271:
"But in a manuscript probably written by his son Sadr al-Din in 1298, based on Nasir al-Din's later thoughts on the subject, there is a new argument based on another hypothesis, also equivalent to Euclid's, [...] The importance of this latter work is that it was published in Rome in 1594 and was studied by European geometers. In particular, it became the starting point for the work of Saccheri and ultimately for the discovery of non-Euclidean geometry."
[15] Boris A. Rosenfeld and Adolf P. Youschkevitch (1996), "Geometry", in Roshdi Rashed, ed., *Encyclopedia of the History of Arabic Science*, Vol. 2, p. 447-494 [469], Routledge, London and New York:
"In *Pseudo-Tusi's Exposition of Euclid*, [...] another statement is used instead of a postulate. It was independent of the Euclidean postulate V and easy to prove. [...] He essentially revised both the Euclidean system of axioms and postulates and the proofs of many propositions from the *Elements*."
[16] MacTutor's Giovanni Girolamo Saccheri
[17] O'Connor, J.J.; Robertson, E.F. "Johann Heinrich Lambert". Retrieved 16 September 2011.
[18] Faber 1983, pg. 161
[19] Heath, T.L., *The thirteen books of Euclid's Elements*, Vol.1, Dover, 1956, pg.309.

[20] Coxeter, H.S.M., *Non-Euclidean Geometry*, 6th Ed., MAA 1998, pg.3
[21] Schopenhauer is referring to Euclid's Common Notion 4: Figures coinciding with one another are equal to one another.

27.7 References

_ Carroll, Lewis, *Euclid and His Modern Rivals*, Dover, ISBN 0-486-22968-8
_ Faber, Richard L. (1983), *Foundations of Euclidean and Non-Euclidean Geometry*, New York: Marcel Dekker Inc., ISBN 0-8247-1748-1
_ Henderson, David W.; Taimiņa, Daina (2005), *Experiencing Geometry: Euclidean and Non-Euclidean with History* (3rd ed.), Upper Saddle River, NJ: Pearson Prentice Hall, ISBN 0-13-143748-8
_ Katz, Victor J. (1998), *History of Mathematics: An Introduction*, Addison-Wesley, ISBN 0-321-01618-1, OCLC 38199387 60154481
_ Rozenfeld, Boris A. (1988), *A History of Non-Euclidean Geometry: Evolution of the Concept of a Geometric Space*, Springer Science+Business Media, ISBN 0-387-96458-4, OCLC 15550634 230166667 230980046 77693662
_ Smith, John D. (1992), *The Remarkable Ibn al-Haytham, The Mathematical Gazette* (Mathematical Association) **76** (475): 189–198, doi:10.2307/3620392, JSTOR 3620392

27.8 External links

_ On Gauss' Mountains
Eder, Michelle (2000), *Views of Euclid's Parallel Postulate in Ancient Greece and in Medieval Islam*, Rutgers University, retrieved 2008-01-23

Chapter 28

Set theory

This article is about the branch of mathematics. For musical set theory, see Set theory (music).
Set theory is the branch of mathematical logic that studies sets, which are collections of objects. Although any type

A *A*∩*B* *B*

A Venn diagram illustrating the intersection of two sets.

of object can be collected into a set, set theory is applied most often to objects that are relevant to mathematics. The language of set theory can be used in the definitions of nearly all mathematical objects.

The modern study of set theory was initiated by Georg Cantor and Richard Dedekind in the 1870s. After the discovery of paradoxes in naive set theory, numerous axiom systems were proposed in the early twentieth century, of which the Zermelo–Fraenkel axioms, with the axiom of choice, are the best-known.

Set theory is commonly employed as a foundational system for mathematics, particularly in the form of Zermelo–Fraenkel set theory with the axiom of choice. Beyond its foundational role, set theory is a branch of mathematics in its own right, with an active research community. Contemporary research into set theory includes a diverse collection of topics, ranging from the structure of the real number line to the study of the consistency of large cardinals.
192
28.1. HISTORY 193

28.1 History

Georg Cantor

Mathematical topics typically emerge and evolve through interactions among many researchers. Set theory, how194
CHAPTER 28. SET THEORY
ever, was founded by a single paper in 1874 by Georg Cantor: "On a Characteristic Property of All Real Algebraic Numbers".[1][2]

Since the 5th century BC, beginning with Greek mathematician Zeno of Elea in the West and early Indian mathematicians in the East, mathematicians had struggled with the concept of infinity. Especially notable is the work of Bernard Bolzano in the first half of the 19th century.[3] Modern understanding of infinity began in 1867–71, with Cantor's work on number theory. An 1872 meeting between Cantor and Richard Dedekind influenced Cantor's thinking and culminated in Cantor's 1874 paper.

Cantor's work initially polarized the mathematicians of his day. While Karl Weierstrass and Dedekind supported Cantor, Leopold Kronecker, now seen as a founder of mathematical constructivism, did not. Cantorian set theory eventually became widespread, due to the utility of Cantorian concepts, such as one-to-one correspondence among sets, his proof that there are more real numbers than integers, and the "infinity of infinities" ("Cantor's paradise") resulting from the power set operation. This utility of set theory led to the article "Mengenlehre" contributed in 1898 by Arthur Schoenflies to Klein's encyclopedia.

The next wave of excitement in set theory came around 1900, when it was discovered that Cantorian set theory gave rise to several contradictions, called antinomies or paradoxes. Bertrand Russell and Ernst Zermelo independently found the simplest and best known paradox, now called Russell's paradox: consider "the set of all sets that are not members of themselves", which leads to a contradiction since it must be a member of itself, and not a member of itself. In 1899 Cantor had himself posed the question "What is the cardinal number of the set of all sets?", and obtained a related paradox. Russell used his paradox as a theme in his 1903 review of continental mathematics in his *The Principles of Mathematics*.

The momentum of set theory was such that debate on the paradoxes did not lead to its abandonment. The work of Zermelo in 1908 and Abraham Fraenkel in 1922 resulted in the set of axioms ZFC, which became the most commonly used set of axioms for set theory. The work of analysts such as Henri Lebesgue demonstrated the great mathematical utility of set theory, which has since become woven into the fabric of modern mathematics. Set theory is commonly used as a foundational system, although in some areas category theory is thought to be a preferred foundation.

28.2 Basic concepts and notation

Main articles: Set (mathematics) and Algebra of sets

Set theory begins with a fundamental binary relation between an object o and a set A. If o is a **member** (or **element**) of A, write $o \in A$. Since sets are objects, the membership relation can relate sets as well.

A derived binary relation between two sets is the subset relation, also called **set inclusion**. If all the members of set A are also members of set B, then A is a **subset** of B, denoted $A \subseteq B$. For example, $\{1,2\}$ is a subset of $\{1,2,3\}$, but $\{1,4\}$ is not. From this definition, it is clear that a set is a subset of itself; for cases where one wishes to rule this out, the term **proper subset** is defined. A is called a **proper subset** of B if and only if A is a subset of B, but B is **not** a subset of A.

Just as arithmetic features binary operations on numbers, set theory features binary operations on sets. The:

_ **Union** of the sets A and B, denoted $A \cup B$, is the set of all objects that are a member of A, or B, or both. The union of $\{1, 2, 3\}$ and $\{2, 3, 4\}$ is the set $\{1, 2, 3, 4\}$.

_ **Intersection** of the sets A and B, denoted $A \cap B$, is the set of all objects that are members of both A and B. The intersection of $\{1, 2, 3\}$ and $\{2, 3, 4\}$ is the set $\{2, 3\}$.

_ **Set difference** of U and A, denoted $U \setminus A$, is the set of all members of U that are not members of A. The set difference $\{1,2,3\} \setminus \{2,3,4\}$ is $\{1\}$, while, conversely, the set difference $\{2,3,4\} \setminus \{1,2,3\}$ is $\{4\}$. When A is a subset of U, the set difference $U \setminus A$ is also called the **complement** of A in U. In this case, if the choice of U is clear from the context, the notation A_c is sometimes used instead of $U \setminus A$, particularly if U is a universal set as in the study of Venn diagrams.

_ **Symmetric difference** of sets A and B, denoted $A \triangle B$ or $A \ominus B$, is the set of all objects that are a member of exactly one of A and B (elements which are in one of the sets, but not in both). For instance, for the sets $\{1,2,3\}$ and $\{2,3,4\}$, the symmetric difference set is $\{1,4\}$. It is the set difference of the union and the intersection, $(A \cup B) \setminus (A \cap B)$ or $(A \setminus B) \cup (B \setminus A)$.

28.3. SOME ONTOLOGY 195

_ **Cartesian product** of A and B, denoted $A \times B$, is the set whose members are all possible ordered pairs (a,b) where a is a member of A and b is a member of B. The cartesian product of $\{1, 2\}$ and $\{red, white\}$ is $\{(1, red), (1, white), (2, red), (2, white)\}$.

_ **Power set** of a set A is the set whose members are all possible subsets of A. For example, the power set of $\{1, 2\}$ is $\{ \{\}, \{1\}, \{2\}, \{1,2\} \}$.

Some basic sets of central importance are the empty set (the unique set containing no elements), the set of natural numbers, and the set of real numbers.

28.3 Some ontology

Main article: von Neumann universe

A set is pure if all of its members are sets, all members of its members are sets, and so on. For example, the set

V_0

V_1

V_2

V_3
V_4
V_5
V_ω
$V_{\omega+1}$
$V_{\omega+2}$
$V_{2\omega}$
$V_{3\omega}$
$V_{\omega*\omega}$

... ...

An initial segment of the von Neumann hierarchy.

{{}} containing only the empty set is a nonempty pure set. In modern set theory, it is common to restrict attention to the **von Neumann universe** of pure sets, and many systems of axiomatic set theory are designed to axiomatize the

pure sets only. There are many technical advantages to this restriction, and little generality is lost, because essentially all mathematical concepts can be modeled by pure sets. Sets in the von Neumann universe are organized into a cumulative hierarchy, based on how deeply their members, members of members, etc. are nested. Each set in this hierarchy is assigned (by transfinite recursion) an ordinal number α, known as its **rank**. The rank of a pure set X is defined to be the least upper bound of all successors of ranks of members of X. For example, the empty set is assigned rank 0, while the set {{}} containing only the empty set is assigned rank 1. For each ordinal α, the set $V\alpha$ is defined to consist of all pure sets with rank less than α. The entire von Neumann universe is denoted V.

28.4 Axiomatic set theory

Elementary set theory can be studied informally and intuitively, and so can be taught in primary schools using Venn diagrams. The intuitive approach tacitly assumes that a set may be formed from the class of all objects satisfying any particular defining condition. This assumption gives rise to paradoxes, the simplest and best known of which are Russell's paradox and the Burali-Forti paradox. Axiomatic set theory was originally devised to rid set theory of such paradoxes.[4]

The most widely studied systems of axiomatic set theory imply that all sets form a cumulative hierarchy. Such systems come in two flavors, those whose ontology consists of:

_ *Sets alone*. This includes the most common axiomatic set theory, **Zermelo–Fraenkel set theory (ZFC)**, which includes the axiom of choice. Fragments of ZFC include:

_ Zermelo set theory, which replaces the axiom schema of replacement with that of separation;

_ General set theory, a small fragment of Zermelo set theory sufficient for the Peano axioms and finite sets;

_ Kripke–Platek set theory, which omits the axioms of infinity, powerset, and choice, and weakens the axiom schemata of separation and replacement.

_ *Sets and proper classes*. These include Von Neumann–Bernays–Gödel set theory, which has the same strength as ZFC for theorems about sets alone, and Morse-Kelley set theory and Tarski–Grothendieck set theory, both of which are stronger than ZFC.

The above systems can be modified to allow **urelements**, objects that can be members of sets but that are not themselves sets and do not have any members.

The systems of **New Foundations NFU** (allowing urelements) and **NF** (lacking them) are not based on a cumulative hierarchy. NF and NFU include a "set of everything," relative to which every set has a complement. In these systems urelements matter, because NF, but not NFU, produces sets for which the axiom of choice does not hold.

Systems of constructive set theory, such as CST, CZF, and IZF, embed their set axioms in intuitionistic logic instead of first order logic. Yet other systems accept standard first order logic but feature a nonstandard membership relation. These include rough set theory and fuzzy set theory, in which the value of an atomic formula embodying the membership relation is not simply **True** or **False**. The Boolean-valued models of ZFC are a related subject.

An enrichment of ZFC called Internal Set Theory was proposed by Edward Nelson in 1977.

28.5 Applications

Many mathematical concepts can be defined precisely using only set theoretic concepts. For example, mathematical structures as diverse as graphs, manifolds, rings, and vector spaces can all be defined as sets satisfying various (axiomatic) properties. Equivalence and order relations are ubiquitous in mathematics, and the theory of mathematical

relations can be described in set theory.
Set theory is also a promising foundational system for much of mathematics. Since the publication of the first volume of *Principia Mathematica*, it has been claimed that most or even all mathematical theorems can be derived using an aptly designed set of axioms for set theory, augmented with many definitions, using first or second order logic. For example, properties of the natural and real numbers can be derived within set theory, as each number system can be identified with a set of equivalence classes under a suitable equivalence relation whose field is some infinite set.

Set theory as a foundation for mathematical analysis, topology, abstract algebra, and discrete mathematics is likewise uncontroversial; mathematicians accept that (in principle) theorems in these areas can be derived from the relevant definitions and the axioms of set theory. Few full derivations of complex mathematical theorems from set theory have been formally verified, however, because such formal derivations are often much longer than the natural language proofs mathematicians commonly present. One verification project, Metamath, includes derivations of more than 10,000 theorems starting from the ZFC axioms and using first order logic.

28.6 Areas of study

Set theory is a major area of research in mathematics, with many interrelated subfields.

28.6.1 Combinatorial set theory

Main article: Infinitary combinatorics
Combinatorial set theory concerns extensions of finite combinatorics to infinite sets. This includes the study of cardinal arithmetic and the study of extensions of Ramsey's theorem such as the Erdős–Rado theorem.

28.6.2 Descriptive set theory

Main article: Descriptive set theory
Descriptive set theory is the study of subsets of the real line and, more generally, subsets of Polish spaces. It begins with the study of pointclasses in the Borel hierarchy and extends to the study of more complex hierarchies such as the projective hierarchy and the Wadge hierarchy. Many properties of Borel sets can be established in ZFC, but proving these properties hold for more complicated sets requires additional axioms related to determinacy and large cardinals.
The field of effective descriptive set theory is between set theory and recursion theory. It includes the study of lightface pointclasses, and is closely related to hyperarithmetical theory. In many cases, results of classical descriptive set theory have effective versions; in some cases, new results are obtained by proving the effective version first and then extending ("relativizing") it to make it more broadly applicable.
A recent area of research concerns Borel equivalence relations and more complicated definable equivalence relations. This has important applications to the study of invariants in many fields of mathematics.

28.6.3 Fuzzy set theory

Main article: Fuzzy set theory
In set theory as Cantor defined and Zermelo and Fraenkel axiomatized, an object is either a member of a set or not. In fuzzy set theory this condition was relaxed by Lotfi A. Zadeh so an object has a *degree of membership* in a set, a number between 0 and 1. For example, the degree of membership of a person in the set of "tall people" is more flexible than a simple yes or no answer and can be a real number such as 0.75.

28.6.4 Inner model theory

Main article: Inner model theory
An **inner model** of Zermelo–Fraenkel set theory (ZF) is a transitive class that includes all the ordinals and satisfies all the axioms of ZF. The canonical example is the constructible universe L developed by Gödel. One reason that the study of inner models is of interest is that it can be used to prove consistency results. For example, it can be shown that regardless of whether a model V of ZF satisfies the continuum hypothesis or the axiom of choice, the inner model L constructed inside the original model will satisfy both the generalized continuum hypothesis and the

axiom of choice. Thus the assumption that ZF is consistent (has at least one model) implies that ZF together with these two principles is consistent.
The study of inner models is common in the study of determinacy and large cardinals, especially when considering axioms such as the axiom of determinacy that contradict the axiom of choice. Even if a fixed model of set theory satisfies the axiom of choice, it is possible for an inner model to fail to satisfy the axiom of choice. For example, the existence of sufficiently large cardinals implies that there is an inner model satisfying the axiom of determinacy (and thus not satisfying the axiom of choice).[5]

28.6.5 Large cardinals

Main article: Large cardinal property
A **large cardinal** is a cardinal number with an extra property. Many such properties are studied, including inaccessible cardinals, measurable cardinals, and many more. These properties typically imply the cardinal number must be very large, with the existence of a cardinal with the specified property unprovable in Zermelo-Fraenkel set theory.

28.6.6 Determinacy

Main article: Determinacy
Determinacy refers to the fact that, under appropriate assumptions, certain two-player games of perfect information are determined from the start in the sense that one player must have a winning strategy. The existence of these strategies has important consequences in descriptive set theory, as the assumption that a broader class of games is determined often implies that a broader class of sets will have a topological property. The axiom of determinacy (AD) is an important object of study; although incompatible with the axiom of choice, AD implies that all subsets of the real line are well behaved (in particular, measurable and with the perfect set property). AD can be used to prove that the Wadge degrees have an elegant structure.

28.6.7 Forcing

Main article: Forcing (mathematics)
Paul Cohen invented the method of forcing while searching for a model of ZFC in which the continuum hypothesis fails, or a model of ZF in which the axiom of choice fails. Forcing adjoins to some given model of set theory additional sets in order to create a larger model with properties determined (i.e. “forced”) by the construction and the original model. For example, Cohen’s construction adjoins additional subsets of the natural numbers without changing any of the cardinal numbers of the original model. Forcing is also one of two methods for proving relative consistency by finitistic methods, the other method being Boolean-valued models.

28.6.8 Cardinal invariants

Main article: Cardinal invariant
A **cardinal invariant** is a property of the real line measured by a cardinal number. For example, a well-studied invariant is the smallest cardinality of a collection of meagre sets of reals whose union is the entire real line. These are invariants in the sense that any two isomorphic models of set theory must give the same cardinal for each invariant. Many cardinal invariants have been studied, and the relationships between them are often complex and related to axioms of set theory.

28.6.9 Set-theoretic topology

Main article: Set-theoretic topology
Set-theoretic topology studies questions of general topology that are set-theoretic in nature or that require advanced methods of set theory for their solution. Many of these theorems are independent of ZFC, requiring stronger axioms for their proof. A famous problem is the normal Moore space question, a question in general topology that was the subject of intense research. The answer to the normal Moore space question was eventually proved to be independent of ZFC.

28.7 Objections to set theory as a foundation for mathematics

From set theory’s inception, some mathematicians have objected to it as a foundation for mathematics. The most common objection to set theory, one Kronecker voiced in set theory’s earliest years, starts from the constructivist view that mathematics is loosely related to computation. If this view is granted, then the treatment of infinite sets, both in naive and in axiomatic set theory, introduces into mathematics methods and objects that are not computable even in principle.
Ludwig Wittgenstein condemned set theory. He wrote that “set theory is wrong”, since it builds on the “nonsense” of fictitious symbolism, has “pernicious idioms”, and that it is nonsensical to talk about “all numbers”.[6] Wittgenstein’s views about the foundations of mathematics were later criticised by Georg Kreisel and Paul Bernays, and investigated by Crispin Wright, among others.
Category theorists have proposed topos theory as an alternative to traditional axiomatic set theory. Topos theory can interpret various alternatives to that theory, such as constructivism, finite set theory, and computable set theory.[7] Topoi also give a natural setting for forcing and discussions of the independence of choice from ZF, as well as providing the framework for pointless topology and Stone spaces.[8]
An active area of research is the univalent foundations arising from homotopy type theory. Here, sets may be defined as classes of types, with universal properties of sets arising from higher inductive types. Principles such as the axiom of choice and the law of the excluded middle appear in a spectrum of different forms, some of which can be proven, others which correspond to the classical notions; this allows for a detailed discussion of the effect of these axioms on

mathematics.[9][10]

28.8 See also

_ Glossary of set theory
_ Category theory
_ List of set theory topics
_ Relational model – borrows from set theory

28.9 Notes

[1] Cantor, Georg (1874), *Ueber eine Eigenschaft des Inbegriffes aller reellen algebraischen Zahlen*, *J. Reine Angew. Math.* **77**: 258–262, doi:10.1515/crll.1874.77.258
[2] Johnson, Philip (1972), *A History of Set Theory*, Prindle, Weber & Schmidt, ISBN 0-87150-154-6
[3] Bolzano, Bernard (1975), Berg, Jan, ed., *Einleitung zur Größenlehre und erste Begriffe der allgemeinen Größenlehre*, Bernard-Bolzano-Gesamtausgabe, edited by Eduard Winter et al., Vol. II, A, 7, Stuttgart, Bad Cannstatt: Friedrich Frommann Verlag, p. 152, ISBN 3-7728-0466-7
[4] In his 1925, John von Neumann observed that "set theory in its first, "naive" version, due to Cantor, led to contradictions. These are the well-known antinomies of the set of all sets that do not contain themselves (Russell), of the set of all transfinte ordinal numbers (Burali-Forti), and the set of all finitely definable real numbers (Richard)." He goes on to observe that two

"tendencies" were attempting to "rehabilitate" set theory. Of the first effort, exemplified by Bertrand Russell, Julius König, Hermann Weyl and L. E. J. Brouwer, von Neumann called the "overall effect of their activity . . . devastating". With regards to the axiomatic method employed by second group composed of Zermelo, Abraham Fraenkel and Arthur Moritz Schoenflies, von Neumann worried that "We see only that the known modes of inference leading to the antinomies fail, but who knows where there are not others?" and he set to the task, "in the spirit of the second group", to "produce, by means of a finite number of purely formal operations . . . all the sets that we want to see formed" but not allow for the antinomies. (All quotes from von Neumann 1925 reprinted in van Heijenoort, Jean (1967, third printing 1976), "From Frege to Gödel: A Source Book in Mathematical Logic, 1979–1931", Harvard University Press, Cambridge MA, ISBN 0-674-32449-8 (pbk). A synopsis of the history, written by van Heijenoort, can be found in the comments that precede von Neumann's 1925.
[5] Jech, Thomas (2003), *Set Theory*, Springer Monographs in Mathematics (Third Millennium ed.), Berlin, New York: Springer-Verlag, p. 642, ISBN 978-3-540-44085-7, Zbl 1007.03002
[6] Wittgenstein, Ludwig (1975). *Philosophical Remarks, §129, §174*. Oxford: Basil Blackwell. ISBN 0631191305.
[7] Ferro, A.; Omodeo, E. G.; Schwartz, J. T. (1980), *Decision procedures for elementary sublanguages of set theory. I. Multilevel syllogistic and some extensions*, *Comm. Pure Appl. Math.* **33** (5): 599–608, doi:10.1002/cpa.3160330503
[8] Saunders Mac Lane and Ieke Moerdijk (1992) *Sheaves in Geometry and Logic: a First Introduction to Topos Theory*. Springer Verlag.
[9] homotopy type theory in *nLab*
[10] *Homotopy Type Theory: Univalent Foundations of Mathematics*. The Univalent Foundations Program. Institute for Advanced Study.

28.10 Further reading

_ Devlin, Keith, 1993. *The Joy of Sets* (2nd ed.). Springer Verlag, ISBN 0-387-94094-4
_ Ferreirós, Jose, 2007 (1999). *Labyrinth of Thought: A history of set theory and its role in modern mathematics*. Basel, Birkhäuser. ISBN 978-3-7643-8349-7
_ Johnson, Philip, 1972. *A History of Set Theory*. Prindle, Weber & Schmidt ISBN 0-87150-154-6
_ Kunen, Kenneth, 1980. *Set Theory: An Introduction to Independence Proofs*. North-Holland, ISBN 0-444-85401-0.
_ Potter, Michael, 2004. *Set Theory and Its Philosophy: A Critical Introduction*. Oxford University Press.
_ Tiles, Mary, 2004 (1989). *The Philosophy of Set Theory: An Historical Introduction to Cantor's Paradise*. Dover Publications. ISBN 978-0-486-43520-6

28.11 External links

_ Foreman, Matthew, Akihiro Kanamori, eds. *Handbook of Set Theory*. 3 vols., 2010. Each chapter surveys some aspect of contemporary research in set theory. Does not cover established elementary set theory, on which see Devlin (1993).
_ Hazewinkel, Michiel, ed. (2001), "Axiomatic set theory", *Encyclopedia of Mathematics*, Springer, ISBN 978-1-55608-010-4
_ Hazewinkel, Michiel, ed. (2001), "Set theory", *Encyclopedia of Mathematics*, Springer, ISBN 978-1-55608-010-4
_ Jech, Thomas (2002). "Set Theory", *Stanford Encyclopedia of Philosophy*.
_ Schoenflies, Arthur (1898). Mengenlehre in Klein's encyclopedia.
_ Online books, and library resources in your library and in other libraries about set theory

Chapter 29

Non-Euclidean geometry

Hyperbolic Euclidean Elliptic

Behavior of lines with a common perpendicular in each of the three types of geometry

In mathematics, **non-Euclidean geometry** consists of two geometries based on axioms closely related to those specifying Euclidean geometry. As Euclidean geometry lies at the intersection of metric geometry and affine geometry, non-Euclidean geometry arises when either the metric requirement is relaxed, or the parallel postulate is replaced with an alternative one. In the latter case one obtains hyperbolic geometry and elliptic geometry, the traditional non-Euclidean geometries. When the metric requirement is relaxed, then there are affine planes associated with the planar algebras which give rise to kinematic geometries that have also been called non-Euclidean geometry.

The essential difference between the metric geometries is the nature of parallel lines. Euclid's fifth postulate, the parallel postulate, is equivalent to Playfair's postulate, which states that, within a two-dimensional plane, for any given line ℓ and a point A, which is not on ℓ, there is exactly one line through A that does not intersect ℓ. In hyperbolic geometry, by contrast, there are infinitely many lines through A not intersecting ℓ, while in elliptic geometry, any line through A intersects ℓ.

Another way to describe the differences between these geometries is to consider two straight lines indefinitely extended in a two-dimensional plane that are both perpendicular to a third line:

_ In Euclidean geometry the lines remain at a constant distance from each other even if extended to infinity, and are known as parallels.

_ In hyperbolic geometry they "curve away" from each other, increasing in distance as one moves further from the points of intersection with the common perpendicular; these lines are often called ultraparallels.

_ In elliptic geometry the lines "curve toward" each other and intersect.

29.1 History

29.1.1 Early history

While Euclidean geometry, named after the Greek mathematician Euclid, includes some of the oldest known mathematics, non-Euclidean geometries were not widely accepted as legitimate until the 19th century.

201

202 *CHAPTER 29. NON-EUCLIDEAN GEOMETRY*

The debate that eventually led to the discovery of the non-Euclidean geometries began almost as soon as Euclid's work *Elements* was written. In the *Elements*, Euclid began with a limited number of assumptions (23 definitions, five common notions, and five postulates) and sought to prove all the other results (propositions) in the work. The most notorious of the postulates is often referred to as "Euclid's Fifth Postulate," or simply the "parallel postulate", which in Euclid's original formulation is:

If a straight line falls on two straight lines in such a manner that the interior angles on the same side are together less than two right angles, then the straight lines, if produced indefinitely, meet on that side on which are the angles less than the two right angles.

Other mathematicians have devised simpler forms of this property. Regardless of the form of the postulate, however, it consistently appears to be more complicated than Euclid's other postulates (which include, for example, "Between any two points a straight line may be drawn").

For at least a thousand years, geometers were troubled by the disparate complexity of the fifth postulate, and believed it could be proved as a theorem from the other four. Many attempted to find a proof by contradiction, including Ibn al-Haytham (Alhazen, 11th century),[1] Omar Khayyám (12th century), Nasīr al-Dīn al-Tūsī (13th century), and Giovanni Girolamo Saccheri (18th century).

The theorems of Ibn al-Haytham, Khayyam and al-Tusi on quadrilaterals, including the Lambert quadrilateral and Saccheri quadrilateral, were "the first few theorems of the hyperbolic and the elliptic geometries." These theorems along with their alternative postulates, such as Playfair's axiom, played an important role in the later development of non-Euclidean geometry. These early attempts at challenging the fifth postulate had a considerable influence on its development among later European geometers, including Witelo, Levi ben Gerson, Alfonso, John Wallis and Saccheri.[2] All of these early attempts made at trying to formulate non-Euclidean geometry however provided flawed proofs of the parallel postulate, containing assumptions that were essentially equivalent to the parallel postulate. These early attempts did, however, provide some early properties of the hyperbolic and elliptic geometries.

Khayyam, for example, tried to derive it from an equivalent postulate he formulated from "the principles of the Philosopher" (Aristotle): "*Two convergent straight lines intersect and it is impossible for two convergent straight lines to diverge in the direction in which they converge.*"[3] Khayyam then considered the three cases right, obtuse, and acute that the summit angles of a Saccheri quadrilateral can take and after proving a number of theorems about them, he correctly refuted the obtuse and acute cases based on his postulate and hence derived the classic postulate of Euclid which he didn't realize was equivalent to his own postulate. Another example is al-Tusi's son, Sadr al-Din (sometimes known as "Pseudo-Tusi"), who wrote a book on the subject in 1298, based on al-Tusi's later thoughts, which presented another hypothesis equivalent to the parallel postulate. "He essentially revised both the Euclidean system of axioms and postulates and the proofs of many propositions from the *Elements*."[4][5] His work was published in Rome in 1594 and was studied by European geometers, including Saccheri[4] who criticised this work as well as that of Wallis.[6]
Giordano Vitale, in his book *Euclide restituo* (1680, 1686), used the Saccheri quadrilateral to prove that if three points are equidistant on the base AB and the summit CD, then AB and CD are everywhere equidistant.
In a work titled *Euclides ab Omni Naevo Vindicatus* (*Euclid Freed from All Flaws*), published in 1733, Saccheri quickly discarded elliptic geometry as a possibility (some others of Euclid's axioms must be modified for elliptic geometry to work) and set to work proving a great number of results in hyperbolic geometry.
He finally reached a point where he believed that his results demonstrated the impossibility of hyperbolic geometry. His claim seems to have been based on Euclidean presuppositions, because no *logical* contradiction was present. In this attempt to prove Euclidean geometry he instead unintentionally discovered a new viable geometry, but did not realize it.
In 1766 Johann Lambert wrote, but did not publish, *Theorie der Parallellinien* in which he attempted, as Saccheri did, to prove the fifth postulate. He worked with a figure that today we call a *Lambert quadrilateral*, a quadrilateral with three right angles (can be considered half of a Saccheri quadrilateral). He quickly eliminated the possibility that the fourth angle is obtuse, as had Saccheri and Khayyam, and then proceeded to prove many theorems under the assumption of an acute angle. Unlike Saccheri, he never felt that he had reached a contradiction with this assumption. He had proved the non-Euclidean result that the sum of the angles in a triangle increases as the area of the triangle decreases, and this led him to speculate on the possibility of a model of the acute case on a sphere of imaginary radius. He did not carry this idea any further.[7]
At this time it was widely believed that the universe worked according to the principles of Euclidean geometry.[8]
29.2. AXIOMATIC BASIS OF NON-EUCLIDEAN GEOMETRY 203

29.1.2 Creation of non-Euclidean geometry

The beginning of the 19th century would finally witness decisive steps in the creation of non-Euclidean geometry. Circa 1813, Carl Friedrich Gauss and independently around 1818, the German professor of law Ferdinand Karl Schweikart[9] had the germinal ideas of non-Euclidean geometry worked out, but neither published any results. Then, around 1830, the Hungarian mathematician János Bolyai and the Russian mathematician Nikolai Ivanovich Lobachevsky separately published treatises on hyperbolic geometry. Consequently, hyperbolic geometry is called Bolyai-Lobachevskian geometry, as both mathematicians, independent of each other, are the basic authors of non-Euclidean geometry. Gauss mentioned to Bolyai's father, when shown the younger Bolyai's work, that he had developed such a geometry several years before,[10] though he did not publish. While Lobachevsky created a non-Euclidean geometry by negating the parallel postulate, Bolyai worked out a geometry where both the Euclidean and the hyperbolic geometry are possible depending on a parameter k. Bolyai ends his work by mentioning that it is not possible to decide through mathematical reasoning alone if the geometry of the physical universe is Euclidean or non-Euclidean; this is a task for the physical sciences.
Bernhard Riemann, in a famous lecture in 1854, founded the field of Riemannian geometry, discussing in particular the ideas now called manifolds, Riemannian metric, and curvature. He constructed an infinite family of geometries which are not Euclidean by giving a formula for a family of Riemannian metrics on the unit ball in Euclidean space. The simplest of these is called elliptic geometry and it is considered to be a non-Euclidean geometry due to its lack of parallel lines.[11]
By formulating the geometry in terms of a curvature tensor, Riemann allowed non-Euclidean geometry to be applied to higher dimensions.

29.1.3 Terminology

It was Gauss who coined the term "non-Euclidean geometry".[12] He was referring to his own work which today we call *hyperbolic geometry*. Several modern authors still consider "non-Euclidean geometry" and "hyperbolic geometry" to be synonyms.
Arthur Cayley noted that distance between points inside a conic could be defined in terms of logarithm and the projective cross-ratio function. The method has become called the Cayley-Klein metric because Felix Klein exploited it to describe the non-euclidean geometries in articles[13] in 1871 and 73 and later in book form. The Cayley-Klein metrics provided working models of hyperbolic and elliptic metric geometries, as well as Euclidean geometry.
Klein is responsible for the terms "hyperbolic" and "elliptic" (in his system he called Euclidean geometry "parabolic",

a term which generally fell out of use[14]). His influence has led to the current usage of the term "non-Euclidean geometry" to mean either "hyperbolic" or "elliptic" geometry.
There are some mathematicians who would extend the list of geometries that should be called "non-Euclidean" in various ways.[15]

29.2 Axiomatic basis of non-Euclidean geometry

Euclidean geometry can be axiomatically described in several ways. Unfortunately, Euclid's original system of five postulates (axioms) is not one of these as his proofs relied on several unstated assumptions which should also have been taken as axioms. Hilbert's system consisting of 20 axioms[16] most closely follows the approach of Euclid and provides the justification for all of Euclid's proofs. Other systems, using different sets of undefined terms obtain the same geometry by different paths. In all approaches, however, there is an axiom which is logically equivalent to Euclid's fifth postulate, the parallel postulate. Hilbert uses the Playfair axiom form, while Birkhoff, for instance, uses the axiom which says that "there exists a pair of similar but not congruent triangles." In any of these systems, removal of the one axiom which is equivalent to the parallel postulate, in whatever form it takes, and leaving all the other axioms intact, produces absolute geometry. As the first 28 propositions of Euclid (in *The Elements*) do not require the use of the parallel postulate or anything equivalent to it, they are all true statements in absolute geometry.[17]
To obtain a non-Euclidean geometry, the parallel postulate (or its equivalent) *must* be replaced by its negation. Negating the Playfair's axiom form, since it is a compound statement (... there exists one and only one ...), can be done in two ways. Either there will exist more than one line through the point parallel to the given line or there will exist no lines through the point parallel to the given line. In the first case, replacing the parallel postulate (or its equivalent)

with the statement "In a plane, given a point P and a line ℓ not passing through P, there exist two lines through P which do not meet ℓ " and keeping all the other axioms, yields hyperbolic geometry.[18] The second case is not dealt with as easily. Simply replacing the parallel postulate with the statement, "In a plane, given a point P and a line ℓ not passing through P, all the lines through P meet ℓ ", does not give a consistent set of axioms. This follows since parallel lines exist in absolute geometry,[19] but this statement says that there are no parallel lines. This problem was known (in a different guise) to Khayyam, Saccheri and Lambert and was the basis for their rejecting what was known as the "obtuse angle case". In order to obtain a consistent set of axioms which includes this axiom about having no parallel lines, some of the other axioms must be tweaked. The adjustments to be made depend upon the axiom system being used. Among others these tweaks will have the effect of modifying Euclid's second postulate from the statement that line segments can be extended indefinitely to the statement that lines are unbounded. Riemann's elliptic geometry emerges as the most natural geometry satisfying this axiom.

29.3 Models of non-Euclidean geometry

For more details on this topic, see Models of non-Euclidean geometry.
Two dimensional Euclidean geometry is modelled by our notion of a "flat plane."
On a sphere, the sum of the angles of a triangle is not equal to 180°. The surface of a sphere is not a Euclidean space, but locally the laws of the Euclidean geometry are good approximations. In a small triangle on the face of the earth, the sum of the angles is very nearly 180°.

29.3.1 Elliptic geometry

Main article: Elliptic geometry

The simplest model for elliptic geometry is a sphere, where lines are "great circles" (such as the equator or the meridians on a globe), and points opposite each other (called antipodal points) are identified (considered to be the same). This is also one of the standard models of the real projective plane. The difference is that as a model of elliptic geometry a metric is introduced permitting the measurement of lengths and angles, while as a model of the projective plane there is no such metric.
In the elliptic model, for any given line ℓ and a point A, which is not on ℓ, all lines through A will intersect ℓ.

29.3.2 Hyperbolic geometry

Main article: Hyperbolic geometry
Even after the work of Lobachevsky, Gauss, and Bolyai, the question remained: "Does such a model exist for hyperbolic geometry?". The model for hyperbolic geometry was answered by Eugenio Beltrami, in 1868, who first showed that a surface called the pseudosphere has the appropriate curvature to model a portion of hyperbolic space and in a second paper in the same year, defined the Klein model which models the entirety of hyperbolic space, and used this to show that Euclidean geometry and hyperbolic geometry were equiconsistent so that hyperbolic geometry was logically consistent if and only if Euclidean geometry was. (The reverse implication follows from the horosphere model of Euclidean geometry.)
In the hyperbolic model, within a two-dimensional plane, for any given line ℓ and a point A, which is not on ℓ, there

are infinitely many lines through A that do not intersect ℓ.
In these models the concepts of non-Euclidean geometries are being represented by Euclidean objects in a Euclidean setting. This introduces a perceptual distortion wherein the straight lines of the non-Euclidean geometry are being represented by Euclidean curves which visually bend. This "bending" is not a property of the non-Euclidean lines, only an artifice of the way they are being represented.

29.3.3 Three-dimensional non-Euclidean geometry

Main article: Thurston geometry
In three dimensions, there are eight models of geometries.[20] There are Euclidean, elliptic, and hyperbolic geometries, as in the two-dimensional case; mixed geometries that are partially Euclidean and partially hyperbolic or spherical; twisted versions of the mixed geometries; and one unusual geometry that is completely anisotropic (i.e. every direction behaves differently).

29.4 Uncommon properties

Euclidean and non-Euclidean geometries naturally have many similar properties, namely those which do not depend upon the nature of parallelism. This commonality is the subject of absolute geometry (also called *neutral geometry*). However, the properties which distinguish one geometry from the others are the ones which have historically received the most attention.
Besides the behavior of lines with respect to a common perpendicular, mentioned in the introduction, we also have the following:
_ A Lambert quadrilateral is a quadrilateral which has three right angles. The fourth angle of a Lambert quadrilateral is acute if the geometry is hyperbolic, a right angle if the geometry is Euclidean or obtuse if the geometry is elliptic. Consequently, rectangles exist (a statement equivalent to the parallel postulate) only in Euclidean geometry.
_ A Saccheri quadrilateral is a quadrilateral which has two sides of equal length, both perpendicular to a side called the *base*. The other two angles of a Saccheri quadrilateral are called the *summit angles* and they have equal measure. The summit angles of a Saccheri quadrilateral are acute if the geometry is hyperbolic, right angles if the geometry is Euclidean and obtuse angles if the geometry is elliptic.

Lambert quadrilateral in hyperbolic geometry
_ The sum of the measures of the angles of any triangle is less than 180° if the geometry is hyperbolic, equal to 180° if the geometry is Euclidean, and greater than 180° if the geometry is elliptic. The *defect* of a triangle is the numerical value (180° - sum of the measures of the angles of the triangle). This result may also be stated as: the defect of triangles in hyperbolic geometry is positive, the defect of triangles in Euclidean geometry is zero, and the defect of triangles in elliptic geometry is negative.

29.5 Importance

Non-Euclidean geometry is an example of a paradigm shift in the history of science.[21] Before the models of a non-Euclidean plane were presented by Beltrami, Klein, and Poincaré, Euclidean geometry stood unchallenged as the mathematical model of space. Furthermore, since the substance of the subject in synthetic geometry was a chief exhibit of rationality, the Euclidean point of view represented absolute authority.
The discovery of the non-Euclidean geometries had a ripple effect which went far beyond the boundaries of mathematics and science. The philosopher Immanuel Kant's treatment of human knowledge had a special role for geometry. It was his prime example of synthetic a priori knowledge; not derived from the senses nor deduced through logic — our knowledge of space was a truth that we were born with. Unfortunately for Kant, his concept of this unalterably true geometry was Euclidean. Theology was also affected by the change from absolute truth to relative truth in mathematics that was a result of this paradigm shift.[22]
The existence of non-Euclidean geometries impacted the intellectual life of Victorian England in many ways[23] and in particular was one of the leading factors that caused a re-examination of the teaching of geometry based on Euclid's Elements. This curriculum issue was hotly debated at the time and was even the subject of a play, *Euclid and his Modern Rivals*, written by Lewis Carroll, the author of Alice in Wonderland.[24]

29.6 Planar algebras

In analytic geometry a plane is described with Cartesian coordinates : $C = \{(x,y) : x, y \text{ in } R\}$. The points are sometimes identified with complex numbers $z = x + y\,\varepsilon$ where the square of ε is in $\{-1, 0, +1\}$. The Euclidean plane corresponds to the case $\varepsilon^2 = -1$ since the modulus of z is given by
zz
_

$$= (x + y\epsilon)(x \square y\epsilon) = x2 + y2$$
and this quantity is the square of the Euclidean distance between z and the origin. For instance, $\{z : z z^* = 1\}$ is the unit circle.

For planar algebra, non-Euclidean geometry arises in the other cases. When $\epsilon2 = +1$, then z is a split-complex number and conventionally j replaces epsilon. Then

zz

$$= (x + yj)(x \square yj) = x2 \square y2$$
and $\{z : z z^* = 1\}$ is the unit hyperbola.

When $\epsilon2 = 0$, then z is a dual number.[25]

This approach to non-Euclidean geometry explains the non-Euclidean angles: the parameters of slope in the dual number plane and hyperbolic angle in the split-complex plane correspond to angle in Euclidean geometry. Indeed, they each arise in polar decomposition of a complex number z.[26]

29.7 Kinematic geometries

Hyperbolic geometry found an application in kinematics with the cosmology introduced by Hermann Minkowski in 1908. Minkowski introduced terms like worldline and proper time into mathematical physics. He realized that the submanifold, of events one moment of proper time into the future, could be considered a hyperbolic space of three dimensions.[27][28] Already in the 1890s Alexander Macfarlane was charting this submanifold through his Algebra of Physics and hyperbolic quaternions, though Macfarlane did not use cosmological language as Minkowski did in 1908. The relevant structure is now called the hyperboloid model of hyperbolic geometry.

The non-Euclidean planar algebras support kinematic geometries in the plane. For instance, the split-complex number $z = e_{aj}$ can represent a spacetime event one moment into the future of a frame of reference of rapidity a. Furthermore, multiplication by z amounts to a Lorentz boost mapping the frame with rapidity zero to that with rapidity a. Kinematic study makes use of the dual numbers $z = x + y\epsilon$; $\epsilon2 = 0$; to represent the classical description of motion in absolute time and space: The equations $x' = x + vt$; $t' = t$ are equivalent to a shear mapping in linear algebra:

$$\begin{pmatrix} x' \\ t' \end{pmatrix} = \begin{pmatrix} 1 & v \\ 0 & 1 \end{pmatrix} \begin{pmatrix} x \\ t \end{pmatrix} :$$

With dual numbers the mapping is $t' + x'\epsilon = (1 + v\epsilon)(t + x\epsilon) = t + (x + vt)\epsilon$: [29]

Another view of special relativity as a non-Euclidean geometry was advanced by E. B. Wilson and Gilbert Lewis in *Proceedings of the American Academy of Arts and Sciences* in 1912. They revamped the analytic geometry implicit in the split-complex number algebra into synthetic geometry of premises and deductions.[30][31]

29.8 Fiction

Non-Euclidean geometry often makes appearances in works of science fiction and fantasy.

Professor James Moriarty, a character in stories written by Sir Arthur Conan Doyle, is a criminal mastermind with a PhD in non-Euclidean geometries.

208 *CHAPTER 29. NON-EUCLIDEAN GEOMETRY*

In 1895 H. G. Wells published the short story "The Remarkable Case of Davidson's Eyes". To appreciate this story one should know how antipodal points on a sphere are identified in a model of the elliptic plane. In the story, in the midst of a thunderstorm, Sidney Davidson sees "Waves and a remarkably neat schooner" while working in an electrical laboratory at Harlow Technical College. At the story's close Davidson proves to have witnessed H.M.S. *Fulmar* off Antipodes Island.

Non-Euclidean geometry is sometimes connected with the influence of the 20th century horror fiction writer H. P. Lovecraft. In his works, many unnatural things follow their own unique laws of geometry: In Lovecraft's Cthulhu Mythos, the sunken city of R'lyeh is characterized by its non-Euclidean geometry. It is heavily implied this is achieved as a side effect of not following the natural laws of this universe rather than simply using an alternate geometric model, as the sheer innate wrongness of it is said to be capable of driving those who look upon it insane.[32]

The main character in Robert Pirsig's *Zen and the Art of Motorcycle Maintenance* mentioned Riemannian Geometry on multiple occasions.
In *The Brothers Karamazov*, Dostoevsky discusses non-Euclidean geometry through his main character Ivan.
Christopher Priest's novel *Inverted World* describes the struggle of living on a planet with the form of a rotating pseudosphere.
Robert Heinlein's *The Number of the Beast* utilizes non-Euclidean geometry to explain instantaneous transport through space and time and between parallel and fictional universes.
Alexander Bruce's *Antichamber* uses non-Euclidean geometry to create a brilliant, minimal, Escher-like world, where geometry and space follow unfamiliar rules.
In the Renegade Legion science fiction setting for FASA's wargame, role-playing-game and fiction, faster-than-light travel and communications is possible through the use of Hsieh Ho's Polydimensional Non-Euclidean Geometry, published sometime in the middle of the twenty-second century.

29.9 See also

_ Hyperbolic space
_ Lénárt sphere
_ Projective geometry

29.10 Notes

[1] Eder, Michelle (2000), *Views of Euclid's Parallel Postulate in Ancient Greece and in Medieval Islam*, Rutgers University, retrieved 2008-01-23
[2] Boris A. Rosenfeld & Adolf P. Youschkevitch, "Geometry", p. 470, in Roshdi Rashed & Régis Morelon (1996), *Encyclopedia of the History of Arabic Science*, Vol. 2, pp. 447–494, Routledge, London and New York:
"Three scientists, Ibn al-Haytham, Khayyam and al-Tusi, had made the most considerable contribution to this branch of geometry whose importance came to be completely recognized only in the nineteenth century. In essence their propositions concerning the properties of quadrangles which they considered assuming that some of the angles of these figures were acute of obtuse, embodied the first few theorems of the hyperbolic and the elliptic geometries. Their other proposals showed that various geometric statements were equivalent to the Euclidean postulate V. It is extremely important that these scholars established the mutual connection between this postulate and the sum of the angles of a triangle and a quadrangle. By their works on the theory of parallel lines Arab mathematicians directly influenced the relevant investigations of their European counterparts. The first European attempt to prove the postulate on parallel lines – made by Witelo, the Polish scientists of the thirteenth century, while revising Ibn al-Haytham's *Book of Optics* (*Kitab al-Manazir*) – was undoubtedly prompted by Arabic sources. The proofs put forward in the fourteenth century by the Jewish scholar Levi ben Gerson, who lived in southern France, and by the above-mentioned Alfonso from Spain directly border on Ibn al-Haytham's demonstration. Above, we have demonstrated that *Pseudo-Tusi's Exposition of Euclid* had stimulated borth J. Wallis's and G. Saccheri's studies of the theory of parallel lines."
[3] Boris A. Rosenfeld & Adolf P. Youschkevitch (1996), "Geometry", p. 467, in Roshdi Rashed & Régis Morelon (1996), *Encyclopedia of the History of Arabic Science*, Vol. 2, pp. 447–494, Routledge, ISBN 0-415-12411-5

[4] Victor J. Katz (1998), *History of Mathematics: An Introduction*, p. 270–271, Addison–Wesley, ISBN 0-321-01618-1:
"But in a manuscript probably written by his son Sadr al-Din in 1298, based on Nasir al-Din's later thoughts on the subject, there is a new argument based on another hypothesis, also equivalent to Euclid's, [...] The importance of this latter work is that it was published in Rome in 1594 and was studied by European geometers. In particular, it became the starting point for the work of Saccheri and ultimately for the discovery of non-Euclidean geometry."
[5] Boris A. Rosenfeld and Adolf P. Youschkevitch (1996), "Geometry", in Roshdi Rashed, ed., *Encyclopedia of the History of Arabic Science*, Vol. 2, p. 447–494 [469], Routledge, London and New York:
"In *Pseudo-Tusi's Exposition of Euclid*, [...] another statement is used instead of a postulate. It was independent of the Euclidean postulate V and easy to prove. [...] He essentially revised both the Euclidean system of axioms and postulates and the proofs of many propositions from the *Elements*."
[6] MacTutor's Giovanni Girolamo Saccheri
[7] O'Connor, J.J.; Robertson, E.F. "Johann Heinrich Lambert". Retrieved 16 September 2011.
[8] A notable exception is David Hume, who as early as 1739 seriously entertained the possibility that our universe was non-Euclidean; see David Hume (1739/1978) *A Treatise of Human Nature*, L.A. Selby-Bigge, ed. (Oxford: Oxford University Press), pp. 51-52.
[9] In a letter of December 1818, Ferdinand Karl Schweikart (1780-1859) sketched a few insights into non-Euclidean geometry. The letter was forwarded to Gauss in 1819 by Gauss's former student Gerling. In his reply to Gerling, Gauss praised Schweikart and mentioned his own, earlier research into non-Euclidean geometry. See:
_ Carl Friedrich Gauss, *Werke* (Leipzig, Germany: B. G. Teubner, 1900), volume 8, pages 180-182.
_ English translations of Schweikart's letter and Gauss's reply to Gerling appear in: Course notes: "Gauss and non-Euclidean geometry", University of Waterloo, Ontario, Canada; see especially pages 10 and 11.
_ Letters by Schweikart and the writings of his nephew Franz Adolph Taurinus (1794-1874), who also was interested in non-Euclidean geometry and who in 1825 published a brief book on the parallel axiom, appear in: Paul Stäckel and

Friedrich Engel, *Die theorie der Parallellinien von Euklid bis auf Gauss, eine Urkundensammlung der nichteuklidischen Geometrie* (The theory of parallel lines from Euclid to Gauss, an archive of non-Euclidean geometry), (Leipzig, Germany: B. G. Teubner, 1895), pages 243 ff.
[10] In the letter to Wolfgang (Farkas) Bolyai of March 6, 1832 Gauss claims to have worked on the problem for thirty or thirty-five years (Faber 1983, pg. 162). In his 1824 letter to Taurinus (Faber 1983, pg. 158) he claimed that he had been working on the problem for over 30 years and provided enough detail to show that he actually had worked out the details. According to Faber (1983, pg. 156) it wasn't until around 1813 that Gauss had come to accept the existence of a new geometry.
[11] However, other axioms besides the parallel postulate must be changed in order to make this a feasible geometry.
[12] Felix Klein, *Elementary Mathematics from an Advanced Standpoint: Geometry*, Dover, 1948 (reprint of English translation of 3rd Edition, 1940. First edition in German, 1908) pg. 176
[13] F. Klein, Über die sogenannte nichteuklidische Geometrie, *Mathematische Annalen*, **4**(1871).
[14] The Euclidean plane is still referred to as "parabolic" in the context of conformal geometry: see Uniformization theorem.
[15] for instance, Manning 1963 and Yaglom 1968
[16] a 21st axiom appeared in the French translation of Hilbert's *Grundlagen der Geometrie* according to Smart 1997, pg. 416
[17] (Smart 1997, pg.366)
[18] while only two lines are postulated, it is easily shown that there must be an infinite number of such lines.
[19] Book I Proposition 27 of Euclid's *Elements*
[20] _ William Thurston. *Three-dimensional geometry and topology. Vol. 1*. Edited by Silvio Levy. Princeton Mathematical Series, 35. Princeton University Press, Princeton, NJ, 1997. x+311 pp. ISBN 0-691-08304-5 (in depth explanation of the eight geometries and the proof that there are only eight)
[21] see Trudeau 1987, p. vii
210 *CHAPTER 29. NON-EUCLIDEAN GEOMETRY*
[22] Imre Toth, "Gott und Geometrie: Eine viktorianische Kontroverse," *Evolutionstheorie und ihre Evolution*, Dieter Henrich, ed. (Schriftenreihe der Universität Regensburg, band 7, 1982) pp. 141–204.
[23] (Richards 1988)
[24] Lewis Carroll, see reference below.
[25] Yaglom 1968
[26] Richard C. Tolman (2004) Theory of Relativity of Motion, page 194, §180 Non-Euclidean angle, §181 Kinematical interpretation of angle in terms of velocity
[27] Hermann Minkowski (1908–9). "Space and Time" (Wikisource).
[28] Scott Walter (1999) Non-Euclidean Style of Special Relativity
[29] Isaak Yaglom (1979) A simple non-Euclidean geometry and its physical basis : an elementary account of Galilean geometry and the Galilean principle of relativity, Springer ISBN 0-387-90332-1
[30] Edwin B. Wilson & Gilbert N. Lewis (1912) "The Space-time Manifold of Relativity. The Non-Euclidean Geometry of Mechanics and Electromagnetics" Proceedings of the American Academy of Arts and Sciences 48:387–507
[31] Synthetic Spacetime, a digest of the axioms used, and theorems proved, by Wilson and Lewis. Archived by WebCite
[32] "The Call of Cthulhu".

29.11 References

_ A'Campo, Norbert and Papadopoulos, Athanase, (2012) *Notes on hyperbolic geometry*, in: Strasbourg Master class on Geometry, pp. 1–182, IRMA Lectures in Mathematics and Theoretical Physics, Vol. 18, Zürich: European Mathematical Society (EMS), 461 pages, SBN ISBN 978-3-03719-105-7, DOI 10.4171/105.
_ Anderson, James W. *Hyperbolic Geometry*, second edition, Springer, 2005
_ Beltrami, Eugenio *Teoria fondamentale degli spazî di curvatura costante*, Annali. di Mat., ser II 2 (1868), 232–255
_ Blumenthal, Leonard M. (1980), *A Modern View of Geometry*, New York: Dover, ISBN 0-486-63962-2
_ Carroll, Lewis *Euclid and His Modern Rivals*, New York: Barnes and Noble, 2009 (reprint) ISBN 978-1-4351-2348-9
_ H. S. M. Coxeter (1942) *Non-Euclidean Geometry*, University of Toronto Press, reissued 1998 by Mathematical Association of America, ISBN 0-88385-522-4.
_ Faber, Richard L. (1983), *Foundations of Euclidean and Non-Euclidean Geometry*, New York: Marcel Dekker, ISBN 0-8247-1748-1
_ Jeremy Gray (1989) *Ideas of Space: Euclidean, Non-Euclidean, and Relativistic*, 2nd edition, Clarendon Press.
_ Greenberg, Marvin Jay *Euclidean and Non-Euclidean Geometries: Development and History*, 4th ed., New York: W. H. Freeman, 2007. ISBN 0-7167-9948-0
_ Morris Kline (1972) *Mathematical Thought from Ancient to Modern Times*, Chapter 36 Non-Euclidean Geometry, pp 861–81, Oxford University Press.
_ Bernard H. Lavenda, (2012) " A New Perspective on Relativity : An Odyssey In Non-Euclidean Geometries", World Scientific, pp. 696, ISBN 9789814340489.
_ Nikolai Lobachevsky (2010) *Pangeometry*, Translator and Editor: A. Papadopoulos, Heritage of European Mathematics Series, Vol. 4, European Mathematical Society.
_ Manning, Henry Parker (1963), *Introductory Non-Euclidean Geometry*, New York: Dover

_ Meschkowski, Herbert (1964), *Noneuclidean Geometry*, New York: Academic Press
29.12. EXTERNAL LINKS 211
_ Milnor, John W. (1982) *Hyperbolic geometry: The first 150 years*, Bull. Amer. Math. Soc. (N.S.) Volume 6, Number 1, pp. 9–24.
_ Richards, Joan L. (1988), *Mathematical Visions: The Pursuit of Geometry in Victorian England*, Boston: Academic Press, ISBN 0-12-587445-6
_ Smart, James R. (1997), *Modern Geometries (5th Ed.)*, Pacific Grove: Brooks/Cole, ISBN 0-534-35188-3
_ Stewart, Ian *Flatterland*. New York: Perseus Publishing, 2001. ISBN 0-7382-0675-X (softcover)
_ John Stillwell (1996) *Sources of Hyperbolic Geometry*, American Mathematical Society ISBN 0-8218-0529-0
.
_ Trudeau, Richard J. (1987), *The Non-Euclidean Revolution*, Boston: Birkhauser, ISBN 0-8176-3311-1
_ Isaak Yaglom (1968) *Complex Numbers in Geometry*, translated by E. Primrose from 1963 Russian original, appendix "Non-Euclidean geometries in the plane and complex numbers", pp 195–219, Academic Press, N.Y.

29.12 External links

_ Roberto Bonola (1912) Non-Euclidean Geometry, Open Court, Chicago.
_ MacTutor Archive article on non-Euclidean geometry
_ Non-euclidean geometry at PlanetMath.org.
_ Non-Euclidean geometries from *Encyclopedia of Math* of European Mathematical Society and Springer Science+ Business Media
_ Synthetic Spacetime, a digest of the axioms used, and theorems proved, by Wilson and Lewis. Archived by WebCite.
_ Annalee Newitz (Oct 30, 2012). "At last, science explains the physics in "Call of Cthulhu"". *i09*. arXiv:1210.8144
212 *CHAPTER 29. NON-EUCLIDEAN GEOMETRY*

A B

D C

A B

D C

D C

right

obtuse

Chapter 30

Argument

This article is about the subject as it is studied in logic and philosophy. For other uses, see Argument (disambiguation).
In logic and philosophy, an **argument** is an attempt to persuade someone of something, by giving reasons for accepting a particular conclusion as evident.[1][2] The general form of an argument in a natural language is that of premises (typically in the form of propositions, statements or sentences) in support of a claim: the conclusion.[3][4][5] The structure of some arguments can also be set out in a formal language, and formally defined "arguments" can be made independently of natural language arguments, as in math, logic and computer science.

In a typical deductive argument, the premises are meant to provide a guarantee of the *truth* of the conclusion, while in an inductive argument, they are thought to provide reasons supporting the conclusion's *probable* truth.[6] The standards for evaluating non-deductive arguments may rest on different or additional criteria than truth, for example, the persuasiveness of so-called "indispensability claims" in transcendental arguments,[7] the quality of hypotheses in

retroduction, or even the disclosure of new possibilities for thinking and acting.[8]
The standards and criteria used in evaluating arguments and their forms of reasoning are studied in logic.[9] Ways of formulating arguments effectively are studied in rhetoric (see also: argumentation theory). An argument in a formal language shows the logical form of the symbolically represented or natural language arguments obtained by its interpretations.

30.1 Formal and informal

Further information: Informal logic and Formal logic
Informal arguments as studied in *informal logic*, are presented in ordinary language and are intended for everyday discourse. Conversely, formal arguments are studied in *formal logic* (historically called *symbolic logic*, more commonly referred to as *mathematical logic* today) and are expressed in a formal language. Informal logic may be said to emphasize the study of argumentation, whereas formal logic emphasizes implication and inference. Informal arguments are sometimes implicit. That is, the rational structure –the relationship of claims, premises, warrants, relations of implication, and conclusion –is not always spelled out and immediately visible and must sometimes be made explicit by analysis.

30.2 Standard types

There are several kinds of arguments in logic, the best-known of which are "deductive" and "inductive." Deductive arguments are sometimes referred to as "truth-preserving" arguments, because the truth of the conclusion follows given that of the premises. A deductive argument asserts that the truth of the conclusion is a logical consequence of the premises. An inductive argument, on the other hand, asserts that the truth of the conclusion is otherwise supported by the premises. Each premise and the conclusion are truth bearers or "truth-candidates", capable of being either true or false (and not both). While statements in an argument are referred to as being either *true* or *false*, arguments are referred to as being *valid* or *invalid* (see logical truth). A deductive argument is valid if and only if the truth of the

conclusion is entailed by (is a logical consequence of) the premises, and its corresponding conditional is therefore a logical truth. A sound argument is a valid argument with true premises; a valid argument may well have false premises under a given interpretation, however, the truth value of a conclusion cannot be determined by an unsound argument.

30.3 Deductive

Main article: Deductive argument
A *deductive argument* is one that, if valid, has a conclusion that is entailed by its premises. In other words, the truth of the conclusion is a logical consequence of the premises—if the premises are true, then the conclusion must be true. It would be self-contradictory to assert the premises and deny the conclusion, because the negation of the conclusion is contradictory to the truth of the premises.

30.3.1 Validity

Main article: Validity
Deductive arguments may be either valid or invalid. If an argument is valid, it is a valid deduction, and if its premises are true, the conclusion must be true: a valid argument cannot have true premises and a false conclusion.
An argument is formally valid if and only if the denial of the conclusion is incompatible with accepting all the premises.
The validity of an argument depends, however, not on the actual truth or falsity of its premises and conclusion, but solely on whether or not the argument has a valid logical form. The validity of an argument is not a guarantee of the truth of its conclusion. Under a given interpretation, a valid argument may have false premises that render it inconclusive: the conclusion of a valid argument with one or more false premises may be either true or false.
Logic seeks to discover the valid forms, the forms that make arguments valid. A form of argument is valid if and only if the conclusion is true under all interpretations of that argument in which the premises are true. Since the validity of an argument depends solely on its form, an argument can be shown to be invalid by showing that its form is invalid. This can be done by giving a counter example of the same form of argument with premises that are true under a given interpretation, but a conclusion that is false under that interpretation. In informal logic this is called a counter argument.
The form of argument can be shown by the use of symbols. For each argument form, there is a corresponding statement form, called a corresponding conditional, and an argument form is valid if and only its corresponding conditional is a logical truth. A statement form which is logically true is also said to be a valid statement form. A statement form is a logical truth if it is true under all interpretations. A statement form can be shown to be a logical truth by either (a) showing that it is a tautology or (b) by means of a proof procedure.
The corresponding conditional of a valid argument is a necessary truth (true *in all possible worlds*) and so the conclusion

necessarily follows from the premises, or follows of logical necessity. The conclusion of a valid argument is not necessarily true, it depends on whether the premises are true. If the conclusion, itself, just so happens to be a necessary truth, it is so without regard to the premises.

Some examples:

_ *Some Greeks are logicians; therefore, some logicians are Greeks.* Valid argument; it would be self-contradictory to admit that *some Greeks are logicians* but deny that *some (any) logicians are Greeks.*

_ *All Greeks are human and all humans are mortal; therefore, all Greeks are mortal.* : Valid argument; if the premises are true the conclusion must be true.

_ *Some Greeks are logicians and some logicians are tiresome; therefore, some Greeks are tiresome.* Invalid argument: the tiresome logicians might all be Romans (for example).

_ *Either we are all doomed or we are all saved; we are not all saved; therefore, we are all doomed.* Valid argument; the premises entail the conclusion. (Remember that this does not mean the conclusion has to be true; it is only true if the premises are true, which they may not be!)

_ *Some men are hawkers. Some hawkers are rich. Therefore, some men are rich.* Invalid argument, what can be easier seen by giving a counter-example with the same argument form:

_ *Some people are herbivores.* Some herbivores are zebras. Therefore, some people are zebras.

In the above last case, the counter-example follows the same logical form as the previous argument, (Premise 1: "Some *X* are *Y*." Premise 2: "Some *Y* are *Z*." Conclusion: "Some *X* are *Z*.") in order to demonstrate that whatever hawkers may be, they may or may not be rich, in consideration of the premises as such. *(See also, existential import).*

The forms of argument that render deductions valid are well-established, however some invalid arguments can also be persuasive depending on their construction (inductive arguments, for example). *(See also, formal fallacy and informal fallacy).*

30.3.2 Soundness

Main article: Soundness

A sound argument is a valid argument whose conclusion follows from its premise(s), and the premise(s) of the argument are true.

30.4 Inductive

Main article: Inductive argument

Non-deductive logic is reasoning using arguments in which the premises support the conclusion but do not entail it. Forms of non-deductive logic include the statistical syllogism, which argues from generalizations true for the most part, and induction, a form of reasoning that makes generalizations based on individual instances. An inductive argument is said to be *cogent* if and only if the truth of the argument's premises would render the truth of the conclusion probable (i.e., the argument is *strong*), and the argument's premises are, in fact, true. Cogency can be considered inductive logic's analogue to deductive logic's "soundness." Despite its name, mathematical induction is not a form of inductive reasoning. The lack of deductive validity is known as the problem of induction.

30.5 Defeasible

An argument is defeasible when additional information (such as new counterreasons) can have the effect that it no longer justifies its conclusion. The term "defeasibility" goes back to the legal theorist H.L.A. Hart, although he focused on concepts instead of arguments. Stephen Toulmin's influential argument model includes the possibility of counterreasons that are characteristic of defeasible arguments, but he did not discuss the evaluation of defeasible arguments. Defeasible arguments give rise to defeasible reasoning.

30.6 By analogy

Argument by analogy may be thought of as argument from the particular to particular. An argument by analogy may use a particular truth in a premise to argue towards a similar particular truth in the conclusion. For example, if A. Plato was mortal, and B. Socrates was like Plato in other respects, then asserting that C. Socrates was mortal is an example of argument by analogy because the reasoning employed in it proceeds from a particular truth in a premise (Plato was mortal) to a similar particular truth in the conclusion, namely that Socrates was mortal.[10]

30.7 Transitional

In epistemology, transitional arguments attempt to show that a particular explanation is better than another because it is able to make sense of a transition from old to new. That is, if explanation *b* can account for the problems that

existed with explanation *a*, but not vice versa, then *b* is regarded to be the more reasonable explanation. A common example in the history of science is the transition from pre-Galilean to Galilean understandings of physical motion.[11]

30.8 Other kinds

Other kinds of arguments may have different or additional standards of validity or justification. For example, Charles Taylor writes that so-called transcendental arguments are made up of a "chain of indispensability claims" that attempt to show why something is necessarily true based on its connection to our experience,[12] while Nikolas Kompridis has suggested that there are two types of "fallible" arguments: one based on truth claims, and the other based on the time-responsive disclosure of possibility (see world disclosure).[13] The late French philosopher Michel Foucault is said to have been a prominent advocate of this latter form of philosophical argument.[14]

30.8.1 In informal logic

Argument is an informal calculus, relating an effort to be performed or sum to be spent,
to possible future gain, either economic or moral.
In informal logic, an argument is a connexion between
a) an *individual* action
b) through which a *generally accepted* good is obtained.
Ex :
1. a) You should marry Jane (individual action, individual decision)
b) because she has the same temper as you. (generally accepted wisdom that marriage is good
in itself, and it is generally accepted that people with the same character get along well).
1. a) You should not smoke (individual action, individual decision)
b) because smoking is harmful (generally accepted wisdom that health is good).
The argument is neither a) *advice* nor b) *moral or economical judgement*, but the connection between the two. An argument always uses the connective **because**. An argument is not an **explanation**. It does not connect two events, cause and effect, which already took place, but a possible individual action and its beneficial outcome. An argument is not a **proof**. A proof is a logical and cognitive concept; an argument is a praxeologic concept. A proof changes our knowledge; an argument compels us to act.

Logical status

Argument does not belong to logic, because it is connected to a real person, a real event, and a real effort to be made.
a) If you, John, will buy this stock, it will become twice as valuable in a year. b) If you, Mary, study dance, you will become a famous ballet dancer.
The value of the argument is connected to the immediate circumstances of the person spoken to. If, in the first case,(a) John has no money, or will die the next year, he will not be interested in buying the stock. If, in the second case (b)she is too heavy, or too old, she will not be interested in studying and becoming a dancer. The argument is not logical, but profitable.

30.8.2 World-disclosing

Main article: World disclosure
World-disclosing arguments are a group of philosophical arguments that are said to employ a disclosive approach, to reveal features of a wider ontological or cultural-linguistic understanding – a "world," in a specifically ontological sense – in order to clarify or transform the background of meaning and "logical space" on which an argument implicitly depends.[15]

30.9 Explanations

Main article: Explanation
While arguments attempt to show that something is, will be, or should be the case, explanations try to show *why* or *how* something is or will be. If Fred and Joe address the issue of *whether* or not Fred's cat has fleas, Joe may state: "Fred, your cat has fleas. Observe, the cat is scratching right now." Joe has made an *argument that* the cat has fleas. However, if Joe asks Fred, "Why is your cat scratching itself?" the explanation, "...because it has fleas." provides understanding.
Both the above argument and explanation require knowing the generalities that a) fleas often cause itching, and b) that one often scratches to relieve itching. The difference is in the intent: an argument attempts to settle whether or not some claim is true, and an explanation attempts to provide understanding of the event. Note, that by subsuming the specific event (of Fred's cat scratching) as an instance of the general rule that "animals scratch themselves when they have fleas", Joe will no longer wonder *why* Fred's cat is scratching itself. Arguments address problems of believe, explanations address problems of understanding. Also note that in the argument above, the statement, "Fred's cat has fleas" is up for debate (i.e. is a claim), but in the explanation, the statement, "Fred's cat has fleas" is assumed to be true (unquestioned at this time) and just needs *explaining*.[16]

Arguments and explanations largely resemble each other in rhetorical use. This is the cause of much difficulty in thinking critically about claims. There are several reasons for this difficulty.
_ People often are not themselves clear on whether they are arguing for or explaining something.
_ The same types of words and phrases are used in presenting explanations and arguments.
_ The terms 'explain' or 'explanation,' et cetera are frequently used in arguments.
_ Explanations are often used within arguments and presented so as to serve *as arguments*.[17]
_ Likewise, "...arguments are essential to the process of justifying the validity of any explanation as there are often multiple explanations for any given phenomenon."[16]
Explanations and arguments are often studied in the field of Information Systems to help explain user acceptance of knowledge-based systems. Certain argument types may fit better with personality traits to enhance acceptance by individuals.[18]

30.10 Fallacies and nonarguments

Main article: Formal fallacy
Fallacies are types of argument or expressions which are held to be of an invalid form or contain errors in reasoning. There is not as yet any general theory of fallacy or strong agreement among researchers of their definition or potential for application but the term is broadly applicable as a label to certain examples of error, and also variously applied to ambiguous candidates.[19]
In Logic types of fallacy are firmly described thus: First the premises and the conclusion must be statements, capable of being true or false. Secondly it must be asserted that the conclusion follows from the premises. In English the words *therefore*, *so*, *because* and *hence* typically separate the premises from the conclusion of an argument, but this is not necessarily so. Thus: *Socrates is a man, all men are mortal therefore Socrates is mortal* is clearly an argument (a valid one at that), because it is clear it is asserted that *Socrates is mortal* follows from the preceding statements. However *I was thirsty and therefore I drank* is NOT an argument, despite its appearance. It is not being claimed that *I drank* is logically entailed by *I was thirsty*. The *therefore* in this sentence indicates *for that reason* not *it follows that*.
Elliptical arguments
Often an argument is invalid because there is a missing premise—the supply of which would render it valid. Speakers and writers will often leave out a strictly necessary premise in their reasonings if it is widely accepted and the writer

does not wish to state the blindingly obvious. Example: *All metals expand when heated, therefore iron will expand when heated.* (Missing premise: iron is a metal). On the other hand, a seemingly valid argument may be found to lack a premise – a 'hidden assumption' – which if highlighted can show a fault in reasoning. Example: A witness reasoned: *Nobody came out the front door except the milkman; therefore the murderer must have left by the back door.* (Hidden assumptions- the milkman was not the murderer, and the murderer has left by the front or back door).

30.11 See also

_ Abductive reasoning
_ Argument map
_ Argumentation theory
_ Argumentative dialogue
_ Belief bias
_ Boolean logic
_ Deductive reasoning
_ Defeasible reasoning
_ Evidence
_ Evidence-based policy
_ Fallacy
_ Dialectic
_ Formal fallacy
_ Inductive reasoning
_ Informal fallacy
_ Inquiry
_ Practical arguments
_ Soundness theorem
_ Soundness
_ Truth
_ Validity

30.12 Notes

[1] "Argument", Internet Encyclopedia of Philosophy." "In everyday life, we often use the word "argument" to mean a verbal dispute or disagreement. This is not the way this word is usually used in philosophy. However, the two uses are related. Normally, when two people verbally disagree with each other, each person attempts to convince the other that his/her viewpoint is the right one. Unless he or she merely results to name calling or threats, he or she typically presents an argument for his or her position, in the sense described above. In philosophy, "arguments" are those statements a person makes in the attempt to convince someone of something, or present reasons for accepting a given conclusion."
[2] Ralph H. Johnson, *Manifest Rationality: A pragmatic theory of argument* (New Jersey: Laurence Erlbaum, 2000), 46-49.
[3] Ralph H. Johnson, *Manifest Rationality: A pragmatic theory of argument* (New Jersey: Laurence Erlbaum, 2000), 46.
[4] The Cambridge Dictionary of Philosophy, 2nd Ed. CUM, 1995 "Argument: a sequence of statements such that some of them (the premises) purport to give reason to accept another of them, the conclusion"

[5] Stanford Enc. Phil., *Classical Logic*
[6] "Deductive and Inductive Arguments," Internet Encyclopedia of Philosophy.
[7] hCharles Taylor, "The Validity of Transcendental Arguments", *Philosophical Arguments* (Harvard, 1995), 20-33. "[Transcendental] arguments consist of a string of what one could call indispensability claims. They move from their starting points to their conclusions by showing that the condition stated in the conclusion is indispensable to the feature identified at the start… Thus we could spell out Kant's transcendental deduction in the first edition in three stages: experience must have an object, that is, be *of* something; for this it must be coherent; and to be coherent it must be shaped by the understanding through the categories."
[8] Kompridis, Nikolas (2006). "World Disclosing Arguments?". *Critique and Disclosure*. Cambridge: MIT Press. pp. 116–124. ISBN 0262277425.
[9] "Argument", Internet Encyclopedia of Philosophy."
[10] Shaw 1922: p. 74.
[11] Charles Taylor, "Explanation and Practical Reasoning", *Philosophical Arguments*, 34-60.
[12] Charles Taylor, "The Validity of Transcendental Arguments", *Philosophical Arguments* (Harvard, 1995), 20-33.
[13] Nikolas Kompridis, "Two Kinds of Fallibilism", *Critique and Disclosure* (Cambridge: MIT Press, 2006), 180-183.
[14] In addition, Foucault said of his own approach that "My role ... is to show people that they are much freer than they feel, that people accept as truth, as evidence, some themes which have been built up at a certain moment during history, and that this so-called evidence can be criticized and destroyed." He also wrote that he was engaged in "the process of putting historico-critical reflection to the *test of concrete practices*… I continue to think that this task requires work on our limits, that is, a patient labor giving form to our impatience for liberty." (emphasis added) Hubert Dreyfus, "Being and Power: Heidegger and Foucault" and Michel Foucault, "What is Enlightenment?"
[15] Nikolas Kompridis, "World Disclosing *Arguments*?" in *Critique and Disclosure*, Cambridge:MIT Press (2006), 118-121.
[16] JONATHAN F. OSBORNE, ALEXIS PATTERSON School of Education, Stanford University, Stanford, CA 94305, USA Received 27 August 2010; revised 22 November 2010; accepted 29 November 2010 DOI 10.1002/sce.20438 Published online 23 May 2011 in Wiley Online Library (wileyonlinelibrary.com)
[17] *Critical Thinking*, Parker and Moore
[18] Justin Scott Giboney, Susan Brown, and Jay F. Nunamaker Jr. (2012). "User Acceptance of Knowledge-Based System Recommendations: Explanations, Arguments, and Fit" 45th Annual Hawaii International Conference on System Sciences, Hawaii, January 5–8.
[19]

30.13 References

_ Shaw, Warren Choate (1922). *The Art of Debate*. Allyn and Bacon. p. 74.
_ Robert Audi, *Epistemology*, Routledge, 1998. Particularly relevant is Chapter 6, which explores the relationship between knowledge, inference and argument.
_ J. L. Austin *How to Do Things With Words*, Oxford University Press, 1976.
_ H. P. Grice, *Logic and Conversation* in *The Logic of Grammar*, Dickenson, 1975.
_ Vincent F. Hendricks, *Thought 2 Talk: A Crash Course in Reflection and Expression*, New York: Automatic Press / VIP, 2005, ISBN 87-991013-7-8
_ R. A. DeMillo, R. J. Lipton and A. J. Perlis, *Social Processes and Proofs of Theorems and Programs*, Communications of the ACM, Vol. 22, No. 5, 1979. A classic article on the social process of acceptance of proofs in mathematics.
_ Yu. Manin, *A Course in Mathematical Logic*, Springer Verlag, 1977. A mathematical view of logic. This book is different from most books on mathematical logic in that it emphasizes the mathematics of logic, as opposed to the formal structure of logic.

_ Ch. Perelman and L. Olbrechts-Tyteca, *The New Rhetoric*, Notre Dame, 1970. This classic was originally published in French in 1958.
_ Henri Poincaré, *Science and Hypothesis*, Dover Publications, 1952
_ Frans van Eemeren and Rob Grootendorst, *Speech Acts in Argumentative Discussions*, Foris Publications, 1984.
_ K. R. Popper *Objective Knowledge; An Evolutionary Approach*, Oxford: Clarendon Press, 1972.
_ L. S. Stebbing, *A Modern Introduction to Logic*, Methuen and Co., 1948. An account of logic that covers the classic topics of logic and argument while carefully considering modern developments in logic.

_ Douglas Walton, *Informal Logic: A Handbook for Critical Argumentation*, Cambridge, 1998
_ Carlos Chesñevar, Ana Maguitman and Ronald Loui, *Logical Models of Argument*, ACM Computing Surveys, vol. 32, num. 4, pp. 337–383, 2000.
_ T. Edward Damer. Attacking Faulty Reasoning, 5th Edition, Wadsworth, 2005. ISBN 0-534-60516-8
_ Charles Arthur Willard, A Theory of Argumentation. 1989.
_ Charles Arthur Willard, Argumentation and the Social Grounds of Knowledge. 1982.

30.14 Further reading

_ Salmon, Wesley C. *Logic*. New Jersey: Prentice-Hall (1963). Library of Congress Catalog Card no. 63-10528.
_ Aristotle, *Prior and Posterior Analytics*. Ed. and trans. John Warrington. London: Dent (1964)
_ Mates, Benson. *Elementary Logic*. New York: OUP (1972). Library of Congress Catalog Card no. 74-166004.
_ Mendelson, Elliot. *Introduction to Mathematical Logic*. New York: Van Nostran Reinholds Company (1964).
_ Frege, Gottlob. *The Foundations of Arithmetic*. Evanston, IL: Northwestern University Press (1980).
_ Martin, Brian. *The Controversy Manual* (Sparsnäs, Sweden: Irene Publishing, 2014).

30.15 External links

_ Argument at PhilPapers
_ Argument at the Indiana Philosophy Ontology Project
_ Argument entry in the *Internet Encyclopedia of Philosophy*

Chapter 31

Formal language

This article is about a technical term in mathematics and computer science. For related studies about natural languages, see Formal semantics (linguistics). For formal modes of speech in natural languages, see Register (sociolinguistics).
In mathematics, computer science, and linguistics, a **formal language** is a set of strings of symbols that may be

S

NP

N'

AdjP

Adj'

Adj

Colorless

N'

N

ideas

N'

AdjP

Adj'

Adj

green
VP
V'
V'
V
sleep
AdvP
Adv'
Adv
furiously

Structure of a syntactically well-formed, although nonsensical English sentence (historical example from Chomsky 1957).

constrained by rules that are specific to it.

The alphabet of a formal language is the set of symbols, letters, or tokens from which the strings of the language may be formed; frequently it is required to be finite.[1] The strings formed from this alphabet are called words, and the

words that belong to a particular formal language are sometimes called *well-formed words* or *well-formed formulas*.

A formal language is often defined by means of a formal grammar such as a regular grammar or context-free grammar, also called its formation rule.

The field of **formal language theory** studies primarily the purely syntactical aspects of such languages—that is, their internal structural patterns. Formal language theory sprang out of linguistics, as a way of understanding the syntactic regularities of natural languages. In computer science, formal languages are used among others as the basis for defining the grammar of programming languages and formalized versions of subsets of natural languages in which the words of the language represent concepts that are associated with particular meanings or semantics. In computational complexity theory, decision problems are typically defined as formal languages, and complexity classes are defined as the sets of the formal languages that can be parsed by machines with limited computational power. In logic and the foundations of mathematics, formal languages are used to represent the syntax of axiomatic systems, and mathematical formalism is the philosophy that all of mathematics can be reduced to the syntactic manipulation of formal languages in this way.

31.1 History

The first formal language is thought be the one used by Gottlob Frege in his *Begriffsschrift* (1879), literally meaning "concept writing", and which Frege described as a "formal language of pure thought."[2]

Axel Thue's early Semi-Thue system which can be used for rewriting strings was influential on formal grammars.

31.2 Words over an alphabet

An **alphabet**, in the context of formal languages, can be any set, although it often makes sense to use an alphabet in the usual sense of the word, or more generally a character set such as ASCII or Unicode. Alphabets can also be infinite; e.g. first-order logic is often expressed using an alphabet which, besides symbols such as $\wedge$, $\neg$, $\forall$ and parentheses, contains infinitely many elements x_0, x_1, x_2, … that play the role of variables. The elements of an alphabet are called its **letters**.

A **word** over an alphabet can be any finite sequence, or string, of characters or letters, which sometimes may include spaces, and are separated by specified word separation characters. The set of all words over an alphabet Σ is usually denoted by Σ^* (using the Kleene star). The length of a word is the number of characters or letters it is composed of. For any alphabet there is only one word of length 0, the *empty word*, which is often denoted by e, ε or λ. By concatenation one can combine two words to form a new word, whose length is the sum of the lengths of the original words. The result of concatenating a word with the empty word is the original word.

In some applications, especially in logic, the alphabet is also known as the *vocabulary* and words are known as *formulas* or *sentences*; this breaks the letter/word metaphor and replaces it by a word/sentence metaphor.

31.3 Definition

A **formal language** *L* over an alphabet Σ is a subset of Σ^*, that is, a set of words over that alphabet. Sometimes the sets of words are grouped into expressions, whereas rules and constraints may be formulated for the creation of 'well-formed expressions'.
In computer science and mathematics, which do not usually deal with natural languages, the adjective "formal" is often omitted as redundant.
While formal language theory usually concerns itself with formal languages that are described by some syntactical rules, the actual definition of the concept "formal language" is only as above: a (possibly infinite) set of finite-length strings composed from a given alphabet, no more nor less. In practice, there are many languages that can be described by rules, such as regular languages or context-free languages. The notion of a formal grammar may be closer to the intuitive concept of a "language," one described by syntactic rules. By an abuse of the definition, a particular formal language is often thought of as being equipped with a formal grammar that describes it.

31.4 Examples

The following rules describe a formal language L over the alphabet Σ = { 0, 1, 2, 3, 4, 5, 6, 7, 8, 9, +, = }:
_ Every nonempty string that does not contain "+" or "=" and does not start with "0" is in L.
_ The string "0" is in L.
_ A string containing "=" is in L if and only if there is exactly one "=", and it separates two valid strings of L.
_ A string containing "+" but not "=" is in L if and only if every "+" in the string separates two valid strings of L.
_ No string is in L other than those implied by the previous rules.
Under these rules, the string "23+4=555" is in L, but the string "=234=+" is not. This formal language expresses natural numbers, well-formed addition statements, and well-formed addition equalities, but it expresses only what they look like (their syntax), not what they mean (semantics). For instance, nowhere in these rules is there any indication that "0" means the number zero, or that "+" means addition.

31.4.1 Constructions

For finite languages one can explicitly enumerate all well-formed words. For example, we can describe a language L as just L = {"a", "b", "ab", "cba"}. The degenerate case of this construction is the **empty language**, which contains no words at all (L = ∅).
However, even over a finite (non-empty) alphabet such as Σ = {a, b} there are infinitely many words: "a", "abb", "ababba", "aaababbbbaab", ... Therefore formal languages are typically infinite, and describing an infinite formal language is not as simple as writing *L* = {"a", "b", "ab", "cba"}. Here are some examples of formal languages:
_ L = Σ^*, the set of *all* words over Σ;
_ L = {"a"}* = {"a"^{n}}, where *n* ranges over the natural numbers and "a"^{n} means "a" repeated *n* times (this is the set of words consisting only of the symbol "a");
_ the set of syntactically correct programs in a given programming language (the syntax of which is usually defined by a context-free grammar);
_ the set of inputs upon which a certain Turing machine halts; or
_ the set of maximal strings of alphanumeric ASCII characters on this line, i.e., the set {"the", "set", "of", "maximal", "strings", "alphanumeric", "ASCII", "characters", "on", "this", "line", "i", "e"}.

31.5 Language-specification formalisms

Formal language theory rarely concerns itself with particular languages (except as examples), but is mainly concerned with the study of various types of formalisms to describe languages. For instance, a language can be given as
_ those strings generated by some formal grammar;
_ those strings described or matched by a particular regular expression;
_ those strings accepted by some automaton, such as a Turing machine or finite state automaton;
_ those strings for which some decision procedure (an algorithm that asks a sequence of related YES/NO questions) produces the answer YES.
Typical questions asked about such formalisms include:
_ What is their expressive power? (Can formalism *X* describe every language that formalism *Y* can describe? Can it describe other languages?)

_ What is their recognizability? (How difficult is it to decide whether a given word belongs to a language described by formalism *X*?)
_ What is their comparability? (How difficult is it to decide whether two languages, one described in formalism *X* and one in formalism *Y*, or in *X* again, are actually the same language?).

Surprisingly often, the answer to these decision problems is "it cannot be done at all", or "it is extremely expensive" (with a characterization of how expensive). Therefore, formal language theory is a major application area of computability theory and complexity theory. Formal languages may be classified in the Chomsky hierarchy based on the expressive power of their generative grammar as well as the complexity of their recognizing automaton. Contextfree grammars and regular grammars provide a good compromise between expressivity and ease of parsing, and are widely used in practical applications.

31.6 Operations on languages

Certain operations on languages are common. This includes the standard set operations, such as union, intersection, and complement. Another class of operation is the element-wise application of string operations.
Examples: suppose L_1 and L_2 are languages over some common alphabet.

_ The *concatenation* L_1L_2 consists of all strings of the form vw where v is a string from L_1 and w is a string from L_2.

_ The *intersection* $L_1 \cap L_2$ of L_1 and L_2 consists of all strings which are contained in both languages

_ The *complement* $\neg L$ of a language with respect to a given alphabet consists of all strings over the alphabet that are not in the language.

_ The Kleene star: the language consisting of all words that are concatenations of 0 or more words in the original language;

_ *Reversal*:

_ Let e be the empty word, then $e^R = e$, and

_ for each non-empty word $w = x_1 \ldots x_n$ over some alphabet, let $w^R = x_\square \ldots x_1$,

_ then for a formal language L, $L^R = \{w^R \mid w \in L\}$.

_ String homomorphism

Such string operations are used to investigate closure properties of classes of languages. A class of languages is closed under a particular operation when the operation, applied to languages in the class, always produces a language in the same class again. For instance, the context-free languages are known to be closed under union, concatenation, and intersection with regular languages, but not closed under intersection or complement. The theory of trios and abstract families of languages studies the most common closure properties of language families in their own right.[3]

31.7 Applications

31.7.1 Programming languages

Main articles: Syntax (programming languages) and Compiler compiler
A compiler usually has two distinct components. A lexical analyzer, generated by a tool like lex, identifies the tokens of the programming language grammar, e.g. identifiers or keywords, which are themselves expressed in a simpler formal language, usually by means of regular expressions. At the most basic conceptual level, a parser, usually

generated by a parser generator like yacc, attempts to decide if the source program is valid, that is if it belongs to the programming language for which the compiler was built. Of course, compilers do more than just parse the source code—they usually translate it into some executable format. Because of this, a parser usually outputs more than a yes/no answer, typically an abstract syntax tree, which is used by subsequent stages of the compiler to eventually generate an executable containing machine code that runs directly on the hardware, or some intermediate code that requires a virtual machine to execute.

31.7.2 Formal theories, systems and proofs

Symbols and
strings of symbols
Well-formed formulas
Theorems

This diagram shows the syntactic divisions within a formal system. Strings of symbols may be broadly divided into nonsense and

well-formed formulas. The set of well-formed formulas is divided into theorems and non-theorems.
Main articles: Theory (mathematical logic) and Formal system
In mathematical logic, a *formal theory* is a set of sentences expressed in a formal language.
A *formal system* (also called a *logical calculus*, or a *logical system*) consists of a formal language together with a deductive apparatus (also called a *deductive system*). The deductive apparatus may consist of a set of transformation rules which may be interpreted as valid rules of inference or a set of axioms, or have both. A formal system is used to derive one expression from one or more other expressions. Although a formal language can be identified with its formulas, a formal system cannot be likewise identified by its theorems. Two formal systems *FS* and *FS′* may have all the same theorems and yet differ in some significant proof-theoretic way (a formula A may be a syntactic consequence of a formula B in one but not another for instance).

A *formal proof* or *derivation* is a finite sequence of well-formed formulas (which may be interpreted as propositions) each of which is an axiom or follows from the preceding formulas in the sequence by a rule of inference. The last sentence in the sequence is a theorem of a formal system. Formal proofs are useful because their theorems can be interpreted as true propositions.
Interpretations and models
Main articles: Formal semantics (logic), Interpretation (logic) and Model theory
Formal languages are entirely syntactic in nature but may be given semantics that give meaning to the elements of the language. For instance, in mathematical logic, the set of possible formulas of a particular logic is a formal language, and an interpretation assigns a meaning to each of the formulas—usually, a truth value.
The study of interpretations of formal languages is called formal semantics. In mathematical logic, this is often done in terms of model theory. In model theory, the terms that occur in a formula are interpreted as mathematical structures, and fixed compositional interpretation rules determine how the truth value of the formula can be derived from the interpretation of its terms; a *model* for a formula is an interpretation of terms such that the formula becomes true.

31.8 See also

_ Combinatorics on words
_ Grammar framework
_ Formal method
_ Mathematical notation
_ Associative array
_ String (computer science)

31.9 References

31.9.1 Citation footnotes

[1] See e.g. Reghizzi, Stefano Crespi (2009), *Formal Languages and Compilation*, Texts in Computer Science, Springer, p. 8, ISBN 9781848820500, "An alphabet is a finite set".
[2] Martin Davis (1995). "Influences of Mathematical Logic on Computer Science". In Rolf Herken. *The universal Turing machine: a half-century survey*. Springer. p. 290. ISBN 978-3-211-82637-9.
[3] Hopcroft & Ullman (1979), Chapter 11: Closure properties of families of languages.

31.9.2 General references

_ A. G. Hamilton, *Logic for Mathematicians*, Cambridge University Press, 1978, ISBN 0-521-21838-1.
_ Seymour Ginsburg, *Algebraic and automata theoretic properties of formal languages*, North-Holland, 1975, ISBN 0-7204-2506-9.
_ Michael A. Harrison, *Introduction to Formal Language Theory*, Addison-Wesley, 1978.
_ John E. Hopcroft and Jeffrey D. Ullman, *Introduction to Automata Theory, Languages, and Computation*, Addison-Wesley Publishing, Reading Massachusetts, 1979. ISBN 81-7808-347-7.
_ Rautenberg, Wolfgang (2010). *A Concise Introduction to Mathematical Logic* (3rd ed.). New York: Springer Science+Business Media. doi:10.1007/978-1-4419-1221-3. ISBN 978-1-4419-1220-6.

_ Grzegorz Rozenberg, Arto Salomaa, *Handbook of Formal Languages: Volume I-III*, Springer, 1997, ISBN 3-540-61486-9.
_ Patrick Suppes, *Introduction to Logic*, D. Van Nostrand, 1957, ISBN 0-442-08072-7.

31.10 External links

_ Hazewinkel, Michiel, ed. (2001), "Formal language", *Encyclopedia of Mathematics*, Springer, ISBN 978-1-

55608-010-4
_ Alphabet at PlanetMath.org.
_ Language at PlanetMath.org.
_ University of Maryland, Formal Language Definitions
_ James Power, "Notes on Formal Language Theory and Parsing", 29 November 2002.
_ Drafts of some chapters in the "Handbook of Formal Language Theory", Vol. 1-3, G. Rozenberg and A. Salomaa (eds.), Springer Verlag, (1997):
_ Alexandru Mateescu and Arto Salomaa, "Preface" in Vol.1, pp. v-viii, and "Formal Languages: An Introduction and a Synopsis", Chapter 1 in Vol. 1, pp.1-39
_ Sheng Yu, "Regular Languages", Chapter 2 in Vol. 1
_ Jean-Michel Autebert, Jean Berstel, Luc Boasson, "Context-Free Languages and Push-Down Automata", Chapter 3 in Vol. 1
_ Christian Choffrut and Juhani Karhumäki, "Combinatorics of Words", Chapter 6 in Vol. 1
_ Tero Harju and Juhani Karhumäki, "Morphisms", Chapter 7 in Vol. 1, pp. 439 - 510
_ Jean-Eric Pin, "Syntactic semigroups", Chapter 10 in Vol. 1, pp. 679-746
_ M. Crochemore and C. Hancart, "Automata for matching patterns", Chapter 9 in Vol. 2
_ Dora Giammarresi, Antonio Restivo, "Two-dimensional Languages", Chapter 4 in Vol. 3, pp. 215 – 267

Chapter 32

Independence (mathematical logic)

In mathematical logic, **independence** refers to the unprovability of a sentence from other sentences.
A sentence σ is **independent** of a given first-order theory *T* if *T* neither proves nor refutes σ; that is, it is impossible to prove σ from *T*, and it is also impossible to prove from *T* that σ is false. Sometimes, σ is said (synonymously) to be ***undecidable*** from *T*; this is not the same meaning of "decidability" as in a decision problem.
A theory *T* is **independent** if each axiom in *T* is not provable from the remaining axioms in *T*. A theory for which there is an independent set of axioms is **independently axiomatizable**.

32.1 Usage note

Some authors say that σ is independent of *T* if *T* simply cannot prove σ, and do not necessarily assert by this that *T* cannot refute σ. These authors will sometimes say "σ is independent of and consistent with *T*" to indicate that *T* can neither prove nor refute σ.

32.2 Independence results in set theory

Many interesting statements in set theory are independent of Zermelo-Fraenkel set theory (ZF). The following statements in set theory are known to be independent of ZF, granting that ZF is consistent:
_ The axiom of choice
_ The continuum hypothesis and the generalised continuum hypothesis
_ The Suslin conjecture
The following statements (none of which have been proved false) cannot be proved in ZFC to be independent of ZFC, even if the added hypothesis is granted that ZFC is consistent. However, they cannot be proved in ZFC (granting that ZFC is consistent), and few working set theorists expect to find a refutation of them in ZFC.
_ The existence of strongly inaccessible cardinals
_ The existence of large cardinals
_ The non-existence of Kurepa trees
The following statements are inconsistent with the axiom of choice, and therefore with ZFC. However they are probably independent of ZF, in a corresponding sense to the above: They cannot be proved in ZF, and few working set theorists expect to find a refutation in ZF. However ZF cannot prove that they are independent of ZF, even with the added hypothesis that ZF is consistent.
228
32.3. SEE ALSO 229
_ The Axiom of determinacy
_ The axiom of real determinacy
_ AD+

32.3 See also

_ List of statements undecidable in ZFC
_ Parallel postulate for an example in geometry
_ Truth

32.4 References

_ Mendelson, Elliott (1997), *An Introduction to Mathematical Logic* (4th ed.), London: Chapman & Hall, ISBN 978-0-412-80830-2
_ Monk, J. Donald (1976), *Mathematical Logic*, Graduate Texts in Mathematics, Berlin, New York: Springer-Verlag, ISBN 978-0-387-90170-1

Chapter 33

Analytic–synthetic distinction

The **analytic–synthetic distinction** (also called the **analytic–synthetic dichotomy**) is a conceptual distinction, used primarily in philosophy to distinguish propositions (in particular, statements that are affirmative subject–predicate judgments) into two types: *analytic propositions* and *synthetic propositions*. **Analytic propositions** are true by virtue of their meaning, while **synthetic propositions** are true by how their meaning relates to the world.[1] However, philosophers have used the terms in very different ways. Furthermore, philosophers have debated whether there is a legitimate distinction.

33.1 Kant

33.1.1 Conceptual containment

The philosopher Immanuel Kant uses the terms "analytic" and "synthetic" to divide propositions into two types. Kant introduces the analytic–synthetic distinction in the Introduction to his *Critique of Pure Reason* (1781/1998, A6–7/B10–11). There, he restricts his attention to statements that are affirmative subject–predicate judgments, and defines "analytic proposition" and "synthetic proposition" as follows:
_ **analytic proposition**: a proposition whose predicate concept is contained in its subject concept
_ **synthetic proposition**: a proposition whose predicate concept is **not** contained in its subject concept but related
Examples of analytic propositions, on Kant's definition, include:
_ "All bachelors are unmarried."
_ "All triangles have three sides."
Kant's own example is:
_ "All bodies are extended," that is, occupy space. (A7/B11)
Each of these statements is an affirmative subject–predicate judgment, and, in each, the predicate concept is *contained* within the subject concept. The concept "bachelor" contains the concept "unmarried"; the concept "unmarried" is part of the definition of the concept "bachelor." Likewise, for "triangle" and "has three sides," and so on.
Examples of synthetic propositions, on Kant's definition, include:
_ "All bachelors are unhappy."
_ "All creatures with hearts have kidneys."
Kant's own example is:
230
33.1. KANT 231
_ "All bodies are heavy," (A7/B11)
As with the previous examples classified as analytic propositions, each of these new statements is an affirmative subject–predicate judgment. However, in none of these cases does the subject concept contain the predicate concept. The concept "bachelor" does not contain the concept "unhappy"; "unhappy" is not a part of the *definition* of "bachelor." The same is true for "creatures with hearts" and "have kidneys"; even if every creature with a heart also has kidneys, the concept "creature with a heart" does not contain the concept "has kidneys."

33.1.2 Kant's version and the *a priori / a posteriori* distinction

Main article: A priori and a posteriori
In the Introduction to the *Critique of Pure Reason*, Kant contrasts his distinction between analytic and synthetic

propositions with another distinction, the distinction between *a priori* and *a posteriori* propositions. He defines these terms as follows:

_ ***a priori*** **proposition**: a proposition whose justification does *not* rely upon experience. Moreover, the proposition can be validated by experience, but is not grounded in experience. Therefore, it is logically necessary.

_ ***a posteriori*** **proposition**: a proposition whose justification does rely upon experience. The proposition is validated by, and grounded in, experience. Therefore, it is logically contingent.

Examples of *a priori* propositions include:

_ "All bachelors are unmarried."

_ "7 + 5 = 12."

The justification of these propositions does not depend upon experience: One need not consult experience to determine whether all bachelors are unmarried, nor whether 7 + 5 = 12. (Of course, as Kant would grant, experience is required to understand the concepts "bachelor," "unmarried," "7", "+" and so forth. However, the *a priori/a posteriori* distinction as employed here by Kant refers not to the *origins* of the concepts but to the *justification* of the propositions. Once we have the concepts, experience is no longer necessary.)

Examples of *a posteriori* propositions include:

_ "All bachelors are unhappy."

_ "Tables exist."

Both of these propositions are *a posteriori*: Any justification of them would require one's experience.

The analytic/synthetic distinction and the *a priori/a posteriori* distinction together yield four types of propositions:

1. analytic *a priori*
2. synthetic *a priori*
3. analytic *a posteriori*
4. synthetic *a posteriori*

Kant says the third type is self-contradictory, so he discusses only the remaining three types as components of his epistemological framework.

33.1.3 The ease of knowing analytic propositions

Part of Kant's argument in the Introduction to the *Critique of Pure Reason* involves arguing that there is no problem figuring out how knowledge of analytic propositions is possible. To know an analytic proposition, Kant argued, one need not consult experience. Instead, one need merely to take the subject and "extract from it, in accordance with the principle of contradiction, the required predicate ..." (A7/B12) In analytic propositions, the predicate concept is contained in the subject concept. Thus, to know an analytic proposition is true, one need merely examine the concept of the subject. If one finds the predicate contained in the subject, the judgment is true.

Thus, for example, one need not consult experience to determine whether "All bachelors are unmarried" is true. One need merely examine the subject concept ("bachelors") and see if the predicate concept "unmarried" is contained in it. And in fact, it is: "unmarried" is part of the definition of "bachelor," and so is contained within it. Thus the proposition "All bachelors are unmarried" can be known to be true without consulting experience.

It follows from this, Kant argued, first: All analytic propositions are *a priori*; there are no *a posteriori* analytic propositions. It follows, second: There is no problem understanding how we can know analytic propositions. We can know them because we just need to consult our concepts in order to determine that they are true.

33.1.4 The possibility of metaphysics

After ruling out the possibility of analytic *a posteriori* propositions, and explaining how we can obtain knowledge of analytic *a priori* propositions, Kant also explains how we can obtain knowledge of synthetic *a posteriori* propositions. That leaves only the question of how knowledge of synthetic *a priori* propositions is possible. This question is exceedingly important, Kant maintains, because all important metaphysical knowledge is of synthetic *a priori* propositions. If it is impossible to determine which synthetic *a priori* propositions are true, he argues, then metaphysics as a discipline is impossible. The remainder of the *Critique of Pure Reason* is devoted to examining whether and how knowledge of synthetic *a priori* propositions is possible.

33.2 Logical positivists

33.2.1 Frege and Carnap revise the Kantian definition

Over a hundred years later, a group of philosophers took interest in Kant and his distinction between analytic and synthetic propositions: the logical positivists.

Part of Kant's examination of the possibility of synthetic *a priori* knowledge involved the examination of mathematical propositions, such as

_ "7 + 5 = 12" (B15–16)

_ "The shortest distance between two points is a straight line." (B16–17)
Kant maintained that mathematical propositions such as these are synthetic *a priori* propositions, and that we know them. That they are synthetic, he thought, is obvious: The concept "12" is not contained within the concept "5," or the concept "7," or the concept "+." And the concept "straight line" is not contained within the concept "the shortest distance between two points." (B15–17) From this, Kant concluded that we have knowledge of synthetic *a priori* propositions. He went on to maintain that it is extremely important to determine how such knowledge is possible.
Frege's notion of analyticity included a number of logical properties and relations beyond containment: symmetry, transitivity, antonymy, or negation and so on. He had a strong emphasis on formality, in particular formal definition, and also emphasized the idea of substitution of synonymous terms. "All bachelors are unmarried" can be expanded out with the formal definition of bachelor as "unmarried man" to form "All unmarried men are unmarried," which is recognizable as tautologous and therefore analytic from its logical form: any statement of the form "All *X* that are (*F* and *G*) are *F*". This expanded idea of analyticity was able to show that all Kant's examples of arithmetical and geometrical truths are analytical *a priori* truths and *not* synthetic *a priori* truths.
"Thanks to Frege's logical semantics, particularly his concept of analyticity, arithmetic truths like "7+5=12" are no longer synthetic *a priori* but analytical *a priori* truths in Carnap's extended sense of

"analytic". Hence logical empiricists are not subject to Kant's criticism of Hume for throwing out mathematics along with metaphysics"[2]
(Here "logical empiricist" is a synonym for "logical positivist")

33.2.2 The origin of the logical positivist's distinction

The logical positivists agreed with Kant that we have knowledge of mathematical truths, and further that mathematical propositions are *a priori*. However, they did not believe that any complex metaphysics, such as the type Kant supplied, are necessary to explain our knowledge of mathematical truths. Instead, the logical positivists maintained that our knowledge of judgments like "all bachelors are unmarried" and our knowledge of mathematics (and logic) are in the basic sense the same: all proceeded from our knowledge of the meanings of terms or the conventions of language.
"Since empiricism had always asserted that *all* knowledge is based on experience, this assertion had to include knowledge in mathematics. On the other hand, we believed that with respect to this problem the rationalists had been right in rejecting the old empiricist view that the truth of "2+2=4" is contingent on the observation of facts, a view that would lead to the unacceptable consequence that an arithmetical statement might possibly be refuted tomorrow by new experiences. Our solution, based upon Wittgenstein's conception, consisted in asserting the thesis of empiricism only for factual truth. By contrast, the truths of logic and mathematics are not in need of confirmation by observations, because they do not state anything about the world of facts, they hold for any possible combination of facts".[3][4]
—Rudolf Carnap, Autobiography: §10: Semantics, p. 64

33.2.3 Logical positivist definitions

Thus the logical positivists drew a new distinction, and, inheriting the terms from Kant, named it the "analytic/synthetic distinction."[5] They provided many different definitions, such as the following:
1. **analytic proposition**: a proposition whose truth depends solely on the meaning of its terms
2. **analytic proposition**: a proposition that is true (or false) by definition
3. **analytic proposition**: a proposition that is made true (or false) solely by the conventions of language
(While the logical positivists believed that the only necessarily true propositions were analytic, they did not define "analytic proposition" as "necessarily true proposition" or "proposition that is true in all possible worlds.")
Synthetic propositions were then defined as:
_ **synthetic proposition**: a proposition that is not analytic
These definitions applied to all propositions, regardless of whether they were of subject–predicate form. Thus, under these definitions, the proposition "It is raining or it is not raining," was classified as analytic, while under Kant's definitions it was neither analytic nor synthetic. And the proposition "7 + 5 = 12" was classified as analytic, while under Kant's definitions it was synthetic.

33.3 Two-dimensionalism

Two-dimensionalism is an approach to semantics in analytic philosophy. It is a theory of how to determine the sense and reference of a word and the truth-value of a sentence. It is intended to resolve a puzzle that has plagued philosophy for some time, namely: How is it possible to discover empirically that a necessary truth is true? Two-dimensionalism provides an analysis of the semantics of words and sentences that makes sense of this possibility. The theory was first developed by Robert Stalnaker, but it has been advocated by numerous philosophers since, including David Chalmers and Berit Brogaard.
Any given sentence, for example, the words,

"Water is H_2O"
is taken to express two distinct propositions, often referred to as a *primary intension* and a *secondary intension*, which together compose its meaning.[6]
The primary intension of a word or sentence is its sense, i.e., is the idea or method by which we find its referent. The primary intension of "water" might be a description, such as *watery stuff*. The thing picked out by the primary intension of "water" could have been otherwise. For example, on some other world where the inhabitants take "water" to mean *watery stuff*, but, where the chemical make-up of watery stuff is not H_2O, it is not the case that water is H_2O for that world.
The *secondary intension* of "water" is whatever thing "water" happens to pick out in *this* world, whatever that world happens to be. So if we assign "water" the primary intension *watery stuff* then the secondary intension of "water" is H_2O, since H_2O is *watery stuff* in this world. The secondary intension of "water" in our world is H_2O, which is H_2O in every world because unlike *watery stuff* it is impossible for H_2O to be other than H_2O. When considered according to its secondary intension, "Water is H_2O" is true in every world.
If two-dimensionalism is workable it solves some very important problems in the philosophy of language. Saul Kripke has argued that "Water is H_2O" is an example of the *necessary a posteriori*, since we had to discover that water was H_2O, but given that it is true (which it is) it cannot be false. It would be absurd to claim that something that is water is not H_2O, for these are known to be *identical*.

33.4 Quine's criticisms

See also: Willard Van Orman Quine § Rejection of the analytic–synthetic distinction and Two Dogmas of Empiricism § Analyticity and circularity
Rudolf Carnap was a strong proponent of the distinction between what he called "internal questions," questions entertained within a "framework" (like a mathematical theory), and "external questions," questions posed outside any framework – posed before the adoption of any framework.[7][8][9] The "internal" questions could be of two types: *logical* (or analytic, or logically true) and *factual* (empirical, that is, matters of observation interpreted using terms from a framework). The "external" questions were also of two types: those that were confused pseudo-questions ("one disguised in the form of a theoretical question") and those that could be re-interpreted as practical, pragmatic questions about whether a framework under consideration was "more or less expedient, fruitful, conducive to the aim for which the language is intended."[7] The adjective "synthetic" was not used by Carnap in his 1950 work: *Empiricism, Semantics, and Ontology*.[7] Carnap did define a "synthetic truth" in his work *Meaning and Necessity*: a sentence that is true, but not simply because "the semantical rules of the system suffice for establishing its truth".[10] The notion of a synthetic truth is something true both because of what it means and because of the way the world is, whereas analytic truths are true in virtue of meaning alone. Thus, what Carnap calls internal *factual* statements (as opposed to internal *logical* statements) could be taken as being also synthetic truths because they require *observations*, but some external statements also could be "synthetic" statements and Carnap would be doubtful about their status. The analytic–synthetic argument therefore is not identical with the internal–external distinction.[11]
In 1951, W.V. Quine published the essay "Two Dogmas of Empiricism" in which he argued that the analytic–synthetic distinction is untenable.[12] The argument at bottom is that there are no "analytic" truths, but all truths involve an empirical aspect. In the first paragraph, Quine takes the distinction to be the following:
_ analytic propositions – propositions grounded in meanings, independent of matters of fact.
_ synthetic propositions – propositions grounded in fact.
Quine's position denying the analytic/synthetic distinction is summarized as follows:
It is obvious that truth in general depends on both language and extralinguistic fact. ... Thus one is tempted to suppose in general that the truth of a statement is somehow analyzable into a linguistic component and a factual component. Given this supposition, it next seems reasonable that in some statements the factual component should be null; and these are the analytic statements. But, for all its *a priori* reasonableness, a boundary between analytic and synthetic statements simply has not been drawn.

That there is such a distinction to be drawn at all is an unempirical dogma of empiricists, a metaphysical article of faith.[13]
—Willard v. O. Quine, Two dogmas of empiricism, p. 64
To summarize Quine's argument, the notion of an analytic proposition requires a notion of synonymy, but establishing synonymy inevitably leads to matters of fact – synthetic propositions. Thus, there is no non-circular (and so no tenable) way to ground the notion of analytic propositions.
While Quine's rejection of the analytic–synthetic distinction is widely known, the precise argument for the rejection and its status is highly debated in contemporary philosophy. However, some (for example, Boghossian[14]) argue that Quine's rejection of the distinction is still widely accepted among philosophers, even if for poor reasons.

33.4.1 Responses

Paul Grice and P.F. Strawson criticized "Two Dogmas" in their (1956) article "In Defense of a Dogma."[15] Among other things, they argue that Quine's skepticism about synonyms leads to a skepticism about meaning. If statements can have meanings, then it would make sense to ask "What does it mean?". If it makes sense to ask "What does it mean?", then synonymy can be defined as follows: Two sentences are synonymous if and only if the true answer of the question "What does it mean?" asked of one of them is the true answer to the same question asked of the other. They also draw the conclusion that discussion about correct or incorrect translations would be impossible given Quine's argument. Four years after Grice and Strawson published their paper, Quine's book *Word and Object* was released. In the book Quine presented his theory of indeterminacy of translation.

In "Speech acts," John R. Searle argues that from the difficulties encountered in trying to explicate analyticity by appeal to specific criteria, it does not follow that the notion itself is void.[16] Considering the way which we would test any proposed list of criteria, which is by comparing their extension to the set of analytic statements, it would follow that any explication of what analyticity means presupposes that we already have at our disposal a working notion of analyticity.

In "'Two Dogmas' revisited," Hilary Putnam argues that Quine is attacking two different notions.[17]

It seems to me there is as gross a distinction between 'All bachelors are unmarried' and 'There is a book on this table' as between any two things in this world, or at any rate, between any two linguistic expressions in the world;[18]

—Hilary Putnam, Philosophical papers, p. 36

Analytic truth defined as a true statement derivable from a tautology by putting synonyms for synonyms is near Kant's account of analytic truth as a truth whose negation is a contradiction. Analytic truth defined as a truth confirmed no matter what, however, is closer to one of the traditional accounts of *a priori*. While the first four sections of Quine's paper concern analyticity, the last two concern a priority. Putnam considers the argument in the two last sections as independent of the first four, and at the same time as Putnam criticizes Quine, he also emphasizes his historical importance as the first top rank philosopher to both reject the notion of a priority and sketch a methodology without it.[19]

Jerrold Katz, a onetime associate of Noam Chomsky's, countered the arguments of *Two Dogmas* directly by trying to define analyticity non-circularly on the syntactical features of sentences.[20][21][22]

In his book *Philosophical Analysis in the Twentieth Century, Volume 1 : The Dawn of Analysis*, Scott Soames has pointed out that Quine's circularity argument needs two of the logical positivists' central theses to be effective:[23]

All necessary (and all *a priori*) truths are analytic

Analyticity is needed to explain and legitimate necessity.

It is only when these two theses are accepted that Quine's argument holds. It is not a problem that the notion of necessity is presupposed by the notion of analyticity if necessity can be explained without analyticity. According to Soames, both theses were accepted by most philosophers when Quine published *Two Dogmas*. Today however, Soames holds both statements to be antiquated. He says: "Very few philosophers today would accept either [of these assertions], both of which now seem decidedly antique."[23]

33.5 See also

- Holophrastic indeterminacy
- Internal–external distinction
- Sense and reference
- Two-dimensionalism

33.6 Footnotes

[1] Rey, Georges. "The Analytic/Synthetic Distinction". *The Stanford Encyclopedia of Philosophy (Winter 2010 Edition)*. Retrieved February 12, 2012.

[2] Jerrold J. Katz (2000). "The epistemic challenge to antirealism". *Realistic Rationalism*. MIT Press. p. 69. ISBN 0262263297.

[3] Reprinted in: Carnap, R. (1999). "Autobiography". In Paul Arthur Schlipp, ed. *The Philosophy of Rudolf Carnap*. Open Court Publishing Company. p. 64. ISBN 0812691539.

[4] This quote is found with a discussion of the differences between Carnap and Wittgenstein in Michael Friedman (1997). "Carnap and Wittgenstein's *Tractatus*". In William W. Tait, Leonard Linsky, eds. *Early Analytic Philosophy: Frege, Russell, Wittgenstein*. Open Court Publishing. p. 29. ISBN 0812693442.

[5] Gary Ebbs (2009). "§51 A first sketch of the pragmatic roots of Carnap's analytic-synthetic distinction". *Rule-Following and Realism*. Harvard University Press. pp. 101 *ff.* ISBN 0674034414.

[6] For a fuller explanation see Chalmers, David. *The Conscious Mind*. Oxford UP: 1996. Chapter 2, section 4.

[7] Rudolf Carnap (1950). "Empiricism, Semantics, and Ontology". *Revue Internationale de Philosophie* **4**: 20–40. Reprinted in the *Supplement to Meaning and Necessity: A Study in Semantics and Modal Logic*, enlarged edition (University of Chicago Press, 1956).

[8] Gillian Russell (2012-11-21). "Analytic/Synthetic Distinction". *Oxford Bibliographies*. Retrieved 2013-05-16.
[9] Mauro Murzi (April 12, 2001). "Rudolf Carnap: §3. Analytic and Synthetic". *Internet Encyclopedia of Philosophy*.
[10] Rudolf Carnap (1947). *Meaning and Necessity: A study in semantics and modal logic* (2nd ed.). University of Chicago. ISBN 0226093476.Google link to Midway reprint.
[11] Stephen Yablo (1998). "Does ontology rest upon a mistake?". *Aristotelian Society Supplementary Volume* **72** (1): 229–262. doi:10.1111/1467-8349.00044. "The usual charge against Carnap's internal/external distinction is one of 'guilt by association with analytic/synthetic'. But it can be freed of this association"
[12] Willard v.O. Quine (1951). "Main Trends in Recent Philosophy: Two Dogmas of Empiricism". *The Philosophical Review* **60**: 20–43. doi:10.2307/2181906. Reprinted in W.V.O. Quine, From a Logical Point of View (Harvard University Press, 1953; second, revised, edition 1961) On-line versions at http://www.calculemus.org and Woodbridge.
[13] Willard v O Quine (1980). "Chapter 2: W.V. Quine: Two dogmas of empiricism". In Harold Morick, ed. *Challenges to empiricism*. Hackett Publishing. p. 60. ISBN 0915144905. Published earlier in *From a Logical Point of View*, Harvard University Press (1953)
[14] Paul Artin Boghossian (August 1996). "Analyticity Reconsidered". *Noûs* **30** (3): 360–391. doi:10.2307/2216275.
[15] H. P. Grice and P. F. Strawson (April 1956). "In Defense of a Dogma". *The Philosophical Review* **65** (2): 41–158. JSTOR 2182828.
[16] Searle, John R. (1969). *Speech Acts: An Essay in the Philosophy of Language*. Cambridge University Press. p. 5. ISBN 052109626X.
[17] Hilary Putnam (1983). *Realism and Reason: Philosophical Papers Volume 3, Realism and Reason*. pp. 87–97. ISBN 9780521246729.
[18] Hilary Putnam (1979). *Philosophical Papers: Volume 2, Mind, Language and Reality*. Harvard University Press. p. 36. ISBN 0521295513.

[19] Putnam, Hilary, "'Two dogmas' revisited." In Gilbert Ryle, *Contemporary Aspects of Philosophy*. Stocksfield: Oriel Press, 1976, 202–213.
[20] Leonard Linsky (October 1970). "Analytical/Synthetic and Semantic Theory". *Synthese* **21** (3/4): 439–448. doi:10.1007/BF00484810. JSTOR 20114738. Reprinted in Donald Davidson, Gilbert Harman, ed. (1973). *Semantics of natural language* (2nd ed.). pp. 473 *ff.* ISBN 9027703043.
[21] Willard v O Quine (February 2, 1967). "On a Suggestion of Katz". *The Journal of Philosophy* **64** (2): 52–54. doi:10.2307/2023770. JSTOR 2023770.
[22] Jerrold J Katz (1974). "Where Things Stand Now with the Analytical/Synthetic Distinction". *Synthese* **28** (3–4): 283–319. doi:10.1007/BF00877579.
[23] Scott Soames (2009). "Evaluating the circularity argument". *'Philosophical Analysis in the Twentieth Century, Volume 1 : The Dawn of Analysis*. Princeton University Press. p. 360. ISBN 1400825792. There are several earlier versions of this work.

33.7 References and further reading

_ Baehr, Jason S. (October 18, 2006). J. Fieser & B. Dowden, eds, ed. "A Priori and A Posteriori". *The Internet Encyclopedia of Philosophy*.
_ Boghossian, Paul. (1996). "Analyticity Reconsidered". *Nous*, Vol. 30, No. 3, pp. 360–391. <http://www.nyu.edu/gsas/dept/philo/faculty/boghossian/papers/AnalyticityReconsidered.html>.
_ Cory Juhl, Eric Loomis (2009). *Analyticity*. Routledge. ISBN 0415773334.
_ Kant, Immanuel. (1781/1998). *The Critique of Pure Reason*. Trans. by P. Guyer and A.W. Wood, Cambridge University Press .
_ Rey, Georges. (2003). "The Analytic/Synthetic Distinction". *The Stanford Encyclopedia of Philosophy*, Edward Zalta (ed.). <http://plato.stanford.edu/entries/analytic-synthetic>
_ Soames, Scott (2009). "Chapter 14: Ontology, Analyticity and Meaning: The Quine-Carnap Dispute". In David John Chalmers, David Manley & Ryan Wasserman, eds. *Metametaphysics: New Essays on the Foundations of Ontology*. Oxford University Press. ISBN 0199546045.
_ Frank X Ryan (2004). "Analytic: Analytic/Synthetic". In John Lachs, Robert B. Talisse, eds. *American Philosophy: An Encyclopedia*. Psychology Press. pp. 36–39. ISBN 020349279X.
_ Quine, W. V. (1951). "Two Dogmas of Empiricism". *Philosophical Review*, Vol.60, No.1, pp. 20–43. Reprinted in *From a Logical Point of View* (Cambridge, MA: Harvard University Press, 1953). <http://www.ditext.com/quine/quine.html>.
_ Robert Hanna (2012). "The return of the analytic-synthetic distinction". *Paradigmi*.

33.8 External links

_ Analytic–synthetic distinction at PhilPapers
_ Analytic–synthetic distinction entry in the *Stanford Encyclopedia of Philosophy*
_ Analytic–synthetic distinction entry in the *Internet Encyclopedia of Philosophy*
_ Analytic–synthetic distinction at the Indiana Philosophy Ontology Project

Chapter 34

Mathematical beauty

Mathematical beauty describes the notion that some mathematicians may derive aesthetic pleasure from their work, and from mathematics in general. They express this pleasure by describing mathematics (or, at least, some aspect of mathematics) as *beautiful*. Mathematicians describe mathematics as an art form or, at a minimum, as a creative activity. Comparisons are often made with music and poetry.

Bertrand Russell expressed his sense of mathematical beauty in these words:

Mathematics, rightly viewed, possesses not only truth, but supreme beauty — a beauty cold and austere, like that of sculpture, without appeal to any part of our weaker nature, without the gorgeous trappings of painting or music, yet sublimely pure, and capable of a stern perfection such as only the greatest art can show. The true spirit of delight, the exaltation, the sense of being more than Man, which is the touchstone of the highest excellence, is to be found in mathematics as surely as poetry.[1]

Paul Erdős expressed his views on the ineffability of mathematics when he said, "Why are numbers beautiful? It's like asking why is Beethoven's Ninth Symphony beautiful. If you don't see why, someone can't tell you. I *know* numbers are beautiful. If they aren't beautiful, nothing is".[2]

34.1 Beauty in method

Mathematicians describe an especially pleasing method of proof as *elegant*. Depending on context, this may mean:

_ A proof that uses a minimum of additional assumptions or previous results.

_ A proof that is unusually succinct.

_ A proof that derives a result in a surprising way (e.g., from an apparently unrelated theorem or collection of theorems.)

_ A proof that is based on new and original insights.

_ A method of proof that can be easily generalized to solve a family of similar problems.

In the search for an elegant proof, mathematicians often look for different independent ways to prove a result—the first proof that is found may not be the best. The theorem for which the greatest number of different proofs have been discovered is possibly the Pythagorean theorem, with hundreds of proofs having been published.[3] Another theorem that has been proved in many different ways is the theorem of quadratic reciprocity—Carl Friedrich Gauss alone published eight different proofs of this theorem.

Conversely, results that are logically correct but involve laborious calculations, over-elaborate methods, very conventional approaches, or that rely on a large number of particularly powerful axioms or previous results are not usually considered to be elegant, and may be called *ugly* or *clumsy*.

238

34.2. BEAUTY IN RESULTS 239

A

B

D C

E

C D A

B

E

An example of "beauty in method"—a simple and elegant proof of the Pythagorean theorem.

34.2 Beauty in results

Some mathematicians[4] see beauty in mathematical results that establish connections between two areas of mathematics that at first sight appear to be unrelated. These results are often described as *deep*.

While it is difficult to find universal agreement on whether a result is deep, some examples are often cited. One is Euler's identity:[5]

e_i_ + 1 = 0 :

This is a special case of Euler's formula, which the physicist Richard Feynman called "our jewel" and "the most remarkable formula in mathematics".[6] Modern examples include the modularity theorem, which establishes an important connection between elliptic curves and modular forms (work on which led to the awarding of the Wolf Prize to Andrew Wiles and Robert Langlands), and "monstrous moonshine", which connects the Monster group to modular functions via string theory for which Richard Borcherds was awarded the Fields Medal.

Starting at $e_0 = 1$*, travelling at the velocity* i *relative to one's position for the length of time π, and adding 1, one arrives at 0. (The diagram is an Argand diagram)*

Other examples of deep results include unexpected insights into mathematical structures. For example, Gauss's Theorema Egregium is a deep theorem which relates a local phenomenon (curvature) to a global phenomenon (area) in a surprising way. In particular, the area of a triangle on a curved surface is proportional to the excess of the triangle and the proportionality is curvature. Another example is the fundamental theorem of calculus (and its vector versions including Green's theorem and Stokes' theorem) which is a wonderfully deep and remarkable insight and is breathtaking in its beauty.

The opposite of *deep* is *trivial*. A trivial theorem may be a result that can be derived in an obvious and straightforward way from other known results, or which applies only to a specific set of particular objects such as the empty set. Sometimes, however, a statement of a theorem can be original enough to be considered deep, even though its proof is fairly obvious.

In his *A Mathematician's Apology*, Hardy suggests that a beautiful proof or result possesses "inevitability", "unexpectedness", and "economy".[7]

Rota, however, disagrees with unexpectedness as a sufficient condition for beauty and proposes a counterexample:

A great many theorems of mathematics, when first published, appear to be surprising; thus for example some twenty years ago [from 1977] the proof of the existence of non-equivalent differentiable structures on spheres of high dimension was thought to be surprising, but it did not occur to anyone to

call such a fact beautiful, then or now.[8]

Perhaps ironically, Monastyrsky writes:

It is very difficult to find an analogous invention in the past to Milnor's beautiful construction of the different differential structures on the seven-dimensional sphere....The original proof of Milnor was not very constructive but later E. Briscorn showed that these differential structures can be described in an extremely explicit and beautiful form.[9]

This disagreement illustrates both the subjective nature of mathematical beauty and its connection with mathematical results: in this case, not only the existence of exotic spheres, but also a particular realization of them.

34.3 Beauty in experience

There is a certain "cold and austere" beauty in this compound of five cubes

Interest in pure mathematics separate from empirical study has been part of the experience of various civilizations, including that of the Ancient Greeks, who "did mathematics for the beauty of it."[10] Mathematical beauty can also be experienced outside the confines of pure mathematics. For example, the aesthetic pleasure that mathematical physicists tend to experience in Einstein's theory of general relativity has been attributed (by Paul Dirac, among others) to its "great mathematical beauty."[11]

Some degree of delight in the manipulation of numbers and symbols is probably required to engage in any mathematics. Given the utility of mathematics in science and engineering, it is likely that any technological society will actively cultivate these aesthetics, certainly in its philosophy of science if nowhere else.

The most intense experience of mathematical beauty for most mathematicians comes from actively engaging in mathematics. It is very difficult to enjoy or appreciate mathematics in a purely passive way—in mathematics there is no real analogy of the role of the spectator, audience, or viewer.[12] Bertrand Russell referred to the *austere beauty* of mathematics.

34.4 Beauty and philosophy

Some mathematicians are of the opinion that the doing of mathematics is closer to discovery than invention, for example:

There is no scientific discoverer, no poet, no painter, no musician, who will not tell you that he found ready made his discovery or poem or picture – that it came to him from outside, and that he did not consciously create it from within.
—William Kingdon Clifford, from a lecture to the Royal Institution titled "Some of the conditions of mental development"

These mathematicians believe that the detailed and precise results of mathematics may be reasonably taken to be true without any dependence on the universe in which we live. For example, they would argue that the theory of the natural numbers is fundamentally valid, in a way that does not require any specific context. Some mathematicians have extrapolated this viewpoint that mathematical beauty is truth further, in some cases becoming mysticism.

Pythagorean mathematicians believed in the literal reality of numbers. The discovery of the existence of irrational numbers was a shock to them, since they considered the existence of numbers not expressible as the ratio of two natural numbers to be a flaw in nature (the Pythagorean world view did not contemplate the limits of infinite sequences of ratios of natural numbers—the modern notion of a real number). From a modern perspective, their mystical approach to numbers may be viewed as numerology.

In Plato's philosophy there were two worlds, the physical one in which we live and another abstract world which contained unchanging truth, including mathematics. He believed that the physical world was a mere reflection of the more perfect abstract world.

Hungarian mathematician Paul Erdős[13] spoke of an imaginary book, in which God has written down all the most beautiful mathematical proofs. When Erdős wanted to express particular appreciation of a proof, he would exclaim "This one's from The Book!" This viewpoint expresses the idea that mathematics, as the intrinsically true foundation on which the laws of our universe are built, is a natural candidate for what has been personified as God by different religious believers.

Twentieth-century French philosopher Alain Badiou claims that ontology is mathematics. Badiou also believes in deep connections between mathematics, poetry and philosophy.

In some cases, natural philosophers and other scientists who have made extensive use of mathematics have made leaps of inference between beauty and physical truth in ways that turned out to be erroneous. For example, at one stage in his life, Johannes Kepler believed that the proportions of the orbits of the then-known planets in the Solar System have been arranged by God to correspond to a concentric arrangement of the five Platonic solids, each orbit lying on the circumsphere of one polyhedron and the insphere of another. As there are exactly five Platonic solids, Kepler's hypothesis could only accommodate six planetary orbits and was disproved by the subsequent discovery of Uranus.

34.5 Beauty and mathematical information theory

In the 1970s, Abraham Moles and Frieder Nake analyzed links between beauty, information processing, and information theory.[14][15] In the 1990s, Jürgen Schmidhuber formulated a mathematical theory of observer-dependent subjective beauty based on algorithmic information theory: the most beautiful objects among subjectively comparable

objects have short algorithmic descriptions (i.e., Kolmogorov complexity) relative to what the observer already knows.[16][17][18] Schmidhuber explicitly distinguishes between beautiful and interesting. The latter corresponds to the first derivative of subjectively perceived beauty: the observer continually tries to improve the predictability and compressibility of the observations by discovering regularities such as repetitions and symmetries and fractal selfsimilarity. Whenever the observer's learning process (possibly a predictive artificial neural network) leads to improved data compression such that the observation sequence can be described by fewer bits than before, the temporary interestingness of the data corresponds to the compression progress, and is proportional to the observer's internal curiosity reward[19][20]

34.6 Mathematics and the arts

Main articles: Mathematics and art and Mathematics and music

34.6.1 Music

Examples of the use of mathematics in music include the stochastic music of Iannis Xenakis, counterpoint of Johann Sebastian Bach, polyrhythmic structures (as in Igor Stravinsky's *The Rite of Spring*), the Metric modulation of Elliott Carter, permutation theory in serialism beginning with Arnold Schoenberg, and application of Shepard tones in Karlheinz Stockhausens *Hymnen*.

34.6.2 Visual arts

Examples of the use of mathematics in the visual arts include applications of chaos theory and fractal geometry to computer-generated art, symmetry studies of Leonardo da Vinci, projective geometries in development of the perspective theory of Renaissance art, grids in Op art, optical geometry in the camera obscura of Giambattista della

Porta, and multiple perspective in analytic cubism and futurism.
The Dutch graphic designer M.C. Escher created mathematically inspired woodcuts, lithographs, and mezzotints. These feature impossible constructions, explorations of infinity, architecture, visual paradoxes and tessellations. British constructionist artist John Ernest created reliefs and paintings inspired by group theory.[21] A number of other British artists of the constructionist and systems schools also draw on mathematics models and structures as a source of inspiration, including Anthony Hill and Peter Lowe. Computer-generated art is based on mathematical algorithms.

34.6.3 Choreography

Shuffling has been applied to choreography as in the *Temple of Rudra* opera.

34.7 See also

- Descriptive science
- Fluency heuristic
- Golden ratio
- Mathematics and architecture
- Normative science
- Philosophy of mathematics
- Processing fluency theory of aesthetic pleasure
- Pythagoreanism

34.8 Notes

[1] Russell, Bertrand (1919). "The Study of Mathematics". *Mysticism and Logic: And Other Essays*. Longman. p. 60. Retrieved 2008-08-22.
[2] Devlin, Keith (2000). "Do Mathematicians Have Different Brains?". *The Math Gene: How Mathematical Thinking Evolved And Why Numbers Are Like Gossip*. Basic Books. p. 140. ISBN 978-0-465-01619-8. Retrieved 2008-08-22.
[3] Elisha Scott Loomis published over 360 proofs in his book Pythagorean Proposition (ISBN 0-873-53036-5).
[4] Rota (1997), *The phenomenology of mathematical beauty*, p. 173
[5] Gallagher, James (13 February 2014). "Mathematics: Why the brain sees maths as beauty". *BBC News online*. Retrieved 13 February 2014.
[6] Feynman, Richard P. (1977). *The Feynman Lectures on Physics* **I**. Addison-Wesley. p. 22-10. ISBN 0-201-02010-6.
[7] Hardy, G.H. "18". Missing or empty |title= (help)
[8] Rota, *The phenomenology of mathematical beautyyear = 1997*, p. 172
[9] Monastyrsky (2001), *Some Trends in Modern Mathematics and the Fields Medal*
[10] Lang, p. 3
[11] Chandrasekhar, p. 148
[12] Phillips, George (2005). "Preface". *Mathematics Is Not a Spectator Sport*. Springer Science+Business Media. ISBN 0-387-25528-1. Retrieved 2008-08-22. ""...there is nothing in the world of mathematics that corresponds to an audience in a concert hall, where the passive listen to the active. Happily, mathematicians are all *doers*, not spectators."
[13] Schechter, Bruce (2000). *My brain is open: The mathematical journeys of Paul Erdős*. New York: Simon & Schuster. pp. 70–71. ISBN 0-684-85980-7.
[14] A. Moles: *Théorie de l'information et perception esthétique*, Paris, Denoël, 1973 (Information Theory and aesthetical perception)
[15] F Nake (1974). Ästhetik als Informationsverarbeitung. (Aesthetics as information processing). Grundlagen und Anwendungen der Informatik im Bereich ästhetischer Produktion und Kritik. Springer, 1974, ISBN 3-211-81216-4, ISBN 978-3-211-81216-7
[16] J. Schmidhuber. Low-complexity art. Leonardo, Journal of the International Society for the Arts, Sciences, and Technology, 30(2):97–103, 1997. http://www.jstor.org/pss/1576418
[17] J. Schmidhuber. Papers on the theory of beauty and low-complexity art since 1994: http://www.idsia.ch/~{}juergen/beauty.html
[18] J. Schmidhuber. Simple Algorithmic Principles of Discovery, Subjective Beauty, Selective Attention, Curiosity & Creativity. Proc. 10th Intl. Conf. on Discovery Science (DS 2007) p. 26-38, LNAI 4755, Springer, 2007. Also in Proc. 18th Intl. Conf. on Algorithmic Learning Theory (ALT 2007) p. 32, LNAI 4754, Springer, 2007. Joint invited lecture for DS 2007 and ALT 2007, Sendai, Japan, 2007. http://arxiv.org/abs/0709.0674
[19] .J. Schmidhuber. Curious model-building control systems. International Joint Conference on Neural Networks, Singapore, vol 2, 1458–1463. IEEE press, 1991
[20] Schmidhuber's theory of beauty and curiosity in a German TV show: http://www.br-online.de/bayerisches-fernsehen/faszination-wissen/schoenheit--aesthetik-wahrnehmung-ID1212005092828.xml
[21] John Ernest's use of mathematics and especially group theory in his art works is analysed in *John Ernest, A Mathematical Artist* by Paul Ernest in Philosophy of Mathematics Education Journal, No. 24 Dec. 2009 (Special Issue on Mathematics and Art): http://people.exeter.ac.uk/PErnest/pome24/index.htm

34.9 References

_ Aigner, Martin, and Ziegler, Gunter M. (2003), *Proofs from THE BOOK,* 3rd edition, Springer-Verlag.
_ Chandrasekhar, Subrahmanyan (1987), *Truth and Beauty: Aesthetics and Motivations in Science,* University of Chicago Press, Chicago, IL.
_ Hadamard, Jacques (1949), *The Psychology of Invention in the Mathematical Field,* 1st edition, Princeton University Press, Princeton, NJ. 2nd edition, 1949. Reprinted, Dover Publications, New York, NY, 1954.
_ Hardy, G.H. (1940), *A Mathematician's Apology*, 1st published, 1940. Reprinted, C.P. Snow (foreword), 1967. Reprinted, Cambridge University Press, Cambridge, UK, 1992.
_ Hoffman, Paul (1992), *The Man Who Loved Only Numbers*, Hyperion.
_ Huntley, H.E. (1970), *The Divine Proportion: A Study in Mathematical Beauty*, Dover Publications, New York, NY.
_ Loomis, Elisha Scott (1968), *The Pythagorean Proposition*, The National Council of Teachers of Mathematics. Contains 365 proofs of the Pythagorean Theorem.
_ Lang, Serge (1985). *The Beauty of Doing Mathematics: Three Public Dialogues*. New York: Springer-Verlag. ISBN 0-387-96149-6.
_ Peitgen, H.-O., and Richter, P.H. (1986), *The Beauty of Fractals*, Springer-Verlag.
_ Reber, R., Brun, M., & Mitterndorfer, K. (2008). The use of heuristics in intuitive mathematical judgment. *Psychonomic Bulletin & Review, 15*, 1174-1178.
_ Strohmeier, John, and Westbrook, Peter (1999), *Divine Harmony, The Life and Teachings of Pythagoras*, Berkeley Hills Books, Berkeley, CA.
_ Rota, Gian-Carlo (1997). "The phenomenology of mathematical beauty". *Synthese* **111** (2): 171–182. doi:10.1023/A:1004930722234.
_ Monastyrsky, Michael (2001). "Some Trends in Modern Mathematics and the Fields Medal". *Can. Math. Soc. Notes* **33** (2 and 3).

34.10 External links

_ Mathematics, Poetry and Beauty
_ Is Mathematics Beautiful?
_ The Beauty of Mathematics
_ Justin Mullins
_ Edna St. Vincent Millay (poet): *Euclid alone has looked on beauty bare*
_ Terence Tao, *What is good mathematics?*
_ Mathbeauty Blog
_ The *Aesthetic Appeal* collection at the Internet Archive [*more*]
_ *A Mathematical Romance* Jim Holt December 5, 2013 issue of The New York Review of Books review of *Love and Math: The Heart of Hidden Reality* by Edward Frenkel

Chapter 35

Paul Erdős

Paul Erdős (Hungarian: *Erdős Pál* [ˈɛrdøːʃ paːl]; 26 March 1913 – 20 September 1996) was a Hungarian mathematician. He was one of the most prolific mathematicians of the 20th century,[2] but also known for his social practice of mathematics (more than 500 collaborators) and eccentric lifestyle (Time Magazine called him *The Oddball's Oddball*).[3]
Erdős pursued problems in combinatorics, graph theory, number theory, classical analysis, approximation theory, set theory, and probability theory.[4]

35.1 Early life, education, life, and death

Paul Erdős was born in Budapest, Hungary, on March 26, 1913.[5] He was the only surviving child of Anna and Lajos Erdős (formerly Engländer).[6] His siblings died before he was born, aged 3 and 5. His parents were both Jewish mathematics teachers from a vibrant intellectual community. His fascination with mathematics developed early—by the age of four, given a person's age, he could calculate, in his head, how many seconds they had lived.[7]
Both of Erdős's parents were high school mathematics teachers, and Erdős received much of his early education from them. Erdős always remembered his parents with great affection. At 16, his father introduced him to two of his lifetime favorite subjects—infinite series and set theory. During high school, Erdős became an ardent solver of the problems proposed each month in KöMaL, the Mathematical and

Physical Monthly for Secondary Schools.[8]
Erdős later published several articles in it about problems in elementary plane geometry.
In 1934, at the age of 21, he was awarded a doctorate in mathematics from the University of Budapest. Erdős's thesis advisor was Leopold Fejér (or Fejér Lipót), who was also the thesis advisor for John von Neumann, George Pólya, and Paul (Pál) Turán.
Much of his family, including two of his aunts, two of his uncles, and his father died in Budapest during the Holocaust. His mother survived in hiding. He was living in America and working at the Princeton Institute for Advanced Study at the time.[9]
On September 20, 1996, at the age of 83, he had a heart attack and died while attending a conference in Warsaw. He never married and had no children. He is buried next to his mother and father in grave 17A-6-29 at Kozma Utcai Temető in Budapest.[10] For his epitaph, he suggested "I've finally stopped getting dumber." (Hungarian: *"Végre nem butulok tovább"*).[11]
His life was documented in the film *N Is a Number: A Portrait of Paul Erdős*, made while he was still alive, and posthumously in the book *The Man Who Loved Only Numbers* (1998).
Erdős's name contains the Hungarian letter "ő" ("o" with double acute accent), but is often incorrectly written as *Erdos* or *Erdös* either "by mistake or out of typographical necessity".[12]
246
35.2. PERSONALITY 247

35.2 Personality

Another roof, another proof.
Paul Erdős[13]
Possessions meant little to Erdős; most of his belongings would fit in a suitcase, as dictated by his itinerant lifestyle. Awards and other earnings were generally donated to people in need and various worthy causes. He spent most of his life as a vagabond, traveling between scientific conferences and the homes of colleagues all over the world. He would typically show up at a colleague's doorstep and announce "my brain is open", staying long enough to collaborate on a few papers before moving on a few days later. In many cases, he would ask the current collaborator about whom to visit next.
His colleague Alfréd Rényi said, "a mathematician is a machine for turning coffee into theorems",[14] and Erdős drank copious quantities. (This quotation is often attributed incorrectly to Erdős,[15] but Erdős himself ascribed it to Rényi.[16]) After 1971 he also took amphetamines, despite the concern of his friends, one of whom (Ron Graham) bet him $500 that he could not stop taking the drug for a month.[17] Erdős won the bet, but complained that during his abstinence, mathematics had been set back by a month: "Before, when I looked at a piece of blank paper my mind was filled with ideas. Now all I see is a blank piece of paper." After he won the bet, he promptly resumed his amphetamine use.
He had his own idiosyncratic vocabulary: Although an agnostic atheist,[18][19] he spoke of "The Book", a visualization of a book in which God had written down the best and most elegant proofs for mathematical theorems.[20] Lecturing in 1985 he said, "You don't have to believe in God, but you should believe in *The Book*." He himself doubted the existence of God, whom he called the "Supreme Fascist" (SF).[21][22] He accused SF of hiding his socks and Hungarian passports, and of keeping the most elegant mathematical proofs to himself. When he saw a particularly beautiful mathematical proof he would exclaim, "This one's from *The Book*!". This later inspired a book entitled *Proofs from the Book*.
Other idiosyncratic elements of Erdős's vocabulary include:[23]
_ Children were referred to as "epsilons" (because in mathematics, particularly calculus, an arbitrarily small positive quantity is commonly denoted by the Greek letter (ε))
_ Women were "bosses"
_ Men were "slaves"
_ People who stopped doing mathematics had "died"
_ People who physically died had "left"
_ Alcoholic drinks were "poison"
_ Music (except classical music) was "noise"
_ People who had married were "captured"
_ People who had divorced were "liberated"
_ To give a mathematical lecture was "to preach"
_ To give an oral exam to a student was "to torture" him/her.
He gave nicknames to many countries, examples being: the U.S. was "samland" (after Uncle Sam), the Soviet Union was "joedom" (after Joseph Stalin), and Israel was "isreal".

35.3 Career

In 1934, he moved to Manchester, England, to be a guest lecturer. In 1938, he accepted his first American position as a scholarship holder at Princeton University. At this time, he began to develop the habit of traveling from campus

to campus. He would not stay long in one place and traveled back and forth among mathematical institutions until his death.

In 1952, the United States Citizenship and Immigration Services denied Erdős, a Hungarian citizen, a re-entry visa into the United States, for reasons that have never been fully explained.[24] Teaching at the University of Notre Dame at the time, Erdős could have chosen to remain in the country. Instead, he packed up and left, albeit requesting reconsideration from the U.S. Immigration Services at periodic intervals.

Erdős, Fan Chung, and her husband Ronald Graham, Japan 1986

Hungary, back then, was under the Warsaw Pact with the Soviet Union. Although the Soviet Union limited the freedom of its own citizens to enter and exit the country, the Soviet Union gave Erdős the exclusive privilege of being allowed to enter and exit Hungary as he pleased in 1956.

The U.S. Immigration Services later on granted a visa in 1963 to Erdős and he resumed including American universities in his teaching and travels. Ten years later, the 60-year-old Erdős left voluntarily from Hungary in 1973.[25]

During the last decades of his life, Erdős received at least fifteen honorary doctorates. He became a member of the scientific academies of eight countries, including the U.S. National Academy of Sciences and the UK Royal Society. Shortly before his death, he renounced his honorary degree from the University of Waterloo over what he considered to be unfair treatment of colleague Adrian Bondy.[26][27]

35.3.1 Mathematical work

Erdős was one of the most prolific publishers of papers in mathematical history, comparable only with Leonhard Euler; Erdős published more papers, mostly in collaboration with other mathematicians, while Euler published more pages, mostly by himself.[28] Erdős wrote around 1,525 mathematical articles in his lifetime,[29] mostly with coauthors. He strongly believed in and practiced mathematics as a social activity,[30] having 511 different collaborators

in his lifetime.[31]

In his mathematical style, Erdős was much more of a "problem solver" than a "theory developer". (See "The Two Cultures of Mathematics"[32] by Timothy Gowers for an in-depth discussion of the two styles, and why problem solvers are perhaps less appreciated.) Joel Spencer states that "his place in the 20th-century mathematical pantheon is a matter of some controversy because he resolutely concentrated on particular theorems and conjectures throughout his illustrious career."[33] Erdős never won the highest mathematical prize, the Fields Medal, nor did he coauthor a paper with anyone who did,[34] a pattern that extends to other prizes.[35] He did win the Wolf Prize, where his contribution is described as "for his numerous contributions to number theory, combinatorics, probability, set theory and mathematical analysis, and for personally stimulating mathematicians the world over".[36] In contrast, the works of the three winners after were recognized as "outstanding", "classic", and "profound", and the three before as "fundamental" or "seminal".

Of his contributions, the development of Ramsey theory and the application of the probabilistic method especially stand out. Extremal combinatorics owes to him a whole approach, derived in part from the tradition of analytic number theory. Erdős found a proof for Bertrand's postulate which proved to be far neater than Chebyshev's original one. He also discovered an elementary proof for the prime number theorem, along with Atle Selberg. However, the circumstances leading up to the proofs, as well as publication disagreements, led to a bitter dispute between Erdős and Selberg.[37][38] Erdős also contributed to fields in which he had little real interest, such as topology, where he is credited as the first person to give an example of a totally disconnected topological space that is not zero-dimensional.[39]

35.3.2 Erdős' problems

Paul Erdős influenced many young mathematicians. In this 1985 photo taken at the University of Adelaide, Erdős explains a problem to Terence Tao — who was 10 years old at the time. Tao received the Fields Medal in 2006, and was elected a Fellow of the Royal Society in 2007. Tao has an Erdős number of 2.

Throughout his career, Erdős would offer payments for solutions to unresolved problems.[40] These ranged from $25 for problems that he felt were just out of the reach of the current mathematical thinking (both his and others), to several thousand dollars for problems that were both difficult to attack and mathematically significant. There are thought to be at least a thousand such unpaid payments, though there is no official or comprehensive list. The offers remain active despite Erdős's death; Ronald Graham is the (informal) administrator of solutions. The solvers can get either the original check signed by Erdős before his death (for memento only, can not be cashed) or a cashable check

from Graham.[41]

Perhaps the most mathematically notable of these problems is the Erdős conjecture on arithmetic progressions:

If the sum of the reciprocals of a sequence of integers diverges, then the sequence contains arithmetic progressions of arbitrary length.

If true, it would solve several other open problems in number theory (although one main implication of the conjecture,

that the prime numbers contain arbitrarily long arithmetic progressions, has since been proved independently as the Green–Tao theorem). The payment for the solution of the problem is currently worth US$5000.[42]
The most familiar problem with an Erdős prize is likely the Collatz conjecture, also called the $3N + 1$ problem. Erdős offered $500 for a solution.

35.3.3 Collaborators

His most frequent collaborators include Hungarian mathematicians András Sárközy (62 papers) and András Hajnal (56 papers), and American mathematician Ralph Faudree (50 papers). Other frequent collaborators were[43]
_ Béla Bollobás (18 papers)
_ Stefan Burr (27 papers)
_ Fan Chung (14 papers)
_ Zoltán Füredi (10 papers)
_ Ron Graham (28 papers)
_ András Gyárfás (15 papers)
_ Richard R. Hall (14 papers)
_ István Joó (12 papers)
_ Eric Charles Milner (15 papers)
_ Melvyn B. Nathanson (19 papers)
_ Jean-Louis Nicolas (19 papers)
_ János Pach (21 papers)
_ George Piranian (14 papers)
_ Carl Pomerance (23 papers)
_ Richard Rado (18 papers)
_ A. R. Reddy (11 papers)
_ Alfréd Rényi (32 papers)
_ Pal Revesz (10 papers)
_ Vojtěch Rödl (11 papers)
_ C. C. Rousseau (35 papers)
_ Richard Schelp (42 papers)
_ John Selfridge (14 papers)
_ Miklós Simonovits (21 papers)
_ Vera Sós (35 papers)

_ Joel Spencer (23 papers)
_ Ernst G. Straus (20 papers)
_ Endre Szemerédi (29 papers)
_ Paul Turán (30 papers)
_ Zsolt Tuza (12 papers)
For other co-authors of Erdős, see the list of people with Erdős number 1 in List of people by Erdős number.

35.4 Erdős number

Main article: Erdős number
Because of his prolific output, friends created the *Erdős number* as a humorous tribute. An Erdős number describes a person's degree of separation from Erdős himself, based on their collaboration with him, or with another who has their own Erdős number. Erdős alone was assigned the Erdős number of 0 (for being himself), while his immediate collaborators could claim an Erdős number of 1, their collaborators have Erdős number at most 2, and so on. Approximately 200,000 mathematicians have an assigned Erdős number,[44] and some have estimated that 90 percent of the world's active mathematicians have an Erdős number smaller than 8 (not surprising in light of the small world phenomenon). Due to collaborations with mathematicians, many scientists in fields such as physics, engineering, biology, and economics have Erdős numbers as well.[45]
Jerry Grossman has written that it could be argued that Baseball Hall of Famer Hank Aaron can be considered to have an Erdős number of 1 because they both autographed the same baseball when Emory University awarded them honorary degrees on the same day.[46] Erdős numbers have also been proposed for an infant, a horse, and several actors.[47]
The Erdős number was most likely first defined by Casper Goffman,[48] an analyst whose own Erdős number is 2.[49] Goffman published his observations about Erdős's prolific collaboration in a 1969 article titled "And what is your Erdős number?"[50]

35.5 Signature

Erdos signed his name “Paul Erdos p g o m”. When he became 60 he added “ld”, at 65 “ad”, at 70 “ld” and at 75 “cd”.
_ pgom means poor great old man
_ ld means living dead
_ ad means archaeological discovery
_ the second ld means legally dead
_ cd means counts dead.[51][52]

35.6 Books about Erdős

A 2013 children’s picture book by Deborah Heligman. *The Boy Who Loved Math; The Improbable Life of Paul Erdős*.[53]

35.7 See also

_ List of topics named after Paul Erdős – including conjectures, numbers, prizes, and theorems

35.8 Notes

[1] “Mathematics Genealogy Project”. Retrieved 13 Aug 2012.
[2] Paul Hoffman (July 8, 2013). “Paul Erdős”. “Encyclopedia Britannica.
[3] Michael D. Lemonick (March 29, 1999). “Paul Erdos: The Oddball’s Oddball”. Time Magazine.
[4] Encyclopædia Britannica article
[5] “Erdos biography”. Gap-system.org. Retrieved 2010-05-29.
[6] Baker, A.; Bollobas, B. (1999). "Paul Erdős 26 March 1913 -- 20 September 1996: Elected For.Mem.R.S. 1989”. *Biographical Memoirs of Fellows of the Royal Society* **45**: 147. doi:10.1098/rsbm.1999.0011.
[7] Hoffman, p. 66.
[8] László Babai. “Paul Erdős just left town”.
[9] Csicsery, George Paul (2005). *N Is a Number: A Portrait of Paul Erdős*. Berlin; Heidelberg: Springer Verlag. ISBN 3-540-22469-6.
[10] grave 17A-6-29
[11] Hoffman, p. 3.
[12] The full quote is “Note the pair of long accents on the "ő," often (even in Erdos’s own papers) by mistake or out of typographical necessity replaced by "ö," the more familiar German umlaut which also exists in Hungarian.”, from Paul Erdős, D. Miklós, Vera T. Sós (1996). *Combinatorics, Paul Erdős is eighty*.
[13] Cited in at least 20 books.
[14] Biography of Alfréd Rényi by J.J. O'Connor and E.F. Robertson
[15] Bruno Schechter (2000), *My Brain is Open: The Mathematical Journeys of Paul Erdős*, p. 155, ISBN 0-684-85980-7
[16] Paul Erdős (1995). “Child Prodigies”. *Mathematics Competitions* **8** (1): 7–15. Retrieved July 17, 2012.
[17] Hill, J. Paul Erdos, Mathematical Genius, Human (In That Order)
[18] Colm Mulcahy (2013-03-26). “Centenary of Mathematician Paul Erdős -- Source of Bacon Number Concept”. Huffington Post. Retrieved 13 April 2013. “In his own words, “I'm not qualified to say whether or not God exists. I kind of doubt He does. Nevertheless, I'm always saying that the SF has this transfinite Book that contains the best proofs of all mathematical theorems, proofs that are elegant and perfect...You don't have to believe in God, but you should believe in the Book.”."
[19] Jack Huberman (2008). *Quotable Atheist: Ammunition for Nonbelievers, Political Junkies, Gadflies, and Those Generally Hell-Bound*. Nation Books. p. 107. ISBN 9781568584195. “I kind of doubt He [exists]. Nevertheless, I'm always saying that the SF has this transfinite Book ... that contains the best proofs of all theorems, proofs that are elegant and perfect.... You don't have to believe in God, but you should believe in the Book.”
[20] Nathalie Sinclair, William Higginson, ed. (2006). *Mathematics and the Aesthetic: New Approaches to an Ancient Affinity*. Springer. p. 36. ISBN 9780387305264. “Erdös, an atheist, named 'the Book' the place where God keeps aesthetically perfect proofs.”
[21] Schechter, Bruce (2000). *My brain is open: The mathematical journeys of Paul Erdős*. New York: Simon & Schuster. pp. 70–71. ISBN 0-684-85980-7.
[22] Varadaraja Raman (2005). *Variety in Religion And Science: Daily Reflections*. iUniverse. p. 256. ISBN 9780595358403.
[23] Hoffman, chapter 1. As included with the New York Times review of the book.
[24] “Erdos biography”. School of Mathematics and Statistics, University of St Andrews, Scotland. January 2000. Retrieved 2008-11-11.
[25] László Babai and Joel Spencer. “Paul Erdős (1913–1996)" (PDF). *Notices of the American Mathematical Society* (American Mathematical Society) **45** (1).
[26] Erdős, Paul (4 June 1996). “Dear President Downey” (PDF). Archived from the original on 15 October 2005. Retrieved 8 July 2014. “With a heavy heart I feel that I have to sever my connections with the University of Waterloo, including resigning my honorary degree which I received from the University in 1981 (which caused me great pleasure). I was very upset by the treatment of Professor Adrian Bondy. I do not maintain that Professor Bondy was innocent, but in view of his accomplishments and distinguished services to the University I feel that 'justice should be tempered with mercy.'"

[27] Transcription of October 2, 1996, article from University of Waterloo Gazette (archive)

[28] Hoffman, p. 42.
[29] Jerry Grossman. "Publications of Paul Erdös". Retrieved 1 Feb 2011.
[30] Charles Krauthammer (September 27, 1996). "Paul Erdos, Sweet Genius". *Washington Post*. p. A25. "?".
[31] "The Erdős Number Project Data Files". Oakland.edu. 2009-05-29. Retrieved 2010-05-29.
[32] This essay is in *Mathematics: Frontiers and Perspectives*, Edited by V. I. Arnold, Michael Atiyah, Peter D. Lax and Barry Mazur, American Mathematical Society, 2000. Available online at .
[33] Joel Spencer, "Prove and Conjecture!", a review of *Mathematics: Frontiers and Perspectives*. *American Scientist*, Volume 88, No. 6 November–December 2000
[34] Paths to Erdös — The Erdös Number Project
[35] From "trails to Erdos", by DeCastro and Grossman, in *The Mathematical Intelligencer*, vol. 21, no. 3 (Summer 1999), 51–63: A careful reading of Table 3 shows that although Erdos never wrote jointly with any of the 42 [Fields] medalists (a fact perhaps worthy of further contemplation)... there are many other important international awards for mathematicians. Perhaps the three most renowned...are the Rolf Nevanlinna Prize, the Wolf Prize in Mathematics, and the Leroy P. Steele Prizes. ... Again, one may wonder why KAPLANSKY is the only recipient of any of these prizes who collaborated with Paul Erdös. (After this paper was written, collaborator Lovasz received the Wolf prize, making 2 in all).
[36] "Wolf Foundation Mathematics Prize Page". Wolffund.org.il. Retrieved 2010-05-29.
[37] Goldfeld, Dorian (2003). "The Elementary Proof of the Prime Number Theorem: an Historical Perspective". *Number Theory: New York Seminar*: 179–192.
[38] Baas, Nils A.; Skau, Christian F. (2008). "The lord of the numbers, Atle Selberg. On his life and mathematics". *Bull. Amer. Math. Soc.* **45** (4): 617–649. doi:10.1090/S0273-0979-08-01223-8
[39] Melvin Henriksen. "Reminiscences of Paul Erdös (1913–1996)". Mathematical Association of America. Retrieved 2008-09-01.
[40] Brent Wittmeier, "Math genius left unclaimed sum," Edmonton Journal, September 28, 2010.
[41] Charles Seife (5 April 2002). "Erdös's Hard-to-Win Prizes Still Draw Bounty Hunters". *Science* **296** (5565): 39–40. doi:10.1126/science.296.5565.39. PMID 11935003.
[42] p. 354, Soifer, Alexander (2008); *The Mathematical Coloring Book: Mathematics of Coloring and the Colorful Life of its Creators*; New York: Springer. ISBN 978-0-387-74640-1
[43] List of collaborators of Erdős by number of joint papers, from the Erdős number project web site.
[44] "From Benford to Erdös". *Radio Lab*. Episode 2009-10-09. 2009-09-30. http://www.wnyc.org/shows/radiolab/episodes/2009/10/09/segments/137643.
[45] Jerry Grossman. "Some Famous People with Finite Erdös Numbers". Retrieved 1 Feb 2011.
[46] Jerry Grossman. "Items of Interest Related to Erdös Numbers".
[47] Extended Erdős Number Project
[48] Michael Golomb's obituary of Paul Erdős
[49] https://files.oakland.edu/users/grossman/enp/ErdosA.html from the Erdos Number Project
[50] Goffman, Casper (1969). "And what is your Erdős number?". *American Mathematical Monthly* **76** (7): 791. doi:10.2307/2317868. JSTOR 2317868.
[51] *My Brain is Open. The Mathematical Journeys of Paul Erdos*, Bruce Schechter, Simon & Schuster, 1998, p.41
[52] Paul Erdös: N is a number on YouTube, a documentary film by George Paul Csicsery, 1991.
[53] Silver, Nate (12 July 2013). [Erdős "Children's Books Beautiful Minds 'The Boy Who Loved Math' and 'On a Beam of Light'"] Check |url= scheme (help). New York Times. Retrieved 29 October 2014.

35.9 References

_ Aigner, Martin; Günther Ziegler (2003). *Proofs from THE BOOK*. Berlin; New York: Springer. ISBN 3-540-40460-0.
_ Hoffman, Paul (1998). *The Man Who Loved Only Numbers: The Story of Paul Erdős and the Search for Mathematical Truth*. London: Fourth Estate Ltd. ISBN 1-85702-811-2.
_ Kolata, Gina (1996-09-24). "Paul Erdos, 83, a Wayfarer In Math's Vanguard, Is Dead". *New York Times*. pp. A1 and B8. Retrieved 2008-09-29.
_ Bruce Schechter (1998). *My Brain is Open: The Mathematical Journeys of Paul Erdős*. Simon & Schuster. ISBN 0-684-84635-7.

35.10 External links

_ Erdős's Scholar Google profile
_ Searchable collection of (almost) all papers of Erdős
_ O'Connor, John J.; Robertson, Edmund F., "Paul Erdős", *MacTutor History of Mathematics archive*, University of St Andrews.
_ Paul Erdős at the Mathematics Genealogy Project
_ Jerry Grossman at Oakland University. *The Erdös Number Project*
_ The Man Who Loved Only Numbers - Royal Society Public Lecture by Paul Hoffman (video)
_ Radiolab: Numbers, with a story on Paul Erdős
_ Fan Chung, "Open problems of Paul Erdős in graph theory"

Chapter 36
Proofs from THE BOOK

Proofs from THE BOOK is a book of mathematical proofs by Martin Aigner and Günter M. Ziegler. The book is dedicated to the mathematician Paul Erdős, who often referred to "The Book" in which God keeps the most elegant proof of each mathematical theorem. During a lecture in 1985, Erdős said, "You don't have to believe in God, but you should believe in The Book."

Proofs from THE BOOK contains 32 sections (40 in the fourth edition), each devoted to one theorem but often containing multiple proofs and related results. It spans a broad range of mathematical fields: number theory, geometry, analysis, combinatorics and graph theory. Erdős himself made many suggestions for the book, but died before its publication. The book is illustrated by Karl Heinrich Hofmann. It has gone through five editions in English, and has been translated into Persian, French, German, Hungarian, Italian, Japanese, Chinese, Polish, Portuguese, Korean, Turkish, Russian and Spanish.

The proofs include:

_ Proof of Bertrand's postulate
_ Proof that e is irrational (also showing the irrationality of certain related numbers)
_ Six proofs of the infinitude of the primes, including Euclid's and Furstenberg's
_ Monsky's theorem (4th edition)

36.1 References

_ *Proofs from the book.* Springer, Berlin 1998, ISBN 3-540-63698-6
_ Aigner, Martin; Ziegler, Günter (2009). *Proofs from THE BOOK* (4th ed.). Berlin, New York: Springer-Verlag. ISBN 978-3-642-00855-9.
_ Table of Contents of the 4th ed.
_ Günter M. Ziegler's homepage, including a list of editions and translations.
_ Shepard, Mary (1999). "Read This! Review of Proofs from THE BOOK". Mathematical Association of America..

255

Chapter 37

Direct proof

In mathematics and logic, a **direct proof** is a way of showing the truth or falsehood of a given statement by a straightforward combination of established facts, usually axioms, existing lemmas and theorems, without making any further assumptions. [1] In order to directly prove a conditional statement of the form "If p, then q", it suffices to consider the situations in which the statement p is true. Logical deduction is employed to reason from assumptions to conclusion. The type of logic employed is almost invariably first-order logic, employing the quantifiers *for all* and *there exists*. Common proof rules used are modus ponens and universal instantiation.

In contrast, an indirect proof may begin with certain hypothetical scenarios and then proceed to eliminate the uncertainties in each of these scenarios until an inescapable conclusion is forced. For example instead of showing directly $p \Rightarrow q$, one proves its contrapositive $\sim q \Rightarrow \sim p$ (one assumes $\sim q$ and shows that it leads to $\sim p$). Since $p \Rightarrow q$ and $\sim q \Rightarrow \sim p$ are equivalent by the principle of transposition (see law of excluded middle), $p \Rightarrow q$ is indirectly proved. Proof methods that are not direct include proof by contradiction, including proof by infinite descent. Direct proof methods include proof by exhaustion and proof by induction.

37.1 Examples

37.1.1 The sum of two even integers equals an even integer

Consider two even integers x and y. Since they are even, they can be written as

$x = 2a$

$y = 2b$
respectively for integers a and b. Then the sum can be written as
$x + y = 2a + 2b = 2(a + b)$
From this it is clear $x + y$ has 2 as a factor and therefore is even, so the sum of any two even integers is even.

37.1.2 Pythagoras Theorem

Observe that we have four right triangles and a square packed into a large square. Each of the triangles has legs a and b and hypotenuse c, just as in the Pythagorean theorem. Of course, on the one hand, the area of the large square is c^2. On the other hand, the area of the large square is the sum of the areas of its component pieces. [2]
Thus we calculate that
$c^2 =$ (area of the large square)
256
37.1. EXAMPLES 257
Diagram of Pythagoras Theorem
We split the large square into the smaller areas and we have that
$c^2 = 4($ area of triangle $) + ($ area of small square $)$
We compute the area of the triangles and area of the inner square so that
$c^2 = 4($
1
2
$ab) + (b \square a)^2$
By symbolic manipulation we have that
$c^2 = 2ab + [a^2 \square 2ab + b^2]$
$c^2 = a^2 + b^2$ ■
258 *CHAPTER 37. DIRECT PROOF*

37.2 References

[1] Cupillari, Antonella. *The Nuts and Bolts of Proofs*. Academic Press, 2001. Page 3.
[2] Krantz, Steven G. *The Proof is the Pudding*. Springer, 2010. Page 43.

37.3 Sources

_ Franklin, J.; A. Daoud (2011). *Proof in Mathematics: An Introduction*. Sydney: Kew Books. ISBN 0-646-54509-4. (Ch. 1.)

Chapter 38

Closure (mathematics)

For other uses, see Closure (disambiguation).
A set has **closure** under an operation if performance of that operation on members of the set always produces a member of the same set; in this case we also say that the set is **closed** under the operation. For example, the integers are closed under subtraction, but the positive integers are not: 1 and 2 are both positive integers, but the result of subtracting 2 from 1 is not a positive integer. Another example is the set containing only the number zero, which is closed under addition, subtraction and multiplication.
Similarly, a set is said to be **closed under a *collection* of operations** if it is closed under each of the operations individually.

38.1 Basic properties

A set that is closed under an operation or collection of operations is said to satisfy a **closure property**. Often a closure property is introduced as an axiom, which is then usually called the **axiom of closure**. Modern set-theoretic definitions usually define operations as maps between sets, so adding closure to a structure as an axiom is superfluous; however in practice operations are often defined initially on a superset of the set in question and a closure proof is required to establish that the operation applied to pairs from that set only produces members of that set. For example, the set of even integers is closed under addition, but the set of odd integers is not.
When a set S is not closed under some operations, one can usually find the smallest set containing S that is closed. This smallest closed set is called the **closure** of S (with respect to these operations). For example, the closure under

subtraction of the set of natural numbers, viewed as a subset of the real numbers, is the set of integers. An important example is that of topological closure. The notion of closure is generalized by Galois connection, and further by monads.

The set S must be a subset of a closed set in order for the closure operator to be defined. In the preceding example, it is important that the reals are closed under subtraction; in the domain of the natural numbers subtraction is not always defined.

The two uses of the word "closure" should not be confused. The former usage refers to the property of being closed, and the latter refers to the smallest closed set containing one that may not be closed. In short, the closure of a set satisfies a closure property.

38.2 Closed sets

A set is closed under an operation if that operation returns a member of the set when evaluated on members of the set. Sometimes the requirement that the operation be valued in a set is explicitly stated, in which case it is known as the **axiom of closure**. For example, one may define a group as a set with a binary product operator obeying several axioms, including an axiom that the product of any two elements of the group is again an element. However the modern definition of an operation makes this axiom superfluous; an n-ary operation on S is just a subset of S_{n+1}. By

259

260 *CHAPTER 38. CLOSURE (MATHEMATICS)*

its very definition, an operator on a set cannot have values outside the set.

Nevertheless, the closure property of an operator on a set still has some utility. Closure on a set does not necessarily imply closure on all subsets. Thus a subgroup of a group is a subset on which the binary product and the unary operation of inversion satisfy the closure axiom.

An operation of a different sort is that of finding the limit points of a subset of a topological space (if the space is first-countable, it suffices to restrict consideration to the limits of sequences but in general one must consider at least limits of nets). A set that is closed under this operation is usually just referred to as a closed set in the context of topology. Without any further qualification, the phrase usually means closed in this sense. Closed intervals like [1,2] = $\{x : 1 \leq x \leq 2\}$ are closed in this sense.

A partially ordered set is **downward closed** (and also called a lower set) if for every element of the set all smaller elements are also in it; this applies for example for the real intervals $(-\infty, p)$ and $(-\infty, p]$, and for an ordinal number p represented as interval $[\,0, p)$; every downward closed set of ordinal numbers is itself an ordinal number. **Upward closed** and upper set are defined similarly.

38.3 *P* closures of binary relations

The notion of a closure can be applied for an arbitrary binary relation $R \subseteq S\times S$, and an arbitrary property P in the following way: the P **closure** of R is the least relation $Q \subseteq S\times S$ that contains R (i.e. $R \subseteq Q$) and for which property P holds (i.e. $P(Q)$ is true). For instance, one can define the **symmetric closure** as the least symmetric relation containing R. This generalization is often encountered in the theory of rewriting systems, where one often uses more "wordy" notions such as the **reflexive transitive closure** $R*$—the smallest preorder containing R, or the **reflexive transitive symmetric closure** R≡—the smallest equivalence relation containing R, and therefore also known as the **equivalence closure**. When considering a particular term algebra, an equivalence relation that is compatible with all operations of the algebra [note 1] is called a congruence relation. The **congruence closure** of R is defined as the smallest congruence relation containing R.

For arbitrary P and R, the P closure of R need not exist. In the above examples, these exist because reflexivity, transitivity and symmetry are closed under arbitrary intersections. In such cases, the P closure can be directly defined as the intersection of all sets with property P containing R.[1]

Some important particular closures can be constructively obtained as follows:

_ cl_{re}□$(R) = R \cup \{ \langle x,x\rangle : x \in S \}$ is the reflexive closure of R,

_ cl□□□$(R) = R \cup \{ \langle y,x\rangle : \langle x,y\rangle \in R \}$ is its symmetry closure,

_ cl□$_r$□$(R) = R \cup \{ \langle x_1,x_n\rangle : n>1 \wedge \langle x_1,x_2\rangle, ..., \langle x_n$□$_1,x_n\rangle \in R \}$ is its transitive closure,

_ cl_e□□,Σ$(R) = R \cup \{ \langle f(x_1,\dots,x_i$□$_1,x_i,x_{i+1},\dots,x_n), f(x_1,\dots,x_i$□$_1,y,x_{i+1},\dots,x_n)\rangle : \langle x_i,y\rangle \in R \wedge f \in \Sigma$ n-ary $\wedge 1 \leq i \leq n \wedge x_1,...,x_n \in S \}$ is its embedding closure with respect to a given set Σ of operations on S, each with a fixed arity.

The relation R is said to have closure under some cl_{xxx}, if $R = cl_{xxx}(R)$; for example R is called symmetric if $R = cl$□□□(R).

Any of these four closures preserves symmetry, i.e., if R is symmetric, so is any $cl_{xxx}(R)$. [note 2] Similarly, all four preserve reflexivity. Moreover, cl□$_r$□ preserves closure under cl_e□□,Σ for arbitrary Σ. As a consequence, the equivalence closure of an arbitrary binary relation R can be obtained as cl□$_r$□$(cl$□□□$(cl_{re}$□$(R)))$, and the congruence closure with respect to some Σ can be obtained as cl□$_r$□$(cl_e$□□,Σ$(cl$□□□$(cl_{re}$□$(R))))$. In the latter case, the nesting order does matter; e.g. if S is the set of terms over $\Sigma = \{ a, b, c, f \}$ and $R = \{ \langle a,b\rangle, \langle f(b),c\rangle \}$, then the pair $\langle f(a),c\rangle$ is contained in the congruence closure cl□$_{\mathrm{r}}$□(cl_{e}□□,Σ(cl□□□(cl_{re}□(R)))) of R, but not in the relation cl_{e}□□,Σ(cl□$_{\mathrm{r}}$□(cl□□□(cl_{re}□(R)))).

38.4 Closure operator

Main article: closure operator

Given an operation on a set X, one can define the closure $C(S)$ of a subset S of X to be the smallest subset closed under that operation that contains S as a subset, if any such subsets exist. Consequently, $C(S)$ is the intersection of all closed sets containing S. For example, the closure of a subset of a group is the subgroup generated by that set.

The closure of sets with respect to some operation defines a **closure operator** on the subsets of X. The closed sets can be determined from the closure operator; a set is closed if it is equal to its own closure. Typical structural properties of all closure operations are: [2]

- The closure is **increasing** or **extensive**: the closure of an object contains the object.
- The closure is **idempotent**: the closure of the closure equals the closure.
- The closure is **monotone**, that is, if X is contained in Y, then also $C(X)$ is contained in $C(Y)$.

An object that is its own closure is called **closed**. By idempotency, an object is closed if and only if it is the closure of some object.

These three properties define an **abstract closure operator**. Typically, an abstract closure acts on the class of all subsets of a set.

If X is contained in a set closed under the operation then every subset of X has a closure.

38.5 Examples

- In topology and related branches, the relevant operation is taking limits. The topological closure of a set is the corresponding closure operator. The Kuratowski closure axioms characterize this operator.
- In linear algebra, the linear span of a set X of vectors is the **closure** of that set; it is the smallest subset of the vector space that includes X and is closed under the operation of linear combination. This subset is a subspace.
- In matroid theory, the closure of X is the largest superset of X that has the same rank as X.
- In set theory, the transitive closure of a set.
- In set theory, the transitive closure of a binary relation.
- In algebra, the algebraic closure of a field.
- In commutative algebra, closure operations for ideals, as integral closure and tight closure.
- In geometry, the convex hull of a set S of points is the smallest convex set of which S is a subset.
- In the theory of formal languages, the Kleene closure of a language can be described as the set of strings that can be made by concatenating zero or more strings from that language.
- In group theory, the conjugate closure or normal closure of a set of group elements is the smallest normal subgroup containing the set.
- In mathematical analysis and in probability theory, the closure of a collection of subsets of X under countably many set operations is called the σ-algebra generated by the collection.

38.6 See also

- Open set
- Clopen set

38.7 Notes

[1] that is, such that e.g. xRy implies $f(x,x_2)\ R\ f(y,x_2)$ and $f(x_1,x)\ R\ f(x_1,y)$ for any binary operation f and arbitrary $x_1,x_2 \in S$

[2] formally: if $R = cl$□□□(R), then $cl_{\mathrm{xxx}}(R) = cl$□□□($cl_{\mathrm{xxx}}(R)$)

38.8 References

[1] Baader, Franz; Nipkow, Tobias (1998). *Term Rewriting and All That*. Cambridge University Press., pp. 8–9

[2] Garrett Birkhoff (1967). *Lattice Theory*. Colloquium Publications **25**. Am. Math. Soc. p. 111.

Chapter 39

Distributive property

"Distributivity" redirects here. It is not to be confused with Distributivism.

In abstract algebra and formal logic, the **distributive property** of binary operations generalizes the **distributive law** from elementary algebra. In propositional logic, **distribution** refers to two valid rules of replacement. The rules allow one to reformulate conjunctions and disjunctions within logical proofs.
For example, in arithmetic:
2 · (1 + 3) = (2 · 1) + (2 · 3), but 2 / (1 + 3) ≠ (2 / 1) + (2 / 3).
In the left-hand side of the first equation, the 2 multiplies the sum of 1 and 3; on the right-hand side, it multiplies the 1 and the 3 individually, with the products added afterwards. Because these give the same final answer (8), we say that multiplication by 2 *distributes* over addition of 1 and 3. Since we could have put any real numbers in place of 2, 1, and 3 above, and still have obtained a true equation, we say that multiplication of real numbers *distributes* over addition of real numbers.

39.1 Definition

Given a set *S* and two binary operators _ and + on *S*, we say that the operation _
_ is *left-distributive* over + if, given any elements *x*, *y*, and *z* of *S*,
x _ (*y* + *z*) = (*x* _ *y*) + (*x* _ *z*)
_ is *right-distributive* over + if, given any elements *x*, *y*, and *z* of *S*:
(*y* + *z*) _ *x* = (*y* _ *x*) + (*z* _ *x*)
_ is *distributive* over + if it is left- and right-distributive.[1]
Notice that when _ is commutative, then the three above conditions are logically equivalent.

39.2 Meaning

The operators used for examples in this section are the binary operations of addition (+) and multiplication (_) of numbers.
There is a distinction between left-distribuitivity and right-distributivity:
263
264 *CHAPTER 39. DISTRIBUTIVE PROPERTY*
a _ (*b* _ *c*) = *a* _ *b* _ *a* _ *c* (left-distributive)
(*a* _ *b*) _ *c* = *a* _ *c* _ *b* _ *c* (right-distributive)
In either case, the distributive property can be described in words as:
To multiply a sum (or difference) by a factor, each summand (or minuend and subtrahend) is multiplied by this factor and the resulting products are added (or subtracted).
If the operation outside the parentheses (in this case, the multiplication) is commutative , then left-distributivity implies right-distributivity and vice-versa.
One example of an operation that is "only" right-distributive is division , which is not commutative:
(*a* _ *b*) _ *c* = *a* _ *c* _ *b* _ *c*
In this case, left-distributivity does not apply:
a _ (*b* _ *c*)≠ *a* _ *b* _ *a* _ *c*
The distributive laws are among the axioms for rings and fields . Examples of structures in which two operations are mutually related to each other by the distributive law are Boolean algebras such as the algebra of sets or the switching algebra . There are also combinations of operations that are not mutually distributive over each other; For example, addition is not distributive over multiplication.
Multiplying sums can be put into words as follows: When a sum is multiplied by a sum, multiply each summand of a sum with each summand of the other sums (keeping track of signs), and then adding up all of the resulting products.

39.3 Examples

39.3.1 Real numbers

In the following examples, the use of the distributive law on the set of real numbers R is illustrated. When multiplication is mentioned in elementary mathematics, it usually refers to this kind of multiplication. From the point of view of algebra , the real numbers form a field , which ensures the validity of the distributive law.
First example (mental and written multiplication)
During mental arithmetic, distributivity is often used unconsciously:
6 _ 16 = 6 _ (10 + 6) = 6 _ 10 + 6 _ 6 = 60 + 36 = 96
Thus, to calculate 6 · 16 in your head, you first multiply 6 · 10 and 6 · 6 and add the intermediate results. Written multiplication is also based on the distributive law.
Second example (with variables)
3*a*2*b* _ (4*a* □ 5*b*) = 3*a*2*b* _ 4*a* □ 3*a*2*b* _ 5*b* = 12*a*3*b* □ 15*a*2*b*2
Third example (with two sums)

(a + b) _ (a □ b) = a _ (a □ b) + b _ (a □ b) = a2 □ ab + ba □ b2 = a2 □ b2
= (a + b) _ a □ (a + b) _ b = a2 + ba □ ab □ b2 = a2 □ b2
Here the distributive law was applied twice and. It does not matter which bracket is first multiplied out.

Fourth Example Here the distributive law is applied the other way around compared to the previous examples. Consider
12a3b2 □ 30a4bc + 18a2b3c2 :
Since the factor 6a2b occurs in all summand, it can be factored out. That is, due to the distributive law one obtains
12a3b2 □ 30a4bc + 18a2b3c2 = 6a2b(2ab □ 5a2c + 3b2c2) :

39.3.2 Matrices

The distributive law is valid for matrix multiplication . More precisely,
(A + B) _ C = A _ C + B _ C
for all l _ m -matrices A;B and m _ n -matrices C , as well as
A _ (B + C) = A _ B + A _ C
for all l _ m -matrices A and m _ n -matrices B;C . Because the commutative property does not hold for matrix multiplication, the second law does not follow from the first law. In this case, they are two different laws.

39.3.3 Other examples

1. Multiplication of ordinal numbers, in contrast, is only left-distributive, not right-distributive.
2. The cross product is left- and right-distributive over vector addition, though not commutative.
3. The union of sets is distributive over intersection, and intersection is distributive over union.
4. Logical disjunction ("or") is distributive over logical conjunction ("and"), and conjunction is distributive over disjunction.
5. For real numbers (and for any totally ordered set), the maximum operation is distributive over the minimum operation, and vice-versa: max(a,min(b,c)) = min(max(a,b),max(a,c)) and min(a,max(b,c)) = max(min(a,b),min(a,c)).
6. For integers, the greatest common divisor is distributive over the least common multiple, and vice-versa: gcd(a,lcm(b,c)) = lcm(gcd(a,b),gcd(a,c)) and lcm(a,gcd(b,c)) = gcd(lcm(a,b),lcm(a,c)).
7. For real numbers, addition distributes over the maximum operation, and also over the minimum operation: a + max(b,c) = max(a+b,a+c) and a + min(b,c) = min(a+b,a+c).

39.4 Propositional logic

39.4.1 Rule of replacement

In standard truth-functional propositional logic, *distribution*[2][3][4] in logical proofs uses two valid rules of replacement to expand individual occurrences of certain logical connectives, within some formula, into separate applications of those connectives across subformulas of the given formula. The rules are:
(P ^ (Q _ R)) , ((P ^ Q) _ (P ^ R))
and
(P _ (Q ^ R)) , ((P _ Q) ^ (P _ R))
where " , ", also written ≡, is a metalogical symbol representing "can be replaced in a proof with" or "is logically equivalent to".

39.4.2 Truth functional connectives

Distributivity is a property of some logical connectives of truth-functional propositional logic. The following logical equivalences demonstrate that distributivity is a property of particular connectives. The following are truth-functional tautologies.
Distribution of conjunction over conjunction (P ^ (Q ^ R)) $ ((P ^ Q) ^ (P ^ R))
Distribution of conjunction over disjunction [5] (P ^ (Q _ R)) $ ((P ^ Q) _ (P ^ R))
Distribution of disjunction over conjunction [6] (P _ (Q ^ R)) $ ((P _ Q) ^ (P _ R))
Distribution of disjunction over disjunction (P _ (Q _ R)) $ ((P _ Q) _ (P _ R))
Distribution of implication (P ! (Q ! R)) $ ((P ! Q) ! (P ! R))
Distribution of implication over equivalence (P ! (Q $ R)) $ ((P ! Q) $ (P ! R))
Distribution of disjunction over equivalence (P _ (Q $ R)) $ ((P _ Q) $ (P _ R))
Double distribution
((P ^ Q) _ (R ^ S)) $ (((P _ R) ^ (P _ S)) ^ ((Q _ R) ^ (Q _ S)))
((P _ Q) ^ (R _ S)) $ (((P ^ R) _ (P ^ S)) _ ((Q ^ R) _ (Q ^ S)))

39.5 Distributivity and rounding

In practice, the distributive property of multiplication (and division) over addition may appear to be compromised or lost because of the limitations of arithmetic precision. For example, the identity ⅓ + ⅓ + ⅓ = (1 + 1 + 1) / 3 appears to fail if the addition is conducted in decimal arithmetic; however, if many significant digits are used, the calculation will result in a closer approximation to the correct results. For example, if the arithmetical calculation takes the form: $0.33333 + 0.33333 + 0.33333 = 0.99999 \neq 1$, this result is a closer approximation than if fewer significant digits had been used. Even when fractional numbers can be represented exactly in arithmetical form, errors will be introduced if those arithmetical values are rounded or truncated. For example, buying two books, each priced at £14.99 before a tax of 17.5%, in two separate transactions will actually save £0.01, over buying them together: £14.99 × 1.175 = £17.61 to the nearest £0.01, giving a total expenditure of £35.22, but £29.98 × 1.175 = £35.23. Methods such as banker's rounding may help in some cases, as may increasing the precision used, but ultimately some calculation errors are inevitable.

39.6 Distributivity in rings

Distributivity is most commonly found in rings and distributive lattices.

A ring has two binary operations (commonly called "+" and "∗"), and one of the requirements of a ring is that ∗ must distribute over +. Most kinds of numbers (example 1) and matrices (example 4) form rings. A lattice is another kind of algebraic structure with two binary operations, ∧ and ∨. If either of these operations (say ∧) distributes over the other (∨), then ∨ must also distribute over ∧, and the lattice is called distributive. See also the article on distributivity (order theory).

Examples 4 and 5 are Boolean algebras, which can be interpreted either as a special kind of ring (a Boolean ring) or a special kind of distributive lattice (a Boolean lattice). Each interpretation is responsible for different distributive laws in the Boolean algebra. Examples 6 and 7 are distributive lattices which are not Boolean algebras.

Failure of one of the two distributive laws brings about near-rings and near-fields instead of rings and division rings respectively. The operations are usually configured to have the near-ring or near-field distributive on the right but not on the left.

Rings and distributive lattices are both special kinds of rigs, certain generalizations of rings. Those numbers in example 1 that don't form rings at least form rigs. Near-rigs are a further generalization of rigs that are left-distributive but not right-distributive; example 2 is a near-rig.

39.7 Generalizations of distributivity

In several mathematical areas, generalized distributivity laws are considered. This may involve the weakening of the above conditions or the extension to infinitary operations. Especially in order theory one finds numerous important variants of distributivity, some of which include infinitary operations, such as the infinite distributive law; others being defined in the presence of only *one* binary operation, such as the according definitions and their relations are given in the article distributivity (order theory). This also includes the notion of a **completely distributive lattice**.

In the presence of an ordering relation, one can also weaken the above equalities by replacing = by either ≤ or ≥. Naturally, this will lead to meaningful concepts only in some situations. An application of this principle is the notion of **sub-distributivity** as explained in the article on interval arithmetic.

In category theory, if (S, μ, η) and (S', μ', η') are monads on a category C, a **distributive law** $S.S' \to S'.S$ is a natural transformation $\lambda : S.S' \to S'.S$ such that (S', λ) is a lax map of monads $S \to S$ and (S, λ) is a colax map of monads $S' \to S'$. This is exactly the data needed to define a monad structure on $S'.S$: the multiplication map is $S'\mu.\mu'S^2.S'\lambda S$ and the unit map is $\eta'S.\eta$. See: distributive law between monads.

A generalized distributive law has also been proposed in the area of information theory.

39.8 Notes

[1] Ayres, p. 20.

[2] Moore and Parker

[3] Copi and Cohen

[4] Hurley

[5] Russell and Whitehead, *Principia Mathematica*

[6] Russell and Whitehead, *Principia Mathematica*

39.9 References

_ Ayres, Frank, *Schaum's Outline of Modern Abstract Algebra*, McGraw-Hill; 1st edition (June 1, 1965). ISBN 0-07-002655-6.

39.10 External links

_ A demonstration of the Distributive Law for integer arithmetic (from cut-the-knot)

Chapter 40

Material conditional

"Logical conditional" redirects here. For other related meanings, see Conditional statement.
Not to be confused with material inference.
The **material conditional** (also known as "**material implication**", "**material consequence**", or simply "**implication**",
Venn diagram of A ! B .
If a member of the set described by this diagram (the red areas) is a member of A , it is in the intersection of A and B , and it therefore is also in B .
"**implies**" or "**conditional**") is a logical connective (or a binary operator) that is often symbolized by a forward arrow "→". The material conditional is used to form statements of the form "$p \to q$" (termed a conditional statement) which is read as "if p then q" and conventionally compared to the English construction "If...then...". But unlike as the English construction may, the conditional statement "$p \to q$" does not specify a causal relationship between *p* and *q* and is to be understood to mean "if *p* is true, then *q* is also true" such that the statement "$p \to q$" is false only when *p* is true and *q* is false.[1] The material conditional is also to be distinguished from logical consequence.
The material conditional is also symbolized using:
268
40.1. DEFINITIONS OF THE MATERIAL CONDITIONAL 269
1. *p _ q* (Although this symbol may be used for the superset symbol in set theory.);
2. *p) q* (Although this symbol is often used for logical consequence (i.e. logical implication) rather than for material conditional.)

With respect to the material conditionals above, *p* is termed the *antecedent*, and *q* the *consequent* of the conditional. Conditional statements may be nested such that either or both of the antecedent or the consequent may themselves be conditional statements. In the example "$(p \to q) \to (r \to s)$" both the antecedent and the consequent are conditional statements.
In classical logic *p ! q* is logically equivalent to *:(p ^ :q)* and by De Morgan's Law logically equivalent to *:p _ q* .[2] Whereas, in minimal logic (and therefore also intuitionistic logic) *p ! q* only logically entails *:(p ^ :q)* ; and in intuitionistic logic (but not minimal logic) *:p _ q* entails *p ! q* .

40.1 Definitions of the material conditional

Logicians have many different views on the nature of material implication and approaches to explain its sense.[3]

40.1.1 As a truth function

In classical logic, the compound $p \to q$ is logically equivalent to the negative compound: not both *p* and not *q*. Thus the compound $p \to q$ is *false* if and only if both *p* is true and *q* is false. By the same stroke, $p \to q$ is *true* if and only if either *p* is false or *q* is true (or both). Thus → is a function from pairs of truth values of the components *p*, *q* to truth values of the compound $p \to q$, whose truth value is entirely a function of the truth values of the components. Hence, this interpretation is called *truth-functional*. The compound $p \to q$ is logically equivalent also to $\neg p \lor q$ (either not *p*, or *q* (or both)), and to $\neg q \to \neg p$ (if not *q* then not *p*). But it is not equivalent to $\neg p \to \neg q$, which is equivalent to $q \to p$.

Truth table

The truth table associated with the material conditional $p \to q$ is identical to that of $\neg p \lor q$ and is also denoted by **Cpq**. It is as follows:
It may also be useful to note that in Boolean algebra, true and false can be denoted as 1 and 0 respectively with an equivalent table.

40.1.2 As a formal connective

The material conditional can be considered as a symbol of a formal theory, taken as a set of sentences, satisfying all the classical inferences involving →, in particular the following characteristic rules:
1. Modus ponens;
2. Conditional proof;
3. Classical contraposition;
4. Classical reductio.

Unlike the truth-functional one, this approach to logical connectives permits the examination of structurally identical propositional forms in various logical systems, where somewhat different properties may be demonstrated. For example, in intuitionistic logic which rejects proofs by contraposition as valid rules of inference, $(p \to q) \Rightarrow \neg p \lor q$ is not a propositional theorem, but the material conditional is used to define negation.
270 *CHAPTER 40. MATERIAL CONDITIONAL*

40.2 Formal properties

When studying logic formally, the material conditional is distinguished from the semantic consequence relation *j*= . We say *A j*= *B* if every interpretation that makes *A* true also makes *B* true. However, there is a close relationship between the two in most logics, including classical logic. For example, the following principles hold:

_ If □ *j*= then ∅ *j*= (*φ*1 ^ _ _ _ ^ *φn* !) for some *φ*1; : : : ; *φn* 2 □ . (This is a particular form of the deduction theorem. In words, it says that if Γ models *ψ* this means that *ψ* can be deduced just from some subset of the theorems in Γ.)

_ The converse of the above

_ Both ! and *j*= are monotonic; i.e., if □ *j*= then Δ *[* □ *j*= , and if *φ* ! then (*φ* ^ _) ! for any *α*, Δ. (In terms of structural rules, this is often referred to as weakening or *thinning*.)

These principles do not hold in all logics, however. Obviously they do not hold in non-monotonic logics, nor do they hold in relevance logics.

Other properties of implication (the following expressions are always true, for any logical values of variables):

_ distributivity: (*s* ! (*p* ! *q*)) ! ((*s* ! *p*) ! (*s* ! *q*))

_ transitivity: (*a* ! *b*) ! ((*b* ! *c*) ! (*a* ! *c*))

_ reflexivity: *a* ! *a*

_ totality: (*a* ! *b*) _ (*b* ! *a*)

_ truth preserving: The interpretation under which all variables are assigned a truth value of 'true' produces a truth value of 'true' as a result of material implication.

_ commutativity of antecedents: (*a* ! (*b* ! *c*)) _ (*b* ! (*a* ! *c*))

Note that *a* ! (*b* ! *c*) is logically equivalent to (*a* ^ *b*) ! *c* ; this property is sometimes called un/currying.

Because of these properties, it is convenient to adopt a right-associative notation for → where *a* ! *b* ! *c* denotes *a* ! (*b* ! *c*) .

Comparison of Boolean truth tables shows that *a* ! *b* is equivalent to *:a* _ *b* , and one is an equivalent replacement for the other in classical logic. See material implication (rule of inference).

40.3 Philosophical problems with material conditional

Outside of mathematics, it is a matter of some controversy as to whether the truth function for material implication provides an adequate treatment of conditional statements in English (a sentence in the indicative mood with a conditional clause attached, i.e., an indicative conditional, or false-to-fact sentences in the subjunctive mood, i.e., a counterfactual conditional).[4] That is to say, critics argue that in some non-mathematical cases, the truth value of a compound statement, "if *p* then *q*", is not adequately determined by the truth values of *p* and *q*.[4] Examples of non-truth-functional statements include: "*q* because *p*", "*p* before *q*" and "it is possible that *p*".[4] "[Of] the sixteen possible truth-functions of *A* and *B*, material implication is the only serious candidate. First, it is uncontroversial that when *A* is true and *B* is false, "If *A*, *B*" is false. A basic rule of inference is modus ponens: from "If *A*, *B*" and *A*, we can infer *B*. If it were possible to have *A* true, *B* false and "If *A*, *B*" true, this inference would be invalid. Second, it is uncontroversial that "If *A*, *B*" is sometimes true when *A* and *B* are respectively (true, true), or (false, true), or (false, false)… Non-truth-functional accounts agree that "If *A*, *B*" is false when *A* is true and *B* is false; and they agree that the conditional is sometimes true for the other three combinations of truth-values for the components; but they deny that the conditional is always true in each of these three cases. Some agree with the truth-functionalist that when *A* and *B* are both true, "If *A*, *B*" must be true. Some do not, demanding a further relation between the facts that *A* and that *B*."[4]

The truth-functional theory of the conditional was integral to Frege's new logic (1879). It was taken up enthusiastically by Russell (who called it "material implication"), Wittgenstein in the *Tractatus*, and the logical positivists, and it is now found in every logic text. It is the first theory of conditionals which students encounter. Typically, it does not strike students as *obviously* correct. It is logic's first surprise. Yet, as the textbooks testify, it does a creditable job in many circumstances. And it has many defenders. It is a strikingly simple theory: "If *A*, *B*" is false when *A* is true and *B* is false. In all other cases, "If *A*, *B*" is true. It is thus equivalent to "~(*A*&~*B*)" and to "~*A* or *B*". "*A* ⊃ *B*" has, by stipulation, these truth conditions.

— Dorothy Edgington, *The Stanford Encyclopedia of Philosophy*, "Conditionals"[4]

The meaning of the material conditional can sometimes be used in the natural language English "if *condition* then *consequence*" construction (a kind of conditional sentence), where *condition* and *consequence* are to be filled with English sentences. However, this construction also implies a "reasonable" connection between the condition (protasis) and consequence (apodosis) (see Connexive logic).

The material conditional can yield some unexpected truths when expressed in natural language. For example, any material conditional statement with a false antecedent is true (see vacuous truth). So the statement "if 2 is odd then 2 is even" is true. Similarly, any material conditional with a true consequent is true. So the statement "if I have a penny in my pocket then Paris is in France" is always true, regardless of whether or not there is a penny in my pocket. These problems are known as the paradoxes of material implication, though they are not really paradoxes in the strict sense; that is, they do not elicit logical contradictions. These unexpected truths arise because speakers of English (and other natural languages) are tempted to equivocate between the material conditional and the indicative conditional, or other conditional statements, like the counterfactual conditional and the material biconditional. It is not surprising that a rigorously defined truth-functional operator does not correspond exactly to all notions of implication or otherwise expressed by 'if...then...' sentences in English (or their equivalents in other natural languages). For an overview of some the various analyses, formal and informal, of conditionals, see the "References" section below.

40.4 See also

40.4.1 Conditionals

_ Counterfactual conditional
_ Indicative conditional
_ Corresponding conditional
_ Strict conditional

40.5 References

[1] Magnus, P.D (January 6, 2012). "forallx: An Introduction to Formal Logic". Creative Commons. p. 25. Retrieved 28 May 2013.
[2] Teller, Paul (January 10, 1989). "A Modern Formal Logic Primer: Sentence Logic Volume 1". Prentice Hall. p. 54. Retrieved 28 May 2013.
[3] Clarke, Matthew C. (March 1996). "A Comparison of Techniques for Introducing Material Implication". Cornell University. Retrieved March 4, 2012.
[4] Edgington, Dorothy (2008). Edward N. Zalta, ed. "Conditionals". *The Stanford Encyclopedia of Philosophy* (Winter 2008 ed.).

40.6 Further reading

_ Brown, Frank Markham (2003), *Boolean Reasoning: The Logic of Boolean Equations*, 1st edition, Kluwer Academic Publishers, Norwell, MA. 2nd edition, Dover Publications, Mineola, NY, 2003.

_ Edgington, Dorothy (2001), "Conditionals", in Lou Goble (ed.), *The Blackwell Guide to Philosophical Logic*, Blackwell.
_ Quine, W.V. (1982), *Methods of Logic*, (1st ed. 1950), (2nd ed. 1959), (3rd ed. 1972), 4th edition, Harvard University Press, Cambridge, MA.
_ Stalnaker, Robert, "Indicative Conditionals", *Philosophia*, **5** (1975): 269–286.

40.7 External links

_ Conditionals entry by Edgington, Dorothy in the *Stanford Encyclopedia of Philosophy*

Chapter 41

Infinite set

In set theory, an **infinite set** is a set that is not a finite set. Infinite sets may be countable or uncountable. Some examples are:

_ the set of all integers, {..., □1, 0, 1, 2, ...}, is a countably infinite set; and
_ the set of all real numbers is an uncountably infinite set.

41.1 Properties

The set of natural numbers (whose existence is postulated by the axiom of infinity) is infinite. It is the only set that is directly required by the axioms to be infinite. The existence of any other infinite set can be proved in Zermelo–Fraenkel set theory (ZFC) only by showing that it follows from the existence of the natural numbers.

A set is infinite if and only if for every natural number the set has a subset whose cardinality is that natural number.

If the axiom of choice holds, then a set is infinite if and only if it includes a countable infinite subset.

If a set of sets is infinite or contains an infinite element, then its union is infinite. The powerset of an infinite set is infinite. Any superset of an infinite set is infinite. If an infinite set is partitioned into finitely many subsets, then at least one of them must be infinite. Any set which can be mapped onto an infinite set is infinite. The Cartesian product of an infinite set and a nonempty set is infinite. The Cartesian product of an infinite number of sets each containing at least two elements is either empty or infinite; if the axiom of choice holds, then it is infinite.

If an infinite set is a well-ordered set, then it must have a nonempty subset that has no greatest element.

In ZF, a set is infinite if and only if the powerset of its powerset is a Dedekind-infinite set, having a proper subset equinumerous to itself. If the axiom of choice is also true, infinite sets are precisely the Dedekind-infinite sets.

If an infinite set is a well-orderable set, then it has many well-orderings which are non-isomorphic.

41.2 History

The first known occurrence of explicitly infinite sets is in Galileo's last book Two New Sciences written while he was under house arrest by the Inquisition.[1]
Galileo argues that the set of squares S =fl; 4; 9; 16; 25; : : :g is the same size as N =fl; 2; 3; 4; 5; : : :g because there is a one-to-one correspondence:
1 $ 1; 2 $ 4; 3 $ 9; 4 $ 16; 5 $ 25; : : :
And yet, as he says, S is a proper subset of N and S even gets less dense as the numbers get larger.
273
274 *CHAPTER 41. INFINITE SET*

41.3 See also

_ Dedekind-infinite set
_ Aleph number
_ Galileo's *Two New Sciences*
_ Infinity

41.4 References

[1] Drake, Stillman, translator (1974). Two New Sciences, University of Wisconsin Press, 1974. ISBN 0-299-06404-2. A new translation including sections on centers of gravity and the force of percussion.

41.5 External links

_ Weisstein, Eric W., "Infinite Set", *MathWorld.*

Chapter 42

Proof by infinite descent

In mathematics, a proof by **infinite descent** is a particular kind of proof by contradiction which relies on the facts that the natural numbers are well ordered and that there are only a finite number of them that are smaller than any given one. One typical application is to show that a given equation has no solutions.
Typically, one shows that if a solution to a problem existed, which in some sense was related to one or more natural numbers, it would necessarily imply that a second solution existed, which was related to one or more 'smaller' natural numbers. This in turn would imply a third solution related to smaller natural numbers, implying a fourth solution, therefore a fifth solution, and so on. However there cannot be an infinity of ever-smaller natural numbers, and therefore by mathematical induction (repeating the same step) the original premise—that any solution exists—must be incorrect. It is disproven because its logical outcome would require a contradiction.
An alternative way to express this is to assume one or more solutions or examples exists. Then there must be a smallest solution or example—a minimal counterexample. We then prove that if a smallest solution exists, it must imply the existence of a smaller solution (in some sense)—which again proves that the existence of any solution would lead to a contradiction.
The method of infinite descent was developed by Fermat, who often used it for Diophantine equations.[1] Two typical examples are showing the non-solvability of the Diophantine equation $r^2 + s^4 = t^4$ and proving Fermat's theorem on sums of two squares, which states that any prime p such that $p \equiv 1 \pmod 4$ can be expressed as a sum of two squares (see proof). In some cases, to a modern eye, what he was using was (in effect) the doubling mapping on an elliptic curve. More precisely, his *method of infinite descent* was an exploitation in particular of the possibility of halving rational points on an elliptic curve *E* by inversion of the doubling formulae. The context is of a hypothetical rational point on *E* with large co-ordinates. Doubling a point on *E* roughly doubles the length of the numbers required to write it (as number of digits): so that a 'halved' point is quite clearly smaller. In this way Fermat was able to show the non-existence of solutions in many cases of Diophantine equations of classical interest (for example, the problem of four perfect squares in arithmetic progression).

42.1 Number theory

In the number theory of the twentieth century, the infinite descent method was taken up again, and pushed to a point where it connected with the main thrust of algebraic number theory and the study of L-functions. The structural result of Mordell, that the rational points on an elliptic curve *E* form a finitely-generated abelian group, used an infinite descent argument based on *E*/2*E* in Fermat's style.
To extend this to the case of an abelian variety *A*, André Weil had to make more explicit the way of quantifying the size of a solution, by means of a height function – a concept that became foundational. To show that *A*(*Q*)/2*A*(*Q*) is finite, which is certainly a necessary condition for the finite generation of the group *A*(*Q*) of rational points of *A*, one must do calculations in what later was recognised as Galois cohomology. In this way, abstractly-defined cohomology groups in the theory become identified with *descents* in the tradition of Fermat. The Mordell–Weil theorem was at the start of what later became a very extensive theory.
275
276 *CHAPTER 42. PROOF BY INFINITE DESCENT*

42.2 Application examples

42.2.1 Irrationality of √2

The proof that the square root of 2 (√2) is irrational (i.e. cannot be expressed as a fraction of two whole numbers) was discovered by the ancient Greeks, and is perhaps the earliest known example of a proof by infinite descent. Pythagoreans discovered that the diagonal of a square is incommensurable with its side, or in modern language, that the square root of two is irrational. Little is known with certainty about the time or circumstances of this discovery, but the name of Hippasus of Metapontum is often mentioned. For a while, the Pythagoreans treated as an official secret the discovery that the square root of two is irrational, and, according to legend, Hippasus was murdered for divulging it.[2][3][4] The square root of two is occasionally called "Pythagoras' number" or "Pythagoras' Constant", for example Conway & Guy (1996).[5]

The ancient Greeks, not having algebra, worked out a geometric proof by infinite descent (John Horton Conway presented another geometric proof (no. 8 ''') by infinite descent that may be more accessible). The following is an algebraic proof along similar lines:-

Suppose that √2 were rational. Then it could be written as

p

2 =

p

q

for two natural numbers, *p* and *q*. Then squaring would give

2 =

p^2

q^2 ;

$2q^2 = p^2$;

so 2 must be a factor of p^2, and therefore 2 must also be a factor of *p* itself (if 2 did not divide *p*, then the prime factorization of *p* (the product of its primes) would contain no 2's. So when one squares *p* by squaring all its factors, there still would be no 2's in the resulting prime factorization of p^2. But since p^2 *has* been found to be divisible by 2, *p* must be divisible by 2 as well.)

As 2 is a factor of *p*, we can now express p as 2 x some number *r*; thus

$p = 2r$

But then

$2q^2 = (2r)^2 = 4r^2$;

$q^2 = 2r^2$;

so 2 must be a factor of q^2, and therefore 2 must also be a factor of *q* itself, and *q* can be written as 2 x *s* for some whole number *s* (same reasoning as above). Therefore *p*/*q* can be written as (2 x *r*)/(2 x *s*), and we find that *p* and *q* are not the smallest natural numbers making √2: we can write √2 as *r*/*s* where *r*<*p* and *s*<*q*. Therefore if √2 could be written as a rational number, it could always be written as a natural number with smaller parts, which itself could be written with yet-smaller parts, *ad infinitum*. But this is impossible in the set of natural numbers. Since √2 is a real number, which can be either rational or irrational, the only option left is for √2 to be irrational.

(Alternatively, this proves that if √2 were rational, no "smallest" representation as a fraction could exist, as any attempt to find a "smallest" representation *p*/*q* would imply a smaller one existed, which is a similar contradiction).

42.2.2 Irrationality of √*k* if it is not an integer

For positive integer *k*, suppose that √k is not an integer, but is rational and can be expressed as *m/n* for natural numbers *m* and *n*, and let *q* be the largest integer no greater than √k. Then

p

k =

m

n

=

m(

p

k □ *q*)

n(

p

k □ *q*)

=

m

p

k □ *mq*

n
p
$k \square nq$
$=$
$nk \square mq$
$m \square nq$
(first the replacing$m$ with numerator the in n
p
k and
p
k with denominator the in m/n)
(The numerator and denominator were each multiplied by a positive expression less than 1, and then simplified independently, to show both products were still integers. Therefore, no matter what natural numbers m and n are used to express √k, there can always be smaller natural numbers $m' < m$ and $n' < n$ that have the same ratio. But infinite descent on the natural numbers is impossible, so this disproves the original assumption that √k can be expressed as a ratio of natural numbers.[6]

42.2.3 Non-solvability of $r^2 + s^4 = t^4$

The non-solvability of $r^2 + s^4 = t^4$ in integers is sufficient to show the non-solvability of $q^4 + s^4 = t^4$ in integers, which is a special case of Fermat's Last Theorem, and the historical proofs of the latter proceeded by more broadly proving the former using infinite descent. The following more recent proof demonstrates both of these impossibilities by proving still more broadly that a Pythagorean triangle cannot have any two of its sides each either a square or twice a square, since there is no smallest such triangle:[7]
Suppose there exists such a Pythagorean triangle. Then it can be scaled down to give a primitive (i.e., with no common factors) Pythagorean triangle with the same property. Primitive Pythagorean triangles' sides can be written as $x = 2ab;\ y = a^2 \square b^2;\ z = a^2 + b^2$, with a and b relatively prime and with $a+b$ odd and hence y and z both odd.
There are three cases, depending on which two sides are postulated to each be a square or twice a square:
_ ***y* and *z***: Neither y nor z, being odd, can be twice a square; if they are both square, the right triangle with legs
p
yz and b^2 and hypotenuse a^2 also would have integer sides including a square leg (b^2) and a square hypotenuse (a^2), and would have a smaller hypotenuse (a^2 compared to $z = a^2 + b^2$).
_ ***y* and *x***: If y is a square and x is a square or twice a square, then each of a and b is a square or twice a square and the integer right triangle with legs b and
p
y and hypotenuse a would have two sides (b and a) each of which
is a square or twice a square, with a smaller hypotenuse than the original triangle (a compared to $z = a^2 + b^2$
).
_ ***z* and *x***: If z is a square and x is a square or twice a square, again each of a and b is a square or twice a square and the integer right triangle with legs a and b and hypotenuse
p
z also would have two sides (a and b) each
of which is a square or twice a square, and a smaller hypotenuse (
p
z compared to z).
In any of these cases, one Pythagorean triangle with two sides each of which is a square or twice a square has led to a smaller one, which in turn would lead to a smaller one, etc.; since such a sequence cannot go on infinitely, the original premise that such a triangle exists must be wrong. This implies that $r^2 + s^4 = t^4$ cannot have a solution, since if it did then r, s^2, and t^2 would be the sides of such a Pythagorean triangle.
For other proofs of this by infinite descent, see[8] and.[9]

42.3 See also

_ Vieta jumping

42.4 References

[1] Weil, André (1984), *Number Theory: An approach through history from Hammurapi to Legendre*, Birkhäuser, pp. 75–79, ISBN 0-8176-3141-0
[2] Stephanie J. Morris, "The Pythagorean Theorem", Dept. of Math. Ed., University of Georgia.
[3] Brian Clegg, "The Dangerous Ratio ...", Nrich.org, November 2004.
[4] Kurt von Fritz, "The discovery of incommensurability by Hippasus of Metapontum", Annals of Mathematics, 1945.
[5] Conway, John H.; Guy, Richard K. (1996), *The Book of Numbers*, Copernicus, p. 25
[6] Sagher, Yoram (February 1988), *What Pythagoras could have done*, *American Mathematical Monthly* **95**: 117, doi:10.2307/2323064

[7] Dolan, Stan, “Fermat’s method of *descente infinie*", *Mathematical Gazette* 95, July 2011, 269–271.
[8] Grant, Mike, and Perella, Malcolm, “Descending to the irrational”, *Mathematical Gazette* 83, July 1999, pp. 263–267.
[9] Barbara, Roy, “Fermat’s last theorem in the case $n = 4$”, *Mathematical Gazette* 91, July 2007, 260–262.

42.5 Other reading

_ Infinite descent at PlanetMath.org.
_ Example of Fermat’s last theorem at PlanetMath.org.

Chapter 43

Square root of 2

“Pythagoras’s constant” redirects here; not to be confused with Pythagoras number

1

2 1

The square root of 2 is equal to the length of the hypotenuse of a right triangle with legs of length 1.

The **square root of 2**, often known as **root 2, radical (rad) 2**, or **Pythagoras’ constant**, and written as √2, is the
279
280 *CHAPTER 43. SQUARE ROOT OF 2*
positive algebraic number that, when multiplied by itself, gives the number 2. Technically, it is called the **principal square root of 2**, to distinguish it from the negative number with the same property.

Geometrically the square root of 2 is the length of a diagonal across a square with sides of one unit of length; this follows from the Pythagorean theorem. It was probably the first number known to be irrational. Its numerical value, truncated to 65 decimal places, is:

1.41421356237309504880168872420969807856967187537694807317667973799... (sequence A002193 in OEIS).

√2

1 2

1 75

41
29

32
17
12
99
70

The square root of 2 (number line not to scale).

The quick approximation 99/70 (≈ 1.41429) for the square root of two is frequently used. Despite having a denominator of only 70, it differs from the correct value by less than 1/10,000 (approx. 7.2 × 10^{-5}). The approximation 665857/470832 is valid to within 1.13 x 10^{12}: its square is 2.0000000000045....

43.1 History

The Babylonian clay tablet YBC 7289 (c. 1800–1600 BC) gives an approximation of √2 in four sexagesimal figures, 1 24 51 10, which is accurate to about six decimal digits,[1] and is the closest possible three-place sexagesimal representation of √2:

1 +
24
60
+
51
60^2 +
10

60^3 =
30547
21600
= 1:41421296:
Another early close approximation is given in ancient Indian mathematical texts, the Sulbasutras (c. 800–200 BC) as follows: *Increase the length [of the side] by its third and this third by its own fourth less the thirty-fourth part of that fourth.*[2] That is,
1 +
1
3
+
1
3 _ 4
□ 1
3 _ 4 _ 34
=
577
408
= 1:4142156862745098039:
This ancient Indian approximation is the seventh in a sequence of increasingly accurate approximations based on the sequence of Pell numbers, that can be derived from the continued fraction expansion of √2. Despite having a smaller denominator, it is only slightly less accurate than the Babylonian approximation.
Pythagoreans discovered that the diagonal of a square is incommensurable with its side, or in modern language, that the square root of two is irrational. Little is known with certainty about the time or circumstances of this discovery, but the name of Hippasus of Metapontum is often mentioned. For a while, the Pythagoreans treated as an official secret the discovery that the square root of two is irrational, and, according to legend, Hippasus was murdered for divulging it.[3][4][5] The square root of two is occasionally called "Pythagoras' number" or "Pythagoras' Constant", for example Conway & Guy (1996).[6]

Babylonian clay tablet YBC 7289 with annotations. Besides showing the square root of 2 in sexagesimal (1 24 51 10), the tablet also gives an example where one side of the square is 30 and the diagonal then is 42 25 35. The sexagesimal digit 30 can also stand for 1/2, in which case 42 25 35 is approximately 0.7071065.

43.2 Computation algorithms

There are a number of algorithms for approximating √2, which in expressions as a ratio of integers or as a decimal can only be approximated. The most common algorithm for this, one used as a basis in many computers and calculators, is the Babylonian method[7] of computing square roots, which is one of many methods of computing square roots. It goes as follows:
First, pick a guess, $a_0 > 0$; the value of the guess affects only how many iterations are required to reach an approximation of a certain accuracy. Then, using that guess, iterate through the following recursive computation:
a_{n+1} =
a_n + 2
a_n
2
=
a_n
2
+
1
a_n
:
The more iterations through the algorithm (that is, the more computations performed and the greater "n"), the better approximation of the square root of 2 is achieved. Each iteration approximately doubles the number of correct digits. Starting with $a_0 = 1$ the next approximations are
_ 3/2 = **1.5**
_ 17/12 = **1.416**...

_ 577/408 = **1.414215**...
_ 665857/470832 = **1.4142135623**746....
The value of √2 was calculated to 137,438,953,444 decimal places by Yasumasa Kanada's team in 1997. In February 2006 the record for the calculation of

p
2 was eclipsed with the use of a home computer. Shigeru Kondo calculated
200,000,000,000 decimal places in slightly over 13 days and 14 hours using a 3.6 GHz PC with 16 GiB of memory.[8] Among mathematical constants with computationally challenging decimal expansions, only π has been calculated more precisely.[9] Such computations aim to check empirically whether such numbers are normal.

43.3 Proofs of irrationality

A short proof of the irrationality of √2 can be obtained from the rational root theorem, that is, if $p(x)$ is a monic polynomial with integer coefficients, then any rational root of $p(x)$ is necessarily an integer. Applying this to the polynomial $p(x) = x^2$ □ 2 , it follows that √2is either an integer or irrational. Because √2 is not an integer (2 is not a perfect square), √2 must therefore be irrational. This proof can be generalized to show that any root of any natural number which is not the square of a natural number is irrational.
See quadratic irrational or infinite descent#Irrationality of √k if it is not an integer for a proof that the square root of any non-square natural number is irrational.

43.3.1 Proof by infinite descent

One proof of the number's irrationality is the following proof by infinite descent. It is also a proof by contradiction, also known as an indirect proof, in that the proposition is proved by assuming that the opposite of the proposition is true and showing that this assumption is false, thereby implying that the proposition must be true.

1. Assume that √2 is a rational number, meaning that there exists a pair of integers whose ratio is √2.
2. If the two integers have a common factor, it can be eliminated using the Euclidean algorithm.
3. Then √2 can be written as an irreducible fraction a/b such that a and b are coprime integers (having no common factor).
4. It follows that $a^2/b^2 = 2$ and $a^2 = 2b^2$. ($(a/b)^n = a^n/b^n$)
5. Therefore a^2 is even because it is equal to $2b^2$. ($2b^2$ is necessarily even because it is 2 times another whole number and multiples of 2 are even.)
6. It follows that a must be even (as squares of odd integers are never even).
7. Because a is even, there exists an integer k that fulfills: $a = 2k$.
8. Substituting $2k$ from step 7 for a in the second equation of step 4: $2b^2 = (2k)^2$ is equivalent to $2b^2 = 4k^2$, which is equivalent to $b^2 = 2k^2$.
9. Because $2k^2$ is divisible by two and therefore even, and because $2k^2 = b^2$, it follows that b^2 is also even which means that b is even.
10. By steps 5 and 8 a and b are both even, which contradicts that a/b is irreducible as stated in step 2.

Q.E.D.

Because there is a contradiction, the assumption (1) that
p
2 is a rational number must be false. This means that
p
2
is not a rational number; i.e.,
p
2 is irrational.
This proof was hinted at by Aristotle, in his *Analytica Priora*, §I.23.[10] It appeared first as a full proof in Euclid's *Elements*, as proposition 117 of Book X. However, since the early 19th century historians have agreed that this proof is an interpolation and not attributable to Euclid.[11]

43.3.2 Proof by unique factorization

An alternative proof uses the same approach with the fundamental theorem of arithmetic which says every integer greater than 1 has a unique factorization into powers of primes.

1. Assume that
p
2 is a rational number. Then there are integers a and b such that a is coprime to b and
p
2 = a/b
. In other words,
p
2 can be written as an irreducible fraction.
2. The value of b cannot be 1 as there is no integer a the square of which is 2.
3. There must be a prime p which divides b and which does not divide a, otherwise the fraction would not be irreducible.
4. The square of a can be factored as the product of the primes into which a is factored but with each power

doubled.
5. Therefore by unique factorization the prime p which divides b, and also its square, cannot divide the square of a.
6. Therefore the square of an irreducible fraction cannot be reduced to an integer.
7. Therefore the square root of 2 cannot be a rational number.

This proof can be generalized to show that if an integer is not an exact kth power of another integer then its kth root is irrational. For a proof of the same result which does not rely on the fundamental theorem of arithmetic, see: quadratic irrational.

43.3.3 Proof by infinite descent, not involving factoring

The following reductio ad absurdum argument showing the irrationality of

p
2 is less well-known. It uses the additional
information 2 >

p
2 > 1 so that 1 >

p
2 □ 1 > 0 .[12]

1. Assume that

p
2 is a rational number. This would mean that there exist positive integers m and n with $n \neq 0$
such that m/n =

p
2 . Then $m = n$

p
2 and m

p
2 = 2n .

2. We may assume that n is the smallest integer so that n

p
2 is an integer. That is, that the fraction m/n is in
lowest terms.

3. Then

p
2 = m
n = m(
p
2□1)
n(
p
2□1)
= 2n□m
m□n

4. Because 1 >

p
2 □ 1 > 0 , it follows that n > n(

p
2 □ 1) = m □ n > 0 .

5. So the fraction m/n for

p
2 , which according to (2) is already in lowest terms, is represented by (3) in strictly
lower terms. This is a contradiction, so the assumption that

p
2 is rational must be false.

43.3.4 Geometric proof

Another reductio ad absurdum showing that

p
2 is irrational is less well-known.[13] It is also an example of proof by
infinite descent. It makes use of classic compass and straightedge construction, proving the theorem by a method similar to that employed by ancient Greek geometers. It is essentially the previous proof viewed geometrically.
Let ABC be a right isosceles triangle with hypotenuse length m and legs n. By the Pythagorean theorem, m/n =

p
2

. Suppose *m* and *n* are integers. Let *m*:*n* be a ratio given in its lowest terms.
Draw the arcs *BD* and *CE* with centre *A*. Join *DE*. It follows that *AB* = *AD*, *AC* = *AE* and the ∠*BAC* and ∠*DAE* coincide. Therefore the triangles *ABC* and *ADE* are congruent by SAS.
Because ∠*EBF* is a right angle and ∠*BEF* is half a right angle, *BEF* is also a right isosceles triangle. Hence $BE = m - n$ implies $BF = m - n$. By symmetry, $DF = m - n$, and *FDC* is also a right isosceles triangle. It also follows that $FC = n - (m - n) = 2n - m$.
Hence we have an even smaller right isosceles triangle, with hypotenuse length $2n - m$ and legs $m - n$. These values are integers even smaller than *m* and *n* and in the same ratio, contradicting the hypothesis that *m*:*n* is in lowest terms. Therefore *m* and *n* cannot be both integers, hence
p
2 is irrational.

A
B C
D
E
F
m−n n
m
m−n 2n−m
m−n
43.3.5 Pythagorean Theorem Proof
Suppose
p
2 is rational.
That means that we can make a right isosceles triangle where the side lengths are natural numbers and the legs and the hypotenuse do not share any common factors (except 1). {1}
Since the legs are equal, so are their squares. So in order for the Pythagorean Theorem to work for this special right

triangle, the square of the hypotenuse has to be an even number (and if we cut it in half once then we have the area of the square of the leg).
Recall that the square of an even number is even and the square of an odd number is odd. So if the square of the hypotenuse is even the hypotenuse is even as well. {2}
Remember that a square is a quadrilateral with 2 pairs of parallel sides which are ***equal in length*** and has 4 right angles. So both sides of the square of the hypotenuse are even.
So the square of the hypotenuse of this right triangle can be cut in half twice and still have integer area. Since we only want to cut it in half once, then we'll get an even number.
So the square of the leg is even. Now according to {2} the leg must be even
This contradicts our assumption at {1} that the leg and hypotenuse have no common factors (except 1). Because if they're both even they share a common factor of 2. So the assumption that
p
2 was rational has to be false. Or in
other words
p
2 is an irrational number. Q. E. D.
43.3.6 Analytic proof
_ Lemma: let _ 2 R+ and *p*1; *p*2; : : : ; *q*1; *q*2; : : : 2 N such that *j_qn* □ *pnj*≠ 0 for all *n* 2 N and

lim $_{n!1}p_n$ = lim $_{n!1}q_n = 1$
lim $_{n!1}$
j_q_n □ $p_nj = 0:$
Then α is irrational.
Proof: suppose $\alpha = a/b$ with $a, b \in \mathbf{N}_+$.
For sufficiently big n,
$0 < j\_q_n$ □ $p_nj <$
1
b
then
$0 < jaq_n/b$ □ $p_nj <$
1
b
$0 < jaq_n$ □ $bp_nj < 1$
but aq_n □ bp_n is an integer, absurd, then α is irrational.

_
p
2 is irrational.
Proof: let $p_1 = q_1 = 1$ and
$p_{n+1} = p^2$
$_n + 2q^2$
$_n$
$q_{n+1} = 2p_nq_n$
for all n 2 **N** .
By induction,
286 *CHAPTER 43. SQUARE ROOT OF 2*
$0 <$

___
p
$2q_n$ □ p_n

___
$<$
1
$2^{2n□1}$
for all n 2 **N** . For $n = 1$,
$0 <$

___
p
$2q_1$ □ p_1

___
$<$
1
2
and if is true for n then is true for $n + 1$. In fact
$0 <$

___
p
$2q_n$ □ p_n

___
$_2$
$<$
1
2^{2n}
$0 <$

___
p
$2(2p_nq_n)$ □ $(p^2$
$_n + 2q^2$
$_n)$

___
$<$

$$0 < \frac{1}{2^{2n}}$$

$$\frac{p}{2q_{n+1} \square p_{n+1}}$$

$$< \frac{1}{2^{2n}}:$$

By application of the lemma,

p
2 is irrational.

43.3.7 Constructive proof

In a constructive approach, one distinguishes between on the one hand not being rational, and on the other hand being irrational (i.e., being quantifiably apart from every rational), the latter being a stronger property. Given positive integers a and b, because the valuation (i.e., highest power of 2 dividing a number) of $2b_2$ is odd, while the valuation of a_2 is even, they must be distinct integers; thus $j2b_2 \square a_2 j _ 1$. Then[14]

$$\frac{\overline{p}}{2} \square a$$
$$b$$

$$= \frac{j2b_2 \square a_2 j}{b_2(}$$

$$\frac{p}{2} + a/b)$$

$$_ \frac{1}{b_2(}$$

$$\frac{p}{2} + a/b)$$

$$_ \frac{1}{3b_2};$$

the latter inequality being true because we assume a
b
$_ 3 \square$
p
2 (otherwise the quantitative apartness can be trivially established). This gives a lower bound of $\frac{1}{3b_2}$ for the difference j

p
2 $\square a/bj$, yielding a direct proof of irrationality not relying on the law of excluded middle; see Errett Bishop (1985, p. 18). This proof constructively exhibits a discrepancy between

p
2 and any rational.

43.4 Properties of the square root of two

One-half of
p
2 , also 1 divided by the square root of 2, approximately 0.70710 67811 86548, is a common quantity in geometry and trigonometry because the unit vector that makes a 45° angle with the axes in a plane has the coordinates

$(p$
2
2
;
p
2
2
)

:
This number satisfies
1
2
p
2 =
√
1
2 =
*p*1
2
= cos(45
°
) = sin(45
°
):
One interesting property of the square root of 2 is as follows:
p 1
2 □ 1
=
p
2 + 1
43.4. PROPERTIES OF THE SQUARE ROOT OF TWO 287
y = ax : a < 1
y = 1/x
y
x
2 sinh(u)
y = x
x 2 + y 2 = 2
2 cos(u)
u
2 sin(u)
2 cosh(u)
Angle size and sector area are the same when the conic radius is square root of two. This diagram illustrates the circular and hyperbolic functions based on sector areas u.
since (
p
2 + 1)(
p
2 □ 1) = 2 □ 1 = 1*:* This is related to the property of silver ratios.
The square root of 2 can also be expressed in terms of the copies of the imaginary unit *i* using only the square root and arithmetic operations:
p
i + *i*
p
i
i and
p
□*i* □ *i*
p
□*i*
□*i*
if the square root symbol is interpreted suitably for the complex numbers *i* and -*i*.

The square root of 2 is also the only real number other than 1 whose infinite tetrate (i.e., infinite exponential tower) is equal to its square. In other words: If for c > 1 we define $x_1 = c$ and $x_{□+1} = c^{x_□}$ for n > 1, we will call the limit of x□ as n → ∞, if this limit exists, by the name f(c). Then sqrt(2) is the only number c > 1 for which $f(c) = c^2$. Or symbolically:

p
2
(
p
2
(
p
2(___
)))
= 2*:*

The square root of 2 appears in Viète's formula for *π*:

2*m*
√
2 □
√
2 +
√
2 + _ _ _ +
p
2 *!* _ as *m ! 1*

for *m* square roots and only one minus sign.[15]

Similar in appearance but with a finite number of terms, the square root of 2 appears in various trigonometric constants:[16]

sin(5 5
8
∘
) =
1
2
√
2 □
√
2 +
√
2 +
p
2;
sin(11 1
4
∘
) =
1
2
√
2 □
√
2 +
p
2;
sin(16 7
8
∘
) =
1
2
√
2 □
√
2 +
√

2 □
p
2;
sin(22 1
2
°
) =
1
2
√
2 □
p
2;
sin(28 1
8
°
) =
1
2
√
2 □
√
2 □
√
2 □
p
2;
sin(33 3
4
°
) =
1
2
√
2 □
√
2 □
p
2;
sin(39 3
8
°
) =
1
2
√
2 □
√
2 □
√
2 +
p
2;
sin(45
°
) =
1
2
p
2;
sin(50 5
8
°

) =
1
2
√
2 +
√
2 □
√
2 +
p
2;
sin(56 1
4
◦
) =
1
2
√
2 +
√
2 □
p
2;
sin(61 7
8
◦
) =
1
2
√
2 +
√
2 □
√
2 □
p
2;
sin(67 1
2
◦
) =
1
2
√
2 +
p
2;
sin(73 1
8
◦
) =
1
2
√
2 +
√
2 +
√
2 □
p
2;
sin(78 3
4

$$^\circ) = \frac{1}{2}\sqrt{2+\sqrt{2+\sqrt{2}}};$$

$$\sin(84\tfrac{3}{8}^\circ) = \frac{1}{2}\sqrt{2+\sqrt{2+\sqrt{2+\sqrt{2}}}}:$$

It is not known whether $\sqrt{2}$ is a normal number, a stronger property than irrationality, but statistical analyses of its binary expansion are consistent with the hypothesis that it is normal to base two.[17]

43.5 Series and product representations

The identity $\cos(\pi/4) = \sin(\pi/4) = 1/\sqrt{2}$, along with the infinite product representations for the sine and cosine, leads to products such as

$$\frac{1}{\sqrt{2}} = \prod_{k=0}^{\infty}\left(1 - \frac{1}{(4k+2)^2}\right) = \left(1-\frac{1}{4}\right)\left(1-\frac{1}{36}\right)\left(1-\frac{1}{100}\right)\cdots$$

and

$$\sqrt{2} = \prod_{k=0}^{\infty}\frac{(4k+2)^2}{(4k+1)(4k+3)} =$$

$$\left(\frac{2\cdot 2}{1\cdot 3}\right)\left(\frac{6\cdot 6}{5\cdot 7}\right)\left(\frac{10\cdot 10}{9\cdot 11}\right)\left(\frac{14\cdot 14}{13\cdot 15}\right)\cdots$$

or equivalently,

$$\sqrt{2}=\prod_{k=0}^{\infty}\left(1+\frac{1}{4k+1}\right)\left(1-\frac{1}{4k+3}\right)=\left(1+\frac{1}{1}\right)\left(1-\frac{1}{3}\right)\left(1+\frac{1}{5}\right)\left(1-\frac{1}{7}\right)\cdots.$$

The number can also be expressed by taking the Taylor series of a trigonometric function. For example, the series for $\cos(\pi/4)$ gives

$$\frac{1}{\sqrt{2}}=\sum_{k=0}^{\infty}\frac{(-1)^k\left(\frac{\pi}{4}\right)^{2k}}{(2k)!}.$$

The Taylor series of $\sqrt{(1+x)}$ with $x = 1$ and using the double factorial $n!!$ gives

$$\sqrt{}$$

$$\sqrt{2} = \sum_{k=0}^{\infty} \frac{(-1)^{k+1}(2k-3)!!}{(2k)!!} = 1 + \frac{1}{2} - \frac{1}{2\cdot 4} + \frac{1\cdot 3}{2\cdot 4\cdot 6} - \frac{1\cdot 3\cdot 5}{2\cdot 4\cdot 6\cdot 8} + \cdots.$$

The convergence of this series can be accelerated with an Euler transform, producing

$$\sqrt{2} = \sum_{k=0}^{\infty} \frac{(2k+1)!}{(k!)^2 2^{3k+1}} = \frac{1}{2} + \frac{3}{8} + \frac{15}{64} + \frac{35}{256} + \frac{315}{4096} + \frac{693}{16384} + \cdots.$$

It is not known whether $\sqrt{2}$ can be represented with a BBP-type formula. BBP-type formulas are known for $\pi\sqrt{2}$ and $\sqrt{2}\ln(1+\sqrt{2})$, however.[18]

43.6 Continued fraction representation

The square root of two has the following continued fraction representation:

$$\sqrt{2} = 1 + \cfrac{1}{2 + \cfrac{1}{2 + \cfrac{1}{\ddots}}}$$

2 +
1
2 + ...
:

The convergents formed by truncating this representation form a sequence of fractions that approximate the square root of two to increasing accuracy, and that are described by the Pell numbers (known as side and diameter numbers to the ancient Greeks because of their use in approximating the ratio between the sides and diagonal of a square). The first convergents are: 1/1, 3/2, 7/5, 17/12, 41/29, 99/70, 239/169, 577/408. The convergent p/q differs from the square root of 2 by almost exactly 1
2q2
p
2 and then the next convergent is $(p + 2q)/(p + q)$.

43.7 Paper size

The approximate aspect ratio of paper sizes under ISO 216 (A4, A0, etc.) is 1:√2. This ratio guarantees that cutting a sheet in half along a line parallel to its short side results in the smaller sheets having the same ratio as the original sheet.

43.8 See also

_ Square root of 3
_ Square root of 5
_ Silver ratio, 1 +
p
2
_ The square root of two is the frequency ratio of a tritone interval in twelve-tone equal temperament music.
_ The square root of two forms the relationship of f-stops in photographic lenses, which in turn means that the ratio of *areas* between two successive apertures is 2.
_ The celestial latitude (declination) of the Sun during a planet's astronomical cross-quarter day points equals the tilt of the planet's axis divided by √2.
_ Viète's formula

43.9 Notes

[1] Fowler and Robson, p. 368.
Photograph, illustration, and description of the *root(2)* tablet from the Yale Babylonian Collection
High resolution photographs, descriptions, and analysis of the *root(2)* tablet (YBC 7289) from the Yale Babylonian Collection
[2] Henderson.
[3] Stephanie J. Morris, "The Pythagorean Theorem", Dept. of Math. Ed., University of Georgia.
[4] Brian Clegg, "The Dangerous Ratio ...", Nrich.org, November 2004.
[5] Kurt von Fritz, "The discovery of incommensurability by Hippasus of Metapontum", Annals of Mathematics, 1945.
[6] Conway, John H.; Guy, Richard K. (1996), *The Book of Numbers*, Copernicus, p. 25
[7] Although the term "Babylonian method" is common in modern usage, there is no direct evidence showing how the Babylonians computed the approximation of √2 seen on tablet YBC 7289. Fowler and Robson offer informed and detailed conjectures.
Fowler and Robson, p. 376. Flannery, p. 32, 158.
[8] "Constants and Records of Computation". Numbers.computation.free.fr. 2010-08-12. Retrieved 2012-09-07.
[9] "Number of known digits". Numbers.computation.free.fr. 2010-08-12. Retrieved 2012-09-07.
[10] All that Aristotle says, while writing about proofs by contradiction, is that "the diagonal of the square is incommensurate with the side, because odd numbers are equal to evens if it is supposed to be commensurate".
[11] The edition of the Greek text of the *Elements* published by E. F. August in Berlin in 1826–1829 already relegates this proof to an Appendix. The same thing occurs with J. L. Heiberg's edition (1883–1888).
[12] Gardner, Martin (2001), *A Gardner's workout: training the mind and entertaining the spirit*, A K Peters, Ltd., ISBN 978-1-56881-120-8, p. 16
[13] Apostol (2000), p. 841
[14] See Katz, Karin Usadi; Katz, Mikhail G. (2011), *Meaning in Classical Mathematics: Is it at Odds with Intuitionism?*, *Intellectica* **56** (2): 223–302 (see esp. Section 2.3, footnote 15), arXiv:1110.5456

[15] Courant, Richard; Robbins, Herbert (1941), *What is mathematics? An Elementary Approach to Ideas and Methods*, London: Oxford University Press, p. 124
[16] Julian D. A. Wiseman Sin and cos in surds
[17] Good & Gover (1967).
[18] http://crd.lbl.gov/~{}dhbailey/dhbpapers/bbp-formulas.pdf

43.10 References

_ Apostol, Tom M. (2000), *Irrationality of the square root of two – A geometric proof*, *American Mathematical Monthly* **107** (9): 841–842, doi:10.2307/2695741, JSTOR 2695741.

_ Aristotle (2007), *Analytica priora*, eBooks@Adelaide
_ Bishop, Errett (1985), Schizophrenia in contemporary mathematics. Errett Bishop: reflections on him and his research (San Diego, Calif., 1983), 1–32, Contemp. Math. 39, Amer. Math. Soc., Providence, RI.
_ Flannery, David (2005), *The Square Root of Two*, Springer-Verlag, ISBN 0-387-20220-X.
_ Fowler, David; Robson, Eleanor (1998), *Square Root Approximations in Old Babylonian Mathematics: YBC 7289 in Context*, *Historia Mathematica* **25** (4): 366–378, doi:10.1006/hmat.1998.2209.
_ Good, I. J.; Gover, T. N. (1967), *The generalized serial test and the binary expansion of √2*, *Journal of the Royal Statistical Society, Series A* **130** (1): 102–107, doi:10.2307/2344040, JSTOR 2344040.
_ Henderson, David W. (2000), "Square roots in the Śulba Sūtras", in Gorini, Catherine A., *Geometry At Work: Papers in Applied Geometry*, Cambridge University Press, pp. 39–45, ISBN 978-0-88385-164-7.

43.11 External links

_ Gourdon, X.; Sebah, P. (2001), "Pythagoras' Constant: √2", *Numbers, Constants and Computation.*
_ Weisstein, Eric W., "Pythagoras's Constant", *MathWorld.*
_ The Square Root of Two to 5 million digits by Jerry Bonnell and Robert Nemiroff. May, 1994.
_ Square root of 2 is irrational, a collection of proofs
_ Grime, James; Bowley, Roger. "The Square Root √2 of Two". *Numberphile.* Brady Haran.

Chapter 44

Contraposition

For contraposition in the field of traditional logic, see Contraposition (traditional logic).
For contraposition in the field of symbolic logic, see Transposition (logic).

In logic, **contraposition** is a law that says that a conditional statement is logically equivalent to its **contrapositive**. The contrapositive of the statement has its antecedent and consequent inverted and flipped: the contrapositive of *P ! Q* is thus *:Q ! :P* . For instance, the proposition "*All bats are mammals*" can be restated as the conditional "*If something is a bat, then it is a mammal*". Now, the law says that statement is identical to the contrapositive "*If something is not a mammal, then it is not a bat.*"

The contrapositive can be compared with three other relationships between conditional statements:

_ **Inversion** (the **inverse**): *:P ! :Q* .
"*If something is not a bat, then it is not a mammal.*" Unlike the contrapositive, the inverse's truth value is not at all dependent on whether or not the original proposition was true, as evidenced here. The inverse here is clearly not true.

_ **Conversion** (the **converse**): *Q ! P* .
"*If something is a mammal, then it is a bat.*" The converse is actually the contrapositive of the inverse and so always has the same truth value as the inverse, which is not necessarily the same as that of the original proposition.

_ **Negation**: *:(P ! Q)* .
"*There exists a bat that is not a mammal.* " If the negation is true, the original proposition (and by extension the contrapositive) is false. Here, of course, the negation is false.

Note that if *P ! Q* is true and we are given that *Q* is false, *:Q* , it can logically be concluded that *P* must be false, *:P* . This is often called the *law of contrapositive*, or the *modus tollens* rule of inference.

44.1 Intuitive explanation

Consider the Euler diagram on the right. According to this diagram, if something is in A, it must be in B as well. So we can interpret "all of A is in B" as:

A ! B

It is also clear that anything that is **not** within B (the white region) **cannot** be within A, either. This statement,

:B ! :A

292

44.2. FORMAL DEFINITION 293

A

B

is the contrapositive. Therefore we can say that

(A ! B) ! (:B ! :A)

Practically speaking, this may make life much easier when trying to prove something. For example, if we want to prove that every girl in the United States (A) is blonde (B), we can either try to directly prove *A ! B* by checking all girls in the United States to see if they are all blonde. Alternatively, we can try to prove *:B ! :A* by checking all non-blonde girls to see if they are all outside the US. This means that if we find at least one non-blonde girl within the US, we will have disproved *:B ! :A* , and equivalently *A ! B* .

To conclude, for any statement where A implies B, then *not B* always implies *not A*. Proving or disproving either one of these statements automatically proves or disproves the other. They are fully equivalent.

44.2 Formal definition

A proposition *Q* is implicated by a proposition *P* when the following relationship holds:

(P ! Q)

This states that, "if *P*, then *Q*", or, "if *Socrates is a man*, then *Socrates is human*." In a conditional such as this, *P* is the antecedent, and *Q* is the consequent. One statement is the **contrapositive** of the other only when its antecedent is the negated consequent of the other, and vice versa. The contrapositive of the example is

(:Q ! :P)

That is, "If not-*Q*, then not-*P*", or, more clearly, "If *Q* is not the case, then *P* is not the case." Using our example, this is rendered "If *Socrates is not human*, then *Socrates is not a man*." This statement is said to be *contraposed* to the original and is logically equivalent to it. Due to their logical equivalence, stating one effectively states the other; when one is true, the other is also true. Likewise with falsity.

Strictly speaking, a contraposition can only exist in two simple conditionals. However, a contraposition may also exist in two complex conditionals, if they are similar. Thus, *8x(Px ! Qx)* , or "All *P*s are *Q*s," is contraposed to *8x(:Qx ! :Px)* , or "All non-*Q*s are non-*P*s."

44.3 Simple proof by definition of a conditional

In first-order logic, the conditional is defined as:

A ! B () :A _ B

We have:

:A _ B () :A _ (::B)

() :(:B) _ :A

() :B ! :A

44.4 Simple proof by contradiction

Let:

(A ! B) ^ :B

It is given that, if A is true, then B is true, and it is also given that B is not true. We can then show that A must not be true by contradiction. For, if A were true, then B would have to also be true (given). However, it is given that B is not true, so we have a contradiction. Therefore, A is not true (assuming that we are dealing with concrete statements that are either true or not true):

(A ! B) ! (:B ! :A)

We can apply the same process the other way round:

(:B ! :A) ^ A

We also know that B is either true or not true. If B is not true, then A is also not true. However, it is given that A is true; so, the assumption that B is not true leads to contradiction and must be false. Therefore, B must be true:

(:B ! :A) ! (A ! B)

Combining the two proved statements makes them logically equivalent:

(A ! B) () (:B ! :A)

44.5 More rigorous proof of the equivalence of contrapositives

Logical equivalence between two propositions means that they are true together or false together. To prove that contrapositives are logically equivalent, we need to understand when material implication is true or false.

(P ! Q)

This is only false when *P* is true and *Q* is false. Therefore, we can reduce this proposition to the statement "False when *P* and not-*Q*" (i.e. "True when it is not the case that *P* and not-*Q*"):
:(P ^ :Q)
The elements of a conjunction can be reversed with no effect (by commutativity):
:(:Q ^ P)
We define *R* as equal to " *:Q* ", and *S* as equal to *:P* (from this, *:S* is equal to *::P* , which is equal to just *P*):
:(R ^ :S)
This reads "It is not the case that (*R* is true and *S* is false)", which is the definition of a material conditional. We can then make this substitution:
(*R ! S*)
When we swap our definitions of *R* and *S*, we arrive at the following:
(*:Q ! :P*)

44.6 Comparisons

44.6.1 Examples

Take the statement "*All red objects have color.*" This can be equivalently expressed as "*If an object is red, then it has color.*"

_ The **contrapositive** is "*If an object does not have color, then it is not red.*" This follows logically from our initial statement and, like it, it is evidently true.
_ The **inverse** is "*If an object is not red, then it does not have color.*" An object which is blue is not red, and still has color. Therefore in this case the inverse is false.
_ The **converse** is "*If an object has color, then it is red.*" Objects can have other colors, of course, so, the converse of our statement is false.
_ The **negation** is "*There exists a red object that does not have color.*" This statement is false because the initial statement which it negates is true.

In other words, the contrapositive is logically equivalent to a given conditional statement, though not sufficient for a biconditional.

Similarly, take the statement "*All quadrilaterals have four sides,*" or equivalently expressed "*If a polygon is a quadrilateral, then it has four sides.*"

296 *CHAPTER 44. CONTRAPOSITION*

_ The **contrapositive** is "*If a polygon does not have four sides, then it is not a quadrilateral.*" This follows logically, and as a rule, contrapositives share the truth value of their conditional.
_ The **inverse** is "*If a polygon is not a quadrilateral, then it does not have four sides.*" In this case, unlike the last example, the inverse of the argument is true.
_ The **converse** is "*If a polygon has four sides, then it is a quadrilateral.*" Again, in this case, unlike the last example, the converse of the argument is true.
_ The **negation** is "*There is at least one quadrilateral that does not have four sides.*" This statement is clearly false.

Since the statement and the converse are both true, it is called a biconditional, and can be expressed as "**A polygon is a quadrilateral *if, and only if,* it has four sides.**" (The phrase *if and only if* is sometimes abbreviated *iff*.) That is, having four sides is both necessary to be a quadrilateral, and alone sufficient to deem it a quadrilateral.

44.6.2 Truth

_ If a statement is true, then its contrapositive is true (and vice versa).
_ If a statement is false, then its contrapositive is false (and vice versa).
_ If a statement's inverse is true, then its converse is true (and vice versa).
_ If a statement's inverse is false, then its converse is false (and vice versa).
_ If a statement's negation is false, then the statement is true (and vice versa).
_ If a statement (or its contrapositive) and the inverse (or the converse) are both true or both false, it is known as a logical biconditional.

44.7 Application

Because the **contrapositive** of a statement always has the same truth value (truth or falsity) as the statement itself, it can be a powerful tool for proving mathematical theorems. A proof by contraposition (contrapositive) is a direct proof of the contrapositive of a statement.[1] However, indirect methods such as proof by contradiction can also be used with contraposition, as, for example, in the proof of the irrationality of the square root of 2. By the definition of a rational number, the statement can be made that "*If*
p
2 *is rational, then it can be expressed as an irreducible*
fraction". This statement is **true** because it is a restatement of a definition. The contrapositive of this statement is "*If*
p

2 *cannot be expressed as an irreducible fraction, then it is not rational*". This contrapositive, like the original statement, is also **true**. Therefore, if it can be proven that

p

2 cannot be expressed as an irreducible fraction, then it must be the case that

p

2 is not a rational number. The latter can be proved by contradiction.

The previous example employed the contrapositive of a definition to prove a theorem. One can also prove a theorem by proving the contrapositive of the theorem's statement. To prove that *if a positive integer* N *is a non-square number, its square root is irrational*, we can equivalently prove its contrapositive, that *if a positive integer* N *has a square root that is rational, then* N *is a square number*. This can be shown by setting $\sqrt{N}$ equal to the rational expression a/b with a and b being positive integers with no common prime factor, and squaring to obtain $N = a2/b2$ and noting that since N is a positive integer $b=1$ so that $N = a2$, a square number.

44.8 See also

_ *Reductio ad absurdum*

44.9 References

[1] Smith, Douglas; Eggen, Maurice; St. Andre, Richard (2001), *A Transition to Advanced Mathematics* (5th ed.), Brooks/Cole, p. 37, ISBN 0-534-38214-2

Chapter 45

Reductio ad absurdum

Reductio ad absurdum (Latin: "reduction to absurdity"; *pl.*: reductiones ad absurdum), also known as *argumentum ad absurdum* (Latin: argument to absurdity), is a common form of argument which seeks to demonstrate that a statement is true by showing that a false, untenable, or absurd result follows from its denial,[1] or in turn to demonstrate that a statement is false by showing that a false, untenable, or absurd result follows from its acceptance. First recognized and studied in classical Greek philosophy (the Latin term derives from the Greek "εις άτοπον απαγωγή" or *eis atopon apagoge*, "reduction to the impossible", for example in Aristotle's *Prior Analytics*),[1] this technique has been used throughout history in both formal mathematical and philosophical reasoning, as well as informal debate.

The "absurd" conclusion of a *reductio ad absurdum* argument can take a range of forms:

_ Rocks have weight, otherwise we would see them floating in the air.

_ Society must have laws, otherwise there would be chaos.

_ There is no smallest positive rational number, because if there were, it could be divided by two to get a smaller one.

_ "if *A* then both *B* and not-*B*, so not-*A*"[2]

_ "if not-*A* then both *B* and not-*B*, so *A*"[2]

The first example above argues that the denial of the assertion would have a ridiculous result that goes against the evidence of our senses. The second argues that the denial would have an untenable result: unacceptable, unworkable or unpleasant for society. The third is a mathematical proof by contradiction, arguing that the denial of the assertion would result in a logical contradiction (there is a smallest positive rational number and yet there is a positive rational number smaller than it).

45.1 Greek philosophy

This technique is used throughout Greek philosophy, beginning with Presocratic philosophers. The earliest Greek example of a *reductio* argument is supposedly in fragments of a satirical poem attributed to Xenophanes of Colophon (c.570 – c.475 BC).[3] Criticizing Homer's attribution of human faults to the Greek gods, he says that humans also believe that the gods' bodies have human form. But if horses and oxen could draw, they would draw the gods with horse and oxen bodies. The gods can't have both forms, so this is a contradiction. Therefore the attribution of other human characteristics to the gods, such as human faults, is also false.

The earlier dialogs of Plato (424 – 348 BC), relating the debates of his teacher Socrates, raised the use of *reductio* arguments to a formal dialectical method (*Elenchus*), now called the *Socratic method*.[4][5] Typically Socrates' opponent would make an innocuous assertion, then Socrates by a step-by-step train of reasoning, bringing in other background assumptions, would make the person admit that the assertion resulted in an absurd or contradictory conclusion, forcing him to abandon his assertion. The technique was also a focus of the work of Aristotle (384 – 322 BC).[5]

297

298 *CHAPTER 45. REDUCTIO AD ABSURDUM*

45.2 The principle of non-contradiction

Aristotle clarified the connection between contradiction and falsity in his principle of non-contradiction.[5] This states that an assertion cannot be both true and false. Therefore if the contradiction of an assertion (not-*P*) can be derived

logically from the assertion (*P*) it can be concluded that a false assumption has been used. This technique, called *proof by contradiction* has formed the basis of *reductio ad absurdum* arguments in formal fields like logic and mathematics.[5]
The principle of non-contradiction has seemed absolutely undeniable to most philosophers.[5] However a few philosophers such as Heraclitus and Hegel have accepted contradictions. The discovery of contradictions at the foundations of mathematics at the beginning of the 20th century, such as Russell's paradox, have led a few philosophers such as Newton da Costa, Walter Carnielli and Graham Priest to reject the principle of non-contradiction, also known as the principle of explosion (Latin: ex falso quodlibet, "from a falsehood, anything follows", or ex contradictione sequitur quodlibet, "from a contradiction, anything follows"), or the principle of Pseudo-Scotus, which are behind the method of argument by Reductio ad absurdum, giving rise to theories such as paraconsistent logic and its particular form, dialethism, which accepts that there exist statements that are both true and false.
Paraconsistent logics usually deny that the principio of explosion holds for all sentences in logic, which amounts to denying that a contradiction entails everything (what is called "deductive explosion"). The Logics of Formal Inconsistency (LFIs) are a family of paraconsistent logics where the notions of contradiction and consistency are not coincident; although the validity of the principle of explosion is not accepted for all sentences, it is accepted for consistent sentences. Most paraconsistent logics, as the LFIs, also reject the principle of non-contradiction.[5]

45.3 Straw man argument

Main article: Straw man
A fallacious argument similar to *reductio ad absurdum* often seen in polemical debate is the *straw man* logical fallacy.[6] A straw man argument attempts to refute a given proposition by showing that a slightly different or inaccurate form of the proposition (the "straw man") has an absurd, unpleasant, or ridiculous consequence, relying on the audience not to notice that the argument does not actually apply to the original proposition. For example, in a 1977 appeal of a U.S. bank robbery conviction, a prosecuting attorney said in his closing argument[7]

I submit to you that if you can't take this evidence and find these defendants guilty on this evidence then we might as well open all the banks and say, "Come on and get the money, boys", because we'll never be able to convict them.

The prosecutor was using this "straw man" to attempt to alarm the appellate judges; the chance that any precedent set by this one particular case would literally make it impossible to convict *any* bank robbers was undoubtedly remote.

45.4 See also

_ Contraposition
_ Mathematical proof
_ Proof by contradiction

45.5 References

[1] Nicholas Rescher. "Reductio ad absurdum". *The Internet Encyclopedia of Philosophy*. Retrieved 21 July 2009.
[2] Read, Stephen (1995). *Thinking About Logic*, p.251. Oxford. ISBN .
[3] Daigle, Robert W. (1991). "The reductio ad absurdum argument prior to Aristotle". *Master's Thesis*. San Jose State Univ. Retrieved August 22, 2012.

[4] Bobzian, Suzanne (2006). "Ancient Logic". *Stanford Encyclopedia of Philosophy*. The Metaphysics Research Lab, Stanford University. Retrieved August 22, 2012.
[5] "Reductio ad absurdum". *New World Encyclopedia*. 2007. Retrieved August 22, 2012.
[6] Lapakko, David (2009). *Argumentation: Critical Thinking in Action*. iUniverse. p. 119. ISBN 1440168385.
[7] Bosanac, Paul (2009). *Litigation Logic: A Practical Guide to Effective Argument*. American Bar Association. p. 393. ISBN 1616327103.

Chapter 46

Coprime integers

In number theory, two integers *a* and *b* are said to be **relatively prime**, **mutually prime**, or **coprime** (also spelled **co-prime**)[1] if the only positive integer that evenly divides both of them is 1. That is, the only common positive factor of the two numbers is 1. This is equivalent to their greatest common divisor being 1.[2] The numerator and denominator of a reduced fraction are coprime. In addition to gcd(a; b) = 1 and (a; b) = 1; the notation *a ? b* is sometimes used to indicate that *a* and *b* are relatively prime.[3]
For example, 14 and 15 are coprime, being commonly divisible by only 1, but 14 and 21 are not, because they are both divisible by 7. The numbers 1 and −1 are coprime to every integer, and they are the only integers to be coprime with 0.
A fast way to determine whether two numbers are coprime is given by the Euclidean algorithm.
The number of integers coprime to a positive integer *n*, between 1 and *n*, is given by Euler's totient function (or Euler's phi function) $\varphi(n)$.
A set of integers can also be called **coprime** if its elements share no common positive factor except 1. A set of

integers is said to be **pairwise coprime** if *a* and *b* are coprime for every pair (*a*, *b*) of different integers in it.

46.1 Properties

9 units

4 units

Figure 1. The numbers 4 and 9 are coprime. Therefore, the diagonal of a 4 x 9 lattice does not intersect any other lattice points

There are a number of conditions which are equivalent to *a* and *b* being coprime:

300

46.2. COPRIMALITY IN SETS 301

_ No prime number divides both *a* and *b*.

_ There exist integers *x* and *y* such that $ax + by = 1$ (see Bézout's identity).

_ The integer *b* has a multiplicative inverse modulo *a*: there exists an integer *y* such that $by \equiv 1 \pmod{a}$. In other words, *b* is a unit in the ring **Z**/*a***Z** of integers modulo *a*.

_ Every pair of congruence relations for an unknown integer *x*, of the form $x \equiv k \pmod{a}$ and $x \equiv l \pmod{b}$, has a solution, as stated by the Chinese remainder theorem; in fact the solutions are described by a single congruence relation modulo *ab*.

_ The least common multiple of *a* and *b* is equal to their product *ab*, i.e. LCM(*a*, *b*) = *ab*.

As a consequence of the third point, if *a* and *b* are coprime and $br \equiv bs \pmod{a}$, then $r \equiv s \pmod{a}$ (because we may "divide by *b*" when working modulo *a*). Furthermore, if b_1 and b_2 are both coprime with *a*, then so is their product b_1b_2 (modulo *a* it is a product of invertible elements, and therefore invertible); this also follows from the first point by Euclid's lemma, which states that if a prime number *p* divides a product *bc*, then *p* divides at least one of the factors *b*, *c*.

As a consequence of the first point, if *a* and *b* are coprime, then so are any powers a^k and b^l.

If *a* and *b* are coprime and *a* divides the product *bc*, then *a* divides *c*. This can be viewed as a generalization of Euclid's lemma.

The two integers *a* and *b* are coprime if and only if the point with coordinates (*a*, *b*) in a Cartesian coordinate system is "visible" from the origin (0,0), in the sense that there is no point with integer coordinates between the origin and (*a*, *b*). (See figure 1.)

In a sense that can be made precise, the probability that two randomly chosen integers are coprime is $6/\pi^2$ (see pi), which is about 61%. See below.

Two natural numbers *a* and *b* are coprime if and only if the numbers $2^a - 1$ and $2^b - 1$ are coprime. As a generalization of this, following easily from Euclidean algorithm in base $n > 1$:

gcd(n^a □ 1; n^b □ 1) = $n^{\gcd(a;b)}$ □ 1:

46.2 Coprimality in sets

A set of integers $S = \{a_1, a_2, ... a_n\}$ can also be called *coprime* or *setwise coprime* if the greatest common divisor of all the elements of the set is 1. If every pair in a (finite or infinite) set of integers is coprime, then the set is said to be *pairwise coprime* (or *pairwise relatively prime*, *mutually coprime* or *mutually relatively prime*). Pairwise coprimality is a stronger condition than setwise coprimality; every pairwise coprime finite set is also setwise coprime, but the reverse is not true. For example, the integers 6, 10, 15 are coprime (because the only positive integer dividing *all* of them is 1), but they are not *pairwise* coprime because the gcd(6, 10) = 2, gcd(10, 15) = 5 and gcd(6, 15) = 3.

The concept of pairwise coprimality is important as a hypothesis in many results in number theory, such as the Chinese remainder theorem.

46.2.1 Infinite set examples

The set of all primes is pairwise coprime, as is the set of elements in Sylvester's sequence, and the set of all Fermat numbers.

46.3 Coprimality in ring ideals

Two ideals *A* and *B* in the commutative ring *R* are called **coprime** (or **comaximal**) if $A + B = R$. This generalizes Bézout's identity: with this definition, two principal ideals (*a*) and (*b*) in the ring of integers **Z** are coprime if and only if *a* and *b* are coprime. If the ideals *A* and *B* of *R* are coprime, then $AB = A \cap B$; furthermore, if *C* is a third

302 *CHAPTER 46. COPRIME INTEGERS*

ideal such that *A* contains *BC*, then *A* contains *C*. The Chinese remainder theorem is an important statement about coprime ideals.

46.4 Cross notation, group

See also: multiplicative group of integers modulo n

If $n \geq 1$ and is an integer, the numbers coprime to *n*, taken modulo *n*, form a group with multiplication as operation; it is written as $(\mathbf{Z}/n\mathbf{Z})^\times$ or **Z**□*.

46.5 Probabilities

Given two randomly chosen integers a and b, it is reasonable to ask how likely it is that a and b are coprime. In this determination, it is convenient to use the characterization that a and b are coprime if and only if no prime number divides both of them (see Fundamental theorem of arithmetic).

Informally, the probability that any number is divisible by a prime (or in fact any integer) p is $1/p$; for example, every 7th integer is divisible by 7. Hence the probability that two numbers are both divisible by p is $1/p^2$, and the probability that at least one of them is not is $1 - 1/p^2$. Any finite collection of divisibility events associated to distinct primes is mutually independent. For example, in the case of two events, a number is divisible by primes p and q if and only if it is divisible by pq; the latter event has probability $1/pq$. If one makes the heuristic assumption that such reasoning can be extended to infinitely many divisibility events, one is led to guess that the probability that two numbers are coprime is given by a product over all primes,

$$\prod_{\text{prime}\,p}\left(1 - \frac{1}{p^2}\right) = \left(\prod_{\text{prime}\,p}\frac{1}{1 - p^{-2}}\right)^{-1} = \frac{1}{\zeta(2)} = \frac{6}{\pi^2} \approx 0.607927102 \approx 61\%.$$

Here ζ refers to the Riemann zeta function, the identity relating the product over primes to $\zeta(2)$ is an example of an Euler product, and the evaluation of $\zeta(2)$ as $\pi^2/6$ is the Basel problem, solved by Leonhard Euler in 1735.

There is no way to choose a positive integer at random so that each positive integer occurs with equal probability, but statements about "randomly chosen integers" such as the ones above can be formalized by using the notion of *natural density*. For each positive integer N, let P_N be the probability that two randomly chosen numbers in $\{1, 2, \ldots, N\}$ are coprime. Although P_N will never equal $6/\pi^2$ exactly, with work[4] one can show that in the limit as $N \to \infty$, the probability P_N approaches $6/\pi^2$.

More generally, the probability of k randomly chosen integers being coprime is $1/\zeta(k)$.

46.6 Generating all coprime pairs

All pairs of positive coprime numbers (m, n) (with $m > n$) can be arranged in two disjoint complete ternary trees, one tree starting from $(2, 1)$ (for even-odd and odd-even pairs),[5] and the other tree starting from $(3, 1)$ (for odd-odd pairs).[6] The children of each vertex (m, n) are generated as follows:

Branch 1: $(2m - n, m)$

Branch 2: $(2m + n, m)$

Branch 3: $(m + 2n, n)$

This scheme is exhaustive and non-redundant with no invalid members.

46.7 See also

- Superpartient number

46.8. REFERENCES 303

0 5 10 15 20 25 30 35 40 45 50 55 60 65 70 75 80

0

5

10

15

20
25
30
35
40

The order of generation of coprime pairs by this algorithm. First node (2,1) is marked red, its three children are shown in orange, third generation is yellow, and so on in the rainbow order.

46.8 References

[1] Eaton, James S. Treatise on Arithmetic. 1872. May be downloaded from: http://archive.org/details/atreatiseonarit05eatogoog
[2] G.H. Hardy; E. M. Wright (2008). *An Introduction to the Theory of Numbers* (6th ed. ed.). Oxford University Press. p. 6. ISBN 978-0-19-921986-5.
[3] Graham, R. L.; Knuth, D. E.; Patashnik, O. (1989), *Concrete Mathematics*, Addison-Wesley
[4] This theorem was proved by Ernesto Cesàro in 1881. For a proof, see G.H. Hardy; E. M. Wright (2008). *An Introduction to the Theory of Numbers* (6th ed. ed.). Oxford University Press. ISBN 978-0-19-921986-5., theorem 332.
[5] Saunders, Robert & Randall, Trevor (July 1994), *The family tree of the Pythagorean triplets revisited*, *Mathematical Gazette* **78**: 190–193.
[6] Mitchell, Douglas W. (July 2001), *An alternative characterisation of all primitive Pythagorean triples*, *Mathematical Gazette* **85**: 273–275, doi:10.2307/3622017.

46.9 Further reading

_ Lord, Nick (March 2008), *A uniform construction of some infinite coprime sequences*, *Mathematical Gazette* **92**: 66–70.

Chapter 47

Constructive proof

In mathematics, a **constructive proof** is a method of proof that demonstrates the existence of a mathematical object by creating or providing a method for creating the object. This is in contrast to a **non-constructive proof** (also known as an *existence proof* or *pure existence theorem*) which proves the existence of a particular kind of object without providing an example.

Some non-constructive proofs show that if a certain proposition is false, a contradiction ensues; consequently the proposition must be true (proof by contradiction). However, the principle of explosion (*ex falso quodlibet*) has been accepted in some varieties of constructive mathematics, including intuitionism.

Constructivism is a mathematical philosophy that rejects all but constructive proofs in mathematics. This leads to a restriction on the proof methods allowed (prototypically, the law of the excluded middle is not accepted) and a different meaning of terminology (for example, the term “or” has a stronger meaning in constructive mathematics than in classical).

Constructive proofs can be seen as defining certified mathematical algorithms: this idea is explored in the Brouwer–Heyting–Kolmogorov interpretation of constructive logic, the Curry–Howard correspondence between proofs and programs, and such logical systems as Per Martin-Löf's Intuitionistic Type Theory, and Thierry Coquand and Gérard Huet's Calculus of Constructions.

47.1 Examples

47.1.1 Non-constructive proofs

First consider the theorem that there are an infinitude of prime numbers. Euclid's proof is constructive. But a common way of simplifying Euclid’s proof postulates that, contrary to the assertion in the theorem, there are only a finite number of them, in which case there is a largest one, denoted n. Then consider the number $n! + 1$ (1 + the product of the first n numbers). Either this number is prime, or all of its prime factors are greater than n. Without establishing a specific prime number, this proves that one exists that is greater than n, contrary to the original postulate.

Now consider the theorem “There exist irrational numbers a and b such that a^b is rational.” This theorem can be proven using a constructive proof, or using a non-constructive proof.

The following 1953 proof by Dov Jarden has been widely used as an example of a non-constructive proof since at least 1970:[1][2]

CURIOSA

339. *A Simple Proof That a Power of an Irrational Number to an Irrational Exponent May Be Rational. p*
2
p
2 is either rational or irrational. If it is rational, our statement is proved. If it is irrational, (

p
2
p
2
)
p
2 =
2 proves our statement.
Dov Jarden Jerusalem
In a bit more detail:
305
306 *CHAPTER 47. CONSTRUCTIVE PROOF*
_ Recall that
p
2 is irrational, and 2 is rational. Consider the number q =
p
2
p
2 . Either it is rational or it is
irrational.
_ If q is rational, then the theorem is true, with a and b both being
p
2 .
_ If q is irrational, then the theorem is true, with a being
p
2
p
2 and b being
p
2 , since
(p
2
p
2
)p
2
=
p
2
(
p
2_
p
2)
=
p
2
2
= 2:
This proof is non-constructive because it relies on the statement "Either q is rational or it is irrational"—an instance of the law of excluded middle, which is not valid within a constructive proof. The non-constructive proof does not construct an example a and b; it merely gives a number of possibilities (in this case, two mutually exclusive possibilities) and shows that one of them—but does not show *which* one—must yield the desired example.
(It turns out that
p
2
p
2 is irrational because of the Gelfond–Schneider theorem, but this fact is irrelevant to the correctness of the non-constructive proof.)

47.1.2 Constructive proofs

A *constructive* proof of the above theorem on irrational powers of irrationals would give an actual example, such as:
a =
p
2 ; b = log2 9 ; ab = 3 :

The square root of 2 is irrational, and 3 is rational. $\log_2 9$ is also irrational: if it were equal to m
n, then, by the
properties of logarithms, 9^n would be equal to 2^m, but the former is odd, and the latter is even.
A more substantial example is the graph minor theorem. A consequence of this theorem is that a graph can be drawn on the torus if, and only if, none of its minors belong to a certain finite set of "forbidden minors". However, the proof of the existence of this finite set is not constructive, and the forbidden minors are not actually specified. They are still unknown.

47.2 Brouwerian counterexamples

In constructive mathematics, a statement may be disproved by giving a counterexample, as in classical mathematics. However, it is also possible to give a **Brouwerian counterexample** to show that the statement is non-constructive. This sort of counterexample shows that the statement implies some principle that is known to be non-constructive. If it can be proved constructively that a statement implies some principle that is not constructively provable, then the statement itself cannot be constructively provable. For example, a particular statement may be shown to imply the law of the excluded middle. An example of a Brouwerian counterexample of this type is Diaconescu's theorem, which shows that the full axiom of choice is non-constructive in systems of constructive set theory, since the axiom of choice implies the law of excluded middle in such systems. The field of constructive reverse mathematics develops this idea further by classifying various principles in terms of "how nonconstructive" they are, by showing they are equivalent to various fragments of the law of the excluded middle.
Brouwer also provided "weak" counterexamples.[3] Such counterexamples do not disprove a statement, however; they only show that, at present, no constructive proof of the statement is known. One weak counterexample begins by taking some unsolved problem of mathematics, such as Goldbach's conjecture. Define a function f of a natural number x as follows:
$f(x) =$
{
0 if Goldbach's conjecture is false
1 if Goldbach's conjecture is true
Although this is a definition by cases, it is still an admissible definition in constructive mathematics. Several facts about f can be proved constructively. However, based on the different meaning of the words in constructive mathematics,

if there is a constructive proof that "$f(0) = 1$ or $f(0) \neq 1$" then this would mean that there is a constructive proof of Goldbach's conjecture (in the former case) or a constructive proof that Goldbach's conjecture is false (in the latter case). Because no such proof is known, the quoted statement must also not have a known constructive proof. However, it is entirely possible that Goldbach's conjecture may have a constructive proof (as we do not know at present whether it does), in which case the quoted statement would have a constructive proof as well, albeit one that is unknown at present. The main practical use of weak counterexamples is to identify the "hardness" of a problem. For example, the counterexample just shown shows that the quoted statement is "at least as hard to prove" as Goldbach's conjecture. Weak counterexamples of this sort are often related to the limited principle of omniscience.

47.3 See also

_ Errett Bishop

47.4 References

[1] J. Roger Hindley, "The Root-2 Proof as an Example of Non-constructivity", unpublished paper, 7 March 2013, full text
[2] Dov Jarden, "A simple proof that a power of an irrational number to an irrational exponent may be rational", *Curiosa* No. 339 in *Scripta Mathematica* **19**:229 (1953)
[3] A. S. Troelstra, *Principles of Intuitionism*, Lecture Notes in Mathematics 95, 1969, p. 102

47.5 Further reading

_ J. Franklin and A. Daoud (2011) *Proof in Mathematics: An Introduction*. Kew Books, ISBN 0-646-54509-4, ch. 4
_ Hardy, G.H. & Wright, E.M. (1979) *An Introduction to the Theory of Numbers* (Fifth Edition). Oxford University Press. ISBN 0-19-853171-0
_ Anne Sjerp Troelstra and Dirk van Dalen (1988) "Constructivism in Mathematics: Volume 1" Elsevier Science. ISBN 978-0-444-70506-8

47.6 External links

_ *Weak counterexamples* by Mark van Atten, Stanford Encyclopedia of Philosophy

Chapter 48

Transcendental number

In mathematics, a **transcendental number** is a real or complex number that is not algebraic—that is, it is not a root of a non-zero polynomial equation with rational coefficients. The most prominent examples of transcendental numbers are π and e. Though only a few classes of transcendental numbers are known (in part because it can be extremely difficult to show that a given number is transcendental), transcendental numbers are not rare. Indeed, almost all real and complex numbers are transcendental, since the algebraic numbers are countable while the sets of real and complex numbers are both uncountable. All real transcendental numbers are irrational, since all rational numbers are algebraic. The converse is not true: not all irrational numbers are transcendental; e.g., the square root of 2 is irrational but not a transcendental number, since it is a solution of the polynomial equation $x^2 - 2 = 0$.

48.1 History

The name "transcendental" comes from Leibniz in his 1682 paper where he proved that sin(x) is not an algebraic function of x.[1][2] Euler was probably the first person to define transcendental *numbers* in the modern sense.[3]
Joseph Liouville first proved the existence of transcendental numbers in 1844,[4] and in 1851 gave the first decimal examples such as the Liouville constant

/Σ

k=1

10

□k! = 0:11000100000000000000000010000 : : :

in which the nth digit after the decimal point is 1 if n is equal to $k!$ (k factorial) for some k and 0 otherwise.[5]
Liouville showed that this number is what we now call a Liouville number; this essentially means that it can be more closely approximated by rational numbers than can any irrational algebraic number. Liouville showed that all Liouville numbers are transcendental.[6]
Johann Heinrich Lambert conjectured that e and π were both transcendental numbers in his 1761 paper proving the number π is irrational. The first number to be proven transcendental without having been specifically constructed for the purpose was e, by Charles Hermite in 1873.
In 1874, Georg Cantor proved that the algebraic numbers are countable and the real numbers are uncountable. He also gave a new method for constructing transcendental numbers.[7] In 1878, Cantor published a construction that proves there are as many transcendental numbers as there are real numbers.[8] Cantor's work established the ubiquity of transcendental numbers.
In 1882, Ferdinand von Lindemann published a proof that the number π is transcendental. He first showed that e to any nonzero algebraic power is transcendental, and since $e^{i\pi} = -1$ is algebraic (see Euler's identity), $i\pi$ and therefore π must be transcendental. This approach was generalized by Karl Weierstrass to the Lindemann–Weierstrass theorem. The transcendence of π allowed the proof of the impossibility of several ancient geometric constructions involving compass and straightedge, including the most famous one, squaring the circle.
In 1900, David Hilbert posed an influential question about transcendental numbers, Hilbert's seventh problem: If a is an algebraic number, that is not zero or one, and b is an irrational algebraic number, is a^b necessarily transcendental?
308
48.2. PROPERTIES 309
The affirmative answer was provided in 1934 by the Gelfond–Schneider theorem. This work was extended by Alan Baker in the 1960s in his work on lower bounds for linear forms in any number of logarithms (of algebraic numbers).[9]

48.2 Properties

The set of transcendental numbers is uncountably infinite. Since the polynomials with integer coefficients are countable, and since each such polynomial has a finite number of zeroes, the algebraic numbers must also be countable. But Cantor's diagonal argument proves that the real numbers (and therefore also the complex numbers) are uncountable; so the set of all transcendental numbers must also be uncountable.
No rational number is transcendental and all real transcendental numbers are irrational. A rational number can be written as p/q, where p and q are integers. Thus, p/q is the root of $qx - p = 0$. However, some irrational numbers are not transcendental. For example, the square root of 2 is irrational and not transcendental (because it is a solution of the polynomial equation $x^2 - 2 = 0$). The same is true for the square root of other non-perfect squares.
Any non-constant algebraic function of a single variable yields a transcendental value when applied to a transcendental argument. For example, from knowing that π is transcendental, we can immediately deduce that numbers such as 5π, $(\pi - 3)/\sqrt{2}$, $(\sqrt{\pi} - \sqrt{3})^8$ and $(\pi^5 + 7)^{1/7}$ are transcendental as well.
However, an algebraic function of several variables may yield an algebraic number when applied to transcendental numbers if these numbers are not algebraically independent. For example, π and $(1 - \pi)$ are both transcendental, but $\pi + (1 - \pi) = 1$ is obviously not. It is unknown whether $\pi + e$, for example, is transcendental, though at least one of $\pi + e$ and πe must be transcendental. More generally, for any two transcendental numbers a and b, at least one of $a + b$ and ab must be transcendental. To see this, consider the polynomial $(x - a)(x - b) = x^2 - (a + b)x + ab$. If $(a + b)$ and ab were both algebraic, then this would be a polynomial with algebraic coefficients. Because algebraic numbers form an algebraically closed field, this would imply that the roots of the polynomial, a and b, must be algebraic. But this is a contradiction, and thus it must be the case that at least one of the coefficients is transcendental.
The non-computable numbers are a strict subset of the transcendental numbers.
All Liouville numbers are transcendental, but not vice versa. Any Liouville number must have unbounded partial

quotients in its continued fraction expansion. Using a counting argument one can show that there exist transcendental numbers which have bounded partial quotients and hence are not Liouville numbers.
Using the explicit continued fraction expansion of *e*, one can show that *e* is not a Liouville number (although the partial quotients in its continued fraction expansion are unbounded). Kurt Mahler showed in 1953 that π is also not a Liouville number. It is conjectured that all infinite continued fractions with bounded terms that are not eventually periodic are transcendental (eventually periodic continued fractions correspond to quadratic irrationals).[10]
A related class of numbers are closed-form numbers, which may be defined in various ways, including rational numbers (and in some definitions all algebraic numbers), but also allow exponentiation and logarithm.

48.3 Numbers proven to be transcendental

Numbers proven to be transcendental:
_ e^a if *a* is algebraic and nonzero (by the Lindemann–Weierstrass theorem).
_ π (by the Lindemann–Weierstrass theorem).
_ e^π, Gelfond's constant, as well as $e^{-\pi/2}=i^i$ (by the Gelfond–Schneider theorem).
_ a^b where *a* is algebraic but not 0 or 1, and *b* is irrational algebraic (by the Gelfond–Schneider theorem), in particular:
2
p
2; the Gelfond–Schneider constant (or Hilbert number).
_ The Continued Fraction Constant, Carl Ludwig Siegel (1929)

1 +
1
2 +
1
3 +
1
4 +
1
5 +
1
6 + ...
_ sin(*a*), cos(*a*) and tan(*a*), and their multiplicative inverses csc(*a*), sec(*a*) and cot(*a*), for any nonzero algebraic number *a* (by the Lindemann–Weierstrass theorem).
_ ln(*a*) if *a* is algebraic and not equal to 0 or 1, for any branch of the logarithm function (by the Lindemann–Weierstrass theorem).
_ *W*(*a*) if *a* is algebraic and nonzero, for any branch of the Lambert W Function (by the Lindemann–Weierstrass theorem).
_ Γ(1/3),[11] Γ(1/4),[12] and Γ(1/6).[12]
_ 0.12345678910111213141516..., the Champernowne constant.[13][14]
_ Ω, Chaitin's constant (since it is a non-computable number).[15]
_ The **Fredholm number**[16][17]
*1*Σ
n=0
2
□2*n*
more generally, any number of the form
*1*Σ
n=0
_2*n*
with 0 < |β| < 1 and β algebraic.[18]
_ The aforementioned Liouville constant
*1*Σ
n=1
10
□*n*!;
more generally any number of the form
*1*Σ
n=1
_*n*!
with 0 < |β| < 1 and β algebraic
_ The Prouhet–Thue–Morse constant.[19][20]

_ Any number for which the digits with respect to some fixed base form a Sturmian word.[21]
_ For $\beta > 1$
$l\Sigma$
$k=0$
10
□$\lfloor_k\rfloor$
;
where _ $7!$ $\lfloor\_\rfloor$ is the floor function.

48.4 Possibly transcendental numbers

Numbers which have yet to be proven to be either transcendental or algebraic:
_ Most sums, products, powers, etc. of the number π and the number e, e.g. $\pi + e$, $\pi - e$, πe, π/e, π^π, e^e, π^e, $\pi^{\sqrt{2}}$, e^{π^2} are not known to be rational, algebraic irrational or transcendental. Notable exceptions are $\pi + e^\pi$, πe^π and $e^{\pi\sqrt{n}}$ (for any positive integer n) which have been proven to be transcendental.[22][23]
_ The Euler–Mascheroni constant γ (which has not even been proven to be irrational).
_ Catalan's constant, also not known to be irrational.
_ Apéry's constant, $\zeta(3)$ (which Apéry proved is irrational)
_ The Riemann zeta function at other odd integers, $\zeta(5)$, $\zeta(7)$, ... (not known to be irrational.)
_ The Feigenbaum constants, δ and α.
_ Mills' constant.
Conjectures:
_ Schanuel's conjecture,
_ Four exponentials conjecture.

48.5 Sketch of a proof that *e* is transcendental

The first proof that the base of the natural logarithms, e, is transcendental dates from 1873. We will now follow the strategy of David Hilbert (1862–1943) who gave a simplification of the original proof of Charles Hermite. The idea is the following:
Assume, for purpose of finding a contradiction, that e is algebraic. Then there exists a finite set of integer coefficients $c_0, c_1, ..., c_n$ satisfying the equation:
$c_0 + c_1 e + c_2 e^2 + \_\_\_ + c_n e^n = 0; \; c_0; \; c_n \neq 0:$
Now for a positive integer k, we define the following polynomial:
$f_k(x) = x^k [(x □ 1) \_\_\_ (x □ n)]^{k+1};$
and multiply both sides of the above equation by
$\int_0^1$
$f_k e$
□x dx;
to arrive at the equation:
c_0
$(\int_0^1$
$f_k e$
□x dx
)
$+ c_1 e$
$(\int_0^1$
$f_k e$
□x dx
)
$+ \_\_\_ + c_n e^n$
$(\int_0^1$
$f_k e$
□x dx
)
$= 0:$
This equation can be written in the form
$P + Q = 0$

where

$$P = c_0\left(\int_0^{\infty} f_k e^{-x}\,dx\right) + c_1 e\left(\int_1^{\infty} f_k e^{-x}\,dx\right) + c_2 e^2\left(\int_2^{\infty} f_k e^{-x}\,dx\right) + \cdots + c_n e^n\left(\int_n^{\infty} f_k e^{-x}\,dx\right)$$

$$Q = c_1 e\left(\int_0^{1} f_k e^{-x}\,dx\right) + c_2 e^2\left(\int_0^{2} f_k e^{-x}\,dx\right) + \cdots + c_n e^n\left(\int_0^{n} f_k e^{-x}\,dx\right)$$

Lemma 1. For an appropriate choice of k, $\frac{P}{k!}$ is a non-zero integer.

Proof. Each term in P is an integer times a sum of factorials, which results from the relation

$$\int_0^{\infty} x^j e^{-x}\,dx = j!$$

which is valid for any positive integer j (consider the Gamma function).

It is non-zero because for every a satisfying $0 < a \leq n$, the integrand in

$$c_a e^a \int_a^{\infty} f_k e^{-x}\,dx$$

is e^{-x} times a sum of terms whose lowest power of x is $k+1$ after substituting x for $x - a$ in the integral. Then this becomes a sum of integrals of the form

$$\int_0^{\infty} x^j e^{-x}\,dx$$

with $k+1 \leq j$, and it is therefore an integer divisible by $(k+1)!$. After dividing by $k!$, we get zero modulo

$(k+1)$. However, we can write:

$$\int_0^1$$

fke

$□x \, dx =$

$$\int_0^1$$

$($

$[(□1)n(n!)]_{k+1}e$

$□xx_k + _ _ _$

$)$

dx

and thus

$$\frac{1}{k!}$$

c_0

$$\int_0^1$$

fke

$□x \, dx = c_0[(□1)n(n!)]_{k+1} \bmod (k+1)$:

By choosing k so that $k+1$ is prime and larger than n and $|c_0|$, we get that $\frac{P}{k!}$ is non-zero modulo $(k+1)$
and is thus non-zero.

Lemma 2.

$$\overline{\frac{Q}{k!}}$$

$\overline{} < 1$ for sufficiently large k.

Proof. Note that

fke

$□x = x_k[(x□1)(x□2) _ _ _ (x□n)]_{k+1}e$

$□x =$

$($

$[x(x □ 1) _ _ _ (x □ n)]_k) ($

$(x □ 1) _ _ _ (x □ n)e$

$□x)$

Using upper bounds G and H for $jx(x □ 1) _ _ _ (x □ n)j$ and $j(x □ 1) _ _ _ (x □ n)e□xj$ on the interval
$[0,n]$ we can infer that

$jQj < G_kH(jc_1je + 2jc_2je_2 + _ _ _ + njc_njen)$

and since

$\lim_{k!1}$

$\dfrac{G_k}{k!}$

$= 0$

it follows that

$\lim_{k!1}$

$\dfrac{Q}{k!}$

$= 0$

which is sufficient to finish the proof of this lemma.

48.6. MAHLER'S CLASSIFICATION 313

Noting that one can choose k so that both Lemmas hold we get the contradiction we needed to prove the transcendence
of e.

48.5.1 The transcendence of π

A similar strategy, different from Lindemann's original approach, can be used to show that the number π is transcendental.
Besides the gamma-function and some estimates as in the proof for e, facts about symmetric polynomials play
a vital role in the proof.

For detailed information concerning the proofs of the transcendence of π and e see the references and external links.

48.6 Mahler's classification

Kurt Mahler in 1932 partitioned the transcendental numbers into 3 classes, called **S**, **T**, and **U**.[24] Definition of these classes draws on an extension of the idea of a Liouville number (cited above).

48.6.1 Measure of irrationality of a real number

One way to define a Liouville number is to consider how small a given real number **x** makes linear polynomials $|qx - p|$ without making them exactly 0. Here p, q are integers with $|p|$, $|q|$ bounded by a positive integer H.

Let $\omega(x, 1, H)$ be the minimum non-zero absolute value these polynomials take and take:

!(x; 1;H) = □log*m*(*x; 1;H*)

log*H*

!(x; 1) = lim sup

H!1

!(x; 1;H):

ω(x, 1) is often called the **measure of irrationality** of a real number x. For rational numbers, ω(x, 1) = 0 and is at least 1 for irrational real numbers. A Liouville number is defined to have infinite measure of irrationality. Roth's theorem says that irrational real algebraic numbers have measure of irrationality 1.

48.6.2 Measure of transcendence of a complex number

Next consider the values of polynomials at a complex number x, when these polynomials have integer coefficients, degree at most n, and height at most H, with n, H being positive integers.

Let ω(x,n,H) be the minimum non-zero absolute value such polynomials take at x and take:

!(x; n;H) = □log*m*(*x; n;H*)

n log*H*

!(x; n) = lim sup

H!1

!(x; n;H):

Suppose this is infinite for some minimum positive integer n. A complex number x in this case is called a **U number** of degree n.

Now we can define

!(x) = lim sup

n!1

!(x; n):

ω(x) is often called the **measure of transcendence** of x. If the ω(x,n) are bounded, then ω(x) is finite, and x is called an **S number**. If the ω(x,n) are finite but unbounded, x is called a **T number**. x is algebraic if and only if ω(x) = 0.

Clearly the Liouville numbers are a subset of the U numbers. William LeVeque in 1953 constructed U numbers of any desired degree.[25][26] The Liouville numbers and hence the U numbers are uncountable sets. They are sets of measure 0.[27]

T numbers also comprise a set of measure 0.[28] It took about 35 years to show their existence. Wolfgang M. Schmidt in 1968 showed that examples exist. It follows that almost all complex numbers are S numbers.[29] Mahler proved that the exponential function sends all non-zero algebraic numbers to S numbers:[30][31] this shows that e is an S number and gives a proof of the transcendence of π. The most that is known about π is that it is not a U number. Many other transcendental numbers remain unclassified.

Two numbers x, y are called **algebraically dependent** if there is a non-zero polynomial P in 2 indeterminates with integer coefficients such that $P(x, y) = 0$. There is a powerful theorem that 2 complex numbers that are algebraically dependent belong to the same Mahler class.[26][32] This allows construction of new transcendental numbers, such as the sum of a Liouville number with e or π.

It is often speculated that S stood for the name of Mahler's teacher Carl Ludwig Siegel and that T and U are just the next two letters.

48.6.3 Koksma's equivalent classification

Jurjen Koksma in 1939 proposed another classification based on approximation by algebraic numbers.[24][33]

Consider the approximation of a complex number x by algebraic numbers of degree ≤ n and height ≤ H. Let α be an algebraic number of this finite set such that $|x - \alpha|$ has the minimum positive value. Define ω*(x,H,n) and ω*(x,n) by:

jx □ *_j* = *H*

□*n!*

(*x;H;n*)□1*:*

!

(*x; n*) = lim sup

H!1

!

(*x; n;H*)*:*

If for a smallest positive integer n, ω*(x,n) is infinite, x is called a **U*-number** of degree n.

If the $\omega^*(x,n)$ are bounded and do not converge to 0, x is called an **S*-number**,
A number x is called an **A*-number** if the $\omega^*(x,n)$ converge to 0.
If the $\omega^*(x,n)$ are all finite but unbounded, x is called a **T*-number**,
Koksma's and Mahler's classifications are equivalent in that they divide the transcendental numbers into the same classes.[33] The A^*-numbers are the algebraic numbers.[29]

48.6.4 LeVeque's construction

Let

_ = 1
3 +
lΣ
k=1
10
□k!

It can be shown that the nth root of λ (a Liouville number) is a U-number of degree n.[34]
This construction can be improved to create an uncountable family of U-numbers of degree n. Let Z be the set consisting of every other power of 10 in the series above for λ. The set of all subsets of Z is uncountable. Deleting any of the subsets of Z from the series for λ creates uncountably many distinct Liouville numbers, whose nth roots are U-numbers of degree n.

48.6.5 Type

The supremum of the sequence $\{\omega(x, n)\}$ is called the **type**. Almost all real numbers are S numbers of type 1, which is minimal for real S numbers. Almost all complex numbers are S numbers of type 1/2, which is also minimal. The claims of almost all numbers were conjectured by Mahler and in 1965 proved by Vladimir Sprindzhuk.[25]

48.7 See also

_ Transcendence theory, the study of questions related to transcendental numbers

48.8 Notes

[1] Gottfried Wilhelm Leibniz, Karl Immanuel Gerhardt, Georg Heinrich Pertz (1858). *Leibnizens mathematische Schriften* **5**. A. Asher & Co. pp. 97–98.
[2] Nicolás Bourbaki (1994). *Elements of the History of Mathematics*. Springer. p. 74.
[3] Paul Erdős, Dudley (December 1943). "Some Remarks and Problems in Number Theory Related to the Work of Euler". *Mathematics Magazine* **76** (5): 292–299. doi:10.2307/2690369. JSTOR 2690369.
[4] Aubrey J. Kempner (October 1916). "On Transcendental Numbers". *Transactions of the American Mathematical Society* (American Mathematical Society) **17** (4): 476–482. doi:10.2307/1988833. JSTOR 1988833.
[5] Weisstein, Eric W. "Liouville's Constant", MathWorld
[6] J. Liouville (1851). "Sur des classes très étendues de quantités dont la valeur n'est ni algébrique, ni même réductible à des irrationnelles algébriques". *J. Math. Pures et Appl.* **16**: 133–142.
[7] Georg Cantor (1874). "Über eine Eigenschaft des Inbegriffes aller reelen algebraischen Zahlen". *J. Reine Angew. Math.* **77**: 258–262.
[8] Georg Cantor (1878). "Ein Beitrag zur Mannigfaltigkeitslehre". *J. Reine Angew. Math.* **84**: 242–258. (Cantor's construction builds a one-to-one correspondence between the set of transcendental numbers and the set of real numbers. In this article, Cantor only applies his construction to the set of irrational numbers. See p. 254.)
[9] J J O'Connor and E F Robertson: Alan Baker. The MacTutor History of Mathematics archive 1998.
[10] Boris Adamczewski and Yann Bugeaud (March 2005). "On the complexity of algebraic numbers, II. Continued fractions". *Acta Mathematica* **195** (1): 1–20. doi:10.1007/BF02588048.
[11] Le Lionnais, F. Les nombres remarquables (ISBN 2-7056-1407-9). Paris: Hermann, p. 46, 1979. via Wolfram Mathworld, Transcendental Number
[12] Chudnovsky, G. V. (1984). *Contributions to the Theory of Transcendental Numbers*. Providence, RI: American Mathematical Society. ISBN 0-8218-1500-8. via Wolfram Mathworld, Transcendental Number
[13] K. Mahler (1937). "Arithmetische Eigenschaften einer Klasse von Dezimalbrüchen". *Proc. Konin. Neder. Akad. Wet. Ser. A.* (40): 421–428.
[14] Mahler (1976) p.12
[15] Calude, Cristian S. (2002). *Information and Randomness: An Algorithmic Perspective*. Texts in Theoretical Computer Science (2nd rev. and ext. ed.). Springer-Verlag. p. 239. ISBN 3-540-43466-6. Zbl 1055.68058.
[16] Allouche & Shallit (2003) pp.385,403
[17] Shallit, Jeffrey (1999). "Number theory and formal languages". In Hejhal, Dennis A.; Friedman, Joel; Gutzwiller, Martin C. et al. *Emerging applications of number theory. Based on the proceedings of the IMA summer program, Minneapolis, MN, USA, July 15-□26, 1996*. The IMA volumes in mathematics and its applications **109**. Springer-Verlag. pp. 547–570. ISBN 0-387-98824-6.
[18] Loxton, J. H. (1988). "13. Automata and transcendence". In Baker, A.. *New Advances in Transcendence Theory*. Cambridge University Press. pp. 215–228. ISBN 0-521-33545-0. Zbl 0656.10032.
[19] Mahler, Kurt (1929). "Arithmetische Eigenschaften der Lösungen einer Klasse von Funktionalgleichungen". *Math. Annalen* **101**: 342–366. doi:10.1007/bf01454845. JFM 55.0115.01.
[20] Allouche & Shallit (2003) p.387

[21] Pytheas Fogg, N. (2002). *Substitutions in dynamics, arithmetics and combinatorics*. Lecture Notes in Mathematics **1794**. Editors Berthé, Valérie; Ferenczi, Sébastien; Mauduit, Christian; Siegel, A. Berlin: Springer-Verlag. ISBN 3-540-44141-7. Zbl 1014.11015.

[22] Weisstein, Eric W., "Irrational Number", *MathWorld*.
[23] Modular functions and transcendence questions, Yu. V. Nesterenko, Sbornik: Mathematics(1996), 187(9):1319
[24] Bugeaud (2012) p.250
[25] Baker (1975) p. 86.
[26] LeVeque (2002) p.II:172
[27] Burger and Tubbs, p. 170.
[28] Burger and Tubbs, p. 172.
[29] Bugeaud (2012) p.251
[30] LeVeque (2002) pp.II:174–186
[31] Burger and Tubbs, p. 182.
[32] Burger and Tubbs, p. 163.
[33] Baker (1975) p.87
[34] Baker(1979), p. 90.

48.9 References

_ David Hilbert, "Über die Transcendenz der Zahlen *e* und _ ", *Mathematische Annalen* **43**:216–219 (1893).
_ A. O. Gelfond, *Transcendental and Algebraic Numbers*, Dover reprint (1960).
_ Baker, Alan (1975). *Transcendental Number Theory*. Cambridge University Press. ISBN 0-521-20461-5. Zbl 0297.10013.
_ Mahler, Kurt (1976). *Lectures on Transcendental Numbers*. Lecture Notes in Mathematics **546**. Springer-Verlag. ISBN 3-540-07986-6. Zbl 0332.10019.
_ Sprindzhuk, Vladimir G. (1979). *Metric theory of Diophantine approximations*. Scripta Series in Mathematics. Transl. from the Russian and ed. by Richard A. Silverman. With a foreword by Donald J. Newman. John Wiley & Sons. Zbl 0482.10047.
_ LeVeque, William J. (2002) [1956]. *Topics in Number Theory, Volumes I and II*. New York: Dover Publications. ISBN 978-0-486-42539-9.
_ Allouche, Jean-Paul; Shallit, Jeffrey (2003). *Automatic Sequences: Theory, Applications, Generalizations*. Cambridge University Press. ISBN 978-0-521-82332-6. Zbl 1086.11015.
_ Burger, Edward B.; Tubbs, Robert (2004). *Making transcendence transparent. An intuitive approach to classical transcendental number theory*. New York, NY: Springer-Verlag. ISBN 0-387-21444-5. Zbl 1092.11031.
_ Peter M Higgins, "Number Story" Copernicus Books, 2008, ISBN 978-1-84800-001-8.
_ Bugeaud, Yann (2012). *Distribution modulo one and Diophantine approximation*. Cambridge Tracts in Mathematics **193**. Cambridge: Cambridge University Press. ISBN 978-0-521-11169-0. Zbl pre06066616.

48.10 External links

_ (English) Proof that *e* is transcendental
_ (English) Proof that the Liouville Constant is transcendental
_ (German) Proof that *e* is transcendental (PDF)
_ (German) Proof that _ is transcendental (PDF)

Chapter 49

Liouville number

In number theory, a **Liouville number** is an irrational number x with the property that, for every positive integer n, there exist integers p and q with $q > 1$ and such that

$$0 < \left| x - \frac{p}{q} \right| < \frac{1}{q^n}.$$

A Liouville number can thus be approximated "quite closely" by a sequence of rational numbers. In 1844, Joseph Liouville showed that all Liouville numbers are transcendental, thus establishing the existence of transcendental numbers for the first time.

49.1 The existence of Liouville numbers (Liouville's constant)

Here we show that Liouville numbers exist by exhibiting a construction that produces such numbers.

For any integer $b \geq 2$, and any sequence of integers (a_1, a_2, ...,), such that $a_k \in \{0, 1, 2, ..., b - 1\}$, $\forall k \in \{1, 2, 3, ...\}$, define the number

$x =$

$_1\Sigma$

$k=1$

a_k

$b_{k!}$

(In the special case when $b = 10$, and $a_k = 1$, $\forall k \in \{1, 2, 3, ...\}$, the resulting number x is called **Liouville's constant**.)

It follows from the definition of x that its base-b representation is

$x = (0{:}a_1a_2000a_3000000000000000000a_4000{:}{:}{:})_b :$

Since this base-b representation is non-repeating it follows that x cannot be rational. Therefore, for any rational number p/q, we have $|x - p/q| > 0$.

Now, for any integer $n \geq 1$, define q_n and p_n as follows:

$q_n = b_{n!}$; $p_n = q_n$

Σ_n

$k=1$

a_k

$b_{k!}$:

Then,

$0 <$

_____

x □ p_n

q_n

_____

$=$

$_1\Sigma$

$k=n+1$

a_k

$b_{k!}$

_

$_1\Sigma$

$k=n+1$

b □ 1

$b_{k!} <$

$_1\Sigma$

$k=(n+1)!$

b □ 1

$b_k =$

b □ 1

$b_{(n+1)!}$

$_1\Sigma$

$k=0$

1

$b_k =$

b □ 1

$b_{(n+1)!}$

_ b

b □ 1

$=$

b

$b_{(n+1)!}$

_ $b_{n!}$

$b_{(n+1)!} =$

1

q_n

n ;

317

318 *CHAPTER 49. LIOUVILLE NUMBER*

...where the last equality follows from the fact that

n _ $n! = n$ _ $n! + n!$ □ $n! = (n + 1)!$ □ $n!$:

Therefore, we conclude that any such x is a Liouville number.

49.2 Irrationality

An equivalent definition to the one given above is that for any positive integer *n*, there exists an *infinite number* of pairs of integers (p, q) obeying the above inequality.

Now we will show that the number $x = c/d$, where c and d are integers and $d > 0$, cannot satisfy the inequalities that define a Liouville number. Since every rational number can be represented as such c/d, we will have proven that **no Liouville number can be rational.**

More specifically, we show that for any positive integer n large enough that $2^{n-1} > d > 0$ (that is, for any integer $n > 1 + \log_2(d)$) no pair of integers (p, q) exists that simultaneously satisfies the two inequalities

$$0 < \left|x - \frac{p}{q}\right| < \frac{1}{q^n}.$$

From this the claimed conclusion follows.

Let p and q be any integers with $q > 1$. Then we have,

$$\left|x - \frac{p}{q}\right| = \left|\frac{c}{d} - \frac{p}{q}\right| = \frac{|cq - dp|}{dq}$$

If $|cq - dp| = 0$, we would have

$$\left|x - \frac{p}{q}\right| = \frac{|cq - dp|}{dq} = 0,$$

meaning that such pair of integers (p, q) would violate the *first* inequality in the definition of a Liouville number, irrespective of any choice of n.

If, on the other hand, $|cq - dp| > 0$, then, since $cq - dp$ is an integer, we can assert the sharper inequality $|cq - dp| \geq 1$. From this it follows that

$$\left|x - \frac{p}{q}\right| = \frac{|cq - dp|}{dq} \geq \frac{1}{dq}$$

Now for any integer $n > 1 + \log_2(d)$, the last inequality above implies

$$\left|x - \frac{p}{q}\right| \geq 1$$

$$\frac{dq}{>}$$
1
$2n\square1q$
$_1$
$q_n:$

Therefore, in the case $|cq - dp| > 0$ such pair of integers (p, q) would violate the *second* inequality in the definition of a Liouville number, for some positive integer n.

We conclude that there is no pair of integers (p, q), with $q > 1$, that would qualify such an $x = c/d$ as a Liouville number.

Hence a Liouville number, if it exists, cannot be rational.

(The section on *Liouville's constant* proves that Liouville numbers exist by exhibiting the construction of one. The proof given in this section implies that this number must be irrational.)

49.3 Uncountability

Consider, for example, the number

3.140001000000000000000000050000....

3.14(3 zeros)1(17 zeros)5(95 zeros)9(599 zeros)2...

where the digits are zero except in positions $n!$ where the digit equals the nth digit following the decimal point in the decimal expansion of π.

As shown in the section on the existence of Liouville numbers, this number, as well as any other non-terminating decimal with its non-zero digits similarly situated, satisfies the definition of a Liouville number. Since the set of all sequences of non-null digits has the cardinality of the continuum, the same thing occurs with the set of all Liouville numbers.

Moreover, the Liouville numbers form a dense subset of the set of real numbers.

49.4 Liouville numbers and measure

From the point of view of measure theory, the set of all Liouville numbers L is small. More precisely, its Lebesgue measure is zero. The proof given follows some ideas by John C. Oxtoby.[1]:8

For positive integers $n > 2$ and $q \geq 2$ set:

$V_{n;q} =$

\bigcup
$p=\square 1$
(
p
q
$\square 1$
$q_n;$
p
q
+
1
q_n
)

we have

$L _$
\bigcup
$q=2$
$V_{n;q}:$

Observe that for each positive integer $n \geq 2$ and $m \geq 1$, we also have

$L \setminus (\square m; m) _$
\bigcup
$q=2$
$V_{n;q} \setminus (\square m; m) _$
\bigcup
$q=2$
\bigcup_{mq}
$p=\square mq$
(
p
q
$\square 1$

q_n;

$$\frac{p}{q} + \frac{1}{q_n})$$

:

Since

$$\overline{(\frac{p}{q} + \frac{1}{q_n})} \Box (\frac{p}{q} \Box \frac{1}{q_n}) \underline{} = \frac{2}{q_n}$$

and $n > 2$ we have

$$m(L\backslash(\Box m; m)) _$$

$$\sum_{q=2}^{l} \sum_{p=\Box mq}^{\Sigma mq} \frac{2}{q_n} =$$

$$\sum_{q=2}^{l} \frac{2(2mq + 1)}{q_n}$$

$$_(4m+1) \overline{\sum_{q=2}^{l} \frac{1}{q_{n\Box 1}}}$$

$$_(4m+1) \overline{\int_{1}^{l} \frac{1}{q_{n\Box 1}} dq}$$

$$_ \frac{4m + 1}{n \Box 2}$$

:

Now

$$\lim_{n!l} \frac{4m + 1}{n \Box 2} = 0$$

and it follows that for each positive integer m, $L \cap (-m, m)$ has Lebesgue measure zero. Consequently, so has L.

In contrast, the Lebesgue measure of the set T of *all* real transcendental numbers is infinite (since T is the complement of a null set).

In fact, the Hausdorff dimension of L is zero, which implies that the Hausdorff measure of L is zero for all dimension $d > 0$.[1] Hausdorff dimension of L under other dimension functions has also been investigated.[2]

49.5 Structure of the set of Liouville numbers

For each positive integer n, set

$$U_n = \bigcup_{q=2}^{\infty} \bigcup_{p=-\infty}^{\infty} \left\{ x \in \mathbf{R} : 0 < \left| x - \frac{p}{q} \right| < \frac{1}{q^n} \right\} = \bigcup_{q=2}^{\infty} \bigcup_{p=-\infty}^{\infty} \left(\frac{p}{q} - \frac{1}{q^n}, \frac{p}{q} + \frac{1}{q^n} \right) \setminus \left\{ \frac{p}{q} \right\}$$

The set of all Liouville numbers can thus be written as

$$L = \bigcap_{n=1}^{\infty} U_n.$$

Each U_n is an open set; as its closure contains all rationals (the {p/q}'s from each punctured interval), it is also a dense subset of real line. Since it is the intersection of countably many such open dense sets, L is comeagre, that is to say, it is a *dense* Gδ set.

Along with the above remarks about measure, it shows that the set of Liouville numbers and its complement decompose the reals into two sets, one of which is meagre, and the other of Lebesgue measure zero.

49.6 Irrationality measure

The **irrationality measure** (or **irrationality exponent** or **approximation exponent** or **Liouville–Roth constant**) of a real number x is a measure of how "closely" it can be approximated by rationals. Generalizing the definition of Liouville numbers, instead of allowing any n in the power of q, we find the least upper bound of the set of *real* numbers μ such that

$$0 < \left| x - \frac{p}{q} \right| < \frac{1}{q^{\mu}}$$

is satisfied by an infinite number of integer pairs (p, q) with $q > 0$. This least upper bound is defined to be the irrationality measure of x.[3] For any value μ less than this upper bound, the infinite set of all rationals p/q satisfying the above inequality yield an approximation of x. Conversely, if μ is greater than the upper bound, then there are at most finitely many (p, q) with $q > 0$ that satisfy the inequality; thus, the opposite inequality holds for all larger values of q. In other words, given the irrationality measure μ of a real number x, whenever a rational approximation $x \cong p/q$, $p,q \in \mathbf{N}$ yields $n + 1$ exact decimal digits, we have

1

$10n$

_

x □ p

q

_ 1

q_

except for at most a finite number of "lucky" pairs (p, q).

For a rational number α the irrationality measure is $\mu(\alpha) = 1$.[3] The Thue–Siegel–Roth theorem states that if α is an algebraic number, real but not rational, then $\mu(\alpha) = 2$.[4]

Almost all numbers have an irrationality measure equal to 2.[3]

Transcendental numbers have irrationality measure 2 or greater. For example, the transcendental number e has $\mu(e) = 2$.[5] The irrationality measure of π is at most 7.60630853: $\mu(\log 2) < 3.57455391$ and $\mu(\log 3) < 5.125$.[6]

The Liouville numbers are precisely those numbers having infinite irrationality measure.[4]

49.7 Liouville numbers and transcendence

All Liouville numbers are transcendental, as will be proven below. Establishing that a given number is a Liouville number provides a useful tool for proving a given number is transcendental. However, not every transcendental number is a Liouville number. The terms in the continued fraction expansion of every Liouville number are unbounded; using a counting argument, one can then show that there must be uncountably many transcendental numbers which are not Liouville. Using the explicit continued fraction expansion of e, one can show that e is an example of a transcendental number that is not Liouville. Mahler proved in 1953 that π is another such example.[7]

The proof proceeds by first establishing a property of irrational algebraic numbers. This property essentially says that irrational algebraic numbers cannot be well approximated by rational numbers. A Liouville number is irrational but does not have this property, so it can't be algebraic and must be transcendental. The following lemma is usually known as **Liouville's theorem (on diophantine approximation)**, there being several results known as Liouville's theorem.

Lemma: If α is an irrational number which is the root of a polynomial f of degree $n > 0$ with integer coefficients, then there exists a real number $A > 0$ such that, for all integers p, q, with $q > 0$,

_ □ p

q

>

A

q^n

Proof of Lemma: Let M be the maximum value of $|f'(x)|$ (the absolute value of the derivative of f) over the interval $[\alpha - 1, \alpha + 1]$. Let $\alpha_1, \alpha_2, ..., \alpha_m$ be the distinct roots of f which differ from α. Select some value $A > 0$ satisfying

$A <$ min

(

1;

1

M

; j_ □ _1j ; j_ □ _2j ; : : : ; j_ □ _mj

)

Now assume that there exist some integers p, q contradicting the lemma. Then

_ □ p

q

_ A

q^n

$$A < \min\left(1, \frac{1}{M}, |\alpha - \alpha_1|, |\alpha - \alpha_2|, \ldots, |\alpha - \alpha_m|\right)$$

Then p/q is in the interval $[\alpha - 1, \alpha + 1]$; and p/q is not in $\{\alpha_1, \alpha_2, \ldots, \alpha_m\}$, so p/q is not a root of f; and there is no root of f between α and p/q.

By the mean value theorem, there exists an x_0 between p/q and α such that

$$f(\alpha) - f\left(\frac{p}{q}\right) = \left(\alpha - \frac{p}{q}\right) f'(x_0)$$

Since α is a root of f but p/q is not, we see that $|f'(x_0)| > 0$ and we can rearrange:

$$\left|\alpha - \frac{p}{q}\right| = \frac{\left|f(\alpha) - f\left(\frac{p}{q}\right)\right|}{|f'(x_0)|} = \left|\frac{f\left(\frac{p}{q}\right)}{f'(x_0)}\right|$$

Now, f is of the form $\sum_{i=0}^{n} c_i x^i$ where each c_i is an integer; so we can express $|f(p/q)|$ as

$$\left|f\left(\frac{p}{q}\right)\right| = \left|\sum_{i=0}^{n} c_i p^i q^{-i}\right| = \frac{1}{q^n}\left|\sum_{i=0}^{n} c_i p^i q^{n-i}\right| \geq 1$$

q^n

the last inequality holding because p/q is not a root of f and the c_i are integers.

Thus we have that $|f(p/q)| \geq 1/q^n$. Since $|f'(x_0)| \leq M$ by the definition of M, and $1/M > A$ by the definition of A, we have that

$$\left|x_0 - \frac{p}{q}\right| = \frac{\left|f\left(\frac{p}{q}\right)\right|}{|f'(x_0)|} \geq \frac{1}{Mq^n} > \frac{A}{q^n} \geq \left|x_0 - \frac{p}{q}\right|$$

which is a contradiction; therefore, no such p, q exist; proving the lemma.

Proof of assertion: As a consequence of this lemma, let x be a Liouville number; as noted in the article text, x is then irrational. If x is algebraic, then by the lemma, there exists some integer n and some positive real A such that for all p, q

$$\left|x - \frac{p}{q}\right| > \frac{A}{q^n}$$

Let r be a positive integer such that $1/(2^r) \leq A$. If we let $m = r + n$, then, since x is a Liouville number, there exists integers a, $b > 1$ such that

$$\left|x - \frac{a}{b}\right| < \frac{1}{b^m} = \frac{1}{b^{r+n}} = \frac{1}{b^r b^n} \leq \frac{1}{2^r}\frac{1}{b^n} \leq \frac{A}{b^n}$$

which contradicts the lemma; therefore x is not algebraic, and is thus transcendental.

49.8 See also

- Diophantine approximation

49.9 References

[1] Oxtoby, John C. (1980). *Measure and Category*. Graduate Texts in Mathematics **2** (2nd ed.). Springer-Verlag. ISBN 0-387-90508-1.
[2] L. Olsen and Dave L. Renfro (February 2006). "On the exact Hausdorff dimension of the set of Liouville numbers. II". *Manuscripta Mathematica* **119** (2): 217–224. doi:10.1007/s00229-005-0604-z.
[3] Bugeaud (2012) p.246
[4] Bugeaud (2012) p.248
[5] Bugeaud (2012) p.185
[6] Zudilin, V.V. (2004). "An essay on the irrationality measure of π and other logarithms". *Chebyshevskii Sbornik* (in Russian) **5** (2(10)): 49–65. Zbl 1140.11036.
[7] The irrationality measure of π does not exceed 7.6304, according to Weisstein, Eric W., "Irrationality Measure", *MathWorld*.
_ Bugeaud, Yann (2012). *Distribution modulo one and Diophantine approximation*. Cambridge Tracts in Mathematics **193**. Cambridge: Cambridge University Press. ISBN 978-0-521-11169-0. Zbl pre06066616.

49.10 External links

_ The Beginning of Transcendental Numbers
_ The least interesting number

Chapter 50

Proof by exhaustion

This article is about the type of mathematical proof. For the method of calculating limits, see Method of exhaustion. "Brute force method" redirects here. For similarly named methods in other disciplines, see Brute force (disambiguation).

Proof by exhaustion, also known as **proof by cases**, **perfect induction**, or the **brute force method**, is a method of mathematical proof in which the statement to be proved is split into a finite number of cases or sets of equivalent cases and each type of case is checked to see if the proposition in question holds.[1] This is a method of direct proof.

A proof by exhaustion contains two stages:

1. A proof that the cases are exhaustive; i.e., that each instance of the statement to be proved matches the conditions of (at least) one of the cases.
2. A proof of each of the cases.

In the Curry–Howard isomorphism, proof by exhaustion and case analysis are related to ML-style pattern matching.

50.1 Example

To prove that every integer that is a perfect cube is a multiple of 9, or is 1 more than a multiple of 9, or is 1 less than a multiple of 9.

Proof:

Each cube number is the cube of some integer *n*. Every integer *n* is either a multiple of 3, or 1 more or 1 less than a multiple of 3. So these 3 cases are exhaustive:

_ Case 1: If $n = 3p$, then $n^3 = 27p^3$, which is a multiple of 9.
_ Case 2: If $n = 3p + 1$, then $n^3 = 27p^3 + 27p^2 + 9p + 1$, which is 1 more than a multiple of 9. For instance, if $n = 4$ then $n^3 = 64 = 9x7 + 1$.
_ Case 3: If $n = 3p - 1$, then $n^3 = 27p^3 - 27p^2 + 9p - 1$, which is 1 less than a multiple of 9. For instance, if $n = 5$ then $n^3 = 125 = 9\times 14 - 1$.■

50.2 Number of cases

There is no upper limit to the number of cases allowed in a proof by exhaustion. Sometimes there are only two or three cases. Sometimes there may be thousands or even millions. For example, rigorously solving an endgame puzzle in chess might involve considering a very large number of possible positions in the game tree of that problem.

323

324 *CHAPTER 50. PROOF BY EXHAUSTION*

The first proof of the four colour theorem was a proof by exhaustion with 1,936 cases. This proof was controversial because the majority of the cases were checked by a computer program, not by hand. The shortest known proof of the four colour theorem today still has over 600 cases.

Mathematicians prefer to avoid proofs with large numbers of cases, as they seem inelegant, and in general the probability of an error in the whole proof increases with the number of cases. A proof with a large number of cases leaves an impression that the theorem is only true by coincidence, and not because of some underlying principle or connection. Other types of proofs—such as proof by induction (mathematical induction)—are considered more elegant. However, there are some important theorems for which no other method of proof has been found, such as

_ The proof that there is no finite projective plane of order 10.
_ The classification of finite simple groups.
_ The Kepler conjecture.

50.3 See also

_ Case analysis

_ Computer-assisted proof

50.4 Notes

[1] Reid, D. A; Knipping, C (2010), *Proof in Mathematics Education: Research, Learning, and Teaching*, Sense Publishers, p. 133, ISBN 978-9460912443.

Chapter 51

Four color theorem

In mathematics, the **four color theorem**, or the **four color map theorem**, states that, given any separation of a plane into contiguous regions, producing a figure called a *map*, no more than four colors are required to color the regions of the map so that no two adjacent regions have the same color. Two regions are called *adjacent* if they share a common boundary that is not a corner, where corners are the points shared by three or more regions.[1] For example, in the map of the United States of America, Utah and Arizona are adjacent, but Utah and New Mexico, which only share a point that also belongs to Arizona and Colorado, are not.

Despite the motivation from coloring political maps of countries, the theorem is not of particular interest to mapmakers. According to an article by the math historian Kenneth May (Wilson 2002, 2), "Maps utilizing only four colors are rare, and those that do usually require only three. Books on cartography and the history of mapmaking do not mention the four-color property."

Three colors are adequate for simpler maps, but an additional fourth color is required for some maps, such as a map in which one region is surrounded by an odd number of other regions that touch each other in a cycle. The five color theorem, which has a short elementary proof, states that five colors suffice to color a map and was proven in the late 19th century (Heawood 1890); however, proving that four colors suffice turned out to be significantly harder. A number of false proofs and false counterexamples have appeared since the first statement of the four color theorem in 1852.

The four color theorem was proven in 1976 by Kenneth Appel and Wolfgang Haken. It was the first major theorem to be proved using a computer. Appel and Haken's approach started by showing that there is a particular set of 1,936 maps, each of which cannot be part of a smallest-sized counterexample to the four color theorem. (If they did appear, you could make a smaller counter-example.) Appel and Haken used a special-purpose computer program to confirm that each of these maps had this property. Additionally, any map that could potentially be a counterexample must have a portion that looks like one of these 1,936 maps. Showing this required hundreds of pages of hand analysis. Appel and Haken concluded that no smallest counterexamples exists because any must contain, yet do not contain, one of these 1,936 maps. This contradiction means there are no counterexamples at all and that the theorem is therefore true. Initially, their proof was not accepted by all mathematicians because the computer-assisted proof was infeasible for a human to check by hand (Swart 1980). Since then the proof has gained wider acceptance, although doubts remain (Wilson 2002, 216–222).

To dispel remaining doubt about the Appel–Haken proof, a simpler proof using the same ideas and still relying on computers was published in 1997 by Robertson, Sanders, Seymour, and Thomas. Additionally in 2005, the theorem was proven by Georges Gonthier with general purpose theorem proving software.

51.1 Precise formulation of the theorem

The intuitive statement of the four color theorem, i.e. 'that given any separation of a plane into contiguous regions, called a map, the regions can be colored using at most four colors so that no two adjacent regions have the same color', needs to be interpreted appropriately to be correct.

First, all corners, points that belong to (technically, are in the closure of) three or more countries, must be ignored. In addition, bizarre maps (using regions of finite area but infinite perimeter) can require more than four colors.[2]

Example of a four-colored map

Second, for the purpose of the theorem, every "country" has to be a connected region, or contiguous. In the real world, this is not true (e.g. the Upper and Lower Peninsula of Michigan, Nakhchivan as part of Azerbaijan, and Kaliningrad as part of Russia are not contiguous). Because all the territory of a particular country must be the same color, four colors may not be sufficient. For instance, consider a simplified map:

In this map, the two regions labeled *A* belong to the same country, and must be the same color. This map then requires five colors, since the two *A* regions together are contiguous with four other regions, each of which is contiguous with

Map of the world using just four colors.

World map colored using the four color theorem, including oceans. This implies that inland seas may not be blue and landlocked countries may be blue.

all the others. If *A* consisted of three regions, six or more colors might be required; one can construct maps that require an arbitrarily high number of colors. A similar construction also applies if a single color is used for all bodies of water, as is usual on real maps.

An easier to state version of the theorem uses graph theory. The set of regions of a map can be represented more abstractly as an undirected graph that has a vertex for each region and an edge for every pair of regions that share a boundary segment. This graph is planar (it is important to note that we are talking about the graphs that have some limitations according to the map they are transformed from only): it can be drawn in the plane without crossings by placing each vertex at an arbitrarily chosen location within the region to which it corresponds, and by drawing the edges as curves that lead without crossing within each region from the vertex location to each shared boundary point of the region. Conversely any planar graph can be formed from a map in this way. In graph-theoretic terminology, the four-color theorem states that the vertices of every planar graph can be colored with at most four colors so that no two adjacent vertices receive the same color, or for short, "every planar graph is four-colorable" (Thomas 1998,

A four-coloring of a map of the states of the United States (ignoring lakes).

Example of a map of Azerbaijan with non-contiguous regions

p. 849; Wilson 2002).

51.2 History

A

A

51.2.1 Early proof attempts

Möbius mentioned the problem in his lectures as early as 1840.[3] The conjecture was first proposed on October 23, 1852 [4] when Francis Guthrie, while trying to color the map of counties of England, noticed that only four different colors were needed. At the time, Guthrie's brother, Frederick, was a student of Augustus De Morgan (the former advisor of Francis) at University College London. Francis inquired with Frederick regarding it, who then took it to De Morgan (Francis Guthrie graduated later in 1852, and later became a professor of mathematics in South Africa).

According to De Morgan:

"A student of mine [Guthrie] asked me to day to give him a reason for a fact which I did not know was a fact — and do not yet. He says that if a figure be any how divided and the compartments differently colored so that figures with any portion of common boundary *line* are differently colored — four colors may be wanted but not more — the following is his case in which four colors *are* wanted. Query cannot a necessity for five or more be invented… " (Wilson 2002, p. 18)

"F.G.", perhaps one of the two Guthries, published the question in *The Athenaeum* in 1854,[5][6] and De Morgan posed the question again in the same magazine in 1860.[7] Another early published reference by Arthur Cayley (1879) in turn credits the conjecture to De Morgan.

There were several early failed attempts at proving the theorem. De Morgan believed that it followed from a simple fact about four regions, though he didn't believe that fact could be derived from more elementary facts.

This arises in the following way. We never need four colors in a neighborhood unless there be four counties, each of which has boundary lines in common with each of the other three. Such a thing cannot happen with four areas unless one or more of them be inclosed by the rest; and the color used for the inclosed county is thus set free to go on with. Now this principle, that four areas cannot each have common boundary with all the other three without inclosure, is not, we fully believe, capable of demonstration upon anything more evident and more elementary; it must stand as a postulate.[7]

One alleged proof was given by Alfred Kempe in 1879, which was widely acclaimed;[3] another was given by Peter Guthrie Tait in 1880. It was not until 1890 that Kempe's proof was shown incorrect by Percy Heawood, and in 1891 Tait's proof was shown incorrect by Julius Petersen—each false proof stood unchallenged for 11 years (Thomas 1998, p. 848).

In 1890, in addition to exposing the flaw in Kempe's proof, Heawood proved the five color theorem (Heawood 1890) and generalized the four color conjecture to surfaces of arbitrary genus—see below.

Tait, in 1880, showed that the four color theorem is equivalent to the statement that a certain type of graph (called a snark in modern terminology) must be non-planar.[8]

In 1943, Hugo Hadwiger formulated the Hadwiger conjecture (Hadwiger 1943), a far-reaching generalization of the four-color problem that still remains unsolved.

51.2.2 Proof by computer

During the 1960s and 1970s German mathematician Heinrich Heesch developed methods of using computers to search for a proof. Notably he was the first to use discharging for proving the theorem, which turned out to be important in the unavoidability portion of the subsequent Appel-Haken proof. He also expanded on the concept of reducibility and, along with Ken Durre, developed a computer test for it. Unfortunately, at this critical juncture, he was unable to procure the necessary supercomputer time to continue his work (Wilson 2002).

Others took up his methods and his computer-assisted approach. While other teams of mathematicians were racing to complete proofs, Kenneth Appel and Wolfgang Haken at the University of Illinois announced, on June 21, 1976,[9] that they had proven the theorem. They were assisted in some algorithmic work by John A. Koch (Wilson 2002).

If the four-color conjecture were false, there would be at least one map with the smallest possible number of regions that requires five colors. The proof showed that such a minimal counterexample cannot exist, through the use of two technical concepts (Wilson 2002; Appel & Haken 1989; Thomas 1998, pp. 852–853):

1. An *unavoidable set* is a set of configurations such that every map that satisfies some necessary conditions for being a minimal non-4-colorable triangulation (such as having minimum degree 5) must have at least one configuration from this set.
2. A *reducible configuration* is an arrangement of countries that cannot occur in a minimal counterexample. If a map contains a reducible configuration, then the map can be reduced to a smaller map. This smaller map has the condition that if it can be colored with four colors, then the original map can also. This implies that if the original map cannot be colored with four colors the smaller map can't either and so the original map is not minimal.

Using mathematical rules and procedures based on properties of reducible configurations, Appel and Haken found an unavoidable set of reducible configurations, thus proving that a minimal counterexample to the four-color conjecture could not exist. Their proof reduced the infinitude of possible maps to 1,936 reducible configurations (later reduced to 1,476) which had to be checked one by one by computer and took over a thousand hours. This reducibility part of the work was independently double checked with different programs and computers. However, the unavoidability part of the proof was verified in over 400 pages of microfiche, which had to be checked by hand (Appel & Haken 1989).

Appel and Haken's announcement was widely reported by the news media around the world, and the math department at the University of Illinois used a postmark stating "Four colors suffice." At the same time the unusual nature of the proof—it was the first major theorem to be proven with extensive computer assistance—and the complexity of the human-verifiable portion, aroused considerable controversy (Wilson 2002).

In the early 1980s, rumors spread of a flaw in the Appel-Haken proof. Ulrich Schmidt at RWTH Aachen examined Appel and Haken's proof for his master's thesis (Wilson 2002, 225). He had checked about 40% of the unavoidability portion and found a significant error in the discharging procedure (Appel & Haken 1989). In 1986, Appel and Haken were asked by the editor of *Mathematical Intelligencer* to write an article addressing the rumors of flaws in their proof. They responded that the rumors were due to a "misinterpretation of [Schmidt's] results" and obliged with a detailed article (Wilson 2002, 225–226). Their magnum opus, *Every Planar Map is Four-Colorable*, a book claiming a complete and detailed proof (with a microfiche supplement of over 400 pages), appeared in 1989 and explained Schmidt's discovery and several further errors found by others (Appel & Haken 1989).

51.2.3 Simplification and verification

Since the proving of the theorem, efficient algorithms have been found for 4-coloring maps requiring only $O(n^2)$ time, where n is the number of vertices. In 1996, Neil Robertson, Daniel P. Sanders, Paul Seymour, and Robin Thomas created a quadratic-time algorithm, improving on a quartic-time algorithm based on Appel and Haken's proof (Thomas 1995; Robertson et al. 1996). This new proof is similar to Appel and Haken's but more efficient because it reduces the complexity of the problem and requires checking only 633 reducible configurations. Both the unavoidability and reducibility parts of this new proof must be executed by computer and are impractical to check by hand (Thomas 1998, pp. 852–853). In 2001, the same authors announced an alternative proof, by proving the snark theorem (Thomas; Pegg et al. 2002).

In 2005, Benjamin Werner and Georges Gonthier formalized a proof of the theorem inside the Coq proof assistant. This removed the need to trust the various computer programs used to verify particular cases; it is only necessary to trust the Coq kernel (Gonthier 2008).

51.3 Summary of proof ideas

The following discussion is a summary based on the introduction to Appel and Haken's book *Every Planar Map is Four Colorable* (Appel & Haken 1989). Although flawed, Kempe's original purported proof of the four color theorem provided some of the basic tools later used to prove it. The explanation here is reworded in terms of the modern graph theory formulation above.

Kempe's argument goes as follows. First, if planar regions separated by the graph are not *triangulated*, i.e. do not have exactly three edges in their boundaries, we can add edges without introducing new vertices in order to make every region triangular, including the unbounded outer region. If this triangulated graph is colorable using four colors or fewer, so is the original graph since the same coloring is valid if edges are removed. So it suffices to prove the four color theorem for triangulated graphs to prove it for all planar graphs, and without loss of generality we assume the graph is triangulated.

Suppose v, e, and f are the number of vertices, edges, and regions (faces). Since each region is triangular and each edge is shared by two regions, we have that $2e = 3f$. This together with Euler's formula, $v - e + f = 2$, can be used to show that $6v - 2e = 12$. Now, the *degree* of a vertex is the number of edges abutting it. If v_n is the number of vertices of degree n and D is the maximum degree of any vertex,

$$6v - 2e = 6\sum_{i=1}^{D} v_i - \sum_{i=1}^{D} i v_i = \sum_{i=1}^{D} (6 - i) v_i = 12.$$

But since $12 > 0$ and $6 - i \leq 0$ for all $i \geq 6$, this demonstrates that there is at least one vertex of degree 5 or less.

If there is a graph requiring 5 colors, then there is a *minimal* such graph, where removing any vertex makes it fourcolorable. Call this graph G. Then G cannot have a vertex of degree 3 or less, because if $d(v) \leq 3$, we can remove v from G, four-color the smaller graph, then add back v and extend the four-coloring to it by choosing a color different from its neighbors.

Kempe also showed correctly that G can have no vertex of degree 4. As before we remove the vertex v and four-color the remaining vertices. If all four neighbors of v are different colors, say red, green, blue, and yellow in clockwise order, we look for an alternating path of vertices colored red and blue joining the red and blue neighbors. Such a path is called a Kempe chain. There may be a Kempe chain joining the red and blue neighbors, and there may be a Kempe chain joining the green and yellow neighbors, but not both, since these two paths would necessarily intersect, and the vertex where they intersect cannot be colored. Suppose it is the red and blue neighbors that are not chained together. Explore all vertices attached to the red neighbor by red-blue alternating paths, and then reverse the colors red and blue on all these vertices. The result is still a valid four-coloring, and v can now be added back and colored red.

This leaves only the case where G has a vertex of degree 5; but Kempe's argument was flawed for this case. Heawood noticed Kempe's mistake and also observed that if one was satisfied with proving only five colors are needed, one could run through the above argument (changing only that the minimal counterexample requires 6 colors) and use Kempe chains in the degree 5 situation to prove the five color theorem.

In any case, to deal with this degree 5 vertex case requires a more complicated notion than removing a vertex. Rather the form of the argument is generalized to considering *configurations*, which are connected subgraphs of G with the degree of each vertex (in G) specified. For example, the case described in degree 4 vertex situation is the configuration consisting of a single vertex labelled as having degree 4 in G. As above, it suffices to demonstrate that if the configuration is removed and the remaining graph four-colored, then the coloring can be modified in such a way that when the configuration is re-added, the four-coloring can be extended to it as well. A configuration for which this is possible is called a *reducible configuration.* If at least one of a set of configurations must occur somewhere in G, that set is called *unavoidable.* The argument above began by giving an unavoidable set of five configurations (a single vertex with degree 1, a single vertex with degree 2, ..., a single vertex with degree 5) and then proceeded to show that the first 4 are reducible; to exhibit an unavoidable set of configurations where every configuration in the set is reducible would prove the theorem.

Because G is triangular, the degree of each vertex in a configuration is known, and all edges internal to the configuration are known, the number of vertices in G adjacent to a given configuration is fixed, and they are joined in a cycle. These vertices form the *ring* of the configuration; a configuration with k vertices in its ring is a k-ring configuration, and the configuration together with its ring is called the *ringed configuration.* As in the simple cases above, one may enumerate all distinct four-colorings of the ring; any coloring that can be extended without modification to a coloring of the configuration is called *initially good.* For example, the single-vertex configuration above with 3 or less neighbors were initially good. In general, the surrounding graph must be systematically recolored to turn the ring's coloring into a good one, as was done in the case above where there were 4 neighbors; for a general configuration with a larger ring, this requires more complex techniques. Because of the large number of distinct four-colorings of the ring, this is the primary step requiring computer assistance.

Finally, it remains to identify an unavoidable set of configurations amenable to reduction by this procedure. The primary method used to discover such a set is the method of discharging. The intuitive idea underlying discharging is to consider the planar graph as an electrical network. Initially positive and negative "electrical charge" is distributed amongst the vertices so that the total is positive.

Recall the formula above:

ΣD

$i=1$

$(6 \square i)v_i = 12$:

Each vertex is assigned an initial charge of 6-deg(v). Then one "flows" the charge by systematically redistributing the charge from a vertex to its neighboring vertices according to a set of rules, the *discharging procedure*. Since charge is preserved, some vertices still have positive charge. The rules restrict the possibilities for configurations of positively-charged vertices, so enumerating all such possible configurations gives an unavoidable set.

As long as some member of the unavoidable set is not reducible, the discharging procedure is modified to eliminate it (while introducing other configurations). Appel and Haken's final discharging procedure was extremely complex

and, together with a description of the resulting unavoidable configuration set, filled a 400-page volume, but the configurations it generated could be checked mechanically to be reducible. Verifying the volume describing the unavoidable configuration set itself was done by peer review over a period of several years.

A technical detail not discussed here but required to complete the proof is *immersion reducibility*.

51.4 False disproofs

The four color theorem has been notorious for attracting a large number of false proofs and disproofs in its long history. At first, *The New York Times* refused as a matter of policy to report on the Appel–Haken proof, fearing that the proof would be shown false like the ones before it (Wilson 2002). Some alleged proofs, like Kempe's and Tait's mentioned above, stood under public scrutiny for over a decade before they were exposed. But many more, authored by amateurs, were never published at all.

Generally, the simplest, though invalid, counterexamples attempt to create one region which touches all other regions. This forces the remaining regions to be colored with only three colors. Because the four color theorem is true, this is always possible; however, because the person drawing the map is focused on the one large region, they fail to notice that the remaining regions can in fact be colored with three colors.

This trick can be generalized: there are many maps where if the colors of some regions are selected beforehand, it becomes impossible to color the remaining regions without exceeding four colors. A casual verifier of the counterexample may not think to change the colors of these regions, so that the counterexample will appear as though it is valid.

Perhaps one effect underlying this common misconception is the fact that the color restriction is not transitive: a region only has to be colored differently from regions it touches directly, not regions touching regions that it touches. If this were the restriction, planar graphs would require arbitrarily large numbers of colors.

Other false disproofs violate the assumptions of the theorem in unexpected ways, such as using a region that consists of multiple disconnected parts, or disallowing regions of the same color from touching at a point.

51.5 Generalizations

1 2 1

3 4 4

5

6 6

7 1 1 2

By joining the single arrows together and the double arrows together, one obtains a torus with seven mutually touching regions; therefore seven colors are necessary

The four-color theorem applies not only to finite planar graphs, but also to infinite graphs that can be drawn without crossings in the plane, and even more generally to infinite graphs (possibly with an uncountable number of vertices) for which every finite subgraph is planar. To prove this, one can combine a proof of the theorem for finite planar graphs with the De Bruijn–Erdős theorem stating that, if every finite subgraph of an infinite graph is k-colorable, then the whole graph is also k-colorable Nash-Williams (1967). This can also be seen as an immediate consequence

This construction shows the torus divided into the maximum of seven regions, each one of which touches every other.

of Kurt Gödel's compactness theorem for first-order logic, simply by expressing the colorability of an infinite graph with a set of logical formulae.

One can also consider the coloring problem on surfaces other than the plane (Weisstein). The problem on the sphere or cylinder is equivalent to that on the plane. For closed (orientable or non-orientable) surfaces with positive genus, the maximum number p of colors needed depends on the surface's Euler characteristic χ according to the formula

$$p = \left\lfloor \frac{7+\sqrt{49-24\chi}}{2} \right\rfloor;$$

where the outermost brackets denote the floor function.

Alternatively, for an orientable surface the formula can be given in terms of the genus of a surface, g:

$$p = \left\lfloor \frac{7+\sqrt{1+48g}}{2} \right\rfloor$$

(Weisstein).

This formula, the Heawood conjecture, was conjectured by P.J. Heawood in 1890 and proven by Gerhard Ringel and J. W. T. Youngs in 1968. The only exception to the formula is the Klein bottle, which has Euler characteristic 0 (hence the formula gives p = 7) and requires 6 colors, as shown by P. Franklin in 1934 (Weisstein).

For example, the torus has Euler characteristic $\chi = 0$ (and genus $g = 1$) and thus $p = 7$, so no more than 7 colors are required to color any map on a torus. Similarly, toroidal polyhedra such as the Szilassi polyhedron all require seven colors.

A Möbius strip requires six colors (Tietze 1910) as do 1-planar graphs (graphs drawn with at most one simple crossing per edge) (Borodin 1984). If both the vertices and the faces of a planar graph are colored, in such a way that no two adjacent vertices, faces, or vertex-face pair have the same color, then again at most six colors are needed (Borodin 1984).

There is no obvious extension of the coloring result to three-dimensional solid regions. By using a set of n flexible rods, one can arrange that every rod touches every other rod. The set would then require n colors, or $n+1$ if you consider the empty space that also touches every rod. The number n can be taken to be any integer, as large as desired. Such examples were known to Fredrick Guthrie in 1880 (Wilson 2002). Even for axis-parallel cuboids (considered to be adjacent when two cuboids share a two-dimensional boundary area) an unbounded number of colors may be necessary (Reed & Allwright 2008; Magnant & Martin (2011)).

51.6 See also

Graph coloring the problem of finding optimal colorings of graphs that are not necessarily planar.

Grötzsch's theorem triangle-free planar graphs are 3-colorable.

Hadwiger–Nelson problem how many colors are needed to color the plane so that no two points at unit distance apart have the same color?

Tietze's subdivision of a Möbius strip into six mutually-adjacent regions, requiring six colors. The vertices and edges of the subdivision form an embedding of Tietze's graph onto the strip.

List of sets of four countries that border one another Contemporary examples of national maps requiring four colors

Apollonian network The planar graphs that require four colors and have exactly one four-coloring

51.7 Notes

[1] Georges Gonthier (December 2008). "Formal Proof—The Four-Color Theorem". *Notices of the AMS* **55** (11): 1382–1393.From this paper: Definitions: A planar map is a set of pairwise disjoint subsets of the plane, called regions. A simple map is one whose regions are connected open sets. Two regions of a map are adjacent if their respective closures have a common point that is not a corner of the map. A point is a corner of a map if and only if it belongs to the closures of at least three regions. Theorem: The regions of any simple planar map can be colored with only four colors, in such a way that any two adjacent regions have different colors.

[2] Hud Hudson (May 2003). "Four Colors Do Not Suffice". *The American Mathematical Monthly* **110** (5): 417–423. JSTOR 3647828.

[3] W. W. Rouse Ball (1960) *The Four Color Theorem*, in Mathematical Recreations and Essays, Macmillan, New York, pp 222-232.

[4] Donald MacKenzie, *Mechanizing Proof: Computing, Risk, and Trust* (MIT Press, 2004) p103

[5] F. G. (June 10, 1854), *Tinting Maps*, *The Athenaeum*: 726.

[6] Brendan D. McKay (2012). "A note on the history of the four-colour conjecture". arXiv:1201.2852.

[7] De Morgan (anonymous), Augustus (April 14, 1860), *The Philosophy of Discovery, Chapters Historical and Critical. By*

W. Whewell., *The Athenaeum*: 501–503
[8] Tait, P. G. (1880), *Remarks on the colourings of maps*, *Proc. R. Soc. Edinburgh* **10**: 729

[9] Gary Chartrand and Linda Lesniak, *Graphs & Digraphs* (CRC Press, 2005) p221

51.8 References

_ Allaire, F. (1997), *Another proof of the four colour theorem—Part I*, *Proceedings, 7th Manitoba Conference on Numerical Mathematics and Computing, Congr. Numer.* **20**: 3–72
_ Appel, Kenneth; Haken, Wolfgang (1977), *Every Planar Map is Four Colorable Part I. Discharging*, *Illinois Journal of Mathematics* **21**: 429–490
_ Appel, Kenneth; Haken, Wolfgang; Koch, John (1977), *Every Planar Map is Four Colorable Part II. Reducibility*, *Illinois Journal of Mathematics* **21**: 491–567
_ Appel, Kenneth; Haken, Wolfgang (October 1977), *Solution of the Four Color Map Problem*, *Scientific American* **237** (4): 108–121, doi:10.1038/scientificamerican1077-108
_ Appel, Kenneth; Haken, Wolfgang (1989), *Every Planar Map is Four-Colorable*, Providence, RI: American Mathematical Society, ISBN 0-8218-5103-9
_ Bernhart, Frank R. (1977), *A digest of the four color theorem.*, *Journal of Graph Theory* **1**: 207–225, doi:10.1002/jgt.3190010305
_ Borodin, O. V. (1984), *Solution of the Ringel problem on vertex-face coloring of planar graphs and coloring of 1-planar graphs*, *Metody Diskretnogo Analiza* (41): 12–26, 108, MR 832128.
_ Cayley, Arthur (1879), *On the colourings of maps*, *Proc. Royal Geographical Society* (Blackwell Publishing) **1** (4): 259–261, doi:10.2307/1799998, JSTOR 1799998
_ Fritsch, Rudolf; Fritsch, Gerda (1998), *The Four Color Theorem: History, Topological Foundations and Idea of Proof*, New York: Springer, ISBN 978-0-387-98497-1
_ Gonthier, Georges (2008), *Formal Proof—The Four-Color Theorem*, *Notices of the American Mathematical Society* **55** (11): 1382–1393
_ Gonthier, Georges (2005), *A computer-checked proof of the four colour theorem*, unpublished
_ Hadwiger, Hugo (1943), *Über eine Klassifikation der Streckenkomplexe*, *Vierteljschr. Naturforsch. Ges. Zürich* **88**: 133–143
_ Heawood, P. J. (1890), *Map-Colour Theorem*, *Quarterly Journal of Mathematics, Oxford* **24**: 332–338
_ Magnant, C.; Martin, D. M. (2011), *Coloring rectangular blocks in 3-space*, *Discussiones Mathematicae Graph Theory* **31** (1): 161–170, doi:10.7151/dmgt.1535
_ Nash-Williams, C. St. J. A. (1967), *Infinite graphs—a survey*, *J. Combinatorial Theory* **3**: 286–301, doi:10.1016/s0021-9800(67)80077-2, MR 0214501.
_ O'Connor; Robertson (1996), *The Four Colour Theorem*, MacTutor archive
_ Pegg, A.; Melendez, J.; Berenguer, R.; Sendra, J. R.; Hernandez; Del Pino, J. (2002), *Book Review: The Colossal Book of Mathematics*, *Notices of the American Mathematical Society* **49** (9): 1084–1086, Bibcode:2002ITED...49.1084A, doi:10.1109/TED.2002.1003756
_ Reed, Bruce; Allwright, David (2008), *Painting the office*, *Mathematics-in-Industry Case Studies* **1**: 1–8
_ Ringel, G.; Youngs, J.W.T. (1968), *Solution of the Heawood Map-Coloring Problem*, *Proc. Natl. Acad. Sci. U.S.A.* **60** (2): 438–445, Bibcode:1968PNAS...60..438R, doi:10.1073/pnas.60.2.438, PMC 225066, PMID 16591648
_ Robertson, Neil; Sanders, Daniel P.; Seymour, Paul; Thomas, Robin (1996), "Efficiently four-coloring planar graphs", *Efficiently four-coloring planar graphs*, STOC'96: Proceedings of the twenty-eighth annual ACM symposium on Theory of computing, ACM Press, pp. 571–575, doi:10.1145/237814.238005
_ Robertson, Neil; Sanders, Daniel P.; Seymour, Paul; Thomas, Robin (1997), *The Four-Colour Theorem*, *J. Combin. Theory Ser. B* **70** (1): 2–44, doi:10.1006/jctb.1997.1750

_ Saaty, Thomas; Kainen, Paul (1986), *The Four Color Problem: Assaults and Conquest*, *Science* (New York: Dover Publications) **202** (4366): 424, Bibcode:1978Sci...202..424S, doi:10.1126/science.202.4366.424, ISBN 0-486-65092-8
_ Swart, ER (1980), *The philosophical implications of the four-color problem*, *American Mathematical Monthly* (Mathematical Association of America) **87** (9): 697–702, doi:10.2307/2321855, JSTOR 2321855
_ Thomas, Robin (1998), *An Update on the Four-Color Theorem*, *Notices of the American Mathematical Society* **45** (7): 848–859
_ Thomas, Robin (1995), *The Four Color Theorem*
_ Tietze, Heinrich (1910), *Einige Bemerkungen zum Problem des Kartenfärbens auf einseitigen Flächen* [*Some remarks on the problem of map coloring on one-sided surfaces*], *DMV Annual Report* **19**: 155–159
_ Thomas, Robin, *Recent Excluded Minor Theorems for Graphs*, p. 14
_ Wilson, Robin (2002), *Four Colors Suffice*, London: Penguin Books, ISBN 0-691-11533-8

51.9 External links

_ Hazewinkel, Michiel, ed. (2001), "Four-colour problem", *Encyclopedia of Mathematics*, Springer, ISBN 978-

1-55608-010-4
_ Weisstein, Eric W., "Blanuša snarks", *MathWorld*.
_ Weisstein, Eric W., "Map coloring", *MathWorld*.

Chapter 52

Probabilistic method

*This article is **not** about interactive proof systems which use probability to convince a verifier that a proof is correct, **nor** about probabilistic algorithms, which give the right answer with high probability but not with certainty, **nor** about Monte Carlo methods, which are simulations relying on pseudo-randomness.*

The **probabilistic method** is a nonconstructive method, primarily used in combinatorics and pioneered by Paul Erdős, for proving the existence of a prescribed kind of mathematical object. It works by showing that if one randomly chooses objects from a specified class, the probability that the result is of the prescribed kind is more than zero. Although the proof uses probability, the final conclusion is determined for *certain*, without any possible error.

This method has now been applied to other areas of mathematics such as number theory, linear algebra, and real analysis, as well as in computer science (e.g. randomized rounding), and information theory.

52.1 Introduction

If every object in a collection of objects fails to have a certain property, then the probability that a random object chosen from the collection has that property is zero. Turning this around, if the probability that the random object has the property is greater than zero, then this proves the existence of at least one object in the collection that has the property. It doesn't matter if the probability is vanishingly small; any positive probability will do.

Similarly, showing that the probability is (strictly) less than 1 can be used to prove the existence of an object that does *not* satisfy the prescribed properties.

Another way to use the probabilistic method is by calculating the expected value of some random variable. If it can be shown that the random variable can take on a value less than the expected value, this proves that the random variable can also take on some value greater than the expected value.

Common tools used in the probabilistic method include Markov's inequality, the Chernoff bound, and the Lovász local lemma.

52.2 Two examples due to Erdős

Although others before him proved theorems via the probabilistic method (for example, Szele's 1943 result that there exist tournaments containing a large number of Hamiltonian cycles), many of the most well known proofs using this method are due to Erdős. Indeed, the Alon-Spencer textbook on the subject has his picture on the cover to highlight the method's association with Erdős. The first example below describes one such result from 1947 that gives a proof of a lower bound for the Ramsey number $R(r, r)$.

52.2.1 First example

Suppose we have a complete graph on n vertices. We wish to show (for small enough values of n) that it is possible to color the edges of the graph in two colors (say red and blue) so that there is no complete subgraph on r vertices

which is monochromatic (every edge colored the same color).

To do so, we color the graph randomly. Color each edge independently with probability 1/2 of being red and 1/2 of being blue. We calculate the expected number of monochromatic subgraphs on r vertices as follows:

For any set S of r vertices from our graph, define the variable $X(S)$ to be 1 if every edge amongst the r vertices is the same color, and 0 otherwise. Note that the number of monochromatic r-subgraphs is the sum of $X(S)$ over all possible subsets. For any S, the expected value of $X(S)$ is simply the probability that all of the
(
r
2
)
edges in S are the same color,
2 _ 2
□(r
2)
(the factor of 2 comes because there are two possible colors).

This holds true for any of the $C(n, r)$ possible subsets we could have chosen, so we have that the sum of $E[X(S)]$ over all S is
(
n

$r)\ 2^{1-\binom{r}{2}}.$

The sum of an expectation is the expectation of the sum (*regardless* of whether the variables are independent), so the expectation of the sum (the expected number of monochromatic r-subgraphs) is

$$\binom{n}{r}2^{1-\binom{r}{2}}.$$

Consider what happens if this value is less than 1. The number of monochromatic r-subgraphs in our random coloring will always be an integer, so at least one coloring must have less than the expected value. But the only integer that satisfies this criterion is 0. Thus if

$$\binom{n}{r} < 2^{\binom{r}{2}-1},$$

(which holds, for example, for n=5 and r=4) then some coloring fits our desired criterion.[1]

By definition of the Ramsey number, this implies that $R(r, r)$ must be bigger than n. In particular, $R(r, r)$ must grow at least exponentially with r.

A peculiarity of this argument is that it is entirely nonconstructive. Even though it proves (for example) that almost every coloring of the complete graph on $(1.1)^r$ vertices contains no monochromatic r-subgraph, it gives no explicit example of such a coloring. The problem of finding such a coloring has been open for more than 50 years.

52.2.2 Second example

A 1959 paper of Erdős (see reference cited below) addressed the following problem in graph theory: given positive integers g and k, does there exist a graph G containing only cycles of length at least g, such that the chromatic number of G is at least k?

It can be shown that such a graph exists for any g and k, and the proof is reasonably simple. Let n be very large and consider a random graph G on n vertices, where every edge in G exists with probability $p = n^{1/g-1}$. It can be shown that with positive probability, the following two properties hold:

Property 1. G contains at most $n/2$ cycles of length less than g.

Proof. Let X be the number cycles of length less than g. Number of cycles of length i in the complete graph on n vertices is

$$\frac{n!}{2\cdot i\cdot(n-i)!} \le \frac{n^i}{2}$$

and each of them is present in G with probability p^i. Hence by Markov's inequality we have

$$\Pr\left(X > \frac{n}{2}\right) \le \frac{2}{n}E[X] \le \frac{1}{n}\sum_{i=3}^{g-1} p^i n^i = \frac{1}{n}\sum_{i=3}^{g-1} n^{\frac{i}{g}}$$

$$\frac{g}{n}$$

n

$g_{\square 1}$

$\frac{g}{\square 1} gn$

$g = o(1)$:

Property 2. G contains no independent set of size $\lceil \frac{n}{2k} \rceil$.

Proof. Let Y be the size of the largest independent set in G. Clearly, we have

$\Pr(Y \geq y) \leq$

$\binom{n}{y}$

$(1 \square p)$
$_{y(y\square 1)}$

$_2 \leq nye$
$_{\square py(y\square 1)}$

$_2 = e$
$_{\square y}$

$_2$

$_{(py\square 2 \ln n\square p)} = o(1)$;

when

$y = \lceil \frac{n}{2k} \rceil$:

Here comes the trick: since G has these two properties, we can remove at most $n/2$ vertices from G to obtain a new graph G' on n' vertices that contains only cycles of length at least g. We can see that this new graph has no independent set of size $\lceil \frac{n'}{k} \rceil$. Hence G' has chromatic number at least k, as chromatic number is lower bounded by 'number of vertices/size of largest independent set'.

This result gives a hint as to why the computation of the chromatic number of a graph is so difficult: even when there are no local reasons (such as small cycles) for a graph to require many colors the chromatic number can still be arbitrarily large.

52.3 See also

_ Random graph
_ Probabilistic proofs of non-probabilistic theorems
_ Method of conditional probabilities
_ Interactive proof system

52.4 References

_ Alon, Noga; Spencer, Joel H. (2000). *The probabilistic method* (2ed). New York: Wiley-Interscience. ISBN 0-471-37046-0.
_ Erdős, P. (1959). "Graph theory and probability". *Canad. J. Math.* **11** (0): 34–38. doi:10.4153/CJM-1959-003-9. MR 0102081.
_ Erdős, P. (1961). "Graph theory and probability, II". *Canad. J. Math.* **13** (0): 346–352. doi:10.4153/CJM-1961-029-9. MR 0120168.
_ J. Matoušek, J. Vondrak. The Probabilistic Method. Lecture notes.
_ Alon, N and Krivelevich, M (2006). Extremal and Probabilistic Combinatorics

52.5 Footnotes

[1] The same fact can be proved without probability, using a simple counting argument:
_ The total number of r-subgraphs is

$\binom{n}{r}$

.

_ Each r-subgraphs has

$\binom{r}{2}$ edges and thus can be colored in $2^{\binom{r}{2}}$ different ways.

_ Of these colorings, only 2 colorings are 'bad' for that subgraph (the colorings in which all vertices are red or all vertices are blue).

_ Hence, the total number of colorings that are bad for *all* subgraphs is at most $2\binom{n}{r}$.

_ Hence, if $2^{\binom{r}{2}} > 2\binom{n}{r}$, there must be at least one coloring which is not 'bad' for any subgraph.

Chapter 53

Probability theory

Probability theory is the branch of mathematics concerned with probability, the analysis of random phenomena.[1] The central objects of probability theory are random variables, stochastic processes, and events: mathematical abstractions of non-deterministic events or measured quantities that may either be single occurrences or evolve over time in an apparently random fashion. If an individual coin toss or the roll of dice is considered to be a random event, then if repeated many times the sequence of random events will exhibit certain patterns, which can be studied and predicted. Two representative mathematical results describing such patterns are the law of large numbers and the central limit theorem.

As a mathematical foundation for statistics, probability theory is essential to many human activities that involve quantitative analysis of large sets of data. Methods of probability theory also apply to descriptions of complex systems given only partial knowledge of their state, as in statistical mechanics. A great discovery of twentieth century physics was the probabilistic nature of physical phenomena at atomic scales, described in quantum mechanics.

53.1 History

The mathematical theory of probability has its roots in attempts to analyze games of chance by Gerolamo Cardano in the sixteenth century, and by Pierre de Fermat and Blaise Pascal in the seventeenth century (for example the "problem of points"). Christiaan Huygens published a book on the subject in 1657[2] and in the 19th century a big work was done by Laplace in what can be considered today as the classic interpretation.[3]

Initially, probability theory mainly considered **discrete** events, and its methods were mainly combinatorial. Eventually, analytical considerations compelled the incorporation of **continuous** variables into the theory.

This culminated in modern probability theory, on foundations laid by Andrey Nikolaevich Kolmogorov. Kolmogorov combined the notion of sample space, introduced by Richard von Mises, and **measure theory** and presented his axiom system for probability theory in 1933. Fairly quickly this became the mostly undisputed axiomatic basis for modern probability theory but alternatives exist, in particular the adoption of finite rather than countable additivity by Bruno de Finetti.[4]

53.2 Treatment

Most introductions to probability theory treat discrete probability distributions and continuous probability distributions separately. The more mathematically advanced measure theory based treatment of probability covers both the discrete, the continuous, any mix of these two and more.

53.2.1 Motivation

Consider an experiment that can produce a number of outcomes. The set of all outcomes is called the *sample space* of the experiment. The *power set* of the sample space is formed by considering all different collections of possible results. For example, rolling an honest die produces one of six possible results. One collection of possible results

corresponds to getting an odd number. Thus, the subset {1,3,5} is an element of the power set of the sample space of die rolls. These collections are called *events*. In this case, {1,3,5} is the event that the die falls on some odd number. If the results that actually occur fall in a given event, that event is said to have occurred.

Probability is a way of assigning every "event" a value between zero and one, with the requirement that the event made up of all possible results (in our example, the event {1,2,3,4,5,6}) be assigned a value of one. To qualify as a probability distribution, the assignment of values must satisfy the requirement that if you look at a collection of mutually exclusive events (events that contain no common results, e.g., the events {1,6}, {3}, and {2,4} are all mutually exclusive), the probability that one of the events will occur is given by the sum of the probabilities of the individual events.[5]

The probability that any one of the events {1,6}, {3}, or {2,4} will occur is 5/6. This is the same as saying that the probability of event {1,2,3,4,6} is 5/6. This event encompasses the possibility of any number except five being rolled. The mutually exclusive event {5} has a probability of 1/6, and the event {1,2,3,4,5,6} has a probability of 1, that is, absolute certainty.

53.2.2 Discrete probability distributions

Main article: Discrete probability distribution

Discrete probability theory deals with events that occur in countable sample spaces.

Examples: Throwing dice, experiments with decks of cards, random walk, and tossing coins

Classical definition: Initially the probability of an event to occur was defined as number of cases favorable for the event, over the number of total outcomes possible in an equiprobable sample space: see Classical definition of probability.

For example, if the event is "occurrence of an even number when a die is rolled", the probability is given by 3

6 = 1

2 ,

since 3 faces out of the 6 have even numbers and each face has the same probability of appearing.

Modern definition: The modern definition starts with a finite or countable set called the **sample space**, which relates to the set of all *possible outcomes* in classical sense, denoted by Ω . It is then assumed that for each element *x 2* Ω , an intrinsic "probability" value *f*(*x*) is attached, which satisfies the following properties:

1. *f*(*x*) *2* [0*;* 1] for all *x 2* Ω;

2.

Σ

*x*2Ω *f*(*x*) = 1 *:*

That is, the probability function *f*(*x*) lies between zero and one for every value of *x* in the sample space *Ω*, and the sum of *f*(*x*) over all values *x* in the sample space *Ω* is equal to 1. An **event** is defined as any subset *E* of the sample space Ω . The **probability** of the event *E* is defined as

P(*E*) =

Σ

x2E

f(*x*) *:*

So, the probability of the entire sample space is 1, and the probability of the null event is 0.

The function *f*(*x*) mapping a point in the sample space to the "probability" value is called a **probability mass function** abbreviated as **pmf**. The modern definition does not try to answer how probability mass functions are obtained; instead it builds a theory that assumes their existence.

53.2.3 Continuous probability distributions

Main article: Continuous probability distribution

Continuous probability theory deals with events that occur in a continuous sample space.

Classical definition: The classical definition breaks down when confronted with the continuous case. See Bertrand's paradox.

344 *CHAPTER 53. PROBABILITY THEORY*

Modern definition: If the outcome space of a random variable *X* is the set of real numbers (R) or a subset thereof, then a function called the **cumulative distribution function** (or **cdf**) *F* exists, defined by *F*(*x*) = *P*(*X* _ *x*) .

That is, *F*(*x*) returns the probability that *X* will be less than or equal to *x*.

The cdf necessarily satisfies the following properties.

1. *F* is a monotonically non-decreasing, right-continuous function;

2. lim*x!*□*1* *F*(*x*) = 0 ;

3. lim*x!1* *F*(*x*) = 1 *:*

If *F* is absolutely continuous, i.e., its derivative exists and integrating the derivative gives us the cdf back again, then the random variable *X* is said to have a **probability density function** or **pdf** or simply **density** *f*(*x*) = *dF*(*x*)

dx :

For a set *E* _ R , the probability of the random variable *X* being in *E* is

P(*X 2 E*) =

∫

x2E

dF(*x*) *:*

In case the probability density function exists, this can be written as

P(*X 2 E*) =

∫

x2E

f(*x*) *dx :*

Whereas the *pdf* exists only for continuous random variables, the *cdf* exists for all random variables (including discrete random variables) that take values in R*:*

These concepts can be generalized for multidimensional cases on R_n and other continuous sample spaces.

53.2.4 Measure-theoretic probability theory

The *raison d'être* of the measure-theoretic treatment of probability is that it unifies the discrete and the continuous cases, and makes the difference a question of which measure is used. Furthermore, it covers distributions that are neither discrete nor continuous nor mixtures of the two.

An example of such distributions could be a mix of discrete and continuous distributions—for example, a random variable that is 0 with probability 1/2, and takes a random value from a normal distribution with probability 1/2. It can still be studied to some extent by considering it to have a pdf of (_[*x*] + *φ*(*x*))/2 , where _[*x*] is the Dirac delta function.

Other distributions may not even be a mix, for example, the Cantor distribution has no positive probability for any single point, neither does it have a density. The modern approach to probability theory solves these problems using measure theory to define the probability space:

Given any set Ω , (also called **sample space**) and a σ-algebra *F* on it, a measure *P* defined on *F* is called a **probability measure** if *P*(Ω) = 1*:*

If *F* is the Borel σ-algebra on the set of real numbers, then there is a unique probability measure on *F* for any cdf, and vice versa. The measure corresponding to a cdf is said to be **induced** by the cdf. This measure coincides with the pmf for discrete variables and pdf for continuous variables, making the measure-theoretic approach free of fallacies.

The *probability* of a set *E* in the σ-algebra *F* is defined as

P(*E*) =

∫

!2E

$_F$ (*d!*)

where the integration is with respect to the measure $_F$ induced by *F :*

Along with providing better understanding and unification of discrete and continuous probabilities, measure-theoretic treatment also allows us to work on probabilities outside R_n , as in the theory of stochastic processes. For example to study Brownian motion, probability is defined on a space of functions.

53.3. PROBABILITY DISTRIBUTIONS 345

53.3 Probability distributions

Main article: Probability distributions

Certain random variables occur very often in probability theory because they well describe many natural or physical processes. Their distributions therefore have gained *special importance* in probability theory. Some fundamental *discrete distributions* are the discrete uniform, Bernoulli, binomial, negative binomial, Poisson and geometric distributions. Important *continuous distributions* include the continuous uniform, normal, exponential, gamma and beta distributions.

53.4 Convergence of random variables

Main article: Convergence of random variables

In probability theory, there are several notions of convergence for random variables. They are listed below in the order of strength, i.e., any subsequent notion of convergence in the list implies convergence according to all of the preceding notions.

Weak convergence: A sequence of random variables $X_1;X_2;$ *: : : ;* converges **weakly** to the random variable *X* if their respective cumulative *distribution functions* $F_1;$ $F_2;$ *: : :* converge to the cumulative distribution function *F* of*X* , wherever *F* is continuous. Weak convergence is also called **convergence in distribution**.

Most common shorthand notation: X_n

D □*!X :*

Convergence in probability: The sequence of random variables$X_1;X_2;$ *: : :* is said to converge towards the random variable *X* **in probability** if lim$_{n!1}$ *P* (*j*X_n □ *Xj* _ ") = 0 for every ε > 0.

Most common shorthand notation: X_n

P □*!*

X

:

Strong convergence: The sequence of random variables $X_1;X_2;$ *: : :* is said to converge towards the random variable *X* **strongly** if *P*(lim$_{n!1}$ X_n = *X*) = 1 . Strong convergence is also known as **almost sure convergence**.

Most common shorthand notation: X_n

□a□*:!s: X :*

As the names indicate, weak convergence is weaker than strong convergence. In fact, strong convergence implies convergence in probability, and convergence in probability implies weak convergence. The reverse statements are not always true.

53.5 Law of large numbers

Main article: Law of large numbers

Common intuition suggests that if a fair coin is tossed many times, then *roughly* half of the time it will turn up *heads*, and the other half it will turn up *tails*. Furthermore, the more often the coin is tossed, the more likely it should be that the ratio of the number of *heads* to the number of *tails* will approach unity. Modern probability provides a formal version of this intuitive idea, known as the **law of large numbers**. This law is remarkable because it is not assumed in the foundations of probability theory, but instead emerges from these foundations as a theorem. Since it links theoretically derived probabilities to their actual frequency of occurrence in the real world, the law of large numbers is considered as a pillar in the history of statistical theory and has had widespread influence.[6]

The **law of large numbers** (LLN) states that the sample average

$X_n =$

1

n

Σ_n

$k=1$

X_k

of a sequence of independent and identically distributed random variables X_k converges towards their common expectation _ , provided that the expectation of jX_kj is finite.

It is in the different forms of convergence of random variables that separates the *weak* and the *strong* law of large numbers

law: Weak X_n

P □!

_

for$n$

!

1

law: Strong X_n

□a□:!s: _ for$n$! *1:*

It follows from the LLN that if an event of probability p is observed repeatedly during independent experiments, the ratio of the observed frequency of that event to the total number of repetitions converges towards p.

For example, if $Y_1; Y_2; :::$ are independent Bernoulli random variables taking values 1 with probability p and 0 with probability 1-p, then E(Y_i) = p for all i, so that _ Y_n converges to p almost surely.

53.6 Central limit theorem

Main article: Central limit theorem

"The central limit theorem (CLT) is one of the great results of mathematics." (Chapter 18 in[7]) It explains the ubiquitous occurrence of the normal distribution in nature.

The theorem states that the average of many independent and identically distributed random variables with finite variance tends towards a normal distribution *irrespective* of the distribution followed by the original random variables.

Formally, let $X_1;X_2; : : :$ be independent random variables with mean _ and variance _2 > 0*:* Then the sequence of random variables

$Z_n =$

Σ_n

$i=1(X_i$ □ _)

_

p

n

converges in distribution to a standard normal random variable.

Notice that for some classes of random variables the classic central limit theorem works rather fast (Berry–Esseen theorem), for example the distributions with finite first, second and third moment from the exponential family, on the other hand for some random variables of the heavy tail and fat tail variety, it works very slow or may not work at all: in such cases one may use the Generalized Central Limit Theorem (GCLT).

53.7 See also

53.8 Notes

[1] "Probability theory, Encyclopaedia Britannica". Britannica.com. Retrieved 2012-02-12.

[2] Grinstead, Charles Miller; James Laurie Snell. "Introduction". *Introduction to Probability*. pp. vii.

[3] Hájek, Alan. "Interpretations of Probability". "The Stanford Encyclopedia of Philosophy". Retrieved 2012-06-20.
[4] ""The origins and legacy of Kolmogorov's Grundbegriffe", by Glenn Shafer and Vladimir Vovk" (PDF). Retrieved 2012-02-12.
[5] Ross, Sheldon. *A First course in Probability*, 8th Edition. Page 26–27.
[6] "Leithner & Co Pty Ltd - Value Investing, Risk and Risk Management - Part I". Leithner.com.au. 2000-09-15. Retrieved 2012-02-12.
[7] David Williams, "Probability with martingales", Cambridge 1991/2008

53.9 References

_ Pierre Simon de Laplace (1812). *Analytical Theory of Probability.*
The first major treatise blending calculus with probability theory, originally in French: *Théorie Analytique des Probabilités.*
_ A. Kolmogoroff (1933). Grundbegriffe der Wahrscheinlichkeitsrechnung.
An English translation by Nathan Morrison appeared under the title *Foundations of the Theory of Probability* (Chelsea, New York) in 1950, with a second edition in 1956.
_ Patrick Billingsley (1979). *Probability and Measure.* New York, Toronto, London: John Wiley and Sons.
_ Olav Kallenberg; *Foundations of Modern Probability,* 2nd ed. Springer Series in Statistics. (2002). 650 pp. ISBN 0-387-95313-2
_ Henk Tijms (2004). *Understanding Probability.* Cambridge Univ. Press.
A lively introduction to probability theory for the beginner.
_ Olav Kallenberg; *Probabilistic Symmetries and Invariance Principles.* Springer -Verlag, New York (2005). 510 pp. ISBN 0-387-25115-4
_ Gut, Allan (2005). *Probability: A Graduate Course.* Springer-Verlag. ISBN 0-387-22833-0.

53.10 External links

_ Animation on YouTube on the probability space of dice.

Chapter 54

Existence theorem

In mathematics, an **existence theorem** is a theorem with a statement beginning 'there exist(s) ..', or more generally 'for all *x*, *y*, ... there exist(s) ...'. That is, in more formal terms of symbolic logic, it is a theorem with a statement involving the existential quantifier. Many such theorems will not do so explicitly, as usually stated in standard mathematical language. For example, the statement that the sine function is continuous; or any theorem written in big O notation. The quantification can be found in the definitions of the concepts used.

A controversy that goes back to the early twentieth century concerns the issue of **pure existence theorems**. Such theorems may depend on non-constructive foundational material such as the axiom of infinity, the axiom of choice, or the law of excluded middle. From a constructivist viewpoint, by admitting them mathematics loses its concrete applicability (see nonconstructive proof). The opposing viewpoint is that abstract methods are far-reaching, in a way that numerical analysis cannot be.

54.1 'Pure' existence results

An existence theorem may be called **pure** if the proof given of it doesn't also indicate a construction of whatever kind of object the existence of which is asserted.

From a more rigorous point of view, this is a problematic concept. This is because it is a tag applied to a *theorem*, but qualifying its *proof*; hence, *pure* is here defined in a way which violates the standard proof irrelevance of mathematical theorems. That is, theorems are statements for which the fact is that a proof exists, without any 'label' depending on the proof: they may be applied without knowledge of the proof, and indeed if that's not the case the statement is faulty. Thus, many constructivist mathematicians work in extended logics (such as intuitionistic logic) where pure existence statements are intrinsically weaker than their constructivist counterparts.

Such pure existence results are in any case ubiquitous in contemporary mathematics. For example, for a linear problem the set of solutions will be a vector space, and some *a priori* calculation of its dimension may be possible. In any case where the dimension is probably at least 1, an existence assertion has been made (that a non-zero solution exists.)

Theoretically, a proof could also proceed by way of a metatheorem, stating that a proof of the original theorem exists (for example, that a proof by exhaustion search for a proof would always succeed). Such theorems are relatively unproblematic when all of the proofs involved are constructive; however, the status of "pure existence metatheorems" is extremely unclear.

54.2 Constructivist ideas

From the other direction there has been considerable clarification of what constructive mathematics is; without the emergence of a 'master theory'. For example according to Errett Bishop's definitions, the continuity of a function

(such as *sin x*) should be proved as a constructive bound on the modulus of continuity, meaning that the existential content of the assertion of continuity is a promise that can always be kept. One could get another explanation from type theory, in which a proof of an existential statement can come only from a *term* (which we can see as the computational content).
348

Chapter 55

Collatz conjecture

The **Collatz conjecture** is a conjecture in mathematics named after Lothar Collatz, who first proposed it in 1937. The conjecture is also known as the **3*n* + 1 conjecture**, the **Ulam conjecture** (after Stanisław Ulam), **Kakutani's problem** (after Shizuo Kakutani), the **Thwaites conjecture** (after Sir Bryan Thwaites), **Hasse's algorithm** (after Helmut Hasse), or the **Syracuse problem**;[1][2] the sequence of numbers involved is referred to as the **hailstone sequence** or **hailstone numbers** (because the values are usually subject to multiple descents and ascents like hailstones in a cloud),[3][4] or as **wondrous numbers**.[5]

Take any natural number n. If n is even, divide it by 2 to get $n / 2$. If n is odd, multiply it by 3 and add 1 to obtain $3n + 1$. Repeat the process (which has been called "Half Or Triple Plus One", or **HOTPO**[6]) indefinitely. The conjecture is that no matter what number you start with, you will always eventually reach 1. The property has also been called **oneness**.[7]

Paul Erdős said about the Collatz conjecture: "Mathematics may not be ready for such problems."[8] He also offered $500 for its solution.[9]

In 1972, J. H. Conway proved that a natural generalization of the Collatz problem is algorithmically undecidable.[10]

55.1 Statement of the problem

Histogram of stopping times for the numbers 1 to 100 million. Stopping time is on the x axis, frequency on the y axis.

Consider the following operation on an arbitrary positive integer:

349

350 *CHAPTER 55. COLLATZ CONJECTURE*

50
100
150
200
250
1,000 3,000 5,000 7,000 9,000

Numbers from 1 to 9999 and their corresponding total stopping time.

_ If the number is even, divide it by two.
_ If the number is odd, triple it and add one.

In modular arithmetic notation, define the function f as follows:

$f(n) =$
{
$n/2$ if n _ 0 (mod 2)
$3n + 1$ if n _ 1 (mod 2):

Now, form a sequence by performing this operation repeatedly, beginning with any positive integer, and taking the result at each step as the input at the next.

In notation:

$a_i =$
{
n for $i = 0$
$f(a_{i□1})$ for $i > 0$

55.2. EXAMPLES 351

(that is: a_i is the value of f applied to n recursively i times; $a_i = f_i(n)$).

The Collatz conjecture is: *This process will eventually reach the number 1, regardless of which positive integer is chosen initially.*

That smallest i such that $ai = 1$ is called the **total stopping time** of n.[11] The conjecture asserts that every n has a well-defined total stopping time. If, for some n, such an i doesn't exist, we say that n has infinite total stopping time and the conjecture is false.

If the conjecture is false, it can only be because there is some starting number which gives rise to a sequence that does not contain 1. Such a sequence might enter a repeating cycle that excludes 1, or increase without bound. No such sequence has been found.

55.2 Examples

For instance, starting with $n = 6$, one gets the sequence 6, 3, 10, 5, 16, 8, 4, 2, 1.
$n = 19$, for example, takes longer to reach 1: 19, 58, 29, 88, 44. 22, 11, 34, 17, 52, 26, 13, 40, 20, 10, 5, 16, 8, 4, 2, 1.
The sequence for $n = 27$, listed and graphed below, takes 111 steps, climbing to 9232 before descending to 1.
(27, 82, 41, 124, 62, 31, 94, 47, 142, 71, 214, 107, 322, 161, 484, 242, 121, 364, 182, 91, 274, 137, 412, 206, 103, 310, 155, 466, 233, 700, 350, 175, 526, 263, 790, 395, 1186, 593, 1780, 890, 445, 1336, 668, 334, 167, 502, 251, 754, 377, 1132, 566, 283, 850, 425, 1276, 638, 319, 958, 479, 1438, 719, 2158, 1079, 3238, 1619, 4858, 2429, 7288, 3644, 1822, 911, 2734, 1367, 4102, 2051, 6154, 3077, ***9232***, 4616, 2308, 1154, 577, 1732, 866, 433, 1300, 650, 325, 976, 488, 244, 122, 61, 184, 92, 46, 23, 70, 35, 106, 53, 160, 80, 40, 20, 10, 5, 16, 8, 4, 2, 1) (sequence A008884 in OEIS)
Numbers with a total stopping time longer than any smaller starting value form a sequence beginning with:
1, 2, 3, 6, 7, 9, 18, 25, 27, 54, 73, 97, 129, 171, 231, 313, 327, 649, 703, 871, 1161, 2223, 2463, 2919, 3711, 6171, ... (sequence A006877 in OEIS).
For the largest number greater than any smaller starting value, they are
1, 2, 3, 7, 15, 27, 255, 447, 639, 703, 1819, 4255, 4591, 9663, 20895, 26623, 31911, 60975, 77671, ... (sequence A006884 in OEIS)

Number of steps for n to reach 1 are
0, 1, 7, 2, 5, 8, 16, 3, 19, 6, 14, 9, 9, 17, 17, 4, 12, 20, 20, 7, 7, 15, 15, 10, 23, 10, 111, 18, 18, 18, 106, 5, 26, 13, 13, 21, 21, 21, 34, 8, 109, 8, 29, 16, 16, 16, 104, 11, 24, 24, ... (sequence A006577 in OEIS)
The longest progression for any initial starting number less than 100 million is 63,728,127, which has 949 steps. For starting numbers less than 1 billion it is 670,617,279, with 986 steps, and for numbers less than 10 billion it is 9,780,657,631, with 1132 steps.[12][13]
The powers of two converge to one quickly because 2^n is halved n times to reach one, and is never increased, but for Mersenne number M_n, they need to increase n times and usually need more steps to reach 1.

55.3 Visualizations

_ Directed graph showing the orbits of small numbers under the Collatz map. The Collatz conjecture is equivalent to the statement that all paths eventually lead to 1.
_ Directed graph showing the orbits of the first 1000 numbers.
_ The x axis represents starting number, the y axis represents the highest number reached during the chain to 1.

55.4 Cycles

Any counterexample to the Collatz conjecture would have to consist either of an infinite divergent trajectory or a cycle different from the trivial (4; 2; 1) cycle. Thus, if one could prove that neither of these types of counterexample could exist, then all natural numbers would have a trajectory that reaches the trivial cycle. Such a strong result is not known, but certain types of cycles have been ruled out.
The type of a cycle may be defined with reference to the "shortcut" definition of the Collatz map, $f(n) = (3n+1)/2$ for odd n and $f(n) = n/2$ for even n.
A *cycle* is a sequence $(a_0; a_1; : : : a_q)$ where $f(a_0) = a_1$, $f(a_1) = a_2$, and so on, up to $f(a_q) = a_0$ in a closed loop.
The only known cycle is (1; 4; 2).
A k-cycle is a cycle that can be partitioned into $2k$ contiguous subsequences: k increasing sequences of odd numbers alternating with k decreasing sequences of even numbers. For instance, if the cycle consists of a single increasing sequence of odd numbers followed by a decreasing sequence of even numbers, it is called a *1-cycle*.[14]
Steiner (1977) proved that there is no 1-cycle other than the trivial (1;2). Simons (2000) used Steiner's method to prove that there is no 2-cycle. Simons & de Weger (2003) extended this proof up to 68-cycles: there is no k-cycle up to $k = 68$.[14] Beyond 68, this method gives upper bounds for the elements in such a cycle: for example, if there is a 75-cycle, then at least one element of the cycle is less than 2385×2^{50}.[14] Therefore as exhaustive computer searches continue, larger cycles may be ruled out.

55.5 Supporting arguments

Although the conjecture has not been proven, most mathematicians who have looked into the problem think the conjecture is true because experimental evidence and heuristic arguments support it.

55.5.1 Experimental evidence

The conjecture has been checked by computer for all starting values up to $5 \times 2^{60} \approx 5.764 \times 10^{18}$.[15] All initial values tested so far eventually end in the repeating cycle (4; 2; 1), which has only three terms. From this lower bound on the starting value, a lower bound can also be obtained for the number of terms a repeating cycle other than (4; 2; 1) must have.[16] When this relationship was established in 1981, the formula gave a lower bound of 35,400 terms.[16]

Such computer evidence is not a proof that the conjecture is true. As shown in the cases of the Pólya conjecture, the

Mertens conjecture and the Skewes' number, sometimes a conjecture's only counterexamples are found when using very large numbers. Since sequentially examining all natural numbers is a process which can never be completed, such an approach can never demonstrate that the conjecture is true, merely that no counterexamples have yet been discovered.

55.5.2 A probabilistic heuristic

If one considers only the *odd* numbers in the sequence generated by the Collatz process, then each odd number is on average 3/4 of the previous one.[17] (More precisely, the geometric mean of the ratios of outcomes is 3/4.) This yields a heuristic argument that every Hailstone sequence should decrease in the long run, although this is not evidence against other cycles, only against divergence. The argument is not a proof because it assumes that Hailstone sequences are assembled from uncorrelated probabilistic events. (It does rigorously establish that the 2-adic extension of the Collatz process has two division steps for every multiplication step for almost all 2-adic starting values.)

55.5.3 Rigorous bounds

Although it is not known rigorously whether all positive numbers eventually reach one according to the Collatz iteration, it is known that many numbers do so. In particular, Krasikov and Lagarias showed that the number of integers in the interval $[1,x]$ that eventually reach one is at least proportional to $x^{0.84}$.[18]

55.6 Other formulations of the conjecture

55.6.1 In reverse

The first 21 levels of the Collatz graph *generated in bottom-up fashion. The graph includes all numbers with an orbit length of 21*

or less.
There is another approach to prove the conjecture, which considers the bottom-up method of growing the so-called *Collatz graph*. The *Collatz graph* is a graph defined by the inverse relation

R(*n*) =
{
*f*2*ng* if*n* _ 0; 1; 2; 3; 5
*f*2*n*; (*n* □ 1)/3*g* if*n* _ 4
(mod 6)*:*

So, instead of proving that all natural numbers eventually lead to 1, we can prove that 1 leads to all natural numbers. For any integer n, $n \equiv 1 \pmod 2$ iff $3n + 1 \equiv 4 \pmod 6$. Equivalently, $(n - 1)/3 \equiv 1 \pmod 2$ iff $n \equiv 4 \pmod 6$. Conjecturally, this inverse relation forms a tree except for the 1–2–4 loop (the inverse of the 4–2–1 loop of the unaltered function *f* defined in the statement of the problem above).

When the relation $3n + 1$ of the function *f* is replaced by the common substitute "shortcut" relation $(3n + 1)/2$, the Collatz graph is defined by the inverse relation,

R(*n*) =
{
*f*2*ng* if*n* _ 0; 1
*f*2*n*; (2*n* □ 1)/3*g* if*n* _ 2
(mod 3)*:*

Conjecturally, this inverse relation forms a tree except for a 1–2 loop (the inverse of the 1–2 loop of the function f(n) revised as indicated above).

55.6.2 As an abstract machine that computes in base two

Repeated applications of the Collatz function can be represented as an abstract machine that handles strings of bits. The machine will perform the following three steps on any odd number until only one "1" remains:

1. Append 1 to the (right) end of the number in binary (giving $2n + 1$);
2. Add this to the original number by binary addition (giving $2n + 1 + n = 3n + 1$);
3. Remove all trailing "0"s (i.e. repeatedly divide by two until the result is odd).

This prescription is plainly equivalent to computing a Hailstone sequence in base two.

Example

The starting number 7 is written in base two as 111. The resulting Hailstone sequence is:
111 1111 10110 10111 100010 100011 110100 11011 101000 1011 10000

55.6.3 As a parity sequence

For this section, consider the Collatz function in the slightly modified form

f(*n*) =
{
n/2 if*n* _ 0
(3*n* + 1)/2 if*n* _ 1*:*
(mod 2)

This can be done because when n is odd, $3n + 1$ is always even.

If P(…) is the parity of a number, that is $P(2n) = 0$ and $P(2n + 1) = 1$, then we can define the Hailstone parity sequence (or parity vector) for a number n as $p_i = P(a_i)$, where $a_0 = n$, and $a_{i+1} = f(a_i)$.

What operation is performed (3n + 1)/2 or n/2 depends on the parity. The parity sequence is the same as the sequence of operations.

Using this form for $f(n)$, it can be shown that the parity sequences for two numbers m and n will agree in the first k terms if and only if m and n are equivalent modulo 2^k. This implies that every number is uniquely identified by its parity sequence, and moreover that if there are multiple Hailstone cycles, then their corresponding parity cycles must be different.[11][19]

Applying the *f* function k times to the number $a \cdot 2^k + b$ will give the result $a \cdot 3^c + d$, where d is the result of applying the *f* function k times to b, and c is how many odd numbers were encountered during that sequence.

55.6.4 As a tag system

For the Collatz function in the form

f(*n*) =
{
n/2 if*n* _ 0
(3*n* + 1)/2 if*n* _ 1*:*
(mod 2)

Hailstone sequences can be computed by the extremely simple 2-tag system with production rules $a \to bc$, $b \to a$, $c \to aaa$. In this system, the positive integer n is represented by a string of n a, and iteration of the tag operation halts on any word of length less than 2. (Adapted from De Mol.)

The Collatz conjecture equivalently states that this tag system, with an arbitrary finite string of *a*'s as the initial word, eventually halts (see *Tag system#Example: Computation of Collatz sequences* for a worked example).

55.7 Extensions to larger domains

55.7.1 Iterating on all integers

An obvious extension is to include all integers, not just positive integers. In this case there are a total of 5 known cycles, which all integers seem to eventually fall into under iteration of *f*. These cycles are listed here, starting with the well-known cycle for positive *n*.

Odd values are listed in bold. Each cycle is listed with its member of least absolute value (which is always odd or zero) first.

The Generalized Collatz Conjecture is the assertion that every integer, under iteration by *f*, eventually falls into one of these five cycles.

55.7.2 Iterating with odd denominators or 2-adic integers

The standard Collatz map can be extended to (positive or negative) rational numbers which have odd denominators when written in lowest terms. The number is taken to be odd or even according to whether its numerator is odd or even. A closely related fact is that the Collatz map extends to the ring of 2-adic integers, which contains the ring of rationals with odd denominators as a subring.

The parity sequences as defined above are no longer unique for fractions. However, it can be shown that any possible parity cycle is the parity sequence for exactly one fraction: if a cycle has length *n* and includes odd numbers exactly *m* times at indices $k_0, \ldots, k_{m-1}$, then the unique fraction which generates that parity cycle is

$$\frac{3^{m-1}2^{k_0} + \cdots + 3^0 2^{k_{m-1}}}{2^n - 3^m}.$$

For example, the parity cycle (1 0 1 1 0 0 1) has length 7 and has 4 odd numbers at indices 0, 2, 3, and 6. The unique fraction which generates that parity cycle is

$$\frac{3^3 2^0 + 3^2 2^2 + 3^1 2^3 + 3^0 2^6}{2^7 - 3^4} = \frac{151}{47};$$

the complete cycle being: 151/47 → 250/47 → 125/47 → 211/47 → 340/47 → 170/47 → 85/47 → 151/47

Although the cyclic permutations of the original parity sequence are unique fractions, the cycle is not unique, each permutation's fraction being the next number in the loop cycle:

$$\frac{3^3 2^1 + 3^2 2^2 + 3^1 2^5 + 3^0 2^6}{2^7 - 3^4} = \frac{250}{47}$$

$$\frac{3^3 2^0 + 3^2 2^1 + 3^1 2^4 + 3^0 2^5}{2^7 - 3^4} = \frac{125}{47}$$

$$\frac{3^3 2^0 + 3^2 2^3 + 3^1 2^4 + 3^0 2^6}{2^7 - 3^4} = \frac{211}{47}$$

$$\frac{3^3 2^2 + 3^2 2^3 + 3^1 2^5 + 3^0 2^6}{2^7 - 3^4} = \frac{340}{47}$$

$$\frac{3^3 2^1 + 3^2 2^2 + 3^1 2^4 + 3^0 2^5}{2^7 - 3^4} = \frac{170}{47}$$

$$\frac{3^3 2^0 + 3^2 2^1 + 3^1 2^3 + 3^0 2^4}{2^7 - 3^4} = \frac{85}{47}$$

Also, for uniqueness, the parity sequence should be "prime", i.e., not partitionable into identical sub-sequences. For example, parity sequence (1 1 0 0 1 1 0 0) can be partitioned into two identical sub-sequences (1 1 0 0)(1 1 0 0). Calculating the 8-element sequence fraction gives

$3_320 + 3_221 + 3_124 + 3_025$

$28 \ \square \ 34 =$

125

175

But when reduced to lowest terms {5/7}, it is the same as that of the 4-element sub-sequence

$3_120 + 3_021$

$24 \ \square \ 32 =$

5

7

:

And this is because the 8-element parity sequence actually represents two circuits of the loop cycle defined by the 4-element parity sequence.

In this context, the Collatz conjecture is equivalent to saying that (0 1) is the only cycle which is generated by positive whole numbers (i.e. 1 and 2).

55.7.3 Iterating on real or complex numbers

Cobweb plot of the orbit 10-5-8-4-2-1-2-1-2-1-etc. in the real extension of the Collatz map (optimized by replacing "3n + 1" with "(3n + 1)/2")

The Collatz map can be viewed as the restriction to the integers of the smooth real and complex map

$f(z) =$

1

2

$z \cos 2$

$($

2

z

$)$

$+ (3z + 1) \sin 2$

$($

2

z

$)$

;

which simplifies to 1

$4 (2 + 7z \ \square \ (2 + 5z) \cos(_z))$:

If the standard Collatz map defined above is optimized by replacing the relation $3n + 1$ with the common substitute "shortcut" relation $(3n + 1)/2$, it can be viewed as the restriction to the integers of the smooth real and complex map

$f(z) =$

1

2

$z \cos 2$

$($

2

z

$)$

$+$

$(3z + 1)$

$2 \sin 2$

$($

2

z

$)$

;

which simplifies to 1

$4 (1 + 4z \ \square \ (1 + 2z) \cos(_z))$.

Collatz fractal

Iterating the above optimized map in the complex plane produces the Collatz fractal.

The point of view of iteration on the real line was investigated by Chamberland (1996), and on the complex plane by Letherman, Schleicher, and Wood (1999).

Collatz map fractal in a neighbourhood of the real line

55.8 Optimizations

55.8.1 Time-space tradeoff

The *As a parity sequence* section above gives a way to speed up simulation of the sequence. To jump ahead k steps on each iteration (using the f function from that section), break up the current number into two parts, b (the k least significant bits, interpreted as an integer), and a (the rest of the bits as an integer). The result of jumping ahead k steps can be found as:

$f^k(a\,2^k + b) = a\,3^{c(b)} + d(b)$.

The c and d arrays are precalculated for all possible k-bit numbers b, where $d(b)$ is the result of applying the f function k times to b, and $c(b)$ is the number of odd numbers encountered on the way.[20] For example, if k=5, you can jump ahead 5 steps on each iteration by separating out the 5 least significant bits of a number and using:

$c(0..31)$ = {0,3,2,2,2,2,2,4,1,4,1,3,2,2,3,4,1,2,3,3,1,1,3,3,2,3,2,4,3,3,4,5}

$d(0..31)$ = {0,2,1,1,2,2,2,20,1,26,1,10,4,4,13,40,2,5,17,17,2,2,20,20,8,22,8,71,26,26,80,242}.

This requires 2^k precomputation and storage to speed up the resulting calculation by a factor of k, a space-time tradeoff.

55.8.2 Modular restrictions

For the special purpose of searching for a counterexample to the Collatz conjecture, this precomputation leads to an even more important acceleration, used by Tomás Oliveira e Silva in his computational confirmations of the Collatz conjecture up to large values of n. If, for some given b and k, the inequality

$f^k(a\,2^k + b) = a\,3^{c(b)} + d(b) < a\,2^k + b$

holds for all a, then the first counterexample, if it exists, cannot be b modulo 2^k.[16] For instance, the first counterexample must be odd because $f(2n) = n$, smaller than $2n$; and it must be 3 mod 4 because $f^2(4n + 1) = 3n + 1$, smaller than $4n + 1$. For each starting value a which is not a counterexample to the Collatz conjecture, there is a k for which such an inequality holds, so checking the Collatz conjecture for one starting value is as good as checking an entire congruence class. As k increases, the search only needs to check those residues b that are not eliminated by lower values of k. Only an exponentially small fraction of the residues survive.[21] For example, the only surviving residues mod 32 are 7, 15, 27, and 31.

55.9 Syracuse function

If k is an odd integer, then $3k + 1$ is even, so $3k + 1 = 2^a k'$, with k' odd and $a \geq 1$. The **Syracuse function** is the function f from the set I of odd integers into itself, for which $f(k) = k'$ (sequence A075677 in OEIS).

Some properties of the Syracuse function are:

- $f(4k + 1) = f(k)$ for all k in I
- For all $p \geq 2$ and h odd, $f^{p-1}(2^p h - 1) = 2 \cdot 3^{p-1} h - 1$ (here, f^{p-1} is function iteration notation)
- For all odd h, $f(2h - 1) \leq (3h - 1)/2$

The Collatz conjecture is equivalent to the statement that, for all k in I, there exists an integer $n \geq 1$ such that $f^n(k) = 1$.

55.10 See also

- Residue-class-wise affine group
- Modular arithmetic
- BOINC
- Distributed computing

55.11 Notes

[1] Maddux, Cleborne D.; Johnson, D. Lamont (1997). *Logo: A Retrospective*. New York: Haworth Press. p. 160. ISBN 0-7890-0374-0. “The problem is also known by several other names, including: Ulam’s conjecture, the Hailstone problem, the Syracuse problem, Kakutani’s problem, Hasse’s algorithm, and the Collatz problem.”

[2] According to Lagarias (1985, p. 4), the name “Syracuse problem” was proposed by Hasse in the 1950s, during a visit to Syracuse University.

[3] Pickover, Clifford A. (2001). *Wonders of Numbers*. Oxford: Oxford University Press. pp. 116–118. ISBN 0-19-513342-0.

[4] “Hailstone Number”. *MathWorld*. Wolfram Research.

[5] Hofstadter, Douglas R. (1979). *Gödel, Escher, Bach*. New York: Basic Books. pp. 400–402. ISBN 0-465-02685-0.

[6] Friendly, Michael (1988). *Advanced Logo: A Language for Learning*. Hillsdale, New Jersey, USA: Lawrence Erlbaum Associates. ISBN 0-89859-933-4.

[7] Bourke, Paul (December 1992). “Decision Procedure for 'Oneness'”. University of West Alabama.

[8] Guy (2004) p. 330

[9] R. K. Guy: Don't try to solve these problems, Amer. Math. Monthly, **90** (1983), 35–41. By this Erdos means that there aren't powerful tools for manipulating such objects.

[10] “J. H. Conway proved the remarkable result that a simple generalization of the problem is algorithmically undecidable.” Quoting Lagarias 1985:

_ Conway, J. H. (1972). "Unpredictable Iterations". *Proceedings of the 1972 Number Theory Conference : University of Colorado, Boulder, Colorado, August 14–18, 1972*. Boulder, Colorado, USA: University of Colorado. pp. 49–52. OCLC 4181683. Zbl 0337.10041.
[11] Lagarias 1985.
[12] Leavens, Gary T.; Vermeulen, Mike (December 1992). "3x+1 Search Programs". *Computers & Mathematics with Applications* **24** (11): 79–99. doi:10.1016/0898-1221(92)90034-F.
[13] Roosendaal, Eric. "3x+1 Delay Records". Retrieved 27 November 2011. (Note: "Delay records" are total stopping time records.)
[14] Simons, J.; de Weger, B.; "Theoretical and computational bounds for *m*-cycles of the $3n + 1$ problem", *Acta Arithmetica*, (on-line version 1.0, November 18, 2003), 2005.
[15] Silva, Tomás Oliveira e Silva. "Computational verification of the 3x+1 conjecture". Retrieved 27 November 2011.
[16] Garner (1981)
[17] Lagarias, 1985, section "A heuristic argument".
[18] Krasikov, Ilia; Lagarias, Jeffrey C. (2003). "Bounds for the $3x + 1$ problem using difference inequalities". *Acta Arithmetica* **109** (3): 237–258. doi:10.4064/aa109-3-4. MR 1980260.
[19] Terras, Riho (1976), *A stopping time problem on the positive integers, Polska Akademia Nauk* **30** (3): 241–252, MR 0568274
[20] Scollo, Giuseppe (2007), "Looking for Class Records in the 3x+1 Problem by means of the COMETA Grid Infrastructure", *Grid Open Days at the University of Palermo*
[21] Lagarias (1985), Theorem D.

55.12 References

55.12.1 Papers

_ Jeffrey C. Lagarias (1985). The 3x + 1 problem and its generalizations. *The American Mathematical Monthly* 92(1): 3-23.
_ Jeffrey C. Lagarias (2001), "Syracuse problem", in Hazewinkel, Michiel, *Encyclopedia of Mathematics*, Springer, ISBN 978-1-55608-010-4.
_ Marc Chamberland. A continuous extension of the $3x + 1$ problem to the real line. Dynam. Contin. Discrete Impuls Systems 2: 4 (1996), 495–509.
_ Garner, Lynn E. (1981). "On the Collatz $3n + 1$ Algorithm". *Proceedings of the American Mathematical Society* **82** (1): 19–22. doi:10.2307/2044308. JSTOR 2044308.
_ Simon Letherman, Dierk Schleicher, and Reg Wood: The $(3n + 1)$-Problem and Holomorphic Dynamics. Experimental Mathematics 8: 3 (1999), 241–252.

_ Eliahou, Shalom, *The 3x+1 problem: new lower bounds on nontrivial cycle lengths*, Discrete Mathematics 118 (1993) p. 45-56; *Le problème 3n+1 : y a-t-il des cycles non triviaux ?, Images des mathématiques* (2011) (French).
_ Andrei, Stefan; Masalagiu, Cristian (1998). "About the Collatz conjecture". *Acta Informatica* **35** (2): 167. doi:10.1007/s002360050117.
_ Van Bendegem, Jean Paul, "The Collatz Conjecture: A Case Study in Mathematical Problem Solving", *Logic and Logical Philosophy*, volume 14 (2005), 7–23.
_ Belaga, Edward G., Mignotte, Maurice, "Walking Cautiously into the Collatz Wilderness: Algorithmically, Number Theoretically, Randomly", Fourth Colloquium on Mathematics and Computer Science : Algorithms, Trees, Combinatorics and Probabilities, September 18–22, 2006, Institut Élie Cartan, Nancy, France.
_ Belaga, Edward G., Mignotte, Maurice, "Embedding the 3x+1 Conjecture in a 3x+d Context", *Experimental Mathematics*, volume 7, issue 2, 1998.
_ Steiner, R. P.; "A theorem on the syracuse problem", *Proceedings of the 7th Manitoba Conference on Numerical Mathematics*, pages 553–559, 1977.
_ Simons, J.; de Weger, B.; "Theoretical and computational bounds for *m*-cycles of the $3n + 1$ problem", *Acta Arithmetica* (on-line version 1.0, November 18, 2003), 2005.
_ Sinyor, J.; "The 3x+1 Problem as a String Rewriting System", *International Journal of Mathematics and Mathematical Sciences*, volume 2010 (2010), Article ID 458563, 6 pages.

55.12.2 Preprints

_ Belaga, Edward G. (1998). "Reflecting on the 3x+1 Mystery". University of Strasbourg. CiteSeerX: 10.1.1.54.483.
_ Bruschi, Mario (2008). "A generalization of the Collatz problem and conjecture". arXiv:0810.5169 [math.NT].
_ De Mol, Liesbeth (January 2008). "Tag systems and Collatz-like functions". *Theoretical Computer Science* **390** (1): 92–101. doi:10.1016/j.tcs.2007.10.020.
_ Jeffrey C. Lagarias (2006). "The $3x + 1$ problem: An annotated bibliography, II (2000–)". arXiv:math.NT/0608208 [math.NT].
_ Ohira, Reiko; Yamashita, Michinori. "A generalization of the Collatz problem" (in Japanese).
_ Sinisalo, Matti K. (June 2003). "On the minimal cycle lengths of the Collatz sequences". University of Oulu. Archived from the original on 2009-10-24.
_ Stadfeld, Paul. "Blueprint for Failure: How to Construct a Counterexample to the Collatz Conjecture".

_ Urata, Toshio. "Some Holomorphic Functions connected with the Collatz Problem". Archived from the original on 2008-04-06.

55.12.3 Books

_ Everest, Graham; van der Poorten, Alf; Shparlinski, Igor; Ward, Thomas (2003). *Recurrence sequences*. Mathematical Surveys and Monographs **104**. Providence, Rhode Island, USA: American Mathematical Society. Chapter 3.4. ISBN 0-8218-3387-1. Zbl 1033.11006.
_ Guy, Richard K. (2004). *Unsolved problems in number theory* (3rd ed.). Springer-Verlag. "E17: Permutation Sequences". ISBN 0-387-20860-7. Zbl 1058.11001. Cf pp. 336–337.
_ Lagarias, Jeffrey C., ed. (2010). *The Ultimate Challenge: the 3x+1 problem*. American Mathematical Society. ISBN 978-0-8218-4940-8. Zbl 1253.11003.
_ Wirsching, Günther J. (1998). *The Dynamical System Generated by the* 3n+1 *Function*. Lecture Notes in Mathematics **1681**. Springer-Verlag. doi:10.1007/BFb0095985. ISBN 978-3-540-63970-1. Zbl 0892.11002.

55.13 External links

_ Keith Matthews' $3x + 1$ page: Review of progress, plus various programs.
_ Distributed computing project that verifies the Collatz conjecture for larger values.
_ An ongoing distributed computing project by Eric Roosendaal verifies the Collatz conjecture for larger and larger values.
_ Another ongoing distributed computing project by Tomás Oliveira e Silva continues to verify the Collatz conjecture (with fewer statistics than Eric Roosendaal's page but with further progress made).
_ An animated implementation that uses arbitrary-precision arithmetic.
_ Weisstein, Eric W., "Collatz Problem", *MathWorld*.
_ Collatz Sequence explanation and exercise.
_ Collatz Problem at PlanetMath.org..
_ Hailstone Patterns • discusses different resonators along with using important numbers in the problem (like 6 and 3₅) to discover patterns.
_ Collatz Iterations on the Ulam Spiral grid on YouTube.
_ Collatz Paths by Jesse Nochella, Wolfram Demonstrations Project.
_ Page allowing to study and to show the guess for a given number.
_ Collatz cycles? About cycles, very basic, contains also some unusual graphs (HTML).
_ Collatz cycles? About loops, compacted text, rather basic (PDF).
_ Collatz sequence for any number up to 500 digits in length.

Chapter 56

Combinatorial proof

In mathematics, the term ***combinatorial proof*** is often used to mean either of two types of mathematical proof:
_ A proof by double counting. A combinatorial identity is proven by counting the number of elements of some carefully chosen set in two different ways to obtain the different expressions in the identity. Since those expressions count the same objects, they must be equal to each other and thus the identity is established.
_ A bijective proof. Two sets are shown to have the same number of members by exhibiting a bijection, i.e. a one-to-one correspondence, between them.
The term "combinatorial proof" may also be used more broadly to refer to any kind of elementary proof in combinatorics. However, as Glass (2003) writes in his review of Benjamin & Quinn (2003) (a book about combinatorial proofs), these two simple techniques are enough to prove many theorems in combinatorics and number theory.

56.1 Example

An archetypal double counting proof is for the well known formula for the number $\binom{n}{k}$ of k-combinations (i.e., subsets of size k) of an n-element set:

$$\binom{n}{k} = \frac{n(n-1)\cdots(n-k+1)}{k(k-1)\cdots 1}$$

Here a direct bijective proof is not possible: because the right-hand side of the identity is a fraction, there is no set *obviously* counted by it (it even takes some thought to see that the denominator always evenly divides the numerator). However its numerator counts the Cartesian product of k finite sets of sizes n, $n - 1$, ..., $n - k + 1$, while its denominator counts the permutations of a k-element set (the set most obviously counted by the denominator would be another Cartesian product k finite sets; if desired one could map permutations to that set by an explicit bijection). Now take S to be the set of sequences of k elements selected from our n-element set without repetition. On one hand, there is an easy bijection of S with the Cartesian product corresponding to the numerator $n(n \,\square\, 1) _ _ _ (n \,\square\, k + 1)$, and on the other hand there is a bijection from the set C of pairs of a k-combination and a permutation σ of k to S, by taking the elements of C in increasing order, and then permuting this sequence by σ to obtain an element of S. The two ways of counting give the equation

$n(n \,\square\, 1) _ _ _ (n \,\square\, k + 1) =$

(
n
k
)
k!;

and after division by $k!$ this leads to the stated formula for

(n
k
)

. In general, if the counting formula involves a division, a similar double counting argument (if it exists) gives the most straightforward combinatorial proof of the identity, but double counting arguments are not limited to situations where the formula is of this form.

362

56.2. THE BENEFIT OF A COMBINATORIAL PROOF 363

56.2 The benefit of a combinatorial proof

Stanley (1997) gives an example of a combinatorial enumeration problem (counting the number of sequences of k subsets S_1, S_2, ... S_k, that can be formed from a set of n items such that the subsets have an empty common intersection) with two different proofs for its solution. The first proof, which is not combinatorial, uses mathematical induction and generating functions to find that the number of sequences of this type is $(2^k - 1)_n$. The second proof is based on the observation that there are $2^k - 1$ proper subsets of the set $\{1, 2, ..., k\}$, and $(2^k - 1)_n$ functions from the set $\{1, 2, ..., n\}$ to the family of proper subsets of $\{1, 2, ..., k\}$. The sequences to be counted can be placed in one-to-one correspondence with these functions, where the function formed from a given sequence of subsets maps each element i to the set $\{j \mid i \in S_j\}$.

Stanley writes, "Not only is the above combinatorial proof much shorter than our previous proof, but also it makes the reason for the simple answer completely transparent. It is often the case, as occurred here, that the first proof to come to mind turns out to be laborious and inelegant, but that the final answer suggests a simple combinatorial proof." Due both to their frequent greater elegance than non-combinatorial proofs and the greater insight they provide into the structures they describe, Stanley formulates a general principle that combinatorial proofs are to be preferred over other proofs, and lists as exercises many problems of finding combinatorial proofs for mathematical facts known to be true through other means.

56.3 The difference between bijective and double counting proofs

Stanley does not clearly distinguish between bijective and double counting proofs, and gives examples of both kinds, but the difference between the two types of combinatorial proof can be seen in an example provided by Aigner & Ziegler (1998), of proofs for Cayley's formula stating that there are n^{n-2} different trees that can be formed from a given set of n nodes. Aigner and Ziegler list four proofs of this theorem, the first of which is bijective and the last of which is a double counting argument. They also mention but do not describe the details of a fifth bijective proof. The most natural way to find a bijective proof of this formula would be to find a bijection between n-node trees and some collection of objects that has n^{n-2} members, such as the sequences of $n - 2$ values each in the range from 1 to n. Such a bijection can be obtained using the Prüfer sequence of each tree. Any tree can be uniquely encoded into a Prüfer sequence, and any Prüfer sequence can be uniquely decoded into a tree; these two results together provide a bijective proof of Cayley's formula.

An alternative bijective proof, given by Aigner and Ziegler and credited by them to André Joyal, involves a bijection between, on the one hand, n-node trees with two designated nodes (that may be the same as each other), and on the other hand, n-node directed pseudoforests. If there are T_n n-node trees, then there are $n^2 T_n$ trees with two designated nodes. And a pseudoforest may be determined by specifying, for each of its nodes, the endpoint of the edge extending outwards from that node; there are n possible choices for the endpoint of a single edge (allowing self-loops) and therefore n^n possible pseudoforests. By finding a bijection between trees with two labeled nodes and

pseudoforests, Joyal's proof shows that $T_n = n^{n-2}$.
Finally, the fourth proof of Cayley's formula presented by Aigner and Ziegler is a double counting proof due to Jim Pitman. In this proof, Pitman considers the sequences of directed edges that may be added to an n-node empty graph to form from it a single rooted tree, and counts the number of such sequences in two different ways. By showing how to derive a sequence of this type by choosing a tree, a root for the tree, and an ordering for the edges in the tree, he shows that there are $T_n n!$ possible sequences of this type. And by counting the number of ways in which a partial sequence can be extended by a single edge, he shows that there are $n^{n-2} n!$ possible sequences. Equating these two different formulas for the size of the same set of edge sequences and cancelling the common factor of $n!$ leads to Cayley's formula.

56.4 Related concepts

_ The principles of double counting and bijection used in combinatorial proofs can be seen as examples of a larger family of combinatorial principles, which include also other ideas such as the pigeonhole principle.
_ Proving an identity combinatorially can be viewed as adding more structure to the identity by replacing numbers by sets; similarly, categorification is the replacement of sets by categories.

56.5 References

_ Aigner, Martin; Ziegler, Günter M. (1998), *Proofs from THE BOOK*, Springer-Verlag, pp. 141–146, ISBN 3-540-40460-0.
_ Benjamin, Arthur T.; Quinn, Jennifer J. (2003), *Proofs that Really Count: The Art of Combinatorial Proof*, Dolciani Mathematical Expositions **27**, Mathematical Association of America, ISBN 978-0-88385-333-7.
_ Glass, Darren (2003), *Read This: Proofs that Really Count*, Mathematical Association of America.
_ Stanley, Richard P. (1997), *Enumerative Combinatorics, Volume I*, Cambridge Studies in Advanced Mathematics **49**, Cambridge University Press, pp. 11–12, ISBN 0-521-55309-1.

Chapter 57

Double counting (proof technique)

In combinatorics, **double counting**, also called **counting in two ways**, is a combinatorial proof technique for showing that two expressions are equal by demonstrating that they are two ways of counting the size of one set. In this technique, which van Lint & Wilson (2001) call "one of the most important tools in combinatorics," one describes a finite set X from two perspectives leading to two distinct expressions for the size of the set. Since both expressions equal the size of the same set, they equal each other.

57.1 Examples

57.1.1 Forming committees

One example of the double counting method counts the number of ways in which a committee can be formed from n people, allowing any number of the people (even zero of them) to be part of the committee. That is, one counts the number of subsets that an n-element set may have. One method for forming a committee is to ask each person to choose whether or not to join it. Each person has two choices – yes or no – and these choices are independent of those of the other people. Therefore there are $2 \times 2 \times \ldots \times 2 = 2^n$ possibilities. Alternatively, one may observe that the size of the committee must be some number between 0 and n. For each possible size k, the number of ways in which a committee of k people can be formed from n people is the binomial coefficient

$$\binom{n}{k}.$$

Therefore the total number of possible committees is the sum of binomial coefficients over $k = 0, 1, 2, \ldots n$. Equating the two expressions gives the identity

$$\sum_{k=0}^{n} \binom{n}{k} = 2^n;$$

a special case of the binomial theorem. A similar double counting method can be used to prove the more general identity

$$\sum_{k=d}^{n}$$

(
n
k
)(
k
d
)
= $2^{n□d}$
(
n
d
)
(Garbano, Malerba & Lewinter 2003; Klavžar 2006).

57.1.2 Handshaking lemma

Another theorem that is commonly proven with a double counting argument states that every undirected graph contains an even number of vertices of odd degree. That is, the number of vertices that have an odd number of incident edges

must be even. In more colloquial terms, in a party of people some of whom shake hands, an even number of people must have shaken an odd number of other people's hands; for this reason, the result is known as the handshaking lemma.

To prove this by double counting, let $d(v)$ be the degree of vertex v. The number of vertex-edge incidences in the graph may be counted in two different ways: by summing the degrees of the vertices, or by counting two incidences for every edge. Therefore

$$\sum_{v} d(v) = 2e$$

where e is the number of edges. The sum of the degrees of the vertices is therefore an even number, which could not happen if an odd number of the vertices had odd degree. This fact, with this proof, appears in the 1736 paper of Leonhard Euler on the Seven Bridges of Königsberg that first began the study of graph theory.

57.1.3 Counting trees

What is the number T_n of different trees that can be formed from a set of n distinct vertices? Cayley's formula gives the answer $T_n = n^{n-2}$. Aigner & Ziegler (1998) list four proofs of this fact; they write of the fourth, a double counting proof due to Jim Pitman, that it is "the most beautiful of them all."

Pitman's proof counts in two different ways the number of different sequences of directed edges that can be added to an empty graph on n vertices to form from it a rooted tree. One way to form such a sequence is to start with one of the T_n possible unrooted trees, choose one of its n vertices as root, and choose one of the $(n - 1)!$ possible sequences in which to add its $n - 1$ edges. Therefore, the total number of sequences that can be formed in this way is $T_n n(n - 1)! = T_n n!$.

Another way to count these edge sequences is to consider adding the edges one by one to an empty graph, and to count the number of choices available at each step. If one has added a collection of $n - k$ edges already, so that the graph formed by these edges is a rooted forest with k trees, there are $n(k - 1)$ choices for the next edge to add: its starting vertex can be any one of the n vertices of the graph, and its ending vertex can be any one of the $k - 1$ roots other than the root of the tree containing the starting vertex. Therefore, if one multiplies together the number of choices from the first step, the second step, etc., the total number of choices is

$$\prod_{k=2}^{n} n(k □ 1) = n^{n□1}(n □ 1)! = n^{n□2} n!:$$

Equating these two formulas for the number of edge sequences results in Cayley's formula:

$$T_n n! = n^{n□2} n!$$

and

$$T_n = n^{n□2}:$$

As Aigner and Ziegler describe, the formula and the proof can be generalized to count the number of rooted forests with k trees, for any k.

57.2 See also

57.2.1 Additional examples

_ Vandermonde's identity, another identity on sums of binomial coefficients that can be proven by double counting.

Cayley's formula implies that there is $1 = 2^{2-2}$ tree on two vertices, $3 = 3^{3-2}$ trees on three vertices, and $16 = 4^{4-2}$ trees on four vertices.

_ Square pyramidal number. The equality between the sum of the first n square numbers and a cubic polynomial can be shown by double counting the triples of numbers x, y, and z where z is larger than either of the other two numbers.
_ Lubell–Yamamoto–Meshalkin inequality. Lubell's proof of this result on set families is a double counting argument on permutations, used to prove an inequality rather than an equality.
_ Proofs of Fermat's little theorem. A divisibility proof by double counting: for any prime p and natural number A, there are $A_p - A$ length-p words over an A-symbol alphabet having two or more distinct symbols. These may be grouped into sets of p words that can be transformed into each other by circular shifts; these sets are

Adding a directed edge to a rooted forest
called necklaces. Therefore, $A_p - A = p \times$ (number of necklaces) and is divisible by p.
_ Proofs of quadratic reciprocity. A proof by Eisenstein derives another important number-theoretic fact by double counting lattice points in a triangle.

57.2.2 Related topics

_ Bijective proof. Where double counting involves counting one set in two ways, bijective proofs involve counting two sets in one way, by showing that their elements correspond one-for-one.
_ The inclusion-exclusion principle, a formula for the size of a union of sets that may, together with another formula for the same union, be used as part of a double counting argument.

57.3 References

_ Aigner, Martin; Ziegler, Günter M. (1998), *Proofs from THE BOOK*, Springer-Verlag. Double counting is described as a general principle on page 126; Pitman's double counting proof of Cayley's formula is on pp. 145–146.
_ Euler, L. (1736), *Solutio problematis ad geometriam situs pertinentis*, *Commentarii Academiae Scientiarum Imperialis Petropolitanae* **8**: 128–140. Reprinted and translated in Biggs, N. L.; Lloyd, E. K.; Wilson, R. J. (1976), *Graph Theory 1736–1936*, Oxford University Press.
_ Garbano, M. L.; Malerba, J. F.; Lewinter, M. (2003), *Hypercubes and Pascal's triangle: a tale of two proofs*, *Mathematics Magazine* **76** (3): 216–217, doi:10.2307/3219324, JSTOR 3219324.
_ Klavžar, Sandi (2006), *Counting hypercubes in hypercubes*, *Discrete Mathematics* **306** (22): 2964–2967, doi:10.1016/j.disc.2005.10.036.
_ van Lint, Jacobus H.; Wilson, Richard M. (2001), *A Course in Combinatorics*, Cambridge University Press, p. 4, ISBN 978-0-521-00601-9.

Chapter 58

Mathematical object

A **mathematical object** is an abstract object arising in philosophy of mathematics and mathematics.
Commonly encountered mathematical objects include numbers, permutations, partitions, matrices, sets, functions, and relations. Geometry as a branch of mathematics has such objects as hexagons, points, lines, triangles, circles, spheres, polyhedra, topological spaces and manifolds. Another branch - Algebra, has groups, rings, fields, grouptheoretic lattices, and order-theoretic lattices. Categories are simultaneously homes to mathematical objects and mathematical objects in their own right.
The ontological status of mathematical objects has been the subject of much investigation and debate by philosophers of mathematics.[1]

58.1 Cantorian framework

One view that emerged around the turn of the 20th century with the work of Cantor is that all mathematical objects can be defined as sets. The set {0,1} is a relatively clear-cut example. On the face of it the group $\mathbf{Z}_2$ of integers mod 2 is also a set with two elements. However, it cannot simply be the set {0,1}, because this does not mention the additional structure imputed to $\mathbf{Z}_2$ by the operations of addition and negation mod 2: how are we to tell which of 0 or 1 is the additive identity, for example? To organize this group as a set it can first be coded as the quadruple ({0,1},+,−,0), which in turn can be coded using one of several conventions as a set representing that quadruple, which in turn entails encoding the operations + and − and the constant 0 as sets.
Sets may include ordered denotation of the particular identities and operations that apply to them, indicating a group, abelian group, ring, field, or other mathematical object. These types of mathematical objects are commonly studied in abstract algebra.

58.2 Foundational paradoxes

If, however, the goal of mathematical ontology is taken to be the internal consistency of mathematics, it is more important that mathematical objects be definable in some uniform way (for example, as sets) regardless of actual practice, in order to lay bare the essence of its paradoxes. This has been the viewpoint taken by foundations of mathematics, which has traditionally accorded the management of paradox higher priority than the faithful reflection of the details of mathematical practice as a justification for defining mathematical objects to be sets.

Much of the tension created by this foundational identification of mathematical objects with sets can be relieved without unduly compromising the goals of foundations by allowing two kinds of objects into the mathematical universe, sets and relations, without requiring that either be considered merely an instance of the other. These form the basis of model theory as the domain of discourse of predicate logic. From this viewpoint, mathematical objects are entities satisfying the axioms of a formal theory expressed in the language of predicate logic.

58.3 Category theory

A variant of this approach replaces relations with operations, the basis of universal algebra. In this variant the axioms often take the form of equations, or implications between equations.

A more abstract variant is category theory, which abstracts sets as objects and the operations thereon as morphisms between those objects. At this level of abstraction mathematical objects reduce to mere vertices of a graph whose edges as the morphisms abstract the ways in which those objects can transform and whose structure is encoded in the composition law for morphisms. Categories may arise as the models of some axiomatic theory and the homomorphisms between them (in which case they are usually concrete, meaning equipped with a faithful forgetful functor to the category **Set** or more generally to a suitable topos), or they may be constructed from other more primitive categories, or they may be studied as abstract objects in their own right without regard for their provenance.

58.4 See also

- Abstract object
- Mathematical structure

58.5 References

[1] Burgess, John, and Rosen, Gideon, 1997. *A Subject with No Object*. Oxford Univ. Press.

- Azzouni, J., 1994. *Metaphysical Myths, Mathematical Practice*. Cambridge University Press.
- Burgess, John, and Rosen, Gideon, 1997. *A Subject with No Object*. Oxford Univ. Press.
- Davis, Philip and Reuben Hersh, 1999 [1981]. *The Mathematical Experience*. Mariner Books: 156-62.
- Gold, Bonnie, and Simons, Roger A., 2008. *Proof and Other Dilemmas: Mathematics and Philosophy*. Mathematical Association of America.
- Hersh, Reuben, 1997. *What is Mathematics, Really?* Oxford University Press.
- Sfard, A., 2000, "Symbolizing mathematical reality into being, Or how mathematical discourse and mathematical objects create each other," in Cobb, P., *et al.*, *Symbolizing and communicating in mathematics classrooms: Perspectives on discourse, tools and instructional design*. Lawrence Erlbaum.
- Stewart Shapiro, 2000. *Thinking about mathematics: The philosophy of mathematics*. Oxford University Press.

58.6 External links

- Stanford Encyclopedia of Philosophy: "Abstract Objects"—by Gideon Rosen.
- Wells, Charles, "Mathematical Objects."
- AMOF: The Amazing Mathematical Object Factory
- Mathematical Object Exhibit

Chapter 59

Rational number

"Rationals" redirects here. For other uses, see Rational (disambiguation).

In mathematics, a **rational number** is any number that can be expressed as the quotient or fraction *p*/*q* of two integers, *p* and *q*, with the denominator *q* not equal to zero.[1] Since *q* may be equal to 1, every integer is a rational number. The set of all rational numbers is usually denoted by a boldface **Q** (or blackboard bold Q , Unicode ℚ); it was thus denoted in 1895 by Peano after *quoziente*, Italian for "quotient".

The decimal expansion of a rational number always either terminates after a finite number of digits or begins to repeat the same finite sequence of digits over and over. Moreover, any repeating or terminating decimal represents a rational number. These statements hold true not just for base 10, but also for binary, hexadecimal, or any other integer base.

A real number that is not rational is called irrational. Irrational numbers include $\sqrt{2}$, π, e, and φ. The decimal expansion of an irrational number continues without repeating. Since the set of rational numbers is countable, and the set of real numbers is uncountable, almost all real numbers are irrational.[1]

The rational numbers can be formally defined as the equivalence classes of the quotient set $(\mathbf{Z} \times (\mathbf{Z} \setminus \{0\})) / \sim$, where the cartesian product $\mathbf{Z} \times (\mathbf{Z} \setminus \{0\})$ is the set of all ordered pairs (m,n) where m and n are integers, n is not 0 ($n \neq 0$), and "~" is the equivalence relation defined by $(m_1,n_1) \sim (m_2,n_2)$ if, and only if, $m_1 n_2 - m_2 n_1 = 0$.

In abstract algebra, the rational numbers together with certain operations of addition and multiplication form a field. This is the archetypical field of characteristic zero, and is the field of fractions for the ring of integers. Finite extensions of **Q** are called algebraic number fields, and the algebraic closure of **Q** is the field of algebraic numbers.[2]
In mathematical analysis, the rational numbers form a dense subset of the real numbers. The real numbers can be constructed from the rational numbers by completion, using Cauchy sequences, Dedekind cuts, or infinite decimals.
Zero divided by any other integer equals zero; therefore, zero is a rational number (but division by zero is undefined).

59.1 Terminology

The term *rational* in reference to the set **Q** refers to the fact that a rational number represents a *ratio* of two integers.
In mathematics, the adjective *rational* often means that the underlying field considered is the field **Q** of rational numbers. Rational polynomial usually, and most correctly, means a polynomial with rational coefficients, also called a "polynomial over the rationals". However, rational function does *not* mean the underlying field is the rational numbers, and a rational algebraic curve is *not* an algebraic curve with rational coefficients.

59.2 Arithmetic

See also: Fraction (mathematics) § Arithmetic with fractions
372
59.2. ARITHMETIC 373

59.2.1 Embedding of integers

Any integer *n* can be expressed as the rational number *n*/1.

59.2.2 Equality

a
b = *c*
d if and only if *ad* = *bc:*

59.2.3 Ordering

Where both denominators are positive:
a
b < *c*
d if and only if *ad* < *bc:*
If either denominator is negative, the fractions must first be converted into equivalent forms with positive denominators, through the equations:
□*a*
□*b*
=
a
b
and
a
□*b*
=
□*a*
b
:

59.2.4 Addition

Two fractions are added as follows:
a
b
+
c
d
=
ad + *bc*
bd
:

59.2.5 Subtraction

a
b
□ *c*
d
=
ad □ *bc*

bd

:

59.2.6 Multiplication

The rule for multiplication is:

a

b

_ *c*

d

=

ac

bd

:

59.2.7 Division

Where $c \neq 0$:

a

b

_ *c*

d

=

ad

bc

:

Note that division is equivalent to multiplying by the reciprocal of the divisor fraction:

ad

bc

=

a

b

_ *d*

c

:

59.2.8 Inverse

Additive and multiplicative inverses exist in the rational numbers:

□

(*a*

b

)

=

□*a*

b

=

a

□*b* and

(*a*

b

)□1

=

b

a if $a \neq 0$:

59.2.9 Exponentiation to integer power

If *n* is a non-negative integer, then

(*a*

b

)*n*

=

an

bn

and (if $a \neq 0$):

$(a$
b
$)^{\square n}$
$=$
b^n
$a^n:$

59.3 Continued fraction representation

Main article: Continued fraction

A **finite continued fraction** is an expression such as

$a_0 +$
1
$a_1 +$
1
$a_2 +$
1
$\ldots +$
1
a_n
$;$

where a_n are integers. Every rational number a/b has two closely related expressions as a finite continued fraction, whose coefficients a_n can be determined by applying the Euclidean algorithm to (a,b).

59.4 Formal construction

Mathematically we may construct the rational numbers as equivalence classes of ordered pairs of integers (m,n), with $n \neq 0$. This space of equivalence classes is the quotient space $(\mathbf{Z} \times (\mathbf{Z} \setminus \{0\})) / \sim$, where $(m_1,n_1) \sim (m_2,n_2)$ if, and only if, $m_1n_2 - m_2n_1 = 0$. We can define addition and multiplication of these pairs with the following rules:

$(m_1; n_1) + (m_2; n_2) \_ (m_1n_2 + n_1m_2; n_1n_2)$
$(m_1; n_1) \_ (m_2; n_2) \_ (m_1m_2; n_1n_2)$

and, if $m_2 \neq 0$, division by

$(m_1; n_1)$
$(m_2; n_2)$
$\_ (m_1n_2; n_1m_2):$

The equivalence relation $(m_1,n_1) \sim (m_2,n_2)$ if, and only if, $m_1n_2 - m_2n_1 = 0$ is a congruence relation, i.e. it is compatible with the addition and multiplication defined above, and we may define $\mathbf{Q}$ to be the quotient set $(\mathbf{Z} \times (\mathbf{Z} \setminus$

A diagram showing a representation of the equivalent classes of pairs of integers

$\{0\})) / \sim$, i.e. we identify two pairs (m_1,n_1) and (m_2,n_2) if they are equivalent in the above sense. (This construction can be carried out in any integral domain: see field of fractions.) We denote by $[(m_1,n_1)]$ the equivalence class containing (m_1,n_1). If $(m_1,n_1) \sim (m_2,n_2)$ then, by definition, (m_1,n_1) belongs to $[(m_2,n_2)]$ and (m_2,n_2) belongs to $[(m_1,n_1)]$; in this case we can write $[(m_1,n_1)] = [(m_2,n_2)]$. Given any equivalence class $[(m,n)]$ there are a countably infinite number of representation, since

$\_\,\_\,\_ = [(\square 2m; \square 2n)] = [(\square m; \square n)] = [(m; n)] = [(2m; 2n)] = \_\,\_\,\_ :$

The canonical choice for $[(m,n)]$ is chosen so that $\gcd(m,n) = 1$, i.e. m and n share no common factors, i.e. m and n are coprime. For example, we would write $[(1,2)]$ instead of $[(2,4)]$ or $[(-12,-24)]$, even though $[(1,2)] = [(2,4)] = [(-12,-24)]$.

We can also define a total order on $\mathbf{Q}$. Let $\wedge$ be the *and*-symbol and $\vee$ be the *or*-symbol. We say that $[(m_1,n_1)] \leq [(m_2,n_2)]$ if:

$(n_1n_2 > 0 \wedge m_1n_2 \_ n_1m_2) \_ (n_1n_2 < 0 \wedge m_1n_2 \_ n_1m_2):$

The integers may be considered to be rational numbers by the embedding that maps m to $[(m,1)]$.

59.5 Properties

2/1

3/1

4/1

1/1
5/1
6/1
7/1
8/1
1/2 1/3 1/4 1/5 1/6 1/7 1/8
2/2
3/2
4/2
5/2
6/2
8/2
7/2
3/3
2/3
4/3
5/3
6/3
7/3
8/3
2/4
3/4
4/4
5/4
6/4

7/4
8/4
2/5
3/5
4/5
5/5
6/5
7/5
8/5
2/6
3/6
4/6
5/6
6/6
7/6
8/6
2/7
3/7
4/7
5/7
6/7
7/7
8/7
2/8
3/8

4/8

5/8

6/8

7/8

8/8 ...

...

...

...

...

...

...

...

...

...

...

...

...

...

... ...

...

A diagram illustrating the countability of the rationals

The set **Q**, together with the addition and multiplication operations shown above, forms a field, the field of fractions of the integers **Z**.

The rationals are the smallest field with characteristic zero: every other field of characteristic zero contains a copy of **Q**. The rational numbers are therefore the prime field for characteristic zero.

The algebraic closure of **Q**, i.e. the field of roots of rational polynomials, is the algebraic numbers.

The set of all rational numbers is countable. Since the set of all real numbers is uncountable, we say that almost all real numbers are irrational, in the sense of Lebesgue measure, i.e. the set of rational numbers is a null set.

The rationals are a densely ordered set: between any two rationals, there sits another one, and, therefore, infinitely many other ones. For example, for any two fractions such that

a

b

$<$
c
d
(where $b; d$ are positive), we have

a
b
$<$
$ad + bc$
$2bd$
$<$
c
d
:

Any totally ordered set which is countable, dense (in the above sense), and has no least or greatest element is order isomorphic to the rational numbers.

59.6 Real numbers and topological properties

The rationals are a dense subset of the real numbers: every real number has rational numbers arbitrarily close to it. A related property is that rational numbers are the only numbers with finite expansions as regular continued fractions. By virtue of their order, the rationals carry an order topology. The rational numbers, as a subspace of the real numbers, also carry a subspace topology. The rational numbers form a metric space by using the absolute difference metric $d(x,y) = |x - y|$, and this yields a third topology on **Q**. All three topologies coincide and turn the rationals into a topological field. The rational numbers are an important example of a space which is not locally compact. The rationals are characterized topologically as the unique countable metrizable space without isolated points. The space is also totally disconnected. The rational numbers do not form a complete metric space; the real numbers are the completion of **Q** under the metric $d(x,y) = |x - y|$, above.

59.7 *p*-adic numbers

See also: P-adic Number

In addition to the absolute value metric mentioned above, there are other metrics which turn **Q** into a topological field:

Let p be a prime number and for any non-zero integer a, let $|a|_p = p^{-n}$, where p^n is the highest power of p dividing a.
In addition set $|0|_p = 0$. For any rational number a/b, we set $|a/b|_p = |a|_p / |b|_p$.
Then $d_p(x,y) = |x - y|_p$ defines a metric on **Q**.

The metric space $(\mathbf{Q}, d_p)$ is not complete, and its completion is the p-adic number field $\mathbf{Q}_p$. Ostrowski's theorem states that any non-trivial absolute value on the rational numbers **Q** is equivalent to either the usual real absolute value or a p-adic absolute value.

59.8 See also

_ Floating point
_ Ford circles
_ Niven's theorem
_ Rational data type

59.9 References

[1] Rosen, Kenneth (2007). *Discrete Mathematics and its Applications* (6th ed.). New York, NY: McGraw-Hill. pp. 105,158–160. ISBN 978-0-07-288008-3.
[2] Gilbert, Jimmie; Linda, Gilbert (2005). *Elements of Modern Algebra* (6th ed.). Belmont, CA: Thomson Brooks/Cole. pp. 243–244. ISBN 0-534-40264-X.

59.10 External links

_ Hazewinkel, Michiel, ed. (2001), "Rational number", *Encyclopedia of Mathematics*, Springer, ISBN 978-1-55608-010-4
_ "Rational Number" From MathWorld – A Wolfram Web Resource

Chapter 60

Statistical proof

Statistical proof is the rational demonstration of degree of certainty for a proposition, hypothesis or theory that is used to convince others subsequent to a statistical test of the supporting evidence and the types of inferences that can be drawn from the test scores. Statistical methods are used to increase the understanding of the facts and the proof demonstrates the validity and logic of inference with explicit reference to a hypothesis, the experimental data, the facts, the test, and the odds. Proof has two essential aims: the first is to convince and the second is to explain the proposition through peer and public review.[1]

The burden of proof rests on the demonstrable application of the statistical method, the disclosure of the assumptions, and the relevance that the test has with respect to a genuine understanding of the data relative to the external world. There are adherents to several different statistical philosophies of inference, such as Bayes theorem versus the likelihood function, or positivism versus critical rationalism. These methods of reason have direct bearing on statistical proof and its interpretations in the broader philosophy of science.[1][2]

A common demarcation between science and non-science is the hypothetico-deductive proof of falsification developed by Karl Popper, which is a well-established practice in the tradition of statistics. Other modes of inference, however, may include the inductive and abductive modes of proof.[3] Scientists do not use statistical proof as a means to attain certainty, but to falsify claims and explain theory. Science cannot achieve absolute certainty nor is it a continuous march toward an objective truth as the vernacular as opposed to the scientific meaning of the term "proof" might imply. Statistical proof offers a kind of proof of a theory's falsity and the means to learn heuristically through repeated statistical trials and experimental error.[2] Statistical proof also has applications in legal matters with implications for the legal burden of proof.[4]

60.1 Axioms

There are two kinds of axioms, 1) conventions that are taken as true that should be avoided because they cannot be tested, and 2) hypotheses.[5] Proof in the theory of probability was built on four axioms developed in the late 17th century:

1. The probability of a hypotheses is a non-negative real number:
{
$Pr(h) \geqq 0$
}
;
2. The probability of necessary truth equals one:
{
$Pr(t) = 1$
}
;
3. If two hypotheses h_1 and h_2 are mutually exclusive, then the sum of their probabilities is equal to the probability of their disjunction:
{
$Pr(h_1) + Pr(h_2) = Pr(h_1 or h_2)$
}
;
4. The conditional probability of h_1 given h_2
{
$Pr(h_1 j h_2)$
}
is equal to the unconditional probability
{
$Pr(h_1$ &
$h_2)$
}
of the conjunction h_1 and h_2, divided by the unconditional probability
{
$Pr(h_2)$
}
of h_2 where that

probability is positive
{
$Pr(h_1 j h_2)=Pr(h_1$ & $h_2)$
$Pr(h_2)$
}
, where
{
$Pr(h_2) > 0$

}

.

The preceding axioms provide the statistical proof and basis for the laws of randomness, or objective chance from where modern statistical theory has advanced. Experimental data, however, can never prove that the hypotheses (h) is true, but relies on an inductive inference by measuring the probability of the hypotheses relative to the empirical data. The proof is in the rational demonstration of using the logic of inference, math, testing, and deductive reasoning of significance.[1][2][6]

60.2 Test and proof

Main article: Statistical tests

The term *proof* descended from its Latin roots (provable, probable, *probare* L.) meaning *to test*.[7][8] Hence, proof is a form of inference by means of a statistical test. Statistical tests are formulated on models that generate probability distributions. Examples of probability distributions might include the binary, normal, or poisson distribution that give exact descriptions of variables that behave according to natural laws of random chance. When a statistical test is applied to samples of a population, the test determines if the sample statistics are significantly different from the assumed null-model. True values of a population, which are unknowable in practice, are called parameters of the population. Researchers sample from populations, which provide estimates of the parameters, to calculate the mean or standard deviation. If the entire population is sampled, then the sample statistic mean and distribution will converge with the parametric distribution.[9]

Using the scientific method of falsification, the probability value that the sample statistic is sufficiently different from the null-model than can be explained by chance alone is given prior to the test. Most statisticians set the prior probability value at 0.05 or 0.1, which means if the sample statistics diverge from the parametric model more than 5 (or 10) times out of 100, then the discrepancy is unlikely to be explained by chance alone and the null-hypothesis is rejected. Statistical models provide exact outcomes of the parametric and estimates of the sample statistics. Hence, the burden of proof rests in the sample statistics that provide estimates of a statistical model. Statistical models contain the mathematical proof of the parametric values and their probability distributions.[10][11]

60.3 Bayes theorem

Main article: Bayes theorem

See also: Evidence under Bayes theorem

Bayesian statistics are based on a different philosophical approach for proof of inference. The mathematical formula for Bayes's theorem is:

$$Pr[ParameterjData] = \frac{Pr[DatajParameter]\_Pr[Parameter]}{Pr[Data]}$$

The formula is read as the probability of the parameter (or hypothesis =*h*, as used in the notation on axioms) "given" the data (or empirical observation), where the horizontal bar refers to "given". The right hand side of the formula calculates the prior probability of a statistical model (Pr [Parameter]) with the likelihood (Pr [Data | Parameter]) to produce a posterior probability distribution of the parameter (Pr [Parameter | Data]). The posterior probability is the likelihood that the parameter is correct given the observed data or samples statistics.[12] Hypotheses can be compared using Bayesian inference by means of the Bayes factor, which is the ratio of the posterior odds to the prior odds. It provides a measure of the data and if it has increased or decreased the likelihood of one hypotheses relative to another.[13]

The statistical proof is the Bayesian demonstration that one hypothesis has a higher (weak, strong, positive) likelihood.[13] There is considerable debate if the Bayesian method aligns with Karl Poppers method of proof of falsification, where some have suggested that "...there is no such thing as "accepting" hypotheses at all. All that one does in science is assign degrees of belief..."[14]:180 According to Popper, hypotheses that have withstood testing and have yet to be falsified are not verified but corroborated. Some researches have suggested that Popper's quest to define corroboration on the premise of probability put his philosophy in line with the Bayesian approach. In this context, the likelihood of

one hypothesis relative to another may be an index of corroboration, not confirmation, and thus statistically proven through rigorous objective standing.[6][15]

60.4 In legal proceedings

Main article: Legal burden of proof

"Where gross statistical disparities can be shown, they alone may in a proper case constitute *prima facie* proof of a pattern or practice of discrimination."[nb 1]:271

Statistical proof in a legal proceeding can be sorted into three categories of evidence:

1. The occurrence of an event, act, or type of conduct,
2. The identity the individual(s) responsible
3. The intent or psychological responsibility[16]

Statistical proof was not regularly applied in decisions concerning United States legal proceedings until the mid 1970's following a landmark jury discrimination case in *Castaneda v. Partida*. The US Supreme Court ruled that gross statistical disparities constitutes "*prima facie* proof" of discrimination, resulting in a shift of the burden of proof from plaintiff to defendant. Since that ruling, statistical proof has been used in many other cases on inequality, discrimination, and DNA evidence.[4][17][18] However, there is not a one-to-one correspondence between statistical proof and the legal burden of proof. "The Supreme Court has stated that the degrees of rigor required in the fact finding processes of law and science do not necessarily correspond."[18]:1533

In an example of a death row sentence (*McCleskey v. Kemp*[nb 2]) concerning racial discrimination, the petitioner, a black man named McCleskey was charged with the murder of a white police officer during a robbery. Expert testimony for McClesky introduced a statistical proof showing that "defendants charged with killing white victims were 4.3 times as likely to receive a death sentence as charged with killing blacks."[19]:595. Nonetheless, the statistics was insufficient "to prove that the decisionmakers in his case acted with discriminatory purpose."[19]:596 It was further argued that there were "inherent limitations of the statistical proof"[19]:596, because it did not refer to the specifics of the individual. Despite the statistical demonstration of an increased probability of discrimination, the legal burden of proof (it was argued) had to be examined on a case by case basis.[19]

60.5 See also

_ Mathematical proof
_ Data analysis

60.6 References

[1] Gold, B.; Simons, R. A. (2008). *Proof and other dilemmas: Mathematics and philosophy*. Mathematics Association of America Inc. ISBN 0-88385-567-4.
[2] Gattei, S. (2008). *Thomas Kuhn's "Linguistic Turn" and the Legacy of Logical Empiricism: Incommensurability, Rationality and the Search for Truth*. Ashgate Pub Co. p. 277. ISBN 0-7546-6160-1.
[3] Pedemont, B. (2007). "How can the relationship between argumentation and proof be analysed?". *Educational Studies in Mathematics* **66** (1): 23–41. doi:10.1007/s10649-006-9057-x.
[4] Meier, P. (1986). "Damned Liars and Expert Witnesses". *Journal of the American Statistical Association* **81** (394): 269–276. doi:10.1080/01621459.1986.10478270.
[5] Wiley, E. O. (1974). "Karl R. Popper, Systematics, and Classification: A Reply to Walter Bock and Other Evolutionary Taxonomists". *Systematic Biology* **24** (2): 233–243. doi:10.1093/sysbio/24.2.233.

[6] Howson, C.; Urbach, P. (1991). "Bayesian reasoning in science". *Nature* **350** (6317): 371–374. doi:10.1038/350371a0.
[7] Sundholm, G. "Proof-Theoretical Semantics and Fregean Identity Criteria for Propositions". *The Monist* **77** (3): 294–314. doi:10.5840/monist199477315.
[8] Bissell, D. (1996). "Statisticians have a Word for it". *Teaching Statistics* **18** (3): 87–89. doi:10.1111/j.1467-9639.1996.tb00300.x.
[9] Sokal, R. R.; Rohlf, F. J. (1995). *Biometry* (3rd ed.). W.H. Freeman & Company. p. 887. ISBN 0-7167-2411-1.
[10] Heath, David (1995). *An introduction to experimental design and statistics for biology*. CRC Press. ISBN 1-85728-132-2.
[11] Hald, Anders (2006). *A History of Parametric Statistical Inference from Bernoulli to Fisher, 1713-1935*. Springer. p. 260. ISBN 0-387-46408-5.
[12] Huelsenbeck, J. P.; Ronquist, F.; Bollback, J. P. (2001). "Bayesian Inference of Phylogeny and Its Impact on Evolutionary Biology". *Science* **294** (5550): 2310=2314. doi:10.1126/science.1065889.
[13] Wade, P. R. (2000). "Bayesian methods in conservation biology". *Conservation Biology* **14** (5): 1308–1316. doi:10.1046/j.1523-1739.2000.99415.x.
[14] Sober, E. (1991). *Reconstructing the Past: Parsimony, Evolution, and Inference*. A Bradford Book. p. 284. ISBN 0-262-69144-2.
[15] Helfenbein, K. G.; DeSalle, R. (2005). "Falsifications and corroborations: Karl Popper's influence on systematics". *Molecular Phylogenetics and Evolution* **35**: 271–280. doi:10.1016/j.ympev.2005.01.003.
[16] Fienberg; Kadane, J. B. "The presentation of Bayesian statistical analyses in legal proceedings". *Journal of the Royal Statistical Society, Series D* **32** (1/2): 88–98. doi:10.2307/2987595.
[17] Garaud, M. C. (1990). "Legal Standards and Statistical Proof in Title VII Litigation: In Search of a Coherent Disparate Impact Model". *University of Pennsylvania Law Review* **139** (2): 455–503.
[18] The Harvard Law Review Association (1995). "Developments in the Law: Confronting the New Challenges of Scientific Evidence". *Harvard Law Review* **108** (7): 1481–1605. doi:10.2307/1341808.
[19] Faigman, D. L. (1991). "Normative Constitutional Fact-Finding": Exploring the Empirical Component of Constitutional Interpretation". *University of Pennsylvania Law Review* **139** (3): 541–613.

60.7 Notes

[1] Supreme Court of the United States *Castaneda v. Partida*, 1977 cited in Meier (1986) Ibid. who states "Thus, in the space of less than half a year, the Supreme Court had moved from the traditional legal disdain for statistical proof to a strong endorsement of it as being capable, on its own, of establishing a prima facie case against a defendant."[4]
[2] 481 U.S. 279 (1987).[19]

Chapter 61

Pure mathematics

An illustration of the Banach–Tarski paradox, a famous result in pure mathematics. Although it is proven that it is possible to convert

one sphere into two using nothing but cuts and rotations, the transformation involves objects that cannot exist in the physical world.
Broadly speaking, **pure mathematics** is mathematics that studies entirely abstract concepts. From the eighteenth century onwards, this was a recognized category of mathematical activity, sometimes characterized as *speculative mathematics*,[1] and at variance with the trend towards meeting the needs of navigation, astronomy, physics, economics, engineering, and so on.
Another insightful view put forth is that *pure mathematics is not necessarily applied mathematics*: it is possible to study abstract entities with respect to their intrinsic nature, and not be concerned with how they manifest in the real world.[2] Even though the pure and applied viewpoints are distinct philosophical positions, in practice there is much overlap in the activity of pure and applied mathematicians.
To develop accurate models for describing the real world, many applied mathematicians draw on tools and techniques that are often considered to be "pure" mathematics. On the other hand, many pure mathematicians draw on natural and social phenomena as inspiration for their abstract research.

61.1 History

61.1.1 Ancient Greece

Ancient Greek mathematicians were among the earliest to make a distinction between pure and applied mathematics. Plato helped to create the gap between "arithmetic", now called number theory, and "logistic", now called arithmetic. Plato regarded logistic (arithmetic) as appropriate for businessmen and men of war who "must learn the art of numbers or [they] will not know how to array [their] troops" and arithmetic (number theory) as appropriate for philosophers "because [they have] to arise out of the sea of change and lay hold of true being."[3] Euclid of Alexandria, when asked by one of his students of what use was the study of geometry, asked his slave to give the student threepence, "since he must needs make gain of what he learns."[4] The Greek mathematician Apollonius of Perga was asked about the usefulness of some of his theorems in Book IV of *Conics* to which he proudly asserted,[5]
They are worthy of acceptance for the sake of the demonstrations themselves, in the same way as we accept many other things in mathematics for this and for no other reason.
383
384 *CHAPTER 61. PURE MATHEMATICS*
And since many of his results were not applicable to the science or engineering of his day, Apollonius further argued in the preface of the fifth book of *Conics* that the subject is one of those that "...seem worthy of study for their own sake."[5]

61.1.2 19th century

The term itself is enshrined in the full title of the Sadleirian Chair, founded (as a professorship) in the mid-nineteenth century. The idea of a separate discipline of *pure* mathematics may have emerged at that time. The generation of Gauss made no sweeping distinction of the kind, between *pure* and *applied*. In the following years, specialisation and professionalisation (particularly in the Weierstrass approach to mathematical analysis) started to make a rift more apparent.

61.1.3 20th century

At the start of the twentieth century mathematicians took up the axiomatic method, strongly influenced by David Hilbert's example. The logical formulation of **pure mathematics** suggested by Bertrand Russell in terms of a quantifier structure of propositions seemed more and more plausible, as large parts of mathematics became axiomatised and thus subject to the simple criteria of *rigorous proof*.
In fact in an axiomatic setting *rigorous* adds nothing to the idea of *proof*. Pure mathematics, according to a view that can be ascribed to the Bourbaki group, is what is proved. **Pure mathematician** became a recognized vocation, achievable through training.

61.2 Generality and abstraction

One central concept in pure mathematics is the idea of generality; pure mathematics often exhibits a trend towards increased generality.
_ Generalizing theorems or mathematical structures can lead to deeper understanding of the original theorems or structures
_ Generality can simplify the presentation of material, resulting in shorter proofs or arguments that are easier to follow.
_ One can use generality to avoid duplication of effort, proving a general result instead of having to prove separate cases independently, or using results from other areas of mathematics.
_ Generality can facilitate connections between different branches of mathematics. Category theory is one area of mathematics dedicated to exploring this commonality of structure as it plays out in some areas of math.
Generality's impact on intuition is both dependent on the subject and a matter of personal preference or learning style. Often generality is seen as a hindrance to intuition, although it can certainly function as an aid to it, especially when it provides analogies to material for which one already has good intuition.
As a prime example of generality, the Erlangen program involved an expansion of geometry to accommodate non-Euclidean geometries as well as the field of topology, and other forms of geometry, by viewing geometry as the

study of a space together with a group of transformations. The study of numbers, called algebra at the beginning undergraduate level, extends to abstract algebra at a more advanced level; and the study of functions, called calculus at the college freshman level becomes mathematical analysis and functional analysis at a more advanced level. Each of these branches of more *abstract* mathematics have many sub-specialties, and there are in fact many connections between pure mathematics and applied mathematics disciplines. A steep rise in abstraction was seen mid 20th century. In practice, however, these developments led to a sharp divergence from physics, particularly from 1950 to 1980. Later this was criticised, for example by Vladimir Arnold, as too much Hilbert, not enough Poincaré. The point does not yet seem to be settled, in that string theory pulls one way, while discrete mathematics pulls back towards proof as central.

61.3 Purism

Mathematicians have always had differing opinions regarding the distinction between pure and applied mathematics. One of the most famous (but perhaps misunderstood) modern examples of this debate can be found in G.H. Hardy's *A Mathematician's Apology*.

It is widely believed that Hardy considered applied mathematics to be ugly and dull. Although it is true that Hardy preferred pure mathematics, which he often compared to painting and poetry, Hardy saw the distinction between pure and applied mathematics to be simply that applied mathematics sought to express *physical* truth in a mathematical framework, whereas pure mathematics expressed truths that were independent of the physical world. Hardy made a separate distinction in mathematics between what he called "real" mathematics, "which has permanent aesthetic value", and "the dull and elementary parts of mathematics" that have practical use.

Hardy considered some physicists, such as Einstein and Dirac, to be among the "real" mathematicians, but at the time that he was writing the *Apology* he also considered general relativity and quantum mechanics to be "useless", which allowed him to hold the opinion that only "dull" mathematics was useful. Moreover, Hardy briefly admitted that—just as the application of matrix theory and group theory to physics had come unexpectedly—the time may come where some kinds of beautiful, "real" mathematics may be useful as well.

Another insightful view is offered by Magid:

I've always thought that a good model here could be drawn from ring theory. In that subject, one has the subareas of commutative ring theory and noncommutative ring theory. An uninformed observer might think that these represent a dichotomy, but in fact the latter subsumes the former: a noncommutative ring is a not necessarily commutative ring. If we use similar conventions, then we could refer to applied mathematics and nonapplied mathematics, where by the latter we *mean not necessarily applied mathematics*... [emphasis added][2]

61.4 Subfields

Analysis is concerned with the properties of functions. It deals with concepts such as continuity, limits, differentiation and integration, thus providing a rigorous foundation for the calculus of infinitesimals introduced by Newton and Leibniz in the 17th century. Real analysis studies functions of real numbers, while complex analysis extends the aforementioned concepts to functions of complex numbers. Functional analysis is a branch of analysis that studies infinite-dimensional vector spaces and views functions as points in these spaces.

Abstract algebra is not to be confused with the manipulation of formulae that is covered in secondary education. It studies sets together with binary operations defined on them. Sets and their binary operations may be classified according to their properties: for instance, if an operation is associative on a set that contains an identity element and inverses for each member of the set, the set and operation is considered to be a group. Other structures include rings, fields, vector spaces and lattices.

Geometry is the study of shapes and space, in particular, groups of transformations that act on spaces. For example, projective geometry is about the group of projective transformations that act on the real projective plane, whereas inversive geometry is concerned with the group of inversive transformations acting on the extended complex plane. Geometry has been extended to topology, which deals with objects known as topological spaces and continuous maps between them. Topology is concerned with the way in which a space is connected and ignores precise measurements of distance or angle.

Number theory is the theory of the positive integers. It is based on ideas such as divisibility and congruence. Its fundamental theorem states that each positive integer has a unique prime factorization. In some ways it is the most accessible discipline in pure mathematics for the general public: for instance the Goldbach conjecture is easily stated (but is yet to be proved or disproved). In other ways it is the least accessible discipline; for example, Wiles' proof that Fermat's equation has no nontrivial solutions requires understanding automorphic forms, which though intrinsic to nature have not found a place in physics or the general public discourse.

61.5 Notes

[1] See for example titles of works by Thomas Simpson from the mid-18th century: *Essays on Several Curious and Useful Subjects in Speculative and Mixed Mathematicks, Miscellaneous Tracts on Some Curious and Very Interesting Subjects in*

Mechanics, Physical Astronomy and Speculative Mathematics.
[2] Andy Magid, Letter from the Editor, in *Notices of the AMS*, November 2005, American Mathematical Society, p.1173.
[3] Boyer, Carl B. (1991). "The age of Plato and Aristotle". *A History of Mathematics* (Second Edition ed.). John Wiley & Sons, Inc. p. 86. ISBN 0-471-54397-7. "Plato is important in the history of mathematics largely for his role as inspirer and director of others, and perhaps to him is due the sharp distinction in ancient Greece between arithmetic (in the sense of the theory of numbers) and logistic (the technique of computation). Plato regarded logistic as appropriate for the businessman and for the man of war, who "must learn the art of numbers or he will not know how to array his troops." The philosopher, on the other hand, must be an arithmetician "because he has to arise out of the sea of change and lay hold of true being.""
[4] Boyer, Carl B. (1991). "Euclid of Alexandria". *A History of Mathematics* (Second Edition ed.). John Wiley & Sons, Inc. p. 101. ISBN 0-471-54397-7. "Evidently Euclid did not stress the practical aspects of his subject, for there is a tale told of him that when one of his students asked of what use was the study of geometry, Euclid asked his slave to give the student threepence, "since he must make gain of what he learns.""
[5] Boyer, Carl B. (1991). "Apollonius of Perga". *A History of Mathematics* (Second Edition ed.). John Wiley & Sons, Inc. p. 152. ISBN 0-471-54397-7. "It is in connection with the theorems in this book that Apollonius makes a statement implying that in his day, as in ours, there were narrow-minded opponents of pure mathematics who pejoratively inquired about the usefulness of such results. The author proudly asserted: "They are worthy of acceptance for the sake of the demonstrations themselves, in the same way as we accept many other things in mathematics for this and for no other reason." (Heath 1961, p.lxxiv).
The preface to Book V, relating to maximum and minimum straight lines drawn to a conic, again argues that the subject is one of those that seem "worthy of study for their own sake." While one must admire the author for his lofty intellectual attitude, it may be pertinently pointed out that s day was beautiful theory, with no prospect of applicability to the science or engineering of his time, has since become fundamental in such fields as terrestrial dynamics and celestial mechanics."

61.6 See also

_ Applied mathematics
_ Logic
_ Metalogic
_ Metamathematics

61.7 External links

_ *What is Pure Mathematics?* Department of Pure Mathematics, University of Waterloo
_ *What is Pure Mathematics?* by Professor P.J. Giblin The University of Liverpool
_ *The Principles of Mathematics* by Bertrand Russell
_ How to Become a Pure Mathematician (or Statistician), a list of undergraduate and basic graduate textbooks and lecture notes, with several comments and links to solutions, companion sites, data sets, errata pages, etc.
_ Pure Mathematics Learning Resources

Chapter 62

Cryptography

"Secret code" redirects here. For the Aya Kamiki album, see Secret Code.
"Cryptology" redirects here. For the David S. Ware album, see Cryptology (album).
Cryptography (or **cryptology**; from Greek κρυπτός *kryptós*, "hidden, secret"; and γράφειν *graphein*, "writing",
German Lorenz cipher machine, used in World War II to encrypt very-high-level general staff messages
or -λογία *-logia*, "study", respectively)[1] is the practice and study of techniques for secure communication in the presence of third parties (called adversaries).[2] More generally, it is about constructing and analyzing protocols that overcome the influence of adversaries[3] and that are related to various aspects in information security such as data confidentiality, data integrity, authentication, and non-repudiation.[4] Modern cryptography intersects the disciplines of mathematics, computer science, and electrical engineering. Applications of cryptography include ATM cards, computer passwords, and electronic commerce.
Cryptography prior to the modern age was effectively synonymous with *encryption*, the conversion of information from a readable state to apparent nonsense. The originator of an encrypted message shared the decoding technique
387
388 *CHAPTER 62. CRYPTOGRAPHY*
needed to recover the original information only with intended recipients, thereby precluding unwanted persons to do the same. Since World War I and the advent of the computer, the methods used to carry out cryptology have become increasingly complex and its application more widespread.
Modern cryptography is heavily based on mathematical theory and computer science practice; cryptographic algorithms are designed around computational hardness assumptions, making such algorithms hard to break in practice by any adversary. It is theoretically possible to break such a system, but it is infeasible to do so by any known practical means. These schemes are therefore termed computationally secure; theoretical advances, e.g., improvements in integer factorization algorithms, and faster computing technology require these solutions to be continually adapted. There exist information-theoretically secure schemes that provably cannot be broken even with unlimited

computing power—an example is the one-time pad—but these schemes are more difficult to implement than the best theoretically breakable but computationally secure mechanisms.
Cryptology-related technology has raised a number of legal issues. In the United Kingdom, additions to the Regulation of Investigatory Powers Act 2000 require a suspected criminal to hand over his or her decryption key if asked by law enforcement. Otherwise the user will face a criminal charge.[5] The Electronic Frontier Foundation (EFF) was involved in a case in the United States which questioned whether requiring suspected criminals to provide their decryption keys to law enforcement is unconstitutional. The EFF argued that this is a violation of the right of not being forced to incriminate oneself, as given in the fifth amendment.[6]

62.1 Terminology

Until modern times cryptography referred almost exclusively to *encryption*, which is the process of converting ordinary information (called plaintext) into unintelligible text (called ciphertext).[7] Decryption is the reverse, in other words, moving from the unintelligible ciphertext back to plaintext. A *cipher* (or *cypher*) is a pair of algorithms that create the encryption and the reversing decryption. The detailed operation of a cipher is controlled both by the algorithm and in each instance by a "key". This is a secret (ideally known only to the communicants), usually a short string of characters, which is needed to decrypt the ciphertext. A "cryptosystem" is the ordered list of elements of finite possible plaintexts, finite possible cyphertexts, finite possible keys, and the encryption and decryption algorithms which correspond to each key. Keys are important, as ciphers without variable keys can be trivially broken with only the knowledge of the cipher used and are therefore useless (or even counter-productive) for most purposes. Historically, ciphers were often used directly for encryption or decryption without additional procedures such as authentication or integrity checks.
In colloquial use, the term "code" is often used to mean any method of encryption or concealment of meaning. However, in cryptography, *code* has a more specific meaning. It means the replacement of a unit of plaintext (i.e., a meaningful word or phrase) with a code word (for example, wallaby replaces attack at dawn). Codes are no longer used in serious cryptography—except incidentally for such things as unit designations (e.g., Bronco Flight or Operation Overlord)—since properly chosen ciphers are both more practical and more secure than even the best codes and also are better adapted to computers.
Cryptanalysis is the term used for the study of methods for obtaining the meaning of encrypted information without access to the key normally required to do so; i.e., it is the study of how to crack encryption algorithms or their implementations.
Some use the terms *cryptography* and *cryptology* interchangeably in English, while others (including US military practice generally) use *cryptography* to refer specifically to the use and practice of cryptographic techniques and *cryptology* to refer to the combined study of cryptography and cryptanalysis.[8][9] English is more flexible than several other languages in which *cryptology* (done by cryptologists) is always used in the second sense above. In the English Wikipedia the general term used for the entire field is *cryptography* (done by cryptographers). RFC 2828[10] advises that steganography is sometimes included in cryptology.
The study of characteristics of languages that have some application in cryptography[11] (or cryptology) (i.e., frequency data, letter combinations, universal patterns, etc.) is called cryptolinguistics.

62.2 History of cryptography and cryptanalysis

Main article: History of cryptography
62.2. HISTORY OF CRYPTOGRAPHY AND CRYPTANALYSIS 389
Before the modern era, cryptography was concerned solely with message confidentiality (i.e., encryption)—conversion of messages from a comprehensible form into an incomprehensible one and back again at the other end, rendering it unreadable by interceptors or eavesdroppers without secret knowledge (namely the key needed for decryption of that message). Encryption was used to (attempt to) ensure secrecy in communications, such as those of spies, military leaders, and diplomats. In recent decades, the field has expanded beyond confidentiality concerns to include techniques for message integrity checking, sender/receiver identity authentication, digital signatures, interactive proofs and secure computation, among others.

62.2.1 Classic cryptography

Reconstructed ancient Greek scytale*, an early cipher device*
The earliest forms of secret writing required little more than writing implements since most people could not read. More literacy, or literate opponents, required actual cryptography. The main classical cipher types are transposition ciphers, which rearrange the order of letters in a message (e.g., 'hello world' becomes 'ehlol owrdl' in a trivially simple rearrangement scheme), and substitution ciphers, which systematically replace letters or groups of letters with other letters or groups of letters (e.g., 'fly at once' becomes 'gmz bu podf' by replacing each letter with the one following it in the Latin alphabet). Simple versions of either have never offered much confidentiality from enterprising opponents. An early substitution cipher was the Caesar cipher, in which each letter in the plaintext was replaced by a letter some fixed number of positions further down the alphabet. Suetonius reports that Julius Caesar used it with a shift of three to communicate with his generals. Atbash is an example of an early Hebrew cipher. The earliest known use of cryptography is some carved ciphertext on stone in Egypt (ca 1900 BCE), but this may have been done for the

amusement of literate observers rather than as a way of concealing information.
The Greeks of Classical times are said to have known of ciphers (e.g., the scytale transposition cipher claimed to have been used by the Spartan military).[12] Steganography (i.e., hiding even the existence of a message so as to keep it confidential) was also first developed in ancient times. An early example, from Herodotus, concealed a message—a tattoo on a slave's shaved head—under the regrown hair.[7] More modern examples of steganography include the use of invisible ink, microdots, and digital watermarks to conceal information.

Ciphertexts produced by a classical cipher (and some modern ciphers) always reveal statistical information about the plaintext, which can often be used to break them. After the discovery of frequency analysis perhaps by the Arab mathematician and polymath Al-Kindi (also known as *Alkindus*) in the 9th century,[13][14] nearly all such ciphers became more or less readily breakable by any informed attacker. Such classical ciphers still enjoy popularity today, though mostly as puzzles (see cryptogram). Al-Kindi wrote a book on cryptography entitled *Risalah fi Istikhraj al-Mu'amma* (*Manuscript for the Deciphering Cryptographic Messages*), which described the first cryptanalysis techniques.[14][15]
16th-century book-shaped French cipher machine, with arms of Henri II of France
Essentially all ciphers remained vulnerable to cryptanalysis using the frequency analysis technique until the development of the polyalphabetic cipher, most clearly by Leon Battista Alberti around the year 1467, though there is some indication that it was already known to Al-Kindi.[15] Alberti's innovation was to use different ciphers (i.e., substitution alphabets) for various parts of a message (perhaps for each successive plaintext letter at the limit). He also invented what was probably the first automatic cipher device, a wheel which implemented a partial realization of his invention.
In the polyalphabetic Vigenère cipher, encryption uses a *key word*, which controls letter substitution depending on which letter of the key word is used. In the mid-19th century Charles Babbage showed that the Vigenère cipher was vulnerable to Kasiski examination, but this was first published about ten years later by Friedrich Kasiski.[16]
Although frequency analysis can be a powerful and general technique against many ciphers, encryption has still often been effective in practice, as many a would-be cryptanalyst was unaware of the technique. Breaking a message without using frequency analysis essentially required knowledge of the cipher used and perhaps of the key involved, thus making espionage, bribery, burglary, defection, etc., more attractive approaches to the cryptanalytically uninformed.
It was finally explicitly recognized in the 19th century that secrecy of a cipher's algorithm is not a sensible nor practical safeguard of message security; in fact, it was further realized that any adequate cryptographic scheme (including ciphers) should remain secure even if the adversary fully understands the cipher algorithm itself. Security of the key used should alone be sufficient for a good cipher to maintain confidentiality under an attack. This fundamental principle was first explicitly stated in 1883 by Auguste Kerckhoffs and is generally called Kerckhoffs's Principle; alternatively and more bluntly, it was restated by Claude Shannon, the inventor of information theory and the fundamentals of theoretical cryptography, as *Shannon's Maxim*—'the enemy knows the system'.
Different physical devices and aids have been used to assist with ciphers. One of the earliest may have been the scytale of ancient Greece, a rod supposedly used by the Spartans as an aid for a transposition cipher (see image above).
In medieval times, other aids were invented such as the cipher grille, which was also used for a kind of steganography.
With the invention of polyalphabetic ciphers came more sophisticated aids such as Alberti's own cipher disk, Johannes Trithemius' tabula recta scheme, and Thomas Jefferson's multi cylinder (not publicly known, and reinvented

Enciphered letter from Gabriel de Luetz d'Aramon, French Ambassador to the Ottoman Empire, after 1546, with partial decipherment
independently by Bazeries around 1900). Many mechanical encryption/decryption devices were invented early in the 20th century, and several patented, among them rotor machines—famously including the Enigma machine used by the German government and military from the late 1920s and during World War II.[17] The ciphers implemented by better quality examples of these machine designs brought about a substantial increase in cryptanalytic difficulty after WWI.[18]

62.2.2 Computer era

Cryptanalysis of the new mechanical devices proved to be both difficult and laborious. In the United Kingdom, cryptanalytic efforts at Bletchley Park during WWII spurred the development of more efficient means for carrying out repetitious tasks. This culminated in the development of the Colossus, the world's first fully electronic, digital, programmable computer, which assisted in the decryption of ciphers generated by the German Army's Lorenz SZ40/42 machine.
Just as the development of digital computers and electronics helped in cryptanalysis, it made possible much more complex ciphers. Furthermore, computers allowed for the encryption of any kind of data representable in any binary format, unlike classical ciphers which only encrypted written language texts; this was new and significant. Computer use has thus supplanted linguistic cryptography, both for cipher design and cryptanalysis. Many computer ciphers can be characterized by their operation on binary bit sequences (sometimes in groups or blocks), unlike classical and mechanical schemes, which generally manipulate traditional characters (i.e., letters and digits) directly. However, computers have also assisted cryptanalysis, which has compensated to some extent for increased cipher complexity.
Nonetheless, good modern ciphers have stayed ahead of cryptanalysis; it is typically the case that use of a quality cipher is very efficient (i.e., fast and requiring few resources, such as memory or CPU capability), while breaking

it requires an effort many orders of magnitude larger, and vastly larger than that required for any classical cipher, making cryptanalysis so inefficient and impractical as to be effectively impossible.

Extensive open academic research into cryptography is relatively recent; it began only in the mid-1970s. In recent times, IBM personnel designed the algorithm that became the Federal (i.e., US) Data Encryption Standard; Whitfield Diffie and Martin Hellman published their key agreement algorithm;[19] and the RSA algorithm was published in Martin Gardner's *Scientific American* column. Since then, cryptography has become a widely used tool in communications, computer networks, and computer security generally. Some modern cryptographic techniques can only keep

their keys secret if certain mathematical problems are intractable, such as the integer factorization or the discrete logarithm problems, so there are deep connections with abstract mathematics. There are no absolute proofs that a cryptographic technique is secure (but see one-time pad); at best, there are proofs that some techniques are secure *if* some computational problem is difficult to solve, or certain assumptions about implementation or practical use are met.

As well as being aware of cryptographic history, cryptographic algorithm and system designers must also sensibly consider probable future developments while working on their designs. For instance, continuous improvements in computer processing power have increased the scope of brute-force attacks, so when specifying key lengths, the required key lengths are similarly advancing.[20] The potential effects of quantum computing are already being considered by some cryptographic system designers; the announced imminence of small implementations of these machines may be making the need for this preemptive caution rather more than merely speculative.[4]

Essentially, prior to the early 20th century, cryptography was chiefly concerned with linguistic and lexicographic patterns. Since then the emphasis has shifted, and cryptography now makes extensive use of mathematics, including aspects of information theory, computational complexity, statistics, combinatorics, abstract algebra, number theory, and finite mathematics generally. Cryptography is also a branch of engineering, but an unusual one since it deals with active, intelligent, and malevolent opposition (see cryptographic engineering and security engineering); other kinds of engineering (e.g., civil or chemical engineering) need deal only with neutral natural forces. There is also active research examining the relationship between cryptographic problems and quantum physics (see quantum cryptography and quantum computer).

62.3 Modern cryptography

The modern field of cryptography can be divided into several areas of study. The chief ones are discussed here; see Topics in Cryptography for more.

62.3.1 Symmetric-key cryptography

Main article: Symmetric-key algorithm

Symmetric-key cryptography refers to encryption methods in which both the sender and receiver share the same key (or, less commonly, in which their keys are different, but related in an easily computable way). This was the only kind of encryption publicly known until June 1976.[19]

Symmetric key ciphers are implemented as either block ciphers or stream ciphers. A block cipher enciphers input in blocks of plaintext as opposed to individual characters, the input form used by a stream cipher.

The Data Encryption Standard (DES) and the Advanced Encryption Standard (AES) are block cipher designs which have been designated cryptography standards by the US government (though DES's designation was finally withdrawn after the AES was adopted).[21] Despite its deprecation as an official standard, DES (especially its still-approved and much more secure triple-DES variant) remains quite popular; it is used across a wide range of applications, from ATM encryption[22] to e-mail privacy[23] and secure remote access.[24] Many other block ciphers have been designed and released, with considerable variation in quality. Many have been thoroughly broken, such as FEAL.[4][25]

Stream ciphers, in contrast to the 'block' type, create an arbitrarily long stream of key material, which is combined with the plaintext bit-by-bit or character-by-character, somewhat like the one-time pad. In a stream cipher, the output stream is created based on a hidden internal state which changes as the cipher operates. That internal state is initially set up using the secret key material. RC4 is a widely used stream cipher; see Category:Stream ciphers.[4] Block ciphers can be used as stream ciphers; see Block cipher modes of operation.

Cryptographic hash functions are a third type of cryptographic algorithm. They take a message of any length as input, and output a short, fixed length hash which can be used in (for example) a digital signature. For good hash functions, an attacker cannot find two messages that produce the same hash. MD4 is a long-used hash function which is now broken; MD5, a strengthened variant of MD4, is also widely used but broken in practice. The US National Security Agency developed the Secure Hash Algorithm series of MD5-like hash functions: SHA-0 was a flawed algorithm that the agency withdrew; SHA-1 is widely deployed and more secure than MD5, but cryptanalysts have identified attacks against it; the SHA-2 family improves on SHA-1, but it isn't yet widely deployed; and the US standards authority thought it "prudent" from a security perspective to develop a new standard to "significantly improve the robustness of NIST's overall hash algorithm toolkit."[26] Thus, a hash function design competition was meant to select a new

Hello
Alice! Encrypt
6EB69570
08E03CE4
Hello
Alice! Decrypt
Secret key
Bob
Alice

Symmetric-key cryptography, where a single key is used for encryption and decryption

U.S. national standard, to be called SHA-3, by 2012. The competition ended on October 2, 2012 when the NIST announced that Keccak would be the new SHA-3 hash algorithm.[27]

Message authentication codes (MACs) are much like cryptographic hash functions, except that a secret key can be used to authenticate the hash value[4] upon receipt.

62.3.2 Public-key cryptography

Main article: Public-key cryptography

Symmetric-key cryptosystems use the same key for encryption and decryption of a message, though a message or group of messages may have a different key than others. A significant disadvantage of symmetric ciphers is the key management necessary to use them securely. Each distinct pair of communicating parties must, ideally, share a different key, and perhaps each ciphertext exchanged as well. The number of keys required increases as the square of the number of network members, which very quickly requires complex key management schemes to keep them all consistent and secret. The difficulty of securely establishing a secret key between two communicating parties, when a secure channel does not already exist between them, also presents a chicken-and-egg problem which is a considerable practical obstacle for cryptography users in the real world.

In a groundbreaking 1976 paper, Whitfield Diffie and Martin Hellman proposed the notion of *public-key* (also, more generally, called *asymmetric key*) cryptography in which two different but mathematically related keys are used—a

K1 K2 K3 K4
K5
K6

One round (out of 8.5) of the patented IDEA cipher, used in some versions of PGP for high-speed encryption of, for instance, e-mail

public key and a *private* key.[28] A public key system is so constructed that calculation of one key (the 'private key') is computationally infeasible from the other (the 'public key'), even though they are necessarily related. Instead, both keys are generated secretly, as an interrelated pair.[29] The historian David Kahn described public-key cryptography as "the most revolutionary new concept in the field since polyalphabetic substitution emerged in the Renaissance".[30]

In public-key cryptosystems, the public key may be freely distributed, while its paired private key must remain secret. In a public-key encryption system, the *public key* is used for encryption, while the *private* or *secret key* is used for decryption. While Diffie and Hellman could not find such a system, they showed that public-key cryptography was indeed possible by presenting the Diffie–Hellman key exchange protocol, a solution that is now widely used in secure communications to allow two parties to secretly agree on a shared encryption key.[19]

Diffie and Hellman's publication sparked widespread academic efforts in finding a practical public-key encryption system. This race was finally won in 1978 by Ronald Rivest, Adi Shamir, and Len Adleman, whose solution has since become known as the RSA algorithm.[31]

The Diffie–Hellman and RSA algorithms, in addition to being the first publicly known examples of high quality publickey algorithms, have been among the most widely used. Others include the Cramer–Shoup cryptosystem, ElGamal encryption, and various elliptic curve techniques. See Category:Asymmetric-key cryptosystems.

To much surprise, a document published in 1997 by the Government Communications Headquarters (GCHQ), a

Hello
Alice!
Alice's
private key
Encrypt
6EB69570
08E03CE4
Hello
Alice! Decrypt
Alice's
public key
Bob
Alice

Public-key cryptography, where different keys are used for encryption and decryption

British intelligence organization, revealed that cryptographers at GCHQ had anticipated several academic developments.[32] Reportedly, around 1970, James H. Ellis had conceived the principles of asymmetric key cryptography. In 1973, Clifford Cocks invented a solution that essentially resembles the RSA algorithm.[32][33] And in 1974, Malcolm J. Williamson is claimed to have developed the Diffie-Hellman key exchange.[34]

Public-key cryptography can also be used for implementing digital signature schemes. A digital signature is reminiscent of an ordinary signature; they both have the characteristic of being easy for a user to produce, but difficult for anyone else to forge. Digital signatures can also be permanently tied to the content of the message being signed; they cannot then be 'moved' from one document to another, for any attempt will be detectable. In digital signature schemes, there are two algorithms: one for *signing*, in which a secret key is used to process the message (or a hash of the message, or both), and one for *verification,* in which the matching public key is used with the message to check the validity of the signature. RSA and DSA are two of the most popular digital signature schemes. Digital signatures are central to the operation of public key infrastructures and many network security schemes (e.g., SSL/TLS, many VPNs, etc.).[25]

Public-key algorithms are most often based on the computational complexity of "hard" problems, often from number theory. For example, the hardness of RSA is related to the integer factorization problem, while Diffie–Hellman and DSA are related to the discrete logarithm problem. More recently, *elliptic curve cryptography* has developed, a system in which security is based on number theoretic problems involving elliptic curves. Because of the difficulty of the underlying problems, most public-key algorithms involve operations such as modular multiplication and exponentiation, which are much more computationally expensive than the techniques used in most block ciphers, especially

Whitfield Diffie and Martin Hellman, authors of the first published paper on public-key cryptography

Padlock icon from the Firefox Web browser, which indicates that TLS, a public-key cryptography system, is in use.

with typical key sizes. As a result, public-key cryptosystems are commonly hybrid cryptosystems, in which a fast high-quality symmetric-key encryption algorithm is used for the message itself, while the relevant symmetric key is sent with the message, but encrypted using a public-key algorithm. Similarly, hybrid signature schemes are often used, in which a cryptographic hash function is computed, and only the resulting hash is digitally signed.[4]

62.3.3 Cryptanalysis

Main article: Cryptanalysis

The goal of cryptanalysis is to find some weakness or insecurity in a cryptographic scheme, thus permitting its subversion or evasion.

It is a common misconception that every encryption method can be broken. In connection with his WWII work at Bell Labs, Claude Shannon proved that the one-time pad cipher is unbreakable, provided the key material is truly random, never reused, kept secret from all possible attackers, and of equal or greater length than the message.[35]
Most ciphers, apart from the one-time pad, can be broken with enough computational effort by brute force attack, but the amount of effort needed may be exponentially dependent on the key size, as compared to the effort needed to make use of the cipher. In such cases, effective security could be achieved if it is proven that the effort required (i.e., "work factor", in Shannon's terms) is beyond the ability of any adversary. This means it must be shown that no efficient method (as opposed to the time-consuming brute force method) can be found to break the cipher. Since no such proof has been found to date, the one-time-pad remains the only theoretically unbreakable cipher.
There are a wide variety of cryptanalytic attacks, and they can be classified in any of several ways. A common distinction turns on what an attacker knows and what capabilities are available. In a ciphertext-only attack, the cryptanalyst has access only to the ciphertext (good modern cryptosystems are usually effectively immune to ciphertext-only attacks). In a known-plaintext attack, the cryptanalyst has access to a ciphertext and its corresponding plaintext (or to many such pairs). In a chosen-plaintext attack, the cryptanalyst may choose a plaintext and learn its corresponding

Variants of the Enigma machine, used by Germany's military and civil authorities from the late 1920s through World War II, implemented a complex electro-mechanical polyalphabetic cipher. Breaking and reading of the Enigma cipher at Poland's Cipher Bureau, for 7 years before the war, and subsequent decryption at Bletchley Park, was important to Allied victory.[7]

ciphertext (perhaps many times); an example is gardening, used by the British during WWII. Finally, in a chosenciphertext attack, the cryptanalyst may be able to *choose* ciphertexts and learn their corresponding plaintexts.[4] Also important, often overwhelmingly so, are mistakes (generally in the design or use of one of the protocols involved; see Cryptanalysis of the Enigma for some historical examples of this).

*Poznań monument (*center*) to Polish cryptologists whose breaking of Germany's Enigma machine ciphers, beginning in 1932, altered the course of World War II*

Cryptanalysis of symmetric-key ciphers typically involves looking for attacks against the block ciphers or stream ciphers that are more efficient than any attack that could be against a perfect cipher. For example, a simple brute force attack against DES requires one known plaintext and 2^{55} decryptions, trying approximately half of the possible keys, to reach a point at which chances are better than even that the key sought will have been found. But this may not be enough assurance; a linear cryptanalysis attack against DES requires 2^{43} known plaintexts and approximately 2^{43} DES operations.[36] This is a considerable improvement on brute force attacks.
Public-key algorithms are based on the computational difficulty of various problems. The most famous of these is integer factorization (e.g., the RSA algorithm is based on a problem related to integer factoring), but the discrete logarithm problem is also important. Much public-key cryptanalysis concerns numerical algorithms for solving these computational problems, or some of them, efficiently (i.e., in a practical time). For instance, the best known algorithms for solving the elliptic curve-based version of discrete logarithm are much more time-consuming than the best known algorithms for factoring, at least for problems of more or less equivalent size. Thus, other things being equal, to achieve an equivalent strength of attack resistance, factoring-based encryption techniques must use larger keys than elliptic curve techniques. For this reason, public-key cryptosystems based on elliptic curves have become popular since their invention in the mid-1990s.
While pure cryptanalysis uses weaknesses in the algorithms themselves, other attacks on cryptosystems are based on actual use of the algorithms in real devices, and are called *side-channel attacks*. If a cryptanalyst has access to, for example, the amount of time the device took to encrypt a number of plaintexts or report an error in a password or PIN character, he may be able to use a timing attack to break a cipher that is otherwise resistant to analysis. An attacker might also study the pattern and length of messages to derive valuable information; this is known as traffic analysis[37] and can be quite useful to an alert adversary. Poor administration of a cryptosystem, such as permitting too short keys, will make any system vulnerable, regardless of other virtues. And, of course, social engineering, and other attacks against the personnel who work with cryptosystems or the messages they handle (e.g., bribery, extortion, blackmail, espionage, torture, ...) may be the most productive attacks of all.

62.3.4 Cryptographic primitives

Much of the theoretical work in cryptography concerns cryptographic *primitives*—algorithms with basic cryptographic properties—and their relationship to other cryptographic problems. More complicated cryptographic tools are then built from these basic primitives. These primitives provide fundamental properties, which are used to develop more complex tools called *cryptosystems* or *cryptographic protocols*, which guarantee one or more high-level security properties. Note however, that the distinction between cryptographic *primitives* and cryptosystems, is quite arbitrary; for example, the RSA algorithm is sometimes considered a cryptosystem, and sometimes a primitive. Typical examples of cryptographic primitives include pseudorandom functions, one-way functions, etc.

62.3.5 Cryptosystems

One or more cryptographic primitives are often used to develop a more complex algorithm, called a cryptographic system, or *cryptosystem*. Cryptosystems (e.g., El-Gamal encryption) are designed to provide particular functionality (e.g., public key encryption) while guaranteeing certain security properties (e.g., chosen-plaintext attack (CPA) security in the random oracle model). Cryptosystems use the properties of the underlying cryptographic primitives to support the system's security properties. Of course, as the distinction between primitives and cryptosystems is somewhat arbitrary, a sophisticated cryptosystem can be derived from a combination of several more primitive cryptosystems. In many cases, the cryptosystem's structure involves back and forth communication among two or more parties in space (e.g., between the sender of a secure message and its receiver) or across time (e.g., cryptographically protected backup data). Such cryptosystems are sometimes called *cryptographic protocols*.

Some widely known cryptosystems include RSA encryption, Schnorr signature, El-Gamal encryption, PGP, etc. More complex cryptosystems include electronic cash[38] systems, signcryption systems, etc. Some more 'theoretical' cryptosystems include interactive proof systems,[39] (like zero-knowledge proofs),[40] systems for secret sharing,[41][42] etc.

Until recently, most security properties of most cryptosystems were demonstrated using empirical techniques or using ad hoc reasoning. Recently, there has been considerable effort to develop formal techniques for establishing the security of cryptosystems; this has been generally called *provable security*. The general idea of provable security is to give arguments about the computational difficulty needed to compromise some security aspect of the cryptosystem (i.e., to any adversary).

The study of how best to implement and integrate cryptography in software applications is itself a distinct field (see Cryptographic engineering and Security engineering).

62.4 Legal issues

See also: Cryptography laws in different nations

62.4.1 Prohibitions

Cryptography has long been of interest to intelligence gathering and law enforcement agencies. Secret communications may be criminal or even treasonous. Because of its facilitation of privacy, and the diminution of privacy attendant on its prohibition, cryptography is also of considerable interest to civil rights supporters. Accordingly, there has been a history of controversial legal issues surrounding cryptography, especially since the advent of inexpensive computers has made widespread access to high quality cryptography possible.

In some countries, even the domestic use of cryptography is, or has been, restricted. Until 1999, France significantly restricted the use of cryptography domestically, though it has since relaxed many of these rules. In China and Iran, a license is still required to use cryptography.[43] Many countries have tight restrictions on the use of cryptography. Among the more restrictive are laws in Belarus, Kazakhstan, Mongolia, Pakistan, Singapore, Tunisia, and Vietnam.[44]

In the United States, cryptography is legal for domestic use, but there has been much conflict over legal issues related to cryptography. One particularly important issue has been the export of cryptography and cryptographic software and hardware. Probably because of the importance of cryptanalysis in World War II and an expectation that cryptography would continue to be important for national security, many Western governments have, at some point,

strictly regulated export of cryptography. After World War II, it was illegal in the US to sell or distribute encryption technology overseas; in fact, encryption was designated as auxiliary military equipment and put on the United States Munitions List.[45] Until the development of the personal computer, asymmetric key algorithms (i.e., public key techniques), and the Internet, this was not especially problematic. However, as the Internet grew and computers became more widely available, high-quality encryption techniques became well known around the globe. As a result, export controls came to be seen to be an impediment to commerce and to research.

62.4.2 Export controls

Main article: Export of cryptography

In the 1990s, there were several challenges to US export regulation of cryptography. After the source code for Philip Zimmermann's Pretty Good Privacy (PGP) encryption program found its way onto the Internet in June 1991, a complaint by RSA Security (then called RSA Data Security, Inc.) resulted in a lengthy criminal investigation of Zimmermann by the US Customs Service and the FBI, though no charges were ever filed.[46][47] Daniel J. Bernstein, then a graduate student at UC Berkeley, brought a lawsuit against the US government challenging some aspects of the restrictions based on free speech grounds. The 1995 case Bernstein v. United States ultimately resulted in a 1999 decision that printed source code for cryptographic algorithms and systems was protected as free speech by the United States Constitution.[48]

In 1996, thirty-nine countries signed the Wassenaar Arrangement, an arms control treaty that deals with the export of arms and "dual-use" technologies such as cryptography. The treaty stipulated that the use of cryptography with short key-lengths (56-bit for symmetric encryption, 512-bit for RSA) would no longer be export-controlled.[49] Cryptography exports from the US became less strictly regulated as a consequence of a major relaxation in 2000;[44] there are no longer very many restrictions on key sizes in US-exported mass-market software. Since this relaxation in US export restrictions, and because most personal computers connected to the Internet include US-sourced web browsers such as Firefox or Internet Explorer, almost every Internet user worldwide has potential access to quality cryptography via their browsers (e.g., via Transport Layer Security). The Mozilla Thunderbird and Microsoft Outlook E-mail client programs similarly can transmit and received emails via TLS, and can send and receive email encrypted with S/MIME. Many Internet users don't realize that their basic application software contains such extensive cryptosystems. These browsers and email programs are so ubiquitous that even governments whose intent is to regulate civilian use of cryptography generally don't find it practical to do much to control distribution or use of cryptography of this quality, so even when such laws are in force, actual enforcement is often effectively impossible.

62.4.3 NSA involvement

See also: Clipper chip

Another contentious issue connected to cryptography in the United States is the influence of the National Security Agency on cipher development and policy. The NSA was involved with the design of DES during its development at IBM and its consideration by the National Bureau of Standards as a possible Federal Standard for cryptography.[50] DES was designed to be resistant to differential cryptanalysis,[51] a powerful and general cryptanalytic technique known to the NSA and IBM, that became publicly known only when it was rediscovered in the late 1980s.[52] According to Steven Levy, IBM discovered differential cryptanalysis,[53] but kept the technique secret at the NSA's request. The technique became publicly known only when Biham and Shamir re-discovered and announced it some years later. The entire affair illustrates the difficulty of determining what resources and knowledge an attacker might actually have.

Another instance of the NSA's involvement was the 1993 Clipper chip affair, an encryption microchip intended to be part of the Capstone cryptography-control initiative. Clipper was widely criticized by cryptographers for two reasons. The cipher algorithm (called Skipjack) was then classified (declassified in 1998, long after the Clipper initiative lapsed). The classified cipher caused concerns that the NSA had deliberately made the cipher weak in order to assist its intelligence efforts. The whole initiative was also criticized based on its violation of Kerckhoffs's Principle, as the scheme included a special escrow key held by the government for use by law enforcement, for example in wiretaps.[47]

62.4.4 Digital rights management

Main article: Digital rights management

Cryptography is central to digital rights management (DRM), a group of techniques for technologically controlling use of copyrighted material, being widely implemented and deployed at the behest of some copyright holders. In 1998, Bill Clinton, President of the United States from 1993 to 2001, signed the Digital Millennium Copyright Act (DMCA), which criminalized all production, dissemination, and use of certain cryptanalytic techniques and technology (now known or later discovered); specifically, those that could be used to circumvent DRM technological schemes.[54] This had a noticeable impact on the cryptography research community since an argument can be made that *any* cryptanalytic research violated, or might violate, the DMCA. Similar statutes have since been enacted in several countries and regions, including the implementation in the EU Copyright Directive. Similar restrictions are called for by treaties signed by World Intellectual Property Organization member-states.

The United States Department of Justice and FBI have not enforced the DMCA as rigorously as had been feared by some, but the law, nonetheless, remains a controversial one. Niels Ferguson, a well-respected cryptography researcher, has publicly stated that he will not release some of his research into an Intel security design for fear of prosecution under the DMCA.[55] Both Alan Cox (longtime number 2 in Linux kernel development) and Edward Felten (and some of his students at Princeton) have encountered problems related to the Act. Dmitry Sklyarov was arrested during a visit to the US from Russia, and jailed for five months pending trial for alleged violations of the DMCA arising from work he had done in Russia, where the work was legal. In 2007, the cryptographic keys responsible for Blu-ray and HD DVD content scrambling were discovered and released onto the Internet. In both cases, the MPAA sent out numerous DMCA takedown notices, and there was a massive Internet backlash[56] triggered by the perceived impact of such notices on fair use and free speech.

62.4.5 Forced disclosure of encryption keys

Main article: Key disclosure law

In the United Kingdom, the Regulation of Investigatory Powers Act gives UK police the powers to force suspects to decrypt files or hand over passwords that protect encryption keys. Failure to comply is an offense in its own right, punishable on conviction by a two-year jail sentence or up to five years in cases involving national security. Successful prosecutions have occurred under the Act; the first in 2009,[57] resulting in a term of 13 months' imprisonment.[58] Similar forced disclosure laws in Australia, Finland, France, and India compel individual suspects under investigation to hand over encryption keys or passwords during a criminal investigation (see Key disclosure law).

In the United States, the federal criminal case of United States v. Fricosu addressed whether a person can be compelled to reveal his or her encryption passphrase or password, despite the U.S. Constitution's Fifth Amendment protection against self-incrimination.[59] In 2012, the court ruled that under the All Writs Act, the defendant was required to produce an unencrypted hard drive.[60]

In many jurisdictions, the legal status of forced disclosure remains unclear.

62.5 See also

_ Category:Cryptographers
_ List of cryptographers
_ Encyclopedia of Cryptography and Security
_ List of important publications in cryptography
_ List of multiple discoveries (see "RSA")
_ List of unsolved problems in computer science
_ Outline of cryptography

_ Global surveillance
_ Strong cryptography
_ Cryptocurrency

62.6 References

[1] Liddell and Scott's Greek-English Lexicon. Oxford University Press. (1984)
[2] Rivest, Ronald L. (1990). "Cryptology". In J. Van Leeuwen. *Handbook of Theoretical Computer Science* **1**. Elsevier.
[3] Bellare, Mihir; Rogaway, Phillip (21 September 2005). "Introduction". *Introduction to Modern Cryptography*. p. 10.
[4] Menezes, A. J.; van Oorschot, P. C.; Vanstone, S. A. *Handbook of Applied Cryptography*. ISBN 0-8493-8523-7.
[5] "UK Data Encryption Disclosure Law Takes Effect". Pcworld.com. 2007-10-01. Retrieved 2012-01-28.
[6] Leyden, John (2011-07-13). "US court test for rights not to hand over crypto keys". Theregister.co.uk. Retrieved 2012-01-28.
[7] David Kahn, *The Codebreakers*, 1967, ISBN 0-684-83130-9.
[8] Oded Goldreich, *Foundations of Cryptography, Volume 1: Basic Tools*, Cambridge University Press, 2001, ISBN 0-521-79172-3
[9] "Cryptology (definition)". *Merriam-Webster's Collegiate Dictionary* (11th ed.). Merriam-Webster. Retrieved 2008-02-01.
[10] https://tools.ietf.org/html/rfc2828
[11] http://staff.neu.edu.tr/~{}fahri/cryptography.html
[12] V. V. IA͡shchenko (2002). "*Cryptography: an introduction*". AMS Bookstore. p.6. ISBN 0-8218-2986-6
[13] Ibrahim A. Al-Kadi, "The origins of cryptology: The Arab contributions." *Cryptologia*, 16(2) (April 1992) pp. 97–126.
[14] Singh, Simon (2000). *The Code Book*. New York: Anchor Books. pp. 14–20. ISBN 9780385495325.
[15] Ibrahim A. Al-Kadi (April 1992), "The origins of cryptology: The Arab contributions", *Cryptologia* **16** (2): 97–126
[16] Schrödel, Tobias (October 2008). "Breaking Short Vigenère Ciphers". *Cryptologia* **32** (4): 334–337. doi:10.1080/01611190802336097.
[17] Hakim, Joy (1995). *A History of Us: War, Peace and all that Jazz*. New York: Oxford University Press. ISBN 0-19-509514-6.
[18] James Gannon, *Stealing Secrets, Telling Lies: How Spies and Codebreakers Helped Shape the Twentieth Century*, Washington, D.C., Brassey's, 2001, ISBN 1-57488-367-4.
[19] Whitfield Diffie and Martin Hellman, "New Directions in Cryptography", IEEE Transactions on Information Theory, vol. IT-22, Nov. 1976, pp: 644–654. (pdf)
[20] Blaze, Matt; Diffie, Whitefield; Rivest, Ronald L.; Schneier, Bruce; Shimomura, Tsutomu; Thompson, Eric; Wiener, Michael (January 1996). "Minimal key lengths for symmetric ciphers to provide adequate commercial security". Fortify. Retrieved 14 October 2011.
[21] FIPS PUB 197: The official Advanced Encryption Standard.
[22] NCUA letter to credit unions, July 2004
[23] RFC 2440 - Open PGP Message Format
[24] SSH at windowsecurity.com by Pawel Golen, July 2004
[25] Bruce Schneier, *Applied Cryptography*, 2nd edition, Wiley, 1996, ISBN 0-471-11709-9.
[26] Archived February 28, 2008 at the Wayback Machine
[27] "NIST Selects Winner of Secure Hash Algorithm (SHA-3) Competition". NIST. October 2, 2012. Retrieved October 2, 2012.

[28] Whitfield Diffie and Martin Hellman, "Multi-user cryptographic techniques" [Diffie and Hellman, AFIPS Proceedings 45, pp109–112, June 8, 1976].
[29] Ralph Merkle was working on similar ideas at the time and encountered publication delays, and Hellman has suggested that the term used should be Diffie–Hellman–Merkle aysmmetric key cryptography.
[30] David Kahn, "Cryptology Goes Public", 58 *Foreign Affairs* 141, 151 (fall 1979), p. 153.
[31] R. Rivest, A. Shamir, L. Adleman. A Method for Obtaining Digital Signatures and Public-Key Cryptosystems at the Wayback Machine (archived November 16, 2001). Communications of the ACM, Vol. 21 (2), pp.120–126. 1978. Previously released as an MIT "Technical Memo" in April 1977, and published in Martin Gardner's *Scientific American* Mathematical recreations column
[32] "British Document Outlines Early Encryption Discovery". *New York Times*. Retrieved 2012-03-27.
[33] Clifford Cocks. A Note on 'Non-Secret Encryption', CESG Research Report, 20 November 1973.
[34] Singh, Simon (1999). *The Code Book*. Doubleday. pp. 279–292.
[35] "Shannon": Claude Shannon and Warren Weaver, *The Mathematical Theory of Communication*, University of Illinois Press, 1963, ISBN 0-252-72548-4
[36] Pascal Junod, "On the Complexity of Matsui's Attack", SAC 2001.
[37] Dawn Song, David Wagner, and Xuqing Tian, "Timing Analysis of Keystrokes and Timing Attacks on SSH", In Tenth USENIX Security Symposium, 2001.

[38] S. Brands, "Untraceable Off-line Cash in Wallets with Observers", In *Advances in Cryptology—Proceedings of CRYPTO*, Springer-Verlag, 1994.
[39] László Babai. "Trading group theory for randomness". *Proceedings of the Seventeenth Annual Symposium on the Theory of Computing*, ACM, 1985.
[40] S. Goldwasser, S. Micali, and C. Rackoff, "The Knowledge Complexity of Interactive Proof Systems", SIAM J. Computing, vol. 18, num. 1, pp. 186–208, 1989.
[41] G. Blakley. "Safeguarding cryptographic keys." In *Proceedings of AFIPS 1979*, volume 48, pp. 313–317, June 1979.
[42] A. Shamir. "How to share a secret." In *Communications of the ACM*, volume 22, pp. 612–613, ACM, 1979.
[43] "Crypto law".
[44] "RSA Laboratories' Frequently Asked Questions About Today's Cryptography". Rsasecurity.com. Retrieved 2011-07-18.
[45] Cryptography & Speech from Cyberlaw
[46] "Case Closed on Zimmermann PGP Investigation", press note from the IEEE.
[47] Levy, Steven (2001). *Crypto: How the Code Rebels Beat the Government—Saving Privacy in the Digital Age*. Penguin Books. p. 56. ISBN 0-14-024432-8. OCLC 244148644 48066852 48846639.
[48] Bernstein v USDOJ, 9th Circuit court of appeals decision.
[49] "The Wassenaar Arrangement on Export Controls for Conventional Arms and Dual-Use Goods and Technologies". Wassenaar.org. Retrieved 2011-07-18.
[50] "The Data Encryption Standard (DES)" from Bruce Schneier's CryptoGram newsletter, June 15, 2000
[51] Coppersmith, D. (May 1994). "The Data Encryption Standard (DES) and its strength against attacks" (PDF). *IBM Journal of Research and Development* **38** (3): 243. doi:10.1147/rd.383.0243.
[52] E. Biham and A. Shamir, "Differential cryptanalysis of DES-like cryptosystems", Journal of Cryptology, vol. 4 num. 1, pp. 3–72, Springer-Verlag, 1991.
[53] Levy 2001, p. 56
[54] "The Digital Millennium Copyright Act of 1998" (PDF). Retrieved 2011-07-18.
[55] Archived December 1, 2001 at the Wayback Machine
[56] "Digg revolt over HD DVD codes". news.com.au. 2 May 2007. Retrieved 2007-05-20.
404 *CHAPTER 62. CRYPTOGRAPHY*
[57] Williams, Christopher Williams (11 August 2009). "Two convicted for refusal to decrypt data • The Register". *theregister.co.uk*. Retrieved 18 October 2012.
[58] Williams, Christopher Williams (24 November 2009). "UK jails schizophrenic for refusal to decrypt files • The Register". *theregister.co.uk*. Retrieved 18 October 2012.
[59] Ingold, John (January 4, 2012). "Password case reframes Fifth Amendment rights in context of digital world". *Denver Post*.
[60] http://www.wired.com/images_blogs/threatlevel/2012/01/decrypt.pdf

62.7 Further reading

Further information: Books on cryptography

_ Becket, B (1988). *Introduction to Cryptology*. Blackwell Scientific Publications. ISBN 0-632-01836-4. OCLC 16832704. Excellent coverage of many classical ciphers and cryptography concepts and of the "modern" DES and RSA systems.
_ *Cryptography and Mathematics* by Bernhard Esslinger, 200 pages, part of the free open-source package CrypTool, PDF download at the Wayback Machine (archived July 22, 2011). CrypTool is the most widespread e-learning program about cryptography and cryptanalysis, open source.
_ *In Code: A Mathematical Journey* by Sarah Flannery (with David Flannery). Popular account of Sarah's awardwinning project on public-key cryptography, co-written with her father.
_ James Gannon, *Stealing Secrets, Telling Lies: How Spies and Codebreakers Helped Shape the Twentieth Century*, Washington, D.C., Brassey's, 2001, ISBN 1-57488-367-4.
_ Oded Goldreich, Foundations of Cryptography, in two volumes, Cambridge University Press, 2001 and 2004.
_ *Introduction to Modern Cryptography* by Jonathan Katz and Yehuda Lindell.
_ *Alvin's Secret Code* by Clifford B. Hicks (children's novel that introduces some basic cryptography and cryptanalysis).
_ Ibrahim A. Al-Kadi, "The Origins of Cryptology: the Arab Contributions," Cryptologia, vol. 16, no. 2 (April 1992), pp. 97–126.
_ Christof Paar, Jan Pelzl, Understanding Cryptography, A Textbook for Students and Practitioners. Springer, 2009. (Slides, online cryptography lectures and other information are available on the companion web site.) Very accessible introduction to practical cryptography for non-mathematicians.
_ *Introduction to Modern Cryptography* by Phillip Rogaway and Mihir Bellare, a mathematical introduction to theoretical cryptography including reduction-based security proofs. PDF download.
_ Johann-Christoph Woltag, 'Coded Communications (Encryption)' in Rüdiger Wolfrum (ed) Max Planck Encyclopedia of Public International Law (Oxford University Press 2009). *"Max Planck Encyclopedia of Public International Law"., giving an overview of international law issues regarding cryptography.
_ Jonathan Arbib & John Dwyer, Discrete Mathematics for Cryptography, 1st Edition ISBN 978-1-907934-01-8.
_ Stallings, William (March 2013). *Cryptography and Network Security: Principles and Practice* (6th ed.). Prentice Hall. ISBN 978-0133354690.

62.8 External links

_ The dictionary definition of cryptography at Wiktionary
_ Media related to Cryptography at Wikimedia Commons
_
_ Cryptography on *In Our Time* at the BBC. (listen now)
_ Crypto Glossary and Dictionary of Technical Cryptography
_ NSA's CryptoKids.
_ Overview and Applications of Cryptology by the CrypTool Team; PDF; 3.8 MB—July 2008
_ A Course in Cryptography by Raphael Pass & Abhi Shelat - offered at Cornell in the form of lecture notes.
_ Cryptocorner.com by Chuck Easttom - A generalized resource on all aspects of cryptology.
_ For more on the use of cryptographic elements in fiction, see: Dooley, John F., William and Marilyn Ingersoll Professor of Computer Science, Knox College (23 August 2012). "Cryptology in Fiction Sorted by Author".
_ The George Fabyan Collection at the Library of Congress has early editions of works of seventeenth-century English literature, publications relating to cryptography.

Chapter 63

Number theory

Not to be confused with Numerology.

Number theory (or **arithmetic**[note 1]) is a branch of pure mathematics devoted primarily to the study of the integers, sometimes called "The Queen of Mathematics" because of its foundational place in the discipline.[1] Number theorists study prime numbers as well as the properties of objects made out of integers (e.g., rational numbers) or defined as generalizations of the integers (e.g., algebraic integers).

Integers can be considered either in themselves or as solutions to equations (Diophantine geometry). Questions in number theory are often best understood through the study of analytical objects (e.g., the Riemann zeta function) that encode properties of the integers, primes or other number-theoretic objects in some fashion (analytic number theory). One may also study real numbers in relation to rational numbers, e.g., as approximated by the latter (Diophantine approximation).

The older term for number theory is *arithmetic*. By the early twentieth century, it had been superseded by "number theory".[note 2] (The word "arithmetic" is used by the general public to mean "elementary calculations"; it has also acquired other meanings in mathematical logic, as in *Peano arithmetic*, and computer science, as in *floating point arithmetic*.) The use of the term *arithmetic* for *number theory* regained some ground in the second half of the 20th century, arguably in part due to French influence.[note 3] In particular, *arithmetical* is preferred as an adjective to *number-theoretic*.

63.1 History

63.1.1 Origins

Dawn of arithmetic

The first historical find of an arithmetical nature is a fragment of a table: the broken clay tablet Plimpton 322 (Larsa, Mesopotamia, ca. 1800 BCE) contains a list of "Pythagorean triples", i.e., integers $(a;b;c)$ such that $a^2+b^2=c^2$. The triples are too many and too large to have been obtained by brute force. The heading over the first column reads: "The *takiltum* of the diagonal which has been subtracted such that the width..."[2]

The table's layout suggests[3] that it was constructed by means of what amounts, in modern language, to the identity

$$\left(\frac{1}{2}\left(x - \frac{1}{x}\right)\right)^2 + 1 = \left(\frac{1}{2}\left(x + \frac{1}{x}\right)\right)^2;$$

which is implicit in routine Old Babylonian exercises.[4] If some other method was used,[5] the triples were first constructed and then reordered by c/a, presumably for actual use as a "table", i.e., with a view to applications.

It is not known what these applications may have been, or whether there could have been any; Babylonian astronomy, for example, truly flowered only later. It has been suggested instead that the table was a source of numerical examples for school problems.[6][note 4]

While Babylonian number theory—or what survives of Babylonian mathematics that can be called thus—consists of this single, striking fragment, Babylonian algebra (in the secondary-school sense of "algebra") was exceptionally well

406

63.1. HISTORY 407

A Lehmer sieve, which is a primitive digital computer once used for finding primes and solving simple Diophantine equations.

408 *CHAPTER 63. NUMBER THEORY*

The Plimpton 322 tablet

developed.[7] Late Neoplatonic sources[8] state that Pythagoras learned mathematics from the Babylonians. Much earlier sources[9] state that Thales and Pythagoras traveled and studied in Egypt.

Euclid IX 21—34 is very probably Pythagorean;[10] it is very simple material ("odd times even is even", "if an odd number measures [= divides] an even number, then it also measures [= divides] half of it"), but it is all that is needed to prove that p
2 is irrational.[11] Pythagorean mystics gave great importance to the odd and the even.[12] The discovery that p
2 is irrational is credited to the early Pythagoreans (pre-Theodorus).[13] By revealing (in modern terms) that numbers could be irrational, this discovery seems to have provoked the first foundational crisis in mathematical history; its proof or its divulgation are sometimes credited to Hippasus, who was expelled or split from the Pythagorean sect.[14] It is only here that we can start to speak of a clear, conscious division between *numbers* (integers and the rationals—the subjects of arithmetic) and *lengths* (real numbers, whether rational or not).

The Pythagorean tradition spoke also of so-called polygonal or figurate numbers.[15] While square numbers, cubic numbers, etc., are seen now as more natural than triangular numbers, square numbers, pentagonal numbers, etc., the study of the sums of triangular and pentagonal numbers would prove fruitful in the early modern period (17th to early 19th century).

We know of no clearly arithmetical material in ancient Egyptian or Vedic sources, though there is some algebra in both. The Chinese remainder theorem appears as an exercise [16] in Sun Zi's *Suan Ching* (also known as *The Mathematical Classic of Sun Zi* (3rd, 4th or 5th century CE.)[17] (There is one important step glossed over in Sun Zi's solution:[note 5] it is the problem that was later solved by Āryabhaṭa's kuṭṭaka – see below.)

There is also some numerical mysticism in Chinese mathematics,[note 6] but, unlike that of the Pythagoreans, it seems to have led nowhere. Like the Pythagoreans' perfect numbers, magic squares have passed from superstition into recreation.

Classical Greece and the early Hellenistic period

Aside from a few fragments, the mathematics of Classical Greece is known to us either through the reports of contemporary non-mathematicians or through mathematical works from the early Hellenistic period.[18] In the case of number theory, this means, by and large, *Plato* and *Euclid*, respectively.

63.1. HISTORY 409

Plato had a keen interest in mathematics, and distinguished clearly between arithmetic and calculation. (By *arithmetic* he meant, in part, theorising on number, rather than what *arithmetic* or *number theory* have come to mean.) It is through one of Plato's dialogues—namely, *Theaetetus* – that we know that Theodorus had proven that p
3;
p
5;:::;
p
17
are irrational. Theaetetus was, like Plato, a disciple of Theodorus's; he worked on distinguishing different kinds of incommensurables, and was thus arguably a pioneer in the study of number systems. (Book X of Euclid's Elements is described by Pappus as being largely based on Theaetetus's work.)

Euclid devoted part of his *Elements* to prime numbers and divisibility, topics that belong unambiguously to number theory and are basic to it (Books VII to IX of Euclid's Elements). In particular, he gave an algorithm for computing the greatest common divisor of two numbers (the Euclidean algorithm; *Elements*, Prop. VII.2) and the first known proof of the infinitude of primes (*Elements*, Prop. IX.20).

In 1773, Lessing published an epigram he had found in a manuscript during his work as a librarian; it claimed to be a letter sent by Archimedes to Eratosthenes.[19][20] The epigram proposed what has become known as Archimedes' cattle problem; its solution (absent from the manuscript) requires solving an indeterminate quadratic equation (which reduces to what would later be misnamed Pell's equation). As far as we know, such equations were first successfully treated by the Indian school. It is not known whether Archimedes himself had a method of solution.

Diophantus

Very little is known about Diophantus of Alexandria; he probably lived in the third century CE, that is, about five hundred years after Euclid. Six out of the thirteen books of Diophantus's *Arithmetica* survive in the original Greek; four more books survive in an Arabic translation. The *Arithmetica* is a collection of worked-out problems where the task is invariably to find rational solutions to a system of polynomial equations, usually of the form f(x;y)=z2 or

$f(x;y;z)=w_2$. Thus, nowadays, we speak of *Diophantine equations* when we speak of polynomial equations to which rational or integer solutions must be found.
One may say that Diophantus was studying rational points — i.e., points whose coordinates are rational — on curves and algebraic varieties; however, unlike the Greeks of the Classical period, who did what we would now call basic algebra in geometrical terms, Diophantus did what we would now call basic algebraic geometry in purely algebraic terms. In modern language, what Diophantus did was to find rational parametrizations of varieties; that is, given an equation of the form (say) $f(x_1;x_2;x_3)=0$, his aim was to find (in essence) three rational functions $g_1;g_2;g_3$ such that, for all values of r and s, setting $x_i=g_i(r;s)$ for $i=1;2;3$ gives a solution to $f(x_1;x_2;x_3)=0:$
Diophantus also studied the equations of some non-rational curves, for which no rational parametrisation is possible. He managed to find some rational points on these curves (elliptic curves, as it happens, in what seems to be their first known occurrence) by means of what amounts to a tangent construction: translated into coordinate geometry (which did not exist in Diophantus's time), his method would be visualised as drawing a tangent to a curve at a known rational point, and then finding the other point of intersection of the tangent with the curve; that other point is a new rational point. (Diophantus also resorted to what could be called a special case of a secant construction.)
While Diophantus was concerned largely with rational solutions, he assumed some results on integer numbers, in particular that every integer is the sum of four squares (though he never stated as much explicitly).

Indian school: Āryabhaṭa, Brahmagupta, Bhāskara

While Greek astronomy probably influenced Indian learning, to the point of introducing trigonometry,[21] it seems to be the case that Indian mathematics is otherwise an indigenous tradition;[22] in particular, there is no evidence that Euclid's Elements reached India before the 18th century.[23]
Āryabhaṭa (476–550 CE) showed that pairs of simultaneous congruences $n\_a_1\ (\mathrm{mod}\ m)_1$, $n\_a_2\ (\mathrm{mod}\ m)_2$ could be solved by a method he called *kuṭṭaka*, or *pulveriser*;[24] this is a procedure close to (a generalisation of) the Euclidean algorithm, which was probably discovered independently in India.[25] Āryabhaṭa seems to have had in mind applications to astronomical calculations.[21]
Brahmagupta (628 CE) started the systematic study of indefinite quadratic equations—in particular, the misnamed Pell equation, in which Archimedes may have first been interested, and which did not start to be solved in the West until the time of Fermat and Euler. Later Sanskrit authors would follow, using Brahmagupta's technical terminology. A general procedure (the chakravala, or "cyclic method") for solving Pell's equation was finally found by Jayadeva (cited in the eleventh century; his work is otherwise lost); the earliest surviving exposition appears in Bhāskara II's Bīja-gaṇita (twelfth century).[26]

Unfortunately, Indian mathematics remained largely unknown in the West until the late eighteenth century;[27] Brahmagupta and Bhāskara's work was translated into English in 1817 by Henry Colebrooke.[28]

Arithmetic in the Islamic golden age

In the early ninth century, the caliph Al-Ma'mun ordered translations of many Greek mathematical works and at least one Sanskrit work (the *Sindhind*, which may [29] or may not[30] be Brahmagupta's Brāhmasphuṭasiddhānta), thus giving rise to the tradition of Islamic mathematics. Diophantus's main work, the *Arithmetica*, was translated into Arabic by Qusta ibn Luqa (820–912). Part of the treatise *al-Fakhri* (by al-Karajī, 953 – ca. 1029) builds on it to some extent. According to Rashed Roshdi, Al-Karajī's contemporary Ibn al-Haytham knew[31] what would later be called Wilson's theorem.

63.2 Western Europe in the Middle Ages

Other than a treatise on squares in arithmetic progression by Fibonacci — who lived and studied in north Africa and Constantinople during his formative years, ca. 1175–1200 — no number theory to speak of was done in western Europe during the Middle Ages. Matters started to change in Europe in the late Renaissance, thanks to a renewed study of the works of Greek antiquity. A catalyst was the textual emendation and translation into Latin of Diophantus's *Arithmetica* (Bachet, 1621, following a first attempt by Xylander, 1575).

63.2.1 Early modern number theory

Fermat

Pierre de Fermat (1601–1665) never published his writings; in particular, his work on number theory is contained almost entirely in letters to mathematicians and in private marginal notes.[32] He wrote down nearly no proofs in number theory; he had no models in the area.[33] He did make repeated use of mathematical induction, introducing the method of infinite descent.
One of Fermat's first interests was perfect numbers (which appear in Euclid, *Elements* IX) and amicable numbers;[note 7] this led him to work on integer divisors, which were from the beginning among the subjects of the correspondence (1636 onwards) that put him in touch with the mathematical community of the day.[34] He had already studied Bachet's edition of Diophantus carefully;[35] by 1643, his interests had shifted largely to Diophantine problems and sums of squares[36] (also treated by Diophantus).
Fermat's achievements in arithmetic include:
_ Fermat's little theorem (1640),[37] stating that, if a is not divisible by a prime p, then $a_{p\square 1}\_1\ (\mathrm{mod}\ p):$ [note 8]

_ If *a* and *b* are coprime, then a^2+b^2 is not divisible by any prime congruent to −1 modulo 4;[38] *and* Every prime congruent to 1 modulo 4 can be written in the form a^2+b^2.[39] These two statements also date from 1640; in 1659, Fermat stated to Huygens that he had proven the latter statement by the method of infinite descent.[40] Fermat and Frenicle also did some work (some of it erroneous or non-rigorous)[41] on other quadratic forms.

_ Fermat posed the problem of solving $x^2\square Ny^2=1$ as a challenge to English mathematicians (1657). The problem was solved in a few months by Wallis and Brouncker.[42] Fermat considered their solution valid, but pointed out they had provided an algorithm without a proof (as had Jayadeva and Bhaskara, though Fermat would never know this.) He states that a proof can be found by descent.

_ Fermat developed methods for (doing what in our terms amounts to) finding points on curves of genus 0 and 1. As in Diophantus, there are many special procedures and what amounts to a tangent construction, but no use of a secant construction.[43]

_ Fermat states and proves (by descent) in the appendix to *Observations on Diophantus* (Obs. XLV)[44] that $x^4+y^4=z^4$ has no non-trivial solutions in the integers. Fermat also mentioned to his correspondents that $x^3+y^3=z^3$ has no non-trivial solutions, and that this could be proven by descent.[45] The first known proof is due to Euler (1753; indeed by descent).[46]

Fermat's claim ("Fermat's last theorem") to have shown there are no solutions to $x^n+y^n=z^n$ for all n_3 (a fact the only known proofs of which were completely beyond his methods) appears only in his annotations on the margin of his copy of Diophantus; he never claimed this to others[47] and thus would have had no need to retract it if he found any mistake in his supposed proof.

Euler

The interest of Leonhard Euler (1707–1783) in number theory was first spurred in 1729, when a friend of his, the amateur[note 9] Goldbach, pointed him towards some of Fermat's work on the subject.[48][49] This has been called the "rebirth" of modern number theory,[35] after Fermat's relative lack of success in getting his contemporaries' attention for the subject.[50] Euler's work on number theory includes the following:[51]

_ *Proofs for Fermat's statements.* This includes Fermat's little theorem (generalised by Euler to non-prime moduli); the fact that $p=x^2+y^2$ if and only if p_1 $mod\ 4$; initial work towards a proof that every integer is the sum of four squares (the first complete proof is by Joseph-Louis Lagrange (1770), soon improved by Euler himself[52]); the lack of non-zero integer solutions to $x^4+y^4=z^2$ (implying the case *n=4* of Fermat's last theorem, the case *n=3* of which Euler also proved by a related method).

_ *Pell's equation*, first misnamed by Euler.[53] He wrote on the link between continued fractions and Pell's equation.[54]

_ *First steps towards analytic number theory.* In his work of sums of four squares, partitions, pentagonal numbers, and the distribution of prime numbers, Euler pioneered the use of what can be seen as analysis (in particular, infinite series) in number theory. Since he lived before the development of complex analysis, most of his work is restricted to the formal manipulation of power series. He did, however, do some very notable (though not fully rigorous) early work on what would later be called the Riemann zeta function.[55]

_ *Quadratic forms.* Following Fermat's lead, Euler did further research on the question of which primes can be expressed in the form x^2+Ny^2, some of it prefiguring quadratic reciprocity.[56] [57][58]

_ *Diophantine equations.* Euler worked on some Diophantine equations of genus 0 and 1.[59][60] In particular, he studied Diophantus's work; he tried to systematise it, but the time was not yet ripe for such an endeavour – algebraic geometry was still in its infancy.[61] He did notice there was a connection between Diophantine problems and elliptic integrals,[61] whose study he had himself initiated.

Lagrange, Legendre and Gauss

Joseph-Louis Lagrange (1736–1813) was the first to give full proofs of some of Fermat's and Euler's work and observations - for instance, the four-square theorem and the basic theory of the misnamed "Pell's equation" (for which an algorithmic solution was found by Fermat and his contemporaries, and also by Jayadeva and Bhaskara II before them.) He also studied quadratic forms in full generality (as opposed to mX^2+nY^2) — defining their equivalence relation, showing how to put them in reduced form, etc.

Adrien-Marie Legendre (1752–1833) was the first to state the law of quadratic reciprocity. He also conjectured what amounts to the prime number theorem and Dirichlet's theorem on arithmetic progressions. He gave a full treatment of the equation $ax^2+by^2+cz^2=0$ [62] and worked on quadratic forms along the lines later developed fully by Gauss.[63] In his old age, he was the first to prove "Fermat's last theorem" for $n = 5$ (completing work by Peter Gustav Lejeune Dirichlet, and crediting both him and Sophie Germain).[64]

In his *Disquisitiones Arithmeticae* (1798), Carl Friedrich Gauss (1777–1855) proved the law of quadratic reciprocity and developed the theory of quadratic forms (in particular, defining their composition). He also introduced some basic notation (congruences) and devoted a section to computational matters, including primality tests.[65] The last section of the *Disquisitiones* established a link between roots of unity and number theory:

The theory of the division of the circle...which is treated in sec. 7 does not belong by itself to arithmetic, but its principles can only be drawn from higher arithmetic.[66]

In this way, Gauss arguably made a first foray towards both Évariste Galois's work and algebraic number theory.

63.2.2 Maturity and division into subfields

Starting early in the nineteenth century, the following developments gradually took place:

- The rise to self-consciousness of number theory (or *higher arithmetic*) as a field of study.[67]
- The development of much of modern mathematics necessary for basic modern number theory: complex analysis, group theory, Galois theory—accompanied by greater rigor in analysis and abstraction in algebra.
- The rough subdivision of number theory into its modern subfields—in particular, analytic and algebraic number theory.

Algebraic number theory may be said to start with the study of reciprocity and cyclotomy, but truly came into its own with the development of abstract algebra and early ideal theory and valuation theory; see below. A conventional starting point for analytic number theory is Dirichlet's theorem on arithmetic progressions (1837),[68] [69] whose proof introduced L-functions and involved some asymptotic analysis and a limiting process on a real variable.[70] The first use of analytic ideas in number theory actually goes back to Euler (1730s),[71] [72] who used formal power series and non-rigorous (or implicit) limiting arguments. The use of *complex* analysis in number theory comes later: the work of Bernhard Riemann (1859) on the zeta function is the canonical starting point;[73] Jacobi's four-square theorem (1839), which predates it, belongs to an initially different strand that has by now taken a leading role in analytic number theory (modular forms).[74]

The history of each subfield is briefly addressed in its own section below; see the main article of each subfield for fuller treatments. Many of the most interesting questions in each area remain open and are being actively worked on.

63.3 Main subdivisions

63.3.1 Elementary tools

The term *elementary* generally denotes a method that does not use complex analysis. For example, the prime number theorem was first proven using complex analysis in 1896, but an elementary proof was found only in 1949 by Erdős and Selberg.[75] The term is somewhat ambiguous: for example, proofs based on complex Tauberian theorems (e.g. Wiener–Ikehara) are often seen as quite enlightening but not elementary, in spite of using Fourier analysis, rather than complex analysis as such. Here as elsewhere, an *elementary* proof may be longer and more difficult for most readers than a non-elementary one.

Number theory has the reputation of being a field many of whose results can be stated to the layperson. At the same time, the proofs of these results are not particularly accessible, in part because the range of tools they use is, if anything, unusually broad within mathematics.[76]

63.3.2 Analytic number theory

Main article: Analytic number theory

Analytic number theory may be defined

- in terms of its tools, as the study of the integers by means of tools from real and complex analysis;[68] or
- in terms of its concerns, as the study within number theory of estimates on size and density, as opposed to identities.[77]

Some subjects generally considered to be part of analytic number theory, e.g., sieve theory,[note 10] are better covered by the second rather than the first definition: some of sieve theory, for instance, uses little analysis,[note 11] yet it does belong to analytic number theory.

The following are examples of problems in analytic number theory: the prime number theorem, the Goldbach conjecture (or the twin prime conjecture, or the Hardy–Littlewood conjectures), the Waring problem and the Riemann Hypothesis. Some of the most important tools of analytic number theory are the circle method, sieve methods and Lfunctions (or, rather, the study of their properties). The theory of modular forms (and, more generally, automorphic forms) also occupies an increasingly central place in the toolbox of analytic number theory.[78]

One may ask analytic questions about algebraic numbers, and use analytic means to answer such questions; it is thus that algebraic and analytic number theory intersect. For example, one may define prime ideals (generalizations of prime numbers living in the field of algebraic numbers) and ask how many prime ideals there are up to a certain size. This question can be answered by means of an examination of Dedekind zeta functions, which are generalizations of the Riemann zeta function, a key analytic object at the roots of the subject.[79] This is an example of a general procedure in analytic number theory: deriving information about the distribution of a sequence (here, prime ideals or prime numbers) from the analytic behavior of an appropriately constructed complex-valued function.[80]

63.3.3 Algebraic number theory

Main article: Algebraic number theory

An *algebraic number* is any complex number that is a solution to some polynomial equation $f(x)=0$ with rational coefficients; for example, every solution x of $x^5+(11/2)x^3-7x^2+9=0$ (say) is an algebraic number. Fields of algebraic numbers are also called *algebraic number fields*, or shortly *number fields*. Algebraic number theory studies algebraic number fields.[81] Thus, analytic and algebraic number theory can and do overlap: the former is defined by its methods,

the latter by its objects of study.
It could be argued that the simplest kind of number fields (viz., quadratic fields) were already studied by Gauss, as the discussion of quadratic forms in *Disquisitiones arithmeticae* can be restated in terms of ideals and norms in quadratic fields. (A *quadratic field* consists of all numbers of the form $a+b$
p
d, where a and b are rational numbers and d
is a fixed rational number whose square root is not rational.) For that matter, the 11th-century chakravala method amounts—in modern terms—to an algorithm for finding the units of a real quadratic number field. However, neither Bhāskara nor Gauss knew of number fields as such.
The grounds of the subject as we know it were set in the late nineteenth century, when *ideal numbers*, the *theory of ideals* and *valuation theory* were developed; these are three complementary ways of dealing with the lack of unique factorisation in algebraic number fields. (For example, in the field generated by the rationals and p
□5 , the number
6 can be factorised both as 6=2_3 and 6=(1+
p
□5)(1□
p
□5) ; all of 2 , 3 , 1+
p
□5 and 1□
p
□5 are irreducible, and thus,
in a naïve sense, analogous to primes among the integers.) The initial impetus for the development of ideal numbers (by Kummer) seems to have come from the study of higher reciprocity laws,[82]i.e., generalisations of quadratic reciprocity.
Number fields are often studied as extensions of smaller number fields: a field L is said to be an *extension* of a field K if L contains K. (For example, the complex numbers C are an extension of the reals R, and the reals R are an extension of the rationals Q.) Classifying the possible extensions of a given number field is a difficult and partially open problem. Abelian extensions—that is, extensions L of K such that the Galois group[note 12] Gal(L/K) of L over K is an abelian group—are relatively well understood. Their classification was the object of the programme of class field theory, which was initiated in the late 19th century (partly by Kronecker and Eisenstein) and carried out largely in 1900—1950.
An example of an active area of research in algebraic number theory is Iwasawa theory. The Langlands program, one of the main current large-scale research plans in mathematics, is sometimes described as an attempt to generalise class field theory to non-abelian extensions of number fields.

63.3.4 Diophantine geometry

Main articles: Diophantine geometry and Glossary of arithmetic and Diophantine geometry
The central problem of *Diophantine geometry* is to determine when a Diophantine equation has solutions, and if it does, how many. The approach taken is to think of the solutions of an equation as a geometric object.
For example, an equation in two variables defines a curve in the plane. More generally, an equation, or system of equations, in two or more variables defines a curve, a surface or some other such object in n-dimensional space. In Diophantine geometry, one asks whether there are any *rational points* (points all of whose coordinates are rationals) or *integral points* (points all of whose coordinates are integers) on the curve or surface. If there are any such points, the next step is to ask how many there are and how they are distributed. A basic question in this direction is: are there finitely or infinitely many rational points on a given curve (or surface)? What about integer points?

An example here may be helpful. Consider the Pythagorean equation $x^2+y^2=1$; we would like to study its rational solutions, i.e., its solutions $(x; y)$ such that x and y are both rational. This is the same as asking for all integer solutions to $a^2+b^2=c^2$; any solution to the latter equation gives us a solution $x=a/c$, $y=b/c$ to the former. It is also the same as asking for all points with rational coordinates on the curve described by $x^2+y^2=1$. (This curve happens to be a circle of radius 1 around the origin.)
The rephrasing of questions on equations in terms of points on curves turns out to be felicitous. The finiteness or not of the number of rational or integer points on an algebraic curve—that is, rational or integer solutions to an equation $f(x; y)=0$, where f is a polynomial in two variables—turns out to depend crucially on the *genus* of the curve. The *genus* can be defined as follows:[note 13] allow the variables in $f(x; y)=0$ to be complex numbers; then $f(x; y)=0$ defines a 2-dimensional surface in (projective) 4-dimensional space (since two complex variables can be decomposed into four real variables, i.e., four dimensions). Count the number of (doughnut) holes in the surface; call this number the *genus* of $f(x; y)=0$. Other geometrical notions turn out to be just as crucial.
There is also the closely linked area of Diophantine approximations: given a number x , how well can it be approximated by rationals? (We are looking for approximations that are good relative to the amount of space that it takes to write the rational: call a/q (with gcd$(a; q)=1$) a good approximation to x if jx□a/qj< 1
qc, where c is large.) This

question is of special interest if x is an algebraic number. If x cannot be well approximated, then some equations do not have integer or rational solutions. Moreover, several concepts (especially that of height) turn out to be crucial both in Diophantine geometry and in the study of Diophantine approximations. This question is also of special interest in transcendence theory: if a number can be better approximated than any algebraic number, then it is a transcendental number. It is by this argument that π and e have been shown to be transcendental.
Diophantine geometry should not be confused with the geometry of numbers, which is a collection of graphical methods for answering certain questions in algebraic number theory. *Arithmetic geometry*, on the other hand, is a contemporary term for much the same domain as that covered by the term *Diophantine geometry*. The term *arithmetic geometry* is arguably used most often when one wishes to emphasise the connections to modern algebraic geometry (as in, for instance, Faltings' theorem) rather than to techniques in Diophantine approximations.

63.4 Recent approaches and subfields

The areas below date as such from no earlier than the mid-twentieth century, even if they are based on older material. For example, as is explained below, the matter of algorithms in number theory is very old, in some sense older than the concept of proof; at the same time, the modern study of computability dates only from the 1930s and 1940s, and computational complexity theory from the 1970s.

63.4.1 Probabilistic number theory

Main article: Probabilistic number theory
Take a number at random between one and a million. How likely is it to be prime? This is just another way of asking how many primes there are between one and a million. Further: how many prime divisors will it have, on average? How many divisors will it have altogether, and with what likelihood? What is the probability that it have many more or many fewer divisors or prime divisors than the average?
Much of probabilistic number theory can be seen as an important special case of the study of variables that are almost, but not quite, mutually independent. For example, the event that a random integer between one and a million be divisible by two and the event that it be divisible by three are almost independent, but not quite.
It is sometimes said that probabilistic combinatorics uses the fact that whatever happens with probability greater than 0 must happen sometimes; one may say with equal justice that many applications of probabilistic number theory hinge on the fact that whatever is unusual must be rare. If certain algebraic objects (say, rational or integer solutions to certain equations) can be shown to be in the tail of certain sensibly defined distributions, it follows that there must be few of them; this is a very concrete non-probabilistic statement following from a probabilistic one.
At times, a non-rigorous, probabilistic approach leads to a number of heuristic algorithms and open problems, notably Cramér's conjecture.

63.4.2 Arithmetic combinatorics

Main articles: Arithmetic combinatorics and Additive number theory
Let A be a set of N integers. Consider the set $A + A = \{ m + n \mid m, n \in A \}$ consisting of all sums of two elements of A. Is $A + A$ much larger than A? Barely larger? If $A + A$ is barely larger than A, must A have plenty of arithmetic structure, for example, does A resemble an arithmetic progression?
If we begin from a fairly "thick" infinite set A , does it contain many elements in arithmetic progression: a , $a + b$, $a + 2b$, $a + 3b$, $: : :$, $a + 10b$, say? Should it be possible to write large integers as sums of elements of A ?
These questions are characteristic of *arithmetic combinatorics*. This is a presently coalescing field; it subsumes *additive number theory* (which concerns itself with certain very specific sets A of arithmetic significance, such as the primes or the squares) and, arguably, some of the *geometry of numbers*, together with some rapidly developing new material. Its focus on issues of growth and distribution accounts in part for its developing links with ergodic theory, finite group theory, model theory, and other fields. The term *additive combinatorics* is also used; however, the sets A being studied need not be sets of integers, but rather subsets of non-commutative groups, for which the multiplication symbol, not the addition symbol, is traditionally used; they can also be subsets of rings, in which case the growth of $A + A$ and $A \cdot A$ may be compared.

63.4.3 Computations in number theory

Main article: Computational number theory
While the word *algorithm* goes back only to certain readers of al-Khwārizmī, careful descriptions of methods of solution are older than proofs: such methods (that is, algorithms) are as old as any recognisable mathematics—ancient Egyptian, Babylonian, Vedic, Chinese—whereas proofs appeared only with the Greeks of the classical period. An interesting early case is that of what we now call the Euclidean algorithm. In its basic form (namely, as an algorithm for computing the greatest common divisor) it appears as Proposition 2 of Book VII in *Elements*, together with a proof of correctness. However, in the form that is often used in number theory (namely, as an algorithm for finding integer solutions to an equation $ax+by=c$, or, what is the same, for finding the quantities whose existence is assured by the Chinese remainder theorem) it first appears in the works of Āryabhaṭa (5th–6th century CE) as an algorithm called *kuṭṭaka* ("pulveriser"), without a proof of correctness.
There are two main questions: "can we compute this?" and "can we compute it rapidly?". Anybody can test whether

a number is prime or, if it is not, split it into prime factors; doing so rapidly is another matter. We now know fast algorithms for testing primality, but, in spite of much work (both theoretical and practical), no truly fast algorithm for factoring.
The difficulty of a computation can be useful: modern protocols for encrypting messages (e.g., RSA) depend on functions that are known to all, but whose inverses (a) are known only to a chosen few, and (b) would take one too long a time to figure out on one's own. For example, these functions can be such that their inverses can be computed only if certain large integers are factorized. While many difficult computational problems outside number theory are known, most working encryption protocols nowadays are based on the difficulty of a few number-theoretical problems.
On a different note — some things may not be computable at all; in fact, this can be proven in some instances. For instance, in 1970, it was proven, as a solution to Hilbert's 10th problem, that there is no Turing machine which can solve all Diophantine equations.[83] In particular, this means that, given a computably enumerable set of axioms, there are Diophantine equations for which there is no proof, starting from the axioms, of whether the set of equations has or does not have integer solutions. (We would necessarily be speaking of Diophantine equations for which there are no integer solutions, since, given a Diophantine equation with at least one solution, the solution itself provides a proof of the fact that a solution exists. We cannot prove, of course, that a particular Diophantine equation is of this kind, since this would imply that it has no solutions.)

63.5 Applications

The number-theorist Leonard Dickson (1874-1954) said "Thank God that number theory is unsullied by any application". Such a view is no longer applicable to number theory.[84] In 1974, Donald Knuth said "...virtually every theorem in elementary number theory arises in a natural, motivated way in connection with the problem of making computers do high-speed numerical calculations".[85] Elementary number theory is taught in discrete mathematics courses for computer scientists; and, on the other hand, number theory also has applications to the continuous in numerical analysis.[86] As well as the well-known applications to cryptography, there are also applications to many other areas of mathematics.[87][88]

63.6 Literature

Two of the most popular introductions to the subject are:
_ G. H. Hardy; E. M. Wright (2008) [1938]. *An introduction to the theory of numbers* (rev. by D. R. Heath-Brown and J. H. Silverman, 6th ed.). Oxford University Press. ISBN 978-0-19-921986-5.
_ Vinogradov, I. M. (2003) [1954]. *Elements of Number Theory* (reprint of the 1954 ed.). Mineola, NY: Dover Publications.
Hardy and Wright's book is a comprehensive classic, though its clarity sometimes suffers due to the authors' insistence on elementary methods.[89] Vinogradov's main attraction consists in its set of problems, which quickly lead to Vinogradov's own research interests; the text itself is very basic and close to minimal. Other popular first introductions are:
_ Ivan M. Niven; Herbert S. Zuckerman; Hugh L. Montgomery (2008) [1960]. *An introduction to the theory of numbers* (reprint of the 5th edition 1991 ed.). John Wiley & Sons. ISBN 978-8-12-651811-1.
_ Kenneth H. Rosen (2010). *Elementary Number Theory* (6th ed.). Pearson Education. ISBN 978-0-32-171775-7.
Popular choices for a second textbook include:
_ Borevich, A. I.; Shafarevich, Igor R. (1966). *Number theory*. Pure and Applied Mathematics **20**. Boston, MA: Academic Press. ISBN 978-0-12-117850-5. MR 0195803.
_ Serre, Jean-Pierre (1996) [1973]. *A course in arithmetic*. Graduate texts in mathematics **7**. Springer. ISBN 978-0-387-90040-7.

63.7 See also

_ Algebraic function field
_ Finite field
_ p-adic number

63.8 Notes

[1] Especially in older sources; see two following notes.
[2] Already in 1921, T. L. Heath had to explain: "By arithmetic, Plato meant, not arithmetic in our sense, but the science which considers numbers in themselves, in other words, what we mean by the Theory of Numbers." (Heath 1921, p. 13)

[3] Take, e.g. Serre 1973. In 1952, Davenport still had to specify that he meant *The Higher Arithmetic*. Hardy and Wright wrote in the introduction to *An Introduction to the Theory of Numbers* (1938): "We proposed at one time to change [the title] to *An introduction to arithmetic*, a more novel and in some ways a more appropriate title; but it was pointed out that this might lead to misunderstandings about the content of the book." (Hardy & Wright 2008)
[4] Robson 2001, p. 201. This is controversial. See Plimpton 322. Robson's article is written polemically (Robson 2001, p.

202) with a view to "perhaps [...] knocking [Plimpton 322] off its pedestal" (Robson 2001, p. 167); at the same time, it settles to the conclusion that

[...] the question "how was the tablet calculated?" does not have to have the same answer as the question "what problems does the tablet set?" The first can be answered most satisfactorily by reciprocal pairs, as first suggested half a century ago, and the second by some sort of right-triangle problems (Robson 2001, p. 202).

Robson takes issue with the notion that the scribe who produced Plimpton 322 (who had to "work for a living", and would not have belonged to a "leisured middle class") could have been motivated by his own "idle curiosity" in the absence of a "market for new mathematics".(Robson 2001, pp. 199–200)

[5] Sun Zi, *Suan Ching*, Ch. 3, Problem 26, in Lam & Ang 2004, pp. 219–220:

[26] Now there are an unknown number of things. If we count by threes, there is a remainder 2; if we count by fives, there is a remainder 3; if we count by sevens, there is a remainder 2. Find the number of things.

Answer: 23.

Method: If we count by threes and there is a remainder 2, put down 140. If we count by fives and there is a remainder 3, put down 63. If we count by sevens and there is a remainder 2, put down 30. Add them to obtain 233 and subtract 210 to get the answer. If we count by threes and there is a remainder 1, put down 70. If we count by fives and there is a remainder 1, put down 21. If we count by sevens and there is a remainder 1, put down 15. When [a number] exceeds 106, the result is obtained by subtracting 105.

[6] See, e.g., Sun Zi, *Suan Ching*, Ch. 3, Problem 36, in Lam & Ang 2004, pp. 223–224:

[36] Now there is a pregnant woman whose age is 29. If the gestation period is 9 months, determine the sex of the unborn child. *Answer*: Male.

Method: Put down 49, add the gestation period and subtract the age. From the remainder take away 1 representing the heaven, 2 the earth, 3 the man, 4 the four seasons, 5 the five phases, 6 the six pitch-pipes, 7 the seven stars [of the Dipper], 8 the eight winds, and 9 the nine divisions [of China under Yu the Great]. If the remainder is odd, [the sex] is male and if the remainder is even, [the sex] is female.

This is the last problem in Sun Zi's otherwise matter-of-fact treatise.

[7] Perfect and especially amicable numbers are of little or no interest nowadays. The same was not true in medieval times – whether in the West or the Arab-speaking world – due in part to the importance given to them by the Neopythagorean (and hence mystical) Nicomachus (ca. 100 CE), who wrote a primitive but influential "Introduction to Arithmetic". See van der Waerden 1961, Ch. IV.

[8] Here, as usual, given two integers *a* and *b* and a non-zero integer *m*, we write $a\_b\ (\mathrm{mod}\ m)$ (read "*a* is congruent to *b* modulo *m*") to mean that *m* divides $a - b$, or, what is the same, *a* and *b* leave the same residue when divided by *m*. This notation is actually much later than Fermat's; it first appears in section 1 of Gauss's Disquisitiones Arithmeticae. Fermat's little theorem is a consequence of the fact that the order of an element of a group divides the order of the group. The modern proof would have been within Fermat's means (and was indeed given later by Euler), even though the modern concept of a group came long after Fermat or Euler. (It helps to know that inverses exist modulo *p* (i.e., given *a* not divisible by a prime *p*, there is an integer *x* such that $xa\_1\ (\mathrm{mod}\ p)$); this fact (which, in modern language, makes the residues mod *p* into a group, and which was already known to Āryabhaṭa; see above) was familiar to Fermat thanks to its rediscovery by Bachet (Weil 1984, p. 7). Weil goes on to say that Fermat would have recognised that Bachet's argument is essentially Euclid's algorithm.

[9] Up to the second half of the seventeenth century, academic positions were very rare, and most mathematicians and scientists earned their living in some other way (Weil 1984, pp. 159, 161). (There were already some recognisable features of professional *practice*, viz., seeking correspondents, visiting foreign colleagues, building private libraries (Weil 1984, pp. 160–161). Matters started to shift in the late 17th century (Weil 1984, p. 161); scientific academies were founded in England (the Royal Society, 1662) and France (the Académie des sciences, 1666) and Russia (1724). Euler was offered a position at this last one in 1726; he accepted, arriving in St. Petersburg in 1727 (Weil 1984, p. 163 and Varadarajan 2006, p. 7). In this context, the term *amateur* usually applied to Goldbach is well-defined and makes some sense: he has been described as a man of letters who earned a living as a spy (Truesdell 1984, p. xv); cited in Varadarajan 2006, p. 9). Notice, however, that Goldbach published some works on mathematics and sometimes held academic positions.

[10] Sieve theory figures as one of the main subareas of analytic number theory in many standard treatments; see, for instance, Iwaniec & Kowalski 2004 or Montgomery & Vaughan 2007

[11] This is the case for small sieves (in particular, some combinatorial sieves such as the Brun sieve) rather than for large sieves; the study of the latter now includes ideas from harmonic and functional analysis.

[12] The Galois group of an extension *K/L* consists of the operations (isomorphisms) that send elements of L to other elements of L while leaving all elements of K fixed. Thus, for instance, *Gal(C/R)* consists of two elements: the identity element (taking every element $x + iy$ of *C* to itself) and complex conjugation (the map taking each element $x + iy$ to $x - iy$). The Galois group of an extension tells us many of its crucial properties. The study of Galois groups started with Évariste Galois; in modern language, the main outcome of his work is that an equation $f(x) = 0$ can be solved by radicals (that is, *x* can be expressed in terms of the four basic operations together with square roots, cubic roots, etc.) if and only if the extension of the rationals by the roots of the equation $f(x) = 0$ has a Galois group that is solvable in the sense of group theory. ("Solvable", in the sense of group theory, is a simple property that can be checked easily for finite groups.)

[13] It may be useful to look at an example here. Say we want to study the curve $y_2 = x_3 + 7$. We allow *x* and *y* to be complex numbers: $(a+bi)_2 = (c+di)_3 + 7$. This is, in effect, a set of two equations on four variables, since both the real and the imaginary part on each side must match. As a result, we get a surface (two-dimensional) in four-dimensional space. After we choose a convenient hyperplane on which to project the surface (meaning that, say, we choose to ignore the coordinate *a*), we can plot the resulting projection, which is a surface in ordinary three-dimensional space. It then becomes clear that the result is a torus, i.e., the surface of a doughnut (somewhat stretched). A doughnut has one hole; hence the genus is 1.

63.9 References

[1] Long (1972, p. 1)
[2] Neugebauer & Sachs 1945, p. 40. The term *takiltum* is problematic. Robson prefers the rendering "The holding-square of the diagonal from which 1 is torn out, so that the short side comes up...".Robson 2001, p. 192
[3] Robson 2001, p. 189. Other sources give the modern formula $(p^2\square q^2;2pq;p^2+q^2)$. Van der Waerden gives both the modern formula and what amounts to the form preferred by Robson.(van der Waerden 1961, p. 79)
[4] van der Waerden 1961, p. 184.
[5] Neugebauer (Neugebauer 1969, pp. 36–40) discusses the table in detail and mentions in passing Euclid's method in modern notation (Neugebauer 1969, p. 39).
[6] Friberg 1981, p. 302.
[7] van der Waerden 1961, p. 43.
[8] Iamblichus, *Life of Pythagoras*,(trans. e.g. Guthrie 1987) cited in van der Waerden 1961, p. 108. See also Porphyry, *Life of Pythagoras*, paragraph 6, in Guthrie 1987 Van der Waerden (van der Waerden 1961, pp. 87–90) sustains the view that Thales knew Babylonian mathematics.
[9] Herodotus (II. 81) and Isocrates (*Busiris* 28), cited in: Huffman 2011. On Thales, see Eudemus ap. Proclus, 65.7, (e.g. Morrow 1992, p. 52) cited in: O'Grady 2004, p. 1. Proclus was using a work by Eudemus of Rhodes (now lost), the *Catalogue of Geometers*. See also introduction, Morrow 1992, p. xxx on Proclus' reliability.
[10] Becker 1936, p. 533, cited in: van der Waerden 1961, p. 108.
[11] Becker 1936.
[12] van der Waerden 1961, p. 109.
[13] Plato, *Theaetetus*, p. 147 B, (e.g. Jowett 1871), cited in von Fritz 2004, p. 212: "Theodorus was writing out for us something about roots, such as the roots of three or five, showing that they are incommensurable by the unit;..." *See also* Spiral of Theodorus.
[14] von Fritz 2004.
[15] Heath 1921, p. 76.
[16] Sun Zi, *Suan Ching*, Chapter 3, Problem 26. This can be found in Lam & Ang 2004, pp. 219–220, which contains a full translation of the *Suan Ching* (based on Qian 1963). See also the discussion in Lam & Ang 2004, pp. 138–140.
[17] The date of the text has been narrowed down to 220–420 AD (Yan Dunjie) or 280–473 AD (Wang Ling) through internal evidence (= taxation systems assumed in the text). See Lam & Ang 2004, pp. 27–28.
[18] Boyer & Merzbach 1991, p. 82.

[19] Vardi 1998, p. 305-319.
[20] Weil 1984, pp. 17–24.
[21] Plofker 2008, p. 119.
[22] Any early contact between Babylonian and Indian mathematics remains conjectural (Plofker 2008, p. 42).
[23] Mumford 2010, p. 387.
[24] Āryabhaṭa, Āryabhatīya, Chapter 2, verses 32–33, cited in: Plofker 2008, pp. 134–140. See also Clark 1930, pp. 42–50. A slightly more explicit description of the kuṭṭaka was later given in Brahmagupta, *Brāhmasphuṭasiddhānta*, XVIII, 3–5 (in Colebrooke 1817, p. 325, cited in Clark 1930, p. 42).
[25] Mumford 2010, p. 388.
[26] Plofker 2008, p. 194.
[27] Plofker 2008, p. 283.
[28] Colebrooke 1817.
[29] Colebrooke 1817, p. lxv, cited in Hopkins 1990, p. 302. See also the preface in Sachau 1888 cited in Smith 1958, pp. 168
[30] Pingree 1968, pp. 97–125, and Pingree 1970, pp. 103–123, cited in Plofker 2008, p. 256.
[31] Rashed 1980, p. 305–321.
[32] Weil 1984, pp. 45–46.
[33] Weil 1984, p. 118. This was more so in number theory than in other areas (remark in Mahoney 1994, p. 284). Bachet's own proofs were "ludicrously clumsy" (Weil 1984, p. 33).
[34] Mahoney 1994, pp. 48, 53–54. The initial subjects of Fermat's correspondence included divisors ("aliquot parts") and many subjects outside number theory; see the list in the letter from Fermat to Roberval, 22.IX.1636, Tannery & Henry 1891, Vol. II, pp. 72, 74, cited in Mahoney 1994, p. 54.
[35] Weil 1984, pp. 1–2.
[36] Weil 1984, p. 53.
[37] Tannery & Henry 1891, Vol. II, p. 209, Letter XLVI from Fermat to Frenicle, 1640, cited in Weil 1984, p. 56
[38] Tannery & Henry 1891, Vol. II, p. 204, cited in Weil 1984, p. 63. All of the following citations from Fermat's *Varia Opera* are taken from Weil 1984, Chap. II. The standard Tannery & Henry work includes a revision of Fermat's posthumous *Varia Opera Mathematica* originally prepared by his son (Fermat 1679).
[39] Tannery & Henry 1891, Vol. II, p. 213.
[40] Tannery & Henry 1891, Vol. II, p. 423.
[41] Weil 1984, pp. 80, 91–92.
[42] Weil 1984, p. 92.
[43] Weil 1984, Ch. II, sect. XV and XVI.
[44] Tannery & Henry 1891, Vol. I, pp. 340–341.
[45] Weil 1984, p. 115.
[46] Weil 1984, pp. 115–116.

[47] Weil 1984, p. 104.
[48] Weil 1984, pp. 2, 172.
[49] Varadarajan 2006, p. 9.
[50] Weil 1984, p. 2 and Varadarajan 2006, p. 37
[51] Varadarajan 2006, p. 39 and Weil 1984, pp. 176–189

[52] Weil 1984, pp. 178–179.
[53] Weil 1984, p. 174. Euler was generous in giving credit to others (Varadarajan 2006, p. 14), not always correctly.
[54] Weil 1984, p. 183.
[55] Varadarajan 2006, pp. 45–55; see also chapter III.
[56] Varadarajan 2006, pp. 44–47.
[57] Weil 1984, pp. 177–179.
[58] Edwards 1983, pp. 285–291.
[59] Varadarajan 2006, pp. 55–56.
[60] Weil 1984, pp. 179–181.
[61] Weil 1984, p. 181.
[62] Weil 1984, pp. 327–328.
[63] Weil 1984, pp. 332–334.
[64] Weil 1984, pp. 337–338.
[65] Goldstein & Schappacher 2007, p. 14.
[66] From the preface of *Disquisitiones Arithmeticae*; the translation is taken from Goldstein & Schappacher 2007, p. 16
[67] See the discussion in section 5 of Goldstein & Schappacher 2007. Early signs of self-consciousness are present already in letters by Fermat: thus his remarks on what number theory is, and how "Diophantus's work [...] does not really belong to [it]" (quoted in Weil 1984, p. 25).
[68] Apostol 1976, p. 7.
[69] Davenport & Montgomery 2000, p. 1.
[70] See the proof in Davenport & Montgomery 2000, section 1
[71] Iwaniec & Kowalski 2004, p. 1.
[72] Varadarajan 2006, sections 2.5, 3.1 and 6.1.
[73] Granville 2008, pp. 322–348.
[74] See the comment on the importance of modularity in Iwaniec & Kowalski 2004, p. 1
[75] Goldfeld 2003.
[76] See, e.g., the initial comment in Iwaniec & Kowalski 2004, p. 1.
[77] Granville 2008, section 1: "The main difference is that in algebraic number theory [...] one typically considers questions with answers that are given by exact formulas, whereas in analytic number theory [...] one looks for *good approximations*."
[78] See the remarks in the introduction to Iwaniec & Kowalski 2004, p. 1: "However much stronger...".
[79] Granville 2008, section 3: "[Riemann] defined what we now call the Riemann zeta function [...] Riemann's deep work gave birth to our subject [...]"
[80] See, e.g., Montgomery & Vaughan 2007, p. 1.
[81] CITEREFMilne2014, p. 2.
[82] Edwards 2000, p. 79.
[83] Davis, Martin; Matiyasevich, Yuri; Robinson, Julia (1976). "Hilbert's Tenth Problem: Diophantine Equations: Positive Aspects of a Negative Solution". In Felix E. Browder. *Mathematical Developments Arising from Hilbert Problems.* Proceedings of Symposia in Pure Mathematics. XXVIII.2. American Mathematical Society. pp. 323–378. ISBN 0-8218-1428-1. Zbl 0346.02026. Reprinted in *The Collected Works of Julia Robinson*, Solomon Feferman, editor, pp.269–378, American Mathematical Society 1996.

[84] "The Unreasonable Effectiveness of Number Theory", Stefan Andrus Burr, George E. Andrews, American Mathematical Soc., 1992, ISBN 9780821855010
[85] Computer science and its relation to mathematics" DE Knuth - The American Mathematical Monthly, 1974
[86] "Applications of number theory to numerical analysis", Lo-keng Hua, Luogeng Hua, Yuan Wang, Springer-Verlag, 1981, ISBN 978-3-540-10382-0
[87] "Practical applications of algebraic number theory". Mathoverflow.net. Retrieved 2012-05-18.
[88] "Where is number theory used in the rest of mathematics?". Mathoverflow.net. 2008-09-23. Retrieved 2012-05-18.
[89] Apostol n.d..

63.10 Sources

_ Apostol, Tom M. (1976). *Introduction to analytic number theory*. Undergraduate texts in mathematics. Springer. ISBN 978-0-387-90163-3.
_ Apostol, Tom M. (n.d.). "An introduction to the theory of numbers". (Review of Hardy & Wright.) Mathematical Reviews (MathSciNet) MR0568909. American Mathematical Society. (Subscription needed)
_ Becker, Oskar (1936). "Die Lehre von Geraden und Ungeraden im neunten Buch der euklidischen Elemente". *Quellen und Studien zur Geschichte der Mathematik, Astronomie und Physik*. Abteilung B:Studien (in German) (Berlin: J. Springer Verlag) **3**: 533–53.
_ Boyer, Carl Benjamin; Merzbach, Uta C. (1991) [1968]. *A History of Mathematics* (2nd ed.). New York: Wiley. ISBN 978-0-471-54397-8. 1968 edition at Google books

_ Clark, Walter Eugene (trans.) (1930). *The Āryabhaṭīya of Āryabhaṭa: An ancient Indian work on mathematics and astronomy*. University of Chicago Press.
_ Colebrooke, Henry Thomas (1817). *Algebra, with arithmetic and mensuration, from the Sanscrit of Brahmegupta and Bháscara.*. London: J. Murray.
_ Davenport, Harold; Montgomery, Hugh L. (2000). *Multiplicative number theory*. Graduate texts in mathematics **74** (revised 3rd ed.). Springer. ISBN 978-0-387-95097-6.
_ Edwards, Harold M. (November 1983). "Euler and quadratic reciprocity". *Mathematics Magazine* (Mathematical Association of America) **56** (5): 285–291. doi:10.2307/2690368. JSTOR 2690368.
_ Edwards, Harold M. (2000) [1977]. *Fermat's Last Theorem: a genetic introduction to algebraic number theory*. Graduate texts in mathematics **50** (reprint of 1977 ed.). Springer Verlag. ISBN 978-0-387-95002-0.
_ Fermat, Pierre de (1679). *Varia Opera Mathematica* (in French & Latin). Toulouse: Joannis Pech.
_ Friberg, Jöran (August 1981). "Methods and traditions of Babylonian mathematics: Plimpton 322, Pythagorean triples and the Babylonian triangle parameter equations". *Historia Mathematica* (Elsevier) **8** (3): 277–318. doi:10.1016/0315-0860(81)90069-0.
_ von Fritz, Kurt (2004). "The discovery of incommensurability by Hippasus of Metapontum". In Christianidis, J. *Classics in the History of Greek Mathematics*. Berlin: Kluwer (Springer). ISBN 978-1-4020-0081-2.
_ Gauss, Carl Friedrich; Waterhouse, William C. (trans.) (1966) [1801]. *Disquisitiones Arithmeticae*. Springer. ISBN 978-0-387-96254-2.
_ Goldfeld, Dorian M. (2003). "Elementary proof of the prime number theorem: a historical perspective".
_ Goldstein, Catherine; Schappacher, Norbert (2007). "A book in search of a discipline". In Goldstein, C.; Schappacher, N.; Schwermer, Joachim. *The Shaping of Arithmetic after Gauss' "Disquisitiones Arithmeticae"*. Berlin & Heidelberg: Springer. pp. 3–66. ISBN 978-3-540-20441-1.
_ Granville, Andrew (2008). "Analytic number theory". In Gowers, Timothy; Barrow-Green, June; Leader, Imre. *The Princeton Companion to to Mathematics*. Princeton University Press. ISBN 978-0-691-11880-2.

_ Porphyry; Guthrie, K. S. (trans.) (1920). *Life of Pythagoras*. Alpine, New Jersey: Platonist Press.
_ Guthrie, Kenneth Sylvan (1987). *The Pythagorean Sourcebook and Library*. Grand Rapids, Michigan: Phanes Press. ISBN 978-0-933999-51-0.
_ Hardy, Godfrey Harold; Wright, E. M. (2008) [1938]. *An Introduction to the Theory of Numbers* (Sixth ed.). Oxford University Press. ISBN 978-0-19-921986-5. MR 2445243.
_ Heath, Thomas L. (1921). *A History of Greek Mathematics, Volume 1: From Thales to Euclid*. Oxford: Clarendon Press.
_ Hopkins, J. F. P. (1990). "Geographical and navigational literature". In Young, M. J. L.; Latham, J. D.; Serjeant, R. B. *Religion, learning and science in the `Abbasid period*. The Cambridge history of Arabic literature. Cambridge University Press. ISBN 978-0-521-32763-3.
_ Huffman, Carl A. (8 August 2011). Zalta, Edward N., ed. "Pythagoras". *Stanford Encyclopaedia of Philosophy* (Fall 2011 ed.). Retrieved 7 February 2012.
_ Iwaniec, Henryk; Kowalski, Emmanuel (2004). *Analytic number theory*. American Mathematical Society Colloquium Publications **53**. Providence, RI,: American Mathematical Society. ISBN 0-8218-3633-1.
_ Plato; Jowett, Benjamin (trans.) (1871). *Theaetetus*.
_ Lam, Lay Yong; Ang, Tian Se (2004). *Fleeting Footsteps: Tracing the conception of arithmetic and algebra in ancient China* (revised ed.). Singapore: World Scientific. ISBN 978-981-238-696-0.
_ Long, Calvin T. (1972). *Elementary Introduction to Number Theory* (2nd ed.). Lexington: D. C. Heath and Company. LCCN 77171950.
_ Mahoney, M. S. (1994). *The mathematical career of Pierre de Fermat, 1601–1665* (Reprint, 2nd ed.). Princeton University Press. ISBN 978-0-691-03666-3.
_ Milne, J. S. (2014). *Algebraic number theory*. Available at www.jmilne.org/math.
_ Montgomery, Hugh L.; Vaughan, Robert C. (2007). *Multiplicative number theory: I, Classical Theory,*. Cambridge University Press. ISBN 978-0-521-84903-6.
_ Morrow, Glenn Raymond (trans., ed.); Proclus (1992). *A commentary on Book 1 of Euclid's Elements*. Princeton University Press. ISBN 978-0-691-02090-7.
_ Mumford, David (March 2010). "Mathematics in India: reviewed by David Mumford". *Notices of the American Mathematical Society* **57** (3): 387. ISSN 1088-9477.
_ Neugebauer, Otto E. (1969). *The exact sciences in antiquity* (corrected reprint of the 1957 ed.). New York: Dover Publications. ISBN 978-0-486-22332-2.
_ Neugebauer, Otto E.; Sachs, Abraham Joseph; Götze, Albrecht (1945). *Mathematical cuneiform texts*. American Oriental Series **29**. American Oriental Society etc.
_ O'Grady, Patricia (September 2004). "Thales of Miletus". The Internet Encyclopaedia of Philosophy. Retrieved 7 February 2012.
_ Pingree, David; Ya'qub, ibn Tariq (1968). "The fragments of the works of Ya'qub ibn Tariq". *Journal of Near*

Eastern Studies (University of Chicago Press) **26**.
_ Pingree, D.; al-Fazari (1970). "The fragments of the works of al-Fazari". *Journal of Near Eastern Studies* (University of Chicago Press) **28**.
_ Plofker, Kim (2008). *Mathematics in India*. Princeton University Press. ISBN 978-0-691-12067-6.
_ Qian, Baocong, ed. (1963). *Suanjing shi shu (Ten mathematical classics)* (in Chinese). Beijing: Zhonghua shuju.
_ Rashed, Roshdi (1980). "Ibn al-Haytham el le théorème de Wilson". *Archive for History of Exact Sciences* **22** (4): 305–321. doi:10.1007/BF00717654.

_ Robson, Eleanor (2001). "Neither Sherlock Holmes nor Babylon: a reassessment of Plimpton 322". *Historia Mathematica* (Elsevier) **28** (28): 167–206. doi:10.1006/hmat.2001.2317.
_ Sachau, Eduard (1888). *Alberuni's India: An account of the religion, philosophy, literature, geography, chronology, astronomy and astrology of India, Vol. 1*. London: Kegan, Paul, Trench, Trübner & Co. |first2= missing |last2= in Authors list (help)
_ Serre, Jean-Pierre (1996) [1973]. *A course in arithmetic*. Graduate texts in mathematics **7**. Springer. ISBN 978-0-387-90040-7.
_ Smith, D. E. (1958). *History of Mathematics, Vol I*. New York: Dover Publications.
_ Tannery, Paul; Henry, Charles (eds.); Fermat, Pierre de (1891). *Oeuvres de Fermat*. (4 Vols.) (in French & Latin). Paris: Imprimerie Gauthier-Villars et Fils. Volume 1 Volume 2 Volume 3 Volume 4 (1912)
_ Iamblichus; Taylor, Thomas (trans.) (1818). *Life of Pythagoras or, Pythagoric life*. London: J. M. Watkins. For other editions, see Iamblichus#List of editions and translations
_ Truesdell, C. A. (1984). "Leonard Euler, supreme geometer". In Hewlett, John (trans.). *Leonard Euler, Elements of Algebra* (reprint of 1840 5th ed.). New York: Springer-Verlag. ISBN 978-0-387-96014-2. This Google books preview of *Elements of algebra* lacks Truesdell's intro, which is reprinted (slightly abridged) in the following book:
_ Truesdell, C. A. (2007). "Leonard Euler, supreme geometer". In Dunham, William. *The Genius of Euler: reflections on his life and work*. Volume 2 of MAA tercentenary Euler celebration. New York: Mathematical Association of America. ISBN 978-0-88385-558-4.
_ Varadarajan, V. S. (2006). *Euler through time: a new look at old themes*. American Mathematical Society. ISBN 978-0-8218-3580-7.
_ Vardi, Ilan (April 1998). "Archimedes' cattle problem". *American Mathematical Monthly* **105** (4): 305–319. doi:10.2307/2589706.
_ van der Waerden, Bartel L.; Dresden, Arnold (trans) (1961). *Science Awakening*. Vol. 1 or Vol 2. New York: Oxford University Press.
_ Weil, André (1984). *Number theory: an approach through history – from Hammurapi to Legendre,*. Boston: Birkhäuser. ISBN 978-0-8176-3141-3.

63.11 External links

_ Hazewinkel, Michiel, ed. (2001), "Number theory", *Encyclopedia of Mathematics*, Springer, ISBN 978-1-55608-010-4
_ Quotations related to Number theory at Wikiquote
_ Number Theory Web

Title page of the 1621 edition of Diophantus' Arithmetica, *translated into Latin by Claude Gaspard Bachet de Méziriac.*

Al-Haytham seen by the West: frontispice of Selenographia, *showing Alhasen [*sic*] representing knowledge through reason, and Galileo representing knowledge through the senses.*

Leonhard Euler

Carl Friedrich Gauss's Disquisitiones Arithmeticae, first edition

Carl Friedrich Gauss

Ernst Kummer

Peter Gustav Lejeune Dirichlet

Riemann zeta function ζ(s) in the complex plane. The color of a point s gives the value of ζ(s): dark colors denote values close to zero and hue gives the value's argument.

The action of the modular group on the upper half plane. The region in grey is the standard fundamental domain.

1 2

$y^2 = x^3 - x$ $y^2 = x^3 - x + 1$

Two examples of an elliptic curve, i.e., a curve of genus 1 having at least one rational point. (Either graph can be seen as a slice of a torus in four-dimensional space.)

Chapter 64

Mathematical statistics

Illustration of linear regression on a data set. Regression analysis is an important part of mathematical statistics.

Mathematical statistics is the application of mathematics to statistics, which was originally conceived as the science of the state — the collection and analysis of facts about a country: its economy, land, military, population, and so forth. Mathematical techniques which are used for this include mathematical analysis, linear algebra, stochastic analysis, differential equations, and measure-theoretic probability theory.[1][2]

64.1 Introduction

Statistical science is concerned with the planning of studies, especially with the design of randomized experiments and with the planning of surveys using random sampling. The initial analysis of the data from properly randomized studies often follows the study protocol.

Of course, the data from a randomized study can be analyzed to consider secondary hypotheses or to suggest new ideas. A secondary analysis of the data from a planned study uses tools from data analysis.

Data analysis is divided into:

_ descriptive statistics - the part of statistics that describes data, i.e. summarises the data and their typical properties.
_ inferential statistics - the part of statistics that draws conclusions from data (using some model for the data):

For example, inferential statistics involves selecting a model for the data, checking whether the data fulfill the conditions of a particular model, and with quantifying the involved uncertainty (e.g. using confidence intervals).

While the tools of data analysis work best on data from randomized studies, they are also applied to other kinds of data --- for example, from natural experiments and observational studies, in which case the inference is dependent on the model chosen by the statistician, and so subjective.[3]

Mathematical statistics has been inspired by and has extended many procedures in applied statistics.

64.2 Topics

The following are some of the important topics in mathematical statistics:[4][5]

64.2.1 Probability distributions

Main article: Probability distribution

A probability distribution assigns a probability to each measurable subset of the possible outcomes of a random experiment, survey, or procedure of statistical inference. Examples are found in experiments whose sample space is non-numerical, where the distribution would be a categorical distribution; experiments whose sample space is encoded by discrete random variables, where the distribution can be specified by a probability mass function; and experiments with sample spaces encoded by continuous random variables, where the distribution can be specified by a probability density function. More complex experiments, such as those involving stochastic processes defined in continuous time, may demand the use of more general probability measures.

A probability distribution can either be univariate or multivariate. A univariate distribution gives the probabilities of a single random variable taking on various alternative values; a multivariate distribution (a joint probability distribution) gives the probabilities of a random vector—a set of two or more random variables—taking on various combinations of values. Important and commonly encountered univariate probability distributions include the binomial distribution, the hypergeometric distribution, and the normal distribution. The multivariate normal distribution is a commonly encountered multivariate distribution.

Special distributions

_ Normal distribution (Gaussian distribution), the most common continuous distribution
_ Bernoulli distribution, for the outcome of a single Bernoulli trial (e.g. success/failure, yes/no)
_ Binomial distribution, for the number of "positive occurrences" (e.g. successes, yes votes, etc.) given a fixed total number of independent occurrences
_ Negative binomial distribution, for binomial-type observations but where the quantity of interest is the number of failures before a given number of successes occurs

_ Geometric distribution, for binomial-type observations but where the quantity of interest is the number of failures before the first success; a special c*Discrete uniform distribution, for a finite set of values (e.g. the outcome of a fair die)
_ Continuous uniform distribution, for continuously distributed values
_ Poisson distribution, for the number of occurrences of a Poisson-type event in a given period of time
_ Exponential distribution, for the time before the next Poisson-type event occurs
_ Gamma distribution, for the time before the next k Poisson-type events occur

_ Chi-squared distribution, the distribution of a sum of squared standard normal variables; useful e.g. for inference regarding the sample variance of normally distributed samples (see chi-squared test)
_ Student's t distribution, the distribution of the ratio of a standard normal variable and the square root of a scaled chi squared variable; useful for inference regarding the mean of normally distributed samples with unknown variance (see Student's t-test)
_ Beta distribution, for a single probability (real number between 0 and 1); conjugate to the Bernoulli distribution and binomial distribution

64.2.2 Statistical inferences

Main article: Statistical inference
Statistical inference is the process of drawing conclusions from data that are subject to random variation, for example, observational errors or sampling variation.[6] Initial requirements of such a system of procedures for inference and induction are that the system should produce reasonable answers when applied to well-defined situations and that it should be general enough to be applied across a range of situations. Inferential statistics are used to test hypotheses and make estimations using sample data. Whereas descriptive statistics describe a sample, inferential statistics infer predictions about a larger population that the sample represents.
The outcome of statistical inference may be an answer to the question "what should be done next?", where this might be a decision about making further experiments or surveys, or about drawing a conclusion before implementing some organizational or governmental policy. For the most part, statistical inference makes propositions about populations, using data drawn from the population of interest via some form of random sampling. More generally, data about a random process is obtained from its observed behavior during a finite period of time. Given a parameter or hypothesis about which one wishes to make inference, statistical inference most often uses:
_ a statistical model of the random process that is supposed to generate the data, which is known when randomization has been used, and
_ a particular realization of the random process; i.e., a set of data.

64.2.3 Regression

Main article: Regression analysis
In statistics, **regression analysis** is a statistical process for estimating the relationships among variables. It includes many techniques for modeling and analyzing several variables, when the focus is on the relationship between a dependent variable and one or more independent variables. More specifically, regression analysis helps one understand how the typical value of the dependent variable (or 'criterion variable') changes when any one of the independent variables is varied, while the other independent variables are held fixed. Most commonly, regression analysis estimates the conditional expectation of the dependent variable given the independent variables – that is, the average value of the dependent variable when the independent variables are fixed. Less commonly, the focus is on a quantile, or other location parameter of the conditional distribution of the dependent variable given the independent variables. In all cases, the estimation target is a function of the independent variables called the **regression function**. In regression analysis, it is also of interest to characterize the variation of the dependent variable around the regression function which can be described by a probability distribution.
Many techniques for carrying out regression analysis have been developed. Familiar methods such as linear regression and ordinary least squares regression are parametric, in that the regression function is defined in terms of a finite number of unknown parameters that are estimated from the data. Nonparametric regression refers to techniques that allow the regression function to lie in a specified set of functions, which may be infinite-dimensional.

64.2.4 Nonparametric statistics

Main article: Nonparametric statistics

Nonparametric statistics are statistics not based on parameterized families of probability distributions. They include both descriptive and inferential statistics. The typical parameters are the mean, variance, etc. Unlike parametric statistics, nonparametric statistics make no assumptions about the probability distributions of the variables being assessed.
Non-parametric methods are widely used for studying populations that take on a ranked order (such as movie reviews receiving one to four stars). The use of non-parametric methods may be necessary when data have a ranking but no clear numerical interpretation, such as when assessing preferences. In terms of levels of measurement, non-parametric methods result in "ordinal" data.

As non-parametric methods make fewer assumptions, their applicability is much wider than the corresponding parametric methods. In particular, they may be applied in situations where less is known about the application in question.
Also, due to the reliance on fewer assumptions, non-parametric methods are more robust.
Another justification for the use of non-parametric methods is simplicity. In certain cases, even when the use of parametric methods is justified, non-parametric methods may be easier to use. Due both to this simplicity and to their greater robustness, non-parametric methods are seen by some statisticians as leaving less room for improper use and misunderstanding.

64.3 Statistics, mathematics, and mathematical statistics

Mathematical statistics has substantial overlap with the discipline of statistics. Statistical theorists study and improve statistical procedures with mathematics, and statistical research often raises mathematical questions. Statistical theory relies on probability and decision theory.
Mathematicians and statisticians like Gauss, Laplace, and C. S. Peirce used decision theory with probability distributions and loss functions (or utility functions). The decision-theoretic approach to statistical inference was reinvigorated by Abraham Wald and his successors,[7][8][9][10][11][12][13] and makes extensive use of scientific computing, analysis, and optimization; for the design of experiments, statisticians use algebra and combinatorics.

64.4 See also

_ Asymptotic theory (statistics)

64.5 References

[1] Lakshmikantham,, ed. by D. Kannan,... V. (2002). *Handbook of stochastic analysis and applications*. New York: M. Dekker. ISBN 0824706609.
[2] Schervish, Mark J. (1995). *Theory of statistics* (Corr. 2nd print. ed.). New York: Springer. ISBN 0387945466.
[3] Freedman, D.A. (2005) *Statistical Models: Theory and Practice*, Cambridge University Press. ISBN 978-0-521-67105-7
[4] Hogg, R. V., A. Craig, and J. W. McKean. "Intro to Mathematical Statistics." (2005).
[5] Larsen, Richard J. and Marx, Morris L. "An Introduction to Mathematical Statistics and Its Applications" (2012). Prentice Hall.
[6] Upton, G., Cook, I. (2008) *Oxford Dictionary of Statistics*, OUP. ISBN 978-0-19-954145-4
[7] Wald, Abraham (1947). *Sequential analysis*. New York: John Wiley and Sons. ISBN 0-471-91806-7. "See Dover reprint, 2004: ISBN 0-486-43912-7"
[8] Wald, Abraham (1950). *Statistical Decision Functions*. John Wiley and Sons, New York.
[9] Lehmann, Erich (1997). *Testing Statistical Hypotheses* (2nd ed.). ISBN 0-387-94919-4.
[10] Lehmann, Erich; Cassella, George (1998). *Theory of Point Estimation* (2nd ed.). ISBN 0-387-98502-6.
[11] Bickel, Peter J.; Doksum, Kjell A. (2001). *Mathematical Statistics: Basic and Selected Topics* **1** (Second (updated printing 2007) ed.). Pearson Prentice-Hall.

[12] Le Cam, Lucien (1986). *Asymptotic Methods in Statistical Decision Theory*. Springer-Verlag. ISBN 0-387-96307-3.
[13] Liese, Friedrich and Miescke, Klaus-J. (2008). *Statistical Decision Theory: Estimation, Testing, and Selection*. Springer.

64.6 Additional reading

_ Borovkov, A. A. (1999). *Mathematical Statistics*. CRC Press. ISBN 90-5699-018-7
_ Virtual Laboratories in Probability and Statistics (Univ. of Ala.-Huntsville)
_ StatiBot, interactive online expert system on statistical tests.

Chapter 65

Computer-assisted proof

A **computer-assisted proof** is a mathematical proof that has been at least partially generated by computer.
Most computer-aided proofs to date have been implementations of large proofs-by-exhaustion of a mathematical theorem. The idea is to use a computer program to perform lengthy computations, and to provide a proof that the result of these computations implies the given theorem. In 1976, the four color theorem was the first major theorem to be verified using a computer program.
Attempts have also been made in the area of artificial intelligence research to create smaller, explicit, new proofs of mathematical theorems from the bottom up using machine reasoning techniques such as heuristic search. Such automated theorem provers have proved a number of new results and found new proofs for known theorems. Additionally, interactive proof assistants allow mathematicians to develop human-readable proofs which are nonetheless formally verified for correctness. Since these proofs are generally human-surveyable (albeit with difficulty, as with the proof of the Robbins conjecture) they do not share the controversial implications of computer-aided proofs-byexhaustion.

65.1 Methods

One method for using computers in mathematical proofs is by means of so-called validated numerics or rigorous numerics. This means computing numerically yet with mathematical rigour. One uses set-valued arithmetic and inclusion principle in order to ensure that the set-valued output of a numerical program encloses the solution of the original mathematical problem. This is done by controlling, enclosing and propagating round-off and truncation errors using for example interval arithmetic. More precisely, one reduces the computation to a sequence of elementary operations, say (+,-,*,/). In a computer, the result of each elementary operation is rounded off by the computer precision. However, one can construct an interval provided by upper and lower bounds on the result of an elementary operation. Then one proceeds by replacing numbers with intervals and performing elementary operations between such intervals of representable numbers.

65.2 Philosophical objections

Computer-assisted proofs are the subject of some controversy in the mathematical world. Some mathematicians believe that lengthy computer-assisted proofs are not, in some sense, 'real' mathematical proofs because they involve so many logical steps that they are not practically verifiable by human beings, and that mathematicians are effectively being asked to replace logical deduction from assumed axioms with trust in an empirical computational process, which is potentially affected by errors in the computer program, as well as defects in the runtime environment and hardware.

Other mathematicians believe that lengthy computer-assisted proofs should be regarded as *calculations*, rather than *proofs*: the proof algorithm itself should be proved valid, so that its use can then be regarded as a mere "verification". Arguments that computer-assisted proofs are subject to errors in their source programs, compilers, and hardware can be resolved by providing a formal proof of correctness for the computer program (an approach which was successfully applied to the four-color theorem in 2005) as well as replicating the result using different programming languages,

different compilers, and different computer hardware.

Another possible way of verifying computer-aided proofs is to generate their reasoning steps in a machine-readable form, and then use an automated theorem prover to demonstrate their correctness. This approach of using a computer program to prove another program correct does not appeal to computer proof skeptics, who see it as adding another layer of complexity without addressing the perceived need for human understanding.

Another argument against computer-aided proofs is that they lack mathematical elegance—that they provide no insights or new and useful concepts. In fact, this is an argument that could be advanced against any lengthy proof by exhaustion.

An additional philosophical issue raised by computer-aided proofs is whether they make mathematics into a quasiempirical science, where the scientific method becomes more important than the application of pure reason in the area of abstract mathematical concepts. This directly relates to the argument within mathematics as to whether mathematics is based on ideas, or "merely" an exercise in formal symbol manipulation. It also raises the question whether, if according to the Platonist view, all possible mathematical objects in some sense "already exist", whether computer-aided mathematics is an observational science like astronomy, rather than an experimental one like physics or chemistry. Interestingly, this controversy within mathematics is occurring at the same time as questions are being asked in the physics community about whether twenty-first century theoretical physics is becoming too mathematical, and leaving behind its experimental roots.

The emerging field of experimental mathematics is confronting this debate head-on by focusing on numerical experiments as its main tool for mathematical exploration.

65.3 Theorems for sale

In 2010, academics at The University of Edinburgh offered people the chance to "buy their own theorem" created through a computer-assisted proof. This new theorem would be named after the purchaser.[1]

65.4 List of theorems proved with the help of computer programs

Inclusion in this list does not imply that a formal computer-checked proof exists, but rather, that a computer program has been involved in some way. See the main articles for details.

_ Four color theorem, 1976
_ Mitchell Feigenbaum's universality conjecture in non-linear dynamics. Proven by O.E. Lanford using rigorous computer arithmetic, 1982.
_ Connect Four, 1988 – a game
_ Non-existence of a finite projective plane of order 10, 1989
_ Robbins conjecture, 1996
_ Kepler conjecture, 1998 – the problem of optimal sphere packing in a box.
_ Lorenz attractor, 2002 - 14th of Smale's problems proved by W. Tucker using interval arithmetic.
_ 17-point case of the Happy Ending problem, 2006
_ NP-hardness of minimum-weight triangulation, 2008
_ Optimal solutions for Rubik's Cube can be obtained in at most 20 face moves, 2010

65.5 See also

_ Mathematical proof

_ Model checking

_ Proof checking
_ Symbolic computation
_ Automated reasoning
_ Formal verification
_ Seventeen or Bust
_ Metamath

65.6 References

[1] "Herald Gazette article on buying your own theorem". *Herald Gazette Scotland*. November 2010.

65.7 Further reading

_ Lenat, D.B., (1976), AM: An artificial intelligence approach to discovery in mathematics as heuristic search, Ph.D. Thesis, STAN-CS-76-570, and Heuristic Programming Project Report HPP-76-8, Stanford University, AI Lab., Stanford, CA.

65.8 External links

_ Oscar E. Lanford; A computer-assisted proof of the Feigenbaum conjectures, "Bull. Amer. Math. Soc.", 1982
_ Edmund Furse; Why did AM run out of steam?
_ Number proofs done by computer might err
_ "A Special Issue on Formal Proof". *Notices of the American Mathematical Society*. December 2008.

Chapter 66

Zermelo–Fraenkel set theory

"ZFC" redirects here. For other uses, see ZFC (disambiguation).

In mathematics, **Zermelo–Fraenkel set theory with the axiom of choice**, named after mathematicians Ernst Zermelo and Abraham Fraenkel and commonly abbreviated **ZFC**, is one of several axiomatic systems that were proposed in the early twentieth century to formulate a theory of sets free of paradoxes such as Russell's paradox. Specifically, ZFC does not allow unrestricted comprehension. Today ZFC is the standard form of axiomatic set theory and as such is the most common foundation of mathematics. The consistency of ZFC cannot be proved within ZFC itself.

ZFC is intended to formalize a single primitive notion, that of a hereditary well-founded set, so that all entities in the universe of discourse are such sets. Thus the axioms of ZFC refer only to sets, not to urelements (elements of sets that are not themselves sets) or classes (collections of mathematical objects defined by a property shared by their members). The axioms of ZFC prevent its models from containing urelements, and proper classes can only be treated indirectly.

Formally, ZFC is a one-sorted theory in first-order logic. The signature has equality and a single primitive binary relation, set membership, which is usually denoted ∈. The formula $a \in b$ means that the set *a* is a member of the set *b* (which is also read, "*a* is an element of *b*" or "*a* is in *b*").

There are many equivalent formulations of the ZFC axioms. Most of the ZFC axioms state the existence of particular sets defined from other sets. For example, the axiom of pairing says that given any two sets *a* and *b* there is a new set $\{a, b\}$ containing exactly *a* and *b*. Other axioms describe properties of set membership. A goal of the ZFC axioms is that each axiom should be true if interpreted as a statement about the collection of all sets in the von Neumann universe (also known as the cumulative hierarchy).

The metamathematics of ZFC has been extensively studied. Landmark results in this area established the independence of the continuum hypothesis from ZFC, and of the axiom of choice from the remaining ZFC axioms.

66.1 History

In 1908, Ernst Zermelo proposed the first axiomatic set theory, Zermelo set theory. However, as first pointed out by Abraham Fraenkel in a 1921 letter to Zermelo, this theory was incapable of proving the existence of certain sets and cardinal numbers whose existence was taken for granted by most set theorists of the time, notably, the cardinal number $\aleph_\omega$ and, where Z_0 is any infinite set and $\wp$ is the power set operation, the set $\{Z_0, \wp(Z_0), \wp(\wp(Z_0)),...\}$ (Ebbinghaus 2007, p. 136). Moreover, one of Zermelo's axioms invoked a concept, that of a "definite" property, whose operational meaning was not clear. In 1922, Fraenkel and Thoralf Skolem independently proposed operationalizing a "definite" property as one that could be formulated as a first order theory whose atomic formulas were limited to set membership and identity. They also independently proposed replacing the axiom schema of specification with the axiom schema of replacement. Appending this schema, as well as the axiom of regularity (first proposed by Dimitry Mirimanoff in 1917), to Zermelo set theory yields the theory denoted by **ZF**. Adding to ZF either the axiom of choice (AC) or a statement that is equivalent to it yields ZFC.

66.2 Axioms

There are many equivalent formulations of the ZFC axioms; for a rich but somewhat dated discussion of this fact, see Fraenkel *et al.* (1973). The following particular axiom set is from Kunen (1980). The axioms per se are expressed in the symbolism of first order logic. The associated English prose is only intended to aid the intuition.
All formulations of ZFC imply that at least one set exists. Kunen includes an axiom that directly asserts the existence of a set, in addition to the axioms given below (although he notes that he does so only "for emphasis" (*ibid.*, p. 10)). Its omission here can be justified in two ways. First, in the standard semantics of first-order logic in which ZFC is typically formalized, the domain of discourse must be nonempty. Hence, it is a logical theorem of first-order logic that something exists — usually expressed as the assertion that something is identical to itself, ∃x(x=x). Consequently, it is a theorem of every first-order theory that something exists. However, as noted above, because in the intended semantics of ZFC there are only sets, the interpretation of this logical theorem in the context of ZFC is that some *set* exists. Hence, there is no need for a separate axiom asserting that a set exists. Second, however, even if ZFC is formulated in so-called free logic, in which it is not provable from logic alone that something exists, the axiom of infinity (below) asserts that an *infinite* set exists. This obviously implies that *a* set exists and so, once again, it is superfluous to include an axiom asserting as much.

66.2.1 1. Axiom of extensionality

Main article: Axiom of extensionality
Two sets are equal (are the same set) if they have the same elements.
8x8y[8z(z 2 x , z 2 y)) x = y]:
The converse of this axiom follows from the substitution property of equality. If the background logic does not include equality "=", *x=y* may be defined as an abbreviation for the following formula (Hatcher 1982, p. 138, def. 1):
8z[z 2 x , z 2 y] ^ 8w[x 2 w , y 2 w]:
In this case, the axiom of extensionality can be reformulated as
8x8y[8z(z 2 x , z 2 y)) 8w(x 2 w , y 2 w)];
which says that if *x* and *y* have the same elements, then they belong to the same sets (Fraenkel *et al.* 1973).

66.2.2 2. Axiom of regularity (also called the *Axiom of foundation*)

Main article: Axiom of regularity
Every non-empty set *x* contains a member *y* such that *x* and *y* are disjoint sets.
8x[9a(a 2 x)) 9y(y 2 x ^ :9z(z 2 y ^ z 2 x))]:
This implies, for example, that no set is an element of itself and that every set has an ordinal rank.

66.2.3 3. Axiom schema of specification (also called the axiom schema of *separation* or of *restricted comprehension*)

Main article: Axiom schema of specification

Subsets are commonly constructed using set builder notation. For example, the even integers can be constructed as the subset of the integers Z satisfying the predicate *x* _ 0 (mod 2) :
fx 2 Z : *x* _ 0 (mod 2)*g:*
In general, the subset of a set *z* obeying a formula *ϕ* (*x*) with one free variable *x* may be written as:
fx 2 z : *ϕ*(*x*)*g:*
The axiom schema of specification states that this subset always exists (it is an axiom *schema* because there is one axiom for each *ϕ*). Formally, let *ϕ*be any formula in the language of ZFC with all free variables among *x; z;w*1*; : : : ;w*n
(*y* is *not* free in *ϕ*). Then:
*8z8w*1*8w*2 *: : : 8w*n*9y8x[x 2 y , (x 2 z ^ ϕ)]:*
Note that the axiom schema of specification can only construct subsets, and does not allow the construction of sets of the more general form:
fx : *ϕ*(*x*)*g:*
This restriction is necessary to avoid Russell's paradox and its variants.
In some other axiomatizations of ZF, this axiom is redundant in that it follows from the axiom schema of replacement.
The axiom of specification can be used to prove the existence of the empty set, denoted ∅ , once at least one set is known to exist (see above). One way to do this is to use a property *ϕ*which no set has. For example, if *w* is any existing set, the empty set can be constructed as
∅ = *fu 2 w j* (*u 2 u*) ^ :(*u 2 u*)*g*
Thus the axiom of the empty set is implied by the nine axioms presented here. The axiom of extensionality implies the empty set is unique (does not depend on *w*). It is common to make a definitional extension that adds the symbol ∅ to the language of ZFC.

66.2.4 4. Axiom of pairing

Main article: Axiom of pairing
If *x* and *y* are sets, then there exists a set which contains *x* and *y* as elements.
8x8y9z(x 2 z ^ y 2 z):

The axiom schema of specification must be used to reduce this to a set with exactly these two elements. This axiom is part of Z, but is redundant in ZF because it follows from the axiom schema of replacement, if we are given a set with at least two elements. The existence of a set with at least two elements is assured by either the axiom of infinity, or by the axiom schema of specification and the axiom of the power set applied twice to any set.

66.2.5 5. Axiom of union

Main article: Axiom of union

The union over the elements of a set exists. For example, the union over the elements of the set *ff*1*; 2g; f*2*; 3gg* is *f*1*; 2; 3g* .

Formally, for any set *F* there is a set *A* containing every element that is a member of some member of *F* :

8F 9A8Y 8x[(*x 2 Y ^ Y 2 F*) *) x 2 A*]*:*

A B

$f(x)$

$f : A \to B$

x

Axiom schema of replacement: the image of the domain set A *under the definable function* f *(i.e. the range of* f*) falls inside a set* B.

66.2.6 6. Axiom schema of replacement

Main article: Axiom schema of replacement

The axiom schema of replacement asserts that the image of a set under any definable function will also fall inside a set.

Formally, let *ϕ*be any formula in the language of ZFC whose free variables are among *x; y; A;w*1*; : : : ;wn* , so that in particular *B* is not free in *ϕ*. Then:

*8A8w*1*8w*2 *: : : 8wn*

[

8x(*x 2 A) 9*!*y ϕ*)*) 9B 8x*

(

x 2 A) 9y(*y 2 B ^ ϕ*)

)]

:

In other words, if the relation *ϕ*represents a definable function *f*, *A* represents its domain, and *f*(*x*) is a set for every *x* in that domain, then the range of *f* is a subset of some set *B* . The form stated here, in which *B* may be larger than strictly necessary, is sometimes called the axiom schema of collection.

66.2.7 7. Axiom of infinity

Main article: Axiom of infinity

Let *S*(*w*) abbreviate *w [fwg* , where *w*is some set (We can see that *fwg* is a valid set by applying the Axiom of Pairing with *x* = *y* = *w*so that the set *z*is *fwg*). Then there exists a set *X* such that the empty set ∅ is a member of *X* and, whenever a set *y* is a member of *X*, then *S*(*y*)is also a member of *X*.

9X [∅ *2 X ^ 8y*(*y 2 X) S*(*y*) *2 X*)] *:*

More colloquially, there exists a set *X* having infinitely many members. The minimal set *X* satisfying the axiom of infinity is the von Neumann ordinal ω, which can also be thought of as the set of natural numbers N .

66.2.8 8. Axiom of power set

Main article: Axiom of power set

By definition a set *z* is a subset of a set *x* if and only if every element of *z* is also an element of *x*:

(*z _ x*) *,* (*8q*(*q 2 z) q 2 x*))*:*

The Axiom of Power Set states that for any set *x*, there is a set *y* that contains every subset of *x*:

8x9y8z[*z _ x) z 2 y*]*:*

The axiom schema of specification is then used to define the power set *P(x)* as the subset of such a *y* containing the subsets of *x* exactly:

P(*x*) = *fz 2 y* : *z _ xg*

Axioms **1–8** define ZF. Alternative forms of these axioms are often encountered, some of which are listed in Jech

(2003). Some ZF axiomatizations include an axiom asserting that the empty set exists. The axioms of pairing, union, replacement, and power set are often stated so that the members of the set x whose existence is being asserted are just those sets which the axiom asserts x must contain.
The following axiom is added to turn ZF into ZFC:

66.2.9 9. Well-ordering theorem

Main article: Well-ordering theorem
For any set X, there is a binary relation R which well-orders X. This means R is a linear order on X such that every nonempty subset of X has a member which is minimal under R.
8X9R(R well-orders *X):*
Given axioms **1–8**, there are many statements provably equivalent to axiom **9**, the best known of which is the axiom of choice (AC), which goes as follows. Let X be a set whose members are all non-empty. Then there exists a function f from X to the union of the members of X, called a "choice function", such that for all $Y \in X$ one has $f(Y) \in Y$.
Since the existence of a choice function when X is a finite set is easily proved from axioms **1–8**, AC only matters for certain infinite sets. AC is characterized as nonconstructive because it asserts the existence of a choice set but says nothing about how the choice set is to be "constructed." Much research has sought to characterize the definability (or lack thereof) of certain sets whose existence AC asserts.

66.3 Motivation via the cumulative hierarchy

One motivation for the ZFC axioms is the cumulative hierarchy of sets introduced by John von Neumann (Shoenfield 1977, sec. 2). In this viewpoint, the universe of set theory is built up in stages, with one stage for each ordinal number. At stage 0 there are no sets yet. At each following stage, a set is added to the universe if all of its elements have been added at previous stages. Thus the empty set is added at stage 1, and the set containing the empty set is added at stage 2; see Hinman (2005, p. 467). The collection of all sets that are obtained in this way, over all the stages, is known as V. The sets in V can be arranged into a hierarchy by assigning to each set the first stage at which that set was added to V.
It is provable that a set is in V if and only if the set is pure and well-founded; and provable that V satisfies all the axioms of ZFC, if the class of ordinals has appropriate reflection properties. For example, suppose that a set x is added at stage α, which means that every element of x was added at a stage earlier than α. Then every subset of x is also added at stage α, because all elements of any subset of x were also added before stage α. This means that any subset of x which the axiom of separation can construct is added at stage α, and that the powerset of x will be added at the next stage after α. For a complete argument that V satisfies ZFC see Shoenfield (1977).
The picture of the universe of sets stratified into the cumulative hierarchy is characteristic of ZFC and related axiomatic set theories such as Von Neumann–Bernays–Gödel set theory (often called NBG) and Morse–Kelley set theory. The cumulative hierarchy is not compatible with other set theories such as New Foundations.
It is possible to change the definition of V so that at each stage, instead of adding all the subsets of the union of the previous stages, subsets are only added if they are definable in a certain sense. This results in a more "narrow" hierarchy which gives the constructible universe L, which also satisfies all the axioms of ZFC, including the axiom of choice. It is independent from the ZFC axioms whether $V = L$. Although the structure of L is more regular and well behaved than that of V, few mathematicians argue that $V = L$ should be added to ZFC as an additional axiom.

66.4 Metamathematics

The axiom schemata of replacement and separation each contain infinitely many instances. Montague (1961) included a result first proved in his 1957 Ph.D. thesis: if ZFC is consistent, it is impossible to axiomatize ZFC using only finitely many axioms. On the other hand, Von Neumann–Bernays–Gödel set theory (NBG) can be finitely axiomatized. The ontology of NBG includes proper classes as well as sets; a set is any class that can be a member of another class. NBG and ZFC are equivalent set theories in the sense that any theorem not mentioning classes and provable in one theory can be proved in the other.
Gödel's second incompleteness theorem says that a recursively axiomatizable system that can interpret Robinson arithmetic can prove its own consistency only if it is inconsistent. Moreover, Robinson arithmetic can be interpreted in general set theory, a small fragment of ZFC. Hence the consistency of ZFC cannot be proved within ZFC itself (unless it is actually inconsistent). Thus, to the extent that ZFC is identified with ordinary mathematics, the consistency of ZFC cannot be demonstrated in ordinary mathematics. The consistency of ZFC does follow from the existence of a weakly inaccessible cardinal, which is unprovable in ZFC if ZFC is consistent. Nevertheless, it is deemed unlikely that ZFC harbors an unsuspected contradiction; it is widely believed that if ZFC were inconsistent, that fact would have been uncovered by now. This much is certain — ZFC is immune to the classic paradoxes of naive set theory: Russell's paradox, the Burali-Forti paradox, and Cantor's paradox.
Abian and LaMacchia (1978) studied a subtheory of ZFC consisting of the axioms of extensionality, union, powerset, replacement, and choice. Using models, they proved this subtheory consistent, and proved that each of the axioms of extensionality, replacement, and power set is independent of the four remaining axioms of this subtheory. If this subtheory is augmented with the axiom of infinity, each of the axioms of union, choice, and infinity is independent

of the five remaining axioms. Because there are non-well-founded models that satisfy each axiom of ZFC except the axiom of regularity, that axiom is independent of the other ZFC axioms.
If consistent, ZFC cannot prove the existence of the inaccessible cardinals that category theory requires. Huge sets of this nature are possible if ZF is augmented with Tarski's axiom (Tarski 1939). Assuming that axiom turns the axioms of infinity, power set, and choice (**7 – 9** above) into theorems.

66.4.1 Independence

Many important statements are independent of ZFC (see list of statements undecidable in ZFC). The independence is usually proved by forcing, whereby it is shown that every countable transitive model of ZFC (sometimes augmented with large cardinal axioms) can be expanded to satisfy the statement in question. A different expansion is then shown to satisfy the negation of the statement. An independence proof by forcing automatically proves independence from arithmetical statements, other concrete statements, and large cardinal axioms. Some statements independent of ZFC can be proven to hold in particular inner models, such as in the constructible universe. However, some statements that are true about constructible sets are not consistent with hypothesized large cardinal axioms.
Forcing proves that the following statements are independent of ZFC:
_ Continuum hypothesis
_ Diamond principle
_ Suslin hypothesis
_ Martin's axiom (which is not a ZFC axiom)
_ Axiom of Constructibility (V=L) (which is also not a ZFC axiom).
Remarks:
_ The consistency of V=L is provable by inner models but not forcing: every model of ZF can be trimmed to become a model of ZFC + V=L.
_ The Diamond Principle implies the Continuum Hypothesis and the negation of the Suslin Hypothesis.
_ Martin's axiom plus the negation of the Continuum Hypothesis implies the Suslin Hypothesis.
_ The constructible universe satisfies the Generalized Continuum Hypothesis, the Diamond Principle, Martin's Axiom and the Kurepa Hypothesis.
_ The failure of the Kurepa hypothesis is equiconsistent with the existence of a strongly inaccessible cardinal.
A variation on the method of forcing can also be used to demonstrate the consistency and unprovability of the axiom of choice, i.e., that the axiom of choice is independent of ZF. The consistency of choice can be (relatively) easily verified by proving that the inner model L satisfies choice. (Thus every model of ZF contains a submodel of ZFC, so that Con(ZF) implies Con(ZFC).) Since forcing preserves choice, we cannot directly produce a model contradicting choice from a model satisfying choice. However, we can use forcing to create a model which contains a suitable submodel, namely one satisfying ZF but not C.
Another method of proving independence results, one owing nothing to forcing, is based on Gödel's second incompleteness theorem. This approach employs the statement whose independence is being examined, to prove the existence of a set model of ZFC, in which case Con(ZFC) is true. Since ZFC satisfies the conditions of Gödel's second theorem, the consistency of ZFC is unprovable in ZFC (provided that ZFC is, in fact, consistent). Hence no statement allowing such a proof can be proved in ZFC. This method can prove that the existence of large cardinals is not provable in ZFC, but cannot prove that assuming such cardinals, given ZFC, is free of contradiction.

66.5 Criticisms

For criticism of set theory in general, see Objections to set theory
ZFC has been criticized both for being excessively strong and for being excessively weak, as well as for its failure to capture objects such as proper classes and the universal set.
Many mathematical theorems can be proven in much weaker systems than ZFC, such as Peano arithmetic and second order arithmetic (as explored by the program of reverse mathematics). Saunders Mac Lane and Solomon Feferman have both made this point. Some of "mainstream mathematics" (mathematics not directly connected with axiomatic

set theory) is beyond Peano arithmetic and second order arithmetic, but still, all such mathematics can be carried out in ZC (Zermelo set theory with choice), another theory weaker than ZFC. Much of the power of ZFC, including the axiom of regularity and the axiom schema of replacement, is included primarily to facilitate the study of the set theory itself.
On the other hand, among axiomatic set theories, ZFC is comparatively weak. Unlike New Foundations, ZFC does not admit the existence of a universal set. Hence the universe of sets under ZFC is not closed under the elementary operations of the algebra of sets. Unlike von Neumann–Bernays–Gödel set theory and Morse–Kelley set theory (MK), ZFC does not admit the existence of proper classes. These ontological restrictions are required for ZFC to avoid Russell's paradox, but critics argue these restrictions make the ZFC axioms fail to capture the informal concept of *set*. A further comparative weakness of ZFC is that the axiom of choice included in ZFC is weaker than the axiom of global choice included in MK.

There are numerous mathematical statements undecidable in ZFC. These include the continuum hypothesis, the Whitehead problem, and the Normal Moore space conjecture. Some of these conjectures are provable with the addition of axioms such as Martin's axiom, large cardinal axioms to ZFC. Some others are decided in ZF+AD where AD is the axiom of determinacy, a strong supposition incompatible with choice. One attraction of large cardinal axioms is that they enable many results from ZF+AD to be established in ZFC adjoined by some large cardinal axiom (see projective determinacy). The Mizar system and Metamath have adopted Tarski–Grothendieck set theory, an extension of ZFC, so that proofs involving Grothendieck universes (encountered in category theory and algebraic geometry) can be formalized.

66.6 See also

_ Foundation of mathematics
_ Inner model
_ Large cardinal axiom
Related axiomatic set theories:
_ Morse–Kelley set theory
_ Von Neumann–Bernays–Gödel set theory
_ Tarski–Grothendieck set theory
_ Constructive set theory
_ Internal set theory

66.7 References

_ Abian, Alexander (1965). *The Theory of Sets and Transfinite Arithmetic*. W B Saunders.
_ ———; LaMacchia, Samuel (1978). "On the Consistency and Independence of Some Set-Theoretical Axioms". *Notre Dame Journal of Formal Logic* **19**: 155–58. doi:10.1305/ndjfl/1093888220.
_ Devlin, Keith (1996) [1984]. *The Joy of Sets*. Springer.
_ Heinz-Dieter Ebbinghaus, 2007. *Ernst Zermelo: An Approach to His Life and Work*. Springer. ISBN 978-3-540-49551-2.
_ Abraham Fraenkel, Yehoshua Bar-Hillel, and Azriel Levy, 1973 (1958). *Foundations of Set Theory*. North-Holland. Fraenkel's final word on ZF and ZFC.
_ Hatcher, William, 1982 (1968). *The Logical Foundations of Mathematics*. Pergamon Press.
_ Peter Hinman, 2005, *Fundamentals of Mathematical Logic*, A K Peters. ISBN 978-1-56881-262-5

_ Thomas Jech, 2003. *Set Theory: The Third Millennium Edition, Revised and Expanded*. Springer. ISBN 3-540-44085-2.
_ Kenneth Kunen, 1980. *Set Theory: An Introduction to Independence Proofs*. Elsevier. ISBN 0-444-86839-9.
_ Richard Montague, 1961, "Semantic closure and non-finite axiomatizability" in *Infinistic Methods*. London: Pergamon Press: 45–69.
_ Patrick Suppes, 1972 (1960). *Axiomatic Set Theory*. Dover reprint. Perhaps the best exposition of ZFC before the independence of AC and the Continuum hypothesis, and the emergence of large cardinals. Includes many theorems.
_ Gaisi Takeuti and Zaring, W M, 1971. *Introduction to Axiomatic Set Theory*. Springer-Verlag.
_ Alfred Tarski, 1939, "On well-ordered subsets of any set,", *Fundamenta Mathematicae* 32: 176-83.
_ Tiles, Mary, 2004 (1989). *The Philosophy of Set Theory*. Dover reprint. Weak on metatheory; the author is not a mathematician.
_ Tourlakis, George, 2003. *Lectures in Logic and Set Theory, Vol. 2*. Cambridge University Press.
_ Jean van Heijenoort, 1967. *From Frege to Gödel: A Source Book in Mathematical Logic, 1879–1931*. Harvard University Press. Includes annotated English translations of the classic articles by Zermelo, Fraenkel, and Skolem bearing on **ZFC**.
_ Zermelo, Ernst (1908). "Untersuchungen über die Grundlagen der Mengenlehre I". *Mathematische Annalen* **65**: 261–281. doi:10.1007/BF01449999. English translation in Heijenoort, Jean van (1967). "Investigations in the foundations of set theory". *From Frege to Gödel: A Source Book in Mathematical Logic, 1879–1931*. Source Books in the History of the Sciences. Harvard University Press. pp. 199–215. ISBN 978-0-674-32449-7.
_ Zermelo, Ernst (1930). "Über Grenzzahlen und Mengenbereiche". *Fundamenta Mathematicae* **16**: 29–47. ISSN 0016-2736.

66.8 External links

_ Hazewinkel, Michiel, ed. (2001), "ZFC", *Encyclopedia of Mathematics*, Springer, ISBN 978-1-55608-010-4
_ Stanford Encyclopedia of Philosophy articles by Thomas Jech:
_ Set Theory;
_ Axioms of Zermelo–Fraenkel Set Theory.
_ Metamath version of the ZFC axioms — A concise and nonredundant axiomatization. The background first order logic is defined especially to facilitate machine verification of proofs.

_ A derivation in Metamath of a version of the separation schema from a version of the replacement schema.
_ Zermelo-Fraenkel Axioms at PlanetMath.org.
_ Weisstein, Eric W., "Zermelo-Fraenkel Set Theory", *MathWorld*.

Chapter 67

Gödel's incompleteness theorems

Gödel's incompleteness theorems are two theorems of mathematical logic that establish inherent limitations of all but the most trivial axiomatic systems capable of doing arithmetic. The theorems, proven by Kurt Gödel in 1931, are important both in mathematical logic and in the philosophy of mathematics. The two results are widely, but not universally, interpreted as showing that Hilbert's program to find a complete and consistent set of axioms for all mathematics is impossible, giving a negative answer to Hilbert's second problem.
The first incompleteness theorem states that no consistent system of axioms whose theorems can be listed by an "effective procedure" (e.g., a computer program, but it could be any sort of algorithm) is capable of proving all truths about the relations of the natural numbers (arithmetic). For any such system, there will always be statements about the natural numbers that are true, but that are unprovable within the system. The second incompleteness theorem, an extension of the first, shows that such a system cannot demonstrate its own consistency.

67.1 Background

Because statements of a formal theory are written in symbolic form, it is possible to verify mechanically that a formal proof from a finite set of axioms is valid. This task, known as automatic proof verification, is closely related to automated theorem proving. The difference is that instead of constructing a new proof, the proof verifier simply checks that a provided formal proof (or, in instructions that can be followed to create a formal proof) is correct. This process is not merely hypothetical; systems such as Isabelle or Coq are used today to formalize proofs and then check their validity.
Many theories of interest include an infinite set of axioms, however. To verify a formal proof when the set of axioms is infinite, it must be possible to determine whether a statement that is claimed to be an axiom is actually an axiom. This issue arises in first order theories of arithmetic, such as Peano arithmetic, because the principle of mathematical induction is expressed as an infinite set of axioms (an axiom schema).
A formal theory is said to be *effectively generated* if its set of axioms is a recursively enumerable set. This means that there is a computer program that, in principle, could enumerate all the axioms of the theory without listing any statements that are not axioms. This is equivalent to the existence of a program that enumerates all the theorems of the theory without enumerating any statements that are not theorems. Examples of effectively generated theories with infinite sets of axioms include Peano arithmetic and Zermelo–Fraenkel set theory.
In choosing a set of axioms, one goal is to be able to prove as many correct results as possible, without proving any incorrect results. A set of axioms is complete if, for any statement in the axioms' language, either that statement or its negation is provable from the axioms. A set of axioms is (simply) consistent if there is no statement such that both the statement and its negation are provable from the axioms. In the standard system of first-order logic, an inconsistent set of axioms will prove every statement in its language (this is sometimes called the principle of explosion), and is thus automatically complete. A set of axioms that is both complete and consistent, however, proves a maximal set of non-contradictory theorems. Gödel's incompleteness theorems show that in certain cases it is not possible to obtain an effectively generated, complete, consistent theory.

67.2 First incompleteness theorem

Gödel's first incompleteness theorem first appeared as "Theorem VI" in Gödel's 1931 paper *On Formally Undecidable Propositions in Principia Mathematica and Related Systems I*.
The formal theorem is written in highly technical language. It may be paraphrased in English as:
Any effectively generated theory capable of expressing elementary arithmetic cannot be both consistent and complete. In particular, for any consistent, effectively generated formal theory that proves certain basic arithmetic truths, there is an arithmetical statement that is true,[1] but not provable in the theory (Kleene 1967, p. 250).
The true but unprovable statement referred to by the theorem is often referred to as "the Gödel sentence" for the theory. The proof constructs a specific Gödel sentence for each consistent effectively generated theory, but there are infinitely many statements in the language of the theory that share the property of being true but unprovable. For example, the conjunction of the Gödel sentence and any logically valid sentence will have this property.
For each consistent formal theory *T* having the required small amount of number theory, the corresponding Gödel sentence *G* asserts: "*G* cannot be proved within the theory *T*". This interpretation of *G* leads to the following informal analysis. If *G* were provable under the axioms and rules of inference of *T*, then *T* would have a theorem, *G*, which effectively contradicts itself, and thus the theory *T* would be inconsistent. This means that if the theory *T* is consistent

then *G* cannot be proved within it, and so the theory *T* is incomplete. Moreover, the claim *G* makes about its own unprovability is correct. In this sense *G* is not only unprovable but true, and provability-within-the-theory-*T* is not the same as truth. This informal analysis can be formalized to make a rigorous proof of the incompleteness theorem, as described in the section "Proof sketch for the first theorem" below. The formal proof reveals exactly the hypotheses required for the theory *T* in order for the self-contradictory nature of *G* to lead to a genuine contradiction.

Each effectively generated theory has its own Gödel statement. It is possible to define a larger theory *T'* that contains the whole of *T*, plus *G* as an additional axiom. This will not result in a complete theory, because Gödel's theorem will also apply to *T'*, and thus *T'* cannot be complete. In this case, *G* is indeed a theorem in *T'*, because it is an axiom. Since *G* states only that it is not provable in *T*, no contradiction is presented by its provability in *T'*. However, because the incompleteness theorem applies to *T'*: there will be a new Gödel statement *G'* for *T'*, showing that *T'* is also incomplete. *G'* will differ from *G* in that *G'* will refer to *T'*, rather than *T*.

To prove the first incompleteness theorem, Gödel represented statements by numbers. Then the theory at hand, which is assumed to prove certain facts about numbers, also proves facts about its own statements, provided that it is effectively generated. Questions about the provability of statements are represented as questions about the properties of numbers, which would be decidable by the theory if it were complete. In these terms, the Gödel sentence states that no natural number exists with a certain, strange property. A number with this property would encode a proof of the inconsistency of the theory. If there were such a number then the theory would be inconsistent, contrary to the consistency hypothesis. So, under the assumption that the theory is consistent, there is no such number.

67.2.1 Meaning of the first incompleteness theorem

Gödel's first incompleteness theorem shows that any consistent effective formal system that includes enough of the theory of the natural numbers is incomplete: there are true statements expressible in its language that are unprovable within the system. Thus no formal system (satisfying the hypotheses of the theorem) that aims to characterize the natural numbers can actually do so, as there will be true number-theoretical statements that that system cannot prove.

This fact is sometimes thought to have severe consequences for the program of logicism proposed by Gottlob Frege and Bertrand Russell, which aimed to define the natural numbers in terms of logic (Hellman 1981, p. 451–468).

Bob Hale and Crispin Wright argue that it is not a problem for logicism because the incompleteness theorems apply equally to first order logic as they do to arithmetic. They argue that only those who believe that the natural numbers are to be defined in terms of first order logic have this problem.

The existence of an incomplete formal system is, in itself, not particularly surprising. A system may be incomplete simply because not all the necessary axioms have been discovered. For example, Euclidean geometry without the parallel postulate is incomplete; it is not possible to prove or disprove the parallel postulate from the remaining axioms. Gödel's theorem shows that, in theories that include a small portion of number theory, a complete and consistent finite list of axioms can *never* be created, nor even an infinite list that can be enumerated by a computer program. Each

time a new statement is added as an axiom, there are other true statements that still cannot be proved, even with the new axiom. If an axiom is ever added that makes the system complete, it does so at the cost of making the system inconsistent.

There *are* complete and consistent lists of axioms for arithmetic that *cannot* be enumerated by a computer program. For example, one might take all true statements about the natural numbers to be axioms (and no false statements), which gives the theory known as "true arithmetic". The difficulty is that there is no mechanical way to decide, given a statement about the natural numbers, whether it is an axiom of this theory, and thus there is no effective way to verify a formal proof in this theory.

Many logicians believe that Gödel's incompleteness theorems struck a fatal blow to David Hilbert's second problem, which asked for a finitary consistency proof for mathematics. The second incompleteness theorem, in particular, is often viewed as making the problem impossible. Not all mathematicians agree with this analysis, however, and the status of Hilbert's second problem is not yet decided (see "Modern viewpoints on the status of the problem").

67.2.2 Relation to the liar paradox

The liar paradox is the sentence "This sentence is false." An analysis of the liar sentence shows that it cannot be true (for then, as it asserts, it is false), nor can it be false (for then, it is true). A Gödel sentence *G* for a theory *T* makes a similar assertion to the liar sentence, but with truth replaced by provability: *G* says "*G* is not provable in the theory *T*." The analysis of the truth and provability of *G* is a formalized version of the analysis of the truth of the liar sentence.

It is not possible to replace "not provable" with "false" in a Gödel sentence because the predicate "Q is the Gödel number of a false formula" cannot be represented as a formula of arithmetic. This result, known as Tarski's undefinability theorem, was discovered independently by Gödel (when he was working on the proof of the incompleteness theorem) and by Alfred Tarski.

67.2.3 Extensions of Gödel's original result

Gödel demonstrated the incompleteness of the theory of *Principia Mathematica*, a particular theory of arithmetic, but a parallel demonstration could be given for any effective theory of a certain expressiveness. Gödel commented on this fact in the introduction to his paper, but restricted the proof to one system for concreteness. In modern statements of the theorem, it is common to state the effectiveness and expressiveness conditions as hypotheses for the incompleteness theorem, so that it is not limited to any particular formal theory. The terminology used to state these conditions was not yet developed in 1931 when Gödel published his results.

Gödel's original statement and proof of the incompleteness theorem requires the assumption that the theory is not just consistent but *ω-consistent*. A theory is **ω-consistent** if it is not ω-inconsistent, and is ω-inconsistent if there is a predicate *P* such that for every specific natural number *m* the theory proves ~*P*(*m*), and yet the theory also proves that there exists a natural number *n* such that *P*(*n*). That is, the theory says that a number with property *P* exists while denying that it has any specific value. The ω-consistency of a theory implies its consistency, but consistency does not imply ω-consistency. J. Barkley Rosser (1936) strengthened the incompleteness theorem by finding a variation of the proof (Rosser's trick) that only requires the theory to be consistent, rather than ω-consistent. This is mostly of technical interest, since all true formal theories of arithmetic (theories whose axioms are all true statements about natural numbers) are ω-consistent, and thus Gödel's theorem as originally stated applies to them. The stronger version of the incompleteness theorem that only assumes consistency, rather than ω-consistency, is now commonly known as Gödel's incompleteness theorem and as the Gödel–Rosser theorem.

67.3 Second incompleteness theorem

Gödel's second incompleteness theorem first appeared as "Theorem XI" in Gödel's 1931 paper *On Formally Undecidable Propositions in Principia Mathematica and Related Systems I.*

Like with the first incompleteness theorem, Gödel wrote this theorem in highly technical formal mathematics. It may be paraphrased in English as:

For any formal effectively generated theory *T* including basic arithmetical truths and also certain truths about formal provability, if *T* includes a statement of its own consistency then *T* is inconsistent.

This strengthens the first incompleteness theorem, because the statement constructed in the first incompleteness theorem does not directly express the consistency of the theory. The proof of the second incompleteness theorem is obtained by formalizing the proof of the first incompleteness theorem within the theory itself.

A technical subtlety in the second incompleteness theorem is how to express the consistency of *T* as a formula in the language of *T*. There are many ways to do this, and not all of them lead to the same result. In particular, different formalizations of the claim that *T* is consistent may be inequivalent in *T*, and some may even be provable. For example, first-order Peano arithmetic (PA) can prove that the largest consistent subset of PA is consistent. But since PA is consistent, the largest consistent subset of PA is just PA, so in this sense PA "proves that it is consistent". What PA does not prove is that the largest consistent subset of PA is, in fact, the whole of PA. (The term "largest consistent subset of PA" is technically ambiguous, but what is meant here is the largest consistent initial segment of the axioms of PA ordered according to specific criteria; i.e., by "Gödel numbers", the numbers encoding the axioms as per the scheme used by Gödel mentioned above).

For Peano arithmetic, or any familiar explicitly axiomatized theory *T*, it is possible to canonically define a formula Con(*T*) expressing the consistency of *T*; this formula expresses the property that "there does not exist a natural number coding a sequence of formulas, such that each formula is either one of the axioms of *T*, a logical axiom, or an immediate consequence of preceding formulas according to the rules of inference of first-order logic, and such that the last formula is a contradiction".

The formalization of Con(*T*) depends on two factors: formalizing the notion of a sentence being derivable from a set of sentences and formalizing the notion of being an axiom of *T*. Formalizing derivability can be done in canonical fashion: given an arithmetical formula A(*x*) defining a set of axioms, one can canonically form a predicate ProvA(*P*), which expresses that a sentence *P* is provable from the set of axioms defined by A(*x*).

In addition, the standard proof of the second incompleteness theorem assumes that ProvA(*P*) satisfies the Hilbert–Bernays provability conditions. Letting #(*P*) represent the Gödel number of a formula *P*, the derivability conditions say:

1. If *T* proves *P*, then T proves ProvA(#(*P*)).
2. *T* proves 1.; that is, *T* proves that if *T* proves *P*, then *T* proves ProvA(#(*P*)). In other words, *T* proves that ProvA(#(*P*)) implies ProvA(#(ProvA(#(P)))).
3. *T* proves that if *T* proves that (*P* → *Q*) and *T* proves *P* then *T* proves *Q*. In other words, *T* proves that ProvA(#(*P* → *Q*)) and ProvA(#(*P*)) imply ProvA(#(*Q*)).

67.3.1 Implications for consistency proofs

Gödel's second incompleteness theorem also implies that a theory T_1 satisfying the technical conditions outlined above cannot prove the consistency of any theory T_2 that proves the consistency of T_1. This is because such a theory T_1 can prove that if T_2 proves the consistency of T_1, then T_1 is in fact consistent. For the claim that T_1 is consistent has form "for all numbers *n*, *n* has the decidable property of not being a code for a proof of contradiction in T_1". If T_1 were in fact inconsistent, then T_2 would prove for some *n* that *n* is the code of a contradiction in T_1. But if T_2 also proved that T_1 is consistent (that is, that there is no such n), then it would itself be inconsistent. This reasoning can be formalized in T_1 to show that if T_2 is consistent, then T_1 is consistent. Since, by second incompleteness theorem, T_1 does not prove its consistency, it cannot prove the consistency of T_2 either.

This corollary of the second incompleteness theorem shows that there is no hope of proving, for example, the consistency of Peano arithmetic using any finitistic means that can be formalized in a theory the consistency of which is provable in Peano arithmetic. For example, the theory of primitive recursive arithmetic (PRA), which is widely accepted as an accurate formalization of finitistic mathematics, is provably consistent in PA. Thus PRA cannot prove the consistency of PA. This fact is generally seen to imply that Hilbert's program, which aimed to justify the use of "ideal" (infinitistic) mathematical principles in the proofs of "real" (finitistic) mathematical statements by giving a finitistic proof that the ideal principles are consistent, cannot be carried out.

The corollary also indicates the epistemological relevance of the second incompleteness theorem. It would actually provide no interesting information if a theory *T* proved its consistency. This is because inconsistent theories prove everything, including their consistency. Thus a consistency proof of *T* in *T* would give us no clue as to whether *T* really is consistent; no doubts about the consistency of *T* would be resolved by such a consistency proof. The interest in consistency proofs lies in the possibility of proving the consistency of a theory *T* in some theory *T'* that is in some sense less doubtful than *T* itself, for example weaker than *T*. For many naturally occurring theories *T* and *T'*, such as

T = Zermelo–Fraenkel set theory and *T'* = primitive recursive arithmetic, the consistency of *T'* is provable in *T*, and thus *T'* can't prove the consistency of *T* by the above corollary of the second incompleteness theorem.

The second incompleteness theorem does not rule out consistency proofs altogether, only consistency proofs that could be formalized in the theory that is proved consistent. For example, Gerhard Gentzen proved the consistency of Peano arithmetic (PA) in a different theory that includes an axiom asserting that the ordinal called ε_0 is wellfounded; see Gentzen's consistency proof. Gentzen's theorem spurred the development of ordinal analysis in proof theory.

67.4 Examples of undecidable statements

See also: List of statements undecidable in ZFC

There are two distinct senses of the word "undecidable" in mathematics and computer science. The first of these is the proof-theoretic sense used in relation to Gödel's theorems, that of a statement being neither provable nor refutable in a specified deductive system. The second sense, which will not be discussed here, is used in relation to computability theory and applies not to statements but to decision problems, which are countably infinite sets of questions each requiring a yes or no answer. Such a problem is said to be undecidable if there is no computable function that correctly answers every question in the problem set (see undecidable problem).

Because of the two meanings of the word undecidable, the term independent is sometimes used instead of undecidable for the "neither provable nor refutable" sense. The usage of "independent" is also ambiguous, however. Some use it to mean just "not provable", leaving open whether an independent statement might be refuted.

Undecidability of a statement in a particular deductive system does not, in and of itself, address the question of whether the truth value of the statement is well-defined, or whether it can be determined by other means. Undecidability only implies that the particular deductive system being considered does not prove the truth or falsity of the statement. Whether there exist so-called "absolutely undecidable" statements, whose truth value can never be known or is illspecified, is a controversial point in the philosophy of mathematics.

The combined work of Gödel and Paul Cohen has given two concrete examples of undecidable statements (in the first sense of the term): The continuum hypothesis can neither be proved nor refuted in ZFC (the standard axiomatization of set theory), and the axiom of choice can neither be proved nor refuted in ZF (which is all the ZFC axioms *except* the axiom of choice). These results do not require the incompleteness theorem. Gödel proved in 1940 that neither of these statements could be disproved in ZF or ZFC set theory. In the 1960s, Cohen proved that neither is provable from ZF, and the continuum hypothesis cannot be proven from ZFC.

In 1973, the Whitehead problem in group theory was shown to be undecidable, in the first sense of the term, in standard set theory.

Gregory Chaitin produced undecidable statements in algorithmic information theory and proved another incompleteness theorem in that setting. Chaitin's incompleteness theorem states that for any theory that can represent enough arithmetic, there is an upper bound *c* such that no specific number can be proven in that theory to have Kolmogorov complexity greater than *c*. While Gödel's theorem is related to the liar paradox, Chaitin's result is related to Berry's paradox.

67.4.1 Undecidable statements provable in larger systems

These are natural mathematical equivalents of the Gödel "true but undecidable" sentence. They can be proved in a larger system which is generally accepted as a valid form of reasoning, but are undecidable in a more limited system such as Peano Arithmetic.

In 1977, Paris and Harrington proved that the Paris-Harrington principle, a version of the Ramsey theorem, is undecidable in the first-order axiomatization of arithmetic called Peano arithmetic, but can be proven in the larger system of second-order arithmetic. Kirby and Paris later showed Goodstein's theorem, a statement about sequences of natural numbers somewhat simpler than the Paris-Harrington principle, to be undecidable in Peano arithmetic. Kruskal's tree theorem, which has applications in computer science, is also undecidable from Peano arithmetic but provable in set theory. In fact Kruskal's tree theorem (or its finite form) is undecidable in a much stronger system codifying the principles acceptable based on a philosophy of mathematics called predicativism. The related but more general graph minor theorem (2003) has consequences for computational complexity theory.

67.5 Limitations of Gödel's theorems

The conclusions of Gödel's theorems are only proven for the formal theories that satisfy the necessary hypotheses. Not all axiom systems satisfy these hypotheses, even when these systems have models that include the natural numbers as a subset. For example, there are first-order axiomatizations of Euclidean geometry, of real closed fields, and of arithmetic in which multiplication is not *provably* total; none of these meet the hypotheses of Gödel's theorems. The key fact is that these axiomatizations are not expressive enough to define the set of natural numbers or develop basic properties of the natural numbers. Regarding the third example, Dan Willard (2001) has studied many weak systems of arithmetic which do not satisfy the hypotheses of the second incompleteness theorem, and which are consistent and capable of proving their own consistency (see self-verifying theories).

Gödel's theorems only apply to effectively generated (that is, recursively enumerable) theories. If all true statements about natural numbers are taken as axioms for a theory, then this theory is a consistent, complete extension of Peano arithmetic (called true arithmetic) for which none of Gödel's theorems apply in a meaningful way, because this theory is not recursively enumerable.

The second incompleteness theorem only shows that the consistency of certain theories cannot be proved from the axioms of those theories themselves. It does not show that the consistency cannot be proved from other (consistent) axioms. For example, the consistency of the Peano arithmetic can be proved in Zermelo–Fraenkel set theory (ZFC), or in theories of arithmetic augmented with transfinite induction, as in Gentzen's consistency proof.

67.6 Relationship with computability

The incompleteness theorem is closely related to several results about undecidable sets in recursion theory.

Stephen Cole Kleene (1943) presented a proof of Gödel's incompleteness theorem using basic results of computability theory. One such result shows that the halting problem is undecidable: there is no computer program that can correctly determine, given any program *P* as input, whether *P* eventually halts when run with a particular given input. Kleene showed that the existence of a complete effective theory of arithmetic with certain consistency properties would force the halting problem to be decidable, a contradiction. This method of proof has also been presented by Shoenfield (1967, p. 132); Charlesworth (1980); and Hopcroft and Ullman (1979).

Franzén (2005, p. 73) explains how Matiyasevich's solution to Hilbert's 10th problem can be used to obtain a proof to Gödel's first incompleteness theorem. Matiyasevich proved that there is no algorithm that, given a multivariate polynomial $p(x_1, x_2, ..., x_{\square})$ with integer coefficients, determines whether there is an integer solution to the equation $p = 0$. Because polynomials with integer coefficients, and integers themselves, are directly expressible in the language of arithmetic, if a multivariate integer polynomial equation $p = 0$ does have a solution in the integers then any sufficiently strong theory of arithmetic *T* will prove this. Moreover, if the theory *T* is ω-consistent, then it will never prove that a particular polynomial equation has a solution when in fact there is no solution in the integers. Thus, if *T* were complete and ω-consistent, it would be possible to determine algorithmically whether a polynomial equation has a solution by merely enumerating proofs of *T* until either "*p* has a solution" or "*p* has no solution" is found, in contradiction to Matiyasevich's theorem. Moreover, for each consistent effectively generated theory *T*, it is possible to effectively generate a multivariate polynomial *p* over the integers such that the equation $p = 0$ has no solutions over the integers, but the lack of solutions cannot be proved in *T* (Davis 2006:416, Jones 1980).

Smorynski (1977, p. 842) shows how the existence of recursively inseparable sets can be used to prove the first incompleteness theorem. This proof is often extended to show that systems such as Peano arithmetic are essentially undecidable (see Kleene 1967, p. 274).

Chaitin's incompleteness theorem gives a different method of producing independent sentences, based on Kolmogorov complexity. Like the proof presented by Kleene that was mentioned above, Chaitin's theorem only applies to theories with the additional property that all their axioms are true in the standard model of the natural numbers. Gödel's incompleteness theorem is distinguished by its applicability to consistent theories that nonetheless include statements that are false in the standard model; these theories are known as ω-inconsistent.

67.7 Proof sketch for the first theorem

Main article: Proof sketch for Gödel's first incompleteness theorem

The proof by contradiction has three essential parts. To begin, choose a formal system that meets the proposed criteria:

1. Statements in the system can be represented by natural numbers (known as Gödel numbers). The significance of this is that properties of statements—such as their truth and falsehood—will be equivalent to determining whether their Gödel numbers have certain properties, and that properties of the statements can therefore be demonstrated by examining their Gödel numbers. This part culminates in the construction of a formula expressing the idea that *"statement S is provable in the system"* (which can be applied to any statement "S" in the system).

2. In the formal system it is possible to construct a number whose matching statement, when interpreted, is selfreferential and essentially says that it (i.e. the statement itself) is unprovable. This is done using a technique called "diagonalization" (so-called because of its origins as Cantor's diagonal argument).
3. Within the formal system this statement permits a demonstration that it is neither provable nor disprovable in the system, and therefore the system cannot in fact be ω-consistent. Hence the original assumption that the proposed system met the criteria is false.

67.7.1 Arithmetization of syntax

The main problem in fleshing out the proof described above is that it seems at first that to construct a statement p that is equivalent to "p cannot be proved", p would somehow have to contain a reference to p, which could easily give rise to an infinite regress. Gödel's ingenious technique is to show that statements can be matched with numbers (often called the arithmetization of syntax) in such a way that *"proving a statement"* can be replaced with *"testing whether a number has a given property"*. This allows a self-referential formula to be constructed in a way that avoids any infinite regress of definitions. The same technique was later used by Alan Turing in his work on the Entscheidungsproblem.
In simple terms, a method can be devised so that every formula or statement that can be formulated in the system gets a unique number, called its **Gödel number**, in such a way that it is possible to mechanically convert back and forth between formulas and Gödel numbers. The numbers involved might be very long indeed (in terms of number of digits), but this is not a barrier; all that matters is that such numbers can be constructed. A simple example is the way in which English is stored as a sequence of numbers in computers using ASCII or Unicode:
_ The word **HELLO** is represented by 72-69-76-76-79 using decimal ASCII, ie the number 7269767679.
_ The logical statement **x=y => y=x** is represented by 120-061-121-032-061-062-032-121-061-120 using octal ASCII, ie the number 120061121032061062032121061120.
In principle, proving a statement true or false can be shown to be equivalent to proving that the number matching the statement does or doesn't have a given property. Because the formal system is strong enough to support reasoning about *numbers in general*, it can support reasoning about *numbers which represent formulae and statements* as well. Crucially, because the system can support reasoning about *properties of numbers*, the results are equivalent to reasoning about *provability of their equivalent statements*.

67.7.2 Construction of a statement about "provability"

Having shown that in principle the system can indirectly make statements about provability, by analyzing properties of those numbers representing statements it is now possible to show how to create a statement that actually does this.
A formula $F(x)$ that contains exactly one free variable x is called a *statement form* or *class-sign*. As soon as x is replaced by a specific number, the statement form turns into a *bona fide* statement, and it is then either provable in the system, or not. For certain formulas one can show that for every natural number n, F(n) is true if and only if it can be proven (the precise requirement in the original proof is weaker, but for the proof sketch this will suffice). In

particular, this is true for every specific arithmetic operation between a finite number of natural numbers, such as "2×3=6".
Statement forms themselves are not statements and therefore cannot be proved or disproved. But every statement form $F(x)$ can be assigned a Gödel number denoted by **G**(F). The choice of the free variable used in the form $F(x)$ is not relevant to the assignment of the Gödel number **G**(F).
Now comes the trick: The notion of provability itself can also be encoded by Gödel numbers, in the following way. Since a proof is a list of statements which obey certain rules, the Gödel number of a proof can be defined. Now, for every statement p, one may ask whether a number x is the Gödel number of its proof. The relation between the Gödel number of p and x, the potential Gödel number of its proof, is an arithmetical relation between two numbers. Therefore there is a statement form Bew(y) that uses this arithmetical relation to state that a Gödel number of a proof of y exists:
Bew(y) = ∃ x (y is the Gödel number of a formula and x is the Gödel number of a proof of the formula encoded by y).
The name **Bew** is short for *beweisbar*, the German word for "provable"; this name was originally used by Gödel to denote the provability formula just described. Note that "Bew(y)" is merely an abbreviation that represents a particular, very long, formula in the original language of T; the string "Bew" itself is not claimed to be part of this language.
An important feature of the formula Bew(y) is that if a statement p is provable in the system then Bew(**G**(p)) is also provable. This is because any proof of p would have a corresponding Gödel number, the existence of which causes Bew(**G**(p)) to be satisfied.

67.7.3 Diagonalization

The next step in the proof is to obtain a statement that says it is unprovable. Although Gödel constructed this statement directly, the existence of at least one such statement follows from the diagonal lemma, which says that for any sufficiently strong formal system and any statement form F there is a statement p such that the system proves
$p \leftrightarrow F(\mathbf{G}(p))$.

By letting F be the negation of Bew(x), we obtain the theorem
$p \leftrightarrow \sim Bew(\mathbf{G}(p))$
and the p defined by this roughly states that its own Gödel number is the Gödel number of an unprovable formula. The statement p is not literally equal to ~Bew($\mathbf{G}(p)$); rather, p states that if a certain calculation is performed, the resulting Gödel number will be that of an unprovable statement. But when this calculation is performed, the resulting Gödel number turns out to be the Gödel number of p itself. This is similar to the following sentence in English:
", when preceded by itself in quotes, is unprovable.", when preceded by itself in quotes, is unprovable.
This sentence does not directly refer to itself, but when the stated transformation is made the original sentence is obtained as a result, and thus this sentence asserts its own unprovability. The proof of the diagonal lemma employs a similar method.
Now, assume that the axiomatic system is ω-consistent, and let p be the statement obtained in the previous section. If p were provable, then Bew($\mathbf{G}(p)$) would be provable, as argued above. But p asserts the negation of Bew($\mathbf{G}(p)$). Thus the system would be inconsistent, proving both a statement and its negation. This contradiction shows that p cannot be provable.
If the negation of p were provable, then Bew($\mathbf{G}(p)$) would be provable (because p was constructed to be equivalent to the negation of Bew($\mathbf{G}(p)$)). However, for each specific number x, x cannot be the Gödel number of the proof of p, because p is not provable (from the previous paragraph). Thus on one hand the system proves there is a number with a certain property (that it is the Gödel number of the proof of p), but on the other hand, for every specific number x,

we can prove that it does not have this property. This is impossible in an ω-consistent system. Thus the negation of p is not provable.
Thus the statement p is undecidable in our axiomatic system: it can neither be proved nor disproved within the system. In fact, to show that p is not provable only requires the assumption that the system is consistent. The stronger assumption of ω-consistency is required to show that the negation of p is not provable. Thus, if p is constructed for a particular system:
_ If the system is ω-consistent, it can prove neither p nor its negation, and so p is undecidable.
_ If the system is consistent, it may have the same situation, or it may prove the negation of p. In the later case, we have a statement ("not p") which is false but provable, and the system is not ω-consistent.
If one tries to "add the missing axioms" to avoid the incompleteness of the system, then one has to add either p or "not p" as axioms. But then the definition of "being a Gödel number of a proof" of a statement changes. which means that the formula Bew(x) is now different. Thus when we apply the diagonal lemma to this new Bew, we obtain a new statement p, different from the previous one, which will be undecidable in the new system if it is ω-consistent.

67.7.4 Proof via Berry's paradox

George Boolos (1989) sketches an alternative proof of the first incompleteness theorem that uses Berry's paradox rather than the liar paradox to construct a true but unprovable formula. A similar proof method was independently discovered by Saul Kripke (Boolos 1998, p. 383). Boolos's proof proceeds by constructing, for any computably enumerable set S of true sentences of arithmetic, another sentence which is true but not contained in S. This gives the first incompleteness theorem as a corollary. According to Boolos, this proof is interesting because it provides a "different sort of reason" for the incompleteness of effective, consistent theories of arithmetic (Boolos 1998, p. 388).

67.7.5 Formalized proofs

Formalized proofs of versions of the incompleteness theorem have been developed by Natarajan Shankar in 1986 using Nqthm (Shankar 1994) and by Russell O'Connor in 2003 using Coq (O'Connor 2005).

67.8 Proof sketch for the second theorem

The main difficulty in proving the second incompleteness theorem is to show that various facts about provability used in the proof of the first incompleteness theorem can be formalized within the system using a formal predicate for provability. Once this is done, the second incompleteness theorem follows by formalizing the entire proof of the first incompleteness theorem within the system itself.
Let p stand for the undecidable sentence constructed above, and assume that the consistency of the system can be proven from within the system itself. The demonstration above shows that if the system is consistent, then p is not provable. The proof of this implication can be formalized within the system, and therefore the statement "p is not provable", or "not $P(p)$" can be proven in the system.
But this last statement is equivalent to p itself (and this equivalence can be proven in the system), so p can be proven in the system. This contradiction shows that the system must be inconsistent.

67.9 Discussion and implications

The incompleteness results affect the philosophy of mathematics, particularly versions of formalism, which use a single system formal logic to define their principles. One can paraphrase the first theorem as saying the following:
An all-encompassing axiomatic system can never be found that is able to prove *all* mathematical truths, but no falsehoods.

On the other hand, from a strict formalist perspective this paraphrase would be considered meaningless because it presupposes that mathematical "truth" and "falsehood" are well-defined in an absolute sense, rather than relative to each formal system.

The following rephrasing of the second theorem is even more unsettling to the foundations of mathematics:

If an axiomatic system can be proven to be consistent from within itself, then it is inconsistent.

Therefore, to establish the consistency of a system S, one needs to use some other system T, but a proof in T is not completely convincing unless T's consistency has already been established without using S.

Theories such as Peano arithmetic, for which any computably enumerable consistent extension is incomplete, are called essentially undecidable or **essentially incomplete**.

67.9.1 Minds and machines

Main article: Mechanism (philosophy) § Gödelian arguments

Authors including the philosopher J. R. Lucas and physicist Roger Penrose have debated what, if anything, Gödel's incompleteness theorems imply about human intelligence. Much of the debate centers on whether the human mind is equivalent to a Turing machine, or by the Church–Turing thesis, any finite machine at all. If it is, and if the machine is consistent, then Gödel's incompleteness theorems would apply to it.

Hilary Putnam (1960) suggested that while Gödel's theorems cannot be applied to humans, since they make mistakes and are therefore inconsistent, it may be applied to the human faculty of science or mathematics in general. Assuming that it is consistent, either its consistency cannot be proved or it cannot be represented by a Turing machine.

Avi Wigderson (2010) has proposed that the concept of mathematical "knowability" should be based on computational complexity rather than logical decidability. He writes that "when *knowability* is interpreted by modern standards, namely via computational complexity, the Gödel phenomena are very much with us."

67.9.2 Paraconsistent logic

Although Gödel's theorems are usually studied in the context of classical logic, they also have a role in the study of paraconsistent logic and of inherently contradictory statements (*dialetheia*). Graham Priest (1984, 2006) argues that replacing the notion of formal proof in Gödel's theorem with the usual notion of informal proof can be used to show that naive mathematics is inconsistent, and uses this as evidence for dialetheism. The cause of this inconsistency is the inclusion of a truth predicate for a theory within the language of the theory (Priest 2006:47). Stewart Shapiro (2002) gives a more mixed appraisal of the applications of Gödel's theorems to dialetheism. Carl Hewitt (2008) has proposed that (inconsistent) paraconsistent logics that prove their own Gödel sentences may have applications in software engineering.

67.9.3 Appeals to the incompleteness theorems in other fields

Appeals and analogies are sometimes made to the incompleteness theorems in support of arguments that go beyond mathematics and logic. Several authors have commented negatively on such extensions and interpretations, including Torkel Franzén (2005); Alan Sokal and Jean Bricmont (1999); and Ophelia Benson and Jeremy Stangroom (2006). Bricmont and Stangroom (2006, p. 10), for example, quote from Rebecca Goldstein's comments on the disparity between Gödel's avowed Platonism and the anti-realist uses to which his ideas are sometimes put. Sokal and Bricmont (1999, p. 187) criticize Régis Debray's invocation of the theorem in the context of sociology; Debray has defended this use as metaphorical (ibid.).

67.9.4 Role of self-reference

Torkel Franzén (2005, p. 46) observes:

67.10. HISTORY 461

Gödel's proof of the first incompleteness theorem and Rosser's strengthened version have given many the impression that the theorem can only be proved by constructing self-referential statements [...] or even that only strange self-referential statements are known to be undecidable in elementary arithmetic. To counteract such impressions, we need only introduce a different kind of proof of the first incompleteness theorem.

He then proposes the proofs based on computability, or on information theory, as described earlier in this article, as examples of proofs that should "counteract such impressions".

67.10 History

After Gödel published his proof of the completeness theorem as his doctoral thesis in 1929, he turned to a second problem for his habilitation. His original goal was to obtain a positive solution to Hilbert's second problem (Dawson 1997, p. 63). At the time, theories of the natural numbers and real numbers similar to second-order arithmetic were known as "analysis", while theories of the natural numbers alone were known as "arithmetic".

Gödel was not the only person working on the consistency problem. Ackermann had published a flawed consistency proof for analysis in 1925, in which he attempted to use the method of ε-substitution originally developed by Hilbert. Later that year, von Neumann was able to correct the proof for a theory of arithmetic without any axioms of induction. By 1928, Ackermann had communicated a modified proof to Bernays; this modified proof led Hilbert to announce his belief in 1929 that the consistency of arithmetic had been demonstrated and that a consistency proof of analysis would likely soon follow. After the publication of the incompleteness theorems showed that Ackermann's modified

proof must be erroneous, von Neumann produced a concrete example showing that its main technique was unsound (Zach 2006, p. 418, Zach 2003, p. 33).
In the course of his research, Gödel discovered that although a sentence which asserts its own falsehood leads to paradox, a sentence that asserts its own non-provability does not. In particular, Gödel was aware of the result now called Tarski's indefinability theorem, although he never published it. Gödel announced his first incompleteness theorem to Carnap, Feigel and Waismann on August 26, 1930; all four would attend a key conference in Königsberg the following week.

67.10.1 Announcement

The 1930 Königsberg conference was a joint meeting of three academic societies, with many of the key logicians of the time in attendance. Carnap, Heyting, and von Neumann delivered one-hour addresses on the mathematical philosophies of logicism, intuitionism, and formalism, respectively (Dawson 1996, p. 69). The conference also included Hilbert's retirement address, as he was leaving his position at the University of Göttingen. Hilbert used the speech to argue his belief that all mathematical problems can be solved. He ended his address by saying,
For the mathematician there is no *Ignorabimus*, and, in my opinion, not at all for natural science either. ... The true reason why [no one] has succeeded in finding an unsolvable problem is, in my opinion, that there is no unsolvable problem. In contrast to the foolish *Ignoramibus*, our credo avers: We must know. We shall know!
This speech quickly became known as a summary of Hilbert's beliefs on mathematics (its final six words, "*Wir müssen wissen. Wir werden wissen!*", were used as Hilbert's epitaph in 1943). Although Gödel was likely in attendance for Hilbert's address, the two never met face to face (Dawson 1996, p. 72).
Gödel announced his first incompleteness theorem at a roundtable discussion session on the third day of the conference. The announcement drew little attention apart from that of von Neumann, who pulled Gödel aside for conversation. Later that year, working independently with knowledge of the first incompleteness theorem, von Neumann obtained a proof of the second incompleteness theorem, which he announced to Gödel in a letter dated November 20, 1930 (Dawson 1996, p. 70). Gödel had independently obtained the second incompleteness theorem and included it in his submitted manuscript, which was received by *Monatshefte für Mathematik* on November 17, 1930.
Gödel's paper was published in the *Monatshefte* in 1931 under the title *Über formal unentscheidbare Sätze der Principia Mathematica und verwandter Systeme I* (On Formally Undecidable Propositions in Principia Mathematica and Related Systems I). As the title implies, Gödel originally planned to publish a second part of the paper; it was never written.

67.10.2 Generalization and acceptance

Gödel gave a series of lectures on his theorems at Princeton in 1933–1934 to an audience that included Church, Kleene, and Rosser. By this time, Gödel had grasped that the key property his theorems required is that the theory must be effective (at the time, the term "general recursive" was used). Rosser proved in 1936 that the hypothesis of ω-consistency, which was an integral part of Gödel's original proof, could be replaced by simple consistency, if the Gödel sentence was changed in an appropriate way. These developments left the incompleteness theorems in essentially their modern form.
Gentzen published his consistency proof for first-order arithmetic in 1936. Hilbert accepted this proof as "finitary" although (as Gödel's theorem had already shown) it cannot be formalized within the system of arithmetic that is being proved consistent.
The impact of the incompleteness theorems on Hilbert's program was quickly realized. Bernays included a full proof of the incompleteness theorems in the second volume of *Grundlagen der Mathematik* (1939), along with additional results of Ackermann on the ε-substitution method and Gentzen's consistency proof of arithmetic. This was the first full published proof of the second incompleteness theorem.

67.10.3 Criticisms

Finsler

Paul Finsler (1926) used a version of Richard's paradox to construct an expression that was false but unprovable in a particular, informal framework he had developed. Gödel was unaware of this paper when he proved the incompleteness theorems (Collected Works Vol. IV., p. 9). Finsler wrote to Gödel in 1931 to inform him about this paper, which Finsler felt had priority for an incompleteness theorem. Finsler's methods did not rely on formalized provability, and had only a superficial resemblance to Gödel's work (van Heijenoort 1967:328). Gödel read the paper but found it deeply flawed, and his response to Finsler laid out concerns about the lack of formalization (Dawson:89). Finsler continued to argue for his philosophy of mathematics, which eschewed formalization, for the remainder of his career.

Zermelo

In September 1931, Ernst Zermelo wrote Gödel to announce what he described as an "essential gap" in Gödel's argument (Dawson:76). In October, Gödel replied with a 10-page letter (Dawson:76, Grattan-Guinness:512-513). But Zermelo did not relent and published his criticisms in print with "a rather scathing paragraph on his young competitor" (Grattan-Guinness:513). Gödel decided that to pursue the matter further was pointless, and Carnap agreed (Dawson:77). Much of Zermelo's subsequent work was related to logics stronger than first-order logic, with which he hoped to show both the consistency and categoricity of mathematical theories.

Wittgenstein

Ludwig Wittgenstein wrote several passages about the incompleteness theorems that were published posthumously in his 1953 *Remarks on the Foundations of Mathematics*. Gödel was a member of the Vienna Circle during the period in which Wittgenstein's early ideal language philosophy and Tractatus Logico-Philosophicus dominated the circle's thinking. Writings in Gödel's Nachlass express the belief that Wittgenstein deliberately misread his ideas.

Multiple commentators have read Wittgenstein as misunderstanding Gödel (Rodych 2003), although Juliet Floyd and Hilary Putnam (2000), as well as Graham Priest (2004) have provided textual readings arguing that most commentary misunderstands Wittgenstein. On their release, Bernays, Dummett, and Kreisel wrote separate reviews on Wittgenstein's remarks, all of which were extremely negative (Berto 2009:208). The unanimity of this criticism caused Wittgenstein's remarks on the incompleteness theorems to have little impact on the logic community. In 1972, Gödel, stated: "Has Wittgenstein lost his mind? Does he mean it seriously?" (Wang 1996:197) And wrote to Karl Menger that Wittgenstein's comments demonstrate a willful misunderstanding of the incompleteness theorems writing:

"It is clear from the passages you cite that Wittgenstein did "not" understand [the first incompleteness theorem] (or pretended not to understand it). He interpreted it as a kind of logical paradox, while in

fact is just the opposite, namely a mathematical theorem within an absolutely uncontroversial part of mathematics (finitary number theory or combinatorics)." (Wang 1996:197)

Since the publication of Wittgenstein's *Nachlass* in 2000, a series of papers in philosophy have sought to evaluate whether the original criticism of Wittgenstein's remarks was justified. Floyd and Putnam (2000) argue that Wittgenstein had a more complete understanding of the incompleteness theorem than was previously assumed. They are particularly concerned with the interpretation of a Gödel sentence for an ω-inconsistent theory as actually saying "I am not provable", since the theory has no models in which the provability predicate corresponds to actual provability. Rodych (2003) argues that their interpretation of Wittgenstein is not historically justified, while Bays (2004) argues against Floyd and Putnam's philosophical analysis of the provability predicate. Berto (2009) explores the relationship between Wittgenstein's writing and theories of paraconsistent logic.

67.11 See also

_ Gödel's completeness theorem
_ Gödel's speed-up theorem
_ Löb's Theorem
_ *Minds, Machines and Gödel*
_ Münchhausen trilemma
_ Non-standard model of arithmetic
_ Provability logic
_ Tarski's undefinability theorem
_ Third Man Argument

67.12 Notes

[1] The word "true" is used disquotationally here: the Gödel sentence is true in this sense because it "asserts its own unprovability and it is indeed unprovable" (Smoryński 1977 p. 825; also see Franzén 2005 pp. 28–33). It is also possible to read "*GT* is true" in the formal sense that primitive recursive arithmetic proves the implication Con(*T*)→*GT*, where Con(*T*) is a canonical sentence asserting the consistency of *T* (Smoryński 1977 p. 840, Kikuchi and Tanaka 1994 p. 403). However, the arithmetic statement in question is *false* in some nonstandard models of arithmetic.

67.13 References

67.13.1 Articles by Gödel

_ 1931, *Über formal unentscheidbare Sätze der Principia Mathematica und verwandter Systeme, I. Monatshefte für Mathematik und Physik 38*: 173-98.
_ 1931, *Über formal unentscheidbare Sätze der Principia Mathematica und verwandter Systeme, I.* and *On formally undecidable propositions of Principia Mathematica and related systems I* in Solomon Feferman, ed., 1986. *Kurt Gödel Collected works, Vol. I.* Oxford University Press: 144-195. The original German with a facing English translation, preceded by a very illuminating introductory note by Kleene.
_ Hirzel, Martin, 2000, *On formally undecidable propositions of Principia Mathematica and related systems I.*. A modern translation by Hirzel.
_ 1951, *Some basic theorems on the foundations of mathematics and their implications* in Solomon Feferman, ed., 1995. *Kurt Gödel Collected works, Vol. III.* Oxford University Press: 304-23.

67.13.2 Translations, during his lifetime, of Gödel's paper into English

None of the following agree in all translated words and in typography. The typography is a serious matter, because Gödel expressly wished to emphasize "those metamathematical notions that had been defined in their usual sense before . . ." (van Heijenoort 1967:595). Three translations exist. Of the first John Dawson states that: "The Meltzer translation was seriously deficient and received a devastating review in the *Journal of Symbolic Logic*; "Gödel also complained about Braithwaite's commentary (Dawson 1997:216). "Fortunately, the Meltzer translation was soon supplanted by a better one prepared by Elliott Mendelson for Martin Davis's anthology *The Undecidable* . . . he found the translation "not quite so good" as he had expected . . . [but because of time constraints he] agreed to its publication" (ibid). (In a footnote Dawson states that "he would regret his compliance, for the published volume was marred throughout by sloppy typography and numerous misprints" (ibid)). Dawson states that "The translation that Gödel favored was that by Jean van Heijenoort" (ibid). For the serious student another version exists as a set of lecture notes recorded by Stephen Kleene and J. B. Rosser "during lectures given by Gödel at to the Institute for Advanced Study during the spring of 1934" (cf commentary by Davis 1965:39 and beginning on p. 41); this version is titled "On Undecidable Propositions of Formal Mathematical Systems". In their order of publication:

_ B. Meltzer (translation) and R. B. Braithwaite (Introduction), 1962. *On Formally Undecidable Propositions of Principia Mathematica and Related Systems*, Dover Publications, New York (Dover edition 1992), ISBN 0-486-66980-7 (pbk.) This contains a useful translation of Gödel's German abbreviations on pp. 33–34. As noted above, typography, translation and commentary is suspect. Unfortunately, this translation was reprinted with all its suspect content by

_ Stephen Hawking editor, 2005. *God Created the Integers: The Mathematical Breakthroughs That Changed History*, Running Press, Philadelphia, ISBN 0-7624-1922-9. Gödel's paper appears starting on p. 1097, with Hawking's commentary starting on p. 1089.

_ Martin Davis editor, 1965. *The Undecidable: Basic Papers on Undecidable Propositions, Unsolvable problems and Computable Functions*, Raven Press, New York, no ISBN. Gödel's paper begins on page 5, preceded by one page of commentary.

_ Jean van Heijenoort editor, 1967, 3rd edition 1967. *From Frege to Gödel: A Source Book in Mathematical Logic, 1879-1931*, Harvard University Press, Cambridge Mass., ISBN 0-674-32449-8 (pbk).[1] van Heijenoort did the translation. He states that "Professor Gödel approved the translation, which in many places was accommodated to his wishes." (p. 595). Gödel's paper begins on p. 595; van Heijenoort's commentary begins on p. 592.

_ Martin Davis editor, 1965, ibid. "On Undecidable Propositions of Formal Mathematical Systems." A copy with Gödel's corrections of errata and Gödel's added notes begins on page 41, preceded by two pages of Davis's commentary. Until Davis included this in his volume this lecture existed only as mimeographed notes.

Citation

[1] van Heijenoort, Jean. "From Frege to Gödel: A Source Book in Mathematical Logic, 1879-1931". *This link goes to the Google Books page for the text.* The original print book was published by Harvard University Press in 1977 and is widely available from booksellers. Retrieved 9 April 2014.

67.13.3 Articles by others

_ George Boolos, 1989, "A New Proof of the Gödel Incompleteness Theorem", *Notices of the American Mathematical Society* v. 36, pp. 388–390 and p. 676, reprinted in Boolos, 1998, *Logic, Logic, and Logic*, Harvard Univ. Press. ISBN 0-674-53766-1

_ Arthur Charlesworth, 1980, "A Proof of Godel's Theorem in Terms of Computer Programs," *Mathematics Magazine*, v. 54 n. 3, pp. 109–121. JStor

_ Martin Davis, "The Incompleteness Theorem", in Notices of the AMS vol. 53 no. 4 (April 2006), p. 414.

_ Jean van Heijenoort, 1963. "Gödel's Theorem" in Edwards, Paul, ed., *Encyclopedia of Philosophy, Vol. 3.* Macmillan: 348-57.

_ Geoffrey Hellman, *How to Gödel a Frege-Russell: Gödel's Incompleteness Theorems and Logicism.* Noûs, Vol. 15, No. 4, Special Issue on Philosophy of Mathematics. (Nov., 1981), pp. 451–468.

_ David Hilbert, 1900, "Mathematical Problems." English translation of a lecture delivered before the International Congress of Mathematicians at Paris, containing Hilbert's statement of his Second Problem.

_ Kikuchi, Makoto; Tanaka, Kazuyuki (1994), *On formalization of model-theoretic proofs of Gödel's theorems*, *Notre Dame Journal of Formal Logic* **35** (3): 403–412, doi:10.1305/ndjfl/1040511346, ISSN 0029-4527, MR 1326122

_ Stephen Cole Kleene, 1943, "Recursive predicates and quantifiers," reprinted from *Transactions of the American Mathematical Society*, v. 53 n. 1, pp. 41–73 in Martin Davis 1965, *The Undecidable* (loc. cit.) pp. 255–287.

_ John Barkley Rosser, 1936, "Extensions of some theorems of Gödel and Church," reprinted from the *Journal of Symbolic Logic* vol. 1 (1936) pp. 87–91, in Martin Davis 1965, *The Undecidable* (loc. cit.) pp. 230–235.

_ John Barkley Rosser, 1939, "An Informal Exposition of proofs of Gödel's Theorem and Church's Theorem", Reprinted from the *Journal of Symbolic Logic*, vol. 4 (1939) pp. 53–60, in Martin Davis 1965, *The Undecidable* (loc. cit.) pp. 223–230

_ C. Smoryński, "The incompleteness theorems", in J. Barwise, ed., *Handbook of Mathematical Logic*, North-Holland 1982 ISBN 978-0-444-86388-1, pp. 821–866.

_ Dan E. Willard (2001), "Self-Verifying Axiom Systems, the Incompleteness Theorem and Related Reflection Principles", *Journal of Symbolic Logic*, v. 66 n. 2, pp. 536–596. doi:10.2307/2695030
_ Zach, Richard (2003), *The Practice of Finitism: Epsilon Calculus and Consistency Proofs in Hilbert's Program*, *Synthese* (Berlin, New York: Springer-Verlag) **137** (1): 211–259, doi:10.1023/A:1026247421383, ISSN 0039-7857
_ Richard Zach, 2005, "Paper on the incompleteness theorems" in Grattan-Guinness, I., ed., *Landmark Writings in Western Mathematics*. Elsevier: 917-25.

67.13.4 Books about the theorems

_ Francesco Berto. *There's Something about Gödel: The Complete Guide to the Incompleteness Theorem* John Wiley and Sons. 2010.
_ Domeisen, Norbert, 1990. *Logik der Antinomien*. Bern: Peter Lang. 142 S. 1990. ISBN 3-261-04214-1. Zentralblatt MATH
_ Torkel Franzén, 2005. *Gödel's Theorem: An Incomplete Guide to its Use and Abuse*. A.K. Peters. ISBN 1-56881-238-8 MR 2007d:03001
_ Douglas Hofstadter, 1979. *Gödel, Escher, Bach: An Eternal Golden Braid*. Vintage Books. ISBN 0-465-02685-0. 1999 reprint: ISBN 0-465-02656-7. MR 80j:03009
_ Douglas Hofstadter, 2007. *I Am a Strange Loop*. Basic Books. ISBN 978-0-465-03078-1. ISBN 0-465-03078-5. MR 2008g:00004
_ Stanley Jaki, OSB, 2005. *The drama of the quantities*. Real View Books.
_ Per Lindström, 1997, *Aspects of Incompleteness*, Lecture Notes in Logic v. 10.
_ J.R. Lucas, FBA, 1970. *The Freedom of the Will*. Clarendon Press, Oxford, 1970.
_ Ernest Nagel, James Roy Newman, Douglas Hofstadter, 2002 (1958). *Gödel's Proof*, revised ed. ISBN 0-8147-5816-9. MR 2002i:03001
_ Rudy Rucker, 1995 (1982). *Infinity and the Mind: The Science and Philosophy of the Infinite*. Princeton Univ. Press. MR 84d:03012
_ Smith, Peter, 2007. *An Introduction to Gödel's Theorems*. Cambridge University Press. MathSciNet

_ N. Shankar, 1994. *Metamathematics, Machines and Gödel's Proof*, Volume 38 of Cambridge tracts in theoretical computer science. ISBN 0-521-58533-3
_ Raymond Smullyan, 1991. *Godel's Incompleteness Theorems*. Oxford Univ. Press.
_ —, 1994. *Diagonalization and Self-Reference*. Oxford Univ. Press. MR 96c:03001
_ Hao Wang, 1997. *A Logical Journey: From Gödel to Philosophy*. MIT Press. ISBN 0-262-23189-1 MR 97m:01090

67.13.5 Miscellaneous references

_ Francesco Berto. "The Gödel Paradox and Wittgenstein's Reasons" *Philosophia Mathematica* (III) 17. 2009.
_ John W. Dawson, Jr., 1997. *Logical Dilemmas: The Life and Work of Kurt Gödel*, A. K. Peters, Wellesley Mass, ISBN 1-56881-256-6.
_ Goldstein, Rebecca, 2005, *Incompleteness: the Proof and Paradox of Kurt Gödel*, W. W. Norton & Company. ISBN 0-393-05169-2
_ Juliet Floyd and Hilary Putnam, 2000, "A Note on Wittgenstein's 'Notorious Paragraph' About the Gödel Theorem", *Journal of Philosophy* v. 97 n. 11, pp. 624–632.
_ Carl Hewitt, 2008, "Large-scale Organizational Computing requires Unstratified Reflection and Strong Paraconsistency", *Coordination, Organizations, Institutions, and Norms in Agent Systems III*, Springer-Verlag.
_ David Hilbert and Paul Bernays, *Grundlagen der Mathematik*, Springer-Verlag.
_ John Hopcroft and Jeffrey Ullman 1979, *Introduction to Automata Theory, Languages, and Computation*, Addison-Wesley, ISBN 0-201-02988-X
_ James P. Jones, *Undecidable Diophantine Equations*, Bulletin of the American Mathematical Society v. 3 n. 2, 1980, pp. 859–862.
_ Stephen Cole Kleene, 1967, *Mathematical Logic*. Reprinted by Dover, 2002. ISBN 0-486-42533-9
_ Russell O'Connor, 2005, "Essential Incompleteness of Arithmetic Verified by Coq", Lecture Notes in Computer Science v. 3603, pp. 245–260.
_ Graham Priest, 2006, *In Contradiction: A Study of the Transconsistent*, Oxford University Press, ISBN 0-19-926329-9
_ Graham Priest, 2004, *Wittgenstein's Remarks on Gödel's Theorem* in Max Kölbel, ed., *Wittgenstein's lasting significance*, Psychology Press, pp. 207–227.
_ Graham Priest, 1984, "Logic of Paradox Revisited", *Journal of Philosophical Logic*, v. 13,` n. 2, pp. 153–179
_ Hilary Putnam, 1960, *Minds and Machines* in Sidney Hook, ed., *Dimensions of Mind: A Symposium*. New York University Press. Reprinted in Anderson, A. R., ed., 1964. *Minds and Machines*. Prentice-Hall: 77.
_ Rautenberg, Wolfgang (2010), *A Concise Introduction to Mathematical Logic* (3rd ed.), New York: Springer Science+Business Media, doi:10.1007/978-1-4419-1221-3, ISBN 978-1-4419-1220-6.

_ Victor Rodych, 2003, "Misunderstanding Gödel: New Arguments about Wittgenstein and New Remarks by Wittgenstein", *Dialectica* v. 57 n. 3, pp. 279–313. doi:10.1111/j.1746-8361.2003.tb00272.x
_ Stewart Shapiro, 2002, "Incompleteness and Inconsistency", *Mind*, v. 111, pp 817–32. doi:10.1093/mind/111.444.817
_ Alan Sokal and Jean Bricmont, 1999, *Fashionable Nonsense: Postmodern Intellectuals' Abuse of Science*, Picador. ISBN 0-312-20407-8
_ Joseph R. Shoenfield (1967), *Mathematical Logic*. Reprinted by A.K. Peters for the Association for Symbolic Logic, 2001. ISBN 978-1-56881-135-2

_ Jeremy Stangroom and Ophelia Benson, *Why Truth Matters*, Continuum. ISBN 0-8264-9528-1
_ George Tourlakis, *Lectures in Logic and Set Theory, Volume 1, Mathematical Logic*, Cambridge University Press, 2003. ISBN 978-0-521-75373-9
_ Wigderson, Avi (2010), "The Gödel Phenomena in Mathematics: A Modern View", *Kurt Gödel and the Foundations of Mathematics: Horizons of Truth*, Cambridge University Press
_ Hao Wang, 1996, *A Logical Journey: From Gödel to Philosophy*, The MIT Press, Cambridge MA, ISBN 0-262-23189-1.
_ Richard Zach, 2006, "Hilbert's program then and now", in *Philosophy of Logic*, Dale Jacquette (ed.), Handbook of the Philosophy of Science, v. 5., Elsevier, pp. 411–447.

67.14 External links

_
_ Godel's Incompleteness Theorems on *In Our Time* at the BBC. (listen now)
_ Stanford Encyclopedia of Philosophy: "Kurt Gödel" — by Juliette Kennedy.
_ MacTutor biographies:
_ Kurt Gödel.
_ Gerhard Gentzen.
_ What is Mathematics:Gödel'{}s Theorem and Around by *Karlis Podnieks*. An online free book.
_ World's shortest explanation of Gödel's theorem using a printing machine as an example.
_ October 2011 RadioLab episode about/including Gödel's Incompleteness theorem
_ Hazewinkel, Michiel, ed. (2001), "Gödel incompleteness theorem", *Encyclopedia of Mathematics*, Springer, ISBN 978-1-55608-010-4

Chapter 68

Experimental mathematics

For the mathematical journal of the same name, see Experimental Mathematics (journal).

Experimental mathematics is an approach to mathematics in which numerical computation is used to investigate mathematical objects and identify properties and patterns.[1] It has been defined as "that branch of mathematics that concerns itself ultimately with the codification and transmission of insights within the mathematical community through the use of experimental (in either the Galilean, Baconian, Aristotelian or Kantian sense) exploration of conjectures and more informal beliefs and a careful analysis of the data acquired in this pursuit."[2]

68.1 History

Mathematicians have always practised experimental mathematics. Existing records of early mathematics, such as Babylonian mathematics, typically consist of lists of numerical examples illustrating algebraic identities. However, modern mathematics, beginning in the 17th century, developed a tradition of publishing results in a final, formal and abstract presentation. The numerical examples that may have led a mathematician to originally formulate a general theorem were not published, and were generally forgotten.

Experimental mathematics as a separate area of study re-emerged in the twentieth century, when the invention of the electronic computer vastly increased the range of feasible calculations, with a speed and precision far greater than anything available to previous generations of mathematicians. A significant milestone and achievement of experimental mathematics was the discovery in 1995 of the Bailey–Borwein–Plouffe formula for the binary digits of π. This formula was discovered not by formal reasoning, but instead by numerical searches on a computer; only afterwards was a rigorous proof found.[3]

68.2 Objectives and uses

The objectives of experimental mathematics are "to generate understanding and insight; to generate and confirm or confront conjectures; and generally to make mathematics more tangible, lively and fun for both the professional researcher and the novice".[4]

The uses of experimental mathematics have been defined as follows:[5]

1. Gaining insight and intuition.
2. Discovering new patterns and relationships.

3. Using graphical displays to suggest underlying mathematical principles.
4. Testing and especially falsifying conjectures.
5. Exploring a possible result to see if it is worth formal proof.
6. Suggesting approaches for formal proof.

7. Replacing lengthy hand derivations with computer-based derivations.
8. Confirming analytically derived results.

68.3 Tools and techniques

Experimental mathematics makes use of numerical methods to calculate approximate values for integrals and infinite series. Arbitrary precision arithmetic is often used to establish these values to a high degree of precision – typically 100 significant figures or more. Integer relation algorithms are then used to search for relations between these values and mathematical constants. Working with high precision values reduces the possibility of mistaking a mathematical coincidence for a true relation. A formal proof of a conjectured relation will then be sought – it is often easier to find a formal proof once the form of a conjectured relation is known.

If a counterexample is being sought or a large-scale proof by exhaustion is being attempted, distributed computing techniques may be used to divide the calculations between multiple computers.

Frequent use is made of general computer algebra systems such as Mathematica, although domain-specific software is also written for attacks on problems that require high efficiency. Experimental mathematics software usually includes error detection and correction mechanisms, integrity checks and redundant calculations designed to minimise the possibility of results being invalidated by a hardware or software error.

68.4 Applications and examples

Applications and examples of experimental mathematics include:

_ Searching for a counterexample to a conjecture

_ Roger Frye used experimental mathematics techniques to find the smallest counterexample to Euler's sum of powers conjecture.

_ The ZetaGrid project was set up to search for a counterexample to the Riemann hypothesis.

_ This project is searching for a counterexample to the Collatz conjecture.

_ Finding new examples of numbers or objects with particular properties

_ The Great Internet Mersenne Prime Search is searching for new Mersenne primes.

_ The distributed.net's OGR project is searching for optimal Golomb rulers.

_ The Riesel Sieve project is searching for the smallest Riesel number.

_ The Seventeen or Bust project is searching for the smallest Sierpinski number.

_ The Sudoku Project is searching for a solution to the minimum Sudoku problem.

_ Finding serendipitous numerical patterns

_ Edward Lorenz found the Lorenz attractor, an early example of a chaotic dynamical system, by investigating anomalous behaviours in a numerical weather model.

_ The Ulam spiral was discovered by accident.

_ Mitchell Feigenbaum's discovery of the Feigenbaum constant was based initially on numerical observations, followed by a rigorous proof.

_ Use of computer programs to check a large but finite number of cases to complete a computer-assisted proof by exhaustion

_ Thomas Hales's proof of the Kepler conjecture.

_ Various proofs of the four colour theorem.

_ Clement Lam's proof of the non-existence of a finite projective plane of order 10.[6]

_ Symbolic validation (via Computer algebra) of conjectures to motivate the search for an analytical proof

_ Solutions to a special case of the quantum three-body problem known as the hydrogen molecule-ion were found standard quantum chemistry basis sets before realizing they all lead to the same unique analytical solution in terms of a *generalization* of the Lambert W function. Related to this work is the isolation of a previously unknown link between gravity theory and quantum mechanics in lower dimensions (see quantum gravity and references therein).

_ In the realm of relativistic many-bodied mechanics, namely the time-symmetric Wheeler–Feynman absorber theory: the equivalence between an advanced Liénard–Wiechert potential of particle *j* acting on particle *i* and the corresponding potential for particle *i* acting on particle *j* was demonstrated exhaustively to order $1/c^{10}$ before being proved mathematically. The Wheeler-Feynman theory has regained interest because of quantum nonlocality.

_ In the realm of linear optics, verification of the series expansion of the envelope of the electric field for ultrashort light pulses travelling in non isotropic media. Previous expansions had been incomplete: the outcome revealed an extra term vindicated by *experiment.*

- Evaluation of infinite series, infinite products and integrals (also see symbolic integration), typically by carrying out a high precision numerical calculation, and then using an integer relation algorithm (such as the Inverse Symbolic Calculator) to find a linear combination of mathematical constants that matches this value. For example, the following identity was first conjectured by Enrico Au-Yeung, a student of Jonathan Borwein using computer search and PSLQ algorithm in 1993:[7]

$$\sum_{k=1}^{\infty}\frac{1}{k^2}\left(1+\frac{1}{2}+\frac{1}{3}+\cdots+\frac{1}{k}\right)^2=\frac{17\pi^4}{360}.$$

- Visual investigations
- In Indra's Pearls, David Mumford and others investigated various properties of Möbius transformation and Schottky group using computer generated images of the groups which: *furnished convincing evidence for many conjectures and lures to further exploration.*[8]

68.5 Plausible but false examples

Main article: mathematical coincidence

Some plausible relations hold to a high degree of accuracy, but are still not true. One example is:

$$\int_0^{\infty}\cos(2x)\prod_{n=1}^{\infty}\cos\left(\frac{x}{n}\right)dx\approx\frac{\pi}{8}.$$

The two sides of this expression only differ after the 42nd decimal place.[9]

Another example is that the maximum height (maximum absolute value of coefficients) of all the factors of $x^n - 1$ appears to be the same as the height of the nth cyclotomic polynomial. This was shown by computer to be true for $n < 10000$ and was expected to be true for all n. However, a larger computer search showed that this equality fails to hold for $n = 14235$, when the height of the nth cyclotomic polynomial is 2, but maximum height of the factors is 3.[10]

68.6 Practitioners

The following mathematicians and computer scientists have made significant contributions to the field of experimental mathematics:

- Fabrice Bellard
- David H. Bailey
- Jonathan Borwein
- David Epstein
- Helaman Ferguson
- Ronald Graham

_ Thomas Callister Hales
_ Donald Knuth
_ Clement Lam
_ Oren Patashnik
_ Simon Plouffe
_ Eric Weisstein
_ Doron Zeilberger
_ A.J. Han Vinck

68.7 See also

_ Borwein integral
_ Computer-aided proof
_ Proofs and Refutations
_ *Experimental Mathematics* (journal)
_ Institute for Experimental Mathematics

68.8 References

[1] Weisstein, Eric W., "Experimental Mathematics", *MathWorld.*
[2] Experimental Mathematics: A Discussion by J. Borwein, P. Borwein, R. Girgensohn and S. Parnes
[3] The Quest for Pi by David H. Bailey, Jonathan M. Borwein, Peter B. Borwein and Simon Plouffe.
[4] Borwein, Jonathan; Bailey, David (2004). *Mathematics by Experiment: Plausible Reasoning in the 21st Century.* A.K. Peters. pp. *vii.* ISBN 1-56881-211-6.
[5] Borwein, Jonathan; Bailey, David (2004). *Mathematics by Experiment: Plausible Reasoning in the 21st Century.* A.K. Peters. p. 2. ISBN 1-56881-211-6.
[6] Clement W. H. Lam (1991). "The Search for a Finite Projective Plane of Order 10". *American Mathematical Monthly* **98** (4): 305–318. doi:10.2307/2323798.
[7] Bailey, David (1997). "New Math Formulas Discovered With Supercomputers". *NAS News* **2** (24).
[8] Mumford, David; Series, Caroline; Wright, David (2002). *Indra's Pearls: The Vision of Felix Klein.* Cambridge. pp. viii. ISBN 0-521-35253-3.
[9] David H. Bailey and Jonathan M. Borwein, Future Prospects for Computer-Assisted Mathematics, December 2005
[10] The height of Φ_{4745} is 3 and 14235 = 3 x 4745. See Sloane sequences A137979 and A160338.

68.9 External links

_ Experimental Mathematics (Journal)
_ Centre for Experimental and Constructive Mathematics (CECM) at Simon Fraser University
_ Collaborative Group for Research in Mathematics Education at University of Southampton
_ Recognizing Numerical Constants by David H. Bailey and Simon Plouffe
_ Psychology of Experimental Mathematics
_ Experimental Mathematics Website (Links and resources)
_ An Algorithm for the Ages: PSLQ, A Better Way to Find Integer Relations (Alternative link)
_ Experimental Algorithmic Information Theory
_ Sample Problems of Experimental Mathematics by David H. Bailey and Jonathan M. Borwein
_ Ten Problems in Experimental Mathematics by David H. Bailey, Jonathan M. Borwein, Vishaal Kapoor, and Eric W. Weisstein
_ Institute for Experimental Mathematics at University of Duisburg-Essen

Chapter 69

Fractal

For other uses, see Fractal (disambiguation).
Figure 1a. The Mandelbrot set illustrates self-similarity. As the image is enlarged, the same pattern re-appears so that it is virtually impossible to determine the scale being examined.
Figure 1b. The same fractal magnified six times.
Figure 1c. The same fractal magnified a hundred times.
Figure 1d. Even at a magnification of 2,000, the Mandelbrot set displays fine detail resembling the full set.

Figure 1e. Another example highlighting how scale is a key feature of a fractal.
A ***fractal*** is a natural phenomenon or a mathematical set that exhibits a repeating pattern that displays at every scale. If the replication is exactly the same at every scale, it is called a self-similar pattern.[1] Fractals can also be nearly the same at different levels. This latter pattern is illustrated in Figure 1.[2][3][4][5] Fractals also includes the idea of a *detailed pattern* that repeats itself.[2]:166; 18[3][6]

Fractals are different from other geometric figures because of the way in which they scale. Doubling the edge lengths of a polygon multiplies its area by four, which is two (the ratio of the new to the old side length) raised to the power of two (the dimension of the space the polygon resides in). Likewise, if the radius of a sphere is doubled, its volume scales by eight, which is two (the ratio of the new to the old radius) to the power of three (the dimension that the sphere resides in). But if a fractal's one-dimensional lengths are all doubled, the spatial content of the fractal scales by a power of two that is not necessarily an integer.[2] This power is called the fractal dimension of the fractal, and it usually exceeds the fractal's topological dimension.[7]

As mathematical equations, fractals are usually nowhere differentiable.[2][5][8] An infinite fractal curve can be conceived of as winding through space differently from an ordinary line, still being a 1-dimensional line yet having a fractal dimension indicating it also resembles a surface.[2]:15[7]:48

The mathematical roots of the idea of fractals have been traced throughout the years as a formal path of published works, starting in the 17th century with notions of recursion, then moving through increasingly rigorous mathematical treatment of the concept to the study of continuous but not differentiable functions in the 19th century, and on to the coining of the word *fractal* in the 20th century with a subsequent burgeoning of interest in fractals and computerbased modelling in the 21st century.[9][10] The term "fractal" was first used by mathematician Benoît Mandelbrot in 1975. Mandelbrot based it on the Latin *frāctus* meaning "broken" or "fractured", and used it to extend the concept of theoretical fractional dimensions to geometric patterns in nature.[2]:405[6]

There is some disagreement amongst authorities about how the concept of a fractal should be formally defined. Mandelbrot himself summarized it as "beautiful, damn hard, increasingly useful. That's fractals."[11] The general consensus is that theoretical fractals are infinitely self-similar, iterated, and detailed mathematical constructs having fractal dimensions, of which many examples have been formulated and studied in great depth.[2][3][4] Fractals are not limited to geometric patterns, but can also describe processes in time.[1][5][12] Fractal patterns with various degrees of self-similarity have been rendered or studied in images, structures and sounds[13] and found in nature,[14][15][16][17][18] technology,[19][20][21][22] art,[23][24][25] and law.[26]

69.1 Introduction

The word "fractal" often has different connotations for laypeople than for mathematicians, where the layperson is more likely to be familiar with fractal art than a mathematical conception. The mathematical concept is difficult to define formally even for mathematicians, but key features can be understood with little mathematical background. The feature of "self-similarity", for instance, is easily understood by analogy to zooming in with a lens or other device that zooms in on digital images to uncover finer, previously invisible, new structure. If this is done on fractals, however, no new detail appears; nothing changes and the same pattern repeats over and over, or for some fractals, nearly the same pattern reappears over and over. Self-similarity itself is not necessarily counter-intuitive (e.g., people have pondered self-similarity informally such as in the infinite regress in parallel mirrors or the homunculus, the little man inside the head of the little man inside the head...). The difference for fractals is that the pattern reproduced must be detailed.[2]:166; 18[3][6]

This idea of being detailed relates to another feature that can be understood without mathematical background: Having a fractional or fractal dimension greater than its topological dimension, for instance, refers to how a fractal scales compared to how geometric shapes are usually perceived. A regular line, for instance, is conventionally understood

to be 1-dimensional; if such a curve is divided into pieces each 1/3 the length of the original, there are always 3 equal pieces. In contrast, consider the curve in Figure 2. It is also 1-dimensional for the same reason as the ordinary line, but it has, in addition, a fractal dimension greater than 1 because of how its detail can be measured. The fractal curve divided into parts 1/3 the length of the original line becomes 4 pieces rearranged to repeat the original detail, and this unusual relationship is the basis of its fractal dimension.

This also leads to understanding a third feature, that fractals as mathematical equations are "nowhere differentiable". In a concrete sense, this means fractals cannot be measured in traditional ways.[2][5][8] To elaborate, in trying to find the length of a wavy non-fractal curve, one could find straight segments of some measuring tool small enough to lay end to end over the waves, where the pieces could get small enough to be considered to conform to the curve in the normal manner of measuring with a tape measure. But in measuring a wavy fractal curve such as the one in Figure 2, one would never find a small enough straight segment to conform to the curve, because the wavy pattern would always re-appear, albeit at a smaller size, essentially pulling a little more of the tape measure into the total length measured each time one attempted to fit it tighter and tighter to the curve. This is perhaps counter-intuitive, but it is how fractals behave.[2]

69.2 History

The history of fractals traces a path from chiefly theoretical studies to modern applications in computer graphics, with several notable people contributing canonical fractal forms along the way.[9][10] According to Pickover, the mathematics behind fractals began to take shape in the 17th century when the mathematician and philosopher Gottfried Leibniz pondered recursive self-similarity (although he made the mistake of thinking that only the straight line was self-similar in this sense).[27] In his writings, Leibniz used the term "fractional exponents", but lamented that

“Geometry” did not yet know of them.[2]:405 Indeed, according to various historical accounts, after that point few mathematicians tackled the issues and the work of those who did remained obscured largely because of resistance to such unfamiliar emerging concepts, which were sometimes referred to as mathematical “monsters”.[8][9][10] Thus, it was not until two centuries had passed that in 1872 Karl Weierstrass presented the first definition of a function with a graph that would today be considered fractal, having the non-intuitive property of being everywhere continuous but nowhere differentiable.[9]:7[10] Not long after that, in 1883, Georg Cantor, who attended lectures by Weierstrass,[10] published examples of subsets of the real line known as Cantor sets, which had unusual properties and are now recognized as fractals.[9]:11–24 Also in the last part of that century, Felix Klein and Henri Poincaré introduced a category of fractal that has come to be called “self-inverse” fractals.[2]:166

One of the next milestones came in 1904, when Helge von Koch, extending ideas of Poincaré and dissatisfied with Weierstrass's abstract and analytic definition, gave a more geometric definition including hand drawn images of a similar function, which is now called the Koch curve (see Figure 2)[9]:25.[10] Another milestone came a decade later in 1915, when Wacław Sierpiński constructed his famous triangle then, one year later, his carpet. By 1918, two French mathematicians, Pierre Fatou and Gaston Julia, though working independently, arrived essentially simultaneously at results describing what are now seen as fractal behaviour associated with mapping complex numbers and iterative functions and leading to further ideas about attractors and repellors (i.e., points that attract or repel other points), which have become very important in the study of fractals (see Figure 3 and Figure 4).[5][9][10] Very shortly after that work was submitted, by March 1918, Felix Hausdorff expanded the definition of “dimension”, significantly for the evolution of the definition of fractals, to allow for sets to have noninteger dimensions.[10] The idea of self-similar curves was taken further by Paul Lévy, who, in his 1938 paper *Plane or Space Curves and Surfaces Consisting of Parts Similar to the Whole* described a new fractal curve, the Lévy C curve.[notes 1]

Different researchers have postulated that without the aid of modern computer graphics, early investigators were limited to what they could depict in manual drawings, so lacked the means to visualize the beauty and appreciate some of the implications of many of the patterns they had discovered (the Julia set, for instance, could only be visualized through a few iterations as very simple drawings hardly resembling the image in Figure 3).[2]:179[8][10] That changed, however, in the 1960s, when Benoît Mandelbrot started writing about self-similarity in papers such as *How Long Is the Coast of Britain? Statistical Self-Similarity and Fractional Dimension*,[28] which built on earlier work by Lewis Fry Richardson. In 1975[6] Mandelbrot solidified hundreds of years of thought and mathematical development in coining the word “fractal” and illustrated his mathematical definition with striking computer-constructed visualizations. These images, such as of his canonical Mandelbrot set pictured in Figure 1, captured the popular imagination; many of them were based on recursion, leading to the popular meaning of the term “fractal”.[29] Currently, fractal studies are essentially exclusively computer-based.[8][9][27]

476 *CHAPTER 69. FRACTAL*

Figure 2a. Koch snowflake, a fractal that begins with an equilateral triangle and then replaces the middle third of every line segment with a pair of line segments that form an equilateral “bump”

69.3 Characteristics

One often cited description that Mandelbrot published to describe geometric fractals is “a rough or fragmented geometric shape that can be split into parts, each of which is (at least approximately) a reduced-size copy of the whole";[2] this is generally helpful but limited. Authorities disagree on the exact definition of *fractal*, but most usually elaborate on the basic ideas of self-similarity and an unusual relationship with the space a fractal is embedded in.[1][2][3][5][30] One point agreed on is that fractal patterns are characterized by fractal dimensions, but whereas these numbers quantify complexity (i.e., changing detail with changing scale), they neither uniquely describe nor specify details of how to construct particular fractal patterns.[31] In 1975 when Mandelbrot coined the word “fractal”, he did so to denote an object whose Hausdorff–Besicovitch dimension is greater than its topological dimension.[6] It has been noted that this dimensional requirement is not met by fractal space-filling curves such as the Hilbert curve.[notes 2]

According to Falconer, rather than being strictly defined, fractals should, in addition to being nowhere differentiable and able to have a fractal dimension, be generally characterized by a gestalt of the following features;[3]

_ Self-similarity, which may be manifested as:

69.3. CHARACTERISTICS 477

Figure 2b. Koch snowflake, a zoom out of the Koch Snowflake

Figure 3. A Julia set, a fractal related to the Mandelbrot set

_ Exact self-similarity: *identical at all scales; e.g. Koch snowflake*

_ Quasi self-similarity: *approximates the same pattern at different scales; may contain small*

478 *CHAPTER 69. FRACTAL*

Figure 4. A strange attractor that exhibits multifractal scaling

copies of the entire fractal in distorted and degenerate forms; e.g., the Mandelbrot set's satellites are approximations of the entire set, but not exact copies, as shown in Figure 1

_ Statistical self-similarity: *repeats a pattern stochastically so numerical or statistical measures are preserved across scales; e.g., randomly generated fractals; the well-known example of the coastline of Britain, for which one would not expect to find a segment scaled*

and repeated as neatly as the repeated unit that defines, for example, the Koch snowflake[5]
_ Qualitative self-similarity: *as in a time series*[12]
_ Multifractal scaling: *characterized by more than one fractal dimension or scaling rule*
_ Fine or detailed structure at arbitrarily small scales. A consequence of this structure is fractals may have emergent properties[32] (related to the next criterion in this list).
_ Irregularity locally and globally that is not easily described in traditional Euclidean geometric language. For images of fractal patterns, this has been expressed by phrases such as "smoothly piling up surfaces" and "swirls upon swirls".[7]
_ Simple and "perhaps recursive" definitions *see Common techniques for generating fractals*
As a group, these criteria form guidelines for excluding certain cases, such as those that may be self-similar without having other typically fractal features. A straight line, for instance, is self-similar but not fractal because it lacks
69.4. BROWNIAN MOTION 479
Uniform Mass Center Triangle Fractal
detail, is easily described in Euclidean language, has the same Hausdorff dimension as topological dimension, and is fully defined without a need for recursion.[2][5]

69.4 Brownian motion

A path generated by a one dimensional Wiener process is a fractal curve of dimension 1.5, and Brownian motion is a finite version of this.[33]

69.5 Common techniques for generating fractals

Images of fractals can be created by fractal generating programs.
_ *Iterated function systems* – use fixed geometric replacement rules; may be stochastic or deterministic;[34] e.g., Koch snowflake, Cantor set, Haferman carpet,[35] Sierpinski carpet, Sierpinski gasket, Peano curve, Harter-Heighway dragon curve, T-Square, Menger sponge
480 *CHAPTER 69. FRACTAL*
Figure 5. Self-similar branching pattern modeled in silico using L-systems principles[18]
_ *Strange attractors* – use iterations of a map or solutions of a system of initial-value differential equations that exhibit chaos (e.g., see multifractal image)
_ *L-systems* - use string rewriting; may resemble branching patterns, such as in plants, biological cells (e.g., neurons and immune system cells[18]), blood vessels, pulmonary structure,[36] etc. (e.g., see Figure 5) or turtle graphics patterns such as space-filling curves and tilings
_ *Escape-time fractals* – use a formula or recurrence relation at each point in a space (such as the complex plane); usually quasi-self-similar; also known as "orbit" fractals; e.g., the Mandelbrot set, Julia set, Burning Ship fractal, Nova fractal and Lyapunov fractal. The 2d vector fields that are generated by one or two iterations of escape-time formulae also give rise to a fractal form when points (or pixel data) are passed through this field repeatedly.
_ *Random fractals* – use stochastic rules; e.g., Lévy flight, percolation clusters, self avoiding walks, fractal landscapes, trajectories of Brownian motion and the Brownian tree (i.e., dendritic fractals generated by modeling diffusion-limited aggregation or reaction-limited aggregation clusters).[5]
_ *Finite subdivision rules* use a recursive topological algorithm for refining tilings[37] and they are similar to the process of cell division.[38] The iterative processes used in creating the Cantor set and the Sierpinski carpet are examples of finite subdivision rules, as is barycentric subdivision.
69.6. SIMULATED FRACTALS 481
A fractal generated by a finite subdivision rule for an alternating link

69.6 Simulated fractals

Fractal patterns have been modeled extensively, albeit within a range of scales rather than infinitely, owing to the practical limits of physical time and space. Models may simulate theoretical fractals or natural phenomena with fractal features. The outputs of the modelling process may be highly artistic renderings, outputs for investigation, or benchmarks for fractal analysis. Some specific applications of fractals to technology are listed elsewhere. Images and other outputs of modelling are normally referred to as being "fractals" even if they do not have strictly fractal characteristics, such as when it is possible to zoom into a region of the fractal image that does not exhibit any fractal properties. Also, these may include calculation or display artifacts which are not characteristics of true fractals.
Modeled fractals may be sounds,[13] digital images, electrochemical patterns, circadian rhythms,[39] etc. Fractal patterns have been reconstructed in physical 3-dimensional space[21]:10 and virtually, often called "in silico" modeling.[36]
Models of fractals are generally created using fractal-generating software that implements techniques such as those outlined above.[5][12][21] As one illustration, trees, ferns, cells of the nervous system,[18] blood and lung vasculature,[36] and other branching patterns in nature can be modeled on a computer by using recursive algorithms and L-systems techniques.[18] The recursive nature of some patterns is obvious in certain examples—a branch from a tree or a frond from a fern is a miniature replica of the whole: not identical, but similar in nature. Similarly, random fractals have been used to describe/create many highly irregular real-world objects. A limitation of modeling fractals is that resemblance

of a fractal model to a natural phenomenon does not prove that the phenomenon being modeled is formed
482 *CHAPTER 69. FRACTAL*
A fractal flame
by a process similar to the modeling algorithms.

69.7 Natural phenomena with fractal features

Further information: Patterns in nature

Approximate fractals found in nature display self-similarity over extended, but finite, scale ranges. The connection between fractals and leaves, for instance, is currently being used to determine how much carbon is contained in trees.[40]

Examples of phenomena known or anticipated to have fractal features are listed below:

69.8 In creative works

Further information: Fractal art

Fractal patterns have been found in the paintings of American artist Jackson Pollock. While Pollock's paintings appear to be composed of chaotic dripping and splattering, computer analysis has found fractal patterns in his work.[25]

Decalcomania, a technique used by artists such as Max Ernst, can produce fractal-like patterns.[49] It involves pressing paint between two surfaces and pulling them apart.

Cyberneticist Ron Eglash has suggested that fractal geometry and mathematics are prevalent in African art, games, divination, trade, and architecture. Circular houses appear in circles of circles, rectangular houses in rectangles of rectangles, and so on. Such scaling patterns can also be found in African textiles, sculpture, and even cornrow hairstyles.[24][50]
69.8. IN CREATIVE WORKS 483
Fractal defrosting patterns, polar Mars. The patterns are formed by sublimation of frozen CO_2. *Width of image is about a kilometer.*
A fractal that models the surface of a mountain (animation)

In a 1996 interview with Michael Silverblatt, David Foster Wallace admitted that the structure of the first draft of *Infinite Jest* he gave to his editor Michael Pietsch was inspired by fractals, specifically the Sierpinski triangle (aka
484 *CHAPTER 69. FRACTAL*
Sierpinski gasket) but that the edited novel is "more like a lopsided Sierpinsky Gasket".[23]

69.9 Applications in technology

Main article: Fractal analysis

69.10 See also

69.10.1 Fractal-generating programs

There are many fractal generating programs available, both free and commercial. Some of the fractal generating programs include:

_ Apophysis - open source software for Microsoft Windows based systems
_ Electric Sheep - open source distributed computing software
_ Fractint - freeware with available source code
_ Sterling - Freeware software for Microsoft Windows based systems
_ SpangFract - For Mac OS
_ Ultra Fractal - A proprietary fractal generator for Microsoft Windows and Mac OS X based systems
_ XaoS - A cross platform open source fractal zooming program
_ Chaotica - a commercial software for Microsoft Windows, Linux and Mac OS
_ Terragen - a fractal terrain generator.

Most of the above programs make two-dimensional fractals, with a few creating three-dimensional fractal objects, such as quaternions, mandelbulbs and mandelboxes.

69.11 Notes

[1] The original paper, Lévy, Paul (1938). "Les Courbes planes ou gauches et les surfaces composées de parties semblables au tout". *Journal de l'École Polytechnique*: 227–247, 249–291., is translated in Edgar, pages 181-239.

[2] The Hilbert curve map is not a homeomorphism, so it does not preserve topological dimension. The topological dimension and Hausdorff dimension of the image of the Hilbert map in $\mathbf{R}^2$ are both 2. Note, however, that the topological dimension of the *graph* of the Hilbert map (a set in $\mathbf{R}^3$) is 1.

69.12 References

[1] Gouyet, Jean-François (1996). *Physics and fractal structures*. Paris/New York: Masson Springer. ISBN 978-0-387-94153-0.

[2] Mandelbrot, Benoît B. (1983). *The fractal geometry of nature*. Macmillan. ISBN 978-0-7167-1186-5. Retrieved 1 February 2012.

[3] Falconer, Kenneth (2003). *Fractal Geometry: Mathematical Foundations and Applications*. John Wiley & Sons, Ltd. xxv. ISBN 0-470-84862-6.

[4] Briggs, John (1992). *Fractals:The Patterns of Chaos*. London, UK: Thames and Hudson. p. 148. ISBN 0-500-27693-5.
69.12. REFERENCES 485

[5] Vicsek, Tamás (1992). *Fractal growth phenomena*. Singapore/New Jersey: World Scientific. pp. 31; 139–146. ISBN 978-981-02-0668-0.
[6] Albers, Donald J.; Alexanderson, Gerald L. (2008). "Benoît Mandelbrot: In his own words". *Mathematical people : profiles and interviews*. Wellesley, MA: AK Peters. p. 214. ISBN 978-1-56881-340-0.
[7] Mandelbrot, Benoît B. (2004). *Fractals and Chaos*. Berlin: Springer. p. 38. ISBN 978-0-387-20158-0. "A fractal set is one for which the fractal (Hausdorff-Besicovitch) dimension strictly exceeds the topological dimension"
[8] Gordon, Nigel (2000). *Introducing fractal geometry*. Duxford, UK: Icon. p. 71. ISBN 978-1-84046-123-7.
[9] Edgar, Gerald (2004). *Classics on Fractals*. Boulder, CO: Westview Press. ISBN 978-0-8133-4153-8.
[10] Trochet, Holly (2009). "A History of Fractal Geometry". *MacTutor History of Mathematics*. Archived from the original on 4 February 2012.
[11] Mandelbrot, Benoit. "24/7 Lecture on Fractals". *2006 Ig Nobel Awards*. Improbable Research.
[12] Peters, Edgar (1996). *Chaos and order in the capital markets : a new view of cycles, prices, and market volatility*. New York: Wiley. ISBN 0-471-13938-6.
[13] Brothers, Harlan J. (2007). "Structural Scaling in Bach's Cello Suite No. 3". *Fractals* **15**: 89–95. doi:10.1142/S0218348X0700337X.
[14] Tan, Can Ozan; Cohen, Michael A.; Eckberg, Dwain L.; Taylor, J. Andrew (2009). "Fractal properties of human heart period variability: Physiological and methodological implications". *The Journal of Physiology* **587** (15): 3929. doi:10.1113/jphysiol.2009.169219.
[15] Buldyrev, Sergey V.; Goldberger, Ary L.; Havlin, Shlomo; Peng, Chung-Kang; Stanley, H. Eugene (1995). "3". In Bunde, Armin; Havlin, Shlomo. "Fractals in Science". Springer.
[16] Liu, Jing Z.; Zhang, Lu D.; Yue, Guang H. (2003). "Fractal Dimension in Human Cerebellum Measured by Magnetic Resonance Imaging". *Biophysical Journal* **85** (6): 4041–4046. doi:10.1016/S0006-3495(03)74817-6. PMC 1303704. PMID 14645092.
[17] Karperien, Audrey L.; Jelinek, Herbert F.; Buchan, Alastair M. (2008). "Box-Counting Analysis of Microglia Form in Schizophrenia, Alzheimer's Disease and Affective Disorder". *Fractals* **16** (2): 103. doi:10.1142/S0218348X08003880.
[18] Jelinek, Herbert F.; Karperien, Audrey; Cornforth, David; Cesar, Roberto; Leandro, Jorge de Jesus Gomes (2002). "MicroMod-an L-systems approach to neural modelling". In Sarker, Ruhul. *Workshop proceedings: the Sixth Australia-Japan Joint Workshop on Intelligent and Evolutionary Systems, University House, ANU,*. University of New South Wales. ISBN 9780731705054. OCLC 224846454. http://researchoutput.csu.edu.au/R/-?func=dbin-jump-full&object_id=6595&local_base=GEN01-CSU01. Retrieved 3 February 2012. "Event location: Canberra, Australia"
[19] Hu, Shougeng; Cheng, Qiuming; Wang, Le; Xie, Shuyun (2012). "Multifractal characterization of urban residential land price in space and time". *Applied Geography* **34**: 161. doi:10.1016/j.apgeog.2011.10.016.
[20] Karperien, Audrey; Jelinek, Herbert F.; Leandro, Jorge de Jesus Gomes; Soares, João V. B.; Cesar Jr, Roberto M.; Luckie, Alan (2008). "Automated detection of proliferative retinopathy in clinical practice". *Clinical ophthalmology (Auckland, N.Z.)* **2** (1): 109–122. doi:10.2147/OPTH.S1579. PMC 2698675. PMID 19668394.
[21] Losa, Gabriele A.; Nonnenmacher, Theo F. (2005). *Fractals in biology and medicine*. Springer. ISBN 978-3-7643-7172-2. Retrieved 1 February 2012.
[22] Vannucchi, Paola; Leoni, Lorenzo (2007). "Structural characterization of the Costa Rica décollement: Evidence for seismically-induced fluid pulsing". *Earth and Planetary Science Letters* **262** (3–4): 413. Bibcode:2007E&PSL.262..413V. doi:10.1016/j.epsl.2007.07.056.
[23] Wallace, David Foster. "Bookworm on KCRW". Kcrw.com. Retrieved 2010-10-17.
[24] Eglash, Ron (1999). "African Fractals: Modern Computing and Indigenous Design". New Brunswick: Rutgers University Press. Retrieved 2010-10-17.
[25] Taylor, Richard; Micolich, Adam P.; Jonas, David. "Fractal Expressionism: Can Science Be Used To Further Our Understanding Of Art?". Phys.unsw.edu.au. Retrieved 2010-10-17.
[26] Stumpff, Andrew (2013). "The Law is a Fractal: The Attempt to Anticipate Everything" **44**. Loyola University Chicago Law Journal. p. 649.
[27] Pickover, Clifford A. (2009). *The Math Book: From Pythagoras to the 57th Dimension, 250 Milestones in the History of Mathematics*. Sterling Publishing Company, Inc. p. 310. ISBN 978-1-4027-5796-9. Retrieved 2011-02-05.

[28] Batty, Michael (1985-04-04). "Fractals - Geometry Between Dimensions". *New Scientist* (Holborn Publishing Group) **105** (1450): 31.
[29] Russ, John C. (1994). *Fractal surfaces* **1**. Springer. p. 1. ISBN 978-0-306-44702-0. Retrieved 2011-02-05.
[30] Edgar, Gerald (2008). *Measure, topology, and fractal geometry*. New York, NY: Springer-Verlag. p. 1. ISBN 978-0-387-74748-4.
[31] Karperien, Audrey (2004). http://www.webcitation.org/65DyLbmF1 *Defining microglial morphology: Form, Function, and Fractal Dimension*. Charles Sturt University. Retrieved 2012-02-05.
[32] Spencer, John; Thomas, Michael S. C.; McClelland, James L. (2009). *Toward a unified theory of development : connectionism and dynamic systems theory re-considered*. Oxford/New York: Oxford University Press. ISBN 978-0-19-530059-8.
[33] Falconer, Kenneth (2013). *Fractals, A Very Short Introduction*. Oxford University Press.
[34] Frame, Angus (3 August 1998). "Iterated Function Systems". In Pickover, Clifford A. *Chaos and fractals: a computer graphical journey : ten year compilation of advanced research*. Elsevier. pp. 349–351. ISBN 978-0-444-50002-1. Retrieved 4 February 2012.
[35] "Haferman Carpet". WolframAlpha. Retrieved 18 October 2012.
[36] Hahn, Horst K.; Georg, Manfred; Peitgen, Heinz-Otto (2005). "Fractal aspects of three-dimensional vascular constructive optimization". In Losa, Gabriele A.; Nonnenmacher, Theo F. *Fractals in biology and medicine*. Springer. pp. 55–66. ISBN 978-3-7643-7172-2.
[37] J. W. Cannon, W. J. Floyd, W. R. Parry. *Finite subdivision rules*. Conformal Geometry and Dynamics, vol. 5 (2001), pp. 153–196.

[38] J. W. Cannon, W. Floyd and W. Parry. *Crystal growth, biological cell growth and geometry*. Pattern Formation in Biology, Vision and Dynamics, pp. 65–82. World Scientific, 2000. ISBN 981-02-3792-8,ISBN 978-981-02-3792-9.
[39] Fathallah-Shaykh, Hassan M. (2011). "Fractal Dimension of the Drosophila Circadian Clock". *Fractals* **19** (4): 423–430. doi:10.1142/S0218348X11005476.
[40] "Hunting the Hidden Dimension." *Nova*. PBS. WPMB-Maryland. 28 October 2008.
[41] Sornette, Didier (2004). *Critical phenomena in natural sciences: chaos, fractals, selforganization, and disorder : concepts and tools*. Springer. pp. 128–140. ISBN 978-3-540-40754-6.
[42] Meyer, Yves; Roques, Sylvie (1993). *Progress in wavelet analysis and applications: proceedings of the International Conference "Wavelets and Applications," Toulouse, France - June 1992*. Atlantica Séguier Frontières. p. 25. ISBN 978-2-86332-130-0. Retrieved 2011-02-05.
[43] Pincus, David (September 2009). "The Chaotic Life: Fractal Brains Fractal Thoughts". *psychologytoday.com*.
[44] Carbone, Alessandra; Gromov, Mikhael; Prusinkiewicz, Przemyslaw (2000). *Pattern formation in biology, vision and dynamics*. World Scientific. p. 78. ISBN 978-981-02-3792-9.
[45] Addison, Paul S. (1997). *Fractals and chaos: an illustrated course*. CRC Press. pp. 44–46. ISBN 978-0-7503-0400-9. Retrieved 2011-02-05.
[46] Ozhovan M.I., Dmitriev I.E., Batyukhnova O.G. Fractal structure of pores of clay soil. Atomic Energy, 74, 241-243 (1993)
[47] Takayasu, H. (1990). *Fractals in the physical sciences*. Manchester: Manchester University Press. p. 36. ISBN 9780719034343.
[48] Jun, Li; Ostoja-Starzewski, Martin. "Saturn's Rings are Fractal". Retrieved 28 June 2014.
[49] Frame, Michael; and Mandelbrot, Benoît B.; *A Panorama of Fractals and Their Uses*
[50] Nelson, Bryn; *Sophisticated Mathematics Behind African Village Designs Fractal patterns use repetition on large, small scale*, San Francisco Chronicle, Wednesday, February 23, 2009
[51] Hohlfeld, Robert G.; Cohen, Nathan (1999). "Self-similarity and the geometric requirements for frequency independence in Antennae". *Fractals* **7** (1): 79–84. doi:10.1142/S0218348X99000098.
[52] Reiner, Richard; Waltereit, Patrick; Benkhelifa, Fouad; Müller, Stefan; Walcher, Herbert; Wagner, Sandrine; Quay, Rüdiger; Schlechtweg, Michael; Ambacher, Oliver; Ambacher, O. (2012). "Fractal structures for low-resistance large area Al-GaN/GaN power transistors". *Proceedings of ISPSD*: 341. doi:10.1109/ISPSD.2012.6229091. ISBN 978-1-4577-1596-9.

[53] Chen, Yanguang (2011). "Modeling Fractal Structure of City-Size Distributions Using Correlation Functions". *PLoS ONE* **6** (9): e24791. doi:10.1371/journal.pone.0024791. PMC 3176775. PMID 21949753.
[54] "Applications". Retrieved 2007-10-21.
[55] Smith, Robert F.; Mohr, David N.; Torres, Vicente E.; Offord, Kenneth P.; Melton III, L. Joseph (1989). "Renal insufficiency in community patients with mild asymptomatic microhematuria". *Mayo Clinic proceedings. Mayo Clinic* **64** (4): 409–414. PMID 2716356.
[56] Landini, Gabriel (2011). "Fractals in microscopy". *Journal of Microscopy* **241** (1): 1–8. doi:10.1111/j.1365-2818.2010.03454.x. PMID 21118245.
[57] Cheng, Qiuming (1997). "Multifractal Modeling and Lacunarity Analysis". *Mathematical Geology* **29** (7): 919–932. doi:10.1023/A:1022355723781.
[58] Chen, Yanguang (2011). "Modeling Fractal Structure of City-Size Distributions Using Correlation Functions". *PLoS ONE* **6** (9): e24791. doi:10.1371/journal.pone.0024791. PMC 3176775. PMID 21949753.
[59] Burkle-Elizondo, Gerardo; Valdéz-Cepeda, Ricardo David (2006). "Fractal analysis of Mesoamerican pyramids". *Nonlinear dynamics, psychology, and life sciences* **10** (1): 105–122. PMID 16393505.
[60] Brown, Clifford T.; Witschey, Walter R. T.; Liebovitch, Larry S. (2005). "The Broken Past: Fractals in Archaeology". *Journal of Archaeological Method and Theory* **12**: 37. doi:10.1007/s10816-005-2396-6.
[61] Saeedi, Panteha; Sorensen, Soren A. "An Algorithmic Approach to Generate After-disaster Test Fields for Search and Rescue Agents". *Proceedings of the World Congress on Engineering 2009*: 93–98. ISBN 978-988-17-0125-1.
[62] Bunde, A.; Havlin, S. (2009). "Fractal Geometry, A Brief Introduction to". "Encyclopedia of Complexity and Systems Science". p. 3700. doi:10.1007/978-0-387-30440-3_218. ISBN 978-0-387-75888-6.

69.13 Further reading

_ Barnsley, Michael F.; and Rising, Hawley; *Fractals Everywhere*. Boston: Academic Press Professional, 1993. ISBN 0-12-079061-0
_ Duarte, German A.; *Fractal Narrative. About the Relationship Between Geometries and Technology and Its Impact on Narrative Spaces*. Bielefeld: Transcript, 2014. ISBN 978-3-8376-2829-6
_ Falconer, Kenneth; *Techniques in Fractal Geometry*. John Wiley and Sons, 1997. ISBN 0-471-92287-0
_ Jürgens, Hartmut; Peitgen, Heins-Otto; and Saupe, Dietmar; *Chaos and Fractals: New Frontiers of Science*. New York: Springer-Verlag, 1992. ISBN 0-387-97903-4
_ Mandelbrot, Benoit B.; *The Fractal Geometry of Nature*. New York: W. H. Freeman and Co., 1982. ISBN 0-7167-1186-9
_ Peitgen, Heinz-Otto; and Saupe, Dietmar; eds.; *The Science of Fractal Images*. New York: Springer-Verlag, 1988. ISBN 0-387-96608-0
_ Pickover, Clifford A.; ed.; *Chaos and Fractals: A Computer Graphical Journey - A 10 Year Compilation of Advanced Research*. Elsevier, 1998. ISBN 0-444-50002-2
_ Jones, Jesse; *Fractals for the Macintosh*, Waite Group Press, Corte Madera, CA, 1993. ISBN 1-878739-46-8.
_ Lauwerier, Hans; *Fractals: Endlessly Repeated Geometrical Figures*, Translated by Sophia Gill-Hoffstadt, Princeton University Press, Princeton NJ, 1991. ISBN 0-691-08551-X, cloth. ISBN 0-691-02445-6 paperback.

"This book has been written for a wide audience..." Includes sample BASIC programs in an appendix.
_ Sprott, Julien Clinton (2003). *Chaos and Time-Series Analysis*. Oxford University Press. ISBN 978-0-19-850839-7.
_ Wahl, Bernt; Van Roy, Peter; Larsen, Michael; and Kampman, Eric; *Exploring Fractals on the Macintosh*, Addison Wesley, 1995. ISBN 0-201-62630-6

_ Lesmoir-Gordon, Nigel; "The Colours of Infinity: The Beauty, The Power and the Sense of Fractals." ISBN 1-904555-05-5 (The book comes with a related DVD of the Arthur C. Clarke documentary introduction to the fractal concept and the Mandelbrot set).
_ Liu, Huajie; *Fractal Art*, Changsha: Hunan Science and Technology Press, 1997, ISBN 9787535722348.
_ Gouyet, Jean-François; *Physics and Fractal Structures* (Foreword by B. Mandelbrot); Masson, 1996. ISBN 2-225-85130-1, and New York: Springer-Verlag, 1996. ISBN 978-0-387-94153-0. Out-of-print. Available in PDF version at."Physics and Fractal Structures" (in French). Jfgouyet.fr. Retrieved 2010-10-17.
_ Bunde, Armin; Havlin, Shlomo (1996). *Fractals and Disordered Systems*. Springer.
_ Bunde, Armin; Havlin, Shlomo (1995). *Fractals in Science*. Springer.
_ ben-Avraham, Daniel; Havlin, Shlomo (2000). *Diffusion and Reactions in Fractals and Disordered Systems*. Cambridge University Press.
_ Falconer, Kenneth (2013). *Fractals, A Very Short Introduction*. Oxford University Press.

69.14 External links

_ Fractals at DMOZ
_ Scaling and Fractals presented by Shlomo Havlin, Bar-Ilan University
_ Hunting the Hidden Dimension, *PBS NOVA*, first aired August 24, 2011
_ Benoit Mandelbrot: Fractals and the Art of Roughness, TED (conference), February 2010
_ Zoom Video in Mandelbox on YouTube (Example of 3D fractal)
_ Video fly through of an animated Mandelbulb world on YouTube
_ Technical Library on Fractals for controlling fluid
_ Equations of self-similar fractal measure based on the fractional-order calculus(2007)

Chapter 70

Proof without words

In mathematics, a **proof without words** is a proof of an identity or mathematical statement which can be demonstrated as self-evident by a diagram without any accompanying explanatory text. Such proofs can be considered more elegant than more formal and mathematically rigorous proofs due to their self-evident nature.[1] When the diagram demonstrates a particular case of a general statement, to be a proof, it must be generalisable.[2]

A proof without words for the sum of odd numbers theorem.

A proof without words for the Pythagorean theorem derived in Zhou Bi Suan Jing.

A graphical proof of Jensen's inequality.

70.1 Examples

70.1.1 Sum of odd numbers

The statement that the sum of all positive odd numbers up to $2n - 1$ is a perfect square—more specifically, the perfect square n^2—can be demonstrated by a proof without words, as shown on the right.[3] The first square is formed by 1 block; 1 is the first square. The next strip, made of white squares, shows how adding 3 more blocks makes another square: four. The next strip, made of black squares, shows how adding 5 more blocks makes the next square. This process can be continued indefinitely.

70.1.2 Pythagorean theorem

The Pythagorean theorem can be proven without words as shown in the second diagram on right. The two different methods for determining the area of the large square give the relation

$$a^2 + b^2 = c^2$$

between the sides. This proof is more subtle than the above, but still can be considered a proof without words.[4]

70.1.3 Jensen's inequality

Jensen's inequality can also be proven graphically, as illustrated on the third diagram. The dashed curve along the X axis is the hypothetical distribution of X, while the dashed curve along the Y axis is the corresponding distribution of Y values. Note that the convex mapping $Y(X)$ increasingly "stretches" the distribution for increasing values of X.[5]

70.2 Usage

The *College Mathematics Journal* runs a regular feature entitled "Proof without words" containing, as the title suggests, proofs without words.[3] The Art of Problem Solving and USAMTS websites run Java applets illustrating proofs without words.[6][7]

70.3 See also

_ Pizza theorem
_ Visual calculus
_ Philosophy of mathematics

70.4 Notes

[1] Dunham 1994, p. 120
[2] Weisstein, Eric W., "Proof without Words", *MathWorld.* Retrieved on 2008-6-20
[3] Dunham 1994, p. 121
[4] Nelsen 1997, p. 3
[5] *Jensen's Inequality, Bulletin of the American Mathematical Society* (American Mathematical Society) **43** (8), 1937: 527
[6] *Gallery of Proofs (AoPS)*, Art of Problem Solving, retrieved 2008-06-20
[7] *Gallery of Proofs (USAMTS)*, Art of Problem Solving

70.5 References

_ Dunham, William (1974), *The Mathematical Universe*, John Wiley and Sons, ISBN 0-471-53656-3
_ Nelsen, Roger B. (1997), *Proofs without Words: Exercises in Visual Thinking*, Mathematical Association of America, p. 160, ISBN 978-0-88385-700-7
_ Nelsen, Roger B. (2000), *Proofs without Words II: More Exercises in Visual Thinking*, Mathematical Association of America, p. 142, ISBN 0-88385-721-9

Chapter 71

Missing square puzzle

Missing square puzzle animation

The **missing square puzzle** is an optical illusion used in mathematics classes to help students reason about geometrical figures, or rather to teach them to not reason using figures, but only using the textual description thereof and the axioms of geometry. It depicts two arrangements made of similar shapes in slightly different configurations. Each apparently forms a 13×5 right-angled triangle, but one has a 1×1 hole in it.

71.1 Solution

The key to the puzzle is the fact that neither of the 13×5 "triangles" is truly a triangle, because what appears to be the hypotenuse is bent. In other words, the "hypotenuse" does not maintain a consistent slope, even though it may appear that way to the human eye. A true 13×5 triangle cannot be created from the given component parts. The four figures (the yellow, red, blue and green shapes) total 32 units of area. The apparent triangles formed from the figures are 13 units wide and 5 units tall, so it appears that the area should be $S = 13\_5$
$2 = 32{:}5$ units. However, the blue triangle
has a ratio of 5:2 (=2.5:1), while the red triangle has the ratio 8:3 (≈2.667:1), so the apparent combined hypotenuse in each figure is actually bent. So with the bent hypotenuse, the first figure actually occupies a combined 32 units, while the second figure occupies 33, including the "missing" square. The amount of bending is approximately 1/28th of a unit (1.245364267°), which is difficult to see on the diagram of this puzzle. Note the grid point where the red and blue triangles in the lower image meet (5 squares to the right and two units up from the lower left corner of the combined figure), and compare it to the same point on the other figure; the edge is slightly under the mark in the upper image, but goes through it in the lower. Overlaying the hypotenuses from both figures results in a very thin parallelogram with an area of exactly one grid square—the same area "missing" from the second figure.

The missing square shown in the lower triangle, where both triangles are in a perfect grid

71.1.1 Principle

According to Martin Gardner,[1] this particular puzzle was invented by a New York City amateur magician, Paul Curry, in 1953. However, the principle of a dissection paradox has been known since the start of the 16th century. The integer dimensions of the parts of the puzzle (2, 3, 5, 8, 13) are successive Fibonacci numbers. Many other geometric dissection puzzles are based on a few simple properties of the Fibonacci sequence.[2]

Missing square puzzle dimensions

71.2 Similar puzzles

8x8=64

5x13=65

5x6+3+5x6=63

Sam Loyd's paradoxical dissection

Sam Loyd's paradoxical dissection. In the “larger” rearrangement, the gaps between the figures have a combined unit square more area than their square gaps counterparts, creating an illusion that the figures there take up more space than those in the square figure. In the “smaller” rearrangement, each quadrilateral needs to overlap the triangle by an area of half a unit for its top/bottom edge to align with a grid line.

Mitsunobu Matsuyama’s “Paradox” uses four congruent quadrilaterals and a small square, which form a larger square. When the quadrilaterals are rotated about their centers they fill the space of the small square, although the total area of the figure seems unchanged. The apparent paradox is explained by the fact that the side of the new large square is a little smaller than the original one. If a is the side of the large square and θ is the angle between two opposing sides in each quadrilateral, then the quotient between the two areas is given by $\sec^2\theta - 1$. For $\theta = 5°$, this is approximately 1.00765, which corresponds to a difference of about 0.8%.

71.3 References

[1] Martin, Gardner (1956). *Mathematics Magic and Mystery*. Dover. pp. 139–150. ISBN 9780486203355.

[2] Weisstein, Eric. “Cassini’s Identity”. Math World.

71.4 External links

_ A printable Missing Square variant with a video demonstration.

_ Curry’s Paradox: How Is It Possible? at cut-the-knot

_ Triangles and Paradoxes at archimedes-lab.org

A variant of Mitsunobu Matsuyama’s “Paradox”

_ The Triangle Problem or What’s Wrong with the Obvious Truth

_ Jigsaw Paradox

_ The Eleven Holes Puzzle

_ Very nice animated Excel workbook of the Missing Square Puzzle

_ A video explaining Curry’s Paradox and Area by James Tanton

Chapter 72

Elementary proof

In mathematics, an **elementary proof** is a mathematical proof that only uses basic techniques. More specifically, the term is used in number theory to refer to proofs that make no use of complex analysis. For some time it was thought that certain theorems, like the prime number theorem, could only be proved using “higher” mathematics. However, over time, many of these results have been reproved using only elementary techniques.

While the meaning has not always been defined precisely, the term is commonly used in mathematical jargon. An elementary proof is not necessarily simple, in the sense of being easy to understand: some elementary proofs can be quite complicated.[1]

72.1 Prime number theorem

The distinction between elementary and non-elementary proofs has been considered especially important in regard to the prime number theorem. This theorem was first proved in 1896 by Jacques Hadamard and Charles Jean de la Vallée-Poussin using complex analysis. Many mathematicians then attempted to construct elementary proofs of the theorem, without success. G. H. Hardy expressed strong reservations; he considered that the essential "depth" of the result ruled out elementary proofs:

No elementary proof of the prime number theorem is known, and one may ask whether it is reasonable to expect one. Now we know that the theorem is roughly equivalent to a theorem about an analytic function, the theorem that Riemann’s zeta function has no roots on a certain line. A proof of such a theorem, not fundamentally dependent on the theory of functions, seems to me extraordinarily unlikely. It is rash to assert that a mathematical theorem *cannot* be proved in a particular way; but one thing seems quite clear. We have certain views about the logic of the theory; we think that some theorems, as we say “lie deep” and others nearer to the surface. If anyone produces an elementary proof of the prime number theorem, he will show that these views are wrong, that the subject does not hang together in the way we have supposed, and that it is time for the books to be cast aside and for the theory to be rewritten.

—G. H. Hardy (1921). Lecture to Mathematical Society of Copenhagen. Quoted in Goldfeld (2003), p. 3

However, in 1948, Atle Selberg produced new methods which led him and Paul Erdős to find elementary proofs of the prime number theorem.[2]

A possible formalization of the notion of "elementary" in connection to a proof of a number-theoretical result is the restriction that the proof can be carried out in Peano arithmetic. Also in that sense, these proofs are elementary.

72.2 Friedman's conjecture

Main article: Grand conjecture

497

498 *CHAPTER 72. ELEMENTARY PROOF*

Harvey Friedman conjectured, "Every theorem published in the *Annals of Mathematics* whose statement involves only finitary mathematical objects (i.e., what logicians call an arithmetical statement) can be proved in elementary arithmetic."[3] The form of elementary arithmetic referred to in this conjecture can be formalized by a small set of axioms concerning integer arithmetic and mathematical induction. For instance, according to this conjecture, Fermat's Last Theorem should have an elementary proof; Wiles' proof of Fermat's Last Theorem is not elementary. However, there are other simple statements about arithmetic such as the existence of iterated exponential functions that cannot be proven in this theory.

72.3 References

[1] Diamond, Harold G. (1982), *Elementary methods in the study of the distribution of prime numbers, Bulletin of the American Mathematical Society* 7 (3): 553–89, doi:10.1090/S0273-0979-1982-15057-1, MR 670132.

[2] Goldfeld, Dorian M. (2003), *The Elementary Proof of the Prime Number Theorem: An Historical Perspective* (PDF), p. 3, retrieved October 31, 2009

[3] Avigad, Jeremy (2003), *Number theory and elementary arithmetic, Philosophia Mathematica* **11** (3): 257, at 258, doi:10.1093/philmat/11.3.257.

Chapter 73

Complex analysis

"Complex analytic" redirects here. For the class of functions often called "complex analytic", see Holomorphic function.

Complex analysis, traditionally known as the **theory of functions of a complex variable**, is the branch of mathematical analysis that investigates functions of complex numbers. It is useful in many branches of mathematics, including algebraic geometry, number theory, applied mathematics; as well as in physics, including hydrodynamics, thermodynamics, nuclear, aerospace, mechanical and electrical engineering.

Murray R. Spiegel described complex analysis as "one of the most beautiful as well as useful branches of Mathematics".

Complex analysis is particularly concerned with analytic functions of complex variables (or, more generally, meromorphic functions). Because the separate real and imaginary parts of any analytic function must satisfy Laplace's equation, complex analysis is widely applicable to two-dimensional problems in physics.

73.1 History

Complex analysis is one of the classical branches in mathematics with roots in the 19th century and just prior. Important mathematicians associated with complex analysis include Euler, Gauss, Riemann, Cauchy, Weierstrass, and many more in the 20th century. Complex analysis, in particular the theory of conformal mappings, has many physical applications and is also used throughout analytic number theory. In modern times, it has become very popular through a new boost from complex dynamics and the pictures of fractals produced by iterating holomorphic functions. Another important application of complex analysis is in string theory which studies conformal invariants in quantum field theory.

73.2 Complex functions

A complex function is one in which the independent variable and the dependent variable are both complex numbers. More precisely, a complex function is a function whose domain and range are subsets of the complex plane.

For any complex function, both the independent variable and the dependent variable may be separated into real and imaginary parts:

$z = x + iy$ and

$w = f(z) = u(x; y) + iv(x; y)$

where $x; y$ 2 R and $u(x; y); v(x; y)$ are real-valued functions.

In other words, the components of the function $f(z)$,

$u = u(x; y)$

499

500 *CHAPTER 73. COMPLEX ANALYSIS*

Plot of the function f(x) = (x2 − 1)(x − 2 − i)2 / (x2 + 2 + 2i). *The hue represents the function argument, while the brightness represents the magnitude.*

$v = v(x; y);$
can be interpreted as real-valued functions of the two real variables, x and y.
The basic concepts of complex analysis are often introduced by extending the elementary real functions (e.g., exponential functions, logarithmic functions, and trigonometric functions) into the complex domain.

73.3 Holomorphic functions

Main article: Holomorphic function
Holomorphic functions are complex functions defined on an open subset of the complex plane that are differentiable. Complex differentiability has much stronger consequences than usual (real) differentiability. For instance, holomorphic functions are infinitely differentiable, whereas some real differentiable functions are not. Most elementary functions, including the exponential function, the trigonometric functions, and all polynomial functions, are holomorphic.
See also: analytic function, holomorphic sheaf and vector bundles.

*The Mandelbrot set, a **fractal**.*

73.4 Major results

One central tool in complex analysis is the line integral. The integral around a closed path of a function that is holomorphic everywhere inside the area bounded by the closed path is always zero; this is the Cauchy integral theorem.
The values of a holomorphic function inside a disk can be computed by a certain path integral on the disk's boundary (Cauchy's integral formula). Path integrals in the complex plane are often used to determine complicated real integrals, and here the theory of residues among others is useful (see methods of contour integration). If a function has a *pole* or isolated singularity at some point, that is, at that point where its values "blow up" and have no finite bound, then one can compute the function's residue at that pole. These residues can be used to compute path integrals involving the function; this is the content of the powerful residue theorem. The remarkable behavior of holomorphic functions near essential singularities is described by Picard's Theorem. Functions that have only poles but no essential singularities are called meromorphic. Laurent series are similar to Taylor series but can be used to study the behavior of functions near singularities.
A bounded function that is holomorphic in the entire complex plane must be constant; this is Liouville's theorem. It can be used to provide a natural and short proof for the fundamental theorem of algebra which states that the field of complex numbers is algebraically closed.
If a function is holomorphic throughout a connected domain then its values are fully determined by its values on any smaller subdomain. The function on the larger domain is said to be analytically continued from its values on the smaller domain. This allows the extension of the definition of functions, such as the Riemann zeta function, which are initially defined in terms of infinite sums that converge only on limited domains to almost the entire complex plane. Sometimes, as in the case of the natural logarithm, it is impossible to analytically continue a holomorphic function to a non-simply connected domain in the complex plane but it is possible to extend it to a holomorphic function on a closely related surface known as a Riemann surface.
All this refers to complex analysis in one variable. There is also a very rich theory of complex analysis in more than one complex dimension in which the analytic properties such as power series expansion carry over whereas most of the geometric properties of holomorphic functions in one complex dimension (such as conformality) do not carry

over. The Riemann mapping theorem about the conformal relationship of certain domains in the complex plane, which may be the most important result in the one-dimensional theory, fails dramatically in higher dimensions.

73.5 See also

_ Complex dynamics
_ List of complex analysis topics
_ Real analysis
_ Runge's theorem
_ Several complex variables
_ Real-valued function
_ Function of a real variable
_ Real multivariable function

73.6 References

_ Ahlfors, L., *Complex Analysis, 3 ed.* (McGraw-Hill, 1979).
_ Carathéodory, C., *Theory of Functions of a Complex Variable* (Chelsea, New York). [2 volumes.]
_ Henrici, P., *Applied and Computational Complex Analysis* (Wiley). [Three volumes: 1974, 1977, 1986.]
_ Kreyszig, E., *Advanced Engineering Mathematics, 10 ed.*, Ch.13-18 (Wiley, 2011).
_ Markushevich, A.I.,*Theory of Functions of a Complex Variable* (Prentice-Hall, 1965). [Three volumes.]
_ Marsden & Hoffman, *Basic Complex Analysis. 3 ed.* (Freeman, 1999).
_ Needham, T., *Visual Complex Analysis* (Oxford, 1997).

_ Rudin, W., *Real and Complex Analysis, 3 ed.* (McGraw-Hill, 1986).
_ Scheidemann, V., *Introduction to complex analysis in several variables* (Birkhauser, 2005)
_ Shaw, W.T., *Complex Analysis with Mathematica* (Cambridge, 2006).
_ Spiegel, Murray R. *Theory and Problems of Complex Variables - with an introduction to Conformal Mapping and its applications* (McGraw-Hill, 1964).
_ Stein & Shakarchi, *Complex Analysis* (Princeton, 2003).

73.7 External links

_ Complex Analysis -- textbook by George Cain
_ Complex analysis course web site by Douglas N. Arnold
_ Example problems in complex analysis
_ A collection of links to programs for visualizing complex functions (and related)
_ Complex Analysis Project by John H. Mathews
_ Hans Lundmark's complex analysis page (many links)
_ Wolfram Research's MathWorld Complex Analysis Page

_ Complex function demos
_ Application of Complex Functions in 2D Digital Image Transformation
_ Complex Visualizer - Java applet for visualizing arbitrary complex functions
_ Complex Map - iOS app for visualizing complex functions and iterations
_ JavaScript complex function graphing tool
_ Earliest Known Uses of Some of the Words of Mathematics: Calculus & Analysis

Chapter 74

Prime number theorem

In number theory, the **prime number theorem** (**PNT**) describes the asymptotic distribution of the prime numbers. The prime number theorem gives a general description of how the primes are distributed amongst the positive integers. It formalizes the intuitive idea that primes become less common as they become larger.

Informally speaking, the prime number theorem states that if a random integer is selected in the range of zero to some large integer *N*, the probability that the selected integer is prime is about 1 / ln(*N*), where ln(*N*) is the natural logarithm of *N*. Consequently, a random integer with at most 2*n* digits (for large enough *n*) is about half as likely to be prime as a random integer with at most *n* digits. For example, among the positive integers of at most 1000 digits, about one in 2300 is prime ($\ln 10^{1000} \approx 2302.6$), whereas among positive integers of at most 2000 digits, about one in 4600 is prime ($\ln 10^{2000} \approx 4605.2$). In other words, the average gap between consecutive prime numbers among the first *N* integers is roughly ln(*N*).[1]

74.1 Statement

1 10^4 10^8 10^{12} 10^{16} 10^{20} 10^{24}

0.9

1.0

1.1

1.2

Graph showing ratio of the prime-counting function π(x) to two of its approximations, x/ln x and Li(x). As x increases (note x axis is logarithmic), both ratios tend towards 1. The ratio for x/ln x converges from above very slowly, while the ratio for Li(x) converges more quickly from below.

1 10^4 10^8 10^{12} 10^{16} 10^{20} 10^{24}

10^4

10^8

10^{12}
10^{16}
10^{20}

Log-log plot showing absolute error of x/ln x *and Li(*x*), two approximations to the prime-counting function π(*x*). Unlike the ratios, the differences increase without bound as* x *increases.*

Let $\pi(x)$ be the prime-counting function that gives the number of primes less than or equal to x, for any real number x. For example, $\pi(10) = 4$ because there are four prime numbers (2, 3, 5 and 7) less than or equal to 10. The prime number theorem then states that $x / \ln(x)$ is a good approximation to $\pi(x)$, in the sense that the limit of the *quotient* of the two functions $\pi(x)$ and $x / \ln(x)$ as x approaches infinity is 1:

lim $_{x!1}$
_(x)
x/ ln(x)
= 1;

known as **the asymptotic law of distribution of prime numbers**. Using asymptotic notation this result can be restated as

_(x) _ x
ln x
:

This notation (and the theorem) does *not* say anything about the limit of the *difference* of the two functions as x approaches infinity. Instead, the theorem states that $x/\ln(x)$ approximates $\pi(x)$ in the sense that the relative error of this approximation approaches 0 as x approaches infinity.

The prime number theorem is equivalent to the statement that the nth prime number p_n satisfies

p_n _ $n \ln(n)$

the asymptotic notation meaning again that the relative error of this approximation approaches 0 as n approaches infinity. For example, the $200 \cdot 10^{15}$th prime number is 8512677386048191063,[2] and $(200 \cdot 10^{15})\ln(200 \cdot 10^{15})$ rounds to 7967418752291744388, a relative error of about 6.8%.

The prime number theorem is also equivalent to lim$_{x!1}$
(x)
x = 1 , and lim$_{x!1}$
#(x)
x = 1 .

74.2 History of the asymptotic law of distribution of prime numbers and its proof

Based on the tables by Anton Felkel and Jurij Vega, Adrien-Marie Legendre conjectured in 1797 or 1798 that $\pi(a)$ is approximated by the function $a/(A \ln(a) + B)$, where A and B are unspecified constants. In the second edition of his book on number theory (1808) he then made a more precise conjecture, with $A = 1$ and $B = -1.08366$. Carl Friedrich Gauss considered the same question at age 15 or 16 "ins Jahr 1792 oder 1793", according to his own recollection in 1849.[3] In 1838 Peter Gustav Lejeune Dirichlet came up with his own approximating function, the logarithmic integral li(x) (under the slightly different form of a series, which he communicated to Gauss). Both Legendre's and Dirichlet's formulas imply the same conjectured asymptotic equivalence of $\pi(x)$ and $x / \ln(x)$ stated above, although it turned out that Dirichlet's approximation is considerably better if one considers the differences instead of quotients.

In two papers from 1848 and 1850, the Russian mathematician Pafnuty L'vovich Chebyshev attempted to prove the asymptotic law of distribution of prime numbers. His work is notable for the use of the zeta function $\zeta(s)$ (for real values of the argument "s", as are works of Leonhard Euler, as early as 1737) predating Riemann's celebrated memoir of 1859, and he succeeded in proving a slightly weaker form of the asymptotic law, namely, that if the limit of $\pi(x)/(x/\ln(x))$ as x goes to infinity exists at all, then it is necessarily equal to one.[4] He was able to prove unconditionally that this ratio is bounded above and below by two explicitly given constants near to 1 for all x.[5] Although Chebyshev's paper did not prove the Prime Number Theorem, his estimates for $\pi(x)$ were strong enough for him to prove Bertrand's postulate that there exists a prime number between n and $2n$ for any integer $n \geq 2$.

An important paper concerning the distribution of prime numbers was Riemann's 1859 memoir *On the Number of Primes Less Than a Given Magnitude*, the only paper he ever wrote on the subject. Riemann introduced new ideas into the subject, the chief of them being that the distribution of prime numbers is intimately connected with the zeros of the analytically extended Riemann zeta function of a complex variable. In particular, it is in this paper of Riemann that the idea to apply methods of complex analysis to the study of the real function $\pi(x)$ originates. Extending the ideas of Riemann, two proofs of the asymptotic law of the distribution of prime numbers were obtained independently by Jacques Hadamard and Charles Jean de la Vallée-Poussin and appeared in the same year (1896). Both proofs used methods from complex analysis, establishing as a main step of the proof that the Riemann zeta function $\zeta(s)$ is non-zero for all complex values of the variable s that have the form $s = 1 + it$ with $t > 0$.[6]

During the 20th century, the theorem of Hadamard and de la Vallée-Poussin also became known as the Prime Number Theorem. Several different proofs of it were found, including the "elementary" proofs of Atle Selberg and Paul Erdős (1949). While the original proofs of Hadamard and de la Vallée-Poussin are long and elaborate, and later proofs have introduced various simplifications through the use of Tauberian theorems but remained difficult to digest, a short proof was discovered in 1980 by American mathematician Donald J. Newman.[7][8] Newman's proof is arguably the simplest known proof of the theorem, although it is non-elementary in the sense that it uses Cauchy's integral theorem from complex analysis.

74.3 Proof methodology

In a lecture on prime numbers for a general audience, Fields medalist Terence Tao described one approach to proving the prime number theorem in poetic terms: listening to the "music" of the primes. We start with a "sound wave" that is "noisy" at the prime numbers and silent at other numbers; this is the von Mangoldt function. Then we analyze its notes or frequencies by subjecting it to a process akin to Fourier transform; this is the Mellin transform. The next and most difficult step is to prove that certain "notes" cannot occur in this music. This exclusion of certain notes leads to the statement of the prime number theorem. According to Tao, this proof yields much deeper insights into the distribution of the primes than the "elementary" proofs.[9]

74.4 Proof sketch

Here is a sketch of the proof referred to in Tao's lecture mentioned above. Like most proofs of the PNT, it starts out by reformulating the problem in terms of a less intuitive, but better-behaved, prime-counting function. The idea is to count the primes (or a related set such as the set of prime powers) with *weights* to arrive at a function with smoother asymptotic behavior. The most common such generalized counting function is the Chebyshev function (x), defined by

$(x) =$

Σ

$pk\_x;$

p prime is

$\log p:$

This is sometimes written as $(x) =$

Σ

n_x $\_(n)$, where $\_(n)$ is the von Mangoldt function, namely

$\_(n) =$

{

$\log p$ if$n = pk$ prime some for p integer and k _ 1;

0 otherwise.

It is now relatively easy to check that the PNT is equivalent to the claim that $\lim_{x!1}$ $(x)/x = 1$. Indeed, this follows from the easy estimates

$(x) =$

Σ

p_x

$\log p$

⌊

$\log x$

$\log p$

⌋

_

Σ

p_x

$\log x = \_(x) \log x$

and (using big O notation) for any $\epsilon > 0$,

(x) _

Σ

$x1\square\epsilon\_p\_x$

$\log p$ _

Σ

$x1\square\epsilon\_p\_x$

$(1 \square \epsilon) \log x = (1 \square \epsilon)(\_(x) + O(x1\square\epsilon)) \log x:$

The next step is to find a useful representation for (x) . Let $\_(s)$ be the Riemann zeta function. It can be shown that $\_(s)$ is related to the von Mangoldt function $\_(n)$, and hence to (x) , via the relation

$\square\_'(s)$

_(s)
=
1Σ
n=1
_(n)n
□s:
A delicate analysis of this equation and related properties of the zeta function, using the Mellin transform and Perron's formula, shows that for non-integer x the equation
(x) = x □
Σ
_
x_
_
□ log(2_)
holds, where the sum is over all zeros (trivial and non-trivial) of the zeta function. This striking formula is one of the so-called explicit formulas of number theory, and is already suggestive of the result we wish to prove, since the term x (claimed to be the correct asymptotic order of (x)) appears on the right-hand side, followed by (presumably) lower-order asymptotic terms.
The next step in the proof involves a study of the zeros of the zeta function. The trivial zeros −2, −4, −6, −8, ... can be handled separately:
1Σ
n=1
1
2n x2n = □1
2 ln
(
1 □ 1
x2
)
;
which vanishes for a large x. The nontrivial zeros, namely those on the critical strip 0 _ ℜ(s) _ 1 , can potentially be of an asymptotic order comparable to the main term x if ℜ(_) = 1 , so we need to show that all zeros have real part strictly less than 1.
To do this, we take for granted that _(s) is meromorphic in the half-plane ℜ(s) > 0 , and is analytic there except for a simple pole at s = 1 , and that there is a product formula _(s) =
Π
p(1 □ p□s)□1 for ℜ(s) > 1: This product
formula follows from the existence of unique prime factorization of integers, and shows that _(s) is never zero in this region, so that its logarithm is defined there and log _(s) = □Σ
p log(1□p□s) =
Σ
p;n p□ns/n: Write s = x+iy
; then
508 CHAPTER 74. PRIME NUMBER THEOREM
j_(x + iy)j = exp(
Σ
n;p
cos ny log p
npnx):
Now observe the identity 3 + 4 cos ϕ + cos 2ϕ = 2(1 + cos ϕ)2 _ 0; so that
j_(x)3_(x + iy)4_(x + 2iy)j = exp
Σ
n;p
3 + 4 cos(ny log p) + cos(2ny log p)
npnx
_ 1
for all x > 1 . Suppose now that _(1 + iy) = 0 . Certainly y is not zero, since _(s) has a simple pole at s = 1
. Suppose that x > 1 and let x tend to 1 from above. Since _(s) has a simple pole at s = 1 and _(x + 2iy) stays analytic, the left hand side in the previous inequality tends to 0 , a contradiction.
Finally, we can conclude that the PNT is "morally" true. To rigorously complete the proof there are still serious technicalities to overcome, due to the fact that the summation over zeta zeros in the explicit formula for (x) does not converge absolutely but only conditionally and in a "principal value" sense. There are several ways around this

problem but many of them require rather delicate complex-analytic estimates that are beyond the scope of this article. Edwards's book[10] provides the details. Another method is to use Ikehara's Tauberian theorem, though this theorem is itself quite hard to prove. D. J. Newman observed that the full strength of Ikehara's theorem is not needed for the prime number theorem, and one can get away with a special case that is much easier to prove.

74.5 Prime-counting function in terms of the logarithmic integral

In a handwritten note on a reprint of his 1838 paper "Sur l'usage des séries infinies dans la théorie des nombres", which he mailed to Carl Friedrich Gauss, Peter Gustav Lejeune Dirichlet conjectured (under a slightly different form appealing to a series rather than an integral) that an even better approximation to $\pi(x)$ is given by the offset logarithmic integral function $\mathrm{Li}(x)$, defined by

$$\mathrm{Li}(x) = \int_2^x \frac{1}{\ln t}\,dt = \mathrm{li}(x) - \mathrm{li}(2).$$

Indeed, this integral is strongly suggestive of the notion that the 'density' of primes around t should be $1/\ln t$. This function is related to the logarithm by the asymptotic expansion

$$\mathrm{Li}(x) \sim \frac{x}{\ln x} \sum_{k=0}^{\infty} \frac{k!}{(\ln x)^k} = \frac{x}{\ln x} + \frac{x}{(\ln x)^2} + \frac{2x}{(\ln x)^3} + \cdots.$$

So, the prime number theorem can also be written as $\pi(x) \sim \mathrm{Li}(x)$. In fact, it follows from the proof of Hadamard and de la Vallée Poussin that

$$\pi(x) = \mathrm{Li}(x) + O\left(x e^{-a\sqrt{\ln x}}\right) \quad \text{as } x \to \infty$$

for some positive constant a, where $O(\ldots)$ is the big O notation. This has been improved to

$$\pi(x) = \mathrm{Li}(x) + O\left(x \exp\left(-\frac{A(\ln x)^{3/5}}{(\ln \ln x)^{1/5}}\right)\right).$$

Because of the connection between the Riemann zeta function and $\pi(x)$, the Riemann hypothesis has considerable importance in number theory: if established, it would yield a far better estimate of the error involved in the prime number theorem than is available today. More specifically, Helge von Koch showed in 1901[11] that, if and only if the Riemann hypothesis is true, the error term in the above relation can be improved to

$$\pi(x) = \mathrm{Li}(x) + O\left(\sqrt{x}\ln x\right).$$

The constant involved in the big O notation was estimated in 1976 by Lowell Schoenfeld:[12] assuming the Riemann hypothesis,

$$|\pi(x) - \mathrm{li}(x)| < \sqrt{}$$

$x \ln x$
8_
for all $x \geq 2657$. He also derived a similar bound for the Chebyshev prime-counting function ψ:
$j(x)$ □ $xj <$
p
$x \ln^2 x$
8_
for all $x \geq 73.2$. This latter bound has been shown to express a variance to mean power law (when regarded as a random function over the integers), $1/f$ noise and to also correspond to the Tweedie compound Poisson distribution.[13] Parenthetically, the Tweedie distributions represent a family of scale invariant distributions that serve as foci of convergence for a generalization of the central limit theorem.[14]
The logarithmic integral $\mathrm{Li}(x)$ is larger than $\pi(x)$ for "small" values of x. This is because it is (in some sense) counting not primes, but prime powers, where a power p^n of a prime p is counted as $1/n$ of a prime. This suggests that $\mathrm{Li}(x)$ should usually be larger than $\pi(x)$ by roughly $\mathrm{Li}(x^{1/2})/2$, and in particular should usually be larger than $\pi(x)$. However, in 1914, J. E. Littlewood proved that this is not always the case. The first value of x where $\pi(x)$ exceeds $\mathrm{Li}(x)$ is probably around $x = 10^{316}$; see the article on Skewes' number for more details.

74.6 Elementary proofs

In the first half of the twentieth century, some mathematicians (notably G. H. Hardy) believed that there exists a hierarchy of proof methods in mathematics depending on what sorts of numbers (integers, reals, complex) a proof requires, and that the prime number theorem (PNT) is a "deep" theorem by virtue of requiring complex analysis.[15] This belief was somewhat shaken by a proof of the PNT based on Wiener's tauberian theorem, though this could be set aside if Wiener's theorem were deemed to have a "depth" equivalent to that of complex variable methods. There is no rigorous and widely accepted definition of the notion of elementary proof in number theory. One definition is "a proof that can be carried out in first order Peano arithmetic." There are number-theoretic statements (for example, the Paris–Harrington theorem) provable using second order but not first order methods, but such theorems are rare to date.
In March 1948, Atle Selberg established, by elementary means, the asymptotic formula
$\#(x)\log(x) +$
Σ
p_x
$\log(p)\,\#$
(
x
p
)
$= 2x\log(x) + O(x)$
where
$\#(x) =$
Σ
p_x
$\log(p)$
for primes p.[16] By July of that year, Selberg and Paul Erdős had each obtained elementary proofs of the PNT, both using Selberg's asymptotic formula as a starting point.[15][17] These proofs effectively laid to rest the notion that the PNT was "deep," and showed that technically "elementary" methods (in other words Peano arithmetic) were more powerful than had been believed to be the case. In 1994, Charalambos Cornaros and Costas Dimitracopoulos proved the PNT using only $I\Delta_0 + \exp$,[18] a formal system far weaker than Peano arithmetic. On the history of the elementary proofs of the PNT, including the Erdős–Selberg priority dispute, see Dorian Goldfeld.[15]

74.7 Computer verifications

In 2005, Avigad *et al.* employed the Isabelle theorem prover to devise a computer-verified variant of the Erdős–Selberg proof of the PNT.[19] This was the first machine-verified proof of the PNT. Avigad chose to formalize the Erdős–Selberg proof rather than an analytic one because while Isabelle's library at the time could implement the notions of limit, derivative, and transcendental function, it had almost no theory of integration to speak of (Avigad et al. p. 19).
In 2009, John Harrison employed HOL Light to formalize a proof employing complex analysis.[20] By developing the necessary analytic machinery, including the Cauchy integral formula, Harrison was able to formalize "a direct, modern and elegant proof instead of the more involved 'elementary' Erdős–Selberg argument."

74.8 Prime number theorem for arithmetic progressions

Let $\_{n;a}(x)$ denote the number of primes in the arithmetic progression $a, a + n, a + 2n, a + 3n, \ldots$ less than x.

Lejeune Dirichlet and Legendre conjectured, and Vallée-Poussin proved, that, if a and n are coprime, then
$\_{n;a}(x)$ _ 1
$\varphi(n)$ Li(x);
where φ is the Euler's totient function. In other words, the primes are distributed evenly among the residue classes $[a]$ modulo n with gcd$(a, n) = 1$. This can be proved using similar methods used by Newman for his proof of the prime number theorem.[21]
The Siegel–Walfisz theorem gives a good estimate for the distribution of primes in residue classes.

74.8.1 Prime number race

Although we have in particular
$\_{4;1}(x)$ _ $\_{4;3}(x)$;
empirically the primes congruent to 3 are more numerous and are nearly always ahead in this "prime number race"; the first reversal occurs at $x = 26,861$.[22]:1–2 However Littlewood showed in 1914[22]:2 that there are infinitely many sign changes for the function
$\_{4;1}(x)$ □ $\_{4;3}(x)$;
so the lead in the race switches back and forth infinitely many times. The phenomenon that $\pi_{4,3}(x)$ is ahead most of the time is called Chebyshev's bias. The prime number race generalizes to other moduli and is the subject of much research; Pál Turán asked whether it is always the case that $\pi(x;a,c)$ and $\pi(x;b,c)$ change places when a and b are coprime to c.[23] Granville and Martin give a thorough exposition and survey.[22]

74.9 Bounds on the prime-counting function

The prime number theorem is an *asymptotic* result. It gives an ineffective bound on $\pi(x)$ as a direct consequence of the definition of the limit: for all $\varepsilon > 0$, there is an S such that for all $x > S$,
(1 □ ")
x
ln x
< _(x) < (1 + ")
x
ln x
:
However, better bounds on $\pi(x)$ are known, for instance Pierre Dusart's

x
ln x
(
1 +
1
ln x
)
< _(x) <
x
ln x
(
1 +
1
ln x
+
2:51
(ln x)2
)
:
The first inequality holds for all $x \geq 599$ and the second one for $x \geq 355991$.[24]
A weaker but sometimes useful bound for $x \geq 55$ is
x
ln x+2 < _(x) < x
ln x□4 .[25]
In Dusart's thesis there are stronger versions of this type of inequality that are valid for larger x.
The proof by de la Vallée-Poussin implies the following. For every $\varepsilon > 0$, there is an S such that for all $x > S$,
x
ln x □ (1 □ ")
< _(x) <
x

$\ln x$ □ $(1 + ″)$
:

74.10 Approximations for the *n*th prime number

As a consequence of the prime number theorem, one gets an asymptotic expression for the *n*th prime number, denoted by p_n:

$$p_n \sim n \ln n.$$

A better approximation is

$$\frac{p_n}{n} = \ln n + \ln \ln n - 1 + \frac{\ln \ln n - 2}{\ln n} - \frac{(\ln \ln n)^2 - 6 \ln \ln n + 11}{2(\ln n)^2} + o\left(\frac{1}{(\ln n)^2}\right).$$ [26]

Again considering the $200 \cdot 10^{15}$ prime number 8512677386048191063, this gives an estimate of 8512681315554715386; the first 5 digits match and relative error is about 0.00005%.

Rosser's theorem states that p_n is larger than $n \ln n$. This can be improved by the following pair of bounds:[27][28]

$$\ln n + \ln \ln n - 1 < \frac{p_n}{n} < \ln n + \ln \ln n \quad \text{for } n \geq 6.$$

74.11 Table of π(*x*), *x* / ln *x*, and li(*x*)

The table compares exact values of $\pi(x)$ to the two approximations $x / \ln x$ and $\mathrm{li}(x)$. The last column, $x / \pi(x)$, is the average prime gap below x.

The value for $\pi(10^{24})$ was originally computed assuming the Riemann hypothesis;[29] it has since been verified unconditionally.[30]

74.12 Analogue for irreducible polynomials over a finite field

There is an analogue of the prime number theorem that describes the "distribution" of irreducible polynomials over a finite field; the form it takes is strikingly similar to the case of the classical prime number theorem.

To state it precisely, let $F = \mathrm{GF}(q)$ be the finite field with q elements, for some fixed q, and let N_n be the number of monic *irreducible* polynomials over F whose degree is equal to n. That is, we are looking at polynomials with

coefficients chosen from F, which cannot be written as products of polynomials of smaller degree. In this setting, these polynomials play the role of the prime numbers, since all other monic polynomials are built up of products of them. One can then prove that

$$N_n \sim \frac{q^n}{n}.$$

If we make the substitution $x = q^n$, then the right hand side is just

$$\frac{x}{\log_q x},$$

which makes the analogy clearer. Since there are precisely q^n monic polynomials of degree n (including the reducible ones), this can be rephrased as follows: if a monic polynomial of degree n is selected randomly, then the probability of it being irreducible is about $1/n$.

One can even prove an analogue of the Riemann hypothesis, namely that

$$N_n = \frac{q^n}{n} + O\left(\frac{q^{n/2}}{n}\right).$$

The proofs of these statements are far simpler than in the classical case. It involves a short combinatorial argument,[31] summarised as follows. Every element of the degree n extension of F is a root of some irreducible polynomial whose degree d divides n; by counting these roots in two different ways one establishes that

$$q_n = \sum_{d|n} dN_d;$$

where the sum is over all divisors d of n. Möbius inversion then yields

$$N_n = \frac{1}{n} \sum_{d|n} \_(n/d)q_d;$$

where $\mu(k)$ is the Möbius function. (This formula was known to Gauss.) The main term occurs for $d = n$, and it is not difficult to bound the remaining terms. The "Riemann hypothesis" statement depends on the fact that the largest proper divisor of n can be no larger than $n/2$.

74.13 See also

_ Abstract analytic number theory for information about generalizations of the theorem.
_ Landau prime ideal theorem for a generalization to prime ideals in algebraic number fields.
_ Riemann hypothesis

74.14 Notes

[1] Hoffman, Paul (1998). *The Man Who Loved Only Numbers*. Hyperion. p. 227. ISBN 0-7868-8406-1.
[2] "Prime Curios!: 8512677386048191063". *Prime Curios!*. University of Tennessee at Martin. 2011-10-09.
[3] C.F. Gauss. Werke, Bd 2, 1st ed, 444-447. Göttingen 1863.
[4] N. Costa Pereira (August–September 1985). "A Short Proof of Chebyshev's Theorem". *American Mathematical Monthly* **92** (7): 494–495. doi:10.2307/2322510. JSTOR 2322510.

[5] M. Nair (February 1982). "On Chebyshev-Type Inequalities for Primes". *American Mathematical Monthly* **89** (2): 126–129. doi:10.2307/2320934. JSTOR 2320934.
[6] Ingham, A.E. (1990). *The Distribution of Prime Numbers*. Cambridge University Press. pp. 2–5. ISBN 0-521-39789-8.
[7] D. J. Newman (1980). "Simple analytic proof of the prime number theorem". *American Mathematical Monthly* **87** (9): 693–696. doi:10.2307/2321853. JSTOR 2321853.
[8] D. Zagier (1997). "Newman's short proof of the prime number theorem". *American Mathematical Monthly* **104** (8): 705–708. doi:10.2307/2975232. JSTOR 2975232.
[9] Video and slides of Tao's lecture on primes, UCLA January 2007.
[10] Edwards, Harold M. (2001). *Riemann's zeta function*. Courier Dover Publications. ISBN 0-486-41740-9.
[11] Helge von Koch (December 1901). "Sur la distribution des nombres premiers". *Acta Mathematica* **24** (1): 159–182. doi:10.1007/BF02403071. (French)
[12] Schoenfeld, Lowell (1976). "Sharper Bounds for the Chebyshev Functions θ(x) and ψ(x). II". *Mathematics of Computation* **30** (134): 337–360. doi:10.2307/2005976. JSTOR 2005976..
[13] Kendal, WS (2013). "Fluctuation scaling and 1/*f* noise: shared origins from the Tweedie family of statistical distributions". *J Basic Appl Phys* **2**: 40–49.
[14] Jørgensen, B; Martinez, JR & Tsao, M (1994). "Asymptotic behaviour of the variance function". *Scandinavian Journal of Statistics* **21**: 223–243.
[15] D. Goldfeld The elementary proof of the prime number theorem: an historical perspective.
[16] Selberg, Atle (Apr 1949). "An Elementary Proof of the Prime-Number Theorem". *Annals of Mathematics*. 2 **50** (2): 305–313. doi:10.2307/1969455.
[17] Baas, Nils A.; Skau, Christian F. (2008). "The lord of the numbers, Atle Selberg. On his life and mathematics". *Bull. Amer. Math. Soc.* **45** (4): 617–649. doi:10.1090/S0273-0979-08-01223-8
[18] Cornaros, Charalambos; Dimitracopoulos, Costas (1994). "The prime number theorem and fragments of *PA*". *Archive for Mathematical Logic* **33** (4): 265–281. doi:10.1007/BF01270626.
[19] Jeremy Avigad, Kevin Donnelly, David Gray, Paul Raff (2005). "A formally verified proof of the prime number theorem". arXiv:cs.AI/0509025 [cs.AI].
[20] "Formalizing an analytic proof of the Prime Number Theorem. (Dedicated to Mike Gordon on the occasion of his 60th birthday)". *Journal of Automated Reasoning*. 2009, volume = 43, pages = 243—261. Check date values in: |date= (help)
[21] Ivan Soprounov (1998). "A short proof of the Prime Number Theorem for arithmetic progressions".
[22] Granville, Andrew; Martin, Greg (January 2006). "Prime Number Races". *American Mathematical Monthly* **113** (1): 1–33. doi:10.2307/27641834. JSTOR 27641834.
[23] Guy, Richard K. (2004). *Unsolved problems in number theory* (3rd ed.). Springer-Verlag. A4. ISBN 978-0-387-20860-2. Zbl 1058.11001.
[24] Dusart, Pierre (1998). *Autour de la fonction qui compte le nombre de nombres premiers. PhD Thesis*. (French)
[25] Barkley Rosser (January 1941). "Explicit Bounds for Some Functions of Prime Numbers". *American Journal of Mathematics* **63** (1): 211–232. doi:10.2307/2371291. JSTOR 2371291.
[26] Ernest Cesàro (1894). "Sur une formule empirique de M. Pervouchine". *Comptes rendus hebdomadaires des séances de l'Académie des sciences* **119**: 848–849. (French)

[27] Eric Bach, Jeffrey Shallit (1996). *Algorithmic Number Theory* **1**. MIT Press. p. 233. ISBN 0-262-02405-5.
[28] Pierre Dusart (1999). "The kth prime is greater than $k(\ln k + \ln \ln k - 1)$ for $k \geq 2$". *Mathematics of Computation* **68**: 411–415.
[29] "Conditional Calculation of pi(10^{24})". Chris K. Caldwell. Retrieved 2010-08-03.
[30] "Computing $\pi(x)$ Analytically)". Retrieved Jul 25, 2012.
[31] Chebolu, Sunil; Ján Mináč (December 2011). "Counting Irreducible Polynomials over Finite Fields Using the Inclusion-Exclusion Principle". *Mathematics Magazine* **84** (5): 369–371. doi:10.4169/math.mag.84.5.369.

74.15 References

_ Hardy, G. H. & Littlewood, J. E. (1916). "Contributions to the Theory of the Riemann Zeta-Function and the Theory of the Distribution of Primes". *Acta Mathematica* **41**: 119–196. doi:10.1007/BF02422942.
_ Granville, Andrew (1995). "Harald Cramér and the distribution of prime numbers". *Scandinavian Actuarial Journal* **1**: 12–28. doi:10.1080/03461238.1995.10413946.

74.16 External links

_ Hazewinkel, Michiel, ed. (2001), "Distribution of prime numbers", *Encyclopedia of Mathematics*, Springer, ISBN 978-1-55608-010-4
_ Table of Primes by Anton Felkel.
_ Short video visualizing the Prime Number Theorem.
_ Prime formulas and Prime number theorem at MathWorld.
_ Prime number theorem at PlanetMath.org.
_ How Many Primes Are There? and The Gaps between Primes by Chris Caldwell, University of Tennessee at Martin.
_ Tables of prime-counting functions by Tomás Oliveira e Silva

Chapter 75

Statistics

More probability density is found as one gets closer to the expected (mean) value in a normal distribution. Statistics used in standardized testing assessment are shown. The scales include standard deviations, cumulative percentages, percentile equivalents, Z-scores, T-scores, standard nines, *and* percentages in standard nines.

Statistics is the study of the collection, analysis, interpretation, presentation, and organization of data.[1] In applying statistics to, e.g., a scientific, industrial, or societal problem, it is necessary to begin with a population or process to be studied. Populations can be diverse topics such as "all persons living in a country" or "every atom composing a crystal". It deals with all aspects of data including the planning of data collection in terms of the design of surveys and experiments.[1]

In case census data cannot be collected, statisticians collect data by developing specific experiment designs and survey samples. Representative sampling assures that inferences and conclusions can safely extend from the sample to the population as a whole. An experimental study involves taking measurements of the system under study, manipulating the system, and then taking additional measurements using the same procedure to determine if the manipulation has modified the values of the measurements. In contrast, an observational study does not involve experimental manipulation.

Two main statistical methodologies are used in data analysis: descriptive statistics, which summarizes data from a sample using indexes such as the mean or standard deviation, and inferential statistics, which draws conclusions from data that are subject to random variation (e.g., observational errors, sampling variation).[2] Descriptive statistics are most often concerned with two sets of properties of a *distribution* (sample or population): *central tendency* (or

Scatter plots are used in descriptive statistics to show the observed relationships between different variables.
location) seeks to characterize the distribution's central or typical value, while *dispersion* (or *variability*) characterizes the extent to which members of the distribution depart from its center and each other. Inferences on mathematical statistics are made under the framework of probability theory, which deals with the analysis of random phenomena. To make an inference upon unknown quantities, one or more estimators are evaluated using the sample.

Standard statistical procedure involve the development of a null hypothesis, a general statement or default position that there is no relationship between two quantities. Rejecting or disproving the null hypothesis is a central task in the modern practice of science, and gives a precise sense in which a claim is capable of being proven false. What statisticians call an alternative hypothesis is simply an hypothesis that contradicts the null hypothesis. Working from a null hypothesis two basic forms of error are recognized: Type I errors (null hypothesis is falsely rejected giving a "false positive") and Type II errors (null hypothesis fails to be rejected and an actual difference between populations is missed giving a "false negative"). A critical region is the set of values of the estimator that leads to refuting the null hypothesis. The probability of type I error is therefore the probability that the estimator belongs to the critical region given that null hypothesis is true (statistical significance) and the probability of type II error is the probability that the estimator doesn't belong to the critical region given that the alternative hypothesis is true. The statistical power of a test is the probability that it correctly rejects the null hypothesis when the null hypothesis is false. Multiple problems have come to be associated with this framework: ranging from obtaining a sufficient sample size to specifying an adequate null hypothesis.

Measurement processes that generate statistical data are also subject to error. Many of these errors are classified as

random (noise) or systematic (bias), but other important types of errors (e.g., blunder, such as when an analyst reports incorrect units) can also be important. The presence of missing data and/or censoring may result in biased estimates and specific techniques have been developed to address these problems. Confidence intervals allow statisticians to express how closely the sample estimate matches the true value in the whole population. Formally, a 95% confidence interval for a value is a range where, if the sampling and analysis were repeated under the same conditions (yielding a different dataset), the interval would include the true (population) value in 95% of all possible cases. Ways to avoid misuse of statistics include using proper diagrams and avoiding bias. In statistics, dependence is any statistical relationship between two random variables or two sets of data. Correlation refers to any of a broad class of statistical relationships involving dependence. If two variables are correlated, they may or may not be the cause of one another. The correlation phenomena could be caused by a third, previously unconsidered phenomenon, called a lurking variable or confounding variable.

Statistics can be said to have begun in ancient civilization, going back at least to the 5th century BC, but it was not until the 18th century that it started to draw more heavily from calculus and probability theory. Statistics continues to be an area of active research, for example on the problem of how to analyze Big data.

75.1 Scope

Statistics is a mathematical body of science that pertains to the collection, analysis, interpretation or explanation, and presentation of data,[3] or as a branch of mathematics.[4] Some consider statistics to be a distinct mathematical science rather than a branch of mathematics.[5][6]

75.1.1 Mathematical statistics

Main article: Mathematical statistics

Mathematical statistics is the application of mathematics to statistics, which was originally conceived as the science of the state — the collection and analysis of facts about a country: its economy, land, military, population, and so forth. Mathematical techniques used for this include mathematical analysis, linear algebra, stochastic analysis, differential equations, and measure-theoretic probability theory.[7][8]

75.2 Overview

In applying statistics to e.g. a scientific, industrial, or societal problem, it is necessary to begin with a population or process to be studied. Populations can be diverse topics such as “all persons living in a country” or “every atom composing a crystal”.

Ideally, statisticians compile data about the entire population (an operation called census). This may be organized by governmental statistical institutes. **Descriptive statistics** can be used to summarize the population data. Numerical descriptors include mean and standard deviation for continuous data types (like income), while frequency and percentage are more useful in terms of describing categorical data (like race).

When a census is not feasible, a chosen subset of the population called a sample is studied. Once a sample that is representative of the population is determined, data is collected for the sample members in an observational or experimental setting. Again, descriptive statistics can be used to summarize the sample data. However, the drawing of the sample has been subject to an element of randomness, hence the established numerical descriptors from the sample are also due to uncertainty. To still draw meaningful conclusions about the entire population, **inferential statistics** is needed. It uses patterns in the sample data to draw inferences about the population represented, accounting for randomness. These inferences may take the form of: answering yes/no questions about the data (hypothesis testing), estimating numerical characteristics of the data (estimation), describing associations within the data (correlation) and modeling relationships within the data (for example, using regression analysis). Inference can extend to forecasting, prediction and estimation of unobserved values either in or associated with the population being studied; it can include extrapolation and interpolation of time series or spatial data, and can also include data mining.

75.3 Data collection

75.3.1 Sampling

In case census data cannot be collected, statisticians collect data by developing specific experiment designs and survey samples. Statistics itself also provides tools for prediction and forecasting the use of data through statistical models. To use a sample as a guide to an entire population, it is important that it truly represent the overall population. Representative sampling assures that inferences and conclusions can safely extend from the sample to the population as a whole. A major problem lies in determining the extent that the sample chosen is actually representative. Statistics offers methods to estimate and correct for any random trending within the sample and data collection procedures. There are also methods of experimental design for experiments that can lessen these issues at the outset of a study, strengthening its capability to discern truths about the population.

Sampling theory is part of the mathematical discipline of probability theory. Probability is used in "mathematical statistics" (alternatively, "statistical theory") to study the sampling distributions of sample statistics and, more generally, the properties of statistical procedures. The use of any statistical method is valid when the system or population under consideration satisfies the assumptions of the method. The difference in point of view between classic probability theory and sampling theory is, roughly, that probability theory starts from the given parameters of a total population to deduce probabilities that pertain to samples. Statistical inference, however, moves in the opposite direction—inductively inferring from samples to the parameters of a larger or total population.

75.3.2 Experimental and observational studies

A common goal for a statistical research project is to investigate causality, and in particular to draw a conclusion on the effect of changes in the values of predictors or independent variables on dependent variables or response. There are two major types of causal statistical studies: experimental studies and observational studies. In both types of studies, the effect of differences of an independent variable (or variables) on the behavior of the dependent variable are observed. The difference between the two types lies in how the study is actually conducted. Each can be very effective. An experimental study involves taking measurements of the system under study, manipulating the system, and then taking additional measurements using the same procedure to determine if the manipulation has modified the values of the measurements. In contrast, an observational study does not involve experimental manipulation. Instead, data are gathered and correlations between predictors and response are investigated. While the tools of data analysis work best on data from randomized studies, they are also applied to other kinds of data – like natural experiments and observational studies[9] – for which a statistician would use a modified, more structured estimation method (e.g., Difference in differences estimation and instrumental variables, among many others) that produce consistent estimators.

Experiments

The basic steps of a statistical experiment are:

1. Planning the research, including finding the number of replicates of the study, using the following information: preliminary estimates regarding the size of treatment effects, alternative hypotheses, and the estimated experimental variability. Consideration of the selection of experimental subjects and the ethics of research is necessary. Statisticians recommend that experiments compare (at least) one new treatment with a standard treatment or control, to allow an unbiased estimate of the difference in treatment effects.
2. Design of experiments, using blocking to reduce the influence of confounding variables, and randomized assignment of treatments to subjects to allow unbiased estimates of treatment effects and experimental error. At this stage, the experimenters and statisticians write the *experimental protocol* that shall guide the performance of the experiment and that specifies the *primary analysis* of the experimental data.
3. Performing the experiment following the experimental protocol and analyzing the data following the experimental protocol.
4. Further examining the data set in secondary analyses, to suggest new hypotheses for future study.
5. Documenting and presenting the results of the study.

Experiments on human behavior have special concerns. The famous Hawthorne study examined changes to the working environment at the Hawthorne plant of the Western Electric Company. The researchers were interested in determining whether increased illumination would increase the productivity of the assembly line workers. The researchers first measured the productivity in the plant, then modified the illumination in an area of the plant and checked if the changes in illumination affected productivity. It turned out that productivity indeed improved (under the experimental conditions). However, the study is heavily criticized today for errors in experimental procedures, specifically for the lack of a control group and blindness. The Hawthorne effect refers to finding that an outcome (in this case, worker productivity) changed due to observation itself. Those in the Hawthorne study became more productive not because the lighting was changed but because they were being observed.[10]

Observational study

An example of an observational study is one that explores the correlation between smoking and lung cancer. This type of study typically uses a survey to collect observations about the area of interest and then performs statistical analysis. In this case, the researchers would collect observations of both smokers and non-smokers, perhaps through a case-control study, and then look for the number of cases of lung cancer in each group.

75.4 Types of data

Main articles: Statistical data type and Levels of measurement

Various attempts have been made to produce a taxonomy of levels of measurement. The psychophysicist Stanley Smith Stevens defined nominal, ordinal, interval, and ratio scales. Nominal measurements do not have meaningful rank order among values, and permit any one-to-one transformation. Ordinal measurements have imprecise differences between consecutive values, but have a meaningful order to those values, and permit any order-preserving transformation. Interval measurements have meaningful distances between measurements defined, but the zero value is arbitrary (as in the case with longitude and temperature measurements in Celsius or Fahrenheit), and permit any linear transformation. Ratio measurements have both a meaningful zero value and the distances between different measurements defined, and permit any rescaling transformation.

Because variables conforming only to nominal or ordinal measurements cannot be reasonably measured numerically, sometimes they are grouped together as categorical variables, whereas ratio and interval measurements are grouped together as quantitative variables, which can be either discrete or continuous, due to their numerical nature. Such distinctions can often be loosely correlated with data type in computer science, in that dichotomous categorical variables may be represented with the Boolean data type, polytomous categorical variables with arbitrarily assigned integers in the integral data type, and continuous variables with the real data type involving floating point computation. But the mapping of computer science data types to statistical data types depends on which categorization of the latter is being implemented.

Other categorizations have been proposed. For example, Mosteller and Tukey (1977)[11] distinguished grades, ranks, counted fractions, counts, amounts, and balances. Nelder (1990)[12] described continuous counts, continuous ratios, count ratios, and categorical modes of data. See also Chrisman (1998),[13] van den Berg (1991).[14]

The issue of whether or not it is appropriate to apply different kinds of statistical methods to data obtained from different kinds of measurement procedures is complicated by issues concerning the transformation of variables and the precise interpretation of research questions. "The relationship between the data and what they describe merely reflects the fact that certain kinds of statistical statements may have truth values which are not invariant under some transformations. Whether or not a transformation is sensible to contemplate depends on the question one is trying to answer" (Hand, 2004, p. 82).[15]

75.5 Terminology and theory of inferential statistics

75.5.1 Statistics, estimators and pivotal quantities

Consider an independent identically distributed (iid) random variables with a given probability distribution: standard statistical inference and estimation theory defines a random sample as the random vector given by the column vector

of these iid variables.[16] The population being examined is described by a probability distribution that may have unknown parameters.

A statistic is a random variable that is a function of the random sample, but *not a function of unknown parameters*. The probability distribution of the statistic, though, may have unknown parameters.

Consider now a function of the unknown parameter: an estimator is a statistic used to estimate such function. Commonly used estimators include sample mean, unbiased sample variance and sample covariance.

A random variable that is a function of the random sample and of the unknown parameter, but whose probability distribution *does not depend on the unknown parameter* is called a pivotal quantity or pivot. Widely used pivots include the z-score, the chi square statistic and Student's t-value.

Between two estimators of a given parameter, the one with lower mean squared error is said to be more efficient. Furthermore, an estimator is said to be unbiased if it's expected value is equal to the true value of the unknown parameter being estimated, and asymptotically unbiased if its expected value converges at the limit to the true value of such parameter.

Other desirable properties for estimators include: UMVUE estimators that have the lowest variance for all possible values of the parameter to be estimated (this is usually an easier property to verify than efficiency) and consistent estimators which converges in probability to the true value of such parameter.

This still leaves the question of how to obtain estimators in a given situation and carry the computation, several methods have been proposed: the method of moments, the maximum likelihood method, the least squares method and the more recent method of estimating equations.

75.5.2 Null hypothesis and alternative hypothesis

Interpretation of statistical information can often involve the development of a null hypothesis in that the assumption is that whatever is proposed as a cause has no effect on the variable being measured.

The best illustration for a novice is the predicament encountered by a jury trial. The null hypothesis, H_0, asserts that the defendant is innocent, whereas the alternative hypothesis, H_1, asserts that the defendant is guilty. The indictment comes because of suspicion of the guilt. The H_0 (status quo) stands in opposition to H_1 and is maintained unless H_1 is supported by evidence "beyond a reasonable doubt". However, "failure to reject H_0" in this case does not imply innocence, but merely that the evidence was insufficient to convict. So the jury does not necessarily *accept* H_0 but *fails to reject* H_0. While one can not "prove" a null hypothesis, one can test how close it is to being true with a power test, which tests for type II errors.

What statisticians call an alternative hypothesis is simply an hypothesis that contradicts the null hypothesis.

75.5.3 Error

Working from a null hypothesis two basic forms of error are recognized:
_ Type I errors where the null hypothesis is falsely rejected giving a "false positive".
_ Type II errors where the null hypothesis fails to be rejected and an actual difference between populations is missed giving a "false negative".
Standard deviation refers to the extent to which individual observations in a sample differ from a central value, such as the sample or population mean, while Standard error refers to an estimate of difference between sample mean and population mean.
A statistical error is the amount by which an observation differs from its expected value, a residual is the amount an observation differs from the value the estimator of the expected value assumes on a given sample (also called prediction).
Mean squared error is used for obtaining efficient estimators, a widely used class of estimators. Root mean square error is simply the square root of mean squared error.
Many statistical methods seek to minimize the residual sum of squares, and these are called "methods of least squares" in contrast to Least absolute deviations. The later gives equal weight to small and big errors, while the former gives
522 *CHAPTER 75. STATISTICS*
A least squares fit: in red the points to be fitted, in blue the fitted line.
more weight to large errors. Residual sum of squares is also differentiable, which provides a handy property for doing regression. Least squares applied to linear regression is called ordinary least squares method and least squares applied to nonlinear regression is called non-linear least squares. Also in a linear regression model the non deterministic part of the model is called error term, disturbance or more simply noise.
Measurement processes that generate statistical data are also subject to error. Many of these errors are classified as random (noise) or systematic (bias), but other important types of errors (e.g., blunder, such as when an analyst reports incorrect units) can also be important. The presence of missing data and/or censoring may result in biased estimates and specific techniques have been developed to address these problems.[17]

75.5.4 Interval estimation

Main article: Interval estimation
Most studies only sample part of a population, so results don't fully represent the whole population. Any estimates obtained from the sample only approximate the population value. Confidence intervals allow statisticians to express how closely the sample estimate matches the true value in the whole population. Often they are expressed as 95% confidence intervals. Formally, a 95% confidence interval for a value is a range where, if the sampling and analysis were repeated under the same conditions (yielding a different dataset), the interval would include the true (population) value in 95% of all possible cases. This does *not* imply that the probability that the true value is in the confidence interval is 95%. From the frequentist perspective, such a claim does not even make sense, as the true value is not a random variable. Either the true value is or is not within the given interval. However, it is true that, before any data are sampled and given a plan for how to construct the confidence interval, the probability is 95% that the yet-to-be*75.5. TERMINOLOGY AND THEORY OF INFERENTIAL STATISTICS* 523

μ

Confidence intervals: the red line is true value for the mean in this example, the blue lines are random confidence intervals for 100 realizations.
calculated interval will cover the true value: at this point, the limits of the interval are yet-to-be-observed random variables. One approach that does yield an interval that can be interpreted as having a given probability of containing the true value is to use a credible interval from Bayesian statistics: this approach depends on a different way of interpreting what is meant by "probability", that is as a Bayesian probability.
In principle confidence intervals can be symmetrical or asymmetrical. An interval can be asymmetrical because it works as lower or upper bound for a parameter (left-sided interval or right sided interval), but it can also be asymmetrical because the two sided interval is built violating symmetry around the estimate. Sometimes the bounds for a confidence interval are reached asymptotically and these are used to approximate the true bounds.

75.5.5 Significance

Main article: Statistical significance
Statistics rarely give a simple Yes/No type answer to the question under analysis. Interpretation often comes down to the level of statistical significance applied to the numbers and often refers to the probability of a value accurately rejecting the null hypothesis (sometimes referred to as the p-value).
The standard approach[16] is to test a null hypothesis against an alternative hypothesis. A critical region is the set of values of the estimator that leads to refuting the null hypothesis. The probability of type I error is therefore the probability that the estimator belongs to the critical region given that null hypothesis is true (statistical significance) and the probability of type II error is the probability that the estimator doesn't belong to the critical region given that the alternative hypothesis is true. The statistical power of a test is the probability that it correctly rejects the null hypothesis when the null hypothesis is false.

Referring to statistical significance does not necessarily mean that the overall result is significant in real world terms. For example, in a large study of a drug it may be shown that the drug has a statistically significant but very small beneficial effect, such that the drug is unlikely to help the patient noticeably.

While in principle the acceptable level of statistical significance may be subject to debate, the p-value is the smallest significance level that allows the test to reject the null hypothesis. This is logically equivalent to saying that the p-value is the probability, assuming the null hypothesis is true, of observing a result at least as extreme as the test statistic. Therefore the smaller the p-value, the lower the probability of committing type I error.

Some problems are usually associated with this framework (See criticism of hypothesis testing):

_ A difference that is highly statistically significant can still be of no practical significance, but it is possible to properly formulate tests in account for this. One response involves going beyond reporting only the significance level to include the *p*-value when reporting whether a hypothesis is rejected or accepted. The p-value, however, does not indicate the size or importance of the observed effect and can also seem to exaggerate the importance of minor differences in large studies. A better and increasingly common approach is to report confidence

524 *CHAPTER 75. STATISTICS*

Set of possible results
Probability density
Observed
data point
More likely ovservations
Very un-likely
observations
P-value
Very un-likely
observations
A p-value (shared green area) is the probability of an
observed (or more extreme) result arising by chance
Important:
Pr (observation | hypothesis) ≠ Pr (hypothesis | observation)
The probability of observing a result given that some bypothesis
is true is *not equivalant* to the probability that a hypothesis is true
given that some rerult has been observed.
Using the p-value as a "score" is commiting an egregious logical error:
the transposed conditional fallacy.

In this graph the black line is probability distribution for the test statistic, the critical region is the set of values to the right of the observed data point (observed value of the test statistic) and the p-value is represented by the green area.

intervals. Although these are produced from the same calculations as those of hypothesis tests or *p*-values, they describe both the size of the effect and the uncertainty surrounding it.

_ Fallacy of the transposed conditional, aka prosecutor's fallacy: criticisms arise because the hypothesis testing approach forces one hypothesis (the null hypothesis) to be favored, since what is being evaluated is probability of the observed result given the null hypothesis and not probability of the null hypothesis given the observed result. An alternative to this approach is offered by Bayesian inference, although it requires establishing a prior probability.[18]

_ Rejecting the null hypothesis does not automatically prove the alternative hypothesis.

_ As everything in inferential statistics it relies on sample size, and therefore under fat tails p-values may be seriously mis-computed.

75.5.6 Examples

Some well-known statistical tests and procedures are:

75.6. MISUSE OF STATISTICS 525

_ Analysis of variance (ANOVA)
_ Chi-squared test
_ Correlation
_ Factor analysis
_ Mann–Whitney *U*
_ Mean square weighted deviation (MSWD)
_ Pearson product-moment correlation coefficient
_ Regression analysis
_ Spearman's rank correlation coefficient
_ Student's *t*-test

_ Time series analysis

75.6 Misuse of statistics

Main article: Misuse of statistics

Misuse of statistics can produce subtle, but serious errors in description and interpretation—subtle in the sense that even experienced professionals make such errors, and serious in the sense that they can lead to devastating decision errors. For instance, social policy, medical practice, and the reliability of structures like bridges all rely on the proper use of statistics.

Even when statistical techniques are correctly applied, the results can be difficult to interpret for those lacking expertise. The statistical significance of a trend in the data—which measures the extent to which a trend could be caused by random variation in the sample—may or may not agree with an intuitive sense of its significance. The set of basic statistical skills (and skepticism) that people need to deal with information in their everyday lives properly is referred to as statistical literacy.

There is a general perception that statistical knowledge is all-too-frequently intentionally misused by finding ways to interpret only the data that are favorable to the presenter.[19] A mistrust and misunderstanding of statistics is associated with the quotation, "There are three kinds of lies: lies, damned lies, and statistics". Misuse of statistics can be both inadvertent and intentional, and the book *How to Lie with Statistics*[19] outlines a range of considerations. In an attempt to shed light on the use and misuse of statistics, reviews of statistical techniques used in particular fields are conducted (e.g. Warne, Lazo, Ramos, and Ritter (2012)).[20]

Ways to avoid misuse of statistics include using proper diagrams and avoiding bias.[21] Misuse can occur when conclusions are overgeneralized and claimed to be representative of more than they really are, often by either deliberately or unconsciously overlooking sampling bias.[22] Bar graphs are arguably the easiest diagrams to use and understand, and they can be made either by hand or with simple computer programs.[21] Unfortunately, most people do not look for bias or errors, so they are not noticed. Thus, people may often believe that something is true even if it is not well represented.[22] To make data gathered from statistics believable and accurate, the sample taken must be representative of the whole.[23] According to Huff, "The dependability of a sample can be destroyed by [bias]... allow yourself some degree of skepticism."[24]

To assist in the understanding of statistics Huff proposed a series of questions to be asked in each case:[24]

_ Who says so? (Does he/she have an axe to grind?)
_ How does he/she know? (Does he/she have the resources to know the facts?)
_ What's missing? (Does he/she give us a complete picture?)
_ Did someone change the subject? (Does he/she offer us the right answer to the wrong problem?)
_ Does it make sense? (Is his/her conclusion logical and consistent with what we already know?)

The confounding variable problem: X *and* Y *may be correlated, not because there is causal relationship between them, but because both depend on a third variable Z. Z is called a confounding factor.*

75.6.1 Misinterpretation: correlation

The concept of correlation is particularly noteworthy for the potential confusion it can cause. Statistical analysis of a data set often reveals that two variables (properties) of the population under consideration tend to vary together, as if they were connected. For example, a study of annual income that also looks at age of death might find that poor people tend to have shorter lives than affluent people. The two variables are said to be correlated; however, they may or may not be the cause of one another. The correlation phenomena could be caused by a third, previously unconsidered phenomenon, called a lurking variable or confounding variable. For this reason, there is no way to immediately infer the existence of a causal relationship between the two variables. (See Correlation does not imply causation.)

75.7 History of statistical science

Main articles: History of statistics and Founders of statistics

Statistical methods date back at least to the 5th century BC.

Some scholars pinpoint the origin of statistics to 1663, with the publication of *Natural and Political Observations upon the Bills of Mortality* by John Graunt.[25] Early applications of statistical thinking revolved around the needs of states to base policy on demographic and economic data, hence its *stat-* etymology. The scope of the discipline of statistics broadened in the early 19th century to include the collection and analysis of data in general. Today, statistics is widely employed in government, business, and natural and social sciences.

Its mathematical foundations were laid in the 17th century with the development of the probability theory by Blaise Pascal and Pierre de Fermat. Mathematical probability theory arose from the study of games of chance, although the concept of probability was already examined in medieval law and by philosophers such as Juan Caramuel.[26] The method of least squares was first described by Adrien-Marie Legendre in 1805.

The modern field of statistics emerged in the late 19th and early 20th century in three stages.[27] The first wave, at the turn of the century, was led by the work of Sir Francis Galton and Karl Pearson, who transformed statistics into a rigorous mathematical discipline used for analysis, not just in science, but in industry and politics as well.

Blaise Pascal, an early pioneer on the mathematics of probability.

Galton's contributions to the field included introducing the concepts of standard deviation, correlation, regression and the application of these methods to the study of the variety of human characteristics – height, weight, eyelash length among others.[28] Pearson developed the Correlation coefficient, defined as a product-moment,[29] the method of moments for the fitting of distributions to samples and the Pearson's system of continuous curves, among many other things.[30] Galton and Pearson founded *Biometrika* as the first journal of mathematical statistics and biometry, and the latter founded the world's first university statistics department at University College London.[31]

The second wave of the 1910s and 20s was initiated by William Gosset, and reached its culmination in the insights of Sir Ronald Fisher, who wrote the textbooks that were to define the academic discipline in universities around the world. Fisher's most important publications were his 1916 seminal paper *The Correlation between Relatives on the Supposition of Mendelian Inheritance* and his classic 1925 work *Statistical Methods for Research Workers*. His paper was the first to use the statistical term, variance. He developed rigorous experimental models and also originated the

Karl Pearson, the founder of mathematical statistics.

concepts of sufficiency, ancillary statistics, Fisher's linear discriminator and Fisher information.[32]

The final wave, which mainly saw the refinement and expansion of earlier developments, emerged from the collaborative work between Egon Pearson and Jerzy Neyman in the 1930s. They introduced the concepts of "Type II" error, power of a test and confidence intervals. Jerzy Neyman in 1934 showed that stratified random sampling was in general a better method of estimation than purposive (quota) sampling.[33]

Today, statistical methods are applied in all fields that involve decision making, for making accurate inferences from a collated body of data and for making decisions in the face of uncertainty based on statistical methodology. The use of modern computers has expedited large-scale statistical computations, and has also made possible new methods that are impractical to perform manually. Statistics continues to be an area of active research, for example on the problem of how to analyze Big data.[34]

75.8 Trivia

75.8.1 Applied statistics, theoretical statistics and mathematical statistics

"Applied statistics" comprises descriptive statistics and the application of inferential statistics.[35] *Theoretical statistics* concerns both the logical arguments underlying justification of approaches to statistical inference, as well encompassing *mathematical statistics*. Mathematical statistics includes not only the manipulation of probability distributions necessary for deriving results related to methods of estimation and inference, but also various aspects of computational statistics and the design of experiments.

75.8.2 Machine learning and data mining

There are two applications for machine learning and data mining: data management and data analysis. Statistics tools are necessary for the data analysis.

Ronald Fisher coined the term "null hypothesis".

75.8.3 Statistics in society

Statistics is applicable to a wide variety of academic disciplines, including natural and social sciences, government, and business. Statistical consultants can help organizations and companies that don't have in-house expertise relevant to their particular questions.

75.8.4 Statistical computing

Main article: Computational statistics

gretl, an example of an open source statistical package

The rapid and sustained increases in computing power starting from the second half of the 20th century have had a substantial impact on the practice of statistical science. Early statistical models were almost always from the class of linear models, but powerful computers, coupled with suitable numerical algorithms, caused an increased interest in nonlinear models (such as neural networks) as well as the creation of new types, such as generalized linear models and multilevel models.

Increased computing power has also led to the growing popularity of computationally intensive methods based on resampling, such as permutation tests and the bootstrap, while techniques such as Gibbs sampling have made use of Bayesian models more feasible. The computer revolution has implications for the future of statistics with new emphasis on "experimental" and "empirical" statistics. A large number of both general and special purpose statistical software are now available.

75.8.5 Statistics applied to mathematics or the arts

Traditionally, statistics was concerned with drawing inferences using a semi-standardized methodology that was "required learning" in most sciences. This has changed with use of statistics in non-inferential contexts. What was once considered a dry subject, taken in many fields as a degree-requirement, is now viewed enthusiastically. Initially

derided by some mathematical purists, it is now considered essential methodology in certain areas.
_ In number theory, scatter plots of data generated by a distribution function may be transformed with familiar tools used in statistics to reveal underlying patterns, which may then lead to hypotheses.
_ Methods of statistics including predictive methods in forecasting are combined with chaos theory and fractal geometry to create video works that are considered to have great beauty.

_ The process art of Jackson Pollock relied on artistic experiments whereby underlying distributions in nature were artistically revealed. With the advent of computers, statistical methods were applied to formalize such distribution-driven natural processes to make and analyze moving video art.
_ Methods of statistics may be used predicatively in performance art, as in a card trick based on a Markov process that only works some of the time, the occasion of which can be predicted using statistical methodology.
_ Statistics can be used to predicatively create art, as in the statistical or stochastic music invented by Iannis Xenakis, where the music is performance-specific. Though this type of artistry does not always come out as expected, it does behave in ways that are predictable and tunable using statistics.

75.9 Specialized disciplines

Main article: List of fields of application of statistics
Statistical techniques are used in a wide range of types of scientific and social research, including: biostatistics, computational biology, computational sociology, network biology, social science, sociology and social research. Some fields of inquiry use applied statistics so extensively that they have specialized terminology. These disciplines include:
_ Actuarial science (assesses risk in the insurance and finance industries)
_ Applied information economics
_ Astrostatistics (statistical evaluation of astronomical data)
_ Biostatistics
_ Business statistics
_ Chemometrics (for analysis of data from chemistry)
_ Data mining (applying statistics and pattern recognition to discover knowledge from data)
_ Demography
_ Econometrics (statistical analysis of economic data)
_ Energy statistics
_ Engineering statistics
_ Epidemiology (statistical analysis of disease)
_ Geography and Geographic Information Systems, specifically in Spatial analysis
_ Image processing
Medical Statistics
_ Psychological statistics
_ Reliability engineering
_ Social statistics
In addition, there are particular types of statistical analysis that have also developed their own specialised terminology and methodology:
_ Bootstrap / Jackknife resampling
_ Multivariate statistics

_ Statistical classification
_ Structured data analysis (statistics)
_ Structural equation modelling
_ Survey methodology
_ Survival analysis
_ Statistics in various sports, particularly baseball - known as 'Sabremetrics' - and cricket
Statistics form a key basis tool in business and manufacturing as well. It is used to understand measurement systems variability, control processes (as in statistical process control or SPC), for summarizing data, and to make data-driven decisions. In these roles, it is a key tool, and perhaps the only reliable tool.

75.10 See also

Main article: Outline of statistics
_ Abundance estimation
_ Glossary of probability and statistics
_ List of academic statistical associations
_ List of important publications in statistics
_ List of national and international statistical services
_ List of statistical packages (software)

_ List of statistics articles
_ List of university statistical consulting centers
_ Notation in probability and statistics
Foundations and major areas of statistics
_ Foundations of statistics
_ List of statisticians
_ Official statistics
_ Multivariate analysis of variance

75.11 References

[1] Dodge, Y. (2006) *The Oxford Dictionary of Statistical Terms*, OUP. ISBN 0-19-920613-9
[2] Lund Research Ltd. "Descriptive and Inferential Statistics". statistics.laerd.com. Retrieved 2014-03-23.
[3] Moses, Lincoln E. (1986) *Think and Explain with Statistics*, Addison-Wesley, ISBN 978-0-201-15619-5 . pp. 1–3
[4] Hays, William Lee, (1973) *Statistics for the Social Sciences*, Holt, Rinehart and Winston, p.xii, ISBN 978-0-03-077945-9

[5] Moore, David (1992). "Teaching Statistics as a Respectable Subject". In F. Gordon and S. Gordon. *Statistics for the Twenty-First Century*. Washington, DC: The Mathematical Association of America. pp. 14–25. ISBN 978-0-88385-078-7.
[6] Chance, Beth L.; Rossman, Allan J. (2005). "Preface". *Investigating Statistical Concepts, Applications, and Methods*. Duxbury Press. ISBN 978-0-495-05064-3.
[7] Lakshmikantham,, ed. by D. Kannan,... V. (2002). *Handbook of stochastic analysis and applications*. New York: M. Dekker. ISBN 0824706609.
[8] Schervish, Mark J. (1995). *Theory of statistics* (Corr. 2nd print. ed.). New York: Springer. ISBN 0387945466.
[9] Freedman, D.A. (2005) *Statistical Models: Theory and Practice*, Cambridge University Press. ISBN 978-0-521-67105-7
[10] McCarney R, Warner J, Iliffe S, van Haselen R, Griffin M, Fisher P (2007). "The Hawthorne Effect: a randomised, controlled trial". *BMC Med Res Methodol* **7**: 30. doi:10.1186/1471-2288-7-30. PMC 1936999. PMID 17608932.
[11] Mosteller, F., & Tukey, J. W. (1977). *Data analysis and regression*. Boston: Addison-Wesley.
[12] Nelder, J. A. (1990). The knowledge needed to computerise the analysis and interpretation of statistical information. In *Expert systems and artificial intelligence: the need for information about data*. Library Association Report, London, March, 23–27.
[13] Chrisman, Nicholas R. (1998). Rethinking Levels of Measurement for Cartography. *Cartography and Geographic Information Science*, vol. 25 (4), pp. 231–242
[14] van den Berg, G. (1991). *Choosing an analysis method*. Leiden: DSWO Press
[15] Hand, D. J. (2004). *Measurement theory and practice: The world through quantification*. London, UK: Arnold.
[16] Piazza Elio, Probabilità e Statistica, Esculapio 2007
[17] Rubin, Donald B.; Little, Roderick J. A.,Statistical analysis with missing data, New York: Wiley 2002
[18] Ioannidis, J. P. A. (2005). "Why Most Published Research Findings Are False". *PLoS Medicine* **2** (8): e124. doi:10.1371/journal.pmed.0020124. PMC 1182327. PMID 16060722.
[19] Huff, Darrell (1954) *How to Lie with Statistics*, WW Norton & Company, Inc. New York, NY. ISBN 0-393-31072-8
[20] Warne, R. Lazo, M., Ramos, T. and Ritter, N. (2012). Statistical Methods Used in Gifted Education Journals, 2006–2010. Gifted Child Quarterly, 56(3) 134–149. doi:10.1177/0016986212444122
[21] Drennan, Robert D. (2008). "Statistics in archaeology". In Pearsall, Deborah M. *Encyclopedia of Archaeology*. Elsevier Inc. pp. 2093–2100. ISBN 978-0-12-373962-9.
[22] Cohen, Jerome B. (December 1938). "Misuse of Statistics". *Journal of the American Statistical Association* (JSTOR) **33** (204): 657–674. doi:10.1080/01621459.1938.10502344.
[23] Freund, J. F. (1988). "Modern Elementary Statistics". *Credo Reference*.
[24] Huff, Darrell; Irving Geis (1954). *How to Lie with Statistics*. New York: Norton. "The dependability of a sample can be destroyed by [bias]... allow yourself some degree of skepticism."
[25] Willcox, Walter (1938) "The Founder of Statistics". *Review of the International Statistical Institute* 5(4):321–328. JSTOR 1400906
[26] J. Franklin, The Science of Conjecture: Evidence and Probability before Pascal,Johns Hopkins Univ Pr 2002
[27] Helen Mary Walker (1975). *Studies in the history of statistical method*. Arno Press.
[28] Galton F (1877) Typical laws of heredity. Nature 15: 492–553
[29] Stigler, S. M. (1989). "Francis Galton's Account of the Invention of Correlation". *Statistical Science* **4** (2): 73–79. doi:10.1214/ss/1177012580.
[30] Pearson, K. (1900). "On the Criterion that a given System of Deviations from the Probable in the Case of a Correlated System of Variables is such that it can be reasonably supposed to have arisen from Random Sampling". *Philosophical Magazine Series 5* **50** (302): 157–175. doi:10.1080/14786440009463897.
[31] "Karl Pearson (1857–1936)". Department of Statistical Science – University College London.

[32] Agresti, Alan; David B. Hichcock (2005). "Bayesian Inference for Categorical Data Analysis". *Statistical Methods & Applications* **14** (14): 298. doi:10.1007/s10260-005-0121-y.
[33] Neyman, J (1934) On the two different aspects of the representative method: The method of stratified sampling and the method of purposive selection. *Journal of the Royal Statistical Society* 97 (4) 557–625 JSTOR 2342192
[34] http://www.santafe.edu/news/item/sfnm-wood-big-data/
[35] Anderson, D.R.; Sweeney, D.J.; Williams, T.A.. (1994) *Introduction to Statistics: Concepts and Applications*, pp. 5–9.

West Group. ISBN 978-0-314-03309-3

Chapter 76

Data analysis

Analysis of data is a process of inspecting, cleaning, transforming, and modeling data with the goal of discovering useful information, suggesting conclusions, and supporting decision-making. Data analysis has multiple facets and approaches, encompassing diverse techniques under a variety of names, in different business, science, and social science domains.

Data mining is a particular data analysis technique that focuses on modeling and knowledge discovery for predictive rather than purely descriptive purposes. Business intelligence covers data analysis that relies heavily on aggregation, focusing on business information. In statistical applications, some people divide data analysis into descriptive statistics, exploratory data analysis (EDA), and confirmatory data analysis (CDA). EDA focuses on discovering new features in the data and CDA on confirming or falsifying existing hypotheses. Predictive analytics focuses on application of statistical models for predictive forecasting or classification, while text analytics applies statistical, linguistic, and structural techniques to extract and classify information from textual sources, a species of unstructured data. All are varieties of data analysis.

Data integration is a precursor to data analysis, and data analysis is closely linked to data visualization and data dissemination. The term *data analysis* is sometimes used as a synonym for data modeling.

76.1 The process of data analysis

Data analysis is a process for obtaining raw data and converting it into information useful for decision-making by users. Data is collected and analyzed to answer questions, test hypotheses or disprove theories.[1]

There are several phases that can be distinguished. The phases are iterative, in that feedback from later phases may result in additional work in earlier phases.[2]

76.1.1 Data requirements

The data necessary as inputs to the analysis are specified based upon the requirements of those directing the analysis or customers who will use the finished product of the analysis. The general type of entity upon which the data will be collected is referred to as an experimental unit (e.g., a person or population of people). Specific variables regarding a population (e.g., age and income) may be specified and obtained. Data may be numerical or categorical (i.e., a text label for numbers).[2]

76.1.2 Data collection

Data is collected from a variety of sources. The requirements may be communicated by analysts to custodians of the data, such as information technology personnel within an organization. The data may also be collected from sensors in the environment, such as traffic cameras, satellites, recording devices, etc. It may also be obtained through interviews, downloads from online sources, or reading documentation.[2]

535

536 *CHAPTER 76. DATA ANALYSIS*

Data science process flowchart

76.1.3 Data processing

Data initially obtained must be processed or organized for analysis. For instance, this may involve placing data into rows and columns in a table format for further analysis, such as within a spreadsheet or statistical software.[2]

76.1.4 Data cleaning

Once processed and organized, the data may be incomplete, contain duplicates, or contain errors. The need for data cleaning will arise from problems in the way that data is entered and stored. Data cleaning is the process of preventing and correcting these errors. Common tasks include record matching, deduplication, and column segmentation.[3]

Such data problems can also be identified through a variety of analytical techniques. For example, with financial information, the totals for particular variables may be compared against separately published numbers believed to be reliable.[4] Unusual amounts above or below pre-determined thresholds may also be reviewed. There are several types of data cleaning that depend on the type of data. Quantitative data methods for outlier detection can be used to get rid of likely incorrectly entered data. Textual data spellcheckers can be used to lessen the amount of mistyped words, but it is harder to tell if the words themselves are correct.[5]

76.1.5 Exploratory data analysis

Once the data is cleaned, it can be analyzed. Analysts may apply a variety of techniques referred to as exploratory data analysis to begin understanding the messages contained in the data.[6] The process of exploration may result in additional data cleaning or additional requests for data, so these activities may be iterative in nature. Descriptive statistics such as the average or median may be generated to help understand the data. Data visualization may also be used to examine the data in graphical format, to obtain additional insight regarding the messages within the data.[2]

76.1. THE PROCESS OF DATA ANALYSIS 537

The phases of the intelligence cycle used to convert raw information into actionable intelligence or knowledge are conceptually similar to the phases in data analysis.

76.1.6 Modeling and algorithms

Mathematical formulas or models called algorithms may be applied to the data to identify relationships among the variables, such as correlation or causation. In general terms, models may be developed to evaluate a particular variable in the data based on other variable(s) in the data, with some residual error depending on model accuracy (i.e., Data = Model + Error).[1]

Inferential statistics includes techniques to measure relationships between particular variables. For example, regression analysis may be used to model whether a change in advertising (independent variable X) explains the variation in sales (dependent variable Y). In mathematical terms, Y (sales) is a function of X (advertising). It may be described as Y = aX + b + error, where the model is designed such that a and b minimize the error when the model predicts Y for a given range of values of X. Analysts may attempt to build models that are descriptive of the data to simplify analysis and communicate results.[1]

76.1.7 Data product

A data product is a computer application that takes data inputs and generates outputs, feeding them back into the environment. It may be based on a model or algorithm. An example is an application that analyzes data about customer purchasing history and recommends other purchases the customer might enjoy.[2]

76.1.8 Communication

Once the data is analyzed, it may be reported in many formats to the users of the analysis to support their requirements. The users may have feedback, which results in additional analysis. As such, much of the analytical cycle is iterative.[2]

When determining how to communicate the results, the analyst may consider data visualization techniques to help clearly and efficiently communicate the message to the audience.

76.2 Quantitative messages

Main article: Data visualization

Author Stephen Few described eight types of quantitative messages that users may attempt to understand or com-

A time series illustrated with a line chart demonstrating trends in U.S. federal spending and revenue over time.

A scatterplot illustrating correlation between two variables (inflation and unemployment) measured at points in time.

municate from a set of data and the associated graphs used to help communicate the message. Customers specifying requirements and analysts performing the data analysis may consider these messages during the course of the process.

1. Time-series: A single variable is captured over a period of time, such as the unemployment rate over a 10-year period. A line chart may be used to demonstrate the trend.
2. Ranking: Categorical subdivisions are ranked in ascending or descending order, such as a ranking of sales performance (the *measure*) by sales persons (the *category*, with each sales person a *categorical subdivision*) during a single period. A bar chart may be used to show the comparison across the sales persons.
3. Part-to-whole: Categorical subdivisions are measured as a ratio to the whole (i.e., a percentage out of 100%). A pie chart or bar chart can show the comparison of ratios, such as the market share represented by competitors in a market.
4. Deviation: Categorical subdivisions are compared again a reference, such as a comparison of actual vs. budget expenses for several departments of a business for a given time period. A bar chart can show comparison of the actual versus the reference amount.
5. Frequency distribution: Shows the number of observations of a particular variable for given interval, such as the number of years in which the stock market return is between intervals such as 0-10%, 11-20%, etc. A histogram, a type of bar chart, may be used for this analysis.
6. Correlation: Comparison between observations represented by two variables (X,Y) to determine if they tend to move in the same or opposite directions. For example, plotting unemployment (X) and inflation (Y) for a sample of months. A scatter plot is typically used for this message.
7. Nominal comparison: Comparing categorical subdivisions in no particular order, such as the sales volume by product code. A bar chart may be used for this comparison.
8. Geographic or geospatial: Comparison of a variable across a map or layout, such as the unemployment rate by state or the number of persons on the various floors of a building. A cartogram is a typical graphic used.[7][8]

76.3 Techniques for analyzing quantitative data

See also: Problem solving

Author Dr. Jonathan Koomey has recommended a series of best practices for understanding quantitative data. These include:

_ Check raw data for anomalies prior to performing your analysis;
_ Re-perform important calculations, such as verifying columns of data that are formula driven;
_ Confirm main totals are the sum of subtotals;

_ Check relationships between numbers that should be related in a predictable way, such as ratios over time;
_ Normalize numbers to make comparisons easier, such as analyzing amounts per person or relative to GDP or as an index value relative to a base year;
_ Break problems into component parts by analyzing factors that led to the results, such as DuPont analysis of return on equity.[4]

For the variables under examination, analysts typically obtain descriptive statistics for them, such as the mean (average), median, and standard deviation. They may also analyze the distribution of the key variables to see how the individual values cluster around the mean.

Consultants at McKinsey and Company named a technique for breaking a quantitative problem down into its component parts called the MECE principle. Each layer can be broken down into its components; each of the subcomponents must be mutually exclusive of each other and collectively add up to the layer above them. The relationship is referred to as "Mutually Exclusive and Collectively Exhaustive" or MECE. For example, profit by definition can be broken down into total revenue and total cost. In turn, total revenue can be analyzed by its components, such as revenue of divisions A, B, and C (which are mutually exclusive of each other) and should add to the total revenue (collectively exhaustive).

Analysts may use robust statistical measurements to solve certain analytical problems. Hypothesis testing is used when a particular hypothesis about the true state of affairs is made by the analyst and data is gathered to determine whether that state of affairs is true or false. For example, the hypothesis might be that "Unemployment has no effect on inflation", which relates to an economics concept called the Phillips Curve. Hypothesis testing involves considering the likelihood of Type I and type II errors, which relate to whether the data supports accepting or rejecting the hypothesis.

Regression analysis may be used when the analyst is trying to determine the extent to which independent variable X affects dependent variable Y (e.g., "To what extent do changes in the unemployment rate (X) affect the inflation rate (Y)?"). This is an attempt to model or fit an equation line or curve to the data, such that Y is a function of X.

76.4 Analytical activities of data users

Users may have particular data points of interest within a data set, as opposed to general messaging outlined above. Such low-level user analytic activities are presented in the following table. The taxonomy can also be organized by three poles of activities: retrieving values, finding data points, and arranging data points.[9][10][11]

76.5 Barriers to effective analysis

Barriers to effective analysis may exist among the analysts performing the data analysis or among the audience. Distinguishing fact from opinion, cognitive biases, and innumeracy are all challenges to sound data analysis.

76.5.1 Confusing fact and opinion

You are entitled to your own opinion, but you are not entitled to your own facts.
Daniel Patrick Moynihan

Effective analysis requires obtaining relevant facts to answer questions, support a conclusion or formal opinion, or test hypotheses. Facts by definition are irrefutable, meaning that any person involved in the analysis should be able to agree upon them. For example, in August 2010, the Congressional Budget Office (CBO) estimated that extending the Bush tax cuts of 2001 and 2003 for the 2011-2020 time period would add approximately $3.3 trillion to the national debt.[12] Everyone should be able to agree that indeed this is what CBO reported; they can all examine the report. This makes it a fact. Whether persons agree or disagree with the CBO is their own opinion.

As another example, the auditor of a public company must arrive at a formal opinion on whether financial statements of publicly traded corporations are "fairly stated, in all material respects." This requires extensive analysis of factual data and evidence to support their opinion. When making the leap from facts to opinions, there is always the possibility that the opinion is erroneous.

76.5.2 Cognitive biases

There are a variety of cognitive biases that can adversely effect analysis. For example, confirmation bias is the tendency to search for or interpret information in a way that confirms one's preconceptions. In addition, individuals may discredit information that does not support their views. Analysts may be trained specifically to be aware of these biases and how to overcome them.

76.5.3 Innumeracy

Effective analysts are generally adept with a variety of numerical techniques. However, audiences may not have such literacy with numbers or numeracy; they are said to be innumerate. Persons communicating the data may also be attempting to mislead or misinform, deliberately using bad numerical techniques.[13]

For example, whether a number is rising or falling may not be the key factor. More important may be the number relative to another number, such as the size of government revenue or spending relative to the size of the economy

(GDP) or the amount of cost relative to revenue in corporate financial statements. This numerical technique is referred to as normalization[14] or common-sizing. There are many such techniques employed by analysts, whether adjusting

for inflation (i.e., comparing real vs. nominal data) or considering population increases, demographics, etc. Analysts apply a variety of techniques to address the various quantitative messages described in the section above.
Analysts may also analyze data under different assumptions or scenarios. For example, when analysts perform financial statement analysis, they will often recast the financial statements under different assumptions to help arrive at an estimate of future cash flow, which they then discount to present value based on some interest rate, to determine the valuation of the company or its stock. Similarly, the CBO analyzes the effects of various policy options on the government's revenue, outlays and deficits, creating alternative future scenarios for key measures.

76.6 Other topics

76.6.1 Analytics and business intelligence

Main article: Analytics
Analytics is the "extensive use of data, statistical and quantitative analysis, explanatory and predictive models, and factbased management to drive decisions and actions." It is a subset of business intelligence, which is a set of technologies and processes that use data to understand and analyze business performance.[15]

76.6.2 Education

Analytic activities of data visualization users
In education, most educators have access to a data system for the purpose of analyzing student data.[16] These data

systems present data to educators in an over-the-counter data format (embedding labels, supplemental documentation, and a help system and making key package/display and content decisions) to improve the accuracy of educators' data analyses.[17]

76.6.3 Nuclear and particle physics

In nuclear and particle physics the data usually originate from the experimental apparatus via a data acquisition system. They are then processed, in a step usually called *data reduction*, to apply calibrations and to extract physically significant information. Data reduction is most often, especially in large particle physics experiments, an automatic, batch-mode operation carried out by software written ad-hoc. The resulting data *n-tuples* are then scrutinized by the physicists, using specialized software tools like ROOT or PAW, comparing the results of the experiment with theory. The theoretical models are often difficult to compare directly with the results of the experiments, so they are used instead as input for Monte Carlo simulation software like Geant4, in order to predict the response of the detector to a given theoretical event, thus producing **simulated events** which are then compared to experimental data.

76.7 Practitioner notes

This section contains rather technical explanations that may assist practitioners but are beyond the typical scope of a Wikipedia article.

76.7.1 Initial data analysis

The most important distinction between the initial data analysis phase and the main analysis phase, is that during initial data analysis one refrains from any analysis that is aimed at answering the original research question. The initial data analysis phase is guided by the following four questions:[18]

Quality of data

The quality of the data should be checked as early as possible. Data quality can be assessed in several ways, using different types of analysis: frequency counts, descriptive statistics (mean, standard deviation, median), normality (skewness, kurtosis, frequency histograms, n: variables are compared with coding schemes of variables external to the data set, and possibly corrected if coding schemes are not comparable.
_ Test for common-method variance.
The choice of analyses to assess the data quality during the initial data analysis phase depends on the analyses that will be conducted in the main analysis phase.[19]

Quality of measurements

The quality of the measurement instruments should only be checked during the initial data analysis phase when this is not the focus or research question of the study. One should check whether structure of measurement instruments corresponds to structure reported in the literature.
There are two ways to assess measurement
_ Analysis of homogeneity (internal consistency), which gives an indication of the reliability of a measurement instrument. During this analysis, one inspects the variances of the items and the scales, the Cronbach's α of the scales, and the change in the Cronbach's alpha when an item would be deleted from a scale.[20]

Initial transformations

After assessing the quality of the data and of the measurements, one might decide to impute missing data, or to perform initial transformations of one or more variables, although this can also be done during the main analysis

phase.[21]
Possible transformations of variables are:[22]
_ Square root transformation (if the distribution differs moderately from normal)

_ Log-transformation (if the distribution differs substantially from normal)
_ Inverse transformation (if the distribution differs severely from normal)
_ Make categorical (ordinal / dichotomous) (if the distribution differs severely from normal, and no transformations help)

Did the implementation of the study fulfill the intentions of the research design?

One should check the success of the randomization procedure, for instance by checking whether background and substantive variables are equally distributed within and across groups.

If the study did not need or use a randomization procedure, one should check the success of the non-random sampling, for instance by checking whether all subgroups of the population of interest are represented in sample.

Other possible data distortions that should be checked are:
_ dropout (this should be identified during the initial data analysis phase)
_ Item nonresponse (whether this is random or not should be assessed during the initial data analysis phase)
_ Treatment quality (using manipulation checks).[23]

Characteristics of data sample

In any report or article, the structure of the sample must be accurately described. It is especially important to exactly determine the structure of the sample (and specifically the size of the subgroups) when subgroup analyses will be performed during the main analysis phase.

The characteristics of the data sample can be assessed by looking at:
_ Basic statistics of important variables
_ Scatter plots
_ Correlations and associations
_ Cross-tabulations[24]

Final stage of the initial data analysis

During the final stage, the findings of the initial data analysis are documented, and necessary, preferable, and possible corrective actions are taken.

Also, the original plan for the main data analyses can and should be specified in more detail or rewritten.

In order to do this, several decisions about the main data analyses can and should be made:
_ In the case of non-normals: should one transform variables; make variables categorical (ordinal/dichotomous); adapt the analysis method?
_ In the case of missing data: should one neglect or impute the missing data; which imputation technique should be used?
_ In the case of outliers: should one use robust analysis techniques?
_ In case items do not fit the scale: should one adapt the measurement instrument by omitting items, or rather ensure comparability with other (uses of the) measurement instrument(s)?
_ In the case of (too) small subgroups: should one drop the hypothesis about inter-group differences, or use small sample techniques, like exact tests or bootstrapping?
_ In case the randomization procedure seems to be defective: can and should one calculate propensity scores and include them as covariates in the main analyses?[25]

544 *CHAPTER 76. DATA ANALYSIS*

Analysis

Several analyses can be used during the initial data analysis phase:[26]
_ Univariate statistics (single variable)
_ Bivariate associations (correlations)
_ Graphical techniques (scatter plots)

It is important to take the measurement levels of the variables into account for the analyses, as special statistical techniques are available for each level:[27]
_ Nominal and ordinal variables
_ Frequency counts (numbers and percentages)
_ Associations
_ circumambulations (crosstabulations)
_ hierarchical loglinear analysis (restricted to a maximum of 8 variables)
_ loglinear analysis (to identify relevant/important variables and possible confounders)
_ Exact tests or bootstrapping (in case subgroups are small)
_ Computation of new variables
_ Continuous variables
_ Distribution
_ Statistics (M, SD, variance, skewness, kurtosis)
_ Stem-and-leaf displays
_ Box plots

Nonlinear analysis

Nonlinear analysis will be necessary when the data is recorded from a nonlinear system. Nonlinear systems can exhibit complex dynamic effects including bifurcations, chaos, harmonics and subharmonics that cannot be analyzed using simple linear methods. Nonlinear data analysis is closely related to nonlinear system identification.[28]

76.7.2 Main data analysis

In the main analysis phase analyses aimed at answering the research question are performed as well as any other relevant analysis needed to write the first draft of the research report.[29]

Exploratory and confirmatory approaches

In the main analysis phase either an exploratory or confirmatory approach can be adopted. Usually the approach is decided before data is collected. In an exploratory analysis no clear hypothesis is stated before analysing the data, and the data is searched for models that describe the data well. In a confirmatory analysis clear hypotheses about the data are tested.

Exploratory data analysis should be interpreted carefully. When testing multiple models at once there is a high chance on finding at least one of them to be significant, but this can be due to a type 1 error. It is important to always adjust the significance level when testing multiple models with, for example, a Bonferroni correction. Also, one should not follow up an exploratory analysis with a confirmatory analysis in the same dataset. An exploratory analysis is used to find ideas for a theory, but not to test that theory as well. When a model is found exploratory in a dataset, then following up that analysis with a confirmatory analysis in the same dataset could simply mean that the results of the confirmatory analysis are due to the same type 1 error that resulted in the exploratory model in the first place. The confirmatory analysis therefore will not be more informative than the original exploratory analysis.[30]

Stability of results

It is important to obtain some indication about how generalizable the results are.[31] While this is hard to check, one can look at the stability of the results. Are the results reliable and reproducible? There are two main ways of doing this:

- Cross-validation: By splitting the data in multiple parts we can check if an analysis (like a fitted model) based on one part of the data generalizes to another part of the data as well.
- Sensitivity analysis: A procedure to study the behavior of a system or model when global parameters are (systematically) varied. One way to do this is with bootstrapping.

Statistical methods

Many statistical methods have been used for statistical analyses. A very brief list of four of the more popular methods is:

- General linear model: A widely used model on which various methods are based (e.g. t test, ANOVA, ANCOVA, MANOVA). Usable for assessing the effect of several predictors on one or more continuous dependent variables.
- Generalized linear model: An extension of the general linear model for discrete dependent variables.
- Structural equation modelling: Usable for assessing latent structures from measured manifest variables.
- Item response theory: Models for (mostly) assessing one latent variable from several binary measured variables (e.g. an exam).

76.8 Free software for data analysis

- Data Applied - an online data mining and data visualization solution.
- DevInfo - a database system endorsed by the United Nations Development Group for monitoring and analyzing human development.
- ELKI - data mining framework in Java with data mining oriented visualization functions.
- KNIME - the Konstanz Information Miner, a user friendly and comprehensive data analytics framework.
- PAW - FORTRAN/C data analysis framework developed at CERN
- SCaViS - a multiplatform (Java-based) data analysis framework from the jWork.ORG community of developers led by Dr. S.Chekanov
- R - a programming language and software environment for statistical computing and graphics.
- ROOT - C++ data analysis framework developed at CERN
- dotplot - cloud based visual designer to create analytic models[32]
- SciPy - A set of Python tools for data analysis http://scipy.org/stackspec.html

76.9 See also

- Analytics
- Business intelligence
- Censoring (statistics)

- Computational physics
- Data acquisition
- Data governance

_ Data mining
_ Data Presentation Architecture
_ Digital signal processing
_ Dimension reduction
_ Early case assessment
_ Exploratory data analysis
_ Fourier analysis
_ Machine learning
_ Multilinear PCA
_ Multilinear subspace learning
_ Multiway Data Analysis
_ Nearest neighbor search
_ nonlinear system identification
_ Predictive analytics
_ Principal component analysis
_ Qualitative research
_ Scientific computing
_ Structured data analysis (statistics)
_ system identification
_ Test method
_ Text analytics
_ Unstructured data
_ Wavelet

76.10 References

76.10.1 Citations

[1] Judd, Charles and, McCleland, Gary (1989). *Data Analysis*. Harcourt Brace Jovanovich. ISBN 0-15-516765-0.
[2] O'Neil, Cathy and, Schutt, Rachel (2014). *Doing Data Science*. O'Reilly. ISBN 978-1-449-35865-5.
[3] "Data Cleaning". Microsoft Research. Retrieved 26 October 2013.
[4] Perceptual Edge-Jonathan Koomey-Best practices for understanding quantitative data-February 14, 2006
[5] Hellerstein, Joseph (27 February 2008). "Quantitative Data Cleaning for Large Databases". *EECS Computer Science Division*: 3. Retrieved 26 October 2013.
[6] Stephen Few-Perceptual Edge-Selecting the Right Graph For Your Message-September 2004

[7] Stephen Few-Perceptual Edge-Selecting the Right Graph for Your Message-2004
[8] Stephen Few-Perceptual Edge-Graph Selection Matrix
[9] Robert Amar, James Eagan, and John Stasko (2005) "Low-Level Components of Analytic Activity in Information Visualization"
[10] William Newman (1994) "A Preliminary Analysis of the Products of HCI Research, Using Pro Forma Abstracts"
[11] Mary Shaw (2002) "What Makes Good Research in Software Engineering?"
[12] "Congressional Budget Office-The Budget and Economic Outlook-August 2010-Table 1.7 on Page 24" (PDF). Retrieved 2011-03-31.
[13] Bloomberg-Barry Ritholz-Bad Math that Passes for Insight-October 28, 2014
[14] Perceptual Edge-Jonathan Koomey-Best practices for understanding quantitative data-February 14, 2006
[15] Davenport, Thomas and, Harris, Jeanne (2007). *Competing on Analytics*. O'Reilly. ISBN 978-1-4221-0332-6.
[16] Aarons, D. (2009). Report finds states on course to build pupil-data systems. *Education Week, 29*(13), 6.
[17] Rankin, J. (2013, March 28). How data Systems & reports can either fight or propagate the data analysis error epidemic, and how educator leaders can help. *Presentation conducted from Technology Information Center for Administrative Leadership (TICAL) School Leadership Summit.*
[18] Adèr, 2008, p. 337.
[19] Adèr, 2008, p. 338-341.
[20] Adèr, 2008, p. 341-3342.
[21] Adèr, 2008, p. 344.
[22] Tabachnick & Fidell, 2007, p. 87-88.
[23] Adèr, 2008, p. 344-345.
[24] Adèr, 2008, p. 345.
[25] Adèr, 2008, p. 345-346.
[26] Adèr, 2008, p. 346-347.
[27] Adèr, 2008, p. 349-353.
[28] Billings S.A. "Nonlinear System Identification: NARMAX Methods in the Time, Frequency, and Spatio-Temporal Domains". Wiley, 2013
[29] Adèr, 2008, p. 363.
[30] Adèr, 2008, p. 361-362.
[31] Adèr, 2008, p. 368-371.
[32] "dotplot designer - Official Website". Retrieved 19 February 2014.

76.10.2 Bibliography

_ Adèr, H.J. (2008). Chapter 14: Phases and initial steps in data analysis. In H.J. Adèr & G.J. Mellenbergh (Eds.) (with contributions by D.J. Hand), Advising on Research Methods: A consultant's companion (pp. 333–356). Huizen, the Netherlands: Johannes van Kessel Publishing.
_ Adèr, H.J. (2008). Chapter 15: The main analysis phase. In H.J. Adèr & G.J. Mellenbergh (Eds.) (with contributions by D.J. Hand), Advising on Research Methods: A consultant's companion (pp. 333–356). Huizen, the Netherlands: Johannes van Kessel Publishing.
_ Tabachnick, B.G. & Fidell, L.S. (2007). Chapter 4: Cleaning up your act. Screening data prior to analysis. In B.G. Tabachnick & L.S. Fidell (Eds.), Using Multivariate Statistics, Fifth Edition (pp. 60–116). Boston: Pearson Education, Inc. / Allyn and Bacon.

76.11 Further reading

_ Adèr, H.J. & Mellenbergh, G.J. (with contributions by D.J. Hand) (2008). Advising on Research Methods: A consultant's companion. Huizen, the Netherlands: Johannes van Kessel Publishing.
_ ASTM International (2002). *Manual on Presentation of Data and Control Chart Analysis*, MNL 7A, ISBN 0-8031-2093-1
_ Juran, Joseph M.; Godfrey, A. Blanton (1999). *Juran's Quality Handbook.* 5th ed. New York: McGraw Hill. ISBN 0-07-034003-X
_ Lewis-Beck, Michael S. (1995). *Data Analysis: an Introduction*, Sage Publications Inc, ISBN 0-8039-5772-6
_ NIST/SEMATEK (2008) *Handbook of Statistical Methods*,
_ Pyzdek, T, (2003). *Quality Engineering Handbook*, ISBN 0-8247-4614-7
_ Richard Veryard (1984). *Pragmatic data analysis.* Oxford : Blackwell Scientific Publications. ISBN 0-632-01311-7
_ Tabachnick, B.G. & Fidell, L.S. (2007). Using Multivariate Statistics, Fifth Edition. Boston: Pearson Education, Inc. / Allyn and Bacon, ISBN 978-0-205-45938-4
_ Vance (September 8, 2011). "Data Analytics: Crunching the Future". Bloomberg Businessweek. Retrieved 26 September 2011.
_ Hair, Joseph (2008). Marketing Research 4th ed. McGraw Hill. *Data Analysis: Testing for Association* ISBN 0-07-340470-5

Chapter 77

Bayesian inference

In statistics, **Bayesian inference** is a method of inference in which Bayes' rule is used to update the probability estimate for a hypothesis as additional evidence is acquired. Bayesian updating is an important technique throughout statistics, and especially in mathematical statistics. For some cases, exhibiting a Bayesian derivation for a statistical method automatically ensures that the method works as well as any competing method. Bayesian updating is especially important in the dynamic analysis of a sequence of data. Bayesian inference has found application in a range of fields including science, engineering, philosophy, medicine and law.
In the philosophy of decision theory, Bayesian inference is closely related to discussions of subjective probability, often called "Bayesian probability". Bayesian probability provides a rational method for updating beliefs.

77.1 Introduction to Bayes' rule

Main article: Bayes' rule
See also: Bayesian probability

77.1.1 Formal

Bayesian inference derives the posterior probability as a consequence of two antecedents, a prior probability and a "likelihood function" derived from a probability model for the data to be observed. Bayesian inference computes the posterior probability according to Bayes' rule:
$P(H j E) =$
$P(E j H) \_ P(H)$
$P(E)$
where
_ *j* denotes a conditional probability; more specifically, it means *given.*
_ *H* stands for any *hypothesis* whose probability may be affected by data (called *evidence* below). Often there are competing hypotheses, from which one chooses the most probable.
_ the *evidence E* corresponds to new data that were not used in computing the prior probability.
_ *P*(*H*) , the *prior probability*, is the probability of*H beforeE* is observed. This indicates one's previous estimate of the probability that a hypothesis is true, before gaining the current evidence.

_ $P(H j E)$, the *posterior probability*, is the probability of *H given E* , i.e., *after E* is observed. This tells us what we want to know: the probability of a hypothesis *given* the observed evidence.
_ $P(E j H)$ is the probability of observing *E given H* . As a function of *E* with *H* fixed, this is the *likelihood.* The likelihood function should **not** be confused with $P(H j E)$ as a function of *H* rather than of *E* . It indicates the compatibility of the evidence with the given hypothesis.
549
550 *CHAPTER 77. BAYESIAN INFERENCE*
_ $P(E)$ is sometimes termed the marginal likelihood or "model evidence". This factor is the same for all possible hypotheses being considered. (This can be seen by the fact that the hypothesis *H* does not appear anywhere in the symbol, unlike for all the other factors.) This means that this factor does not enter into determining the relative probabilities of different hypotheses.
Note that, for different values of *H* , only the factors $P(H)$ and $P(E j H)$ affect the value of $P(H j E)$. As both of these factors appear in the numerator, the posterior probability is proportional to both. In words:
_ (more exactly) *The posterior probability of a hypothesis is determined by a combination of the inherent likeliness of a hypothesis (the prior) and the compatibility of the observed evidence with the hypothesis (the likelihood).*
_ (more concisely) *Posterior is proportional to likelihood times prior.*
Note that Bayes' rule can also be written as follows:
$P(H j E) =$
$P(E j H)$
$P(E)$
_ $P(H)$
where the factor $P(EjH)$
$P(E)$ represents the impact of *E* on the probability of *H* .

77.1.2 Informal

Rationally, Bayes' rule makes a great deal of sense. If the evidence does not match up with a hypothesis, one should reject the hypothesis. But if a hypothesis is extremely unlikely *a priori*, one should also reject it, even if the evidence does appear to match up.
For example, imagine that I have various hypotheses about the nature of a newborn baby of a friend, including:
_ H_1 : the baby is a brown-haired boy.
_ H_2 : the baby is a blond-haired girl.
_ H_3 : the baby is a dog.
Then consider two scenarios:
1. I'm presented with evidence in the form of a picture of a blond-haired baby girl. I find this evidence supports H_2 and opposes H_1 and H_3 .
2. I'm presented with evidence in the form of a picture of a baby dog. Although this evidence, treated in isolation, supports H_3 , my prior belief in this hypothesis (that a human can give birth to a dog) is extremely small, so the posterior probability is nevertheless small.
The critical point about Bayesian inference, then, is that it provides a principled way of combining new evidence with prior beliefs, through the application of Bayes' rule. (Contrast this with frequentist inference, which relies only on the evidence as a whole, with no reference to prior beliefs.) Furthermore, Bayes' rule can be applied iteratively: after observing some evidence, the resulting posterior probability can then be treated as a prior probability, and a new posterior probability computed from new evidence. This allows for Bayesian principles to be applied to various kinds of evidence, whether viewed all at once or over time. This procedure is termed "Bayesian updating".

77.1.3 Bayesian updating

Bayesian updating is widely used and computationally convenient. However, it is not the only updating rule that might be considered "rational".
Ian Hacking noted that traditional "Dutch book" arguments did not specify Bayesian updating: they left open the possibility that non-Bayesian updating rules could avoid Dutch books. Hacking wrote[1] "And neither the Dutch
77.2. FORMAL DESCRIPTION OF BAYESIAN INFERENCE 551
book argument, nor any other in the personalist arsenal of proofs of the probability axioms, entails the dynamic assumption. Not one entails Bayesianism. So the personalist requires the dynamic assumption to be Bayesian. It is true that in consistency a personalist could abandon the Bayesian model of learning from experience. Salt could lose its savour."
Indeed, there are non-Bayesian updating rules that also avoid Dutch books (as discussed in the literature on "probability kinematics" following the publication of Richard C. Jeffrey's rule, which applies Bayes' rule to the case where the evidence itself is assigned a probability.[2] The additional hypotheses needed to uniquely require Bayesian updating have been deemed to be substantial, complicated, and unsatisfactory.[3]

77.2 Formal description of Bayesian inference

77.2.1 Definitions

_ x , a data point in general. This may in fact be a vector of values.
_ _ , the parameter of the data point's distribution, i.e., x _ $p(x j$ _) . This may in fact be a vector of parameters.
_ _ , the hyperparameter of the parameter, i.e., _ _ $p($_ j _) . This may in fact be a vector of hyperparameters.
_ **X** , a set of n observed data points, i.e., $x_1; : : : ; x_n$.
_ ~x , a new data point whose distribution is to be predicted.

77.2.2 Bayesian inference

_ The prior distribution is the distribution of the parameter(s) before any data is observed, i.e. $p($_ j _) .
_ The prior distribution might not be easily determined. In this case, we can use the Jeffreys prior to obtain the posterior distribution before updating them with newer observations.
_ The sampling distribution is the distribution of the observed data conditional on its parameters, i.e. $p($**X** j _) . This is also termed the likelihood, especially when viewed as a function of the parameter(s), sometimes written L(_ ; **X**) = $p($**X** j _) .
_ The marginal likelihood (sometimes also termed the *evidence*) is the distribution of the observed data marginalized over the parameter(s), i.e. $p($**X** j _) =
∫
_ $p($**X** j _)$p($_ j _) d_ .
_ The posterior distribution is the distribution of the parameter(s) after taking into account the observed data. This is determined by Bayes' rule, which forms the heart of Bayesian inference:
$p($_ j **X**; _) =
$p($**X** j _)$p($_ j _)
$p($**X** j _)
/ $p($**X** j _)$p($_ j _)
Note that this is expressed in words as "posterior is proportional to likelihood times prior", or sometimes as "posterior = likelihood times prior, over evidence".

77.2.3 Bayesian prediction

_ The posterior predictive distribution is the distribution of a new data point, marginalized over the posterior:
$p($~x j **X**; _) =
∫
_
$p($~x j _)$p($_ j **X**; _) d_
_ The prior predictive distribution is the distribution of a new data point, marginalized over the prior:

$p($~x j _) =
∫
_
$p($~x j _)$p($_ j _) d_
Bayesian theory calls for the use of the posterior predictive distribution to do predictive inference, i.e., to predict the distribution of a new, unobserved data point. That is, instead of a fixed point as a prediction, a distribution over possible points is returned. Only this way is the entire posterior distribution of the parameter(s) used. By comparison, prediction in frequentist statistics often involves finding an optimum point estimate of the parameter(s)—e.g., by maximum likelihood or maximum a posteriori estimation (MAP)—and then plugging this estimate into the formula for the distribution of a data point. This has the disadvantage that it does not account for any uncertainty in the value of the parameter, and hence will underestimate the variance of the predictive distribution.
(In some instances, frequentist statistics can work around this problem. For example, confidence intervals and prediction intervals in frequentist statistics when constructed from a normal distribution with unknown mean and variance are constructed using a Student's t-distribution. This correctly estimates the variance, due to the fact that (1) the average of normally distributed random variables is also normally distributed; (2) the predictive distribution of a normally distributed data point with unknown mean and variance, using conjugate or uninformative priors, has a student's t-distribution. In Bayesian statistics, however, the posterior predictive distribution can always be determined exactly—or at least, to an arbitrary level of precision, when numerical methods are used.)
Note that both types of predictive distributions have the form of a compound probability distribution (as does the marginal likelihood). In fact, if the prior distribution is a conjugate prior, and hence the prior and posterior distributions come from the same family, it can easily be seen that both prior and posterior predictive distributions also come from the same family of compound distributions. The only difference is that the posterior predictive distribution uses the updated values of the hyperparameters (applying the Bayesian update rules given in the conjugate prior article), while the prior predictive distribution uses the values of the hyperparameters that appear in the prior distribution.

77.3 Inference over exclusive and exhaustive possibilities

If evidence is simultaneously used to update belief over a set of exclusive and exhaustive propositions, Bayesian inference may be thought of as acting on this belief distribution as a whole.

77.3.1 General formulation

Suppose a process is generating independent and identically distributed events E_n, but the probability distribution is unknown. Let the event space Ω represent the current state of belief for this process. Each model is represented by event M_m. The conditional probabilities $P(E_n j\ M_m)$ are specified to define the models. $P(M_m)$ is the degree of belief in M_m. Before the first inference step, $fP(M_m)g$ is a set of *initial prior probabilities*. These must sum to 1, but are otherwise arbitrary.
Suppose that the process is observed to generate $E\ 2\ fE_ng$. For each $M\ 2\ fM_mg$, the prior $P(M)$ is updated to the posterior $P(M j\ E)$. From Bayes' theorem:[4]

$P(M j\ E) =$
$P(E j\ \Sigma\ M)$
$_m\ P(E j\ M_m)P(M_m)$
$\_\ P(M)$

Upon observation of further evidence, this procedure may be repeated.

77.3.2 Multiple observations

For a set of independent and identically distributed observations $\mathbf{E} = fe_1; : : : ; e_ng$, it may be shown that repeated application of the above is equivalent to

$P(M j\ \mathbf{E}) =$
$\Sigma\ P(\mathbf{E} j\ M)$
$_m\ P(\mathbf{E} j\ M_m)P(M_m)$
$\_\ P(M)$

Where

P(M) ? P(M) ?
P(M) ?
1 2
3
P(E | M1)
P(E | M1)
P(E | M1)
1
2
3
P(E | M2)
P(E | M2)
P(E | M2)
1
2
3
Ω

• • •

• • • • • •

P(E | M3)

P(E | M3)

P(E | M3)

1

2

3

• • •

Diagram illustrating event space Ω in general formulation of Bayesian inference. Although this diagram shows discrete models and events, the continuous case may be visualized similarly using probability densities.

P(**E** j M) =
Π
k
P(ek j M):

This may be used to optimize practical calculations.

77.3.3 Parametric formulation

By parametrizing the space of models, the belief in all models may be updated in a single step. The distribution of belief over the model space may then be thought of as a distribution of belief over the parameter space. The distributions in this section are expressed as continuous, represented by probability densities, as this is the usual situation. The technique is however equally applicable to discrete distributions.

Let the vector _ span the parameter space. Let the initial prior distribution over _ be p(_ j _) , where _ is a set of parameters to the prior itself, or *hyperparameters*. Let **E** = fe1; : : : ; eng be a set of independent and identically distributed event observations, where all ei are distributed as p(e j _) for some _ . Bayes' theorem is applied to find the posterior distribution over _ :

p(_ j **E**; _) =
p(**E** j _; _)
p(**E** j _)
_ p(_ j _)
=
p(**E** j ∫ _; _)
_ p(**E**j _; _)p(_ j _) d_
_ p(_ j _)

Where

p(**E** j _; _) =
Π
k
p(ek j _)

77.4 Mathematical properties

77.4.1 Interpretation of factor

P(EjM)
P(E) > 1) P(E j M) > P(E) . That is, if the model were true, the evidence would be more likely than is predicted by the current state of belief. The reverse applies for a decrease in belief. If the belief does not change,
P(EjM)
P(E) = 1) P(E j M) = P(E) . That is, the evidence is independent of the model. If the model were true, the evidence would be exactly as likely as predicted by the current state of belief.

77.4.2 Cromwell's rule

Main article: Cromwell's rule

If $P(M) = 0$ then $P(M j E) = 0$. If $P(M) = 1$, then $P(MjE) = 1$. This can be interpreted to mean that hard convictions are insensitive to counter-evidence.

The former follows directly from Bayes' theorem. The latter can be derived by applying the first rule to the event "not M " in place of " M ", yielding "if 1 □ $P(M) = 0$, then 1 □ $P(M j E) = 0$ ", from which the result immediately follows.

77.4.3 Asymptotic behaviour of posterior

Consider the behaviour of a belief distribution as it is updated a large number of times with independent and identically distributed trials. For sufficiently nice prior probabilities, the Bernstein-von Mises theorem gives that in the limit of infinite trials, the posterior converges to a Gaussian distribution independent of the initial prior under some conditions firstly outlined and rigorously proven by Joseph L. Doob in 1948, namely if the random variable in consideration has a finite probability space. The more general results were obtained later by the statistician David A. Freedman who published in two seminal research papers in 1963 and 1965 when and under what circumstances the asymptotic behaviour of posterior is guaranteed. His 1963 paper treats, like Doob (1949), the finite case and comes to a satisfactory conclusion. However, if the random variable has an infinite but countable probability space (i.e., corresponding to a die with infinite many faces) the 1965 paper demonstrates that for a dense subset of priors the Bernstein-von Mises theorem is not applicable. In this case there is almost surely no asymptotic convergence. Later in the 1980s and 1990s Freedman and Persi Diaconis continued to work on the case of infinite countable probability spaces.[5] To summarise, there may be insufficient trials to suppress the effects of the initial choice, and especially for large (but finite) systems the convergence might be very slow.

77.4.4 Conjugate priors

Main article: Conjugate prior

In parameterized form, the prior distribution is often assumed to come from a family of distributions called conjugate priors. The usefulness of a conjugate prior is that the corresponding posterior distribution will be in the same family, and the calculation may be expressed in closed form.

77.4.5 Estimates of parameters and predictions

It is often desired to use a posterior distribution to estimate a parameter or variable. Several methods of Bayesian estimation select measurements of central tendency from the posterior distribution.

For one-dimensional problems, a unique median exists for practical continuous problems. The posterior median is attractive as a robust estimator.[6]

If there exists a finite mean for the posterior distribution, then the posterior mean is a method of estimation.

~_ = E[_] =
∫

_p(_j **X**; _) d_

Taking a value with the greatest probability defines maximum *a posteriori* (MAP) estimates:

*f*_MAPg _ arg max _

p(_j **X**; _):

There are examples where no maximum is attained, in which case the set of MAP estimates is empty.

There are other methods of estimation that minimize the posterior *risk* (expected-posterior loss) with respect to a loss function, and these are of interest to statistical decision theory using the sampling distribution ("frequentist statistics").

The posterior predictive distribution of a new observation ~*x* (that is independent of previous observations) is determined by

p(~*xj***X**; _) =
∫

p(~*x*; _ *j* **X**; _) *d*_ =
∫

p(~*x j* _)*p*(_ *j* **X**; _) *d*_:

77.5 Examples

77.5.1 Probability of a hypothesis

Suppose there are two full bowls of cookies. Bowl #1 has 10 chocolate chip and 30 plain cookies, while bowl #2 has 20 of each. Our friend Fred picks a bowl at random, and then picks a cookie at random. We may assume there is no reason to believe Fred treats one bowl differently from another, likewise for the cookies. The cookie turns out to be a plain one. How probable is it that Fred picked it out of bowl #1?

Intuitively, it seems clear that the answer should be more than a half, since there are more plain cookies in bowl #1. The precise answer is given by Bayes' theorem. Let H_1 correspond to bowl #1, and H_2 to bowl #2. It is given that

the bowls are identical from Fred's point of view, thus $P(H_1) = P(H_2)$, and the two must add up to 1, so both are equal to 0.5. The event E is the observation of a plain cookie. From the contents of the bowls, we know that $P(E j H_1) = 30/40 = 0{:}75$ and $P(E j H_2) = 20/40 = 0{:}5$. Bayes' formula then yields
$P(H_1 j E) =$
$P(E j H_1) P(H_1)$
$P(E j H_1) P(H_1) + P(E j H_2) P(H_2)$
=
0:75 _ 0:5
0:75 _ 0:5 + 0:5 _ 0:5
= 0:6
Before we observed the cookie, the probability we assigned for Fred having chosen bowl #1 was the prior probability, $P(H_1)$, which was 0.5. After observing the cookie, we must revise the probability to $P(H_1 j E)$, which is 0.6.

77.5.2 Making a prediction

Example results for archaeology example. This simulation was generated using c=15.2.
An archaeologist is working at a site thought to be from the medieval period, between the 11th century to the 16th century. However, it is uncertain exactly when in this period the site was inhabited. Fragments of pottery are found, some of which are glazed and some of which are decorated. It is expected that if the site were inhabited during the early medieval period, then 1% of the pottery would be glazed and 50% of its area decorated, whereas if it had been inhabited in the late medieval period then 81% would be glazed and 5% of its area decorated. How confident can the archaeologist be in the date of inhabitation as fragments are unearthed?
The degree of belief in the continuous variable C (century) is to be calculated, with the discrete set of events
fGD;G_D; _ GD; _G
_D
g as evidence. Assuming linear variation of glaze and decoration with time, and that these variables are independent,
P(E = GD j C = c) = (0:01 + 0:16(c □ 11))(0:5 □ 0:09(c □ 11))
P(E = G_D j C = c) = (0:01 + 0:16(c □ 11))(0:5 + 0:09(c □ 11))
P(E = _ GD j C = c) = (0:99 □ 0:16(c □ 11))(0:5 □ 0:09(c □ 11))
P(E = _G
_D
j C = c) = (0:99 □ 0:16(c □ 11))(0:5 + 0:09(c □ 11))
Assume a uniform prior of fC(c) = 0:2 , and that trials are independent and identically distributed. When a new fragment of type e is discovered, Bayes' theorem is applied to update the degree of belief for each c :
fC(c j E = e) = P(E=ejC=c)
P(E=e) fC(c) = P(E=ej ∫ C=c) 16
11 P(E=ejC=c)fC(c)dc
fC(c)
A computer simulation of the changing belief as 50 fragments are unearthed is shown on the graph. In the simulation, the site was inhabited around 1420, or $c = 15{:}2$. By calculating the area under the relevant portion of the graph for 50 trials, the archaeologist can say that there is practically no chance the site was inhabited in the 11th and 12th centuries, about 1% chance that it was inhabited during the 13th century, 63% chance during the 14th century and 36% during the 15th century. Note that the Bernstein-von Mises theorem asserts here the asymptotic convergence to the "true" distribution because the probability space corresponding to the discrete set of events fGD;G_D ; _ GD; _G
_D
g
is finite (see above section on asymptotic behaviour of the posterior).

77.6 In frequentist statistics and decision theory

A decision-theoretic justification of the use of Bayesian inference was given by Abraham Wald, who proved that every Bayesian procedure is admissible. Conversely, every admissible statistical procedure is either a Bayesian procedure or a limit of Bayesian procedures.[7]
Wald characterized admissible procedures as Bayesian procedures (and limits of Bayesian procedures), making the Bayesian formalism a central technique in such areas of frequentist inference as parameter estimation, hypothesis testing, and computing confidence intervals.[8] For example:
_ "Under some conditions, all admissible procedures are either Bayes procedures or limits of Bayes procedures (in various senses). These remarkable results, at least in their original form, are due essentially to Wald. They are useful because the property of being Bayes is easier to analyze than admissibility."[7]
_ "In decision theory, a quite general method for proving admissibility consists in exhibiting a procedure as a unique Bayes solution."[9]

_ “In the first chapters of this work, prior distributions with finite support and the corresponding Bayes procedures were used to establish some of the main theorems relating to the comparison of experiments. Bayes procedures with respect to more general prior distributions have played a very important role in the development of statistics, including its asymptotic theory.” “There are many problems where a glance at posterior distributions, for suitable priors, yields immediately interesting information. Also, this technique can hardly be avoided in sequential analysis.”[10]

_ “A useful fact is that any Bayes decision rule obtained by taking a proper prior over the whole parameter space must be admissible”[11]

_ “An important area of investigation in the development of admissibility ideas has been that of conventional sampling-theory procedures, and many interesting results have been obtained.”[12]

77.6.1 Model selection

See Bayesian model selection

77.7 Applications

77.7.1 Computer applications

Bayesian inference has applications in artificial intelligence and expert systems. Bayesian inference techniques have been a fundamental part of computerized pattern recognition techniques since the late 1950s. There is also an ever growing connection between Bayesian methods and simulation-based Monte Carlo techniques since complex models cannot be processed in closed form by a Bayesian analysis, while a graphical model structure *may* allow for efficient simulation algorithms like the Gibbs sampling and other Metropolis–Hastings algorithm schemes.[13] Recently Bayesian inference has gained popularity amongst the phylogenetics community for these reasons; a number of applications allow many demographic and evolutionary parameters to be estimated simultaneously.

As applied to statistical classification, Bayesian inference has been used in recent years to develop algorithms for identifying e-mail spam. Applications which make use of Bayesian inference for spam filtering include CRM114, DSPAM, Bogofilter, SpamAssassin, SpamBayes, and Mozilla. Spam classification is treated in more detail in the article on the naive Bayes classifier.

Solomonoff’s Inductive inference is the theory of prediction based on observations; for example, predicting the next symbol based upon a given series of symbols. The only assumption is that the environment follows some unknown but computable probability distribution. It is a formal inductive framework that combines two well-studied principles of inductive inference: Bayesian statistics and Occam’s Razor.[14] Solomonoff’s universal prior probability of any prefix p of a computable sequence x is the sum of the probabilities of all programs (for a universal computer) that

compute something starting with p. Given some p and any computable but unknown probability distribution from which x is sampled, the universal prior and Bayes’ theorem can be used to predict the yet unseen parts of x in optimal fashion.[15][16]

77.7.2 In the courtroom

Bayesian inference can be used by jurors to coherently accumulate the evidence for and against a defendant, and to see whether, in totality, it meets their personal threshold for 'beyond a reasonable doubt'.[17][18][19] Bayes’ theorem is applied successively to all evidence presented, with the posterior from one stage becoming the prior for the next. The benefit of a Bayesian approach is that it gives the juror an unbiased, rational mechanism for combining evidence. It may be appropriate to explain Bayes’ theorem to jurors in odds form, as betting odds are more widely understood than probabilities. Alternatively, a logarithmic approach, replacing multiplication with addition, might be easier for a jury to handle.

Adding up evidence.

If the existence of the crime is not in doubt, only the identity of the culprit, it has been suggested that the prior should be uniform over the qualifying population.[20] For example, if 1,000 people could have committed the crime, the

prior probability of guilt would be 1/1000.

The use of Bayes’ theorem by jurors is controversial. In the United Kingdom, a defence expert witness explained Bayes’ theorem to the jury in *R v Adams*. The jury convicted, but the case went to appeal on the basis that no means of accumulating evidence had been provided for jurors who did not wish to use Bayes’ theorem. The Court of Appeal upheld the conviction, but it also gave the opinion that “To introduce Bayes’ Theorem, or any similar method, into a criminal trial plunges the jury into inappropriate and unnecessary realms of theory and complexity, deflecting them from their proper task.”

Gardner-Medwin[21] argues that the criterion on which a verdict in a criminal trial should be based is *not* the probability of guilt, but rather the *probability of the evidence, given that the defendant is innocent* (akin to a frequentist p-value). He argues that if the posterior probability of guilt is to be computed by Bayes’ theorem, the prior probability of guilt must be known. This will depend on the incidence of the crime, which is an unusual piece of evidence to consider in a criminal trial. Consider the following three propositions:

A The known facts and testimony could have arisen if the defendant is guilty

B The known facts and testimony could have arisen if the defendant is innocent
C The defendant is guilty.
Gardner-Medwin argues that the jury should believe both A and not-B in order to convict. A and not-B implies the truth of C, but the reverse is not true. It is possible that B and C are both true, but in this case he argues that a jury should acquit, even though they know that they will be letting some guilty people go free. See also Lindley's paradox.

77.7.3 Bayesian epistemology

Bayesian epistemology is an epistemological movement that uses techniques of Bayesian inference as a means of justifying the rules of inductive logic.
Karl Popper and David Miller have rejected the alleged rationality of Bayesianism, i.e. using Bayes rule to make epistemological inferences:[22] It is prone to the same vicious circle as any other justificationist epistemology, because it presupposes what it attempts to justify. According to this view, a rational interpretation of Bayesian inference would see it merely as a probabilistic version of falsification, rejecting the belief, commonly held by Bayesians, that high likelihood achieved by a series of Bayesian updates would prove the hypothesis beyond any reasonable doubt, or even with likelihood greater than 0.

77.7.4 Other

_ The scientific method is sometimes interpreted as an application of Bayesian inference. In this view, Bayes' rule guides (or should guide) the updating of probabilities about hypotheses conditional on new observations or experiments.[23]
_ Bayesian search theory is used to search for lost objects.
_ Bayesian inference in phylogeny
_ Bayesian tool for methylation analysis

77.8 Bayes and Bayesian inference

The problem considered by Bayes in Proposition 9 of his essay, "An Essay towards solving a Problem in the Doctrine of Chances", is the posterior distribution for the parameter *a* (the success rate) of the binomial distribution.

77.9 History

Main article: History of statistics § Bayesian statistics

The term *Bayesian* refers to Thomas Bayes (1702–1761), who proved a special case of what is now called Bayes' theorem. However, it was Pierre-Simon Laplace (1749–1827) who introduced a general version of the theorem and used it to approach problems in celestial mechanics, medical statistics, reliability, and jurisprudence.[24] Early Bayesian inference, which used uniform priors following Laplace's principle of insufficient reason, was called "inverse probability" (because it infers backwards from observations to parameters, or from effects to causes[25]). After the 1920s, "inverse probability" was largely supplanted by a collection of methods that came to be called frequentist statistics.[25]
In the 20th century, the ideas of Laplace were further developed in two different directions, giving rise to *objective* and *subjective* currents in Bayesian practice. In the objective or "non-informative" current, the statistical analysis depends on only the model assumed, the data analyzed,[26] and the method assigning the prior, which differs from one objective Bayesian to another objective Bayesian. In the subjective or "informative" current, the specification of the prior depends on the belief (that is, propositions on which the analysis is prepared to act), which can summarize information from experts, previous studies, etc.
In the 1980s, there was a dramatic growth in research and applications of Bayesian methods, mostly attributed to the discovery of Markov chain Monte Carlo methods, which removed many of the computational problems, and an increasing interest in nonstandard, complex applications.[27] Despite growth of Bayesian research, most undergraduate teaching is still based on frequentist statistics.[28] Nonetheless, Bayesian methods are widely accepted and used, such as for example in the field of machine learning.[29]

77.10 See also

_ Bayes' theorem
_ Bayesian hierarchical modeling
_ Bayesian Analysis, the journal of the ISBA
_ Inductive probability
_ International Society for Bayesian Analysis (ISBA)

77.11 Notes

[1] Hacking (1967, Section 3, p. 316), Hacking (1988, p. 124)
[2] "Bayes' Theorem (Stanford Encyclopedia of Philosophy)". Plato.stanford.edu. Retrieved 2014-01-05.
[3] van Fraassen, B. (1989) *Laws and Symmetry*, Oxford University Press. ISBN 0-19-824860-1
[4] Gelman, Andrew; Carlin, John B.; Stern, Hal S.; Dunson, David B.;Vehtari, Aki; Rubin, Donald B. (2013). *Bayesian Data Analysis*, Third Edition. Chapman and Hall/CRC. ISBN 978-1-4398-4095-5.
[5] Larry Wasserman et alia, JASA 2000.
[6] Sen, Pranab K.; Keating, J. P.; Mason, R. L. (1993). *Pitman's measure of closeness: A comparison of statistical estimators.*

Philadelphia: SIAM.
[7] Bickel & Doksum (2001, p. 32)
[8] _ Kiefer, J. and Schwartz, R. (1965). "Admissible Bayes Character of T_2-, R_2-, and Other Fully Invariant Tests for Multivariate Normal Problems". *Annals of Mathematical Statistics* **36**: 747–770. doi:10.1214/aoms/1177700051.
_ Schwartz, R. (1969). "Invariant Proper Bayes Tests for Exponential Families". *Annals of Mathematical Statistics* **40**: 270–283. doi:10.1214/aoms/1177697822.
_ Hwang, J. T. and Casella, George (1982). "Minimax Confidence Sets for the Mean of a Multivariate Normal Distribution". *Annals of Statistics* **10**: 868–881. doi:10.1214/aos/1176345877.
[9] Lehmann, Erich (1986). *Testing Statistical Hypotheses* (Second ed.). (see p. 309 of Chapter 6.7 "Admissibilty", and pp. 17–18 of Chapter 1.8 "Complete Classes"

[10] Le Cam, Lucien (1986). *Asymptotic Methods in Statistical Decision Theory*. Springer-Verlag. ISBN 0-387-96307-3. (From "Chapter 12 Posterior Distributions and Bayes Solutions", p. 324)
[11] Cox, D. R. and Hinkley, D.V (1974). *Theoretical Statistics*. Chapman and Hall. ISBN 0-04-121537-0. page 432
[12] Cox, D. R. and Hinkley, D. V. (1974). *Theoretical Statistics*. Chapman and Hall. ISBN 0-04-121537-0. p. 433)
[13] Jim Albert (2009). *Bayesian Computation with R, Second edition*. New York, Dordrecht, etc.: Springer. ISBN 978-0-387-92297-3.
[14] Samuel Rathmanner and Marcus Hutter. "A Philosophical Treatise of Universal Induction". *Entropy*, 13(6):1076–1136, 2011.
[15] "The Problem of Old Evidence", in §5 of "On Universal Prediction and Bayesian Confirmation", M. Hutter - Theoretical Computer Science, 2007 - Elsevier
[16] "Raymond J. Solomonoff", Peter Gacs, Paul M. B. Vitanyi, 2011 cs.bu.edu
[17] Dawid, A. P. and Mortera, J. (1996) "Coherent Analysis of Forensic Identification Evidence". *Journal of the Royal Statistical Society*, Series B, 58, 425–443.
[18] Foreman, L. A.; Smith, A. F. M., and Evett, I. W. (1997). "Bayesian analysis of deoxyribonucleic acid profiling data in forensic identification applications (with discussion)". *Journal of the Royal Statistical Society*, Series A, 160, 429–469.
[19] Robertson, B. and Vignaux, G. A. (1995) *Interpreting Evidence: Evaluating Forensic Science in the Courtroom*. John Wiley and Sons. Chichester. ISBN 978-0-471-96026-3
[20] Dawid, A. P. (2001) "Bayes' Theorem and Weighing Evidence by Juries"; http://128.40.111.250/evidence/content/dawid-paper.pdf
[21] Gardner-Medwin, A. (2005) "What Probability Should the Jury Address?". *Significance*, 2 (1), March 2005
[22] David Miller: *Critical Rationalism*
[23] Howson & Urbach (2005), Jaynes (2003)
[24] Stigler, Stephen M. (1986). "Chapter 3". *The History of Statistics*. Harvard University Press.
[25] Fienberg, Stephen E. (2006). "When did Bayesian Inference Become 'Bayesian'?". *Bayesian Analysis* **1** (1): 1–40 [p. 5]. doi:10.1214/06-ba101.
[26] Bernardo, José-Miguel (2005). "Reference analysis". *Handbook of statistics* **25**. pp. 17–90.
[27] Wolpert, R. L. (2004). "A Conversation with James O. Berger". *Statistical Science* **19** (1): 205–218. doi:10.1214/088342304000000053. MR 2082155.
[28] Bernardo, José M. (2006). "A Bayesian mathematical statistics primer". *ICOTS-7*.
[29] Bishop, C. M. (2007). *Pattern Recognition and Machine Learning*. New York: Springer. ISBN 0387310738.

77.12 References

_ Aster, Richard; Borchers, Brian, and Thurber, Clifford (2012). *Parameter Estimation and Inverse Problems*, Second Edition, Elsevier. ISBN 0123850487, ISBN 978-0123850485
_ Bickel, Peter J. and Doksum, Kjell A. (2001). *Mathematical Statistics, Volume 1: Basic and Selected Topics* (Second (updated printing 2007) ed.). Pearson Prentice–Hall. ISBN 0-13-850363-X.
_ Box, G. E. P. and Tiao, G. C. (1973) *Bayesian Inference in Statistical Analysis*, Wiley, ISBN 0-471-57428-7
_ Edwards, Ward (1968). "Conservatism in Human Information Processing". In Kleinmuntz, B. *Formal Representation of Human Judgment*. Wiley.
_ Edwards, Ward (1982). "Conservatism in Human Information Processing (excerpted)". In Daniel Kahneman, Paul Slovic and Amos Tversky. *Judgment under uncertainty: Heuristics and biases*. Cambridge University Press.

_ Jaynes E. T. (2003) *Probability Theory: The Logic of Science*, CUP. ISBN 978-0-521-59271-0 (Link to Fragmentary Edition of March 1996).
_ Howson, C. and Urbach, P. (2005). *Scientific Reasoning: the Bayesian Approach* (3rd ed.). Open Court Publishing Company. ISBN 978-0-8126-9578-6.
_ Phillips, L. D.; Edwards, Ward (October 2008). "Chapter 6: Conservatism in a Simple Probability Inference Task (*Journal of Experimental Psychology* (1966) 72: 346-354)". In Jie W. Weiss and David J. Weiss. *A Science of Decision Making:The Legacy of Ward Edwards*. Oxford University Press. p. 536. ISBN 978-0-19-532298-9.

77.13 Further reading

77.13.1 Elementary

The following books are listed in ascending order of probabilistic sophistication:
_ Stone, JV (2013), "Bayes' Rule: A Tutorial Introduction to Bayesian Analysis", Download first chapter here, Sebtel Press, England.
_ Colin Howson and Peter Urbach (2005). *Scientific Reasoning: The Bayesian Approach* (3rd ed.). Open Court Publishing Company. ISBN 978-0-8126-9578-6.
_ Berry, Donald A. (1996). *Statistics: A Bayesian Perspective*. Duxbury. ISBN 0-534-23476-3.
_ Morris H. DeGroot and Mark J. Schervish (2002). *Probability and Statistics* (third ed.). Addison-Wesley. ISBN 978-0-201-52488-8.
_ Bolstad, William M. (2007) *Introduction to Bayesian Statistics*: Second Edition, John Wiley ISBN 0-471-27020-2
_ Winkler, Robert L (2003). *Introduction to Bayesian Inference and Decision* (2nd ed.). Probabilistic. ISBN 0-9647938-4-9. Updated classic textbook. Bayesian theory clearly presented.
_ Lee, Peter M. *Bayesian Statistics: An Introduction*. Fourth Edition (2012), John Wiley ISBN 978-1-1183-3257-3
_ Carlin, Bradley P. and Louis, Thomas A. (2008). *Bayesian Methods for Data Analysis, Third Edition*. Boca Raton, FL: Chapman and Hall/CRC. ISBN 1-58488-697-8.
_ Gelman, Andrew; Carlin, John B.; Stern, Hal S.; Dunson, David B.; Vehtari, Aki; Rubin, Donald B. (2013). *Bayesian Data Analysis, Third Edition*. Chapman and Hall/CRC. ISBN 978-1-4398-4095-5.

77.13.2 Intermediate or advanced

_ Berger, James O (1985). *Statistical Decision Theory and Bayesian Analysis*. Springer Series in Statistics (Second ed.). Springer-Verlag. ISBN 0-387-96098-8.
_ Bernardo, José M.; Smith, Adrian F. M. (1994). *Bayesian Theory*. Wiley.
_ DeGroot, Morris H., *Optimal Statistical Decisions*. Wiley Classics Library. 2004. (Originally published (1970) by McGraw-Hill.) ISBN 0-471-68029-X.
_ Schervish, Mark J. (1995). *Theory of statistics*. Springer-Verlag. ISBN 0-387-94546-6.
_ Jaynes, E. T. (1998) *Probability Theory: The Logic of Science*.
_ O'Hagan, A. and Forster, J. (2003) *Kendall's Advanced Theory of Statistics*, Volume 2B: *Bayesian Inference*. Arnold, New York. ISBN 0-340-52922-9.
_ Robert, Christian P (2001). *The Bayesian Choice – A Decision-Theoretic Motivation* (second ed.). Springer. ISBN 0-387-94296-3.

_ Glenn Shafer and Pearl, Judea, eds. (1988) *Probabilistic Reasoning in Intelligent Systems*, San Mateo, CA: Morgan Kaufmann.
_ Pierre Bessière et al. (2013), "Bayesian Programming", CRC Press. ISBN 9781439880326

77.14 External links

_ Hazewinkel, Michiel, ed. (2001), "Bayesian approach to statistical problems", *Encyclopedia of Mathematics*, Springer, ISBN 978-1-55608-010-4
_ Bayesian Statistics from Scholarpedia.
_ Introduction to Bayesian probability from Queen Mary University of London
_ Mathematical Notes on Bayesian Statistics and Markov Chain Monte Carlo
_ Bayesian reading list, categorized and annotated by Tom Griffiths
_ A. Hajek and S. Hartmann: Bayesian Epistemology, in: J. Dancy et al. (eds.), A Companion to Epistemology. Oxford: Blackwell 2010, 93-106.
_ S. Hartmann and J. Sprenger: Bayesian Epistemology, in: S. Bernecker and D. Pritchard (eds.), Routledge Companion to Epistemology. London: Routledge 2010, 609-620.
_ *Stanford Encyclopedia of Philosophy*: "Inductive Logic"
_ Bayesian Confirmation Theory
_ What Is Bayesian Learning?

Chapter 78

Probability

For probability in mathematics, see Probability theory.

Probability is the measure of the likeliness that an event will occur.[1]

Probability is used to quantify an attitude of mind towards some proposition of whose truth we are not certain.[2] The proposition of interest is usually of the form "Will a specific event occur?" The attitude of mind is of the form "How certain are we that the event will occur?" The certainty we adopt can be described in terms of a numerical measure and this number, between 0 and 1 (where 0 indicates impossibility and 1 indicates certainty), we call probability.[3] Thus the higher the probability of an event, the more certain we are that the event will occur. A simple example would be the toss of a fair coin. Since the 2 outcomes are deemed equiprobable, the probability of "heads" equals the probability of "tails" and each probability is 1/2 or equivalently a 50% chance of either "heads" or "tails". These concepts have been given an axiomatic mathematical formalization in probability theory (see probability axioms), which is used widely in such areas of study as mathematics, statistics, finance, gambling, science (in particular physics), artificial intelligence/machine learning, computer science, and philosophy to, for example, draw inferences about the expected frequency of events. Probability theory is also used to describe the underlying mechanics and regularities of complex systems.[4]

78.1 Interpretations

Main article: Probability interpretations

When dealing with experiments that are random and well-defined in a purely theoretical setting (like tossing a fair coin), probabilities can be numerically described by the statistical number of outcomes considered favorable divided by the total number of all outcomes (tossing a fair coin twice will yield head-head with probability 1/4, because the four outcomes head-head, head-tails, tails-head and tails-tails are equally likely to occur). When it comes to practical application however there are two major competing categories of **probability interpretations**, whose adherents possess different views about the fundamental nature of probability:

1. Objectivists assign numbers to describe some objective or physical state of affairs. The most popular version of objective probability is frequentist probability, which claims that the probability of a random event denotes the *relative frequency of occurrence* of an experiment's outcome, when repeating the experiment. This interpretation considers probability to be the relative frequency "in the long run" of outcomes.[5] A modification of this is propensity probability, which interprets probability as the tendency of some experiment to yield a certain outcome, even if it is performed only once.
2. Subjectivists assign numbers per subjective probability, i.e., as a degree of belief.[6] The degree of belief has been interpreted as, "the price at which you would buy or sell a bet that pays 1 unit of utility if E, 0 if not E."[7] The most popular version of subjective probability is Bayesian probability, which includes expert knowledge as well as experimental data to produce probabilities. The expert knowledge is represented by some (subjective) prior probability distribution. The data is incorporated in a likelihood function. The product of the prior and

564

78.2. ETYMOLOGY 565

the likelihood, normalized, results in a posterior probability distribution that incorporates all the information known to date.[8] Starting from arbitrary, subjective probabilities for a group of agents, some Bayesians claim that all agents will eventually have sufficiently similar assessments of probabilities, given enough evidence (see Cromwell's rule).

78.2 Etymology

The word *probability* derives from the Latin *probabilitas*, which can also mean "probity", a measure of the authority of a witness in a legal case in Europe, and often correlated with the witness's nobility. In a sense, this differs much from the modern meaning of *probability*, which, in contrast, is a measure of the weight of empirical evidence, and is arrived at from inductive reasoning and statistical inference.[9]

78.3 History

Main article: History of probability

The scientific study of probability is a modern development. Gambling shows that there has been an interest in quantifying the ideas of probability for millennia, but exact mathematical descriptions arose much later. There are reasons of course, for the slow development of the mathematics of probability. Whereas games of chance provided the impetus for the mathematical study of probability, fundamental issues are still obscured by the superstitions of gamblers.[10]

According to Richard Jeffrey, "Before the middle of the seventeenth century, the term 'probable' (Latin *probabilis*) meant *approvable*, and was applied in that sense, univocally, to opinion and to action. A probable action or opinion was one such as sensible people would undertake or hold, in the circumstances."[11] However, in legal contexts especially, 'probable' could also apply to propositions for which there was good evidence.[12]

The sixteenth century polymath Gerolamo Cardano demonstrated the efficacy of defining odds as the ratio of favourable to unfavourable outcomes (which implies that the probability of an event is given by the ratio of favourable outcomes to the total number of possible outcomes [13]). Aside from the elementary work by Cardano, the doctrine of probabilities dates to the correspondence of Pierre de Fermat and Blaise Pascal (1654). Christiaan Huygens (1657) gave the earliest known scientific treatment of the subject.[14] Jakob Bernoulli's *Ars Conjectandi* (posthumous, 1713) and Abraham de Moivre's *Doctrine of Chances* (1718) treated the subject as a branch of mathematics.[15] See Ian Hacking's *The Emergence of Probability*[9] and James Franklin's *The Science of Conjecture* for histories of the early development of the very concept of mathematical probability.

The theory of errors may be traced back to Roger Cotes's *Opera Miscellanea* (posthumous, 1722), but a memoir prepared by Thomas Simpson in 1755 (printed 1756) first applied the theory to the discussion of errors of observation. The reprint (1757) of this memoir lays down the axioms that positive and negative errors are equally probable, and that certain assignable limits define the range of all errors. Simpson also discusses continuous errors and describes a probability curve.

The first two laws of error that were proposed both originated with Pierre-Simon Laplace. The first law was published in 1774 and stated that the frequency of an error could be expressed as an exponential function of the numerical magnitude of the error, disregarding sign. The second law of error was proposed in 1778 by Laplace and stated that the frequency of the error is an exponential function of the square of the error.[16] The second law of error is called the normal distribution or the Gauss law. "It is difficult historically to attribute that law to Gauss, who in spite of his well-known precocity had probably not made this discovery before he was two years old."[16]

Daniel Bernoulli (1778) introduced the principle of the maximum product of the probabilities of a system of concurrent errors.

Adrien-Marie Legendre (1805) developed the method of least squares, and introduced it in his *Nouvelles méthodes pour la détermination des orbites des comètes* (*New Methods for Determining the Orbits of Comets*). In ignorance of Legendre's contribution, an Irish-American writer, Robert Adrain, editor of "The Analyst" (1808), first deduced the law of facility of error,

566 *CHAPTER 78. PROBABILITY*

Christiaan Huygens probably published the first book on probability

$$\phi(x) = ce^{-h^2x^2};$$

where *h* is a constant depending on precision of observation, and *c* is a scale factor ensuring that the area under the curve equals 1. He gave two proofs, the second being essentially the same as John Herschel's (1850). Gauss gave

78.3. HISTORY 567

Carl Friedrich Gauss

the first proof that seems to have been known in Europe (the third after Adrain's) in 1809. Further proofs were given by Laplace (1810, 1812), Gauss (1823), James Ivory (1825, 1826), Hagen (1837), Friedrich Bessel (1838), W. F. Donkin (1844, 1856), and Morgan Crofton (1870). Other contributors were Ellis (1844), De Morgan (1864), Glaisher (1872), and Giovanni Schiaparelli (1875). Peters's (1856) formula for *r*, the probable error of a single observation, is well known.

In the nineteenth century authors on the general theory included Laplace, Sylvestre Lacroix (1816), Littrow (1833), Adolphe Quetelet (1853), Richard Dedekind (1860), Helmert (1872), Hermann Laurent (1873), Liagre, Didion, and Karl Pearson. Augustus De Morgan and George Boole improved the exposition of the theory.

Andrey Markov introduced the notion of Markov chains (1906), which played an important role in stochastic processes theory and its applications. The modern theory of probability based on the measure theory was developed by Andrey Kolmogorov (1931).

On the geometric side (see integral geometry) contributors to *The Educational Times* were influential (Miller, Crofton, McColl, Wolstenholme, Watson, and Artemas Martin).

568 *CHAPTER 78. PROBABILITY*

Further information: History of statistics

78.4 Theory

Main article: Probability theory

Like other theories, the theory of probability is a representation of probabilistic concepts in formal terms—that is, in terms that can be considered separately from their meaning. These formal terms are manipulated by the rules of mathematics and logic, and any results are interpreted or translated back into the problem domain.

There have been at least two successful attempts to formalize probability, namely the Kolmogorov formulation and the Cox formulation. In Kolmogorov's formulation (see probability space), sets are interpreted as events and probability itself as a measure on a class of sets. In Cox's theorem, probability is taken as a primitive (that is, not further analyzed) and the emphasis is on constructing a consistent assignment of probability values to propositions. In both cases, the laws of probability are the same, except for technical details.

There are other methods for quantifying uncertainty, such as the Dempster–Shafer theory or possibility theory, but those are essentially different and not compatible with the laws of probability as usually understood.

78.5 Applications

Probability theory is applied in everyday life in risk assessment and in trade on financial markets. Governments apply probabilistic methods in environmental regulation, where it is called pathway analysis. A good example is the effect of the perceived probability of any widespread Middle East conflict on oil prices—which have ripple effects in the economy as a whole. An assessment by a commodity trader that a war is more likely vs. less likely sends prices up or down, and signals other traders of that opinion. Accordingly, the probabilities are neither assessed independently nor necessarily very rationally. The theory of behavioral finance emerged to describe the effect of such groupthink on pricing, on policy, and on peace and conflict.[17]

The discovery of rigorous methods to assess and combine probability assessments has changed society. It is important for most citizens to understand how probability assessments are made, and how they contribute to decisions.

Another significant application of probability theory in everyday life is reliability. Many consumer products, such as automobiles and consumer electronics, use reliability theory in product design to reduce the probability of failure. Failure probability may influence a manufacturer's decisions on a product's warranty.[18]

The cache language model and other statistical language models that are used in natural language processing are also examples of applications of probability theory.

78.6 Mathematical treatment

See also: Probability axioms

Consider an experiment that can produce a number of results. The collection of all results is called the sample space of the experiment. The power set of the sample space is formed by considering all different collections of possible results. For example, rolling a dice can produce six possible results. One collection of possible results gives an odd number on the dice. Thus, the subset {1,3,5} is an element of the power set of the sample space of dice rolls. These collections are called "events." In this case, {1,3,5} is the event that the dice falls on some odd number. If the results that actually occur fall in a given event, the event is said to have occurred.

A probability is a way of assigning every event a value between zero and one, with the requirement that the event made up of all possible results (in our example, the event {1,2,3,4,5,6}) is assigned a value of one. To qualify as a probability, the assignment of values must satisfy the requirement that if you look at a collection of mutually exclusive events (events with no common results, e.g., the events {1,6}, {3}, and {2,4} are all mutually exclusive), the probability that at least one of the events will occur is given by the sum of the probabilities of all the individual

events.[19]

The probability of an event *A* is written as *P*(*A*), *p*(*A*) or Pr(*A*).[20] This mathematical definition of probability can extend to infinite sample spaces, and even uncountable sample spaces, using the concept of a measure.

The *opposite* or *complement* of an event *A* is the event [not *A*] (that is, the event of *A* not occurring); its probability is given by *P*(not *A*) = 1 − *P*(*A*).[21] As an example, the chance of not rolling a six on a six-sided die is 1 – (chance of rolling a six) = 1 □ 1

6 = 5

6 . See Complementary event for a more complete treatment.

If two events *A* and *B* occur on a single performance of an experiment, this is called the intersection or joint probability of *A* and *B*, denoted as *P*(*A* \ *B*) .

78.6.1 Independent events

If two events, *A* and *B* are independent then the joint probability is

P(*A* and *B*) = *P*(*A* \ *B*) = *P*(*A*)*P*(*B*);

for example, if two coins are flipped the chance of both being heads is 1

2

_ 1

2 = 1

4 : [22]

78.6.2 Mutually exclusive events

If either event *A* or event *B* or both events occur on a single performance of an experiment this is called the union of the events *A* and *B* denoted as *P*(*A* [*B*) . If two events are mutually exclusive then the probability of either occurring is

P(*A* or *B*) = *P*(*A* [*B*) = *P*(*A*) + *P*(*B*):

For example, the chance of rolling a 1 or 2 on a six-sided die is *P*(1 or 2) = *P*(1) + *P*(2) = 1

6 + 1

6 = 1

3 :

78.6.3 Not mutually exclusive events

If the events are not mutually exclusive then

P (*A* or *B*) = *P* (*A*) + *P* (*B*) □ *P* (*A* and *B*) :

For example, when drawing a single card at random from a regular deck of cards, the chance of getting a heart or a face card (J,Q,K) (or one that is both) is 13

52 + 12

52

□ 3

52 = 11

26 , because of the 52 cards of a deck 13 are hearts, 12 are face cards, and 3 are both: here the possibilities included in the "3 that are both" are included in each of the "13 hearts" and the "12 face cards" but should only be counted once.

78.6.4 Conditional probability

Conditional probability is the probability of some event *A*, given the occurrence of some other event *B*. Conditional probability is written *P(A j B)* , and is read "the probability of *A*, given *B*". It is defined by[23]

P(A j B) =
P(A \ B)
P(B)
:

If *P(B)* = 0 then *P(A j B)* is formally undefined by this expression. However, it is possible to define a conditional probability for some zero-probability events using a σ-algebra of such events (such as those arising from a continuous random variable).

For example, in a bag of 2 red balls and 2 blue balls (4 balls in total), the probability of taking a red ball is 1/2 ; however, when taking a second ball, the probability of it being either a red ball or a blue ball depends on the ball previously taken, such as, if a red ball was taken, the probability of picking a red ball again would be 1/3 since only 1 red and 2 blue balls would have been remaining.

78.6.5 Inverse probability

In probability theory and applications, **Bayes' rule** relates the odds of eventA_1 to eventA_2 , before (prior to) and after (posterior to) conditioning on another event *B* . The odds on A_1 to event A_2 is simply the ratio of the probabilities of the two events. When arbitrarily many events *A* are of interest, not just two, the rule can be rephrased as **posterior is proportional to prior times likelihood**, *P(AjB) / P(A)P(BjA)* where the proportionality symbol means that the left hand side is proportional to (i.e., equals a constant times) the right hand side as *A* varies, for fixed or given *B* (Lee, 2012; Bertsch McGrayne, 2012). In this form it goes back to Laplace (1774) and to Cournot (1843); see Fienberg (2005). See Inverse probability and Bayes' rule.

78.6.6 Summary of probabilities

78.7 Relation to randomness

Main article: Randomness

In a deterministic universe, based on Newtonian concepts, there would be no probability if all conditions were known (Laplace's demon), (but there are situations in which sensitivity to initial conditions exceeds our ability to measure them, i.e. know them). In the case of a roulette wheel, if the force of the hand and the period of that force are known, the number on which the ball will stop would be a certainty (though as a practical matter, this would likely be true only of a roulette wheel that had not been exactly levelled — as Thomas A. Bass' Newtonian Casino revealed). Of course, this also assumes knowledge of inertia and friction of the wheel, weight, smoothness and roundness of the ball, variations in hand speed during the turning and so forth. A probabilistic description can thus be more useful than Newtonian mechanics for analyzing the pattern of outcomes of repeated rolls of a roulette wheel. Physicists face the same situation in kinetic theory of gases, where the system, while deterministic *in principle*, is so complex (with the number of molecules typically the order of magnitude of Avogadro constant $6.02 \cdot 10^{23}$) that only a statistical description of its properties is feasible.

Probability theory is required to describe quantum phenomena.[24] A revolutionary discovery of early 20th century physics was the random character of all physical processes that occur at sub-atomic scales and are governed by the laws of quantum mechanics. The objective wave function evolves deterministically but, according to the Copenhagen interpretation, it deals with probabilities of observing, the outcome being explained by a wave function collapse when an observation is made. However, the loss of determinism for the sake of instrumentalism did not meet with universal approval. Albert Einstein famously remarked in a letter to Max Born: "I am convinced that God does not play dice".[25] Like Einstein, Erwin Schrödinger, who discovered the wave function, believed quantum mechanics is a statistical approximation of an underlying deterministic reality.[26] In modern interpretations, quantum decoherence accounts for subjectively probabilistic behavior.

78.8 See also

Main article: Outline of probability

_ Chance (disambiguation)
_ Class membership probabilities
_ Equiprobability
_ Heuristics in judgment and decision-making
_ Probability theory
_ Statistics

78.9 Notes

[1] "Probability". *Webster's Revised Unabridged Dictionary*. G & C Merriam, 1913

[2] "Kendall's Advanced Theory of Statistics, Volume 1: Distribution Theory", Alan Stuart and Keith Ord, 6th Ed, (2009), ISBN 9780534243128

[3] William Feller, "An Introduction to Probability Theory and Its Applications", (Vol 1), 3rd Ed, (1968),Wiley ,ISBN 0-471-

25708-7
[4] Probability Theory The Britannica website
[5] Hacking, Ian (1965). *The Logic of Statistical Inference*. Cambridge University Press. ISBN 0-521-05165-7.
[6] Finetti, Bruno de (1970). "Logical foundations and measurement of subjective probability". *Acta Psychologica* **34**: 129–145. doi:10.1016/0001-6918(70)90012-0.
[7] Hájek, Alan. "Interpretations of Probability". *The Stanford Encyclopedia of Philosophy (Winter 2012 Edition), Edward N. Zalta (ed.)*. Retrieved 22 April 2013.
[8] Hogg, Robert V.; Craig, Allen; McKean, Joseph W. (2004). *Introduction to Mathematical Statistics* (6th ed.). Upper Saddle River: Pearson. ISBN 0-13-008507-3.
[9] Hacking, I. (2006) *The Emergence of Probability: A Philosophical Study of Early Ideas about Probability, Induction and Statistical Inference*, Cambridge University Press, ISBN 978-0-521-68557-3
[10] Freund, John. (1973) *Introduction to Probability*. Dickenson ISBN 978-0822100782 (p. 1)
[11] Jeffrey, R.C., *Probability and the Art of Judgment,* Cambridge University Press. (1992). pp. 54-55 . ISBN 0-521-39459-7
[12] Franklin, J. (2001) *The Science of Conjecture: Evidence and Probability Before Pascal,* Johns Hopkins University Press. (pp. 22, 113, 127)
[13] *Some laws and problems in classical probability and how Cardano anticipated them* Gorrochum, P. *Chance* magazine 2012
[14] Abrams, William, *A Brief History of Probability*, Second Moment, retrieved 2008-05-23
[15] Ivancevic, Vladimir G.; Ivancevic, Tijana T. (2008). *Quantum leap : from Dirac and Feynman, across the universe, to human body and mind.* Singapore ; Hackensack, NJ: World Scientific. p. 16. ISBN 978-981-281-927-7.
[16] Wilson EB (1923) "First and second laws of error". Journal of the American Statistical Association, 18, 143
[17] Singh, Laurie (2010) "Whither Efficient Markets? Efficient Market Theory and Behavioral Finance". The Finance Professionals' Post, 2010.
[18] Gorman, Michael (2011) "Management Insights". *Management Science*
[19] Ross, Sheldon. *A First course in Probability*, 8th Edition. Page 26-27.
[20] Olofsson (2005) Page 8.
[21] Olofsson (2005), page 9
[22] Olofsson (2005) page 35.
[23] Olofsson (2005) page 29.
[24] Burgi, Mark (2010) "Interpretations of Negative Probabilities", p. 1. arXiv:1008.1287v1
[25] *Jedenfalls bin ich überzeugt, daß der Alte nicht würfelt.* Letter to Max Born, 4 December 1926, in: Einstein/Born Briefwechsel 1916-1955.
[26] Moore, W.J. (1992). *Schrödinger: Life and Thought*. Cambridge University Press. p. 479. ISBN 0-521-43767-9.

78.10 Bibliography

_ Kallenberg, O. (2005) *Probabilistic Symmetries and Invariance Principles*. Springer -Verlag, New York. 510 pp. ISBN 0-387-25115-4
_ Kallenberg, O. (2002) *Foundations of Modern Probability,* 2nd ed. Springer Series in Statistics. 650 pp. ISBN 0-387-95313-2
_ Olofsson, Peter (2005) *Probability, Statistics, and Stochastic Processes*, Wiley-Interscience. 504 pp ISBN 0-471-67969-0.

78.11 External links

_ Virtual Laboratories in Probability and Statistics (Univ. of Ala.-Huntsville)

_
_ Probability on *In Our Time* at the BBC. (/In_Our_Time_Probability listen now)
_ Probability and Statistics EBook
_ Edwin Thompson Jaynes. *Probability Theory: The Logic of Science*. Preprint: Washington University, (1996).
— HTML index with links to PostScript files and PDF (first three chapters)
_ People from the History of Probability and Statistics (Univ. of Southampton)
_ Probability and Statistics on the Earliest Uses Pages (Univ. of Southampton)
_ Earliest Uses of Symbols in Probability and Statistics on Earliest Uses of Various Mathematical Symbols
_ A tutorial on probability and Bayes' theorem devised for first-year Oxford University students
_ pdf file of An Anthology of Chance Operations (1963) at UbuWeb
_ Introduction to Probability - eBook, by Charles Grinstead, Laurie Snell Source *(GNU Free Documentation License)*
_ (English) (Italian) Bruno de Finetti, *Probabilità e induzione*, Bologna, CLUEB, 1993. ISBN 88-8091-176-7 (digital version)

Chapter 79

Mathematical physics

An example of mathematical physics: solutions of Schrödinger's equation for quantum harmonic oscillators (left) with their amplitudes (right).

Mathematical physics refers to development of mathematical methods for application to problems in physics. The *Journal of Mathematical Physics* defines the field as "the application of mathematics to problems in physics and the development of mathematical methods suitable for such applications and for the formulation of physical theories".[1]
573
574 *CHAPTER 79. MATHEMATICAL PHYSICS*

79.1 Scope

There are several distinct branches of mathematical physics, and these roughly correspond to particular historical periods.

79.1.1 Classical mechanics

Main articles: Lagrangian mechanics and Hamiltonian mechanics

The rigorous, abstract and advanced re-formulation of Newtonian mechanics adopting the Lagrangian mechanics and the Hamiltonian mechanics even in the presence of constraints. Both formulations are embodied in the so-called analytical mechanics. It leads, for instance, to discover the deep interplay of the notion of symmetry and that of conserved quantities during the dynamical evolution, stated within the most elementary formulation of Noether's theorem. These approaches and ideas can be and, in fact, have been extended to other areas of physics as statistical mechanics, continuum mechanics, classical field theory and quantum field theory. Moreover they have provided several examples and basic ideas in differential geometry (e.g. the theory of vector bundles and several notions in symplectic geometry).

79.1.2 Partial differential equations

Main article: Partial differential equations

The theory of partial differential equations (and the related areas of variational calculus, Fourier analysis, potential theory, and vector analysis) are perhaps most closely associated with mathematical physics. These were developed intensively from the second half of the eighteenth century (by, for example, D'Alembert, Euler, and Lagrange) until the 1930s. Physical applications of these developments include hydrodynamics, celestial mechanics, continuum mechanics, elasticity theory, acoustics, thermodynamics, electricity, magnetism, and aerodynamics.

79.1.3 Quantum theory

Main article: Quantum mechanics

The theory of atomic spectra (and, later, quantum mechanics) developed almost concurrently with the mathematical fields of linear algebra, the spectral theory of operators, operator algebras and more broadly, functional analysis. Nonrelativistic quantum mechanics includes Schrödinger operators, and it has connections to atomic and molecular physics. Quantum information theory is another subspecialty.

79.1.4 Relativity and Quantum Relativistic Theories

Main articles: Theory of relativity and Quantum field theory

The special and general theories of relativity require a rather different type of mathematics. This was group theory, which played an important role in both quantum field theory and differential geometry. This was, however, gradually supplemented by topology and functional analysis in the mathematical description of cosmological as well as quantum field theory phenomena. In this area both homological algebra and category theory are important nowadays.

79.1.5 Statistical mechanics

Main article: Statistical mechanics

Statistical mechanics forms a separate field, which includes the theory of phase transitions. It relies upon the Hamiltonian mechanics (or its quantum version) and it is closely related with the more mathematical ergodic theory and some parts
79.2. USAGE 575
of probability theory. There are increasing interactions between combinatorics and physics, in particular statistical physics.

79.2 Usage

The usage of the term "mathematical physics" is sometimes idiosyncratic. Certain parts of mathematics that initially arose from the development of physics are not, in fact, considered parts of mathematical physics, while other closely related fields are. For example, ordinary differential equations and symplectic geometry are generally viewed as purely mathematical disciplines, whereas dynamical systems and Hamiltonian mechanics belong to mathematical physics.

79.2.1 Mathematical vs. theoretical physics

The term "mathematical physics" is sometimes used to denote research aimed at studying and solving problems inspired by physics or thought experiments within a mathematically rigorous framework. In this sense, mathematical physics covers a very broad academic realm distinguished only by the blending of pure mathematics and physics. Although related to theoretical physics,[2] mathematical physics in this sense emphasizes the mathematical rigour of the same type as found in mathematics.

On the other hand, theoretical physics emphasizes the links to observations and experimental physics, which often requires theoretical physicists (and mathematical physicists in the more general sense) to use heuristic, intuitive, and approximate arguments.[3] Such arguments are not considered rigorous by mathematicians. Arguably, rigorous mathematical

physics is closer to mathematics, and theoretical physics is closer to physics. This is reflected institutionally: mathematical physicists are often members of the mathematics department.
Such mathematical physicists primarily expand and elucidate physical theories. Because of the required level of mathematical rigour, these researchers often deal with questions that theoretical physicists have considered to already be solved. However, they can sometimes show (but neither commonly nor easily) that the previous solution was incomplete, incorrect, or simply, too naive. Issues about attempts to infer the second law of thermodynamics

from statistical mechanics are examples. Other examples concerns all the subtleties involved with synchronisation procedures in special and general relativity (Sagnac effect and Einstein synchronisation)
The effort to put physical theories on a mathematically rigorous footing has inspired many mathematical developments. For example, the development of quantum mechanics and some aspects of functional analysis parallel each other in many ways. The mathematical study of quantum mechanics, quantum field theory and quantum statistical mechanics has motivated results in operator algebras. The attempt to construct a rigorous quantum field theory has also brought about progress in fields such as representation theory. Use of geometry and topology plays an important role in string theory.

79.3 Prominent mathematical physicists

79.3.1 Before Newton

The roots of mathematical physics can be traced back to the likes of Archimedes in Greece, Ptolemy in Egypt, Alhazen in Iraq, and Al-Biruni in Persia.
In the first decade of the 16th century, amateur astronomer Nicolaus Copernicus proposed heliocentrism, and published a treatise on it in 1543. Not quite radical, Copernicus merely sought to simplify astronomy and achieve orbits of more perfect circles, stated by Aristotelian physics to be the intrinsic motion of Aristotle's fifth element—the quintessence or universal essence known in Greek as *aither* for the English *pure air*—that was the pure substance beyond the sublunary sphere, and thus was celestial entities' pure composition. The German Johannes Kepler [1571–1630], Tycho Brahe's assistant, modified Copernican orbits to *ellipses*, however, formalized in the equations of Kepler's laws of planetary motion.
An enthusiastic atomist, Galileo Galilei in his 1623 book *The Assayer* asserted that the "book of nature" is written in mathematics.[4] His 1632 book, upon his telescopic observations, supported heliocentrism.[5] Having introducing experimentation, Galileo then refuted geocentric cosmology by refuting Aristotelian physics itself. Galilei's 1638 book *Discourse on Two New Sciences* established law of equal free fall as well as the principles of inertial motion, founding the central concepts of what would become today's classical mechanics.[5] By the Galilean law of inertia as well as the principle Galilean invariance, also called Galilean relativity, for any object experiencing inertia, there is empirical justification of knowing only its being at *relative* rest or *relative* motion—rest or motion with respect to another object.
René Descartes adopted Galilean principles and developed a complete system of heliocentric cosmology, anchored on the principle of vortex motion, Cartesian physics, whose widespread acceptance brought demise of Aristotelian physics. Descartes sought to formalize mathematical reasoning in science, and developed Cartesian coordinates for geometrically plotting locations in 3D space and marking their progressions along the flow of time.[6]

79.3.2 Newtonian and post Newtonian

Isaac Newton [1642–1727] developed new mathematics, including calculus and several numerical methods such as Newton's method to solve problems in physics. Newton's theory of motion, published in 1687, modeled three Galilean laws of motion along with Newton's law of universal gravitation on a framework of absolute space—hypothesized by Newton as a physically real entity of Euclidean geometric structure extending infinitely in all directions—while presuming absolute time, supposedly justifying knowledge of absolute motion, the object's motion with respect to absolute space. The principle Galilean invariance/relativity was merely implicit in Newton's theory of motion. Having ostensibly reduced Keplerian celestial laws of motion as well as Galilean terrestrial laws of motion to a unifying force, Newton achieved great mathematic rigor if theoretical laxity.[7]
In the 18th century, the Swiss Daniel Bernoulli [1700–1782] made contributions to fluid dynamics, and vibrating strings. The Swiss Leonhard Euler [1707–1783] did special work in variational calculus, dynamics, fluid dynamics, and other areas. Also notable was the Italian-born Frenchman, Joseph-Louis Lagrange [1736–1813] for work in analytical mechanics (he formulated the so-called Lagrangian mechanics) and variational methods. A major contribution to the formulation of Analytical Dynamics called Hamiltonian Dynamics was also made by the Irish physicist, astronomer and mathematician, William Rowan Hamilton [1805-1865]. Hamiltonian Dynamics had played an important role in the formulation of modern theories in physics including field theory and quantum mechanics. The French mathematical physicist Joseph Fourier [1768 – 1830] introduced the notion of Fourier series to solve the heat equation giving rise to a new approach to handle partial differential equations by means of integral transforms.

Into the early 19th century, the French Pierre-Simon Laplace [1749–1827] made paramount contributions to mathematical astronomy, potential theory, and probability theory. Siméon Denis Poisson [1781–1840] worked in analytical

mechanics and potential theory. In Germany, Carl Friedrich Gauss [1777–1855] made key contributions to the theoretical foundations of electricity, magnetism, mechanics, and fluid dynamics.
A couple of decades ahead of Newton's publication of a particle theory of light, the Dutch Christiaan Huygens [1629–1695] developed the wave theory of light, published in 1690. By 1804, Thomas Young's double-slit experiment revealed an interference pattern as though light were a wave, and thus Huygens's wave theory of light, as well as Huygens's inference that that light waves were vibrations of the luminiferous aether was accepted. Jean-Augustin Fresnel modeled hypothetical behavior of the aether. Michael Faraday introduced the theoretical concept of a field—not action at a distance. Mid-19th century, the Scottish James Clerk Maxwell [1831–1879] reduced electricity and magnetism to Maxwell's electromagnetic field theory, whittled down by others to the four Maxwell's equations. Initially, optics was found consequent of Maxwell's field. Later, radiation and then today's known electromagnetic spectrum were found also consequent of this electromagnetic field.
The English physicist Lord Rayleigh [1842–1919] worked on sound. The Irishmen William Rowan Hamilton [1805–1865], George Gabriel Stokes [1819–1903] and Lord Kelvin [1824–1907] did a lot of major work: Stokes was a leader in optics and fluid dynamics; Kelvin made substantial discoveries in thermodynamics; Hamilton did notable work on analytical mechanics finding out a new and powerful approach nowadays known as Hamiltonian mechanics. Very relevant contributions to this approach are due to his German colleague Carl Gustav Jacobi [1804–1851] in particular referring to the so-called canonical transformations. The German Hermann von Helmholtz [1821–1894] is greatly contributed to electromagnetism, waves, fluids, and sound. In the United States, the pioneering work of Josiah Willard Gibbs [1839–1903] became the basis for statistical mechanics. Fundamental theoretical results in this area were achieved by the German Ludwig Boltzmann [1844-1906]. Together, these individuals laid the foundations of electromagnetic theory, fluid dynamics, and statistical mechanics.

79.3.3 Relativistic

By the 1880s, prominent was the paradox that an observer within Maxwell's electromagnetic field measured it at approximately constant speed regardless of the observer's speed relative to other objects within the electromagnetic field. Thus, although the observer's speed was continually lost relative to the electromagnetic field, it was preserved relative to other objects *in* the electromagnetic field. And yet no violation of Galilean invariance within physical interactions among objects was detected. As Maxwell's electromagnetic field was modeled as oscillations of the aether, physicists inferred that motion within the aether resulted in aether drift, shifting the electromagnetic field, explaining the observer's missing speed relative to it. Physicists' mathematical process to translate the positions in one reference frame to predictions of positions in another reference frame, all plotted on Cartesian coordinates, had been the Galilean transformation, which was newly replaced with Lorentz transformation, modeled by the Dutch Hendrik Lorentz [1853–1928].
In 1887, experimentalists Michelson and Morley failed to detect aether drift, however. It was hypothesized that motion *into* the aether prompted aether's shortening, too, as modeled in the Lorentz contraction. Hypotheses at the aether thus kept Maxwell's electromagnetic field aligned with the principle Galilean invariance across all inertial frames of reference, while Newton's theory of motion was spared.
In the 19th century, Gauss's contributions to non-Euclidean geometry, or geometry on curved surfaces, laid the groundwork for the subsequent development of Riemannian geometry by Bernhard Riemann [1826–1866]. Austrian theoretical physicist and philosopher Ernst Mach criticized Newton's postulated absolute space. Mathematician Jules-Henri Poincaré [1854–1912] questioned even absolute time. In 1905, Pierre Duhem published a devastating criticism of the foundation of Newton's theory of motion.[7] Also in 1905, Albert Einstein [1879–1955] published special theory of relativity, newly explaining both the electromagnetic field's invariance and Galilean invariance by discarding all hypotheses at aether, including aether itself. Refuting the framework of Newton's theory—absolute space and absolute time—special relativity states *relative space* and *relative time*, whereby *length* contracts and *time* dilates along the travel pathway of an object experiencing kinetic energy.
In 1908, Einstein's former professor Hermann Minkowski modeled 3D space together with the 1D axis of time by treating the temporal axis like a fourth spatial dimension—altogether 4D spacetime—and declared the imminent demise of the separation of space and time. Einstein initially called this "superfluous learnedness", but later used Minkowski spacetime to great elegance in general theory of relativity,[8] extending invariance to all reference frames—whether perceived as inertial or as accelerated—and thanked Minkowski, by then deceased. General relativity replaces Cartesian coordinates with Gaussian coordinates, and replaces Newton's claimed empty yet Euclidean

space traversed instantly by Newton's vector of hypothetical gravitational force—an instant action at a distance—with a gravitational *field*. The gravitational field is Minkowski spacetime itself, the 4D topology of Einstein aether modeled on a Lorentzian manifold that "curves" geometrically, according to the Riemann curvature tensor, in the vicinity of either mass or energy. (By special relativity—a special case of general relativity—even massless energy exerts gravitational effect by its mass equivalence locally "curving" the geometry of the four, unified dimensions of space and time.)

79.3.4 Quantum

Another revolutionary development of the twentieth century has been quantum theory, which emerged from the seminal contributions of Max Planck [1856–1947] (on black body radiation) and Einstein's work on the photoelectric effect. This was, at first, followed by a heuristic framework devised by Arnold Sommerfeld [1868–1951] and Niels Bohr [1885–1962], but this was soon replaced by the quantum mechanics developed by Max Born [1882–1970], Werner Heisenberg [1901–1976], Paul Dirac [1902–1984], Erwin Schrödinger [1887–1961], Satyendra Nath Bose [1894 –1974], and Wolfgang Pauli [1900–1958]. This revolutionary theoretical framework is based on a probabilistic interpretation of states, and evolution and measurements in terms of self-adjoint operators on an infinite dimensional vector space. That is the so-called Hilbert space, introduced in its elementary form by David Hilbert [1862–1943] and Frigyes Riesz [1880-1956], and rigorously defined within the axiomatic modern version by John von Neumann in his celebrated book on mathematical foundations of quantum mechanics, where he built up a relevant part of modern functional analysis on Hilbert spaces, the spectral theory in particular. Paul Dirac used algebraic constructions to produce a relativistic model for the electron, predicting its magnetic moment and the existence of its antiparticle, the positron.

79.3.5 List of important mathematical physicists in the 20th century

Prominent contributors to the 20th century's mathematical physics (although the list contains some typical theoretical, not mathematical, physicists and leaves many, many contributors out) include (ordered by birth date) Arnold Sommerfeld [1868–1951], Albert Einstein [1879–1955], Max Born [1882–1970], Niels Bohr [1885–1962], Hermann Weyl [1885–1955], Satyendra Nath Bose [1894–1974], Wolfgang Pauli [1900–1958], Werner Heisenberg [1901–1976], Paul Dirac [1902–1984], Eugene Wigner [1902–1995], John von Neumann [1903–1957], Sin-Itiro Tomonaga [1906–1979], Hideki Yukawa [1907–1981], Lev Landau [1908-1968], Nikolay Bogolyubov [1909–1992], Mark Kac [1914–1984], Julian Schwinger [1918–1994], Richard Feynman [1918–1988], Arthur Strong Wightman [1922–2013], Chen-Ning Yang [1922–], Rudolf Haag [1922–], Freeman Dyson [1923–], Martin Gutzwiller [1925–2014], Abdus Salam [1926–1996], Jürgen Moser [1928–1999], Peter Higgs [1929–], Michael Atiyah [1929–], Joel Lebowitz [1930–], Roger Penrose [1931–], Elliott H. Lieb [1932–], Sheldon Lee Glashow [1932–], Steven Weinberg [1933–], Ludvig D. Faddeev [1934–], David Ruelle [1935–], Yakov G. Sinai [1935–], Vladimir Arnold [1937–2010], Arthur Jaffe [1937–], Roman Jackiw [1939–], Leonard Susskind [1940–], Rodney J. Baxter [1940–], Stephen Hawking [1942–], Alexander M. Polyakov [1945–], Barry Simon [1946–], John L. Cardy [1947–], Edward Witten [1951–], and Juan M. Maldacena [1968–].

79.4 See also

_ International Association of Mathematical Physics
_ Notable publications in mathematical physics

79.5 Notes

[1] Definition from the *Journal of Mathematical Physics*. http://jmp.aip.org/jmp/staff.jsp
[2] Quote: " ... a negative definition of the theorist refers to his inability to make physical experiments, while a positive one.. implies his encyclopaedic knowledge of physics combined with possessing enough mathematical armament. Depending on the ratio of these two components, the theorist may be nearer either to the experimentalist or to the mathematician. In the latter case, he is usually considered as a specialist in mathematical physics.", Ya. Frenkel, as related in A.T. Filippov, *The Versatile Soliton*, pg 131. Birkhauser, 2000.

[3] Quote: "Physical theory is something like a suit sewed for Nature. Good theory is like a good suit. ... Thus the theorist is like a tailor." Ya. Frenkel, as related in Filippov (2000), pg 131.
[4] Peter Machamer "Galileo Galilei"—sec 1 "Brief biography", in Zalta EN, ed, *The Stanford Encyclopedia of Philosophy*, Spring 2010 edn
[5] Antony G Flew, *Dictionary of Philosophy*, rev 2nd edn (New York: St Martin's Press, 1984), p 129
[6] Antony G Flew, *Dictionary of Philosophy*, rev 2nd edn (New York: St Martin's Press, 1984), p 89
[7] Imre Lakatos, auth, Worrall J & Currie G, eds, *The Methodology of Scientific Research Programmes: Volume 1: Philosophical Papers* (Cambridge: Cambridge University Press, 1980), pp 213-4, 220
[8] Salmon WC & Wolters G, eds, *Logic, Language, and the Structure of Scientific Theories* (Pittsburgh: University of Pittsburgh Press, 1994), p 125

79.6 References

_ Zalsow, Eric (2005), *Physmatics*, arXiv:physics/0506153, Bibcode:2005physics...6153Z

79.7 Further reading

79.7.1 The Classics

_ Abraham, Ralph; Marsden, Jerrold E. (2008), *'Foundations of mechanics: a mathematical exposition of classical mechanics with an introduction to the qualitative theory of dynamical systems'* (2nd ed.), Providence, [RI.]: AMS Chelsea Pub., ISBN 978-0-8218-4438-0
_ Arnold, Vladimir I.; Vogtmann, K.; Weinstein, A. (tr.) (1997), *'Mathematical methods of classical mechanics / [Matematicheskie metody klassicheskoĭ mekhaniki]'* (2nd ed.), New York, [NY.]: Springer-Verlag, ISBN 0-387-96890-3
_ Courant, Richard; Hilbert, David (1989), *Methods of mathematical physics*, New York, [NY.]: Interscience Publishers

_ Glimm, James; Jaffe, Arthur (1987), *'Quantum physics: a functional integral point of view'* (2nd ed.), New York, [NY.]: Springer-Verlag, ISBN 0-387-96477-0 (pbk.)
_ Haag, Rudolf (1996), *'Local quantum physics: fields, particles, algebras'* (2nd rev. & enl. ed.), Berlin, [Germany] ; New York, [NY.]: Springer-Verlag, ISBN 3-540-61049-9 (softcover)
_ Hawking, Stephen W.; Ellis, George F. R. (1973), *'The large scale structure of space-time'*, Cambridge, [England]: Cambridge University Press, ISBN 0-521-20016-4
_ Kato, Tosio (1995), *'Perturbation theory for linear operators'* (2nd repr. ed.), Berlin, [Germany]: Springer-Verlag, ISBN 3-540-58661-X (This is a reprint of the second (1980) edition of this title.)
_ Margenau, Henry; Murphy, George Moseley (1976), *'The mathematics of physics and chemistry'* (2nd repr. ed.), Huntington, [NY.]: R. E. Krieger Pub. Co., ISBN 0-88275-423-8 (This is a reprint of the 1956 second edition.)
_ Morse, Philip McCord; Feshbach, Herman (1999), *'Methods of theoretical physics'* (repr. ed.), Boston, [Mass.]: McGraw Hill, ISBN 0-07-043316-X (This is a reprint of the original (1953) edition of this title.)
_ von Neumann, John; Beyer, Robert T. (tr.) (1955), *'Mathematical foundations of quantum mechanics'*, Princeton, [NJ.]: Princeton University Press
_ Reed, Michael C.; Simon, Barry (1972–1977), *Methods of modern mathematical physics* **4**, New York City: Academic Press, ISBN 0-12-585001-8
_ Titchmarsh, Edward Charles (1939), *'The theory of functions'* (2nd ed.), London, [England]: Oxford University Press (This tome was reprinted in 1985.)
_ Thirring, Walter E.; Harrell, Evans M. (tr.) (1978–1983), *'A course in mathematical physics / [Lehrbuch der mathematischen Physik] (4 vol.)'*, New York, [NY.]: Springer-Verlag

_ Weyl, Hermann; Robertson, H. P. (tr.) (1931), *'The theory of groups and quantum mechanics / [Gruppentheorie und Quantenmechanik]'*, London, [England]: Methuen & Co.
_ Whittaker, Edmund Taylor; Watson, George Neville (1927), *'A course of modern analysis: an introduction to the general theory of infinite processes and of analytic functions, with an account of the principal transcendental functions'* (1st AMS ed.), Cambridge: Cambridge University Press, ISBN 978-0-521-58807-2

79.7.2 Textbooks for undergraduate studies

_ Arfken, George B.; Weber, Hans J. (1995), *'Mathematical methods for physicists'* (4th ed.), San Diego, [CA.]: Academic Press, ISBN 0-12-059816-7 (pbk.)
_ Boas, Mary L. (2006), *'Mathematical Methods in the Physical Sciences'* (3rd ed.), Hoboken, [NJ.]: John Wiley & Sons, ISBN 978-0-471-19826-0
_ Butkov, Eugene (1968), *'Mathematical physics'*, Reading, [Mass.]: Addison-Wesley
_ Jeffreys, Harold; Swirles Jeffreys, Bertha (1956), *'Methods of mathematical physics'* (3rd rev. ed.), Cambridge, [England]: Cambridge University Press
_ Kusse, Bruce R. (2006), *'Mathematical Physics: Applied Mathematics for Scientists and Engineers'* (2nd ed.), [Germany]: Wiley-VCH, ISBN 3-527-40672-7
_ Joos, Georg; Freeman, Ira M. (1987), *Theoretical Physics*, Dover Publications, ISBN 0-486-65227-0
_ Mathews, Jon; Walker, Robert L. (1970), *'Mathematical methods of physics'* (2nd ed.), New York, [NY.]: W. A. Benjamin, ISBN 0-8053-7002-1
_ Menzel, Donald Howard (1961), *Mathematical Physics*, Dover Publications, ISBN 0-486-60056-4
_ Stakgold, Ivar (c. 2000), *'Boundary value problems of mathematical physics (2 vol.)'*, Philadelphia, [PA.]: Society for Industrial and Applied Mathematics, ISBN 0-89871-456-7 (set : pbk.)

79.7.3 Textbooks for graduate studies

_ Hassani, Sadri (1999), *'Mathematical Physics: A Modern Introduction to Its Foundations'*, Berlin, [Germany]: Springer-Verlag, ISBN 0-387-98579-4
_ Reed, M.; Simon, B. (1972–1977). *Methods of Mathematical Physics*. Vol 1-4. Academic Press.
_ Teschl, G. (2009). *Mathematical Methods in Quantum Mechanics; With Applications to Schrödinger Operators*. Providence: American Mathematical Society. ISBN 978-0-8218-4660-5.
_ Moretti, V. (2013). *Spectral Theory and Quantum Mechanics; With an Introduction to the Algebraic Formulation*. Berlin, Milan: Springer. ISBN 978-88-470-2834-0.

79.7.4 Other specialised subareas

_ Aslam, Jamil; Hussain, Faheem (2007), *'Mathematical physics' Proceedings of the 12th Regional Conference, Islamabad, Pakistan, 27 March – 1 April 2006*, Singapore: World Scientific, ISBN 978-981-270-591-4
_ Baez, John C.; Muniain, Javier P. (1994), *'Gauge fields, knots, and gravity'*, Singapore ; River Edge, [NJ.]: World Scientific, ISBN 981-02-2034-0 (pbk.)

_ Geroch, Robert (1985), *'Mathematical physics'*, Chicago, [IL.]: University of Chicago Press, ISBN 0-226-28862-5 (pbk.)
_ Polyanin, Andrei D. (2002), *'Handbook of linear partial differential equations for engineers and scientists'*, Boca Raton, [FL.]: Chapman & Hall / CRC Press, ISBN 1-58488-299-9
_ Polyanin, Alexei D.; Zaitsev, Valentin F. (2004), *'Handbook of nonlinear partial differential equations'*, Boca Raton, [FL.]: Chapman & Hall / CRC Press, ISBN 1-58488-355-3
_ Szekeres, Peter (2004), *'A course in modern mathematical physics: groups, Hilbert space and differential geometry'*, Cambridge, [England]; New York, [NY.]: Cambridge University Press, ISBN 0-521-53645-6 (pbk.)

_ Yndurain, Francisco J (2006), *'Theoretical and Mathematical Physics. The Theory of Quark and Gluon Interactions'*, Berlin, [Germany]: Springer, ISBN 978-3642069741 (pbk.)

Chapter 80

Bayesian probability

Bayesian probability is one interpretation of the concept of probability. The Bayesian interpretation of probability can be seen as an extension of propositional logic that enables reasoning with hypotheses, i.e., the propositions whose truth or falsity is uncertain.

Bayesian probability belongs to the category of evidential probabilities; to evaluate the probability of a hypothesis, the Bayesian probabilist specifies some prior probability, which is then updated in the light of new, relevant data (evidence).[1] The Bayesian interpretation provides a standard set of procedures and formulae to perform this calculation.

In contrast to interpreting probability as the "frequency" or "propensity" of some phenomenon, Bayesian probability is a quantity that we assign for the purpose of representing a state of knowledge,[2] or a state of belief.[3] In the Bayesian view, a probability is assigned to a hypothesis, whereas under the frequentist view, a hypothesis is typically tested without being assigned a probability.

The term "Bayesian" refers to the 18th century mathematician and theologian Thomas Bayes, who provided the first mathematical treatment of a non-trivial problem of Bayesian inference.[4] Mathematician Pierre-Simon Laplace pioneered and popularised what is now called Bayesian probability.[5]

Broadly speaking, there are two views on Bayesian probability that interpret the *probability* concept in different ways. According to the *objectivist view*, the rules of Bayesian statistics can be justified by requirements of rationality and consistency and interpreted as an extension of logic.[2][6] According to the *subjectivist view*, probability quantifies a "personal belief".[3] Many modern machine learning methods are based on objectivist Bayesian principles.[7]

80.1 Bayesian methodology

Bayesian methods are characterized by the following concepts and procedures:

_ The use of random variables, or, more generally, unknown quantities,[8] to model all sources of uncertainty in statistical models. This also includes uncertainty resulting from lack of information (see also the aleatoric and epistemic uncertainty).
_ The need to determine the *prior probability distribution* taking into account the available (prior) information.
_ The *sequential use of the Bayes' formula*: when more data becomes available, calculate the *posterior distribution* using the Bayes' formula; subsequently, the posterior distribution becomes the next prior.
_ For the frequentist a hypothesis is a proposition (which must be either true or false), so that the frequentist probability of a hypothesis is either one or zero. In Bayesian statistics, a probability can be assigned to a hypothesis that can differ from 0 or 1 if the truth value is uncertain.

80.2 Objective and subjective Bayesian probabilities

Broadly speaking, there are two views on Bayesian probability that interpret the 'probability' concept in different ways. For **objectivists**, *probability* objectively measures the plausibility of propositions, i.e. the probability of a proposition corresponds to a reasonable belief everyone (even a "robot") sharing the same knowledge should share in accordance with the rules of Bayesian statistics, which can be justified by requirements of rationality and consistency.[2][6] For **subjectivists**, probability corresponds to a 'personal belief'.[3] For subjectivists, rationality and coherence constrain the probabilities a subject may have, but allow for substantial variation within those constraints. The objective and subjective variants of Bayesian probability differ mainly in their interpretation and construction of the prior probability.

80.3 History

Main article: History of statistics § Bayesian statistics

The term *Bayesian* refers to Thomas Bayes (1702–1761), who proved a special case of what is now called Bayes' theorem in a paper titled "An Essay towards solving a Problem in the Doctrine of Chances".[9] In that special case, the prior and posterior distributions were Beta distributions and the data came from Bernoulli trials. It was Pierre-Simon Laplace (1749–1827) who introduced a general version of the theorem and used it to approach problems in celestial mechanics, medical statistics, reliability, and jurisprudence.[10] Early Bayesian inference, which used uniform priors following Laplace's principle of insufficient reason, was called "inverse probability" (because it infers backwards from observations to parameters, or from effects to causes).[11] After the 1920s, "inverse probability" was largely supplanted by a collection of methods that came to be called frequentist statistics.[11]

In the 20th century, the ideas of Laplace were further developed in two different directions, giving rise to *objective* and *subjective* currents in Bayesian practice. In the objectivist stream, the statistical analysis depends on only the model assumed and the data analysed.[12] No subjective decisions need to be involved. In contrast, "subjectivist" statisticians deny the possibility of fully objective analysis for the general case.

In the 1980s, there was a dramatic growth in research and applications of Bayesian methods, mostly attributed to the discovery of Markov chain Monte Carlo methods, which removed many of the computational problems, and an increasing interest in nonstandard, complex applications.[13] Despite the growth of Bayesian research, most undergraduate teaching is still based on frequentist statistics.[14] Nonetheless, Bayesian methods are widely accepted and used, such as in the field of machine learning.[7]

80.4 Justification of Bayesian probabilities

The use of Bayesian probabilities as the basis of Bayesian inference has been supported by several arguments, such as the Cox axioms, the Dutch book argument, arguments based on decision theory and de Finetti's theorem.

80.4.1 Axiomatic approach

Richard T. Cox showed that[6] Bayesian updating follows from several axioms, including two functional equations and a controversial hypothesis of differentiability. It is known that Cox's 1961 development (mainly copied by Jaynes) is non-rigorous, and in fact a counterexample has been found by Halpern.[15] The assumption of differentiability or even continuity is questionable since the Boolean algebra of statements may only be finite.[8] Other axiomatizations have been suggested by various authors to make the theory more rigorous.[8]

80.4.2 Dutch book approach

The Dutch book argument was proposed by de Finetti, and is based on betting. A Dutch book is made when a clever gambler places a set of bets that guarantee a profit, no matter what the outcome of the bets. If a bookmaker follows the rules of the Bayesian calculus in the construction of his odds, a Dutch book cannot be made.

However, Ian Hacking noted that traditional Dutch book arguments did not specify Bayesian updating: they left open the possibility that non-Bayesian updating rules could avoid Dutch books. For example, Hacking writes[16] "And neither the Dutch book argument, nor any other in the personalist arsenal of proofs of the probability axioms, entails the dynamic assumption. Not one entails Bayesianism. So the personalist requires the dynamic assumption to be Bayesian. It is true that in consistency a personalist could abandon the Bayesian model of learning from experience. Salt could lose its savour."

In fact, there are non-Bayesian updating rules that also avoid Dutch books (as discussed in the literature on "probability kinematics" following the publication of Richard C. Jeffreys' rule, which is itself regarded as Bayesian [17]). The additional hypotheses sufficient to (uniquely) specify Bayesian updating are substantial, complicated, and unsatisfactory.[18]

80.4.3 Decision theory approach

A decision-theoretic justification of the use of Bayesian inference (and hence of Bayesian probabilities) was given by Abraham Wald, who proved that every admissible statistical procedure is either a Bayesian procedure or a limit of Bayesian procedures.[19] Conversely, every Bayesian procedure is admissible.[20]

80.5 Personal probabilities and objective methods for constructing priors

Following the work on expected utility theory of Ramsey and von Neumann, decision-theorists have accounted for rational behavior using a probability distribution for the agent. Johann Pfanzagl completed the *Theory of Games and Economic Behavior* by providing an axiomatization of subjective probability and utility, a task left uncompleted by von Neumann and Oskar Morgenstern: their original theory supposed that all the agents had the same probability distribution, as a convenience.[21] Pfanzagl's axiomatization was endorsed by Oskar Morgenstern: "Von Neumann and I have anticipated" the question whether probabilities "might, perhaps more typically, be subjective and have stated specifically that in the latter case axioms could be found from which could derive the desired numerical utility together with a number for the probabilities (cf. p. 19 of The Theory of Games and Economic Behavior). We did not carry this out; it was demonstrated by Pfanzagl ... with all the necessary rigor".[22]

Ramsey and Savage noted that the individual agent's probability distribution could be objectively studied in experiments. The role of judgment and disagreement in science has been recognized since Aristotle and even more clearly with Francis Bacon. The objectivity of science lies not in the psychology of individual scientists, but in the process of science and especially in statistical methods, as noted by C. S. Peirce.[23] Recall that the objective methods for falsifying propositions about personal probabilities have been used for a half century, as noted previously. Procedures for testing hypotheses about probabilities (using finite samples) are due to Ramsey (1931) and de Finetti (1931,

1937, 1964, 1970). Both Bruno de Finetti and Frank P. Ramsey acknowledge their debts to pragmatic philosophy, particularly (for Ramsey) to Charles S. Peirce.

The "Ramsey test" for evaluating probability distributions is implementable in theory, and has kept experimental psychologists occupied for a half century.[24] This work demonstrates that Bayesian-probability propositions can be falsified, and so meet an empirical criterion of Charles S. Peirce, whose work inspired Ramsey. (This falsifiabilitycriterion was popularized by Karl Popper.[25][26])

Modern work on the experimental evaluation of personal probabilities uses the randomization, blinding, and Booleandecision procedures of the Peirce-Jastrow experiment.[27] Since individuals act according to different probability judgments, these agents' probabilities are "personal" (but amenable to objective study).

Personal probabilities are problematic for science and for some applications where decision-makers lack the knowledge or time to specify an informed probability-distribution (on which they are prepared to act). To meet the needs of science and of human limitations, Bayesian statisticians have developed "objective" methods for specifying prior probabilities.

Indeed, some Bayesians have argued the prior state of knowledge defines *the* (unique) prior probability-distribution for "regular" statistical problems; cf. well-posed problems. Finding the right method for constructing such "objective" priors (for appropriate classes of regular problems) has been the quest of statistical theorists from Laplace to John Maynard Keynes, Harold Jeffreys, and Edwin Thompson Jaynes: These theorists and their successors have suggested several methods for constructing "objective" priors:

_ Maximum entropy

_ Transformation group analysis

_ Reference analysis

Each of these methods contributes useful priors for "regular" one-parameter problems, and each prior can handle some challenging statistical models (with "irregularity" or several parameters). Each of these methods has been useful in Bayesian practice. Indeed, methods for constructing "objective" (alternatively, "default" or "ignorance") priors have been developed by avowed subjective (or "personal") Bayesians like James Berger (Duke University) and José-Miguel Bernardo (Universitat de València), simply because such priors are needed for Bayesian practice, particularly in science.[28] The quest for "the universal method for constructing priors" continues to attract statistical theorists.[28]

Thus, the Bayesian statistician needs either to use informed priors (using relevant expertise or previous data) or to choose among the competing methods for constructing "objective" priors.

80.6 See also

_ Bertrand's paradox — a paradox in classical probability, solved by E.T. Jaynes in the context of Bayesian probability

_ De Finetti's game — a procedure for evaluating someone's subjective probability

_ QBism — a controversial application of Bayesian probabilities to quantum mechanics

_ Uncertainty

_ *An Essay towards solving a Problem in the Doctrine of Chances*

80.7 References

[1] Paulos, John Allen. *The Mathematics of Changing Your Mind,* New York Times (US). August 5, 2011; retrieved 2011-08-06

[2] Jaynes, E.T. "Bayesian Methods: General Background." In Maximum-Entropy and Bayesian Methods in Applied Statistics, by J. H. Justice (ed.). Cambridge: Cambridge Univ. Press, 1986

[3] de Finetti, B. (1974) *Theory of probability* (2 vols.), J. Wiley & Sons, Inc., New York

[4] Stigler, Stephen M. (1986) *The history of statistics.* Harvard University press. pg 131.

[5] Stigler, Stephen M. (1986) *The history of statistics.*, Harvard University press. pp97-98, 131.

[6] Cox, Richard T. *Algebra of Probable Inference*, The Johns Hopkins University Press, 2001

[7] Bishop, C.M. *Pattern Recognition and Machine Learning.* Springer, 2007

[8] Dupré, Maurice J., Tipler, Frank T. *New Axioms For Bayesian Probability*, Bayesian Analysis (2009), Number 3, pp. 599-606

[9] McGrayne, Sharon Bertsch. (2011). The Theory That Would Not Die, *p. 10.*, p. 10, at Google Books

[10] Stigler, Stephen M. (1986) *The history of statistics.* Harvard University press. Chapter 3.

[11] Fienberg, Stephen. E. (2006) *When did Bayesian Inference become "Bayesian"? Bayesian Analysis*, 1 (1), 1–40. See page 5.

[12] Bernardo, J.M. (2005), *Reference analysis, Handbook of statistics*, 25, 17–90

[13] Wolpert, R.L. (2004) *A conversation with James O. Berger*, Statistical science, 9, 205–218

[14] Bernardo, José M. (2006) *A Bayesian mathematical statistics primer.* ICOTS-7

[15] Halpern, J. *A counterexample to theorems of Cox and Fine*, Journal of Artificial Intelligence Research, 10: 67-85.

[16] Hacking (1967, Section 3, page 316), Hacking (1988, page 124)

[17] http://plato.stanford.edu/entries/bayes-theorem/

[18] van Frassen, B. (1989) *Laws and Symmetry*, Oxford University Press. ISBN 0-19-824860-1
[19] Wald, Abraham. *Statistical Decision Functions.* Wiley 1950.
[20] Bernardo, José M., Smith, Adrian F.M. *Bayesian Theory.* John Wiley 1994. ISBN 0-471-92416-4.
[21] Pfanzagl (1967, 1968)
[22] Morgenstern (1976, page 65)
[23] Stigler, Stephen M. (1978). "Mathematical statistics in the early States". *Annals of Statistics* **6** (March): 239–265 esp. p. 248. doi:10.1214/aos/1176344123. JSTOR 2958876. MR 483118.
[24] Davidson et al. (1957)
[25] "Karl Popper" in *Stanford Encyclopedia of Philosophy*
[26] Popper, Karl. (2002) *The Logic of Scientific Discovery* 2nd Edition, Routledge ISBN 0-415-27843-0 (Reprint of 1959 translation of 1935 original) Page 57.
[27] Peirce & Jastrow (1885)
[28] Bernardo, J. M. (2005). *Reference Analysis. Handbook of Statistics* 25 (D. K. Dey and C. R. Rao eds). Amsterdam: Elsevier, 17-90

80.8 Bibliography

_ Berger, James O. (1985). *Statistical Decision Theory and Bayesian Analysis*. Springer Series in Statistics (Second ed.). Springer-Verlag. ISBN 0-387-96098-8.
_ Bessière, Pierre; Mazer, E., Ahuacatzin, J-M, Mekhnacha, K. (2013). *Bayesian Programming.* CRC Press. ISBN 9781439880326.
_ Bernardo, José M.; Smith, Adrian F. M. (1994). *Bayesian Theory.* Wiley. ISBN 0-471-49464-X.
_ Bickel, Peter J.; Doksum, Kjell A. (2001). *Mathematical statistics, Volume 1: Basic and selected topics* (Second (updated printing 2007) of the Holden-Day 1976 ed.). Pearson Prentice–Hall. ISBN 0-13-850363-X. MR 443141.
_ Davidson, Donald; Suppes, Patrick; Siegel, Sidney (1957). *Decision-Making: An Experimental Approach.* Stanford University Press.
_ de Finetti, Bruno. "Probabilism: A Critical Essay on the Theory of Probability and on the Value of Science," (translation of 1931 article) in *Erkenntnis,* volume 31, September 1989.
_ de Finetti, Bruno (1937) "La Prévision: ses lois logiques, ses sources subjectives," Annales de l'Institut Henri Poincaré,
_ de Finetti, Bruno. "Foresight: its Logical Laws, Its Subjective Sources," (translation of the 1937 article in French) in H. E. Kyburg and H. E. Smokler (eds), *Studies in Subjective Probability,* New York: Wiley, 1964.
_ de Finetti, Bruno (1974–5). *Theory of Probability. A Critical Introductory Treatment*, (translation by A.Machi and AFM Smith of 1970 book) 2 volumes. Wiley ISBN 0-471-20141-3, ISBN 0-471-20142-1
_ DeGroot, Morris (2004) *Optimal Statistical Decisions.* Wiley Classics Library. (Originally published 1970.) ISBN 0-471-68029-X.

_ Hacking, Ian (December 1967). "Slightly More Realistic Personal Probability". *Philosophy of Science* **34** (4): 311–325. doi:10.1086/288169. JSTOR 186120. Partly reprinted in: Gärdenfors, Peter and Sahlin, Nils-Eric. (1988) *Decision, Probability, and Utility: Selected Readings*. 1988. Cambridge University Press. ISBN 0-521-33658-9
_ Hajek, A. and Hartmann, S. (2010): "Bayesian Epistemology", in: Dancy, J., Sosa, E., Steup, M. (Eds.) (2001) *A Companion to Epistemology*, Wiley. ISBN 1-4051-3900-5 Preprint
_ Hald, Anders (1998). *A History of Mathematical Statistics from 1750 to 1930*. New York: Wiley. ISBN 0-471-17912-4.
_ Hartmann, S. and Sprenger, J. (2011) "Bayesian Epistemology", in: Bernecker, S. and Pritchard, D. (Eds.) (2011) *Routledge Companion to Epistemology*. Routledge. ISBN 978-0-415-96219-3 (Preprint)
_ Hazewinkel, Michiel, ed. (2001), "Bayesian approach to statistical problems", *Encyclopedia of Mathematics*, Springer, ISBN 978-1-55608-010-4
_ Howson, C.; Urbach, P. (2005). *Scientific Reasoning: the Bayesian Approach* (3rd ed.). Open Court Publishing Company. ISBN 978-0-8126-9578-6.
_ Jaynes E.T. (2003) *Probability Theory: The Logic of Science*, CUP. ISBN 978-0-521-59271-0 (Link to Fragmentary Edition of March 1996).
_ McGrayne, SB. (2011). *The Theory That Would Not Die: How Bayes' Rule Cracked The Enigma Code, Hunted Down Russian Submarines, & Emerged Triumphant from Two Centuries of Controversy.* New Haven: Yale University Press. 13-ISBN 9780300169690/10-ISBN 0300169698; OCLC 670481486
_ Morgenstern, Oskar (1978). "Some Reflections on Utility". In Andrew Schotter. *Selected Economic Writings of Oskar Morgenstern.* New York University Press. pp. 65–70. ISBN 978-0-8147-7771-8.
_ Peirce, C.S. and Jastrow J. (1885). "On Small Differences in Sensation". *Memoirs of the National Academy of Sciences* **3**: 73–83.
_ Pfanzagl, J (1967). "Subjective Probability Derived from the Morgenstern-von Neumann Utility Theory". In Martin Shubik. *Essays in Mathematical Economics In Honor of Oskar Morgenstern.* Princeton University Press.

pp. 237–251.
_ Pfanzagl, J. in cooperation with V. Baumann and H. Huber (1968). "Events, Utility and Subjective Probability". *Theory of Measurement*. Wiley. pp. 195–220.
_ Ramsey, Frank Plumpton (1931) "Truth and Probability" (PDF), Chapter VII in *The Foundations of Mathematics and other Logical Essays*, Reprinted 2001, Routledge. ISBN 0-415-22546-9,
_ Stigler, SM. (1990). *The History of Statistics: The Measurement of Uncertainty before 1900*. Belknap Press/Harvard University Press. ISBN 0-674-40341-X.
_ Stigler, SM. (1999) *Statistics on the Table: The History of Statistical Concepts and Methods*. Harvard University Press. ISBN 0-674-83601-4
_ Stone, JV (2013). Download chapter 1 of book "Bayes' Rule: A Tutorial Introduction to Bayesian Analysis", Sebtel Press, England.
_ Winkler, RL (2003). *Introduction to Bayesian Inference and Decision* (2nd ed.). Probabilistic. ISBN 0-9647938-4-9. Updated classic textbook. Bayesian theory clearly presented.

Chapter 81

Psychologism

Psychologism is a generic type of position in philosophy according to which psychology plays a central role in grounding or explaining some other, non-psychological type of fact or law. "The view or doctrine that a theory of psychology or ideas forms the basis of an account of metaphysics, epistemology, or meaning; (sometimes) spec. the explanation or derivation of mathematical or logical laws in terms of psychological facts."[1] The most common types of psychologism are logical psychologism and mathematical psychologism.

Logical psychologism is a position in logic (or the philosophy of logic) according to which logical laws and mathematical laws are grounded in, derived from, explained or exhausted by psychological facts (or laws). Psychologism in the philosophy of mathematics is the position that mathematical concepts and/or truths are grounded in, derived from or explained by psychological facts (or laws).

John Stuart Mill seems to have been an advocate of a type of logical psychologism (although his rejection of a static ontology arguably makes his psychologism flexible enough to accommodate its detractors' criticisms), as were many nineteenth-century German logicians such as Sigwart and Erdmann as well as a number of psychologists, past and present: for example, Gustave Le Bon. Psychologism was famously criticized by Frege in his *The Foundations of Arithmetic*, and many of his works and essays, including his review of Husserl's *Philosophy of Arithmetic*. Edmund Husserl, in the first volume of his *Logical Investigations*, called "The Prolegomena of Pure Logic", criticized psychologism thoroughly and sought to distance himself from it. The "Prolegomena" is considered a more concise, fair, and thorough refutation of psychologism than the criticisms made by Frege, and also it is considered today by many as being a memorable refutation for its decisive blow to psychologism. Psychologism was also criticized by Charles Sanders Peirce and Maurice Merleau-Ponty.

In "Psychologism and Behaviorism," Ned Block takes psychologism as the position that "whether behavior is intelligent behavior depends on the character of the internal information processing that produces it." This is in contrast to a behavioral view which would state that intelligence can be ascribed to a being solely via observing its behavior. This type of behavioral view is strongly associated with the Turing test.

81.1 See also

_ Anti-psychologism
_ Naturalized epistemology
_ Blockhead

81.2 External links

_ Kusch, Martin (Nov 7, 2011). "Psychologism". The Stanford Encyclopedia of Philosophy. Retrieved May 18, 2012.
_ Husserl's Criticism of Psychologism. Link broken, page preserved most recently from October 22, 2009 at Internet Archive: Eprint. From *Diwatao*, (apparently former) online journal of the philosophy department of San Beda College, Manila, the Philippines.
588
81.2. EXTERNAL LINKS 589
_ The Turing test entry in the *Stanford Encyclopedia of Philosophy*
_ Block, Ned (1981), *Psychologism and Behaviorism*, *The Philosophical Review* (Duke University Press) **90** (1): 5–43, doi:10.2307/2184371, JSTOR 2184371.

Chapter 82

Language of thought hypothesis

In philosophy of mind, the **language of thought hypothesis** (**LOTH**) put forward by American philosopher Jerry Fodor describes thoughts as represented in a "language" (sometimes known as *mentalese*) that allows complex thoughts to be built up by combining simpler thoughts in various ways. In its most basic form the theory states that thought follows the same rules as language: thought has syntax.

Using empirical data drawn from linguistics and cognitive science to describe mental representation from a philosophical vantage-point, the hypothesis states that thinking takes place in a language of thought (LOT): cognition and cognitive processes are only 'remotely plausible' when expressed as a system of representations that is "tokened" by a linguistic or semantic structure and operated upon by means of a combinatorial syntax.[1] Linguistic tokens used in mental language describe elementary concepts which are operated upon by logical rules establishing causal connections to allow for complex thought. Syntax as well as semantics have a causal effect on the properties of this system of mental representations.

These mental representations are not present in the brain in the same way as symbols are present on paper; rather, the LOT is supposed to exist at the cognitive level, the level of thoughts and concepts. The LOTH has wide-ranging significance for a number of domains in cognitive science. It relies on a version of functionalist materialism, which holds that mental representations are actualized and modified by the individual holding the propositional attitude, and it challenges eliminative materialism and connectionism. It implies a strongly rationalist model of cognition in which many of the fundamentals of cognition are innate.

82.1 Presentation

The hypothesis applies to thoughts that have propositional content, and is not meant to describe everything that goes on in the mind. It appeals to the representational theory of thought to explain what those tokens actually are and how they behave. There must be a mental representation that stands in some unique relationship with the subject of the representation and has specific content. Complex thoughts get their semantic content from the content of the basic thoughts and the relations that they hold to each other. Thoughts can only relate to each other in ways that do not violate the syntax of thought. The syntax by means of which these two sub-parts are combined can be expressed in first-order predicate calculus.

The thought "John is tall" is clearly composed of two sub-parts, the concept of John and the concept of tallness, combined in a manner that may be expressed in first-order predicate calculus as a predicate 'T' ("is tall") that holds of the entity 'j' (John). A fully articulated proposal for what a LOT would have to take into account greater complexities such as quantification and propositional attitudes (the various attitudes people can have towards statements; for example I might *believe* or *see* or merely *suspect* that John is tall).

82.1.1 Precepts

1. There can be no higher cognitive processes without mental representation. The only plausible psychological models represent higher cognitive processes as representational and computational thought needs a representational system as an object upon which to compute. We must therefore attribute a representational system to organisms for cognition and thought to occur.

2. There is causal relationship between our intentions and our actions. Because mental states are structured in a way that causes our intentions to manifest themselves by what we do, there is a connection between how we view the world and ourselves and what we do.

82.2 Reception

Some philosophers have argued that our public language is our mental language, that a person who speaks English *thinks* in English. Others contend that people who do not know a public language (e.g. babies, aphasics) *can* think, and that therefore some form of mentalese must be present innately.

The notion that mental states are causally efficacious diverges from behaviorists like Gilbert Ryle, who held that there is no break between cause of mental state and effect of behavior. Rather, Ryle proposed that people act in some way because they are in a disposition to act in that way, that these causal mental states are representational. An objection to this point comes from John Searle in the form of biological naturalism, a nonrepresentational theory of mind that accepts the causal efficacy of mental states. Searle divides intentional states into low-level brain activity and high-level mental activity. The lower-level, nonrepresentational neurophysiological processes have causal power in intention and behavior rather than some higher-level mental representation.

Tim Crane, in his book *The Mechanical Mind*,[2] states that, while he agrees with Fodor, his reason is very different. A logical objection challenges LOTH's explanation of how sentences in natural languages get their meaning. That is the view that "Snow is white" is TRUE if and only if P is TRUE in the LOT, where P means the same thing in LOT as "Snow is white" means in the natural language. Any symbol manipulation is in need of some way of deriving what those symbols mean.[2] If the meaning of sentences is explained in terms of sentences in the LOT, then the meaning of sentences in LOT must get their meaning from somewhere else. There seems to be an infinite regress of sentences getting their meaning. Sentences in natural languages get their meaning from their users (speakers, writers).[2] Therefore sentences in mentalese must get their meaning from the way in which they are used by thinkers and so on *ad infinitum*. This regress is often called the homunculus regress.[2]

Daniel Dennett accepts that homunculi may be explained by other homunculi and denies that this would yield an infinite regress of homunculi. Each explanatory homunculus is "stupider" or more basic than the homunculus it explains but this regress is not infinite but bottoms out at a basic level that is so simple that it does not need interpretation.[2] John Searle points out that it still follows that the bottom-level homunculi are manipulating some sorts of symbols.

LOTH implies that the mind has some tacit knowledge of the logical rules of inference and the linguistic rules of syntax (sentence structure) and semantics (concept or word meaning).[2] If LOTH cannot show that the mind knows that it is following the particular set of rules in question then the mind is not computational because it is not governed by computational rules.[2][3] Also, the apparent incompleteness of this set of rules in explaining behavior is pointed out. Many conscious beings behave in ways that are contrary to the rules of logic. Yet this irrational behavior is not accounted for by any rules, showing that there is at least some behavior that does not act in accordance with this set of rules.[2]

Another objection within representational theory of mind has to do with the relationship between propositional attitudes and representation. Dennett points out that a chess program can have the attitude of "wanting to get its queen out early," without having a representation or rule that explicitly states this. A multiplication program on a computer computes in the computer language of 1's and 0's, yielding representations that do not correspond with any propositional attitude.[3]

Susan Schneider has recently developed a version of LOT that departs from Fodor's approach in numerous ways. She argues that Fodor's pessimism about the success of cognitive science is misguided, and she outlines an approach to LOT that integrates LOT with neuroscience. She also stresses that LOT that is not wedded to the extreme view that all concepts are innate. She fashions a new theory of mental symbols, and a related two-tiered theory concepts, in which a concept's nature is determined by their LOT symbol type and their meaning.[4]

82.3 Relation to connectionism

Connectionism is a recent applied approach to artificial intelligence that often accepts a lot of the same theoretical framework that LOTH accepts; that mental states are computational and causally efficacious and very often that they are representational. However, connectionism stresses the possibility of thinking machines, most often realized as neural networks, an inter-connectional set of nodes, and describes mental states as able to create memory by modifying

the strength of these connections over time. Some popular types of neural networks are interpretations of units, and learning algorithm. "Units" can be interpreted as neurons or groups of neurons. A learning algorithm is such that, over time, a change in connection weight is possible, allowing networks to modify their connections. Connectionist neural networks are able to change over time via their activation. An activation is a numerical value that represents any aspect of a unit that a neural network has at any time. Activation spreading is the spreading or taking over of other over time of the activation to all other units connected to the activated unit.

Since connectionist models can change over time, supporters of connectionism claim that it can solve the problems that LOTH brings to classical AI. These problems are those that show that machines with a LOT syntactical framework very often are much better at solving problems and storing data than human minds, yet much worse at things that the human mind is quite adept at such as recognizing facial expressions, objects in photographs and understanding nuanced gestures.[2] Fodor defends LOTH by arguing that a connectionist model is just some realization or implementation of the classical computational theory of mind and therein necessarily employs a symbol-manipulating LOT.

Fodor and Zenon Pylyshyn use the notion of cognitive architecture in their defense. Cognitive architecture is the set of basic functions of an organism with representational input and output. They argue that it is a law of nature that cognitive capacities are productive, systematic and inferentially coherent - they have the ability to produce and understand sentences of a certain structure if they can understand one sentence of that structure.[5] A cognitive model must have a cognitive architecture that explains these laws and properties in some way that is compatible with the scientific method. Fodor and Pylyshyn say that cognitive architecture can only explain the property of systematicity by appealing to a system of representations and that connectionism either employs a cognitive architecture of representations or else does not. If it does, then connectionism uses LOT. If it does not then it is empirically false.[3]

Connectionists have responded to Fodor and Pylyshyn by denying that connectionism uses LOT, by denying that cognition is essentially a function that uses representational input and output or denying that systematicity is a law of nature that rests on representation.

82.4 Empirical testing

Since LOTH came to be it has been empirically tested. Not all experiments have confirmed the hypothesis;

_ In 1971, Roger Shepard and Jacqueline Metzler tested Pylyshyn's particular hypothesis that all symbols are understood by the mind in virtue of their fundamental mathematical descriptions.[6] Shepard and Metzler's experiment consisted of showing a group of subjects a 2-D line drawing of a 3-D object, and then that same object at some rotation. According to Shepard and Metzler, if Pylyshyn were correct, then the amount of time it took to identify the object as the same object would not depend on the degree of rotation of the object. Their finding that the time taken to recognize the object was proportional to its rotation contradicts this hypothesis.

_ There may be a connection between prior knowledge of what relations hold between objects in the world and the time it takes subjects to recognize the same objects. For example, it is more likely that subjects will not recognize a hand that is rotated in such a way that it would be physically impossible for an actual hand.

It has since also been empirically tested and supported that the mind might better manipulate mathematical descriptions in topographical wholes. These findings have illuminated what the mind is not doing in terms of how it manipulates symbols.

82.5 See also

_ Private language argument
_ Universal grammar
_ Psycholinguistics
_ Psychological nativism

82.6 References

[1] Stanford Encyclopedia of Philosophy http://plato.stanford.edu/entries/language-thought/

[2] Crane, Tim (2005). *The mechanical mind : a philosophical introduction to minds, machines and mental representation* (2nd, repr. ed.). London: Routledge. ISBN 978-0-415-29031-9.

[3] Murat Aydede (2004-07-27). "The Language of Thought Hypothesis".

[4] Schneider, Susan (2011). *The Language of Thought: a New Direction.* Boston: Mass: MIT Press.

[5] James Garson (2010-07-27). "Connectionism".

[6] Shepard, Roger N.; Metzler, Jacqueline (1971-02-19). "Mental Rotation of Three-Dimensional Objects". *Science* **171** (3972): 701–703.

_ Ravenscroft, Ian, *Philosophy of mind.* Oxford University press, 2005. pp 91.
_ Fodor, Jerry A., *The Language Of Thought*. Crowell Press, 1975. pp 214.
_ John R. Searle (June 29, 1972). "Chomsky's Revolution in Linguistics". *New York Review of Books.*

82.7 External links

_ The Language of Thought Hypothesis at The Internet Encyclopedia of Philosophy.
_ The Language of Thought Hypothesis at The Stanford Encyclopedia of Philosophy.
_ Language of Thought - By Larry Kaye.
_ Revealing The Language Of Thought - By Brent Silby
_ Jerry Fodor Homepage
_ The Language Of Thought Hypothesis: State Of The Art - By Murat Aydede

Chapter 83

Science

This article is about the general term. For other uses, see Science (disambiguation).

Science (from Latin *scientia*, meaning "knowledge"[1]) is a systematic enterprise that builds and organizes knowledge in the form of testable explanations and predictions about the universe.[2][3][4] In an older and closely related meaning, "science" also refers to a body of knowledge itself, of the type that can be rationally explained and reliably applied. A practitioner of science is known as a scientist.

In classical antiquity, science as a type of knowledge was closely linked to philosophy. During the Islamic Golden Age, the foundation for the scientific method was laid, which emphasized experimental data and reproducibility of its results.[5][6][7][8][9][10][11][12] In the West during the early modern period the words "science" and "philosophy of nature" were sometimes used interchangeably,[13] And not until the 17th century, natural philosophy (which is today called "natural science") was considered a separate branch of philosophy in the West.[14]

In modern usage, "science" most often refers to a way of pursuing knowledge, not only the knowledge itself. It is also often restricted to those branches of study that seek to explain the phenomena of the material universe.[15] In the 17th and 18th centuries scientists increasingly sought to formulate knowledge in terms of *laws of nature* such as Newton's laws of motion. And over the course of the 19th century, the word "science" became increasingly associated with the scientific method itself, as a disciplined way to study the natural world, including physics, chemistry, geology and biology. It is in the 19th century also that the term *scientist* was created by the naturalist-theologian William Whewell to distinguish those who sought knowledge on nature from those who sought other types of knowledge.[16]

However, "science" has also continued to be used in a broad sense to denote reliable and teachable knowledge about a topic, as reflected in modern terms like library science or computer science. This is also reflected in the names of some areas of academic study such as "social science" or "political science".

83.1 History

Main article: History of science

Science in a broad sense existed before the modern era, and in many historical civilizations,[19] but modern science is so distinct in its approach and successful in its results that it now defines what science is in the strictest sense of the term.[3] Much earlier than the modern era, another important turning point was the development of classical natural philosophy in the ancient Greek-speaking world.

83.1.1 Pre-philosophical

Science in its original sense is a word for a type of knowledge (Latin *scientia*, Ancient Greek *epistemē*), rather than a specialized word for the pursuit of such knowledge. In particular it is one of the types of knowledge which people can communicate to each other and share. For example, knowledge about the working of natural things was gathered long before recorded history and led to the development of complex abstract thinking. This is shown by the construction of complex calendars, techniques for making poisonous plants edible, and buildings such as the pyramids. However no consistent conscientious distinction was made between knowledge of such things which are true in every community
594
83.1. HISTORY 595
The scale of the universe mapped to the branches of science and the hierarchy of science.[17]
Further information: Outline of science
and other types of communal knowledge, such as mythologies and legal systems.

83.1.2 Philosophical study of nature

See also: Nature (philosophy)

Before the invention or discovery of the concept of "nature" (Ancient Greek *phusis*), by the Pre-Socratic philosophers, the same words tend to be used to describe the *natural* "way" in which a plant grows,[20] and the "way" in which, for example, one tribe worships a particular god. For this reason it is claimed these men were the first philosophers in the strict sense, and also the first people to clearly distinguish "nature" and "convention".[21] Science was therefore distinguished as the knowledge of nature, and the things which are true for every community, and the name of the specialized pursuit of such knowledge was philosophy — the realm of the first philosopher-physicists. They were mainly speculators or theorists, particularly interested in astronomy. In contrast, trying to use knowledge of nature to imitate nature (artifice or technology, Greek *technē*) was seen by classical scientists as a more appropriate interest for lower class artisans.[22]

83.1.3 Philosophical turn to human things

A major turning point in the history of early philosophical science was the controversial but successful attempt by Socrates to apply philosophy to the study of human things, including human nature, the nature of political communities, and human knowledge itself. He criticized the older type of study of physics as too purely speculative, and lacking in self-criticism. He was particularly concerned that some of the early physicists treated nature as if it could be assumed that it had no intelligent order, explaining things merely in terms of motion and matter. The study of human things had been the realm of mythology and tradition, and Socrates was executed.[23] Aristotle later created
596 *CHAPTER 83. SCIENCE*
Both Aristotle and Kuan Tzu (4th century BCE), in an example of simultaneous scientific discovery, mention that some marine animals were subject to a lunar cycle, and increase and decrease in size with the waxing and waning of the moon. Aristotle was referring specifically to the sea urchin, pictured above.[18]
a less controversial systematic programme of Socratic philosophy, which was teleological, and human-centred. He rejected many of the conclusions of earlier scientists. For example in his physics the sun goes around the earth, and many things have it as part of their nature that they are for humans. Each thing has a formal cause and final cause and a role in the rational cosmic order. Motion and change is described as the actualization of potentials already in things, according to what types of things they are. While the Socratics insisted that philosophy should be used to consider the practical question of the best way to live for a human being (a study Aristotle divided into ethics and political philosophy), they did not argue for any other types of applied science.

Aristotle maintained the sharp distinction between science and the practical knowledge of artisans, treating theoretical speculation as the highest type of human activity, practical thinking about good living as something less lofty, and the knowledge of artisans as something only suitable for the lower classes. In contrast to modern science, Aristotle's influential emphasis was upon the "theoretical" steps of deducing universal rules from raw data, and did not treat the gathering of experience and raw data as part of science itself.[24]

83.1.4 Medieval science, and foundations for scientific method

During late antiquity and the early Middle Ages, the Aristotelian approach to inquiries on natural phenomenon was used. Some ancient knowledge was lost, or in some cases kept in obscurity, during the fall of the Roman Empire and periodic political struggles. However, the general fields of science, or natural philosophy as it was called, and much of the general knowledge from the ancient world remained preserved though the works of the early Latin encyclopedists
83.1. HISTORY 597
Ibn al-Haytham (Alhazen), 965–1039 Iraq. The Muslim scholar who is considered by some to be the first Scientist due to his emphasis on experimental data and reproducibility of its results.[6][7]
like Isidore of Seville. Also, in the Byzantine empire, many Greek science texts were preserved in Syriac translations done by groups such as Nestorians and Monophysites.[25] Many of these were translated later on into Arabic under

Caliphate, during which many types of classical learning were preserved and in some cases improved upon.[25] During the Islamic Golden Age, a Muslim scholar Ibn al-Haytham, who is considered by some to be the father of modern scientific method, argued for it by emphasizing experimental data and reproducibility of its results.[6][7] The House of Wisdom, considered to be the first university in the world, was established in Abbasid-era Baghdad, Iraq.[26] It is considered to have been a major intellectual center during the Islamic Golden Age. In the later medieval period, as science in Byzantium and the Islamic world waned, Western Europeans began collecting ancient texts from the Mediterranean, not only in Latin, but also in Greek, Arabic, and Hebrew. Knowledge of ancient researchers such as Aristotle, Ptolemy, Euclid, amongst Catholic scholars, were recovered with renewed interest in diverse aspects of

natural phenomenon. In Europe, men like Roger Bacon in England argued for more experimental science. By the late Middle Ages, a synthesis of Catholicism and Aristotelianism known as Scholasticism was flourishing in Western Europe, which had become a new geographic center of science.

83.1.5 Renaissance, and early modern science

Main article: Scientific revolution

By the late Middle Ages, especially in Italy there was an influx of Greek texts and scholars from the collapsing

Galileo is considered one of the fathers of modern science.[27]

Byzantine empire. Copernicus formulated a heliocentric model of the solar system unlike the geocentric model of Ptolemy's Almagest. All aspects of scholasticism were criticized in the 15th and 16th centuries; one author who was notoriously persecuted was Galileo, who made innovative use of experiment and mathematics. However the persecution began after Pope Urban VIII blessed Galileo to write about the Copernican system. Galileo had used arguments from the Pope and put them in the voice of the simpleton in the work "Dialogue Concerning the Two Chief World Systems" which caused great offense to him.[28]

In Northern Europe, the new technology of the printing press was widely used to publish many arguments including some that disagreed with church dogma. René Descartes and Francis Bacon published philosophical arguments in favor of a new type of non-Aristotelian science. Descartes argued that mathematics could be used in order to study nature, as Galileo had done, and Bacon emphasized the importance of experiment over contemplation. Bacon questioned the Aristotelian concepts of formal cause and final cause, and promoted the idea that science should study the laws of "simple" natures, such as heat, rather than assuming that there is any specific nature, or "formal cause", of each complex type of thing. This new modern science began to see itself as describing "laws of nature". This updated approach to studies in nature was seen as mechanistic. Bacon also argued that science should aim for the first time at practical inventions for the improvement of all human life.

1 2 3 4 5

700

800

900

1000

Experiment No.

Speed of light (km/s minus 299,000)

true speed

Data from the famous Michelson–Morley experiment that refuted 19th century theory of light-bearing aether as the medium that had to be a fluid in order to fill space, more rigid than steel in order to support the high frequencies of light waves, as well as massless and without viscosity or it would visibly affect the orbits of planets.

83.1.6 Age of Enlightenment

In the 17th and 18th centuries, the project of modernity, as had been promoted by Bacon and Descartes, led to rapid scientific advance and the successful development of a new type of natural science, mathematical, methodically experimental, and deliberately innovative. Newton and Leibniz succeeded in developing a new physics, now referred to as Newtonian physics, which could be confirmed by experiment and explained using mathematics. Leibniz also incorporated terms from Aristotelian physics, but now being used in a new non-teleological way, for example "energy" and "potential" (modern versions of Aristotelian "*energeia* and *potentia*"). In the style of Bacon, he assumed that different types of things all work according to the same general laws of nature, with no special formal or final causes for each type of thing.

It is during this period that the word "science" gradually became more commonly used to refer to a *type of pursuit* of a type of knowledge, especially knowledge of nature — coming close in meaning to the old term "natural philosophy".

83.1.7 19th century

Both John Herschel and William Whewell systematized methodology: the latter coined the term scientist. When Charles Darwin published *On the Origin of Species* he established descent with modification as the prevailing evolutionary explanation of biological complexity. His theory of natural selection provided a natural explanation of how species originated, but this only gained wide acceptance a century later. John Dalton developed the idea of atoms. The laws of Thermodynamics and the electromagnetic theory were also established in the 19th century, which raised new questions which could not easily be answered using Newton's framework.

83.1.8 20th century and beyond

Einstein's Theory of Relativity and the development of quantum mechanics led to the replacement of Newtonian physics with a new physics which contains two parts, that describe different types of events in nature. The extensive use of scientific innovation during the wars of this century, led to the space race, increased life expectancy, and the Nuclear arms race, giving a widespread public appreciation of the importance of modern science. More recently it has been argued that the ultimate purpose of science is to make sense of human beings and our nature – for example in his book *Consilience*, EO Wilson said "The human condition is the most important frontier of the natural sciences."[29]

83.2 Philosophy of science

Main article: Philosophy of science

Working scientists usually take for granted a set of basic assumptions that are needed to justify the scientific method: (1) that there is an objective reality shared by all rational observers; (2) that this objective reality is governed by natural laws; (3) that these laws can be discovered by means of systematic observation and experimentation.[3] Philosophy of science seeks a deep understanding of what these underlying assumptions mean and whether they are valid.

The belief that scientific theories should and do represent metaphysical reality is known as realism. It can be contrasted with anti-realism, the view that the success of science does not depend on it being accurate about unobservable entities such as electrons. One form of anti-realism is idealism, the belief that the mind or consciousness is the most basic essence, and that each mind generates its own reality.[30] In an idealistic world view, what is true for one mind need not be true for other minds.

There are different schools of thought in philosophy of science. The most popular position is empiricism,[31] which claims that knowledge is created by a process involving observation and that scientific theories are the result of generalizations from such observations.[32] Empiricism generally encompasses inductivism, a position that tries to explain the way general theories can be justified by the finite number of observations humans can make and the hence finite amount of empirical evidence available to confirm scientific theories. This is necessary because the number of predictions those theories make is infinite, which means that they cannot be known from the finite amount of evidence using deductive logic only. Many versions of empiricism exist, with the predominant ones being bayesianism[33] and the hypothetico-deductive method.[34]

Empiricism has stood in contrast to rationalism, the position originally associated with Descartes, which holds that

83.2. PHILOSOPHY OF SCIENCE 601

Karl Popper c. 1980s

knowledge is created by the human intellect, not by observation.[35] Critical rationalism is a contrasting 20th-century approach to science, first defined by Austrian-British philosopher Karl Popper. Popper rejected the way that empiricism describes the connection between theory and observation. He claimed that theories are not generated by observation, but that observation is made in the light of theories and that the only way a theory can be affected by observation is when it comes in conflict with it.[36] Popper proposed replacing verifiability with falsifiability as the landmark of scientific theories, and replacing induction with falsification as the empirical method.[37] Popper further claimed that there is actually only one universal method, not specific to science: the negative method of criticism, trial and error.[38] It covers all products of the human mind, including science, mathematics, philosophy, and art.[39]

602 *CHAPTER 83. SCIENCE*

Another approach, instrumentalism, colloquially termed "shut up and calculate", emphasizes the utility of theories as instruments for explaining and predicting phenomena.[40] It views scientific theories as black boxes with only their input (initial conditions) and output (predictions) being relevant. Consequences, theoretical entities and logical structure are claimed to be something that should simply be ignored and that scientists shouldn't make a fuss about (see interpretations of quantum mechanics). Close to instrumentalism is constructive empiricism, according to which the main criterion for the success of a scientific theory is whether what it says about observable entities is true.

Paul K Feyerabend advanced the idea of epistemological anarchism, which holds that there are no useful and exceptionfree methodological rules governing the progress of science or the growth of knowledge, and that the idea that science can or should operate according to universal and fixed rules is unrealistic, pernicious and detrimental to science itself.[41] Feyerabend advocates treating science as an ideology alongside others such as religion, magic and mythology, and considers the dominance of science in society authoritarian and unjustified. He also contended (along with Imre Lakatos) that the demarcation problem of distinguishing science from pseudoscience on objective grounds is not possible and thus fatal to the notion of science running according to fixed, universal rules.[41] Feyerabend also stated that science does not have evidence for its philosophical precepts, particularly the notion of Uniformity of Law and the Uniformity of Process across time and space.[42]

Finally, another approach often cited in debates of scientific skepticism against controversial movements like "scientific creationism", is methodological naturalism. Its main point is that a difference between natural and supernatural explanations should be made, and that science should be restricted methodologically to natural explanations.[43] That the restriction is merely methodological (rather than ontological) means that science should not consider supernatural explanations itself, but should not claim them to be wrong either. Instead, supernatural explanations should be left a matter of personal belief outside the scope of science. Methodological naturalism maintains that proper science requires strict adherence to empirical study and independent verification as a process for properly developing and evaluating explanations for observable phenomena.[44] The absence of these standards, arguments from authority, biased observational studies and other common fallacies are frequently cited by supporters of methodological naturalism as characteristic of the non-science they criticize.

83.2.1 Certainty and science

A scientific theory is empirical,[31] and is always open to falsification if new evidence is presented. That is, no theory is ever considered strictly certain as science accepts the concept of fallibilism.[45] The philosopher of science Karl Popper sharply distinguishes truth from certainty. He writes that scientific knowledge "consists in the search for truth", but it "is not the search for certainty ... All human knowledge is fallible and therefore uncertain."[46]

New scientific knowledge rarely results in vast changes in our understanding. According to psychologist Keith Stanovich, it may be the media's overuse of words like "breakthrough" that leads the public to imagine that science is constantly proving everything it thought was true to be false.[47]:119–138 While there are such famous cases as the theory of relativity that required a complete reconceptualization, these are extreme exceptions. Knowledge in science is gained by a gradual synthesis of information from different experiments, by various researchers, across different branches of science; it is more like a climb than a leap.[48]:123 Theories vary in the extent to which they have been tested and verified, as well as their acceptance in the scientific community.[49] For example, heliocentric theory, the theory of evolution, relativity theory, and germ theory still bear the name "theory" even though, in practice, they are considered factual.[50] Philosopher Barry Stroud adds that, although the best definition for "knowledge" is contested, being skeptical and entertaining the *possibility* that one is incorrect is compatible with being correct. Ironically then, the scientist adhering to proper scientific approaches will doubt themselves even once they possess the truth.[51] The fallibilist C. S. Peirce argued that inquiry is the struggle to resolve actual doubt and that merely quarrelsome, verbal, or hyperbolic doubt is fruitless[52]—but also that the inquirer should try to attain genuine doubt rather than resting uncritically on common sense.[53] He held that the successful sciences trust, not to any single chain of inference (no stronger than its weakest link), but to the cable of multiple and various arguments intimately connected.[54]

Stanovich also asserts that science avoids searching for a "magic bullet"; it avoids the single-cause fallacy. This means a scientist would not ask merely "What is *the* cause of ...", but rather "What *are* the most significant *causes* of ...". This is especially the case in the more macroscopic fields of science (e.g. psychology, cosmology).[55]:141–147 Of course, research often analyzes few factors at once, but these are always added to the long list of factors that are most important to consider.[55]:141–147 For example: knowing the details of only a person's genetics, or their history and upbringing, or the current situation may not explain a behaviour, but a deep understanding of all these variables combined can be very predictive.

83.2.2 Pseudoscience, fringe science, and junk science

An area of study or speculation that masquerades as science in an attempt to claim a legitimacy that it would not otherwise be able to achieve is sometimes referred to as pseudoscience, fringe science, or junk science.[56] Physicist Richard Feynman coined the term "cargo cult science" for cases in which researchers believe they are doing science because their activities have the outward appearance of science but actually lack the "kind of utter honesty" that allows their results to be rigorously evaluated.[57] Various types of commercial advertising, ranging from hype to fraud, may fall into these categories.

There also can be an element of political or ideological bias on all sides of scientific debates. Sometimes, research may be characterized as "bad science", research that may be well-intentioned but is actually incorrect, obsolete, incomplete, or over-simplified expositions of scientific ideas. The term "scientific misconduct" refers to situations such as where researchers have intentionally misrepresented their published data or have purposely given credit for a discovery to the wrong person.[58]

83.3 Scientific practice

"If a man will begin with certainties, he shall end in doubts; but if he will be content to begin with doubts, he shall end in certainties." —Francis Bacon (1605) *The Advancement of Learning*, Book 1, v, 8

A skeptical point of view, demanding a method of proof, was the practical position taken as early as 1000 years ago, with Alhazen, *Doubts Concerning Ptolemy*, through Bacon (1605), and C. S. Peirce (1839–1914), who note that a community will then spring up to address these points of uncertainty. The methods of inquiry into a problem have been known for thousands of years,[59] and extend beyond theory to practice. The use of measurements, for example, is a practical approach to settle disputes in the community.

John Ziman points out that intersubjective pattern recognition is fundamental to the creation of all scientific knowledge.[60] Ziman shows how scientists can identify patterns to each other across centuries: Needham 1954 (illustration facing page 164) shows how today's trained Western botanist can identify *Artemisia alba* from images taken from a 16thcentury

Chinese pharmacopeia,[61] and Ziman refers to this ability as 'perceptual consensibility'.[62] Ziman then makes consensibility, leading to consensus, the touchstone of reliable knowledge.[63]

83.3.1 The scientific method

Main article: Scientific method

The scientific method seeks to explain the events of nature in a reproducible way.[64] An explanatory thought experiment or hypothesis is put forward, as explanation, using principles such as parsimony (also known as "Occam's Razor") and are generally expected to seek consilience—fitting well with other accepted facts related to the phenomena.[65] This new explanation is used to make falsifiable predictions that are testable by experiment or observation. The predictions are to be posted before a confirming experiment or observation is sought, as proof that no tampering has occurred. Disproof of a prediction is evidence of progress.[66][67] This is done partly through observation of natural phenomena, but also through experimentation, that tries to simulate natural events under controlled conditions, as appropriate to the discipline (in the observational sciences, such as astronomy or geology, a predicted observation might take the place of a controlled experiment). Experimentation is especially important in science to help establish causal relationships (to avoid the correlation fallacy).

When a hypothesis proves unsatisfactory, it is either modified or discarded.[68] If the hypothesis survived testing, it may become adopted into the framework of a scientific theory. This is a logically reasoned, self-consistent model or framework for describing the behavior of certain natural phenomena. A theory typically describes the behavior of much broader sets of phenomena than a hypothesis; commonly, a large number of hypotheses can be logically bound together by a single theory. Thus a theory is a hypothesis explaining various other hypotheses. In that vein, theories are formulated according to most of the same scientific principles as hypotheses. In addition to testing hypotheses, scientists may also generate a model based on observed phenomena. This is an attempt to describe or depict the phenomenon in terms of a logical, physical or mathematical representation and to generate new hypotheses that can be tested.[69]

While performing experiments to test hypotheses, scientists may have a preference for one outcome over another, and so it is important to ensure that science as a whole can eliminate this bias.[70][71] This can be achieved by careful experimental design, transparency, and a thorough peer review process of the experimental results as well as any conclusions.[72][73] After the results of an experiment are announced or published, it is normal practice for independent researchers to double-check how the research was performed, and to follow up by performing similar experiments to determine how dependable the results might be.[74] Taken in its entirety, the scientific method allows for highly creative problem solving while minimizing any effects of subjective bias on the part of its users (namely the confirmation bias).[75]

83.3.2 Mathematics and formal sciences

Main article: Mathematics

Mathematics is essential to the sciences. One important function of mathematics in science is the role it plays in the expression of scientific models. Observing and collecting measurements, as well as hypothesizing and predicting, often require extensive use of mathematics. Arithmetic, algebra, geometry, trigonometry and calculus, for example, are all essential to physics. Virtually every branch of mathematics has applications in science, including "pure" areas such as number theory and topology.

Statistical methods, which are mathematical techniques for summarizing and analyzing data, allow scientists to assess the level of reliability and the range of variation in experimental results. Statistical analysis plays a fundamental role in many areas of both the natural sciences and social sciences.

Computational science applies computing power to simulate real-world situations, enabling a better understanding of scientific problems than formal mathematics alone can achieve. According to the Society for Industrial and Applied Mathematics, computation is now as important as theory and experiment in advancing scientific knowledge.[76]

Whether mathematics itself is properly classified as science has been a matter of some debate. Some thinkers see mathematicians as scientists, regarding physical experiments as inessential or mathematical proofs as equivalent to experiments. Others do not see mathematics as a science, since it does not require an experimental test of its theories and hypotheses. Mathematical theorems and formulas are obtained by logical derivations which presume axiomatic systems, rather than the combination of empirical observation and logical reasoning that has come to be known as the scientific method. In general, mathematics is classified as formal science, while natural and social sciences are classified as empirical sciences.[77]

83.3.3 Basic and applied research

Distinguished Men of Science.[1] Use your cursor to see who is who.[2]

1. ^ Engraving after 'Men of Science Living in 1807-8', John Gilbert engraved by George Zobel and William Walker, ref. NPG 1075a, National Portrait Gallery, London, accessed February 2010
2. ^ Smith, HM (May 1941). "Eminent men of science living in 1807-8". *J. Chem. Educ* **18** (5): 203. doi:10.1021/ed018p203.

Although some scientific research is applied research into specific problems, a great deal of our understanding comes from the curiosity-driven undertaking of basic research. This leads to options for technological advance that were not planned or sometimes even imaginable. This point was made by Michael Faraday when, allegedly in response to the question "what is the *use* of basic research?" he responded "Sir, what is the use of a new-born child?".[78] For example, research into the effects of red light on the human eye's rod cells did not seem to have any practical purpose; eventually, the discovery that our night vision is not troubled by red light would lead search and rescue teams (among others) to adopt red light in the cockpits of jets and helicopters.[79]:106–110 In a nutshell: Basic research is the search for knowledge. Applied research is the search for solutions to practical problems using this knowledge. Finally, even basic research can take unexpected turns, and there is some sense in which the scientific method is built to harness luck.

83.3.4 Research in practice

Due to the increasing complexity of information and specialization of scientists, most of the cutting-edge research today is done by well funded groups of scientists, rather than individuals.[80] D.K. Simonton notes that due to the breadth of very precise and far reaching tools already used by researchers today and the amount of research generated so far, creation of new disciplines or revolutions within a discipline may no longer be possible as it is unlikely that some phenomenon that merits its own discipline has been overlooked. Hybridizing of disciplines and finessing knowledge is, in his view, the future of science.[80]

83.3.5 Practical impacts of scientific research

Discoveries in fundamental science can be world-changing. For example:

83.4 Scientific community

Main article: Scientific community

The scientific community is the group of all interacting scientists. It includes many sub-communities working on particular scientific fields, and within particular institutions; interdisciplinary and cross-institutional activities are also significant.

83.4.1 Branches and fields

Main article: Branches of science

Scientific fields are commonly divided into two major groups: natural sciences, which study natural phenomena (including biological life), and social sciences, which study human behavior and societies. These groupings are empirical sciences, which means the knowledge must be based on observable phenomena and capable of being tested for its validity by other researchers working under the same conditions.[82] There are also related disciplines that are grouped into interdisciplinary and applied sciences, such as engineering and medicine. Within these categories are specialized scientific fields that can include parts of other scientific disciplines but often possess their own nomenclature and expertise.[83]

Mathematics, which is classified as a formal science,[84][85] has both similarities and differences with the empirical sciences (the natural and social sciences). It is similar to empirical sciences in that it involves an objective, careful and systematic study of an area of knowledge; it is different because of its method of verifying its knowledge, using *a priori* rather than empirical methods.[86] The formal sciences, which also include statistics and logic, are vital to the empirical sciences. Major advances in formal science have often led to major advances in the empirical sciences. The formal sciences are essential in the formation of hypotheses, theories, and laws,[87] both in discovering and describing how things work (natural sciences) and how people think and act (social sciences).

83.4.2 Institutions

Learned societies for the communication and promotion of scientific thought and experimentation have existed since the Renaissance period.[88] The oldest surviving institution is the Italian *Accademia dei Lincei* which was established in 1603.[89] The respective National Academies of Science are distinguished institutions that exist in a number of countries, beginning with the British Royal Society in 1660[90] and the French *Académie des Sciences* in 1666.[91] International scientific organizations, such as the International Council for Science, have since been formed to promote cooperation between the scientific communities of different nations. Many governments have dedicated agencies to support scientific research. Prominent scientific organizations include, the National Science Foundation in the

U.S., the National Scientific and Technical Research Council in Argentina, the academies of science of many nations, CSIRO in Australia, Centre national de la recherche scientifique in France, Max Planck Society and Deutsche Forschungsgemeinschaft in Germany, and in Spain, CSIC.

83.4.3 Literature

Main article: Scientific literature

An enormous range of scientific literature is published.[92] Scientific journals communicate and document the results of research carried out in universities and various other research institutions, serving as an archival record of science. The first scientific journals, *Journal des Sçavans* followed by the *Philosophical Transactions*, began publication in 1665. Since that time the total number of active periodicals has steadily increased. In 1981, one estimate for the number of scientific and technical journals in publication was 11,500.[93] The United States National Library of Medicine currently indexes 5,516 journals that contain articles on topics related to the life sciences. Although the journals are in 39 languages, 91 percent of the indexed articles are published in English.[94]

Most scientific journals cover a single scientific field and publish the research within that field; the research is normally expressed in the form of a scientific paper. Science has become so pervasive in modern societies that it is generally considered necessary to communicate the achievements, news, and ambitions of scientists to a wider populace. Science magazines such as *New Scientist*, *Science & Vie*, and *Scientific American* cater to the needs of a much wider readership and provide a non-technical summary of popular areas of research, including notable discoveries and advances in certain fields of research. Science books engage the interest of many more people. Tangentially, the science fiction genre, primarily fantastic in nature, engages the public imagination and transmits the ideas, if not the methods, of science.

Recent efforts to intensify or develop links between science and non-scientific disciplines such as Literature or, more specifically, Poetry, include the *Creative Writing Science* resource developed through the Royal Literary Fund.[95]

83.5 Science and society

83.5.1 Women in science

Main article: Women in science

Science has traditionally been a male-dominated field, with some notable exceptions.[96] Women historically faced considerable discrimination in science, much as they did in other areas of male-dominated societies, such as frequently being passed over for job opportunities and denied credit for their work.[97] For example, Christine Ladd (1847-1930) was able to enter a Ph.D. program as 'C. Ladd'; Christine "Kitty" Ladd completed the requirements in 1882, but was awarded her degree only in 1926, after a career which spanned the algebra of logic (see truth table), color vision, and psychology. Her work preceded notable researchers like Ludwig Wittgenstein and Charles Sanders Peirce. The achievements of women in science have been attributed to their defiance of their traditional role as laborers within the domestic sphere.[98]

In the late 20th century, active recruitment of women and elimination of institutional discrimination on the basis of sex greatly increased the number of female scientists, but large gender disparities remain in some fields; over half of new biologists are female, while 80% of PhDs in physics are given to men. Feminists claim this is the result of culture rather than an innate difference between the sexes, and some experiments have shown that parents challenge and explain more to boys than girls, asking them to reflect more deeply and logically.[99] In the early part of the 21st century, in America, women earned 50.3% bachelor's degrees, 45.6% master's degrees, and 40.7% of PhDs in science and engineering fields with women earning more than half of the degrees in three fields: Psychology (about 70%), Social Sciences (about 50%), and Biology (about 50-60%). However, when it comes to the Physical Sciences, Geosciences, Math, Engineering, and Computer Science; women earned less than half the degrees.[100] However, lifestyle choice also plays a major role in female engagement in science; women with young children are 28% less likely to take tenure-track positions due to work-life balance issues,[101] and female graduate students' interest in careers in research declines dramatically over the course of graduate school, whereas that of their male colleagues remains unchanged.[102]

83.5.2 Science policy

Main articles: Science policy, History of science policy, Funding of science and Economics of science

Science policy is an area of public policy concerned with the policies that affect the conduct of the scientific enterprise, including research funding, often in pursuance of other national policy goals such as technological innovation to promote commercial product development, weapons development, health care and environmental monitoring. Science policy also refers to the act of applying scientific knowledge and consensus to the development of public policies. Science policy thus deals with the entire domain of issues that involve the natural sciences. In accordance with public policy being concerned about the well-being of its citizens, science policy's goal is to consider how science and technology can best serve the public.

State policy has influenced the funding of public works and science for thousands of years, dating at least from the time of the Mohists, who inspired the study of logic during the period of the Hundred Schools of Thought, and the study of defensive fortifications during the Warring States period in China. In Great Britain, governmental approval of the Royal Society in the 17th century recognized a scientific community which exists to this day. The professionalization of science, begun in the 19th century, was partly enabled by the creation of scientific organizations such as the National Academy of Sciences, the Kaiser Wilhelm Institute, and State funding of universities of their respective nations. Public policy can directly affect the funding of capital equipment, intellectual infrastructure for industrial research, by providing tax incentives to those organizations that fund research. Vannevar Bush, director of the Office of Scientific Research and Development for the United States government, the forerunner of the National Science Foundation, wrote in July 1945 that "Science is a proper concern of government".[103]

Science and technology research is often funded through a competitive process, in which potential research projects are evaluated and only the most promising receive funding. Such processes, which are run by government, corporations or foundations, allocate scarce funds. Total research funding in most developed countries is between 1.5% and 3% of GDP.[104] In the OECD, around two-thirds of research and development in scientific and technical fields is carried out by industry, and 20% and 10% respectively by universities and government. The government funding proportion in certain industries is higher, and it dominates research in social science and humanities. Similarly, with some exceptions (e.g. biotechnology) government provides the bulk of the funds for basic scientific research. In commercial research and development, all but the most research-oriented corporations focus more heavily on near-term commercialisation possibilities rather than "blue-sky" ideas or technologies (such as nuclear fusion).

83.5.3 Media perspectives

The mass media face a number of pressures that can prevent them from accurately depicting competing scientific claims in terms of their credibility within the scientific community as a whole. Determining how much weight to give different sides in a scientific debate may require considerable expertise regarding the matter.[105] Few journalists have real scientific knowledge, and even beat reporters who know a great deal about certain scientific issues may be ignorant about other scientific issues that they are suddenly asked to cover.[106][107]

83.5.4 Political usage

See also: Politicization of science

Many issues damage the relationship of science to the media and the use of science and scientific arguments by politicians. As a very broad generalisation, many politicians seek certainties and *facts* whilst scientists typically offer probabilities and caveats. However, politicians' ability to be heard in the mass media frequently distorts the scientific understanding by the public. Examples in Britain include the controversy over the MMR inoculation, and the 1988 forced resignation of a Government Minister, Edwina Currie for revealing the high probability that battery farmed eggs were contaminated with *Salmonella*.[108]

John Horgan, Chris Mooney, and researchers from the US and Canada have described Scientific Certainty Argumentation Methods (SCAMs), where an organization or think tank makes it their only goal to cast doubt on supported science because it conflicts with political agendas.[109][110][111][112] Hank Campbell and microbiologist Alex Berezow have described "feel-good fallacies" used in politics, where politicians frame their positions in a way that makes people feel good about supporting certain policies even when scientific evidence shows there is no need to worry or there is no need for dramatic change on current programs.[113]

83.6 See also

- Outline of science
- Outline of natural science
- Outline of physical science
- Outline of earth science
- Outline of formal science
- Outline of social science
- Outline of applied science
- Antiquarian science books
- Research
- Criticism of science
- Protoscience
- Science wars
- Sociology of scientific knowledge

83.7 Notes

[1] "science". Online Etymology Dictionary. Retrieved 2014-09-20.

[2] Wilson, Edward O. (1998). *Consilience: The Unity of Knowledge* (1st ed.). New York, NY: Vintage Books. pp. 49–71. ISBN 0-679-45077-7.

[3] "... modern science is a discovery as well as an invention. It was a discovery that nature generally acts regularly enough to be described by laws and even by mathematics; and required invention to devise the techniques, abstractions, apparatus, and organization for exhibiting the regularities and securing their law-like descriptions." —p.vii, J. L. Heilbron, (2003, editor-in-chief). *The Oxford Companion to the History of Modern Science*. New York: Oxford University Press. ISBN 0-19-511229-6.

[4] "science". *Merriam-Webster Online Dictionary*. Merriam-Webster, Inc. Retrieved 2011-10-16. "**3 a:** knowledge or a system of knowledge covering general truths or the operation of general laws especially as obtained and tested through scientific method **b:** such knowledge or such a system of knowledge concerned with the physical world and its phenomena"

[5] http://bioinfo.aizeonpublishers.net/content/2014/1/bioinfo321-326.pdf

[6] Jim Al-Khalili (4 January 2009). "The 'first true scientist'". BBC News.

[7] Tracey Tokuhama-Espinosa (2010). *Mind, Brain, and Education Science: A Comprehensive Guide to the New Brain-Based Teaching*. W. W. Norton & Company. p. 39. ISBN 9780393706079. "Alhazen (or Al-Haytham; 965–1039 C.E.) was perhaps one of the greatest physicists of all times and a product of the Islamic Golden Age or Islamic Renaissance (7th–13th centuries). He made significant contributions to anatomy, astronomy, engineering, mathematics, medicine, ophthalmology, philosophy, physics, psychology, and visual perception and is primarily attributed as the inventor of the scientific method, for which author Bradley Steffens (2006) describes him as the "first scientist"."

[8] El-Bizri, Nader, "A Philosophical Perspective on Alhazen's Optics", *Arabic Sciences and Philosophy* **15** (2005-08-05), 189–218

[9] Malik, Kenan (2010-10-22). "Pathfinders: The Golden Age of Arabic Science, By Jim Al-Khalili". *The Independent*. Retrieved 2014-10-22.
[10] Haq, Syed (2009). "Science in Islam". *Oxford Dictionary of the Middle Ages*. ISSN 1703-7603. Retrieved 2014-10-22.
[11] Lindberg, D. C., Theories of Vision from al-Kindi to Kepler, (Chicago, Univ. of Chicago Pr., 1976), pp. 60–7.
[12] Sabra, A. I. (1989). *The Optics of Ibn al-Haytham. Books I–II–III: On Direct Vision*. London: The Warburg Institute, University of London. pp. 25–29. ISBN 0-85481-072-2.

[13] David C. Lindberg (2007), *The beginnings of Western science: the European Scientific tradition in philosophical, religious, and institutional context*, Second ed. Chicago: Univ. of Chicago Press ISBN 978-0-226-48205-7, p. 3
[14] Isaac Newton's Philosophiae Naturalis Principia Mathematica (1687), for example, is translated "Mathematical Principles of Natural Philosophy", and reflects the then-current use of the words "natural philosophy", akin to "systematic study of nature"
[15] Oxford English Dictionary
[16] The *Oxford English Dictionary* dates the origin of the word "scientist" to 1834.
[17] Feynman, *Lectures in Physics*, Vol.1, Chap.1.
[18] Needham 1954, p. 150
[19] "The historian ... requires a very broad definition of "science" — one that ... will help us to understand the modern scientific enterprise. We need to be broad and inclusive, rather than narrow and exclusive ... and we should expect that the farther back we go [in time] the broader we will need to be." — David Pingree (1992), "Hellenophilia versus the History of Science" *Isis* **83** 554–63, as cited on p.3, David C. Lindberg (2007), *The beginnings of Western science: the European Scientific tradition in philosophical, religious, and institutional context*, Second ed. Chicago: Univ. of Chicago Press ISBN 978-0-226-48205-7
[20] See the quotation in Homer (8th century BCE) *Odyssey* 10.302–3
[21] "Progress or Return" in An Introduction to Political Philosophy: Ten Essays by Leo Strauss. (Expanded version of Political Philosophy: Six Essays by Leo Strauss, 1975.) Ed. Hilail Gilden. Detroit: Wayne State UP, 1989.
[22] Strauss and Cropsey eds. History of Political Philosophy, Third edition, p.209.
[23] Plato, *Apology* 30e
[24] "... [A] man knows a thing scientifically when he possesses a conviction arrived at in a certain way, and when the first principles on which that conviction rests are known to him with certainty—for unless he is more certain of his first principles than of the conclusion drawn from them he will only possess the knowledge in question accidentally." — Aristotle, *Nicomachean Ethics* **6** (H. Rackham, ed.) Aristot. Nic. Eth. 1139b
[25] Grant, Edward (2007). *A History of Natural Philosophy: From the Ancient World to the Nineteenth Century*. Cambridge University Press. pp. 62–67. ISBN 978-0-521-68957-1.
[26] The ʿAbbāsid Caliphate. *Encyclopædia Britannica*.
[27] "Galileo and the Birth of Modern Science, by Stephen Hawking, American Heritage's Invention & Technology, Spring 2009, Vol. 24, No. 1, p. 36
[28] "Galileo Project – Pope Urban VIII Biography".
[29] Wilson, EO. 1998. *Consilience: The unity of knowledge*. New York. Alfred A. Knopf. p334
_ Jeremy Griffith (2011) *What is Science?*. In *The Book of Real Answers to Everything!*, ISBN 978-1-74129-007-3. http://www.worldtransformation.com/what-is-science/ accessed November 20, 2012.
[30] This realization is the topic of intersubjective verifiability, as recounted, for example, by Max Born (1949, 1965) *Natural Philosophy of Cause and Chance*, who points out that all knowledge, including natural or social science, is also subjective. p. 162: "Thus it dawned upon me that fundamentally everything is subjective, everything without exception. That was a shock."
[31] In his investigation of the law of falling bodies, Galileo (1638) serves as example for scientific investigation: *Two New Sciences* "A piece of wooden moulding or scantling, about 12 cubits long, half a cubit wide, and three finger-breadths thick, was taken; on its edge was cut a channel a little more than one finger in breadth; having made this groove very straight, smooth, and polished, and having lined it with parchment, also as smooth and polished as possible, we rolled along it a hard, smooth, and very round bronze ball. Having placed this board in a sloping position, by lifting one end some one or two cubits above the other, we rolled the ball, as I was just saying, along the channel, noting, in a manner presently to be described, the time required to make the descent. We . . . now rolled the ball only one-quarter the length of the channel; and having measured the time of its descent, we found it precisely one-half of the former. Next we tried other distances, comparing the time for the whole length with that for the half, or with that for two-thirds, or three-fourths, or indeed for any fraction; in such experiments, repeated many, many, times." Galileo solved the problem of time measurement by weighing a jet of water collected during the descent of the bronze ball, as stated in his *Two New Sciences*.
[32] "... [T]he logical empiricists thought that the great aim of science was to discover and establish *generalizations*." —Godfrey-Smith 2003, p. 41

[33] "Bayesianism tries to understand evidence using probability theory." —Godfrey-Smith 2003, p. 203
[34] Godfrey-Smith 2003, p. 236
[35] Godfrey-Smith 2003, p. 20
[36] Godfrey-Smith 2003, pp. 63–7
[37] Godfrey-Smith 2003, p. 68
[38] Popper called this *Conjecture and Refutation* Godfrey-Smith 2003, pp. 117–8
[39] Karl Popper: *Objective Knowledge* (1972)
[40] Newton-Smith, W. H. (1994). *The Rationality of Science*. London: Routledge. p. 30. ISBN 0-7100-0913-5.
[41] Feyerabend 1993.

[42] Feyerabend, Paul (1987). *Farewell To Reason.* Verso. p. 100. ISBN 0-86091-184-5.
[43] Godfrey-Smith 2003, p. 151 credits Willard Van Orman Quine (1969) "Epistemology Naturalized" *Ontological Relativity and Other Essays* New York: Columbia University Press, as well as John Dewey, with the basic ideas of naturalism — Naturalized Epistemology, but Godfrey-Smith diverges from Quine's position: according to Godfrey-Smith, "A naturalist can think that science can contribute to *answers* to philosophical questions, without thinking that philosophical questions can be replaced by science questions.".
[44] Brugger, E. Christian (2004). "Casebeer, William D. Natural Ethical Facts: Evolution, Connectionism, and Moral Cognition". *The Review of Metaphysics* **58** (2).
[45] "No amount of experimentation can ever prove me right; a single experiment can prove me wrong." —Albert Einstein, noted by Alice Calaprice (ed. 2005) *The New Quotable Einstein* Princeton University Press and Hebrew University of Jerusalem, ISBN 0-691-12074-9 p. 291. Calaprice denotes this not as an exact quotation, but as a paraphrase of a translation of A. Einstein's "Induction and Deduction". *Collected Papers of Albert Einstein* **7** Document 28. Volume 7 is *The Berlin Years: Writings, 1918-1921.* A. Einstein; M. Janssen, R. Schulmann, et al., eds.
[46] Popper 1996, p. 4.
[47] Stanovich 2007
[48] Stanovich 2007
[49] Fleck, Ludwik (1979). Trenn, Thaddeus J.; Merton, Robert K, eds. *Genesis and Development of a Scientific Fact.* Chicago: University of Chicago Press. ISBN 0-226-25325-2. Claims that before a specific fact "existed", it had to be created as part of a social agreement within a community. Steven Shapin (1980) "A view of scientific thought" *Science* ccvii (7 Mar 1980) 1065–66 states "[To Fleck,] facts are invented, not discovered. Moreover, the appearance of scientific facts as discovered things is itself a social construction: a *made* thing. "
[50] Dawkins, Richard; Coyne, Jerry (2005-09-02). "One side can be wrong". *The Guardian* (London).
[51] "Barry Stroud on Scepticism". philosophy bites. 2007-12-16. Retrieved 2012-02-05.
[52] Peirce (1877), "The Fixation of Belief", Popular Science Monthly, v. 12, pp. 1–15, see §IV on p. 6–7. Reprinted *Collected Papers* v. 5, paragraphs 358–87 (see 374–6), *Writings* v. 3, pp. 242–57 (see 247–8), *Essential Peirce* v. 1, pp. 109–23 (see 114–15), and elsewhere.
[53] Peirce (1905), "Issues of Pragmaticism", *The Monist*, v. XV, n. 4, pp. 481–99, see "Character V" on p. 491. Reprinted in *Collected Papers* v. 5, paragraphs 438–63 (see 451), *Essential Peirce* v. 2, pp. 346–59 (see 353), and elsewhere.
[54] Peirce (1868), "Some Consequences of Four Incapacities", *Journal of Speculative Philosophy* v. 2, n. 3, pp. 140–57, see p. 141. Reprinted in *Collected Papers*, v. 5, paragraphs 264–317, *Writings* v. 2, pp. 211–42, *Essential Peirce* v. 1, pp. 28–55, and elsewhere.
[55] Stanovich 2007
[56] "*Pseudoscientific – pretending to be scientific, falsely represented as being scientific*", from the *Oxford American Dictionary*, published by the Oxford English Dictionary; Hansson, Sven Ove (1996)."Defining Pseudoscience", Philosophia Naturalis, 33: 169–176, as cited in "Science and Pseudo-science" (2008) in Stanford Encyclopedia of Philosophy. The Stanford article states: "Many writers on pseudoscience have emphasized that pseudoscience is non-science posing as science. The foremost modern classic on the subject (Gardner 1957) bears the title Fads and Fallacies in the Name of Science. According to Brian Baigrie (1988, 438), "[w]hat is objectionable about these beliefs is that they masquerade as genuinely scientific ones." These and many other authors assume that to be pseudoscientific, an activity or a teaching has to satisfy the following two criteria (Hansson 1996): (1) it is not scientific, and (2) its major proponents try to create the impression that it is scientific".
83.7. NOTES 611
_ For example, Hewitt et al. *Conceptual Physical Science* Addison Wesley; 3 edition (July 18, 2003) ISBN 0-321-05173-4, Bennett et al. *The Cosmic Perspective* 3e Addison Wesley; 3 edition (July 25, 2003) ISBN 0-8053-8738-2; *See also*, e.g., Gauch HG Jr. *Scientific Method in Practice* (2003).
_ A 2006 National Science Foundation report on Science and engineering indicators quoted Michael Shermer's (1997) definition of pseudoscience: '"claims presented so that they appear [to be] scientific even though they lack supporting evidence and plausibility"(p. 33). In contrast, science is "a set of methods designed to describe and interpret observed and inferred phenomena, past or present, and aimed at building a testable body of knowledge open to rejection or confirmation"(p. 17)'.Shermer M. (1997). *Why People Believe Weird Things: Pseudoscience, Superstition, and Other Confusions of Our Time.* New York: W. H. Freeman and Company. ISBN 0-7167-3090-1. as cited by National Science Board. National Science Foundation, Division of Science Resources Statistics (2006). "Science and Technology: Public Attitudes and Understanding". *Science and engineering indicators 2006.*
_ "A pretended or spurious science; a collection of related beliefs about the world mistakenly regarded as being based on scientific method or as having the status that scientific truths now have," from the *Oxford English Dictionary*, second edition 1989.
[57] Cargo Cult Science by Feyman, Richard. Retrieved 2011-07-21.
[58] "Coping with fraud" (PDF). *The COPE Report 1999*: 11–18. Archived from the original on 2007-09-28. Retrieved 2011-07-21. "It is 10 years, to the month, since Stephen Lock ... Reproduced with kind permission of the Editor, The Lancet."
[59] In mathematics, Plato's *Meno* demonstrates that it is possible to know logical propositions, such as the Pythagorean theorem, and even to prove them, as cited by Crease 2009, pp. 35–41
[60] Ziman cites Polanyi 1958 chapter 12, as referenced in Ziman 1978, p. 44
[61] Ziman 1978, pp. 46–47
[62] Ziman 1978, p. 46
[63] Ziman 1978, p. 104.
[64] di Francia 1976, p. 13: "The amazing point is that for the first time since the discovery of mathematics, a method has been introduced, the results of which have an intersubjective value!" *(Author's punctuation)*
[65] Wilson, Edward (1999). *Consilience: The Unity of Knowledge.* New York: Vintage. ISBN 0-679-76867-X

[66] di Francia 1976, pp. 4–5: "One learns in a laboratory; one learns how to make experiments only by experimenting, and one learns how to work with his hands only by using them. The first and fundamental form of experimentation in physics is to teach young people to work with their hands. Then they should be taken into a laboratory and taught to work with measuring instruments — each student carrying out real experiments in physics. This form of teaching is indispensable and cannot be read in a book."
[67] Fara 2009, p. 204: "Whatever their discipline, scientists claimed to share a common scientific method that ... distinguished them from non-scientists."
[68] Nola & Irzik 2005, p. 208.
[69] Nola & Irzik 2005, pp. 199–201.
[70] van Gelder, Tim (1999). ""Heads I win, tails you lose": A Foray Into the Psychology of Philosophy" (PDF). University of Melbourne. Archived from the original on 2008-04-09. Retrieved 2008-03-28.
[71] Pease, Craig (September 6, 2006). "Chapter 23. Deliberate bias: Conflict creates bad science". *Science for Business, Law and Journalism*. Vermont Law School. Archived from the original on 19 June 2010.
[72] Shatz, David (2004). *Peer Review: A Critical Inquiry*. Rowman & Littlefield. ISBN 0-7425-1434-X. OCLC 54989960.
[73] Krimsky, Sheldon (2003). *Science in the Private Interest: Has the Lure of Profits Corrupted the Virtue of Biomedical Research.* Rowman & Littlefield. ISBN 0-7425-1479-X. OCLC 185926306.
[74] Bulger, Ruth Ellen; Heitman, Elizabeth; Reiser, Stanley Joel (2002). *The Ethical Dimensions of the Biological and Health Sciences* (2nd ed.). Cambridge University Press. ISBN 0-521-00886-7. OCLC 47791316.
[75] Backer, Patricia Ryaby (October 29, 2004). "What is the scientific method?". San Jose State University. Retrieved 2008-03-28.
[76] Graduate Education for Computational Science and Engineering, SIAM Working Group on CSE Education. Retrieved 2008-04-27.

[77] Bunge, Mario Augusto (1998). *Philosophy of Science: From Problem to Theory*. Transaction Publishers. p. 24. ISBN 0-7658-0413-1.
[78] "To Live at All Is Miracle Enough — Richard Dawkins". RichardDawkins.net. 2006-05-10. Retrieved 2012-02-05.
[79] Stanovich 2007
[80] Simonton, Dean Keith (2013). "After Einstein: Scientific genius is extinct". *Nature* **493** (7434): 602–602. doi:10.1038/493602a.
[81] Evicting Einstein, March 26, 2004, NASA. *"Both [relativity and quantum mechanics] are extremely successful. The Global Positioning System (GPS), for instance, wouldn't be possible without the theory of relativity. Computers, telecommunications, and the Internet, meanwhile, are spin-offs of quantum mechanics."*
[82] Popper 2002, p. 20.
[83] See: Editorial Staff (March 7, 2008). "Scientific Method: Relationships among Scientific Paradigms". Seed magazine. Retrieved 2007-09-12.
[84] Tomalin, Marcus (2006). *Linguistics and the Formal Sciences*. Cambridge.org. doi:10.2277/0521854814. Retrieved 2012-02-05.
[85] Benedikt Löwe (2002) "The Formal Sciences: Their Scope, Their Foundations, and Their Unity"
[86] Popper 2002, pp. 10–11.
[87] Popper 2002, pp. 79–82.
[88] Parrott, Jim (August 9, 2007). "Chronicle for Societies Founded from 1323 to 1599". Scholarly Societies Project. Retrieved 2007-09-11.
[89] "Accademia Nazionale dei Lincei" (in Italian). 2006. Retrieved 2007-09-11.
[90] "History of the Royal Society". The Royal Society. Retrieved 2011-10-16.
[91] Meynell, G.G. "The French Academy of Sciences, 1666–91: A reassessment of the French Académie royale des sciences under Colbert (1666–83) and Louvois (1683–91)". Retrieved 2011-10-13.
[92] Ziman, J.M. (1980). "The proliferation of scientific literature: a natural process". *Science* **208** (4442): 369–371. doi:10.1126/science.7367863. PMID 7367863.
[93] Subramanyam, Krishna; Subramanyam, Bhadriraju (1981). *Scientific and Technical Information Resources.* CRC Press. ISBN 0-8247-8297-6. OCLC 232950234.
[94] "MEDLINE Fact Sheet". Washington DC: United States National Library of Medicine. Retrieved 2011-10-15.
[95] Petrucci, Mario. "Creative Writing – Science". Retrieved 2008-04-27.
[96] Women in science have included:
_ Hypatia (c. 350–415 CE), of the Library of Alexandria.
_ Trotula of Salerno, a physician c. 1060 CE.
_ Caroline Herschel one of the first professional astronomers of the 18th and 19th centuries.
_ Christine Ladd-Franklin, a doctoral student of C. S. Peirce, who published Wittgenstein's proposition 5.101 in her dissertation, 40 years before Wittgenstein's publication of Tractatus Logico-Philosophicus.
_ Henrietta Leavitt, a professional human computer and astronomer, who first published the significant relationship between the luminosity of Cepheid variable stars and their distance from Earth. This allowed Hubble to make the discovery of the expanding universe, which led to the Big Bang theory.
_ Emmy Noether, who proved the conservation of energy and other constants of motion in 1915.
_ Marie Curie, who made discoveries relating to radioactivity along with her husband, and for whom Curium is named.
_ Rosalind Franklin, who worked with x-ray diffraction.
[97] Nina Byers,Contributions of 20th Century Women to Physics which details and 83 female physicists of the 20th century, By 1976, more women were physicists, and the 83 who were detailed were joined by other women in noticeably larger numbers.
[98] Bonnie Spanier, From Molecules to Brains, Normal Science Supports Sexist Beliefs About Differences, The Gender and

Science Reader (New York: Routledge 2001)

[99] Crowley, K. Callanan, M.A., Tenenbaum, H. R., & Allen, E. (2001). Parents explain more often to boys than to girls during shared scientific thinking. Psychological Science, 258–261.
[100] Rosser, Sue V. *Breaking into the Lab : Engineering Progress for Women in Science*. New York: New York University Press. p. 7. ISBN 9780814776452.
[101] Goulden et al. 2009. Center for American Progress
[102] Royal Society of Chemistry. 2009. Change of Heart;
[103] "Vannevar Bush (July 1945), "Science, the Endless Frontier"". Nsf.gov. Retrieved 2012-02-05.
[104] "Main Science and Technology Indicators – 2008-1" (PDF). OECD. Retrieved 20 April 2012. 50.8 KB
[105] Dickson, David (October 11, 2004). "Science journalism must keep a critical edge". Science and Development Network. Archived from the original on 21 June 2010.
[106] Mooney, Chris (Nov–Dec 2004). "Blinded By Science, How 'Balanced' Coverage Lets the Scientific Fringe Hijack Reality" **43** (4). Columbia Journalism Review. Retrieved 2008-02-20.
[107] McIlwaine, S.; Nguyen, D. A. (2005). "Are Journalism Students Equipped to Write About Science?". *Australian Studies in Journalism* **14**: 41–60. Retrieved 2008-02-20.
[108] "1988: Egg industry fury over salmonella claim", "On This Day," BBC News, December 3, 1988.
[109] "Original "Doubt is our product ..." memo". University of California, San Francisco. 21 August 1969. Retrieved 3 October 2012. The memo reads "Doubt is our product since it is the best means of competing with the 'body of fact' that exists in the mind of the general public. It is also the means of establishing a controversy."
[110] "'THE REPUBLICAN WAR ON SCIENCE,' BY CHRIS MOONEY", Political Science, Review by JOHN HORGAN, Published: December 18 2005
[111] Mooney, Chris (2005). *The Republican War on Science*. Basic Books. ISBN 0-465-04676-2.
[112] William R. Freudenburg, Robert Gramling, Debra J. Davidson (2008) "Scientific Certainty Argumentation Methods (SCAMs): Science and the politics of doubt". *Sociological Inquiry*. Vol. **78**, No. 1. 2–38
[113] Hank Campbell, Alex Berezow,. *Science Left Behind : Feel-good Fallacies and the Rise of the Anti-Scientific Left* (1st ed.). New York: PublicAffairs. ISBN 978-1-61039-164-1.

83.8 References

_ Crease, Robert P. (2009). *The Great Equations*. New York: W.W. Norton. p. 317. ISBN 978-0-393-06204-5
_ Crease, Robert P. (2011). *World in the Balance: the historic quest for an absolute system of measurement*. New York: W.W. Norton. p. 317. ISBN 978-0-393-07298-3.
_ di Francia, Giuliano Toraldo (1976). *The Investigation of the Physical World*. Cambridge: Cambridge University Press. ISBN 0-521-29925-X Originally published in Italian as *L'Indagine del Mondo Fisico* by Giulio Einaudi editore 1976; first published in English by Cambridge University Press 1981.
_ Fara, Patricia (2009). *Science : a four thousand year history*. Oxford: Oxford University Press. p. 408. ISBN 978-0-19-922689-4.
_ Feyerabend, Paul (1993). *Against Method* (3rd ed.). London: Verso. ISBN 0-86091-646-4.
_ Feyerabend, Paul (2005). *Science, history of the philosophy*, as cited in Honderich, Ted (2005). *The Oxford companion to philosophy*. Oxford Oxfordshire: Oxford University Press. ISBN 0-19-926479-1. OCLC 173262485.
_ Godfrey-Smith, Peter (2003). *Theory and Reality*. Chicago 60637: University of Chicago. p. 272. ISBN 0-226-30062-5
_ Feynman, R.P. (1999). *The Pleasure of Finding Things Out: The Best Short Works of Richard P. Feynman*. Perseus Books Group. ISBN 0-465-02395-9. OCLC 181597764.

_ Needham, Joseph (1954). "Science and Civilisation in China: Introductory Orientations" **1**. Cambridge University Press
_ Nola, Robert; Irzik, Gürol (2005). *Philosophy, science, education and culture*. Science & technology education library **28**. Springer. ISBN 1-4020-3769-4.
_ Papineau, David. (2005). *Science, problems of the philosophy of.*, as cited in Honderich, Ted (2005). *The Oxford companion to philosophy*. Oxford Oxfordshire: Oxford University Press. ISBN 0-19-926479-1. OCLC 173262485.
_ Parkin, D. (1991). "Simultaneity and Sequencing in the Oracular Speech of Kenyan Diviners". In Philip M. Peek. *African Divination Systems: Ways of Knowing*. Indianapolis, IN: Indiana University Press..
_ Polanyi, Michael (1958). *Personal Knowledge: Towards a Post-Critical Philosophy*. University of Chicago Press. ISBN 0-226-67288-3
_ Popper, Karl Raimund (1996) [1984]. *In search of a better world: lectures and essays from thirty years*. New York, NY: Routledge. ISBN 0-415-13548-6.
_ Popper, Karl R. (2002) [1959]. *The Logic of Scientific Discovery*. New York, NY: Routledge Classics. ISBN 0-415-27844-9. OCLC 59377149.
_ Stanovich, Keith E. (2007). *How to Think Straight About Psychology*. Boston: Pearson Education. ISBN 978-0-205-68590-5.
_ Ziman, John (1978). *Reliable knowledge: An exploration of the grounds for belief in science*. Cambridge: Cambridge University Press. p. 197. ISBN 0-521-22087-4

83.9 Further reading

_ Augros, Robert M., Stanciu, George N., "The New Story of Science: mind and the universe", Lake Bluff, Ill.:

Regnery Gateway, c1984. ISBN 0-89526-833-7
_ Becker, Ernest (1968). *The structure of evil; an essay on the unification of the science of man.* New York: G. Braziller.
_ Cole, K. C., *Things your teacher never told you about science: Nine shocking revelations* Newsday, Long Island, New York, March 23, 1986, pg 21+
_ Feynman, Richard "Cargo Cult Science"
_ Gaukroger, Stephen (2006). *The Emergence of a Scientific Culture: Science and the Shaping of Modernity 1210–1685.* Oxford: Oxford University Press. ISBN 0-19-929644-8.
_ Gopnik, Alison, "Finding Our Inner Scientist", Daedalus, Winter 2004.
_ Krige, John, and Dominique Pestre, eds., *Science in the Twentieth Century*, Routledge 2003, ISBN 0-415-28606-9
_ Levin, Yuval (2008). *Imagining the Future: Science and American Democracy.* New York, Encounter Books. ISBN 1-59403-209-2
_ Kuhn, Thomas, *The Structure of Scientific Revolutions*, 1962.
_ William F., McComas (1998). "The principal elements of the nature of science: Dispelling the myths". In McComas, William F. *The nature of science in science education: rationales and strategies.* Springer. ISBN 978-0-7923-6168-8
_ Obler, Paul C.; Estrin, Herman A. (1962). *The New Scientist: Essays on the Methods and Values of Modern Science.* Anchor Books, Doubleday.
_ Russell, Bertrand (1985) [1952]. *The Impact of Science on Society.* London: Unwin. ISBN 0-04-300090-8.

_ Rutherford, F. James; Ahlgren, Andrew (1990). *Science for all Americans.* New York, NY: American Association for the Advancement of Science, Oxford University Press. ISBN 0-19-506771-1.
_ Thurs, Daniel Patrick (2007). *Science Talk: Changing Notions of Science in American Popular Culture.* New Brunswick, NJ: Rutgers University Press. pp. 22–52. ISBN 978-0-8135-4073-3.

83.10 External links

Publications

_ "*GCSE Science textbook*". Wikibooks.org

News

_ Nature News. Science news by the journal *Nature*
_ New Scientist. An weekly magazine published by Reed Business Information
_ ScienceDaily
_ Science Newsline
_ Sciencia
_ Discover Magazine
_ Irish Science News from Discover Science & Engineering
_ Science Stage Scientific Videoportal and Community

Resources

_ Euroscience:
_ "ESOF: Euroscience Open Forum". Archived from the original on 10 June 2010.
_ Science Development in the *Latin American docta*
_ Classification of the Sciences in *Dictionary of the History of Ideas.* (Dictionary's new electronic format is badly botched, entries after "Design" are inaccessible. *Internet Archive* old version).
_ "Nature of Science" *University of California Museum of Paleontology*
_ United States Science Initiative Selected science information provided by US Government agencies, including research & development results
_ How science works *University of California Museum of Paleontology*

DNA determines the genetic structure of all known life

Astronomy became much more accurate after Tycho Brahe devised his scientific instruments for measuring angles between two celestial bodies, before the invention of the telescope. Brahe's observations were the basis for Kepler's laws.

A scientist's lab book entry

Johannes Hevelius and wife Elisabetha making observations, 1673. The Royal Society numbers Hevelius among its first foreign members.

The Meissner effect causes a magnet to levitate above a superconductor

Louis XIV visiting the Académie des sciences in 1671

Vera Rubin, the first astronomer to infer galactic clumping from astronomical data in 1953, was not allowed to use the telescope at Palomar until 1965, with the given reason that the facility did not have a women's restroom.

President Clinton meets the 1998 U.S. Nobel Prize winners in the White House.

Chapter 84
Baruch Spinoza

"Spinoza" redirects here. For other uses, see Spinoza (disambiguation).

Baruch Spinoza (/bəˈruːkspɪˈnoʊzə/; born Benedito de Espinosa; 24 November 1632 – 21 February 1677, later **Benedict de Spinoza**) was a Dutch philosopher.[2] The breadth and importance of Spinoza's work was not fully realized until many years after his death. By laying the groundwork for the 18th-century Enlightenment[3] and modern biblical criticism,[4] including modern conceptions of the self and, arguably, the universe,[5] he came to be considered one of the great rationalists of 17th-century philosophy.[6] His magnum opus, the posthumous *Ethics*, in which he opposed Descartes's mind–body dualism, has earned him recognition as one of Western philosophy's most important thinkers. In the *Ethics*, "Spinoza wrote the last indisputable Latin masterpiece, and one in which the refined conceptions of medieval philosophy are finally turned against themselves and destroyed entirely."[7] Philosopher Georg Wilhelm Friedrich Hegel said of all contemporary philosophers, "You are either a Spinozist or not a philosopher at all."[8]

Spinoza's given name in different languages is Hebrew: ברוך שפינוזה *Baruch Spinoza*, Portuguese: *Benedito or Bento de Espinosa* and Latin: *Benedictus de Spinoza*; in all these languages, the given name means "the Blessed". Spinoza was raised in the Portuguese Jewish community in Amsterdam. He developed highly controversial ideas regarding the authenticity of the Hebrew Bible and the nature of the Divine. The Jewish religious authorities issued a *cherem* (Hebrew: חרם , a kind of ban, shunning, ostracism, expulsion, or excommunication) against him, effectively excluding him from Jewish society at age 23. His books were also later put on the Catholic Church's *Index of Forbidden Books*. Spinoza lived an outwardly simple life as a lens grinder, turning down rewards and honors throughout his life, including prestigious teaching positions. The family inheritance he gave to his sister. His philosophical accomplishments and moral character prompted 20th-century philosopher Gilles Deleuze to name him "the 'prince' of philosophers".[9] Spinoza died at the age of 44 allegedly of a lung illness, perhaps tuberculosis or silicosis exacerbated by fine glass dust inhaled while grinding optical lenses. Spinoza is buried in the churchyard of the Christian Nieuwe Kerk in The Hague.[10]

84.1 Biography

84.1.1 Family and community origins

Spinoza's ancestors were of Sephardic Jewish descent, and were a part of the community of Portuguese Jews that had settled in the city of Amsterdam in the wake of the Alhambra Decree in Spain (1492) and the Portuguese Inquisition (1536), which had resulted in forced conversions and expulsions from the Iberian peninsula.[11]

Attracted by the Decree of Toleration issued in 1579 by the Union of Utrecht, Portuguese "conversos" first sailed to Amsterdam in 1593 and promptly reconverted to Judaism.[12] In 1598 permission was granted to build a synagogue, and in 1615 an ordinance for the admission and government of the Jews was passed.[13] As a community of exiles, the Portuguese Jews of Amsterdam were highly proud of their identity.[13]

The Spinoza family ("Espinosa" in Portuguese) probably had its origins in Espinosa de los Monteros, near Burgos, or in Espinosa de Cerrato, near Palencia, both in Northern Castile, Spain. The family was expelled from Spain in 1492

and fled to Portugal. Portugal compelled them to convert to Catholicism in 1498.[14][15]

Spinoza's father was born roughly a century after this forced conversion in the small Portuguese city of Vidigueira, near Beja in Alentejo. When Spinoza's father was still a child, Spinoza's grandfather, Isaac de Spinoza (who was from Lisbon), took his family to Nantes in France. They were expelled in 1615 and moved to Rotterdam, where Isaac died in 1627.

Spinoza's father, Miguel (Michael), and his uncle, Manuel, then moved to Amsterdam where they resumed the practice of Judaism. Miguel was a successful merchant and became a warden of the synagogue and of the Amsterdam Jewish school.[13] He buried three wives and three of his six children died before reaching adulthood.[16]

84.1.2 17th-century Holland

Amsterdam and Rotterdam operated as important cosmopolitan centers where merchant ships from many parts of the world brought people of various customs and beliefs. This flourishing commercial activity made for a culture relatively tolerant of the play of new ideas, sheltered from the censorious hand of ecclesiastical authority. Not by chance the philosophical works of both Descartes and Spinoza were developed in the cultural and intellectual background of the Dutch Republic in the 17th century.[17] Spinoza may have had access to a circle of friends who were unconventional in terms of social tradition, including members of the Collegiants.[18] One of the people he knew was Niels Stensen, a brilliant Danish student in Leiden;[19] others included Albert Burgh, with whom Spinoza is known to have corresponded.[20]

84.1.3 Early life

Map by Balthasar Florisz van Berckenrode (1625) with the present location of the Moses and Aaron church in white, but also the spot where Spinoza grew up.[21]

Baruch de Espinoza was born on 24 November 1632 in the Jodenbuurt in Amsterdam, Netherlands. He was the second son of Miguel de Espinoza, a successful, although not wealthy, Portuguese Sephardic Jewish merchant in Amsterdam.[23] His mother, Ana Débora, Miguel's second wife, died when Baruch was only six years old.[24] Spinoza's

Spinoza lived where the Moses and Aaron Church is located now, and there is strong evidence that he may have been born there.[22] mother tongue was Portuguese, although he also knew Hebrew, Spanish, Dutch, perhaps French, and later Latin.[25] Although he wrote in Latin, Spinoza learned Latin late in his youth.

Spinoza had a traditional Jewish upbringing, attending the Keter Torah yeshiva of the Amsterdam Talmud Torah

congregation headed by the learned and traditional senior Rabbi Saul Levi Morteira. His teachers also included the less traditional Rabbi Manasseh ben Israel, "a man of wide learning and secular interests, a friend of Vossius, Grotius, and Rembrandt".[26] While presumably a star pupil, and perhaps considered as a potential rabbi, Spinoza never reached the advanced study of the Torah in the upper levels of the curriculum.[27] Instead, at the age of 17, after the death of his elder brother, Isaac, he cut short his formal studies in order to begin working in the family importing business.[27]

In 1653, at age 20, Spinoza began studying Latin with Frances van den Enden (Franciscus van den Enden), a notorious free thinker, former Jesuit, and radical democrat who likely introduced Spinoza to scholastic and modern philosophy, including that of Descartes.[28] (A decade later, in the early 1660s, Van den Enden was considered to be a Cartesian and atheist,[29] and his books were put on the Catholic Index of Banned Books.)

Spinoza's father, Miguel, died in 1654 when Spinoza was 21. He duly recited Kaddish, the Jewish prayer of mourning, for eleven months as required by Jewish law.[30] When his sister Rebekah disputed his inheritance, he took her to court to establish his claim, won his case, but then renounced his claim in her favor.[31]

Spinoza adopted the Latin name Benedictus de Spinoza,[32] began boarding with Van den Enden, and began teaching in his school.[33] Following an anecdote in an early biography by Johannes Corelus,[34] he is said to have fallen in love with his teacher's daughter, Clara, but she rejected him for a richer student. (This story has been discounted on the basis that Clara Maria van den Enden was born in 1643 and would have been no more than about 18 years old when Spinoza left Amsterdam.[25] In 1671 she married Dirck Kerckring.)

During this period Spinoza also became acquainted with the Collegiants, an anti-clerical sect of Remonstrants with tendencies towards rationalism, and with the Mennonites who had existed for a century but were close to the Remonstrants.[35] Many of his friends belonged to dissident Christian groups which met regularly as discussion groups and which typically rejected the authority of established churches as well as traditional dogmas.[2]

Spinoza's break with the prevailing dogmas of Judaism, and particularly the insistence on non-Mosaic authorship of the Pentateuch, was not sudden; rather, it appears to have been the result of a lengthy internal struggle: "If anyone thinks my criticism [regarding the authorship of the Bible] is of too sweeping a nature and lacking sufficient foundation, I would ask him to undertake to show us in these narratives a definite plan such as might legitimately be imitated by historians in their chronicles... If he succeeds, I shall at once admit defeat, and he will be my mighty Apollo. For I confess that all my efforts over a long period have resulted in no such discovery. Indeed, I may add that I write nothing here that is not the fruit of lengthy reflection; and although I have been educated from boyhood in the accepted beliefs concerning Scripture, I have felt bound in the end to embrace the views I here express."[36]

Nevertheless, once branded as a heretic, Spinoza's clashes with authorities became more pronounced. For example, questioned by two members of his synagogue, Spinoza apparently responded that God has a body and nothing in scripture says otherwise.[37] He was later attacked on the steps of the synagogue by a knife-wielding assailant shouting "Heretic!" He was apparently quite shaken by this attack and for years kept (and wore) his torn cloak, unmended, as a souvenir.[38]

After his father's death in 1654, Spinoza and his younger brother Gabriel (Abraham).[30] ran the family importing business. The business ran into serious financial difficulties, however, perhaps as a result of the First Anglo-Dutch War. In March 1656, Spinoza filed suit with the Amsterdam municipal authorities to be declared an orphan in order to escape his father's business debts and so that he could inherit his mother's estate (which at first was incorporated into his father's estate) without it being subject to his father's creditors.[39] In addition, after having made substantial contributions to the Talmud Torah synagogue in 1654 and 1655, he reduced his December 1655 contribution and his March 1656 pledge to nominal amounts (and the March 1656 pledge was never paid).[40]

Spinoza was eventually able to relinquish responsibility for the business and its debts to his younger brother, Gabriel, and devote himself chiefly to the study of philosophy, especially the system expounded by Descartes, and to optics.

84.1.4 Expulsion from the Jewish community

On 27 July 1656, the Talmud Torah congregation of Amsterdamissued a writ of *cherem* (Hebrew: חרם , a kind of ban, shunning, ostracism, expulsion, or excommunication) against the 23-year-old Spinoza.[41] The following document translates the official record of the censure:[42]

The Lords of the ma'amad, having long known of the evil opinions and acts of Baruch de Espinoza, have endeavored by various means and promises, to turn him from his evil ways. But having failed to make him mend his wicked ways, and, on the contrary, daily receiving more and more serious informa
tion about the abominable heresies which he practiced and taught and about his monstrous deeds, and having for this numerous trustworthy witnesses who have deposed and born witness to this effect in the presence of the said Espinoza, they became convinced of the truth of the matter; and after all of this has been investigated in the presence of the honorable chachamin, they have decided, with their consent, that the said Espinoza should be excommunicated and expelled from the people of Israel. By the decree of the angels, and by the command of the holy men, we excommunicate, expel, curse and damn Baruch de Espinoza, with the consent of God, Blessed be He, and with the consent of all the Holy Congregation, in front of these holy Scrolls with the six-hundred-and-thirteen precepts which are written therein, with the excommunication with which Joshua banned Jericho, with the curse with which Elisha cursed the boys, and with all the curses which are written in the Book of the Law. Cursed be he by day and cursed

be he by night; cursed be he when he lies down, and cursed be he when he rises up; cursed be he when he goes out, and cursed be he when he comes in. The Lord will not spare him; the anger and wrath of the Lord will rage against this man, and bring upon him all the curses which are written in this book, and the Lord will blot out his name from under heaven, and the Lord will separate him to his injury from all the tribes of Israel with all the curses of the covenant, which are written in the Book of the Law. But you who cleave unto the Lord God are all alive this day. We order that no one should communicate with him orally or in writing, or show him any favor, or stay with him under the same roof, or within four ells of him, or read anything composed or written by him.

The Talmud Torah congregation issued censure routinely, on matters great and small, so such an edict was not unusual.[43]

The language of Spinoza's censure is unusually harsh, however, and does not appear in any other censure known to have been issued by the Portuguese Jewish community in Amsterdam.[44] The exact reason for expelling Spinoza is not stated.[45] The censure refers only to the "abominable heresies that he practiced and taught," to his "monstrous deeds," and to the testimony of witnesses "in the presence of the said Espinoza." There is no record of such testimony, but there appear to have been several likely reasons for the issuance of the censure.

First, there were Spinoza's radical theological views that he was apparently expressing in public. As philosopher and Spinoza biographer Steven Nadler puts it: "No doubt he was giving utterance to just those ideas that would soon appear in his philosophical treatises. In those works, Spinoza denies the immortality of the soul; strongly rejects the notion of a providential God—the God of Abraham, Isaac and Jacob; and claims that the Law was neither literally given by God nor any longer binding on Jews. Can there be any mystery as to why one of history's boldest and most radical thinkers was sanctioned by an orthodox Jewish community?"[46]

Second, there is ample basis to assume that the Amsterdam Jewish community, largely comprising former "conversos" having within the last century fled from the Portuguese Inquisition (and their children and grandchildren), must have been concerned to protect its reputation from any association with Spinoza lest his controversial views provide the basis for their own possible persecution or expulsion.[47] There is little or no evidence that the Amsterdam municipal authorities were directly involved in Spinoza's censure itself. But "in 1619, the town council expressly ordered [the Portuguese Jewish community] to regulate their conduct and ensure that the members of the community kept to a strict observance of Jewish law";[48] and other evidence, such as bans adopted by the synagogue itself on public wedding or funeral processions and on discussing religious matters with Christians, lest such activity might "disturb the liberty we enjoy,"[49] makes it clear that the danger of upsetting the civil authorities was never far from mind. Thus, the issuance of Spinoza's censure was almost certainly, in part, an exercise in self-censorship by the Portuguese Jewish community in Amsterdam.[50]

Third, it appears likely that Spinoza himself had already taken the initiative to separate himself from the Talmud Torah congregation and was vocally expressing his hostility to Judaism itself. He had probably stopped attending services at the synagogue either after the lawsuit with his sister or after the knife attack on its steps. He might already have been voicing the view expressed later, in his *Theological-Political Treatise*, that the civil authorities should suppress Judaism as harmful to the Jews themselves. Either for financial or other reasons,[51] he had in any case effectively stopped contributing to the synagogue by March 1656. He had also committed the "monstrous deed," contrary to the regulations of the synagogue and the views of certain rabbinical authorities (including Maimonides), of filing suit in a civil court rather than with the synagogue authorities[52]—to renounce his father's heritage, no less. Upon being notified of the issuance of the censure, he is reported to have said: "Very well; this does not force me to do anything that I would not have done of my own accord, had I not been afraid of a scandal."[53] Thus, unlike most of the censure issued routinely by the Amsterdam congregation to discipline its members, the censure issued against Spinoza did not lead to repentance and so was never withdrawn.

After the censure, Spinoza is said to have addressed an "Apology" (defense), written in Spanish, to the elders of the synagogue, "in which he defended his views as orthodox, and condemned the rabbis for accusing him of 'horrible practices and other enormities' merely because he had neglected ceremonial observances."[54] This "Apology" does not survive, but some of its contents may later have been included in his *Theological-Political Treatise*.[55] For example, he cited a series of cryptic statements by medieval biblical commentator Abraham Ibn Ezra intimating that certain apparently anachronistic passages of the Pentateuch (i.e., "[t]he Canaanite was then in the land," Genesis 12:6, which Ibn Ezra called a "mystery" and exhorted those "who understand[] it keep silent") were not of Mosaic authorship as proof that his own views had valid historical precedent.[36]

The most remarkable aspect of the censure may be not so much its issuance, or even Spinoza's refusal to submit, but the fact that Spinoza's expulsion from the Jewish community did not lead to his conversion to Christianity.[56] Spinoza kept the Latin (and so implicitly Christian) name Benedict de Spinoza, maintained a close association with the Collegiants, a Christian sect, even moved to a town near the Collegiants' headquarters, and was buried in a Christian graveyard—but there is no evidence or suggestion that he ever accepted baptism or participated in a Christian mass. Thus, by default, Baruch de Espinoza became the first secular Jew of modern Europe.[56]

In September 2012, the Portugees-Israëlietische Gemeente te Amsterdam asked the chief rabbi of their community Haham Pinchas Toledano to reconsider the cherem after consulting several Spinoza experts. However he declined to remove it, citing Spinoza's "preposterous ideas, where he was tearing apart the very fundaments of our religion", and stating that Judaism did not share the modern concept of free speech.[57]

84.1.5 Later life and career

Spinoza spent his remaining 21 years writing and studying as a private scholar.[2]

Spinoza believed in a "Philosophy of tolerance and benevolence"[58] and actually lived the life he preached. He was criticized and ridiculed during his life and afterwards for his alleged atheism. However even those who were against

him "had to admit he lived a saintly life".[58] Besides the religious controversies, nobody really had much bad to say about Spinoza other than, "he sometimes enjoyed watching spiders chase flies".[58]
After the cherem, the Amsterdam municipal authorities, "responding to the appeals of the rabbis, and also of the Calvinist clergy, who had been vicariously offended by the existence of a free thinker in the synagogue,"[59] promptly expelled Spinoza from Amsterdam. He spent a brief time in or near the village of Ouderkerk aan de Amstel, but returned soon afterwards to Amsterdam and lived there quietly for several years, giving private philosophy lessons and grinding lenses, before leaving the city in 1660 or 1661.[60]
During this time in Amsterdam, Spinoza wrote his *Short Treatise on God, Man, and His Well-Being*, "of which two Dutch translations survive, discovered about 1810."[61]
Spinoza moved around 1660 or 1661 from Amsterdam to Rijnsburg, (near Leiden), the headquarters of the Collegiants.[62] In Rijnsburg he began work on his *Descartes' "Principles of Philosophy"* as well as on his masterpiece, the *Ethics*. In 1663 he returned briefly to Amsterdam, where he finished and published *Descartes's "Principles of Philosophy"* (the only work published in his lifetime under his own name), and then moved the same year to Voorburg.[63]

84.1.6 Voorburg

In Voorburg, Spinoza continued work on the *Ethics* and corresponded with scientists, philosophers, and theologians across Europe.[64] He also wrote and in 1670 published his *Theological Political Treatise* in defense of secular and constitutional government—and in support of Jan de Witt, the Grand Pensionary of the Netherlands, against the Stadholder, the Prince of Orange.[65] Leibniz, who visited Spinoza, claimed that Spinoza's life was in danger when supporters of the Prince of Orange murdered de Witt in 1672.[66] While published anonymously, the work did not long remain so, and de Witt's enemies characterized it as "forged in Hell by a renegade Jew and the Devil, and issued with the knowledge of Jan de Witt."[67] It was condemned in 1673 by the Synod of the Reformed Church and formally banned in 1674.[68]

84.1.7 Lens-grinding and optics

Spinoza earned a modest living from lens-grinding and instrument making, yet while living in Voorburg through correspondence and friendships with scientist Christiaan Huygens and mathematician Johannes Hudde he was involved

in important optical investigations of the day including debate over microscope design with Huygens, favoring small objectives[69] and collaborating on calculations for a prospective 40 ft telescope which would have been one of the largest in Europe at the time.[70] The quality of Spinoza's lenses was much praised by Christiaan Huygens among others[71] in fact his technique and instruments were so esteemed Constantijn Huygens in 1687 ground a "clear and bright" 42 ft. telescope lens from one of Spinoza's grinding dishes 10 years after his death.[72] The exact type of lenses Spinoza made are not known, but very likely included lenses for both the microscope and telescope. He was said by anatomist Theodor Kerckring to have produced an "excellent" microscope, the quality of which was the foundation of Kerckring's anatomy claims.[73] During his time as a lens and instrument maker he was also supported by small, but regular, donations from close friends.[2]

84.1.8 The Hague

In 1670 Spinoza moved to The Hague, where he lived on a small pension from Jan de Witt and a small annuity from the brother of his dead friend, Simon de Vries.[74] He worked on the *Ethics*, wrote an unfinished Hebrew grammar, began his *Political Treatise*, wrote two scientific essays ("On the Rainbow" and "On the Calculation of Chances"), and began a Dutch translation (that he later destroyed) of the Bible.[75]
Spinoza chose the Latin word "caute" (be cautious), inscribed beneath a rose, itself a symbol of secrecy, as his device.[7] "For, having chosen to write in a language that was so widely intelligible, he was compelled to hide what he had written."[7]
Spinoza was offered the chair of philosophy at the University of Heidelberg, but he refused it because of the possibility that it might in some way curb his complete freedom of thought.
In 1676, Spinoza met with Leibniz at The Hague for a discussion of his principal philosophical work, *Ethics*, which had been completed in 1676. This meeting was described in Matthew Stewart's *The Courtier and the Heretic*.[76]
Spinoza's health began to fail in 1676, and he died on 20 February 1677, at the age of 44.[77] His premature death was said to be due to lung illness, possibly silicosis as a result of breathing in glass dust from the lenses he ground. Later, a shrine was made of his home in The Hague.[3]
Textbooks and encyclopedias often depict Spinoza as a solitary soul who eked out a living as a lens grinder; in reality, he had many friends but kept his needs to a minimum.[2] He preached a philosophy of tolerance and benevolence. Anthony Gottlieb described him as living "a saintly life."[2] The reviewer M. Stuart Phelps noted "No one has ever come nearer to the ideal life of the philosopher than Spinoza."[78] Another reviewer, Harold Bloom, wrote: "As a teacher of reality, he practiced his own wisdom, and was surely one of the most exemplary human beings ever to have lived."[79] According to the New York Times "In outward appearance he was unpretending, but not careless. His way of living was exceedingly modest and retired; often he did not leave his room for many days together. He was likewise almost incredibly frugal; his expenses sometimes amounted only to a few pence a day."[80] According to Harold Bloom and the Chicago Tribune "He appears to have had no sexual life."[79][81] Spinoza also corresponded with Peter Serrarius, a radical Protestant and millennarian merchant. Serrarius was a patron to Spinoza after Spinoza left the Jewish community, and even had letters sent and received for the philosopher to and from third parties. Spinoza and Serrarius maintained their relationship until Serrarius' death in 1669.[82] By the beginning of the 1660s, Spinoza's name became more widely known, and eventually Gottfried Leibniz[76] and Henry Oldenburg paid him visits, as stated in Matthew Stewart's *The Courtier and the Heretic*.[76] Spinoza corresponded with Oldenburg for the rest of his short life.

84.1.9 Writings and correspondence

The writings of René Descartes have been described as "Spinoza's starting point."[79] Spinoza's first publication was his geometric exposition (proofs using the geometric method on the model of Euclid with definitions, axioms, etc.) of Descartes's Parts I and II of *Principles of Philosophy* (1663). Spinoza has been associated with Leibniz and Descartes as "rationalists" in contrast to "empiricists".[83]

From December 1664 to June 1665, Spinoza engaged in correspondence with Willem van Blijenbergh, an amateur Calvinist theologian, who questioned Spinoza on the definition of evil. Later in 1665, Spinoza notified Oldenburg that he had started to work on a new book, the *Theologico-Political Treatise*, published in 1670. Leibniz disagreed harshly with Spinoza in Leibniz's own later published manuscript "Refutation of Spinoza,"[84] but he is also known to have met with Spinoza on at least one occasion[76][83] (as mentioned above), and his own work bears some striking resemblances to specific important parts of Spinoza's philosophy (see: Monadology).

When the public reactions to the anonymously published *Theologico-Political Treatise* were extremely unfavourable to his brand of Cartesianism, Spinoza was compelled to abstain from publishing more of his works. Wary and independent, he wore a signet ring which he used to mark his letters and which was engraved with a rose and the word "caute" (Latin for "cautiously").[85]

The *Ethics* and all other works, apart from the *Descartes' Principles of Philosophy* and the *Theologico-Political Treatise*, were published after his death, in the *Opera Posthuma* edited by his friends in secrecy to avoid confiscation and destruction of manuscripts. The *Ethics* contains many still-unresolved obscurities and is written with a forbidding mathematical structure modeled on Euclid's geometry[2] and has been described as a "superbly cryptic masterwork."[79]

84.2 Philosophy

84.2.1 Substance, attributes and modes

Main article: Philosophy of Spinoza

These are the fundamental concepts with which Spinoza sets forth a vision of Being, illuminated by his awareness of God. They may seem strange at first sight. To the question "What is?" he replies: "Substance, its attributes, and modes".

— Karl Jaspers[86]

Spinoza argued that God exists and is abstract and impersonal.[2] Spinoza's system imparted order and unity to the tradition of radical thought, offering powerful weapons for prevailing against "received authority." He contended that everything that exists in Nature (i.e., everything in the Universe) is one Reality (substance) and there is only one set of rules governing the whole of the reality which surrounds us and of which we are part. Spinoza viewed God and Nature as two names for the same reality,[79] namely a single, fundamental substance (meaning "that which stands beneath" rather than "matter") that is the basis of the universe and of which all lesser "entities" are actually modes or modifications, that all things are determined by Nature to exist and cause effects, and that the complex chain of cause and effect is understood only in part. His identification of God with nature was more fully explained in his posthumously published *Ethics*.[2] Spinoza's main contention with Cartesian mind–body dualism was that, if mind and body were truly distinct, then it is not clear how they can coordinate in any manner. That humans presume themselves to have free will, he argues, is a result of their awareness of appetites which affect their minds while being unable to understand the reasons why they want and act as they do. Spinoza has been described by one writer as an "Epicurean materialist,"[79] although to call Spinoza a materialist (as the Epicureans were) would be misleading as he treats both thought (the realm of the mind and thought) and extension (physical reality) as attributes of an ultimate, infinite substance, "*Deus sive Natura*" ("God or Nature"),[87] which has infinite attributes and modes (though these terms are used in quite technical ways by Spinoza).

Spinoza contends that "*Deus sive Natura*" is a being of infinitely many attributes, of which thought and extension are two. His account of the nature of reality, then, seems to treat the physical and mental worlds as intertwined, causally related, and deriving from the same substance. It is important to note here that, in Parts 3 through 4 of the *Ethics*, Spinoza describes how the human mind is affected by both mental and physical factors. He directly contests dualism. The universal substance emanates both body and mind; while they are different attributes, there is no fundamental difference between these aspects. This formulation is a historically significant solution to the mind–body problem known as neutral monism. Spinoza's system also envisages a God that does not rule over the universe by Providence in which God can make changes, but a God which itself is the deterministic system of which everything in nature is a part. Spinoza argues that "things could not have been produced by God in any other way or in any other order than is the case,";[88] he directly challenges a transcendental God which actively responds to events in the universe. Everything that has and will happen is a part of a long chain of cause and effect which, at a metaphysical level, humans are unable to change. No amount of prayer or ritual will sway God. Only knowledge of God, or the existence which humans inhabit, allows them to best respond to the world around them. Not only is it impossible for two infinite substances to exist (two infinities being absurd),[89] God—being the ultimate substance—cannot be affected by anything else, or else it would be affected by something else, and not be the fundamental substance.

Spinoza was a thoroughgoing determinist who held that absolutely everything that happens occurs through the operation of necessity. For him, even human behaviour is fully determined, with freedom being our capacity to know we are determined and to understand *why* we act as we do. By forming more "adequate" ideas about what we do and our

emotions or affections, we become the adequate cause of our effects (internal or external), which entails an increase in activity (versus passivity). This means that we become both more free and more like God, as Spinoza argues in the Scholium to Prop. 49, Part II. However, Spinoza also held that everything must necessarily happen the way that it does. Therefore, humans have no free will. They believe, however, that their will is free. This illusionary perception of freedom stems from our human consciousness, experience, and indifference to prior natural causes. Humans think they are free but they "dream with their eyes open". For Spinoza, our actions are guided entirely by natural impulses. In his letter to G. H. Schuller (Letter 58), he wrote: "men are conscious of their desire and unaware of the causes by which [their desires] are determined."[90]

This picture of Spinoza's determinism is ever more illuminated through reading this famous quote in *Ethics*: "the infant believes that it is by free will that it seeks the breast; the angry boy believes that by free will he wishes vengeance; the timid man thinks it is with free will he seeks flight; the drunkard believes that by a free command of his mind he speaks the things which when sober he wishes he had left unsaid. ... All believe that they speak by a free command of the mind, whilst, in truth, they have no power to restrain the impulse which they have to speak."[91] Thus for Spinoza morality and ethical judgment like choice is predicated on an illusion. For Spinoza, "Blame" and "Praise" are non existent human ideals only fathomable in the mind because we are so acclimatized to human consciousness interlinking with our experience that we have a false ideal of choice predicated upon this.

Spinoza's philosophy has much in common with Stoicism inasmuch as both philosophies sought to fulfill a therapeutic role by instructing people how to attain happiness. However, Spinoza differed sharply from the Stoics in one important respect: he utterly rejected their contention that reason could defeat emotion. On the contrary, he contended, an emotion can only be displaced or overcome by a stronger emotion. For him, the crucial distinction was between active and passive emotions, the former being those that are rationally understood and the latter those that are not. He also held that knowledge of true causes of passive emotion can transform it to an active emotion, thus anticipating one of the key ideas of Sigmund Freud's psychoanalysis.[92]

84.2.2 Ethical philosophy

Encapsulated at the start in his *Treatise on the Improvement of the Understanding* (*Tractatus de intellectus emendatione*) is the core of Spinoza's ethical philosophy, what he held to be the true and final good. Spinoza held good and evil to be relative concepts, claiming that nothing is intrinsically good or bad except relative to a particularity. Things that had classically been seen as good or evil, Spinoza argued, were simply good or bad for humans. Spinoza believes in a deterministic universe in which "All things in nature proceed from certain [definite] necessity and with the utmost perfection." Nothing happens by chance in Spinoza's world, and nothing is contingent.

84.2.3 Spinoza's *Ethics*

Main article: Ethics (book)

In the universe anything that happens comes from the essential nature of objects, or of God/Nature. According to Spinoza, reality is perfection. If circumstances are seen as unfortunate it is only because of our inadequate conception of reality. While components of the chain of cause and effect are not beyond the understanding of human reason, human grasp of the infinitely complex whole is limited because of the limits of science to empirically take account of the whole sequence. Spinoza also asserted that sense perception, though practical and useful, is inadequate for discovering truth. His concept of "conatus" states that human beings' natural inclination is to strive toward preserving an essential being and an assertion that virtue/human power is defined by success in this preservation of being by the guidance of reason as one's central ethical doctrine. According to Spinoza, the highest virtue is the intellectual love or knowledge of God/Nature/Universe.

In the final part of the "Ethics", his concern with the meaning of "true blessedness", and his explanation of how emotions must be detached from external cause and so master them, foreshadow psychological techniques developed in the 1900s. His concept of three types of knowledge—opinion, reason, intuition—and his assertion that intuitive knowledge provides the greatest satisfaction of mind, lead to his proposition that the more we are conscious of ourselves and Nature/Universe, the more perfect and blessed we are (in reality) and that only intuitive knowledge is eternal.

Given Spinoza's insistence on a completely ordered world where "necessity" reigns, Good and Evil have no absolute meaning. The world as it exists looks imperfect only because of our limited perception.

84.3 History of reception

84.3.1 Pantheist, panentheist, or atheist?

Main article: Spinozism

See also: Pantheism controversy

It is a widespread belief that Spinoza equated God with the material universe. He has therefore been called the "prophet"[93] and "prince"[94] and most eminent expounder of pantheism. More specifically, in a letter to Henry Oldenburg he states, "as to the view of certain people that I identify God with Nature (taken as a kind of mass or corporeal matter), they are quite mistaken".[95] For Spinoza, our universe (cosmos) is a *mode* under two *attributes* of Thought and Extension. God has infinitely many other attributes which are not present in our world.

According to German philosopher Karl Jaspers, when Spinoza wrote "*Deus sive Natura*" (God or Nature) Spinoza meant God was *Natura naturans* not *Natura naturata*, and Jaspers believed that Spinoza, in his philosophical system, did not mean to say that God and Nature are interchangeable terms, but rather that God's transcendence was attested by his infinitely many attributes, and that two attributes known by humans, namely Thought and Extension, signified God's *immanence*.[96] Even God under the attributes of thought and extension cannot be identified strictly with our world. That world is of course "divisible"; it has parts. But Spinoza insists that "no attribute of a substance can be truly conceived from which it follows that the substance can be divided" (Which means that one cannot conceive an attribute in a way that leads to division of substance), and that "a substance which is absolutely infinite is indivisible" (Ethics, Part I, Propositions 12 and 13).[97] Following this logic, our world should be considered as a mode under two attributes of thought and extension. Therefore, according to Jaspers, the pantheist formula "One and All" would apply to Spinoza only if the "One" preserves its transcendence and the "All" were not interpreted as the totality of

finite things.[96]
Martial Guéroult suggested the term "panentheism", rather than "pantheism" to describe Spinoza's view of the relation between God and the world. The world is not God, but it is, in a strong sense, "in" God. Not only do finite things have God as their cause; they cannot be conceived without God.[97] In other words, the world is a subset of God. However, American panentheist philosopher Charles Hartshorne insisted on the term Classical Pantheism to describe Spinoza's view.[98]
In 1785, Friedrich Heinrich Jacobi published a condemnation of Spinoza's pantheism, after Lessing was thought to have confessed on his deathbed to being a "Spinozist", which was the equivalent in his time of being called an atheist. Jacobi claimed that Spinoza's doctrine was pure materialism, because all Nature and God are said to be nothing but extended substance. This, for Jacobi, was the result of Enlightenment rationalism and it would finally end in absolute atheism. Moses Mendelssohn disagreed with Jacobi, saying that there is no actual difference between theism and pantheism. The issue became a major intellectual and religious concern for European civilization at the time.
The attraction of Spinoza's philosophy to late 18th-century Europeans was that it provided an alternative to materialism, atheism, and deism. Three of Spinoza's ideas strongly appealed to them:
_ the unity of all that exists;
_ the regularity of all that happens;
_ the identity of spirit and nature.
By 1879, Spinoza's pantheism was praised by many, but was considered by some to be alarming and dangerously inimical.[99]
Spinoza's "God or Nature" (*Deus sive Natura*) provided a living, natural God, in contrast to the Newtonian mechanical "First Cause" or the dead mechanism of the French "Man Machine". Coleridge and Shelley saw in Spinoza's philosophy a *religion of nature*.[2] Novalis called him the "God-intoxicated man".[79][100] Spinoza inspired the poet Shelley to write his essay "The Necessity of Atheism".[79]
Spinoza was considered to be an atheist because he used the word "God" (Deus) to signify a concept that was different from that of traditional Judeo–Christian monotheism. "Spinoza expressly denies personality and consciousness to God; he has neither intelligence, feeling, nor will; he does not act according to purpose, but everything follows necessarily from his nature, according to law...."[101] Thus, Spinoza's cool, indifferent God[102] is the antithesis to the concept of an anthropomorphic, fatherly God who cares about humanity.
According to the Stanford Encyclopedia of Philosophy: Spinoza's God is an "infinite intellect", (Ethics 2p11c) all knowing, (2p3) and capable of loving both himself—and us, insofar as we are part of his perfection. (5p35c) And

if the mark of a personal being is that it is one towards which we can entertain personal attitudes, then we should note too that Spinoza recommends *amor intellectualist dei* (the intellectual love of God) as the supreme good for man. (5p33) However, the matter is complex. Spinoza's God does not have free will (1p32c1), he does not have purposes or intentions (1apendix), and Spinoza insists that "neither intellect nor will pertain to the nature of God" (1p17s1). Moreover, while we may love God, we need to remember that God is really not the kind of being who could ever love us back. "He who loves God cannot strive that God should love him in return," says Spinoza (5p19).[103]
Steven Nadler suggests that settling the question of Spinoza's atheism or pantheism depends on an analysis of attitudes. If pantheism is associated with religiosity, then Spinoza is not a pantheist, since Spinoza believes that the proper stance to take towards God is not one of reverence or religious awe, but instead one of objective study and reason, since taking the religious stance would leave one open to the possibility of error and superstition.[104]

84.3.2 Comparison to Eastern philosophies

Similarities between Spinoza's philosophy and Eastern philosophical traditions have been discussed by many authors. The 19th-century German Sanskritist Theodore Goldstücker was one of the early figures to notice the similarities between Spinoza's religious conceptions and the Vedanta tradition of India, writing that Spinoza's thought was "... a western system of philosophy which occupies a foremost rank amongst the philosophies of all nations and ages, and which is so exact a representation of the ideas of the Vedanta, that we might have suspected its founder to have borrowed the fundamental principles of his system from the Hindus, did his biography not satisfy us that he was wholly unacquainted with their doctrines... We mean the philosophy of Spinoza, a man whose very life is a picture of that moral purity and intellectual indifference to the transitory charms of this world, which is the constant longing of the true Vedanta philosopher... comparing the fundamental ideas of both we should have no difficulty in proving that, had Spinoza been a Hindu, his system would in all probability mark a last phase of the Vedanta philosophy."[105][106]
Max Muller, in his lectures, noted the striking similarities between Vedanta and the system of Spinoza, saying "the Brahman, as conceived in the Upanishads and defined by Sankara, is clearly the same as Spinoza's 'Substantia'."[107]
Helena Blavatsky, a founder of the Theosophical Society also compared Spinoza's religious thought to Vedanta, writing in an unfinished essay "As to Spinoza's Deity—natura naturans—conceived in his attributes simply and alone; and the same Deity—as natura naturata or as conceived in the endless series of modifications or correlations, the direct outflowing results from the properties of these attributes, it is the Vedantic Deity pure and simple."[108]

84.3.3 Spinoza's reception in the 20th century

Late 20th-century Europe demonstrated a greater philosophical interest in Spinoza, often from a left-wing or Marxist perspective. Karl Marx liked Spinoza's account of the universe, interpreting it as materialistic.[2] Notable philosophers Louis Althusser, Gilles Deleuze, Antonio Negri and Étienne Balibar have each drawn upon Spinoza's philosophy. Deleuze's doctoral thesis, published in 1968, refers to him as "the prince of philosophers."[109] Other philosophers heavily influenced by Spinoza include Constantin Brunner and John David Garcia. Stuart Hampshire wrote a major English language study of Spinoza, though H. H. Joachim's work is equally valuable. Unlike most philosophers, Spinoza was highly regarded by Nietzsche.

Spinoza was an important philosophical inspiration for George Santayana. When Santayana graduated from college, he published an essay, "The Ethical Doctrine of Spinoza", in *The Harvard Monthly*.[110] Later, he wrote an introduction to *Spinoza's Ethics and "De intellectus emendatione"*.[111] In 1932, Santayana was invited to present an essay (published as "Ultimate Religion")[112] at a meeting at The Hague celebrating the tricentennial of Spinoza's birth. In Santayana's autobiography, he characterized Spinoza as his "master and model" in understanding the naturalistic basis of morality.[113]

84.3.4 Spinoza's religious criticism and its effect on the philosophy of language

Philosopher Ludwig Wittgenstein evoked Spinoza with the title (suggested to him by G. E. Moore) of the English translation of his first definitive philosophical work, *Tractatus Logico-Philosophicus*, an allusion to Spinoza's *Tractatus Theologico-Politicus*. Elsewhere, Wittgenstein deliberately borrowed the expression *sub specie aeternitatis* from Spinoza (*Notebooks, 1914-16*, p. 83). The structure of his *Tractatus Logico-Philosophicus* does have some structural affinities with Spinoza's *Ethics* (though, admittedly, not with the latter's own *Tractatus*) in erecting complex philosophical arguments upon basic logical assertions and principles. Furthermore, in propositions 6.4311 and 6.45 he alludes to a Spinozian understanding of eternity and interpretation of the religious concept of eternal life, stating

that "If by eternity is understood not eternal temporal duration, but timelessness, then he lives eternally who lives in the present." (6.4311) "The contemplation of the world sub specie aeterni is its contemplation as a limited whole." (6.45)

Leo Strauss dedicated his first book ("Spinoza's Critique of Religion") to an examination of the latter's ideas. In the book, Strauss identified Spinoza as part of the tradition of Enlightenment rationalism that eventually produced Modernity. Moreover, he identifies Spinoza and his works as the beginning of Jewish Modernity.[79] More recently Jonathan Israel, Professor of Modern European History at The Institute for Advanced Study, Princeton, has made a detailed case that from 1650 to 1750 Spinoza was "the chief challenger of the fundamentals of revealed religion, received ideas, tradition, morality, and what was everywhere regarded, in absolutist and non-absolutist states alike, as divinely constituted political authority."[114]

84.3.5 Spinoza in literature and popular culture

Spinoza has had influence beyond the confines of philosophy.

_ On the Chair's table in the Dutch Parliament, Spinoza's Tractatus theologico-politicus is one of three books, thought to be most representative of the beliefs and ethics of the Dutch people; the other two are the Bible and the Quran.

_ The 19th century novelist George Eliot produced her own translation of the *Ethics*, the first known English translation of it. Eliot liked Spinoza's vehement attacks on superstition.[2]

_ In his autobiography "From My Life: Poetry and Truth", Goethe recounts the way in which Spinoza's *Ethics* calmed the sometimes unbearable emotional turbulence of his youth. Goethe later displayed his grasp of Spinoza's metaphysics in a fragmentary elucidation of some Spinozist ontological principles entitled *Study After Spinoza*.[115] Moreover, he cited Spinoza alongside Shakespeare and Carl Linnaeus as one of the three strongest influences on his life and work.[116]

_ The 20th century novelist W. Somerset Maugham alluded to one of Spinoza's central concepts with the title of his novel *Of Human Bondage*.

_ In the early Star Trek episode, "Where No Man Has Gone Before", the antagonist, Gary Mitchell is seen reading Spinoza and the dialogue implies that Captain Kirk also may have read him as part of his studies at Starfleet Academy.

_ Albert Einstein named Spinoza as the philosopher who exerted the most influence on his world view (Weltanschauung). Spinoza equated God (infinite substance) with Nature, consistent with Einstein's belief in an impersonal deity. In 1929, Einstein was asked in a telegram by Rabbi Herbert S. Goldstein whether he believed in God. Einstein responded by telegram: "I believe in Spinoza's God who reveals himself in the orderly harmony of what exists, not in a God who concerns himself with the fates and actions of human beings."[117][118]

_ Spinoza's pantheism has also influenced environmental theory; Arne Næss, the father of the deep ecology movement, acknowledged Spinoza as an important inspiration.

_ The Argentine writer Jorge Luis Borges was greatly influenced by Spinoza's world view. Borges makes allusions to the philosopher's work in many of his poems and short stories, as does Isaac Bashevis Singer in his short story *The Spinoza of Market Street*.[119]

_ The title character of *Hoffman's Hunger*, the fifth novel by the Dutch novelist Leon de Winter, reads and comments upon the *Tractatus de Intellectus Emendatione* over the course of the novel.

_ Spinoza has been the subject of numerous biographies and scholarly treatises.[100][120][121][122]

_ Spinoza is an important historical figure in the Netherlands, where his portrait was featured prominently on the Dutch 1000-guilder banknote, legal tender until the euro was introduced in 2002. The highest and most prestigious scientific award of the Netherlands is named the *Spinozaprijs* (Spinoza prize). Spinoza was included in a 50 theme canon that attempts to summarise the history of the Netherlands.[123]

_ Spinoza's life has been honored by educators.[124]

_ In the sequel to Eric Flint's alternate-history novel, 1632, a Jewish man and his wife are killed during an attack on Amsterdam, leaving behind a less-than-year-old son. The identity of the child is quickly revealed to be the infant Spinoza himself.

_ The 2008 play "New Jerusalem," by David Ives, is based on the cherem (ban, shunning, ostracism, expulsion or excommunication) issued against Spinoza by the Talmud Torah congregation in Amsterdam in 1656, and events leading to it.[125]

84.4 Bibliography

- c. 1660. *Korte Verhandeling van God, de mensch en deszelvs welstand* (*A Short Treatise on God, Man and His Well-Being*).
- 1662. *Tractatus de Intellectus Emendatione* (*On the Improvement of the Understanding*).
- 1663. *Principia philosophiae cartesianae* (*The Principles of Cartesian Philosophy*, translated by Samuel Shirley, with an Introduction and Notes by Steven Barbone and Lee Rice, Indianapolis, 1998). Gallica (in Latin).
- 1670. *Tractatus Theologico-Politicus* (A Theologico-Political Treatise).
- 1675–76. *Tractatus Politicus* (unfinished) (PDF version)
- 1677. *Ethica Ordine Geometrico Demonstrata* (*The Ethics*, finished 1674, but published posthumously)
- 1677. *Compendium grammatices linguae hebraeae* (Hebrew Grammar).[126]
- Morgan, Michael L. (ed.), 2002. *Spinoza: Complete Works*, (Indianapolis/Cambridge: Hackett Publishing Company). ISBN 978-0-87220-620-5.
- Spruit, Leen 2011. *The Vatican Manuscript of Spinoza's Ethica*, Leiden: Brill 2011.

84.5 See also

- Criticism of Judaism
- Pantheism
- Philosophy of Spinoza
- Plane of immanence
- Spinozism

84.6 References

[1] Scruton 1986 (2002 ed.), ch. 2: "Through the works of Moses Maimonides and the commentaries of the Arab Averroës, Spinoza would have become acquainted with Aristotle."

[2] Anthony Gottlieb. "God Exists, Philosophically (review of "Spinoza: A Life" by Steven Nadler)". The New York Times, Books. 18 July 1999. Retrieved 7 September 2009.

[3] Yalom, Irvin (21 February 2012). "The Spinoza Problem". *The Washington Post.* Retrieved 7 March 2013.

[4] Yovel, Yirmiyahu, *Spinoza and Other Heretics: The Adventures of Immanence* (Princeton University Press, 1992), p. 3

[5] "Destroyer and Builder". *The New Republic.* 3 May 2012. Retrieved 7 March 2013.

[6] Scruton 1986 (2002 ed.), ch. 2, p.26

[7] Scruton 1986 (2002 ed.), ch. 1, p.32.

636 *CHAPTER 84. BARUCH SPINOZA*

[8] *Hegel's History of Philosophy.* Google Books. Archived from the original on 13 May 2011. Retrieved 2 May 2011.

[9] quoted in the translator's preface of Deleuze *Expressionism in Philosophy: Spinoza* (1990).

[10] de Spinoza, Benedictus; Hessing, Siegfried (1977). *Speculum Spinozanum, 1677-1977.* Routledge & Kegan Paul. p. 828., Snipped view of page 828

[11] Magnusson 1990.

[12] Scruton 1986 (2002 ed.), ch. 1, p.15.

[13] Scruton 1986 (2002 ed.), ch. 1, p.19.

[14] Javier Muguerza in his *Desde la perplejidad*

[15] Ben-Menahem, Ari, *Historical Encyclopedia of Natural and Mathematical Sciences, Volume 1* (Springer, 2009), p. 1095.

[16] Scruton 1986 (2002 ed.), ch. 1, p.20. (Scruton states that only Baruch and Rebekah reached adulthood, but Baruch's younger brother Gabriel apparently did as well.)

[17] Israel, J. (1998), The Dutch Republic: Its Rise, Greatness, and Fall, 1477–1806, Oxford, Oxford University Press, p. 4, p. 583, p. 677, p. 917.

[18] De Dijn, Herman, *Spinoza: The Way to Wisdom* (Purdue University Press, 1996), pp. 3 & 4.

[19] Nadler, Steven, *Spinoza: A Life* (Cambridge University Press, 2001), p. 195.

[20] Curley, Edwin, "Spinoza's exchange with Albert Burgh", in Melamed & Rosenthal (eds.), *Spinoza's* Theological-Political Treatise*: A Critical Guide* (Cambridge University Press, 2010), pp. 11–28.

[21] Historische Gids van Amsterdam, opnieuw bewerkt door Mr H.F. Wijnman, p. 205; Vaz Dias A.M. & W.G. van der Tak (1932) Spinoza, Merchant & autodidact, p. 140, 174-175. Reprint in: Studia Resenthaliana. Vol. XVI, number 2, 1982.

[22] Die Lebensgeschichte Spinozas. Zweite, stark erweiterte und vollständig neu kommentierte Auflage der Ausgabe von Jakob Freudenthal 1899. M. e. Bibliographie hg. v. Manfred Walther unter Mitarbeit v. Michael Czelinski. 2 Bde. Stuttgart-Bad Canstatt: frommann-holzboog, 2006. (Specula 4,1 – 4,2.) Erläuterungen. S. 98, 119.

[23] See Nadler 2001, ch.1, p.1.

[24] Nadler 2001, ch.2, p.23 (his mother's death when he was six years old).

[25] S. Nadler, Spinoza: A Life, Cambridge University Press, 1999, p. 47

[26] Scruton 1986 (2002 ed.), ch. 1, p.8.

[27] See Nadler 2001, ch.1, p.1

[28] Scruton 1986 (2002 ed.), ch. 1, pp.20-21; Nadler 2001, ch.2, p.27, n.27, p.189.

[29] Frank Mertens, Ghent University (30 June 2009). "Franciscus van den Enden/Biography". Retrieved 7 October 2011.

[30] Nadler 2001, ch.1, p.1.

[31] Scruton 1986 (2002 ed.), ch.1, p.21.

[32] Strathern, Paul (25 September 1998). *Spinoza in 90 Minutes.* Ivan R. Dee. pp. 24–25. ISBN 978-1-56663-215-7.

[33] Scruton 1986 (2002 ed.), ch.1, p.21; Nadler 2001 ch.2 p.27, n.27, p.189.

[34] Johannes Colerus, The Life of Benedict de Sponisa (London: Benjamin Bragg, 1706) 4

[35] Scruton 1986 (2002 ed.), ch. 1, p.20.
[36] Benedictus de Spinoza, Tractatus Theologico-Politicus: Gebhardt Edition (E.J. Brill 1989), p. 179 (available at http: //books.google.com/books?id=y3Cd7Yd73esC&pg=PA186&source=gbs_toc_r&cad=4#v=onepage&q&f=false).
[37] Scruton 1986 (2002 ed.), ch.1, p.21.
[38] Scruton 1986 (2002 ed.), ch. 1, p.21.
[39] Nadler 2001, ch.2, p.25.
[40] Nadler 2001, ch.2, pp.26-27.

[41] Scruton 1986 (2002 ed.), ch. 1, p.21.
[42] Steven Nadler, Spinoza: A Life, Cambridge University Press; 1 edition (23 April 2001), ISBN 978-0-521-00293-6, Page: 120
[43] Yitzhak Melamed, Associate Professor of Philosophy, Johns Hopkins University, speaking at an Artistic Director's Roundtable, Theater J, Washington D.C., 18 March 2012. See also Nadler 2001, ch. 1, p.7.
[44] Nadler 2001, ch.1, p.2.
[45] Steven B. Smith, *Spinoza's book of life: freedom and redemption in the Ethics*, Yale University Press (1 December 2003), p.xx/Introduction google books
[46] Steven Nadler, Baruch Spinoza, Stanford Encyclopedia of Philosophy, First published Fri 29 June 2001; substantive revision Mon 1 December 2008, plato.standord.eu
[47] Nadler 2001, ch. 2, pp. 17–22.
[48] Nadler 2001, ch.2, p.19.
[49] Nadler 2001, ch.2, p.20.
[50] See Nadler 2001, ch.2, pp.19-21.
[51] See Nadler 2001, ch.2, p.28, n.28, p.189.
[52] Nadler 2001, ch.2, pp.25-25.
[53] Scruton 1986 (2002 ed.), ch.1, p.22.
[54] Scruton 1986 (2002 ed.), ch.1, p.22.
[55] Scruton 1986 (2002 ed.), ch.1, p.22.
[56] Yitzhak Melamed, Associate Professor of Philosophy, Johns Hopkins University, speaking at an Artistic Director's Roundtable, Theater J, Washington D.C., 18 March 2012.
[57] Simon Rocker (August 28, 2014). "Why Baruch Spinoza is still excommunicated". *The Jewish Chronicle Online*.
[58] GOTTLIEB, ANTHONY. "God Exists, Philosophically". *The New York Times*. The New York Times. Retrieved 14 July 2014.
[59] Scruton 1986 (2002 ed.), ch.1, p.22.
[60] Scruton 1986 (2002 ed.), ch.1, p.22.
[61] Scruton 1986 (2002 ed.), ch.1, p.22.
[62] Scruton 1986 (2002 ed.), ch. 1, p.23.
[63] Scruton 1986 (2002 ed.), ch. 1, p.24.
[64] Scruton 1986 (2002 ed.), ch. 1, p.25.
[65] Scruton 1986 (2002 ed.), ch. 1, p.25-26.
[66] "...he [Spinoza] told me [Leibniz] he had a strong desire, on the day of the massacre of Mess. De Witt, to sally forth at night, and put up somewhere, near the place of the massacre, a paper with the words *Ultimi barbarorum* [ultimate barbarity]. But his host had shut the house to prevent his going out, for he would have run the risk of being torn to pieces." (*A Refutation Recently Discovered of Spinoza by Leibnitz*, "Remarks on the Unpublished Refutation of Spinoza by Leibnitz," Edinburg: Thomas Constable and Company, 1855. Page 70. Available on Google Books
[67] Scruton 1986 (2002 ed.), ch. 1, p.26.
[68] Scruton 1986 (2002 ed.), ch. 1, p.26.
[69] Christiaan Huygens, Oeuvres completes, Letter No. 1638, 11 Maj 1668
[70] Christiaan Huygens, Oeuvres completes, letter to his brother 23 September 1667
[71] Stephen Nadler, Spinoza: A Life (2001) p.183
[72] Christiaan Huygens, Oeuvres completes vol. XXII, p. 732, footnote

[73] Theodore Kerckring, "Spicilegium Anatomicum" Observatio XCIII (1670)
[74] Scruton 1986 (2002 ed.), ch. 1, p.26.
[75] Scruton 1986 (2002 ed.), ch. 1, p.26.
[76] Lucas, 1960.
[77] Scruton 1986 (2002 ed.), ch. 1, p.29.
[78] Phelps, M. Stuart (21 February 1877). "Spinoza. Oration by M. Ernest Renan, delivered at the Hague, February 21, 1877 by Translated by M. Stuart Phelps [pp. 763-776]". New Englander and Yale Review Volume 0037 Issue 147 (November 1878). Retrieved 8 September 2009.
[79] Harold Bloom (book reviewer) (16 June 2006). "Deciphering Spinoza, the Great Original -- Book review of "Betraying Spinoza. The Renegade Jew Who Gave Us Modernity." By Rebecca Goldstein". *The New York Times*. Retrieved 8 September 2009.
[80] "How Spinoza lived". *The New York Times*. 17 March 1878. Retrieved 8 September 2009.
[81] "New Light on Spinoza -- Joseph Freudenthal's Book, Published in German, Gives Facts.". *The Chicago Tribune*. 19 November 1899. Retrieved 8 September 2009.
[82] Popkin, Richard H., "Spinoza de Spinoza" in *The Columbia History of Western Philosophy* (Columbia University Press, 1999), p. 381.

[83] Lisa Montanarelli (book reviewer) (8 January 2006). "Spinoza stymies 'God's attorney' -- Stewart argues the secular world was at stake in Leibniz face off". *San Francisco Chronicle*. Retrieved 8 September 2009.
[84] see Refutation of Spinoza
[85] Stewart, Matthew, *The Courtier and the Heretic* (W. W. Norton & Company, 2006), p. 106.
[86] *Spinoza*, Karl Jaspers p.9
[87] *Ethics*, Part IV, preface: "*Deus seu Natura*".
[88] Baruch Spinoza. *Ethics*, in *Spinoza: Complete Works*, trans. by Samuel Shirley and ed. by Michael L. Morgan (Indianapolis: Hackett Publishing, 2002), see Part I, Proposition 33.
[89] *Ethics*, Part I, Proposition 6.
[90] *Ethics*, Pt. I, Prop. XXXVI, Appendix: "[M]en think themselves free inasmuch as they are conscious of their volitions and desires, and never even dream, in their ignorance, of the causes which have disposed them so to wish and desire."
[91] *Ethics*, Part III, Proposition 2.
[92] Roger Scruton, *Spinoza, A very Short Introduction*, p.86
[93] Picton, J. Allanson, "Pantheism: Its Story and Significance", 1905
[94] Fraser, Alexander Campbell "Philosophy of Theism", William Blackwood and Sons, 1895, p 163
[95] Correspondence of Benedict de Spinoza, Wilder Publications (26 March 2009), ISBN 978-1-60459-156-9, letter 73
[96] Karl Jaspers, Spinoza (Great Philosophers), Harvest Books (23 October 1974), ISBN 978-0-15-684730-8, Pages: 14 and 95
[97] Genevieve Lloyd, Routledge Philosophy GuideBook to Spinoza and The Ethics (Routledge Philosophy Guidebooks), Routledge; 1 edition (2 October 1996), ISBN 978-0-415-10782-2, Page: 40
[98] Charles Hartshorne and William Reese, "Philosophers Speak of God," Humanity Books, 1953 ch 4
[99] "The Pantheism of Spinoza Dr. Smith regarded as the most dangerous enemy of Christianity, and as he announced his conviction that it had gained the control of the schools, press and pulpit of the Old World [Europe], and was rapidly gaining the same control of the New [United States], his alarm and indignation sometimes rose to the eloquence of genuine passion." *Memorial of the Rev. Henry Smith, D.D., LL D., Professor of Sacred Rhetoric and Pastoral Theology in Lane Theological Seminary, Consisting of Addresses on Occasion of the Anniversary of the Seminary, May 8th, 1879, Together with Commemorative Resolutions*, p. 26.

[100] Hutchison, Percy (20 November 1932). "Spinoza, "God-Intoxicated Man"; Three Books Which Mark the Three Hundredth Anniversary of the Philosopher's Birth Blessed Spinoza. A Biography. By Lewis Browne. 319 pp. New York: The Macmillan Company. $4. Spinoza . Liberator of God and Man. By Benjamin De Casseres, 145pp. New York: E. Wickham Sweetland. $2. *Spinoza the Biospher Pinoza*. By Frederick Kettner. Introduc- tion by Nicholas Roerich, New Era Library. 255 pp. New York: Roerich Museum Press. $2.50. Spinoza". *The New York Times*. Retrieved 8 September 2009.
[101] Frank Thilly, *A History of Philosophy*, § 47, Holt & Co., New York, 1914
[102] "I believe in Spinoza's God who reveals himself in the orderly harmony of what exists, not in a God who concerns himself with fates and actions of human beings." These words were spoken by Albert Einstein, upon being asked if he believed in God by Rabbi Herbert Goldstein of the Institutional Synagogue, New York, 24 April 1921, published in the New York Times, 25 April 1929; from *Einstein: The Life and Times* Ronald W. Clark, New York: World Publishing Co., 1971, p. 413; also cited as a telegram to a Jewish newspaper, 1929, Einstein Archive 33-272, from Alice Calaprice, ed., *The Expanded Quotable Einstein*, Princeton, NJ: Princeton University
[103] "Pantheism (Stanford Encyclopedia of Philosophy)". plato.stanford.edu. Retrieved 3 October 2014.
[104] "Baruch Spinoza (Stanford Encyclopedia of Philosophy)". Plato.stanford.edu. Retrieved 24 December 2011.
[105] Literary Remains of the Late Professor Theodore Goldstucker, W. H. Allen, 1879. p32.
[106] The Westminster Review, Volumes 78-79, Baldwin, Cradock, and Joy, 1862. p1862
[107] Three Lectures on the Vedanta Philosophy. F. Max Muller. Kessinger Publishing, 2003. p123
[108] H.P Blavatsky's Collected Writings, Volume 13, pages 308-310. Quest Books
[109] Deleuze, 1968.
[110] George Santayana, "The Ethical Doctrine of Spinoza", *The Harvard Monthly*, 2 (June 1886: 144–52)
[111] George Santayana, "Introduction", in *Spinoza's Ethics and "De intellectus emendatione"*(London: Dent, 1910, vii–xxii)
[112] George Santayana, "Ultimate Religion", in *Obiter Scripta*, eds. Justus Buchler and Benjamin Schwartz (New York and London: Charles Scribner's Sons, 1936) 280-297.
[113] George Santayana, *Persons and Places* (Cambridge, MA and London: MIT Press, 1986) 233–36.
[114] Israel, J. (2001) Radical Enlightenment; Philosophy and the Making of Modernity 1650–1750, Oxford, Oxford University Press, p159.
[115] "Goethe: Studie nach Spinoza - Aufsätze und Rezensionen". Textlog.de. 30 October 2007. Archived from the original on 13 May 2011. Retrieved 2 May 2011.
[116] "Linné on line - What people have said about Linnaeus". Linnaeus.uu.se. Archived from the original on 8 June 2011. Retrieved 2 May 2011.
[117] "Einstein believes in "Spinoza's God"; Scientist Defines His Faith in Reply, to Cablegram From Rabbi Here. Sees a Divine Order But Says Its Ruler Is Not Concerned "Wit Fates and Actions of Human Beings."". *The New York Times*. 25 April 1929. Retrieved 8 September 2009.
[118] "Einstein's Third Paradise, by Gerald Holton". Aip.org. Archived from the original on 22 May 2011. Retrieved 2 May 2011.
[119] *Spinoza of Market Street and Other ... - Google Books*. Google Books. Archived from the original on 13 May 2011. Retrieved 2 May 2011.
[120] "Spinoza's First Biography Is Recovered; The oldest biography of Spinoza. Edited with Translations, Introduction, Annotations,

&c., by A. Wolf. 196 pp. New York: Lincoln Macveagh. The Dial Press.". *The New York Times*. 11 December 1927. Retrieved 8 September 2009.
[121] Irwin Edman (22 July 1934). "The Unique and Powerful Vision of Baruch Spinoza; Professor Wolfson's Long-Awaited Book Is a Work of Illuminating Scholarship. (Book review) *The Philosophy of Spinoza*. By Henry Austryn Wolfson". *The New York Times*. Retrieved 8 September 2009.
[122] Cummings, M E (8 September 1929). "Roth Evaluates Spinoza". *Los Angeles Times*. Retrieved 8 September 2009.
[123] "Entoen.nu". Entoen.nu. Archived from the original on 13 May 2011. Retrieved 2 May 2011.
[124] Richard H. Popkin (2004). "Spinoza".

[125] Ives 2009.
[126] See G. Licata, "Spinoza e la cognitio universalis dell'ebraico. Demistificazione e speculazione grammaticale nel Compendio di grammatica ebraica", Giornale di Metafisica, 3 (2009), pp. 625–661.

84.7 Sources

_ Albiac, Gabriel, 1987. *La sinagoga vacía: un estudio de las fuentes marranas del espinosismo*. Madrid: Hiperión D.L. ISBN 978-84-7517-214-9
_ Balibar, Étienne, 1985. *Spinoza et la politique* ("Spinoza and politics") Paris: PUF.
_ Bennett, Jonathan, 1984. *A Study of Spinoza's Ethics*. Hackett.
_ Boucher, Wayne I., 1999. *Spinoza in English: A Bibliography from the Seventeenth Century to the Present*. 2nd edn. Thoemmes Press.
_ Boucher, Wayne I., ed., 1999. *Spinoza: Eighteenth and Nineteenth-Century Discussions*. 6 vols. Thoemmes Press.
_ Carlisle, Claire. "Questioning Transcendence, Teleology and Truth" in *Kierkegaard and the Renaissance and Modern Traditions* (ed. Jon Stewart. Farnham: Ashgate Publishing, 2009).
_ Damásio, António, 2003. Looking for Spinoza: Joy, Sorrow, and the Feeling Brain, Harvest Books, ISBN 978-0-15-602871-4
_ Deleuze, Gilles, 1968. *Spinoza et le problème de l'expression*. Trans. "Expressionism in Philosophy: Spinoza" Martin Joughin (New York: Zone Books).
_ ———, 1970. *Spinoza: Philosophie pratique*. Transl. "Spinoza: Practical Philosophy".
_ ———, 1990. *Negotiations* trans. Martin Joughin (New York: Columbia University Press).
_ Della Rocca, Michael. 1996. *Representation and the Mind-Body Problem in Spinoza*. Oxford University Press. ISBN 978-0-19-509562-3
_ Garrett, Don, ed., 1995. *The Cambridge Companion to Spinoza*. Cambridge Uni. Press.
_ Gatens, Moira, and Lloyd, Genevieve, 1999. *Collective imaginings : Spinoza, past and present*. Routledge. ISBN 978-0-415-16570-9, ISBN 978-0-415-16571-6
_ Goldstein, Rebecca, 2006. *Betraying Spinoza: The Renegade Jew Who Gave Us Modernity*. Schocken. ISBN 978-0-8052-1159-7
_ Goode, Francis, 2012. *Life of Spinoza*. Smashwords edition. ISBN 978-1-4661-3399-0
_ Gullan-Whur, Margaret, 1998. *Within Reason: A Life of Spinoza*. Jonathan Cape. ISBN 978-0-224-05046-3
_ Hampshire, Stuart, 1951. Spinoza and Spinozism, OUP, 2005 ISBN 978-0-19-927954-8
_ Hardt, Michael, trans., University of Minnesota Press. Preface, in French, by Gilles Deleuze, available here: "Multitudes Web - 01. Préface à L'Anomalie sauvage de Negri". Multitudes.samizdat.net. Archived from the original on 11 June 2011. Retrieved 2 May 2011.
_ Israel, Jonathan, 2001. *The Radical Enlightenment*, Oxford: Oxford University Press.
_ ———, 2006. *Enlightenment Contested: Philosophy, Modernity, and the Emancipation of Man 1670–1752*, (ISBN 978-0-19-927922-7 hardback)
_ Ives 2009: Ives, David, "New Jerusalem: The Interrogation of Baruch de Spinoza at Talmud Torah Congregation: Amsterdam, July 27, 1656," 2009 (Dramatists Play Service, Inc., New York, ISBN 978-0-8222-2385-6).
_ Kasher, Asa, and Shlomo Biderman. "Why Was Baruch de Spinoza Excommunicated?"

_ Kayser, Rudolf, 1946, with an introduction by Albert Einstein. *Spinoza: Portrait of a Spiritual Hero*. New York: The Philosophical Library.
_ Lloyd, Genevieve, 1996. *Spinoza and the Ethics*. Routledge. ISBN 978-0-415-10781-5, ISBN 978-0-415-10782-2
_ LeBuffe, Michael. 2010. *Spinoza and Human Freedom*. Oxford University Press.
_ Lucas, P. G., 1960. "Some Speculative and Critical Philosophers", in I. Levine (ed.), *Philosophy* (London: Odhams)
_ Lovejoy, Arthur O., 1936. "Plenitude and Sufficient Reason in Leibniz and Spinoza" in his *The Great Chain of Being*. Harvard University Press: 144-82 (ISBN 978-0-674-36153-9). Reprinted in Frankfurt, H. G., ed., 1972. *Leibniz: A Collection of Critical Essays*. Anchor Books.
_ Macherey, Pierre, 1977. *Hegel ou Spinoza*, Maspéro (2nd ed. La Découverte, 2004).
_ ———, 1994-98. *Introduction à l'Ethique de Spinoza*. Paris: PUF.
_ Magnusson 1990: Magnusson, M (ed.), *Spinoza, Baruch*, Chambers Biographical Dictionary, Chambers 1990, ISBN 978-0-550-16041-6.
_ Matheron, Alexandre, 1969. *Individu et communauté chez Spinoza*, Paris: Minuit.
_ Montag, Warren. *Bodies, Masses, Power: Spinoza and his Contemporaries*. (London: Verso, 2002).
_ Moreau, Pierre-François, 2003, *Spinoza et le spinozisme*, PUF (Presses Universitaires de France)

_ Nadler 1999: Nadler, Steven, *Spinoza: A Life*, 1999 (Cambridge University Press, Cambridge England, ISBN 978-0-521-55210-3).
_ Nadler 2001: Nadler, Steven, *Spinoza's Heresy: Immortality and the Jewish Mind*, 2001 (Oxford University Press, Oxford England, New York NY, reprinted 2004, ISBN 0-19-926887-8).
_ Nadler 2006: Nadler, Steven, *Spinoza's Ethics: An Introduction*, 2006 (Cambridge University Press, Cambridge England, ISBN 978-0-521-83620-3).
_ Nadler 2011: Nadler, Steven, *A Book Forged in Hell: Spinoza's Scandalous Treatise and the Birth of the Secular Age*, 2011 (Princeton University Press, Princeton NJ, ISBN 978-0-691-13989-0).
_ Negri, Antonio, 1991. *The Savage Anomaly: The Power of Spinoza's Metaphysics and Politics.*
_ ———, 2004. *Subversive Spinoza: (Un)Contemporary Variations).*
_ Popkin, R. H., 2004. *Spinoza* (Oxford: One World Publications)
_ Prokhovnik, Raia (2004). *Spinoza and republicanism.* Houndmills, Basingstoke, Hampshire New York: Palgrave Macmillan. ISBN 0333733908.
_ Ratner, Joseph, 1927. *The Philosophy of Spinoza* (The Modern Library: Random House)
_ Scruton 1986: Scruton, Roger, *Spinoza: A Very Short Introduction*, 1986 (Oxford University Press, Oxford England), 2002 (reprinted as A Very Short Introduction, Oxford University Press, Oxford England, ISBN 0-19-280316-6).
_ Stewart, Matthew. *The Courtier and the Heretic: Leibniz, Spinoza and the Fate of God.* 2006. W. W. Norton
_ Stolze, Ted and Warren Montag (eds.), *The New Spinoza*; Minneapolis: University of Minnesota Press, 1997.
_ Strauss, Leo. *Persecution and the Art of Writing.* Glencoe, Ill.: Free Press, 1952. Reprint. Chicago: University of Chicago Press, 1988.
_ ———ch. 5, "How to Study Spinoza's *Tractus Theologico-Politicus*;" reprinted in Strauss, *Jewish Philosophy and the Crisis of Modernity*, ed. Kenneth Hart Green (Albany, N.Y.: SUNY Press, 1997), 181-233.
_ ———*Spinoza's Critique of Religion.* New York: Schocken Books, 1965. Reprint. University of Chicago Press, 1996.

_ ———, "Preface to the English Translation" reprinted as "Preface to Spinoza's Critique of Religion," in Strauss, *Liberalism Ancient and Modern* (New York: Basic Books, 1968, 224-59; also in Strauss, *Jewish Philosophy and the Crisis of Modernity*, 137-77).
_ Smilevski, Goce. *Conversation with SPINOZA.* Chicago: Northwestern University Press, 2006.
_ Williams, David Lay. 2010. "Spinoza and the General Will," *The Journal of Politics*, Vol. 72 (April): 341-56.
_ Wolfson, Henry A. "The Philosophy of Spinoza". 2 vols. Harvard University Press.
_ Yalom, I. (2012) The Spinoza Problem: A Novel. New York: Basic Books.
_ Yovel, Yirmiyahu, "Spinoza and Other Heretics, Vol. 1: The Marrano of Reason." Princeton, Princeton University Press, 1989.
_ Yovel, Yirmiyahu, "Spinoza and Other Heretics, Vol. 2: The Adventures of Immanence." Princeton, Princeton University Press, 1989.
_ Vinciguerra, Lorenzo *Spinoza in French Philosophy Today.* Philosophy Today, Vol. 53, No. 4, Winter 2009.

84.8 External links

_ Stanford Encyclopedia of Philosophy:
_ "Spinoza" by Steven Nadler.
_ "Spinoza's Psychological Theory" by Michael LeBuffe.
_ "Spinoza's Physical Theory" by Richard Manning.
_ "Spinoza's Political Philosophy" by Justin Steinberg.
_ Benedict De Spinoza entry by Blake Dutton in the *Internet Encyclopedia of Philosophy*
_ Bulletin Spinoza of the journal Archives de philosophie
_ Susan James on Spinoza on the Passions, *Philosophy Bites* podcast
_ Spinoza and Spinozism - BDSweb
_ Spinoza, the Moral Heretic by Matthew J. Kisner
_ Immortality in Spinoza
_ BBC Radio 4 In Our Time programme on Spinoza
_ Spinoza: Mind of the Modern - audio from Radio Opensource
_ Infography about Baruch Spinoza
_ Spinoza Csack's Website (with pdf files of Spinoza's works)
_ Spinoza's grave in The Hague
_ The Escamoth stating Spinoza's excommunication
_ Gilles Deleuze's lectures about Spinoza (1978-1981)
_ Spinoza in the Jewish Encyclopedia
_ Audio interview with Steven Nadler on Spinoza - Minerva podcast
_ Video lecture on Baruch Spinoza by Dr. Henry Abramson
Works
_ Works by Baruch Spinoza at Project Gutenberg

_ Refutation of Spinoza by Leibniz In full at Google Books
_ More easily readable versions of the Correspondence, Ethics Demonstrated in Geometrical Order and Treatise on Theology and Politics

_ EthicaDB Hypertextual and multilingual publication of Ethics
_ A Theologico-Political Treatise - English Translation
_ Theological-Political Treatise - English Translation
_ A Theologico-Political Treatise - English Translation (at sacred-texts.com)
_ A letter from Spinoza to Albert Burgh
_ Ethica Ordine Geometrico Demonstrata et in quinque partes distincta, in quibus agetur
_ *Opera posthuma* - Amsterdam 1677. Complete photographic reproduction, ed. by F. Mignini (Quodlibet publishing house website)
644 *CHAPTER 84. BARUCH SPINOZA*
Statue of Spinoza, near his house on the Paviljoensgracht in The Hague.
84.8. EXTERNAL LINKS 645
Spinoza's house in Rijnsburg from 1661 to 1663, now a museum
Study room of Spinoza
646 *CHAPTER 84. BARUCH SPINOZA*
The opening page of Spinoza's magnum opus, Ethics
84.8. EXTERNAL LINKS 647
An unfavorable engraving depiction of philosopher Spinoza, captioned in Latin, "A Jew and an Atheist".
648 *CHAPTER 84. BARUCH SPINOZA*
Tractatus Theologico-Politicus

Chapter 85

Certainty

For statistical certainty, see Probability. For the film, see Certainty (film).
"Certain" redirects here. For the French footballer, see François Certain.
Certainty is perfect knowledge that has total security from error, or the mental state of being without doubt. Objectively defined, certainty is total continuity and validity of all foundational inquiry, to the highest degree of precision. Something is certain only if no skepticism can occur. Philosophy (at least historical Cartesian philosophy) seeks this state.
It is widely held that certainty about the real world is a failed historical enterprise (that is, beyond deductive truths, tautology, etc.).[1] This is in large part due to the power of David Hume's problem of induction. Physicist Carlo Rovelli adds that certainty, in real life, is useless or often damaging (the idea is that "total security from error" is impossible in practice, and a complete "lack of doubt" is undesirable).[2]

85.1 History

85.1.1 Pyrrho – ancient Greece

Main article: Pyrrho
Pyrrho is credited as being the first Skeptic philosopher. The main principle of Pyrrho's thought is expressed by the word acatalepsia, which denotes the ability to withhold assent from doctrines regarding the truth of things in their own nature; against every statement its contradiction may be advanced with equal justification. Secondly, it is necessary in view of this fact to preserve an attitude of intellectual suspense, or, as Timon expressed it, no assertion can be known to be better than another.

85.1.2 Al-Ghazali – Islamic theologian

Main article: Al-Ghazali
Al-Ghazali was a professor of philosophy in the 11th century. His book titled *The Incoherence of the Philosophers* marks a major turn in Islamic epistemology, as Ghazali effectively discovered philosophical skepticism that would not be commonly seen in the West until Averroes, René Descartes, George Berkeley and David Hume. He described the necessity of proving the validity of reason—independently from reason. He attempted this and failed. The doubt that he introduced to his foundation of knowledge could not be reconciled using philosophy. Taking this very seriously, he resigned from his post at the university, and suffered serious psychosomatic illness. It was not until he became a religious sufi that he found a solution to his philosophical problems, which are based on Islamic religion; this encounter with skepticism led Ghazali to embrace a form of theological occasionalism, or the belief that all causal events and interactions are not the product of material conjunctions but rather the immediate and present will of God.
649
650 *CHAPTER 85. CERTAINTY*

85.1.3 Ibn-Rushd - Averroes

Main article: Ibn Rushd
Latinized name **Averroës**
Averroes was a defender of Aristotelian philosophy against Ash'ari theologians led by Al-Ghazali. Averroes' philosophy was considered controversial in Muslim circles.[3] Averroes had a greater impact on Western European circles and he has been described as the "founding father of secular thought in Western Europe".

85.1.4 Descartes – 17th century

Descartes' *Meditations on First Philosophy* is a book in which Descartes first discards all belief in things which are not absolutely certain, and then tries to establish what can be known for sure. Although the phrase "Cogito, ergo sum" is often attributed to Descartes' *Meditations on First Philosophy*, it is actually put forward in his *Discourse on*

Method. Due to the implications of inferring the conclusion within the predicate, however, he changed the argument to “I think, I exist"; this then became his first certainty.

85.1.5 Ludwig Wittgenstein – 20th century

On Certainty is a series of notes made by Ludwig Wittgenstein just prior to his death. The main theme of the work is that context plays a role in epistemology. Wittgenstein asserts an anti-foundationalist message throughout the work: that every claim can be doubted but certainty is possible in a framework. “The function [propositions] serve in language is to serve as a kind of framework within which empirical propositions can make sense”.[4]

85.2 Degrees of certainty

See also: Inductive reasoning, Probability interpretations and Philosophy of statistics

Physicist Lawrence M. Krauss suggests that identifying degrees of certainty is under-appreciated in various domains, including policy making and the understanding of science. This is because different goals require different degrees of certainty—and politicians are not always aware of (or do not make it clear) how much certainty we are working with.[5]

Rudolf Carnap viewed certainty as a matter of degree (degrees of certainty) which could be objectively measured, with degree one being certainty. Bayesian analysis derives degrees of certainty which are interpreted as a measure of subjective psychological belief.

Alternatively, one might use the legal degrees of certainty. These standards of evidence ascend as follows: no credible evidence, some credible evidence, a preponderance of evidence, clear and convincing evidence, beyond reasonable doubt, and beyond any shadow of a doubt (i.e. *undoubtable*—recognized as an impossible standard to meet—which serves only to terminate the list).

85.3 Foundational crisis of mathematics

Main article: Foundations of mathematics

The *foundational crisis of mathematics* was the early 20th century’s term for the search for proper foundations of mathematics.

After several schools of the philosophy of mathematics ran into difficulties one after the other in the 20th century, the assumption that mathematics had any foundation that could be stated within mathematics itself began to be heavily challenged.

One attempt after another to provide unassailable foundations for mathematics was found to suffer from various paradoxes (such as Russell’s paradox) and to be inconsistent.

Various schools of thought on the right approach to the foundations of mathematics were fiercely opposing each other. The leading school was that of the formalist approach, of which David Hilbert was the foremost proponent, culminating in what is known as Hilbert’s program, which sought to ground mathematics on a small basis of a formal system proved sound by metamathematical finitistic means. The main opponent was the intuitionist school, led by L.E.J. Brouwer, which resolutely discarded formalism as a meaningless game with symbols. The fight was acrimonious. In 1920 Hilbert succeeded in having Brouwer, whom he considered a threat to mathematics, removed from the editorial board of *Mathematische Annalen*, the leading mathematical journal of the time.

Gödel’s incompleteness theorems, proved in 1931, showed that essential aspects of Hilbert’s program could not be attained. In Gödel's first result he showed how to construct, for any sufficiently powerful and consistent finitely axiomatizable system—such as necessary to axiomatize the elementary theory of arithmetic—a statement that can be shown to be true, but that does not follow from the rules of the system. It thus became clear that the notion of mathematical truth can not be reduced to a purely formal system as envisaged in Hilbert’s program. In a next result Gödel showed that such a system was not powerful enough for proving its own consistency, let alone that a simpler system could do the job. This dealt a final blow to the heart of Hilbert’s program, the hope that consistency could be established by finitistic means (it was never made clear exactly what axioms were the “finitistic” ones, but whatever axiomatic system was being referred to, it was a *weaker* system than the system whose consistency it was supposed to prove). Meanwhile, the intuitionistic school had failed to attract adherents among working mathematicians, and floundered due to the difficulties of doing mathematics under the constraint of constructivism.

In a sense, the crisis has not been resolved, but faded away: most mathematicians either do not work from axiomatic systems, or if they do, do not doubt the consistency of Zermelo–Fraenkel set theory, generally their preferred axiomatic system. In most of mathematics as it is practiced, the various logical paradoxes never played a role anyway, and in those branches in which they do (such as logic and category theory), they may be avoided.

85.4 Quotes

85.5 See also

- Uncertainty
- Almost surely
- Fideism
- Gut feeling
- Infallibility
- Justified true belief
- Neuroethological innate behavior, instinct
- Pascal’s Wager
- Pragmatism
- Skeptical hypothesis

_ As contrary concepts
_ Fallibilism
_ Indeterminism
_ Multiverse

85.6 References

[1] Peat, F. David (2002). *From Certainty to Uncertainty: The Story of Science and Ideas in the Twentieth Century*. National Academies Press. ISBN 978-0-309-09620-1.
[2] edge.org
[3] "Averroës (Ibn Rushd) > By Individual Philosopher > Philosophy". Philosophybasics.com. Retrieved 2012-10-13.
[4] Wittgenstein, Ludwig. "On Certainty". SparkNotes.
[5] "question center, SHAs – cognitive tools". edge.com.

85.7 External links

_ "Certitude". *Catholic Encyclopedia*. New York: Robert Appleton Company. 1913.
_ certainty, The American Heritage Dictionary of the English Language. Bartleby.com
_ "certainty vs. doubt". *About.com*. Retrieved 2008-02-23.
_ Certainty entry by Baron Reed in the *Stanford Encyclopedia of Philosophy*
_ certainty.co.uk – The UK's National Will Register

Chapter 86

Cogito ergo sum

Cogito ergo sum[lower-alpha 1] (/ˈkoʊɡɨtoʊˈɜrɡoʊˈsʊm/, also /ˈkɒɡɨtoʊ/, /ˈsʌm/; Classical Latin: [ˈkoːɡitoː ˈɛrɡoː ˈsʊm], "I think, therefore I am", or better "I am thinking, therefore I exist") is a philosophical proposition by René Descartes. The simple meaning of the Latin phrase is that thinking about one's existence proves—in and of itself—that an "I" exists to do the thinking; or, as Descartes explains, "[W]e cannot doubt of our existence while we doubt … ."

This proposition became a fundamental element of Western philosophy, as it was perceived to form a foundation for all knowledge. While other knowledge could be a figment of imagination, deception or mistake, the very act of doubting one's own existence arguably serves as proof of the reality of one's own existence, or at least of one's thought.

Descartes' original phrase, *je pense, donc je suis* (French pronunciation: [ʒə pɑ̃s dɔ̃k ʒə sɥi]), appeared in his *Discourse on the Method* (1637), which was written in French rather than Latin to reach a wider audience in his country than scholars.[1] He used the Latin *cogito ergo sum* in the later *Principles of Philosophy* (1644).

The argument is popularly known in the English speaking world as "the *cogito ergo sum* argument" or, more briefly, as "the *cogito*".

86.1 In Descartes' writings

Descartes first wrote the phrase in French in his 1637 *Discours De la Méthode*. He referred to it in Latin without explicitly stating the familiar form of the phrase in his 1641 *Meditationes de Prima Philosophia.* The earliest written record of the phrase in Latin is in his 1644 *Principia Philosophiae*, where he also provides a clear explanation of his intent in a margin note. Fuller forms of the phrase are due to other authors. [Formatting note: *cogito* variants in this section are highlighted in **boldface** to facilitate comparison; *italics* only as in originals.]

86.1.1 In *Discours de la Méthode* (1637)

The phrase first appeared (in French) in Descartes' 1637 *Discours de la Méthode* (full title in English: *Discourse on the Method of Rightly Conducting the Reason, and Seeking Truth in the Sciences*). From the first paragraph of Part IV:

French: "… Ainsi, à cause que nos sens nous trompent quelquefois, je voulus supposer qu'il n'y avoit aucune chose qui fût telle qu'ils nous la font imaginer ; et parce qu'il y a des hommes qui se méprennent en raisonnant, même touchant les plus simples matières de géométrie, et y font des paralogismes, jugeant que j'étois sujet à faillir autant qu'aucun autre, je rejetai comme fausses toutes les raisons que j'avois prises auparavant pour démonstrations ; et enfin, considérant que toutes les mêmes pensées que nous avons étant éveillés nous peuvent aussi venir quand nous dormons, sans qu'il y en ait aucune pour lors qui soit vraie, je me résolus de feindre que toutes les choses qui m'étoient jamais entrées en l'esprit n'étoient non plus vraies que les illusions de mes songes. Mais aussitôt après je pris garde que, pendant que je voulois ainsi penser que tout étoit faux, il falloit nécessairement que moi qui le pensois fusse quelque chose ; et remarquant que cette vérité, ***je pense, donc je suis*** [italics in original], étoit si ferme et si assurée, que toutes les plus extravagantes suppositions des sceptiques n'étoient pas capables de l'ébranler, je jugeai que je pouvois la recevoir sans scrupule pour le premier principe de la philosophie que je cherchois."

English: "… Accordingly, seeing that our senses sometimes deceive us, I was willing to suppose that there existed nothing really such as they presented to us; and because some men err in reasoning, and fall into paralogisms, even on the simplest matters of geometry, I, convinced that I was as open to error as any other, rejected as false all the reasonings I had hitherto taken for demonstrations; and finally, when I considered that the very same thoughts (presentations) which we experience when awake may also be experienced when we are asleep, while there is at that time not one of them true, I supposed that all the objects (presentations) that had ever entered into my mind when awake, had in them no more truth than the illusions of my dreams. But immediately upon this I observed that, whilst I thus wished to think that all was false, it was absolutely necessary that I, who thus thought, should be somewhat; and as I observed that this truth, ***I think, therefore I am***, was so certain and of such evidence that no ground of doubt, however extravagant, could be alleged by the sceptics capable of shaking it, I concluded that I might, without scruple, accept it as the first principle of the philosophy of which I was in search."[lower-alpha 2][lower-alpha 3]

86.1.2 In *Meditationes de Prima Philosophia* (1641)

In 1641, Descartes published (in Latin) *Meditationes de Prima Philosophia* (English: *Meditations on first philosophy*) in which he referred to the proposition, though not explicitly as "cogito ergo sum" in Meditation II:

Latin: "... hoc pronuntiatum: **ego sum, ego existo**, quoties a me profertur, vel mente concipitur, necessario esse verum."

English: "... this proposition: **I am, I exist**, whenever it is uttered from me, or conceived by the mind, necessarily is true."[lower-alpha 4]

86.1.3 In *Principia Philosophiae* (1644)

In 1644, Descartes published (in Latin), *Principia Philosophiae* (English: Principles of Philosophy) where the phrase "ego cogito, ergo sum" appears in Part 1, article 7:

Latin: "Sic autem rejicientes illa omnia, de quibus aliquo modo possumus dubitare, ac etiam, falsa esse fingentes, facilè quidem, supponimus nullum esse Deum, nullum coelum, nulla corpora; nosque etiam ipsos, non habere manus, nec pedes, nec denique ullum corpus, non autem ideò nos qui talia cogitamus nihil esse: repugnat enim ut putemus id quod cogitat eo ipso tempore quo cogitat non existere. Ac proinde haec cognitio, ***ego cogito, ergo sum*** [italics in original], est omnium prima & certissima, quae cuilibet ordine philosophanti occurrat."

English: "While we thus reject all of which we can entertain the smallest doubt, and even imagine that it is false, we easily indeed suppose that there is neither God, nor sky, nor bodies, and that we ourselves even have neither hands nor feet, nor, finally, a body; but we cannot in the same way suppose that we are not while we doubt of the truth of these things; for there is a repugnance in conceiving that what thinks does not exist at the very time when it thinks. Accordingly, the knowledge, ***I think, therefore I am***, is the first and most certain that occurs to one who philosophizes orderly."[lower-alpha 5]

Descartes' margin note for the above paragraph is:

Latin: "Non posse à nobis dubitari, quin existamus dum dubitamus: at que hoc esse primum quod ordine philosophando cognoscimus."

English: "That we cannot doubt of our existence while we doubt, and that this is the first knowledge we acquire when we philosophize in order."[lower-alpha 5]

86.1.4 Other forms

The proposition is sometimes given as "**dubito, ergo cogito, ergo sum**". This fuller form was penned by the eloquent French literary critic, Antoine Léonard Thomas, in an award-winning 1765 essay in praise of Descartes, where it

appeared as "Puisque je doute, je pense ; puisque je pense, j'existe." In English, this is "Since I doubt, I think; since I think I exist"; with rearrangement and compaction, "I doubt, therefore I think, therefore I am", or in Latin, "dubito, ergo cogito, ergo sum".[lower-alpha 6]

A further expansion, "**dubito, ergo cogito, ergo sum—res cogitans**" ("...—a thinking thing") extends the *cogito* with Descartes' statement in the subsequent Meditation, "Ego sum res cogitans, id est dubitans, affirmans, negans, pauca intelligens, multa ignorans, volens, nolens, imaginans etiam et sentiens ...", or, in English, "I am a thinking (conscious) thing, that is, a being who doubts, affirms, denies, knows a few objects, and is ignorant of many ...".[lower-alpha 7] This has been referred to as "the expanded *cogito*".[11]

86.2 Interpretation

The phrase *cogito ergo sum* is not used in Descartes' *Meditations on First Philosophy* but the term "the *cogito*" is used to refer to an argument from it. In the Meditations, Descartes phrases the conclusion of the argument as "that the proposition, *I am, I exist,* is necessarily true whenever it is put forward by me or conceived in my mind." (*Meditation* II)

At the beginning of the second meditation, having reached what he considers to be the ultimate level of doubt — his argument from the existence of a deceiving god — Descartes examines his beliefs to see if any have survived the doubt. In his belief in his own existence, he finds that it is impossible to doubt that he exists. Even if there were a deceiving god (or an evil demon), one's belief in their own existence would be secure, for there is no way one could be deceived unless one existed in order to be deceived.

But I have convinced myself that there is absolutely nothing in the world, no sky, no earth, no minds, no bodies. Does it now follow that I, too, do not exist? No. If I convinced myself of something [or thought anything at all], then I certainly existed. But there is a deceiver of supreme power and cunning who deliberately and constantly deceives me. In that case, I, too, undoubtedly exist, if he deceives me; and let him deceive me as much as he can, he will never bring it about that I am nothing, so long as I think that I am something. So, after considering everything very thoroughly, I must finally conclude that the proposition, *I am, I exist,* is necessarily true whenever it is put forward by me or conceived in my mind. (AT VII 25; CSM II 16–17)

There are three important notes to keep in mind here. First, he claims only the certainty of *his own* existence from the first-person point of view — he has not proved the existence of other minds at this point. This is something that has to be thought through by each of us for ourselves, as we follow the course of the meditations. Second, he does not say that his existence is necessary; he says that *if he thinks*, then necessarily he exists (see the instantiation principle). Third, this proposition "I am, I exist" is held true not based on a deduction (as mentioned above) or on empirical induction but on the clarity and self-evidence of the proposition. Descartes does not use this first certainty, the *cogito*, as a foundation upon which to build further knowledge; rather, it is the firm ground upon which he can stand as he works to restore his beliefs. As he puts it:

Archimedes used to demand just one firm and immovable point in order to shift the entire earth; so I too can hope for great things if I manage to find just one thing, however slight, that is certain and unshakable. (AT VII 24; CSM II 16)

According to many of Descartes' specialists, including Étienne Gilson, the goal of Descartes in establishing this first truth is to demonstrate the capacity of his criterion — the immediate clarity and distinctiveness of self-evident propositions — to establish true and justified propositions despite having adopted a method of generalized doubt. As a consequence of this demonstration, Descartes considers science and mathematics to be justified to the extent that their proposals are established on a similarly immediate clarity, distinctiveness, and self-evidence that presents itself to the mind. The originality of Descartes' thinking, therefore, is not so much in expressing the cogito — a feat accomplished by other predecessors, as we shall see — but on using the cogito as demonstrating the most fundamental epistemological principle, that science and mathematics are justified by relying on clarity, distinctiveness, and selfevidence.

Baruch Spinoza in "*Principia philosophiae cartesianae*" at its *Prolegomenon* identified "cogito ergo sum" the "*ego sum cogitans*" (I am a thinking being) as the thinking substance with his ontological interpretation. It can also be considered that *Cogito ergo sum* is needed before any living being can go further in life".[12]

86.3 Predecessors

Although the idea expressed in *cogito ergo sum* is widely attributed to Descartes, he was not the first to mention it. Plato spoke about the "knowledge of knowledge" (Greek *νόησις νοήσεως* - *nóesis noéseos*) and Aristotle explains the idea in full length:

But if life itself is good and pleasant (...) and if one who sees is conscious that he sees, one who hears that he hears, one who walks that he walks and similarly for all the other human activities there is a faculty that is conscious of their exercise, so that whenever we perceive, we are conscious that we perceive, and whenever we think, we are conscious that we think, and to be conscious that we are perceiving or thinking is to be conscious that we exist... (*Nicomachean Ethics*, 1170a25 ff.)

Augustine of Hippo in *De Civitate Dei* writes *Si [...] fallor, sum* ("If I am mistaken, I am") (book XI, 26), and also anticipates modern refutations of the concept. Furthermore, in the *Enchiridion* Augustine attempts to refute skepticism by stating, "[B]y not positively affirming that they are alive, the skeptics ward off the appearance of error in themselves, yet they do make errors simply by showing themselves alive; one cannot err who is not alive. That we live is therefore not only true, but it is altogether certain as well" (Chapter 7 section 20). Another predecessor was Avicenna's "Floating Man" thought experiment on human self-awareness and self-consciousness.[13]

The 8th Century Hindu philosopher Adi Shankara wrote in a similar fashion, No one thinks, 'I am not', arguing that one's existence cannot be doubted, as there must be someone there to doubt.[14]

Further information: Gómez Pereira

86.4 Criticisms

There have been a number of criticisms of the argument. One concerns the nature of the step from "I am thinking" to "I exist." The contention is that this is a syllogistic inference, for it appears to require the extra premise: "Whatever has the property of thinking, exists", a premise Descartes did not justify. In fact, he conceded that there would indeed be an extra premise needed, but denied that the *cogito* is a syllogism (see below).

To argue that the *cogito* is not a syllogism, one may call it self-evident that "Whatever has the property of thinking, exists". In plain English, it seems incoherent to actually doubt that one exists and is doubting. Strict skeptics maintain that only the property of 'thinking' is indubitably a property of the meditator (presumably, they imagine it possible that a thing thinks but does not exist). This countercriticism is similar to the ideas of Jaakko Hintikka, who offers a nonsyllogistic interpretation of *cogito ergo sum*. He claimed that one simply cannot doubt the proposition "I exist". To be mistaken about the proposition would mean something impossible: I do not exist, but I am still wrong.

Perhaps a more relevant contention is whether the "I" to which Descartes refers is justified. In *Descartes, The Project of Pure Enquiry*, Bernard Williams provides a history and full evaluation of this issue. Apparently, the first scholar who raised the problem was Pierre Gassendi. He *points out that recognition that one has a set of thoughts does not imply that one is a particular thinker or another. Were we to move from the observation that there is thinking occurring to the attribution of this thinking to a particular agent, we would simply assume what we set out to prove, namely, that there exists a particular person endowed with the capacity for thought* . In other words, *the only claim that is indubitable here is the agent-independent claim that there is cognitive activity present*[15] The objection, as presented by Georg Lichtenberg, is that rather than supposing an entity that is thinking, Descartes should have said: "thinking is occurring." That is, whatever the force of the *cogito*, Descartes draws too much from it; the existence of a thinking thing, the reference of the "I," is more than the *cogito* can justify. Friedrich Nietzsche criticized the phrase in that it presupposes that there is an "I", that there is such an activity as "thinking", and that "I" know what "thinking" is. He suggested a more appropriate phrase would be "it thinks." In other words the "I" in "I think" could be similar to the "It" in "It is raining." David Hume claims that the philosophers who argue for a self that can be found using reason are confusing "similarity" with "identity". This means that the similarity of our thoughts and the continuity of them in this similarity do not mean that we can identify ourselves as a self but that our thoughts are similar.

86.4.1 Williams' argument in detail

In addition to the preceding two arguments against the *cogito*, other arguments have been advanced by Bernard Williams. He claims, for example, that what we are dealing with when we talk of thought, or when we say "I am thinking," is something conceivable from a third-person perspective; namely objective "thought-events" in the former case, and an objective thinker in the latter.

Williams provides a meticulous and exhaustive examination of this objection. He argues, first, that it is impossible to make sense of "there is thinking" without relativizing it to *something*. However, this something cannot be Cartesian egos, because it is impossible to differentiate objectively between things just on the basis of the pure content of consciousness.

The obvious problem is that, through introspection, or our experience of consciousness, we have no way of moving to conclude the existence of any third-personal fact, to conceive of which would require something above and beyond just the purely subjective contents of the mind.

86.4.2 Søren Kierkegaard's critique

The Danish philosopher Søren Kierkegaard provided a critical response to the *cogito*.[16] Kierkegaard argues that the *cogito* already presupposes the existence of "I", and therefore concluding with existence is logically trivial.

Kierkegaard's argument can be made clearer if one extracts the premise "I think" into two further premises:

"x" thinks

I am that "x"

Therefore I think

Therefore I am

Where "x" is used as a placeholder in order to disambiguate the "I" from the thinking thing.[17]

Here, the *cogito* has already assumed the "I"'s existence as that which thinks. For Kierkegaard, Descartes is merely "developing the content of a concept", namely that the "I", which already exists, thinks.[18]

Kierkegaard argues that the value of the *cogito* is not its logical argument, but its psychological appeal: a thought must have something that exists to think the thought. It is psychologically difficult to think "I do not exist". But as Kierkegaard argues, the proper logical flow of argument is that existence is already assumed or presupposed in order for thinking to occur, not that existence is concluded from that thinking.[19]

86.4.3 John Macmurray's Rejection

The Scottish philosopher John Macmurray rejects the *cogito* outright in order to place action at the center of a philosophical system. "We must reject this, both as standpoint and as method. If this be philosophy, then philosophy is a bubble floating in an atmosphere of unreality." [20] The reliance on thought creates an irreconcilable dualism between thought and action in which the unity of experience is lost. In order to formulate a more adequate *cogito*, Macmurray proposes the substitution of "I do" for "I think".

86.4.4 Skepticism

Many philosophical skeptics and particularly radical skeptics would say that indubitable knowledge does not exist, is impossible, or has not been found yet, and would apply this criticism to the assertion that the "cogito" is beyond doubt.

86.5 See also

_ Solipsism

_ Gómez Pereira

86.6 Notes

[1] The *cogito ergo sum* phrase was not capitalized by Descartes in his *Principia Philosophiae*.[21]

[2] This translation, from *Discours de la méthode* at Project Gutenberg, inserted the uppercase Latin phrase "COGITO ERGO SUM" in parentheses after the "I think, therefore I am". However, as this was not in the original French, it has been removed here.

[3] The 1637 *Discours* was translated to Latin in the 1644 *Specimina Philosophiae*[2] but this is not referenced here because of issues raised regarding translation quality.[3]

[4] This combines, for clarity and to retain phrase ordering, the translations of Cress[4] and Haldane.[5]

[5] Translation from *The Principles of Philosophy* at Project Gutenberg.

[6] The 1765 work, *Éloge de René Descartes*,[6] by Antoine Léonard Thomas, was awarded the 1765 Le Prix De L'académie Française and republished in the 1826 compilation of Descartes' work, *Oeuvres de Descartes*[7] by Victor Cousin. The French text is available in more accessible format at Project Gutenberg. The compilation by Cousin is credited with a revival of interest in Descartes.[8][9]

[7] This translation by Veitch[10] is the first English translation from Descartes as "I am a thinking thing".

86.7 References

[1] Burns, William E. (2001). *The scientific revolution: an encyclopedia*. Santa Barbara, California: ABC-CLIO. p. 84. ISBN 0-87436-875-8.

[2] Descartes, René (1644). *Specimina philosophiae*.

[3] Vermeulen, Corinna Lucia (2006). *René Descartes,* Specimina philosophiae. *Introduction and Critical Edition* (Dissertation, Utrecht University).

[4] Descartes, René (1986). *Discourse on Method and Meditations on First Philosophy*. Translated by Donald A. Cress. p. 65. ISBN 978-1-60384-551-9.

[5] Descartes, René (1960). *Meditations on first philosophy*. Translated by Elizabeth S. Haldane. p. 29. ISBN 978-1-61536-

207-3.
[6] Thomas, Antoine Léonard (1765). *Éloge de René Descartes.*
[7] Cousin, Victor (1824). *Oeuvres de Descartes.*
[8] *The Edinburgh Review for July, 1890 ... October, 1890.* Leonard Scott Publication Co. 1890. p. 469.
[9] Descartes, René (2007). *The Correspondence between Princess Elisabeth of Bohemia and René Descartes.* Translated by Lisa Shapiro. University of Chicago Press. p. 5. ISBN 978-0226204420.
[10] Veitch, John (1880). *The Method, Meditations and Selections from the Principles of René Descartes* (7th ed.). Edinburgh: William Blackwood and Sons. p. 115.
[11] Kline, George L. (1967). "Randall's Interpretation of the Philosophies of Descartes, Spinoza, and Leibniz". In Randall, Jr., John P. *Naturalism and Historical Understanding.* SUNY, Albany. p. 85. Unknown parameter |DUPLICATE_editor1-last= ignored (help); |first2= missing |last2= in Editors list (help)
[12] Vesey, Nicholas (2011). *Developing Consciousness.* United Kingdom: O-Books. p. 16. ISBN 978-1-84694-461-1.
[13] Nasr, Seyyed Hossein and Leaman, Oliver (1996), *History of Islamic Philosophy*, p. 315, Routledge, ISBN 0-415-13159-6.
[14] Radhakrishnan, S. (1948), *Indian Philosophy, vol II*, p. 476, George Allen & Unwin Ltd,
[15] Fisher, Saul. "Pierre Gassendi". Retrieved 2010-11-02. from Stanford Encyclopedia of Philosophy
[16] Kierkegaard, Søren. *Philosophical Fragments.* Trans. Hong, Princeton, 1985. p. 38-42.
[17] Schönbaumsfeld, Genia. *A Confusion of the Spheres.* Oxford, 2007. p.168-170.
[18] Kierkegaard, Søren. *Philosophical Fragments.* Trans. Hong, Princeton, 1985. p. 40.

[19] Archie, Lee C., "Søren Kierkegaard, *God's Existence Cannot Be Proved*". Philosophy of Religion. Lander Philosophy, 2006.
[20] Macmurray, John. *The Self as Agent.* Humanity books, 1991. p. 78.
[21] Descartes, René (1644). *Principia Philosophiae.*

86.8 Further reading

_ Abraham, W.E. "Disentangling the Cogito", *Mind* 83:329 (1974)
_ Boufoy-Bastick, Z. *Introducing 'Applicable Knowledge' as a Challenge to the Attainment of Absolute Knowledge , Sophia Journal of Philosophy*, VIII (2005), pp 39–52.
_ Descartes, R. (translated by John Cottingham), *Meditations on First Philosophy*, in *The Philosophical Writings of Descartes* vol. II (edited Cottingham, Stoothoff, and Murdoch; Cambridge University Press, 1984) ISBN 0-521-28808-8
_ Hatfield, G. *Routledge Philosophy Guidebook to Descartes and the Meditations* (Routledge, 2003) ISBN 0-415-11192-7
_ Kierkegaard, S. *Concluding Unscientific Postscript* (Princeton, 1985) ISBN 978-0-691-02081-5
_ Kierkegaard, S. *Philosophical Fragments* (Princeton, 1985) ISBN 978-0-691-02036-5
_ Williams, B. *Descartes, The Project of Pure Enquiry* (Penguin, 1978) OCLC 4025089
_ Baird, Forrest E.; Walter Kaufmann (2008). *From Plato to Derrida.* Upper Saddle River, New Jersey: Pearson Prentice Hall. ISBN 0-13-158591-6.
_ Macmurray, John. "The Self as Agent" 1951

86.9 External links

_ See External Links for Descartes' 1637 Discourse on the Method
_ See External Links for Descartes' 1641 Meditations on First Philosophy
_ See External Links for Descartes' 1644 Principles of Philosophy
_ Descartes' Epistemology entry in the *Stanford Encyclopedia of Philosophy*
_ Routledge Encyclopedia of Philosophy: Descartes — The Cogito Argument

Chapter 87

Automated theorem proving

Argonne National Laboratory was a leader in automated theorem proving from the 1960s to the 2000s

Automated theorem proving (also known as **ATP** or **automated deduction**) is a subfield of automated reasoning and mathematical logic dealing with proving mathematical theorems by computer programs. Automated reasoning over mathematical proof was a major impetus for the development of computer science.

87.1 Logical foundations

While the roots of formalised logic go back to Aristotle, the end of the 19th and early 20th centuries saw the development of modern logic and formalised mathematics. Frege's *Begriffsschrift* (1879) introduced both a complete propositional calculus and what is essentially modern predicate logic.[1] His *Foundations of Arithmetic*, published 1884,[2] expressed (parts of) mathematics in formal logic. This approach was continued by Russell and Whitehead

in their influential *Principia Mathematica*, first published 1910–1913,[3] and with a revised second edition in 1927.[4] Russell and Whitehead thought they could derive all mathematical truth using axioms and inference rules of formal logic, in principle opening up the process to automatisation. In 1920, Thoralf Skolem simplified a previous result by Leopold Löwenheim, leading to the Löwenheim–Skolem theorem and, in 1930, to the notion of a Herbrand universe and a Herbrand interpretation that allowed (un)satisfiability of first-order formulas (and hence the validity of a

theorem) to be reduced to (potentially infinitely many) propositional satisfiability problems.[5]
In 1929, Mojżesz Presburger showed that the theory of natural numbers with addition and equality (now called Presburger arithmetic in his honor) is decidable and gave an algorithm that could determine if a given sentence in the language was true or false.[6][7] However, shortly after this positive result, Kurt Gödel published *On Formally Undecidable Propositions of Principia Mathematica and Related Systems* (1931), showing that in any sufficiently strong axiomatic system there are true statements which cannot be proved in the system. This topic was further developed in the 1930s by Alonzo Church and Alan Turing, who on the one hand gave two independent but equivalent definitions of computability, and on the other gave concrete examples for undecidable questions.

87.2 First implementations

Shortly after World War II, the first general purpose computers became available. In 1954, Martin Davis programmed Presburger's algorithm for a JOHNNIAC vacuum tube computer at the Princeton Institute for Advanced Study. According to Davis, "Its great triumph was to prove that the sum of two even numbers is even".[7][8] More ambitious was the Logic Theory Machine, a deduction system for the propositional logic of the *Principia Mathematica*, developed by Allen Newell, Herbert A. Simon and J. C. Shaw. Also running on a JOHNNIAC, the Logic Theory Machine constructed proofs from a small set of propositional axioms and three deduction rules: modus ponens, (propositional) variable substitution, and the replacement of formulas by their definition. The system used heuristic guidance, and managed to prove 38 of the first 52 theorems of the *Principia*.[7]
The "heuristic" approach of the Logic Theory Machine tried to emulate human mathematicians, and could not guarantee that a proof could be found for every valid theorem even in principle. In contrast, other, more systematic algorithms achieved, at least theoretically, completeness for first-order logic. Initial approaches relied on the results of Herbrand and Skolem to convert a first-order formula into successively larger sets of propositional formulae by instantiating variables with terms from the Herbrand universe. The propositional formulas could then be checked for unsatisfiability using a number of methods. Gilmore's program used conversion to disjunctive normal form, a form in which the satisfiability of a formula is obvious.[7][9]

87.3 Decidability of the problem

Depending on the underlying logic, the problem of deciding the validity of a formula varies from trivial to impossible. For the frequent case of propositional logic, the problem is decidable but Co-NP-complete, and hence only exponential-time algorithms are believed to exist for general proof tasks. For a first order predicate calculus, Gödel's completeness theorem states that the theorems (provable statements) are exactly the logically valid well-formed formulas, so identifying valid formulas is recursively enumerable: given unbounded resources, any valid formula can eventually be proven. However, *invalid* formulas (those that are *not* entailed by a given theory), cannot always be recognized.
The above applies to first order theories, such as Peano Arithmetic. However, for a specific model that may be described by a first order theory, some statements may be true but undecidable in the theory used to describe the model. For example, by Gödel's incompleteness theorem, we know that any theory whose proper axioms are true for the natural numbers cannot prove all first order statements true for the natural numbers, even if the list of proper axioms is allowed to be infinite enumerable. It follows that an automated theorem prover will fail to terminate while searching for a proof precisely when the statement being investigated is undecidable in the theory being used, even if it is true in the model of interest. Despite this theoretical limit, in practice, theorem provers can solve many hard problems, even in models that are not fully described by any first order theory (such as the integers).

87.4 Related problems

A simpler, but related, problem is **proof verification**, where an existing proof for a theorem is certified valid. For this, it is generally required that each individual proof step can be verified by a primitive recursive function or program, and hence the problem is always decidable.
Since the proofs generated by automated theorem provers are typically very large, the problem of proof compression is crucial and various techniques aiming at making the prover's output smaller, and consequently more easily understandable and checkable, have been developed.
Proof assistants require a human user to give hints to the system. Depending on the degree of automation, the prover can essentially be reduced to a proof checker, with the user providing the proof in a formal way, or significant proof tasks can be performed automatically. Interactive provers are used for a variety of tasks, but even fully automatic systems have proved a number of interesting and hard theorems, including at least one that has eluded human mathematicians for a long time, namely the Robbins conjecture.[10][11] However, these successes are sporadic, and work on hard problems usually requires a proficient user.
Another distinction is sometimes drawn between theorem proving and other techniques, where a process is considered to be theorem proving if it consists of a traditional proof, starting with axioms and producing new inference steps using rules of inference. Other techniques would include model checking, which, in the simplest case, involves bruteforce enumeration of many possible states (although the actual implementation of model checkers requires much cleverness, and does not simply reduce to brute force).
There are hybrid theorem proving systems which use model checking as an inference rule. There are also programs which were written to prove a particular theorem, with a (usually informal) proof that if the program finishes with a certain result, then the theorem is true. A good example of this was the machine-aided proof of the four color theorem, which was very controversial as the first claimed mathematical proof which was essentially impossible to verify by humans due to the enormous size of the program's calculation (such proofs are called non-surveyable proofs). Another example would be the proof that the game Connect Four is a win for the first player.

87.5 Industrial uses

Commercial use of automated theorem proving is mostly concentrated in integrated circuit design and verification. Since the Pentium FDIV bug, the complicated floating point units of modern microprocessors have been designed with extra scrutiny. AMD, Intel and others use automated theorem proving to verify that division and other operations are correctly implemented in their processors.

87.6 First-order theorem proving

First-order theorem proving is one of the most mature subfields of automated theorem proving. The logic is expressive enough to allow the specification of arbitrary problems, often in a reasonably natural and intuitive way. On the other hand, it is still semi-decidable, and a number of sound and complete calculi have been developed, enabling *fully* automated systems. More expressive logics, such as higher order logics, allow the convenient expression of a wider range of problems than first order logic, but theorem proving for these logics is less well developed.

87.7 Benchmarks and competitions

The quality of implemented systems has benefited from the existence of a large library of standard benchmark examples — the Thousands of Problems for Theorem Provers (TPTP) Problem Library[12] — as well as from the CADE ATP System Competition (CASC), a yearly competition of first-order systems for many important classes of first-order problems.

Some important systems (all have won at least one CASC competition division) are listed below.

- E is a high-performance prover for full first-order logic, but built on a purely equational calculus, developed primarily in the automated reasoning group of Technical University of Munich.

- Otter, developed at the Argonne National Laboratory, is based on first-order resolution and paramodulation. Otter has since been replaced by Prover9, which is paired with Mace4.
- SETHEO is a high-performance system based on the goal-directed model elimination calculus. It is developed in the automated reasoning group of Technical University of Munich. E and SETHEO have been combined (with other systems) in the composite theorem prover E-SETHEO.
- Vampire is developed and implemented at Manchester University by Andrei Voronkov and Krystof Hoder, formerly also by Alexandre Riazanov. It has won the CADE ATP System Competition in the most prestigious CNF (MIX) division for eleven years (1999, 2001–2010).
- Waldmeister is a specialized system for unit-equational first-order logic. It has won the CASC UEQ division for the last fourteen years (1997–2010).
- SPASS is a first order logic theorem prover with equality. This is developed by the research group Automation of Logic, Max Planck Institute for Computer Science.

87.8 Popular techniques

- First-order resolution with unification
- Lean theorem proving
- Model elimination
- Method of analytic tableaux
- Superposition and term rewriting
- Model checking
- Mathematical induction
- Binary decision diagrams
- DPLL
- Higher-order unification

87.9 Comparison

See also: Proof assistant#Comparison and Category:Theorem proving software systems

87.9.1 Free software

- Alt-Ergo
- Automath
- CVC
- E
- Gödel-machines
- iProver
- IsaPlanner
- KED theorem prover
- LCF

- LoTREC
- MetaPRL
- NuPRL
- Paradox
- Simplify (GPL'ed since 5/2011)
- Twelf
- SPARK (programming language)

87.9.2 Proprietary software

_ Acumen RuleManager (commercial product)
_ ALLIGATOR
_ CARINE
_ KIV
_ Prover Plug-In (commercial proof engine product)
_ ProverBox
_ ResearchCyc
_ Spear modular arithmetic theorem prover

87.10 Notable people

_ Leo Bachmair, co-developer of the superposition calculus.
_ Woody Bledsoe, artificial intelligence pioneer.
_ Robert S. Boyer, co-author of the Boyer-Moore theorem prover, co-recipient of the Herbrand Award 1999.
_ Alan Bundy, University of Edinburgh, meta-level reasoning for guiding inductive proof, proof planning and recipient of 2007 IJCAI Award for Research Excellence, Herbrand Award, and 2003 Donald E. Walker Distinguished Service Award.
_ William McCune Argonne National Laboratory, author of Otter, the first high-performance theorem prover. Many important papers, recipient of the Herbrand Award 2000.
_ Hubert Comon, CNRS and now ENS Cachan. Many important papers.
_ Robert Lee Constable, Cornell University. Important contributions to type theory, NuPRL.
_ Martin Davis, author of the "Handbook of Artificial Reasoning", co-inventor of the DPLL algorithm, recipient of the Herbrand Award 2005.
_ Branden Fitelson University of California at Berkeley. Work in automated discovery of shortest axiomatic bases for logic systems.
_ Harald Ganzinger, co-developer of the superposition calculus, head of the MPI Saarbrücken, recipient of the Herbrand Award 2004 (posthumous).
_ Michael Genesereth, Stanford University professor of Computer Science.
_ Keith Goolsbey chief developer of the Cyc inference engine.
_ Michael J. C. Gordon led the development of the HOL theorem prover.

_ Gérard Huet Term rewriting, HOL logics, Herbrand Award 1998.
_ Robert Kowalski developed the connection graph theorem-prover and SLD resolution, the inference engine that executes logic programs.
_ Donald W. Loveland Duke University. Author, co-developer of the DPLL-procedure, developer of model elimination, recipient of the Herbrand Award 2001.
_ Norman Megill, developer of Metamath, and maintainer of its site at metamath.org, an online database of automatically verified proofs.
_ J Strother Moore, co-author of the Boyer–Moore theorem prover, co-recipient of the Herbrand Award 1999.
_ Robert Nieuwenhuis University of Barcelona. Co-developer of the superposition calculus.
_ Tobias Nipkow of the Technical University of Munich, contributions to (higher-order) rewriting, co-developer of the Isabelle proof assistant
_ Ross Overbeek Argonne National Laboratory. Founder of The Fellowship for Interpretation of Genomes
_ Lawrence C. Paulson of the University of Cambridge, work on higher-order logic system, co-developer of the Isabelle Theorem Prover
_ David Plaisted University of North Carolina at Chapel Hill. Complexity results, contributions to rewriting and completion, instance-based theorem proving.
_ John Rushby Program Director – SRI International[13]
_ J. Alan Robinson Syracuse University. Developed original resolution and unification based first order theorem proving, co-editor of the "Handbook of Automated Reasoning", recipient of the Herbrand Award 1996
_ Jürgen Schmidhuber Work on Gödel Machines: Self-Referential Universal Problem Solvers Making Provably Optimal Self-Improvements
_ Stephan Schulz, E theorem Prover.
_ Natarajan Shankar SRI International, work on decision procedures, *little engines of proof*, co-developer of PVS.
_ Mark Stickel SRI International. Recipient of the Herbrand Award 2002.
_ Geoff Sutcliffe University of Miami. Maintainer of the TPTP collection, an organizer of the CADE annual contest.
_ Dolph Ulrich Purdue, Work on automated discovery of shortest axiomatic bases for systems.
_ Robert Veroff University of New Mexico. Many important papers.
_ Andrei Voronkov Developer of Vampire and Co-Editor of the "Handbook of Automated Reasoning"
_ Larry Wos Argonne National Laboratory. (Otter) Many important papers. Very first Herbrand Award winner (1992)
_ Wen-Tsun Wu Work in geometric theorem proving: Wu's method, Herbrand Award 1997
_ Christoph Weidenbach, author of SPASS, automated theorem prover.

87.11 See also

_ Symbolic computation

_ Computer-aided proof
_ Automated reasoning
_ Formal verification

_ Logic programming
_ Proof checking
_ Model checking
_ Proof complexity
_ Computer algebra system
_ Program analysis (computer science)
_ General Problem Solver
_ Metamath language for formalized mathematics

87.12 Notes

[1] Frege, Gottlob (1879). *Begriffsschrift*. Verlag Louis Neuert.
[2] Frege, Gottlob (1884). *Die Grundlagen der Arithmetik*. Breslau: Wilhelm Kobner.
[3] Bertrand Russell; Alfred North Whitehead (1910–1913). *Principia Mathematica* (1st ed.). Cambridge University Press.
[4] Bertrand Russell; Alfred North Whitehead (1927). *Principia Mathematica* (2nd ed.). Cambridge University Press.
[5] Herbrand, Jaques (1930). *Recherches sur la théorie de la démonstration.*
[6] Presburger, Mojżesz (1929). "Über die Vollständigkeit eines gewissen Systems der Arithmetik ganzer Zahlen, in welchem die Addition als einzige Operation hervortritt". *Comptes Rendus du I congrès de Mathématiciens des Pays Slaves* (Warszawa): 92–101.
[7] Davis, Martin (2001), "The Early History of Automated Deduction", in Robinson, Alan; Voronkov, Andrei, *Handbook of Automated Reasoning* **1**, Elsevier)
[8] Bibel, Wolfgang (2007). "Early History and Perspectives of Automated Deduction". *KI 2007*. LNAI (Springer) (4667): 2–18. Retrieved 2 September 2012.
[9] Gilmore, Paul (1960). "A proof procedure for quantification theory: its justification and realisation". *IBM Journal of Research and Development* **4**: 28–35. doi:10.1147/rd.41.0028.
[10] W.W. McCune (1997). "Solution of the Robbins Problem". *Journal of Automated Reasoning* **19** (3).
[11] Gina Kolata (December 10, 1996). "Computer Math Proof Shows Reasoning Power". The New York Times. Retrieved 2008-10-11.
[12] Sutcliffe, Geoff. "The TPTP Problem Library for Automated Theorem Proving". Retrieved 8 September 2012.
[13] "SRI International Computer Science Laboratory – John Rushby". SRI International. Retrieved 22 September 2012.

87.13 References

_ Chin-Liang Chang; Richard Char-Tung Lee (1973). *Symbolic Logic and Mechanical Theorem Proving*. Academic Press.
_ Loveland, Donald W. (1978). *Automated Theorem Proving: A Logical Basis. Fundamental Studies in Computer Science Volume 6*. North-Holland Publishing.
_ Gallier, Jean H. (1986). *Logic for Computer Science: Foundations of Automatic Theorem Proving*. Harper & Row Publishers (Available for free download).
_ Duffy, David A. (1991). *Principles of Automated Theorem Proving*. John Wiley & Sons.
_ Wos, Larry; Overbeek, Ross; Lusk, Ewing; Boyle, Jim (1992). *Automated Reasoning: Introduction and Applications* (2nd ed.). McGraw–Hill.

_ Alan Robinson and Andrei Voronkov (eds.), ed. (2001). *Handbook of Automated Reasoning Volume I & II*. Elsevier and MIT Press.
_ Fitting, Melvin (1996). *First-Order Logic and Automated Theorem Proving* (2nd ed.). Springer.

Chapter 88

Mathematical fallacy

In mathematics, certain kinds of mistaken proof are often exhibited, and sometimes collected, as illustrations of a concept of **mathematical fallacy**. There is a distinction between a simple *mistake* and a *mathematical fallacy* in a proof: a mistake in a proof leads to an **invalid proof** just in the same way, but in the best-known examples of mathematical fallacies, there is some concealment in the presentation of the proof. For example, the reason validity fails may be a division by zero that is hidden by algebraic notation. There is a striking quality of the mathematical fallacy: as typically presented, it leads not only to an absurd result, but does so in a crafty or clever way.[1] Therefore these fallacies, for pedagogic reasons, usually take the form of spurious proofs of obvious contradictions. Although the proofs are flawed, the errors, usually by design, are comparatively subtle, or designed to show that certain steps are conditional, and should not be applied in the cases that are the exceptions to the rules.
The traditional way of presenting a mathematical fallacy is to give an invalid step of deduction mixed in with valid steps, so that the meaning of fallacy is here slightly different from the logical fallacy. The latter applies normally to a form of argument that is not a genuine rule of logic, where the problematic mathematical step is typically a correct rule applied with a tacit wrong assumption. Beyond pedagogy, the resolution of a fallacy can lead to deeper insights into a subject (such as the introduction of Pasch's axiom of Euclidean geometry).[2] *Pseudaria*, an ancient lost book of false proofs, is attributed to Euclid.[3]

Mathematical fallacies exist in many branches of mathematics. In elementary algebra, typical examples may involve a step where division by zero is performed, where a root is incorrectly extracted or, more generally, where different values of a multiple valued function are equated. Well-known fallacies also exist in elementary Euclidean geometry and calculus.

88.1 Howlers

Examples exist of *mathematically **correct** results derived by **incorrect** lines of reasoning*. Such an argument, however true the conclusion, is mathematically invalid and is commonly known as a **howler**. Consider for instance the calculation:

16
64
=
16/
6/4
=
1
4
:

Although the conclusion 16
64 = 1
4 is correct, there is a fallacious, invalid cancellation in the middle step. Bogus
proofs, calculations, or derivations constructed to produce a correct result in spite of incorrect logic or operations were termed *howlers* by Maxwell.[4] Outside the field of mathematics the term "*howler*" has various meanings, generally less specific.

88.2 Division by zero

The division-by-zero fallacy has many variants.

88.2.1 All numbers equal all other numbers

The following example uses division by zero to "prove" that 2 = 1, but can be modified to prove that any number equals any other number.

1. Let *a* and *b* be equal non-zero quantities

a = *b*

2. Multiply by *a*

*a*2 = *ab*

3. Subtract *b*2

*a*2 □ *b*2 = *ab* □ *b*2

4. Factor both sides

(*a* □ *b*)(*a* + *b*) = *b*(*a* □ *b*)

5. Divide out (*a* □ *b*)

a + *b* = *b*

6. Observing that *a* = *b*

b + *b* = *b*

7. Combine like terms on the left

2*b* = *b*

8. Divide by the non-zero *b*

2 = 1

Q.E.D.[5]

The fallacy is in line 5: the progression from line 4 to line 5 involves division by $a - b$, which is zero since *a* equals *b*. Since division by zero is undefined, the argument is invalid.

88.3 Multivalued functions

Many functions do not have a unique inverse. For instance squaring a number gives a unique value, but there are two possible square roots of a positive number. The square root is multivalued. One value can be chosen by convention as the principal value, in the case of the square root the non-negative value is the principal value, but there is no guarantee that the square root function given by this principal value of the square of a number will be equal to the original number, e.g. the square root of the square of −2 is 2.

88.4 Calculus

Calculus as the mathematical study of infinitesimal change and limits can lead to mathematical fallacies if the properties of integrals and differentials are ignored. For instance, a naive use of integration by parts can be used to give a false proof that 0 = 1.[6] Letting *u* = 1
log *x* and *dv* = *dx*
x , we may write:
∫
1

$$\frac{}{x \log x}\,dx = 1 + \int \frac{1}{x \log x}\,dx$$

after which the antiderivatives may be cancelled yielding 0 = 1. The problem is that antiderivatives are only defined up to a constant and shifting them by 1 or indeed any number is allowed. The error really comes to light when we introduce arbitrary integration limits a and b.

$$\int_a^b \frac{1}{x \log x}\,dx = 1\big|_a^b + \int_a^b \frac{1}{x \log x}\,dx = 0 + \int_a^b \frac{1}{x \log x}\,dx = \int_a^b \frac{1}{x \log x}\,dx$$

Since the difference between two values of a constant function vanishes, the same definite integral appears on both sides of the equation.

88.5 Power and root

Fallacies involving disregarding the rules of elementary arithmetic through an incorrect manipulation of the radical. For complex numbers the failure of power and logarithm identities has led to many fallacies.

88.5.1 Positive and negative roots

Invalid proofs utilizing powers and roots are often of the following kind:[7]

$$1 = \sqrt{1} = \sqrt{(-1)(-1)} = \sqrt{-1}\sqrt{-1} = i \cdot i = -1.$$

The fallacy is that the rule $\sqrt{xy} = \sqrt{x}\sqrt{y}$ is generally valid only if both x and y are positive (when dealing with real numbers), which is not the case here.

Although the fallacy is easily detected here, sometimes it is concealed more effectively in notation. For instance,[8] consider the equation

$$\cos^2 x = 1 - \sin^2 x$$

which holds as a consequence of the Pythagorean theorem. Then, by taking a square root,

$$\cos x = (1 - \sin^2 x)^{\frac{1}{2}}$$

so that

$$1 + \cos x = 1 + (1 - \sin^2 x)^{\frac{1}{2}}.$$

But evaluating this when $x = \pi$ implies

$$1 - 1 = 1 + (1 - 0)^{\frac{1}{2}}$$

or

$0 = 2$

which is incorrect.

The error in each of these examples fundamentally lies in the fact that any equation of the form

$x^2 = a^2$

has two solutions, provided $a \neq 0$,

$x = \_a$

and it is essential to check which of these solutions is relevant to the problem at hand.[9] In the above fallacy, the square root that allowed the second equation to be deduced from the first is valid only when $\cos x$ is positive. In particular, when x is set to π, the second equation is rendered invalid.

Another example of this kind of fallacy, where the error is immediately detectable, is the following invalid proof that $-2 = 2$. Letting $x = -2$, and then squaring gives

$x^2 = 4$

whereupon taking a square root implies

$x =$

p

$4 = 2;$

so that $x = -2 = 2$, which is absurd. Clearly when the square root was extracted, it was the *negative* root -2, rather than the *positive* root, that was relevant for the particular solution in the problem.

Alternatively, imaginary roots are obfuscated in the following:

p

□1 = (□1)

2

4= ((□1)2)

1

4= 1

1

4= 1

The error here lies in the last equality, where we are ignoring the other fourth roots of 1,[10] which are -1, i and $-i$ (where i is the imaginary unit). Seeing as we have squared our figure and then taken roots, we cannot always assume that all the roots will be correct. So the correct fourth are i and $-i$, which are p the imaginary numbers defined to be □1 .

88.5.2 Complex exponents

When a number is raised to a complex power, the result is not uniquely defined (see Failure of power and logarithm identities). If this property is not recognized, then errors such as the following can result:

$e2\_i = 1$

$(e2\_i)i = 1i$

e

□2_ = 1

The error here is that the rule of multiplying exponents as when going to the third line does not apply unmodified with complex exponents, even if when putting both sides to the power i only the principal value is chosen. When treated as multivalued functions, both sides produce the same set of values, being $\{e2\pi n \mid n \in \mathbb{Z}\}$.

88.6 Geometry

Many mathematical fallacies in geometry arise from using in an additive equality involving oriented quantities (such adding vectors along a given line or adding oriented angles in the plane) a valid identity, but which fixes only the absolute value of (one of) these quantities. This quantity is then incorporated into the equation with the wrong orientation, so as to produce an absurd conclusion. This wrong orientation is usually suggested implicitly by supplying an imprecise diagram of the situation, where relative positions of points or lines are chosen in a way that is actually impossible under the hypotheses of the argument, but non-obviously so. Such a fallacy is easy to expose by drawing a precise picture of the situation, in which some relative positions will be different form those in the provided diagram. In order to avoid such fallacies, a correct geometric argument using addition or subtraction of distances or angles should always prove that quantities are being incorporated with their correct orientation.

88.6.1 Fallacy of the isosceles triangle

The fallacy of the isosceles triangle, from (Maxwell 1959, Chapter II, § 1), purports to show that every triangle is isosceles, meaning that two sides of the triangle are congruent. This fallacy has been attributed to Lewis Carroll.[11]

Given a triangle △ABC, prove that AB = AC:

1. Draw a line bisecting ∠A
2. Draw the perpendicular bisector of segment BC, which bisects BC at a point D
3. Let these two lines meet at a point O.
4. Draw line OR perpendicular to AB, line OQ perpendicular to AC
5. Draw lines OB and OC
6. By RHS, △RAO ≅ △QAO (∠ORA = ∠OQA = 90; AO=AO (COMMON SIDE); ∠RAO = ∠QAO)
7. By RHS,[12] △ROB ≅ △QOC
8. Thus, AR = AQ, RB = QC, and AB = AR + RB = AQ + QC = AC

Q.E.D.
As a corollary, one can show that all triangles are equilateral, by showing that AB = BC and AC = BC in the same way.
The error in the proof is the assumption in the diagram that the point O is *inside* the triangle. In fact, O always lies at the circumcircle of the △ABC (except for isosceles and equilateral triangles where AO and OD coincides . Furthermore, it can be shown that, if AB is longer than AC, then R will lie *within* AB, while Q will lie *outside* of AC (and vice versa). (Any diagram drawn with sufficiently accurate instruments will verify the above two facts.) Because of this, AB is still AR + RB, but AC is actually AQ − QC; and thus the lengths are not necessarily the same.

88.7 Proof by induction

There exist several fallacious proofs by induction in which one of the components, basis case or inductive step, is incorrect. Intuituvely, proofs by induction work by arguing that, if a statement is true in one case, it is true in the next case, and hence by repeatedly applying this it can be shown to be true for all cases. This "proof" shows that all horses are the same colour.
1. Let us say that any group of N horses is all of the same colour.
2. If we remove a horse from the group, we have a group of N - 1 horses of the same colour. If we add another horse, we have another group of N horses. By our previous assumption, all the horses are of the same colour in this new group, since it is a group of N horses.
3. Thus we have constructed two groups of N horses all of the same colour, with N - 1 horses in common. Since these two groups have some horses in common, the two groups must be of the same colour as each other.
4. Therefore combining all the horses used, we have a group of N + 1 horses of the same colour.
5. Thus if any N horses are all the same colour, any N + 1 horses are the same colour.
6. This is clearly true for N = 1 (i.e. one horse is a group where all the horses are the same colour). Thus, by induction, N horses are the same colour for any positive integer N. i.e. all horses are the same colour.

The fallacy in this proof arises in line 3. For N = 1, the two groups of horses have N − 1 = 0 horses in common, and thus are not necessarily the same colour as each other, so the group of N + 1 = 2 horses is not necessarily all of the same colour. The implication "Every N horses are of the same color, then N+1 horses are of the same color" works for any N greater than one, but fails to be true when N=1. The basis case is correct, but the induction step has a fundamental flaw.

88.8 See also

- List of incomplete proofs
- Paradox
- Proof by intimidation

88.9 Notes

[1] Maxwell 1959, p. 9
[2] Maxwell 1959
[3] Heath & Helberg 1908, Chapter II, §I
[4] Maxwell 1959
[5] Harro Heuser: *Lehrbuch der Analysis - Teil 1*, 6th edition, Teubner 1989, ISBN 978-3-8351-0131-9, page 51 (German).
[6] Barbeau, Ed (1990). "Fallacies, Flaws and Flimflam #19: Dolt's Theorem". *The College Mathematics Journal* **21** (3): 216–218.
[7] Maxwell 1959, Chapter VI, §I.2
[8] Maxwell 1959, Chapter VI, §I.1
[9] Maxwell 1959, Chapter VI, §II
[10] In general, the expression $\sqrt[n]{1}$ evaluates to *n* complex numbers, called the *n*th roots of unity.
[11] Robin Wilson (2008). *Lewis Carroll in Numberland*. Penguin Books. pp. 169–170. ISBN 978-0-14-101610-8.
[12] Hypotenuse-leg congruence

88.10 References

- Barbeau, Edward J. (2000), *Mathematical fallacies, flaws, and flimflam*, MAA Spectrum, Mathematical Association of America, ISBN 978-0-88385-529-4, MR 1725831.
- Bunch, Bryan (1997), *Mathematical fallacies and paradoxes*, New York: Dover Publications, ISBN 978-0-486-29664-7, MR 1461270.
- Heath, Sir Thomas Little; Heiberg, Johan Ludvig (1908), *The thirteen books of Euclid's Elements, Volume 1*, The University Press.
- Maxwell, E. A. (1959), *Fallacies in mathematics*, Cambridge University Press, ISBN 0-521-05700-0, MR 0099907.

88.11 External links

- Invalid proofs at Cut-the-knot (including literature references)
- Classic fallacies with some discussion
- More invalid proofs from AhaJokes.com
- Math jokes including an invalid proof

Chapter 89

List of incomplete proofs

This page lists notable examples of incomplete published mathematical proofs. Most of these were accepted as correct for several years but later discovered to contain gaps. There are both examples where a complete proof was later found and where the alleged result turned out to be false.

Lecat (1935) is a list over a hundred pages long of errors made by mathematicians.

89.1 Legend

_ Result is correct and was later rigorously proved.
_ Result was wrong as stated, but a modified version was later rigorously proved.
_ Status of the result is unclear
_ Result is wrong and is unfixable

89.2 Examples

The examples are arranged roughly in order of the publication date of the incomplete proof. Several of the examples on the list were taken from answers to questions on the MathOverflow site, listed in the external links below.

_ Euclid's Elements. Euclid's proofs are essentially correct, but strictly speaking sometimes contain gaps because he tacitly uses some unstated assumptions, such as the existence of intersection points. In 1899 Hilbert gave a complete set of (second order) axioms for Euclidean geometry, called Hilbert's axioms, and between 1926 and 1959 Tarski gave some complete sets of first order axioms, called Tarski's axioms.

_ Infinitesimals. In the 18th century there was widespread use of infinitesimals in calculus, though these were not really well defined. Calculus was put on firm foundations in the 19th century, and Robinson put infinitesimals in a rigorous basis with the introduction of nonstandard analysis in the 20th century.

_ In 1803, Gian Francesco Malfatti claimed to prove that a certain arrangement of three circles would cover the maximum possible area inside a right triangle. However, to do so he made certain unwarranted assumptions about the configuration of the circles. It was shown in 1930 that circles in a different configuration could cover a greater area, and in 1967 that Malfatti's configuration was *never* optimal. See Malfatti circles.

_ In 1806 André-Marie Ampère claimed to prove that a continuous function is differentiable at most points, but in 1872 Weierstrass gave an example of a continuous function that was not differentiable anywhere: The Weierstrass function.

_ Uniform convergence. In his *Cours d'analyse* of 1821, Cauchy "proved" that if a sum of continuous functions converges pointwise, then its limit is also continuous. However, Abel observed three years later that this is not the case. For the conclusion to hold, "pointwise convergence" must be replaced with "uniform convergence".[1] There are many counterexamples. For example, a Fourier series of sine and cosine functions, all continuous, may converge to a discontinuous function such as a step function.

_ Intersection theory. In 1848 Steiner claimed that the number of conics tangent to 5 given conics is 7776 = 6^5, but later realized this was wrong. The correct number 3264 was found by Berner in 1865 and by de Jonquieres around 1859 and by Chasles in 1864 using his theory of characteristics. However these results, like many others in classical intersection theory, do not seem to have been given complete proofs until the work of Fulton and Macpherson in about 1978.

_ Dirichlet's principle. This was used by Riemann in 1851, but Weierstrass found a counterexample to one version of this principle in 1870, and Hilbert stated and proved a correct version in 1900.

_ In 1879, Alfred Kempe published a purported proof of the four-color map theorem, whose validity as a proof was accepted for eleven years before it was refuted. The proof did, however, suffice to show the weaker five-color map theorem. The four-color theorem was eventually proved in 1976.[2]

_ Jordan curve theorem. There has been some controversy about whether Jordan's original proof of this in 1887 contains gaps. Oswald Veblen in 1905 claimed that Jordan's proof is incomplete, but in 2007 Hales said that the gaps are minor and that Jordan's proof is essentially complete.

_ Vahlen (1891) published a purported example of an algebraic curve in 3-dimensional projective space that could not be defined as the zeros of 3 polynomials, but in 1941 Perron found 3 equations defining Vahlen's curve. In 1961 Kneser showed that any algebraic curve in projective 3-space can be given as the zeros of 3 polynomials.[3]

_ In 1898 Miller published a paper incorrectly claiming to prove that the Mathieu group M_{24} does not exist, though in 1900 he pointed out that his proof was wrong.

_ In 1905 Lebesgue tried to prove the (correct) result that a function implicitly defined by a Baire function is Baire, but his proof incorrectly assumed that the projection of a Borel set is Borel. Suslin pointed out the error and was inspired by it to define analytic sets as continuous images of Borel sets.

_ Dehn's lemma. Dehn published an attempted proof in 1910, but Kneser found a gap in 1929. It was finally proven in 1956 by Christos Papakyriakopoulos.

_ Italian school of algebraic geometry. Most gaps in proofs are caused either by a subtle technical oversight, or before the 20th century by a lack of precise definitions. A major exception to this is the Italian school of algebraic geometry in the first half of the 20th century, where lower standards of rigor gradually became acceptable. The result was that there are many papers in this area where the proofs are incomplete, or the

theorems are not stated precisely. This list contains a few representative examples, where the result was not just incompletely proved but also hopelessly wrong.
_ Perko pair, a pair of knots listed as distinct in tables for many years until Perko discovered in 1974 that they were the same. This gives a counterexample to a theorem claimed by Little in 1900 that the writhe of a reduced knot diagram is an invariant.
_ Hilbert's sixteenth problem. Henri Dulac published a partial solution to this problem in 1923, but in about 1980 Écalle and Ilyashenko independently found a serious gap, and fixed it in about 1991.[4]
_ Hilbert's twenty-first problem. In 1908 Plemelj claimed to have shown the existence a Fuchsian differential equations with any given monodromy group, but in 1989 Bolibruch discovered a counterexample.

_ Kurt Gödel proved in 1932 that the truth of a certain class of sentences of first-order arithmetic, known in the literature as $[\exists^*\forall^2\exists^*, all, (0)]$, was decidable. That is, there was a method for deciding correctly whether any statement of that form was true. In the final sentence of that paper, he asserted that the same proof would work for the decidability of the larger class $[\exists^*\forall^2\exists^*, all, (0)]_=$, which also includes formulas that contain an equality predicate. However, in the mid-1960s, Stål Aanderaa showed that Gödel's proof would *not* go through for the larger class, and in 1982 Warren Goldfarb showed that validity of formulas from the larger class was in fact undecidable.[5][6]
_ Grunwald–Wang theorem. Wilhelm Grunwald published an incorrect proof in 1933 of an incorrect theorem, and Whaples later published another incorrect proof. Shianghao Wang found a counterexample in 1948 and published a corrected version of the theorem in 1950.
_ In 1934 Severi claimed that the space of rational equivalence classes of cycles on an algebraic surface is finite-dimensional, but Mumford (1968) showed that this is false for surfaces of positive geometric genus.
_ Littlewood–Richardson rule. Robinson published an incomplete proof in 1938, though the gaps were not noticed for many years. The first complete proofs were given by Schützenberger in 1977 and Thomas in 1974.
_ Jacobian conjecture. Keller asked this as a question in 1939, and in the next few years there were several published incomplete proofs, including 3 by B. Segre, but Vitushkin found gaps in many of them. The Jacobian conjecture is (as of 2010) an open problem, and more incomplete proofs are regularly announced. Hyman Bass, Edwin H. Connell, and David Wright (1982) discuss the errors in some of these incomplete proofs.
_ One of many examples from algebraic geometry in the first half of the 20th century: Severi (1946) claimed that that a degree-n surface in 3-dimensional projective space has at most $\binom{n+2}{3}-4$ nodes, B. Segre pointed out that this was wrong; for example, for degree 6 the maximum number of nodes is 65, achieved by the Barth sextic, which is more than the maximum of 52 claimed by Severi.
_ Rokhlin invariant. Rokhlin (1951) incorrectly claimed that the third stable stem of the homotopy groups of spheres is of order 12. In 1952 he discovered his error: it is in fact cyclic of order 24. The difference is crucial as it results in the existence of the Rokhlin invariant, a fundamental tool in the theory of 3- and 4-dimensional manifolds.
_ Class numbers of imaginary quadratic fields. In 1952 Heegner published a solution to this problem. His paper was not accepted as a complete proof as it contained a gap, and the first complete proofs were given in about 1967 by Baker and Stark. In 1969 Stark showed how to fill the gap in Heegner's paper.
_ Hilbert's sixteenth problem. In the 1950s, Evgenii Landis and Ivan Petrovsky published a purported solution, but it was shown wrong in the early 1960s.[4]
_ Nielsen realization problem. Kravetz claimed to solve this in 1959 by first showing that Teichmuller space is negatively curved, but in 1974 Masur showed that it is not negatively curved. The Nielsen realization problem was finally solved in 1980 by Kerskhoff.
_ Yamabe problem. Yamabe claimed a solution in 1960, but Trudinger discovered a gap in 1968, and a complete proof was not given until 1984.
_ In 1961, Jan-Erik Roos published an incorrect theorem about the vanishing of the first derived functor of the inverse limit functor under certain general conditions.[7] However, over forty years later, Amnon Neeman and Pierre Deligne constructed a counterexample.[8]
_ Mordell conjecture over function fields. Manin published a proof in 1963, but Coleman (1990) found and corrected a gap in the proof.

_ The Schur multiplier of the Mathieu group M_{22} is particularly notorious as it was miscalculated more than once: Burgoyne & Fong (1966) first claimed it had order 3, then in a 1968 correction claimed it had order 6; its order is in fact (currently believed to be) 12. This caused an error in the title of Janko's paper *A new finite simple group of order 86,775,570,046,077,562,880 which possesses M_{24} and the full covering group of M_{22} as subgroup* on J_4: it does not have the full covering group as a subgroup, as the full covering group is larger than was realized at the time.
_ Complex structures on the 6-sphere. In 1969 Alfred Adler published a paper in the American Journal of Mathematics claiming that the 6-sphere has no complex structure. His argument was incomplete, and this is (as of 2011) still a major open problem.
_ In 1973 Britton published a 282 page attempted solution of Burnside's problem. In his proof he assumed the existence of a set of parameters satisfying some inequalities, but Adian pointed out that these inequalities were inconsistent. Novikov and Adian had previously found a correct solution around 1968.
_ Closed geodesics. In 1978 Wilhelm Klingenberg published a proof that smooth compact manifolds without boundary have infinitely many closed geodesics. His proof was controversial, and there is currently (as of 2011) no consensus on whether his proof is complete.

_ Classification of finite simple groups. In 1983, Gorenstein announced that the proof of the classification had been completed, but he had been misinformed about the status of the proof of classification of quasithin groups, which had a serious gap in it. A complete proof for this case was published by Aschbacher and Smith in 2004.
_ Kepler conjecture. Hsiang published an incomplete proof of this in 1993. Hales later published a proof (currently believed to be correct) depending on some very long computer calculations.
_ Fermat's Last Theorem. In the words of mathematical historian Howard Eves, "Fermat's Last Theorem has the peculiar distinction of being the mathematical problem for which the greatest number of incorrect proofs have been published."[9] In June 1993 Andrew Wiles presented his proof of Fermat's Last Theorem. However, it became apparent during peer review that a critical point in the proof was incorrect. It took Wiles more than a year and collaboration with a co-author before the proof was fixed.[10]:128–130
_ Busemann–Petty problem. Zhang published two papers in the Annals of Mathematics in 1994 and 1999, in the first of which he proved that the Busemann–Petty problem in $\mathbf{R}^4$ has a negative solution, and in the second of which he proved that it has a positive solution.
_ Algebraic stacks. The book Laumon & Moret-Bailly (2000) on algebraic stacks mistakenly claimed that morphisms of algebraic stacks induce morphisms of lisse-étale topoi. The results depending on this were repaired by Olsson (2007).
_ Matroid bundles. In 2003 Biss published a paper in the Annals of Mathematics claiming to show that matroid bundles are equivalent to real vector bundles, but in 2009 published a correction pointing out a serious gap in the proof.

89.3 See also

_ List of long proofs

89.4 References

[1] Porter, Roy (2003). *The Cambridge History of Science*. Cambridge University Press. p. 476. ISBN 0-521-57199-5.

[2] Thomas L. Saaty and Paul C. Kainen (1986). *The Four-Color Problem: Assaults and Conquest*. Dover Publications. ISBN 978-0-486-65092-0.
[3] http://mathoverflow.net/questions/35476
[4] Yulij Ilyashenko (2002). "Centennial History of Hilbert's 16th problem". *Bulletin of the AMS* **39** (3): 301–354. doi:10.1090/s0273-0979-02-00946-1.
[5] Boerger, Egon; Grädel, Erich; Gurevich, Yuri (1997). *The Classical Decision Problem*. Springer. p. 188. ISBN 3-540-42324-9.
[6] Goldfarb, Warren (1986). Feferman, Solomon, ed. *Kurt Gödel: Collected Works* **1**. Oxford University Press. pp. 229–231. ISBN 0-19-503964-5.
[7] Roos, Jan-Erik (1961). "Sur les foncteurs dérivés de lim. Applications.". *C. R. Acad. Sci. Paris* **252**: 3702–3704. MR 0132091.
[8] Neeman, Amnon (2002). "A counterexample to a 1961 "theorem" in homological algebra (with an appendix by P. Deligne)". *Inv. Math.* **148** (2): 397–420. doi:10.1007/s002220100197. MR 1906154.
[9] Koshy T (2001). *Elementary number theory with applications*. New York: Academic Press. p. 544. ISBN 978-0-12-421171-1.
[10] Aczel, Amir (30 September 1996). *Fermat's Last Theorem: Unlocking the Secret of an Ancient Mathematical Problem*. Four Walls Eight Windows. ISBN 978-1-56858-077-7.
_ Bass, Hyman; Connell, Edwin H.; Wright, David (1982), *The Jacobian conjecture: reduction of degree and formal expansion of the inverse, American Mathematical Society. Bulletin. New Series* **7** (2): 287–330, doi:10.1090/S0273-0979-1982-15032-7, ISBN 978-1-982150-32-7, MR 663785
_ Burgoyne, N.; Fong, Paul (1966), *The Schur multipliers of the Mathieu groups, Nagoya Mathematical Journal* **27**: 733–745, ISSN 0027-7630, MR 0197542
_ Coleman, Robert F. (1990), *Manin's proof of the Mordell conjecture over function fields, L'Enseignement Mathématique. Revue Internationale. IIe Série* **36** (3): 393–427, ISSN 0013-8584, MR 1096426
_ Laumon, Gérard; Moret-Bailly, Laurent (2000), *Champs algébriques*, Ergebnisse der Mathematik und ihrer Grenzgebiete. 3. Folge. A Series of Modern Surveys in Mathematics [Results in Mathematics and Related Areas. 3rd Series. A Series of Modern Surveys in Mathematics] **39**, Berlin, New York: Springer-Verlag, ISBN 978-3-540-65761-3, MR 1771927
_ Lecat, Maurice (1935), *Erreurs de mathématiciens des origines à nos jours*, Bruxelles - Louvain: Librairie Castaigne - Ém. Desbarax
_ Mumford, David (1968), *Rational equivalence of 0-cycles on surfaces, Journal of Mathematics of Kyoto University* **9**: 195–204, ISSN 0023-608X, MR 0249428
_ Olsson, Martin (2007), *Sheaves on Artin stacks, Journal für die reine und angewandte Mathematik* **603**: 55–112, doi:10.1515/CRELLE.2007.012, ISSN 0075-4102, MR 2312554
_ Rohlin, V. A. (1951), *Classification of mappings of an (n+3)-dimensional sphere into an n-dimensional one, Doklady Akad. Nauk SSSR (N.S.)* **81**: 19–22, MR 0046043
_ Severi, Francesco (1946), *Sul massimo numero di nodi di una superficie di dato ordine dello spazio ordinario o di una forma di un operspazio, Annali di Matematica Pura ed Applicata. Serie Quarta* **25**: 1–41, doi:10.1007/bf02418077, ISSN 0003-4622
_ Vahlen, K. T. (1891), *Bemerkung zur vollställndigen Darstellung algebraischer Raumkurven, J. Reine Angew. Math.* **108**: 346–347

89.5 External links

_ David Mumford email about the errors of the Italian algebraic geometry school under Severi

89.5.1 MathOverflow questions

_ Ilya Nikokoshev, Most interesting mathematics mistake?
_ Kevin Buzzard what mistakes did the Italian algebraic geometers actually make?
_ Will Jagy, Widely accepted mathematical results that were later shown wrong?
_ John Stillwell, What are some correct results discovered with incorrect (or no) proofs?

Chapter 90

List of long mathematical proofs

This is a list of unusually long mathematical proofs.
As of 2011, the longest mathematical proof, measured by number of published journal pages, is the classification of finite simple groups with well over 10000 pages. There are several proofs that would be far longer than this if the details of the computer calculations they depend on were published in full.

90.1 Long proofs

The length of unusually long proofs has increased with time. As a rough rule of thumb, 100 pages in 1900, or 200 pages in 1950, or 500 pages in 2000 is unusually long for a proof.
_ 1799 The Abel–Ruffini theorem was nearly proved by Paolo Ruffini, but his proof, spanning 500 pages, was mostly ignored and later, in 1824, Niels Henrik Abel published a proof that required just six pages
_ 1890 Killing's classification of simple complex Lie algebras, including his discovery of the exceptional Lie algebras, took 180 pages in 4 papers.
_ 1894 The ruler-and-compass construction of a polygon of 65537 sides by Johann Gustav Hermes took over 200 pages.
_ 1905 Lasker–Noether theorem Emanuel Lasker's original proof took 98 pages, but has since been simplified: modern proofs are less than a page long.
_ 1963 Odd order theorem This was 255 pages long, which at the time was over 10 times as long as what had previously been considered a long paper in group theory.
_ 1964 Resolution of singularities Hironaka's original proof was 216 pages long; it has since been simplified considerably down to about 10 or 20 pages.
_ 1966 Abyhankar's proof of resolution of singularities for 3-folds in characteristic greater than 6 covered about 500 pages in several papers. (In 2009 Cutkosky simplified this to about 40 pages.)
_ 1966 Discrete series representations of Lie groups. Harish-Chandra's construction of these involved a long series of papers totaling around 500 pages. His later work on the Plancherel theorem for semisimple groups added another 150 pages to these.
_ 1968 the Novikov-Adian proof solving Burnside's problem on finitely generated infinite groups with finite exponents negatively. The three-part original paper is more than 300 pages long. (Britton later published a 282 page paper attempting to solve the problem, but his paper contained a serious gap.)
_ 1960–1970 Fondements de la Géometrie Algébrique, Éléments de géométrie algébrique and Séminaire de géométrie algébrique. Grothendieck's work on the foundations of algebraic geometry covers many thousands of pages. Although this is not a proof of a single theorem, there are several theorems in it whose proofs depend on hundreds of earlier pages.

_ 1974 N-group theorem Thompson's classification of N-groups used 6 papers totaling about 400 pages, but also used earlier results of his such as the odd order theorem, which bring to total length up to more than 700 pages.
_ 1974 Ramanujan conjecture and the Weil conjectures. While Deligne's final paper proving these was "only" about 30 pages long, it depended on background results in algebraic geometry and étale cohomology that Deligne estimated to be about 2000 pages long.
_ 1974 4-color theorem. Appel and Haken's proof of this took 139 pages, and also depended on long computer calculations.
_ 1974 The Gorenstein–Harada theorem classifying finite groups of sectional 2-rank at most 4 was 464 pages long.
_ 1976 Eisenstein series Langlands's proof of the functional equation for Eisenstein series was 337 pages long.
_ 1983 Trichotomy theorem Gorenstein and Lyons's proof for the case of rank at least 4 was 731 pages long, and Aschbacher's proof of the rank 3 case adds another 159 pages, for a total of 890 pages.
_ 1983 Selberg trace formula Hejhal's proof of a general form of the Selberg trace formula consisted of 2 volumes with a total length of 1322 pages.
_ Arthur–Selberg trace formula. Arthur's proofs of the various versions of this cover several hundred pages spread over many papers.
_ 2000 Almgren's regularity theorem Almgren's proof was 955 pages long.
_ 2000 Lafforgue's theorem on the Langlands conjecture for the general linear group over function fields. Laurent Lafforgue's proof of this was about 600 pages long, not counting many pages of background results.

_ 2003 Poincaré conjecture, Geometrization theorem, Geometrization conjecture. Perelman's original proofs of the Poincaré conjecture and the Geometrization conjecture were not lengthy, but were rather sketchy. Several other mathematicians have published proofs with the details filled in, which come to several hundred pages.
_ 2004 Quasi-thin groups The classification of the simple quasi-thin groups by Aschbacher and Smith was 1221 pages long, one of the longest single papers ever written.
_ 2004 Classification of finite simple groups. The proof of this is spread out over hundreds of journal articles which makes it hard to estimate its total length, which is probably around 10000 to 20000 pages.
_ 2004 Robertson–Seymour theorem. The proof takes about 500 pages spread over about 20 papers.
_ 2005 Kepler conjecture Hales's proof of this involves several hundred pages of published arguments, together with several gigabytes of computer calculations.
_ 2006 the strong perfect graph theorem, by Maria Chudnovsky, Neil Robertson, Paul Seymour, and Robin Thomas. 180 pages in the Annals of Mathematics.
_ 2012 Inter-universal Teichmüller theory Mochizuki's work on this covers many hundreds of pages spread over several long papers.

90.2 Long computer calculations

There are many mathematical theorems that have been checked by long computer calculations. If these were written out as proofs many would be far longer than most of the proofs above. There is not really a clear distinction between computer calculations and proofs, as several of the proofs above, such as the 4-color theorem and the Kepler conjecture, use long computer calculations as well as many pages of mathematical argument. For the computer calculations in this section, the mathematical arguments are only a few pages long, and the length is due to long but routine calculations. Some typical examples of such theorems include:
_ Several proofs of the existence of sporadic simple groups, such as the Lyons group, originally used computer calculations with large matrices or with permutations on billions of symbols. In most cases, such as the baby monster group, the computer proofs were later replaced by shorter proofs avoiding computer calculations. Similarly the calculation of the maximal subgroups of the larger sporadic groups uses a lot of computer calculations.

_ 2004 Verification of the Riemann hypothesis for the first 10^{13} zeros of the Riemann zeta function.
_ 2007 Verification that Checkers is a draw.
_ 2008 Proofs that various Mersenne numbers with around ten million digits are prime.
_ Calculations of large numbers of digits of π.
_ 2010 Showing that Rubik's Cube can be solved in 20 moves.
_ 2012 Showing that Sudoku needs at least 17 clues .
_ 2013 Ternary Goldbach conjecture: Every odd number greater than 5 can be expressed as the sum of three primes.
_ 2014 Proof of Erdős discrepancy conjecture for particular case C=2: every ±1-sequence of the length 1161 has a discrepancy at least 3, original proof generated by a SAT solver had a size of 13 gigabytes, it has been reduced later to 850 megabytes.

90.3 Long proofs in mathematical logic

Main article: Gödel's speed-up theorem
Kurt Gödel showed how to find explicit examples of statements in formal systems that are provable in that system but whose shortest proof is absurdly long. For example, the statement:
"This statement cannot be proved in Peano arithmetic in less than a googolplex symbols"
is provable in Peano arithmetic but the shortest proof has at least a googolplex symbols. It has a short proof in a more powerful system: in fact it is easily provable in Peano arithmetic together with the statement that Peano arithmetic is consistent (which cannot be proved in Peano arithmetic by Gödel's incompleteness theorem).
In this argument, Peano arithmetic can be replaced by any more powerful consistent system, and a googolplex can be replaced by any number that can be described concisely in the system.
Harvey Friedman found some explicit natural examples of this phenomenon, giving some explicit statements in Peano arithmetic and other formal systems whose shortest proofs are ridiculously long (Smoryński 1982). For example, the statement that
"there is an integer n such that if there is a sequence of rooted trees T_1, T_2, ..., T_n such that T_k has at most k+10 vertices, then some tree can be homeomorphically embedded in a later one"
is provable in Peano arithmetic, but the shortest proof has length at least $A(1000)$, where $A(0)=1$ and $A(n+1)=2^{A(n)}$.
The statement is a special case of Kruskal's theorem and has a short proof in second order arithmetic.

90.4 See also

_ List of incomplete proofs

90.5 References

_ Krantz, Steven G. (2011), *The proof is in the pudding. The changing nature of mathematical proof*, Berlin, New York: Springer-Verlag, ISBN 978-0-387-48908-7, MR 2789493
_ Smoryński, C. (1982), *The varieties of arboreal experience*, *Math. Intelligencer* **4** (4): 182–189, doi:10.1007/bf03023553, MR 0685558

Chapter 91

List of mathematical proofs

A list of articles with mathematical proofs:

91.1 Theorems of which articles are primarily devoted to proving them

See also: Category:Article proofs

_ Bertrand's postulate and a proof
_ Estimation of covariance matrices
_ Fermat's little theorem and some proofs
_ Gödel's completeness theorem and its original proof
_ Mathematical induction and a proof
_ Proof that 0.999... equals 1
_ Proof that 22/7 exceeds π
_ Proof that e is irrational
_ Proof that π is irrational
_ Proof that the sum of the reciprocals of the primes diverges

91.2 Articles devoted to theorems of which a (sketch of a) proof is given

See also: Category:Articles containing proofs

_ Banach fixed point theorem
_ Banach–Tarski paradox
_ Basel problem
_ Bolzano–Weierstrass theorem
_ Brouwer fixed point theorem
_ Buckingham π theorem (proof in progress)
_ Burnside's lemma
_ Cantor's theorem

_ Cantor–Bernstein–Schroeder theorem
_ Cayley's formula
_ Cayley's theorem
_ Clique problem (to do)
_ Compactness theorem (very compact proof)
_ Erdős–Ko–Rado theorem
_ Euler's formula
_ Euler's four-square identity
_ Euler's theorem
_ Five color theorem
_ Five lemma
_ Fundamental theorem of arithmetic
_ Gauss–Markov theorem (brief pointer to proof)
_ Gödel's incompleteness theorem
_ Gödel's first incompleteness theorem
_ Gödel's second incompleteness theorem
_ Goodstein's theorem
_ Green's theorem (to do)
_ Green's theorem when D is a simple region
_ Heine–Borel theorem
_ Intermediate value theorem
_ Itō's lemma
_ König's lemma
_ König's theorem (set theory)
_ König's theorem (graph theory)
_ Lagrange's theorem
_ Liouville's theorem (brief pointer to proof)
_ Markov's inequality (proof of a generalization)
_ Mean value theorem
_ Multivariate normal distribution (to do)
_ Holomorphic functions are analytic
_ Pythagorean theorem
_ Quadratic equation
_ Quotient rule
_ Ramsey's theorem
_ Rao–Blackwell theorem

_ Rice's theorem

_ Rolle’s theorem
_ Splitting lemma
_ squeeze theorem
_ Sum rule in differentiation
_ Sum rule in integration
_ Sylow theorems
_ Transcendence of e and π (as corollaries of Lindemann–Weierstrass)
_ Tychonoff’s theorem (to do)
_ Ultrafilter lemma
_ Ultraparallel theorem
_ Urysohn’s lemma
_ Van der Waerden’s theorem
_ Wilson’s theorem
_ Zorn’s lemma

91.3 Articles devoted to algorithms in which their correctness is proven

_ Bellman–Ford algorithm (to do)
_ Euclidean algorithm
_ Kruskal’s algorithm
_ Gale–Shapley algorithm
_ Prim’s algorithm
_ Shor’s algorithm (incomplete)

91.4 Articles where example statements are proven

See also: Category:Articles containing proofs

_ Basis (linear algebra)
_ Burrows–Abadi–Needham logic
_ Direct proof
_ Generating a vector space
_ Linear independence
_ Polynomial
_ Proof
_ Pumping lemma
_ Simpson’s rule

91.5 Other articles containing proofs

See also: Category:Articles containing proofs

_ Addition in N
_ associativity of addition in N
_ commutativity of addition in N
_ uniqueness of addition in N
_ Algorithmic information theory
_ Boolean ring
_ commutativity of a boolean ring
_ Boolean satisfiability problem
_ NP-completeness of the Boolean satisfiability problem
_ Cantor’s diagonal argument
_ set is smaller than its power set
_ uncountability of the real numbers
_ Cantor’s first uncountability proof
_ uncountability of the real numbers
_ Combinatorics
_ Combinatory logic
_ Co-NP
_ Coset
_ Countable
_ countability of a subset of a countable set (to do)
_ Counter
_ Angle of parallelism
_ Galois group
_ Fundamental theorem of Galois theory (to do)
_ Gödel number
_ Gödel’s incompleteness theorem
_ Group (mathematics)
_ Halting problem
_ insolubility of the halting problem
_ Harmonic series (mathematics)

_ divergence of the (standard) harmonic series
_ Highly composite number
_ Area of hyperbolic sector, basis of hyperbolic angle
_ Infinite series

_ convergence of the geometric series with first term 1 and ratio 1/2
_ Integer partition
_ Irrational number
_ irrationality of $\log_2 3$
_ irrationality of the square root of 2
_ Limit point
_ Mathematical induction
_ sum identity
_ Power rule
_ differential of x^n
_ Product and Quotiont Rules
_ Derivation of Product and Quotient rules for differentiating.
_ Prime number
_ Infinitude of the prime numbers
_ Primitive recursive function
_ Principle of bivalence
_ no propositions are neither true nor false in intuitionistic logic
_ Recursion
_ Relational algebra (to do)
_ Solvable group
_ Square root of 2
_ Tetris
_ Algebra of sets
_ idempotent laws for set union and intersection

91.6 Articles which mention dependencies of theorems

_ Cauchy's integral formula
_ Cauchy integral theorem
_ Computational geometry
_ Fundamental theorem of algebra
_ Lambda calculus
_ Invariance of domain
_ Minkowski inequality
_ Nash embedding theorem
_ Open mapping theorem (functional analysis)
_ Product topology
_ Riemann integral
_ Time hierarchy theorem
_ Deterministic time hierarchy theorem

91.7 Articles giving mathematical proofs within a physical model

_ No cloning theorem
_ Torque

91.8 See also

_ Gödel's ontological proof
_ Invalid proof
_ List of theorems
_ List of incomplete proofs
_ List of long proofs

Chapter 92

Proof by intimidation

Proof by intimidation (or argumentum verbosium) is a jocular phrase used mainly in mathematics to refer to a style of presenting a purported mathematical proof by giving an argument loaded with jargon and appeal to obscure results, so that the audience is simply obliged to accept it, lest they have to admit their ignorance and lack of understanding.[1]
The phrase is also used when the author is an authority in his field presenting his proof to people who respect *a priori* his insistence that the proof is valid or when the author claims that his statement is true because it is trivial or because he simply says so. Usage of this phrase is for the most part in good humour, though it also appears in serious criticism.[2]

More generally, "proof by intimidation" has also been used by critics of junk science to describe cases in which scientific evidence is thrown aside in favour of a litany of tragic individual cases presented to the public by articulate advocates who pose as experts in their field.[3]
Gian-Carlo Rota claimed in a memoir that the expression "proof by intimidation" was coined by Mark Kac to describe a technique used by William Feller in his lectures.[4]

92.1 See also

_ *Ad nauseam*
_ Argument from authority
_ Chewbacca defense
_ Gish Gallop
_ Handwaving
_ Obscurantism
_ Sophism

92.2 References

[1] Michael H. F. Wilkinson. "Cogno-Intellectualism, Rhetorical Logic, and the Craske-Trump Theorem". *Annals of Improbable Research* **6** (5): 15–16. Retrieved 2008-02-22.
[2] Tony Hey (1999). "Richard Feynman and computation". *Contemporary Physics* **40** (4): 257–265. doi:10.1080/001075199181459. Retrieved 2008-02-22.
[3] Marjorie K. Jeffcoat (July 2003). "Junk science: Appearances can be deceiving". *Journal of the American Dental Association* **134** (7): 802–803. doi:10.14219/jada.archive.2003.0268. PMID 12892436. Retrieved 2008-02-22.

[4] *He took umbrage when someone interrupted his lecturing by pointing out some glaring mistake. He became red in the face and raised his voice, often to full shouting range. It was reported that on occasion he had asked the objector to leave the classroom. The expression "proof by intimidation" was coined after Feller's lectures (by Mark Kac). During a Feller lecture, the hearer was made to feel privy to some wondrous secret, one that often vanished by magic as he walked out of the classroom at the end of the period. Like many great teachers, Feller was a bit of a con man.* Proof by intimidation was also referenced in xkcd: http://xkcd.com/982/. Gian-Carlo Rota (1996). *Indiscrete Thoughts*. Boston: Birkhäuser. ISBN 0-8176-3866-0.

Chapter 93

Termination analysis

In computer science, a **termination analysis** is program analysis which attempts to determine whether the evaluation of a given program will definitely terminate. Because the halting problem is undecidable, termination analysis cannot be total. The aim is to find the answer "program does terminate" (or "program does not terminate") whenever this is possible. Without success the algorithm (or human) working on the termination analysis may answer with "maybe" or continue working infinitely long.

93.1 Termination proof

A *termination proof* is a type of mathematical proof that plays a critical role in formal verification because total correctness of an algorithm depends on termination.
A simple, general method for constructing termination proofs involves associating a **measure** with each step of an algorithm. The measure is taken from the domain of a well-founded relation, such as from the ordinal numbers. If the measure "decreases" according to the relation along every possible step of the algorithm, it must terminate, because there are no infinite descending chains with respect to a well-founded relation.
Some types of termination analysis can automatically generate or imply the existence of a termination proof.

93.2 Example

An example of a programming language construct which may or may not terminate is a loop, as they can be run repeatedly. Loops implemented using a counter variable as typically found in data processing algorithms will usually terminate, demonstrated by the pseudocode example below:
i := 0 **loop** until i = SIZE_OF_DATA process_data(data[i])) //process the data chunk at position i i := i + 1 //move to the next chunk of data to be processed
If the value of *SIZE_OF_DATA* is non-negative, fixed and finite, the loop will eventually terminate, assuming *process_data* terminates too.
Some loops can be shown to always terminate or never terminate, through human inspection. For example, even a non-programmer should see that, in theory, the following never stops (but it may halt on physical machines due to arithmetic overflow):
i := 1 **loop** until i = 0 i := i + 1
In termination analysis one may also try to determine the termination behaviour of some program depending on some unknown input. The following example illustrates this problem.
i := 1 **loop** until i = UNKNOWN i := i + 1
Here the loop condition is defined using some value UNKNOWN, where the value of UNKNOWN is not known (e.g. defined by the user's input when the program is executed). Here the termination analysis must take into account all possible values of UNKNOWN and find out that in the possible case of UNKNOWN = 0 (as in the original example)

the termination cannot be shown.
There is, however, no general procedure for determining whether an expression involving looping instructions will halt, even when humans are tasked with the inspection. The theoretical reason for this is the undecidability of the Halting Problem: there cannot exist some algorithm which determines whether any given program stops after finitely many computation steps.
In practice one fails to show termination (or non-termination) because every algorithm works with a finite set of methods being able to extract relevant information out of a given program. A method might look at how variables change with respect to some loop condition (possibly showing termination for that loop), other methods might try to transform the program's calculation to some mathematical construct and work on that, possibly getting information about the termination behaviour out of some properties of this mathematical model. But because each method is only able to "see" some specific reasons for (non)termination, even through combination of such methods one cannot cover all possible reasons for (non)termination.
Recursive functions and loops are equivalent in expression; any expression involving loops can be written using recursion, and vice versa. Thus the termination of recursive expressions is also undecidable in general. Most recursive expressions found in common usage (i.e. not pathological) can be shown to terminate through various means, usually depending on the definition of the expression itself. As an example, the function argument in the recursive expression for the factorial function below will always decrease by 1; from the well-ordering property on natural numbers, the argument will eventually reach 1 and the recursion will terminate.
function factorial (argument **as** natural number) **if** argument = 0 **or** argument = 1 **return** 1 **otherwise return** argument * factorial(argument - 1)

93.3 Dependent types

Termination check is very important in dependently typed programming language and theorem proving systems like Coq and Agda. These systems use Curry-Howard isomorphism between programs and proofs. Proofs over inductively defined data types were traditionally described using induction and recursion principles which are in fact, primitive recursion. However, it was found later, that describing a program via a recursively defined function with pattern matching is more natural way of proving than using induction principle directly. Unfortunately, allowing arbitrary, including non terminating definitions, leads to possibility of logical inconsistencies in type theories. That's why Agda and Coq have termination checkers built-in.

93.3.1 Sized types

One of the approaches to termination checking in dependently typed programming languages are sized types. The main idea is to annotate the types over which we can recurse with size annotations and allow recursive calls only on smaller arguments. Sized types are implemented in Agda as a syntactic extension.

93.4 Current Research

There are several research teams that work on new methods that can show (non)termination. Many researchers include these methods into programs[1] that try to analyze the termination behavior automatically (so without human interaction). An on-going aspect of research is to allow the existing methods to be used to analyze termination behavior of programs written in "real world" programming languages. For declarative languages like Haskell, Mercury and Prolog, many results exist[2][3][4] (mainly because of the strong mathematical background of these languages). The research community also works on new methods to analyze termination behavior of programs written in imperative languages like C and Java.
Because of the undecidability of the Halting Problem research in this field cannot reach completeness. One can always think of new methods that find new (complicated) reasons for termination.

93.5 See also

_ Complexity analysis — the problem of estimating the time needed to terminate
_ Loop variant
_ Total functional programming — a programming paradigm that restricts the range of programs to those that are provably terminating
_ Walther recursion

93.6 References

[1] Tools at termination-portal.org
[2] Giesl, J. and Swiderski, S. and Schneider-Kamp, P. and Thiemann, R. "Automated Termination Analysis for Haskell: From Term Rewriting to Programming Languages (invited lecture)" (postscript). In Pfenning, F. "Term Rewriting and Applications, 17th Int. Conf., RTA-06". LNCS **4098**. pp. 297–312.
[3] Compiler options for termination analysis in Mercury
[4] http://verify.rwth-aachen.de/giesl/papers/lopstr07-distribute.pdf
Research papers on automated program termination analysis include:
_ Christoph Walther (1988). "Argument-Bounded Algorithms as a Basis for Automated Termination Proofs". *Proc. 9th Conference on Automated Deduction*. LNAI **310**. Springer. pp. 602–621.
_ Christoph Walther (1991). "On Proving the Termination of Algorithms by Machine". *Artificial Intelligence* **70** (1).
_ Xi, Hongwei (1998). "Towards Automated Termination Proofs through *Freezing*". In Tobias Nipkow. *Rewriting*

Techniques and Applications, 9th Int. Conf., RTA-98. LNCS **1379**. Springer. pp. 271–285.
_ Jürgen Giesl; Christoph Walther; Jürgen Brauburger (1998). "Termination Analysis for Functional Programs" (postscript). In W. Bibel; P. Schmitt. *Automated Deduction - A Basis for Applications* **3**. Dordrecht: Kluwer Academic Publishers. pp. 135–164.
_ Christoph Walther (2000). "Criteria for Termination" (postscript). In S. Hölldobler. *Intellectics and Computational Logic*. Dordrecht: Kluwer Academic Publishers. pp. 361–386.
_ Christoph Walther; Stephan Schweitzer (2005). "Automated Termination Analysis for Incompletely Defined Programs". In Franz Baader; Andrei Voronkov. *Proc. 11th Int. Conf. on Logic for Programming, Artificial Intelligence and Reasoning (LPAR)*. LNAI **3452**. Springer. pp. 332–346.
_ Adam Koprowski; Johannes Waldmann (2008). "Arctic Termination ...Below Zero". In Andrei Voronkov. *Rewriting Techniques and Applications, 19th Int. Conf., RTA-08*. Lecture Notes in Computer Science **5117**. Springer. pp. 202–216. ISBN 978-3-540-70588-8.
System descriptions of automated termination analysis tools include:
_ Giesl, J. (1995). "Generating Polynomial Orderings for Termination Proofs (system description)" (postscript). In Hsiang, Jieh. *Rewriting Techniques and Applications, 6th Int. Conf., RTA-95*. LNCS **914**. Springer. pp. 426–431.
_ Ohlebusch, E.; Claves, C.; Marché, C. (2000). "TALP: A Tool for the Termination Analysis of Logic Programs (system description)" (compressed postscript). In Bachmair, Leo. *Rewriting Techniques and Applications, 11th Int. Conf., RTA-00*. LNCS **1833**. Springer. pp. 270–273.

_ Hirokawa, N.; Middeldorp, A. (2003). "Tsukuba Termination Tool (system description)". In Nieuwenhuis, R. *Rewriting Techniques and Applications, 14th Int. Conf., RTA-03*. LNCS **2706**. Springer. pp. 311–320.
_ Giesl, J.; Thiemann, R.; Schneider-Kamp, P.; Falke, S. (2004). "Automated Termination Proofs with AProVE (system description)". In van Oostrom, V. *Rewriting Techniques and Applications, 15th Int. Conf., RTA-04*. LNCS **3091**. Springer. pp. 210–220. ISBN 3-540-22153-0.
_ Hirokawa, N.; Middeldorp, A. (2005). "Tyrolean Termination Tool (system description)". In Giesl, J. *Term Rewriting and Applications, 16th Int. Conf., RTA-05*. LNCS **3467**. Springer. pp. 175–184. ISBN 978-3-540-25596-3.
_ Koprowski, A. (2006). "TPA: Termination Proved Automatically (system description)". In Pfenning, F. *Term Rewriting and Applications, 17th Int. Conf., RTA-06*. LNCS **4098**. Springer. pp. 257–266.
_ Marché, C.; Zantema, H. (2007). "The Termination Competition (system description)". In Baader, F. *Term Rewriting and Applications, 18th Int. Conf., RTA-07*. LNCS **4533**. Springer. pp. 303–313.

93.7 External links

_ Termination Analysis of Higher-Order Functional Programs
_ Termination Tools mailing list
_ Termination Competition — see Marché, Zantema (2007) for a description
_ Termination Portal

Chapter 94

What the Tortoise Said to Achilles

"**What the Tortoise Said to Achilles**", written by Lewis Carroll in 1895 for the philosophical journal *Mind*, is a brief dialogue which problematises the foundations of logic. The title alludes to one of Zeno's paradoxes of motion, in which Achilles could never overtake the tortoise in a race. In Carroll's dialogue, the tortoise challenges Achilles to use the force of logic to make him accept the conclusion of a simple deductive argument. Ultimately, Achilles fails, because the clever tortoise leads him into an infinite regression.

94.1 Summary of the dialogue

The discussion begins by considering the following logical argument:
_ *A*: "Things that are equal to the same are equal to each other" (Euclidean relation, a weakened form of the transitive property)
_ *B*: "The two sides of this triangle are things that are equal to the same"
_ Therefore *Z*: "The two sides of this triangle are equal to each other"
The Tortoise asks Achilles whether the conclusion logically follows from the premises, and Achilles grants that it obviously does. The Tortoise then asks Achilles whether there might be a reader of Euclid who grants that the argument is *logically valid*, as a *sequence*, while denying that *A* and *B* are true. Achilles accepts that such a reader might exist, and that he would hold that *if A* and *B* are true, *then Z* must be true, while not yet accepting that *A* and *B are* true. (A reader who denies the premises.)
The Tortoise then asks Achilles whether a second kind of reader might exist, who accepts that *A* and *B are* true, but who does *not* yet accept the principle that *if A* and *B* are both true, *then Z* must be true. Achilles grants the Tortoise that this second kind of reader might also exist. The Tortoise, then, asks Achilles to treat the Tortoise as a reader of this second kind. Achilles must now logically compel the Tortoise to accept that *Z* must be true. (The tortoise is a reader who denies the argument itself; the syllogism's conclusion, structure, or validity.)
After writing down *A*, *B*, and *Z* in his notebook, Achilles asks the Tortoise to accept the hypothetical:
_ *C*: "If *A* and *B* are true, *Z* must be true"
The Tortoise agrees to accept *C*, if Achilles will write down what it has to accept in his notebook, making the new

argument:
_ *A*: "Things that are equal to the same are equal to each other"
_ *B*: "The two sides of this triangle are things that are equal to the same"
_ *C*: "If *A* and *B* are true, *Z* must be true"
_ Therefore *Z*: "The two sides of this triangle are equal to each other"

But now that the Tortoise accepts premise *C*, it still refuses to accept the expanded argument. When Achilles demands that "If you accept *A* and *B* and *C*, you must accept *Z*," the Tortoise remarks that that's *another* hypothetical proposition, and suggests even if it accepts *C*, it could still fail to conclude *Z* if it did not see the truth of:
_ *D*: "If *A* and *B* and *C* are true, *Z* must be true"
The Tortoise continues to accept each hypothetical premise once Achilles writes it down, but denies that the conclusion necessarily follows, since each time it denies the hypothetical that if all the premises written down so far are true, *Z* must be true:
"And at last we've got to the end of this ideal racecourse! Now that you accept *A* and *B* and *C* and *D*, *of course* you accept *Z*."
"Do I?" said the Tortoise innocently. "Let's make that quite clear. I accept *A* and *B* and *C* and *D*. Suppose I *still* refused to accept *Z*?"
"Then Logic would take you by the throat, and *force* you to do it!" Achilles triumphantly replied. "Logic would tell you, 'You can't help yourself. Now that you've accepted *A* and *B* and *C* and *D*, you must accept *Z*!' So you've no choice, you see."
"Whatever Logic is good enough to tell me is worth *writing down*," said the Tortoise. "So enter it in your notebook, please. We will call it
(*E*) If *A* and *B* and *C* and *D* are true, *Z* must be true.
Until I've granted that, of course I needn't grant *Z*. So it's quite a necessary step, you see?"
"I see," said Achilles; and there was a touch of sadness in his tone.
Thus, the list of premises continues to grow without end, leaving the argument always in the form:
_ (1): "Things that are equal to the same are equal to each other"
_ (2): "The two sides of this triangle are things that are equal to the same"
_ (3): (1) and (2) ⇒ (Z)
_ (4): (1) and (2) and (3) ⇒ (Z)
_ ...
_ (*n*): (1) and (2) and (3) and (4) and ... and (*n* − 1) ⇒ (*Z*)
_ Therefore (*Z*): "The two sides of this triangle are equal to each other"
At each step, the Tortoise argues that even though he accepts all the premises that have been written down, there is some further premise (that if all of (1)–(*n*) are true, then (*Z*) must be true) that it still needs to accept before it is compelled to accept that (*Z*) is true.

94.2 Explanation

Lewis Carroll was showing that there is a regress problem that arises from modus ponens deductions.
P ! Q; P
) Q
The regress problem arises because, in order to explain the logical principle, we have to propose a prior principle. And, once we explain *that* principle, we have to introduce *another* principle to explain *that* principle. Thus, if the causal chain is to continue, we are to fall into infinite regress. However, if we introduce a formal system where modus ponens is simply an axiom, then we are to abide by it simply, because it is so. For example, in a chess game there are particular rules, and the rules simply go without question. As players of the chess game, we are to simply follow the

rules. Likewise, if we are engaging in a formal system of logic, then we are to simply follow the rules without question. Hence, introducing the formal system of logic stops the infinite regression—that is, because the regress would stop at the axioms or rules, per se, of the given game, system, etc. Though, it does also state that there are problems with this as well, because, within the system, no proposition or variable carries with it any semantic content. So, the moment you add to any proposition or variable semantic content, the problem arises again, because the propositions and variables *with* semantic content run outside the system. Thus, if the solution is to be said to work, then it is to be said to work solely within the given formal system, and not otherwise.
Some logicians (Kenneth Ross, Charles Wright) draw a firm distinction between the conditional connective (the syntactic sign "→"), and the implication relation (the formal object denoted by the double arrow symbol "⇒"). These logicians use the phrase *not p or q* for the conditional connective and the term *implies* for the implication relation. Some explain the difference by saying that the conditional is the *contemplated* relation while the implication is the *asserted* relation. In most fields of mathematics, it is treated as a variation in the usage of the single sign "⇒," not requiring two separate signs. Not all of those who use the sign "→" for the conditional connective regard it as a sign that denotes any kind of object, but treat it as a so-called *syncategorematic sign*, that is, a sign with a purely syntactic function. For the sake of clarity and simplicity in the present introduction, it is convenient to use the two-sign notation, but allow the sign "→" to denote the boolean function that is associated with the truth table of the material conditional. These considerations result in the following scheme of notation.
p ! q p) q
not *p* or *q p* implies *q*

The paradox ceases to exist the moment we replace informal logic with propositional logic. The Tortoise and Achilles don't agree on any definition of logical implication. In propositional logic the logical implication is defined as follows: $P \Rightarrow Q$ if and only if the proposition $P \rightarrow Q$ is a tautology.

Hence de modus ponens $[P \wedge (P \rightarrow Q)] \Rightarrow Q$, is a valid logical implication according to the definition of logical implication just stated. There is no need to recurse since the logical implication can be translated into symbols, and propositional operators such as $\rightarrow$. Demonstrating the logical implication simply translates into verifying that the compound truth table is producing a tautology.

94.3 Discussion

Several philosophers have tried to resolve Carroll's paradox. Bertrand Russell discussed the paradox briefly in § 38 of *The Principles of Mathematics* (1903), distinguishing between *implication* (associated with the form "if *p*, then *q*"), which he held to be a relation between *unasserted* propositions, and *inference* (associated with the form "*p*, therefore *q*"), which he held to be a relation between *asserted* propositions; having made this distinction, Russell could deny that the Tortoise's attempt to treat *inferring Z* from *A* and *B* is equivalent to, or dependent on, agreeing to the *hypothetical* "If *A* and *B* are true, then *Z* is true."

The Wittgensteinian philosopher Peter Winch discussed the paradox in *The Idea of a Social Science and its Relation to Philosophy* (1958), where he argued that the paradox showed that "the actual process of drawing an inference, which is after all at the heart of logic, is something which cannot be represented as a logical formula ... Learning to infer is not just a matter of being taught about explicit logical relations between propositions; it is learning *to do* something" (p. 57). Winch goes on to suggest that the moral of the dialogue is a particular case of a general lesson, to the effect that the proper *application* of rules governing a form of human activity cannot itself be summed up with a set of *further* rules, and so that "a form of human activity can never be summed up in a set of explicit precepts" (p. 53).

According to Penelope Maddy,[1] Carroll's dialogue is apparently the first description of an obstacle to Conventionalism about logical truth, then reworked in more sober philosophical terms by W. O. Quine.[2]

94.4 See also

- Deduction theorem
- Homunculus argument
- Münchhausen trilemma

- Paradox
- Regress argument
- Rule of inference

94.5 References

[1] Maddy, P. (December 2012). "The philosophy of logic". *Bulletin of Symbolic Logic* **18** (4): 481–504. doi:10.2178/bsl.1804010. JSTOR 23316289.

[2] Quine, W.V.O. (1976). *The ways of paradox, and other essays*. Cambridge, MA: Havard University Press. ISBN 9780674948358. OCLC 185411480.

94.6 Sources

- Carroll, Lewis (1995). "What the Tortoise Said to Achilles". *Mind* **104** (416): 691–693. doi:10.1093/mind/104.416.691. JSTOR 2254477.
- Hofstadter, Douglas. *Gödel, Escher, Bach: an Eternal Golden Braid.* See the second dialogue, entitled "Two-Part Invention". Hofstadter appropriated the characters of Achilles and the Tortoise for other, original, dialogues in the book which alternate contrapuntally with prose chapters. Hofstadter's Tortoise is of the male sex, though the Tortoise's sex is never specified by Carroll. The French translation of the book rendered the Tortoise's name as "Madame Tortue".
- A number of websites, including "What the Tortoise Said to Achilles" at the Lewis Carroll Society of North America, "What the Tortoise Said to Achilles" at Digital Text International, and "What the Tortoise Said to Achilles" at Fair Use Repository.

94.7 External links

- Works related to What the Tortoise Said to Achilles at Wikisource

Chapter 95

Combinatorics

Not to be confused with combinatoriality.

Combinatorics is a branch of mathematics concerning the study of finite or countable discrete structures. Aspects of combinatorics include counting the structures of a given kind and size (enumerative combinatorics), deciding when certain criteria can be met, and constructing and analyzing objects meeting the criteria (as in combinatorial designs and matroid theory), finding "largest", "smallest", or "optimal" objects (extremal combinatorics and combinatorial optimization), and studying combinatorial structures arising in an algebraic context, or applying algebraic techniques to combinatorial problems (algebraic combinatorics).

Combinatorial problems arise in many areas of pure mathematics, notably in algebra, probability theory, topology, and geometry,[1] and combinatorics also has many applications in mathematical optimization, computer science, ergodic

theory and statistical physics. Many combinatorial questions have historically been considered in isolation, giving an *ad hoc* solution to a problem arising in some mathematical context. In the later twentieth century, however, powerful and general theoretical methods were developed, making combinatorics into an independent branch of mathematics in its own right. One of the oldest and most accessible parts of combinatorics is graph theory, which also has numerous natural connections to other areas. Combinatorics is used frequently in computer science to obtain formulas and estimates in the analysis of algorithms.
A mathematician who studies combinatorics is called a **combinatorialist** or a **combinatorist**.

95.1 History

Main article: History of combinatorics
Basic combinatorial concepts and enumerative results appeared throughout the ancient world. In 6th century BCE, ancient Indian physician Sushruta asserts in Sushruta Samhita that 63 combinations can be made out of 6 different tastes, taken one at a time, two at a time, etc., thus computing all $2^6 - 1$ possibilities. Greek historian Plutarch discusses an argument between Chrysippus (3rd century BCE) and Hipparchus (2nd century BCE) of a rather delicate enumerative problem, which was later shown to be related to Schröder numbers.[2][3] In the *Ostomachion*, Archimedes (3rd century BCE) considers a tiling puzzle.
In the Middle Ages, combinatorics continued to be studied, largely outside of the European civilization. The Indian mathematician Mahāvīra (c. 850) provided formulae for the number of permutations and combinations,[4][5] and these formulas may have been familiar to Indian mathematicians as early as the 6th century CE.[6] The philosopher and astronomer Rabbi Abraham ibn Ezra (c. 1140) established the symmetry of binomial coefficients, while a closed formula was obtained later by the talmudist and mathematician Levi ben Gerson (better known as Gersonides), in 1321.[7] The arithmetical triangle— a graphical diagram showing relationships among the binomial coefficients— was presented by mathematicians in treatises dating as far back as the 10th century, and would eventually become known as Pascal's triangle. Later, in Medieval England, campanology provided examples of what is now known as Hamiltonian cycles in certain Cayley graphs on permutations.[8]
During the Renaissance, together with the rest of mathematics and the sciences, combinatorics enjoyed a rebirth.

Works of Pascal, Newton, Jacob Bernoulli and Euler became foundational in the emerging field. In modern times, the works of J. J. Sylvester (late 19th century) and Percy MacMahon (early 20th century) laid the foundation for enumerative and algebraic combinatorics. Graph theory also enjoyed an explosion of interest at the same time, especially in connection with the four color problem.
In the second half of 20th century, combinatorics enjoyed a rapid growth, which led to establishment of dozens of new journals and conferences in the subject.[9] In part, the growth was spurred by new connections and applications to other fields, ranging from algebra to probability, from functional analysis to number theory, etc. These connections shed the boundaries between combinatorics and parts of mathematics and theoretical computer science, but at the same time led to a partial fragmentation of the field.

95.2 Approaches and subfields of combinatorics

95.2.1 Enumerative combinatorics

Main article: Enumerative combinatorics
Enumerative combinatorics is the most classical area of combinatorics, and concentrates on counting the number of certain combinatorial objects. Although counting the number of elements in a set is a rather broad mathematical problem, many of the problems that arise in applications have a relatively simple combinatorial description. Fibonacci numbers is the basic example of a problem in enumerative combinatorics. The twelvefold way provides a unified framework for counting permutations, combinations and partitions.

95.2.2 Analytic combinatorics

Main article: Analytic combinatorics
Analytic combinatorics concerns the enumeration of combinatorial structures using tools from complex analysis and probability theory. In contrast with enumerative combinatorics which uses explicit combinatorial formulae and generating functions to describe the results, analytic combinatorics aims at obtaining asymptotic formulae.

95.2.3 Partition theory

Main article: Partition theory
Partition theory studies various enumeration and asymptotic problems related to integer partitions, and is closely related to q-series, special functions and orthogonal polynomials. Originally a part of number theory and analysis, it is now considered a part of combinatorics or an independent field. It incorporates the bijective approach and various tools in analysis, analytic number theory, and has connections with statistical mechanics.

95.2.4 Graph theory

Main article: Graph theory
Graphs are basic objects in combinatorics. The questions range from counting (e.g., the number of graphs on n vertices with k edges) to structural (e.g., which graphs contain Hamiltonian cycles) to algebraic questions (e.g., given a graph G and two numbers x and y, does the Tutte polynomial $T_G(x,y)$ have a combinatorial interpretation?). It should be noted that while there are very strong connections between graph theory and combinatorics, these two are sometimes thought of as separate subjects.[10]

95.2.5 Design theory

Main article: Combinatorial design
Design theory is a study of combinatorial designs, which are collections of subsets with certain intersection properties. Block designs are combinatorial designs of a special type. This area is one of the oldest parts of combinatorics, such as in Kirkman's schoolgirl problem proposed in 1850. The solution of the problem is a special case of a Steiner system, which systems play an important role in the classification of finite simple groups. The area has further connections to coding theory and geometric combinatorics.

95.2.6 Finite geometry

Main article: Finite geometry
Finite geometry is the study of geometric systems having only a finite number of points. Structures analogous to those found in continuous geometries (Euclidean plane, real projective space, etc.) but defined combinatorially are the main items studied. This area provides a rich source of examples for Design theory. It should not be confused with Discrete geometry (Combinatorial geometry).

95.2.7 Order theory

Main article: Order theory
Order theory is the study of partially ordered sets, both finite and infinite. Various examples of partial orders appear in algebra, geometry, number theory and throughout combinatorics and graph theory. Notable classes and examples of partial orders include lattices and Boolean algebras.

95.2.8 Matroid theory

Main article: Matroid theory
Matroid theory abstracts part of geometry. It studies the properties of sets (usually, finite sets) of vectors in a vector space that do not depend on the particular coefficients in a linear dependence relation. Not only the structure but also enumerative properties belong to matroid theory. Matroid theory was introduced by Hassler Whitney and studied as a part of the order theory. It is now an independent field of study with a number of connections with other parts of combinatorics.

95.2.9 Extremal combinatorics

Main article: Extremal combinatorics
Extremal combinatorics studies extremal questions on set systems. The types of questions addressed in this case are about the largest possible graph which satisfies certain properties. For example, the largest triangle-free graph on $2n$ vertices is a complete bipartite graph $K_{n,n}$. Often it is too hard even to find the extremal answer $f(n)$ exactly and one can only give an asymptotic estimate.
Ramsey theory is another part of extremal combinatorics. It states that any sufficiently large configuration will contain some sort of order. It is an advanced generalization of the pigeonhole principle.

95.2.10 Probabilistic combinatorics

Main article: Probabilistic method

In probabilistic combinatorics, the questions are of the following type: what is the probability of a certain property for a random discrete object, such as a random graph? For instance, what is the average number of triangles in a random graph? Probabilistic methods are also used to determine the existence of combinatorial objects with certain prescribed properties (for which explicit examples might be difficult to find), simply by observing that the probability of randomly selecting an object with those properties is greater than 0. This approach (often referred to as *the* probabilistic method) proved highly effective in applications to extremal combinatorics and graph theory. A closely related area is the study of finite Markov chains, especially on combinatorial objects. Here again probabilistic tools are used to estimate the mixing time.
Often associated with Paul Erdős, who did the pioneer work on the subject, probabilistic combinatorics was traditionally viewed as a set of tools to study problems in other parts of combinatorics. However, with the growth of applications to analysis of algorithms in computer science, as well as classical probability, additive and probabilistic number theory, the area recently grew to become an independent field of combinatorics.

95.2.11 Algebraic combinatorics

Main article: Algebraic combinatorics
Algebraic combinatorics is an area of mathematics that employs methods of abstract algebra, notably group theory and representation theory, in various combinatorial contexts and, conversely, applies combinatorial techniques to problems in algebra. Algebraic combinatorics is continuously expanding its scope, in both topics and techniques, and can be seen as the area of mathematics where the interaction of combinatorial and algebraic methods is particularly strong and significant.

95.2.12 Combinatorics on words

Main article: Combinatorics on words
Combinatorics on words deals with formal languages. It arose independently within several branches of mathematics, including number theory, group theory and probability. It has applications to enumerative combinatorics, fractal analysis, theoretical computer science, automata theory and linguistics. While many applications are new, the classical Chomsky–Schützenberger hierarchy of classes of formal grammars is perhaps the best known result in the field.

95.2.13 Geometric combinatorics

Main article: Geometric combinatorics
Geometric combinatorics is related to convex and discrete geometry, in particular polyhedral combinatorics. It asks, for example, how many faces of each dimension can a convex polytope have. Metric properties of polytopes play an

important role as well, e.g. the Cauchy theorem on rigidity of convex polytopes. Special polytopes are also considered, such as permutohedra, associahedra and Birkhoff polytopes. We should note that combinatorial geometry is an old fashioned name for discrete geometry.

95.2.14 Topological combinatorics

Main article: Topological combinatorics

Combinatorial analogs of concepts and methods in topology are used to study graph coloring, fair division, partitions, partially ordered sets, decision trees, necklace problems and discrete Morse theory. It should not be confused with combinatorial topology which is an older name for algebraic topology.

95.2.15 Arithmetic combinatorics

Main article: Arithmetic combinatorics

Arithmetic combinatorics arose out of the interplay between number theory, combinatorics, ergodic theory and harmonic analysis. It is about combinatorial estimates associated with arithmetic operations (addition, subtraction, multiplication, and division). *Additive combinatorics* refers to the special case when only the operations of addition and subtraction are involved. One important technique in arithmetic combinatorics is the ergodic theory of dynamical systems.

95.2.16 Infinitary combinatorics

Main article: Infinitary combinatorics

Infinitary combinatorics, or combinatorial set theory, is an extension of ideas in combinatorics to infinite sets. It is a part of set theory, an area of mathematical logic, but uses tools and ideas from both set theory and extremal combinatorics.

Gian-Carlo Rota used the name *continuous combinatorics*[11] to describe probability and measure theory, since there are many analogies between *counting* and *measure*.

95.3 Related fields

95.3.1 Combinatorial optimization

Combinatorial optimization is the study of optimization on discrete and combinatorial objects. It started as a part of combinatorics and graph theory, but is now viewed as a branch of applied mathematics and computer science, related to operations research, algorithm theory and computational complexity theory.

95.3.2 Coding theory

Coding theory started as a part of design theory with early combinatorial constructions of error-correcting codes. The main idea of the subject is to design efficient and reliable methods of data transmission. It is now a large field of study, part of information theory.

95.3.3 Discrete and computational geometry

Discrete geometry (also called combinatorial geometry) also began a part of combinatorics, with early results on convex polytopes and kissing numbers. With the emergence of applications of discrete geometry to computational geometry, these two fields partially merged and became a separate field of study. There remain many connections with geometric and topological combinatorics, which themselves can be viewed as outgrowths of the early discrete geometry.

95.3.4 Combinatorics and dynamical systems

Combinatorial aspects of dynamical systems is another emerging field. Here dynamical systems can be defined on combinatorial objects. See for example graph dynamical system.

95.3.5 Combinatorics and physics

There are increasing interactions between combinatorics and physics, particularly statistical physics. Examples include an exact solution of the Ising model, and a connection between the Potts model on one hand, and the chromatic and Tutte polynomials on the other hand.

95.4 See also

_ Combinatorial biology
_ Combinatorial chemistry
_ Combinatorial data analysis
_ Combinatorial game theory
_ Combinatorial group theory
_ List of combinatorics topics
_ Phylogenetics

95.5 Notes

[1] Björner and Stanley, p. 2

[2] Stanley, Richard P.; "Hipparchus, Plutarch, Schröder, and Hough", *American Mathematical Monthly* **104** (1997), no. 4, 344–350.

[3] Habsieger, Laurent; Kazarian, Maxim; and Lando, Sergei; "On the Second Number of Plutarch", *American Mathematical Monthly* **105** (1998), no. 5, 446.

[4] O'Connor, John J.; Robertson, Edmund F., "Combinatorics", *MacTutor History of Mathematics archive*, University of St Andrews.

[5] Puttaswamy, Tumkur K. (2000), "The Mathematical Accomplishments of Ancient Indian Mathematicians", in Selin, Helaine, *Mathematics Across Cultures: The History of Non-Western Mathematics*, Netherlands: Kluwer Academic Publishers, p. 417, ISBN 978-1-4020-0260-1
[6] Biggs, Norman L.; "The Roots of Combinatorics", *Historia Mathematica* 6 (1979), 109–136.
[7] History of Combinatorics, chapter in a textbook.
[8] White, Arthur T.; "Ringing the Cosets", *American Mathematical Monthly*, **94** (1987), no. 8, 721–746; White, Arthur T.; "Fabian Stedman: The First Group Theorist?", *American Mathematical Monthly*, **103** (1996), no. 9, 771–778.
[9] See Journals in Combinatorics and Graph Theory
[10] Sanders, Daniel P.; *2-Digit MSC Comparison*
[11] *Continuous and profinite combinatorics*

95.6 References

_ Björner, Anders; and Stanley, Richard P.; (2010); *A Combinatorial Miscellany*
_ Bóna, Miklós; (2011); *A Walk Through Combinatorics (3rd Edition)*. ISBN 978-981-4335-23-2, ISBN 978-981-4460-00-2(pbk)
_ Graham, Ronald L.; Groetschel, Martin; and Lovász, László; eds. (1996); *Handbook of Combinatorics*, Volumes 1 and 2. Amsterdam, NL, and Cambridge, MA: Elsevier (North-Holland) and MIT Press. ISBN 0-262-07169-X
_ Lindner, Charles C.; and Rodger, Christopher A.; eds. (1997); *Design Theory*, CRC-Press; 1st. edition (October 31, 1997). ISBN 0-8493-3986-3.
_ Riordan, John (1958); *An Introduction to Combinatorial Analysis*, New York, NY: Wiley & Sons (republished)
_ Stanley, Richard P. (1997, 1999); *Enumerative Combinatorics*, Volumes 1 and 2, Cambridge University Press. ISBN 0-521-55309-1, ISBN 0-521-56069-1
_ van Lint, Jacobus H.; and Wilson, Richard M.; (2001); *A Course in Combinatorics*, 2nd Edition, Cambridge University Press. ISBN 0-521-80340-3

95.7 External links

_ Hazewinkel, Michiel, ed. (2001), "Combinatorial analysis", *Encyclopedia of Mathematics*, Springer, ISBN 978-1-55608-010-4
_ Combinatorial Analysis – an article in Encyclopædia Britannica Eleventh Edition
_ Combinatorics, a MathWorld article with many references.
_ Combinatorics, from a *MathPages.com* portal.
_ The Hyperbook of Combinatorics, a collection of math articles links.
_ The Two Cultures of Mathematics by W. T. Gowers, article on problem solving vs theory building.

An example of change ringing (with six bells), one of the earliest nontrivial results in Graph Theory.

Five binary trees on three vertices, an example of Catalan numbers.
A plane partition.

Petersen graph.

{x,y,z}
{x,y} {x,z} {y,z}
{x} {y} {z}
Ø

Hasse diagram of the powerset of {x,y,z} ordered by inclusion.

Self-avoiding walk in a square grid graph.

Young diagram of a partition (5,4,1).
Construction of a Thue–Morse infinite word.

An icosahedron.

Splitting a necklace with two cuts.

Kissing spheres are connected to both coding theory and discrete geometry.

Chapter 96

Graph theory

This article is about sets of vertices connected by edges. For graphs of mathematical functions, see Graph of a function. For other uses, see Graph (disambiguation).
In mathematics and computer science, **graph theory** is the study of *graphs*, which are mathematical structures used

1

3 2

4 5

6

A drawing of a graph

to model pairwise relations between objects. A "graph" in this context is made up of "vertices" or "nodes" and lines called *edges* that connect them. A graph may be *undirected*, meaning that there is no distinction between the two vertices associated with each edge, or its edges may be *directed* from one vertex to another; see graph (mathematics) for more detailed definitions and for other variations in the types of graph that are commonly considered. Graphs are one of the prime objects of study in discrete mathematics.
Refer to the glossary of graph theory for basic definitions in graph theory.

96.1 Definitions

Definitions in graph theory vary. The following are some of the more basic ways of defining graphs and related mathematical structures.

96.1.1 Graph

In the most common sense of the term,[1] a **graph** is an ordered pair $G = (V, E)$ comprising a set V of **vertices** or **nodes** together with a set E of **edges** or **lines**, which are 2-element subsets of V (i.e., an edge is related with two vertices, and the relation is represented as an unordered pair of the vertices with respect to the particular edge). To avoid ambiguity, this type of graph may be described precisely as undirected and simple.
Other senses of *graph* stem from different conceptions of the edge set. In one more generalized notion,[2] E is a set together with a relation of **incidence** that associates with each edge two vertices. In another generalized notion, E is a multiset of unordered pairs of (not necessarily distinct) vertices. Many authors call this type of object a multigraph or pseudograph.
All of these variants and others are described more fully below.
The vertices belonging to an edge are called the **ends**, **endpoints**, or **end vertices** of the edge. A vertex may exist in a graph and not belong to an edge.
V and E are usually taken to be finite, and many of the well-known results are not true (or are rather different) for **infinite graphs** because many of the arguments fail in the infinite case. The **order** of a graph is jVj (the number of vertices). A graph's **size** is jEj , the number of edges. The **degree** of a vertex is the number of edges that connect to it, where an edge that connects to the vertex at both ends (a loop) is counted twice.
For an edge $\{u, v\}$, graph theorists usually use the somewhat shorter notation uv.

96.2 Applications

Graphs can be used to model many types of relations and processes in physical, biological,[4] social and information systems. Many practical problems can be represented by graphs.
In computer science, graphs are used to represent networks of communication, data organization, computational devices, the flow of computation, etc. For instance, the link structure of a website can be represented by a directed graph, in which the vertices represent web pages and directed edges represent links from one page to another. A similar approach can be taken to problems in travel, biology, computer chip design, and many other fields. The development of algorithms to handle graphs is therefore of major interest in computer science. The transformation of graphs is often formalized and represented by graph rewrite systems. Complementary to graph transformation systems focusing on rule-based in-memory manipulation of graphs are graph databases geared towards transactionsafe, persistent storing and querying of graph-structured data.
Graph-theoretic methods, in various forms, have proven particularly useful in linguistics, since natural language often lends itself well to discrete structure. Traditionally, syntax and compositional semantics follow tree-based structures, whose expressive power lies in the principle of compositionality, modeled in a hierarchical graph. More contemporary

approaches such as head-driven phrase structure grammar model the syntax of natural language using typed feature structures, which are directed acyclic graphs. Within lexical semantics, especially as applied to computers, modeling word meaning is easier when a given word is understood in terms of related words; semantic networks are therefore important in computational linguistics. Still other methods in phonology (e.g. optimality theory, which uses lattice graphs) and morphology (e.g. finite-state morphology, using finite-state transducers) are common in the analysis of language as a graph. Indeed, the usefulness of this area of mathematics to linguistics has borne organizations such as TextGraphs, as well as various 'Net' projects, such as WordNet, VerbNet, and others.

Graph theory is also used to study molecules in chemistry and physics. In condensed matter physics, the threedimensional structure of complicated simulated atomic structures can be studied quantitatively by gathering statistics on graph-theoretic properties related to the topology of the atoms. In chemistry a graph makes a natural model for a molecule, where vertices represent atoms and edges bonds. This approach is especially used in computer processing of molecular structures, ranging from chemical editors to database searching. In statistical physics, graphs can represent

ar
bg
ca
cs
da
de
en
es
fa
fr fi
he
hu
id
it
ja
ko
nl
no
pl
pt
ro
ru
sv
tr
uk
zh

The network graph formed by Wikipedia editors (edges) contributing to different Wikipedia language versions (nodes) during one month in summer 2013.[3]

local connections between interacting parts of a system, as well as the dynamics of a physical process on such systems. Graphs are also used to represent the micro-scale channels of porous media, in which the vertices represent the pores and the edges represent the smaller channels connecting the pores.

Graph theory is also widely used in sociology as a way, for example, to measure actors' prestige or to explore rumor spreading, notably through the use of social network analysis software. Under the umbrella of social networks are many different types of graphs:[5] Acquaintanceship and friendship graphs describe whether people know each other.

Influence graphs model whether certain people can influence the behavior of others. Finally, collaboration graphs model whether two people work together in a particular way, such as acting in a movie together.
Likewise, graph theory is useful in biology and conservation efforts where a vertex can represent regions where certain species exist (or habitats) and the edges represent migration paths, or movement between the regions. This information is important when looking at breeding patterns or tracking the spread of disease, parasites or how changes to the movement can affect other species.
In mathematics, graphs are useful in geometry and certain parts of topology such as knot theory. Algebraic graph theory has close links with group theory.
A graph structure can be extended by assigning a weight to each edge of the graph. Graphs with weights, or weighted graphs, are used to represent structures in which pairwise connections have some numerical values. For example if a graph represents a road network, the weights could represent the length of each road.

96.3 History

The Königsberg Bridge problem

The paper written by Leonhard Euler on the *Seven Bridges of Königsberg* and published in 1736 is regarded as the first paper in the history of graph theory.[6] This paper, as well as the one written by Vandermonde on the *knight problem,* carried on with the *analysis situs* initiated by Leibniz. Euler's formula relating the number of edges, vertices, and faces of a convex polyhedron was studied and generalized by Cauchy[7] and L'Huillier,[8] and is at the origin of topology.
More than one century after Euler's paper on the bridges of Königsberg and while Listing introduced topology, Cayley was led by the study of particular analytical forms arising from differential calculus to study a particular class of graphs, the *trees*.[9] This study had many implications in theoretical chemistry. The involved techniques mainly concerned the enumeration of graphs having particular properties. Enumerative graph theory then rose from the results of Cayley and the fundamental results published by Pólya between 1935 and 1937 and the generalization of these by De Bruijn in 1959. Cayley linked his results on trees with the contemporary studies of chemical composition.[10] The fusion of the ideas coming from mathematics with those coming from chemistry is at the origin of a part of the standard terminology of graph theory.
In particular, the term "graph" was introduced by Sylvester in a paper published in 1878 in *Nature*, where he draws an analogy between "quantic invariants" and "co-variants" of algebra and molecular diagrams:[11]
"[...] Every invariant and co-variant thus becomes expressible by a *graph* precisely identical with a Kekuléan diagram or chemicograph. [...] I give a rule for the geometrical multiplication of graphs, *i.e.* for constructing a *graph* to the product of in- or co-variants whose separate graphs are given. [...]" (italics as in the original).
The first textbook on graph theory was written by Dénes Kőnig, and published in 1936.[12] Another book by Frank Harary, published in 1969, was "considered the world over to be the definitive textbook on the subject",[13] and enabled mathematicians, chemists, electrical engineers and social scientists to talk to each other. Harary donated all of the royalties to fund the Pólya Prize.[14]

One of the most famous and stimulating problems in graph theory is the four color problem: "Is it true that any map drawn in the plane may have its regions colored with four colors, in such a way that any two regions having a common border have different colors?" This problem was first posed by Francis Guthrie in 1852 and its first written record is in a letter of De Morgan addressed to Hamilton the same year. Many incorrect proofs have been proposed, including those by Cayley, Kempe, and others. The study and the generalization of this problem by Tait, Heawood, Ramsey and Hadwiger led to the study of the colorings of the graphs embedded on surfaces with arbitrary genus. Tait's reformulation generated a new class of problems, the *factorization problems*, particularly studied by Petersen and Kőnig. The works of Ramsey on colorations and more specially the results obtained by Turán in 1941 was at the origin of another branch of graph theory, *extremal graph theory*.
The four color problem remained unsolved for more than a century. In 1969 Heinrich Heesch published a method for solving the problem using computers.[15] A computer-aided proof produced in 1976 by Kenneth Appel and Wolfgang Haken makes fundamental use of the notion of "discharging" developed by Heesch.[16][17] The proof involved checking the properties of 1,936 configurations by computer, and was not fully accepted at the time due to its complexity.
A simpler proof considering only 633 configurations was given twenty years later by Robertson, Seymour, Sanders and Thomas.[18]
The autonomous development of topology from 1860 and 1930 fertilized graph theory back through the works of Jordan, Kuratowski and Whitney. Another important factor of common development of graph theory and topology came from the use of the techniques of modern algebra. The first example of such a use comes from the work of the physicist Gustav Kirchhoff, who published in 1845 his Kirchhoff's circuit laws for calculating the voltage and current in electric circuits.
The introduction of probabilistic methods in graph theory, especially in the study of Erdős and Rényi of the asymptotic probability of graph connectivity, gave rise to yet another branch, known as *random graph theory*, which has been a fruitful source of graph-theoretic results.

96.4 Graph drawing

Main article: Graph drawing
Graphs are represented visually by drawing a dot or circle for every vertex, and drawing an arc between two vertices if they are connected by an edge. If the graph is directed, the direction is indicated by drawing an arrow.
A graph drawing should not be confused with the graph itself (the abstract, non-visual structure) as there are several ways to structure the graph drawing. All that matters is which vertices are connected to which others by how many

edges and not the exact layout. In practice it is often difficult to decide if two drawings represent the same graph. Depending on the problem domain some layouts may be better suited and easier to understand than others.
The pioneering work of W. T. Tutte was very influential in the subject of graph drawing. Among other achievements, he introduced the use of linear algebraic methods to obtain graph drawings.
Graph drawing also can be said to encompass problems that deal with the crossing number and its various generalizations. The crossing number of a graph is the minimum number of intersections between edges that a drawing of the graph in the plane must contain. For a planar graph, the crossing number is zero by definition.
Drawings on surfaces other than the plane are also studied.

96.5 Graph-theoretic data structures

Main article: Graph (abstract data type)
There are different ways to store graphs in a computer system. The data structure used depends on both the graph structure and the algorithm used for manipulating the graph. Theoretically one can distinguish between list and matrix structures but in concrete applications the best structure is often a combination of both. List structures are often preferred for sparse graphs as they have smaller memory requirements. Matrix structures on the other hand provide faster access for some applications but can consume huge amounts of memory.
List structures include the incidence list, an array of pairs of vertices, and the adjacency list, which separately lists

the neighbors of each vertex: Much like the incidence list, each vertex has a list of which vertices it is adjacent to.
Matrix structures include the incidence matrix, a matrix of 0's and 1's whose rows represent vertices and whose columns represent edges, and the adjacency matrix, in which both the rows and columns are indexed by vertices. In both cases a 1 indicates two adjacent objects and a 0 indicates two non-adjacent objects. The Laplacian matrix is a modified form of the adjacency matrix that incorporates information about the degrees of the vertices, and is useful in some calculations such as Kirchhoff's theorem on the number of spanning trees of a graph. The distance matrix, like the adjacency matrix, has both its rows and columns indexed by vertices, but rather than containing a 0 or a 1 in each cell it contains the length of a shortest path between two vertices.

96.6 Problems in graph theory

96.6.1 Enumeration

There is a large literature on graphical enumeration: the problem of counting graphs meeting specified conditions. Some of this work is found in Harary and Palmer (1973).

96.6.2 Subgraphs, induced subgraphs, and minors

A common problem, called the subgraph isomorphism problem, is finding a fixed graph as a subgraph in a given graph. One reason to be interested in such a question is that many graph properties are *hereditary* for subgraphs, which means that a graph has the property if and only if all subgraphs have it too. Unfortunately, finding maximal subgraphs of a certain kind is often an NP-complete problem.

- Finding the largest complete graph is called the clique problem (NP-complete).

A similar problem is finding induced subgraphs in a given graph. Again, some important graph properties are hereditary with respect to induced subgraphs, which means that a graph has a property if and only if all induced subgraphs also have it. Finding maximal induced subgraphs of a certain kind is also often NP-complete. For example,

- Finding the largest edgeless induced subgraph, or independent set, called the independent set problem (NPcomplete).

Still another such problem, the *minor containment problem*, is to find a fixed graph as a minor of a given graph. A minor or **subcontraction** of a graph is any graph obtained by taking a subgraph and contracting some (or no) edges. Many graph properties are hereditary for minors, which means that a graph has a property if and only if all minors have it too. A famous example:

- A graph is planar if it contains as a minor neither the complete bipartite graph $K_{3;3}$ (See the Three-cottage problem) nor the complete graph K_5 .

Another class of problems has to do with the extent to which various species and generalizations of graphs are determined by their *point-deleted subgraphs*, for example:

- The reconstruction conjecture.

96.6.3 Graph coloring

Many problems have to do with various ways of coloring graphs, for example:

- The four-color theorem
- The strong perfect graph theorem
- The Erdős–Faber–Lovász conjecture(unsolved)

- The total coloring conjecture, also called Behzad's conjecture) (unsolved)
- The list coloring conjecture (unsolved)
- The Hadwiger conjecture (graph theory) (unsolved)

96.6.4 Subsumption and unification

Constraint modeling theories concern families of directed graphs related by a partial order. In these applications, graphs are ordered by specificity, meaning that more constrained graphs—which are more specific and thus contain a greater amount of information—are subsumed by those that are more general. Operations between graphs include evaluating the direction of a subsumption relationship between two graphs, if any, and computing graph unification. The unification of two argument graphs is defined as the most general graph (or the computation thereof) that is consistent with (i.e. contains all of the information in) the inputs, if such a graph exists; efficient unification algorithms

are known.
For constraint frameworks which are strictly compositional, graph unification is the sufficient satisfiability and combination function. Well-known applications include automatic theorem proving and modeling the elaboration of linguistic structure.

96.6.5 Route problems

- Hamiltonian path and cycle problems
- Minimum spanning tree
- Route inspection problem (also called the "Chinese Postman Problem")
- Seven Bridges of Königsberg
- Shortest path problem
- Steiner tree
- Three-cottage problem
- Traveling salesman problem (NP-hard)

96.6.6 Network flow

There are numerous problems arising especially from applications that have to do with various notions of flows in networks, for example:

- Max flow min cut theorem

96.6.7 Visibility problems

- Museum guard problem

96.6.8 Covering problems

Covering problems in graphs are specific instances of subgraph-finding problems, and they tend to be closely related to the clique problem or the independent set problem.

- Set cover problem
- Vertex cover problem

96.6.9 Decomposition problems

Decomposition, defined as partitioning the edge set of a graph (with as many vertices as necessary accompanying the edges of each part of the partition), has a wide variety of question. Often, it is required to decompose a graph into subgraphs isomorphic to a fixed graph; for instance, decomposing a complete graph into Hamiltonian cycles. Other problems specify a family of graphs into which a given graph should be decomposed, for instance, a family of cycles, or decomposing a complete graph K_n into $n - 1$ specified trees having, respectively, 1, 2, 3, ..., $n - 1$ edges.
Some specific decomposition problems that have been studied include:

- Arboricity, a decomposition into as few forests as possible
- Cycle double cover, a decomposition into a collection of cycles covering each edge exactly twice
- Edge coloring, a decomposition into as few matchings as possible
- Graph factorization, a decomposition of a regular graph into regular subgraphs of given degrees

96.6.10 Graph classes

Many problems involve characterizing the members of various classes of graphs. Some examples of such questions are below:

- Enumerating the members of a class
- Characterizing a class in terms of forbidden substructures
- Ascertaining relationships among classes (e.g., does one property of graphs imply another)
- Finding efficient algorithms to decide membership in a class
- Finding representations for members of a class.

96.7 See also

- Gallery of named graphs
- Glossary of graph theory
- List of graph theory topics
- Publications in graph theory

96.7.1 Related topics

- Algebraic graph theory
- Citation graph
- Conceptual graph
- Data structure
- Disjoint-set data structure
- Entitative graph
- Existential graph
- Graph algebras
- Graph automorphism

- Graph coloring
- Graph database
- Graph data structure
- Graph drawing

_ Graph equation
_ Graph rewriting
_ Graph sandwich problem
_ Graph property
_ Intersection graph
_ Logical graph
_ Loop
_ Network theory
_ Null graph
_ Pebble motion problems
_ Percolation
_ Perfect graph
_ Quantum graph
_ Random regular graphs
_ Semantic networks
_ Spectral graph theory
_ Strongly regular graphs
_ Symmetric graphs
_ Transitive reduction
_ Tree data structure

96.7.2 Algorithms

_ Bellman–Ford algorithm
_ Dijkstra's algorithm
_ Ford–Fulkerson algorithm
_ Kruskal's algorithm
_ Nearest neighbour algorithm
_ Prim's algorithm
_ Depth-first search
_ Breadth-first search

96.7.3 Subareas

_ Algebraic graph theory
_ Geometric graph theory
_ Extremal graph theory
_ Probabilistic graph theory
_ Topological graph theory

96.7.4 Related areas of mathematics

_ Combinatorics
_ Group theory
_ Knot theory
_ Ramsey theory

96.7.5 Generalizations

_ Hypergraph
_ Abstract simplicial complex

96.7.6 Prominent graph theorists

_ Alon, Noga
_ Berge, Claude
_ Bollobás, Béla
_ Bondy, Adrian John
_ Brightwell, Graham
_ Chudnovsky, Maria
_ Chung, Fan
_ Dirac, Gabriel Andrew
_ Erdős, Paul
_ Euler, Leonhard
_ Faudree, Ralph
_ Golumbic, Martin
_ Graham, Ronald
_ Harary, Frank
_ Heawood, Percy John
_ Kotzig, Anton
_ Kőnig, Dénes
_ Lovász, László

_ Murty, U. S. R.

_ Nešetřil, Jaroslav
_ Rényi, Alfréd
_ Ringel, Gerhard
_ Robertson, Neil
_ Seymour, Paul
_ Szemerédi, Endre
_ Thomas, Robin
_ Thomassen, Carsten
_ Turán, Pál
_ Tutte, W. T.
_ Whitney, Hassler

96.8 Notes

[1] See, for instance, Iyanaga and Kawada, **69 J**, p. 234 or Biggs, p. 4.
[2] See, for instance, Graham et al., p. 5.
[3] Hale, Scott A. (2013). "Multilinguals and Wikipedia Editing". arXiv:1312.0976 [cs.CY].
[4] Mashaghi, A.; *et al.* (2004). "Investigation of a protein complex network". *European Physical Journal B* **41** (1): 113–121. doi:10.1140/epjb/e2004-00301-0.
[5] Rosen, Kenneth H. *Discrete mathematics and its applications* (7th ed.). New York: McGraw-Hill. ISBN 978-0-07-338309-5.
[6] Biggs, N.; Lloyd, E. and Wilson, R. (1986), *Graph Theory, 1736-1936*, Oxford University Press
[7] Cauchy, A.L. (1813), *Recherche sur les polyèdres - premier mémoire, Journal de l'École Polytechnique*, 9 (Cahier 16): 66–86.
[8] L'Huillier, S.-A.-J. (1861), *Mémoire sur la polyèdrométrie, Annales de Mathématiques* **3**: 169–189.
[9] Cayley, A. (1857), *On the theory of the analytical forms called trees, Philosophical Magazine*, Series IV **13** (85): 172–176, doi:10.1017/CBO9780511703690.046.
[10] Cayley, A. (1875), *Ueber die Analytischen Figuren, welche in der Mathematik Bäume genannt werden und ihre Anwendung auf die Theorie chemischer Verbindungen, Berichte der deutschen Chemischen Gesellschaft* **8** (2): 1056–1059, doi:10.1002/cber.18750080252.
[11] John Joseph Sylvester (1878), *Chemistry and Algebra.* Nature, volume 17, page 284. doi:10.1038/017284a0. Online version. Retrieved 2009-12-30.
[12] Tutte, W.T. (2001), *Graph Theory*, Cambridge University Press, p. 30, ISBN 978-0-521-79489-3.
[13] Gardner, Martin (1992), *Fractal Music, Hypercards, and more...Mathematical Recreations from Scientific American*, W. H. Freeman and Company, p. 203.
[14] Society for Industrial and Applied Mathematics (2002), "The George Polya Prize", *Looking Back, Looking Ahead: A SIAM History*, p. 26.
[15] Heinrich Heesch: Untersuchungen zum Vierfarbenproblem. Mannheim: Bibliographisches Institut 1969.
[16] Appel, K. and Haken, W. (1977), *Every planar map is four colorable. Part I. Discharging, Illinois J. Math.* **21**: 429–490.
[17] Appel, K. and Haken, W. (1977), *Every planar map is four colorable. Part II. Reducibility, Illinois J. Math.* **21**: 491–567.
[18] Robertson, N.; Sanders, D.; Seymour, P. and Thomas, R. (1997), *The four color theorem, Journal of Combinatorial Theory Series B* **70**: 2–44, doi:10.1006/jctb.1997.1750.

96.9 References

_ Berge, Claude (1958), *Théorie des graphes et ses applications*, Collection Universitaire de Mathématiques **II**, Paris: Dunod. English edition, Wiley 1961; Methuen & Co, New York 1962; Russian, Moscow 1961; Spanish, Mexico 1962; Roumanian, Bucharest 1969; Chinese, Shanghai 1963; Second printing of the 1962 first English edition, Dover, New York 2001.
_ Biggs, N.; Lloyd, E.; Wilson, R. (1986), *Graph Theory, 1736–1936*, Oxford University Press.
_ Bondy, J.A.; Murty, U.S.R. (2008), *Graph Theory*, Springer, ISBN 978-1-84628-969-9.
_ Bondy, Riordan, O.M (2003), *Mathematical results on scale-free random graphs in "Handbook of Graphs and Networks" (S. Bornholdt and H.G. Schuster (eds)), Wiley VCH, Weinheim, 1st ed.*.
_ Chartrand, Gary (1985), *Introductory Graph Theory*, Dover, ISBN 0-486-24775-9.
_ Gibbons, Alan (1985), *Algorithmic Graph Theory*, Cambridge University Press.
_ Reuven Cohen, Shlomo Havlin (2010), *Complex Networks: Structure, Robustness and Function*, Cambridge University Press
_ Golumbic, Martin (1980), *Algorithmic Graph Theory and Perfect Graphs*, Academic Press.
_ Harary, Frank (1969), *Graph Theory*, Reading, MA: Addison-Wesley.
_ Harary, Frank; Palmer, Edgar M. (1973), *Graphical Enumeration*, New York, NY: Academic Press.
_ Mahadev, N.V.R.; Peled, Uri N. (1995), *Threshold Graphs and Related Topics*, North-Holland.
_ Mark Newman (2010), *Networks: An Introduction*, Oxford University Press.

96.10 External links

_ Graph theory with examples
_ Hazewinkel, Michiel, ed. (2001), "Graph theory", *Encyclopedia of Mathematics*, Springer, ISBN 978-1-55608-010-4
_ Graph theory tutorial
_ A searchable database of small connected graphs
_ Image gallery: graphs at the Wayback Machine (archived February 6, 2006)

_ Concise, annotated list of graph theory resources for researchers
_ rocs — a graph theory IDE
_ The Social Life of Routers — non-technical paper discussing graphs of people and computers
_ Graph Theory Software — tools to teach and learn graph theory
_ Online books, and library resources in your library and in other libraries about graph theory

96.10.1 Online textbooks

_ Phase Transitions in Combinatorial Optimization Problems, Section 3: Introduction to Graphs (2006) by Hartmann and Weigt
_ Digraphs: Theory Algorithms and Applications 2007 by Jorgen Bang-Jensen and Gregory Gutin
_ Graph Theory, by Reinhard Diestel

Chapter 97

Proof of Bertrand's postulate

In mathematics, Bertrand's postulate (actually a theorem) states that for each n _ 1 there is a prime p such that $n < p$ _ $2n$. It was first proven by Pafnuty Chebyshev, and a short but advanced proof was given by Srinivasa Ramanujan.[1] The gist of the following elementary proof is due to Paul Erdős. The basic idea of the proof is to show that a certain central binomial coefficient needs to have a prime factor within the desired interval in order to be large enough. This is made possible by a careful analysis of the prime factorization of central binomial coefficients.
The main steps of the proof are as follows. First, one shows that every prime power factor p^r that enters into the prime decomposition of the central binomial coefficient $\binom{2n}{n} := \frac{(2n)!}{(n!)^2}$ is at most $2n$. In particular, every prime larger than $\sqrt{2n}$ can enter at most once into this decomposition; that is, its exponent r is at most one. The next step is to prove that $\binom{2n}{n}$ has no prime factors at all in the gap interval $\left(\frac{2n}{3}; n\right)$. As a consequence of these two bounds, the contribution to the size of $\binom{2n}{n}$ coming from all the prime factors that are at most n grows asymptotically as $O($_$^n)$ for some _ < 4 . Since the asymptotic growth of the central binomial coefficient is at least $4^n/2n$, one concludes that for n large enough the binomial coefficient must have another prime factor, which can only lie between n and $2n$. Indeed, making these estimates quantitative, one obtains that this argument is valid for all $n > 468$. The remaining smaller values of n are easily settled by direct inspection, completing the proof of the Bertrand's postulate.

97.1 Lemmas and computation

97.1.1 Lemma 1: A lower bound on the central binomial coefficients

Lemma: For any integer $n > 0$, we have

$$\frac{4^n}{2n} \text{ \_ } \binom{2n}{n} .$$

Proof: Applying the binomial theorem,

$$4^n = (1+1)^{2n} = \sum_{k=0}^{2n} \binom{2n}{k}$$

$$) = 2 + 2\sum_{k=1}^{n-1}\binom{2n}{k} \le 2n\binom{2n}{n};$$

since $\binom{2n}{n}$ is the largest term in the sum in the right-hand side, and the sum has $2n$ terms (including the initial two outside the summation).

97.1.2 Lemma 2: An upper bound on prime powers dividing central binomial coefficients

For a fixed prime p , define $R(p; n)$ to be the largest natural number r such that p^r divides $\binom{2n}{n}$.

Lemma: For any prime p , $p^{R(p;n)} \le 2n$.

Proof: The exponent of p in $n!$ is (see Factorial#Number theory):

$$\sum_{j=1}^{\infty}\left\lfloor \frac{n}{p^j} \right\rfloor;$$

so

$$R(p; n) = \sum_{j=1}^{\infty}\left\lfloor \frac{2n}{p^j} \right\rfloor - 2\sum_{j=1}^{\infty}\left\lfloor \frac{n}{p^j} \right\rfloor = \sum_{j=1}^{\infty}\left(\left\lfloor \frac{2n}{p^j} \right\rfloor - 2\left\lfloor \frac{n}{p^j} \right\rfloor\right):$$

But each term of the last summation can either be zero (if $n/p^j \bmod 1 < 1/2$) or 1 (if $n/p^j \bmod 1 \ge 1/2$) and all terms with $j > \log_p(2n)$ are zero. Therefore

$$R(p; n) \le \log_p(2n);$$

and

pR(*p*;*n*) _ *p*log*p* 2*n* = 2*n*:

This completes the proof of the lemma.

97.1.3 Lemma 3: The exact power of a large prime in a central binomial coefficient

Lemma: If *p* is odd and 2*n*

3 < *p* _ *n* , then *R*(*p*; *n*) = 0:

Proof: The factors of *p* in the numerator come from the terms *p* and 2*p* , and in the denominator from two factors of *p* . These cancel since *p* is odd.

97.1.4 Lemma 4: An upper bound on the primorial

We estimate the primorial function,

x# =

Π

p _ *x*

p;

where the product is taken over all *prime* numbers *p* less than or equal to the real number *x* .

Lemma: For all real numbers *x* _ 3 , *x*# < 22*x*□3 [2]

Proof: Since

(2*n*

n

)

is an integer and all the primes *n* + 1 _ *p* _ 2*n* □ 1 appears in its numerator (2*x* □ 1)#/(*x*)# _

(2*n*

n

)

< 22*x*□2 holds. The proof is by mathematical induction.

_ *n* = 3 : *n*# = 6 < 8:

_ *n* = 4 : *n*# = 6 < 32:

_ 2*m* □ 1# < 22(2*m*□1)□3

_ 2*m*# < 22(2*m*□3)

_ *x*# = ⌊*x*⌋# < 22*x*□3

Thus the lemma is proven.

97.2 Proof of Bertrand's Postulate

Assume there is a counterexample: an integer $n \geq 2$ such that there is no prime p with $n < p < 2n$.

If $2 \leq n < 468$, then p can be chosen from among the prime numbers 3, 5, 7, 13, 23, 43, 83, 163, 317, 631 (each being less than twice its predecessor) such that $n < p < 2n$. Therefore $n \geq 468$.

There are no prime factors *p* of

(2*n*

n

)

such that:

_ $2n < p$, because every factor must divide $(2n)!$;

_ $p = 2n$, because $2n$ is not prime;

_ $n < p < 2n$, because we assumed there is no such prime number;

_ $2n / 3 < p \leq n$: by Lemma 3.

Therefore, every prime factor p satisfies $p \leq 2n/3$.

When *p* >

p

2*n*; the number

(2*n*

n

)

has at most one factor of *p*. By Lemma 2, for any prime *p* we have *pR*(*p*,*n*) ≤ 2*n*, so the product of the *pR*(*p*,*n*) over the primes less than or equal to

p

2*n* is at most (2*n*)

p

2*n* . Then, starting with Lemma

1 and decomposing the right-hand side into its prime factorization, and finally using Lemma 4, these bounds give:

4*n*

2*n*

_

(

2*n*

n

)

=

$$\left(\prod_{p\le\sqrt{2n}} p^{R(p,n)}\right)\left(\prod_{\sqrt{2n}<p\le\frac{2n}{3}} p^{R(p,n)}\right) < (2n)^{\sqrt{2n}} \prod_{1<p\le\frac{2n}{3}} p = (2n)^{\sqrt{2n}} \left(\frac{2n}{3}\right)\# \le (2n)^{\sqrt{2n}} 4^{2n/3}.$$

Taking logarithms yields to

$$\frac{\log 4}{3} n \le (\sqrt{2n} + 1) \log 2n.$$

By concavity of the right-hand side as a function of n, the last inequality is necessarily verified on an interval. Since it holds true for $n=467$ and it does not for $n=468$, we obtain

$$n < 468.$$

But these cases have already been settled, and we conclude that no counterexample to the postulate is possible.

97.2.1 Proof by Shigenori Tochiori

Using Lemma 4, Tochiori refined Erdos's method and proved if there exists a positive integer $n \ge 5$ such that there is no prime number $n < p \le 2n$ then $n < 64$. [3]

First, refine lemma 1 to:

Lemma 1': For any integer $n \ge 4$, we have

$$\frac{4^n}{n} < \binom{2n}{n}.$$

Proof: By induction: $\frac{4^4}{4} = 64 < 70 = \binom{8}{4}$; and assuming the truth of the lemma for $n - 1$,

$$\binom{2n}{n} = 2\frac{2n-1}{n}\binom{2(n-1)}{n-1}$$

n □ 1
)
> 2
2n □ 1
n
4n□1
n □ 1
> 2 _ 2
4n□1
n
=
4n
n
:
97.3. REFERENCES 731
Then, refine the estimate of the product of all small primes via a better estimate on _(x) (the number of primes at most n):
Lemma 5: For any natural number n , we have
_(n) _ 1
3
n + 2:
Proof: Except for p = 2; 3 , every prime number has p _ 1 or p _ 5 (mod 6) . Thus _(n) is upper bounded by the number of numbers with k _ 1 or k _ 5 (mod 6) , plus one (since this counts 1 and misses 2; 3). Thus
_(n) _
⌊
n + 5
6
⌋
+
⌊
n + 1
6
⌋
+ 1 _ n + 5
6
+
n + 1
6
+ 1 =
1
3
n + 2:
Now, calculating the binomial coefficient as in the previous section, we can use the improved bounds to get (for n _ 5 , which implies
p
2n _ 3 so that
p
2n# _ 3# = 6):
4n
n
_
(
2n
n
)
=
Π
p_
p
2n
pR(p;n) _
Π
p
2n<p_2n
3
pR(p;n)
< (2n)_(
p
2n)

Π
p
$2n<p$ _ $2n$
3
$p = (2n)$
1
3
p
$2n+2$ $(2pn/3)\#$
$2n\#$
$< (2n)$
1
3
p
$2n+2$ 22_ $2n/3$□3
6
$< (2n)$
1
3
p
$2n+2$24$n/3$□5:
Taking logarithms to get
2
3
$n \log 2 <$
1
3
p
$2n \log 2n + 3 \log n$
2
and dividing both sides by 2
$3n$:
$\log 2 <$
p
2 _ log
p
p n
n
+
9
4
log n
2
n
2
+ $p$log 2
$2n$
_ $f(n)$:
Now the function $g(x) = \log x$
x is decreasing for x _ e , so $f(n)$ is decreasing when n _ $e2 > 2e$. But
$f(26)$
log 2
=
p
2 _ 3
8
+
9
4
_ 5
32
+
p
2
16
= 0:97 _ _ _ < 1 <
$f(n)$
log 2
;
so $n < 26 = 64$. The remaining cases are proven by an explicit list of primes, as above.

97.3 References

[1] Ramanujan, S. (1919), *A proof of Bertrand's postulate*, *Journal of the Indian Mathematical Society* **11**: 181–182
[2] http://www.chart.co.jp/subject/sugaku/suken_tsushin/76/76-8.pdf
[3] http://www.chart.co.jp/subject/sugaku/suken_tsushin/76/76-8.pdf
_ Aigner, Martin, G., Günter M. Ziegler, Karl H. Hofmann, *Proofs from THE BOOK*, Fourth edition, Springer, 2009. ISBN 978-3-642-00855-9.

Chapter 98

Proof that e is irrational

The number e was introduced by Jacob Bernoulli in 1683. More than half a century later, Euler, who had been a student of Jacob's younger brother Johann, proved that e is irrational, that is, that it can not be expressed as the quotient of two integers.

98.1 Euler's proof

Euler wrote the first proof of the fact that e is irrational in 1737 (but the text was only published seven years later).[1][2][3] He computed the representation of e as a simple continued fraction, which is

$$e = [2; 1, 2, 1, 1, 4, 1, 1, 6, 1, 1, 8, 1, 1, \ldots, 2n, 1, 1, \ldots].$$

Since this continued fraction is infinite, e is irrational. A short proof of the previous equality is known.[4] Since the simple continued fraction of e is not periodic, this also proves that e is not a root of second degree polynomial with rational coefficients; in particular, e^2 is irrational.

98.2 Fourier's proof

The most well-known proof is Joseph Fourier's proof by contradiction,[5] which is based upon the equality

$$e = \sum_{n=0}^{\infty} \frac{1}{n!}.$$

Initially e is assumed to be a rational number of the form a/b. Note that b couldn't be equal to one as e is not an integer. It can be shown using the above equality that e is strictly between 2 and 3.

$$\frac{1}{1} + \frac{1}{1} < e = \frac{1}{1} + \frac{1}{1} + \frac{1}{1 \cdot 2} + \frac{1}{1 \cdot 2 \cdot 3} + \cdots < \frac{1}{1} + \frac{1}{1} + \frac{1}{1 \cdot 2} + \frac{1}{1 \cdot 2 \cdot 2} + \cdots = 3$$

We then analyze a blown-up difference x of the series representing e and its strictly smaller b th partial sum, which approximates the limiting value e. By choosing the magnifying factor to be the factorial of b, the fraction a/b and the b th partial sum are turned into integers, hence x must be a positive integer. However, the fast convergence of the series representation implies that the magnified approximation error x is still strictly smaller than 1. From this contradiction we deduce that e is irrational.

Suppose that e is a rational number. Then there exist positive integers a and b such that $e = a/b$. Define the number

$$x = b!\left(e - \sum_{n=0}^{b} \frac{1}{n!}\right)$$

To see that if e is rational, then x is an integer, substitute $e = a/b$ into this definition to obtain

$$x = b!\left(\frac{a}{b} - \sum_{n=0}^{b} \frac{1}{n!}\right) = a(b - 1)! - \sum_{n=0}^{b} \frac{b!}{n!} :$$

The first term is an integer, and every fraction in the sum is actually an integer because $n \leq b$ for each term. Therefore x is an integer.

We now prove that $0 < x < 1$. First, to prove that x is strictly positive, we insert the above series representation of e into the definition of x and obtain

$$x = b!\left(\sum_{n=0}^{\infty} \frac{1}{n!} - \sum_{n=0}^{b} \frac{1}{n!}\right) = \sum_{n=b+1}^{\infty} \frac{b!}{n!} > 0 ;$$

because all the terms are strictly positive.

We now prove that $x < 1$. For all terms with $n \geq b + 1$ we have the upper estimate

$$\frac{b!}{n!} = \frac{1}{(b+1)(b+2) \cdots (b+(n-b))} < \frac{1}{(b+1)^{n-b}} :$$

This inequality is strict for every $n \geq b + 2$. Changing the index of summation to $k = n - b$ and using the formula for the infinite geometric series, we obtain

$$x = \sum_{n=b+1}^{\infty} \frac{b!}{n!} < \sum_{n=b+1}^{\infty} \frac{1}{\ldots}$$

$$(b+1)n\,\square\,b = \sum_{k=1}^{\infty}\frac{1}{(b+1)k} = \frac{1}{b+1}\left(\frac{1}{1-\frac{1}{b+1}}\right) = \frac{1}{b} < 1;$$

Since there is no integer strictly between 0 and 1, we have reached a contradiction, and so e must be irrational. Q.E.D.

98.3 Alternate proofs

Another proof[6] can be obtained from the previous one by noting that

$$(b+1)x = 1 + \frac{1}{b+2} + \frac{1}{(b+2)(b+3)} + ___ < 1 + \frac{1}{b+1} + \frac{1}{(b+1)(b+2)} + ___ = 1 + x;$$

and this inequality is equivalent to the assertion that $bx < 1$. This is impossible, of course, since b and x are natural numbers.

Still another proof[7] can be obtained from the fact that

$$\frac{1}{e} = e^{\square 1} = \sum_{n=0}^{\infty}\frac{(\square 1)n}{n!}$$

98.4 Generalizations

In 1840, Liouville published a proof of the fact that $e2$ is irrational[8] followed by a proof that $e2$ is not a root of a second degree polynomial with rational coefficients.[9] This last fact implies that $e4$ is irrational. His proofs are similar to Fourier's proof of the irrationality of e. In 1891, Hurwitz explained how it is possible to prove along the same line of ideas that e is not a root of a third degree polynomial with rational coefficients.[10] In particular, $e3$ is irrational. More generally, eq is irrational for any non-zero rational q.[11]

98.5 See also

_ Characterizations of the exponential function
_ Transcendental number, including a proof that e is transcendental
_ Lindemann–Weierstrass theorem

98.6 References

[1] Euler, Leonhard (1744). "De fractionibus continuis dissertatio" [A dissertation on continued fractions]. *Commentarii academiae scientiarum Petropolitanae* **9**: 98–137.

[2] Euler, Leonhard (1985). "An essay on continued fractions". *Mathematical Systems Theory* **18**: 295–398. doi:10.1007/bf01699475.

[3] Sandifer, C. Edward (2007). "Chapter 32: Who proved e is irrational?". *How Euler did it*. Mathematical Association of America. pp. 185–190. ISBN 978-0-88385-563-8. LCCN 2007927658.

[4] Cohn, Henry (2006). "A short proof of the simple continued fraction expansion of e". *American Mathematical Monthly* (Mathematical Association of America) **113** (1): 57–62. JSTOR 27641837.

[5] de Stainville, Janot (1815). *Mélanges d'Analyse Algébrique et de Géométrie* [A mixture of Algebraic Analysis and Geometry]. Veuve Courcier. pp. 340–341.

[6] MacDivitt, A. R. G.; Yanagisawa, Yukio (1987), *An elementary proof that* e *is irrational*, *The Mathematical Gazette* (London: Mathematical Association) **71** (457): 217, JSTOR 3616765
[7] Penesi, L. L. (1953). "Elementary proof that *e* is irrational". *American Mathematical Monthly* (Mathematical Association of America) **60** (7): 474. JSTOR 2308411.
[8] Liouville, Joseph (1840). "Sur l'irrationalité du nombre *e* = 2,718...". *Journal de Mathématiques Pures et Appliquées*. 1 (in French) **5**: 192.
[9] Liouville, Joseph (1840). "Addition à la note sur l'irrationnalité du nombre *e*". *Journal de Mathématiques Pures et Appliquées*. 1 (in French) **5**: 193–194.
[10] Hurwitz, Adolf (1933) [1891]. "Über die Kettenbruchentwicklung der Zahl *e*". *Mathematische Werke* (in German) **2**. Basel: Birkhäuser. pp. 129–133.
[11] Aigner, Martin; Ziegler, Günter M. (1998), *Proofs from THE BOOK* (4th ed.), Berlin, New York: Springer-Verlag, pp. 27–36, doi:10.1007/978-3-642-00856-6, ISBN 978-3-642-00855-9.

Chapter 99

Prime number

"Prime" redirects here. For other uses, see Prime (disambiguation).

A **prime number** (or a **prime**) is a natural number greater than 1 that has no positive divisors other than 1 and itself. A natural number greater than 1 that is not a prime number is called a composite number. For example, 5 is prime because 1 and 5 are its only positive integer factors, whereas 6 is composite because it has the divisors 2 and 3 in addition to 1 and 6. The fundamental theorem of arithmetic establishes the central role of primes in number theory: any integer greater than 1 can be expressed as a product of primes that is unique up to ordering. The uniqueness in this theorem requires excluding 1 as a prime because one can include arbitrarily many instances of 1 in any factorization, e.g., 3, 1 × 3, 1 × 1 × 3, etc. are all valid factorizations of 3.

The property of being prime (or not) is called primality. A simple but slow method of verifying the primality of a given number n is known as trial division. It consists of testing whether n is a multiple p of any integer between 2 and n . Algorithms much more efficient than trial division have been devised to test the primality of large numbers. Particularly fast methods are available for numbers of special forms, such as Mersenne numbers. As of April 2014, the largest known prime number has 17,425,170 decimal digits.

There are infinitely many primes, as demonstrated by Euclid around 300 BC. There is no known useful formula that sets apart all of the prime numbers from composites. However, the distribution of primes, that is to say, the statistical behaviour of primes in the large, can be modelled. The first result in that direction is the prime number theorem, proven at the end of the 19th century, which says that the probability that a given, randomly chosen number n is prime is inversely proportional to its number of digits, or to the logarithm of n.

Many questions regarding prime numbers remain open, such as Goldbach's conjecture (that every even integer greater than 2 can be expressed as the sum of two primes), and the twin prime conjecture (that there are infinitely many pairs of primes whose difference is 2). Such questions spurred the development of various branches of number theory, focusing on analytic or algebraic aspects of numbers. Primes are used in several routines in information technology, such as public-key cryptography, which makes use of properties such as the difficulty of factoring large numbers into their prime factors. Prime numbers give rise to various generalizations in other mathematical domains, mainly algebra, such as prime elements and prime ideals.

99.1 Definition and examples

A natural number (i.e. 1, 2, 3, 4, 5, 6, etc.) is called a **prime number** (or a **prime**) if it has exactly two positive divisors, 1 and the number itself.[1] Natural numbers greater than 1 that are not prime are called *composite*.

Among the numbers 1 to 6, the numbers 2, 3, and 5 are the prime numbers, while 1, 4, and 6 are not prime. 1 is excluded as a prime number, for reasons explained below. 2 is a prime number, since the only natural numbers dividing it are 1 and 2. Next, 3 is prime, too: 1 and 3 do divide 3 without remainder, but 3 divided by 2 gives remainder 1. Thus, 3 is prime. However, 4 is composite, since 2 is another number (in addition to 1 and 4) dividing 4 without remainder:

$4 = 2 \cdot 2.$

735

736 *CHAPTER 99. PRIME NUMBER*

The number 12 is not a prime, as 12 items can be placed into 3 equal-size columns of 4 each (among other ways). 11 items cannot be all placed into several equal-size columns of more than 1 item each without some extra items leftover (a remainder). Therefore the number 11 is a prime.

5 is again prime: none of the numbers 2, 3, or 4 divide 5. Next, 6 is divisible by 2 or 3, since

$6 = 2 \cdot 3.$

Hence, 6 is not prime. The image at the right illustrates that 12 is not prime: $12 = 3 \cdot 4$. No even number greater than 2 is prime because by definition, any such number n has at least three distinct divisors, namely 1, 2, and n. This implies that n is not prime. Accordingly, the term *odd prime* refers to any prime number greater than 2. In a similar vein, all prime numbers bigger than 5, written in the usual decimal system, end in 1, 3, 7, or 9, since even numbers are multiples of 2 and numbers ending in 0 or 5 are multiples of 5.

If n is a natural number, then 1 and n divide n without remainder. Therefore, the condition of being a prime can also be restated as: a number is prime if it is greater than one and if none of

$2, 3, \ldots, n - 1$

divides n (without remainder). Yet another way to say the same is: a number $n > 1$ is prime if it cannot be written as a product of two integers a and b, both of which are larger than 1:
$n = a \cdot b$.
In other words, n is prime if n items cannot be divided up into smaller equal-size groups of more than one item.
The set of all primes is often denoted by **P**.
The first 168 prime numbers (all the prime numbers less than 1000) are:

2, 3, 5, 7, 11, 13, 17, 19, 23, 29, 31, 37, 41, 43, 47, 53, 59, 61, 67, 71, 73, 79, 83, 89, 97, 101, 103, 107, 109, 113, 127, 131, 137, 139, 149, 151, 157, 163, 167, 173, 179, 181, 191, 193, 197, 199, 211, 223, 227, 229, 233, 239, 241, 251, 257, 263, 269, 271, 277, 281, 283, 293, 307, 311, 313, 317, 331, 337, 347, 349, 353, 359, 367, 373, 379, 383, 389, 397, 401, 409, 419, 421, 431, 433, 439, 443, 449, 457, 461, 463, 467, 479, 487, 491, 499, 503, 509, 521, 523, 541, 547, 557, 563, 569, 571, 577, 587, 593, 599, 601, 607, 613, 617, 619, 631, 641, 643, 647, 653, 659, 661, 673, 677, 683, 691, 701, 709, 719, 727, 733, 739, 743, 751, 757, 761, 769, 773, 787, 797, 809, 811, 821, 823, 827, 829, 839, 853, 857, 859, 863, 877, 881, 883, 887, 907, 911, 919, 929, 937, 941, 947, 953, 967, 971, 977, 983, 991, 997 (sequence A000040 in OEIS).

99.2 Fundamental theorem of arithmetic

Main article: Fundamental theorem of arithmetic
The crucial importance of prime numbers to number theory and mathematics in general stems from the *fundamental theorem of arithmetic*, which states that every integer larger than 1 can be written as a product of one or more primes in a way that is unique except for the order of the prime factors.[2] Primes can thus be considered the "basic building blocks" of the natural numbers. For example:
As in this example, the same prime factor may occur multiple times. A decomposition:
$n = p_1 \cdot p_2 \cdot ... \cdot p_t$
of a number n into (finitely many) prime factors p_1, p_2, ... to p_t is called *prime factorization* of n. The fundamental theorem of arithmetic can be rephrased so as to say that any factorization into primes will be identical except for the order of the factors. So, albeit there are many prime factorization algorithms to do this in practice for larger numbers, they all have to yield the same result.
If p is a prime number and p divides a product ab of integers, then p divides a or p divides b. This proposition is known as Euclid's lemma.[3] It is used in some proofs of the uniqueness of prime factorizations.

99.2.1 Primality of one

Most early Greeks did not even consider 1 to be a number,[4] and so they did not consider it a prime. In the 19th century, however, many mathematicians did consider the number 1 a prime. For example, Derrick Norman Lehmer's list of primes up to 10,006,721, reprinted as late as 1956,[5] started with 1 as its first prime.[6] Henri Lebesgue is said to be the last professional mathematician to call 1 prime.[7]
Although a large body of mathematical work would still be valid when calling 1 a prime, the fundamental theorem of arithmetic (mentioned above) would not hold as stated. For example, the number 15 can be factored as $3 \cdot 5$ and $1 \cdot 3 \cdot 5$; if 1 were admitted as a prime, these two presentations would be considered different factorizations of 15 into prime numbers, so the statement of that theorem would have to be modified. Similarly, the sieve of Eratosthenes would not work correctly if 1 were considered a prime: a modified version of the sieve that considers 1 as prime would eliminate all multiples of 1 (that is, all numbers) and produce as output only the single number 1. Furthermore, the prime numbers have several properties that the number 1 lacks, such as the relationship of the number to its corresponding value of Euler's totient function or the sum of divisors function.[8][9]

99.3 History

There are hints in the surviving records of the ancient Egyptians that they had some knowledge of prime numbers: the Egyptian fraction expansions in the Rhind papyrus, for instance, have quite different forms for primes and for composites. However, the earliest surviving records of the explicit study of prime numbers come from the Ancient

The Sieve of Eratosthenes is a simple algorithm for finding all prime numbers up to a specified integer. It was created in the 3rd century BC by Eratosthenes, an ancient Greek mathematician.
Greeks. Euclid's Elements (circa 300 BC) contain important theorems about primes, including the infinitude of primes and the fundamental theorem of arithmetic. Euclid also showed how to construct a perfect number from a Mersenne prime. The Sieve of Eratosthenes, attributed to Eratosthenes, is a simple method to compute primes, although the large primes found today with computers are not generated this way.
After the Greeks, little happened with the study of prime numbers until the 17th century. In 1640 Pierre de Fermat stated (without proof) Fermat's little theorem (later proved by Leibniz and Euler). Fermat also conjectured that all numbers of the form $2^{2^n} + 1$ are prime (they are called Fermat numbers) and he verified this up to $n = 4$ (or $2^{16} + 1$). However, the very next Fermat number $2^{32} + 1$ is composite (one of its prime factors is 641), as Euler discovered later, and in fact no further Fermat numbers are known to be prime. The French monk Marin Mersenne looked at primes of the form $2^p - 1$, with p a prime. They are called Mersenne primes in his honor.
Euler's work in number theory included many results about primes. He showed the infinite series 1/2 + 1/3 + 1/5 + 1/7 + 1/11 + … is divergent. In 1747 he showed that the even perfect numbers are precisely the integers of the form $2^{p-1}(2^p - 1)$, where the second factor is a Mersenne prime.
At the start of the 19th century, Legendre and Gauss independently conjectured that as x tends to infinity, the number of primes up to x is asymptotic to $x/\ln(x)$, where $\ln(x)$ is the natural logarithm of x. Ideas of Riemann in his 1859

paper on the zeta-function sketched a program that would lead to a proof of the prime number theorem. This outline was completed by Hadamard and de la Vallée Poussin, who independently proved the prime number theorem in 1896.
Proving a number is prime is not done (for large numbers) by trial division. Many mathematicians have worked on primality tests for large numbers, often restricted to specific number forms. This includes Pépin's test for Fermat numbers (1877), Proth's theorem (around 1878), the Lucas–Lehmer primality test (originated 1856),[10] and the generalized Lucas primality test. More recent algorithms like APRT-CL, ECPP, and AKS work on arbitrary numbers but remain much slower.

For a long time, prime numbers were thought to have extremely limited application outside of pure mathematics.[11] This changed in the 1970s when the concepts of public-key cryptography were invented, in which prime numbers formed the basis of the first algorithms such as the RSA cryptosystem algorithm.
Since 1951 all the largest known primes have been found by computers. The search for ever larger primes has generated interest outside mathematical circles. The Great Internet Mersenne Prime Search and other distributed computing projects to find large primes have become popular in the last ten to fifteen years, while mathematicians continue to struggle with the theory of primes.

99.4 Number of prime numbers

Main article: Euclid's theorem
There are infinitely many prime numbers. Another way of saying this is that the sequence
2, 3, 5, 7, 11, 13, ...
of prime numbers never ends. This statement is referred to as *Euclid's theorem* in honor of the ancient Greek mathematician Euclid, since the first known proof for this statement is attributed to him. Many more proofs of the infinitude of primes are known, including an analytical proof by Euler, Goldbach's proof based on Fermat numbers,[12] Furstenberg's proof using general topology,[13] and Kummer's elegant proof.[14]

99.4.1 Euclid's proof

Euclid's proof (Book IX, Proposition 20[15]) considers any finite set S of primes. The key idea is to consider the product of all these numbers plus one:

$$N = 1 + \prod_{p \in S} p.$$

Like any other natural number, N is divisible by at least one prime number (it is possible that N itself is prime).
None of the primes by which N is divisible can be members of the finite set S of primes with which we started, because dividing N by any one of these leaves a remainder of 1. Therefore the primes by which N is divisible are additional primes beyond the ones we started with. Thus any finite set of primes can be extended to a larger finite set of primes.
It is often erroneously reported that Euclid begins with the assumption that the set initially considered contains all prime numbers, leading to a contradiction, or that it contains precisely the n smallest primes rather than any arbitrary finite set of primes.[16] Today, the product of the smallest n primes plus 1 is conventionally called the nth Euclid number.

99.4.2 Euler's analytical proof

Euler's proof uses the sum of the reciprocals of primes,

$$S(p) = \frac{1}{2} + \frac{1}{3} + \frac{1}{5} + \frac{1}{7} + \cdots + \frac{1}{p}.$$

This sum becomes larger than any arbitrary real number provided that p is big enough.[17] This shows that there are infinitely many primes, since otherwise this sum would grow only until the biggest prime p is reached. The growth of $S(p)$ is quantified by Mertens' second theorem.[18] For comparison, the sum

$$\frac{1}{1^2} + \frac{1}{2^2} + \frac{1}{3^2} + \cdots +$$

$$\frac{1}{n^2} = \sum_{i=1}^{n} \frac{1}{i^2}$$

does not grow to infinity as n goes to infinity. In this sense, prime numbers occur more often than squares of natural numbers. Brun's theorem states that the sum of the reciprocals of twin primes,

$$\left(\frac{1}{3}+\frac{1}{5}\right)+\left(\frac{1}{5}+\frac{1}{7}\right)+\left(\frac{1}{11}+\frac{1}{13}\right)+\cdots=\sum_{\substack{p \text{ prime},\\ p+2 \text{ prime}}}\left(\frac{1}{p}+\frac{1}{p+2}\right);$$

is finite.

99.5 Testing primality and integer factorization

There are various methods to determine whether a given number n is prime. The most basic routine, trial division, is of little practical use because of its slowness. One group of modern primality tests is applicable to arbitrary numbers, while more efficient tests are available for particular numbers. Most such methods only tell whether n is prime or not. Routines also yielding one (or all) prime factors of n are called factorization algorithms.

99.5.1 Trial division

The most basic method of checking the primality of a given integer n is called *trial division*. This routine consists of dividing n by each integer m that is greater than 1 and less than or equal to the square root of n. If the result of any of these divisions is an integer, then n is not a prime, otherwise it is a prime. Indeed, if $n = ab$ is composite (with a and $b \neq 1$) then one of the factors a or b is necessarily at most $\sqrt{n}$. For example, for $n = 37$, the trial divisions are by $m = 2, 3, 4, 5$, and 6. None of these numbers divides 37, so 37 is prime. This routine can be implemented more efficiently if a complete list of primes up to $\sqrt{n}$ is known—then trial divisions need to be checked only for those m that are prime. For example, to check the primality of 37, only three divisions are necessary ($m = 2, 3$, and 5), given that 4 and 6 are composite.

While a simple method, trial division quickly becomes impractical for testing large integers because the number of possible factors grows too rapidly as n increases. According to the prime number theorem explained below, the number of prime numbers less than

p

n is approximately given by
p
$n/ \ln($
p
$n)$, so the algorithm may need up to
this number of trial divisions to check the primality of n. For $n = 10^{20}$, this number is 450 million—too large for many practical applications.

99.5.2 Sieves

An algorithm yielding all primes up to a given limit, such as required in the trial division method, is called a prime number sieve. The oldest example, the sieve of Eratosthenes (see above) is useful for relatively small primes. The modern sieve of Atkin is more complicated, but faster when properly optimized. Before the advent of computers, lists of primes up to bounds like 10^7 were also used.[19]

99.5.3 Primality testing versus primality proving

Modern primality tests for general numbers n can be divided into two main classes, probabilistic (or "Monte Carlo") and deterministic algorithms. Deterministic algorithms provide a way to tell **for sure** whether a given number is prime or not. For example, trial division is a deterministic algorithm because, if it performed correctly, it will always identify a prime number as prime and a composite number as composite. Probabilistic algorithms are normally faster, but do not completely prove that a number is prime. These tests rely on testing a given number in a partly random way. For example, a given test might pass all the time if applied to a prime number, but pass only with probability p if applied to a composite number. If we repeat the test n times and pass every time, then the probability that our number is composite is $1/(1-p)^n$, which decreases exponentially with the number of tests, so we can be as sure as we like (though never perfectly sure) that the number is prime. On the other hand, if the test ever fails, then we know that the number is composite.

A particularly simple example of a probabilistic test is the Fermat primality test, which relies on the fact (Fermat's little theorem) that $n^p \equiv n \pmod{p}$ for any n if p is a prime number. If we have a number b that we want to test for primality,

then we work out $n^b \pmod{b}$ for a random value of n as our test. A flaw with this test is that there are some composite numbers (the Carmichael numbers) that satisfy the Fermat identity even though they are not prime, so the test has no way of distinguishing between prime numbers and Carmichael numbers. Carmichael numbers are substantially rarer than prime numbers, though, so this test can be useful for practical purposes. More powerful extensions of the Fermat primality test, such as the Baillie-PSW, Miller-Rabin, and Solovay-Strassen tests, are guaranteed to fail at least some of the time when applied to a composite number.

Deterministic algorithms do not erroneously report composite numbers as prime. In practice, the fastest such method is known as elliptic curve primality proving. Analyzing its run time is based on heuristic arguments, as opposed to the rigorously proven complexity of the more recent AKS primality test. Deterministic methods are typically slower than probabilistic ones, so the latter ones are typically applied first before a more time-consuming deterministic routine is employed.

The following table lists a number of prime tests. The running time is given in terms of n, the number to be tested and, for probabilistic algorithms, the number k of tests performed. Moreover, ε is an arbitrarily small positive number, and log is the logarithm to an unspecified base. The big O notation means that, for example, elliptic curve primality proving requires a time that is bounded by a factor (not depending on n, but on ε) times $\log^{5+\varepsilon}(n)$.

99.5.4 Special-purpose algorithms and the largest known prime

Further information: List of prime numbers

In addition to the aforementioned tests applying to any natural number n, a number of much more efficient primality tests is available for special numbers. For example, to run Lucas' primality test requires the knowledge of the prime factors of $n - 1$, while the Lucas–Lehmer primality test needs the prime factors of $n + 1$ as input. For example, these tests can be applied to check whether

$n! \pm 1 = 1 \cdot 2 \cdot 3 \cdot \ldots \cdot n \pm 1$

are prime. Prime numbers of this form are known as factorial primes. Other primes where either $p + 1$ or $p - 1$ is of a particular shape include the Sophie Germain primes (primes of the form $2p + 1$ with p prime), primorial primes, Fermat primes and Mersenne primes, that is, prime numbers that are of the form $2^p - 1$, where p is an arbitrary prime. The Lucas–Lehmer test is particularly fast for numbers of this form. This is why the largest *known* prime has almost always been a Mersenne prime since the dawn of electronic computers.

Fermat primes are of the form

$F_k = 2^{2^k} + 1,$

with k an arbitrary natural number. They are named after Pierre de Fermat who conjectured that all such numbers F_k are prime. This was based on the evidence of the first five numbers in this series—3, 5, 17, 257, and 65,537—being prime. However, F_5 is composite and so are all other Fermat numbers that have been verified as of 2011. A regular n-gon is constructible using straightedge and compass if and only if

$n = 2^i \cdot m$

where m is a product of any number of distinct Fermat primes and i is any natural number, including zero.

The following table gives the largest known primes of the mentioned types. Some of these primes have been found using distributed computing. In 2009, the Great Internet Mersenne Prime Search project was awarded a US$100,000 prize for first discovering a prime with at least 10 million digits.[20] The Electronic Frontier Foundation also offers $150,000 and $250,000 for primes with at least 100 million digits and 1 billion digits, respectively.[21] Some of the largest primes not known to have any particular form (that is, no simple formula such as that of Mersenne primes) have been found by taking a piece of semi-random binary data, converting it to a number n, multiplying it by 256^k for some positive integer k, and searching for possible primes within the interval $[256^k n + 1, 256^k(n + 1) - 1]$.

99.5.5 Integer factorization

Main article: Integer factorization

Construction of a regular pentagon. 5 is a Fermat prime.

Given a composite integer n, the task of providing one (or all) prime factors is referred to as *factorization* of n. Elliptic curve factorization is an algorithm relying on arithmetic on an elliptic curve.

99.6 Distribution

In 1975, number theorist Don Zagier commented that primes both[25]

The distribution of primes in the large, such as the question how many primes are smaller than a given, large threshold, is described by the prime number theorem, but no efficient formula for the n-th prime is known.

There are arbitrarily long sequences of consecutive non-primes, as for every positive integer n the n consecutive integers from $(n + 1)! + 2$ to $(n + 1)! + n + 1$ (inclusive) are all composite (as $(n + 1)! + k$ is divisible by k for k between 2 and $n + 1$).

Dirichlet's theorem on arithmetic progressions, in its basic form, asserts that linear polynomials

$p(n) = a + bn$

with coprime integers a and b take infinitely many prime values. Stronger forms of the theorem state that the sum of

the reciprocals of these prime values diverges, and that different such polynomials with the same b have approximately the same proportions of primes.

The corresponding question for quadratic polynomials is less well-understood.

99.6.1 Formulas for primes

Main article: formulas for primes

There is no known efficient formula for primes. For example, Mills' theorem and a theorem of Wright assert that there are real constants $A>1$ and μ such that

⌊

$A3_n$

⌋

and

⌊

2:::22_ ⌋

are prime for any natural number n. Here ⌊□⌋ represents the floor function, i.e., largest integer not greater than the number in question. The latter formula can be shown using Bertrand's postulate (proven first by Chebyshev), which states that there always exists at least one prime number p with $n < p < 2n - 2$, for any natural number $n > 3$. However, computing A or μ requires the knowledge of infinitely many primes to begin with.[26] Another formula is based on Wilson's theorem and generates the number 2 many times and all other primes exactly once.

There is no non-constant polynomial, even in several variables, that takes *only* prime values. However, there is a set of Diophantine equations in 9 variables and one parameter with the following property: the parameter is prime if and only if the resulting system of equations has a solution over the natural numbers. This can be used to obtain a single formula with the property that all its *positive* values are prime.

99.6.2 Number of prime numbers below a given number

Main articles: Prime number theorem and Prime-counting function

The prime counting function $\pi(n)$ is defined as the number of primes not greater than n. For example $\pi(11) = 5$,

20 000 40 000 60 000 80 000 100 000

2000

4000

6000

8000

A chart depicting π(n) (blue), n / ln (n) (green) and Li(n) (red)

since there are five primes less than or equal to 11. There are known algorithms to compute exact values of $\pi(n)$ faster

than it would be possible to compute each prime up to n. The *prime number theorem* states that $\pi(n)$ is approximately given by

_(n) _ n

ln n

;

in the sense that the ratio of $\pi(n)$ and the right hand fraction approaches 1 when n grows to infinity. This implies that the likelihood that a number less than n is prime is (approximately) inversely proportional to the number of digits in n. A more accurate estimate for $\pi(n)$ is given by the offset logarithmic integral

Li(n) =

∫ n

2

dt

ln t

:

The prime number theorem also implies estimates for the size of the n-th prime number pn (i.e., $p1 = 2$, $p2 = 3$, etc.): up to a bounded factor, pn grows like $n \log(n)$.[27] In particular, the prime gaps, i.e. the differences $pn - pn_{-1}$ of two consecutive primes, become arbitrarily large. This latter statement can also be seen in a more elementary way by noting that the sequence $n! + 2, n! + 3, \ldots, n! + n$ (for the notation $n!$ read factorial) consists of $n - 1$ composite numbers, for any natural number n.

99.6.3 Arithmetic progressions

An arithmetic progression is the set of natural numbers that give the same remainder when divided by some fixed number q called modulus. For example,

3, 12, 21, 30, 39, ...,

is an arithmetic progression modulo $q = 9$. Except for 3, none of these numbers is prime, since $3 + 9n = 3(1 + 3n)$ so that the remaining numbers in this progression are all composite. (In general terms, all prime numbers above q are of the form $q\#\cdot n + m$, where $0 < m < q\#$, and m has no prime factor $\leq q$.) Thus, the progression

$a, a + q, a + 2q, a + 3q, \ldots$

can have infinitely many primes only when a and q are coprime, i.e., their greatest common divisor is one. If this necessary condition is satisfied, *Dirichlet's theorem on arithmetic progressions* asserts that the progression contains infinitely many primes. The picture below illustrates this with $q = 9$: the numbers are "wrapped around" as soon as a multiple of 9 is passed. Primes are highlighted in red. The rows (=progressions) starting with $a = 3$, 6, or 9 contain at most one prime number. In all other rows ($a = 1$, 2, 4, 5, 7, and 8) there are infinitely many prime numbers. What is more, the primes are distributed equally among those rows in the long run—the density of all primes congruent a modulo 9 is 1/6.

The Green–Tao theorem shows that there are arbitrarily long arithmetic progressions consisting of primes.[28] An odd prime p is expressible as the sum of two squares, $p = x2 + y2$, exactly if p is congruent 1 modulo 4 (Fermat's theorem on sums of two squares).

99.6.4 Prime values of quadratic polynomials

Euler noted that the function

$n2 + n + 41$

gives prime numbers for $0 \leq n < 40$,[29][30] a fact leading into deep algebraic number theory, more specifically Heegner numbers. For bigger n, it does take composite values. The Hardy-Littlewood conjecture F makes an asymptotic prediction about the density of primes among the values of quadratic polynomials (with integer coefficients a, b, and c)

Prime numbers (highlighted in red) in arithmetic progression modulo 9.

$f(n) = ax2 + bx + c$

in terms of $\mathrm{Li}(n)$ and the coefficients a, b, and c. However, progress has proved hard to come by: no quadratic polynomial (with $a \neq 0$) is known to take infinitely many prime values. The Ulam spiral depicts all natural numbers in a spiral-like way. Surprisingly, prime numbers cluster on certain diagonals and not others, suggesting that some quadratic polynomials take prime values more often than other ones.

99.7 Open questions

99.7.1 Zeta function and the Riemann hypothesis

Main article: Riemann hypothesis

The Riemann zeta function $\zeta(s)$ is defined as an infinite sum

_(s) =

$I\Sigma$

$n=1$

1

ns ;

where s is a complex number with real part bigger than 1. It is a consequence of the fundamental theorem of arithmetic that this sum agrees with the infinite product

Π

$p$prime

1

1 □ p□s :

The zeta function is closely related to prime numbers. For example, the aforementioned fact that there are infinitely many primes can also be seen using the zeta function: if there were only finitely many primes then $\zeta(1)$ would have a finite value. However, the harmonic series 1 + 1/2 + 1/3 + 1/4 + ... diverges (i.e., exceeds any given number), so there must be infinitely many primes. Another example of the richness of the zeta function and a glimpse of modern algebraic number theory is the following identity (Basel problem), due to Euler,

_(2) =

Π

p

1

1 □ p□2 =

_2

6

:

The Ulam spiral. Red pixels show prime numbers. Primes of the form $4n^2 - 2n + 41$ *are highlighted in blue.*
The reciprocal of ζ(2), $6/\pi^2$, is the probability that two numbers selected at random are relatively prime.[31][32]
The unproven *Riemann hypothesis*, dating from 1859, states that except for $s = -2, -4, \ldots$, all zeroes of the ζ-function have real part equal to 1/2. The connection to prime numbers is that it essentially says that the primes are as regularly distributed as possible. From a physical viewpoint, it roughly states that the irregularity in the distribution of primes only comes from random noise. From a mathematical viewpoint, it roughly states that the asymptotic distribution of primes (about x/log *x* of numbers less than *x* are primes, the prime number theorem) also holds for much shorter intervals of length about the square root of *x* (for intervals near *x*). This hypothesis is generally believed to be correct. In particular, the simplest assumption is that primes should have no significant irregularities without good reason.

99.7.2 Other conjectures

Further information: Category:Conjectures about prime numbers
In addition to the Riemann hypothesis, many more conjectures revolving about primes have been posed. Often having an elementary formulation, many of these conjectures have withstood a proof for decades: all four of Landau's problems from 1912 are still unsolved. One of them is Goldbach's conjecture, which asserts that every even integer *n* greater than 2 can be written as a sum of two primes. As of February 2011, this conjecture has been verified

Plot of the zeta function ζ(s). At s=1, the function has a pole, that is to say, it tends to infinity.
for all numbers up to $n = 2 \cdot 10^{17}$.[33] Weaker statements than this have been proven, for example Vinogradov's theorem says that every sufficiently large odd integer can be written as a sum of three primes. Chen's theorem says that every sufficiently large even number can be expressed as the sum of a prime and a semiprime, the product of two primes. Also, any even integer can be written as the sum of six primes.[34] The branch of number theory studying such questions is called additive number theory.
Other conjectures deal with the question whether an infinity of prime numbers subject to certain constraints exists. It is conjectured that there are infinitely many Fibonacci primes[35] and infinitely many Mersenne primes, but not Fermat primes.[36] It is not known whether or not there are an infinite number of Wieferich primes and of prime Euclid numbers.
A third type of conjectures concerns aspects of the distribution of primes. It is conjectured that there are infinitely many twin primes, pairs of primes with difference 2 (twin prime conjecture). Polignac's conjecture is a strengthening of that conjecture, it states that for every positive integer *n*, there are infinitely many pairs of consecutive primes that differ by $2n$.[37] It is conjectured there are infinitely many primes of the form $n^2 + 1$.[38] These conjectures are special cases of the broad Schinzel's hypothesis H. Brocard's conjecture says that there are always at least four primes between the squares of consecutive primes greater than 2. Legendre's conjecture states that there is a prime number between n^2 and $(n + 1)^2$ for every positive integer *n*. It is implied by the stronger Cramér's conjecture.

99.8 Applications

For a long time, number theory in general, and the study of prime numbers in particular, was seen as the canonical example of pure mathematics, with no applications outside of the self-interest of studying the topic with the exception of use of prime numbered gear teeth to distribute wear evenly. In particular, number theorists such as British mathematician G. H. Hardy prided themselves on doing work that had absolutely no military significance.[39] However, this vision was shattered in the 1970s, when it was publicly announced that prime numbers could be used as the basis for the creation of public key cryptography algorithms. Prime numbers are also used for hash tables and pseudorandom number generators.
Some rotor machines were designed with a different number of pins on each rotor, with the number of pins on any one rotor either prime, or coprime to the number of pins on any other rotor. This helped generate the full cycle of possible rotor positions before repeating any position.
The International Standard Book Numbers work with a check digit, which exploits the fact that 11 is a prime.

99.8.1 Arithmetic modulo a prime and finite fields

Main article: Modular arithmetic
Modular arithmetic modifies usual arithmetic by only using the numbers
f0; 1; 2; : : : ; n □ 1*g;*
where *n* is a fixed natural number called modulus. Calculating sums, differences and products is done as usual, but whenever a negative number or a number greater than $n - 1$ occurs, it gets replaced by the remainder after division by *n*. For instance, for $n = 7$, the sum 3 + 5 is 1 instead of 8, since 8 divided by 7 has remainder 1. This is referred to by saying "3 + 5 is congruent to 1 modulo 7" and is denoted
3 + 5 _ 1 (mod 7)*:*
Similarly, $6 + 1 \equiv 0 \pmod 7$, $2 - 5 \equiv 4 \pmod 7$, since $-3 + 7 = 4$, and $3 \cdot 4 \equiv 5 \pmod 7$ as 12 has remainder 5. Standard properties of addition and multiplication familiar from the integers remain valid in modular arithmetic. In the parlance of abstract algebra, the above set of integers, which is also denoted **Z**/*n***Z**, is therefore a commutative ring for any *n*. Division, however, is not in general possible in this setting. For example, for $n = 6$, the equation
3 _ *x* _ 2 (mod 6)*;*
a solution *x* of which would be an analogue of 2/3, cannot be solved, as one can see by calculating $3 \cdot 0, \ldots, 3 \cdot 5$ modulo 6. The distinctive feature of prime numbers is the following: division *is* possible in modular arithmetic if and only if *n* is a prime. Equivalently, *n* is prime if and only if all integers *m* satisfying $2 \le m \le n - 1$ are *coprime* to *n*, i.e. their only common divisor is one. Indeed, for $n = 7$, the equation
3 _ *x* _ 2 (mod 7)*;*
has a unique solution, $x = 3$. Because of this, for any prime *p*, **Z**/*p***Z** (also denoted $\mathbf{F}_p$) is called a field or, more

specifically, a finite field since it contains finitely many, namely p, elements.
A number of theorems can be derived from inspecting **F***p* in this abstract way. For example, Fermat's little theorem, stating
ap□1 _ 1(mod *p*)
for any integer *a* not divisble by *p*, may be proved using these notions. This implies

Σ*p*□1
a=1
ap□1 _ (*p* □ 1) _ 1 _ □1 (mod *p*):
Giuga's conjecture says that this equation is also a sufficient condition for *p* to be prime. Another consequence of Fermat's little theorem is the following: if *p* is a prime number other than 2 and 5, 1/*p* is always a recurring decimal, whose period is $p - 1$ or a divisor of $p - 1$. The fraction 1/*p* expressed likewise in base *q* (rather than base 10) has similar effect, provided that *p* is not a prime factor of *q*. Wilson's theorem says that an integer $p > 1$ is prime if and only if the factorial $(p - 1)! + 1$ is divisible by *p*. Moreover, an integer $n > 4$ is composite if and only if $(n - 1)!$ is divisible by *n*.

99.8.2 Other mathematical occurrences of primes

Many mathematical domains make great use of prime numbers. An example from the theory of finite groups are the Sylow theorems: if *G* is a finite group and *pn* is the highest power of the prime *p* that divides the order of *G*, then *G* has a subgroup of order *pn*. Also, any group of prime order is cyclic (Lagrange's theorem).

99.8.3 Public-key cryptography

Main article: Public key cryptography
Several public-key cryptography algorithms, such as RSA and the Diffie–Hellman key exchange, are based on large prime numbers (for example 512 bit primes are frequently used for RSA and 1024 bit primes are typical for Diffie–Hellman.). RSA relies on the assumption that it is much easier (i.e., more efficient) to perform the multiplication of two (large) numbers *x* and *y* than to calculate *x* and *y* (assumed coprime) if only the product *xy* is known. The Diffie–Hellman key exchange relies on the fact that there are efficient algorithms for modular exponentiation, while the reverse operation the discrete logarithm is thought to be a hard problem.

99.8.4 Prime numbers in nature

The evolutionary strategy used by cicadas of the genus *Magicicada* make use of prime numbers.[40] These insects spend most of their lives as grubs underground. They only pupate and then emerge from their burrows after 7, 13 or 17 years, at which point they fly about, breed, and then die after a few weeks at most. The logic for this is believed to be that the prime number intervals between emergences make it very difficult for predators to evolve that could specialize as predators on *Magicicadas*.[41] If *Magicicadas* appeared at a non-prime number intervals, say every 12 years, then predators appearing every 2, 3, 4, 6, or 12 years would be sure to meet them. Over a 200-year period, average predator populations during hypothetical outbreaks of 14- and 15-year cicadas would be up to 2% higher than during outbreaks of 13- and 17-year cicadas.[42] Though small, this advantage appears to have been enough to drive natural selection in favour of a prime-numbered life-cycle for these insects.
There is speculation that the zeros of the zeta function are connected to the energy levels of complex quantum systems.[43]

99.9 Generalizations

The concept of prime number is so important that it has been generalized in different ways in various branches of mathematics. Generally, "prime" indicates minimality or indecomposability, in an appropriate sense. For example, the prime field is the smallest subfield of a field *F* containing both 0 and 1. It is either **Q** or the finite field with *p* elements, whence the name.[44] Often a second, additional meaning is intended by using the word prime, namely that any object can be, essentially uniquely, decomposed into its prime components. For example, in knot theory, a prime knot is a knot that is indecomposable in the sense that it cannot be written as the knot sum of two nontrivial knots. Any knot can be uniquely expressed as a connected sum of prime knots.[45] Prime models and prime 3-manifolds are other examples of this type.

99.9.1 Prime elements in rings

Main articles: Prime element and Irreducible element
Prime numbers give rise to two more general concepts that apply to elements of any commutative ring *R*, an algebraic structure where addition, subtraction and multiplication are defined: *prime elements* and *irreducible elements*. An element *p* of *R* is called prime element if it is neither zero nor a unit (i.e., does not have a multiplicative inverse) and satisfies the following requirement: given *x* and *y* in *R* such that *p* divides the product *xy*, then *p* divides *x* or *y*. An element is irreducible if it is not a unit and cannot be written as a product of two ring elements that are not units. In the ring **Z** of integers, the set of prime elements equals the set of irreducible elements, which is
f: : : ;□11;□7;□5;□3;□2; 2; 3; 5; 7; 11; : : : *g* :
In any ring *R*, any prime element is irreducible. The converse does not hold in general, but does hold for unique factorization domains.
The fundamental theorem of arithmetic continues to hold in unique factorization domains. An example of such a domain is the Gaussian integers **Z**[*i*], that is, the set of complex numbers of the form $a + bi$ where *i* denotes the imaginary unit and *a* and *b* are arbitrary integers. Its prime elements are known as Gaussian primes. Not every prime (in **Z**) is a Gaussian prime: in the bigger ring **Z**[*i*], 2 factors into the product of the two Gaussian primes $(1 + i)$ and $(1 - i)$. Rational primes (i.e. prime elements in **Z**) of the form $4k + 3$ are Gaussian primes, whereas rational primes

of the form $4k + 1$ are not.

99.9.2 Prime ideals

Main article: Prime ideals

In ring theory, the notion of number is generally replaced with that of ideal. *Prime ideals*, which generalize prime elements in the sense that the principal ideal generated by a prime element is a prime ideal, are an important tool and object of study in commutative algebra, algebraic number theory and algebraic geometry. The prime ideals of the ring of integers are the ideals (0), (2), (3), (5), (7), (11), ... The fundamental theorem of arithmetic generalizes to the Lasker–Noether theorem, which expresses every ideal in a Noetherian commutative ring as an intersection of primary ideals, which are the appropriate generalizations of prime powers.[46]

Prime ideals are the points of algebro-geometric objects, via the notion of the spectrum of a ring.[47] Arithmetic geometry also benefits from this notion, and many concepts exist in both geometry and number theory. For example, factorization or ramification of prime ideals when lifted to an extension field, a basic problem of algebraic number theory, bears some resemblance with ramification in geometry. Such ramification questions occur even in numbertheoretic questions solely concerned with integers. For example, prime ideals in the ring of integers of quadratic number fields can be used in proving quadratic reciprocity, a statement that concerns the solvability of quadratic equations

$x2\_p \ (\bmod q);$

where x is an integer and p and q are (usual) prime numbers.[48] Early attempts to prove Fermat's Last Theorem climaxed when Kummer introduced regular primes, primes satisfying a certain requirement concerning the failure of unique factorization in the ring consisting of expressions

$a0 + a1\_ + \_\_\_ + ap□1\_p□1 ;$

where $a0$, ..., $ap-1$ are integers and ζ is a complex number such that $\zeta p = 1$.[49]

99.9.3 Valuations

Valuation theory studies certain functions from a field K to the real numbers **R** called valuations.[50] Every such valuation yields a topology on K, and two valuations are called equivalent if they yield the same topology. A *prime*

of K (sometimes called a *place of K*) is an equivalence class of valuations. For example, the p-adic valuation of a rational number q is defined to be the integer $vp(q)$, such that

$q = pvp(q) \ r$

s

;

where both r and s are not divisible by p. For example, $v3(18/7) = 2$. The p-adic norm is defined as[nb 1]

jqj

$p := p$

$□vp(q):$

In particular, this norm gets smaller when a number is multiplied by p, in sharp contrast to the usual absolute value (also referred to as the infinite prime). While completing **Q** (roughly, filling the gaps) with respect to the absolute value yields the field of real numbers, completing with respect to the p-adic norm $|-|p$ yields the field of p-adic numbers.[51] These are essentially all possible ways to complete **Q**, by Ostrowski's theorem. Certain arithmetic questions related to **Q** or more general global fields may be transferred back and forth to the completed (or local) fields. This local-global principle again underlines the importance of primes to number theory.

99.10 In the arts and literature

Prime numbers have influenced many artists and writers. The French composer Olivier Messiaen used prime numbers to create ametrical music through "natural phenomena". In works such as *La Nativité du Seigneur* (1935) and *Quatre études de rythme* (1949–50), he simultaneously employs motifs with lengths given by different prime numbers to create unpredictable rhythms: the primes 41, 43, 47 and 53 appear in the third étude, "Neumes rythmiques". According to Messiaen this way of composing was "inspired by the movements of nature, movements of free and unequal durations".[52]

In his science fiction novel *Contact*, NASA scientist Carl Sagan suggested that prime numbers could be used as a means of communicating with aliens, an idea that he had first developed informally with American astronomer Frank Drake in 1975.[53] In the novel *The Curious Incident of the Dog in the Night-Time* by Mark Haddon, the narrator arranges the sections of the story by consecutive prime numbers.[54]

Many films, such as *Cube*, *Sneakers*, *The Mirror Has Two Faces* and *A Beautiful Mind* reflect a popular fascination with the mysteries of prime numbers and cryptography.[55] Prime numbers are used as a metaphor for loneliness and isolation in the Paolo Giordano novel *The Solitude of Prime Numbers*, in which they are portrayed as "outsiders" among integers.[56]

99.11 See also

_ Adleman–Pomerance–Rumely primality test
_ Bonse's inequality
_ Brun sieve
_ Burnside theorem
_ Chebotarev's density theorem
_ Chinese remainder theorem
_ Cullen number
_ Illegal prime

_ List of prime numbers
_ Mersenne prime
_ Multiplicative number theory

_ Number field sieve
_ Pepin's test
_ Prime k-tuple
_ Primon gas
_ Quadratic residuosity problem
_ RSA number
_ Smooth number
_ Super-prime
_ Woodall number

99.12 Notes

[1] Some sources also put *jqj*
p := *e*
□*vp*(*q*): .

[1] Dudley, Underwood (1978), *Elementary number theory* (2nd ed.), W. H. Freeman and Co., ISBN 978-0-7167-0076-0, p. 10, section 2
[2] Dudley 1978, Section 2, Theorem 2
[3] Dudley 1978, Section 2, Lemma 5
[4] See, for example, David E. Joyce's commentary on Euclid's Elements, Book VII, definitions 1 and 2.
[5] Riesel 1994, p. 36
[6] Conway & Guy 1996, pp. 129–130
[7] Derbyshire, John (2003), "The Prime Number Theorem", *Prime Obsession: Bernhard Riemann and the Greatest Unsolved Problem in Mathematics*, Washington, D.C.: Joseph Henry Press, p. 33, ISBN 978-0-309-08549-6, OCLC 249210614
[8] "“Arguments for and against the primality of 1".
[9] "Why is the number one not prime?"
[10] The Largest Known Prime by Year: A Brief History Prime Curios!: 17014...05727 (39-digits)
[11] For instance, Beiler writes that number theorist Ernst Kummer loved his ideal numbers, closely related to the primes, "because they had not soiled themselves with any practical applications", and Katz writes that Edmund Landau, known for his work on the distribution of primes, "loathed practical applications of mathematics", and for this reason avoided subjects such as geometry that had already shown themselves to be useful. Beiler, Albert H. (1966), *Recreations in the Theory of Numbers: The Queen of Mathematics Entertains*, Dover, p. 2, ISBN 9780486210964. Katz, Shaul (2004), *Berlin roots—Zionist incarnation: the ethos of pure mathematics and the beginnings of the Einstein Institute of Mathematics at the Hebrew University of Jerusalem, Science in Context* **17** (1-2): 199–234, doi:10.1017/S0269889704000092, MR 2089305.
[12] Letter in Latin from Goldbach to Euler, July 1730.
[13] Furstenberg 1955
[14] Ribenboim 2004, p. 4
[15] James Williamson (translator and commentator), *The Elements of Euclid, With Dissertations*, Clarendon Press, Oxford, 1782, page 63, English translation of Euclid's proof
[16] Hardy, Michael; Woodgold, Catherine (2009). "Prime Simplicity". *Mathematical Intelligencer* **31** (4): 44–52. doi:10.1007/s00283-009-9064-8.
[17] Apostol, Tom M. (1976), *Introduction to Analytic Number Theory*, Berlin, New York: Springer-Verlag, ISBN 978-0-387-90163-3, Section 1.6, Theorem 1.13
[18] Apostol 1976, Section 4.8, Theorem 4.12

[19] (Lehmer 1909).
[20] "Record 12-Million-Digit Prime Number Nets $100,000 Prize". Electronic Frontier Foundation. October 14, 2009. Retrieved 2010-01-04.
[21] "EFF Cooperative Computing Awards". Electronic Frontier Foundation. Retrieved 2010-01-04.
[22] Chris K. Caldwell. "The Top Twenty: Factorial". Primes.utm.edu. Retrieved 2013-02-05.
[23] Chris K. Caldwell. "The Top Twenty: Primorial". Primes.utm.edu. Retrieved 2013-02-05.
[24] Chris K. Caldwell. "The Top Twenty: Twin Primes". Primes.utm.edu. Retrieved 2013-02-05.
[25] Havil 2003, p. 171
[26] http://books.google.com/books?id=oLKlk5o6WroC&pg=PA13#v=onepage&q&f=false p. 15
[27] (Tom M. Apostol 1976), Section 4.6, Theorem 4.7
[28] (Ben Green & Terence Tao 2008).
[29] Hua (2009), pp. 176–177"
[30] See list of values, calculated by Wolfram Alpha
[31] Caldwell, Chris. "What is the probability that gcd(n,m)=1?". *The Prime Pages*. Retrieved 2013-09-06.
[32] C. S. Ogilvy & J. T. Anderson *Excursions in Number Theory*, pp. 29–35, Dover Publications Inc., 1988 ISBN 0-486-25778-9
[33] Tomás Oliveira e Silva (2011-04-09). "Goldbach conjecture verification". Ieeta.pt. Retrieved 2011-05-21.
[34] Ramaré, O. (1995), *On šnirel'man's constant, Annali della Scuola Normale Superiore di Pisa. Classe di Scienze. Serie IV* **22** (4): 645–706, retrieved 2008-08-22.

[35] Caldwell, Chris, *The Top Twenty: Lucas Number* at The Prime Pages.
[36] E.g., see Guy 1981, problem A3, pp. 7–8
[37] Tattersall, J.J. (2005), *Elementary number theory in nine chapters*, Cambridge University Press, ISBN 978-0-521-85014-8, p. 112
[38] Weisstein, Eric W., "Landau's Problems", *MathWorld.*
[39] Hardy 1940 "No one has yet discovered any warlike purpose to be served by the theory of numbers or relativity, and it seems unlikely that anyone will do so for many years."
[40] Goles, E.; Schulz, O.; Markus, M. (2001). "Prime number selection of cycles in a predator-prey model". *Complexity* **6** (4): 33–38. doi:10.1002/cplx.1040.
[41] Paulo R. A. Campos, Viviane M. de Oliveira, Ronaldo Giro, and Douglas S. Galvão. (2004), *Emergence of Prime Numbers as the Result of Evolutionary Strategy, Physical Review Letters* **93** (9): 098107, arXiv:q-bio/0406017, Bibcode:2004PhRvL..93i8107C, doi:10.1103/PhysRevLett.93.098107.
[42] "Invasion of the Brood". *The Economist.* May 6, 2004. Retrieved 2006-11-26.
[43] Ivars Peterson (June 28, 1999). "The Return of Zeta". *MAA Online*. Retrieved 2008-03-14.
[44] Lang, Serge (2002), *Algebra*, Graduate Texts in Mathematics **211**, Berlin, New York: Springer-Verlag, ISBN 978-0-387-95385-4, MR 1878556, Section II.1, p. 90
[45] Schubert, H. "Die eindeutige Zerlegbarkeit eines Knotens in Primknoten". *S.-B Heidelberger Akad. Wiss. Math.-Nat. Kl.* 1949 (1949), 57–104.
[46] Eisenbud 1995, section 3.3.
[47] Shafarevich, Basic Algebraic Geometry volume 2 (Schemes and Complex Manifolds), p. 5, section V.1
[48] Neukirch, Algebraic Number theory, p. 50, Section I.8
[49] Neukirch, Algebraic Number theory, p. 38, Section I.7
[50] Endler, Valuation Theory, p. 1
754 *CHAPTER 99. PRIME NUMBER*
[51] Gouvea: p-adic numbers: an introduction, Chapter 3, p. 43
[52] Hill, ed. 1995
[53] Carl Pomerance, Prime Numbers and the Search for Extraterrestrial Intelligence, Retrieved on December 22, 2007
[54] Mark Sarvas, Book Review: *The Curious Incident of the Dog in the Night-Time*, at The Modern Word, Retrieved on March 30, 2012
[55] The music of primes, Marcus du Sautoy's selection of films featuring prime numbers.
[56] "Introducing Paolo Giordano". Books Quarterly.

99.13 References

_ Apostol, Thomas M. (1976), *Introduction to Analytic Number Theory*, New York: Springer, ISBN 0-387-90163-9
_ Conway, John Horton; Guy, Richard K. (1996), *The Book of Numbers*, New York: Copernicus, ISBN 978-0-387-97993-9
_ Crandall, Richard; Pomerance, Carl (2005), *Prime Numbers: A Computational Perspective* (2nd ed.), Berlin, New York: Springer-Verlag, ISBN 978-0-387-25282-7
_ Derbyshire, John (2003), *Prime obsession*, Joseph Henry Press, Washington, DC, ISBN 978-0-309-08549-6, MR 1968857
_ Eisenbud, David (1995), *Commutative algebra*, Graduate Texts in Mathematics **150**, Berlin, New York: Springer-Verlag, ISBN 978-0-387-94268-1, MR 1322960
_ Fraleigh, John B. (1976), *A First Course In Abstract Algebra* (2nd ed.), Reading: Addison-Wesley, ISBN 0-201-01984-1
_ Furstenberg, Harry (1955), *On the infinitude of primes, The American Mathematical Monthly* (Mathematical Association of America) **62** (5): 353, doi:10.2307/2307043, JSTOR 2307043
_ Green, Ben; Tao, Terence (2008), *The primes contain arbitrarily long arithmetic progressions, Annals of Mathematics* **167** (2): 481–547, arXiv:math.NT/0404188, doi:10.4007/annals.2008.167.481
_ Gowers, Timothy (2002), *Mathematics: A Very Short Introduction*, Oxford University Press, ISBN 978-0-19-285361-5
_ Guy, Richard K. (1981), *Unsolved Problems in Number Theory*, Berlin, New York: Springer-Verlag, ISBN 978-0-387-90593-8
_ Havil, Julian (2003), *Gamma: Exploring Euler's Constant*, Princeton University Press, ISBN 978-0-691-09983-5
_ Hardy, Godfrey Harold (1908), *A Course of Pure Mathematics*, Cambridge University Press, ISBN 978-0-521-09227-2
_ Hardy, Godfrey Harold (1940), *A Mathematician's Apology*, Cambridge University Press, ISBN 978-0-521-42706-7
_ Herstein, I. N. (1964), *Topics In Algebra*, Waltham: Blaisdell Publishing Company, ISBN 978-1114541016
_ Hill, Peter Jensen, ed. (1995), *The Messiaen companion*, Portland, Or: Amadeus Press, ISBN 978-0-931340-95-6
_ Hua, L. K. (2009), *Additive Theory of Prime Numbers*, Translations of Mathematical Monographs **13**, AMS Bookstore, ISBN 978-0-8218-4942-2
_ Lehmer, D. H. (1909), *Factor table for the first ten millions containing the smallest factor of every number not divisible by 2, 3, 5, or 7 between the limits 0 and 10017000*, Washington, D.C.: Carnegie Institution of Washington

_ McCoy, Neal H. (1968), *Introduction To Modern Algebra, Revised Edition*, Boston: Allyn and Bacon, LCCN 68-15225
_ Narkiewicz, Wladyslaw (2000), *The development of prime number theory: from Euclid to Hardy and Littlewood*, Springer Monographs in Mathematics, Berlin, New York: Springer-Verlag, ISBN 978-3-540-66289-1
_ Ribenboim, Paulo (2004), *The little book of bigger primes*, Berlin, New York: Springer-Verlag, ISBN 978-0-387-20169-6
_ Riesel, Hans (1994), *Prime numbers and computer methods for factorization*, Basel, Switzerland: Birkhäuser, ISBN 978-0-8176-3743-9
_ Sabbagh, Karl (2003), *The Riemann hypothesis*, Farrar, Straus and Giroux, New York, ISBN 978-0-374-25007-2, MR 1979664
_ du Sautoy, Marcus (2003), *The music of the primes*, HarperCollins Publishers, ISBN 978-0-06-621070-4, MR 2060134

99.13.1 Further references

_ Kelly, Katherine E., ed. (2001), *The Cambridge companion to Tom Stoppard*, Cambridge University Press, ISBN 978-0-521-64592-8
_ Stoppard, Tom (1993), *Arcadia*, London: Faber and Faber, ISBN 978-0-571-16934-4

99.14 External links

_ Hazewinkel, Michiel, ed. (2001), "Prime number", *Encyclopedia of Mathematics*, Springer, ISBN 978-1-55608-010-4
_ Caldwell, Chris, The Prime Pages at primes.utm.edu.
_
_ Prime Numbers on *In Our Time* at the BBC.
_ An Introduction to Analytic Number Theory, by Ilan Vardi and Cyril Banderier
_ Plus teacher and student package: prime numbers from Plus, the free online mathematics magazine produced by the Millennium Mathematics Project at the University of Cambridge.

99.14.1 Prime number generators and calculators

_ Prime Number Checker identifies the smallest prime factor of a number.
_ Fast Online primality test with factorization makes use of the Elliptic Curve Method (up to thousand-digits numbers, requires Java).
_ Prime Number Generator generates a given number of primes above a given start number.
_ Huge database of prime numbers
_ Prime Numbers up to 1 trillion

Chapter 100

Euclid's theorem

Euclid's theorem is a fundamental statement in number theory that asserts that there are infinitely many prime numbers. There are several well-known proofs of the theorem.

100.1 Euclid's proof

Euclid offered the following proof published in his work *Elements* (Book IX, Proposition 20),[1] which is paraphrased here.

Consider any finite list of prime numbers $p_1, p_2, ..., p_n$. It will be shown that at least one additional prime number not in this list exists. Let P be the product of all the prime numbers in the list: $P = p_1p_2...p_n$. Let $q = P + 1$. Then q is either prime or not:

_ If q is prime, then there is at least one more prime than is in the list.
_ If q is not prime, then some prime factor p divides q. If this factor p were on our list, then it would divide P (since P is the product of every number on the list); but p divides $P + 1 = q$. If p divides P and q, then p would have to divide the difference[2] of the two numbers, which is $(P + 1) - P$ or just 1. Since no prime number divides 1, this would be a contradiction and so p cannot be on the list. This means that at least one more prime number exists beyond those in the list.

This proves that for every finite list of prime numbers there is a prime number not on the list, and therefore there must be infinitely many prime numbers.

Euclid is often erroneously reported to have proved this result by contradiction, beginning with the assumption that the set initially considered contains all prime numbers, or that it contains precisely the n smallest primes, rather than any arbitrary finite set of primes.[3] Although the proof as a whole is not by contradiction (it does not assume that only finitely many primes exist), a proof by contradiction is within it, which is that none of the initially considered primes can divide the number q above.

100.2 Euler's proof

Another proof, by the Swiss mathematician Leonhard Euler, relies on the fundamental theorem of arithmetic: that every integer has a unique prime factorization. If P is the set of all prime numbers, Euler wrote that:

Π
$p2P$
1

1 □ 1/*p*
=
Π
p2P
Σ
*k*_0
1
pk =
Σ
n
1
n
:
The first equality is given by the formula for a geometric series in each term of the product. To show the second equality, distribute the product over the sum:

Π
p2P
Σ
*k*_0
1
pk =
Σ
*k*_0
1
2*k*
_
Σ
*k*_0
1
3*k*
_
Σ
*k*_0
1
5*k*
_
Σ
*k*_0
1
7*k*
_ _ _ _ =
Σ
*k;l;m;n;*____0
1
2*k*3*l*5*m*7*n* _ _ _ =
Σ
n
1
n
in the result, every product of primes appears exactly once and so by the fundamental theorem of arithmetic the sum is equal to the sum over all integers.
The sum on the right is the harmonic series, which diverges. Thus the product on the left must also diverge. Since each term of the product is finite, the number of terms must be infinite; therefore, there is an infinite number of primes.

100.3 Erdős's proof

Paul Erdős gave a third proof that relies on the fundamental theorem of arithmetic. First note that every integer n can be uniquely written as
rs2
where r is *square-free*, or not divisible by any square numbers (let *s2* be the largest square number that divides n and then let $r = n/s2$). Now suppose that there are only finitely many prime numbers and call the number of prime numbers k.
Fix a positive integer N and try to count the number of integers between 1 and N. Each of these numbers can be written as *rs2* where r is square-free and r and *s2* are both less than N. By the fundamental theorem of arithmetic, there are only 2k square-free numbers r (see Combination#Number of k-combinations for all k) as each of the prime numbers factorizes r at most once, and we must have $s < \sqrt{N}$. So the total number of integers less than N is at most 2*k*√*N*; i.e.:
2*k*

p
$N \_ N$

Since this inequality does not hold for N sufficiently large, there must be infinitely many primes.

100.4 Furstenberg's proof

In the 1950s, Hillel Furstenberg introduced a proof using point-set topology. See Furstenberg's proof of the infinitude of primes.

100.5 Some recent proofs

100.5.1 Pinasco

Juan Pablo Pinasco has written the following proof.[4]

Let $p_1, \ldots, p_N$ be the smallest N primes. Then by the inclusion–exclusion principle, the number of positive integers less than or equal to x that are divisible by one of those primes is

$$1 + \sum_i \left\lfloor \frac{x}{p_i} \right\rfloor - \sum_{i<j} \left\lfloor \frac{x}{p_i p_j} \right\rfloor + \sum_{i<j<k} \left\lfloor \frac{x}{p_i p_j p_k} \right\rfloor - \cdots \pm (-1)^{N+1} \left\lfloor \frac{x}{p_1 \cdots p_N} \right\rfloor. \quad (1)$$

Dividing by x and letting $x \to \infty$ gives

$$\sum_i \frac{1}{p_i} - \sum_{i<j} \frac{1}{p_i p_j} + \sum_{i<j<k} \frac{1}{p_i p_j p_k} - \cdots \pm (-1)^{N+1} \frac{1}{p_1 \cdots p_N}. \quad (2)$$

This can be written as

$$1 - \prod_{i=1}^{N} \left(1 - \frac{1}{p_i} \right). \quad (3)$$

If no other primes than $p_1, \ldots, p_N$ exist, then the expression in (1) is equal to $\lfloor x \rfloor$ and the expression in (2) is equal to 1, but clearly the epression in (3) exceeds 1. Therefore there must be more primes than $p_1, \ldots, p_N$.

100.5.2 Whang

In 2010, Junho Peter Whang published the following proof by contradiction.[5] Let k be any positive integer. Then according to de Polignac's formula (actually due to Legendre)

$$k! = \prod_{p \text{ prime}} p^{f(p,k)}$$

where

$$f(p,k) = \left\lfloor \frac{k}{p} \right\rfloor + \left\lfloor \frac{k}{p^2} \right\rfloor + \cdots.$$

$$f(p,k) < \frac{k}{p} + \frac{k}{p^2} + \cdots = \frac{k}{p-1} \le k.$$

But if only finitely many primes exist, then

$$\lim_{k \to \infty} \frac{\left(\prod_p p\right)^k}{k!} = 0,$$

(the numerator of the fraction would grow singly exponentially while by Stirling's approximation the denominator grows more quickly than singly exponentially), contradicting the fact that for each k the numerator is greater than or equal to the denominator.

100.6 Proof using the irrationality of π

Representing the Leibniz formula for π as an Euler product gives[6]

$$\frac{\pi}{4} = \frac{3}{4} \cdot \frac{5}{4} \cdot \frac{7}{8} \cdot \frac{11}{12} \cdot \frac{13}{12} \cdot \frac{17}{16} \cdot \frac{19}{20} \cdot \frac{23}{24} \cdot \frac{29}{28} \cdot \frac{31}{32} \cdots$$

The numerators of this product are the odd prime numbers, and each denominator is the multiple of four nearest to the numerator.
If there were finitely many primes this formula would show that π is rational, contradicting the fact that π is actually irrational.

100.7 See also

_ Dirichlet's theorem on arithmetic progressions
_ Prime number theorem

100.8 Notes and references

[1] James Williamson (translator and commentator), *The Elements of Euclid, With Dissertations*, Clarendon Press, Oxford, 1782, page 63.
[2] In general, for any integers a, b, c if *a j b* and *a j c* , then *a j* (*b* □ *c*) . For more information, see Divisibility.
[3] Michael Hardy and Catherine Woodgold, "Prime Simplicity", *Mathematical Intelligencer*, volume 31, number 4, fall 2009, pages 44–52.
[4] Juan Pablo Pinasco, "New Proofs of Euclid's and Euler's theorems", *American Mathematical Monthly*, volume 116, number 2, February, 2009, pages 172–173.
[5] Junho Peter Whang, "Another Proof of the Infinitude of the Prime Numbers", *American Mathematical Monthly*, volume 117, number 2, February 2010, page 181.
[6] Debnath, Lokenath (2010), *The Legacy of Leonhard Euler: A Tricentennial Tribute*, World Scientific, p. 214, ISBN 9781848165267.

100.9 External links

_ Weisstein, Eric W., "Euclid's Theorem", *MathWorld.*
_ Euclid's Elements, Book IX, Prop. 20 (Euclid's proof, on David Joyce's website at Clark University)

Chapter 101

Furstenberg's proof of the infinitude of primes

In number theory, **Hillel Furstenberg's proof of the infinitude of primes** is a celebrated topological proof that the integers contain infinitely many prime numbers. When examined closely, the proof is less a statement about topology than a statement about certain properties of arithmetic sequences.[1] Unlike Euclid's classical proof, Furstenberg's proof is a proof by contradiction. The proof was published in 1955 in the *American Mathematical Monthly* while Furstenberg was still an undergraduate student at Yeshiva University.

101.1 Furstenberg's proof

Define a topology on the integers **Z**, called the evenly spaced integer topology, by declaring a subset $U \subseteq \mathbf{Z}$ to be an open set if and only if it is either the empty set, ∅, or it is a union of arithmetic sequences $S(a, b)$ (for $a \neq 0$), where
$S(a; b) = fan + b j n 2 Zg = aZ + b:$
In other words, U is open if and only if every $x \in U$ admits some non-zero integer a such that $S(a, x) \subseteq U$. The axioms for a topology are easily verified:
_ By definition, ∅ is open; **Z** is just the sequence $S(1, 0)$, and so is open as well.
_ Any union of open sets is open: for any collection of open sets U_i and x in their union U, any of the numbers a_i for which $S(a_i, x) \subseteq U_i$ also shows that $S(a_i, x) \subseteq U$.
_ The intersection of two (and hence finitely many) open sets is open: let U_1 and U_2 be open sets and let $x \in U_1 \cap U_2$ (with numbers a_1 and a_2 establishing membership). Set a to be the lowest common multiple of a_1 and a_2. Then $S(a, x) \subseteq S(a_i, x) \subseteq U_i$.
This topology has two notable properties:
1. Since any non-empty open set contains an infinite sequence, a finite set cannot be open; put another way, the complement of a finite set cannot be a closed set.
2. The basis sets $S(a, b)$ are both open and closed: they are open by definition, and we can write $S(a, b)$ as the complement of an open set as follows:
$S(a; b) = Z n$
a∪□1
j=1
$S(a; b + j):$
760

The only integers that are not integer multiples of prime numbers are −1 and +1, i.e.
$Z n f$□1; +1g =
∪
p prime
$S(p; 0):$
By the first property, the set on the left-hand side cannot be closed. On the other hand, by the second property, the

sets $S(p, 0)$ are closed. So, if there were only finitely many prime numbers, then the set on the right-hand side would be a finite union of closed sets, and hence closed. This would be a contradiction, so there must be infinitely many prime numbers.

101.2 Notes

[1] Mercer, Idris D. (2009). "On Furstenberg's Proof of the Infinitude of Primes". *American Mathematical Monthly* **116**: 355–356. doi:10.4169/193009709X470218.

101.3 References

_ Aigner, Martin; Ziegler, Günter M. (1998). "Proofs from The Book". Berlin, New York: Springer-Verlag.
_ Furstenberg, Harry (1955). "On the infinitude of primes". *American Mathematical Monthly* (Mathematical Association of America) **62** (5): 353. doi:10.2307/2307043. JSTOR 2307043. MR 0068566

101.4 External links

_ Furstenberg's proof that there are infinitely many prime numbers at Everything2
_ Fürstenberg's proof of the infinitude of primes at PlanetMath.org.

Chapter 102

Monsky's theorem

In geometry, **Monsky's theorem** states that it is not possible to dissect a square into an odd number of triangles of equal area.[1] In other words, a square does not have an odd equidissection.

The problem was posed by Fred Richman in the American Mathematical Monthly in 1965, and was proved by Paul Monsky in 1970.[2][3][4]

102.1 Proof

Monsky's proof combines combinatorial and algebraic techniques, and in outline is as follows:

1. Take the square to be the unit square with vertices at (0,0), (0,1), (1,0) and (1,1). If there is a dissection into n triangles of equal area then the area of each triangle is $1/n$.
2. Colour each point in the square with one of three colours, depending on the 2-adic valuation of its coordinates.
3. Show that a straight line can contain points of only two colours.
4. Use Sperner's lemma to show that every triangulation of the square into triangles meeting edge-to-edge must contain at least one triangle whose vertices have three different colours.
5. Conclude from the straight-line property that a tricolored triangle must also exist in every dissection of the square into triangles, not necessarily meeting edge-to-edge.
6. Use Cartesian geometry to show that the 2-adic valuation of the area of a triangle whose vertices have three different colours is greater than 1. So every dissection of the square into triangles must contain at least one triangle whose area has a 2-adic valuation greater than 1.
7. If n is odd then the 2-adic valuation of $1/n$ is 1, so it is impossible to dissect the square into triangles all of which have area $1/n$.[5]

102.2 Generalizations

Main article: Equidissection

The theorem can be generalized to higher dimensions: an n-dimensional hypercube can only be divided into simplices of equal volume, if the number of simplices is a multiple of $n!$.[2]

102.3 References

[1] Aigner, Martin; Ziegler, Günter M. (2010). "One square and an odd number of triangles". *Proofs from The Book* (4th ed.). Berlin: Springer-Verlag. pp. 131–138. doi:10.1007/978-3-642-00856-6_20.

[2] Sperner's Lemma, Moor Xu

[3] Monsky, P. (1970). "On Dividing a Square into Triangles". *The American Mathematical Monthly* **77** (2): 161–164. doi:10.2307/2317329. MR 0252233.

[4] Kleber, M.; Vakil, R.; Stein, S. (2004). "Cutting a Polygon into Triangles of Equal Areas". *The Mathematical Intelligencer* **26**: 17. doi:10.1007/BF02985395.

[5] Dissecting a square into triangles

102.4 Text and image sources, contributors, and licenses

102.4.1 Text

_ **Mathematical proof** *Source:* http://en.wikipedia.org/wiki/Mathematical%20proof?oldid=634309344 *Contributors:* AxelBoldt, Robert Merkel, The Anome, Toby Bartels, Youandme, Edward, TeunSpaans, Michael Hardy, Spazlink, TakuyaMurata, LittleDan, Pratyeka, Glenn, Tim Retout, Hashar, Revolver, Charles Matthews, Dysprosia, Jitse Niesen, Bjh21, Markhurd, Booya, Bcorr, Denelson83, Fredrik, Romanm, Lowellian, Gandalf61, Paul G, Hadal, Giftlite, Markus Krötzsch, Yekrats, CSTAR, Karl-Henner, Tyler McHenry, Hkpawn, Petershank, Peter Kwok, Andreas Kaufmann, PhotoBox, Jason Carreiro, Leibniz, Paul August, Robertbowerman, Syp, El C, Skeppy, Jojit fb, Nk, ACW, Jumbuck, Kanie, Arthena, Itsmine, Oleg Alexandrov, Jslu, Mindmatrix, Rodrigo Rocha, Bkkbrad, Pol098, Jeff3000, Tygar, Stevey7788, Seyon, Rjwilmsi, Salix alba, R.e.b., Alexb@cut-the-knot.com, KarlFrei, Jrtayloriv, BMF81, Chobot, Jersey Devil, DVdm, Borgx, Trovatore, Cheeser1, Boivie, FF2010, PTSE, Josh3580, HereToHelp, Willtron, Curpsbot-unicodify, Brentt, □□□□ robot,

SmackBot, KnowledgeOfSelf, Bomac, Jagged 85, Diegotorquemada, Gracenotes, Can't sleep, clown will eat me, Rantingsteve, Rrburke, Drackap, MichaelBillington, Occultations, Luke Gustafson, Byelf2007, Lambiam, Eliyak, Scottie 000, Heimstern, Loadmaster, Mets501, ZodoJats, Stephen B Streater, Newone, Mrdthree, Braddodson, CRGreathouse, CBM, Larrywcusick, Dgw, MarsRover, Myasuda, Gregbard, Cydebot, NotQuiteEXPComplete, Thijs!bot, Epbr123, Nnn9245, Escarbot, AntiVandalBot, Joasiak, Nikolas Karalis, JAnDbot, Leuko, Hut 8.5, Four Dog Night, VoABot II, Soulbot, Singularity, Craw-daddy, Johnbibby, Seberle, JJ Harrison, David Eppstein, Bernard Hurley, Rohan Ghatak, JaimeLesMaths, Naveed ahmad14382, Wiki Raja, J.delanoy, Leon math, Numbo3, Maurice Carbonaro, Sefog, Vvitor, Smeira, He is a shithead, Lupussy, M.M.S., Tparameter, Policron, Jonjesbuzz, Dessources, Vanished user 39948282, Idiomabot, VolkovBot, JohnBlackburne, Jimmaths, Greatwalk, TXiKiBoT, Gentlemath, Anonymous Dissident, Ocolon, AlleborgoBot, Symane, EmxBot, SieBot, YonaBot, Phe-bot, RiskAverse, RJaguar3, Triwbe, X-Fi6, LeadSongDog, 360 Degree, Barliner, Radon210, Byrialbot, Thehotelambush, DesolateReality, Nusumareta, Randomblue, Myrvin, Tautologist, Athenean, ClueBot, Justin W Smith, Dobermanji, Lozersk, Adrianwn, ChandlerMapBot, I am a violinist, Wikiimee, DragonBot, Alexbot, La Pianista, Humanengr, Pichpich, Stickee, Gerhardvalentin, Kwjbot, WikiDao, EEng, Addbot, Betterusername, Glane23, AndersBot, Roux, ChenzwBot, Jaydec, Numbo3-bot, BOOLE1847, Legobot, Luckas-bot, Yobot, Kan8eDie, JorgeFierro, Texhausballa, AnomieBOT, JRB-Europe, Are you ready for IPv6?, La comadreja, ArthurBot, LilHelpa, MauritsBot, Xqbot, Sourceholder, Almabot, J04n, Geometryfan, FrescoBot, LucienBOT, Pinethicket, FriedrickMILBarbarossa, Foodimentary, Di1000, Jauhienij, Leasnam, FoxBot, TobeBot, DixonDBot, Robert hoffman, EmausBot, AmigoDoPaulo, Josve05a, Sven Manguard, Jijo925, ClueBot NG,Wikigold96, CocuBot, Melville88,Widr, ساجد امجد ساجد, Keetanii, Helpful Pixie Bot, Joolsa123, Wasbeer, Khonkhortisan, Kiewbra, Hebert Peró, Was123ification, Dexbot, Mogism, Brirush, Epicgenius, Gdaniel111, Jamesmcmahon0, Purnendu Karmakar, Schopenhauerswille, Blackbombchu, Seansmoove27 and Anonymous: 182

_ **Proposition** *Source:* http://en.wikipedia.org/wiki/Proposition?oldid=634053896 *Contributors:* AxelBoldt, Mav, Toby Bartels, Zoe, Stevertigo, K.lee, Michael Hardy, Zeno Gantner, TakuyaMurata, Minesweeper, Evercat, Sethmahoney, Conti, Reddi, Greenrd, Markhurd, Hyacinth, Banno, RedWolf, Ojigiri, Timrollpickering, Tobias Bergemann, Giftlite, Jason Quinn, Stevietheman, Antandrus, Superborsuk, Sebbe, Amicuspublilius, Martpol, Hapsiainen, Vanished user lp09qa86ft, Chalst, Phiwum, Duesentrieb, Bobo192, Larry V, MPerel, Helix84, V2Blast, Ish ishwar, Emvee, RJFJR, Bobrayner, Philthecow, Velho, Woohookitty, Kzollman, Isnow, Patl, Brolin Empey, Lakitu, Fresheneesz, Bornhj, YurikBot, Hairy Dude, Rick Norwood, Wknight94, Finell, SmackBot, Evanreyes, Ignacioerrico, Bluebot, Jaymay, DHN-bot, Cybercobra, Richard001, Lacatosias, Jon Awbrey, Vina-iwbot, Byelf2007, Harryboyles, SilkTork, Ckatz, 16@r, Grumpyyoungman01, Stwalkerster, Caiaffa, Levineps, Iridescent, JoeBot, Gveret Tered, Eastlaw, CRGreathouse, CBM, Sdorrance, Andkore, Gregbard, Juansempere, Yesterdog, Thijs!bot, Barticus88, Kredal, AllenFerguson, Voyaging, NSH001, JAnDbot, MER-C, Leolaursen, Bookinvestor, Connormah, VoABot II, WhatamIdoing, Pomte, Tgeairn, J.delanoy, Ali, Ginsengbomb, Katalaveno, Coppertwig, Nieske, Funandtrvl, King Lopez, ABF, TXiKiBoT, Philogo, Tracerbullet11, Cnilep, Barkeep, SieBot, Legion fi, Oxymoron83, OKBot, ClueBot, The Thing That Should Not Be, Mate2code, Estirabot, Hans Adler, Hugo Herbelin, DumZiBoT, Makotoy, Crazy Boris with a red beard, Dthomsen8, Dwnelson, SilvonenBot, Good Olfactory, Addbot, Andrewghutchison, LAAFan, Luckas-bot, TheSuave, Denyss, THEN WHO WAS PHONE?, Ehuss, KamikazeBot, AnomieBOT, E235, Yalckram, Wortafad, ArthurBot, Luis Felipe Schenone, Omnipaedista, BrideOfKripkenstein, Motomuku, Pinethicket, A8UDI, Monkeymanman, Gamewizard71, FoxBot, Lotje, Daliot, EmausBot, John of Reading, Eekerz, Honestrosewater, Bollyjeff, Coasterlover1994, Chewings72, ClueBot NG, MelbourneStar, Satellizer, Masssly, Helpful Pixie Bot, Hans-Jürgen Streicher, زكريا, ChrisGualtieri, Jochen Burghardt, Eyesnore, Purnendu Karmakar, DetectiveKraken, SanketDash, Ashika Bieber, Eavestn and Anonymous: 157

_ **Argument-deduction-proof distinctions** *Source:* http://en.wikipedia.org/wiki/Argument-deduction-proof%20distinctions?oldid=607599544 *Contributors:* Discospinster, Rjwilmsi, Gregbard, BOOLE1847, WikiDan61, FrescoBot, NABRASA, POLY1956 and JHU1959

_ **Theorem** *Source:* http://en.wikipedia.org/wiki/Theorem?oldid=633670840 *Contributors:* AxelBoldt, Mav, Zundark, The Anome, Tarquin, Tbackstr, XJaM, Aldie, Michael Hardy, Zeno Gantner, TakuyaMurata, Bagpuss, Glenn, Tim Retout, Rotem Dan, Andres, Charles Matthews, Dcoetzee, Bemoeial, Hyacinth, Traroth, SirPeebles, Moriel, Josh Cherry, Fredrik, MathMartin, Ojigiri, Timrollpickering, Hadal, Alan Liefting, Snobot, Ancheta Wis, Tosha, Giftlite, Monedula, Fropuff, Fishal, Chowbok, Alaz, MarkSweep, Karol Langner, Jacob grace, Pmanderson, Tyler McHenry, Hkpawn, Tzanko Matev, Joyous!, EugeneZelenko, Discospinster, Rich Farmbrough, Paul August, Bender235, Gauge, Tompw, El C, Edwinstearns, Billymac00, Sasquatch, Alansohn, Gary, Sciurinæ, Joriki, Igny, Ruud Koot, Erasmus, Mekong Bluesman, Tslocum, Graham87, BD2412, Gmelli, Sdornan, Salix alba, Jrtayloriv, AndriuZ, M7bot, Chobot, MithrandirMage, Sbrools, DVdm, Algebraist, Roboto de Ajvol, Borgx, Chaos, Trovatore, Dbfirs, Bota47, Tomisti, Arthur Rubin, Kier07, Anclation, Curpsbot-unicodify, Erudy, Finell, Sardanaphalus, SmackBot, RDBury, Bomac, Skizzik, Fuzzform, MalafayaBot, DHN-bot, Sholto Maud, Cybercobra, Acdx, SashatoBot, Lambiam, IronGargoyle, Craigblock, Lalaith, Autonova, Mike Fikes, Zero sharp, JRSpriggs, CRGreathouse, Hi.ro, CBM, Gregbard, Cydebot, R Harris, Moxmalin, Thijs!bot, Epbr123, Mrcs, James086, Nick Number, AntiVandal-Bot, Thenub314, .anacondabot, Magioladitis, Dvptl, Animum, Gwern, Stephenchou0722, Pomte, Hippasus the Younger, Fcsuper, Jeepday, Nznancy, Coppertwig, Haseldon, Tparameter, Fjbfour, Dessources, DavidCBryant, Caiodnh, Austinmohr, VolkovBot, Psmythirl, Am Fiosaigear, TXiKiBoT, Rei-bot, IKiddo, Voorlandt, Philogo, Geometry guy, Graymornings, Dmcq, HiDrNick, DestroyerofDreams, Hthoreau2, Newbyguesses, SieBot, Respir, Huzzah018, Oxymoron83, SimonTrew, OKBot, Kumioko (renamed), DesolateReality, Wjemather, Loren.wilton, ClueBot, Blanchardb, Excirial, Alexbot, TheSnacks, Hans Adler, Muzz 2008, MonoBot, XLinkBot, Burkaja, Marc van Leeuwen, SilvonenBot, Badgernet, Addbot, DrThunder88, Some jerk on the Internet, CanadianLinuxUser, CarsracBot, Dr. Universe, Favonian, Numbo3-bot, Flatfish89, Stepfordswife, Zorrobot, Legobot, Cote d'Azur, Luckas-bot, Yobot, II MusLiM HyBRiD II, Kristen Eriksen, LilHelpa, Xqbot, Doezxcty, Capricorn42, Almabot, Miym, GrouchoBot, Jubb-green, RibotBOT, FrescoBot, Sirtywell, Haeinous, Heptadecagon, BigDwiki, RedBot, Eric wisniewski, EmausBot, ZéroBot, The Nut, Xzenu, GZ-Bot, H3llBot, D.Lazard, ChuispastonBot, ClueBot NG, Helpful Pixie Bot, Howald, Jibun, bukiyou desu kara, Khazar2, Oracions, Wywin, Yamaha5 and Anonymous:

110

_ **Axiom** *Source:* http://en.wikipedia.org/wiki/Axiom?oldid=630172920 *Contributors:* AxelBoldt, LC, Brion VIBBER, Mav, Zundark, The Anome, Youssefsan, XJaM, Stevertigo, Michael Hardy, JakeVortex, Nixdorf, Graue, Glenn, Tim Retout, Nikai, Rotem Dan, Andres, Hectorthebat, EdH, Rob Hooft, Mxn, Charles Matthews, Dysprosia, Greenrd, Markhurd, Hyacinth, CecilTyme, Robbot, Josh Cherry, RedWolf, Altenmann, Jeronim, Flauto Dolce, Saforrest, Wikibot, Benc, Tobias Bergemann, Ancheta Wis, Tosha, Giftlite, Mshonle, Gene Ward Smith, Dissident, Alterego, Neilc, Andycjp, Knutux, Alberto da Calvairate, Antandrus, Dunks58, Sam Hocevar, Guppyfinsoup, Mormegil, EugeneZelenko, Rich Farmbrough, Iainscott, Mani1, Paul August, El C, Rgdboer, Irr^a^ti^o^nal, Bobo192, Babomb, Mintywalker, ParticleMan, Nk, Rje, Obradovic Goran, Haham hanuka, Jumbuck, Alansohn, 119, Jeltz, Hu, Simplebrain, Stephan Leeds, HenryLi, Killing Vector, Oleg Alexandrov, Natalya, Roland2, Kzollman, Ruud Koot, Gimboid13, Xeonx, Mandarax, Qwertyus, FreplySpang, Salix alba, Nmegill, FlaBot, Matharvest, Mathbot, Alvin-cs, Chobot, YurikBot, Wavelength, Splash, Chaos, Trovatore, Dbmag9, VinnyCee, Bucketsofg, Arthur Rubin, Vicarious, David Biddulph, Nahaj, SmackBot, McGeddon, Od Mishehu, SaxTeacher, Bomac, KocjoBot, Alksub, RobotJcb, Bluebot, Adam M. Gadomski, DMS, MalafayaBot, SchfiftyThree, Spellchecker, Zhuravskij, Xiner, Xyzzyplugh, EPM, Pissant, Eddiesegoura, Jon Awbrey, Vina-iwbot, Byelf2007, SashatoBot, Lambiam, DA3N, Dialectic, Physis, Atoll, 16@r, Loadmaster, MTSbot, Hu12, Dreftymac, Lenoxus, Bruno321, Happy-melon, Bharatveer, CRGreathouse, CaveBat, CBM, RoddyYoung, Gregbard,

Cydebot, Peterdjones, Julian Mendez, Gproud, PamD, Fourchette, Ken Burns, CharlotteWebb, X96lee15, AlefZet, Escarbot, Quintote, Readro, Steelpillow, JAnDbot, Oxinabox, MER-C, Four Dog Night, Magioladitis, VoABot II, JamesBWatson, Swpb, Amorette, Daarznieks, David Eppstein, R'n'B, Trusilver, MistyMorn, Maurice Carbonaro, Amritasen7, Cpiral, SlowJog, McSly, Daniel5Ko, Juliancolton, Dorftrottel, VolkovBot, Hotfeba, Jeff G., LokiClock, Philip Trueman, TXiKiBoT, AlexDenney, Dmcq, Submachinegum, Sfmammamia, LegoAxiom1007, SieBot, Dawn Bard, JerrySteal, MiNombreDeGuerra, IdNotFound, BrianGo28, Denisarona, DEMcAdams, ClueBot, Uncle Milty, Mate2code, Jotterbot, Eloifigueiredo, Aitias, Blackhawkjxx7, GAT27, YouRang?, JKeck, Marc van Leeuwen, Gnowor, Seshadri105, Addbot, Aelisabeths, NjardarBot, Numbo3-bot, Dayewalker, Tide rolls, Legobot, Luckas-bot, Yobot, Fraggle81, TaBOT-zerem, KamikazeBot, Goldhawk08, Jalal0, AnomieBOT, Arjun G. Menon, Jim1138, Piano non troppo, LlywelynII, Bob Burkhardt, ArthurBot, Xqbot, LordDelta, J JMesserly, GrouchoBot, The Wiki ghost, Die4cheese5, Chjoaygame, Paine Ellsworth, I dream of horses, RedBot, Xiglofre, Meier99, Gamewizard عقيل كاشف , 71 , Lotje, Dinamik-bot, 777sms, Duoduoduo, Reaper Eternal, Minimac, EmausBot, Najeeb1010, GoingBatty, Wikipelli, AvicBot, MonoAV, NTox, ClueBot NG, Masssly, Helpful Pixie Bot, Calabe1992, Frog23, Davidiad, Glacialfox, HumanNaturOriginal, Jeremy112233, ChrisGualtieri, Br'er Rabbit, Spectral sequence, Josell2, Lakshya08, Yolow1, Cthulhu is love cthulhu is life and Anonymous: 266

_ **Deductive reasoning** *Source:* http://en.wikipedia.org/wiki/Deductive%20reasoning?oldid=630179231 *Contributors:* The Cunctator, Toby Bartels, Youandme, Mrwojo, DennisDaniels, Michael Hardy, TakuyaMurata, BenKovitz, EdH, DesertSteve, Charles Matthews, Dtgm, Hyacinth, Lumos3, Robbot, R3m0t, Romanm, AceMyth, Blainster, Tobias Bergemann, Ancheta Wis, Giftlite, Lethe, Guanaco, Bovlb, Archenzo, Jason Quinn, Piotrus, Karol Langner, Aequo, Stepp-Wulf, EricBright, TedPavlic, Kevinb, Stbalbach, PhilHibbs, Causa sui, Flammifer, Jumbuck, Ryanmcdaniel, CyberSkull, Nasukaren, Garrisonroo, SidP, Kenyon, Tariqabjotu, Stephen, Velho, OwenX, Mindmatrix, TheNightFly, Ruud Koot, Jon Harald Søby, ZephyrAnycon, Teemu Leisti, BD2412, Nightscream, Koavf, Gmelli, Jweiss11, Tangotango, YAZASHI, Ggfevans, DirkvdM, FlaBot, Fresheneesz, Skc0001, Alphachimp, Chobot, YurikBot, Borgx, Erachima, DTRY, Rick Norwood, Holycharly, TriGen, EEMIV, Bota47, Shadro, Mjhy0926, Allens, Infinity0, GrinBot, DVD R W, Sardanaphalus, SmackBot, Aim Here, KocjoBot, Thunderboltz, Stephensuleeman, WookieInHeat, Ieopo, The great kawa, Gilliam, Q1w2e3, Mhss, Psiphiorg, Bluebot, ViolinGirl, MalafayaBot, George Rodney Maruri Game, Octahedron80, DHN-bot, Javalenok, Chlewbot, Mr.Z-man, ConMan, Cybercobra, Jon Awbrey, RossF18, Byelf2007, The Ungovernable Force, SashatoBot, Nishkid64, Dbtfz, JoseREMY, Extremophile, Penagate, Comicist, Quaeler, Iridescent, K, Zarex, Van helsing, ChristineD, Neelix, Gregbard, Julian Mendez, Thijs!bot, LactoseTI, Marek69, Kborer, Noaa, AntiVandalBot, MaTT, AaronY, IrishPete, Oliver Simon, BenMcLean, JAnDbot, Skomorokh, Agentnj, Hewinsj, Gurch-Bot, Probios, Djradon, Kirrages, Rupes, VoABot II, Arno Matthias, Snowded, Oxford Dictionary, Illspirit, Vanished user ty12kl89jq10, Cathalwoods, HemRaj Singh, Pbroks13, Pomte, Stjeanp, N4nojohn, J.delanoy, Trusilver, Shawn in Montreal, OAC, Tparameter, Jaxha, CompuChip, Heyitspeter, DorganBot, Vinsfan368, Lallallal, Jonwilliamsl, Pasixxxx, MARVINWAGNER, Rucha58, Leoremy, TXiKi-BoT, Technopat, A4bot, Msviolone, Philogo, Broadbot, Abdullais4u, Dprust, Andrewaskew, Graymornings, Lova Falk, Kusyadi, MCTales, Cnilep, Sfmammamia, SieBot, Paradoctor, Meldor, Mankar Camoran, Svick, DesolateReality, Mygerardromance, Escape Orbit, Troy 07, Kenji000, De728631, ClueBot, R000t, Philosophy.dude, Bfeylia, Neverquick, Excirial, Rohit nit, GoldenGoose100, PaulKincaidSmith, SpikeToronto, Ember of Light, GlasGhost, Thingg, Vanished User 1004, Zenwhat, XLinkBot, BodhisattvaBot, Kwork2, Gerhardvalentin, Saeed.Veradi, WikHead, Qwertytastic, Ewger, Addbot, CanadianLinuxUser, H92Bot, Glane23, GlobalSequence, Tide rolls, Lightbot, ScienceApe, سعی , Legobot, Luckas-bot, Yobot, 2D, Azcolvin429, Rubinbot, PresMan, Flewis, Prbclj25, ArthurBot, Xqbot, Doezxcty, S h i v a (Visnu), Lord Archer, Capricorn42, Xephras, Hartkiller, Hjurgen, Ordning, Lord Bane, Ruby.red.roses, RowanEvans, RibotBOT, SEASONnmr, FrescoBot, Beclp, Rushbugled13, Mohehab, JimRech, NeuroBells123, Humble Rat, Difu Wu, Whisky drinker, Badelmann, Tesseract2, DASHBot, EmausBot, Hedonistbot4000, Mo ainm, Tommy2010, TheGeomaster, JSquish, Fæ, Oncenawhile, MindShifts, Foreverlove642, Wayne Slam, Donner60, Tziemer991, Jimmynudes2, ClueBot NG, Drdoug5, Kimberleyporter, Fauzan, Jj1236, Tideflat, Amr.rs, Dictabeard, O.Koslowski, Masssly, Widr, Chillllls, Helpful Pixie Bot, Nichole773, Hallows AG, Wiki13, Luke13579, Richard84041, Ninjagoat, Sopidex, Dhruv-NJITWILL, Sgilmore10, Courtneysfoster, ChrisGualtieri, GoShow, Oligocene, ShangTsung87, Lugia2453, 8ty3hree, MostlyListening, BreakfastJr, Strikingstar, VogelsangLorenzo, Ginsuloft, Mauriziogeri2013, Monkbot and Anonymous: 356

_ **Inductive reasoning** *Source:* http://en.wikipedia.org/wiki/Inductive%20reasoning?oldid=631721302 *Contributors:* AxelBoldt, The Cunctator, Ryguasu, DennisDaniels, Michael Hardy, Earth, Owl, Voidvector, BoNoMoJo (old), Jfitzg, Andres, Evercat, EdH, DesertSteve, Timwi, Trontonian, Bemoeial, Ike9898, Wolfgang Kufner, Radiojon, Markhurd, Peregrine981, Banno, Nufy8, Nurg, Romanm, Ojigiri, Mikiher, Tobias Bergemann, Filemon, Ancheta Wis, Giftlite, Zigger, PeruvianIlama, Bovlb, Jason Quinn, Jackol, ELApro, Guppyfinsoup, Lucidish, Archer3, Discospinster, Freestylefrappe, Ivan Bajlo, Bender235, Kbh3rd, El C, Aaronbrick, David Crawshaw, Bobo192, QTxVi4bEMRbrNqOorWBV, Flammifer, Samohyl Jan, Yuckfoo, Mikeo, Recury, Nightstallion, Kinema, Kazvorpal, Kenyon, Hq3473, Velho, Mindmatrix, Kzollman, Ruud Koot, Alfakim, Andrea.gf, Rjwilmsi, Jweiss11, Strake, Bryan H Bell, Reinis, Matt Deres, Chris Pressey, Latka, RexNL, Fresheneesz, NotJackhorkheimer, Spencerk, King of Hearts, Chobot, Dresdnhope, YurikBot, RussBot, Gaius Cornelius, Grafen, Holycharly, SAE1962, 24ip, Pkearney, Roy Brumback, Bota47, Shadro, Tomisti, Sethery, Fram, Curpsbot-unicodify, Teply, Infinity0, Bo Jacoby, Jer ome, Sardanaphalus, SmackBot, RedHouse18, David Kernow, Rtc, McGeddon, Istvan, Eskimbot, Klokie, Gilliam, Duke Ganote, Ohnoitsjamie, Betacommand, Bluebot, Anthonzi, LaggedOnUser, DHN-bot, DoctorStrangelove, Can't sleep, clown will eat me, Go For It, Avb, Edivorce, Mr.Z-man, Jmnbatista, Richard001, Kalexander, Jon Awbrey, Neshatian, Andeggs, Vina-iwbot, Byelf2007, Jonrgrover, Normalityrelief, RichMorin, Antonielly, Aleenf1, Lukemcgrath, Grumpyyoungman01, Domokato, Levineps, Iridescent, K, Wjejskenewr, FleetCommand, CWY2190, Indigenius, El aprendelenguas, TMN, Gregbard, Slazenger, Peterdjones, Khromatikos, Gogo Dodo, Wikipediarules2221, Miguel de Servet, Letranova, Gacggt, Ucanlookitup, Second Quantization, Danny

766 *CHAPTER 102. MONSKY'S THEOREM*

Reese, Defeatedfear, Fotomatt, AntiVandalBot, Luna Santin, Minhtung91, Spencer, Salgueiro, JAnDbot, Davewho2, Dmar198, Coffee2theorems, Magioladitis, Bongwarrior, Equinexus, Hasek is the best, Arno Matthias, Farquaadhnchmn, DAGwyn, Snowded, Moopiefoof, Cathalwoods, MetsBot, Chrisdone, WLU, Mommyzbrat, STBot, Dionysiaca, Pomte, OttoMäkelä, LordAnubisBOT, Mahewa, Touisiau, Chiswick Chap, Heyitspeter, Pianoman55, MetsFan76, STBotD, Andy Marchbanks, Straw Cat, Zach425, VolkovBot, Thewolf37, Pasixxxx, Hotfeba, Shinju, Jimmaths, Tiktuk, Deleet, Katoa, Jazzwick, Philogo, Abdullais4u, Jackfork, PDFbot, Anarchangel, Jor344, Shifter95, Cnilep, Harmonicemundi, PhysPhD, Jammycaketin, AlleborgoBot, Newbyguesses, Dwandelt, Matthew Yeager, Mark Klamberg, Flyer22, Bobklahn, Oxymoron83, Vanished user oij8h435jweih3, MiNombreDeGuerra, Bagatelle, Sunrise, Linkboyz, Melcombe, Oneforlogic, ClueBot, Farras Octara, Eric Schoettle, Niceguyedc, Vandalometer, Rbakels, Excirial, Jusdafax, Kikilamb, Estirabot, ChrisKalt, Hazzzzzz12, Lx 121, XLinkBot, Fastily, Gerhardvalentin, Tegiap, Saeed.Veradi, Skarebo, WikHead, Kwjbot, Kbdankbot, Tayste, Addbot, Tanhabot, Jtradke, Numbo3-bot, Tide rolls, ScienceApe, KUSSOMAK, Legobot, Luckas-bot, Oilstone, THEN WHO WAS PHONE?, AnomieBOT, Doingmorestuffonline, Vanakaris, Bob Burkhardt, Parthian Scribe, Xqbot, Lord Bane, Hanberke, A157247, F-22 Raptored, Omnipaedista, RibotBOT, Delbertpeach, Alialiac, FieldOperative, Paine Ellsworth, SBA1870, Machine Elf 1735, Pinethicket, Kiefer.Wolfowitz, Mavit, A8UDI, Cleon7177, NeuroBells123, Gamewizard71, TobeBot, Jonkerz, Miracle Pen, Dbmikus, Hyperbytev2, Ripchip Bot, Elspru, NerdyScienceDude, George Richard Leeming, EmausBot, Elanguescence, Grjoni88, Logical Cowboy, Gfoley4, RenamedUser01302013, Mo ainm, ZéroBot, Leminh91, CanonLawJunkie, Oncenawhile, Wagino 20100516, Erianna, EricWesBrown, L Kensington, 28bot, ClueBot NG, Gareth Griffith-Jones, MelbourneStar, Ek65, Millermk, Schicagos, Tsunamicharlie,

Alberttttt, Thepigdog, Masssly, Widr, Helpful Pixie Bot, HMSSolent, Curb Chain, BG19bot, Wiki13, MusikAnimal, Jander80, Wandwiki, BlueMist, Will.Oliver, Trailspark, RichardMills65, Ctasa221, Fangli997376557, ChrisGualtieri, Lhu720, Hagrid da fifth, Watchpup32, Neurocitizen, Oligocene, Moonstroller-2, Jochen Burghardt, M strat17, 90b56587, Reid12345, Londomollari42, Cauzality, Aubreybardo, Liz, Logicman2, Hoffoholi, Superploro, Temprack5446, Hellerrr and Anonymous: 390

_ **Empirical evidence** *Source:* http://en.wikipedia.org/wiki/Empirical%20evidence?oldid=632184066 *Contributors:* Stevertigo, Michael Hardy, Eurleif, Glenn, Kaihsu, Raven in Orbit, Markhurd, Martinphi, Robbot, ShaunMacPherson, Taak, Just Another Dan, Chowbok, Piotrus, Mvuijlst, Jameshowison, Mdd, Gary, Atlant, Joepzander, Kenyon, PeregrineAY, BD2412, Michaelbluejay, Jrtayloriv, Phoenix2, VolatileChemical, RL0919, Scolaire, SmackBot, Jagged 85, Thumperward, Normxxx, Memming, Hammer1980, Madbull, Tktktk, Bjankuloski06en, A. Parrot, Frederlck, Khanssen, Levineps, K, Ncosmob, Danneckar, WeggeBot, Apterygial, Gregbard, Travelbird, ST47, TheM62Manchester, Ruaraidh-dobson, Runch, ThomasPusch, AntiVandalBot, Ingolfson, Ioeth, JAnDbot, Cgingold, Astragale, Geboy, Removepam, STBot, J.delanoy, Filll, ChrisfromHouston, Tdadamemd, Mikael Häggström, Fainites, Legoland12342, Jeff G., DianaGaleM, Poopsmith21, Jmath666, Falcon8765, Rurik3, Newbyguesses, Nihil novi, Gerakibot, Sunrise, Melcombe, Twinsday, Elassint, ClueBot, Gorillasapiens, Desellliers, Carolyn 3300, Bala990, Skarebo, Addbot, Ronhjones, Jubeidono, Yobot, Fmrauch, AnomieBOT, Tryptofish, Trabucogold, Wortafad, Apollo, Xqbot, LogoX, Mikejens, Srich32977, Al Azif, GrouchoBot, Speednat, Pinethicket, Edderso, TobeBot, Ale And Quail, Dark Lord of the Sith, Theo10011, Reaper Eternal, GrindtXX, Wayne Slam, Donner60, Atlantictire, ClueBot NG, Mcgee123, Kathryn Whitney, JordanSeymore, Wbm1058, Maestro814, Malerooster, Mark viking, Theemathas, Aubreybardo, Mohammad sohag khan, Loraof and Anonymous: 109

_ **Logic** *Source:* http://en.wikipedia.org/wiki/Logic?oldid=634041395 *Contributors:* AxelBoldt, The Anome, Toby Bartels, Ryguasu, Hirzel, Dwheeler, Stevertigo, Edward, Patrick, Chas zzz brown, Michael Hardy, Lexor, TakuyaMurata, Bagpuss, Looxix, Ahoerstemeier, Notheruser, BigFatBuddha, Александър, Glenn, Marco Krohn, Rossami, Tim Retout, Rotem Dan, Evercat, EdH, DesertSteve, Caffelice, Mxn, Michael Voytinsky, Renamed user 4, Rzach, Charles Matthews, Dcoetzee, Paul Stansifer, Dysprosia, Jitse Niesen, Xiaodai, Markhurd, MikeS, Flyingbird, SEWilco, Samsara, JMD, Shizhao, Olathe, Jusjih, Ldo, Banno, Chuunen Baka, Robbot, Iwpg, Fredrik, R3m0t, Altenmann, MathMartin, Rorro, Rholton, Saforrest, Borislav, Robertoalencar, Michael Snow, Raeky, Guy Peters, Jooler, Filemon, Ancheta Wis, Exploding Boy, Giftlite, Recentchanges, Inter, Wolfkeeper, Lee J Haywood, COMPATT, Everyking, Rookkey, Malyctenar, Andris, Bovlb, Jason Quinn, Sundar, Siroxo, Deus Ex, Rheun, LiDaobing, Roachgod, Quadell, Starbane, Piotrus, Ludimer, Karol Langner, CSTAR, Rdsmith4, APH, JimWae, OwenBlacker, Kntg, Mysidia, Pmanderson, Eduardoporcher, Eliazar, Grunt, Guppyfinsoup, Mike Rosoft, Freakofnurture, Ultratomio, Lorenzo Martelli, Discospinster, KillerChihuahua, Rhobite, Guanabot, Leibniz, Hippojazz, Vsmith, Raistlinjones, Slipstream, ChadMiller, Paul August, Bender235, El C, Chalst, Mwanner, Tverbeek, Bobo192, Cretog8, Johnkarp, Shenme, Amerindianarts, Passw0rd, Knucmo2, Storm Rider, Red Winged Duck, Alansohn, Shadikka, Rh, Chira, ABCD, Kurt Shaped Box, SlimVirgin, Batmanand, Yummifruitbat, Shinjiman, Velella, Sciurinæ, MIT Trekkie, Alai, CranialNerves, Velho, Mel Etitis, Mindmatrix, Camw, Kokoriko, Kzollman, Ruud Koot, Orz, MONGO, Apokrif, Jok2000, Wikiklrsc, CharlesC, MarcoTolo, DRHansen, Gerbrant, Tslocum, Alienus, BD2412, Porcher, Rjwilmsi, Mayumashu, Саша Стефановић, GOD, Brucelee, Salix alba, Crazynas, Ligulem, Baryonic Being, Titoxd, FlaBot, Kwhittingham, Latka, Mathbot, Twipley, RexNL, AndriuZ, Quuxplusone, Celendin, Influence, R Lee E, JegaPRIME, Malhonen, Spencerk, Chobot, DVdm, EamonnPKeane, Roboto de Ajvol, Wavelength, Deeptrivia, KSmrq, Endgame, Polyvios, CambridgeBayWeather, KSchutte, NawlinWiki, Rick Norwood, SEWilcoBot, Mipadi, Brimstone, LaszloWalrus, AJHalliwell, Trovatore, Pontifexmaximus, Chunky Rice, Cleared as filed, Nick, Darkfred, Wjwma, Googl, Mendicott, StuRat, Open2universe, ChrisGriswold, Nikkimaria, OEMCUST, Nahaj, Extreme Unction, Allens, Sardanaphalus, Johndc, SmackBot, Lestrade, InverseHypercube, Pschelden, Jim62sch, Jagged 85, WookieInHeat, Josephprymak, Timotheus Canens, Srnec, LonesomeDrifter, Collingsworth, Gilliam, Skizzik, RichardClarke, Heliostellar, Chris the speller, Jaymay, Da nuke, Unbreakable MJ, MK8, Andrew Parodi, Kevin Hanse, MalafayaBot, Clconway, Sciyoshi, Go for it!, Mikker, Zsinj, Can't sleep, clown will eat me, Misgnomer, Grover cleveland, Fuhghettaboutit, Cybercobra, Jiddisch, Richard001, MEJ119, Kabain52, Lacatosias, Jon Awbrey, DMacks, Henning Makholm, Ged UK, Ceoil, Byelf2007, SashatoBot, Lambiam, Dbtfz, Deaconse, UberCryxic, FrozenMan, Heimstern, Shlomke, Shadowlynk, F15 sanitizing eagle, Prince153, WithstyleCMC, Hvn0413, Meco, RichardF, Novangelis, Vagary, Pamplmoose, KJS77, Hu12, Levineps, BananaFiend, K, Lottamiata, Catherineyronwode, Mrdthree, Igoldste, Themanofnines, Adambiswanger1, Satarnion, Tawkerbot2, Galex, SkyWalker, CRGreathouse, CBM, Editorius, Rubberchix, Gregbard, Kpossin, Cydebot, Pce3@ij.net, Jasperdoomen, Samuell, Quinnculver, Peterdjones, Travelbird, Pv2b, Drksl, Julian Mendez, Dancter, Tawkerbot4, Shirulashem, Dougweller, DumbBOT, Garik, Progicnet, Mattisse, Letranova, Thijs!bot, Epbr123, Kredal, Smee, Marek69, AgentPeppermint, OrenBochman, Escarbot, Eleuther, Mentifisto, Vafthrudnir, Anti-VandalBot, Peoppenheimer, Majorly, Gioto, Hidayat ullah, GeePriest, Dougher, Sluzzelin, JAnDbot, Narssarssuaq, MER-C, The Transhumanist, Avaya1, Zizon, Frankie816, Savant13, Dr mindbender, LittleOldMe, Bongwarrior, VoABot II, SDas, JNW, Arno Matthias, Appraiser, Gamkiller, Smihael, Caesarjbsquitti, Midgrid, Bubba hotep, Moopiefoof, GeorgeFThomson, Virtlink, Epsilon0, DerHexer, Waninge, Exbuzz, MartinBot, Wylve, CommonsDelinker, EdBever, C.R.Selvakumar, J.delanoy, Trusilver, Jbessie, Fictionpuss, Cpiral, RJMalko, McSly, Lightest, Classicalsubjects, Mrg3105, Daniel5Ko, The Transhumanist (AWB), Policron, MetsFan76, Kenneth M Burke, Steel1943, Idioma-bot, Spellcast, WraithM, VolkovBot, Cireshoe, Rucha58, Macedonian, Hotfeba, Jennavecia, Fundamental metric tensor, Jimmaths, Djhmoore, Aesopos, Rei-bot, Llamabr, Ontoraul, Philogo, Leafyplant, Sanfranman59, Abdullais4u, Cullowheean, Maxim, Myscience, LIBLAHLIBLAHTIMMAH, Synthebot, Rurik3, Koolo, Nagy, Symane, PGWG, W4chris, Prom2008, NHRHS2010, Radagast3, Demmy, JonnyJD, Newbyguesses, Linguist1, SieBot, StAnselm, Maurauth, Gerakibot, RJaguar3, Yintan, Bjrslogii, Soler97, Til Eulenspiegel, Flyer22, DanEdmonds, Undead Herle King, Crowstar, Redmarkviolinist, Spinethetic, Thelogicthinker, DancingPhiloso

pher, Svick, Valeria.depaiva, Adhawk, Sginc, Tognopop, CBM2, 3rdAlcove, PsyberS, Francvs, Classicalecon, Athenean, Atif.t2, Martarius, ClueBot, Andrew Nutter, Snigbrook, The Thing That Should Not Be, Taroaldo, Ukabia, TheOldJacobite, Boing! said Zebedee, Niceguyedc, Blanchardb, DragonBot, Jessieslame, Excirial, Alexbot, Jusdafax, Mate2code, AENAON, NuclearWarfare, Arjayay, Thingg, JDPhD, Budelberger, Skunkboy74, Gerhardvalentin, Duncan, Saeed.Veradi, Mcgauley08, NellieBly, Noctibus, Aunt Entropy, Jjfuller123, Spidz, Addbot, Rdanneskjold, Proofreader77, Atethnekos, Sully111, Logicist, Vitruvius3, Glane23, Uber WoMensch!, Chzz, Favonian, LinkFA-Bot, AgadaUrbanit, Numbo3-bot, Ehrenkater, Tide rolls, Lightbot, Macro Shell, Zorrobot, JEN9841, Aarsalankhalid, Arcvirgos 08, Jammie101, Francos22, Azcolvin429, MassimoAr, AnomieBOT, Jim1138, IRP, AdjustShift, Melune, Nick UA, Neurolysis, ArthurBot, Gemtpm, Blueeyedbombshell, Junho7391, Xqbot, Gilo1969, The Land Surveyor, Tyrol5, A157247, Petropoxy (Lithoderm Proxy), Uarrin, GrouchoBot, Hifcelik, Omnipaedista, , جم شد يدى مهدى 78.26 פריימן ,, Tales23, GhalyBot, Aaron Kauppi, GliderMaven, FrescoBot, Liridon, D'Artagnol, Tobby72, D'ohBot, Mewulwe, Itisnotme, Rhalah, Citation bot 1, Chenopodiaceous, AstaBOTh15, Gus the mouse, Pinethicket, Vicenarian, A8UDI, Ninjasaves, Seryred123, Maokart444, Gamewizard71, FoxBot, TobeBot, Burritoburritoburrito, Mysticcooperfox, Lotje, Gregkaye, Vrenator, Duoduoduo, גרשון בן , Merlinsorca, Literateur, Jarpup, Whisky drinker, Mean as custard, Rlnewma, TjBot, Walkinxyz, EmausBot, Orphan Wiki, The Kytan Apprentice, Pologic, Jedstamas, Wham Bam Rock II, Solarra, ZéroBot, Leminh91, Josve05a, Shuipzv3, Mar4d, Wayne Slam, Frigotoni, Rcsprinter123, FrankFlanagan, L Kensington, Danmuz, Eranderson, Donner60, Chewings72, GKaczinsky, ChuispastonBot, NTox, Poopnubblet, Xanchester, Rememberway, ClueBot NG, W.Kaleem, Jack Greenmaven, SusikMkr, Quantamflux, Validlessness, Bastianperrot, Wdchk, Snotbot, Masssly, Widr, Lawsonstu, ESL75, Helpful Pixie Bot, Anav2221, BG19bot, Darekaon, Norma Romm, PTJoshua, Northamerica1000, Geegeeg, JohnChrysostom,

Frze, Chjohnson39, Alex.Ramek, CJMacalister, CitationCleanerBot, Jilliandivine, Flosfa, Chrisct1993, Brad7777, Mewhho18, A.coolmcfly, Compulogger, Cyberbot II, Roger Smalling, The Illusive Man, NanishaOpaenyak, Rhlozier, EagerToddler39, Dexbot, Marius siuram, Табалдыев Ысламбек, Omanchandy007, RideLightning, Jochen Burghardt, Wieldthespade, Hippocamp, Wickid123, Matticusmadness, JMCF125, NIXONDIXON, CsDix, I am One of Many, ניר גלעד, Tentinator, EvergreenFir, Ugog Nizdast, Melody Lavender, JustBerry, Skansi.sandro, Ginsuloft, Robf00f1235, Calvinator8, The Annoyed Logician, Liz, GreyWinterOwl, ByDash, Jbob13, Henniepenny, Matthew Derick B Cruz, Filedelinkerbot, Sherlock502, Fvdedphill, Norwo037, Karnaoui, Pat132, The Expedia, Sbcdave, Muneeb Masoud, Jacksplay, Esicam, Ntuser123, Cthulhu is love cthulhu is life, Adamrobson28, Josmust222 and Anonymous: 736

_ **Natural language** *Source:* http://en.wikipedia.org/wiki/Natural%20language?oldid=632919021 *Contributors:* Damian Yerrick, Chuck Smith, Lee Daniel Crocker, Tox, Stevertigo, Michael Hardy, Kku, Plasticlax, Nanshu, Grin, Timwi, Furrykef, Taxman, Nricardo, Wikibot, Ruakh, Aphaia, Micru, Jason Quinn, Prosfilaes, Edcolins, Andycjp, Ryan524, Beland, OverlordQ, Vina, Marcos, Burschik, Poccil, Leibniz, Mrevan, Gadykozma, Rama, Dbachmann, Ntennis, El C, Szyslak, Kwamikagami, Jordan123, Carbon Caryatid, Ray Foster, Dejvid, Angr, Ruud Koot, WadeSimMiser, Dolfrog, Dpr, Sjö, Quiddity, Armandeh, FlaBot, Windchaser, Paperflowergirl, Chobot, Krishnavedala, Peterl, YurikBot, Tristatestar, NTBot, Red Slash, Bhny, Akamad, Maunus, Pooryorick, Netrapt, Minur, Mabisa, Sardanaphalus, SmackBot, Ccalvin, PiKeeper, MalafayaBot, Hibernian, Maxsonbd, AdeMiami, Cookie90, SashatoBot, Autoterm, Guyjohnston, MagnaMopus, JorisvS, NongBot, Drork, 16@r, Icez, Peter Horn, JMK, Joseph Solis in Australia, IvanLanin, Gregbard, Davius, Jessicanr, Trident13, Garik, Mattisse, Headbomb, Odoncaoa, WikiRuler03, Luna Santin, Kauczuk, Bongwarrior, JamesBWatson, Edward321, Jackson Peebles, Nono64, AgarwalSumeet, Mange01, Mike.lifeguard, LittleHow, Cal Evans, Maghnus, TXiKiBoT, ElinorD, Srikipedia, Alphaios, Triplejumper, Mjs072, UnitedStatesian, Lova Falk, AlleborgoBot, Ivan Štambuk, Purbo T, Jonathan.Wickens, Mr. Stradivarius, Linforest, Tanvir Ahmmed, B J Bradford, ClueBot, Pi zero, Pompeufabra, Mild Bill Hiccup, ChandlerMapBot, Cenarium, Bonewah, JasonAQuest, Thingg, BendersGame, Saeed.Veradi, Addbot, Yolgnu, Fgnievinski, DVassallo, MXVN, FiriBot, Zorrobot, TaBOT-zerem, Pcap, AnomieBOT, 1exec1, Citation bot, ArthurBot, Xqbot, Jburlinson, Omnipaedista, Danebell, Феникс, The Wiki ghost, A. di M., FrescoBot, Nirmos, Dinamik-bot, Dmwpowers, EmausBot, Alvatov, Accents, AvicBot, Sodmy, Stephanos21, MerlynCooper, BioPupil, ClueBot NG, Movses-bot, Snotbot, Ryan Vesey, MerlIwBot, Helpful Pixie Bot, Mr. Credible, Smith felix, Kyoakoa, Smcg8374, DewBmtn, JPaestpreornJeolhlna, Cumar1, Kahtar, Artist robert jones, Jochpoch, Malistomailie and Anonymous: 106

_ **Informal logic** *Source:* http://en.wikipedia.org/wiki/Informal%20logic?oldid=632692969 *Contributors:* Chris Q, DennisDaniels, Kwertii, Pcb21, Docu, Julesd, Dysprosia, Markhurd, Furrykef, Hyacinth, Tomchiukc, Leonard G., CSTAR, Paulscrawl, Guppyfinsoup, Chalst, Femto, A.t.bruland, Velho, Simetrical, Mindmatrix, Rjwilmsi, Gurch, Mattpeck, Tsch81, NawlinWiki, NostinAdrek, Closedmouth, GraemeL, SmackBot, Stev0, Bluebot, Colonies Chris, Jon Awbrey, Byelf2007, Grumpyyoungman01, Meco, JohnsonRalph, Moreschi, Gregbard, Kpossin, Steel, Al Lemos, Escarbot, Doremítzwr, WinBot, Deeplogic, Clan-destine, Cathalwoods, Gomm, Gwern, R'n'B, RickardV, Adedayoojo, Homo logos, Nburden, Igglebop, Guillaume2303, Ontoraul, GlassFET, Kumioko (renamed), ClueBot, Napzilla, Staticshakedown, Addbot, SpBot, Lightbot, Luckas-bot, Luce nordica, AnomieBOT, Omnipaedista, Gerald Roark, FrescoBot, LucienBOT, Machine Elf 1735, Abductive, Wotnow, John of Reading, ZéroBot, Tijfo098, ClueBot NG, Helpful Pixie Bot, GabeIglesia and Anonymous: 38

_ **Formal proof** *Source:* http://en.wikipedia.org/wiki/Formal%20proof?oldid=630249488 *Contributors:* Charles Matthews, Hyacinth, Timrollpickering, Pmanderson, EmilJ, Dfranke, BD2412, Salix alba, Gaius Cornelius, Seegoon, Chrylis, Gregbard, Nick Number, Philogo, VanishedUserABC, Radagast3, Ndenison, Mild Bill Hiccup, Hans Adler, HexaChord, Addbot, Numbo3-bot, AnomieBOT, MattTait, The Wiki ghost, Disinvented and Anonymous: 8

_ **Proof theory** *Source:* http://en.wikipedia.org/wiki/Proof%20theory?oldid=630512528 *Contributors:* Mav, Bryan Derksen, The Anome, Toby Bartels, Youandme, Michael Hardy, Llywrch, Dominus, Rotem Dan, Charles Matthews, Dysprosia, Markhurd, Hyacinth, Jni, Giftlite, Markus Krötzsch, Jorend, Kntg, Leibniz, Pj.de.bruin, Luqui, Number 0, Brian0918, Chalst, Nortexoid, Msh210, Krappie, Wtmitchell, Gene Nygaard, Oleg Alexandrov, Jtauber, Porcher, Mathbot, Comiscuous, Tillmo, Chobot, YurikBot, Hairy Dude, Tong, Arthur Rubin, Sardanaphalus, SmackBot, Cabe6403, Jahiegel, Vina-iwbot, Byelf2007, Lambiam, Dbtfz, Kuru, Rizome, Dicklyon, JRSpriggs, CBM, Gregbard, Nick Number, Escarbot, LaForge, David Eppstein, Yonaa, J.delanoy, Policron, Alan U. Kennington, JohnBlackburne, Hqb, Qxz, Magmi, VanishedUserABC, Radagast3, OKBot, Kumioko, Anchor Link Bot, CBM2, ClueBot, Mannypabla, Thisthat12345, Beach drifter, Good Olfactory, Addbot, Matěj Grabovský, Dima125, Yobot, Ptbotgourou, JRB-Europe, MattTait, Citation bot, Tbvdm, Sa'y, Foobarnix, Gamewizard71, Dbmikus, WildBot, EmausBot, ZéroBot, Donner60, ClueBot NG, Rezabot, Helpful Pixie Bot, Brad7777, Rsmbf, Begadkepat, Kkval0, JHU1959 and Anonymous: 46

_ **Mathematical practice** *Source:* http://en.wikipedia.org/wiki/Mathematical%20practice?oldid=605493760 *Contributors:* AxelBoldt, Michael Hardy, Dcljr, Revolver, Charles Matthews, OmegaMan, Khalid hassani, Mathbot, AKeen, Closedmouth, Robofish, M a s, AntiVandalBot, R'n'B, ConcernedScientist, Alexbot, Addbot, Yobot, John of Reading, Mark viking and Anonymous: 17

_ **Quasi-empiricism in mathematics** *Source:* http://en.wikipedia.org/wiki/Quasi-empiricism%20in%20mathematics?oldid=549790325 *Contributors:* Robert Merkel, Michael Hardy, Minesweeper, Cyan, Charles Matthews, Vespristiano, Everyking, Jason Quinn, PDH, Andreas Kaufmann, PhotoBox, M0rph, Diego Moya, CAD6DEE2E8DAD95A, RussBot, Rbarreira, Skullfission, SmackBot, Chris the

speller, JMSwtlk, Akriasas, Jon Awbrey, Sparkleyone, Lambiam, Grumpyyoungman01, Gregbard, Cydebot, M a s, Pneumonoultramicroscopicsilicovolcanoconiosis, Matthew Fennell, Nowletsgo, Filll, Pleasantville, The Tetrast, Coffee, Editor2020, Legobot, ZéroBot, Alpha Quadrant (alt), ClueBot NG and Anonymous: 16

_ **Mathematical folklore** *Source:* http://en.wikipedia.org/wiki/Mathematical%20folklore?oldid=622942454 *Contributors:* AxelBoldt, The Anome, SimonP, Chas zzz brown, Michael Hardy, Mic, Islandboy99, Rotem Dan, Charles Matthews, Gandalf61, Macrakis, Chameleon, JRR Trollkien, Pt, Rgdboer, Diego Moya, Linas, Smmurphy, YurikBot, SmackBot, Nbarth, Can't sleep, clown will eat me, Kingdon, BigrTex, Cnilep, Tautologist, Lightbot, Yobot, DelSarto, Suslindisambiguator and Anonymous: 14

_ **Philosophy of mathematics** *Source:* http://en.wikipedia.org/wiki/Philosophy%20of%20mathematics?oldid=632366477 *Contributors:* Damian Yerrick, AxelBoldt, Matthew Woodcraft, Derek Ross, LC, Bryan Derksen, Zundark, The Anome, Hhanke, SimonP, Zadcat, Ryguasu, Cwitty, IanS, The hanged man, Michael Hardy, Nixdorf, BoNoMoJo (old), Gabbe, Chinju, GTBacchus, Dori, Eric119, Ahoerstemeier, Snoyes, Darkwind, Cyan, Tim Retout, Rotem Dan, Andres, EdH, Schneelocke, Renamed user 4, Charles Matthews, RickK, Ww, Dtgm, Markhurd, Maximus Rex, Unknown, Robbot, Romanm, Gandalf61, Tim Ivorson, MathMartin, OmegaMan, Hadal, JohannesMarat, Tobias Bergemann, Adam78, Giftlite, Lee J Haywood, Lethe, Monedula, Everyking, Esap, WHEELER, Just Another Dan, JRR Trollkien, Stevietheman, Alberto da Calvairate, Karol Langner, Rdsmith4, Pmanderson, Robin klein, Random account 47, Shahab, D6, Splatty, Rich Farmbrough, Wclark, YUL89YYZ, Paul August, Pban92, Elwikipedista, Syp, Lycurgus, Rgdboer, Episcopo, Pokereth, Root4(one), Pearle, IonNerd, Jumbuck, Vesal, Droob, Eagleamn, Ossiemanners, Sligocki, Mr. Hyde, Hu, Pernest2002, Bookandcoffee, Euphrosyne, Angr, Boothy443, Mel Etitis, Woohookitty, Linas, Barrylb, Jok2000, Al E., Dfranke, DaveApter, BD2412, Qwertyus, Gigapixel, Rjwilmsi, Salix alba, Mathbot, Nihiltres, Echeneida, RexNL, Mark J, David H Braun (1964), JamesLee, YurikBot, Wavelength, Shonk, Hairy Dude, Conscious, Hydrargyrum, Gaius Cornelius, Chaos, KSchutte, Aldux, Hakeem.gadi, Tomisti, Igiffin, Arthur Rubin, Skullfission, LeonardoRob0t, Canadianism, Meegs, Infinity0, Brentt, Sardanaphalus, JJL, SmackBot, Reedy, Tom Lougheed, Pokipsy76, Eskimbot, Gilliam, Ghosts&empties, Chris the speller, Bluebot, Trebor, JMSwtlk, Helder Ribeiro, Go for it!, Lesnail, Cybercobra,

John wesley, Jon Awbrey, Just plain Bill, Bidabadi, Vina-iwbot, FlyHigh, Clicketyclack, Byelf2007, Igrant, Dbtfz, JHunterJ, Mets501, Ryulong, Iridescent, K, Aeternus, Rhetth, Tawkerbot2, 8754865, CRGreathouse, CmdrObot, CBM, Pan Camel, Gregbard, Logicus, Danman3459, Peterdjones, M a s, JamesAM, Old port, Thijs!bot, W3asal, 271828182, Headbomb, Marek69, Nick Number, Klausness, Vodello, Rjmars97, GeePriest, Wayiran, Leuqarte, Tayl1257, Ophion, Huphelmeyer, Andrewthomas10, Wlod, Lucaas, Revery, Seberle, David Eppstein, Martynas Patasius, Nowletsgo, Ynotds, CommonsDelinker, Filll, Altes, Maurice Carbonaro, NerdyNSK, Dispenser, Mikael Häggström, Rnest2002, Infarom, Milogardner, AlnoktaBOT, Jimmaths, TXiKiBoT, Aleph42, The Tetrast, Philogo, Broadbot, Geometry guy, Popopp, Finalfantasy2012, Wenli, Tomaxer, Sapphic, Dmcq, Newbyguesses, SieBot, Darrell Wheeler, Lightmouse, Roran659, DesolateReality, Classicalecon, ClueBot, DFRussia, Rockfang, CohesionBot, Azadeh.a, Sun Creator, Vegetator, JKeck, Marc van Leeuwen, Gerhardvalentin, Zodon, Addbot, Fyrael, With goodness in mind, MrOllie, Imtg5102, ProfessorThunderlips, Drdonzi, Nallimbot, Synchronism, AnomieBOT, Materialscientist, Citation bot, Xqbot, Capricorn42, Crzer07, Uarrin, J04n, Freddyfirre, Omnipaedista, Taekwandean, Aaron Kauppi, Constructive editor, Mr fabs, Hugetim, FrescoBot, Mark Renier, EricAndrewWallace, Bride-OfKripkenstein, Alboran, Steve Quinn, CESSMASTER, Machine Elf 1735, Mary rose arias, Tkuvho, Pinethicket, Pollinosisss, Jdapayne, Vrenator, Hueyha, Jowa fan, Merehap, EmausBot, John of Reading, ZéroBot, Amacfiew, RaptureBot, HarmoniousMembrane, BartlebytheScrivener, RockMagnetist, Logicalgregory, Anita5192, E. Fokker, ClueBot NG, Alexander E Ross, Deer*lake, Helpful Pixie Bot, Adriaan Joubert, Brian Tomasik, Harizotoh9, Brad7777, Gibbja, Dexbot, Jochen Burghardt, Mark viking, Stara729, Workstern and Anonymous: 218

_ **Language of mathematics** *Source:* http://en.wikipedia.org/wiki/Language%20of%20mathematics?oldid=632056343 *Contributors:* Bdesham, Snoyes, Nikai, Charles Matthews, Bevo, Gandalf61, Giftlite, Waltpohl, Marcos, Andreas Kaufmann, Leibniz, Xezbeth, Gauge, CanisRufus, Rgdboer, QTxVi4bEMRbrNqOorWBV, Mdd, Diego Moya, Oleg Alexandrov, Woohookitty, Splintax, SixWingedSeraph, Kbdank71, Bgwhite, Katieh5584, SmackBot, Bluebot, A Geek Tragedy, Rrburke, Lambiam, CRGreathouse, Gregbard, Sluzzelin, Cic, David Eppstein, SuneJ, DesolateReality, Foxj, Rjd0060, CohesionBot, Muhandes, Dmoursund, Addbot, Jarble, AnomieBOT, Xqbot, Jebdm, Kierkkadon, TRBP, GoingBatty, Fixblor, Bluefairyturnedred and Anonymous: 27

_ **Arithmetic** *Source:* http://en.wikipedia.org/wiki/Arithmetic?oldid=633703384 *Contributors:* LC, Tarquin, Christian List, Toby Bartels, Nonenmac, Juuitchan, TeunSpaans, Michael Hardy, JakeVortex, Meekohi, Ixfd64, Dcljr, Dori, J-Wiki, Angela, Julesd, AugPi, Rotem Dan, Mikue, Pizza Puzzle, Hashar, Revolver, Charles Matthews, Dysprosia, Jitse Niesen, Selket, Markhurd, Grendelkhan, Robbot, Red-Wolf, Gandalf61, Henrygb, OmegaMan, Rebrane, Wikibot, Fuelbottle, Mandel, Guy Peters, Mfc, Centrx, Giftlite, Gene Ward Smith, Herbee, Monedula, Michael Devore, Siroxo, Bobblewik, Chowbok, Alexf, Beland, Smallstraw, Rdsmith4, APH, Histrion, Iantresman, Joyous!, Frenchwhale, Jh51681, CliffordEW, Zondor, Trevor MacInnis, Freakofnurture, Skal, Rich Farmbrough, Guanabot, Gadykozma, Rama, Paul August, ESkog, Andrejj, Blotwell, Jojit fb, Haham hanuka, Mdd, Ogress, Jumbuck, Msh210, Arthena, Diego Moya, Hippophaë, Mrholybrain, Caesura, Velella, Computerjoe, Freyr, Kenyon, Oleg Alexandrov, Mindmatrix, Kzollman, ^demon, MFH, Amikeco, KingsleyIdehen, Waldir, Palica, BD2412, Josh Parris, Sdornan, Salix alba, Brighterorange, FlavrSavr, Yamamoto Ichiro, FlaBot, Mathbot, Greg321, RexNL, Gurch, Ayla, Brendan Moody, BradBeattie, Chobot, YurikBot, SpikeJones, RobotE, Red Slash, KSmrq, RadioFan, NawlinWiki, Rick Norwood, DryaUnda, Jhinman, Brisvegas, Lt-wiki-bot, Xaxafrad, DmitriyV, Sardanaphalus, SmackBot, YellowMonkey, Melchoir, Jagged 85, BiT, Timotheus Canens, CrypticBacon, Gilliam, G O T R, IMacWin95, Bluebot, Keegan, Pimrietbroek, Oli Filth, Adam Lewis, Frap, SundarBot, Khoikhoi, Acepectif, Jiddisch, StephenMacmanus, Ruwanraj, Kalathalan, Bpeel, Via strass, SashatoBot, Wideangle, Goodnightmush, Asdfv, 16@r, Mets501, Citicat, Robertwb, Captainj, Vanisaac, Stifynsemons, DWarrior, Mikeliuk, Patchouli, Ninetyone, Timichal, MarsRover, Nilfanion, Equendil, Cydebot, Richardguk, Flowerpotman, Dr.enh, Tdvance, PlanetCoder, Dougweller, M a s, Daven200520, Emmett5, Kansas Sam, Thijs!bot, Pmagyar, Headbomb, AndresV, Marek69, Missvain, JustAGal, Dfrg.msc, Escarbot, Quintote, Edokter, Doktor Who, Danger, JAnDbot, Sangwinc, MER-C, The Transhumanist, 100110100, VoABot II, Mclay1, Akgupta, Redaktor, Schwarzbichler, CountingPine, Grape Soda, Ryeterrell, Seberle, David Eppstein, Spellmaster, JaGa, Khalid Mahmood, Misibacsi, MartinBot, Nono64, Roelvandijk, Uncle Dick, Katalaveno, Mathwhizx2, Merceris, The Transhumanist (AWB), Bobianite, Milogardner, Jazzbruce, Treisijs, Idioma-bot, Ottershrew, VolkovBot, JohnBlackburne, Am Fiosaigear, DoorsAjar, TXiKi-BoT, Anonymous Dissident, Ocolon, Yk Yk Yk, Meters, Synthebot, Enviroboy, Dmcq, AlleborgoBot, Symane, NHRHS2010, SieBot, Gerakibot, Keilana, Flyer22, LDCutter, Oxymoron83, Janfri, Nic bor, UKe-CH, ClueBot, PipepBot, Justin W Smith, The Thing That Should Not Be, Cliff, VsBot, Ignorance is strength, JRD RockS, Rejka, Dekanfari, Muro Bot, Workman63, Workman64, Workman65, Workman100, DumZiBoT, Fastily, Pichpich, WikHead, Addbot, Jojhutton, Olli Niemitalo, Istvánka, NjardarBot, MrOllie, LaaknorBot, CarsracBot, AndersBot, VASANTH S.N., Tide rolls, Lightbot, Jarble, Luckas-bot, Yobot, II MusLiM HyBRiD II, AnomieBOT, Jim1138, Nick UA, Materialscientist, Bob Burkhardt, Xqbot, Timir2, Capricorn42, Johnferrer, Isheden, Almabot, Novonium, GrouchoBot, Omnipaedista, Charvest, Aashaa, Schekinov Alexey Victorovich, Aaron Kauppi, FrescoBot, Tobby72, Aleksa Lukic, MacMed, Pinethicket, Niaz632, Martinvl, Shiva Khanal, MastiBot, SpaceFlight89, FoxBot, TobeBot, Yunshui, Lotje, Zvn, January, Joodeak, Reach Out to the Truth, RjwilmsiBot, Slon02, Mr. Anon515, EmausBot, John of Reading, WikitanvirBot, Lipsio, Dewritech, Gfuy, Fayimora, Savh,

AvicBot, PBS-AWB, Fæ, Simulations, Skipper per, D.Lazard, Lightbeamrider55, ChuispastonBot, Jordibuma, Strangely Real, ClueBot NG, Wcherowi, AnthonyNotes, Movses-bot, Quantamflux, Yster76, Frietjes, Braincricket, Lincoln Josh, Helpful Pixie Bot, Jerrydeanrsmith, Vagobot, Mrjohncummings, Furkaocean, DrTechDaddy, Brad7777, Cky2250, Helloimjustintimefordinner, Arcandam, Dexbot, Saehry, Lugia2453, Frosty, Konstantin.bay, Kevin12xd, JustAMuggle, NHCLS, 1canuckbuck, Ramanujan srinivasa, LateralMoraine, Galois2718, Whoppabang, Luram, Kylieh10 and Anonymous: 261

_ **Algebra** *Source:* http://en.wikipedia.org/wiki/Algebra?oldid=628789233 *Contributors:* AxelBoldt, Brion VIBBER, Mav, Bryan Derksen, Zundark, Tarquin, Ed Poor, Andre Engels, Oliverkroll, David spector, Heron, Netesq, Stevertigo, Spiff, Frecklefoot, Boud, Xavic69, Michael Hardy, Bertfried Fauser, DopefishJustin, Dominus, Nixdorf, Dineshjk, Chinju, Dcljr, Sannse, TakuyaMurata, Karada, Delirium, Egil, Cyp, Haakon, J'raxis, Samuelsen, Angela, Dietary Fiber, AugPi, Poor Yorick, Nikai, Andres, Mxn, Pizza Puzzle, Charles Matthews, Dcoetzee, Dysprosia, Jitse Niesen, Doug Hogg, Quoth, Secretlondon, Aenar, Robbot, Fredrik, Chris 73, R3m0t, Romanm, Rebrane, Wikibot, Casito, Pengo, Giftlite, Gene Ward Smith, Wikilibrarian, Fastfission, Ayman, Ido50, Everyking, No Guru, Brona, Michael Devore, Gilgamesh, Guanaco, Ptk, Siroxo, Python eggs, Bobblewik, Wmahan, Stinerman, Vivero, Utcursch, Fred Fury, Zarvok, Ryan524, Antandrus, Beland, MarkSweep, Yafujifide, Jossi, Vina, Rdsmith4, Mzajac, Gauss, Sam Hocevar, Neonstarlight, Iantresman, Barnaby dawson, Stdarg, Corti, DanielCD, Rich Farmbrough, Batkins, Emilva, Dbachmann, Paul August, Bender235, ESkog, Zaslav, JoeSmack, Ben Standeven, Gauge, Brian0918, Tompw, El C, Lycurgus, Edward Z. Yang, Shanes, Triona, One-dimensional Tangent, Jlin, Causa sui, Bobo192, Meggar, Smalljim, Maurreen, Jordgubbe, La goutte de pluie, Nk, Nintendomon74, Obradovic Goran, HasharBot, Jumbuck, Danski14, Kuratowski's Ghost, Msh210, Gary, Anthony Appleyard, Hi332211-, Rgclegg, Chiapr, Riana, Mysdaao, Snowolf, Baron-Larf, Garzo, Docboat, Evil Monkey, Endersdouble, RainbowOfLight, Sciurinæ, Pwqn, Drbreznjev, Mattbrundage, Zereshk, Kazvorpal, Dan100, Oleg Alexandrov, DarTar, Snowmanmelting, Tbsmith, Boothy443, DealPete, Starblind, Woohookitty, Linas, Guardian of Light, Ma Baker, Ruud Koot, Jeff3000, Miss Madeline, Terence, Smmurphy, Coolfire276, Prashanthns, LinkTiger, Palica, Dysepsion, Graham87, BD2412, Qwertyus, Pranathi, Jclemens, Mayumashu, JVz, Bgohla, Vary, Amire80, Tangotango, Sdornan, Salix alba, MZMcBride, Kierah, Tdowling, Sohmc, Bfigura, Yamamoto Ichiro, Titoxd, FlaBot, Jeff Fries, Latka, Mathbot, Nivix, Dantecubed, RexNL, Mark J, Jrtayloriv, Didre, Ichudov, Fresheneesz, Alphachimp, Malhonen, SpectrumDT, Imnotminkus, MichaelCaricofe, Jidan, Chobot,

Dark Dragon, The Rambling Man, YurikBot, Angus Lepper, I need a name, Hairy Dude, Vertaloni, RussBot, Michael Slone, Grubber, Manop, Gaius Cornelius, Nis81, Shanel, NawlinWiki, Rick Norwood, Wiki alf, Lesotho, Orioneight, Badagnani, Trovatore, ONEder Boy, Cleared as filed, Irishguy, Abb3w, Aldux, Matthew0028, Farmanesh, Crasshopper, Zwobot, Dbfirs, BOT-Superzerocool, DeadEyeArrow, Bota47, Aristotle2600, Alpha 4615, WCX, Ms2ger, Googl, Tista, E-Dogg, Johndburger, Lt-wiki-bot, Ninly, Chase me ladies, I'm the Cavalry, Closedmouth, Sean Whitton, GraemeL, SidJ, Mikus, Shastra, TLSuda, Meegs, Benandorsqueaks, Stumps, That Guy, From That Show!, Sardanaphalus, SmackBot, FocalPoint, YellowMonkey, Selfworm, Incnis Mrsi, Rose Garden, InverseHypercube, KnowledgeOf-Self, AndyZ, Masonprof, Fheo, KocjoBot, Jagged 85, Onebravemonkey, Swerdnaneb, HalfShadow, Sebesta, Commander Keane bot, PeterSymonds, Macintosh User, Gilliam, G O T R, Llanowan, Grokmoo, Bluebot, TheDarkArchon, DVader, Miquonranger03, MalafayaBot, Domthedude001, Silly rabbit, SchfiftyThree, Akanemoto, Kevin Ryde, Go for it!, Baa, Yanksox, Weierstraß, Can't sleep, clown will eat me, HLwiKi, Gnp, Yorick8080, TheKMan, Edivorce, Arab Hafez, Khoikhoi, DavidStern, Nibuod, Jiddisch, SP angel 1, Sljaxon, Ruwanraj, RayGates, Kukini, Ugur Basak Bot, SashatoBot, Lambiam, Mukadderat, ArglebargleIV, AThing, Kuru, John, Ascend, Cronholm144, Ubertoaster, Sayama, NongBot, MarkSutton, VooDooChild, Slakr, Kirbytime, Mr Stephen, Unknownroad4, Mets501, Mathsci, Ryulong, Tonybrown100, Reccanboy, Miladsafa, Squirepants101, NinjaCharlie, Keycard, Iridescent, Fatima irshad, Nilamdoc, Mas0090, Westfall, Mrdthree, Cbrown1023, Fsotrain09, Tony Fox, Tawkerbot2, AbsolutDan, Tifego, Geomprof, CmdrObot, Fumblebruschi, Van helsing, Scohoust, Xanthoxyl, Amir1981, Circuit dreamer, GHe, THF, Dgw, MarsRover, MrFish, Captmog, AndrewHowse, Yaris678, Doctormatt, Cydebot, Kanags, Reywas92, SyntaxError55, Gogo Dodo, Karafias, Tawkerbot4, DumbBOT, Ameliorate!, Kozuch, Daven200520, Xantharius, Starship Trooper, JodyB, Thijs!bot, Epbr123, LeeG, Kilva, Yboord028, Wmgan, SilverSurf, Cowsnatcher27, Oliver202, Amitprabhakar, Stiltskin, RobHar, Benstam, EdJohnston, Platinum Knight, Klausness, JohnSteinbeck, Natalie Erin, Icep, Escarbot, KrakatoaKatie, AntiVandalBot, Luna Santin, Widefox, KopiKat, Mk*, Hagrinas, Mack2, Math Teacher, Sbarnard, Jaredroberts, Gökhan, Res2216firestar, JAnDbot, Sangwinc, The Transhumanist, Instinct, Kingnosis, Hut 8.5, YK Times, Howsthatfordamage25, Acroterion, Mardavich, Lost-theory, Meeples, Magioladitis, Bongwarrior, VoABot II, CiteCop, JamesBWatson, Think outside the box, Dorum, WODUP, Balloonguy, Jmartinsson, SwiftBot, WhatamIdoing, Bcherkas, Adrian J. Hunter, 28421u2232nfenfcenc, Aziz1005, Canyouhearmenow, DerHexer, JaGa, Edward321, Fantastic4boy, Hbent, Bbowenjr, Palestine48, Cliff smith, MartinBot, EyeSerene, BetBot, Gnayshkr3020, Kaspg, Trickmyster, Alfred Legrand, Ccmolik, Thefutureschannel, AlphaEta, J.delanoy, The dark lord trombonator, Trusilver, Bogey97, Maurice Carbonaro, LarsTheViking, Tcop, Lantonov, RedKlonoa, Ncmvocalist, Nymphomaniac, Nemo bis, Samtheboy, TylerDorsanoTJmans, Gurchzilla, AntiSpamBot, (jarbarf), The Transhumanist (AWB), Arms & Hearts, NewEnglandYankee, SJP, Imawsome, Fullmetal123321, Shoessss, DavidCBryant, Burzmali, Ross Fraser, Gtg204y, Useight, Xiahou, Idarin, Lights, Nikthestunned, PeaceNT, Deor, VolkovBot, Ramah71, ABF, JGHowes, A.Ou, Jeff G., Jennavecia, JohnBlackburne, LokiClock, AlnoktaBOT, DancingMan, Nousernamesleft, TXiKiBoT, BuickCenturyDriver, Hacker300, Miranda, Rei-bot, Anonymous Dissident, Vanished user ikijeirw34iuaeolaseriffic, Jddphd, Zimbardo Cookie Experiment, LeaveSleaves, Aiowe ;nhg8ohegpo8haeog, Bored461, Geometry guy, Prb4, MearsMan, CrackdownSamSung, Fivelittlemonkeys, Nery00, Kmhkmh, Blurpeace, Finngall, Commator, Synthebot, Burntsauce, Dib14, Monty845, Twooars, AlleborgoBot, Symane, Jirt, ERHSHaxor, Ken Kuniyuki, GirasoleDE, Demmy100, Markdraper, GoonerDP, SieBot, Ivan Štambuk, Dusti, Strawberryx7, VVVBot, Viskonsas, Caltas, Uhomako, Triwbe, Waynebrady 77, Cdthedude, Keilana, Oda Mari, Ezh, Paolo.dL, Ur ugly face, JuanFox, Oxymoron83, Editor91, Weston.pace, JackSchmidt, RSStockdale, Hornetchuck, Skull eyes, Macy, Wormdoggy, Hippie Metalhead, Vice regent, Pfffff, Eriksensei, Raptorz, Randomblue, Mountainofdoom, 3rdAlcove, Rdobet, Artistspace222, Ncfriel, Jamesfranklingresham, Madden king, ClueBot, Artichoker, PipepBot, C1932, TrigWorks, Danaghafari, Supertouch, Mr Rofl, Xenon54, DaveReece, Monkeyboy1268, Neverquick, BlueAmethyst, Inflate, Puchiko, Suliu, MindstormsKid, Aua, DragonBot, Alexbot, Aligebra, Erikp 92, Francine 21, Robert impey, StarcraftBuff, Edgesnext, Leonard^Bloom, Mumia-w-18, Zaharous, Vv22, Cenarium, TJTheRealist36, Singhalawap, Hans Adler, Airplanesrus, Diaa abdelmoneim, Rmsgrey, Calor, Iraq-Irak, BANMENOWPLZ, Thingg, Franklin.vp, Puranjan Dev, Heatmizer323, BlueDevil, NERIC-Security, Masterjake1295, Arc112, Sgunteratparamus. k12.nj.us, Pichpich, WikHead, Makmac22, RyanCross, Markerdancerboy, Nabuchadnessar, Flatbush756, Addbot, Morriswa, Metagraph, SpellingBot, Aaronthegr8, Kpthanuman, CanadianLinuxUser, Eleinax0x0, WFPM, Sweetkidjoey, Bumdah, Joel Butts, CarsracBot, Vega2, Z. Patterson, Tumadreescaliente, Saycgthm999, AgadaUrbanit, Youlittlehandsomeguy, VASANTH S.N., Gail, Zorrobot, TeH nOmInAtOr, Jarble, Kamek900, Jim, Undeadwarlock, Luckas-bot, Yobot, 2D, Senator Palpatine, Kan8eDie, THEN WHO WAS PHONE?, Dale S. Satre, AnomieBOT, Manouchehr78, Rjanag, Galoubet, JackieBot, AllUrBseRB3l0ngToM3, Nick UA, Thebrat132, The High Fin Sperm Whale, Citation bot, Lolololh4x, Stargazer84, Dagger20, Xqbot, TinucherianBot II, Timir2, S56k, Drilnoth, Gigemag76, Poetaris, Bubbles16 22, Xedret, Srich32977, GrouchoBot, Gott wisst, GhalyBot, Zytroft, GliderMaven, Tobby72, Citation bot 1, Iterate, Ebony Jackson, Historyscholar123, 1to0to-1, Jujutacular, AustralianMelodrama, Gamewizard71, FoxBot, TobeBot, Buddy23Lee, Alokprasad, JLincoln, Diannaa, RjwilmsiBot, EmausBot, John of Reading, GoingBatty, ZxxZxxZ, AOC25, Quondum, D.Lazard, EdoBot, Wcherowi, JimsMaher, SusikMkr, Yourmomblah, Rurik the Varangian, Lincoln Josh, Helpful Pixie Bot, Mrjohn770

cummings, Northamerica1000, Mysterytrey, ShellPond, Teika kazura, Brad7777, DrLewisphd, Cengime, Class Avesta, Mark L MacDonald, Hmainsbot1, Jamietwells, Tary123, YvelinesFrance, William2001, HistoryofIran, Wikifan2744, Jackmcbarn, Omphalosskeptic, Raymond37, JMP EAX and Anonymous: 608

_ **Irrational number** *Source:* http://en.wikipedia.org/wiki/Irrational%20number?oldid=633673343 *Contributors:* AxelBoldt, Zundark, The Anome, Jan Hidders, Youssefsan, Christian List, Toby Bartels, PierreAbbat, Mjb, BL, B4hand, Edward, Bdesham, Michael Hardy, Chuck SMITH, TakuyaMurata, Minesweeper, Card, UserGoogol, Panoramix, Mxn, Pizza Puzzle, Charles Matthews, Nohat, Dysprosia, Furrykef, Hyacinth, AndrewKepert, Quoth, Daran, Robert2957, Robbot, Fredrik, Schutz, Romanm, Henrygb, Saforrest, Tobias Bergemann, Psb777, Giftlite, Gene Ward Smith, Recentchanges, Ævar Arnfjörð Bjarmason, JeffBobFrank, Pne, LiDaobing, Antandrus, Zfr, Sam Hocevar, Starx, Gscshoyru, Kutulu, Spiffy sperry, Discospinster, Guanabot, Paul August, Dmr2, ESkog, Ben Standeven, El C, JHarris, Grue, Wood Thrush, Smalljim, SpeedyGonsales, Obradovic Goran, C-4, Mdd, Alansohn, Andrewpmk, Crobzub, Falcorian, Oleg Alexandrov, Nuno Tavares, Kelly Martin, Simetrical, Woohookitty, Shreevatsa, Igny, LOL, Chaosmotic, Mreult, Howabout1, GregorB, Sudhee26, Dysepsion, Yurik, Josh Parris, Salix alba, Alexb@cut-the-knot.com, Ysangkok, Nihiltres, RexNL, Gurch, Fresheneesz, Glenn L, JonathanFreed, Haonhien, CiaPan, Chobot, Roboto de Ajvol, YurikBot, Wavelength, Hairy Dude, KSmrq, Grubber, NawlinWiki, Dialectric, Schmock, DYLAN LENNON, Moe Epsilon, Luke-Jr, Bota47, Crisco 1492, FF2010, Lt-wiki-bot, Arthur Rubin, Jogers, Shawnc, Katieh5584, Finell, AndrewWTaylor, SmackBot, RDBury, FocalPoint, Selfworm, Adam majewski, Diggers2004, Melchoir, Shoy, Jagged 85, AustinKnight, BiT, Gilliam, Dauto, Raja Hussain, MalafayaBot, Akanemoto, DHN-bot, Can't sleep, clown will eat me, Dharmabum420, Khukri, Flyingspuds, Bowlhover, RandomP, Kleuske, Fuzzypeg, DMacks, Bidabadi, Drunken Pirate, John Reid, Thejerm, SashatoBot, Lambiam, Johncatsoulis, Maverick starstrider, Kuru, Khazar, Btg2290, Bjankuloski06en, IronGargoyle, Loadmaster, Androl, Mr Stephen, EdC, Jdahm, Quaeler, BranStark, Madmath789, Newone, Catherineyronwode, Zero sharp, Captainj, Az1568, JForget, Vaughan Pratt, CRGreathouse, Dycedarg, Smiloid, Retrovirus, SuperMidget, Odie5533, Christian75, Robertinventor, Xantharius, Thijs!bot, Epbr123, Koeplinger, AbcXyz, Natalie Erin, AntiVandalBot, Seaphoto, Blue Tie, Opelio, Shirt58, Dylan Lake, Rsocol, JAnDbot, CosineKitty, Sonicsuns, Dsp13, T. Moitie, Rothorpe, Acroterion, Xororaz, Connormah, LordFoom, VoABot II, Transcendence, Dwineman, Catgut, Loonymonkey, DerHexer, Kayau, Hdt83, MartinBot, R'n'B, Pbroks13, Nono64, J.delanoy, Numbo3, Chrisfrom-Houston, TimDoster, LordAnubisBOT, Trumpet marietta 45750, Krasniy, Beremiz, Potatoswatter, Cometstyles, STBotD, DorganBot,

CardinalDan, Pgb23, VolkovBot, TallNapoleon, Ktalon, Barneca, TXiKiBoT, AstroWiki, Shahkent, Plclark, Martin451, Mouse is back, Wolfrock, Enviroboy, Bilal.alsallakh, Dmcq, Demmy, GirasoleDE, SieBot, Jsc83, Caltas, S711, Keilana, Flyer22, Tiptoety, Masgatotkaca, Janopus, Momo san, Oxymoron83, Faradayplank, SimonTrew, Manway, Macy, Hatster301, CharlesGillingham, Randomblue, Escape Orbit, Athenean, Atif.t2, ClueBot, PipepBot, The Thing That Should Not Be, Vonbontee, Cptmurdok, Drmies, Mild Bill Hiccup, Skäpperöd, Alexbot, Jusdafax, SchreiberBike, Aitias, Johnuniq, Thymaridas, Nitsua60, Feinoha, Theonlydavewilliams, Beach drifter, Jaan Vajakas, D.M. from Ukraine, Addbot, DOI bot, Betterusername, Captain-tucker, Ronhjones, Glane23, LinkFA-Bot, Tide rolls, Jarble, Luckas-bot, Yobot, 2D, Fraggle81, Nallimbot, A Stop at Willoughby, გიგა, AnomieBOT, Jim1138, 9258fahsflkh917fas, Sz-iwbot, MaybeJesusMaybeNot, Materialscientist, Citation bot, ArthurBot, Xqbot, TinucherianBot II, Zoso96, RJGray, Isheden, ProtectionTaggingBot, VladimirReshetnikov, Wheelingcrows, 78.26, Joshalex88, Universalss, Brfettig, Stirred-not-shaken, Amazing Steve, Tobby72, Ulrich Müller, Worldrimroamer, Citation bot 1, Tkuvho, MacMed, Pinethicket, HRoestBot, Supreme Deliciousness, Franci.cariati, MastiBot, Steel Tortoise, Juanr2099, White Shadows, Double sharp, TobeBot, Popovvk, Vrenator, Duoduoduo, Setitup, Danny5000, DASHBot, Whywhenwhohow, EmausBot, Sir Arthur Williams, Njaikrishna, Ibbn, Khalidmathematics, Wham Bam Rock II, Slightsmile, Tommy2010, Slawekb, Thecheesykid, JSquish, Traxs7, TestSubject528491, Wayne Slam, Shreeradha, Paulmiko, IGeMiNix, L Kensington, Colm Keogh, Donner60, Ego White Tray, Orange Suede Sofa, Some1new4ya, ResearchRave, ClueBot NG, Wcherowi, Kawaikx15, Etioquaesitor, Masssly, Widr, JayShaw TT, Kavir1698, Iatemath, Helpful Pixie Bot, BG19bot, Hallows AG, ElphiBot, Dan653, Sbarner, CitationCleanerBot, Snow Blizzard, Cliff12345, Anbu121, ChrisGualtieri, Dexbot, Lugia2453, N8 Toe, Rkaup, Nitpicking polish, Micahqueen, Unlikelyuser, Zararsiz, X catching clouds x, B14709, Ginsuloft, OtakuWiz, Tombycrap and Anonymous: 465

_ **Mathematical induction** *Source:* http://en.wikipedia.org/wiki/Mathematical%20induction?oldid=631273770 *Contributors:* AxelBoldt, Bryan Derksen, Zundark, Tarquin, Jan Hidders, XJaM, Mjb, Youandme, Bdesham, Michael Hardy, Wshun, David Martland, Dominus, Kku, Meekohi, Sannse, TakuyaMurata, Ahoerstemeier, Snoyes, Александър, Charles Matthews, Dcoetzee, Sbloch, Dysprosia, Jitse Niesen, Selket, Markhurd, Furrykef, Sabbut, Bloodshedder, .mau., Aenar, Robbot, Iwpg, Altenmann, Gandalf61, Bruceq, Aetheling, Ruakh, Tobias Bergemann, Tosha, Giftlite, Nickptar, Fintor, Peter Kwok, Kousu, PhotoBox, Lucidish, Shipmaster, Ericamick, Paul August, Bender235, Elwikipedista, Zenohockey, CXI, Spoon!, Aplusbi, NetBot, Obradovic Goran, Polylerus, Jumbuck, Msh210, Arthena, Dirac1933, Bsadowski1, Oleg Alexandrov, Zntrip, Mindmatrix, StradivariusTV, Acone, Drostie, Ruud Koot, Mpatel, Sartak, M412k, Ryan Reich, Palica, Mandarax, Chun-hian, JIP, Rjwilmsi, SMC, FlaBot, Maitch, Mathbot, Chobot, YurikBot, Borgx, Hairy Dude, Russ-Bot, Hyad, KSmrq, Debroglie, TEB728, Trovatore, Nick, Jstrater, Schmock, Pyroclastic, Froth, Gadget850, Bota47, Addps4cat, Mgnbar, Brisvegas, Jwissick, Reyk, SmackBot, RDBury, Selfworm, Melchoir, Pgk, Bomac, Pokipsy76, Jagged 85, Chris the speller, Fintler, JCSantos, Trebor, Nbarth, Nixeagle, ConMan, Radagast83, Acepectif, Lambiam, Wvbailey, J. Finkelstein, Pliny, Mets501, MissingNOOO, Dreftymac, Blackhawk charlie2003, Zero sharp, JRSpriggs, CRGreathouse, Wafulz, CBM, WeggeBot, Gregbard, Cydebot, Zahlentheorie, JFreeman, BillWeiss, Headbomb, Pcbene, AntiVandalBot, Vic226, Gramby, Sekky, Alphachimpbot, MER-C, Wisnuops, David Eppstein, DerHexer, SquidSK, Stephenchou0722, MartinBot, JCraw, Maurice Carbonaro, TomS TDotO, Paulecoyote, OttoMäkelä, It Is Me Here, LordAnubisBOT, Tparameter, Policron, DavidCBryant, Omegamormegil, Fbarton, VolkovBot, AlnoktaBOT, Jimmaths, TXiKiBoT, Anonymous Dissident, Philogo, PaulTanenbaum, Gilisa, Pboulus, SieBot, Paradoctor, Alexsmail, MiNombreDeGuerra, Ellamosi, Ngriffeth, DesolateReality, Altzinn, Amahoney, AutoFire, Classicalecon, Gjakovit, ClueBot, Justin W Smith, The Thing That Should Not Be, Jdgilbey, Mathemajor, J8079s, DragonBot, Muro Bot, Tired time, SoxBot III, Dyhan81, Brentsmith101, Kbdankbot, Addbot, DOI bot, Riyuky, CanadianLinuxUser, NjardarBot, MrVanBot, Chamal N, CarsracBot, LinkFA-Bot, Lightbot, Dminkovsky, Jarble, Estudiarme, Jasonschock, AnomieBOT, 1exec1, MattTait, Citation bot, ArthurBot, TaySpace, Poetaris, DSisyphBot, J04n, RibotBOT, D'ohBot, Citation bot 1, Tkuvho, Kiefer.Wolfowitz, Unnachamois, Bozo the bear, Gabrielgmendonca, Mrhota, Vrenator, EdEveridge, Duoduoduo, One15969, Jowa fan, Whywhenwhohow, EmausBot, Vasanthloganathan, Slawekb, Ebrambot, Noodleki, ClueBot NG, Moneysorter, Tideflat, Meingbg, Helpful Pixie Bot, DBigXray, FutureTrillionaire, Alpert1, CitationCleanerBot, Glacialfox, Justincheng12345-bot, Dexbot, Jochen Burghardt, I am One of Many, TI. Gracchus, Glasbys, Beneficii, Monkbot, Swashski, Dsacf1234567 and Anonymous: 224

_ **Arithmetic progression** *Source:* http://en.wikipedia.org/wiki/Arithmetic%20progression?oldid=629533732 *Contributors:* AxelBoldt, Tarquin, William Avery, Patrick, Chas zzz brown, Michael Hardy, UserGoogol, Andres, Charles Matthews, Dcoetzee, Kbk, Hyacinth, McKay, Fredrik, Altenmann, Nikitadanilov, Giftlite, Jackol, CryptoDerk, Tbjablin, Discospinster, Murtasa, Goochelaar, Bobo192, Robotje, C S, .:Ajvol:., Haham hanuka, Msh210, Rh, Snowolf, Terrible tony, Oleg Alexandrov, Hyperfusion, Sanjaymjoshi, Shreevatsa, GregorB, Cornince, Tbone, Salvatore Ingala, Chobot, Bgwhite, Siddhant, YurikBot, Wavelength, Hairy Dude, Icedemon, KSmrq, Bill-on-the-Hill, Petter Strandmark, Nick, Googl, Zzuuzz, Closedmouth, RDBury, Selfworm, InverseHypercube, Melchoir, Ohnoitsjamie, PrimeHunter, Octahedron80, DHN-bot, Rrburke, Sebo.PL, Attys, Catapult, Aleenf1, Ckatz, 16@r, Euphrates, MTSbot, ST47,

Goldencako, Marek69, John254, Majorly, JAnDbot, Ricardo sandoval, VoABot II, JNW, David Eppstein, Nguyễn Hữu Dung, Ulfalizer, DarwinPeacock, Daniel5Ko, Policron, Milogardner, Jamesontai, Steel1943, VolkovBot, Johan1298, TXiKiBoT, Anonymous Dissident, Ravig sagi, SieBot, Rlendog, Flyer22, Masgatotkaca, Harry, Xhackeranywhere, Denisarona, Loren.wilton, Tai Chi Tech, ClueBot, DR23, Mild Bill Hiccup, Sohail555, DumZiBoT, XLinkBot, Gonzonoir, Kal-El-Bot, Addbot, Macarse, Delaszk, SamatBot, Numbo3-bot, Tide rolls, Legobot, Luckas-bot, Yobot, TaBOT-zerem, Mmxx, Writer on wiki, გიგა, AnomieBOT, Ciphers, Rubinbot, Jim1138, 9258fahsflkh917fas, Flewis, Materialscientist, Citation bot, Donanayath, Xqbot, Öncel Acar, RibotBOT, SassoBot, LuisVillegas, Sophus Bie, A.amitkumar, Dougofborg, RTFVerterra, Robo37, Pinethicket, SpaceFlight89, RandomStringOfCharacters, Dude1818, FoxBot, DixonDBot, Duoduoduo, KurtSchwitters, Jowa fan, EmausBot, Maschen, Nikunj Pandya, ClueBot NG, Widr, Princetct.007, Krenair, Walrus068, Kingsbrook, Sparkie82, Brad7777, Williamdemeo, Arpitkjain, Justincheng12345-bot, Scientific Alan 2, Jochen Burghardt, Razibot, Vshender, Zorch713, Chacho39, AtoiyonTayib and Anonymous: 213

_ **Binomial theorem** *Source:* http://en.wikipedia.org/wiki/Binomial%20theorem?oldid=632378487 *Contributors:* AxelBoldt, 0, SimonP, Michael Hardy, TakuyaMurata, Eric119, Poor Yorick, Andres, Charles Matthews, Dfeuer, Carbuncle, JorgeGG, Robbot, Cdang, Robinh, Tobias Bergemann, Giftlite, Mikez, Lupin, MSGJ, Duncharris, Antandrus, MarkSweep, Wroscel, Adashiel, Shahab, Discospinster, Rich Farmbrough, .:Ajvol:., SpeedyGonsales, La goutte de pluie, Nk, Haham hanuka, Alansohn, Borisblue, Pion, Blaxthos, Kenyon, Oleg Alexandrov, Matevzk, Tbsmith, Linas, Kbdank71, Sjö, R.e.b., Mohawkjohn, MarnetteD, Nomet, Joe056, RexNL, Gurch, Pevernagie, Sodin, Glenn L, Salvatore Ingala, Chobot, YurikBot, Wavelength, Michael Slone, Wiki alf, Saric, Arthur Rubin, Gesslein, Bo Jacoby, Zvika, SmackBot, BeteNoir, InverseHypercube, Jagged 85, BiT, Gilliam, Skizzik, Nbarth, Bsodmike, Rrburke, LouScheffer, -Ozone-, Bombshell, Acdx, Vina-iwbot, Zchenyu, Cronholm144, MvH, Jim.belk, Loadmaster, Mets501, ILovePlankton, Iridescent, Bharatveer, TNeloms, JForget, CRGreathouse, Stebulus, Gregbard, MC10, Goldencako, Verdy p, RickDC, Thijs!bot, Opabinia regalis, Nonagonal Spider, Jojan, Urdutext, Crabula, JAnDbot, MER-C, RebelRobot, Charibdis, Jay Gatsby, Nyttend, SWNut, Myrizio, Akulo, Leyo, J.delanoy, Ulfalizer, Gombang, Warut, Policron, PTP2009, TopGun, VolkovBot, Pleasantville, Chuchogl, TXiKiBoT, The absolute real deal, Anonymous Dissident, JhsBot, Geometry guy, Madhero88, Kmhkmh, Dirkbb, Oceanwitch, Dmcq, PericlesofAthens, Quietbritishjim, Nirvana89, SieBot, Meldor, Caltas, Phorgan1, Keilana, Cpoynton, Cffk, Android Mouse Bot 3, AlanUS, DEMcAdams, FlamingSilmaril, ClueBot, Justin W Smith, The Thing That Should Not Be, Thetruthseer, PMDrive1061, Carusus, Jusdafax, Subash.chandran007, RMFan1, Kiensvay, Marc van Leeuwen, Kal-El-Bot, Badgernet, Chemninja, Addbot, Binary TSO, Nerak99, WardenWalk, CarsracBot, Hawk8103, Wings Upon My Feet, Zorrobot, Luckas-bot, TheSuave, Yobot, Marshall2389, THEN WHO

WAS PHONE?, KamikazeBot, Greg Holden 08, AmritasyaPutra, Kristen Eriksen, Rubinbot, Kingpin13, Giants27, Aj.sujit, Danno uk, ArthurBot, Gsmgm, FactSpewer, Xqbot, Poetaris, Molotron, Jmundo, NOrbeck, Omnipaedista, RibotBOT, Kloddant, Sławomir Biały, Ebony Jackson, Elockid, RedBot, Meaghan, Tcnuk, 777sms, DARTH SIDIOUS 2, RjwilmsiBot, EcneicsFlogCitanaf, EmausBot, KIMWOONGJI, Knight1993, Sbealing, Bineapple, AManWithNoPlan, Babababoshka, ChuispastonBot, DASHBotAV, ClueBot NG, Wcherowi, GSZaum, Jbr326, Delusion23, O.Koslowski, Minthellen, StandardizerII, Vagobot, Brad7777, 069952497a, Epicgenius, EditGuy101, Hollylilholly, WikiOjos, Wsvkforever, Peiffers, Nozitall, Bcalden and Anonymous: 303

_ **Proof by contradiction** *Source:* http://en.wikipedia.org/wiki/Proof%20by%20contradiction?oldid=629745038 *Contributors:* AxelBoldt, Magnus Manske, Lee Daniel Crocker, Vicki Rosenzweig, Mav, Zundark, The Anome, Tarquin, Larry Sanger, Andre Engels, Roadrunner, FvdP, B4hand, Patrick, Chas zzz brown, Michael Hardy, Oliver Pereira, DopefishJustin, Kidburla, Dominus, Dcljr, Skysmith, Andrewa, Scott, EdH, Ideyal, Revolver, Charles Matthews, Populus, Fibonacci, Leonariso, Robbot, Mohan ravichandran, Altenmann, Romanm, Chancemill, Sam Spade, Bkell, UtherSRG, Centrx, Giftlite, Barbara Shack, Everyking, Cortina, 20040302, Chrismear, Jason Quinn, Chameleon, Gubbubu, Toytoy, Michaelcarraher, Rdsmith4, Grossdomestic, Peter Kwok, Neale Monks, Lacrimosus, Hydrox, Roybb95, Paul August, Brian0918, DrewRobinson, Cretog8, Dungodung, Man vyi, Jonsafari, Nsaa, Rd232, Burn, Bart133, Omphaloscope, Out180, Chamaeleon, Axeman89, Kazvorpal, Harvestdancer, Mindmatrix, Rodrigo Rocha, Hdante, Graham87, BD2412, Chun-hian, Patrick Zanon, X1011, Mathbot, Markkbilbo, Mattman00000, Lemuel Gulliver, Glenn L, Chobot, Reetep, Algebraist, YurikBot, Severa, Zafiroblue05, Kerowren, Dkg11hu, Doctor Whom, TERdON, Bota47, Andersersej, MagicOgre, SV Resolution, PurplePlatypus, Seanjacksontc, SmackBot, Pgk, BiT, Wje, Psiphiorg, Qwasty, ScottForschler, Sampi, Nbarth, Kobayen, Simpsons contributor, Aerobird, Ioscius, Mhym, Byelf2007, Lambiam, Dbtfz, Pliny, Loodog, Robofish, Mgiganteus1, Grumpyyoungman01, Xiaphias, Dr.K., Emx, Ncosmob, Wafulz, CBM, Gregbard, Mattbuck, Cydebot, Shirulashem, MarcelLionheart, Epbr123, Kajisol, Pampas Cat, Flarity, CZeke, Hopiakuta, JAnDbot, Pedro, LookingGlass, NeighborTotoro, Electriceel, TheBusiness, JuanPaBJ16, Neonguru, AltiusBimm, Colincbn, Pmbcomm, Kesal, DorganBot, VolkovBot, Jimmaths, TXiKiBoT, Liko81, Melsaran, Synthebot, GlassFET, Katzmik, Arkwatem, The Evil Spartan, Luciengav, Marc van Leeuwen, SilvonenBot, NjardarBot, MrVanBot, CarsracBot, Nikie42, AgadaUrbanit, Tide rolls, Alexander.mitsos, Legobot, Yobot, Cflm001, Majestic-chimp, AnomieBOT, OpenFuture, RavShimon, Dave3457, LegendFSL, Duoduoduo, PetroniusArb, DexDor, Jdl22, Bastian964, Makecat, Mehdi, Aflyhorse, Helpful Pixie Bot, Vesta Zenobia, Vclam068, Jochen Burghardt, Password is DOB, Xin-Xin W. and Anonymous: 146

_ **Euclidean geometry** *Source:* http://en.wikipedia.org/wiki/Euclidean%20geometry?oldid=632544128 *Contributors:* AxelBoldt, Brion VIBBER, Zundark, The Anome, Gareth Owen, Youssefsan, Toby Bartels, SimonP, Boleslav Bobcik, Nealmcb, Michael Hardy, Nixdorf, Bcrowell, Sannse, Fwappler, Eric119, Angela, Александър, LittleDan, Rl, Charles Matthews, Guaka, Dino, Joshuabowman, Dysprosia, Kbk, Peregrine981, Bwefler, Merriam, Robbot, Mattblack82, Chancemill, Lowellian, Rholton, Meelar, Timrollpickering, Jleedev, Filemon, Tosha, Giftlite, Ananda, Scaevola, Lupin, Michael Devore, Mike40033, Ezhiki, Eequor, OverlordQ, Oscar, DragonflySixtyseven, Elroch, Icairns, Discospinster, Vsmith, Paul August, Chewie, El C, Rgdboer, Shanes, Bobo192, Func, Che090572, BrokenSegue, Rje, Msh210, Theaterfreak64, Borisblue, Demi, SteinbDJ, Agutie, Talkie tim, Bookandcoffee, Oleg Alexandrov, Zntrip, Simetrical, Woohookitty, Linas, Consequencefree, Jimbryho, Eelvex, JATerg, Dionyziz, Isnow, Noetica, Audiovideo, Gerbrant, DaveApter, Marudubshinki, Graham87, Magister Mathematicae, BD2412, Qwertyus, Rjwilmsi, Salix alba, Makaristos, Oo64eva, FlaBot, VKokielov, Mathbot, Chobot, Russmack, DVdm, Gwernol, The Rambling Man, YurikBot, Wavelength, Borgx, Darkstar949, Michael Slone, KSmrq, DanMS, RadioFan, Stephenb, B-Con, Joelr31, Daniel Mietchen, Raven4x4x, EEMIV, Cinik, PyroGamer, Tetracube, Saric, Bayerischermann, Closedmouth, Ray Chason, Vicarious, Badgettrg, Cjfsyntropy, Whouk, Finell, Sardanaphalus, SmackBot, Fanblade, Nihonjoe, Lestrade, Proficient, Jagged 85, Srnec, Gilliam, Mhss, Chris the speller, Autarch, Concerned cynic, Dreg743, Thumperward, Chesaguy, Rrburke, Anfernyjohnsun, Addshore, Grover cleveland, Akriasas, Astarael8, Q Hill, John Reid, Blahm, SashatoBot, Dannytee, MagnaMopus, Mets501, Newone, Aeons, JRSpriggs, Jh12, Stifynsemons, Vaughan Pratt, CRGreathouse, Dycedarg, Aherunar, Gregbard, Equendil, Flowerpotman, Frozenport, Headbomb, Davidhorman, Dawnseeker2000, Grouchy Chris, Quintote, Leevclarke, Exo Kopaka, Yancyfry jr, JAnDbot, MER-C, Thenub314, 100110100, Bongwarrior, Avjoska, Chevinki, Greg park avenue, Catgut, EagleFan, Seberle, DerHexer, Hbent, TheRanger, Simon perez2, Shaffer193, MartinBot, J.delanoy, Nev1, Leon math, Katalaveno, NewEnglandYankee, Policron, D4RK-L3G10N, WarFox, Pdcook, Gertrudethetramp, JohnBlackburne, Sxcrunner0402, Philip Trueman, TXiKiBoT, Hqb, Someguy1221, Anna Lincoln, Jackfork, Geometry guy, Blurpeace, Y, Falcon8765, Dmcq, Monty845, Arcfrk, Logan, ChrisMiddleton, SieBot, AS, Mod Herman, RJaguar3, JerrySteal, Flyer22, Wilson44691, Taemyr, Yerpo, Weston.pace, 3rdAlcove, Xiscobernal,

Mr. Granger, Martarius, ClueBot, Ideal gas equation, The Thing That Should Not Be, ArdClose, Supertouch, JuPitEer, Blanchardb, Brews ohare, Bercant, Domni, LieAfterLie, SoxBot III, Pichpich, Mifter, Addbot, Kongr43gpen, ILovePie91, Ironholds, Ahmedettaf, D0762, MrOllie, Pmod, Glane23, LinkFA-Bot, RoadieRich, Shukyking, Lightbot, Bermicourt, Luckas-bot, Yobot, Kartano, Truereplica, Ytiugibma, Newportm, AnomieBOT, Piano non troppo, BipedalP, Jeremiah-360, Materialscientist, Citation bot, Bean49, Ayda D, Xqbot, J04n, Omnipaedista, Charvest, IShadowed, In fact, Griffinofwales, Undsoweiter, Csk444, Louperibot, Tkuvho, Pinethicket, Ambarsande, Fama Clamosa, Standardfact, Reaper Eternal, Surabhijimii, DARTH SIDIOUS 2, The Utahraptor, RjwilmsiBot, Balph Eubank, DASHBot, EmausBot, Eekerz, Wham Bam Rock II, JSquish, Aeonx, D.Lazard, Md haris4u, Donner60, Jordanyoung17, Weimer, Thuytnguyen48, LM2000, ClueBot NG, Wcherowi, Johnblittle512, Jtoland1, Schuylerreid, SusikMkr, *half-moon* bubba, Mesoderm, 123Hedgehog456, MillingMachine, Helpful Pixie Bot, Vagobot, AvocatoBot, Mm32pc, Falkirks, NotWith, Brad7777, Morning Sunshine, Fylbecatulous, BattyBot, Riley Huntley, ChrisGualtieri, YFdyh-bot, Aszlisna, Harsh 2580, Dexbot, Jamesx12345, Dsumms, Ugog Nizdast, Mooseandbruce1, Glassbreaker, Jianhui67, Francois-Pier, MewisPLipkin, DarthRyn, EuclidGeometry, Dzmanto, Je.est.un.autre, Swagata Halder, Richik das, Loraof and Anonymous: 410

_ **Parallel postulate** *Source:* http://en.wikipedia.org/wiki/Parallel%20postulate?oldid=634158219 *Contributors:* The Anome, Tarquin, B4hand, Michael Hardy, Grizzly, Dcljr, Minesweeper, Charles Matthews, John Cross, Joshuabowman, Francs2000, Robbot, Lowellian, Stewartadcock, Filemon, Giftlite, Smjg, Gene Ward Smith, Herbee, Waltpohl, DragonflySixtyseven, Gauss, Icairns, Sonett72, Qef, TheObtuseAngleOfDoom, Mani1, Paul August, ESkog, Ben Standeven, Rgdboer, .:Ajvol:., Haham hanuka, Sligocki, Geckogrover, Ilse@, Kay Dekker, Oleg Alexandrov, Joriki, Linas, Ruud Koot, Dodiad, Someone42, Mathbot, John Z, Chobot, Wiki alf, Twin Bird, Alecmconroy, Bob Hu, Carlosguitar, SmackBot, RDBury, Lestrade, Jagged 85, Ruanperia, Gilliam, Skizzik, Bluebot, Jprg1966, DroEsperanto, DHN-bot, Nedlum, Ustadny, Ianmacm, Jon Awbrey, Chwech, SashatoBot, Libertyblues, TwistOfCain, CBM, Setsuden, Casper2k3, Myasuda, Gregbard, Ntsimp, Juansempere, Thijs!bot, AbcXyz, Flarity, Magioladitis, Laur2ro, Jukeboxlord, Master shepherd, PatriciaJH, Kunjaan, JohnBlackburne, Philip Trueman, TXiKiBoT, Anonymous Dissident, Finlux, Qxz, Ferengi, AlleborgoBot, NHRHS2010, SieBot, EoGuy, J8079s, Brews ohare, Euclidmathproject-wjhs, Quark-Bomb, Mikaey, Kyslyi, 51kwad, Addbot, DOI bot, Laaknor-Bot, ChenzwBot, Bob K31416, Lightbot, Uroboros, ВиКо, Luckas-bot, Disgustingangel, Dickdock, AnomieBOT, Rubinbot, Nifky?, Mkotrba2, Machodog, RibotBOT, Abuk SABUK, FrescoBot, Tobby72, Citation bot 1, MarcelB612, Vkgfx, TobeBot, DARTH SIDIOUS 2, AvicBot, ZéroBot, U+003F, ChuispastonBot, ClueBot NG, Wcherowi, Widr, Helpful Pixie Bot, Brad7777, BattyBot, Dexbot, Jbeyerl, ბზულა, YogiBuddhaTao69 and Anonymous: 89

_ **Set theory** *Source:* http://en.wikipedia.org/wiki/Set%20theory?oldid=630090454 *Contributors:* AxelBoldt, Bryan Derksen, Zundark, The Anome, Christian List, Toby Bartels, Enchanter, Michael Hardy, Karada, William M. Connolley, Plaudite, Andres, Evercat, Marcosantonio,

Dysprosia, ThomasStrohmann, HappyDog, Hyacinth, VeryVerily, Robbot, Fredrik, Peak, Romanm, Tobias Bergemann, Giftlite, Gene Ward Smith, Bogdanb, Mintleaf, Lethe, Siroxo, Python eggs, Andycjp, Crawdaddio, Sam Hocevar, Marcos, Mubor, David Sneek, EugeneZelenko, Discospinster, Zaheen, Vsmith, Mani1, Harriv, Paul August, Bender235, *drew, Rgdboer, Randall Holmes, Mike Schwartz, Giraffedata, Obradovic Goran, Msh210, Tablizer, Pinar, Max Naylor, Dirac1933, Itsmine, Feezo, Slgrandson, Chun-hian, Mendaliv, Search4Lancer, Rjwilmsi, Koavf, Salix alba, R.e.b., JonnyR, Gurch, BMF81, Chobot, Pip2andahalf, RussBot, CarlHewitt, Rick Norwood, Joth, Dysmorodrepanis, Trovatore, PrologFan, Jpbowen, Crasshopper, Wheelybrook, Haemo, Googl, Arthur Rubin, Modify, Sardanaphalus, SmackBot, Unschool, Jagged 85, Brick Thrower, Skizzik, Dugas, Jprg1966, MalafayaBot, Darth Panda, Aremith, Nick Levine, Glloq, KevM, Lesnail, Grover cleveland, Jiddisch, Chrylis, Wvbailey, Rundquist, Bjankuloski06en, Jim.belk, Dfass, Mr. Vernon, Loadmaster, Mets501, Iridescent, Quantum Burrito, Clarityfiend, Tophtucker, RekishiEJ, Supertigerman, Mjohnrussell, Spdegabrielle, CRGreathouse, CBM, MarsRover, Gregbard, Dragonflare82, Thijs!bot, J. W. Love, RichardVeryard, Klausness, Urdutext, Escarbot, AntiVandalBot, Emeraldcityserendipity, LibLord, Salgueiro, Perelaar, Knotwork, JAnDbot, MER-C, Bongwarrior, Soulbot, Animum, David Eppstein, Kope, Khalid Mahmood, Musictheorist, Pharaoh of the Wizards, Maurice Carbonaro, NewEnglandYankee, Policron, Cometstyles, Tellerman, DorganBot, Treisijs, Idioma-bot, Alan U. Kennington, ABF, JohnBlackburne, LokiClock, AlnoktaBOT, TXiKiBoT, Red Act, Rei-bot, Anonymous Dissident, Digby Tantrum, Kmitchell19, Synthebot, VanishedUserABC, AlleborgoBot, Jtarr, YohanN7, Dogah, SieBot, BotMultichill, Gerakibot, Viskonsas, Disneyfreek, Mnlkpo, Billyg, Yoda of Borg, CBM2, Neurophysics, B J Bradford, ClueBot, Smithpith, Cliff, DragonBot, BearMachine, Cenarium, Njardarlogar, 7&6=thirteen, Hans Adler, Palnot, BodhisattvaBot, Jin-Jian, Multipundit, Deineka, Addbot, Manuel Trujillo Berges, Blake de doosten, Betterusername, Renamed user 5, Topology Expert, Ronhjones, NjardarBot, MrOllie, Download, Dyaa, Ozob, Bob K31416, Numbo3-bot, Tide rolls, TeH nOmInAtOr, Rrmsjp, DaveChild, Ben Ben, Legobot, Tartarus, Yobot, Fraggle81, TaBOT-zerem, Pcap, Synchronism, Galoubet, Faizulhasan, OllieFury, Tbvdm, Xqbot, RJGray, Qualitydemise, Solphusion, Point-set topologist, WissensDürster, La Mejor Ratonera, IShadowed, Red van man, Joxemai, Bekus, FrescoBot, Tobby72, Mark Renier, Dagme, Citation bot 1, FirmBenevolence, Tkuvho, Boulaur, Jschnur, Foobarnix, MPeterHenry, JumpDiscont, DARTH SIDIOUS 2, EmausBot, Set theorist, Jmencisom, Kimartz, Wikipelli, Wayne Slam, Computationalverb, ChuispastonBot, ClueBot NG, SusikMkr, Movses-bot, The Master of Mayhem, MerlIwBot, Ftonti, Helpful Pixie Bot, MKar, Hallows AG, Xosé Antonio, Ingmar.lippert, Knwlgc, YatharthROCK, Brad7777, Sofia karampataki, Spasoev, Justincheng12345-bot, Alfasst, Deltahedron, Nandanchoudhary05, Mark viking, Flimflam97, William2001, Tentinator, NeapleBerlina, SakeUPenn, K401sTL3, Mario Castelán Castro, Galileo9 and Anonymous: 207

_ **Non-Euclidean geometry** *Source:* http://en.wikipedia.org/wiki/Non-Euclidean%20geometry?oldid=633094563 *Contributors:* AxelBoldt, Uriyan, AstroNomer, Gareth Owen, Deb, Formulax, Michael Hardy, Llywrch, Kku, MartinHarper, Deadstar, Bcrowell, Chinju, Sannse, Looxix, Evercat, Revolver, Dave1999, Charles Matthews, Ww, Joshuabowman, Dysprosia, Populus, Fibonacci, Robert2957, Robbot, RedWolf, Gandalf61, Henrygb, Rholton, Auric, Lzur, Diberri, Tosha, Connelly, Giftlite, Gene Ward Smith, Mintleaf, BenFrantzDale, Lupin, Noone, Waltpohl, Sidar, Jackol, Alexf, Dmaftei, Tomruen, Icairns, Esquire, Metahacker, Mike Rosoft, Helohe, Rich Farmbrough, Vsmith, Bender235, Elwikipedista, Davidw, BenjBot, Rgdboer, RoyBoy, Mentatus, C S, Foobaz, Kevin Lamoreau, Msh210, Alansohn, Proteus71, ChristopherWillis, GenkiNeko, Wtmitchell, Ascetic, Falcorian, Oleg Alexandrov, Velho, OwenX, Woohookitty, Graham87, BD2412, Rjwilmsi, FAR, Koavf, Eyu100, Wragge, FlaBot, Chobot, Bgwhite, Siddhant, YurikBot, Sceptre, Hairy Dude, Mathlaura, Jpetry, CarlHewitt, NawlinWiki, Mipadi, Grafen, Predawn, 2over0, Zzuuzz, SFGiants, Nentuaby, Vicarious, Badgettrg, Geoffrey.landis, Spliffy, Cjfsyntropy, SkerHawx, Finell, Sardanaphalus, SmackBot, Lestrade, Incnis Mrsi, Jagged 85, Jipcy, BiT, Commander Keane bot, Ohnoitsjamie, Chris the speller, TimBentley, Nbarth, DHN-bot, Berland, Whpq, Taggart Transcontinental, Michaelrccurtis, Byelf2007, Battlemonk, Nharipra, Ckatz, Loadmaster, Korovioff, Ryulong, Tsolosmi, K, Clarityfiend, Newone, Bonti, Aeons, Phoenixrod, JRSpriggs, CRGreathouse, Mattbr, Myasuda, Fph, Gregbard, Arrataz, SuperGerbil, Jessealama, Quirex, Eubulide, Thijs!bot, Epbr123, Coelacan, John254, Paquitotrek, Mentifisto, AntiVandalBot, Exteray, Salgueiro, Acroterion, VoABot II, JamesBWatson, Xanderill, Dave Z Coates, Zip123, Burnedthru, Mycroft7, Pbroks13, TomyDuby, Robert Illes, Mjg0, Theowoll, Jance, Sarregouset, PatriciaJH, Varletcard, Wikieditor06, Deor, VolkovBot, Flyingidiot, JohnBlackburne, AlnoktaBOT, Kokpit, TXiKiBoT, Erik the Red 2, Kiyabg, Prb4, RadiantRay, Lamro, Bacchio, Falcon8765, Bathroomboy99, Cwkmail, Sings-With-Spirits, Oxymoron83, MiNombreDeGuerra, KathrynLybarger, OKBot, Anchor Link Bot, Charum, Tomasz Prochownik, ClueBot, SummerWithMorons, Mild Bill Hiccup, Blanchardb, Cpt-CutLess, Mad2Physicist, Sorpresa03, PixelBot, Wprlh, 7&6=thirteen, JDPhD, ZooFari, MystBot, Addbot, ERK, Supphavit, Metagraph,

MrOllie, LaaknorBot, Pmod, CarsracBot, Peti610botH, Tassedethe, Quaristice, Lightbot, Vasil', Legobot, Luckas-bot, Yobot, Amicon, AnomieBOT, KDS4444, Rubinbot, Citation bot, LilHelpa, Xqbot, Ubcule, NOrbeck, FrescoBot, Elcidia, Dylanayer, TobeBot, Brian the Editor, PleaseStand, DARTH SIDIOUS 2, Rockyourbod77, Mattmatt1987, EmausBot, Heathdj, Tommy2010, Wikipelli, JSquish, AOC25, L Kensington, Nightly Scribbles, Tablethree, VictorianMutant, ClueBot NG, Derfel73, Wcherowi, Alexmarsella0828, Widr, Helpful Pixie Bot, Turkeybob777, Koertefa, BG19bot, Kmukayed, MusikAnimal, Bolerno, Kmc74, Brad7777, Glacialfox, The1337gamer, Judiakok1985, Scgam3rz, SantoshBot, Mdpacer, Elephantplot, Brirush, Badbadboy25, Pdecalculus, Tentinator, Pseudo-toadfish, Ugog Nizdast and Anonymous: 244

_ **Argument** *Source:* http://en.wikipedia.org/wiki/Argument?oldid=633660793 *Contributors:* AxelBoldt, Gaurav, Ronz, Markhurd, Hyacinth, Jjshapiro, Robbot, Michael Devore, Discospinster, Paul August, Bender235, Bobo192, Nsaa, Kata Alreshim, Velella, RainbowOfLight, Bjones, BD2412, Nightscream, Koavf, Chobot, FrankTobia, Stephenb, Kimchi.sg, DRosenbach, Josh3580, Trickstar, SmackBot, Od Mishehu, Jmendez, Gilliam, Chris the speller, MartinPoulter, Fuzzform, DHN-bot, Hallenrm, Darth Panda, Scwlong, Rrburke, Mr.Z-man, Cybercobra, Dreadstar, Richard001, Trbdavies, Lambiam, Khazar, Dwpaul, Shlomke, Ckatz, CRGreathouse, Penbat, Gregbard, Gogo Dodo, DumbBOT, Alaibot, Letranova, Northumbrian, JAnDbot, Bongwarrior, VoABot II, Kevinmon, Sonawin, Kateshortforbob, J.delanoy, Rgoodermote, Maurice Carbonaro, Johnfos, Jimmaths, TXiKiBoT, Philogo, Go2slash, Graymornings, Synthebot, Lova Falk, Cnilep, Dmcq, Botev, SieBot, Yintan, Enti342, Steven Zhang, Techman224, Correogsk, Msrasnw, Mhnin0, Sokari, Twinsday, ClueBot, The Thing That Should Not Be, Pelle Ohlander, Niceguyedc, Mack-the-random, Leonard^Bloom, Peachypoh, Johnuniq, DumZiBoT, Gerhardvalentin, Little Mountain 5, NellieBly, Rexroad2, Quaint and curious, Addbot, Cxz111, Grayfell, Tcncv, Ronhjones, Download, Favonian, 5 albert square, Bfigura's puppy, Shulha, Luckas-bot, TaBOT-zerem, Suntag, AnomieBOT, Grolltech, Flewis, E235, Citation bot, Maxis ftw, Obersachsebot, Xqbot, Capricorn42, Lord Bane, The Evil IP address, GrouchoBot, Omnipaedista, Linkubus, Wikiresearchman, Sky Attacker, Steelstring, D'ohBot, Machine Elf 1735, Pinethicket, Momergil, Oshnutz, TobeBot, Tangoleader, Gregkaye, Walkinxyz, EmausBot, Wikipelli, Zacchro, Shadfurman, ZéroBot, LiterateTiger, Staszek Lem, TyA, Tijfo098, Considering..., ClueBot NG, Lisnabreeny, Wesniel, Masssly, Widr, Bluemax711, Bigjonstuff, Titodutta, DonBogdan, Yumyum36, Infinite Loop-Maker, Wannabemodel, Teammm, Pratyya Ghosh, ChrisGualtieri, Alanajames, 5000tut, Mogism, Harlequinjelly, Metcalm, MarcusSchmitz, WHAZZUP204, Laleluuuuuuuu, Smr0151993, Thomn12, BubbleFoot and Anonymous: 202

_ **Formal language** *Source:* http://en.wikipedia.org/wiki/Formal%20language?oldid=630173077 *Contributors:* LC, Jan Hidders, Andre Engels, Youandme, Rp, Ahoerstemeier, Nanshu, Schneelocke, Charles Matthews, Bemoeial, Hyacinth, Cleduc, Spikey, Robbot, Jaredwf, Millosh, Tobias Bergemann, Connelly, Giftlite, Kim Bruning, Eequor, ConradPino, MarkSweep, Mukerjee, Tyler McHenry, Discospinster, Rich Farmbrough, Dbachmann, Ntennis, Chalst, Nk, Jonsafari, Obradovic Goran, Waku, Helix84, Mdd, Jumbuck, MrTree, Stephan Leeds, Nuno Tavares, Linas, Ruud Koot, Flamingspinach, ThomasOwens, Sdornan, LKenzo, FlaBot, Eubot, PlatypeanArchcow, Pexatus,

Chobot, MithrandirMage, YurikBot, Wavelength, RussBot, Hede2000, Archelon, Rick Norwood, Muu-karhu, Bkil, Aaron Schulz, BOTSuperzerocool, Bota47, Saric, Ripper234, Arthur Rubin, Jogers, GrinBot, TuukkaH, Finell, SmackBot, Wic2020, Incnis Mrsi, Unyoyega, Od Mishehu, Fikus, Jpvinall, Jerome Charles Potts, J. Spencer, Frap, Cerebralpayne, Jon Awbrey, FelisLeo, SashatoBot, Astuishin, Mike Fikes, Iridescent, Dreftymac, Adriatikus, DBooth, CRGreathouse, CBM, Ezrakilty, Gregbard, Alaibot, NERIUM, Nick Number, Cyclonenim, Vantelimus, Danny lost, VictorAnyakin, Hermel, AndriesVanRenssen, Nyq, JNW, Tedickey, Cic, David Eppstein, J.delanoy, Trusilver, Brest, Daniele.tampieri, Policron, Bonadea, Halukakin, Idioma-bot, Philomathoholic, VolkovBot, AlnoktaBOT, Philogo, ASHPvanRenssen, AlleborgoBot, Newbyguesses, MiNombreDeGuerra, Fratrep, OKBot, CBM2, Classicalecon, ClueBot, Alpha Beta Epsilon, Excirial, Quercus basaseachicensis, Alexbot, NuclearWarfare, Hans Adler, Addbot, Cuaxdon, LemmeyBOT, OlEnglish, Zorrobot, Jarble, JakobVoss, Legobot, Luckas-bot, Yobot, Pcap, AnomieBOT, Jim1138, JackieBot, Hahahaha4, Citation bot, Clickey, Xqbot, GrouchoBot, Charvest, Wei.cs, Serberimor, MastiBot, RobinK, Full-date unlinking bot, ActivExpression, Gzorg, Lokentaren, LoStrangolatore, Mean as custard, Ankog, Dziadgba, Architectchao, Tijfo098, ClueBot NG, Helpful Pixie Bot, Garsd, Sff9, ChrisGualtieri, Dmunene, Chunliang Lyu, Dulaambaw, Akerbos, Jochen Burghardt, Isthatmoe and Anonymous: 86

_ **Independence (mathematical logic)** *Source:* http://en.wikipedia.org/wiki/Independence%20(mathematical%20logic)?oldid=630176020 *Contributors:* Charles Matthews, Hyacinth, Thue, Gandalf61, LX, Barnaby dawson, ESkog, Oleg Alexandrov, Roboto de Ajvol, YurikBot, Trovatore, CRGreathouse, CBM, Gregbard, JAnDbot, Pavel Jelínek, DesolateReality, IsleLaMotte, Addbot, Theking17825, Yobot, Pcap, Erik9bot, EffeX2, LucienBOT, Ebony Jackson, RA0808, Masssly, Daysrr, Nathanielfirst and Anonymous: 11

_ **Analytic–synthetic distinction** *Source:* http://en.wikipedia.org/wiki/Analytic%E2%80%93synthetic%20distinction?oldid=626148466 *Contributors:* Renamed user 4, Charles Matthews, Markhurd, Banno, Rursus, Giftlite, RayBirks, Zenohockey, Causa sui, Bart133, Velho, SDC, BD2412, Rjwilmsi, Koavf, BMF81, Wavelength, NawlinWiki, Tomisti, Drboisclair, Arthur Rubin, NielsenGW, SmackBot, Jaymay, Drphilharmonic, Byelf2007, Wvbailey, Mgiganteus1, Falk Lieder, Sdorrance, Gregbard, Cydebot, Peterdjones, Rt3368, Thijs!bot, Epbr123, Kilva, HBinswanger, AntiVandalBot, Seaphoto, Antique Rose, Ran4, Empyrius, Sluzzelin, JAnDbot, Pmooremath, Anarchia, R'n'B, Herbsewell, AceNZ, Maurice Carbonaro, Colchicum, Mlsquirrel, Kyle the bot, Scottglloyd, Philogo, Optigan13, Insanity Incarnate, Tuscon, SieBot, Flyer22, ClueBot, Leopedia2, SomeGuy11112, Brews ohare, Gavinbeeker, SchreiberBike, Addbot, Atethnekos, JEN9841, Luckas-bot, Yobot, Ptbotgourou, Denispir, AnomieBOT, Ikant, Srich32977, Aaron Kauppi, Sémaphore, Bachoui, Blubro, DixonDBot, Maestrelli, EmausBot, Bollyjeff, Gbsrd, MisterDub, Angelo Mascaro, ClueBot NG, Braincricket, Helpful Pixie Bot, BG19bot, Objectivesea2, PhnomPencil, Qetuth, Justincheng12345-bot, Deletes living, Aelephant, M.wurtenberger and Anonymous: 74

_ **Mathematical beauty** *Source:* http://en.wikipedia.org/wiki/Mathematical%20beauty?oldid=622802368 *Contributors:* Tarquin, Miguel, Michael Hardy, Nixdorf, MartinHarper, GTBacchus, Karada, Ahoerstemeier, DavidWBrooks, Jimfbleak, Angela, Nikai, Nikola Smolenski, Revolver, Dysprosia, KRS, Ann O'nyme, Jmartinezot, Jose Ramos, Bevo, Aleph4, Gandalf61, Hadal, Randomness, Ancheta Wis, Giftlite, Dbenbenn, Lupin, Bfinn, WHEELER, Dav4is, JRR Trollkien, Andycjp, Alexf, Elroch, Sam Hocevar, Gscshoyru, ChaTo, Thorwald, Paul August, Billymac00, C S, Andrewbadr, IonNerd, Orimosenzon, Critical, Linas, Mindmatrix, LOL, Ruud Koot, Tedneeman, Ryan Reich, Lawrence King, Grammarbot, Hack-Man, Salix alba, Slac, R.e.b., Nihiltres, Jersey Devil, DVdm, YurikBot, Wolfmankurd, Wikinick, NawlinWiki, Arichnad, Raven4x4x, EEMIV, Sandstein, Cullinane, Gesslein, Allens, SmackBot, Stepa, PeterSymonds, Kithburd, SMP, Silly rabbit, Nbarth, DHN-bot, Scalene, Rrelf, Armend, Jon Awbrey, DMacks, Xiutwel, Lambiam, Howdoesthiswo, Amenzix, RandomCritic, Waggers, Mets501, E-Kartoffel, Aeternus, Sakurambo, CRGreathouse, Wafulz, A civilian, WeggeBot, Gregbard, Nauticashades, Tsenapathy, MC10, M a s, Int3gr4te, Thijs!bot, Kilva, Liquid-aim-bot, Mhaitham.shammaa, Narssarssuaq, HarmonicFeather, Xeno, Ineffable3000, Maurice Carbonaro, Nigholith, Rnest2002, Ogranut, Funandtrvl, Hammersoft, Chitownmack, SteveStrummer, Broadbot, Lambyte, Psyche825, Everything counts, Mouse is back, Billinghurst, Adam.J.W.C., Euryalus, Anchor Link Bot, Tautologist, ClueBot, KarenSutherland, MonoBot, XLinkBot, Addbot, Idbelange, Guffydrawers, Tassedethe, Yobot, Fleabox, AnomieBOT, Citation bot, ProtectionTaggingBot, Kurosuke88, Ron Aharoni, Citation bot 1, Daclyff, EdEveridge, PPdd, WikitanvirBot, Kiatdd, ZéroBot, Thewhyman, Mixedberries17, ZeroCool4ta, Fuzzy artist, Miegoreng, Rmashhadi, ClueBot NG, Frietjes, Delusion23, Luckimg, Helpful Pixie Bot, BG19bot, Brad7777, Pankaj Jyoti Mahanta, Mathbeauty, Ashorocetus, Nigellwh, Lamaballa, LZNQBD and Anonymous: 111

_ **Paul Erdős** *Source:* http://en.wikipedia.org/wiki/Paul%20Erd%C5%91s?oldid=634059198 *Contributors:* The Anome, XJaM, Arvindn, Tox, Stevertigo, Michael Hardy, Lexor, Isomorphic, Jschrempp, Karada, Tregoweth, CesarB, Julesd, Bogdangiusca, Eszett, Charles Matthews, Berteun, Tpbradbury, Grendelkhan, Itai, Zero0000, McKay, EldKatt, Carlossuarez46, Dougefresh42, Kevinatilusa, Robbot, Friedo, Jaredwf, Fredrik, Sanders muc, Saforrest, JerryFriedman, Lzur, Wile E. Heresiarch, Adam78, Centrx, Giftlite, Gwalla, Dbenbenn, Vir4030, BenFrantzDale, Cobaltbluetony, Marcika, Curps, DO'Neil, Andris, Daniel Brockman, Fak119, Matt Crypto, Rjyanco, Neilc, Gubbubu, Ato, Andycjp, Farside, Gaul, DragonflySixtyseven, Thincat, Two Bananas, Urhixidur, Robin klein, Klemen Kocjancic, Trilobite, Eduardoporcher, Alperen, Thorwald, Corti, Grstain, PhotoBox, D6, AgentSteel, Blanchette, Zaheen, Rich Farmbrough, Guanabot, Paul August, Bender235, Cyclopia, Jnestorius, BACbKA, Billlion, Commonbrick, Aranel, Livajo, Indil, Kwamikagami, EmilJ, Deanos, Jpgordon, Causa sui, Grue, Mike Schwartz, Mikemsd, Viriditas, Cmdrjameson, Rajah, ריינהארט לערי, Obradovic Goran, Haham hanuka, Knucmo2, Jumbuck, Msh210, TheParanoidOne, PaulHanson, Borisblue, Keenan Pepper, Ricky81682, Hégésippe Cormier, Burn, Ross Burgess, VivaEmilyDavies, Aquae, BlastOButter42, Gene Nygaard, Igorpak, Bookandcoffee, Kazvorpal, Ceyockey, Kenyon, Mcsee, Bobrayner, Angr, DealPete, Jak86, Shreevatsa, Jacobolus, Wikiklrsc, Kvetch, Marudubshinki, Emerson7, Graham87, Kbdank71, Reisio, Kane5187, Rjwilmsi, Koavf, Sorenr, Zbxgscqf, Lockley, Hack-Man, Quiddity, Salix alba, Arbor, Kalogeropoulos, DoubleBlue, MarnetteD, Anurup, Kerowyn, JYOuyang, RexNL, Redwolf24, Goudzovski, Haonhien, CJLL Wright, Chobot, Jared Preston, Commander Nemet, Adoniscik, EamonnPKeane, JPD, YurikBot, Jamesmorrison, RobotE, Swerty, RussBot, Michael Slone, KSmrq, Lar, David Woodward, Emiellaiendiay, Bachrach44, LaszloWalrus, JocK, Daniel Mietchen, Pnrj, Syrthiss, Cheeser1, Bota47, Avraham, Ott2, Pegship, Ozaru, Kompik, Saric, Arthur Rubin, Ydam, Nealeyoung, RobertBorgersen, T. Anthony, Curpsbot-unicodify, Garion96, Canadianism, Kgf0, Tom Morris, Vulturell, Marquez, SmackBot, TheBilly, Tigerghost, Jclerman, Roger Hui, InverseHypercube, CRKingston, C.Fred, D'n, KocjoBot, Senordingdong, Stifle, Gilliam, Duke Ganote, Ohnoitsjamie, JCSantos, Shawn M. O'Hare, Thumperward, Snori, Jayanta Sen, Taxipom, RyanEberhart, Laura47, Mhym, LouScheffer, Squigish, Khoikhoi, Cyberevil, Eeyore22, Sashato-Bot, Lambiam, Cold Light, Vanished user 9i39j3, Thlayli, Bo99, JoshuaZ, Goelano, BillFlis, Optimale, Dicklyon, Doczilla, Cbuckley, Quaeler, Dan Gluck, Iridescent, Stotr, Robbie Cook, Twas Now, Rhetth, JForget, Markjoseph125, Patchouli, ZICO, Wafulz, W guice, Gritzko, Kowalmistrz, Erik Kennedy, Myasuda, Icarus of old, Dr Zak, Pit-yacker, Cydebot, Korky Day, Ntsimp, Mycroft.Holmes, Garik, Thijs!bot, Biruitorul, Jaqo, Joe in Australia, K. Lastochka, Bunzil, Hires an editor, AnAj, Atavi, Vanjagenije, Danger, GodGell, Paulelastic, Qwerty Binary, Kuteni, Aldousdj, Dreaded Walrus, JAnDbot, Turgidson, Sophie means wisdom, Jce358, Magioladitis, Hroðulf, Vanish2, Singularity, NicoSan, David Eppstein, Emw, Thibbs, Kope, Turtlens, Nocklas, Ben MacDui, CommonsDelinker, Johnpacklambert, Pomte, J.delanoy, SureFire, SuperGirl, Ttwo, Maurice Carbonaro, Unimaginative Username, Cpiral, Salih, Jon Ascton, Anonywiki, SteveChervitzTrutane, Alais4, Beremiz, Serenthia, Swanny18, Bnynms, VolkovBot, TXiKiBoT, Paulburnett, Mathwhiz 29, Anonymous Dissident, Combatentropy, Oxfordwang, Voiceofreason01, Ichigo9559, Duncan.Hull, Gilisa, Kmhkmh, Lamro, Hopson-Road, Ceranthor, Gbjbaanb, Quietbritishjim, JulesN, SieBot, Gerakibot, Hamster X, Mathosaur, Utternutter, Flyer22, Elakhna, Svick, Aumnamahashiva, Vaarky, Sitush, SageMab, RS1900, ClueBot, SummerWithMorons, Perroazul, Tintinobelisk, All Hallow's Wraith, WriterListener, ImperfectlyInformed, Barnwani, Trivialist, Flaming, Masterpiece2000, Excirial, Roomsmight, Eeekster, Estirabot, Aurora2698,

Tnxman307, Mickey gfss2007, Doprendek, Dyrob, Johnuniq, DumZiBoT, 51kwad, Koraki, Pioneer42, Luwilt, Addbot, Out of Phase User, C6541, DOI bot, Betterusername, Ronhjones, Gzhanstong, KöMaL314, LaaknorBot, Proton donor, Favonian, LinkFABot, Feketekave, Luckas-bot, Yobot, Zoe17, Amirobot, AnomieBOT, Jim1138, Mintrick, Citation bot, Xqbot, The Elves Of Dunsimore, Locos epraix, U664003803, Omnipaedista, Andypar, Touchatou, FrescoBot, Surv1v4l1st, Lothar von Richthofen, PorkoltLover60, Citation bot 1, PrBeacon, Kiefer.Wolfowitz, Plucas58, Skyerise, Mikewarbz, Lineslarge, Bgpaulus, Oracleofottawa, 777sms, Minimac, RjwilmsiBot, In ictu oculi, EmausBot, Grinxen, Chimpionspeak, WikiDude1776, Leslie.Hetherington, Bbeamer007, Ida Shaw, Infinte loop, H3llBot, ClueBot NG, Gregerdos, HumbleAbelard, Subrahmanya Hegde, LJosil, Frietjes, Joel B. Lewis, Mark Hymann-Adler, Helpful Pixie Bot, Hopeandreason, Kendall-K1, Exercisephys, CitationCleanerBot, Slushy9, The Uncyclopedian, Dysrhythmia, Batty-Bot, Yonoteam, Jeremy112233, Smasongarrison, Ninmacer20, Micronation132, Purdygb, Sergeyuzilov, JYBot, Qwertymany, Mogism, Kephir, Everything Is Numbers, Aloneinthewild, Jkuodo, Vanamonde93, K9re11, Dutch mastiff lover, MisterHeadroom, Matplotlib, Monkbot, AntiqueReader, Pieter202, ShulMaven, Garfield Garfield, Krajjam, Leeuwe and Anonymous: 283

_ **Proofs from THE BOOK** *Source:* http://en.wikipedia.org/wiki/Proofs%20from%20THE%20BOOK?oldid=629106042 *Contributors:* Kpjas, Michael Hardy, Markhurd, Gandalf61, Giftlite, Marcika, DragonflySixtyseven, Mani1, PaulHanson, Jeff3000, Salix alba, Bubba73, Jaraalbe, CecilWard, SmackBot, Wdvorak, E-Kartoffel, CBM, Cydebot, Skomorokh, Matthew Komorowski, David Eppstein, TXiKiBoT, PaulTanenbaum, Lamro, AlleborgoBot, Palnot, Addbot, Luckas-bot, Citation bot, Twri, Citation bot 1, Peepo36, Andreschulz, Zmwangx and Anonymous: 16

_ **Direct proof** *Source:* http://en.wikipedia.org/wiki/Direct%20proof?oldid=634208463 *Contributors:* Michael Hardy, TakuyaMurata, Kaihsu, Korath, Nifboy, Peter Kwok, Andreas Kaufmann, Robertbowerman, Lachatdelarue, Gauge, Amerindianarts, Daranz, MarSch, GregAsche, Splintercellguy, Soltras, SmackBot, Luke Gustafson, Lambiam, Pliny, Bjankuloski06en, 16@r, CBM, Gregbard, Cydebot, Thijs!bot, David Eppstein, STBot, VolkovBot, Jimmaths, Anonymous Dissident, Synthebot, ClueBot, Estirabot, Addbot, Luckas-bot, La comadreja, GrouchoBot, RibotBOT, Erik9bot, MastiBot, Igor N Dimovski, Duoduoduo, EmausBot, ClueBot NG, Hoorayforturtles, Attempt4, Widr, Helpful Pixie Bot, CitationCleanerBot, Monkbot, CS104G19, Unaizalakain and Anonymous: 12

_ **Closure (mathematics)** *Source:* http://en.wikipedia.org/wiki/Closure%20(mathematics)?oldid=629747640 *Contributors:* Arvindn, Patrick, Michael Hardy, Dominus, David Shay, Populus, Tobias Bergemann, Giftlite, Lethe, Neilc, ArnoldReinhold, Paul August, Spayrard, R. S. Shaw, Obradovic Goran, Kazvorpal, Oleg Alexandrov, Unixxx, Linas, Jftsang, Staecker, Salix alba, FlaBot, Mathbot, RexNL, Vonkje, Exe, Chobot, Wingchi, YurikBot, Bhny, Gwaihir, Occono, Googl, Kompik, JuJube, CIreland, SmackBot, RaulMiller, Mhss, PJTraill, Octahedron80, Nbarth, Sct72, Can't sleep, clown will eat me, NeilFraser, Lambiam, ILikeThings, Gregbard, Mr Gronk, Thijs!bot, .anacondabot, Sodabottle, David Eppstein, JaGa, Trusilver, Sapien2, U.S.A.U.S.A.U.S.A., VolkovBot, Skylarkmichelle, Flyer22, Yoda of Borg, Classicalecon, ClueBot, Razimantv, Iranway, Addbot, Jarble, Ettrig, Pcap, AnomieBOT, Eumolpo, Capricorn42, I dream of horses, PrincessofLlyr, Kallikanzarid, DARTH SIDIOUS 2, Donner60, Puffin, ClueBot NG, Wcherowi, Hofmic, Momobeauty, The1337gamer, Assassin Sorcerer, Jochen Burghardt, Brirush, Tentinator, Darencline and Anonymous: 53

_ **Distributive property** *Source:* http://en.wikipedia.org/wiki/Distributive%20property?oldid=632024107 *Contributors:* AxelBoldt, Tarquin, Youssefsan, Toby Bartels, Patrick, Xavic69, Michael Hardy, Andres, Ideyal, Dysprosia, Malcohol, Andrewman327, Shizhao, PuzzletChung, Romanm, Chris Roy, Wikibot, Tobias Bergemann, Giftlite, Markus Krötzsch, Dissident, Nodmonkey, Mike Rosoft, Smimram, Discospinster, Paul August, ESkog, Rgdboer, EmilJ, Bobo192, Robotje, Smalljim, Jumbuck, Arthena, Keenan Pepper, Mykej, Bsadowski1, Blaxthos, Linas, Evershade, Isnow, Marudubshinki, Salix alba, Vegaswikian, Nneonneo, Bfigura, FlaBot, Alexb@cut-theknot.com, Mathbot, Andy85719, Ichudov, DVdm, YurikBot, Michael Slone, Grafen, Trovatore, Bota47, Banus, Melchoir, Bluebot, Ladislav the Posthumous, Octahedron80, UNV, Jiddisch, Khazar, FrozenMan, Bando26, 16@r, Dicklyon, EdC, Engelec, Exzakin, Jokes Free4Me, Simeon, Gregbard, Thijs!bot, Barticus88, Marek69, Nezzadar, Escarbot, Mhaitham.shammaa, Salgueiro, JAnDbot, Onkel Tuca, Acroterion, Numbo3, Katalaveno, AntiSpamBot, GaborLajos, Lyctc, Idioma-bot, Janice Vian, Montchav, TXiKiBoT,

Anonymous Dissident, Dictouray, Oxfordwang, Martin451, Skylarkmichelle, Jackfork, Enviroboy, AlleborgoBot, Gerakibot, Bentogoa, Radon210, Hello71, ClueBot, The Thing That Should Not Be, Cliff, Goldkingtut5, Excirial, Jusdafax, NuclearWarfare, NERICSecurity, Pichpich, Mm40, Addbot, Jojhutton, Ronhjones, Zarcadia, Favonian, Squandermania, Jarble, Ben Ben, Legobot, Luckasbot, Materialscientist, NFD9001, False vacuum, RibotBOT, Pinethicket, I dream of horses, MastiBot, Andrea105, Slon02, Saul34, J36miles, Davejohnsan, Orphan Wiki, Super48paul, Sp33dyphil, Slawekb, Quondum, BrokenAnchorBot, TyA, Donner60, Chewings72, DASHBotAV, AlecJansen, ClueBot NG, Wcherowi, IfYouDoIfYouDon't, Dreth, O.Koslowski, Asukite, Vibhijain, Helpful Pixie Bot, Pmi1924, BG19bot, TCN7JM, Dan653, Forkloop, CallofDutyboy9, EuroCarGT, Sandeep.ps4, Christian314, Ivashikhmin, Makecatbot, Lugia2453, Gphilip, Brirush, Wywin, BB-GUN101, ElHef, DavidLeighEllis, Shaun9876, Pkramer2021, Kitkat1234567880, Kcolemantwin3, Gracecandy1143, David88063, Abruce123412, Amortias, Jj 1213 wiki, Dalangster, Iwamwickham, 123456me123456 and Anonymous: 233

_ **Material conditional** *Source:* http://en.wikipedia.org/wiki/Material%20conditional?oldid=634271622 *Contributors:* William Avery, Dcljr, AugPi, Charles Matthews, Dcoetzee, Doradus, Cholling, Giftlite, Jason Quinn, Nayuki, TedPavlic, Elwikipedista, Nortexoid, Vesal, Eric Kvaalen, BD2412, Kbdank71, Martin von Gagern, Joel D. Reid, Fresheneesz, Vonkje, NevilleDNZ, RussBot, KSchutte, NawlinWiki, Trovatore, Avraham, Closedmouth, Arthur Rubin, SyntaxPC, Fctk, SmackBot, Amcbride, Incnis Mrsi, Pokipsy76, BiT, Mhss, Jaymay, Tisthammerw, Sholto Maud, Robma, Cybercobra, Jon Awbrey, Oceanofperceptions, Byelf2007, Grumpyyoungman01, Clark Mobarry, Beefyt, Rory O'Kane, Dreftymac, Eassin, JRSpriggs, Gregbard, FilipeS, Cydebot, Julian Mendez, Thijs!bot, Egriffin, Jojan, Escarbot, Applemeister, WinBot, Salgueiro, JAnDbot, Olaf, Alastair Haines, Arno Matthias, JaGa, Santiago Saint James, Pharaoh of the Wizards, Pyrospirit, SFinside, Anonymous Dissident, The Tetrast, Cnilep, Radagast3, Newbyguesses, Lightbreather, Paradoctor, Iamthedeus, Soler97, Francvs, Classicalecon, Josang, Ruy thompson, Mate2code, Hans Adler, Djk3, Marc van Leeuwen, Tbsdy lives, Addbot, Melab-1, Fyrael, Morriswa, SpellingBot, CarsracBot, Chzz, Jarble, Meisam, Luckas-bot, AnomieBOT, Sonia, Pnq, Bearnfæder, FrescoBot, Greyfriars, Machine Elf 1735, RedBot, MoreNet, Beyond My Ken, John of Reading, Hgetnet, Hibou57, ClueBot NG, Movses-bot, Jiri 1984, Masssly, Dooooot, Noobnubcakes, Hanlon1755, Leif Czerny, CarrieVS, Jochen Burghardt, Lukekfreeman, ServiceableVillain, NickDragonRyder, Indomitavis, Rathkirani, AnotherPseudonym, Xerula, Matthew Kastor, Mathematical Truth and Anonymous: 69

_ **Infinite set** *Source:* http://en.wikipedia.org/wiki/Infinite%20set?oldid=632950519 *Contributors:* The Anome, Toby Bartels, Dennis-Daniels, Charles Matthews, David Shay, Bkell, Tobias Bergemann, Giftlite, Paul August, Rgdboer, Benji22210, Zerofoks, Salix alba, FlaBot, VKokielov, Chobot, 4C, Grubber, Trovatore, Maksim-e, Addshore, Bidabadi, Vina-iwbot, Lambiam, StevenPatrickFlynn, Bjankuloski06en, Fell Collar, JRSpriggs, CRGreathouse, CBM, Escarbot, JAnDbot, Olaf, Ttwo, Maurice Carbonaro, Doug, DFRussia, Cliff, JP.Martin-Flatin, Alexbot, Hans Adler, Hatsoff, Addbot, Favonian, Luckas-bot, TaBOT-zerem, AnomieBOT, JackieBot, Materialscientist, VladimirReshetnikov, Erik9bot, Nicolas Perrault III, BenzolBot, Tkuvho, Pinethicket, SkyMachine, TobeBot, Beyond My Ken, Wgunther, Tommy2010, ZéroBot, Donner60, ClueBot NG, Widr, Juro2351, Magic6ball, Sauood07, YFdyh-bot, Blackbombchu, Centralpanic and Anonymous: 37

_ **Proof by infinite descent** *Source:* http://en.wikipedia.org/wiki/Proof%20by%20infinite%20descent?oldid=625994360 *Contributors:* Zundark, Michael Hardy, Charles Matthews, Dcoetzee, Giftlite, Dissident, Silenteuphony, Rich Farmbrough, KneeLess, FT2, Bender235, Mailer diablo, Joriki, Rjwilmsi, VKokielov, Mathbot, Maxal, Don Gosiewski, Salvatore Ingala, King of Hearts, David Pierce, SamuelRiv,

JCipriani, Flowersofnight, Mattmm, SmackBot, Mgreenbe, Bh3u4m, JCSantos, MalafayaBot, Mecrazywong, Sohale, TPIRFanSteve, CRGreathouse, Cydebot, Asmeurer, David Eppstein, Joshua Issac, This, that and the other, Azumashii, M4gnum0n, BOTarate, Rinpoche, MystBot, Addbot, Lightbot, דניאל ב., Yobot, PMLawrence, LilHelpa, GeometryGirl, Bibekmaths, Erik9bot, RedBot, Taylor561, Duoduoduo, DexDor, WikitanvirBot, ZéroBot, Fred Gandt, ChuispastonBot, Pastafarianist, Makecat-bot, Ginsuloft, DeathOfBalance, Cklasky, Monkbot and Anonymous: 26

_ **Square root of 2** *Source:* http://en.wikipedia.org/wiki/Square%20root%20of%202?oldid=634098373 *Contributors:* Vicki Rosenzweig, The Anome, XJaM, Stevertigo, D, Michael Hardy, Chinju, Komap, Delirium, Charles Matthews, Crissov, MatrixFrog, Dcoetzee, Dysprosia, Jitse Niesen, Prumpf, Hyacinth, Daran, Dale Arnett, Josh Cherry, Fredrik, Chocolateboy, Lowellian, Henrygb, Bkell, Superm401, Giftlite, Smjg, Anton Mravcek, Dmmaus, Tonymaric, DragonflySixtyseven, Drrosslg, Uaxuctum, Zondor, ArnoldReinhold, D-Notice, Paul August, Bender235, Zaslav, MisterSheik, El C, Rgdboer, EmilJ, Touriste, Alansohn, Caesura, Oleg Alexandrov, Japanese Searobin, Woohookitty, Mindmatrix, Pluvius, Sympleko, Qaddosh, Pranathi, Rjwilmsi, Tizio, Nneonneo, Alexb@cut-the-knot.com, DevastatorIIC, Kri, Glenn L, King of Hearts, Chobot, Bgwhite, PointedEars, Siddhant, MathiasRav, Chuck Carroll, Wimt, DavidConrad, Mipadi, VIGNERON, Sahodaran, Super Rad!, Closedmouth, Arthur Rubin, Finell, AndrewWTaylor, SmackBot, RDBury, Mitchan, Melchoir, C.Fred, Jdmt, DTM, Eskimbot, Gene Thomas, Kmarinas86, JCSantos, Silly rabbit, TotalTommyTerror, Hgrosser, Foxjwill, Chlewbot, Daqu, Jbergquist, John Reid, Jim.belk, Dicklyon, A-cai, Hodge Star, RekishiEJ, Amakuru, Eric Le Bigot, INVERTED, Carifio24, Saintrain, Thijs!bot, Timo3, Jojan, Davidhorman, Dgies, WinBot, Edokter, PeteS, Kaobear, Albmont, Soulbot, Usien6, Nyttend, Animum, David Eppstein, N.Nahber, JaGa, Jtir, Arithmonic, R'n'B, Pbroks13, Hippasus, Fowler&fowler, Aqwis, Johnbod, Trumpet marietta 45750, Milogardner, TreasuryTag, Barneca, Gentlemath, Anonymous Dissident, Ask123, Shuhab, Luqqe, Yk Yk Yk, Tomaxer, VanBuren, Dmcq, Arcfrk, Radagast3, EnJx, Portalian, PlanetStar, Messagetolove, Spaglia, Iameukarya, Skeptical scientist, S2000magician, Randomblue, AllHailZeppelin, Kortaggio, ClueBot, Htkootc2362, Justin W Smith, Blanchardb, Calimo, Gtstricky, Aw.rootbeer1, NuclearWarfare, 3d vector, ChrisHodgesUK, Thehelpfulone, Rinpoche, Anty munt, Djk3, Goodvac, TimothyRias, MystBot, Knowland, Thatguyflint, Addbot, DOI bot, Pete Bevin, Download, LinkFA-Bot, Matěj Grabovský, خالد حسني, Yobot, Amirobot, Paepaok, FredrikMeyer, Citation bot, Srinivas, LilHelpa, Xqbot, Anna Frodesiak, Srich32977, A. di M., Maxidater, FrescoBot, Sławomir Biały, DelphinidaeZeta, Robo37, Hexagon70, Citation bot 1, Tkuvho, Pinethicket, Phil1881, Jean-François Clet, Double sharp, Itu, Ckrp, Lotje, Vrenator, Extra999, Duoduoduo, EmausBot, Orphan Wiki, Temi4-hik, AlanSiegrist, Cvsoft, Chewings72, Orange Suede Sofa, C0617470r, ClueBot NG, AmosClecure, LutherVinci, ElectricRazor, Fauzan, Thanaaoesat, Kiwislushie, Frietjes, *half-moon* bubba, Karlbonner1982, Helpful Pixie Bot, Adamhgolding, Benzband, Pratyush Sarkar, HotdogPi, YFdyh-bot, La marts boys, Dylanfrinkled, Dexbot, Thewikicontributor, BeaumontTaz, Gautamh, Sharno, William2001, Salmaislambd, Wkkim11, Monkbot, Friendsofmath and Anonymous: 156

_ **Contraposition** *Source:* http://en.wikipedia.org/wiki/Contraposition?oldid=619535304 *Contributors:* Edward, Ixfd64, Stevenj, Stismail, Amerindianarts, Dasunst3r, BD2412, ScottJ, FlaBot, Kwammi, Chobot, Roboto de Ajvol, Bamgooly, Widdma, Red Slash, Jaymax, Big Brother 1984, Googl, Carabinieri, HereToHelp, Otto ter Haar, SmackBot, Javalenok, HLwiKi, NickPenguin, Byelf2007, John, DouglasCalvert, Iridescent, Harold f, Dycedarg, CBM, WeggeBot, Andkore, Simeon, Gregbard, Blindman shady, Gimmetrow, Thijs!bot, Epbr123, Raeven0, Jheiv, Prgrmr@wrk, Okloster, McSly, Shinju, Jimmaths, Deleet, Synthebot, AlleborgoBot, SieBot, Wilson44691, Fratrep, Pointylittlethingy, XDanielx, The Thing That Should Not Be, Metaprimer, Mild Bill Hiccup, Aquillyne, SchreiberBike, DumZiBoT, Addbot, Tide rolls, AtiwH, Luckas-bot, Ptbotgourou, Burning.flamer, AnomieBOT, Makeswell, Dendropithecus, Elliottwolf, Erik9bot, FrescoBot, Dashed, Katovatzschyn, Antonone, JSquish, ClueBot NG, Wcherowi, Widr, Henneyj, BlueMist, Jaysn1 and Anonymous: 66

_ **Reductio ad absurdum** *Source:* http://en.wikipedia.org/wiki/Reductio%20ad%20absurdum?oldid=632663998 *Contributors:* Michael Hardy, Dominus, Julesd, Hyacinth, Xanzzibar, Centrx, Giftlite, Suspekt, 20040302, Paul August, Richard Arthur Norton (1958-), Mindmatrix, TheAlphaWolf, Mandarax, Spezied, Mendaliv, XP1, Theinsomniac4life, Savethemooses, Diza, Newagelink, Thnidu, HereToHelp, Williamjacobs, SmackBot, InverseHypercube, Bomac, Bmearns, BenAveling, Hibbleton, Occultations, Byelf2007, Loodog, Gobonobo, Wizard191, Chetvorno, N2e, Gregbard, Jordan Brown, PamD, Sensemaker, Gapooh007, JAnDbot, Asnac, Matthew Fennell, JamesBWatson, Cardamon, ForestAngel, David Eppstein, Nev1, Uranium grenade, Ohms law, GDW13, Treisijs, Anarchangel, Y, KirbenS, Iamthedeus, Hello71, Yoda of Borg, Martarius, Reydeyo, Alexbot, SpikeToronto, Zebrasil, Tired time, HumphreyW, Gerhardvalentin, Ronhjones, AgadaUrbanit, Numbo3-bot, Yobot, Majestic-chimp, Angel ivanov angelov, Nallimbot, Tojasonharris, Ciphers, OpenFuture, Transity, Dave3457, Much noise, Thedarkknight491, Pappaj333, Dashed, Pollinosisss, Jordgette, LilyKitty, Diyan.boyanov, BertSeghers, EmausBot, Jrjspencer, Ὁ οἶστρος, Sudozero, TheConduqtor, Aflyhorse, BlueMist, E1618978, Snow Blizzard, Cengime, JSWIFT13, Lugia2453, Jochen Burghardt, You Can Act Like A Man, Scythemantic, Yamaha5, Lance Chance, Immanuel Thoughtmaker and Anonymous: 77

_ **Coprime integers** *Source:* http://en.wikipedia.org/wiki/Coprime%20integers?oldid=626320084 *Contributors:* AxelBoldt, Bryan Derksen, Zundark, Tarquin, Andre Engels, XJaM, Bdesham, Patrick, Michael Hardy, SGBailey, TakuyaMurata, Ahoerstemeier, Stevenj, Dod1, Revolver, Timwi, AC, Glimz, Hyacinth, AnthonyQBachler, PuzzletChung, MathMartin, Tobias Bergemann, Filemon, Giftlite, DNewhall, Tsemii, Paul August, Bender235, EmilJ, Jumbuck, Oleg Alexandrov, Graham87, Rjwilmsi, Salix alba, Maxim Razin, FlaBot, Mathbot, Chobot, YurikBot, RobotE, Dmharvey, Wolfmankurd, Dantheox, Trovatore, Netrapt, Digfarenough, Lapointe64, HereToHelp, Pred, Maksim-e, BiT, Psiphiorg, Kostmo, AdamSmithee, Dreadstar, N Shar, Andrei Stroe, Bezenek, CRGreathouse, Domanix, Funnyfarmofdoom, Ntsimp, Benzi455, Kazubon, JAnDbot, Vanish2, JamesBWatson, David Eppstein, DirkEn, DorganBot, Matiasholte, X!, Jeff G., TXiKiBoT, Anonymous Dissident, Ferengi, Doug, SieBot, Thehotelambush, Cnotgate, OKBot, Martarius, ClueBot, Lartoven, Computer97, Jwpat7, Perchy22, Marc van Leeuwen, Virginia-American, ماني, Loupeter, Legobot, Luckas-bot, Yobot, Ptbotgourou, Jgmoxness, AnomieBOT, GB fan, Frankenpuppy, Xqbot, Cryptography project, Anne Bauval, GrouchoBot, Grinevitski, Ebony Jackson, Stpasha, FoxBot, Duoduoduo, WillNess, RjwilmsiBot, D.Lazard, Chewings72, MSDousti, Forever Dusk, JonRichfield, Anita5192, ClueBot NG, Joel B. Lewis, Pluma, Helpful Pixie Bot, BG19bot, Consorveyapaaj2048394, Glacialfox, CodeTom, Amdoskarr, Bryanrutherford0, Monkbot, Erickspedian, GeoffreyT2000 and Anonymous: 80

_ **Constructive proof** *Source:* http://en.wikipedia.org/wiki/Constructive%20proof?oldid=627596487 *Contributors:* Michael Hardy, Gandalf61, Tobias Bergemann, Giftlite, Macrakis, Peter Kwok, Txa, Pol098, Chenxlee, Jameshfisher, Gurch, Quuxplusone, Bgwhite, Jayme, Gaius Cornelius, SmackBot, Bluebot, Lambiam, Lim Wei Quan, CBM, Pierre de Lyon, Cydebot, Salgueiro, RebelRobot, Dr Caligari, Jimmaths, Dmcq, SieBot, Classicalecon, Mahue, Leberbaum, Addbot, Unzerlegbarkeit, Yobot, Obscuranym, AnomieBOT, JRBEurope, Twri, Noamz, FrescoBot, Steve2011, Thomassteinke, DARTH SIDIOUS 2, Slawekb, Chricho, ClueBot NG, Helpful Pixie Bot, Khonkhortisan, Jochen Burghardt, Acetotyce, LudicrousTripe and Anonymous: 25

_ **Transcendental number** *Source:* http://en.wikipedia.org/wiki/Transcendental%20number?oldid=632937781 *Contributors:* AxelBoldt, Eloquence, The Anome, Youssefsan, XJaM, PierreAbbat, Stevertigo, Bdesham, Chas zzz brown, Michael Hardy, Dante Alighieri, Dominus, Dcljr, TakuyaMurata, GTBacchus, Eric119, Pde, Kosebamse, Iulianu, Glenn, Ideyal, Loren Rosen, Charles Matthews, Dino, Markhurd, Hyacinth, VeryVerily, AnonMoos, Jerzy, Robbot, Fredrik, Voyager640, Lowellian, Henrygb, Academic Challenger, Humus sapiens, Tobias Bergemann, Giftlite, Lethe, Herbee, Mboverload, Macrakis, Gadfium, LiDaobing, Profvk, DragonflySixtyseven, Kahkonen, Mschlindwein, JasticE, Jfpierce, Ericamick, Bender235, Ben Standeven, Plugwash, Crisófilax, Billymac00, Jung dalglish, SpeedyGonsales,

Rajah, Haham hanuka, Richard Taytor, Suruena, Oleg Alexandrov, Linas, Mindmatrix, GregorB, SDC, Gisling, Chenxlee, Rjwilmsi, X1011, YurikBot, Petiatil, Dkostic, Gaius Cornelius, Ihope127, Rick Norwood, Joth, Bmdavll, Crasshopper, Bota47, Hakeem.gadi, Arthur Rubin, Cojoco, HereToHelp, Ethan Mitchell, GrinBot, SmackBot, Tom Lougheed, Slashme, Alan McBeth, BiT, IronSwallow, SMP, Cavie82, Nbarth, Calc rulz, Kotra, Can't sleep, clown will eat me, Joerite, Davezarzycki, CorbinSimpson, Monguin61, John Reid, Modus Vivendi, JzG, Ksn, MSchmahl, Turanyuksel, MTSbot, Mike Fikes, Dan Gluck, UnnotableWorldFigure, Lottamiata, Newone, Me lkjhgfdsa, IanOfNorwich, Eeben, CRGreathouse, SuperMidget, Doctormatt, Ntsimp, Asymptote, Gravitroid, Mon4, Falcon9x5.com, Michael C Price, Xantharius, Julian.cancino, Hanche, Thijs!bot, Koeplinger, Woody, AbcXyz, Northumbrian, JAnDbot, Oxinabox, Nadalle, Scott Tillinghast, Houston TX, .anacondabot, Vanish2, Johnbibby, Pagw, Capeyork, Prokofiev2, Ttwo, Maurice Carbonaro, CompuChip, WikiGrrrl, PMajer, TXiKiBoT, Nxavar, George Moromisato, Anonymous Dissident, Liko81, Colourplay, Spur, SieBot, Cwkmail, Arknascar44, Mr. Granger, ClueBot, Rotational, Hans Adler, Europe22, SilvonenBot, Virginia-American, WikiDao, Addbot, DOI bot, AkhtaBot, CarsracBot, 5 albert square, KALZOID-73-20METER, Lightbot, Jarble, Legobot, Luckas-bot, Yobot, Roviury, Ptbotgourou, AnomieBOT, Citation bot, ArthurBot, Xqbot, Alanonala, TinucherianBot II, Nasnema, RJGray, Isheden, Gap9551, Jhbdel, VladimirReshetnikov, IShadowed, Qm2008q, FrescoBot, Majopius, Steve Quinn, Citation bot 1, I dream of horses, Franci.cariati, The.megapode, RedBot, Taylor561, MondalorBot, Jujutacular, Mstempin, JokerXtreme, Pbrower2a, Gegege13, RjwilmsiBot, Emaus-Bot, High Tinker, Vvceph, Slawekb, Thecheesykid, ZéroBot, Quondum, ManU0710, ClueBot NG, Chicken Wing999, MelbourneStar, Anon5791, Discordianchao5, BG19bot, Consorveyapaaj2048394, Minitech.me, EdvinW, Spabble, Rhaycock, Deltahedron, Vitaash, Spectral sequence, Cvictor34, Zimboras, Cerellon, Suelru, Second Skin, Cosmia Nebula, Monkbot and Anonymous: 161

_ **Liouville number** *Source:* http://en.wikipedia.org/wiki/Liouville%20number?oldid=613986832 *Contributors:* AxelBoldt, PierreAbbat, Edward, Chas zzz brown, Michael Hardy, Snoyes, Charles Matthews, Jitse Niesen, Markhurd, Nnh, Tobias Bergemann, Tosha, Giftlite, Tubular, Rick Block, Rjyanco, Icairns, Eranb, Kate, Gauge, EmilJ, HannsEwald, StephanCom, Maxal, 4C, Zvika, SmackBot, Bluebot, JCSantos, Gutworth, SerialJaywalker, Tesseran, Ohconfucius, Cronholm144, OzOz, CRGreathouse, Tex, Scott Tillinghast, Houston TX, R'n'B, Ale2006, Lechatjaune, AlleborgoBot, OV-CA, Thehotelambush, Addbot, Uncia, Lightbot, דניאל ב., PV=nRT, Legobot, Yobot, Kilom691, AnomieBOT, C2equalA2plusB2, Mr.gondolier, VladimirReshetnikov, JovanCormac, Citation bot 1, RjwilmsiBot, Jowa fan, TheodoreYou, JordiGH, Fenderbear, Wim Nobel, Lojano, Deltahedron, Josephk, AHusain314, Monkbot and Anonymous: 38

_ **Proof by exhaustion** *Source:* http://en.wikipedia.org/wiki/Proof%20by%20exhaustion?oldid=627479159 *Contributors:* Michael Hardy, Dcljr, Karada, Charles Matthews, Gandalf61, Ian Maxwell, Peter Kwok, PhotoBox, Jnestorius, DimaDorfman, Dfeldmann, Anthony Appleyard, StradivariusTV, Bubba73, Eubot, YurikBot, Hakeem.gadi, NYKevin, Cybercobra, Desmond71, Dbtfz, Robofish, Mets501, ShelfSkewed, Gregbard, Cydebot, Eurobas, Edward321, EnOreg, Classicalecon, Tautologist, Addbot, Yobot, Idealboat, EmausBot, Jumpythehat, Mario Castelán Castro and Anonymous: 13

_ **Four color theorem** *Source:* http://en.wikipedia.org/wiki/Four%20color%20theorem?oldid=633835990 *Contributors:* Damian Yerrick, AxelBoldt, LC, Brion VIBBER, Vicki Rosenzweig, Bryan Derksen, Tarquin, Koyaanis Qatsi, Andre Engels, XJaM, PierreAbbat, SJK, Heron, Patrick, Chas zzz brown, Minimax, Michael Hardy, Dominus, SGBailey, Wapcaplet, Ixfd64, Cyde, Tomi, Eric119, CesarB, Mdebets, Ahoerstemeier, Docu, Kingturtle, Darkwind, Peter Kaminski, Ciphergoth, TraxPlayer, Rotem Dan, Ideyal, Loren Rosen, A5, Charles Matthews, Dcoetzee, Lfh, Joshuabowman, Jitse Niesen, Tpbradbury, McKay, Joy, Sandman, EldKatt, David.Monniaux, Twang,

Sjorford, ZimZalaBim, Lowellian, Mayooranathan, MathMartin, Timrollpickering, Jleedev, Tosha, Decrypt3, Giftlite, Smjg, Marnanel, Gene Ward Smith, Lupin, Jacob1207, Dratman, JeffBobFrank, Andris, Avsa, Mobius, Neilc, LiDaobing, HorsePunchKid, Blazotron, SimonLyall, Zfr, Kutulu, Sam, Arminius, Poccil, Rich Farmbrough, Dbachmann, Martpol, Paul August, ZeroOne, Closeapple, Ben Standeven, Elwikipedista, Tompw, Syp, Pt, PhilHibbs, Art LaPella, Bobo192, Aplusbi, Illuvatar,, John Vandenberg, C S, Viriditas, Guido del Confuso, Apyule, Flammifer, Treborbassett, Haham hanuka, Tsirel, Jason Davies, Uogl, Keenan Pepper, Doopokko, MKultra, Podstawko, Velho, Shreevatsa, Georgia guy, Sburke, Miaow Miaow, GregorB, Graham87, Belhelvie, BD2412, Zzedar, Rjwilmsi, Quiddity, Salix alba, Makaristos, Bubba73, Mkehrt, SystemBuilder, Wikihuh, Ian Pitchford, Chris Pressey, Mathbot, Margosbot, Harmil, JYOuyang, Ewlyahoocom, Gurch, Pevernagie, Sodin, Chobot, Jersey Devil, Gdrbot, Algebraist, ColdFeet, YurikBot, Jonrock, Dmharvey, Michael Slone, Red Slash, Sasuke Sarutobi, Ori.livneh, Robertvan1, Pyrotec, Froth, Zwobot, Asbl, Ott2, Tetracube, Arthur Rubin, Mursel, Pred, Dpotop, GrinBot, Auroranorth, Erik Sandberg, SmackBot, BeteNoir, Nihonjoe, KocjoBot, NathanHess, Zanetu, Aivazovsky, GraemeMcRae, Alsandro, The Famous Movie Director, DanPope, MK8, MalafayaBot, Taxipom, Nbarth, Hgrosser, Shalom Yechiel, Mhym, Lapisphil, LouScheffer, Cybercobra, Chrylis, SpiderJon, Bidabadi, Bejnar, Lambiam, ArglebargleIV, OliverTwist, Fanx, Svartkell, Barabum, JLeander, Bless sins, Mets501, Cbuckley, Quaeler, Pqrstuv, Master Yugin, RJChapman, Credner, Twas Now, Tawkerbot2, Jrmski, Devourer09, JForget, Fsswsb, CRGreathouse, Ale jrb, MatthewMain, WeggeBot, MaxEnt, Oo7565, Reywas92, A876, DumbBOT, Epbr123, Elfred, R'nway, Andyjsmith, Colin Rowat, Headbomb, Chet nc, RobHar, Luna Santin, Opelio, Storkk, GromXXVII, AlphaPhoenixDown, Alexandermiller, Rich257, WODUP, David Eppstein, Kope, DerHexer, Projectstann, AgarwalSumeet, Kguirnela, J.delanoy, Tlim7882, SiliconDioxide, Policron, KylieTastic, Majolo, ReddyVarun, Idioma-bot, VolkovBot, YB3, Jimmaths, Ptrillian, Philip Trueman, Walor, Agricola44, Sean D Martin, Fizzackerly, RiverStyx23, Brianga, ZBrannigan, SieBot, Dusti, Mandsford, Taemyr, JSpung, Oxymoron83, Faradayplank, Steven Zhang, SimonTrew, Davidjmarcus, Hatster301, OKBot, Anchor Link Bot, Classicalecon, ClueBot, Justin W Smith, Ndenison, Jrezaieb, JuPitEer, Razimantv, DragonBot, Tim32, Relata refero, Bender2k14, J.Gowers, Doucthec, Lkruijsw, FsswsbX, FiachDubh, FsswsbY, GlasGhost, BahTab, Dthomsen8, Ost316, D.M. from Ukraine, Addbot, Off we go, Ronhjones, Ozone009, Chzz, Favonian, Ehrenkater, Lightbot, Loupeter, Zorrobot, Legobot, Luckas-bot, Yobot, OrgasGirl, Charleswallingford, DBoffey, IW.HG, Dinesh smita, Пика Пика, Citation bot, ArthurBot, Tomwsulcer, FrescoBot, Dosilasol, Sławomir Biały, BenzolBot, Citation bot 1, Redrose64, Kiefer.Wolfowitz, Giovanni.ramirez, TobeBot, Duoduoduo, RjwilmsiBot, NameIsRon, Bert-Seghers, Afteread, Louis van Appeven, WikitanvirBot, K6ka, ZéroBot, Cobaltcigs, ClueBot NG, Hazhk, O.Koslowski, Mouse20080706, Lilmonty94, Rks.9690, Coltonmagnant, Bibcode Bot, Lowercase sigmabot, BG19bot, Yucong Duan, PhnomPencil, QuarkyPi, Batty-Bot, Electricmuffin11, Dexbot, Kjanow, Darcourse, Ffages, Thechickenontheroof, Marchino61, □□□□, Glaisher, Anrnusna, Monkbot, Ischeinfeld, Ameoba5, Մակար Ղազարյան, Adam Iragaël and Anonymous: 256

_ **Probabilistic method** *Source:* http://en.wikipedia.org/wiki/Probabilistic%20method?oldid=633204306 *Contributors:* Michael Hardy, Dominus, Charles Matthews, Viz, Kevinatilusa, Giftlite, Andris, Peter Kwok, Qutezuce, Paul August, Ryan Reich, Adking80, Mathbot, Maxal, Michael Slone, Shell Kinney, Ott2, Nealeyoung, Pierre de Lyon, Cydebot, Headbomb, Lantonov, Nagy, Melcombe, ClueBot, Alexbot, M4gnum0n, Addbot, Dyaa, DemocraticLuntz, Erel Segal, Psdey1, Buenasdiaz, Miym, Charvest, Citation bot 1, Agrinshp and Anonymous: 18

_ **Probability theory** *Source:* http://en.wikipedia.org/wiki/Probability%20theory?oldid=631752070 *Contributors:* Lee Daniel Crocker, Bryan Derksen, Zundark, The Anome, Miguel, Boleslav Bobcik, Patrick, Michael Hardy, Goatasaur, Den fjättrade ankan, Jonik, Bjcairns, Charles Matthews, Dysprosia, Gutsul, Johannes Hüsing, MH, Robbot, MathMartin, TMLutas, Hadal, Borislav, Tobias Bergemann, Weialawaga, Tosha, Giftlite, Raymond Meredith, ShaunMacPherson, Lethe, Fastfission, Niteowlneils, ChicXulub, Utcursch, Knutux, Ynh, Beland, APH, Maximaximax, Gbr3, Vivacissamamente, Rich Farmbrough, Paul August, Bender235, RJHall, MisterSheik, El C, Zenohockey, AvidDismantler, Hayabusa future, Bobo192, Cretog8, Zwilson, Flammifer, Obradovic Goran, Cyrillic, Mdd, Tsirel, Msh210, Rgclegg, PAR, KingTT, Frankman, Suruena, Jheald, RainbowOfLight, Oleg Alexandrov, INic, Btyner, Graham87, Porcher,

Rjwilmsi, Mayumashu, Tizio, Salix alba, FutureNJGov, Nguyen Thanh Quang, Mathbot, RexNL, Krun, Malhonen, Chobot, Wavelength, Sceptre, Spudbeach, Michael Slone, Lenthe, TheMandarin, ENeville, Robertetaylor, Trovatore, Srinivasasha, User27091, Wknight94, Hirak 99, Pb30, JoanneB, Kungfuadam, Sardanaphalus, SmackBot, Snielsen, Arjay369, Aastrup, PJTraill, Chris the speller, Bluebot, Jprg1966, MalafayaBot, DHN-bot, Wynand.winterbach, Gala.martin, Decltype, Skiminki, Andeggs, Ncmathsadist, Lambiam, Dbtfz, Bjankuloski06en, Levineps, Spebudmak, Tawkerbot2, Kurtan, Devourer09, Sleeping123, CRGreathouse, Unionhawk, Ali Obeid, Myasuda, Leroytirebiter, Omicronpersei8, Progicnet, Thijs!bot, Josephbrophy, Urdutext, AntiVandalBot, DarkAudit, Coyets, Dylan Lake, Britcruise, JAnDbot, MER-C, Drizzd, SiobhanHansa, Freshacconci, Ensign beedrill, David Eppstein, Fantastic4boy, Pax:Vobiscum, MartinBot, Numbo3, Cpiral, McSly, Roman V. Odaisky, Gill110951, JayJasper, Treisijs, JohnBlackburne, TXiKiBoT, Anonymous Dissident, Liko81, Magmi, JhsBot, Raymondwinn, Christopher Connor, Geometry guy, Arcfrk, Givegains, SieBot, Tiddly Tom, Jason Patton, Calabraxthis, Oda Mari, Cyfal, Melcombe, Elassint, ClueBot, Josang, GorillaWarfare, Justin W Smith, Lbertolotti, Excirial, PixelBot, Eeekster, Zaharous, Estirabot, 1ForTheMoney, Qwfp, XLinkBot, Gerhardvalentin, Wyatt915, Tayste, Addbot, Some jerk on the Internet, Betterusername, DutchDevil, Fluffernutter, LaaknorBot, Dyaa, Debresser, LinkFA-Bot, Booniesyeo, Tide rolls, Luckas-bot, Yobot, Ptbotgourou, TaBOT-zerem, Hairer, JackieBot, Materialscientist, Citation bot, ArthurBot, MauritsBot, Xqbot, TinucherianBot II, Capricorn42, XZeroBot, Борис Пряха, GrouchoBot, Abce2, RibotBOT, Peto1994, FrescoBot, GreenRoot, Rhalah, Goldfinger 93, Pinethicket, MastiBot, Jauhienij, DARTH SIDIOUS 2, EmausBot, PrisonerOfIce, Felix Hoffmann, K6ka, ZéroBot, ClueBot NG, Mathstat, CocuBot, Reidbaileyrocks, Bsmath802, MerllwBot, Helpful Pixie Bot, Brad7777, BattyBot, ChrisGualtieri, Illia Connell, Mogism, Crystalfile, Lugia2453, Matty.007, Sboysel, Rcehy, Monkbot and Anonymous: 220

_ **Existence theorem** *Source:* http://en.wikipedia.org/wiki/Existence%20theorem?oldid=622076234 *Contributors:* Charles Matthews, Dysprosia, Jitse Niesen, Murray Langton, Giftlite, Lupin, FunnyMan3595, Waltpohl, Ntmatter, Oleg Alexandrov, Marudubshinki, SmackBot, LeaChim, JMSwtlk, Clconway, Gmw, Cydebot, Whatever1111, Spiffyman, Classicalecon, Ideal gas equation, Addbot, Lightbot, OlEnglish, Jarble, Erik9bot, Tkuvho, Brad7777 and Anonymous: 6

_ **Collatz conjecture** *Source:* http://en.wikipedia.org/wiki/Collatz%20conjecture?oldid=633355380 *Contributors:* AxelBoldt, The Anome, Dze27, Alex.tan, Awaterl, XJaM, Edward, Michael Hardy, Oliver Pereira, Pnm, SGBailey, GTBacchus, Jocago, Schneelocke, WolfgangRieger, Timwi, Dysprosia, Twang, Phil Boswell, Fredrik, BitwiseMan, Rorro, Danjel, Giftlite, Smjg, ShaunMacPherson, MSGJ, Fropuff, Anton Mravcek, Dratman, Jason Quinn, Sietse, Chowbok, LiDaobing, Mukerjee, Pmanderson, Tyler McHenry, Urhixidur, Mschlindwein, Thorwald, Oskar Sigvardsson, Rich Farmbrough, Paul August, Bender235, ZeroOne, Sunborn, El C, Crisófilax, EmilJ, Fuzzytess, Kevin Lamoreau, ACW, 99of9, Jumbuck, Danski14, Keenan Pepper, Sligocki, Pontus, Greg Kuperberg, Hlg, Allen McC., Drbreznjev, Oleg Alexandrov, Japanese Searobin, Simetrical, OwenX, David Haslam, Oliphaunt, Miss Madeline, Wikiklrsc, Ryan Reich, Mensanator, Yoghurt, Graham87, Qwertyus, Rjwilmsi, TerrorBite, Kinu, Eyu100, Nneonneo, Bubba73, Matt Cook, Volfy, Ian Pitchford, Hashproduct, Stephantom, Akuma2k5, Bgwhite, Algebraist, Cirne, YurikBot, Laurentius, Hairy Dude, Jfeise, Stephenb, Ihope127, GeeJo, R.e.s., Hv, Zwobot, Kyle Barbour, Elkman, Ott2, StuRat, Arthur Rubin, Cedar101, Reyk, Mario23, HereToHelp, Sbyrnes321, AndrewWTaylor, SmackBot, RDBury, Herostratus, Melchoir, Pokipsy76, Fulldecent, GraemeMcRae, Brianski, Ohnoitsjamie, Betacommand, Garfieldsk,

Thumperward, PrimeHunter, Modest Genius, Samrawlins, Tamfang, Vanished User 0001, HLwiKi, Afrozenator, LouScheffer, Jgoulden, Jmnbatista, Robma, Daqu, DRLB, Paddy3118, Stefan Kohl, Kleuske, Adamarthurryan, Druseltal2005, Bejnar, Vgy7ujm, Gobonobo, JoshuaZ, Deadcode, Metao, Stephen B Streater, Supereddy, Newyorkbrad, CRGreathouse, Neelix, Reywas92, Mon4, Jibbist, Goldencako, Agamir, Elfred, Headbomb, P.gibellini, Davidhorman, Hannes Eder, Bhamv, Marcus Rowland, Holshy, Magioladitis, WolfmanSF, JamesBWatson, Swpb, Ling.Nut, Singularity, David Eppstein, Kope, JaGa, Ztobor, RaitisMath, Dukeofalba, Pedrito, Aqwis, JonMcLoone, Thisma, Eckolicious, DASonnenfeld, 28bytes, Austinmohr, Pleasantville, Jeff G., Anonymous Dissident, Aymatth2, Voorlandt, Ocolon, Rubseb, Don4of4, Jmath666, Cdehaan, Math-kika, Rjgodoy, Dmcq, Alexmagnus2, Thamaki, Sicksideburns33, LeadSongDog, Dirk P Broer, Byrialbot, Jdcox1999, Sean.hoyland, ClueBot, Paralda, Joulesm, Nnemo, Drmies, Rudiwiki1234, Dkf11, DragonBot, Mate2code, Armin Rigo, Sun Creator, AmirOnWiki, Hans Adler, Manco Capac, DumZiBoT, XLinkBot, Jed 20012, Skorpionek, SilvonenBot, Wanderingjew, Arulo1984, Addbot, Pbruckman, Debresser, Numbo3-bot, Yggdradhyl, Mps, Luckas-bot, Yobot, Legobot II, Nalfeindo, AnomieBOT, Letuño, Jim1138, Sz-iwbot, Kimsey0, Citation bot, Xqbot, Freebirth Toad, OPi, Nitrxgen, Kays666, Druiffic, Anne Bauval, Gap9551, Nippashish, Kaushiks.nitt, Shirik, Mn monkey, Shadowjams, NoldorinElf, Fredericksgary, Nicolas Perrault III, Nageh, Mirughaz, Citation bot 1, Pinethicket, Supplican, Turian, Niri.M, Gamewizard71, Unbitwise, Skrio, Bento00, ButOnMethItIs, Ludovicusmagister, Leedm777, Eloquant, Mussermaster, Twild1990, Kungfuj35u5, Glittalogik, Savh, Rodriguez-Torres, Dennis714, Jsayre64, Dmitry123456, ChuispastonBot, Nanotube09, Superboy777, ClueBot NG, David C Bailey, Deer*lake, Snotbot, Joseph Sinyor, Helpful Pixie Bot, The engineer37, Francesco.fragnito, Plantdrew, BG19bot, Adriaan Joubert, Guitarman19, TricksterWolf, Compfreak7, Ryanmcnamara10132, Moncef Elmoumen, Bingwiki08, Phoenixia1177, Mleeds12, Donald6309, ChrisGualtieri, Rluketich, Dameniphel, Deltahedron, Profeshermyguad, Данило Дубінін, MrShoe, Frosty, Jc86035, Ashley345, Contributeur62, K9re11, Franz Scheerer (Olbers), Suelru, Monkbot, Bariskanber and Anonymous: 330

_ **Combinatorial proof** *Source:* http://en.wikipedia.org/wiki/Combinatorial%20proof?oldid=573037719 *Contributors:* Michael Hardy, Mxn, Charles Matthews, Zarvok, Peter Kwok, Oleg Alexandrov, Will Orrick, Rjwilmsi, Alexb@cut-the-knot.com, Mhym, Bwsulliv, Stebulus, Sopoforic, Cydebot, MER-C, David Eppstein, Lantonov, PaulTanenbaum, Hffman, Arjayay, Marc van Leeuwen, Addbot, Citation bot 1, EmausBot, John of Reading, Joel B. Lewis, Helpful Pixie Bot, Brad7777 and Anonymous: 4

_ **Double counting (proof technique)** *Source:* http://en.wikipedia.org/wiki/Double%20counting%20(proof%20technique)?oldid=634183739 *Contributors:* Arvindn, Heron, Tomo, Michael Hardy, Dominus, Charles Matthews, Dysprosia, Populus, Henrygb, Giftlite, Almit39, Gauge, Grutness, Yiliu60, Igorpak, Shreevatsa, Btyner, Salix alba, King of Hearts, DaGizza, Jurriaan, Shyam, Melchoir, Voulouza, Eskimbot, Mhym, SirIsaacBrock, Stebulus, Cydebot, JAnDbot, David Eppstein, Leon math, Lantonov, PaulTanenbaum, Kmhkmh, Plastikspork, PixelBot, BOTarate, Marc van Leeuwen, Addbot, Twri, Anne Bauval, MathsPoetry, Citation bot 1, Wcherowi, Helpful Pixie Bot, Samoon, CitationCleanerBot, Pugiator and Anonymous: 14

_ **Mathematical object** *Source:* http://en.wikipedia.org/wiki/Mathematical%20object?oldid=599602597 *Contributors:* Michael Hardy, Andycjp, Rich Farmbrough, SixWingedSeraph, Reverendgraham, SmackBot, Byelf2007, Bn, Vaughan Pratt, Gregbard, EagleFan, Robin S, Maurice Carbonaro, Tautologist, Certes, Addbot, OlEnglish, JEN9841, Luckas-bot, Yobot, LGB, JRB-Europe, Twri, ArthurBot, FrescoBot, Tkuvho, GreenGrammarian, TobeBot, Slawekb, ZéroBot, Super-real dance, ClueBot NG, Rjs.swarnkar, Brad7777, Bg9989 and Anonymous: 19

_ **Rational number** *Source:* http://en.wikipedia.org/wiki/Rational%20number?oldid=634228447 *Contributors:* AxelBoldt, Brion VIBBER, Bryan Derksen, Zundark, Tarquin, Jan Hidders, Andre Engels, XJaM, Christian List, Toby Bartels, PierreAbbat, Roadrunner, FvdP, Stevertigo, Patrick, Michael Hardy, Wshun, MartinHarper, Ixfd64, TakuyaMurata, Mdebets, Ciphergoth, AugPi, Andres, Evercat, Panoramix, Pizza Puzzle, Hashar, Hawthorn, Revolver, Charles Matthews, Dcoetzee, Dysprosia, Jitse Niesen, Greenrd, Hyacinth, Thue, McKay, Guppy, Denelson83, Robbot, Romanm, Mayooranathan, Thunderbolt16, Henrygb, Borislav, Fuelbottle, Lupo, Tobias Bergemann, Giftlite, Gene Ward Smith, Ævar Arnfjörð Bjarmason, Lethe, Dissident, Fropuff, Dratman, Guanaco, Bovlb, Jorge Stolfi, Nayuki, Tagishsimon, Rheun, Ato, Antandrus, MarkSweep, Bob.v.R, Vina, Scott Burley, Ehamberg, Lostchicken, Mormegil, Discospinster, Notinasnaid, Paul August, El C, EmilJ, Deanos, Bobo192, Elipongo, Jung dalglish, Blotwell, Deryck Chan, Obradovic Goran,

Jumbuck, Msh210, Alansohn, Ncik, Silver86, Wtmitchell, Velella, L33th4x0rguy, Mikeo, Btornado, Oleg Alexandrov, Linas, StradivariusTV, Prashanthns, Graham87, BD2412, SixWingedSeraph, Yurik, Zzedar, Jshadias, Josh Parris, Sdornan, Salix alba, The wub, Bhadani, Yamamoto Ichiro, David H Braun (1964), DVdm, YurikBot, Wavelength, Pseudomonas, NawlinWiki, Rick Norwood, E rulez, Hennobrandsma, Charlie Wiederhold, Lt-wiki-bot, Gesslein, GrinBot, Crystallina, Hydrogen Iodide, Zerida, Unyoyega, Yamaguchi□□, Gilliam, Skizzik, Persian Poet Gal, Raymond arritt, Raja Hussain, MalafayaBot, Akanemoto, DHN-bot, Can't sleep, clown will eat me, Shunpiker, Grover cleveland, Daqu, Nakon, Dreadstar, NickPenguin, Salamurai, Bidabadi, Vina-iwbot, Ck lostsword, SashatoBot, Lambiam, Nishkid64, Btritchie, Kuru, CorvetteZ51, Cronholm144, Gobonobo, Jim.belk, Ekrub-ntyh, Loadmaster, Dr Greg, Mets501, Lee Carre, Quaeler, Jazzcello, Majora4, ILikeThings, JForget, CRGreathouse, Wafulz, Penbat, SuperMidget, Cydebot, Worthingtonse, Boardhead, Epbr123, Koeplinger, Martin Hogbin, Marek69, Wmasterj, AbcXyz, Escarbot, Ju66l3r, AntiVandalBot, Vvidetta, Edokter, Braindrain0000, JAnDbot, MER-C, Smiddle, .anacondabot, Connormah, Bongwarrior, VoABot II, Twsx, WODUP, Avicennasis, Seberle, JoergenB, DerHexer, GermanX, Aschmitz, Tokidoki27, MartinBot, Kostisl, Jarhed, J.delanoy, Katalaveno, Stwitzel, SJP, Policron, CompuChip, Angular, DavidCBryant, Wikieditor06, VolkovBot, Johan1298, TallNapoleon, AlnoktaBOT, VasilievVV, TXiKiBoT, Maximillion Pegasus, Dendodge, Broadbot, Hrundi Bakshi, Maxim, Wolfrock, Synthebot, Allan1114, Monty845, AlleborgoBot, LuigiManiac, EmxBot, Omerks, Demmy, SieBot, Legion fi, Yintan, Bentogoa, Flyer22, Tiptoety, Macy, OKBot, Diego Grez, Angielaj, Randomblue, Dolphin51, Troy 07, Joe Photon, ClueBot, Rumping, PipepBot, Fyyer, The Thing That Should Not Be, Cliff, Jpcs, Dylan620, Paritybit, Jusdafax, Eeekster, Cenarium, Jotterbot, Aitias, Katanada, SoxBot III, Crazy Boris with a red beard, AgnosticPreachersKid, Spitfire, Crapme, Zrs 12, NellieBly, Dnvrfantj, RyanCross, HexaChord, Addbot, Proofreader77, ConCompS, Friginator, Ronhjones, Tutter-Mouse, Fieldday-sunday, Skyezx, LaaknorBot, LinkFA-Bot, Jaydec, AgadaUrbanit, Numbo3-bot, Tide rolls, MZaplotnik, Teles, LuK3, Luckas-bot, Yobot, Tempodivalse, 9258fahsflkh917fas, Kingpin13, Materialscientist, Erikekahn, OllieFury, Xelnx, ArthurBot, Xqbot, TinucherianBot II, Capricorn42, Doctor Rosenberg, Isheden, Hackabhihack, RibotBOT, The Wiki Octopus, Aaron Kauppi, Stlrams22, Lothar von Richthofen, DivineAlpha, Tkuvho, Pinethicket, ShadowRangerRIT, I dream of horses, Adlerbot, MarcelB612, BigDwiki, Jujutacular, Reconsider the static, Tim1357, ItsZippy, Jonkerz, Dinamik-bot, Vrenator, JuanGabrielRobalino, Stroppolo, Luhar1997, TjBot, Bento00, TomT0m, EmausBot, Khalidmathematics, Slightsmile, TuHan-Bot, Wikipelli, ReySquared, Alpha Quadrant, Quondum, Mburdis, Wayne Slam, L Kensington, Donner60, Chewings72, Orange Suede Sofa, 28bot, ClueBot NG, This lousy T-shirt, MikuMiku-Cookie, Dfarrell07, Mpaa, Ichliebepferde, Kasirbot, Widr, WikiPuppies, Calabe1992, Kelsi1122, Sheilds, Darksonn, Ihatechickens214, Mercrutio, Nick white03, Glacialfox, Bodema, Kishugoyal, Teammm, Pratyya Ghosh, Dexbot, FoCuSandLeArN, Mcash001, Scharan09, Lugia2453, Epicgenius, Encyclopedia 12, Harry styles5555554, Imperial Marshmallow, AmaryllisGardener, Maxtheaxe1999, Ray Lightyear, Zenibus, JDiala, Wikiwonka7777, Moorelife, Programmer 112, Wilson Widyadhana, Anirudh Babu, DemonKiller3527, Engilukol albert and Anonymous: 578

102.4. TEXT AND IMAGE SOURCES, CONTRIBUTORS, AND LICENSES 779

_ **Statistical proof** *Source:* http://en.wikipedia.org/wiki/Statistical%20proof?oldid=628383160 *Contributors:* Michael Hardy, Wtmitchell, Btyner, Rjwilmsi, Vegaswikian, Piet Delport, SmackBot, G716, AndrewHowse, Conquistador2k6, AlasdairBailey, Melcombe, Tautologist, Mild Bill Hiccup, SchreiberBike, Qwfp, Tayste, Ettrig, Are you ready for IPv6?, J04n, Thompsma, Pollinosisss, RjwilmsiBot, Helpful Pixie Bot, Qetuth, Illia Connell and Anonymous: 3

_ **Pure mathematics** *Source:* http://en.wikipedia.org/wiki/Pure%20mathematics?oldid=632388056 *Contributors:* Derek Ross, Michael Hardy, Ahoerstemeier, Charles Matthews, Jitse Niesen, Markhurd, Bevo, Finlay McWalter, Sander123, Altenmann, MathMartin, Giftlite, Jareha, Marsvin, Paul August, Jpgordon, Dirac1933, Critical, E=MC^2, Isnow, BD2412, Zbxgscqf, Salix alba, Volfy, Dougluce, Yurik-Bot, TexasAndroid, Annabel, Larry laptop, EEMIV, DeadEyeArrow, JMBrust, Smurrayinchester, Canadianism, Meegs, Sardanaphalus, JJL, Selfworm, Mmernex, Eskimbot, Cazort, MalafayaBot, Cybercobra, Diocles, Bidabadi, Cronholm144, Eastlaw, DonkeyKong64, Rjms, Gregbard, Purplenite, Chrislk02, Thijs!bot, N5iln, Mojo Hand, Nick Number, Delataur, AntiVandalBot, Salgueiro, VoABot II, Systemlover, Rajpaj, Manticore, Trusilver, Suenm, Maurice Carbonaro, Smeira, DavidCBryant, Mplourde, JohnBlackburne, Barneca, TXiKiBoT, Anonymous Dissident, Someguy1221, ClueBot, Polyamorph, Excirial, Prancibaldfpants, Dekisugi, Sandrobt, XLinkBot, Nepenthes, Leonini, Addbot, Fgnievinski, AkhtaBot, LaaknorBot, Favonian, ChenzwBot, Numbo3-bot, Tide rolls, Lightbot, Pugno, Luckasbot, Yobot, Eadaless, Ptbotgourou, Virgilius Eremite, Obersachsebot, Xqbot, Anna Frodesiak, Naylor101, Jandalhandler, Atkins450, RjwilmsiBot, EmausBot, WikitanvirBot, Anirudh Emani, Chricho, Sara cuttie, ClueBot NG, Satellizer, Bcapetta, TCN7JM, Brad7777, Carliitaeliza, Makecat-bot, Dingle665, AmaryllisGardener, Purnendu Karmakar, Hierarchivist and Anonymous: 104

_ **Cryptography** *Source:* http://en.wikipedia.org/wiki/Cryptography?oldid=633623023 *Contributors:* AxelBoldt, WojPob, LC, Brion VIBBER, Mav, Uriyan, Zundark, The Anome, Taw, Ap, Tao, Ted Longstaffe, Dachshund, Arvindn, Gianfranco, PierreAbbat, Ortolan88, Roadrunner, Boleslav Bobcik, Maury Markowitz, Imran, Graft, Heron, Sfdan, Stevertigo, Nevilley, Patrick, Chas zzz brown, Michael Hardy, GABaker, Dante Alighieri, Liftarn, Ixfd64, Cyde, TakuyaMurata, Karada, Dori, (, Goatasaur, Card, Ahoerstemeier, DavidWBrooks, ZoeB, Theresa knott, Cferrero, Jdforrester, Julesd, Glenn, Kylet, Nikai, Andres, Cimon Avaro, Evercat, Delifisek, Dgreen34, Schneelocke, Norwikian, Revolver, Novum, Htaccess, Timwi, Wikiborg, Dmsar, Ww, Dysprosia, Jitse Niesen, Phr, The Anomebot, Greenrd, Dtgm, Tpbradbury, GimmeFuel, K1Bond007, Tempshill, Ed g2s, Raul654, Rbellin, Pakaran, Jeffq, Ckape, Robbot, Fredrik, Chris 73, RedWolf, Donreed, Altenmann, Kuszi, Securiger, Georg Muntingh, MathMartin, Jsdeancoearthlink.net, Academic Challenger, Meelar, Timrollpickering, Rasmus Faber, Cyrius, Superm401, Ludraman, Tobias Bergemann, Dave6, Snobot, Giftlite, Dbenbenn, Jacoplane, HippoMan, Wolfkeeper, Netoholic, Farnik, PeruvianIlama, Michael Devore, Yekrats, Per Honor et Gloria, Sietse, Mboverload, Ferdinand Pienaar, Matt Crypto, Mobius, Neilc, Gubbubu, Geni, CryptoDerk, Antandrus, Beland, Vanished user 1234567890, Pale blue dot, Rdsmith4, APH, Mzajac, Euphoria, SimonLyall, Oiarbovnb, TiMike, Ta bu shi da yu, Freakofnurture, Monkeyman, Blokhead, Heryu, Mark Zinthefer, Moverton, Discospinster, Rich Farmbrough, Guanabot, MaxMad, ArnoldReinhold, YUL89YYZ, Ivan Bajlo, Paul August, DcoetzeeBot, Bender235, TerraFrost, Surachit, JRM, Prsephone1674, Bobo192, Harley peters, AnyFile, John Vandenberg, Myria, Jericho4.0, Davidgothberg, Slipperyweasel, Wrs1864, ClementSeveillac, M5, Stephen G. Brown, LoganK, Msh210, Wereldburger758, Alansohn, JYolkowski, Dhar, Mo0, Fg, Seamusandrosy, Complex01, ABCD, Logologist, InShaneee, Avenue, Snowolf, Super-Magician, Saga City, Zyarb, Daedelus, Egg, H2g2bob, Vadim Makarov, Richwales, Oleg Alexandrov, Zntrip, Woohookitty, Mindmatrix, Starwiz, Deeahbz, Jacobolus, Madchester, E=MC^2, Brentdax, Duncan.france, Nfearnley, Shmitra, Jok2000, Wikiklrsc, Mangojuice, SDC, Plrk, DarkBard, Cedrus-Libani, Stefanomione, Turnstep, Jimgawn, Tslocum, Graham87, Abach, FreplySpang, Vyse, JIP, Sinar, Jorunn, Sjakkalle, Ner102, Rjwilmsi, Demian12358, Adjusting, MarSch, Mike Segal, Edggar, Miserlou, HappyCamper, Brighterorange, The wub, DoubleBlue, Volfy, CBR1kboy, Vuong Ngan Ha, RobertG, Mathbot, Gouldja, PleaseSendMoneyToWikipedia, Crazycomputers, Jameshfisher, RobyWayne, KFP, King of Hearts, Chobot, Manscher, Roboto de Ajvol, Siddhant, Wavelength, Laurentius, Auyongcheemeng, Mukkakukaku, RussBot, Lpmusix, Pigman, Manop, The1physicist, Gaius Cornelius, Chaos, Zeno of Elea, Nawlin-Wiki, Welsh, Joel7687, Exir Kamalabadi, Proidiot, ONEder Boy, Schlafly, DavidJablon, Thiseye, Dhollm, Peter Delmonte, Misza13, Grafikm fr, Xompanthy, Deckiller, BOT-Superzerocool, Jeremy Visser, FF2010, 21655, Papergrl, Closedmouth, Nemu, CharlesHBennett, Aeon1006, Peyna, Bernd Paysan, Echartre, Anclation, Wbrameld, Who-is-me, MagneticFlux, Crazyquesadilla, Endymi0n, Dr1819, DVD R W, ChemGardener, Yakudza, A bit iffy, SmackBot, Sean.nobles, Mmernex, Nihonjoe, 1dragon, Impaciente, Uncle Lemon, Jacek Kendysz, Jagged 85, Jrockley, David G Brault, BiT, JohnMac777, Mauls, Peter Isotalo, Gilliam, Ohnoitsjamie, Hmains, Skizzik, Chaojoker, Lakshmin, Chris the speller, Ciacchi, Agateller, Hibbleton, Thumperward, Delfeye, Snori, Alan smithee, PrimeHunter, Iago4096,

NYKevin, DevSolar, Vkareh, ZachPruckowski, DrDnar, Wes!, Rashad9607, Alieseraj, Kazov, Wonderstruck, Maxt, DRLB, OutRIAAge, Sovietmah, Bidabadi, Chungc, Andrewrabbott, Harryboyles, Dr. Sunglasses, Molerat, Fatespeaks, Ksn, Sidmow, JoshuaZ, Minna Sora no Shita, ManiF, Michael miceli, Jacopo, Ryanwammons, Slayemin, Chrisd87, Eltzermay, Meco, TastyPoutine, Dhp1080, Serlin, DeathLoofah, Drink666, Hectorian, DouglasCalvert, RudyB, Judgesurreal777, Pegasus1138, Detach, Shenron, Nightswatch, Gilabrand, Tawkerbot2, Chetvorno, Jafet, Powerslide, Sansbras, CRGreathouse, Hermitage17, Crownjewel82, GeorgeLouis, Thehockeydude44, CWY2190, Saoirse11, Raghunath88, Blackvault, Grandexandi, Cydebot, Ntsimp, Mblumber, John Yesberg, Gogo Dodo, Corpx, Tawkerbot4, XP105, Kozuch, Brad101, Omicronpersei8, Robertsteadman, Antura, Pallas44, Saber Cherry, Oerjan, Mojo Hand, Lotte Monz, Dgies, DPdH, Scircle, AntiVandalBot, Luna Santin, Jj137, Dylan Lake, Oddity-, G Rose, JAnDbot, Monkeymonkey11, Komponisto, WPIsFlawed, Hut 8.5, GurchBot, SCCC, Jahoe, Richard Burr, Acroterion, Koolkirby, Calcton, Hong ton po, MoleRat, CrazyComputers, Heinze, MooCowz69, Connormah, Bongwarrior, VoABot II, Michi.bo, Nyttend, Homunq, KConWiki, David Eppstein, NoychoH, Havanafreestone, JaGa, Mmustafa, BetBot, Rettetast, Speck-Made, David Nicoson, Glrx, CommonsDelinker, Artaxiad, J.delanoy, Hans Dunkelberg, Maurice Carbonaro, Syphertext, Cadence-, Darth Mike, Salih, MezzoMezzo, Touisiau, AntiSpamBot, SJP, Wilson.canadian, Chandu iiet, R Math, Treisijs, Ross Fraser, Adam7117, Remi0o, Reddy212, Cralar, Tw mama, Mrstoltz, VolkovBot, Thomas.W, Macedonian, DSRH, JohnBlackburne, Jimmaths, Greatdebtor, Mercurish, TXiKiBoT, GimmeBot, Xnquist, Qxz, David-Saff, Ocolon, TedColes, Praveen pillay, Abdullais4u, Msanford, LeaveSleaves, Geometry guy, Bkassay, Rich5411, Symane, Legoktm, NHRHS2010, Radagast3, Botev, SieBot, TJRC, Nihil novi, Moonriddengirl, James Banogon, Caltas, Yintan, Browner87, Mayevski, Yob kejor, Branger, Enti342, WannabeAmatureHistorian, Lightmouse, Skippydo, StaticGull, Hamiltondaniel, Secrefy, PerryTachett, Tom Reedy, Joel Rennie, Dlrohrer2003, Leranedo, WikipedianMarlith, ClueBot, Binksternet, The Thing That Should Not Be, JuPitEer, Niceguyedc, Mspraveen, Sv1xv, Excirial, Infomade, Ziko, Lunchscale, Jpmelos, Kakofonous, Unmerklich, Aitias, Johnuniq, MasterOfHisOwnDomain, Skunkboy74, Bletchley, Hotcrocodile, IAMTrust, Bill431412, Kbdankbot, IsmaelLuceno, B Fizz, Addbot, Ghettoblaster, Some jerk on the Internet, DOI bot, Mabdul, CL, Madmax8712, Blethering Scot, TutterMouse, Gus Buonafalce, Fieldday-sunday, D0762, Bte99, Leszek Jańczuk, Harrymph, MrOllie, Protonk, AndersBot, Porkolt60, Maslen, 5 albert square, Hollerme, Tide rolls, Artusstormwind, Luckas-bot, Yobot, 2D, MarioS, Amirobot, Anypodetos, Maxí, AnomieBOT, BeEs1, Rubinbot, Jim1138, Galoubet, AdjustShift, Gowr, Wiki5d, Materialscientist, Rohitnwg, Citation bot, Clickey, Xtremejames183, Xqbot, Tomasz Dolinowski, Cluckkid, Capricorn42, Permethius, Jessicag12, ProtectionTaggingBot, Omnipaedista, Shirik, Brandon5485, Kernel.package, Smallman12q, Aaron Kauppi, WhatisFeelings?, StevieNic, 00mitpat, FrescoBot, Dogposter, Tobby72, Nageh, Krj373, Mark Renier, D'ohBot, Mohdavary, LaukkuTheGreit, HamburgerRadio, Citation bot 1, Geoffreybernardo, Quartekoen, Pinethicket, Jonesey95, Hoo man, Rochdalehornet, Pbsouthwood, Strigoides, Pezanos, Lightlowemon, FoxBot,Wsu-dm-a, عقيل كاشف, Lotje, PPerviz, Vrenator, Aoidh, Diannaa, Social780

workerking, Sonam.r.88, Dienlei, Episcopus, RjwilmsiBot, VernoWhitney, Church074, Skamecrazy123, EmausBot, John of Reading, Immunize, Udopr, Japs 88, GoingBatty, Slightsmile, Beleary, MithrandirAgain, Akerans, DanDao, OnePt618, Msaied75, FrankFlanagan, Donner60, Dev-NJITWILL, Herk1955, Jramio, Rememberway, ClueBot NG, Wcherowi, Frapter, Nikola1891, Lord Roem, Ap375-NJITWILL, Braincricket, Widr, Mvoorzanger, Kapanidze, Dzu33, Sprishi, BG19bot, 2pem, Hdrugge, Chrisbx1, Wiki13, Anubhab91, Mm32pc, ZipoBibrok5x10^8, Drift chambers, Difbobatl, Brad7777, Sciguystfm, Giacomo.vacca, Winston Chuen-Shih Yang, Melenc, Sam Edward c, OldishTim, Dexbot, Kushalbiswas777, Denis Fadeev, Numbermaniac, Bobanobahoba, Hamerbro, WiHkibew, JustAMuggle, Fshtea, Faizan, Sachin Hariharan1992, Ac130195, Tentinator, Hendrick 99, Raseman, R00stare, Alidad1261, NorthBySouthBaranof, Orhanozkilic, Jianhui67, Whizz محمد علي عراقي ال, 40, ₫b, GrantWishes, Joineir, Ninja1123, JohnDoe4000, Claw of Slime, Monkbot, Cayelr, Jordanbailey123456789, Shammie23, Ephemeratta, Phayzfaustyn, Je.est.un.autre, Redkilla007, Whikie, TD-Linux, Bagulbol, Crypto Funcault, Sizzy1337, Pcfan500 and Anonymous: 672

_ **Number theory** *Source:* http://en.wikipedia.org/wiki/Number%20theory?oldid=631521192 *Contributors:* AxelBoldt, Derek Ross, Calypso, Brion VIBBER, Mav, Zundark, Tarquin, XJaM, Michael Shulman, Christian List, Miguel, Twilsonb, Stevertigo, TeunSpaans, Michael Hardy, Booyabazooka, Ixfd64, Dcljr, GTBacchus, Delirium, Minesweeper, Ams80, Ahoerstemeier, Nikai, Rotem Dan, Iorsh, [212], Schneelocke, Hashar, Markb, Revolver, Charles Matthews, Timwi, Dcoetzee, Dysprosia, Jitse Niesen, The Anomebot, Xiaodai, Tpbradbury, Sabbut, Jose Ramos, Qianfeng, Finlay McWalter, Bearcat, Robbot, Jaredwf, Fredrik, Romanm, Lowellian, Gandalf61, Fuelbottle, Lupo, PrimeFan, Jleedev, Ancheta Wis, Giftlite, Recentchanges, Pretzelpaws, Lethe, Fropuff, Everyking, Gubbubu, Gadfium, LiDaobing, Antandrus, Beland, Robert Brockway, Bob.v.R, Khaosworks, Karol Langner, APH, Stefan64, Tsemii, Xmlizer, Rich Farmbrough, Paul August, Bender235, ESkog, Ben Standeven, Appleboy, Tompw, El C, Jpgordon, Mysteronald, Maurreen, Hagerman, Pearle, Storm Rider, Msh210, Alansohn, Arthena, Neonumbers, Diego Moya, Velella, SidP, Evil Monkey, CloudNine, Dirac1933, Igorpak, HenryLi, Oleg Alexandrov, Mcsee, Richard Arthur Norton (1958-), Linas, Unixer, Jimbryho, Ruud Koot, Wikiklrsc, GregorB, Dionyziz, Graham87, Dpv, Mayumashu, R.e.b., The wub, DoubleBlue, Sango123, Vuong Ngan Ha, FlaBot, Mathbot, Malhonen, Scythe33, Haonhien, Chobot, Digitalme, YurikBot, Wavelength, Lexi Marie, Lenthe, JabberWok, Stassats, Joth, Welsh, Ino5hiro, Ms2ger, Kompik, Lt-wiki-bot, Arthur Rubin, Willtron, GrinBot, That Guy, From That Show!, Marquez, Sardanaphalus, SmackBot, Mmernex, Rebollo fr, McGeddon, Jagged 85, Rouenpucelle, AustinKnight, Hmains, Anastasios, Chris the speller, Bluebot, ChuckHG, Prime-Hunter, MalafayaBot, Spellchecker, DHN-bot, Colonies Chris, Sct72, Modest Genius, Ianmacm, Lwassink, Bidabadi, Andrew Dalby, SashatoBot, Lambiam, ArglebargleIV, UberCryxic, Evildictaitor, Tdudkowski, Jim.belk, Gary13579, Stwalkerster, Childzy, Mets501, Mathsci, Kripkenstein, Joseph Solis in Australia, LDH, Zero sharp, Igoldste, Courcelles, Tawkerbot2, Stifynsemons, CRGreathouse, CmdrObot, Mikeliuk, Sdorrance, Chrisahn, Ken Gallager, Myasuda, Kronecker, Doctormatt, Gogo Dodo, Karl-H, Thijs!bot, Epbr123, Atmd, O, Bhowmickr, Marek69, Woody, RobHar, Sherbrooke, Escarbot, Luna Santin, Marquess, Qwerty Binary, Normanzhang, JAnDbot, MER-C, Hut 8.5, Magioladitis, JamesBWatson, Wlod, Usien6, Kroposky, Bubba hotep, Systemlover, NJR ZA, Kope, DerHexer, Khalid Mahmood, TheRanger, Vandermude, R'n'B, Nono64, J.delanoy, Trusilver, Numbo3, Maurice Carbonaro, Smeira, Tarotcards, Policron, CompuChip, Milogardner, CombFan, Treisijs, VolkovBot, JohnBlackburne, Philip Trueman, TXiKiBoT, Hotjava, Anna Lincoln, Plclark, Magmi, Pleaseee, Broadbot, Kmhkmh, Blurpeace, Joseph A. Spadaro, Symane, GirasoleDE, SieBot, Calliopejen1, Groove-Dog, Iames, KoenDelaere, S2000magician, Amahoney, ClueBot, MeowMeow163, Justin W Smith, Paulsavala, Drmies, Mild Bill Hiccup, DragonBot, PixelBot, Bercant, Cenarium, Arjayay, Jotterbot, H.Marxen, Crowsnest, XLinkBot, Marc van Leeuwen, Killthesteel, Ajcheema, Virginia-American, Addbot, Betterusername, Ronhjones, SpillingBot, Download, Uncia, Ausefi1900, Feketekave, Bluebusy, TeH nOmInAtOr, Luckas-bot, Yobot, MinorProphet, Zhouhaigang, Xylune, AnomieBOT, Rubinbot, Royote, Materialscientist, Citation bot, Maxis ftw, ArthurBot, Gypsydave5, Xqbot, Anne Bauval, Tyrol5, Vaywatch, GrouchoBot, Charvest, Raulshc, Aprogrammer, Lexy-lou, Bekus, DanRawsthorne, FrescoBot, Imbalzanog, LucienBOT, Tobby72, BrideOfKripkenstein, Machine Elf 1735, Pinethicket, MarcelB612, Artorio, Garald, Jauhienij, Tim1357, FoxBot, Dinamik-bot, Vrenator, Reach Out to the Truth, Korepin, KurtSchwitters, Ccrazymann, EmausBot, Lollipopweare, Fly by Night, EleferenBot, Jmencisom, Wikipelli, Bethnim, ZéroBot, Knight1993, D.Lazard, Bobdylan1234567, Donner60, Nobrook, Unga Khan, Anita5192, ClueBot NG, Satellizer, Helpful Pixie Bot, PhnomPencil, JohnChrysostom, AvocatoBot, Brad7777, Weierstrass1, Ducknish, JYBot, Dexbot, Deltahedron, Apdenum, Jamesx12345, Prem nath singh, Jodosma, SakeUPenn, Syferion, Programmingcaffeine, ProfessorMoriarty1811, Canto55 and Anonymous: 278

_ **Mathematical statistics** *Source:* http://en.wikipedia.org/wiki/Mathematical%20statistics?oldid=631469117 *Contributors:* Dcljr, Cvore, Den fjättrade ankan, Charles Matthews, Dysprosia, Creidieki, HedgeHog, Rich Farmbrough, Dallashan, Joolz, Avenue, Igny, Acerperi,

NeoUrfahraner, Joe Decker, Algebraist, Neilbeach, Number 57, Bota47, Sardanaphalus, JJL, SmackBot, Boris Barowski, MalafayaBot, Dreadstar, Richard001, G716, Bjankuloski06en, Andrew Davidson, CBM, Panda17, Lehalle, Chrislk02, Thijs!bot, Soulviver, JAnDbot, Ugajin, R'n'B, J.delanoy, VolkovBot, JohnBlackburne, AlleborgoBot, Melcombe, ClueBot, Turbojet, Qwfp, Addbot, Luckas-bot, Yobot, AnomieBOT, Jim1138, Materialscientist, Obersachsebot, Xqbot, Capricorn42, GrouchoBot, AstaBOTh15, DrilBot, Kiefer.Wolfowitz, H.ehsaan, EmausBot, ZéroBot, ChuispastonBot, ClueBot NG, Helpful Pixie Bot, Marcocapelle, Brad7777, Illia Connell, APerson, Dexbot, Brirush, Seppi333 and Anonymous: 26

_ **Computer-assisted proof** *Source:* http://en.wikipedia.org/wiki/Computer-assisted%20proof?oldid=625595169 *Contributors:* Poor Yorick, Dcoetzee, Furrykef, Stephan Schulz, Giftlite, Beland, Bender235, Rgdboer, John Vandenberg, Ultra megatron, Guthrie, Alai, Salix alba, Mallocks, Zvika, Bird of paradox, Nbarth, RyanEberhart, JonHarder, Germandemat, Acdx, Barabum, Skapur, CRGreathouse, Matthew Treder, Roccorossi, Hilgerdenaar, Oddity-, David Eppstein, MaD70, VanishedUserABC, Educres, Alik Kirillovich, Classicalecon, H.Marxen, Dthomsen8, MystBot, Addbot, Lightbot, Ptbotgourou, AnomieBOT, GnawnBot, Thosjleep, Mikrosam Akademija 3, Afteread, Helpful Pixie Bot, Hellachaz, Limit-theorem, Jmio17 and Anonymous: 22

_ **Zermelo–Fraenkel set theory** *Source:* http://en.wikipedia.org/wiki/Zermelo%E2%80%93Fraenkel%20set%20theory?oldid=632950757 *Contributors:* AxelBoldt, Matthew Woodcraft, Zundark, Tarquin, Toby Bartels, Dwheeler, Patrick, Michael Hardy, MartinHarper, Bcrowell, Chinju, Haakon, Habj, Tim Retout, Schneelocke, Charles Matthews, Dcoetzee, Dysprosia, Hyacinth, VeryVerily, Fibonacci, JohnH, Aleph4, Rursus, Tobias Bergemann, Giftlite, Smjg, Dratman, CyborgTosser, Mellum, Jorend, Ajgorhoe, Tarantoga, David Sneek, Bender235, Elwikipedista, Peter M Gerdes, Nortexoid, Obradovic Goran, Msh210, Suruena, TXlogic, Gible, Oleg Alexandrov, Joriki, OwenX, Drostie, Ma Baker, Hdante, Esben, Dionyziz, MarSch, Salix alba, R.e.b., STarry, Chobot, Karch, YurikBot, Hairy Dude, Michael Slone, Piet Delport, Ksnortum, Ogai, Trovatore, Twin Bird, Expensivehat, Insipid, Jpbowen, Crasshopper, Wknight94, Arthur Rubin, Josh3580, Banus, Otto ter Haar, Schizobullet, A bit iffy, SmackBot, Fulldecent, Mhss, Darth Panda, Foxjwill, Tsca.bot, Miguel1626, TKD, Allan McInnes, Grover cleveland, Acepectif, Jon Awbrey, Meni Rosenfeld, Stefano85, Vina-iwbot, Noegenesis, Rainwarrior, Dicklyon, Tophtucker, JRSpriggs, CRGreathouse, CBM, Myasuda, Gregbard, Awmorp, Thijs!bot, Whooooooknows, Odoncaoa, Jirka6, VictorAnyakin, JAnDbot, Quentar, Giler, Mathfreq, Omicron18, JustinRosenstein, Diroth, The Real Marauder, Numbo3, Ttwo, Trumpet marietta 45750, Policron, JavierMC, The enemies of god, Pasixxxx, Magmi, Bistromathic, Henry Delforn (old), Jjepfl, C xong, JP.Martin-Flatin, Alexbot, Iohannes Animosus, Palnot, Marc van Leeuwen, Addbot, Matěj Grabovský, Yobot, AnomieBOT, Materialscientist, La comadreja, Control.valve, VladimirReshetnikov, Nicolas Perrault III, Andrewjameskirk, NSH002, Tkuvho, Zdorovo, ClueBot

NG, Chetrasho, Snotbot, Helpful Pixie Bot, Brad7777, Daysrr, Khazar2, Jochen Burghardt, Mark viking and Anonymous: 112

_ **Gödel's incompleteness theorems** *Source:* http://en.wikipedia.org/wiki/G%C3%B6del's%20incompleteness%20theorems?oldid=633209113 *Contributors:* AxelBoldt, Joao, Chenyu, LC, Lee Daniel Crocker, Mav, Bryan Derksen, The Anome, Tarquin, Jan Hidders, Andre Engels, Danny, MadSurgeon, SimonP, Camembert, Genneth, Olivier, Michael Hardy, Tim Starling, Kwertii, Dominus, Lousyd, MartinHarper, Wapcaplet, Chinju, TakuyaMurata, GTBacchus, Eric119, CesarB, HarmonicSphere, ThirdParty, Ootachi, Suisui, Den fjättrade ankan, Mark Foskey, Александър, BuzzB, AugPi, Tim Retout, Rotem Dan, Evercat, Madir, Gamma, Rzach, Charles Matthews, Timwi, Dcoetzee, Reddi, Dysprosia, Wikid, Doradus, Markhurd, Lfwlfw, Hyacinth, Bevo, Joseaperez, .mau., Gakrivas, Pakaran, Ldo, David.Monniaux, Dmytro, Owen, Aleph4, Gandalf61, MathMartin, Bethenco, Rasmus Faber, Bkell, Mjscud, Ruakh, Tobias Bergemann, Jimpaz, Ancheta Wis, Tosha, Matt Gies, Giftlite, Lethe, Lupin, Fropuff, Mellum, Waltpohl, Geoffroy, Wikiwikifast, Andris, Bovlb, Sundar, Prosfilaes, Siroxo, Khalid hassani, C17GMaster, Neilc, Utcursch, Andycjp, Sigfpe, Toytoy, Gdr, SarekOfVulcan, Gdm, Lightst, Antandrus, Beland, Amoss, APH, DragonflySixtyseven, Mike Storm, Elroch, Sam Hocevar, Asbestos, Karl Dickman, Gazpacho, D6, Sysy, 4pq1injbok, Guanabot, FT2, Cacycle, Smyth, Francis Davey, Maksym Ye., Paul August, Chalst, Crisófilax, EmilJ, Lance Williams, DG, AshtonBenson, Jumbuck, Keenan Pepper, Hu, Wtmitchell, Simplebrain, Cal 1234, GabrielF, Gene Nygaard, MIT Trekkie, TXlogic, Oleg Alexandrov, Revived, Billhpike, OwenX, Rodrigo Rocha, Pol098, Ruud Koot, Apokrif, Frungi, Waldir, Marudubshinki, MACherian, Mandarax, Graham87, Qwertyus, Rjwilmsi, Tim!, Kinu, WoodenTaco, RCSB, Pasky, R.e.b., Magidin, FlaBot, VKokielov, Docbug, Mathbot, Jrtayloriv, Celendin, Sperxios, Xelloss, Eric.dane, Chobot, YurikBot, Wavelength, Hairy Dude, Dmharvey, Rcaetano, Me and, Ksnortum, Jabber-Wok, KSmrq, Bergsten, SpuriousQ, Thoreaulylazy, Vibritannia, Ytcracker, Aatu, Trovatore, Długosz, JoeBruno, Dogcow, Nick, Anetode, Philosofool, Guruparan18, Hakeem.gadi, Liyang, Arthur Rubin, Curpsbot-unicodify, GrinBot, Finell, SolarMcPanel, SmackBot, RDBury, Selfworm, BeteNoir, InverseHypercube, Melchoir, SaxTeacher, Pokipsy76, Mandelum, Biedermann, Brick Thrower, Srnec, The Rhymesmith, Chris the speller, MagnusW, Irving Anellis, RoyArne, MalafayaBot, Kostmo, Jdthood, Charles Moss, Grover cleveland, AndySimpson, Trifon Triantafillidis, John wesley, Kid A, LoveMonkey, Rhkramer, Byelf2007, Zchenyu, Lambiam, Wvbailey, Evildictaitor, Illythr, Mets501, ASKingquestions, Sgutkind, Nbhatla, Quaeler, Dan Gluck, Jason.grossman, Dreftymac, Joseph Solis in Australia, Zero sharp, Wfructose, Matthew Kornya, JRSpriggs, CRGreathouse, Geremia, Diegueins, CBM, Nicolaennio, Gregbard, Nilfanion, Sopoforic, Cydebot, Pce3@ij.net, Steel, Peterdjones, TicketMan, Mon4, Blaisorblade, Michael C Price, DumbBOT, Robertinventor, Morgaladh, Clickheretologin, Thijs!bot, Headbomb, Tamalet, Towopedia, Turkeyphant, Kborer, Stan the fisher, Joegoodbud, Gioto, Blue Tie, Ad88110, NBeale, Husond, Avaya1, Gavia immer, JamesBWatson, Renosecond, Tedickey, Baccyak4H, K95, David Eppstein, Exiledone, Pavel Jelínek, Infovarius, Franp9am, Eliko, Abecedare, Fredeaker, Warut, CeilingCrash, DadaNeem, Policron, Mad7777, DavidCBryant, Merzul, Quux0r, Germanium, VolkovBot, DDSaeger, AlnoktaBOT, TXiKiBoT, Sacramentis, Lou2261, CaptinJohn, Broadbot, David in DC, Lifeisfun0007, Latulla, Popopp, Gbawden, Davebuehler, YohanN7, SieBot, Simplifier, Iamthedeus, Gerakibot, Noaqiyeum, Likebox, Flyer22, Scouto2, OKBot, Valeria.depaiva, D14C050, IsleLaMotte, Anchor Link Bot, S2000magician, Tesi1700, CBM2, ReluctantPhilosopher, Beeblebrox, ClueBot, Admiral Norton, DFRussia, Juustensson, Razimantv, Niceguyedc, Libett, Ajoykt, Ademh, Byates5637, Bender2k14, Sun Creator, Coinmanj, AmirOnWiki, Tnxman307, Snacks, Hans Adler, Muro Bot, Darkicebot, Palnot, Gerhardvalentin, TravisAF, Xamce, Addbot, Yousou, Harttwood47, RPHv, Mpholmes, EjsBot, Vishnava, LaaknorBot, דניאל ב., Legobot, Luckas-bot, Yobot, Ht686rg90, TaBOT-zerem, Nallimbot, Third Merlin, FeydHuxtable, AnomieBOT, 9258fahsflkh917fas, Materialscientist, Citation bot, Twiceuponatime, Markworthen, LilHelpa, Obersachsebot, Xqbot, Psyoptix, False vacuum, VladimirReshetnikov, Shaun.mk, Kristjan.Jonasson, Andrewjameskirk, Mark Renier, Citation bot 1, Þjóðólfr, Sptzimas, Tkuvho, Aunin, MarcelB612, MPeterHenry, Standardfact, H.ehsaan, RjwilmsiBot, Fictionalist, Kozation, EmausBot, Bijuro, AvicAWB, Negyek, Dominique.devriese, Herk1955, Llightex, CasualUser1729, ClueBot NG, Kevin Gorman, Helpful Pixie Bot, Wbm1058, Lowercase sigmabot, BG19bot, Modelpractice, Cyberpower678, Minsbot, Marktoiii0, Kiewbra, Cyberbot II, Deltahedron, Fernandodelucia, Harri jensen, LFOlsnes-Lea, Paul1andrews, EvergreenFir, Dinadineke, 22merlin, Zeus000000, Latinosopher and Anonymous: 357

_ **Experimental mathematics** *Source:* http://en.wikipedia.org/wiki/Experimental%20mathematics?oldid=632604295 *Contributors:* Patrick, Karada, Loren Rosen, Jitse Niesen, Gandalf61, Henrygb, (:Julien:), AsianAstronaut, Giftlite, Alberto da Calvairate, Abdull, Thorwald, Rich Farmbrough, Billymac00, C S, Jamiemichelle, Avenue, Oleg Alexandrov, Kelisi, Porcher, Rjwilmsi, Salix alba, Maxal, Googl, JJL, Berland, Cybercobra, BryanG, Big Paradox, Quibik, Aqwis, JohnBlackburne, Nusumareta, Gaz v pol, DFRussia, Nsk92, Aleksd, Libcub, Virginia-American, Addbot, Balabiot, Legobot, Citation bot, Obersachsebot, LoKiLeCh, 2ndjpeg, Thehelpfulbot, FrescoBot, TonyMath, Ben Tillman, ClueBot NG, Frietjes, Brad7777, Cnt vdw, Pimp slap the funk, Monkbot and Anonymous: 24

_ **Fractal** *Source:* http://en.wikipedia.org/wiki/Fractal?oldid=631960163 *Contributors:* AxelBoldt, Lee Daniel Crocker, Brion VIBBER, Archibald Fitzchesterfield, Mav, Zundark, Piotr Gasiorowski, Mirwin, XJaM, PierreAbbat, The Ostrich, Hhanke, Miguel, SimonP, Edemaine,

Heron, Aafuss, Arj, Youandme, JDG, Olivier, Frecklefoot, Lir, Infrogmation, Michael Hardy, Jdandr2, Wapcaplet, Ixfd64, Firebirth, TakuyaMurata, Arpingstone, Card, Ahoerstemeier, William M. Connolley, Theresa knott, Ijon, DropDeadGorgias, Glenn, Aragorn2, Llull, Evercat, Lancevortex, Schneelocke, Flajann, Revolver, Charles Matthews, Nostrum, Dino, Jitse Niesen, WhisperToMe, Timc, CBDunkerson, Tpbradbury, Maximus Rex, Hyacinth, Morwen, Saltine, Bevo, Xyb, Shizhao, Khym Chanur, Raul654, AnonMoos, Jeffq, Robbot, Ke4roh, Fredrik, Chris 73, Matt me, RedWolf, Bkalafut, Lowellian, Gandalf61, Chopchopwhitey, Sverdrup, Texture, Timrollpickering, Paul Murray, Fuelbottle, Seth Ilys, Gregorsamsa11, Superm401, MikeCapone, Tobias Bergemann, David Gerard, Cedars, Stirling Newberry, Matthew Stannard, Giftlite, Elf, Wolfkeeper, Everyking, Anville, Wikibob, Frencheigh, DO'Neil, Andris, Avsa, Eequor, Macrakis, Foobar, Gzornenplatz, Matt Crypto, Chameleon, SWAdair, Bobblewik, PhiloVivero, Utcursch, Shibboleth, Antandrus, Rajasekaran Deepak, ClockworkLunch, Kaldari, Joizashmo, MacGyverMagic, IYY, APH, MarkBurnett, Maximaximax, Kevin B12, Jawed, Eranb, Eric B. and Rakim, Urhixidur, Joyous!, Grunt, Mike Rosoft, Shiftchange, Mormegil, DanielCD, Discospinster, Haruki, Rich Farmbrough, TedPavlic, Pak21, Michal Jurosz, Solkoll, User2004, AlexKepler, Jamadagni, Michael Zimmermann, Nard the Bard, Paul August, Pban92, Bender235, ESkog, Kaisershatner, Neko-chan, Brian0918, EastNile, El C, Shanes, Mkosmul, RoyBoy, ~K, Bobo192, Reinyday, Cohesion, Sidjaggi, QTxVi4bEMRbrNqOorWBV, Blotwell, Bert Hickman, Rambus, Davidgothberg, MPerel, Gsklee, Nsaa, Mdd, Danski14, Sswn, Alansohn, Julesruis, Gargaj, Arthena, Rgclegg, Darrenthebaron, Craigy144, Weezer Ohio, ABCD, Hu, Snowolf, Malber, Yogi de, DRJacobson, Twisp, JCSP, Danhash, Uffish, Dirac1933, Lerdsuwa, DV8 2XL, Redvers, Oleg Alexandrov, Levan, JordanSamuels, Lkinkade, Novacatz, Woohookitty, TigerShark, Camw, LOL, TheNightFly, Duncan.france, Kelisi, SDC, Plrk, DocRuby, Cshirky, Mekong Bluesman, DaveApter, Kobi L, Aidje, Runis57, Graham87, Magister Mathematicae, Electricmoose, MC MasterChef, FreplySpang, Grammarbot, Saperaud, JVz, MarSch, Jamesrskemp, TheRingess, Salix alba, SpNeo, Mbutts, Feil0014, Titoxd, FlaBot, Mathbot, ZoneSeek, Crazycomputers, Who, Fragglet, Gparker, RexNL, Gurch, Ayla, Nimur, Physchim62, Chobot, Visor, ShadowHntr, WriterHound, Gwernol, The Rambling Man, YurikBot, Wavelength, Sceptre, Adamhauner, RussBot, Sarranduin, AVM, Hellbus, RadioFan, Stephenb, Gaius Cornelius, Chaos, Wimt, Wiki alf, Raven4x4x, Moe Epsilon, Charron, Tony1, Supten, Cheeser1, Cat2020, BazookaJoe, Sandstein, JonathanD, Zzuuzz, RDF, Theda, Arthur Rubin, Josh3580, Ianbolland, Reyk, BorgQueen, Paul Erik, Bo Jacoby, Roke, Choi9999, Yakudza, SmackBot, RDBury, GBarnett, Mujahideen, Unschool, Nadimghaznavi, Nihonjoe, Ravage386, KnowledgeOfSelf, Vald, Chairman S., Mdd4696, Jonathan Karlsson, BiT, Gilliam, Benjaminevans82, Ohnoitsjamie, Skizzik,

RobertM525, Qtoktok, Full Shunyata, RDBrown, MalafayaBot, Ida noeman, DHN-bot, Colonies Chris, Gbok, Max David, OrphanBot, Pevarnj, Gunjankg, Nakon, TheLimbicOne, Aelffin, Infovoria, Dreadstar, Kirils, Dr. Gabriel Gojon, Acdx, Kukini, SashatoBot, Lambiam, Wvbailey, Harryboyles, Lakinekaki, Rigadoun, Ocee, Rijkbenik, Sir Nicholas de Mimsy-Porpington, Ivanip, Johnsen953, JoshuaZ, Noegenesis, Samirdmonte, Thegreatdr, Ben Moore, Hvgard, 16@r, Grumpyyoungman01, Fractalmichel, Tasc, GilbertoSilvaFan, Waggers, Dhp1080, Farzaad, H, AntOnTrack, Hu12, Kevin R Johnson, Jason7825, Pegasus1138, Aeternus, Tó campos, Neoking, FakeTango, Courcelles, Padvi, Experiment123, Tawkerbot2, Lasserempe, Generalcp702, Fdot, JForget, InvisibleK, Edward Vielmetti, Ninetyone, Basawala, ShelfSkewed, David Traver, Karenjc, Nauticashades, Badseed, Slazenger, Doctormatt, Cydebot, MC10, SyntaxError55, Tawkerbot4, Christian75, Optimist on the run, Junglerolf, Victoriaedwards, PamD, Ael 2, Riojajar, Letranova, Epbr123, Barticus88, Wikid77, Ante Aikio, Hervegirod, Dafydd Williams, Marek69, John254, Drewboy64, Bunzil, Big Bird, Elert, AntiVandalBot, Seaphoto, Quite-Unusual, Quintote, Danger, Tillman, Alphachimpbot, Derouch, Byrgenwulf, Qwerty Binary, Myanw, Kaini, JAnDbot, MER-C, Fetchcomms, Dizzydog11235, Plm209, Beaumont, LittleOldMe, S0uj1r0, Penubag, Bongwarrior, VoABot II, Santisan, Nitku, Bubba hotep, Catgut, Indon, Animum, Dobi, Seberle, David Eppstein, Gregly, Martynas Patasius, GermanX, Vishvax, Ineffable3000, Denis tarasov, MartinBot, Geometricarts, NAHID, Reguiieee, Gcranston, EdwardHades, CommonsDelinker, Alexnevzorov, Fconaway, Jwoehr, Lilac Soul, LedgendGamer, Erkan Yilmaz, Pravirmalik, J.delanoy, Prokofiev2, Gandreas, Rayquaza11, Uncle Dick, McFarty, Maurice Carbonaro, Semajdraehs, Power Gear, Bot-Schafter, 5theye, Silver The Slammer, Nemo bis, C quest000, Chiswick Chap, WHeimbigner, Nwbeeson, SJP, Robertgreer, Submanifold, Gtg204y, Idioma-bot, Funandtrvl, Signalhead, Mbheyman, VolkovBot, Thedjatclubrock, Kriplozoik, Nikhil Varma, Sssmok1, TXiKiBoT, Zurishaddai, Arnon Chaffin, Voorlandt, Anna Lincoln, Una Smith, Auaug, Everything counts, Modocc, Hristos, Simzer, Kmhkmh, Emc2rocks, Synthebot, Murrogh, White Witch of Narnia, Lampica, Koldito, Chilti, Duckmackay, Logan, Ishboyfay, GirasoleDE, Dark Jackalope, SieBot, Bjtaylor01, Gopher292, Tresiden, Pengyanan, Caltas, Paul beaulieu, Kkrouni, Nikos.salingaros, LeadSongDog, Soler97, Andersmusician, SiegeLord, Keilana, Maphyche, Zucchini Marie, Oda Mari, Arbor to SJ, Rolandnine, Rhanyeia, Poop78432, BlueCerinthe, Oxymoron83, Faradayplank, Hobartimus, Svick, Eglash, Jons63, Arkixml, WikipedianMarlith, Casp7, ClueBot, Rumping, Fyyer, Foxj, The Thing That Should Not Be, JASONQUANTUM1, Hadrianheugh, JuPitEer, Meekywiki, Ryoutou, Lbertolotti, Edo 555, Patalbwil, Tamaratrouts, Wavedoc1, Sun Creator, Mr45acp, Francisco Albani, Jackrm, Sarsaparilla, ChrisHodgesUK, Ori benjamin, Aitias, Vanished User 1004, Jeflecastin, XLinkBot, Gonzonoir, Rortaruiter, XalD, Ehsan Nikooee, Addbot, DOI bot, Betterusername, Landon1980, SunDragon34, Twsntwsntwsm, Download, CarsracBot, Delaszk, LinkFA-Bot, Tide rolls, ScAvenger, Luckas Blade, Zorrobot, David0811, Jarble, Tinsol, Legobot, Luckas-bot, Yobot, Pink!Teen, Squish7, Synchronism, AnomieBOT, Accuruss, AdjustShift, Aditya, Kingpin13, Jacksonroberts25, Citation bot, Dromioofephesus, LilHelpa, PavelSolin, Xqbot, Mgroover, Hayley Tales, Hip2BSquare, Ptrf, Puvircho, Spw0766, Xenodream, Abce2, Web420, RibotBOT, FillerBrushMan, Locobot, Baba11111, WaysToEscape, Samwb123, Sesu Prime, Prari, FrescoBot, LucienBOT, Polynomial123, Harry f seldon, Citation bot 4, Pinethicket, Arctic Night, Hamtechperson, Moreinfopleasenow, Xxslorexx, Sumant Sethi, Mikejbau, Jandalhandler, Keri, Dr. Snew, Tatasz, Jordgette, Willihans, Clarkcj12, □□□, Percyzhang, Satdeep gill, DARTH SIDIOUS 2, Mean as custard, RjwilmsiBot, Utrytr, Calcyman, Beyond My Ken, BertSeghers, DSP-user, Afteread, Skamecrazy123, Der Künstler, Spennyize, Rusfuture, Scotty-Berg, RA0808, Awall2012, Ochristi, FatPope, Tommy2010, Wikipelli, K6ka, Cegalegolog99, TheLunarFrog, Josve05a, Empty Buffer, Fred Gandt, Muffiewrites, Dubravka2, Fractus-1, TyA, Gut Monk, James Krug, Maschen, Scientific29, Orange Suede Sofa, Foldedwater, FeatherPluma, Dllu, Signalizing, ClueBot NG, Elcubano91, A520, -sche, Austinlittle93, Teichlersmith, Widr, Calumcjr, Helpful Pixie Bot, Ahmad.829, Yiliangchen0113, DBigXray, BG19bot, DrJimothyCatface, Papadim.G, M0rphzone, Smitty121981, MSa1, MaxxyXD, Zhenxinghua, Ура-ура12, Lekro, Tigris35711, Achowat, Akarpe, Ezhu94, Josep m batlle, BattyBot, Justincheng12345-bot, HueSatLum, Khazar2, Dexbot, Cwobeel, DeniseKShull, Czech is Cyrillized, Brirush, Seaman4516, FairyTale'sEnd, Polytope4d, Fenniakaidan, Vanamonde93, Josep m batlle2, GailTheOx, Frac2012, Ignaciokriche, Ben christianson, QuahogClamMan, Anupama Srinivas, Wamiq, Ashorocetus, NorthBySouthBaranof, Revolverc24, Jackmcbarn, Francois-Pier, Metingle, Dkapetansky, Antideregister, Moogooshoe, Ali salsa, BethNaught, Shastamist, AKFS Editor, Garfield Garfield, Shira201, Loraof and Anonymous: 805

_ **Proof without words** *Source:* http://en.wikipedia.org/wiki/Proof%20without%20words?oldid=613705537 *Contributors:* Michael Hardy, Gandalf61, Postdlf, Giftlite, Shreevatsa, Graham87, Rsrikanth05, Royalbroil, Froth, Attilios, Nbarth, John, Jim.belk, Cydebot, AstroHurricane001, Nousernamesleft, Dmcq, Anyeverybody, Addbot, Uncia, Tkuvho, RedBot, TobeBot, EdoBot, Hmainsbot1 and Anonymous: 4

_ **Missing square puzzle** *Source:* http://en.wikipedia.org/wiki/Missing%20square%20puzzle?oldid=618466598 *Contributors:* Ed Poor, Minesweeper, Aarchiba, Evercat, Charles Matthews, Hadal, Wizzy, Kapow, Herbee, SarekOfVulcan, Noe, ALE!, DragonflySixtyseven, MCBastos, Paul August, Violetriga, Wfisher, Art LaPella, Kaganer, Jjron, Alexb@cut-the-knot.com, Mathbot, DVdm, Whosasking, Hydrargyrum, Nebogipfel, Zirland, PTSE, SmackBot, Hux, McGeddon, Nachra, Aldaron, Henning Makholm, N Shar, NeMewSys, Bjankuloski06en, Michael miceli, 041744, Stratadrake, Yvesnimmo, Pro bug catcher, Reywas92, Alvesgaspar, DavidRF, Omicronpersei8,

Tewapack, Thijs!bot, Salgueiro, David Eppstein, Gphoto, Rehrenberg, VolkovBot, Victorrocha, A4bot, Aymatth2, Trekky0623, MTHarden, Hatster301, WikiLaurent, Drmies, Piledhigheranddeeper, Bigup sim, Tired time, XLinkBot, Wikiuser100, MystBot, Addbot, Moosehadley, Mohamedhp, Download, LaaknorBot, SpBot, Tide rolls, Gribozavr, Luckas-bot, Fraggle81, Legobot II, Amirobot, Ayda D, Xqbot, Jayarathina, Khajidha, Vedabit, Aaron Kauppi, FrescoBot, Reaper Eternal, RjwilmsiBot, DASHBot, WTM, Rahatf, Tommy2010, AsceticRose, Traxs7, Nomalas, Jothikumar Rathinamoorthy, ClueBot NG, Widr, Chaseed530, Cleverdie, Cw.pondering, Gwickwire, Morning Sunshine, KSCgrace, BattyBot, PPMBrouwers, Nanera, Guilhermeappolinario, Ginsuloft, Andyni123 and Anonymous: 85

_ **Elementary proof** *Source:* http://en.wikipedia.org/wiki/Elementary%20proof?oldid=474298901 *Contributors:* Michael Hardy, Charles Matthews, Sam nead, Paul August, Alansohn, Greg Kuperberg, Oleg Alexandrov, BD2412, R.e.b., KarlFrei, Star trooper man, Finell, Timotheus Canens, Kukini, Lambiam, JoshuaZ, Braddodson, Myasuda, Doctormatt, Cydebot, Ntsimp, Headbomb, Dalassa, David Eppstein, Atheuz, ClueBot, Citation bot 1 and Anonymous: 2

_ **Complex analysis** *Source:* http://en.wikipedia.org/wiki/Complex%20analysis?oldid=633818680 *Contributors:* AxelBoldt, Michael Hardy, TakuyaMurata, Ahoerstemeier, Schneelocke, Charles Matthews, Dysprosia, Jitse Niesen, Taxman, Robbot, Fredrik, Sverdrup, AndreasB, Diberri, Alan Liefting, Rock69, Giftlite, Dratman, Gareth Wyn, Mboverload, Daniel Brockman, Neilc, Karl-Henner, Abdull, Discospinster, Rich Farmbrough, Paul August, El C, Sjoerd visscher, 3mta3, Obradovic Goran, Hesperian, HasharBot, Msh210, ABCD, Velella, Oleg Alexandrov, Ruud Koot, Isnow, Mana Excalibur, Koavf, Vvelup, Mathbot, Bgwhite, YurikBot, Sceptre, Michael Slone, Piet Delport, Thatoneguy, Raven4x4x, Crayolacrime, Brian Tvedt, Paul D. Anderson, Adam majewski, Blindsuperhero, MalafayaBot, Jonatan Swift, Jóna Þórunn, Jim.belk, CJauff, Ft1, Tony Fox, Blehfu, AlsatianRain, Tawkerbot2, Ali Obeid, Myasuda, Nauticashades, Thijs!bot, Divide-ByZero14, MCrawford, RobHar, BigJohnHenry, Liquid-aim-bot, Edokter, JAnDbot, Fiona CS, MSBOT, Penubag, Magioladitis, Ineffable3000, Maurice Carbonaro, Rocchini, Robert Illes, DavidCBryant, LokiClock, Anonymous Dissident, Ocolon, Digby Tantrum, Maxim, Haiviet, SieBot, Cwkmail, Vice regent, Anchor Link Bot, ClueBot, Excirial, Jusdafax, Aitias, Addbot, Rockyrackoon, CarsracBot, AnnaFrance, Favonian, שי דוד, Jarble, Tayoun7, Legobot, Luckas-bot, TaBOT-zerem, AnomieBOT, Speller26, 9258fahsflkh917fas, Citation

bot, Xqbot, Point-set topologist, RibotBOT, D'ohBot, Bigsigma1, MastiBot, Jkimath, Didactik, Gamewizard71, FoxBot, Duoduoduo, DARTH SIDIOUS 2, Laurent MAYER, TjBot, Jowa fan, StringTheory11, Allforrous, Maschen, Mentibot, ClueBot NG, Aarishshaheen, Helpful Pixie Bot, Mohammad kalanaki, Brad7777, Kephir, Cerabot, Mark viking, Samreid94 and Anonymous: 117

_ **Prime number theorem** *Source:* http://en.wikipedia.org/wiki/Prime%20number%20theorem?oldid=625621346 *Contributors:* Axel-Boldt, Bryan Derksen, XJaM, Bernfarr, Chas zzz brown, Michael Hardy, Kidburla, TakuyaMurata, Looxix, Wael Ellithy, Charles Matthews, Dcoetzee, Dysprosia, Jitse Niesen, Robbot, TMC1221, Gandalf61, Wereon, JensG, JerryFriedman, Tobias Bergemann, Decrypt3, Giftlite, Dbenbenn, Lupin, Herbee, Python eggs, Fred Stober, Rich Farmbrough, Guanabot, Roybb95, Paul August, Dmr2, Bender235, Jnestorius, EmilJ, Billymac00, John Vandenberg, Sligocki, Skimaxpower, Mcsee, Linas, Lucienve, Xiong Chiamiov, Reddwarf2956, Ketiltrout, Rjwilmsi, HappyCamper, R.e.b., Bubba73, FlaBot, Mathbot, Maxal, Alexjohnc3, Enon, Sstrader, Diza, Sodin, Scythe33, Glenn L, Chobot, ScottAlanHill, YurikBot, Wavelength, Dmharvey, KSmrq, LMSchmitt, DYLAN LENNON, Bbaumer, Kompik, Arthur Rubin, GrinBot, RDBury, BeteNoir, Dahn, PrimeHunter, MalafayaBot, Viebel, Ck lostsword, Khazar, Ninjagecko, ZAB, Statsone, JoshuaZ, WAREL, CRGreathouse, CmdrObot, Myasuda, Gregbard, Doctormatt, Mon4, Robertinventor, Karl-H, Thijs!bot, Headbomb, Nuesken, Noel Bush, AbcXyz, Mdotley, Salgueiro, Olaf, JamesBWatson, Zooloo, David Eppstein, Kope, Nono64, Rrostrom, Maurice Carbonaro, Tarotcards, Policron, Dessources, DavidCBryant, Inwind, Eric Ng, TXiKiBoT, Nxavar, Anonymous Dissident, Hagman, Arcfrk, GirasoleDE, SieBot, Murraythemathgeek, Justin W Smith, DragonBot, Mate2code, Jsondow, DumZiBoT, Katsushi, Nicolae Coman, Addbot, DOI bot, Uncia, LinkFA-Bot, SPat, Charleswallingford, II MusLiM HyBRiD II, Bab dz, AnomieBOT, Citation bot, ArthurBot, Xqbot, Shirik, Raulshc, D'ohBot, Motomuku, Citation bot 1, PrBeacon, Pinethicket, Primalbeing, Mikewarbz, TobeBot, Duoduoduo, Thomassteinke, RjwilmsiBot, EmausBot, John of Reading, Vincent Semeria, RA0808, Slawekb, Endlessoblivion, Toshio Yamaguchi, Maschen, Sapphorain, Pottermagic, ClueBot NG, Helpful Pixie Bot, Asmallworld, Brad7777, BattyBot, Boeing720, Dexbot, Deltahedron, AppliedMathematics, Mark viking, Monkbot, Whyes19 and Anonymous: 111

_ **Statistics** *Source:* http://en.wikipedia.org/wiki/Statistics?oldid=634274955 *Contributors:* Brion VIBBER, Mav, The Anome, Tarquin, Stephen Gilbert, Ap, Larry Sanger, Eclecticology, Saikat, Youssefsan, Christian List, Enchanter, Miguel, SimonP, Peterlin, Ben-Zin, Hefaistos, Waveguy, Heron, Rsabbatini, Camembert, Marekan, Olivier, Stevertigo, Edward, Boud, Michael Hardy, GABaker, Fred Bauder, Lexor, Nixdorf, Shyamal, Kku, Tannin, Dcljr, Tomi, CesarB, Looxix, Ahoerstemeier, DavidWBrooks, Ronz, BevRowe, Snoyes, Salsa Shark, Netsnipe, Big iron, Jfitzg, Cherkash, Samuel, Mxn, Schneelocke, Hike395, Guaka, Vanished user 5zariu3jisj0j4irj, Wikiborg, Dysprosia, Jitse Niesen, Quux, Jake Nelson, Maximus Rex, Wakka, Wernher, Optim, Rbellin, Secretlondon, Noeckel, Phil Boswell, Robbot, Jakohn, Benwing, ZimZalaBim, Gandalf61, Tim Ivorson, RossA, Henrygb, Hemanshu, Gidonb, Borislav, Ianml, Roozbeh, Dhodges, SoLando, Wile E. Heresiarch, Cutler, Dave6, Aomarks, Ancheta Wis, Matthew Stannard, Tophcito, Giftlite, Sj, Wikilibrarian, Netoholic, Lethe, Tom harrison, Meursault2004, Everyking, Maha ts, Curps, Dmb000006, Muzzle, Jfdwolff, BrendanH, Maarten van Vliet, Guanaco, Skagedal, Eequor, Mdb, SWAdair, Brazuca, Hereticam, Andycjp, Mats Kindahl, Antandrus, MarkSweep, Piotrus, Ampre, L353a1, Sean Heron, CSTAR, APH, Oneiros, Gsociology, PFHLai, Bodnotbod, Mysidia, Icairns, Simoneau, Sam Hocevar, Jeremykemp, Howardjp, Zondor, Bluemask, Drchris, Richardelainechambers, Moverton, Discospinster, Rich Farmbrough, Guanabot, Michal Jurosz, IlyaHaykinson, Paul August, Bender235, Kbh3rd, Brian0918, El C, Lycurgus, Zenohockey, Art LaPella, RoyBoy, 2005, Bobo192, Janna Isabot, O18, Gianlu, Smalljim, Maurreen, 3mta3, Minghong, Mdd, Passw0rd, Drf5n, Schissel, Jigen III, Msh210, Alansohn, Gary, Anthony Appleyard, Kanie, Rgclegg, Avenue, Evil Monkey, Oleg Alexandrov, AustinZ, Waabu, Linas, Karnesky, LOL, Before My Ken, WadeSimMiser, Acerperi, Wikiklrsc, Sengkang, BlaiseFEgan, Gimboid13, Mr Anthem, Marudubshinki, Graham87, Ilya, Galwhaa, Chun-hian, FreplySpang, Dragoneye776, Dpr, Tlroche, Jorunn, Koolkao, Rjwilmsi, Mayumashu, Amire80, Carbonite, Salix alba, Jb-adder, Willetjo, Crazynas, Jeffmcneill, Zero0w, FlaBot, Chocolatier, RobertG, Windchaser, Dibowen5, Latka, Mathbot, Airumel, Nivix, Celestianpower, RexNL, Gurch, AndriuZ, Pete.Hurd, Mathieumcguire, Shaile, Malhonen, BradBeattie, CiaPan, Chobot, Nagytibi, DVdm, Simesa, Adoniscik, Gwernol, Wavelength, Phantomsteve, Loom91, Cswrye, Epolk, Donwarnersaklad, Hydrargyrum, Stephenb, Manop, Chaos, NawlinWiki, Wiki alf, Grafen, Tailpig, ONEder Boy, TCrossland, Johndarrington, Isolani, D. Wu, Alex43223, BOT-Superzerocool, Mgnbar, Tigershrike, Saric, Closedmouth, Terfgiu, Modify, Beaker342, GraemeL, AGToth, Whouk, NeilN, DVD R W, Sardanaphalus, Veinor, JJL, SmackBot, YellowMonkey, Twerges, Unschool, Honza Záruba, Stux, Hydrogen Iodide, McGeddon, CommodiCast, Timotheus Canens, Dhochron, Gilliam, Brotherbobby, Skizzik, ERcheck, Chris the speller, Bychan, Bluebot, Keegan, Jjalexand, DocKrin, Wikisamh, Silly rabbit, Ekalin, RayAYang, Klnorman, Dlohcierekim's sock, Robth, Zven, John Reaves, Scwlong, Chendy, SLC1, Iwaterpolo, PierreAnoid, Can't sleep, clown will eat me, DéRahier, Asarko, Hve, Terry Oldberg, Addshore, Kcordina, Amazins490, Mosca, SundarBot, Jmlk17, Aldaron, ConMan, Valenciano, Krexer, Chadmbol, Richard001, Nrcprm2026, Mini-Geek, G716, Photoleif, GumbyProf, Fschoonj, Wybot, Zeamays, SashatoBot, Lambiam, Arodb, Derek farn, Harryboyles, Chocolateluvr88, Sina2, Archimerged, Kuru, MagnaMopus, Lapaz, Soumyasch, Tim bates, SpyMagician, Deviathan, Ckatz, RandomCritic, 16@r, Beetstra, Santa Sangre, Daphne A, Mets501, Spiel496, Ctacmo, RichardF, Roderickmunro, Hu12, Levineps, BranStark, Joseph Solis in Australia, Wjejskenewr, Mangesh.dashpute, Chris53516, Igoldste, Tawkerbot2, Daniel5127, Filelakeshoe, Kevin Murray, Kendroche, JForget, Robertdamron, CRGreathouse, CmdrObot, Dycedarg, Dexter inside, Requestion, MarsRover, Neelix, Hingenivrutti,

Nnp, Art10, MrFish, Myasuda, Mct mht, Slack---line, Mjhoy, Arauzo, Ramitmahajan, Gogo Dodo, Anonymi, Bornsommer, Odie5533, Christian75, DumbBOT, Richard416282, Englishnerd, Optimist on the run, Lindsay658, Finn krogstad, FrancoGG, Mattisse, Talgalili, Sarvesh85@gmail.com, Epbr123, Jrl306, LeeG, Pattern86, Willworkforicecream, N5iln, Marek69, John254, Escarbot, Dainis, Mentifisto, Wikiwilly, AntiVandalBot, Luna Santin, Seaphoto, Memset, Zappernapper, Mack2, Sbarnard, Gökhan, Golgofrinchian, MikeLynch, JAnDbot, Ldc, Markbold, The Transhumanist, Db099221, BenB4, PhilKnight, IamHope, SiobhanHansa, Magioladitis, Bongwarrior, VoABot II, Jeff Dahl, JamesBWatson, Hubbardaie, Ranger2006, Trugster, Skew-t, Recurring dreams, Ddr, Caesarjbsquitti, Avicennasis, Nevvers, KConWiki, Catgut, Animum, Depressedrobot, Johnbibby, Robotman1974, Boffob, Bobby H. Heffley, Xerxes minor, JoergenB, DerHexer, JaGa, Khalid Mahmood, AllenDowney, Apdevries, Pax:Vobiscum, Gjd001, Rustyfence, Cliff smith, MartinBot, Vigyani, BetBot, Jim.henderson, R'n'B, Lilac Soul, Mausy5043, J.delanoy, Trusilver, Rlsheehan, Numbo3, Mthibault, Ulyssesmsu, Yannick56, TheSeven, Cpiral, Gzkn, M C Y 1008, Luntertun, It Is Me Here, Noschool3, Macrolizard, Bmilicevic, The Transhumanist (AWB), Kenneth M Burke, DavidCBryant, Tiggerjay, Afv2006, HyDeckar, WinterSpw, Ron shelf, Tanyawade, Idioma-bot, Funandtrvl, Wikieditor06, Lights, VolkovBot, DrMicro, ABF, JohnBlackburne, Paxcoder, Jimmaths, Barneca, Philip Trueman, DoorsAjar, TXiKi-BoT, Ranajeet, Jacob Lundberg, Wikipediatoperfection, Tomsega, ElinorD, Qxz, Arpabr, The Tetrast, Seanstock, Jackfork, Christopher Connor, Onore Baka Sama, Manik762007, Careercornerstone, Wikidan829, Richard redfern, Skarz, Dmcq, Symane, EmxBot, Kolmorogoff, Demmy, Thefellswooper, SieBot, BotMultichill, Katonal, Triwbe, Toddst1, Flyer22, Tiptoety, JD554, Ireas, Jt512, Free Software Knight, Strife911, Oxymoron83, Faradayplank, Boromir123, Hinaaa, BenoniBot, Emesee, OKBot, Melcombe, Yhkhoo, Nn123645, Superbeecat, Digisus, Richard David Ramsey, Escape Orbit, Maniac2910, Tautologist, XDanielx, WikipedianMarlith, ClueBot, Rumping, Fyyer, John ellenberger, DesertAngel, Gaia Octavia Agrippa, Giusippe, Turbojet, Uncle Milty, Niceguyedc, LizardJr8, Morten Münchow, Chickenman78, Lbertolotti, DragonBot, Pumpmeup, Jusdafax, Three-quarter-ten, Rwilli13, Adamjslund, Livius3, Stathope17, Notteln, Precanalytics, Diaa abdelmoneim, Dekisugi, Gundersen53, BOTarate, Aitias, ShawnAGaddy, Dbenzvi, JDPhD, FinnMan, Qwfp, Antonwg, Ano-User, GKantaris, Editorofthewiki, Helixweb, XLinkBot, Avoided, WikHead, Alexius08, Tayste, Addbot, Proofreader77,

Hgberman, DOI bot, Captain-tucker, Atethnekos, Johnjohn83, Kwanesum, Br1z, Bte99, CanadianLinuxUser, MrOllie, Chamal N, Glane23, Delaszk, Glass Sword, Debresser, Favonian, Quercus solaris, Aitambong, Ssschhh, Tide rolls, Lightbot, Kiril Simeonovski, Teles, MuZemike, TeH nOmInAtOr, LuK3, Megaman en m, Nbeltz, Jim, Luckas-bot, Yobot, Notizy1251, OrgasGirl, Fraggle81, Vimalp, DisillusionedBitterAndKnackered, Mathinik, Gobbleswoggler, THEN WHO WAS PHONE?, ECEstats, Brougham96, Mhmolitor, AnomieBOT, DemocraticLuntz, VX, Cavarrone, Galoubet, Dwayne, Piano non troppo, Youkbam, Templatehater, Walter Grassroot, Htim, Materialscientist, The High Fin Sperm Whale, Citation bot, OllieFury, Markmagdy, Sweeraha, GB fan, Apollo, Neurolysis, Arthur-Bot, Herreradavid33, Xqbot, TinucherianBot II, Class ruiner, Kenz0402, Drilnoth, Fishiface, Locos epraix, Spetzznaz, AbigailAbernathy, Clear range, Coretheapple, GrouchoBot, Ute in DC, SassoBot, Loizbec, 78.26, Rstatx, Stynyr, Doulos Christos, Chen-Pan Liao, N.j.hansen, Shadowjams, Brennan41292, FrescoBot, Tobby72, Hallway916, Shadowpsi, HJ Mitchell, Winterswift, Citation bot 1, PrBeacon, Boxplot, Yuanfangdelang, Pinethicket, Kiefer.Wolfowitz, Stpasha, Brian Everlasting, Île flottante, Bwana2009, Dee539, Florendobe, White Shadows, Gamewizard71, FoxBot, Mjs1991, Ruzihm, TobeBot, LAUD, Arfgab, Decstop, MrX, Spegali, Keepitup.sid, Sourishdas, Tbhotch, Drivi86, Sandman888, DARTH SIDIOUS 2, Chrisrayner, Whisky drinker, Mean as custard, Updatehelper, TjBot, Kastchei, Karlheinz037, Becritical, Elitropia, Jordan.brayanov, EmausBot, Orphan Wiki, Gfoley4, Racerx11, Hiamy, Tommy2010, Kellylautt, Tuxedo junction, Daonguyen95, Fæ, Josve05a, Bollyjeff, Tastewrong1234, WeijiBaikeBianji, Cbratsas, JA(000)Davidson, Access Denied, Dylthaavatar, Kgwet, SporkBot, Jorjulio, GrindtXX, Makecat, Sak11sl, Future ahead, Anglais1, Sunur7, Mr. Kenan Bek, Noodleki, Donner60, Agatecat2700, NTox, DemonicPartyHat, 28bot, Petrb, ClueBot NG, MelbourneStar, Chrisminter, Dvsbmx, BarrelProof, Bped1985, Andreas.Persson, Shawnluft, Cntras, Braincricket, ScottSteiner, Ryan Vesey, Amircrypto, Helpful Pixie Bot, Xandrox, Mishnadar, Ldownss00, Calabe1992, KLBot2, Lowercase sigmabot, BG19bot, Juro2351, Northamerica1000, Absconded Northerner, Muhehej1000, MusikAnimal, Marcocapelle, Stalve, EmadIV, Rm1271, Htrkaya, Omiswiki, Manoguru, Kittipatv, Meclee, Brad7777, Glacialfox, Anbu121, Europeancentralbank, Bsutradhar, Ca3tki, Kodiologist, Codeh, Gr khan veroana kharal, Markk waugh, Illia Connell, Dexbot, Ubertook, Mogism, Wikignome1213, Princessandthepi, Lugia2453, Brownstat, Norazoey, Speakel, 069952497a, Faizan, RG57, FallingGravity, AmericanLemming, Tentinator, Beasarah, Butter7938, Seppi333, SpuriousTwist, Ginsuloft, Sean4424, Sarwan khan, Adirlanz, AddWittyNameHere, Narasandraprabhakara, Science.philosophy.arts, Akuaku123, Mendisar Esarimar Desktrwaimar, Mconnolly17, Zib2542, Therealthings, MelaniePS, Monkbot, Poepkop, Soon Son Simps, Vieque, Majormuesli, Trackteur, Andri Kuawko, Romelthomas, Umkan, Amortias, NQ, Morgantaschuk, Sumonratin, Thebearedguy, Mj3322, Rainamagdalena and Anonymous: 1188

_ **Data analysis** *Source:* http://en.wikipedia.org/wiki/Data%20analysis?oldid=633898590 *Contributors:* Edward, Michael Hardy, Ronz, Altenmann, Giftlite, Khalid hassani, Pgan002, OverlordQ, Piotrus, Gscshoyru, Discospinster, Tavkhelidzem, JoeHenzi, Art LaPella, Bobo192, Mdd, Alansohn, SteinbDJ, Oleg Alexandrov, Camw, Mandarax, Piet Delport, Lijealso, Arthur Rubin, Dontaskme, Curpsbotunicodify, Mhkay, SmackBot, McGeddon, Jtneill, WookieInHeat, CommodiCast, Gilliam, Chris the speller, A. B., Sergio.ballestrero, Shalom Yechiel, 16@r, Zgemignani, ShelfSkewed, Jmswisher, Funnyfarmofdoom, Steel, Gogo Dodo, Chasingsol, Barticus88, Hcberkowitz, Dawnseeker2000, Schekanov, AuburnPilot, Elringo, Nyttend, Shim'on, R'n'B, Rlsheehan, Singularitarian, Maurice Carbonaro, TheSeven, TacoBelly, STBotD, Sheliak, Sergivs-en, 28bytes, VolkovBot, Zenstat, Kyle the bot, Saddy Dumpington, Philip Trueman, Udufruduhu, 3bi, Ericmelse, Sue Rangell, Farcaster, Ttonyb1, Loveless2, JHollingsworth, Hello71, Tesi1700, Melcombe, PerryTachett, ClueBot, Apptrain, Xkrebstarx, Supertouch, Excirial, M4gnum0n, Mleusink, Lostraven, Inspector 34, Jthetzel, Qwfp, Apparition11, XLinkBot, Rror, Nepenthes, SilvonenBot, PL290, Addbot, Delaszk, Tide rolls, Tedtoal, Luckas-bot, Yobot, Fraggle81, Aboalbiss, AnomieBOT, Kristen Eriksen, Materialscientist, Jeffrey Mall, Thosjleep, Shirik, Matthew Blake, Mark Schierbecker, SassoBot, Doulos Christos, Smallman12q, A. di M., Endothermic, Mark Renier, Er.piyushkp, Pinethicket, Analyst246, Dmitry St, TobeBot, Sijit21, Lotje, Duoduoduo, Mike78465, Marksenizer, Valeropedro, Slon02, Helwr, EmausBot, Kellylautt, Thecheesykid, Hhhippo, Wayne Slam, Cit helper, Stormymountain, Calle lund LTH, ClueBot NG, Bayes Puppy, Hulbert88, MethAdvice2010, Frietjes, WikiMSL, Lawsonstu, Helpful Pixie Bot, Tobbilla83, Cbrick77, Mrjohncummings, Muhehej1000, Mark Arsten, Glacialfox, Timbrooker, BattyBot, Teammm, Mdann52, Dexbot, Gbrouce, Stevebillings, Municca, Druzdzel, Cmartines, Saxondigital, Tami Allen, Melonkelon, Sudipta456, Ogretmem, Jenny Rankin, Prussonyc, Garretvanderboom, Auditing2, Wccsnow, Csjacobs24, Narvan2013, RICHA TARAR, JaconaFrere, Mgt88drcr, SWB456, JayOnDione, Beth.Alex123, Jonathan Pan and Anonymous: 209

_ **Bayesian inference** *Source:* http://en.wikipedia.org/wiki/Bayesian%20inference?oldid=631706712 *Contributors:* The Anome, Fubar Obfusco, DavidSJ, Jinian, Edward, JohnOwens, Michael Hardy, Lexor, Karada, Ronz, Suisui, Den fjättrade ankan, LouI, EdH, Jonik, Hike395, Novum, Timwi, WhisperToMe, Selket, SEWilco, Jose Ramos, Insightaction, Banno, Robbot, Kiwibird, Benwing, Meduz, Henrygb, AceMyth, Wile E. Heresiarch, Ancheta Wis, Giftlite, DavidCary, Dratman, Leonard G., JimD, Wmahan, Pcarbonn, MarkSweep, L353a1, FelineAvenger, APH, Sam Hocevar, Perey, Discospinster, Rich Farmbrough, Bender235, ZeroOne, Donsimon, MisterSheik, El C, Edward Z. Yang, DimaDorfman, Cje, John Vandenberg, LeonardoGregianin, Jung dalglish, Hooperbloob, Landroni, Arcenciel, Nurban, Avenue, Cburnett, Jheald, Facopad, Sjara, Oleg Alexandrov, Roylee, Joriki, Mindmatrix, BlaiseFEgan, Btyner, Magister Mathematicae, Tlroche, Rjwilmsi, Ravik, Jeffmcneill, Billjefferys, FlaBot, Brendan642, Kri, Chobot, Reetep, Gdrbot, Adoniscik, Wavelength, Pacaro, Gaius Cornelius, ENeville, Dysmorodrepanis, Snek01, BenBildstein, Modify, Mastercampbell, Nothlit, NielsenGW, Mebden, Bo Jacoby, Boggie, Harthacnut, SmackBot, Mmernex, Rtc, Mcld, Cunya, Gilliam, DoctorW, Nbarth, G716, Jbergquist, Bejnar, Gh02t,

Wyxel, Josephsieh, JeonghunNoh, Thermochap, BoH, Basar, TheRegicider, Farzaneh, Lindsay658, Tdunning, Helgus, EdJohnston, Jvstone, Mack2, Lfstevens, Makohn, Stephanhartmannde, Comrade jo, Ph.eyes, Coffee2theorems, Ling.Nut, Charlesbaldo, DAGwyn, User A1, Tercer, STBot, Tobyr2, LittleHow, Policron, Jeffbadge, Bhepburn, Robcalver, James Kidd, VolkovBot, Thedjatclubrock, Maghnus, TXiKiBoT, Andrewaskew, GirasoleDE, SieBot, Doctorfree, Natta.d, Anchor Link Bot, Melcombe, Kvihill, Rfinchdavis, Smithpith, GeneCallahan, Krogstadt, Reovalis, Hussainshafqat, Charledl, ERosa, Qwfp, Tdslk, XLinkBot, Erreip, Addbot, K-MUS, Metagraph, LaaknorBot, Ozob, Legobot, Yobot, Gongshow, AnomieBOT, Citation bot, Shadak, Danielshin, VladimirReshetnikov, KingScot, Jon-DePlume, Thehelpfulbot, FrescoBot, WhatWasDone, Haeinous, JFK0502, Kiefer.Wolfowitz, 124Nick, Night Jaguar, Scientist2, Trappist the monk, Gnathan87, Jonkerz, Jowa fan, EmausBot, Blumehua, Montgolfière, Moswento, McPastry, Bagrowjp, Willy.pregliasco, Floombottle, Epdeloso, ClueBot NG, Mathstat, Bayes Puppy, Jj1236, Albertttt, Thepigdog, Helpful Pixie Bot, Michael.d.larkin, Whyking thc, Intervallic, CitationCleanerBot, DaleSpam, Kaseton, Simonsm21, Danielribeirosilva, ChrisGualtieri, Alialamifard, Yongli Han, 90b56587, MittensR, Mark viking, Boomx09, Waynechew87, Hamoudafg, Promise her a definition, Abacenis, Engheta, Avehtari and Anonymous: 218

_ **Probability** *Source:* http://en.wikipedia.org/wiki/Probability?oldid=634491852 *Contributors:* Bryan Derksen, Zundark, Koyaanis Qatsi, Ap, Enchanter, Deb, Heron, Youandme, Patrick, D, Chas zzz brown, Michael Hardy, Fred Bauder, DopefishJustin, David Martland, Wapcaplet, Ixfd64, Minesweeper, Looxix, ArnoLagrange, Ronz, Snoyes, Suisui, Evercat, Clausen, Jonik, Ideyal, Bjcairns, Lit-sci, Charles Matthews, Reddi, Dysprosia, Jitse Niesen, DJ Clayworth, Markhurd, Hyacinth, Wetman, Johnleemk, Jni, Robbot, Jwpurple, Altenmann, MathMartin, Henrygb, Hadal, Wxlfsr, Wile E. Heresiarch, Centrx, Giftlite, Recentchanges, Lee J Haywood, Bfinn, DaveBrondsema, Solipsist, Bobblewik, Wmahan, LiDaobing, Fangz, Antandrus, MarkSweep, CSTAR, APH, Maximaximax, JamesTeterenko, Vivacissamamente, Grstain, Mike Rosoft, Brianjd, JTN, Discospinster, Paul August, Bender235, ESkog, Kaisershatner, MisterSheik, CanisRufus,

Bobo192, Robotje, John Vandenberg, Jung dalglish, NickSchweitzer, Flammifer, Nsaa, Googie man, Localhost00, Msh210, Alansohn, Mac Davis, Avenue, Fourthords, Jheald, Wyatts, Kazvorpal, INic, Velho, OwenX, Mindmatrix, TigerShark, Kzollman, Lifung, Acerperi, Pdn, Bhound89, Sengkang, Btyner, RuM, Graham87, BD2412, Qwertyus, FreplySpang, NatusRoma, Jake Wartenberg, Salix alba, FlaBot, Sky Harbor, Mathbot, Vandal B, Demnevanni, RexNL, Alexjohnc3, Cyclone49, Chobot, Stephen Compall, Cactus.man, Gwernol, YurikBot, Wavelength, Spacepotato, RussBot, Bhny, Ogai, Vibritannia, Wimt, NawlinWiki, RattleMan, Haduong, Trovatore, Srinivasasha, Mortein, Jpbowen, Dbfirs, Hirak 99, 21655, Takeda, Jeffw57, Vicarious, QmunkE, Garion96, Ybbor, TLSuda, Mebden, Capitalist, Marquez, Eog1916, Sycthos, Sardanaphalus, Amalthea, JJL, SmackBot, FishSpeaker, Rtc, Reedy, Stux, Jacek Kendysz, Cessator, Edgar181, Ohnoitsjamie, Rmosler2100, Guess Who, Jprg1966, SonOfNothing, MartinPoulter, SchfiftyThree, Octahedron80, Whispering, DHN-bot, Putgeminmouth, Ladislav Mecir, Suicidalhamster, NYKevin, KG6YKN, Rrburke, Parent5446, SundarBot, Stevenmitchell, Jmlk17, Robma, Richard001, G716, LoveMonkey, Andeggs, Mlpkr, Valodim, Lambiam, Doug Bell, Sina2, Dbtfz, Demicx, Hoomank, Chetan.Panchal, Nijdam, Santa Sangre, Ehheh, Mets501, Dl2000, Hu12, Levineps, Capitan Obvio, Shoeofdeath, Amakuru, Lenoxus, CapitalR, Happy-melon, Tawkerbot2, Braddodson, AbsolutDan, Kurtan, Philtr, Dhammapal, Markjoseph125, Wafulz, CBM, Matthew Auger, Dgw, ShelfSkewed, Requestion, Myasuda, Gregbard, Buttonius, Captmog, AndrewHowse, Gogo Dodo, Lugnuts, Leobh, Ceannaideachd, Clovis Sangrail, Dougweller, DumbBOT, Omicronpersei8, Ebraminio, Progicnet, Mattisse, Epbr123, Btball, David from Downunder, PaperTruths, Scfencer, Philippe, CharlotteWebb, Nick Number, SusanLesch, Hmrox, QuiteUnusual, Readro, Jj137, Dylan Lake, Leuqarte, Sluzzelin, .alyn.post., JAnDbot, MER-C, The Transhumanist, Instinct, Fetchcomms, Zalle, Hut 8.5, PhilKnight, Beaumont, SiobhanHansa, Acroterion, Connormah, Jaysweet, Bongwarrior, VoABot II, Swpb, Twisted86, Ranger2006, King Mir, Aka042, FrF, Giggy, Khalid Mahmood, Fantastic4boy, Pax:Vobiscum, Arenarax, JosephCampisi, MartinBot, Arjun01, Tgeairn, J.delanoy, Abby, Uncle Dick, Nigholith, Ignacio Icke, Zifnab25, DarkFalls, Ncmvocalist, McSly, Infarom, Coppertwig, NewEnglandYankee, Mycroft80, Treisijs, Pdcook, Ja 62, VolkovBot, GoldenPi, Jim.Callahan,Orlando, Jeff G., JohnBlackburne, VasilievVV, Jimmaths, Classical geographer, Philip Trueman, Sagittarian Milky Way, Asarlaí, Drestros power, Someguy1221, Wordsmith, Jackfork, Psyche825, United-Statesian, Cremepuff222, Onore Baka Sama, FrankSanMiguel, CO, Craphouse, Paulcd2000, Larklight, Eliotwiki, Billinghurst, Mabsjenbu123, Richardajohns, Monty845, Arcfrk, Klapper, XKL, Symane, Desteg, SieBot, Mikemoral, Hasanbay, YourEyesOnly, Caltas, Soler97, Wikistudent 1, CouldOughta, Flyer22, Free Software Knight, Yerpo, Oxymoron83, Vericuester, Wormdoggy, Randomblue, Melcombe, Dabomb87, Denisarona, Disneycat, Tautologist, DEMcAdams, Varnesavant, ClueBot, Jackollie, The Thing That Should Not Be, Arakunem, Drmies, Sevilledade, Mild Bill Hiccup, Yamakiri, Puchiko, DragonBot, Excirial, CrazyChemGuy, Mate2code, L.tak, Cenarium, Peter.C, Razorflame, Dekisugi, Thingg, Aitias, ShawnAGaddy, Footballfan190, Kaksag, Qwfp, Johnuniq, Vanished User 1004, XLinkBot, Tarheel95, Gerhardvalentin, Badgernet, Noctibus, Menthaxpiperita, Rogimoto, Krantz2, Tayste, Addbot, Brumski, Metagraph, Ronhjones, CanadianLinuxUser, Leszek Jańczuk, Cst17, MrOllie, Glane23, Peterjhlee, Favonian, Ebsith, Aliyah4499, Kevmus, Tide rolls, Lightbot, Luckas Blade, Gail, Ettrig, Packersfannn101, Liang9993, Ben Ben, Aubrey, Luckas-bot, Indicom217, Fraggle81, II MusLiM HyBRiD II, Taxisfolder, KamikazeBot, AnomieBOT, Valueyou, Ciphers, Jim1138, Kingpin13, RVS, Xqbot, Bdmy, Thesoxlost, Capricorn42, Kbodouhi, El Caro, GrouchoBot, Riotrocket8676,Mario777Zelda, Omnipaedista, פריימן, Shadowjams, Ricklethickets, Carricko, FrescoBot, Paine Ellsworth, Danno12345, Tobby72, Machine Elf 1735, William915, Alberg15, AstaBOTh15, DrilBot, Tintenfischlein, Pinethicket, I dream of horses, Boulaur, 10metreh, Aizquier, BRUTE, RedBot, Ezhuttukari, Jauhienij, Debator of mathematics, 0zlw, Vrenator, Bluefist, Reaper Eternal, Antonwalter, Jeffrd10, Fastilysock, Drivi86, Zach1994, Regancy42, Emaus-Bot, NoisyJinx, Learnhead, Abby1019, RabinZhao, Wikipelli, Dirtytedd, John Cline, Daonguyen95, Fæ, טוב ,נו, Hasihfiadhfoiahsio, Brendo4, TheGreenCarrot, SporkBot, Cymru.lass, Pointless.FF59F5C9, Donner60, Ali cheif, Bomazi, Herebo, Kushalneo, ChuispastonBot, NTox, Herk1955, Virgilian, DASHBotAV, Rocketrod1960, Petrb, ClueBot NG, Mathstat, Tillander, Chester Markel, Baseball Watcher, JohnsonL623, T.Montague19-NJITWILL, Abcd1234 manutd, ScottSteiner, Go Phightins!, Widr, Helpful Pixie Bot, Strike Eagle, BG19bot, Bigbird195, Pine, Pacerier, TeamRocketPikachu, Zmoney918, Ann12h, The Determinator, Jawabiit, Alijamal14, Jranes, Brad7777, JZCL, CurtisKime, EuroCarGT, Alunaifuseng, Illia Connell, Dexbot, Profeshermyguad, Narender khola, Epicgenius, I am One of Many, Oh Goes the Waterholes, EvergreenFir, Nigellwh, Mrm7171, Ginsuloft, XXEniiGmAx, Bryanrutherford0, Banzai6666, Mendisar Esarimar Desktrwaimar, Michaelalc, Raman2608, Jolifer, Yuri716, Soon Son Simps, Scarlettail, Occults, Chancemreed, Nelsonalp, Sachinnarula77 and Anonymous: 674

_ **Mathematical physics** *Source:* http://en.wikipedia.org/wiki/Mathematical%20physics?oldid=633443396 *Contributors:* CYD, Ed Poor, Andre Engels, Michael Hardy, Dgrant, Joy, Robbot, Gandalf61, Ilya (usurped), Tobias Bergemann, Ancheta Wis, Giftlite, Mintleaf, Bkonrad, Python eggs, CryptoDerk, Karol Langner, APH, Elroch, Lumidek, Eduardoporcher, Igorivanov, Richard Boyer, Maksym Ye., Paul August, Bender235, Jung dalglish, Helix84, Sswn, Jason Davies, Joolz, Burn, Jheald, Oleg Alexandrov, Linas, Karnesky, Ruud Koot, Acerperi, Bluemoose, BD2412, Porcher, Rjwilmsi, Koavf, MarSch, Andrei Polyanin, R.e.b., Fred Bradstadt, Kasparov, FlaBot, Margosbot, Srleffler, Tdoune, Bgwhite, YurikBot, DanMS, Salsb, JocK, E2mb0t, Iztok.jeras, NBS525, Sardanaphalus, JJL, Jagged 85, Ajt, Spellchecker, RyanC., JustUser, Jhausauer, Pax85, Cybercobra, SashatoBot, Cronholm144, IvanLanin, Albertod4, CRGreathouse, WeggeBot, Myasuda, Mct mht, Quibik, Islescape, Headbomb, BehnamFarid, D.H, Seaphoto, Robsavoie, VoABot II, PeterStJohn, Hans Lundmark, Etale, Ludvikus, JaGa, Matqkks, Maurice Carbonaro, Hodja Nasreddin, DavidCBryant, Idioma-bot, VolkovBot, ABF, John-Blackburne, Hyperlinker, TXiKiBoT, Clarince63, Arcfrk, SieBot, Henry Delforn (old), Thehotelambush, StewartMH, ClueBot, Jan1nad, Razimantv, Ammarsakaji, J8079s, Awickert, Muro Bot, HumphreyW, Boethius65, DumZiBoT, TimothyRias, XLinkBot, Dthomsen8,

Saeed.Veradi, Addbot, DOI bot, R physicist, AndersBot, Lightbot, TeH nOmInAtOr, Yndurain, Luckas-bot, Yobot, Materialscientist, ArthurBot, Xqbot, J04n, SassoBot, Charvest, Inasilentway, CES1596, FrescoBot, Ironboy11, Haeinous, G2kdoe, Citation bot 1, I dream of horses, Jonesey95, RedBot, Kallikanzarid, Mthlwn, EmausBot, Sajadsandilo, Somerwind, Hhhippo, Donner60, GKaczinsky, Bomazi, Mathuvw, RockMagnetist, YtivarG, Liuthar, ClueBot NG, Aarishshaheen, Musamanyama, Helpful Pixie Bot, Bibcode Bot, Rderdwien, Brad7777, Randomguess, BattyBot, Ema--or, RFWerner, Donn300, Makecat-bot, Lugia2453, BeaumontTaz, Brirush, Moretti.valter9, Mfb, Airwoz, EliDika, Almaionescu, Vieque, Baba i deda and Anonymous: 122

_ **Bayesian probability** *Source:* http://en.wikipedia.org/wiki/Bayesian%20probability?oldid=633432598 *Contributors:* AxelBoldt, Matthew Woodcraft, The Anome, Taw, Ap, Andre Engels, Cable Hills, Stevertigo, Edward, JohnOwens, Michael Hardy, Fred Bauder, Martin-Harper, Dcljr, Arthur3030, Ahoerstemeier, Snoyes, Den fjättrade ankan, Cyan, Poor Yorick, Jonik, Jm34harvey, AC, Pheon, Populus, MH, Henrygb, KellyCoinGuy, Bkell, Wile E. Heresiarch, SpellBott, Unfree, Snobot, Giftlite, Jao, BenFrantzDale, Bfinn, Cathy Linton, Dratman, Duncharris, Macrakis, Pcarbonn, Quadell, L353a1, Gene s, Miorea, BlairZajac, Discospinster, Smyth, Xezbeth, Nybbles, El C, Cretog8, O18, Sebastianlutz, Mcdonaldsguy, Jung dalglish, Flammifer, KarlHallowell, Larry V, Hooperbloob, ClementSeveillac, Zachlipton, Diego Moya, Moanzhu, John Quiggin, Monado, Avenue, Schaefer, Samohyl Jan, Jheald, Count Iblis, RainbowOfLight, Oleg Alexandrov, Brookie, INic, Tomlillis, GregorB, BlaiseFEgan, Marudubshinki, Graham87, Rjwilmsi, Commander, MarSch, Ravik, Billjefferys, Wragge, Mathbot, RexNL, Valor, Exelban, Fresheneesz, Chobot, Jdannan, Beanyk, MacMog, Jules.LT, Che829, Daniel roy, Finell, Capitalist, Roydanroy, SmackBot, Tomyumgoong, Incnis Mrsi, Cazort, Aaadddaaammm, MartinPoulter, Snori, Nbarth, Jdthood, Ladislav Mecir, Trekphiler, Jahiegel, BenE, Cybercobra, Dwchin, G716, OverInsured, Harryboyles, Aroundthewayboy, Nijdam, RichardF, AdjustablePliers, Hetar, Emote, DavidGSDavies, Alfpooh, Link2009, Panda17, N2e, ShelfSkewed, Requestion, Moreschi, Basar, Vizier,

Gregbard, FilipeS, Logicus, Hebrides, Anthonyhcole, CNMIN, Daa89563, Helgus, Mr pand, EdJohnston, Jvstone, Tomixdf, Mack2, Storkk, Knotwork, Stephanhartmannde, Coffee2theorems, Avjoska, Ranger2006, Freddie McPhyll, Robin S, Topagae, Gwern, CommonsDelinker, Nono64, AgarwalSumeet, Rlsheehan, Aleksandr Grigoryev, Gill110951, Coppertwig, LittleHow, Normanfenton, Policron, Robcalver, JohnBlackburne, The Tetrast, Econterms, Lambyte, Andrewaskew, Enkyo2, PlanetStar, Doctorfree, GentDave, Mateat, Janopus, Ddxc, Melcombe, ClueBot, Reovalis, Viviannevilar, Pot, Charledl, ERosa, Qwfp, Tdslk, Erreip, Gerhardvalentin, Imperial Star Destroyer, Hossdave, WMdeMuynck, NjardarBot, Lihaas, LemmeyBOT, Tassedethe, Legobot, PlankBot, Yobot, Mindbuilder, UNSEENUNHEARD, AnomieBOT, Materialscientist, Citation bot, Glenn Stokowski, Jockocampbell, Shadowjams, Borkert, X7q, Argumzio, Eurdem, DrilBot, Kiefer.Wolfowitz, Trappist the monk, Gnathan87, Richardcherron, Dinamik-bot, Arrowzf, EmausBot, JA(000)Davidson, Matteo.taiana, AManWithNoPlan, GreenMachine86, Peter M. Brown, Udaya.s.k, Iratheclimber, Bayes Puppy, Habil zare, Helpful Pixie Bot, Scochran4, Dasonk, Lolapellicer, Lifeformnoho, Joydeep, Ellewarren, CeraBot, Acuppert, ChrisGualtieri, Dexbot, Monkbot, Sennsationalist and Anonymous: 185

_ **Psychologism** *Source:* http://en.wikipedia.org/wiki/Psychologism?oldid=634360293 *Contributors:* Prosario 2000, Kku, Jm34harvey, Wclark, Chalst, Oop, Knucmo2, Keenan Pepper, Oleg Alexandrov, FlaBot, Pigman, Nicke L, M3taphysical, Tomisti, TransUtopian, Jules.LT, Bluebot, JoshuaZ, Gregbard, Cydebot, Thijs!bot, William Knorpp, JAnDbot, Xeno, Pomte, Stdbrouw, DorganBot, VolkovBot, The Tetrast, Billinghurst, Jackbars, Myrvin, Hazzzzzz12, Addbot, SpBot, AnomieBOT, Dr Oldekop, GrouchoBot, Omnipaedista, Updatehelper, Bluszczokrzew, EmausBot, ZéroBot, Gibbja, BattyBot, Rodney.k.b.parker and Anonymous: 11

_ **Language of thought hypothesis** *Source:* http://en.wikipedia.org/wiki/Language%20of%20thought%20hypothesis?oldid=620957436 *Contributors:* Edward, Cadr, Banno, Diberri, Eequor, Mporch, D6, Posiduck, Gary, Keenan Pepper, Velho, Ylem, Mandarax, Rjwilmsi, KYPark, Spencerk, Ncsaint, SmackBot, Srnec, WikiPedant, OrphanBot, JBel, StN, Gregbard, Peterdjones, Quibik, Thrapper, Ninjakannon, Harborsparrow, Nono64, VanishedUserABC, Newbyguesses, Jojalozzo, Ddgromit, Krzysztofgajewski, EPadmirateur, Alexbot, Brews ohare, Lucky Bottlecap, Earcanal, Addbot, Redheylin, Yobot, MMPedwardg, Unara, FrescoBot, Abductive, Jemoore31688, Dmanzelmann, Helpful Pixie Bot, 220 of Borg, Cgeggie, Jochen Burghardt, Aubreybardo, Philosopher of Mind and Anonymous: 28

_ **Science** *Source:* http://en.wikipedia.org/wiki/Science?oldid=633316216 *Contributors:* AxelBoldt, Eloquence, Zundark, The Anome, Stephen Gilbert, Malcolm Farmer, Ed Poor, RK, Andre Engels, Eclecticology, Vignaux, XJaM, Fredbauder, PierreAbbat, Fubar Obfusco, William Avery, SimonP, Anthere, KF, Hephaestos, JDG, ChrisSteinbach, Twilsonb, Stevertigo, Edward, Lir, Infrogmation, Michael Hardy, Fred Bauder, Lexor, Grizzly, BoNoMoJo (old), Tannin, Mic, Ixfd64, Lquilter, Tango, Dcljr, GTBacchus, Pagingmrherman, Ahoerstemeier, Arwel Parry, Snoyes, Angela, Den fjättrade ankan, JWSchmidt, Glenn, RadRafe, Cyan, Poor Yorick, Big iron, Rotem Dan, Andres, Kaihsu, Evercat, Sethmahoney, Mxn, Hemmer, Ec5618, Charles Matthews, Guaka, Pingchen, Wikiborg, Ed Cormany, Reddi, Terse, Fuzheado, Greenrd, Lord Kenneth, DJ Clayworth, Markhurd, Tpbradbury, Ksolway, Maximus Rex, Morwen, Saltine, Martinphi, SEWilco, Paul-L, Omegatron, Buridan, Fvw, Stormie, Wetman, Gakrivas, Secretlondon, Pilaf, Flockmeal, Banno, Francs2000, Owen, PuzzletChung, Phil Boswell, Gentgeen, Robbot, Chrism, Fredrik, Gwrede, Zandperl, R3m0t, RedWolf, Goethean, Altenmann, Modulatum, Lowellian, Gkochanowsky, Stewartadcock, Rholton, Rursus, Texture, Davodd, Hadal, UtherSRG, Robinh, HaeB, Guy Peters, Spell-Bott, Dina, Alan Liefting, Enochlau, Leighxucl, Vaoverland, Ancheta Wis, Matthew Stannard, Centrx, Giftlite, Christopher Parham, Pretzelpaws, Inter, Lee J Haywood, Tom harrison, Lysis, Brian Kendig, Fastfission, Aphaia, Hokanomono, Everyking, Curps, Bensaccount, Wikibob, Jorend, Jfdwolff, Avsa, Ezhiki, Mboverload, Prosfilaes, AlistairMcMillan, Solipsist, Costyn, SWAdair, AdamJacobMuller, Tagishsimon, Wmahan, OldakQuill, James Crippen, Gadfium, Andycjp, R. fiend, Quadell, Stephan Leclercq, Antandrus, The Singing Badger, Beland, Estel, OverlordQ, PDH, Jossi, Karol Langner, Rdsmith4, APH, Maximaximax, Bodnotbod, Huwr, Pethan, DanielDemaret, Sam Hocevar, Rlcantwell, Popadopolis, Darksun, Iantresman, Neutrality, Joyous!, Ukexpat, Jh51681, Frau Holle, Karl Dickman, Eduardoporcher, Grm wnr, Deglr6328, Gleet, Zondor, Adashiel, Grunt, Bluemask, Mike Rosoft, Brianjd, SimonEast, Reinthal, Juan Ponderas, Haiduc, EugeneZelenko, Discospinster, ElTyrant, Rich Farmbrough, KillerChihuahua, Rhobite, Guanabot, FT2, Schuetzm, Vsmith, HeikoEvermann, Smyth, Dave souza, 1pezguy, Paul August, SpookyMulder, BBB, Bender235, ESkog, Melamed, Eric Forste, Brian0918, TFK, RJHall, El C, Cap'n Refsmmat, Lycurgus, Mwanner, QuartierLatin1968, Skeppy, Aude, Shanes, Art LaPella, Riyehn, Adambro, Guettarda, Shoujun, Bobo192, Flxmghvgvk, Jung dalglish, Ziggurat, Guiltyspark, Greenleaf, Jeffreyn, Dzou, VBGFscJUn3, Malcolm rowe, Vanished user 19794758563875, John Fader, MPerel, Pharos, Pearle, Nsaa, Mdd, Ekhalom, HasharBot, ADM, Jumbuck, Beyondthislife, Poweroid, Alansohn, JYolkowski, Eleland, Polarscribe, Arthena, Atlant, Paleorthid, Plumbago, JoaoRicardo, Logologist, Riana, AzaToth, Lightdarkness, Eukesh, Malo, Titanium Dragon, Avenue, Caesura, Cortonin, Cugel, Velella, SidP, Tycho, Rick Sidwell, Gaussianzz, Knowledge Seeker, Suruena, Garzo, Evil Monkey, Omphaloscope, Dmccabe, Harej, Tony Sidaway, Amorymeltzer, Grenavitar, CloudNine, Sciurinæ, Mikeo, BlastOButter42, Redvers, Kenyon, Mullet, FrancisTyers, The JPS, Simetrical, Mel Etitis, OwenX, Woohookitty, Mindmatrix, N1r4v, Pmberry, Georgia guy, Consequencefree, Swamp Ig, Daniel Case, Brunnock, Before My Ken, Ruud Koot, MONGO, Eleassar777, Tygar, Friarslantern, Wikiklrsc, Ledouche, Terence, Striver, Sengkang, GregorB, Kralizec!, Noetica, Wayward, Joke137, Gimboid13, MarcoTolo, Phlebas, Allen3, LexCorp, GSlicer, DavidParfitt, Raguks, Graham87, Alienus, Magister Mathematicae, BD2412, Kbdank71, FreplySpang, Malangthon, Island, Icey, Sjö, Drbogdan, Sjakkalle, Rjwilmsi, Mayumashu, Dpark, Jake Wartenberg, Commander, Vary, Bob A, Mll, MarSch, Quiddity, Jiohdi, Xosé, Rschen7754, HolyApocalypse, MZMcBride, Mork the delayer, Tawker, Plotinus, Mm35173, Bhadani, Dar-Ape, GregAsche, Jesus Is Love, Cassowary, Tommy Kronkvist, Falphin, Titoxd, Ian Pitchford, RobertG, Airumel, Nihiltres, MethodicEvolution, SouthernNights, Nivix, Chanting Fox, Hottentot, Andy85719, Pathoschild, RexNL, Alexjohnc3, AndriuZ, Agesilaus II, Diza, Malhonen, Daycd, Snailwalker, Imnotminkus, King of Hearts, Chobot,

Mhking, VolatileChemical, Skraz, Gwernol, Roboto de Ajvol, Summalogicae, YurikBot, Wavelength, TexasAndroid, Sceptre, Sarranduin, WhatPotato?, Jlc46, Pigman, Cswrye, Markus Schmaus, Netscott, SpuriousQ, Ansell, Matt Fitzpatrick, Akamad, Stephenb, Grubber, Cate, Gaius Cornelius, CambridgeBayWeather, Alex Bakharev, KSchutte, Ergzay, Ugur Basak, MosheA, Shanel, NawlinWiki, Rick Norwood, Nowa, Wiki alf, Hwasungmars, Deskana, Jaxl, Johann Wolfgang, InformationalAnarchist, Ino5hiro, Jfsaiya, Dureo, Nick, Ragesoss, Brythain, Banes, Daniel Mietchen, Rmky87, Raven4x4x, Stevenwmccrary58, Alex43223, Nate1481, RonCram, PrimeCupEevee, Mysid, DeadEyeArrow, ThreePD, Haemo, Xpclient, Werdna, Efbrazil, Wknight94, Pooryorick, FF2010, Enormousdude, Rolf-Peter Wille, Zzuuzz, Snotface, Andrew Lancaster, Mike Dillon, Closedmouth, The Son of Oink, Arthur Rubin, KGasso, Dspradau, Jake Spooky, BorgQueen, GraemeL, JoanneB, CWenger, Cjwright79, HereToHelp, ArielGold, CKelly, Ilmari Karonen, Kungfuadam, Junglecat, Pfistermeister, Aeosynth, Meegs, Bsod2, Benandorsqueaks, Paul Erik, Asterion, MansonP, DVD R W, Algae, Jdcompguy, Luk, Sardanaphalus, Xygtshadow, A bit iffy, BonsaiViking, SmackBot, NSLE, Rtc, Zazaban, Brianyoumans, Prodego, KnowledgeOfSelf, Test-Pilot, Hydrogen Iodide, Od Mishehu, Vald, Rokfaith, Bomac, Jagged 85, Gabi bart, Anastrophe, Delldot, Alan McBeth, Hardyplants, RobotJcb, Canthusus, Chauncey27, Shamalyguy, Jpvinall, Edgar181, Commander Keane bot, M stone, Sloman, Portillo, Ohnoitsjamie, Hmains, Skizzik, Ppntori, Jwgraham, Kmarinas86, Lindosland, Anwar saadat, Wigren, Bluebot, Geneb1955, Samosa Poderosa, Bartimaeus, Persian Poet Gal, NCurse, Tito4000, Bduke, Cattus, Stubblyhead, MartinPoulter, Lddnhan, Fplay, Silly rabbit, Papa November, Ryan Paddy, SchfiftyThree, Deli nk, Mark7-2, J. Spencer, Go for it!, CMacMillan, DHN-bot, Sbharris, Hallenrm, A. B., Reaper X, D-Rock, Can't sleep, clown will eat me, MisterHand, Nick Levine, Милан Јелисавчић, Danielkueh, Frap, Sommers, Aelsi, Darthgriz98,

Voyajer, Xiner, Run!, Addshore, Kcordina, Meepster, SundarBot, Zophar1, Barkman34, Fuhghettaboutit, Cybercobra, Wapp, Bowlhover, Nakon, Theodore7, Jiddisch, Nick125, Rezecib, Dacoutts, Salt Yeung, BullRangifer, Hgilbert, Polonium, Jon Awbrey, Adrigon, Hammer1980, Jklin, Wizardman, Just plain Bill, Metamagician3000, Mystaker1, Sadi Carnot, Ck lostsword, Pilotguy, Kensor, Kukini, Dkusic, Ged UK, Byelf2007, Chwech, Lambiam, Mchavez, ArglebargleIV, Rory096, Orbicle, Giovanni33, Paaerduag, Zahid Abdassabur, Kuru, UberCryxic, Vgy7ujm, Nzgabriel, JoshuaZ, Chodorkovskiy, JorisvS, Dumelow, Mgiganteus1, CaptainVindaloo, Singh.vish, Runningfridgesrule, IKR1, AdAdAdAd, 16@r, Cjackb, Stwalkerster, Martinp23, Laogeodritt, Wstomv, Mr Stephen, Bendzh, Z E U S, Waggers, Icez, Dhp1080, Tuspm, Anonymous anonymous, RichardF, Jose77, RHB, Osame, Travia21, Snezzy, ShakingSpirit, Phuzion, Hu12, Stephen B Streater, Tawkerbot, Ginkgo100, Levineps, Kevlar992, K, Dekaels, Paul venter, David Little, TWIS, J Di, Mrdthree, Andrew Hampe, Shoreranger, Lenoxus, Secretpizaparty, AGK, Az1568, Courcelles, Audiosmurf, Peteweez, Nkayesmith, Secos5, Tawkerbot2, MarylandArtLover, Blueracer6, Kurtan, Lahiru k, Trubkozub, MightyWarrior, Firefly111, Farnyl, Efrum, JForget, Wolfdog, Ken McRitchie, Phillip J, CmdrObot, Wafulz, Insanephantom, Van helsing, Makeemlighter, Enselic, Page Up, Taylorhewitt, Dan2119, GHe, Dark24spn, Dgw, Yarnalgo, Ballista, Ezrakilty, ButterApple, Neelix, GregW, Longshot.222, Andrew Delong, Murderd2death, Funnyfarmofdoom, TJDay, Slazenger, Bkessler23, Don.atreides, Mualphachi, Mato, Mortus Est, Michaelas10, Gogo Dodo, JFreeman, Chasingsol, Eu.stefan, Dancter, He Who Is, Tawkerbot4, Shirulashem, DumbBOT, FastLizard4, NaLalina, Crana, IComputerSaysNo, Joe11miles, Omicronpersei8, Dyanega, Wexcan, Trev M, Casliber, FrancoGG, BetacommandBot, Smellyk, Alquri, Epbr123, Barticus88, Wikid77, Btball, Qwyrxian, Goods21, N5iln, Marek69, Joeprempeh, John254, Tapir Terrific, Second Quantization, Tellyaddict, Sturm55, BlytheG, Random Tree, SusanLesch, Beezle1999, DblGkid, AlefZet, Eleuther, Danielfolsom, Porqin, WikiSlasher, AntiVandalBot, RobotG, Luna Santin, Guy Macon, Why My Fleece?, Wenisboy111, Emeraldcityserendipity, TimVickers, Vicè, Geogeogeo, Pheoinixflame, Danger, Lperez2029, Gdo01, Spencer, Zidane tribal, Myanw, Res2216firestar, Ioeth, Mikenorton, JAnDbot, Narssarssuaq, Husond, Poga, Jen-Louise, MER-C, Plantsurfer, The Transhumanist, Gtorell, Hello32020, Arturo 7, Plm209, Andonic, Dcooper, Hut 8.5, Lirter, Tstrobaugh, Acroterion, Magioladitis, Pedro, Bennybp, Bongwarrior, VoABot II, Dekimasu, Wikidudeman, Yandman, JamesBWatson, CattleGirl, A10brown, Genedoug, Avicennasis, Wolfram.Tungsten, Cardamon, Rachita Sephiroth, D-rew, Viconpr, Dinohunter, User86654, Spacegoat, Bloodredrover, Ciaccona, Biokinetica, Allstarecho, Chivista, Cpl Syx, Gomm, MCG, Vssun, Sabedon, Glen, DerHexer, LW77, WLU, Calltech, Akhil999in, Lightnin Boltz, 0612, Weiojranwie v5a, MartinBot, BetBot, It334, Arjun01, Tvoz, Jessoupe, Rettetast, Ottantotto, R'n'B, Snozzer, Boston, Erkan Yilmaz, Artaxiad, Gizmo321, J.delanoy, Pharaoh of the Wizards, Trusilver, Bogey97, Psycho Kirby, Maurice Carbonaro, Nigholith, Ginsengbomb, Eliz81, Jason-rule, Kimhyunwoo, Taop, G. Campbell, Josisb, Heat023.robes, Jkaplan, SharkD, Dontrustme, Nnamdinwokoro, Smeira, Jeepday, Olithal, Gabe mayne, Pyrospirit, Rossenglish, The Transhumanist (AWB), Arms & Hearts, NewEnglandYankee, Antony-22, Philmacrackin, Luctor IV, Mrfriedchicken, Daerg, Unknownguy123456789, Jjdukejj403, Olegwiki, Shoessss, 2help, Miaferron, Juliancolton, Cometstyles, WJBscribe, Remember the dot, Zara1709, C-word, Tae Guk Gi, Diego, Khargas, Andy Marchbanks, The Fat Guy, Useight, Axle12693, Markguitar333, Nigger1234567, Nwanda, Xiahou, CardinalDan, Cromoser, Idioma-bot, Wikieditor06, DaDawg22, Vranak, Deor, VolkovBot, TreasuryTag, Swfcowls, ABF, Somebodyreallycool, Arialboundaries123, Alexandria, AlnoktaBOT, Bacchus87, Mugander, Fences and windows, Miguelzinho, Voronwae, Soap Poisoning, Sześćsetsześćdziesiątsześć, Tzetzes, Eedo Bee, Philip Trueman, Nerm12, TXiKiBoT, Rollo44, KateBerry, Cosmic Latte, The Original Wildbear, Red Act, A4bot, Hqb, Scilit, Applerw, Joel Kincaid, Jazzwick, GDonato, Gerrish, Ask123, Figgisfiggisfiggis, Qxz, Taimaster, PolarBearoughey, Forrest1966, Indy 900, Retiono Virginian, Sciencegrl101, Littlealien182, H2ono2, Seth103, Kiwi1234, Explosiv, Dendodge, Gjgarrett, The Tetrast, Martin451, Blacktriangle10k, RedAndr, Rexeken, Abdullais4u, Fbs. 13, LeaveSleaves, Supernerd 10, Themcman1, PDFbot, 1yesfan, StillTrill, Koolkatie, Colin stuart, Songrit, Aphilo, Mwilso24, Lethalraptor, Blurpeace, Simonwerner, Lerdthenerd, Farkas János, Synthebot, Lainer21193, Falcon8765, Enviroboy, Sam1993, SeizeThe Dayy, Thanatos666, Insanity Incarnate, Sebastjanmm, Pjoef, AlleborgoBot, Roadcreature, S4ndp4perm4n, TheXenocide, Riverwaste, TimProof, Randula, Roberdor, SieBot, Tiddly Tom, Nihil novi, Graham Beards, ElphabaThropp95, Scarian, Lemonflash, Elmllama, Parhamr, Dawn Bard, Mammamiamania, Whiteghost.ink, Breakyunit, GlassCobra, Keilana, Bobcrankins, Ujjwol, Elvissenthil, Tiptoety, Radon210, Michaelgerety, Nopetro, Oiws, Copperwing0, JSpung, Elmacenderesi, Oxymoron83, Faradayplank, Yoshimad123, Lightmouse, LaidOff, Jimmycleveland, RyanParis, AMackenzie, Sunrise, DancingPhilosopher, Mygerardromance, Fox red star, Markster2000, Vig vimarsh, Productof-Society, Neurophysics, Amahoney, Verdadero, JL-Bot, TracySurya, Onemado, Myrvin, Asher196, Ainlina, TheCatalyst31, Afiya27, Kleinhev, Martarius, Tanvir Ahmmed, Elassint, ClueBot, PipepBot, Jackollie, Jncc0, Panoptik, The Thing That Should Not Be, Chocoforfriends, Rjd0060, Papa Smurf11, Tractorboy60, Pwitham, Ukabia, Drmies, VQuakr, Polyamorph, Boing! said Zebedee, Counter-VandalismBot, Themully, Ryan1182, The mullisk, Briankohl, Turbo566, Neverquick, Jasonssmith94, Puchiko, MindstormsKid, DragonBot, Excirial, Anne Prouse, Keithbowden, SkE, Vanisheduser12345, Jjvikingsfan, Lingo pen, Estirabot, Mit027*, Carau, Cenarium, JoelDick, Jotterbot, Medos2, Ttpp7, Tnxman307, Singhalawap, B-man79210, Laughitup2, Dekisugi, Banime, SchreiberBike, Spykodemon, Casualpsycho, Xme, Polly, Dpthurs, GFHandel, Invisibill, Thingg, Aitias, JDPhD, Ertemplin, Versus22, LieAfterLie, Macderv15h, Relly Komaruzaman, JKeck, Dakrismeno, Against the current, XLinkBot, Nicholiser, PseudoOne, Jovianeye, Dsgdfshfdshdsfh, Little Mountain 5, Rreagan007, Whisky it Up!, SilvonenBot, Pogipogi, Mifter, Sikig, Tannerthegreat, Badgernet, HarlandQPitt, Navy Blue, The Rationalist, Branrile09, HexaChord, Rmiddl, Pamejudd, Loueiler, Addbot, Cxz111, Willking1979, Fireheart7397, Nickenge, Bobafett29, DOI bot, Jojhutton, Captain-tucker, Binary TSO, DougsTech, EliteAthlete, Chris19910, Ronhjones, TutterMouse, Screwdis, Wikiwizzard123, Camarinha, Scient, Ilya-108, Leszek Jańczuk, Kapaleev, Devrit, Looie496, MrOllie, Chamal N, CarsracBot, Glane23, Dizzle13, Debresser, NittyG, Dr. Universe, Favonian, LemmeyBOT, AtheWeatherman, Azurefox, Connect1, Numbo3-bot, Tide rolls, Lightbot, Jan eissfeldt, Avono, Tenth Plague, Luckas Blade, Gail, Greyhood, Trotter, Quantumobserver, Luckas-bot, ZX81, Finbob83, Yobot, 2D, ALL OF YOU ST, Ajh16, THEN WHO WAS PHONE?, Runinbraces12, Nallimbot, Thehappymoustaches, Jimmysevolution,

KamikazeBot, Sumail, 2008CM, Alexkin, Tempodivalse, Rlogan2, Licor, AnomieBOT, Tryptofish, Kristen Eriksen, Rubinbot, VX, Jim1138, Short Brigade Harvester Boris, Piano non troppo, PhaseChanger, AdjustShift, Quispiam, Ulric1313, Flewis, Materialscientist, Limideen, The High Fin Sperm Whale, Citation bot, Alkhowarizmi, François Pichette, GB fan, ArthurBot, Xqbot, Timir2, Marko Grobelnik, Intelati, Cureden, JimVC3, Raphyortanez, Capricorn42, Dsoconno, A455bcd9, TechBot, Jeffrey Mall, Stsang, Prettygirlswebshow, Fancy steve, Grim23, Edward Luva, Jmundo, Jakwra, Locos epraix, The Evil IP address, Billybob131, Aa77zz, Miguel in Portugal, Ewindward, Bakaw69, Srich32977, NEDM4EVER, Nicholas.a.chambers, ScreaminXD, Almabot, IntellectToday, ClareZeBearZe, J04n, Shuzo123456789, Elitefart505, Flaucinausihilipilifacation, Corruptcopper, Biggedawg, Rockmasterdan, XXIsuelXx, Yuhh, KEEHAM, Redpanda900, Omnipaedista, Robert froste, Tclgb, Yugolervan, Earlypsychosis, Gott wisst, Voheezy, Thogan3, Robert leon, Beatson121, Jamierobb893, 14albeev, Sexyz, Sports36, Doulos Christos, XxHolyDiverxx, Persontgssbdx, Nanana180, RFC posting script, Hamamelis, AlimanRuna, SchnitzelMannGreek, Doodoocacapeepee, Sicronet, Legobot III, Hugetim, Nagualdesign, FrescoBot, Liridon, Paine Ellsworth, Tobby72, Sky Attacker, Alberttruong, Aleksa Lukic, Machine Elf 1735, Drew R. Smith, Xhaoz, ClickRick, Citation bot 1, Killian441, Gravityguy, Tnt uncontested, Skunen1, Pinethicket, Kiefer.Wolfowitz, Per Ardua, Therustinator132, MJ94, Calmer Waters, MTDinoHunter, Shan3coley, Fancynancy1244521, Elnene15, Georgiaham, RedBot, Lalalllala, Tensil, SpaceFlight89, Meaghan, I am a ginger bread man, EdoDodo, Gilbeysjame, Holybassist, Saayiit, Anticent, Dude1818, Max Duchess, FoxBot, TobeBot, DixonDBot, Kellymaj, Sznax, WebEdHC, Jovenmae, Gfdfgshdhdhdfhfghgfhfh, Mrgarcia94, Jdavie, DragonofFire, Toniiiix, Suffusion of Yellow,

788 *CHAPTER 102. MONSKY'S THEOREM*

Tbhotch, Deanmullen09, Bricaniwi, RjwilmsiBot, TjBot, DHooke1973, Tesseract2, DASHBot, Ficz44, Ejamtiger, EmausBot, Proud Liberal 6, Dasher246, WikitanvirBot, Sciencenerdphd, Gfoley4, Dominus Vobisdu, Chloeey97, Farpre, GoingBatty, Riggr Mortis, Oceans and oceans, Hhhippo, JSquish, ZéroBot, MithrandirAgain, Andyman1125, The Nut, AshforkAZ, Aeonx, H3llBot, GrindtXX, Henesuri, L Kensington, Lilgas52, Phronetic, Neuberliner, KyleAraujo, Imagoofygooberyeah, Freecie1, Kiesewetter, RockMagnetist, ClamDip, Sharonmil, U3964057, Teapeat, WordDiver, Helpsome, ClueBot NG, W.Kaleem, Another n00b, Movses-bot, Nahiyan8, Schicagos, Frietjes, Delusion23, Fbarrera09, IvoryMeerkat, Skiles1611, Helpful Pixie Bot, Calabe1992, BG19bot, Ryker-Smith, Hallows AG, ElphiBot, AvocatoBot, Ramos1990, Mthoodhood, Mbotee, I am huge liar listen, Thestickman91, Jakakla, Soerfm, CitationCleanerBot, Samyriup, Hairandfashion142, Seanpkenny, Smsagro, In11Chaudri, BattyBot, Warheadpat, Kenixkil, Karanhbhatt, Tow, IjonTichyIjonTichy, Linkman22, Dexbot, Yash!, Mogism, Numbermaniac, ਹਰਜੋਤ ਸਿੰਘ, Sowlos, Josophie, Alexis1812w, MarchOrDie, Zjrong, Thepooj9, Theo's Little Bot, Jp4gs, Globalcooling400, Melonkelon, Rakomwolvesbane, Praemonitus, AKYF, Arnlodg, Sssssss340, Sol1, Mrm7171, Hansmuller, Jackmcbarn, VeryCrocker, Man of Steel 85, Avelez00, Monkbot, Teaksmitty, ShawntheGod, UglowT, Batsgasps, Jack Pepa, BeUtkarsh and Anonymous: 1277

_ **Baruch Spinoza** *Source:* http://en.wikipedia.org/wiki/Baruch%20Spinoza?oldid=633523168 *Contributors:* Damian Yerrick, AxelBoldt, Magnus Manske, Kpjas, Marj Tiefert, Eloquence, Mav, Clasqm, Tsja, Heron, Olivier, Rickyrab, Edward, Michael Hardy, Kwertii, BoNoMoJo (old), Gabbe, Damnedkingdom, Mic, IZAK, Shanemac, Paul A, Ellywa, Snoyes, Den fjättrade ankan, Glenn, Poor Yorick, Sethmahoney, Adam Conover, RodC, Charles Matthews, Radgeek, Dtgm, Markhurd, Tpbradbury, YeshivaResearch2, Zero0000, Anon-Moos, JorgeGG, Robbot, Goethean, Sparky, Gidonb, Blainster, Diderot, Wikibot, JerryFriedman, Marc Venot, Stirling Newberry, Andries, Wilfried Derksen, Karn, Dratman, Wikiwikifast, Gilgamesh, Ferdinand Pienaar, Edcolins, Mporch, Chowbok, James Crippen, Gadfium, Pgan002, Scottryan, Quadell, Antandrus, Evertype, Mikko Paananen, Pmanderson, Jareha, Neutrality, Marcus2, Robin klein, Esperant, Kate, Lucidish, D6, Jayjg, Sfeldman, Varada, Discospinster, Brianhe, Rich Farmbrough, KillerChihuahua, Guanabot, FranksValli, ArnoldReinhold, Bender235, CanisRufus, Causa sui, NetBot, Whosyourjudas, Meggar, Olve Utne, SpeedyGonsales, Tresoldi, (aeropagitica), Pharos, HasharBot, Knucmo2, Jumbuck, Szczels, Danski14, Alansohn, Ricky81682, Logologist, Wikidea, Ksnow, Bbsrock, Garzo, Bsadowski1, Drbreznjev, Iustinus, Velho, Woohookitty, TheGoblin, Kzollman, Ruud Koot, JeremyA, Kelisi, Pdn, Ferg2k, Prashanthns, Gerbrant, Marudubshinki, Lawrence King, Graham87, Magister Mathematicae, Wraybm1, Qwertyus, Elvey, FurciferRNB, FreplySpang, Porcher, Rjwilmsi, Koavf, Zbxgscqf, KungFuMonkey, Golden Eternity, Dudegalea, Kalogeropoulos, Lairor, Brighterorange, The wub, FayssalF, Goclenius, FlaBot, Hottentot, Str1977, TeaDrinker, Bmicomp, Rynogertie, King of Hearts, Chobot, DVdm, Random user 39849958, Gwernol, Wavelength, Kinneyboy90, Ssimsekler, Wester, Kafziel, Manicsleeper, Anonymous editor, Warshy, Pigman, SpuriousQ, Lusanaherandraton, MosheA, NawlinWiki, The Ogre, Grafen, BlackAndy, BirgitteSB, Peter Delmonte, Moe Epsilon, AdiJapan, Ospalh, Larsobrien, Gadget850, Morgan Leigh, Yesselman, C i d, Jules.LT, Nikkimaria, Arthur Rubin, Fram, Mavaddat, Allens, Kungfuadam, Roke, DVD R W, Ryūkotsusei, C mon, Sardanaphalus, Attilios, SmackBot, Amcbride, Lestrade, Jdoniach, Inverse-Hypercube, McGeddon, Lukeasrodgers, Serte, Bozartas, Sludgehaichoi, Hmains, Nmacri, David Ludwig, Chris the speller, Michbich, Rex Germanus, Kaliz, Snori, Drzax, Go for it!, DHN-bot, MerricMaker, Jammus, WikiPedant, Can't sleep, clown will eat me, Милан Јелисавчић, Eliyahu S, Cophus, Writtenright, Snowmanradio, AndySimpson, Stevenmitchell, Bbarlavi, Sei Shonagon, Jiddisch, Pipifaxa, Richard001, MEJ119, Hgilbert, Diocles, Vblanton, Epf, Will Beback, SashatoBot, Ser Amantio di Nicolao, Lapaz, Ocanter, Bo99, Mgiganteus1, Antireconciler, RandomCritic, Comicist, Rkmlai, Noah Salzman, Macellarius, Plattler01, TheOtherStephan, H, Noleander, Kripkenstein, Chgwheeler, Hkd2029, Hu12, DwightKingsbury, Norm mit, Thinkingfreely, Missionary, Woxie Ninian, Shoeofdeath, Ivan-Lanin, Bryan nelson, Bifgis, Mthomas1776, Jatrius, INkubusse, CmdrObot, BryanWoody, CWY2190, Jokes Free4Me, Neelix, Gregbard, Aovechkin, Logicus, AndrewHowse, Jane023, TheRegicider, Cydebot, Jasperdoomen, Mikebrand, Reywas92, Evenmadderjon, Hebrides, Kahananite, ST47, Studerby, Krator, DBaba, Dleisawitz, Grison, AVIosad, Nishidani, Mamalujo, Thijs!bot, Barticus88, Mime, Mayor Pez, Headbomb, Pacific PanDeist, Jamesfrost, Peterrr, Top.Squark, Pemboid, SusanLesch, Rlitwin, Suriak, AntiVandalBot, Akradecki, Antique Rose, Anacrolix, Ruwel, Zachwoo, Fayenatic london, Schaheb, TheRepairMan, Sluzzelin, Atlas Mugged, JAnDbot, Tigga, MER-C, Skomorokh, Dsp13, Ericoides, Matthew Fennell, Sightbeyondsite, Détruire, Kmaguir1, Anthony Krupp, Be1981, Judejones, Taksen, Demophon, Promking, Hurmata, Magioladitis, Unused0029, Dr.Crawboney, Tito-, S-fury, Chesdovi, DBWikis, Toddcs, Fang 23, Philg88, Okcomp333, MartinBot, Nandt1, Healkids, Autocratique, CommonsDelinker, J.delanoy, Svetovid, Lizrael, Peter Chastain, Shannondale, Lhynard, Dextrase, Skier Dude, Mattximus, SteveChervitzTrutane, Jay ryann, Engelo, Vanished user 39948282, Rcrath, SuW, Inwind, Useight, Quejlfaspasma, Dynnik, RJASE1, Idioma-bot, N.B. Miller, RaulCovita, VolkovBot, Hce1132, Shakesphere17, OakMt, EchoBravo, TXiKiBoT, Java7837, BertSen, Fl 2007, Miguel Chong, Woodsstock, Ontoraul, Gekritzl, Don4of4, Wassermann, Lisa, Gilisa, Gavin.collins, 75tpickupsx1983, Ninjatacoshell, Joelleabirached, Fleurstigter, Enviroboy, Oldwes, Ignacio Bibcraft, AlleborgoBot, Struway, Abcdwiki, Category Mistake, Austriacus, Hucksterling, Ahmadiskandarshah, Demmy, Gaelen S., SieBot, Kwork, Tresiden, Taffaplatzel, Anklefear, RJaguar3, S711, Nite-Sirk, Tiptoety, Surferhere, Yone Fernandes, Lightmouse, Epikouros, Sanya3, BenoniBot, Anchor Link Bot, Leranedo, Martarius, ClueBot, Jchatter, Mc2000, PipepBot, All Hallow's Wraith, Kendo70133, Pi zero, Gregcaletta, Saddhiyama, Mild Bill Hiccup, Oxvox, TypoBoy, RafaAzevedo, Eversman, BlueAmethyst, DragonBot, Alexbot, Random user iooi23ialdjkjk4, Sreifa01, 10FingerJoe, Bauer 1046, Rhododendrites, Sun Creator, Jotterbot, Stefano510, Orlando098, Gjames017, MickCallaghan, ThegreatLolofchina, Alkra1, Genesiswinter, Crowsnest, DumZiBoT, Vandort1, XLinkBot, User2102, Jan D. Berends, BodhisattvaBot, Kwork2, Dthomsen8, SilvonenBot, Good Olfactory, Kbdankbot, Addbot, Houthakker, Manuel Trujillo Berges, Atethnekos, Toyokuni3, LightSpectra, Cognatus, Groundsquirrel13, NjardarBot, Bertramhp, LinkFA-Bot, Bentaura, Tassedethe, Numbo3-bot, Rjaf29, Taketa, Faunas, BennyQuixote, Margin1522, Luckas-bot, Yobot, Stevenpinker, Ptbotgourou, TaBOT-zerem, קרול ישראל, Nallimbot, Matanya, Jimjilin, Astonzia, Fffffffffffffffffffffffff, AnomieBOT, 1exec1, BlessedButThorny, JackieBot, Sz-iwbot, Materialscientist,

Wandering Courier, Bob Burkhardt, Xqbot, 613kpiggy, მოცარტი, Stuthehistoryguy, Ekwos, St.nerol, Drilnoth, Davshul, Tomwsulcer, GrouchoBot, APassionCane, Omnipaedista, SassoBot, JonDePlume, BoomerAB, FreeKnowledgeCreator, Rotnerl, Msolow, FrescoBot, Platonykiss, Arouet lj, Zlatno Pile, Jamesooders, Kwiki, Naturalistic, Wikitza, MikeGurlitz, Fixer88, Nednednerb, Jandalhandler, Molly-RoseCopyEdit, Bgpaulus, Jauhienij, Hart, aber ungerecht, FoxBot, Marcobale, Sensantius, TobeBot, DixonDBot, Lotje, Marielle H, Bgmax2, Joey1978, 777sms, Milad10us1985, Jaunda, RjwilmsiBot, Bhawani Gautam, DASHBot, EmausBot, Pooya1312, ImprovingWiki, John of Reading, Atwarwiththem, Rarevogel, Everything Else Is Taken, Evanh2008, AvicBot, Kkm010, ZéroBot, Mrcrumplar, Bollyjeff, Naviguessor, Bustakey, Ghione, Gbsrd, Erianna, Δ, Vasio, Jeffeux, Polisher of Cobwebs, Chewings72, Abacatabacaxi, Jerryfrancis, Gilquentin, ChuispastonBot, AgentSniff, Anoop24, Rocketrod1960, Turmerick, Xanchester, Helpsome, ClueBot NG, Lhimec, B.Harrus, Caute AF, ATX-NL, EricWR, Joefromrandb, Philosopher12, LeTechnogoat, Thorsten Wiesmann, Editør, L69, Morris Saunders, Rezabot, Widr, Morgan Riley, Oxford73, Meotian, Helpful Pixie Bot, Edisonqv, BG19bot, Arch8887, JohnChrysostom, Darouet, Cold Season, Allecher, OttawaAC, Ostera65, The Almightey Drill, Dobrich, Harizotoh9, Jafdc, Striking13, Clakhi, Anthrophilos, SergeantHippyZombie, Francisgoode, Khazar2, Archer47, Dexbot, Belisariusgroup, NaturaNaturans, Tishtolo, The Vintage Feminist, VIAFbot, Mousetext, 420mysteryman69, Skippco, G8r Scott, Epicgenius, HumphreyBurke, Fontao24, Brokal, Liza Llewellyn, RaphaelQS, Torah-Cafe, Gavelboy, WalterStevenson, BenEsq, Mustagnom, Sasha.gazmori, Liz, Jennifer Lost the War, Azovdelt, Francisco.j.gonzalez, Vlad the implorer, Guy355, JeanLuc Discard, ErikSummerville, Palestinewillbefree and Anonymous: 639

_ **Certainty** *Source:* http://en.wikipedia.org/wiki/Certainty?oldid=623714729 *Contributors:* Edward, Michael Hardy, Kku, Ixfd64, JASpencer, Lumos3, Cholling, Dbenbenn, Gyrofrog, Fishal, Andycjp, Sonjaaa, Thorsten1, D6, Zy26, Cherry blossom tree, Wayfarer, Zachlipton, Melaen, DonQuixote, Woohookitty, Jeff3000, Lawrence King, Rjwilmsi, Wragge, Captwheeler, Spencerk, Jayme, YurikBot, Bhny, MightyGiant, Aldux, Josh3580, SmackBot, Elonka, Od Mishehu, NGC6254, Egsan Bacon, RedHillian, 16@r, Macellarius, Clarityfiend, DeadCow, George100, Gregbard, Nauticashades, Dr.enh, Mattisse, Teh tennisman, Nimakha, AntiVandalBot, Matthew Fennell, Yahel Guhan, Tsinoyboi, Passanger, Slash, Someman6, Tamabec, STBotD, Squids and Chips, Jeff G., Bovineboy2008, Philogo, Insanity Incarnate, Tautologist, PipepBot, SamuelTheGhost, Jusdafax, S19991002, SchreiberBike, TwiLighT1126, Natdudeuk, Addbot, Elfstones, Lightbot, Kookyunii, Unara, Jonathan321, Captainnipples, Abillionistoomany, ThurstonMoore123, Aaron Kauppi, FrescoBot, Rkr1991, Dawgboy47, Tom.Reding, MastiBot, Yutsi, Redbeanpaste, Lotje, Miracle Pen, Some Wiki Editor, Wiggalama, Tesseract2, Ngc0202, AvicBot, Ὁ οἶστρος, ClueBot NG, Amr.rs, Rezabot, BattyBot, AmericanLemming and Anonymous: 72

_ **Cogito ergo sum** *Source:* http://en.wikipedia.org/wiki/Cogito%20ergo%20sum?oldid=632489663 *Contributors:* Tobias Hoevekamp, Mav, The Anome, Someone else, Stevertigo, Edward, Bdesham, Michael Hardy, Kwertii, Liftarn, Minesweeper, Typhoon, Snoyes, Evercat, Conti, Emperorbma, Charles Matthews, Timwi, Andrewman327, Haukurth, Furrykef, David Shay, Saltine, Buridan, Pigsonthewing, Fredrik, Fifelfoo, Ajd, Academic Challenger, Gbog, Andrew Levine, Wereon, Giftlite, Monedula, Piquan, Eequor, Lucky 6.9, Wmahan, Toytoy, SarekOfVulcan, Zeimusu, Quadell, JoJan, Jossi, Phil Sandifer, Tothebarricades.tk, Sam Hocevar, Publunch, Neutrality, Lacrimosus, Lucidish, Rfl, Freakofnurture, Discospinster, Rich Farmbrough, Guanabot, FranksValli, Vague Rant, Ericamick, Paul August, Horsten, Schmeitgeist, CanisRufus, Maclean25, El C, Kwamikagami, Spearhead, Art LaPella, Jpgordon, Bjardine, TheProject, Sam Korn, Jumbuck, Mark Lewis, Alansohn, Diego Moya, Andrewpmk, R Calvete, Lightdarkness, Clarahamster, Velella, Immanuel Giel, Saga City, Cburnett, LFaraone, Iustinus, Oleg Alexandrov, Kelly Martin, Mel Etitis, Woohookitty, MattGiuca, Jeff3000, Nick Drake, Bkwillwm, Tydaj, Gerbrant, Edsmilde, BD2412, Reisio, Dpv, Rjwilmsi, Tizio, Harry491, Bill37212, Bhadani, Cassowary, Yamamoto Ichiro, Nihiltres, TheMidnighters, Nivix, Backflash001, Mark Yen, Chobot, Voodoom, Adoniscik, YurikBot, Hede2000, Chaser, Gaius Cornelius, ENeville, Holycharly, Trovatore, ScottyWZ, Denihilonihil, Tomisti, Scott.stratford, MarsJenkar, CharlesHBennett, Pollo318, SmackBot, Larvatus, Hydrogen Iodide, Jagged 85, ViewFromNowhere, Eskimbot, Frymaster, BiT, Edgar181, Septegram, The Famous Movie Director, The monkeyhate, Schmiteye, Kaliz, Roscelese, Baa, DHN-bot, Mladifilozof, Sct72, WikiPedant, Mooncow, Can't sleep, clown will eat me, Metroid dragon, Harnad, Bowlhover, Jklin, Wybot, Vina-iwbot, Kukini, Byelf2007, Visium, Bando26, The Man in Question, Grumpyyoungman01, DragonWR12LB, Hu12, CzarB, Ariel Pontes, Roobydo, Non organ, S ried, BoH, Aherunar, Sdorrance, Bybbyy, Gregbard, Cydebot, Jasperdoomen, Reywas92, JFreeman, Arthurian Legend, Eu.stefan, Meol, Atcevik, Cyclonenim, AntiVandalBot, Dylan Lake, Danny lost, Minhtung91, JAnDbot, Skomorokh, Awien, Gaeddal, RainbowCrane, Michael Keats, Cockfag69, RebelRobot, Yahel Guhan, Magioladitis, Gatorinvancouver, Celithemis, VoABot II, Adamumansky, Sam Medany, Hiplibrarianship, Mtd2006, Spontini, Edward321, Hdt83, Ugajin, Anarchia, Cmulvaney, Dionysiaca, Subjectivist, J.delanoy, 0207848m, Raistlin11325, Maurice Carbonaro, Blew1500, Tdadamemd, Dispenser, 04cah, Ilikenuts, Tarotcards, Pygenot, Kvdveer, Straw Cat, Izno, Bane II, Deor, VolkovBot, Sokoljan, 0-Jenny-0, Eeldrop, Hqb, Rei-bot, Qxz, Fredvdp, Broadbot, PDFbot, Popopp, Beverley12, AlleborgoBot, GeertHa, SieBot, Guevara27, VVVBot, Alexbook, Andersmusician, Andersas, Drax89, Quest for Truth, Fratrep, Vice regent, Grisabre, BRHutchins, XDanielx, WikipedianMarlith, ClueBot, SummerWithMorons, Ve4ernik, Mild Bill Hiccup, DragonBot, Naleh, Abrech, Requires1GB, Editor2020, Humanengr, Tarlneustaedter, Pfhorrest, Feinoha, Good Olfactory, Tulli, 9step10, LightSpectra, The C of E, AndersBot, AZDub, West.andrew.g, Artaxus, Abiyoyo, Tide rolls, Fragtion, Legobot, Luckas-bot, Yobot, Amirobot, NakuruAngel, KeelNar, AnomieBOT, Mauro Lanari, Trevithj, Jeff Muscato, ImperatorExercitus, Citation bot, Johan11131982, Racconish, Javier666, ArthurBot, Brighamhb, Xqbot, Vangen, Stho002, Armbrust, Omnipaedista, DillonLarson, Machine Elf 1735, 77persons, Hluup, Buddy23Lee, Gregkaye, Vistascan, LilyKitty, Mariadelcarmenpatricia, Science001, VernoWhitney, Tesseract2, Zujine, DASHBot, Sk8erboy70, Ajoones, Italia2006, Mz7, ZéroBot, Leminh91, Knight1993, Zap Rowsdower, Tolly4bolly, Inswoon, WMC, ClueBot NG, Philclasstoday1234, Adrianmanjarrez92, O.Koslowski, Widr, Jacobburns99, Helpful Pixie Bot, DrJimothyCatface, LMPorter0000, TCN7JM, ZMD123, W.andrea, Marly1929==, Teivtaoht, Sephiroth4923, Thelivingparadox, Khazar2, Kolega2357, Lfdder, Brontologique, DavidLeighEllis, Apophaticlogos, Chickenfrend, Monkbot and Anonymous: 405

_ **Automated theorem proving** *Source:* http://en.wikipedia.org/wiki/Automated%20theorem%20proving?oldid=628958083 *Contributors:* AxelBoldt, The Anome, Taw, Dwheeler, Cwitty, Michael Hardy, JakeVortex, Karada, Rotem Dan, Charles Matthews, Dcoetzee, Dysprosia, Michaeln, Jimbreed, Stephan Schulz, TittoAssini, Wikibot, Lumingz, Tobias Bergemann, Ancheta Wis, Giftlite, Bfinn, Bobblewik, Beland, TheosThree, Tc, Rich Farmbrough, Rama, Peter M Gerdes, AshtonBenson, Diego Moya, Nealcardwell, Krischik, Pontus, Alai, LunaticFringe, Oleg Alexandrov, Geoffgeoffgeoff3, Paul Haroun, Ruud Koot, JosefUrban, Graham87, Qwertyus, Grammarbot, Rjwilmsi, Tizio, MZMcBride, Brighterorange, Mathbot, Ysangkok, CarolGray, JYOuyang, Jrtayloriv, GreyCat, Nehalem, Algebraist, Ksyrie, Grafen, Jpbowen, Arthur Rubin, Saeed Jahed, Zmoboros, PhS, Nahaj, Zvika, Pintman, InverseHypercube, McGeddon, Bluebot, Clconway, MovGP0, Haberg, Slawekk, Akriasas, Jon Awbrey, Byelf2007, Ramykaram, Disavian, Michael Bednarek, Antonielly, Lancebledsoe, Loadmaster, JHalcomb, ILikeThings, CRGreathouse, CBM, Pgr94, Ezrakilty, Blaisorblade, Agent1, Magioladitis, Detla, Cic, David Eppstein, Epsilon0, Zacchiro, Laurusnobilis, Nattfodd, Uwe Bubeck, MaD70, BotKung, Brian Huffman, Synthebot, Newbyguesses, Arapajoe, DaYZman, Vanished user oij8h435jweih3, PaulBrinkley, BSoD, DainDwarf, Tatzelworm, Logperson, JohnAspinall, Eusebius, Adrianwn, Simon04, Dkf11, D.scain.farenzena, Tim32, Bracton, Ceilican, WikHead, Tassedethe, Lightbot, Jarble, Legobot, Adelpine, Linket, AnomieBOT, Citation bot, Doezxcty, FrescoBot, Wikinglouis, Iislucas, Kishmakov, Trappist the monk, Vincent Aravantinos, Jfmantis, BertSeghers, Mekeor, Chricho, Peskoj, Thetna123, Sonic7406, Arrandale, Frietjes, BattyBot, ChrisGualtieri, Khazar2, Y256, Jochen Burghardt, Pimp slap the funk, Ragerdl, Monkbot, SiddMahen, NQ, Kyle1009 and Anonymous: 141

_ **Mathematical fallacy** *Source:* http://en.wikipedia.org/wiki/Mathematical%20fallacy?oldid=634098720 *Contributors:* AxelBoldt, Bryan

Derksen, Zundark, Arvindn, Michael Hardy, Dominus, Eric119, Minesweeper, Ijon, LittleDan, UserGoogol, Schneelocke, Charles Matthews, Timwi, Dcoetzee, Dysprosia, Jitse Niesen, Wik, Wiwaxia, Fredrik, Altenmann, Gandalf61, Merovingian, Henrygb, Giftlite, Wolfkeeper, Paul Pogonyshev, Guanaco, Matt Crypto, Mdob, Chowbok, Bongbang, Starx, Peter Kwok, Gdabski, Gazpacho, The-James, Paul August, ESkog, Lankiveil, Spoon!, Nandhp, JRM, Wood Thrush, QTxVi4bEMRbrNqOorWBV, Martinultima, Tsirel, JYolkowski, Anders Kaseorg, SurrealWarrior, Splat, Mikeo, Drbreznjev, Axeman89, Feezo, StradivariusTV, Apokrif, Waldir, BD2412, Jshadias, Josh Parris, Eyu100, R.e.b., Tomtheman5, Tedd, Alexb@cut-the-knot.com, Mathbot, Celestianpower, King of Hearts, Sbrools, DVdm, X42bn6, RussBot, IanManka, Gaius Cornelius, Pnrj, Cheeser1, Simxp, Syko, Brentt, KnightRider, SmackBot, Incnis Mrsi, Melchoir, Rokfaith, Fulldecent, Bluebot, Thumperward, SchfiftyThree, Tavianator, RyanEberhart, Calbaer, Fuhghettaboutit, Pwjb, Turms, Louisng114, Henning Makholm, Ged UK, Byelf2007, Zchenyu, Lambiam, Polihale, Omnedon, Illythr, Cstella23, Kpengboy, Mets501, Hyperwiz, Dr.K., AlsatianRain, JRSpriggs, George100, Whyareall, CRGreathouse, CBM, JPadron, Cydebot, Reywas92, WillowW, MC10, Steel, Crossmr, Carifio24, Kacie Jane, Odie5533, Krzysiu Jarzyna, Englishnerd, Yurell, Kilva, Pallas44, Jojan, AntiVandal-Bot, Seaphoto, Joe Schmedley, MarvinCZ, Dylan Lake, Husond, Oxinabox, MER-C, Boleslaw, Drhlajos, Some Guy123, Timanderso, Albmont, Email4mobile, Catgut, MetsBot, Error792, Cpl Syx, Dravick, Patstuart, Connor Behan, Ztobor, MartinBot, Ariel., Xoran99,
790 *CHAPTER 102. MONSKY'S THEOREM*
Arjun01, Comperr, Anaxial, J.delanoy, AstroHurricane001, Uncle Dick, Paidgenius, GEWilker, Soccersabo, Useight, TWiStErRob, RJASE1, Kimandy, Science4sail, Jennavecia, Nousernamesleft, Wannger27, Anonymous Dissident, Amahdy, Kmhkmh, Geoffreyfishing, Secretss, Dmcq, Sue Rangell, Misha Mullov-Abbado, Oboeboy, Paradoctor, Phe-bot, Keilana, Dragnmn, Taemyr, 0rrAvenger, Kudret abi, □□□□□, Kortaggio, Tuntable, ClueBot, EoGuy, Mild Bill Hiccup, Xenon54, Oxnard27, Doloco, Fletcher17, Lartoven, Ykhwong, H.Marxen, Djk3, XLinkBot, Marc van Leeuwen, Fastily, Pichpich, Tongrongtian, Gwandoya, Gerhardvalentin, Charles Sturm, AlexFekken, Luca Antonelli, Nickolai kazimir, Addbot, JoeMoron2000, 067012732s, CanadianLinuxUser, Fluffernutter, Barak Sh, 3qwertbbb7, Calculuslover, Tide rolls, NKapustin, Yobot, TaBOT-zerem, Timeroot, Spenalzo, AnomieBOT, Pkukiss, Jim1138, AdjustShift, Terminatore, Georgepowell2008, Xqbot, TechBot, The Evil IP address, Point-set topologist, POTUS270, Jetpackboy14, Pottersson, Pinethicket, Number Googol, Patwotrik, Tcnuk, Jujutacular, Barras, Double sharp, Trappist the monk, Niketmalik, Phatency, Le Docteur, Thewriter006, Tbhotch, Sideways713, Whisky drinker, Martianpackets, Mr. Anon515, EmausBot, John of Reading, Slawekb, Anoop.dixith, Derekleungtszhei, Quondum, L Kensington, JonRichfield, ClueBot NG, Rtucker913, Helpful Pixie Bot, Hguy, BG19bot, Hawkwindeb, Mocky3497, Nicolae-boicu, Avengingbandit, Rupert'sscribe, Lugia2453, Mmitchell10, Yehianumb, Gtklocker, Bilorv, That kiwi guy, Stishuk.hf and Anonymous: 277

_ **List of incomplete proofs** *Source:* http://en.wikipedia.org/wiki/List%20of%20incomplete%20proofs?oldid=629758049 *Contributors:* AxelBoldt, Zundark, Michael Hardy, Dominus, Schneelocke, Giftlite, Jason Quinn, Paul August, Kzollman, Rjwilmsi, R.e.b., Wavelength, Spacepotato, Geraschenko, SmackBot, RDBury, Espresso Addict, PrimeHunter, Scwlong, Tamfang, Makyen, Harlekeyn, JRSpriggs, CmdrObot, Cydebot, MC10, Blicher, Quibik, BetacommandBot, Headbomb, RobHar, Hermel, Fabrictramp, David Eppstein, DGG, Hasteur, Cold Phoenix, Pichpich, Yobot, Kilom691, Citation bot 1, Kiefer.Wolfowitz, Xnn, RjwilmsiBot, Helpful Pixie Bot and Anonymous: 6

_ **List of long mathematical proofs** *Source:* http://en.wikipedia.org/wiki/List%20of%20long%20mathematical%20proofs?oldid=626965714 *Contributors:* Edward, TakuyaMurata, Charles Matthews, Dfeldmann, Apokrif, Rjwilmsi, R.e.b., Drone5, Myasuda, Cydebot, David Eppstein, Kope, Olsonist, R'n'B, Yobot, The Evil IP address, Trappist the monk, GoingBatty, Brandmeister, EefeG0hi, A.lisitsa and Anonymous: 3

_ **List of mathematical proofs** *Source:* http://en.wikipedia.org/wiki/List%20of%20mathematical%20proofs?oldid=583470132 *Contributors:* Manning Bartlett, Edward, Michael Hardy, Dominus, Revolver, Pfortuny, Marc Venot, Tosha, Giftlite, Dbenbenn, Dissident, Gro-Tsen, Golbez, Zarvok, Peter Kwok, Shahab, Paul August, ZeroOne, ABCD, Oleg Alexandrov, Jacobolus, Salix alba, R.e.b., Mathbot, YurikBot, IanManka, Grubber, Buster79, Figaro, Googl, Zvika, Silly rabbit, Syrcatbot, Cydebot, Neko244, Weixifan, Leon math, JohnBlackburne, Jimmaths, Synthebot, Dmcq, OlEnglish, RJGray, Set theorist, Joemkhan and Anonymous: 9

_ **Proof by intimidation** *Source:* http://en.wikipedia.org/wiki/Proof%20by%20intimidation?oldid=606933925 *Contributors:* The Anome, Michael Hardy, Dominus, Altenmann, Rheun, Nerdfiles, Florian Blaschke, Linas, Rjwilmsi, Finell, SmackBot, Bill3000, Lambiam, Antonielly, MessedRobot, Cydebot, MC10, EdJohnston, Magioladitis, Bearian, JackSchmidt, Nnemo, Pichpich, Addbot, DOI bot, Citation bot, Xqbot, Toa Nidhiki05, DrilBot, PowerWiki112233, Lotje, Scientific29, KlappCK and Anonymous: 16

_ **Termination analysis** *Source:* http://en.wikipedia.org/wiki/Termination%20analysis?oldid=629301084 *Contributors:* Palmcluster, Mdd, Diego Moya, Ruud Koot, Robert A West, Robertbyrne, Garion96, SmackBot, Derek farn, Wvbailey, Jafet, Konstantin.Solomatov, Bazzargh, AnAj, Faizhaider, Yobot, AnomieBOT, ClueBot NG, Braincricket, CarrieVS, Jochen Burghardt, Nomoteretes and Anonymous: 8

_ **What the Tortoise Said to Achilles** *Source:* http://en.wikipedia.org/wiki/What%20the%20Tortoise%20Said%20to%20Achilles?oldid=627280560 *Contributors:* AxelBoldt, Magnus Manske, Toby Bartels, Enchanter, Ryguasu, Ahoerstemeier, ThirdParty, AugPi, Charles Matthews, Paul Stansifer, Radgeek, David Shay, AnonMoos, BenRG, Aleph4, Superm401, Marc Venot, Everyking, Savant1984, Czrisher, Floorsheim, Cretog8, Pyrosim, Algorithm, Alan Canon, Porcher, Rjwilmsi, Tim!, Koavf, Cfortunato, KSchutte, Saric, Mais oui!, SmackBot, Bluebot, MeekSaffron, TKD, Andeggs, Ohconfucius, Byelf2007, Ace Frahm, Gregbard, Cydebot, JPG-GR, Rxtreme, Cognita, Llygadebrill, DrDentz, Bawm79, Mikemoral, Sun Creator, Hans Adler, Addbot, Hisarmwasherleg, Favonian, TriniMuñoz, Narayan, Ptbotgourou, AnomieBOT, Dromioofephesus, Omnipaedista, Aaron Kauppi, Wikiain, AATroop, TjBot, Zafar142003, ZéroBot, Jenks24, Стюарт Радзинский, Accelerometer, Wbm1058, BG19bot, Paolo Lipparini, Necocco123, BreakfastJr, Monkbot, Trackteur and Anonymous: 36

_ **Combinatorics** *Source:* http://en.wikipedia.org/wiki/Combinatorics?oldid=618008581 *Contributors:* AxelBoldt, Zundark, Stevertigo, Chas zzz brown, Michael Hardy, Kku, Mcarling, Pcb21, Goatasaur, Ejrh, Mxn, Rodney Topor, RodC, Charles Matthews, Viz, Dysprosia, Peregrine981, Zudu29, Furrykef, Hyacinth, McKay, Traroth, Tjdw, Robbot, Naddy, Lowellian, Ianb, MathMartin, Lesonyrra, AceMyth, Aniu, JesseW, Marc Venot, Centrx, Giftlite, Dbenbenn, DocWatson42, Paul Richter, Eran, Lethe, Tom harrison, Dratman, CyborgTosser, Bobblewik, Zarvok, Alberto da Calvairate, Antandrus, Gunnar Larsson, APH, Almit39, Chadernook, Urhixidur, Camipco, Robin klein, ELApro, Shahab, Mormegil, Wfaulk, Ralph Corderoy, Guanabot, Sam Derbyshire, Paul August, Zaslav, Spoon!, Causa sui, Bobo192, R. S. Shaw, Haham hanuka, Jumbuck, Msh210, Stack, Cowsandmilk, Kusma, Igorpak, Ultramarine, Oleg Alexandrov, Japanese Searobin, Alkarex, Dryguy, Woohookitty, Rocastelo, Will Orrick, Oliphaunt, Mondhir, Chochopk, Btyner, Palica, Graham87, Magister Mathematicae, FreplySpang, Kitarak, Rjwilmsi, Nneonneo, Nandesuka, DrBozzball, FlaBot, SchuminWeb, Mathbot, Malhonen, Masnevets, Chobot, Jersey Devil, DVdm, Nehalem, Gwernol, YurikBot, Wavelength, Hairy Dude, Deeptrivia, Michael Slone, KSmrq, Dkostic, Dbmag9, Hakkinen, Nathan11g, LarryLACa, Oakwood, TomJF, Arcades, Anclation, Kaicarver, Memodude, GrEp, Infinity0, Crni-Bombarder!!!, Finell, Capitalist, Sardanaphalus, SmackBot, Peterven, Ttzz, Sticky Parkin, Davidhand, InverseHypercube, Jagged 85, Mcld, Brianv, PrimeHunter, AndrewBuck, Taxipom, Sisodia, RyanEberhart, Mhym, Nonforma, DRLB, -Ozone-, G716, Kensor, Doug Bell, Teutanic, 16@r, Macarion, Dicklyon, Kvng, GoCooL, EZio, Mulder416sBot, CRGreathouse, Wafulz, JRavn, Vyznev Xnebara, ShelfSkewed, WeggeBot, Some P. Erson, Myasuda, Ntsimp, Mblumber, Zahlentheorie, Red Director, Aajaja, Thijs!bot, King Bee, Gryspnik, Tocharianne, Urdutext, Fedayee, Hermel, JAnDbot, MER-C, VoABot II, Ling.Nut, David Eppstein, DirkOliverTheis, FANSTARbot, Koko90, Lantonov, CombAuc, Indeed123, Aresch, Nwbeeson, Policron, Ultra two, FurnaldHall, Carter, JohnBlackburne, LokiClock,

Ptrillian, TXiKiBoT, Malsaqer, Anonymous Dissident, PaulTanenbaum, Peter Alan McAllister, Mickster810, Kmhkmh, Bbukh, Arcfrk, Symane, EuPhyte, Gatinha, Gerakibot, Matthew Yeager, Tiptoety, Thehotelambush, Mike2vil, Anchor Link Bot, S2000magician, Athenean, Wantnot, ClueBot, Rks22, Bob1960evens, Thegeneralguy, PixelBot, Bracton, Thingg, Heyzeuss, Erodium, RMFan1, Vadimvadim, XLinkBot, Marc van Leeuwen, SilvonenBot, Willking1979, Betterusername, Ron B. Thomson, Tomthecool, NjardarBot, MrOllie, Delaszk, Nassrat, Zorrobot, Jarble, Noumenon, Legobot, Luckas-bot, Yobot, KamikazeBot, AnomieBOT, Ciphers, □□, Citation bot, Timir2, Thore Husfeldt, GrouchoBot, Arid Zkwelty, Charvest, BoomerAB, Alexisastupidnoob, Tobby72, HRoestBot, RedBot, Fparnon, TjBot, EmausBot, Jmencisom, Bethnim, Phoenixthebird, Sourabh Katagade, Ahughes6, Devanshuhpandey, OnePt618, Agatecat2700, ClueBot NG, Wcherowi, CocuBot, Joel B. Lewis, Widr, Helpful Pixie Bot, Leonxlin, Ijgt, Rajathsbhat, Will Gladstone, Brad7777, Dtotoo, JYBot, RichardMarioFratini, Mark viking, Purnendu Karmakar, SakeUPenn, Jorgelin10 and Anonymous: 227

102.4. TEXT AND IMAGE SOURCES, CONTRIBUTORS, AND LICENSES 791

_ **Graph theory** *Source:* http://en.wikipedia.org/wiki/Graph%20theory?oldid=629736426 *Contributors:* AxelBoldt, Kpjas, LC, Robert Merkel, Zundark, Taw, Jeronimo, BlckKnght, Dze27, Oskar Flordal, Andre Engels, Karl E. V. Palmen, Shd, XJaM, JeLuF, Arvindn, Gianfranco, Matusz, PierreAbbat, Miguel, Boleslav Bobcik, FvdP, Camembert, Hirzel, Tomo, Patrick, Chas zzz brown, Michael Hardy, Wshun, Booyabazooka, Glinos, Meekohi, Jakob Voss, TakuyaMurata, GTBacchus, Grog, Pcb21, Dgrant, CesarB, Looxix, Ellywa, Ams80, Ronz, Nanshu, Gyan, Nichtich, Mark Foskey, Александър, Poor Yorick, Caramdir, Mxn, Charles Matthews, Berteun, Almi, Hbruhn, Dysprosia, Daniel Quinlan, Gutza, Doradus, Zoicon5, Roachmeister, Populus, Zero0000, Doctorbozzball, McKay, Shizhao, Optim, Robbot, Brent Gulanowski, Fredrik, Altenmann, Dittaeva, Gandalf61, MathMartin, Sverdrup, Puckly, KellyCoinGuy, Thesilverbail, Bkell, Paul Murray, Fuelbottle, ElBenevolente, Aknxy, Dina, Tobias Bergemann, Giftlite, Dbenbenn, Thv, The Cave Troll, Elf, Lupin, Brona, Pashute, Duncharris, Andris, Jorge Stolfi, Tyir, Sundar, GGordonWorleyIII, Alan Au, Bact, Knutux, APH, Tyler McHenry, Naerbnic, Peter Kwok, Robin klein, Ratiocinate, Andreas Kaufmann, Chmod007, Madewokherd, Discospinster, Solitude, Guanabot, Qutezuce, Mani1, Paul August, Bender235, Zaslav, Tompw, Diego UFCG, Chalst, Shanes, Renice, C S, Csl77, Jojit fb, Photonique, Jonsafari, Obradovic Goran, Jumbuck, Msh210, Alansohn, Liao, Mailer diablo, Marianocecowski, Aquae, Blair Azzopardi, Oleg Alexandrov, Youngster68, Linas, LOL, Ruud Koot, Tckma, Astrophil, Davidfstr, GregorB, SCEhardt, Stochata, Xiong, Graham87, Magister Mathematicae, SixWingedSeraph, Rjwilmsi, Gmelli, George Burgess, Eugeneiiim, Arbor, Kalogeropoulos, Fred Bradstadt, FayssalF, FlaBot, PaulHoadley, RexNL, Vonkje, Chobot, Jinma, YurikBot, Wavelength, Michael Slone, Gaius Cornelius, Alex Bakharev, Morphh, SEWilcoBot, Jaxl, Ino5hiro, Xdenizen, Daniel Mietchen, Shepazu, Rev3nant, Lt-wiki-bot, Jwissick, LeonardoRob0t, Agro1986, Eric.weigle, Allens, Sardanaphalus, Melchoir, Brick Thrower, Ohnoitsjamie, Oli Filth, OrangeDog, Taxipom, DHN-bot, Tsca.bot, Onorem, GraphTheoryPwns, Lpgeffen, Jon Awbrey, Henning Makholm, Mlpkr, SashatoBot, Whyfish, Disavian, MynameisJayden, Idiosyncratic-bumblebee, Dicklyon, Quaeler, Lanem, Tawkerbot2, Ylloh, Mahlerite, CRGreathouse, Dycedarg, Requestion, Bumbulski, Myasuda, RUVARD, The Isiah, Ntsimp, Abeg92, Corpx, DumbBOT, Anthonynow12, Thijs!bot, Jheuristic, King Bee, Pstanton, Hazmat2, Headbomb, Marek69, Eleuther, AntiVandalBot, Whiteknox, Hannes Eder, Spacefarer, Myanw, JAnDbot, MER-C, Igodard, Restname, Sangak, Tmusgrove, Feeeshboy, Usien6, Ldecola, David Eppstein, Kope, DerHexer, Oroso, MartinBot, R'n'B, Uncle Dick, Joespiff, Ignatzmice, Shikhar1986, Tarotcards, Policron, XxjwuxX, Yecril, JohnBlackburne, Anonymous Dissident, Alcidesfonseca, Anna Lincoln, Ocolon, Magmi, PaulTanenbaum, Geometry guy, Fivelittlemonkeys, Sacredmint, Spitfire8520, Radagast3, SieBot, Dawn Bard, Toddst1, Jon har, Bananastalktome, Titanic4000, Beda42, Maxime.Debosschere, Damien Karras, ClueBot, DFRussia, PipepBot, Justin W Smith, Vacio, Wraithful, Garyzx, Mild Bill Hiccup, DragonBot, Fchristo, Hans Adler, Dafyddg, Razorflame, Rmiesen, Kruusamägi, Pugget, Darkicebot, BodhisattvaBot, Tangi-tamma, Addbot, Dr.S.Ramachandran, Cerber, DOI bot, Ronhjones, Low-frequency internal, CanadianLinuxUser, Protonk, LaaknorBot, Smoke73, Delaszk, Favonian, Maurobio, Lightbot, Jarble, Ettrig, Luckas-bot, Yobot, Kilom691, Trinitrix, Jean.julius, AnomieBOT, Womiller99, Sonia, Jim1138, Piano non troppo, Gragragra, RandomAct, Citation bot, Ayda D, Xqbot, Capricorn42, Nasnema, Miym, GiveAFishABone, RibotBOT, Jalpar75, Aaditya 7, Ankitbhatt, FrescoBot, Mark Renier, SlumdogAramis, Citation bot 1, Sibian, Pinethicket, RobinK, Wsu-dm-jb, D75304, Wsu-f, Xnn, Obankston, Andrea105, RjwilmsiBot, TjBot, Powerthirst123, Aaronzat, EmausBot, Domesticenginerd, EleferenBot, Jmencisom, Slawekb, Akutagawa10, D.Lazard, Netha Hussain, Tolly4bolly, Chuispaston-Bot, EdoBot, ClueBot NG, Watersmeetfreak, Matthiaspaul, MelbourneStar, Outraged duck, OMurgo, Bazuz, Aks1521, Masssly, Joel B. Lewis, Johnsopc, HMSSolent, 4368a, BG19bot, Ajweinstein, Канеюку, MusikAnimal, AvocatoBot, Bereziny, Brad7777, Sofia karampataki, ChrisGualtieri, GoShow, Dexbot, Cerabot, Omgigotanaccount, Wikiisgreat123, Faizan, Maxwell bernard, Bg9989, Zsoftua, Sake-UPenn, Yloreander, StaticElectricity, Gold4444, Cyborgbadger and Anonymous: 374

_ **Proof of Bertrand's postulate** *Source:* http://en.wikipedia.org/wiki/Proof%20of%20Bertrand's%20postulate?oldid=633955252 *Contributors:* AxelBoldt, XJaM, Michael Hardy, Eric119, Mydogategodshat, Charles Matthews, Fredrik, Dbenbenn, Dissident, Porges, Supercoop, Wikiacc, Paul August, Gauge, EmilJ, Marco Polo, Gene Nygaard, Cgibbard, Woohookitty, Linas, Ryan Reich, Salix alba, FlaBot, Bgwhite, Dmharvey, Buster79, Hv, Zvika, SmackBot, Eskimbot, PrimeHunter, Richard L. Peterson, Hiiiiiiiiiiiiiiiiiiiii, CRGreathouse, Gihanuk, Icek, Jasperdoomen, Magioladitis, Vernanimalcula, David Eppstein, PMajer, Addbot, Touchatou, FrescoBot, Motomuku, SpaceFlight89, PleaseStand, Goldfruits, Danieljohngoldstein, Kodip, Kannan Nambiar th3an0maly, TakuyaMishiba and Anonymous: 24

_ **Proof that e is irrational** *Source:* http://en.wikipedia.org/wiki/Proof%20that%20e%20is%20irrational?oldid=629561637 *Contributors:* Gareth Owen, Stevertigo, Michael Hardy, Paul A, Smack, Wooster, Revolver, Charles Matthews, Pmineault, Centrx, Giftlite, Ikitakoja, Icairns, Pt, EmilJ, R6MaY89, Woohookitty, Linas, Shreevatsa, Starwiz, Rjwilmsi, Salix alba, Mathbot, Bgwhite, WriterHound, YurikBot, Lofty, Schmock, Arthur Rubin, Mordacil, HereToHelp, Zvika, Meshach, BeteNoir, Eskimbot, JCSantos, Oli Filth, Nbarth, AeroSpace, Cybercobra, Kirbytime, Tobias Pfanner, Myasuda, RohanDhruva, Nyttend, David Eppstein, Jimothy 46, Synthebot, This, that and the other, Jsondow, Cmr08, Addbot, J.alzaili, 9258fahsflkh917fas, Ssola, Capricorn42, Citation bot 1, Jowa fan, GabKBel, Wackywill1001, Vaselli, Pirokiazuma, Kavigupta and Anonymous: 36

_ **Prime number** *Source:* http://en.wikipedia.org/wiki/Prime%20number?oldid=632231512 *Contributors:* Damian Yerrick, AxelBoldt, LC, Lee Daniel Crocker, Mav, Bryan Derksen, Zundark, Tarquin, Hajhouse, XJaM, Arvindn, Christian List, The Ostrich, Boleslav Bobcik, FvdP, Imran, Karl Palmen, Olivier, Paul Ebermann, Jim McKeeth, Bdesham, Patrick, Michael Hardy, Pwlfong, Wshun, Dougmerritt, Dominus, Ixfd64, Dcljr, TakuyaMurata, GTBacchus, Eric119, Tengai, Pcb21, Egil, Ahoerstemeier, Stevenj, Docu, Jpatokal, Samuelsen, Snoyes, 5ko, Angela, Kingturtle, Lupinoid, Nikai, Rotem Dan, Cimon Avaro, Jacquerie27, Schneelocke, Jengod, Revolver, Charles Matthews, Timwi, Dcoetzee, Dysprosia, Malcohol, Fuzheado, Evgeni Sergeev, Tpbradbury, Furrykef, Taxman, VeryVerily, Fibonacci, Paul-L, Darkhorse, Sabbut, Bevo, Shizhao, Warofdreams, Optim, Pakaran, Johnleemk, AnthonyQBachler, Garo, Jeffq, Gromlakh, Robbot, Sander123, Chrism, Fredrik, Chris 73, Schutz, Chocolateboy, Calmypal, Lowellian, Gandalf61, Georg Muntingh, Merovingian, Henrygb, Bkell, Intangir, Mark Krueger, Saforrest, JackofOz, Kent Wang, PrimeFan, Seth Ilys, Tobias Bergemann, Giftlite, Dbenbenn, JamesMLane, Abigail-II, Reub2000, Hagedis, MSGJ, Herbee, Spencer195, Monedula, Numerao, Anton Mravcek, Mboverload, Pascal666, Eequor, Macrakis, SWAdair, ALargeElk, Haggis, Tonymaric, CryptoDerk, Gdr, Knutux, LucasVB, Antandrus, OverlordQ, Profvk, CSTAR, Gauss, Martin Wisse, Thincat, Elektron, Pmanderson, Elroch, Icairns, Lumidek, Marcos, Gscshoyru, Tzanko Matev, Creidieki, Topaz, Mschlindwein, Neale Monks, Azuredu, Kasreyn, Eisnel, Mike Rosoft, Oskar Sigvardsson, D6, Mormegil, CALR, Discospinster, E. abu Filumena, Helohe, Rich Farmbrough, Guanabot, Cacycle, Andros 1337, Qutezuce, Antaeus Feldspar, Paul August, Jelammers, Bender235, Petrus, Bombcar, Cyclopia, El C, Lankiveil, Crisófilax, Shanes, RoyBoy, EmilJ, Vdm, Billymac00, Smalljim,

Redlentil, John Vandenberg, C S, BrokenSegue, Drange net, Sasquatch, Larry V, Obradovic Goran, Haham hanuka, Krellis, Pearle, Juanpabl, Jonathunder, Eddideigel, A2Kafir, Jumbuck, Schissel, Msh210, Alansohn, Keenan Pepper, Water Bottle, Sligocki, Zyqqh, In-Shaneee, Bart133, Hohum, Wtmitchell, Velella, Bsadowski1, Gunter, BDD, Blaxthos, Cxxl, HenryLi, Jefflundberg, Oleg Alexandrov, Mahanga, Arbol01, Gatewaycat, Cyclotronwiki, Simetrical, OwenX, Woohookitty, Linas, Mindmatrix, Georgia guy, Uncle G, Aaron McDaid, Oliphaunt, Unixer, Potatojunkie, ^demon, Hdante, MONGO, Zingi, Easyas12c, GregorB, Frankie1969, □□□□□□, Dedalus, Gerbrant, Alofferman, Dysepsion, Kakashi-sensei, Graham87, GoldRingChip, Ilya, Jclemens, MichelleG, BorgHunter, Rjwilmsi, Tim!, Matt.whitby, WCFrancis, Strait, John Nixon, QuickFox, Bubba73, Dianelos, Erkcan, Afterwriting, Amelio Vázquez, FlaBot, Windchaser, Sean Gray, Mathbot, Crazycomputers, Tezh, Pathoschild, Gurch, Intgr, Goudzovski, Glenn L, Imnotminkus, Chobot, Crosstimer,

Faseidman, SujinYH, Roboto de Ajvol, Siddhant, YurikBot, Wavelength, Jeugeorge, Neitherday, Ilanpi, Jimp, Vecter, Hyad, DMahalko, JabberWok, GLaDOS, Stephenb, JohnJSal, Gaius Cornelius, Ewx, NawlinWiki, Hillcino368, Wiki alf, Msikma, Robertvan1, Arichnad, Welsh, Stompbox, UVW, Thiseye, Aaron Brenneman, Xdenizen, EverettColdwell, Raven4x4x, Moe Epsilon, STufaro, Alex43223, Zwobot, Arr jay, Cheeser1, BOT-Superzerocool, Jangid, Noosfractal, Hirak 99, FF2010, 21655, Super Rad!, StuRat, Jwissick, Arthur Rubin, KGasso, Redgolpe, Gulliveig, Th1rt3en, GraemeL, Haddock420, JoanneB, CWenger, Kevin, Anclation, Jaranda, Garion96, Gesslein, John Broughton, Kle0012, DVD R W, Eigenlambda, SmackBot, NickyMcLean, RDBury, Cubs Fan, Krychek, InverseHypercube, Hydrogen Iodide, Imsaguy, Melchoir, Unyoyega, Lifebaka, Nickst, Scifiintel, Delldot, Eskimbot, RobotJcb, Iph, HalfShadow, Xaosflux, Betacommand, Skizzik, Chaojoker, Rmosler2100, LinguistAtLarge, Geneb1955, MK8, B00P, Oli Filth, PrimeHunter, MalafayaBot, SchfiftyThree, CSWarren, Octahedron80, Epastore, Colonies Chris, Reaper X, Scwlong, Can't sleep, clown will eat me, Timothy Clemans, Dingler, Shalom Yechiel, Crasic, Rrburke, Parent5446, Rashad9607, Gogino, Bolonium, Korako, Ianmacm, Lhf, Nakon, Jiddisch, Kntrabssi, Mini-Geek, Drphilharmonic, Jon Awbrey, Tanyakh, LavosBaconsForgotHisPassword, Xiutwel, Panchitaville, Bidabadi, Pilotguy, SashatoBot, Lambiam, ArglebargleIV, Kuru, Richard L. Peterson, Vgy7ujm, Ekpyrotic Architect, Ksn, Statsone, Mr.K., Misosoup, Ocatecir, IronGargoyle, Psmsis, Sohale, Loadmaster, Slakr, TheHYPO, Lukis100, Ehheh, Dicklyon, Mets501, Mathsci, Novangelis, Avant Guard, FinnG, Rtcvb32, Amitch, Asyndeton, Lee Carre, Hu12, Stephen B Streater, Iridescent, Criticofpurereason, Bio rules, Abel Cavaşi, WAREL, Dreftymac, Madmath789, JoeBot, Shoeofdeath, StephenBuxton, Twas Now, CapitalR, Parbiter, Newyorkbrad, Marysunshine, Ulfben, Tawkerbot2, Gco, JForget, Wolfdog, CRGreathouse, CmdrObot, ZICO, Wafulz, PIXTOM, Olaf Davis, Prlsmith, Drinibot, NickW557, McVities, Moreschi, Bumnut, Myasuda, Ketorin, Flammingo, Doctormatt, Peripitus, Cvindustries, Gogo Dodo, Pimpsolo, Corpx, Mon4, ST47, Tawkerbot4, Clovis Sangrail, DumbBOT, Dipics, Optimist on the run, Jc42, Ryan Gittins, Omicronpersei8, Lo2u, Tewapack, Cheveyo, Malleus Fatuorum, Epbr123, Barticus88, O, Timo3, Mojo Hand, Mungomba, Headbomb, Marek69, John254, Neil916, Tellyaddict, Brett Dunbar, RobHar, CielProfond, DaveJ7, Michael A. White, Rotundo, AbcXyz, Dugwiki, Openlander, Jomoal99, Escarbot, Hmrox, AntiVandalBot, Gioto, Luna Santin, Mfiorentino, Mnp, Leuqarte, DOSGuy, Asmeurer, Husond, Roman à clef, Oxinabox, MER-C, Arch dude, QuantumEngineer, Dreamster, Hut 8.5, Savant13, AOL account, Alexandre Vassalotti, Sunrise.it, FaerieInGrey, Penubag, Magioladitis, Freedomlinux, Bongwarrior, VoABot II, Vanish2, JNW, David Cat, JamesBWatson, Kinston eagle, Jerome Kohl, Wlod, Jakob.scholbach, Jim Douglas, Jrssr5, Baccyak4H, Ryeterrell, Avicennasis, Mwalimu59, Catgut, Indon, Johnbibby, Bpcrao, Sullivan.t.j, David Eppstein, Cpl Syx, Just James, Craig Mayhew, Kope, Glen, DerHexer, JaGa, Excesses, RaitisMath, B9 hummingbird hovering, MartinBot, Franp9am, Rakscyn, Rubasov, Anaxial, Tbone55, R'n'B, CommonsDelinker, AlexiusHoratius, Pbroks13, YassirLaCama, Ash, Coolkie, Laljag, Worldedixor, AlphaEta, J.delanoy, Pharaoh of the Wizards, Lordbonzion, Hdam59, Trusilver, Bogey97, UBeR, Plutophanes, ChrisfromHouston, Uncle Dick, Maurice Carbonaro, Marek Wolf, Karthixinbox, Speed8ump, Acalamari, Shawn in Montreal, Abhijitsathe, Nemo bis, (jarbarf), Daniel5Ko, NewEnglandYankee, Labachevskij, Policron, Cometstyles, Toyentory, Vanished user 39948282, Mike V, Vcpandya, Permarbor0, Lcawte, Pdcook, JavierMC, CardinalDan, Idioma-bot, Fainites, VolkovBot, ABF, Iosef, Pleasantville, Kelapstick, JohnBlackburne, Kevinkor2, James Callahan, Akwdb, Soliloquial, Alexkorn, Ryan032, Pescofish, CameronPG, Philip Trueman, Chrishepner, TXiKiBoT, JorgeAranda, Zidonuke, Malsaqer, Anonymous Dissident, Qxz, Someguy1221, Monkey Bounce, Anna Lincoln, Plclark, Mmbabies, Martin451, Adiamas, Jackfork, LeaveSleaves, Vornez, Doug, Greenerturtle, Illumini85, Insanity Incarnate, Koalorka, Zmohd2, Brianga, Dmcq, AlleborgoBot, Symane, Portia327, MathPerson, Superdeterminism, NHRHS2010, Ajonnet, RJSprengnether, Hmwith, Newbyguesses, Demmy100, SieBot, 99nintynine, Andlima, TJRC, TYLER, K. Annoyomous, Scarian, Juniuswikia, Kerplunk83, Dawn Bard, Caltas, X-Fi6, Aeuoah, Calabraxthis, Amin Morshed, Srushe, Garde, Arda Xi, Keilana, Flyer22, Radon210, Jimmie24, Smiertelnik, Taemyr, Oxymoron83, Falcifer, Ehccheehcche, Avnjay, Harry, MiNombre-DeGuerra, Android Mouse Bot 3, PhiEaglesfan712, MrWikiMiki, Taggard, BenoniBot, Oliver Kent, JonyFredUnit, Dereklid, Svick, Mori Riyo, Randomblue, Contestcen, Bkumartvm, Dabomb87, Georg Friedrich Bernhard Riemann, Wahrmund, Mxlyons, C0nanPayne, Startswithj, Martarius, Sarahtheawesome, ClueBot, DeaconJohnFairfax, DFRussia, Justin W Smith, Kennvido, Cliff, Flimsy.twiddle, Threshold Pilot, CarlLamb, Maniac18, VQuakr, Mild Bill Hiccup, Darkmoon802, Joshi1983, Zeus000, Ivan-Davey, Boing! said Zebedee, Futurefaust, Blanchardb, Neverquick, Cirt, Geopol, Salutmoncon, Wesaq, Excirial, Jusdafax, He7d3r, Bender2k14, Timmyp320, Wikitumnus, KClick91, Conical Johnson, Feline Hymnic, Bradchristie, SpikeToronto, J.Gowers, Jotterbot, Clayton-Jai, Kaiba, ChrisHodgesUK, ChrisHamburg, Aprock, Aitias, Versus22, Ubardak, Un Piton, DumZiBoT, Darkicebot, XLinkBot, Megankerr, Forbes72, Ovis23, Dark Mage, Stickee, Little Mountain 5, Anturiaethwr, Skarebo, Badgernet, Alexius08, ZooFari, Vy0123, Ironwill02, Addbot, Hubertsimson, Roentgenium111, Quaoarp, Some jerk on the Internet, DOI bot, Thomas888b, Haruth, Motboylol, WardenWalk, Ronhjones, Frobitz, Mattiedebest, Shirtwaist, MrOllie, CarsracBot, AndersBot, Favonian, SpBot, Doniago, Farmercarlos, LinkFA-Bot, Ozob, 5 albert square, Senseoften, Delibebek, Numbo3-bot, Ehrenkater, Bigzteve, Emdrgreg, AsphyxiateDrake, Tide rolls, Weplayit, Vasil', Mjquinn id, Teles, Timmygoud, CarlHinton, Joebastone, Narutolovehinata5, Legobot, Toriboo96, Luckas-bot, Yobot, OrgasGirl, Stocker741, Bunnyhop11, Zoe17, Fraggle81, Charleswallingford, Marino10, THEN WHO WAS PHONE?, A Stop at Willoughby, KamikazeBot, IW.HG, Synchronism, AnomieBOT, Countercouper, Jbsjbs, Killiondude, Jim1138, Freond, AdjustShift, Silverks, Ulric1313, Tekhnolyze2, Bocutadriansebastian, Bluerasberry, Materialscientist, Citation bot, Maxis ftw, ArthurBot, Xqbot, Biglad2k8mc, Intelati, Mtanderson.cam, Capricorn42, Jeffrey Mall, Rattenkrieg, Gap9551, Kamrul2010, William452, Tarantulae, Whomping willow, Herocksmyworld94, Doulos Christos, Wickedauthor, GhalyBot, Ksureshbabu, Shadowjams, BertoRich, Jgillespiecsc, Us441, A.amitkumar, Kigore, Imbalzanog, Supergeek345, LucienBOT, Wikipe-tan, Afrey82, Majopius, Agemon, JAboy, Sae1962, Citation bot 1, Amplitude101, Ez leviathan, Pinethicket, I dream of horses, 10metreh, Calmer Waters, JasonAJensen, Achim1999, A8UDI, Jusses2, RedBot, MastiBot, Foobarnix, Dposedi, Docmarkc1, Bcmia, Lkhiger, TobeBot, Fractalcrazy, Gonzagalaier, Comet Tuttle, Vrenator, Xkyve, Reaper Eternal, Davish Krail, Gold Five, Kudda080893, Innotata, Tbhotch, Wonky the Worm, WillNess, Xnn, Dexter Nextnumber, Updatehelper, Ripchip Bot, BjörnBergman, Hajatvrc, Jowa fan, Grondemar, Matsgranvik, DASHBot, Urbank, EmausBot, Sir Arthur Williams, Energy Dome, John of Reading, WikitanvirBot, Immunize, Az29, Dreamkxd, Roier, ScottyBerg, Huck42, NotAnonymous0, Illogicalpie, Tommy2010, Lollypop123456789, Anirudh Emani, Slawekb, Endlessoblivion, Vvvhellovvv, ZéroBot, Nerfshots911, Александр Федянинов, MithrandirAgain, East of Borschov, Babahadi, Corpbeast Jr, Adolfbatman, TurilCronburg, Quondum, Rspence1234, Gpounder, Wayne Slam, Eswanhorst, Signsamongafter, RMPK, L Kensington, Awsdert, Donner60, Morgan TG, Wrbodine, Camooseman, Chewings72, TurtleMelody, OisinisiO, AndyTheGrump, Primetravel, Senator2029, Tawarama, Hemiboso, Xprettyxgirlx, DASHBotAV, 28bot, Ebehn, Sriwantha, Anita5192, Petrb, Brianbicknell, ClueBot NG, AznBurger, Primenlight, FlavioMattos, Satellizer, Baseball Watcher, Frietjes, Cntras, PeggyCummins, Rezabot, Joel B. Lewis, Widr, AlexGTV, Helpful Pixie Bot, HMSSolent, Bibcode Bot, 2001:db8, Wikisian, Jacks1881,

Corpsecreate, Utacity, Cyberpower678, Wiki13, Inquisitor1323, Nachhattardhammu, Rderdwien, M hariprasad, Gallagher783, Harizotoh9, Snow Blizzard, יהודה שמחה ולדמן, Rebecca G, Glacialfox, David Bent, Ecallow, Jdlrobson, Mdann52, The Illusive Man, GoShow, AsK x RaNDoMz, ClearShadow, Dexbot, Deltahedron, Timmywosere, Joeharrycarter, Spectral sequence, Lugia2453, Skanik, RudiPo, Jaxxmaster11, ISkeetrainbows, Faizan, Myname111111, Technicwriter, CsDix, Kaka jason, Acetotyce, Septimus.stevens, ColinRW, Tilly 123, Escspeed, LudicrousTripe, Hd n3h, Hippetty, Golyho, Blackbombchu, The Herald, CarringtonEnglish, Mimo, Jackmcbarn, Csikoszi, JDiala, Derekdoth, Byravcev, K9re11 and Anonymous: 1543

_ **Euclid's theorem** *Source:* http://en.wikipedia.org/wiki/Euclid's%20theorem?oldid=631418696 *Contributors:* Zundark, Michael Hardy,

Giftlite, DragonflySixtyseven, Satori, Teorth, RobertStar20, Oleg Alexandrov, Tabletop, Reddwarf2956, Reetep, ThanksForTheFish, Pouchkidium, RDBury, Kaiwen1, Old-fool, JCSantos, PrimeHunter, RekishiEJ, CRGreathouse, Olaf Davis, Myasuda, Mattbuck, Gproud, Thijs!bot, David Eppstein, Wikiwikiwildwest, Shaan4uall, Mmanganello, Jeff G., JohnBlackburne, Yugsdrawkcabeht, Anonymous Dissident, Shakko, Dobermanji, Mwasheim, J.Gowers, DumZiBoT, CàlculIntegral, Addbot, דניאל ב., Luckas-bot, AnomieBOT, Ciphers, Jim1138, Xqbot, -), Mox Hox, Brambleclawx, Rhansakk, Slawekb, Dvqr, Tawarama, ClueBot NG, Wcherowi, Sam X, Mesoderm, Widr, Parsiad.azimzadeh, Duxwing, Ca2james, Chicken345 and Anonymous: 43

_ **Furstenberg's proof of the infinitude of primes** *Source:* http://en.wikipedia.org/wiki/Furstenberg's%20proof%20of%20the%20infinitude%20of%20primes?oldid=*Contributors:* Michael Hardy, Giftlite, Bender235, Woohookitty, Shreevatsa, Salix alba, Sodin, PrimeHunter, RekishiEJ, CRGreathouse, Erzbischof, Sullivan.t.j, David Eppstein, Borat fan, VolkovBot, Lechatjaune, Thehotelambush, Malatinszky, Cenarium, MystBot, Addbot, DOI bot, Topology Expert, Favonian, Yobot, Citation bot, Point-set topologist, Elseif, Citation bot 1, Fly by Night, CitationCleanerBot and Anonymous: 8

_ **Monsky's theorem** *Source:* http://en.wikipedia.org/wiki/Monsky's%20theorem?oldid=588036430 *Contributors:* Michael Hardy, Gandalf61, Robin klein, Naraht, Melchoir, David Eppstein, Addbot, Darkdieuguerre, ZéroBot and Anonymous: 1

102.4.2 Images

_ **File:10_DM_Serie4_Vorderseite.jpg** *Source:* http://upload.wikimedia.org/wikipedia/commons/0/0d/10_DM_Serie4_Vorderseite.jpg *License:* Public domain *Contributors:* http://www.bundesbank.de/Redaktion/DE/Standardartikel/Kerngeschaeftsfelder/Bargeld/dm_banknoten.html#doc18118bodyText2 *Original artist:* Deutsche Bundesbank, Frankfurt am Main, Germany

_ **File:16th_century_French_cypher_machine_in_the_shape_of_a_book_with_arms_of_Henri_II.jpg** *Source:* http://upload.wikimedia.org/wikipedia/commons/a/a2/16th_century_French_cypher_machine_in_the_shape_of_a_book_with_arms_of_Henri_II.jpg *License:* CCBY-SA-3.0 *Contributors:* Own work, photographed at Musee d'Ecouen *Original artist:* Uploadalt

_ **File:1919_eclipse_negative.jpg** *Source:* http://upload.wikimedia.org/wikipedia/commons/d/da/1919_eclipse_negative.jpg *License:* Public domain *Contributors:* F. W. Dyson, A. S. Eddington, and C. Davidson, "A Determination of the Deflection of Light by the Sun's Gravitational Field, from Observations Made at the Total Eclipse of May 29, 1919" *Philosophical Transactions of the Royal Society of London. Series A, Containing Papers of a Mathematical or Physical Character* (1920): 291-333, on 332. *Original artist:* F. W. Dyson, A. S. Eddington, and C. Davidson

_ **File:2008-09_Kaiserschloss_Kryptologen.JPG** *Source:* http://upload.wikimedia.org/wikipedia/commons/a/ad/2008-09_Kaiserschloss_Kryptologen.JPG *License:* CC-BY-SA-3.0-2.5-2.0-1.0 *Contributors:* Own work *Original artist:* Ziko

_ **File:4CT_Inadequacy_Example.svg** *Source:* http://upload.wikimedia.org/wikipedia/commons/b/b5/4CT_Inadequacy_Example.svg *License:* CC-BY-SA-3.0 *Contributors:* Based on a this raster image by Wapcaplet on en.wikipedia. *Original artist:* Inductiveload

_ **File:4CT_Non-Counterexample_1.svg** *Source:* http://upload.wikimedia.org/wikipedia/commons/a/a6/4CT_Non-Counterexample_1.svg *License:* Public domain *Contributors:* Based on a this raster image by Dmharvey on en.wikipedia. *Original artist:* Inductiveload

_ **File:4CT_Non-Counterexample_2.svg** *Source:* http://upload.wikimedia.org/wikipedia/commons/7/7a/4CT_Non-Counterexample_2.svg *License:* Public domain *Contributors:* Based on a this raster image by Dmharvey on en.wikipedia. *Original artist:* Inductiveload

_ **File:6n-graf.svg** *Source:* http://upload.wikimedia.org/wikipedia/commons/5/5b/6n-graf.svg *License:* Public domain *Contributors:* Image:6n-graf.png simlar input data *Original artist:* User:AzaToth

_ **File:ANL-E_aerial_22037k4.jpg** *Source:* http://upload.wikimedia.org/wikipedia/commons/0/07/ANL-E_aerial_22037k4.jpg *License:* Public domain *Contributors:* Argonne National Laboratory photo [1] *Original artist:* Argonne National Laboratory

_ **File:Académie_des_Sciences_1671.jpg** *Source:* http://upload.wikimedia.org/wikipedia/commons/7/78/Acad%C3%A9mie_des_Sciences_1671.jpg *License:* Public domain *Contributors:* ? *Original artist:* ?

_ **File:Algebraic_equation_notation.svg** *Source:* http://upload.wikimedia.org/wikipedia/commons/b/be/Algebraic_equation_notation.svg *License:* CC-BY-SA-3.0 *Contributors:* PC generated image *Original artist:* Iantresman / Iantresman at English Wikipedia

_ **File:Ambox_content.png** *Source:* http://upload.wikimedia.org/wikipedia/en/f/f4/Ambox_content.png *License:* ? *Contributors:* Derived from Image:Information icon.svg *Original artist:*
El T (original icon); David Levy (modified design); Penubag (modified color)

_ **File:Animated_fractal_mountain.gif** *Source:* http://upload.wikimedia.org/wikipedia/commons/6/6d/Animated_fractal_mountain.gif *License:* Public domain *Contributors:* self made based in own JAVA animation *Original artist:* António Miguel de Campos - en:User:Tó campos

_ **File:Antiguo-Artículo_bueno.svg** *Source:* http://upload.wikimedia.org/wikipedia/commons/f/f1/Antiguo-Art%C3%ADculo_bueno.svg *License:* Public domain *Contributors:* Circle taken from Image:Symbol support vote.svg. *Original artist:* Paintman y Chabacano

_ **File:Apophysis-100303-104.jpg** *Source:* http://upload.wikimedia.org/wikipedia/en/1/10/Apophysis-100303-104.jpg *License:* PD *Contributors:* I (Gut Monk (talk)) created this work entirely by myself. *Original artist:*
Gut Monk (talk)

_ **File:Archimedes_sphere_and_cylinder.svg** *Source:* http://upload.wikimedia.org/wikipedia/commons/7/70/Archimedes_sphere_and_cylinder.svg *License:* CC-BY-SA-2.5 *Contributors:*
_ Archimedes_sphere_and_cylinder.png *Original artist:*
_ derivative work: Pbroks13 (talk)

_ **File:Aristotle_Altemps_Inv8575.jpg** *Source:* http://upload.wikimedia.org/wikipedia/commons/a/ae/Aristotle_Altemps_Inv8575.jpg *License:* Public domain *Contributors:* Jastrow (2006) *Original artist:* Copy of Lysippus

_ **File:Arithmetic_symbols.svg** *Source:* http://upload.wikimedia.org/wikipedia/commons/a/a3/Arithmetic_symbols.svg *License:* Public domain *Contributors:* Own work *Original artist:* This vector image was created with Inkscape by Elembis, and then manually replaced.

_ **File:Artículo_bueno.svg** *Source:* http://upload.wikimedia.org/wikipedia/commons/e/e5/Art%C3%ADculo_bueno.svg *License:* Public domain *Contributors:* Circle taken from Image:Symbol support vote.svg *Original artist:* Paintman y Chabacano

_ **File:Banach-Tarski_Paradox.svg** *Source:* http://upload.wikimedia.org/wikipedia/commons/7/74/Banach-Tarski_Paradox.svg *License:* Public domain *Contributors:* Based upon en:Image:Tarksi.png by en:User:Sean Kelly. This version created by bdesham in Inkscape. *Original artist:* Benjamin D. Esham (bdesham)

_ **File:Baruch_de_Spinoza_cover_portrait.jpg** *Source:* http://upload.wikimedia.org/wikipedia/commons/8/88/Baruch_de_Spinoza_cover_portrait.jpg *License:* Public domain *Contributors:* New York Public Library Archives *Original artist:* Historical and Public Figures Archives

_ **File:Bayes_icon.svg** *Source:* http://upload.wikimedia.org/wikipedia/commons/e/ed/Bayes_icon.svg *License:* CC0 *Contributors:* <a href='http://validator.w3.org/' data-x-rel='nofollow'><img alt='W3C' src='//upload.wikimedia.org/wikipedia/commons/thumb/1/1a/Valid_SVG_1.1_%28green%29.svg/88px-Valid_SVG_1.1_%28green%29.svg.png' width='88' height='30' style='vertical-align: top' srcset='//upload.wikimedia.org/wikipedia/commons/thumb/1/1a/Valid_SVG_1.1_%28green%29.svg/132px-Valid_SVG_1.1_%28green%29.svg.png 1.5x, //upload.wikimedia.org/wikipedia/commons/thumb/1/1a/Valid_SVG_1.1_%28green%29.svg/176px-Valid_SVG_1.1_%28green%29.svg.png 2x' data-file-width='91' data-file-height='31' /></a>iThe source code of this SVG is <a data-x-rel='nofollow' class='external text' href='http://validator.w3.org/check?uri=http%3A%2F%2Fcommons.wikimedia.org%2Fwiki%2FSpecial%3AFilepath%2FBayes_icon.svg,<span>,&,</span>,ss=1#source'>valid</a>.
Original artist: Mikhail Ryazanov

_ **File:Bayesian_inference_archaeology_example.jpg** *Source:* http://upload.wikimedia.org/wikipedia/commons/6/6d/Bayesian_inference_archaeology_example.jpg *License:* CC0 *Contributors:* Own work *Original artist:* Gnathan87

_ **File:Bayesian_inference_event_space.svg** *Source:* http://upload.wikimedia.org/wikipedia/commons/a/ad/Bayesian_inference_event_space.svg *License:* CC0 *Contributors:* Own work *Original artist:* Gnathan87

_ **File:Bendixen_-_Carl_Friedrich_Gauß,_1828.jpg** *Source:* http://upload.wikimedia.org/wikipedia/commons/3/33/Bendixen_-_Carl_Friedrich_Gau%C3%9F%2C_1828.jpg *License:* Public domain *Contributors:* published in “Astronomische Nachrichten” 1828 *Original artist:* Siegfried Detlev Bendixen

_ **File:BinomialTheorem.png** *Source:* http://upload.wikimedia.org/wikipedia/commons/0/0f/BinomialTheorem.png *License:* Public domain *Contributors:* Danilo Guanabara Fernandes *Original artist:* Danilo Guanabara Fernandes

_ **File:Blaise_Pascal_Versailles.JPG** *Source:* http://upload.wikimedia.org/wikipedia/commons/9/98/Blaise_Pascal_Versailles.JPG *License:* CC-BY-3.0 *Contributors:* Own work *Original artist:* unknown; a copy of the painture of François II Quesnel, which was made for Gérard Edelinck en 1691.

_ **File:Brain.png** *Source:* http://upload.wikimedia.org/wikipedia/commons/7/73/Nicolas_P._Rougier%27s_rendering_of_the_human_brain.png *License:* GPL *Contributors:* http://www.loria.fr/~{}rougier *Original artist:* Nicolas Rougier

_ **File:CSIRO_ScienceImage_11171_CSIRO_lab_book_entry.jpg** *Source:* http://upload.wikimedia.org/wikipedia/commons/8/8b/CSIRO_ScienceImage_11171_CSIRO_lab_book_entry.jpg *License:* CC-BY-3.0 *Contributors:* http://www.scienceimage.csiro.au/image/11171 *Original artist:* Morgan Brown, CSIRO

_ **File:Casa_espinoza.jpg** *Source:* http://upload.wikimedia.org/wikipedia/commons/f/f8/Casa_espinoza.jpg *License:* CC-BY-SA-3.0 *Contributors:* ? *Original artist:* ?

_ **File:Catalan_4_leaves_binary_tree_example.svg** *Source:* http://upload.wikimedia.org/wikipedia/commons/b/b4/Catalan_4_leaves_binary_tree_example.svg *License:* Public domain *Contributors:* Originally from es.wikipedia; description page is/was here. *Original artist:* Original uploader was Jasampler at es.wikipedia

_ **File:Cayley'{}s_formula_2-4.svg** *Source:* http://upload.wikimedia.org/wikipedia/commons/e/ee/Cayley%27s_formula_2-4.svg *License:* CC-BY-SA-3.0 *Contributors:* Fully drawn by myself, after Cayley_graph_formula_2_4.gif by Rocchini *Original artist:* Júlio Reis

_ **File:Chinese_pythagoras.jpg** *Source:* http://upload.wikimedia.org/wikipedia/commons/c/c3/Chinese_pythagoras.jpg *License:* Public domain *Contributors:* ? *Original artist:* ?

_ **File:Christiaan_Huygens-painting.jpeg** *Source:* http://upload.wikimedia.org/wikipedia/commons/a/a4/Christiaan_Huygens-painting.jpeg *License:* Public domain *Contributors:* http://ressources2.techno.free.fr/informatique/sites/inventions/inventions.html *Original artist:* Caspar Netscher (circa 1639–1684)

_ **File:Circular_and_hyperbolic_angle.svg** *Source:* http://upload.wikimedia.org/wikipedia/commons/c/c6/Circular_and_hyperbolic_angle.svg *License:* CC0 *Contributors:* Own work *Original artist:* Maschen

_ **File:Clinton&1998NobelLaureates.jpg** *Source:* http://upload.wikimedia.org/wikipedia/commons/9/9d/Clinton%261998NobelLaureates.jpg *License:* Public domain *Contributors:* http://clinton3.nara.gov/WH/EOP/OSTP/html/19981125.html *Original artist:* The White House

_ **File:CobwebCollatz2.PNG** *Source:* http://upload.wikimedia.org/wikipedia/commons/6/65/CobwebCollatz2.PNG *License:* Public domain *Contributors:* ? *Original artist:* ?

_ **File:Codomain2_A_B.SVG** *Source:* http://upload.wikimedia.org/wikipedia/commons/d/d1/Codomain2_A_B.SVG *License:* Public domain *Contributors:* Own work (Original text: *I created this work entirely by myself.*) *Original artist:* Damien Karras (talk)

_ **File:Collatz-stopping-time.svg** *Source:* http://upload.wikimedia.org/wikipedia/commons/b/b9/Collatz-stopping-time.svg *License:* Public domain *Contributors:* self-made, based on plot by Goldencako *Original artist:* Cirne

_ **File:Collatz-tree,_depth=20.svg** *Source:* http://upload.wikimedia.org/wikipedia/commons/0/0e/Collatz-tree%2C_depth%3D20.svg *License:* CC-BY-SA-3.0 *Contributors:* Own work *Original artist:* Kleuske

_ **File:Collatz5.svg** *Source:* http://upload.wikimedia.org/wikipedia/commons/9/96/Collatz5.svg *License:* Public domain *Contributors:* Own work *Original artist:* Pokipsy76

_ **File:CollatzFractal.png** *Source:* http://upload.wikimedia.org/wikipedia/commons/1/1c/CollatzFractal.png *License:* Public domain *Contributors:* English wikipedia *Original artist:* Pokipsy76

_ **File:CollatzStatistic100million.png** *Source:* http://upload.wikimedia.org/wikipedia/commons/2/26/CollatzStatistic100million.png *License:* CC-BY-SA-3.0 *Contributors:* Own work *Original artist:* AllenMcC.

102.4. TEXT AND IMAGE SOURCES, CONTRIBUTORS, AND LICENSES 795

_ **File:Color_complex_plot.jpg** *Source:* http://upload.wikimedia.org/wikipedia/commons/d/d3/Color_complex_plot.jpg *License:* CCBY-2.5 *Contributors:* Own work *Original artist:* Claudio Rocchini

_ **File:Commons-logo.svg** *Source:* http://upload.wikimedia.org/wikipedia/en/4/4a/Commons-logo.svg *License:* ? *Contributors:* ? *Original artist:* ?

_ **File:Complex_zeta.jpg** *Source:* http://upload.wikimedia.org/wikipedia/commons/1/1b/Complex_zeta.jpg *License:* Public domain *Contributors:* made with Mathematica, own work. See User:Jan_Homann/Mathematics for an explanation of how to generate images like these with Mathematica. *Original artist:* Jan Homann

_ **File:Compound_of_five_cubes.png** *Source:* http://upload.wikimedia.org/wikipedia/commons/0/03/Compound_of_five_cubes.png *License:* Attribution *Contributors:* Transferred from en.wikipedia to Commons by User:Sreejithk2000 using CommonsHelper. *Original artist:* Tomruen at en.wikipedia

_ **File:Congruent_triangles.svg** *Source:* http://upload.wikimedia.org/wikipedia/commons/7/72/Congruent_triangles.svg *License:* Public domain *Contributors:* Own work *Original artist:* Ilmari Karonen

_ **File:Congruentie.svg** *Source:* http://upload.wikimedia.org/wikipedia/commons/b/bc/Congruentie.svg *License:* Public domain *Contributors:* Originally from nl.wikipedia; description page is/was here. *Original artist:* Original uploader was MADe at nl.wikipedia

_ **File:Coprime-lattice.svg** *Source:* http://upload.wikimedia.org/wikipedia/commons/5/55/Coprime-lattice.svg *License:* Public domain *Contributors:* Transferred from en.wikipedia; transfer was stated to be made by User:JamesR. *Original artist:* Original uploader was

Dmharvey at en.wikipedia
_ **File:Coprime8.svg** *Source:* http://upload.wikimedia.org/wikipedia/commons/2/26/Coprime8.svg *License:* CC-BY-SA-3.0 *Contributors:* Own work *Original artist:* Jgmoxness
_ **File:Crypto_key.svg** *Source:* http://upload.wikimedia.org/wikipedia/commons/6/65/Crypto_key.svg *License:* CC-BY-SA-3.0 *Contributors:* Own work based on image:Key-crypto-sideways.png by MisterMatt originally from English Wikipedia *Original artist:* MesserWoland
_ **File:DNA_Overview2_crop.png** *Source:* http://upload.wikimedia.org/wikipedia/commons/4/4c/DNA_Overview2_crop.png *License:* CC-BY-SA-3.0 *Contributors:* Own work *Original artist:* Soerfm
_ **File:Data_visualization_process_v1.png** *Source:* http://upload.wikimedia.org/wikipedia/en/b/ba/Data_visualization_process_v1.png *License:* CC-BY-SA-3.0 *Contributors:*
Flowchart created from data visualization process descriptions
Original artist:
Farcaster
_ **File:Dedekind_cut_sqrt_2.svg** *Source:* http://upload.wikimedia.org/wikipedia/commons/7/72/Dedekind_cut_sqrt_2.svg *License:* Public domain *Contributors:* Own work *Original artist:* User:Jafet
_ **File:Diagonal_argument.svg** *Source:* http://upload.wikimedia.org/wikipedia/commons/8/85/Diagonal_argument.svg *License:* CC-BYSA-3.0 *Contributors:* Own work *Original artist:* Cronholm144
_ **File:Diagram_of_Pythagoras_Theorem_simplified.png** *Source:* http://upload.wikimedia.org/wikipedia/commons/6/6e/Diagram_of_Pythagoras_Theorem_simplified.png *License:* CC-BY-SA-4.0 *Contributors:* Own work *Original artist:* Unaizalakain
_ **File:Diffie_and_Hellman.jpg** *Source:* http://upload.wikimedia.org/wikipedia/commons/8/88/Diffie_and_Hellman.jpg *License:* CCBY-SA-3.0 *Contributors:* ? *Original artist:* ?
_ **File:Diophantus-cover.jpg** *Source:* http://upload.wikimedia.org/wikipedia/commons/6/60/Diophantus-cover.jpg *License:* Public domain *Contributors:* ? *Original artist:* ?
_ **File:Disqvisitiones-800.jpg** *Source:* http://upload.wikimedia.org/wikipedia/commons/e/e3/Disqvisitiones-800.jpg *License:* Public domain *Contributors:* ? *Original artist:* ?
_ **File:Dominoeffect.png** *Source:* http://upload.wikimedia.org/wikipedia/commons/9/92/Dominoeffect.png *License:* CC-BY-SA-3.0 *Contributors:* ? *Original artist:* ?
_ **File:ECClines-3.svg** *Source:* http://upload.wikimedia.org/wikipedia/commons/d/d0/ECClines-3.svg *License:* CC-BY-SA-3.0-2.5-2.0-1.0 *Contributors:* Own work based on Image:ECexamples01.png by en:User:Image:Dake and Image:ECClines-2.svg by SuperManu. *Original artist:* <a href='//commons.wikimedia.org/wiki/User:YassineMrabet/Gallery' title='User:YassineMrabet/Gallery'>**G**</a><a href='//commons.wikimedia.org/wiki/User:YassineMrabet' title='User:YassineMrabet'>YassineMrabet</a><a href='//commons.wikimedia.org/wiki/User_talk:YassineMrabet' title='User talk:YassineMrabet'>Talk</a><a class='external text' href='http://commons.wikipedia.org/w/index.php?title=User_talk:YassineMrabet,<span>,&,</span>,action=edit,<span>,&,</span>,section=new'>✉</a>
_ **File:Ebits2c.png** *Source:* http://upload.wikimedia.org/wikipedia/commons/2/2c/Ebits2c.png *License:* GFDL *Contributors:* Own work *Original artist:* P. Fraundorf
_ **File:Edit-clear.svg** *Source:* http://upload.wikimedia.org/wikipedia/en/f/f2/Edit-clear.svg *License:* ? *Contributors:* The *Tango! Desktop Project*. *Original artist:*
The people from the Tango! project. And according to the meta-data in the file, specifically: "Andreas Nilsson, and Jakub Steiner (although minimally)."
_ **File:Encoded_letter_of_Gabriel_Luetz_d_Aramon_after_1546_with_partial_deciphering.jpg** *Source:* http://upload.wikimedia.org/wikipedia/commons/b/b8/Encoded_letter_of_Gabriel_Luetz_d_Aramon_after_1546_with_partial_deciphering.jpg *License:* CC-BY-SA-3.0 *Contributors:* Own work, photographed at Ecouen Museum *Original artist:* Uploadalt
_ **File:Enigma.jpg** *Source:* http://upload.wikimedia.org/wikipedia/commons/a/ae/Enigma.jpg *License:* Public domain *Contributors:* *Original artist:*
_ **File:Erdos_head_budapest_fall_1992.jpg** *Source:* http://upload.wikimedia.org/wikipedia/commons/f/f1/Erdos_head_budapest_fall_1992.jpg *License:* CC-BY-3.0 *Contributors:* Own work *Original artist:* Topsy Kretts
_ **File:ErnstKummer.jpg** *Source:* http://upload.wikimedia.org/wikipedia/en/2/2f/ErnstKummer.jpg *License:* ? *Contributors:* ? *Original artist:* ?

_ **File:Espinoza_estatua.jpg** *Source:* http://upload.wikimedia.org/wikipedia/commons/b/b0/Espinoza_estatua.jpg *License:* CC-BY-SA-3.0 *Contributors:* ? *Original artist:* ?
_ **File:Estudio_espinoza.jpg** *Source:* http://upload.wikimedia.org/wikipedia/commons/2/28/Estudio_espinoza.jpg *License:* CC-BY-SA-3.0 *Contributors:* ? *Original artist:* ?
_ **File:Euclid-proof.svg** *Source:* http://upload.wikimedia.org/wikipedia/commons/b/b3/Euclid-proof.svg *License:* GPL *Contributors:* _ Euclid-proof.jpg *Original artist:* Euclid-proof.jpg: Original uploader was Bcrowell at en.wikipedia
_ **File:Euclidian_and_non_euclidian_geometry.png** *Source:* http://upload.wikimedia.org/wikipedia/commons/5/5e/Euclidian_and_non_euclidian_geometry.png *License:* CC-BY-SA-3.0 *Contributors:* ? *Original artist:* ?
_ **File:Euler'{}s_formula.svg** *Source:* http://upload.wikimedia.org/wikipedia/commons/7/71/Euler%27s_formula.svg *License:* CC-BYSA-3.0 *Contributors:* Drawn by en User:Gunther, modified by others. *Original artist:* Originally created by gunther using xfig, recreated in Inkscape by Wereon, italics fixed by lasindi.
_ **File:EulerIdentity2.svg** *Source:* http://upload.wikimedia.org/wikipedia/commons/3/3b/EulerIdentity2.svg *License:* CC-BY-SA-3.0 *Contributors:* ? *Original artist:* ?
_ **File:Fallacy_of_the_isosceles_triangle2.svg** *Source:* http://upload.wikimedia.org/wikipedia/commons/a/ab/Fallacy_of_the_isosceles_triangle2.svg *License:* CC-BY-SA-4.0 *Contributors:* Own work *Original artist:* Turms
_ **File:Fibonacci_spiral_34.svg** *Source:* http://upload.wikimedia.org/wikipedia/commons/9/93/Fibonacci_spiral_34.svg *License:* Public domain *Contributors:* self-drawn in Inkscape *Original artist:* User:Dicklyon
_ **File:Finite_subdivision_of_a_radial_link.png** *Source:* http://upload.wikimedia.org/wikipedia/commons/d/d6/Finite_subdivision_of_a_radial_link.png *License:* CC-BY-SA-3.0 *Contributors:* Own work *Original artist:* Brirush
_ **File:Fisher_iris_versicolor_sepalwidth.svg** *Source:* http://upload.wikimedia.org/wikipedia/commons/4/40/Fisher_iris_versicolor_sepalwidth.svg *License:* CC-BY-SA-3.0 *Contributors:* en:Image:Fisher iris versicolor sepalwidth.png *Original artist:* en:User:Qwfp (original); Pbroks13 (talk) (redraw)
_ **File:Flag_of_Amsterdam.svg** *Source:* http://upload.wikimedia.org/wikipedia/commons/6/6d/Flag_of_Amsterdam.svg *License:* Public domain *Contributors:* ? *Original artist:* ?
_ **File:Folder_Hexagonal_Icon.svg** *Source:* http://upload.wikimedia.org/wikipedia/en/4/48/Folder_Hexagonal_Icon.svg *License:* ? *Contributors:* ? *Original artist:* ?
_ **File:Formal_languages.svg** *Source:* http://upload.wikimedia.org/wikipedia/commons/d/da/Formal_languages.svg *License:* CC-BY-SA-

3.0 *Contributors:* Own work based on: en:Image:Formal languages.png by Gregbard. *Original artist:* MithrandirMage
_ **File:Four_Colour_Map_Example.svg** *Source:* http://upload.wikimedia.org/wikipedia/commons/8/8a/Four_Colour_Map_Example.svg *License:* CC-BY-SA-3.0 *Contributors:* Based on a this raster image by chas zzz brown on en.wikipedia. *Original artist:* Inductiveload
_ **File:Four_Colour_Planar_Graph.svg** *Source:* http://upload.wikimedia.org/wikipedia/commons/a/a4/Four_Colour_Planar_Graph.svg *License:* CC-BY-SA-3.0 *Contributors:* Based on a this raster image by chas zzz brown on en.wikipedia. *Original artist:* Inductiveload
_ **File:Fractal_Broccoli.jpg** *Source:* http://upload.wikimedia.org/wikipedia/commons/4/4f/Fractal_Broccoli.jpg *License:* Public domain *Contributors:* http://pdphoto.org/PictureDetail.php?mat=pdef&pg=8232 *Original artist:* Jon Sullivan
_ **File:Fractal_defrosting_patterns_on_Mars.jpg** *Source:* http://upload.wikimedia.org/wikipedia/commons/1/1c/Fractal_defrosting_patterns_on_Mars.jpg *License:* Public domain *Contributors:* http://hirise.lpl.arizona.edu/ESP_029614_1105 *Original artist:* NASA HiRISE camera, Mars Reconnaissance Orbiter.
_ **File:Fractal_fern_explained.png** *Source:* http://upload.wikimedia.org/wikipedia/commons/4/4b/Fractal_fern_explained.png *License:* Public domain *Contributors:* Own work *Original artist:* António Miguel de Campos
_ **File:Frans_Hals_-_Portret_van_René_Descartes.jpg** *Source:* http://upload.wikimedia.org/wikipedia/commons/7/73/Frans_Hals_-_Portret_van_Ren%C3%A9_Descartes.jpg *License:* Public domain *Contributors:* André Hatala [e.a.] (1997) *De eeuw van Rembrandt*, Bruxelles: Crédit communal de Belgique, ISBN 2-908388-32-4. *Original artist:* After Frans Hals (1582/1583–1666)
_ **File:Frost_patterns_2.jpg** *Source:* http://upload.wikimedia.org/wikipedia/commons/0/0a/Frost_patterns_2.jpg *License:* CC-BY-SA-3.0 *Contributors:* Own work *Original artist:* Schnobby
_ **File:Georg_Cantor_1894.jpg** *Source:* http://upload.wikimedia.org/wikipedia/commons/7/73/Georg_Cantor_1894.jpg *License:* Public domain *Contributors:* Transferred from de.wikipedia; transferred to Commons by User:Ireas using CommonsHelper.
Original artist: unbekannt. Original uploader was Taxiarchos228 at de.wikipedia
_ **File:Gerolamo_Cardano_(colour).jpg** *Source:* http://upload.wikimedia.org/wikipedia/en/2/2f/Gerolamo_Cardano_%28colour%29.jpg *License:* ? *Contributors:* ? *Original artist:* ?
_ **File:Glue1_800x600.jpg** *Source:* http://upload.wikimedia.org/wikipedia/commons/6/69/Glue1_800x600.jpg *License:* CC-BY-SA-3.0 *Contributors:* ? *Original artist:* ?
_ **File:Graph.tree._Cayley'{}s_formula.png** *Source:* http://upload.wikimedia.org/wikipedia/commons/8/83/Graph.tree._Cayley%27s_formula.png *License:* CC0 *Contributors:* made with Microsoft paint,own work. *Original artist:* □□□□
_ **File:Greek_letter_uppercase_Phi.svg** *Source:* http://upload.wikimedia.org/wikipedia/commons/9/96/Greek_letter_uppercase_Phi.svg *License:* ? *Contributors:* A character from the font Linux Libertine. *Original artist:*
_ SVG by Tryphon
_ **File:Gretl_screenshot.png** *Source:* http://upload.wikimedia.org/wikipedia/commons/b/b9/Gretl_screenshot.png *License:* GPL *Contributors:* ? *Original artist:* ?
_ **File:Hasse_diagram_of_powerset_of_3.svg** *Source:* http://upload.wikimedia.org/wikipedia/commons/e/ea/Hasse_diagram_of_powerset_of_3.svg *License:* CC-BY-SA-3.0 *Contributors:* self-made using graphviz's dot. *Original artist:* KSmrq
_ **File:Hevelius_Selenographia_frontispiece.png** *Source:* http://upload.wikimedia.org/wikipedia/commons/3/32/Hevelius_Selenographia_frontispiece.png *License:* Public domain *Contributors:* Johannes Hevelius, Selenographia *Original artist:* file: myself; artwork drawn by Adolph Boÿ, engraved by Jeremias Falck

_ **File:Hevelius_and_wife.jpg** *Source:* http://upload.wikimedia.org/wikipedia/commons/d/db/Hevelius_and_wife.jpg *License:* Public domain *Contributors:* ? *Original artist:* ?
_ **File:Hilbert.jpg** *Source:* http://upload.wikimedia.org/wikipedia/commons/7/79/Hilbert.jpg *License:* Public domain *Contributors:* ? *Original artist:* ?
_ **File:Ibn_al-Haytham.png** *Source:* http://upload.wikimedia.org/wikipedia/commons/3/36/Ibn_al-Haytham.png *License:* CC-BY-SA-3.0 *Contributors:*
_ Iraqi_Dinar_1982_Ibn_al-Haitham.jpg *Original artist:*
_ derivative work: Wronkiew (<a href='//commons.wikimedia.org/wiki/User_talk:Wronkiew' title='User talk:Wronkiew'>talk</a>)
_ **File:Icosahedron.svg** *Source:* http://upload.wikimedia.org/wikipedia/commons/b/b7/Icosahedron.svg *License:* CC-BY-SA-3.0 *Contributors:* Vectorisation of Image:Icosahedron.jpg *Original artist:* User:DTR
_ **File:Image-Al-Kitāb_al-muḫtaṣar_fī_ḥisāb_al-ğabr_wa-l-muqābala.jpg** *Source:* http://upload.wikimedia.org/wikipedia/commons/2/23/Image-Al-Kit%C4%81b_al-mu%E1%B8%ABta%E1%B9%A3ar_f%C4%AB_%E1%B8%A5is%C4%81b_al-%C4%9Fabr_wa-l-muq%C4%81bala.jpg *License:* Public domain *Contributors:* John L. Esposito. *The Oxford History of Islam*. Oxford University Press. ISBN 0195107993. *Original artist:* Muhammad ibn Musa al-Khwarizmi
_ **File:International_Data_Encryption_Algorithm_InfoBox_Diagram.svg** *Source:* http://upload.wikimedia.org/wikipedia/commons/a/af/International_Data_Encryption_Algorithm_InfoBox_Diagram.svg *License:* Public domain *Contributors:* Transfered from en.wikipedia *Original artist:* Original uploader was Surachit at en.wikipedia
_ **File:Irrationality_of_sqrt2.svg** *Source:* http://upload.wikimedia.org/wikipedia/commons/f/ff/Irrationality_of_sqrt2.svg *License:* Public domain *Contributors:* Own work *Original artist:* Smjg
_ **File:Jensen_graph.png** *Source:* http://upload.wikimedia.org/wikipedia/commons/b/b4/Jensen_graph.png *License:* Public domain *Contributors:* Transferred from en.wikipedia; transfer was stated to be made by User:Nousernamesleft. *Original artist:* Original uploader was Artur adib at en.wikipedia Later version(s) were uploaded by Nousernamesleft at en.wikipedia.
_ **File:Jodenbreestraat_Amsterdam_from_a_map_of_1625.png** *Source:* http://upload.wikimedia.org/wikipedia/commons/0/0d/Jodenbreestraat_Amsterdam_from_a_map_of_1625.png *License:* Public domain *Contributors:* Balthasar_Florisz._van_Berckenrode_-_Amsterdam_(1625)_2-1.jpg <a href='//commons.wikimedia.org/wiki/File:Balthasar_Florisz._van_Berckenrode_-_Amsterdam_(1625)_2-1.jpg' class='image'><img alt='Balthasar Florisz. van Berckenrode - Amsterdam (1625) 2-1.jpg' src='//upload.wikimedia.org/wikipedia/commons/thumb/4/47/Balthasar_Florisz._van_Berckenrode_-_Amsterdam_%281625%29_2-1.jpg/50px-Balthasar_Florisz._van_Berckenrode_-_Amsterdam_%281625%29_2-1.jpg' width='50' height='42' srcset='//upload.wikimedia.org/wikipedia/commons/thumb/4/47/Balthasar_Florisz._van_Berckenrode_-_Amsterdam_%281625%29_2-1.jpg/75px-Balthasar_Florisz._van_Berckenrode_-_Amsterdam_%281625%29_2-1.jpg 1.5x, //upload.wikimedia.org/wikipedia/commons/thumb/4/47/Balthasar_Florisz._van_Berckenrode_-_Amsterdam_%281625%29_2-1.jpg/100px-Balthasar_Florisz._van_Berckenrode_-_Amsterdam_%281625%29_2-1.jpg 2x' data-file-width='2500' data-file-height='2123' /></a>
Original artist: Balthasar_Florisz._van_Berckenrode_-_Amsterdam_(1625)_2-1.jpg:
_ **File:Julia_set_(indigo).png** *Source:* http://upload.wikimedia.org/wikipedia/commons/b/bd/Julia_set_%28indigo%29.png *License:* Public domain *Contributors:* ? *Original artist:* ?
_ **File:Justus_Sustermans_-_Portrait_of_Galileo_Galilei,_1636.jpg** *Source:* http://upload.wikimedia.org/wikipedia/commons/d/d4/Justus_Sustermans_-_Portrait_of_Galileo_Galilei%2C_1636.jpg *License:* Public domain *Contributors:* http://www.nmm.ac.uk/mag/pages/mnuExplore/PaintingDetail.cfm?ID=BHC2700 *Original artist:* Justus Sustermans

_ **File:Karl_Popper.jpg** *Source:* http://upload.wikimedia.org/wikipedia/commons/4/43/Karl_Popper.jpg *License:* ? *Contributors:* http://www.flickr.com/photos/lselibrary/3833724834/in/set-72157623156680255/ *Original artist:* LSE library
_ **File:KarperienFractalBranch.jpg** *Source:* http://upload.wikimedia.org/wikipedia/commons/b/b7/KarperienFractalBranch.jpg *License:* CC-BY-SA-3.0 *Contributors:* Own work *Original artist:* Audrey Karperien
_ **File:Karperien_Strange_Attractor_200.gif** *Source:* http://upload.wikimedia.org/wikipedia/commons/5/5e/Karperien_Strange_Attractor_200.gif *License:* CC-BY-SA-3.0 *Contributors:* Own work *Original artist:* Akarpe
_ **File:Kissing-3d.png** *Source:* http://upload.wikimedia.org/wikipedia/commons/8/8f/Kissing-3d.png *License:* CC-BY-SA-3.0 *Contributors:* Transferred from en.wikipedia; transfer was stated to be made by User:Yelm. *Original artist:* Original uploader was Robertwb at en.wikipedia
_ **File:Koch_Snowflake_Animated_Fractal.gif** *Source:* http://upload.wikimedia.org/wikipedia/en/e/e7/Koch_Snowflake_Animated_Fractal.gif *License:* CC-BY-SA-3.0 *Contributors:*
I created this using Xaos and the Gimp
Original artist:
Nadimghaznavi
_ **File:Konigsberg_bridges.png** *Source:* http://upload.wikimedia.org/wikipedia/commons/5/5d/Konigsberg_bridges.png *License:* CCBY-SA-3.0 *Contributors:* Public domain (PD), based on the image
_ <a href='//commons.wikimedia.org/wiki/File:Image-Koenigsberg,_Map_by_Merian-Erben_1652.jpg' class='image'><img alt='Image-Koenigsberg, Map by Merian-Erben 1652.jpg' src='//upload.wikimedia.org/wikipedia/commons/thumb/1/15/Image-Koenigsberg%2C_Map_by_Merian-Erben_1652.jpg/120px-Image-Koenigsberg%2C_Map_by_Merian-Erben_1652.jpg' width='120' height='84' data-file-width='628' data-file-height='437' /></a>
Original artist: Bogdan Giuşcă
_ **File:Lambert_quadrilateral.svg** *Source:* http://upload.wikimedia.org/wikipedia/commons/8/84/Lambert_quadrilateral.svg *License:* Public domain *Contributors:* Image:Lambert quadrilateral.png *Original artist:* Tosha (converted by King of Hearts)
_ **File:Lattice_of_the_divisibility_of_60.svg** *Source:* http://upload.wikimedia.org/wikipedia/commons/5/51/Lattice_of_the_divisibility_of_60.svg *License:* CC-BY-SA-3.0 *Contributors:* ? *Original artist:* ?__

www.ingramcontent.com/pod-product-compliance
Lightning Source LLC
Chambersburg PA
CBHW080755180526
45168CB00006B/2220